Space Sciences Series of ISSI

Volume 91

The Space Sciences Series of ISSI books are coherent reports of the findings, discussions, and ideas that result from international scientific workshops regularly held at the International Space Science Institute (ISSI) in Bern, Switzerland. ISSI's main task is to contribute to the achievement of a deeper understanding of the results from space-research missions, adding value to those results through multi-disciplinary research in an atmosphere of international cooperation. The books are reprints of special issues in the Space Science Reviews journal and occasionally of special issues in the Surveys in Geophysics journal.

Rumi Nakamura • James L. Burch
Editors

Magnetic Reconnection

Explosive Energy Conversion in Space Plasmas

Previously published in *Space Science Reviews, Collection "Magnetic Reconnection"*

Editors
Rumi Nakamura
Space Research Inst
Austrian Academy of Sciences
Graz, Austria

James L. Burch
Space Science and Engineering Division
Southwest Research Institute
San Antonio, TX, USA

ISSN 1385-7525
Space Sciences Series of ISSI
ISBN 978-94-024-2360-0

Cover illustration: Left Caption: Artist image of a black hole (Credit: Richard Menchaca)
Middle Caption: Artist image of the Earth's magnetosphere (blue region with magnetic field) along with 3D simulation of magnetic reconnection and four MMS spacecraft (Image credit: NASA/Goddard/Conceptual Image Lab and Bill Daughton)
Right Caption: Artist image of the Sun (Credit: Richard Menchaca)

This Springer imprint is published by the registered company Springer Nature B.V.
The registered company address is: Van Godewijckstraat 30, 3311 GX Dordrecht, The Netherlands

If disposing of this product, please recycle the paper.

Contents

VI. Data Analysis Methods and Simulation Models

VII. Summary and Outlook

Space Science Reviews (2025) 221:88
https://doi.org/10.1007/s11214-025-01220-3

Magnetic Reconnection: Explosive Energy Conversion in Space Plasmas – Editorial

J.L. Burch[1] · Rumi Nakamura[2,3]

Received: 10 September 2025 / Accepted: 10 September 2025 / Published online: 22 September 2025

Magnetic reconnection is a powerful energy-conversion process that operates anywhere in the universe where there is a magnetized plasma. Magnetic reconnection converts magnetic energy to charged particle thermal and kinetic energy and results in magnetic linkage between adjacent plasma regions, which can transfer momentum from one region to the other. Results of magnetic reconnection can be spectacular and also harmful to technological systems and humans as they include auroras, geomagnetic storms, solar and stellar flares, beams of energetic particles and radio bursts resulting from magnetic interactions between accretion disks and the neutron stars and black holes they surround.

Research on magnetic reconnection has proceeded since the 1950s through theory, plasma simulations and spacecraft measurements including imaging of the solar corona and in-situ measurements within planetary magnetospheres and the solar wind. While great progress had been made, up until 2015 the in-situ measurements were limited to the fluid and ion scales, and it was recognized that further rapid progress would depend on the availability of measurements of plasmas and fields at the electron scale along with 3D measurements of electric and magnetic fields. This step was made with the NASA Magnetospheric Multiscale (MMS) mission, which has tested many of the theoretical predictions about magnetic reconnection, measured the reconnection rate, determined the source of the reconnection electric field, and discovered reconnection in current sheets embedded in the bow shock, magnetosheath, and Kelvin–Helmholtz vortices. MMS also discovered electron-only reconnection, which is not embedded in larger-scale ion reconnection regions as assumed in all previous theories of the phenomenon.

Following the launch of the NASA Magnetospheric Multiscale (MMS) mission in 2015 a rapid increase of published results on magnetic reconnection occurred. The explosion of knowledge on this important energy-conversion process in plasmas was due to the unprecedentedly high temporal resolution of the MMS plasma measurements together with the electron-scale tetrahedral formation of its four spacecraft. During the same period, steady advances in kinetic plasma simulation continued, laboratory investigations of reconnection accelerated, and Parker Solar Probe started reporting on reconnection in the solar corona. Several years after the launch of MMS it became timely and even urgent to review progress

✉ J.L. Burch
jburch@swri.edu

[1] Southwest Research Institute, San Antonio, TX, USA

[2] Space Research Institute, Austrian Academy of Sciences, Graz, Austria

[3] International Space Science Institute, Bern, Switzerland

in research on magnetic reconnection and relevant processes in space plasma based on the recent in-situ multi-point observations and theoretical simulations, and to discuss its astrophysical context. A proposal was accepted by the International Space Science Institute (ISSI) to hold a workshop on the subject, and the workshop was held at the International Space Science Institute (ISSI) in Bern, Switzerland. This topical collection contains papers that resulted from discussions during the workshop, which was attended by about 65 leaders of reconnection research including six early-career scientists. The inevitable Covid restrictions caused the delay of the workshop and for it to be hybrid with a few remote participants although most attended in person. During the workshop held on June 27 – July 1, 2022, an outline of a topical collection of *Space Science Reviews* was developed with papers mainly authored by workshop participants. While most of the papers concern theory and MMS measurements of reconnection at Earth's magnetosphere, relevant results from other measurements in geospace such as Cluster and THEMIS are also highlighted. To provide broad context of the magnetic reconnection research in the MMS era, there are individual papers on laboratory, solar, astrophysical and planetary reconnection. We believe that this special issue will be a good resource for graduate students and early-career scientists who are interested in the latest status of magnetic reconnection research and for more senior scientists to put their own work in perspective.

1 Introduction

The paper by Jim Burch and Rumi Nakamura (*Magnetic Reconnection in Space: An Introduction*) introduces the topic of magnetic reconnection and the challenges faced by the MMS mission. They summarize the capabilities of MMS and how they have led to new understanding of the microphysics of the reconnection process. They briefly introduce the results of the topical collection of papers including the effects of reconnection in geospace, planetary and astrophysical environments, and in the laboratory.

2 Micro-Scale Physics of Magnetic Reconnection

The Kevin Genestreti et al. paper (*Structure of the Electron Diffusion Region During Magnetic Reconnection*) discusses recent advances made in understanding the dynamics of reconnection at the smallest scale: the electron kinetic scale. They focus on the electron-scale region known as the electron diffusion region or EDR, wherein the magnetic field topology changes and magnetic energy is converted to heat and charged particle kinetic energy. The Yi-Hsin Liu et al. paper (*Ohm's Law, the Reconnection Rate, and Energy Conversion in Collisionless Magnetic Reconnection*) discusses the origin of the reconnection electric field and how it sustains the current and pressure in thin current sheets, thereby determining the rate of reconnection. They also review key diagnostics and modeling of energy conversion around the reconnection diffusion region, seeking insights from recently developed theories. The Cecilia Norgren et al. paper (*Particle Dynamics in Reconnection Diffusion Regions*) discusses charged-particle dynamics within the reconnection diffusion region and the surrounding environment including the formation of outflow jets. They further discuss the effects of guide fields in symmetric magnetotail reconnection as well as in asymmetric magnetopause reconnection and in Kelvin-Helmholtz vortices. Details on the structure and dynamics of both the electron and ion diffusion regions are provided.

3 Waves and Turbulence

The Daniel Graham et al. paper (*Kinetic Instabilities and Waves in Collisionless Magnetic Reconnection*) discusses the wide variety of kinetic waves that are generated by magnetic reconnection via wave-particle interactions and how they contribute to anomalous resistivity, particle diffusion and heating and transfer of energy between particle populations. The Julia Stawarz et al. paper (*The Interplay Between Collisionless Magnetic Reconnection and Turbulence*) discusses the importance of turbulence as a fundamental nonlinear plasma phenomenon that plays a key role in energy transport and conversion in space and astrophysical plasmas. They trace the interplay between turbulence and magnetic reconnection and the new insights provided by MMS measurements.

4 Context and Consequence of Magnetic Reconnection in Geospace

The Stephen Fuselier et al. paper (*Global-Scale Processes and Effects of Magnetic Reconnection on the Geospace Environment*) discusses recent advances in the global structure and consequences of reconnection including the location and steadiness of reconnection at the dayside magnetopause, the importance of multiple plasma sources in the global circulation, and reconnection consequences in the magnetotail. The Joo Hwang et al. paper (*Cross-Scale Processes of Magnetic Reconnection*) reviews multi-scale and cross-scale aspects of magnetic reconnection revealed in near-Earth space beyond the general global-scale features and magnetospheric circulation organized by the Dungey Cycle. The Mitsuo Oka et al. paper (*Particle Acceleration by Magnetic Reconnection in Geospace*) reviews recent progress in the understanding of particle acceleration by magnetic reconnection in Earth's magnetosphere. They discuss outstanding problems that remain regarding energy partition between thermal and non-thermal components and the precise role of turbulence in the particle acceleration process.

5 Magnetic Reconnection Beyond Geospace

The Jim Drake et al. paper (*Magnetic Reconnection in Solar Flares and the Near-Sun Solar Wind*) presents an overview of current understanding and open questions related to magnetic reconnection in solar flares and the near-Sun solar wind. They discuss mechanisms that facilitate fast energy release and that control flare onset, electron energization, ion energization and abundance enhancement, electron and ion transport, and flare-driven heating. The Dan Gershman et al. paper (*Magnetic Reconnection at Planetary Bodies and Astrospheres*) presents the latest understanding of reconnection throughout the solar system and beyond. They specifically discuss how reconnection can couple magnetized obstacles to both sub- and super-magnetosonic upstream flows and discuss how reconnection energy conversion scales throughout the solar system. The Fan Guo et al. paper (*Magnetic Reconnection and Associated Particle Acceleration in High-Energy Astrophysics*) discusses how magnetic reconnection can be a mechanism to explain a broad range of high-energy astrophysical phenomena and radiation signatures including high-energy radiation flares, emissions from accretion flows and fast radio bursts. The Hantao Ji et al. paper (*Laboratory Study of Collisionless Magnetic Reconnection*) discusses results from laboratory experiments on collisionless magnetic reconnection and their relationship to space measurements. Highlights include spatial structures of electromagnetic fields in ion and electron diffusion regions as a function of upstream symmetry and guide field strength, energy conversion and partitioning from magnetic field to ions and electrons including particle acceleration, electrostatic and electromagnetic kinetic plasma waves, and plasmoid-mediated multiscale reconnection.

6 Data Analysis Techniques and Simulation Models

The Hiroshi Hasegawa et al. paper (*Advanced Methods for Analyzing in-Situ Observations of Magnetic Reconnection*) provides an overview of a variety of single- and multi-spacecraft data analysis techniques that are key to revealing the context of in-situ observations of magnetic reconnection in space and for detecting and analyzing the diffusion regions where ions and/or electrons are demagnetized. Included are tools for magnetic reconstruction of the reconnection regions. The Mike Shay et al. paper (*Simulation Models for Exploring Magnetic Reconnection*) reviews a wide range of simulation methods to study particular aspects and consequences of magnetic reconnection. The methods include magnetohydrodynamics (MHD), Hall MHD, hybrid, kinetic particle-in-cell (PIC), kinetic Vlasov, fluid models with embedded PIC, fluid models with direct feedback from energetic particle populations, and the Rice Convection Model.

7 Outlook & Summary

The Rumi Nakamura et al. paper (*Outstanding Questions and Future Research on Magnetic Reconnection*) reviews outstanding questions concerning the complex dynamics and structures in the diffusion region, cross-scale and regional couplings, the onset of magnetic reconnection, and the details of particle energization. They also discuss future directions for magnetic reconnection research, including new observations, new simulations, and interdisciplinary approaches.

Acknowledgements The editors would like to thank the authors for their dedicated efforts to prepare the papers that are included in this collection. We are grateful to the other convenors — Barbara Giles, Jim Drake, Michael Hesse, Masahiro Hoshino, Roy Torbert, Ruedi von Steiger and Ruedi's retirement replacement Maurizio Falanga, who passed away in 2025. We greatly appreciate the work of the session coordinators and all of the workshop participants. The friendly and always helpful assistance by the staff of ISSI before, during, and after the workshop was an invaluable asset for the participants and the conveners. We also thank the editorial staff of *Space Science Reviews* for the smooth organization of the publishing process and efficient help with technical issues. We are very grateful to ISSI for generous financial support in organizing the workshop and for publication of the book.

Declarations

Competing Interests The authors declare no competing interests.

I. Introduction

Space Science Reviews (2025) 221:19
https://doi.org/10.1007/s11214-025-01145-x

Magnetic Reconnection in Space: An Introduction

J.L. Burch[1] · Rumi Nakamura[2,3]

Received: 19 November 2024 / Accepted: 24 January 2025 / Published online: 12 February 2025

Abstract
An International Space Science Institute (ISSI) workshop was convened to assess recent rapid advances in studies of magnetic reconnection made possible by the NASA Magnetospheric Multiscale (MMS) mission and to place them in context with concurrent advances in solar physics by the Parker Solar Probe, astrophysics, planetary science and laboratory plasma physics. The review papers resulting from this study focus primarily on results obtained by MMS, and these papers are complemented by reports of advances in magnetic reconnection physics in these other plasma environments. This paper introduces the topical collection "Magnetic Reconnection: Explosive Energy Conversion in Space Plasmas", in particular introducing the new capabilities of the MMS mission used in majority of the articles in the collection and briefly summarizing the advances obtained from MMS.

1 Context and Purpose

The study of magnetic reconnection is a relatively young branch of plasma physics, which has recently garnered great interest as its importance in the laboratory, the Earth's magnetosphere, the solar photosphere and corona and objects such as accretion disks surrounding neutron stars and black holes and supernova remnants has become widely recognized. The common occurrence of magnetic reconnection and its frequent explosive nature make it one of the most important agents of energy transfer throughout the universe. A recent commentary by Hesse and Cassak (2020) describes the universal importance of reconnection along with prospects for its further understanding.

An International Space Science Institute (ISSI) workshop was convened to assess recent rapid advances in this field made possible by the NASA Magnetospheric Multiscale (MMS) mission (Burch et al. 2016a) and to place them in context with concurrent advances in solar physics by the Parker Solar Probe (Drake et al. 2025, this collection), astrophysics (Guo et al. 2024, this collection), planetary science (Gershman et al. 2024, this collection) and

✉ J.L. Burch
jburch@swri.edu

✉ R. Nakamura
Rumi.Nakamura@oeaw.ac.at

1 Southwest Research Institute, San Antonio, TX, USA

2 Space Research Institute, Austrian Academy of Sciences, Graz, Austria

3 International Space Science Institute, Bern, Switzerland

laboratory plasma physics (Ji et al. 2023b, this collection). The review papers resulting from this study focus to a large extent on the results obtained by MMS, and these papers are complemented by reports of advances in magnetic reconnection physics in these other plasma environments.

Leading up to the launch of MMS in March 2015, significant progress in understanding magnetic reconnection at the MHD and ion scales was made by the European Space Agency Cluster mission (e.g., Eastwood et al. 2010) while theoretical predictions of electron-scale phenomena by plasma simulations helped set the stage for future measurements (e.g., Hesse et al. 2014; Bessho et al. 2014). As described by Burch and Drake (2009), four great mysteries of magnetic reconnection were: (1) What produces dissipation in a collisionless plasma, allowing reconnection to occur? (2) What determines the aspect ratio of the dissipation region and the rate of release of magnetic energy? (3) What is the "spark" that causes the magnetic energy that has built up over a period of time to be suddenly released? and (4) What is the mechanism for efficient conversion of magnetic energy into the kinetic energy of charged particles? The unprecedented temporal and spatial resolution of the MMS measurements (Burch et al. 2016a) have led to significant progress toward solving these four mysteries while at the same time revealing many unanticipated aspects of magnetic reconnection in the boundary regions of the Earth's magnetosphere. Note that the above mysteries of magnetic reconnection apply to many other plasma environments as discussed in Ji et al. (2023a). Outstanding questions of magnetic reconnection in different plasma environments are further discussed in Nakamura et al. (2025, this collection).

2 New Capabilities of MMS

By 2005 MMS had become the next NASA Solar-Terrestrial Probe. This event followed the ever-increasing scientific focus on magnetic reconnection that had occurred in the previous 25 years. The International Sun-Earth Explorer mission (ISEE) had made the first measurement of the reconnection ion exhaust at the dayside magnetopause (Paschmann et al. 1979), thereby establishing the existence of magnetic reconnection as an important mechanism of energy-transfer from the solar wind to the Earth's magnetosphere. Subsequent investigations by the WIND, Polar, Cluster and THEMIS missions formed a comprehensive picture of how magnetic reconnection and other associated plasma processes operate at the fluid and ion scales in the boundary regions of the magnetosphere (e.g., Øieroset et al. 2001; Mozer et al. 2002; Eastwood et al. 2010; Angelopoulos et al. 2008). The new frontier for MMS was recognized to be the electron diffusion region, where magnetic fields in adjacent regions (magnetosheath-magnetosphere, north-south tail lobes) become interlinked. The primary advance that was needed was to increase the measurement cadence of the three-dimensional electron distribution function by at least a factor of 100. This requirement was reached through a simple consideration of the following estimates that were made at the time: The electron diffusion region has a width of a few km; the magnetopause moves radially inward and outward at speeds of a few 10s of km/s; and at least three measurements of the distribution function within the diffusion region are needed to characterize it. These numbers led to a round-figure minimum time resolution for electrons of 30 ms. Because of the larger size of the ion diffusion region, the time resolution for ions was relaxed to 150 ms. This major advance in time resolution required a new approach to measuring plasmas over 4π steradians (Pollock et al. 2016). Most previous missions used the spacecraft rotation to sample the full sky with a 2D angular array of sensors, but the required time resolution for

Table 1 Summary of measurements made on each MMS spacecraft.

Measurement	Time Resolution	Sensitivity/Energy Resolution
DC B	3D to ~1 ms	<0.1 nT
DC E	3D to ~1 ms SC Potential (proxy for n_e) to ~1 ms	<0.5 mV/m
AC B	Waveform, spectra to 6 kHz	$<2 \times 10^{-5}$ nT/Hz$^{1/2}$ @ 1 kHz
AC E	Waveform, spectra to 100 kHz	$<1 \times 10^{7}$ V/m-Hz$^{1/2}$ @ 10 kHz
Plasma Electrons	3D $f_e(v)$ every 30 ms; 1 eV to 30 keV Electron beams at single E to 1 ms	20% ΔE/E resolution Poisson Statistics limited
Plasma Ions	3D $f_i(v)$ every 150 ms; 1 eV to 20 keV/q	20% ΔE/E resolution Poisson Statistics limited
Plasma Ion Composition	3D $f_i(v)$ every 10 s 10 eV/q to 30 keV/q for H^+, He^{++}, He^+, O^+	20% ΔE/E resolution
Energetic Ions/Electrons	3D $f(v)$ every 10 s; 20 keV to 1000 keV	35% ΔE/E resolution
Energetic Ions	3D $f_i(v)$ every 20 s; 40 keV to 1000 keV For H^+, He^{++}, He^+, O^+	35% ΔE/E resolution

MMS led to spacecraft rotation speeds that were not technically achievable. The new approach taken by MMS was to use multiple 2D arrays (tophats) with ±22.5° electrostatic deflection of the azimuthal fields of view. This approach required accurate intercalibration of eight electron and eight ion instruments on each spacecraft and across the four spacecraft.

Other key requirements were (1) to determine the reconnection electric field with high-resolution three-axis electric field measurements, which had been difficult or impossible to achieve in the previous missions (Torbert et al. 2016a; Ergun et al. 2016); (2) to measure plasma composition accurately in the presence of the high proton fluxes at the magnetopause, which required a new instrument design (Burch et al. 2005; Young et al. 2016); (3) to measure energetic electrons and ions produced by magnetic reconnection and associated processes (Mauk et al. 2016); (4) to control the spacecraft potential to <4 V, as was done on Cluster (Torkar et al. 2016); and (5) to maintain a tetrahedron of four spacecraft with separations in the range from 10 to 160 km for dayside and nightside observations of reconnection (Tooley et al. 2016). The measurements made by the four MMS spacecraft are summarized in Table 1.

Calibrated Level-2 data in physical units are provided to the public in common data format (CDF) with 30-day latency through the MMS Science Data Center at https://lasp.colorado.edu/mms/sdc/public/. To aid in MMS data plotting and analysis, an extensive library of IDL and Python codes are maintained in the Space Physics Environment Data Analysis System (SPEDAS) (Angelopoulos et al. 2019).

3 Summary of Results from MMS

Regarding reconnection mystery (1), the plasma simulations of Hesse et al. (2014) suggested that accelerated electrons could rapidly carry energy from the magnetic field out of the diffusion region from laminar reconnection without the need for turbulence or the associated anomalous resistivity. Early MMS measurements (Burch et al. 2016b) confirmed the existence of the crescent-type distributions predicted by Hesse et al. (2014) and by Bessho et al.

(2014) while also confirming the prediction of Cassak and Shay (2007) that the dissipation occurs Earthward of the X line for asymmetric reconnection at the magnetopause. Torbert et al. (2016b) showed further that the non-ideal electric field in generalized Ohm's law was produced mainly by the divergence of the electron pressure tensor with a much smaller contribution from electron inertia and a significant residual, which could represent anomalous resistivity. The primary conclusion from these early studies is that at the magnetopause the reconnection electric field is produced by non-isotropic electron pressure, that the electron crescent distributions carry the out-of-plane current and the dissipation, as measured by $\mathbf{J} \cdot \mathbf{E}$, is caused completely by the electron dynamics. Norgren et al. (2025, this collection) provide a detailed discussion of electron dynamics in the electron diffusion region (EDR), while further results on the generalized Ohm's Law and reconnection rate from theory and measurements are reviewed by Liu et al. (2025, this collection).

As noted by Burch and Drake (2009), there is the possibility that "with turbulence, the field lines would be strongly twisted so that multiple ones could reconnect simultaneously, vastly increasing the reconnection rate." The Earth's magnetosheath is a region that is often turbulent by virtue of the dynamical interaction of solar-wind plasma with the bow shock. While reconnecting magnetosheath current sheets have been observed by Cluster (Retino et al. 2007) and THEMIS (Øieroset et al. 2017), the electron dynamics were not observable owing to their 4-s and 3-s time resolutions, respectively. With MMS, turbulent reconnection is observed routinely in the magnetosheath at the electron scale with the significant result that often there is only an electron diffusion region, which is not embedded within an ion diffusion region as occurs in the standard reconnection model (Phan et al. 2018).

In the MMS era, the magnetosheath has become an important laboratory for the study of kinetic plasma turbulence and its relationship with magnetic reconnection (Wilder et al. 2017; Chasapis et al. 2018; Bandyopadhyay et al. 2021). Stawarz et al. (2024, this collection) cover the latest advances in turbulent reconnection by MMS in the magnetosheath and bow shock as well as in Kelvin-Helmholtz vortices in the flank magnetopause. In the magnetosheath, MMS revealed that reconnection between interlinked flux tubes originating from multiple X-lines leads to the formation of flux ropes or flux transfer events (FTEs). This 3D FTE formation process occurs predominantly when the interplanetary magnetic field has a strong east-west component, a condition that leads magnetic field lines to collide and reconnect. These phenomena are covered in detail by Hwang et al. (2023, this collection).

Regarding reconnection mystery (2), the rate of release of magnetic energy, or the reconnection rate, can be measured by determining the aspect ratio of the electron diffusion region, by measuring the inflow rate of electrons into the diffusion region, and by measuring the reconnection electric field. Theoretical estimates summarized by Cassak et al. (2017) are that the normalized reconnection rate is near 0.1. MMS data have been used for all three approaches with the aspect ratio being determined for a magnetotail reconnection event by Nakamura et al. (2019); the electron inflow velocity being measured by Burch et al. (2020, 2022); and the reconnection electric field being determined by Nakamura et al. (2018) and Genestreti et al. (2018). These measurements have yielded normalized reconnection rates in the range of 0.05 to 0.25 with the highest value being for the Phan et al. (2018) electron-only reconnection event. Further discussion of the latest theoretical and experimental advances on the reconnection rate appears in Liu et al. (2025, this collection).

For reconnection mystery (3), the magnetotail is an opportune region for determining what initiates reconnection because there is a growth phase during which the magnetic flux is transferred from the dayside to the nightside magnetosphere leading eventually to the rapid onset of magnetic reconnection across the tail neutral sheet and the initiation of a magnetospheric substorm (Angelopoulos et al. 2008). A recent result by Genestreti et al.

(2023) showed how a thinning magnetotail current sheet triggered by a solar-wind pressure pulse develops multiple X lines with low reconnection rates until one of them reaches lobe field lines. At this time, rapid reconnection is initiated, a primary X line is formed, and the other X lines are swept down the tail within the exhaust of the primary X line. This observation of the initiation of rapid reconnection and its implications are discussed further by Nakamura et al. (2025, this collection).

Mystery (4) stems from the observation of electrons and ions with energies of hundreds of keV during magnetotail reconnection events. These energies exceed the full potential drop across the magnetotail and far exceed the energies that can result from the reconnection electric field of a few mV/m or the energy of reconnection outflows, which proceed at near the Alfvén speed. While some energetic particles, particularly on the day side of the magnetosphere can be explained by leakage of radiation-belt particles through the magnetopause, this process does not apply in the mid to distant magnetotail. Betatron and Fermi acceleration, along with acceleration by parallel electric fields have been predicted and observed by MMS to be important in the Earthward flows, reconnection exhaust and separatrices, respectively (Turner et al. 2016; Ma et al. 2022). Particularly intriguing is the turbulent acceleration of ions and electrons near the reconnection X line and the surrounding diffusion region (Ergun et al. 2022). These results are discussed in detail by Oka et al. (2023, this collection).

An important and unexpected result from MMS is the ubiquitous occurrence of reconnection wherever thin current sheets co-exist with either laminar or turbulent magnetic-field reversals. Examples include thin current sheets embedded within the vortices of Kelvin-Helmholtz MHD instability, which often occur along the flanks of the magnetotail (Eriksson et al. 2016; Hwang et al. 2023, this collection); and the bow shock, which also has been found by MMS to contain reconnecting thin current sheets on a regular basis (Stawarz et al. 2024, this collection).

The reconnection diffusion region has been found to be the site of a wide variety of plasma waves with frequencies from below the ion cyclotron frequency to above the electron plasma frequency, which often develop from plasma instabilities that result from the accumulation of free energy. Advances from MMS concerning these waves and the instabilities that cause them are reviewed by Graham et al. (2025, this collection).

During the time over which rapid advances in reconnection physics have been made by MMS, similar advances have been made in the plasma simulations that predicted many of the experimental results and were later used to provide deeper theoretical understanding of unexpected measurements. The development of plasma simulations during the MMS era is summarized by Shay et al. (2025, this collection). Important advances have also been made in methods for analyzing the multi-spacecraft MMS data including reconstruction of the magnetic-field surrounding reconnection sites. A comprehensive review of these analysis methods is provided by Hasegawa et al. (2024, this collection).

4 Advances in Theory and Modeling

As reviewed by Shay et al. (2025, this collection), a wide variety of plasma simulation techniques have been developed to explore the kinetic physics of reconnection diffusion regions, the transfer of particles and energy to the region closely surrounding the diffusion regions, as well as the macroscale and global effects of reconnection. Because of limitations on computer speed and memory, complete simulation of kinetic physics in the diffusion region is still not feasible at realistic ion-to-electron mass ratios or with time-resolution sufficient to track electron-scale wave phenomena. However, evolution of computer capabilities is steadily improving this situation.

Before the launch of MMS, plasma simulation was somewhat more advanced than the measurements because they could access electron kinetic scales with the most common technique being 2.5-D particle-in-cell (PIC) simulations. With this technique electric and magnetic fields and plasma distribution functions are resolved in 3D but are only allowed to vary along a 2D grid. While limited in scope, these simulations nevertheless made important predictions, some of which have been verified with MMS data. A classic example is the electron crescent distribution, which is described by Genestreti et al. (2025, this collection).

As noted by Shay et al. (2025, this collection), mesoscale, macroscale and up to global simulation can be handled by embedding PIC codes into MHD codes. A similar approach is used in solar flare physics where explicit particle-acceleration codes are embedded in MHD codes as described by Drake et al. (2025, this collection).

5 Summary and Conclusions

Considering the many advances in magnetic reconnection physics that have been made by MMS, it is possible to identify further advances that can be made by this remarkable mission or by subsequent missions that may focus on specific targets for further study. These potential advances are discussed in the outlook paper by Nakamura et al. (2025, this collection). One example is the investigation of cross-scale coupling through which the many microscale reconnection phenomena identified by MMS propagate to the successively larger meso and macro scales. Early attempts at understanding the first stages of cross-scale coupling are being made by MMS through the use of asymmetric tetrahedron and string-of-pearls formations. Ultimately, the determination of how magnetic reconnection affects global magnetospheric dynamics will require coordinated measurements at the various scales from micro to global. Such a large undertaking would no doubt require significant multi-national cooperation, which would be well justified based on the extensive advances made at the microscale by MMS. Outstanding questions about the coupling of the MMS-observed kinetic processes and large-scale magnetospheric processes are reviewed by Fuselier et al. (2024, this collection).

Acknowledgements The authors are indebted to the International Space Science Institute for supporting the workshop on which this collection of review papers is based, to their authors, and to all who participated in the discussion. The authors thank J. F. Drake, B. L. Giles, M. Hesse, M. Hoshino, B. Lavraud, and R. B. Torbert for co-convening the workshop. We acknowledge the important contributions of Rudolf von Steiger and Maurizio Falanga of ISSI.

Funding Open access funding provided by Österreichische Akademie der Wissenschaften. The work of JLB was supported by NASA Contract NNG04EB99C at SwRI. The work of RN was supported by Austrian Science Fund (FWF): https://doi.org/10.55776/P32175.

Declarations

Competing Interests The first author (JLB) is a member of the Space Science Reviews Editorial Board. There are no other competing interests to report.

References

Angelopoulos V, McFadden JP, Larson D, Carlson CW, Mende SG, Frey H, Phan T, Sibeck DG, Glassmeier K-H, Auster U, Donovan W, Mann IR, Ray IJ, Russell CT, Runov A, Zhou X-Z, Kepko L (2008) Tail reconnection triggering substorm onset. Science 321(5891):931. https://doi.org/10.1126/science.1160495

Angelopoulos V, Cruce P, Drozdov A, Grimes EW, Hatzigeorgiu N, King DA, Larson D, Lewis JW, McTiernan JM, Roberts DA, Russell CT, Hori T, Kasahara Y, Kumamoto A, Matsuoka A, Miyashita Y, Miyoshi Y, Shinohara I, Teramoto M, Faden JB, Halford AJ, McCarthy M, Millan RM, Sample JG, Smith DM, Woodger LA, Masson A, Narock AA, Asamura K, Chang TF, Chiang C-Y, Kazama Y, Keika K, Matsuda S, Segawa T, Seki K, Shoji M, Tam SWY, Umemura N, Wang B-J, Wang S-Y, Redmon R, Rodriguez JV, Singer HJ, Vandegriff J, Abe S, Nose M, Shinbori A, Tanaka Y-M, Ueno S, Andersson L, Dunn P, Fowler C, Halekas JS, Hara T, Harada Y, Lee CO, Lillis R, Mitchell DL, Argall MR, Bromund K, Burch JL, Cohen IJ, Galloy M, Giles B, Jaynes AN, Le Contel O, Oka M, Phan TD, Walsh BM, Westlake J, Wilder FD, Bale SD, Livi R, Pulupa M, Whittlesey P, DeWolfe A, Harter B, Lucas E, Auster U, Bonnell JW, Cully CM, Donovan E, Ergun RE, Frey HU, Jackel B, Keiling A, Korth H, McFadden JP, Nishimura Y, Plaschke F, Robert P, Turner DL, Weygand JM, Candey RM, Johnson RC, Kovalick T, Liu MH, McGuire RE, Breneman A, Kersten K, Schroeder P (eds) (2019) The Space Physics Environment Data Analysis System (SPEDAS). Space Sci Rev 215(1):9. https://doi.org/10.1007/s11214-018-0576-4

Bandyopadhyay R, Chasapis A, Matthaeus WH, Parashar TN, Haggerty CC, Shay MA, Gershman DJ, Giles BL, Burch JL (2021) Energy dissipation in turbulent reconnection. Phys Plasmas 28:112305. https://doi.org/10.1063/5.0071015

Bessho N, Chen L-J, Shuster JR, Wang S (2014) Electron distribution functions in the electron diffusion region of magnetic reconnection: physics behind the fine structures. Geophys Res Lett 41:8688–8695. https://doi.org/10.1002/2014GL062034

Burch JL, Drake JF (2009) Reconnecting magnetic fields. Am Sci 97:392–399. https://doi.org/10.1511/2009.80.392

Burch JL, Miller GP, De Los Santos A, Pollock CJ, Pope SE, Valek PW, Young DT (2005) Technique for increasing dynamic range of space-borne ion composition instruments. Rev Sci Instrum 76:103301. https://doi.org/10.1063/1.2084867

Burch JL, Moore TE, Torbert RB, Giles BL (2016a) Magnetospheric Multiscale overview and science objectives. Space Sci Rev 199:5–21. https://doi.org/10.1007/s11214-015-0164-9

Burch JL, Torbert RB, Phan TD, Chen L-J, Moore TE, Ergun RE, et al (2016b) Electron-scale measurements of magnetic reconnection in space. Science 352(6290):aaf2939. https://doi.org/10.1126/science.aaf2939

Burch JL, Webster JM, Hesse M, Genestreti KJ, Denton RE, Phan TD, Hasegawa H, Cassak PA, Torbert RB, Giles BL, Gershman DJ, Ergun RE, Russell CT, Strangeway RJ, Le Contel O, Pritchard KR, Marshall AT, Hwang K-J, Dokgo K, Fuselier SA, Chen L-J, Yamada M, Wang S, Swisdak M, Drake JF, Argall MR, Trattner KJ, Paschmann G (2020) Electron inflow velocities and reconnection rates at Earth's magnetopause and magnetosheath. Geophys Res Lett 47:e2020GL089082. https://doi.org/10.1029/2020GL089082

Burch JL, Hesse M, Webster JM, Genestreti KJ, Torbert RB, Denton RE, Ergun RE, Giles BL, Gershman DJ, Russell CT, Wang S, Chen L-J, Dokgo K, Hwang K-J, Pollock CJ (2022) The EDR inflow region of a reconnecting current sheet in the geomagnetic tail. Phys Plasmas 29:052903. https://doi.org/10.1063/5.0083169

Cassak PA, Shay MA (2007) Scaling of asymmetric magnetic reconnection: general theory and collisional simulations. Phys Plasmas 14(10):102114. https://doi.org/10.1063/1.2795630

Cassak PA, Liu Y-H, Shay MA (2017) A review of the 0.1 reconnection rate problem. J Plasma Phys 83:715830501. https://doi.org/10.1017/S0022377817000666

Chasapis A, Yang Y, Matthaeus WH, Parashar TN, Haggerty CC, Burch JL, Moore TE, Pollock CJ, Dorelli J, Gershman DJ, Torbert RB, Russell CT (2018) Energy conversion and collisionless plasma dissipation channels in the turbulent magnetosheath observed by the magnetospheric multiscale mission. Astrophys J 862:32. https://doi.org/10.3847/1538-4357/aac775

Drake JF, Antiochos S, Bale S, Chen B, Cohen C, Dahlin J, Glesener L, Guo F, Hoshino M, Imada S, Oka M, Phan T, Reeves K, Swisdak M (2025) Magnetic reconnection in solar flares and the near-Sun solar wind. Space Sci Rev 221

Eastwood JP, Phan TD, Øieroset M, Shay MA (2010) Average properties of the magnetic reconnection ion diffusion region in the Earth's magnetotail: the 2001-2005 Cluster observations and comparison with simulations. J Geophys Res Space Phys 115. https://doi.org/10.1029/2009JA014962

Ergun RE, Tucker S, Westfall J, Goodrich KA, Malaspina DM, Summers D, Wallace J, Karlsson M, Mack J, Brennan N, Pyke B, Withnell P, Torbert R, Macri J, Rau D, Dors I, Needell J, Lindqvist P-A, Olsson G, Cully CM (2016) The axial double probe and fields signal processing for the MMS mission. Space Sci Rev 199(1–4):167–188. https://doi.org/10.1007/s11214-014-0115-x

Ergun RE, Pathak N, Usanova ME, Qi Y, Vo T, Burch JL, Schwartz SJ, Torbert RB, Ahmadi N, Wilder FD, Chasipis A, Newman DL, Stawarz JE, Hesse M, Turner DL, Gershman D (2022) Observation of magnetic reconnection in a region of strong turbulence. Astrophys J Lett 935:L8. https://doi.org/10.3847/2041-8213/ac81d4

Eriksson S, Lavraud B, Wilder FD, Stawarz JE, Giles BL, Burch JL, Baumjohann W, Ergun RE, Lindqvist P-A, Magnes W, Pollock CJ, Russell CT, Saito Y, Strangeway RJ, Torbert RB, Gershman DJ, Khotyaintsev YV, Dorelli JC, Schwartz SJ, Avanov L, Grimes E, Vernisse Y, Sturner AP, Phan TD, Marklund GT, Moore TE, Paterson WR, Goodrich KA (2016) Magnetospheric Multiscale observations of magnetic reconnection associated with Kelvin-Helmholtz waves. Geophys Res Lett 43:5606–5615. https://doi.org/10.1002/2016GL068783

Fuselier SA, Petrinec SM, Reiff PH, Birn J, Baker DN, Cohen IJ, Nakamura R, Sitnov MI, Stephens GK, Hwang K-J, Lavraud B, Moore TE, Trattner KJ, Giles BL, Gershman DJ, Toledo-Redondo S, Eastwood JP (2024) Global-scale processes and effects of magnetic reconnection on the geospace environment. Space Sci Rev 220:34. This collection. https://doi.org/10.1007/s11214-024-01067-0

Genestreti KJ, Nakamura TKM, Nakamura R, Denton RE, Torbert RB, Burch JL, Plaschke F, Fuselier SA, Ergun RE, Giles BL, Torbert RB, Russell CT (2018) How accurately can we measure the reconnection rate for the MMS diffusion region event of 11 July 2017? J Geophys Res Space Phys 123(11):9130–9149. https://doi.org/10.1029/2018JA025711

Genestreti KJ, Farrugia CJ, Lu S, Vines SK, Reiff PH, Phan T, Baker DN, Leonard TW, Burch JL, Bingham ST, Cohen IJ, Shuster JR, Gershman DJ, Mouikis CG, Rogers AJ, Torbert RB, Trattner KJ, Webster JM, Chen L-J, Giles BL, Ahmadi N, Ergun RE, Russell CT, Strangeway RJ, Nakamura R, Turner DL (2023) Multi-scale observation of magnetotail reconnection onset: 2. Microscopic dynamics. J Geophys Res Space Phys 128:e2023JA031760. https://doi.org/10.1029/2023JA031760

Genestreti KJ, Nakamura R, Burch JL, Liu Y-H, Norgren C, Shuster J, Hesse M, Torbert RB (2025) Structure of the electron diffusion region during magnetic reconnection. Space Sci Rev 221

Gershman DJ, Fuselier SA, Cohen I, Turner D, Liu Y-H, Chen L-J, Phan T, Stawarz JE, DiBraccio G, Masters A, Ebert R, Sun W, Harada Y, Swisdak M (2024) Magnetic reconnection at planetary bodies and astrospheres. Space Sci Rev 220:7. https://doi.org/10.1007/s11214-023-01017-2

Graham DB, Cozzani G, Khotyaintsev Y, Wilder F, Holmes J, Nakamura R, Büchner J, Dokgo K, Richard L, Steinvall K, Norgren C, Chen L-J, Ji H, Drake JF, Stawarz JE, Eriksson (2025) The role of kinetic instabilities and waves in collisionless magnetic reconnection. Space Sci Rev 221. https://doi.org/10.1007/s11214-024-01133-7

Guo F, Liu Y-H, Zenitani S, Hoshino M (2024) Magnetic reconnection and associated particle acceleration in high-energy astrophysics. Space Sci Rev 220:43. https://doi.org/10.1007/s11214-024-01073-2

Hasegawa H, Argall MR, Aunai N, Bandyopadhyay R, Bessho N, Cohen IJ, Denton RE, Dorelli JC, Egedal J, Fuselier SA, Garnier P, Génot V, Graham DB, Hwang K-J, Khotyaintsev YV, Korovinskiy DB, Lavraud B, Lenouvel Q, Li TC, Liu Y-H, Michotte de Welle B, Nakamura TKM, Payne DS, Petrinec SM, Qi Y, Rager AC, Reiff PH, Schroeder JM, Shuster JR, Sitnov MI, Stephens GK, Swisdak M, Tian AM, Torbert RB, Trattner KJ, Zenitani S (2024) Advanced methods for analyzing in-situ observations of magnetic reconnection. Space Sci Rev 220:68. https://doi.org/10.1007/s11214-024-01095-w

Hesse M, Cassak PA (2020) Magnetic reconnection in the space sciences: past, present, and future. J Geophys Res Space Phys 125:e2018JA025935. https://doi.org/10.1029/2018JA025935

Hesse M, Aunai N, Sibeck D, Birn J (2014) On the electron diffusion region in planar, asymmetric, systems. Geophys Res Lett 41:8673–8680. https://doi.org/10.1002/2014GL061586

Hwang K-J, Nakamura R, Eastwood JP, Fuselier SA, Hasegawa H, Nakamura T, Lavraud B, Dokgo K, Turner DL, Ergun RE, Reiff PH (2023) Cross-scale processes of magnetic reconnection. Space Sci Rev 219:71. https://doi.org/10.1007/s11214-023-01010-9

Ji H, Karpen J, Alt A, Bellan P, Begelman M, Beresnyak A, Blackman E, Bose S, Brown M, Burch J, Carter T, Cassak P, Chen B, Chen L-J, Cheung M, Comisso L, Dahlin J, Daughton W, DeLuca E, Dong C, Dorfman S, Drake J, Ebrahimi F, Egedal J, Forest C, Froula D, Fujimoto K, Gao L, Genestreti K, Gibson S, Guo F, Hoshino M, Hu Q, Huang Y-M, Karimabadi H, Kepco L, Klimchuk J, Kunz M, Kusano K,

Lazarian A, Lebedev S, Li H, Li X, Lin Y, Linton M, Liu Y-H, Loureiro N, Majeski S, Matthaeus W, McLaughlin J, Murphy N, Ono Y, Opher M, Qiu J, Rempel M, Ren Y, Rosner R, Roytershteyn V, Savcheva A, Schoeffier K, Scime E, Shi P, Sironi L, Stanier A, TenBarge J, Vaivads A, Wang H, Yamada M, Yokoyama T, Yoo J, Zenitani S, Zhang J, Zweibel E (2023a) Major scientific challenges and opportunities in understanding magnetic reconnection and related explosive phenomena in heliophysics and beyond. Bull Am Astron Soc 55:192. https://doi.org/10.3847/25c2cfeb.e22a8d1f

Ji H, Yoo J, Fox W, Yamada M, Argall M, Egedal J, Liu Y-H, Wilder F, Eriksson S, Daughton W, Bergstedt K, Bose S, Burch J, Torbert R, Ng J, Chen L-J (2023b) Laboratory study of collisionless magnetic reconnection. Space Sci Rev 219:76. https://doi.org/10.1007/s11214-023-01024-3

Liu Y-H, Hesse M, Genestreti KJ, Nakamura R, Burch JL, Cassak PA, Bessho N, Eastwood JP, Phan T, Swisdak M, Toledo-Redondo S, Hoshino M, Norgren C, Ji H, Nakamura TKM, et al (2025) Ohm's Law, reconnection rate, and energy conversion in collisionless magnetic reconnection. Space Sci Rev 221. https://doi.org/10.1007/s11214-025-01142-0

Ma W, Zhou M, Zhong Z, Deng X (2022) Contrasting the mechanisms of reconnection-driven electron acceleration with in situ observations from MMS in the terrestrial magnetotail. Astrophys J 931:135. https://doi.org/10.3847/1538-4357/ac6be6

Mauk BH, Blake JB, Baker DN, Clemmons JH, Reeves GD, Spence HE, Jaskulek SE, Schlemm CE, Brown LE, Cooper SA, Craft JV, Fennell JF, Gurnee RS, Hammock CM, Hayes JR, Hill PA, Ho GC, Hutcheson JC, Jacques AD, Kerem S, Mitchell DG, Nelson KS, Paschalidis NP, Rossano E, Stokes MR, Westlake JH (2016) The Energetic Particle Detector (EPD) Investigation and the Energetic Ion Spectrometer (EIS) for the Magnetospheric Multiscale (MMS) mission. Space Sci Rev 199:471–514. https://doi.org/10.1007/s11214-014-0055-5

Mozer FS, Bale SD, Phan TD (2002) Evidence of diffusion regions at a subsolar magnetopause crossing. Phys Rev Lett 89(1):015002. https://doi.org/10.1103/PhysRevLett.89.015002

Nakamura TKM, Genestreti KJ, Liu Y-H, Nakamura R, Teh W-L, Hasegawa H, Daughton W, Hesse M, Torbert RB, Burch JL, Giles BL (2018) Measurement of the magnetic reconnection rate in the Earth's magnetotail. J Geophys Res Space Phys 123:9150–9168. https://doi.org/10.1029/2018JA025713

Nakamura R, Genestreti KJ, Nakamura T, Baumjohann W, Varsani A, Nagai T, Bessho N, Burch JL, Denton RE, Eastwood JP, Ergun RE, Gershman DJ, Giles BL, Hasegawa H, Hesse M, Lindqvist P-A, Russell CT, Stawarz JE, Strangeway RJ, Torbert RB (2019) Structure of the current sheet in the 11 July 2017 electron diffusion region event. J Geophys Res Space Phys 124(2):1173–1186. https://doi.org/10.1029/2018JA026028

Nakamura R, Burch JL, Birn J, Chen L-J, Graham DB, Guo F, Hwang K-J, Ji H, Khotyaintsev Y, Liu Y-H, Oka M, Payne D, Sitnov MI, Swisdak M, Zenitani S, Drake JF, Fuselier SA, Genestreti KJ, Gershman DJ, Hasegawa H, Hoshino M, Norgren C, Shay MA, Shuster JR, Stawarz JE (2025) Outstanding questions and future research of magnetic reconnection. Space Sci Rev 221. https://doi.org/10.1007/s11214-025-01143-z

Norgren C, Chen L-J, Graham DB, Bessho N, Egedal J, Richard L, Khotyaintsev YV, Shuster J, Toledo-Redondo S, Lavraud B, Hasegawa H, Eastwood JP, Hesse M, Liu YH, Holmes JC, Argall M (2025) Electron and ion dynamics in reconnection diffusion regions. Space Sci Rev 221

Øieroset M, Phan TD, Fujimoto M, Lin RP, Lepping RP (2001) In situ detection of collisionless reconnection in the Earth's magnetotail. Nature 412(6845):414–417. https://doi.org/10.1038/35086520

Øieroset M, Phan TD, Shay MA, Haggerty CC, Fujimoto M, Angelopoulos V, Eastwood JP, Mozer FS (2017) THEMIS multispacecraft observations of a reconnecting magnetosheath current sheet with symmetric boundary conditions and a large guide field. Geophys Res Lett 44:7598–7606. https://doi.org/10.1002/2017GL074196

Oka M, Birn J, Egedal J, Guo F, Ergun RE, Turner DL, Khotyaintsev Y, Hwang K-J, Cohen IJ, Drake JF (2023) Particle acceleration by magnetic reconnection in geospace. Space Sci Rev 219:75. https://doi.org/10.1007/s11214-023-01011-8

Paschmann G, Sckopke N, Haerendel G, Sonnerup Ö BU, Bame SJ, Asbridge JR, Gosling JT, Russell CT, Elphic RC (1979) Plasma acceleration at the Earth's magnetopause - evidence for reconnection. Nature 282(5736):243–246. https://doi.org/10.1038/282243a0

Phan TD, Eastwood JP, Shay MA, et al (2018) Electron magnetic reconnection without ion coupling in Earth's turbulent magnetosheath. Nature 557:202–206. https://doi.org/10.1038/s41586-018-0091-5

Pollock CJ, Moore T, Jacques A, Burch J, Gliese U, Saito Y, Omoto T, Avanov L, Barrie A, Coffey V, Dorelli J, Gershman D, Giles B, Rosnack T, Salo C, Yokota S, Adrian M, Aoustin C, Auletti C, Aung S, Bigio V, Cao N, Chandler M, Chornay D, Christian K, Clark G, Collinson G, Corris T, De Los Santos A, Devlin R, Diaz T, Dickerson T, Dickson C, Diekmann A, Diggs F, Duncan C, Figueroa-Vinas A, Firman C, Freeman M, Galassi N, Garcia K, Goodhart G, Guererro D, Hageman J, Hanley J, Hemminger E, Holland M, Hutchins M, James T, Jones W, Kreisler S, Kujawski J, Lavu V, Lobell J, LeCompte E, Lukemire A, MacDonald E, Mariano A, Mukai T, Narayanan K, Nguyan Q, Onizuka M, Paterson W,

Persyn S, Piepgrass B, Cheney F, Rager A, Raghuram T, Ramil A, Reichenthal L, Rodriguez H, Rouzaud J, Rucker A, Saito Y, Samara M, Sauvaud J-A, Schuster D, Shappirio M, Shelton K, Sher D, Smith D, Smith K, Smith S, Steinfeld D, Szymkiewicz R, Tanimoto K, Taylor J, Tucker C, Tull K, Uhl A, Vloet J, Walpole P, Weidner S, White D, Winkert G, Yeh P-S, Zeuch M (2016) Fast plasma investigation for Magnetospheric Multiscale. Space Sci Rev 199:331–406. https://doi.org/10.1007/s11214-016-0245-4

Retino A, Sundkvist D, Vaivads A, Mozer F, André M, Owen CJ (2007) In situ evidence of magnetic reconnection in turbulent plasma. Nat Phys 3:235–238. https://doi.org/10.1038/nphys574

Shay MA, Adhikari S, Beesho N, Birn J, Büchner J, Cassak PA, Chen L-J, Chen Y, Cozzani G, Drake JF, Guo F, Hesse M, Jain N, Pfau-Kempf Y, Lin Y, Liu Y-H, Oka M, Omelchenko Y, Palmroth M, Pezzi1 O, Reiff PH, Swisdak M, Toffoletto F, Toth G, Wolf RA (2025) Simulation models for exploring magnetic reconnection. Space Sci Rev 221

Stawarz JE, Munõz PA, Bessho N, Bandyopadhyay R, Nakamura TKM, Eriksson S, Graham D, Büchner J, Chasapis A, Drake JF, Shay MA, Ergun RE, Hasegawa H, Khotyaintsev YV, Swisdak M, Wilder F (2024) The interplay between collisionless magnetic reconnection and turbulence. Space Sci Rev 220:90. https://doi.org/10.1007/s11214-024-01124-8

Tooley CR, Black RK, Robertson BP, Stone JM, Pope SE, Davis GT (2016) The Magnetospheric Multiscale constellation. Space Sci Rev 199:23–76. https://doi.org/10.1007/s11214-015-0220-5

Torbert RB, Burch JL, Giles BL, Gershman DJ, et al (2016b) Estimates of terms in Ohm's law during an encounter with an electron diffusion region. Geophys Res Lett 43:5918–5925. https://doi.org/10.1002/2016GL069553

Torbert RB, Russell CT, Magnes W, Ergun RE, Lindqvist P-A, Le Contel O, Vaith H, Macri J, Myers S, Rau D, Needell J, King B, Granoff M, Chutter M, Dors I, Olsson G, Khotyaintsev YV, Eriksson A, Kletzing CA, Bounds S, Anderson B, Baumjohann W, Steller M, Bromund K, Le G, Nakamura R, Strangeway RJ, Leinweber HK, Tucker S, Westfall J, Fischer D, Plaschke F, Porter J, Lappalainen K (2016a) The FIELDS instrument suite on MMS: scientific objectives, measurements, and data products. Space Sci Rev 199:105–135. https://doi.org/10.1007/s11214-014-0109-8

Torkar K, Nakamura R, Tajmar M, Scharlemann C, Jeszenszky H, Laky G, Fremuth G, Escoubet CP, Svenes K (2016) Active spacecraft potential control investigation. Space Sci Rev 199:515–544. https://doi.org/10.1007/s11214-014-0049-3

Turner DL, Fennell JF, Blake JB, Clemmons JH, Mauk BH, Cohen IJ, Jaynes AN, Craft JV, Wilder FD, Baker DN, Reeves GD, Gershman DJ, Avanov LA, Dorelli JC, Giles BL, Pollock CJ, Schmid D, Nakamura R, Strangeway RJ, Russell CT, Artemyev AV, Runov A, Angelopoulos V, Spence HE, Torbert RB, Burch JL (2016) Energy limits of electron acceleration in the plasma sheet during substorms: a case study with the Magnetospheric Multiscale (MMS) mission. Geophys Res Lett 43:7785–7794. https://doi.org/10.1002/2016GL069691

Wilder FD, Ergun RE, Eriksson S, Phan TD, Burch JL, Ahmadi N, Goodrich KA, Newman DL, Trattner KJ, Torbert RB, Giles BL, Strangeway RJ, Magnes W, Lindqvist P-A, Khotyaintsev Y-V (2017) Multipoint measurements of the electron jet of symmetric magnetic reconnection with a moderate guide field. Phys Rev Lett 118:265101. https://doi.org/10.1103/PhysRevLett.118.265101

Young DT, Burch JL, Gomez RG, De Los Santos A, Miller GP, Wilson P, Paschalidis N, Fuselier SA, Pickens K, Hertzberg E, Pollock CJ, Scherrer J, Wood PB, Donald ET, Aaron D, Furman J, George D, Gurnee RS, Hourani RS, Jacques A, Johnson T, Orr T, Pan KS, Persyn S, Pope S, Roberts J, Stokes MR, Trattner KJ, Webster JM (2016) Hot plasma composition analyzer for the Magnetospheric Multiscale mission. Space Sci Rev 199(1–4):407–470. https://doi.org/10.1007/s11214-014-0119-6

II. Micro-scale Physics of Magnetic Reconnection

Space Science Reviews (2025) 221:59
https://doi.org/10.1007/s11214-025-01188-0

Structure of the Electron Diffusion Region During Magnetic Reconnection

K.J. Genestreti[1] · R. Nakamura[2] · Y.-H. Liu[3] · J.L. Burch[4] · C. Norgren[5] · J. Shuster[6] · M. Hesse[7] · R.B. Torbert[6,1] · L.-J. Chen[8] · S.V. Heuer[6]

Received: 12 February 2025 / Accepted: 10 June 2025 / Published online: 27 June 2025

Abstract
This article reviews recent developments in our understanding of the electron diffusion region (EDR) of magnetic reconnection, focusing on how the structure and dynamics of the EDR vary with the background plasma conditions. In particular, we highlight results from the Magnetospheric Multiscale (MMS) mission, a four-satellite NASA mission that measures plasma particles and electromagnetic fields in and around Earth's magnetosphere. Theoretical and modeling-driven predictions regarding the structure of the EDR and the key processes that enable reconnection are compared and contrasted with observational studies using in-situ MMS measurements.

1 Introduction

1.1 Background and Scope of This Review

Magnetized plasmas permeate the universe, being the most common form of matter by an overwhelming majority. In many settings, the magnetic field energy density exceeds the thermal energy of the plasma. Magnetic reconnection is the primary mechanism for releasing energy stored in the magnetic field. Reconnection occurs between plasmas with sheared

✉ K.J. Genestreti
kevin.genestreti@swri.org

✉ R. Nakamura
rumi.nakamura@oeaw.ac.at

1 Earth Oceans and Space, Southwest Research Institute, Durham, NH, USA

2 Space Research Institute, Austrian Academy of Sciences, Graz, Austria

3 Department of Physics and Astronomy, Dartmouth College, Hanover, NH, USA

4 Space Science and Engineering, Southwest Research Institute, San Antonio, TX, USA

5 Space Plasma Physics Research Programme, Swedish Institute for Space Physics, Uppsala, Sweden

6 Space Science Center, University of New Hampshire, Durham, NH, USA

7 Ames Research Center, National Aeronautics and Space Administration, Moffet Field, CA, USA

8 Goddard Space Flight Center, National Aeronautics and Space Administration, Greenbelt, MD, USA

magnetic fields, which drift together, merge, and annihilate, expending a portion of their energy toward accelerating and expelling nearby plasma in fast jets. This process can continue to operate as long as particles and magnetic flux convected away in the jets are replenished by inflows. Reconnection impacts its surroundings by accelerating particles, driving convection and redistributing mass and magnetic flux, and changing the magnetic topology of the reconnecting boundary, which allows previously separated plasmas to mix along merged field lines.

Reconnection drives explosive phenomena in plasmas throughout the universe and has therefore garnered attention from scientists with a broad spectrum of interests. Reconnection heats the solar corona and drives coronal mass ejections and flares (Drake et al. 2025, this collection), powers geomagnetic storms and aurorae at Earth and in planetary magnetospheres (Gershman et al. 2024, this collection; Fuselier et al. 2024, this collection), and drives energy conversion in pulsar nebulae and black holes (Koide et al. 2006; Uzdensky et al. 2011; Guo et al. 2024, this collection). While these examples illustrate reconnection's potential to impact the dynamics of a variety of systems on a global scale, it is enabled by plasma dynamics that play out over comparably microscopic kinetic scales.

This paper details recent advances made in understanding the dynamics of reconnection at the smallest scale: the electron-kinetic scale. In particular, we focus on the electron-scale region known as the electron diffusion region or EDR, wherein the magnetic field topology changes, as is defined in Sect. 1.2. Many of these advances were made using in-situ plasma and field data from NASA's Magnetospheric Multiscale (MMS) mission, which is described in Sect. 1.3. We contrast the electron-kinetic dynamics in and near EDRs in differing plasma environments (Sect. 2) and, where possible, connect the microphysics to the larger-scale impact reconnection has on its environment. Finally, in Sect. 3, the state-of-the-art understanding is summarized and open questions are identified. Topics introduced here are expanded upon in Liu et al. (2025) and Norgren et al. (2025), which focus on the fluid and kinetic aspects of reconnection, respectively.

1.2 Substructures of Reconnection and Their Roles

A simplified diagram of magnetic reconnection is shown in Fig. 1. Figure 1a shows the large-scale structure of reconnection relative to the commonly-used LMN coordinate system. The inflow regions (green) supply oppositely-directed magnetic fields to the diffusion region (grey), where the inflowing fields are interconnected. Reconnection magnetic fields and accelerated particles are transported away from the diffusion region in the outflow jets (orange). Particles can cross from the inflow to outflow either through the diffusion region or across the separatrices (blue, dashed), which divide the inflow and outflow regions. The three-dimensional extension of this 2-dimensional picture, assuming translational invariance in the third (M) dimension, is illustrated in Fig. 1c.

The diffusion region is, in reality, a nested set of ion and electron diffusion regions (IDR and EDR respectively), as illustrated in Figures 1a and 1b (cf., Sonnerup 1979). Particles crossing from inflowing to reconnected magnetic fields in the outflow require a curled electric field in the plasma rest frame, i.e., $\nabla \times \left(\vec{E} + \vec{u} \times \vec{B}\right) \neq 0$ such that $\partial \vec{B}/\partial t \neq 0$ in the rest frame, where $\vec{E}$ is the electric field, $\vec{u}$ is the bulk velocity, and $\vec{B}$ is the magnetic field. Particle motions become demagnetized when the magnetic field gradient scale length is comparable to or smaller than the thermal gyroradius, hence the larger size of the IDR relative to the EDR (Figures 1a and 1b). In the IDR ions are demagnetized while electrons remain magnetized. In the EDR, the frozen-in condition is violated for all species and field

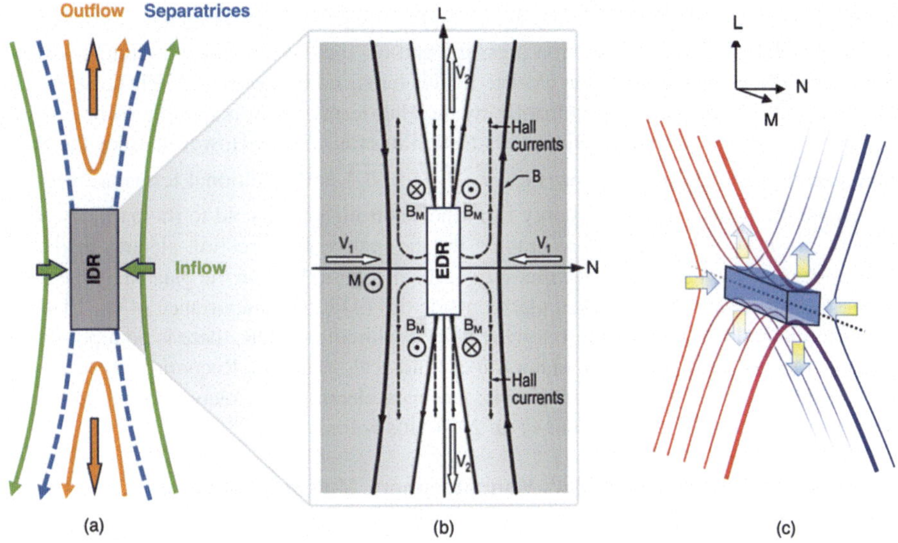

Fig. 1 (a) and (b) illustrate the fluid and kinetic-scale features of reconnection in the L-N plane (see coordinate axes in panel b), which contains the inflow and outflow directions. Panel (c) illustrates how the simple schematic of (a) extends in the third dimension. ((a), (b): Adapted from Burch et al. 2016a)

lines from the inflow regions can slip through the plasma and merge. The EDR is sometimes broken into an "inner" and "outer" EDR (Phan et al. 2007). The outer EDR is characterized by super-ion-Alfvénic electron jets that flow outward from the inner EDR, i.e., the perpendicular-to-B portion of the Hall current system. While both the inner and outer EDRs have non-ideal electric fields, only those in the inner EDR change the magnetic topology. Hence, we refer to the "inner EDR" as the EDR and the "outer EDR" as the electron outflow throughout this review. Typically, most inflowing plasma crosses into the exhausts along separatrices (rather than through the diffusion regions alone), which span the entire length of the exhausts, meaning that the small sizes of the diffusion regions (Fig. 1b) do not limit the pace of the much larger convection cycle driven by reconnection (Fig. 1a) (cf. Kulsrud 2001). In a $\geq$3-mass-species fluid and during reconnection with cold and hot populations of the same species, additional diffusion regions may develop at the characteristic scale lengths of the particles' motions; for simplicity, we limit the scope of this paper to the two-species case and direct interested readers to Liu et al. (2025).

At all scales, the pattern of inflow and outflow is supported by an electric field E_{M}, called the reconnection electric field (cf., Hesse et al. 2018). Further insight into the nature of the reconnection electric field is provided by generalized Ohm's law, which is derived from the electron momentum equation in a collisionless plasma as

$$\vec{E} = -\vec{u} \times \vec{B} + \frac{1}{en}\vec{J} \times \vec{B} - \frac{1}{en}\nabla \cdot \overline{P}_e + \frac{m_e}{e^2 n}\left(\nabla \cdot \left(\vec{u}\vec{J} + \vec{J}\vec{u} - \frac{\vec{J}\vec{J}}{en}\right) + \frac{\partial \vec{J}}{\partial t}\right), \tag{1}$$

where $\vec{J}$ is the current density, $\vec{u}$ is the center-of-mass bulk velocity, $\overline{P}_e$ is the electron pressure tensor, and n is the plasma density, equivalent to n_e and $\sum_s q_s n_{i,s}/e$ for a quasineutral plasma. From left to right, the terms on the right-hand-side of equation (1) are the (1) ideal-MHD, (2) Hall, (3) electron pressure, and (4) inertial terms. In the fluid-scale inflow and outflow, the reconnection electric field is balanced by the ideal MHD term. In the IDR, the field is balanced predominantly by the Hall term, which is thought to be crucial for enabling fast reconnection (Birn et al. 2001; Liu et al. 2022). However, since the Hall force cannot do work on a plasma, i.e., $\vec{J} \cdot \left(\vec{J} \times \vec{B}\right) = 0$, additional terms are needed to mediate the rapid transfer of energy from the electromagnetic field to the particles. The primary function of the EDR is to provide the dissipative reconnection electric field that enables reconnection; however, the majority of the net work done on the plasma by the field during reconnection typically takes place outside the EDR. The importance of the EDR to reconnection is illustrated by the following points: without an EDR, there is no topological rearrangement of the magnetic field nor ion or fluid-scale exhausts. Reconnection may also occur without the full participation of ions; however, electron-only reconnection is beyond the scope of this review (see Stawarz et al. 2024, this volume).

1.3 Science and Data from NASA's Magnetospheric Multiscale Mission

As early as the 1960s, reconnection was used to explain reconfigurations of the magnetospheric field and magnetospheric and ionospheric convection patterns (Dungey 1961; Sonnerup and Cahill 1967; Hones 1979; Paschmann et al. 1979). Observational studies on Hall physics of magnetic reconnection started with the availability of full three-dimensional particle distribution functions with cadences sufficient to resolve ion-scale current sheets. Geotail made the first in-situ observations of the Hall current system of magnetic reconnection by identifying the ion-electron decoupling and resolving inflowing electrons toward the X-line and outflowing electron jets (Nagai et al. 1998, 2001). Additional single-spacecraft observations of the Hall electromagnetic fields were made with data from the Wind and Polar missions (Øieroset et al. 2001; Mozer et al. 2002). Cluster, which is a four-spacecraft mission, provided the first quantitative analyses of the reconnection current sheet by determining the 3-d gradients and thereby successfully confirmed the expected Hall-current system in the ion-diffusion region of X-line (Runov et al. 2003; Nakamura et al. 2006; Eastwood et al. 2010). However, gradients (determined from Cluster) and particle distribution functions (from Cluster and Geotail) were measured at the time and spatial scales of the ions. Hence, investigations of kinetic effects using data from these prior missions were limited predominantly to studies of ion kinetics. Furthermore, the electric field measurements were performed using a spin-plane double probe that did not provide the full 3-d vector, which is essential in determining the reconnection electric field.

The majority of the works reviewed in this paper use plasma and field measurements made by NASA's Magnetospheric Multiscale (MMS) mission, four satellites designed to study the kinetic-scale physics of diffusion regions in the boundary regions of Earth's magnetosphere (Burch et al. 2016a). The identical spacecraft each carry a complement of instruments for measuring the three-dimensional electromagnetic field and velocity-space distributions of ions and electrons. The unique capabilities of MMS have enabled a rapid development in our understanding of diffusion regions in the nearly ten years since its launch in March 2015. First, MMS particle distributions are measured at a much faster cadence than previous missions (e.g., ions and electron distributions are measured $\sim$30 and $\sim$130 times faster than ESA's Cluster mission). The high cadence of MMS's fast plasma instrument (FPI) was required to resolve the particle dynamics in small (several to hundreds km) diffusion regions that often move rapidly (tens to hundreds km/s) across the satellites (Pollock et al. 2016). Fluid moments derived from FPI velocity distribution functions permit

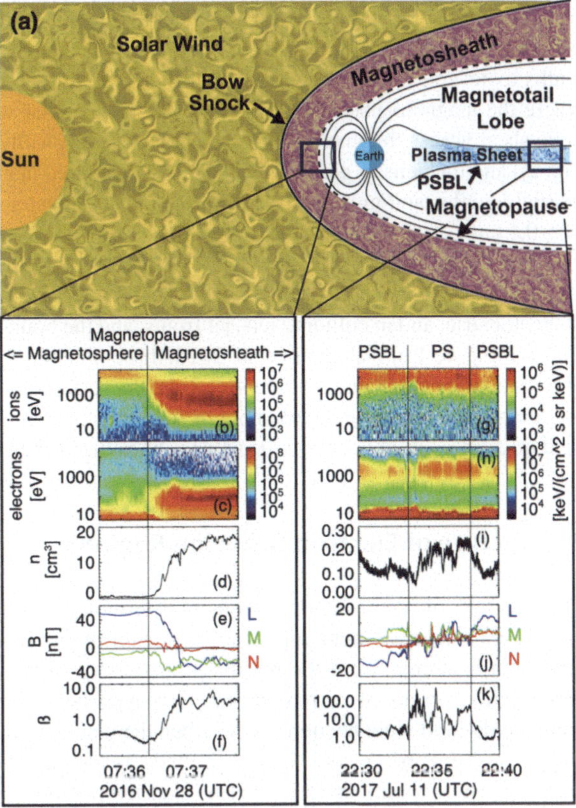

Fig. 2 (a) side-on view of the magnetosphere, adapted from Stawarz and Genestreti (2023), with example MMS data during reconnection events observed (b)-(f) at the dayside magnetopause and (g)-(k) in the magnetotail. The time intervals corresponding to the current sheets and their boundary layers are labeled above (b) and (g), where PS and PSBL are plasma sheet and plasma sheet boundary layer, respectively. (b)/(g): ion and (c)/(h): electron energy fluxes, (d)/(i): number densities, (e)/(j): magnetic field vector components B_L, B_M and B_N in blue, green and red, respectively, and (f)/(k): plasma beta

the accurate calculation of the current density as $\vec{J} = en\left(\vec{u}_i - \vec{u}_e\right)$ at each of the four spacecraft (Phan et al. 2016; Lavraud et al. 2016); together with the 3-d electric field (Ergun et al. 2016; Lindqvist et al. 2016), this allows for the calculation of the energy conversion rate $\vec{J} \cdot \vec{E}$ at each of the four spacecraft. In combination with the 3-d magnetic field vector (Torbert et al. 2016a), these data are used to evaluate a broad spectrum of local plasma parameters, including electron pressure non-gyrotropy, and associated phase space density structures (e.g., Burch et al. 2016b). The constellation has by-and-large been maintained in a tight, electron-scale tetrahedron. Comparative analysis and differencing of the multi-satellite data therefore allows for detailed investigations of the spatial structure of diffusion regions and the estimation of gradients of measured quantities (e.g., Paschmann and Daly 2000; Hasegawa et al. 2024, this collection). These aforementioned parameters, which are used to diagnose the electron-kinetic physics that enables reconnection, were not possible to calculate reliably before MMS.

The orbit of MMS was optimized to study reconnection at the dayside magnetopause and magnetotail plasma sheet (see Fig. 2a) (Fuselier et al. 2016). The magnetopause is the outer boundary of the magnetosphere that separates the dense magnetosheath (shocked solar wind) and sparse magnetospheric plasmas (Fig. 2b-d); sharp gradients in the density (Fig. 2d) and plasma beta (Fig. 2f) are common. The magnetic shear at the magnetopause varies both temporally, with the direction of the interplanetary magnetic field (IMF), and spatially, as the IMF drapes around the magnetospheric field (e.g., the shear angle from Fig. 2e is $\sim$130°). Typical densities, proton inertial lengths, and proton Alfvén speeds in the dayside magnetosheath are on the order of 10's of cm^{-3}, 10's km, 100's km/s, respectively; the magnetospheric plasma near the dayside magnetopause is typically less dense, with densities of tenths to several cm^{-3}. The magnetotail is the elongated nightside of the magnetosphere, where the magnetospheric field is stretched by the solar wind flow (Fig. 2g-k). The magnetic shear across the magnetotail current sheet is typically very large (close to 180°, as in Fig. 2j) and the density gradient is typically considered negligible. Typical densities, proton inertial lengths, and proton Alfvén speeds in the magnetotail plasma sheet are on the order of $\sim$0.1 cm^{-3}, $\sim$100 km, and $\sim$100 km/s, respectively.

Reconnection also occurs and is observed by MMS in the solar wind, in turbulent current sheets at the bow shock and magnetosheath, in twisted vortices that develop along the flank magnetopause, and in colliding jets, jet fronts, and the boundary layers of other reconnection sites (cf. Figure 1 of Nakamura et al. 2025, this collection; Hwang et al. 2023). In laboratory observations or simulations of reconnection, the structure of a current sheet is known or can be inferred for a wide area/volume. For in-situ measurements, like MMS', the plasma and field quantities are only determined along the paths of the satellites.

2 Structure of Electron Diffusion Regions

The structure of reconnection regions is determined to a large degree by the background plasma and magnetic field conditions upstream of the reconnecting current sheet. A current sheet is symmetric if the plasma properties on either side are identical. Anti-parallel reconnection occurs at current sheets with oppositely-directed upstream magnetic fields, while guide-field reconnection occurs when the magnetic shear is $<$180°, e.g., the upstream $B_M \neq 0$. Asymmetric reconnection is defined as having differing upstream magnetic fields and/or plasma conditions (e.g., density and temperature). While myriads of other inflow conditions are known or expected to impact the diffusion region structure, the scope of the following subsections is limited to the impact of the symmetry and magnetic shear. Readers are directed to Liu et al. (2025), Norgren et al. (2025), and Hwang et al. (2023) for additional information.

2.1 Basic Concepts, Theory and Modeling

2.1.1 Symmetric and Anti-Parallel Reconnection

Since the momenta and magnetic stresses of the two inflow regions are identical in symmetric inflow regions, plasma flows stagnate ($U_{eL} = U_{eN} = 0$) where the magnetic field vanishes ($B_L = B_N = 0$), i.e., the inflow stagnation (S) and X-type magnetic field lines coincide. By Ampere's law, a strong out-of-plane current J_M is required to balance the difference of the strong magnetic field gradient $\partial B_L/\partial N$ across the current sheet and weaker gradient $\partial B_N/\partial L$ along the current sheet (see Fig. 3). The current in the EDR is carried by electrons trapped in the weak magnetic field region around the X-line, which bounce around the magnetic field reversal in so-called Speiser orbits (Speiser 1965).

The EDR is defined as the region around an X-line where electron-kinetic forces balance the reconnection electric field force and violate the frozen-in flux condition (i.e., $\nabla \times (\vec{E} + \vec{u}_e \times \vec{B}) \neq 0$), enabling the change of magnetic topology. In the absence of strong out-of-plane gradients ∂/∂M and time variations, these bouncing electrons also provide the only viable mechanism for balancing the reconnection electric field E_M at the inflow stagnation

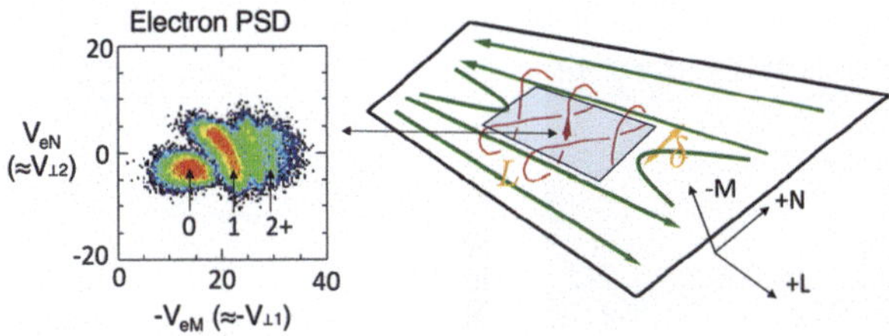

Fig. 3 (left) Simulated electron velocity space distribution at the X-line showing the characteristic striated pattern resulting from acceleration by the reconnection electric field. Distinct populations are labeled numerically by the number of orbits they have taken through the EDR. (right) Magnetic field structure (green) and bouncing orbits of trapped electrons (red) for a symmetric, anti-parallel EDR (shaded region); L and δ denote the half-width and half-height of the EDR, respectively. Adapted from Shuster et al. (2015) (left) and Hesse et al. (2011) (right)

line (which is coincident with the X-line in the symmetric case); all other terms on the right-hand side of Eq. (1) vanish except the gradient of the off-diagonal elements of the electron pressure tensor (e.g., Hesse et al. 2011). The half-thickness of the EDR δ can be defined as the average bounce-length of an electron trapped in the B_L reversal (see Fig. 3) (Schindler and Biskamp 1971, Hesse et al. 1999, 2011). Similarly, the half-length L is defined for the bounce-length in the B_N reversal. The aspect ratio δ/L is a close proxy for the normalized reconnection rate in the small aspect ratio limit, i.e., $\delta/L \lesssim 0.1$ (Liu et al. 2017; Heuer et al. 2022; Liu et al. 2025).

The work rate per unit volume of the non-ideal-MHD electric field ($\vec{J} \cdot (\vec{E} + \vec{u}_e \times \vec{B}) \equiv \vec{J} \cdot \vec{E}'$) is used to diagnose the kinetic physics driving energy conversion (Zenitani et al. 2011). Unsurprisingly the key physics that drives $\vec{J} \cdot \vec{E}' > 0$ in symmetric, anti-parallel EDRs are related to bouncing and trapped electrons. With each bounce, electrons get a 'kick' from the reconnection electric field E_M and increase in energy. The net result is a striated and highly non-gyrotropic distribution of electrons in velocity space (Ng et al. 2012; Shuster et al. 2015). The nth stripe of the straited distribution corresponds to the population of electrons that have bounced n times and received n 'kicks' in energy from the reconnection electric field. In the distribution function shown in Fig. 3, this energization can be seen by comparing the discrete bands in phase space of the inflowing (labeled "0"), 1-bounce, and $\geq$2-bounce populations. Eventually, electrons escape into the exhausts where they re-magnetize and thermalize (Ng et al. 2012; Shuster et al. 2015).

2.1.2 Guide-Field Reconnection

The introduction of a guide field changes two key features of the EDR, relative to the anti-parallel case: (1) the magnetic field at the X-line and (2) the Lorentz force term $u_{eL}B_M$ downstream of the X-line no longer vanish.

Resulting from (1), the electron current near the X-line becomes increasingly aligned with the magnetic field as the ratio of the guide and reconnecting magnetic field components (B_M/B_L) in the inflow regions increases. The guide field may substantially alter the structure of the EDR when the gyroradius about the guide field is smaller than the bounce length of thermal electrons in the B_L reversal (Hesse et al. 2011, 2016) such that bounce motions for electrons within the thermal energy range are prohibited; such strong guide fields can magnetize electrons and inhibit meandering bounce motions at the current sheet center, which provide the non-gyrotropic pressure forces required to violate the frozen-in flux

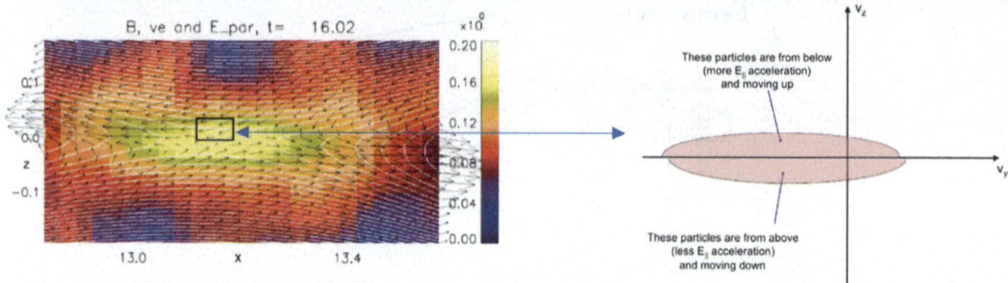

Fig. 4 (left) 2.5-dimensional PIC simulation of guide field reconnection showing the parallel electric field (color), magnetic field (white lines), and electron velocity field (black arrows). (right) A cartoon of a non-gyrotropic electron velocity distribution near an X-line of guide field reconnection. Adapted from Hesse et al. (2011). Here, X, Y and Z correspond to L, M and N, respectively

condition in the anti-parallel case. However, in the quasi-2d steady case, a non-gyrotropic electron pressure force is still required to violate the frozen-in flux condition regardless of the guide field (Hesse et al. 2011, 2016). Even for extremely large guide fields ($B_G \sim 100$, which is more than an order of magnitude larger than what MMS has thus far observed), non-gyrotropic pressures are predicted within a narrow gyro-scale layer of the EDR based on 2-d simulations (Liu et al. 2014). Non-gyrotropic current layers must become thinner with increasing guide field strength, such that the meandering bounce length remains smaller than the gyro-radius about the guide field. It is predicted that during guide field reconnection, the thickness of the non-gyrotropic layer scales with the electron gyro-radius, rather than the electron inertial length, as in the anti-parallel case (Cassak et al. 2017).

The non-gyrotropy may still be generated due to acceleration by the reconnection electric field, however the mechanism may differ from the anti-parallel case. As illustrated in Fig. 4, electron distributions above (or below) the current sheet are composed of a population that has crossed the current sheet and undergone more acceleration by E_M and a population that has just entered the current sheet and undergone less acceleration. If the quasi-2-d condition is relaxed, i.e., $\partial/\partial M \neq 0$, variations of the gyrotropic electron pressure in the current direction may violate the frozen-in condition (see Eq. 9 Hesse et al. 2011; Liu et al. 2013).

Resulting from (2), the electron jets, which carry the perpendicular portion of the Hall current loop, are diverted away from the center of the current sheet toward one of the separatrices (Goldman et al. 2011) as they propagate downstream from the EDR. The resulting deflection of the jet leads asymmetries in the jet structure (Pritchett and Mozer 2009) and heating (Eastwood et al. 2018) across the jet, and modifies the dissipation and thermalization process (Barbhuiya et al. 2022).

2.1.3 Asymmetric Reconnection

Unlike its symmetric counterpart, there can be a net current J_N across the X-line during asymmetric reconnection, even in the simplest (steady state and quasi-two-dimensional) configurations. During density-asymmetric reconnection, the inflow stagnation line (S-line) – where the momenta of the two inflowing populations balance and the normal component of the bulk velocity vanishes – is displaced toward the inflow region with lower density. During B_L-asymmetric reconnection, the X-line is displaced toward the inflow region with the weaker B_L. The separation between the X- and S-lines therefore depends on the asymmetries in B_L and n, as predicted by Cassak and Shay (2007) and illustrated in Fig. 5a. Additionally,

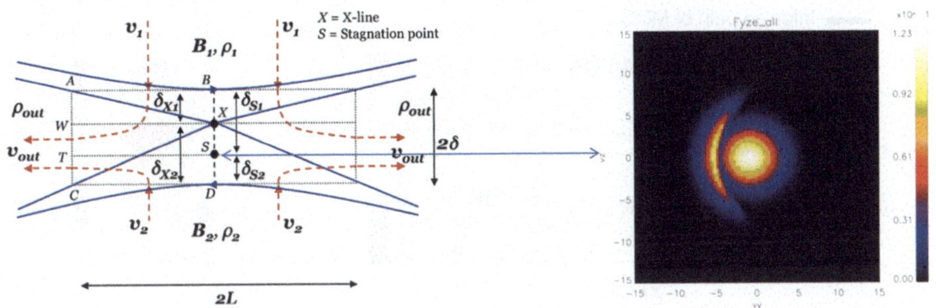

Fig. 5 (left) Schematic diagram of asymmetric reconnection, where blue lined represent the magnetic field, red dashed lines represent the plasma flow, and subscripts "1" and "2" represent parameters associated with the inflow regions and subscript "out" indicates outflow quantities. "S" and "X" mark the inflow stagnation- and X-lines, respectively. (Right) A simulated electron velocity distribution shown in the velocity plane perpendicular to the magnetic field near the S-line. (Adapted from Cassak and Shay 2007 and Hesse et al. 2011)

the stagnation lines of higher-mass ions (S_i) are displaced further from the center of the X-line than the electron stagnation line (S_e), leading to a strong normal-directed polarization electric field (Malakit et al. 2013).

Regardless of the asymmetry, the reconnection electric field at the S_e-line must be balanced by pressure forces associated with electron non-gyrotropy forces for steady and quasi-two-dimensional reconnection, as discussed in Sects. 2.1.1 and 2.1.2. However, at the X-line of asymmetric reconnection, the electron inertial (Hesse et al. 2014) and gyrotropic electron pressure forces (Shay et al. 2016) are non-vanishing when the X- and S_e-lines are not collocated.

The current density in asymmetric EDRs is maximized between the X- and S_e-lines (Cassak et al. 2017), where B_L is non-vanishing. The current is carried by electrons moving under a combination of (1) $E\text{x}B$-drift from the normal electric field E_N and reconnecting field B_L, (2) meandering motion, comparable to the symmetric case, and (3) diamagnetic drifts associated with differences in the pressures of the inflowing electrons. The meandering population, which is visible as the crescent-shaped feature in the velocity-space distribution function of Fig. 5, is responsible for the non-gyrotropic pressure (Hesse et al. 2014). (Note that similar crescent-shaped distributions are expected in non-reconnecting diamagnetic current sheets, but can be distinguished from those in an EDR by the former's lack of a non-gyrotropic pressure force (Egedal et al. 2016; Rager et al. 2018)). Work is done on the electrons by both the normal and reconnection electric fields, E_N and E_M respectively. For comparison, the work rate in symmetric EDRs is controlled by E_M and thus scales with the reconnection rate.

2.2 Observations

2.2.1 Symmetric and Anti-Parallel Reconnection

MMS observations of an anti-parallel and symmetric EDR is shown in Fig. 6c-d along with a fully-kinetic particle-in-cell (PIC) simulation of the event in Fig. 6a-b. Context for this event is provided in a number of papers (Torbert et al. 2018; Genestreti et al. 2018a; R. Nakamura et al. 2019). The estimated path of MMS through the current sheet is shown in black in Fig. 6a, while field and plasma quantities along that path are shown in Fig. 6b. The EDR is identified in Fig. 6a as the yellow-colored region at the center with a non-ideal

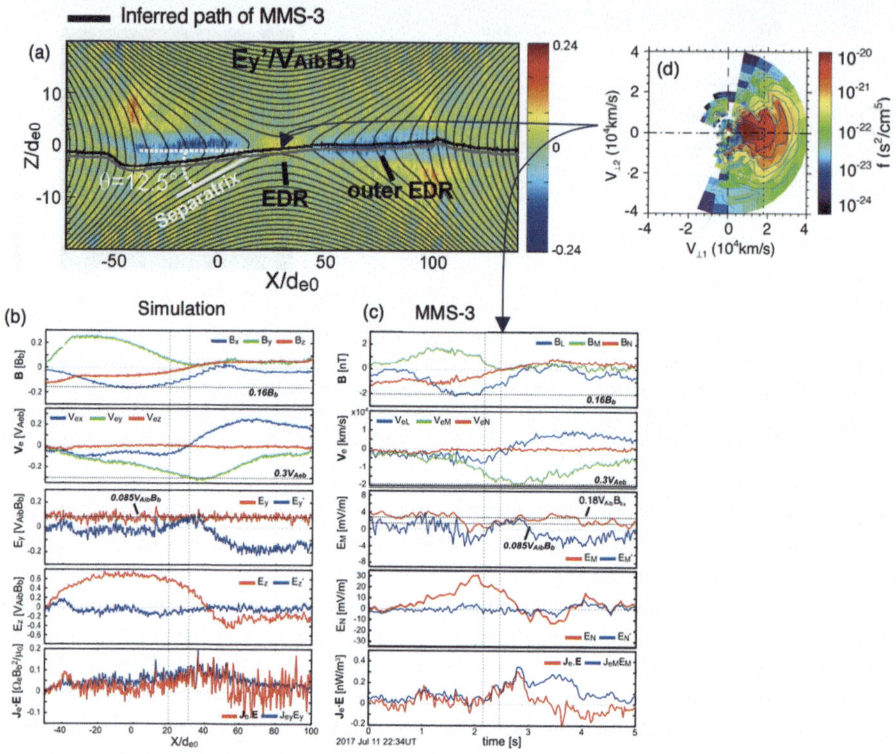

Fig. 6 (a) Simulation of a symmetric anti-parallel EDR observed by MMS on 11 July 2017 at 22:34 coordinated universal time (UTC). The thick black line is the inferred path of MMS, thin black lines are the magnetic field, and X, Y, and Z correspond to L, M, and N, respectively. (b) Cuts of the simulation along the path of MMS showing (top to bottom) the magnetic field vector, electron bulk velocity, out-of-plane electric field, normal electric field, and energy conversion rate. (c) MMS data, where the quantities are identical to (b). (d) Electron velocity space distribution measured by MMS at the closest approach to the X-line, where $V_{\perp 1}\approx$-M and $V_{\perp 2}\approx$N. Adapted from TKM Nakamura et al. (2018) and Torbert et al. (2018). ((a) Note that the outer EDR, described in Sect. 1.2, is not a focus of this review)

electric field $E'_y > 0$. MMS transited the EDR laterally through the current sheet from the $U_{eL} < 0$ to $U_{eL} > 0$ jets. The closest approach to the X-line is marked by vertical dashed lines in Fig. 6b-c and is identified by: (1) a reversal of B_N concurrent with (2) a reversal in the electron outflow U_{eL}, (3) small B_L, approximately 8% of the inflow magnetic field (not pictured), (4) a strong out-of-plane electron flow U_{eM} carried by (5) bouncing, trapped electrons and (6) a positive work rate of the non-ideal electric field $\vec{J} \cdot \vec{E}' > 0$ (cf. Torbert et al. 2018).

Figure 6b and 6c show the reconnection electric field (E_M) and non-ideal out-of-plane electric field (E_M') in the simulation and data, respectively. The reconnection electric field is fairly steady across the whole EDR with a normalized value (corresponding to the normalized reconnection rate) of 0.18 ± 0.035 (Nakamura et al. 2018; Genestreti et al. 2018a). At the closest approach to the X-line MMS observed $E'_M \approx E_M$, i.e., a reconnection electric field supported by non-ideal terms in generalized Ohm's law, in excellent agreement with the simulation. Immediately before and after the X-line, the spacecraft were upstream ($\Delta N < 0$) then downstream ($\Delta L > 0$) of the EDR. A cut of the electron velocity space distribution in the M-N plane near the X-line is shown in Fig. 6d. The core electron popu-

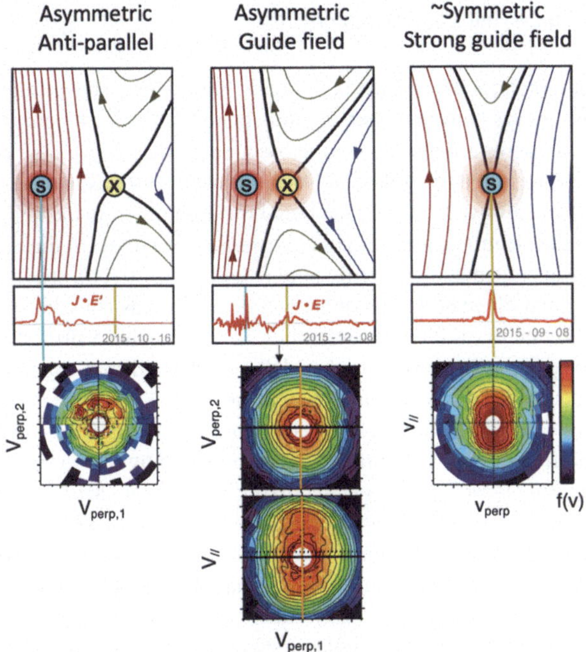

Fig. 7 Three example EDR events arranged from left to right in order of guide field strength. The top row shows cartoons of the field line geometry, the middle row shows MMS-observed work rates of the non-ideal electric field, and the bottom row shows the characteristic electron velocity distributions in the EDR current. For the anti-parallel case (left), current density and energy conversion peak near the electron stagnation line (blue "s"); the current there is carried by non-gyrotropic meandering electrons. For moderate guide fields (middle column), both perpendicular and parallel acceleration are observed between the X and Se-lines, with markedly weaker signatures of gyration visible compared to the anti-parallel case. Lastly, for strong guide fields (right column), parallel acceleration dominates almost completely. (Adapted from Genestreti et al. 2017)

lation appears in red while multiple stripes of higher phase space density, corresponding to electrons that have undergone one or multiple bounces through the EDR, appear at larger velocities V_{eM}, consistent with predictions (e.g., Fig. 3). The energy gain is a result of energization from the reconnection electric field, as evidenced by $J.E \approx J_M E_M$ (Fig. 6b-c). (See Liu et al. 2025 for details of the electron dynamics and the changes in the velocity distributions around the EDR of this event). The reconnection electric field can be independently estimated from the electron distributions alone and was determined to be ~4 mV/m ± 1-to-2 mV/m, corresponding to a normalized value of ~0.2 (Bessho et al. 2018).

Lastly, the half-thickness ($\delta \approx 1\ d_{e0}$) and half-length ($L \approx 4\ d_{e0}$) of the EDR in Fig. 6 were determined to be in good agreement with the estimated bounce length of electrons trapped in the $\partial B_L/\partial N$ and $\partial B_N/\partial L$ field reversals, respectively (Nakamura et al. 2019). The aspect ratio of the EDR was determined independently in a number of studies (Torbert et al. 2018; Nakamura et al. 2019; Heuer et al. 2022) and corresponded to a normalized reconnection rate of ~0.17 ± 0.01 (Heuer et al. 2022).

2.2.2 Impact of Guide Fields and Asymmetries

Various EDR configurations are presented in Fig. 7, highlighting the differences in the electron dynamics and energy conversion when guide fields and/or asymmetries are introduced. In general, the peak current density – and subsequently, the energy conversion – are displaced from the X-line when the inflow plasmas are asymmetric. Electrons flow across the X-line and undergo a half gyration, with net bulk flow pointing from the high-density to low-density inflow; as the electrons turn from the inflow direction to the M direction, they are accelerated by a strong perpendicular electric field (Burch et al. 2016b), specifically, by a combination of the normal-directed and reconnection electric fields (Bessho et al. 2016; Burch et al. 2018a,b; Cozzani et al. 2019; Pritchard et al. 2019). These strongly non-gyrotropic electrons carry the primary J_M current in the EDR and, as predicted, manifest as a crescent-shaped feature in cuts of the velocity distribution function in the plane perpendicular to the magnetic field (Burch et al. 2016b), as in the left-most example of Fig. 7. Also as predicted, the frozen-in condition in the EDR current sheet is violated by primarily by the divergence of the electron pressure tensor (Torbert et al. 2016b; Rager et al. 2018).

The introduction of a guide field, as shown in the middle example EDR event of Fig. 7, reduces the electron non-gyrotropy substantially (Genestreti et al. 2017, 2018b; Zhou et al. 2019; Eriksson et al. 2016). The event of Fig. 7 had a guide field approximately half as large as the upstream reconnecting magnetic field component. The current sheet thickness ($\sim$3 d_e), determined from the magnetic field gradient scale length, was $\sim$5-to-6 times larger than the thermal electron gyroradius about the guide field. Non-gyrotropic "crescent"-like features were not observed in the thermal energy range. Electrons with sufficiently large perpendicular energies (corresponding to gyroradii larger than the current sheet thickness) did exhibit these crescent-like features, which were associated with a weak non-gyrotropy, as the phase space density of this higher energy population was fairly small. The current was carried by strongly anisotropic, parallel-heating electrons, which stream along the guide field in the M direction. As the guide field increases, the current density and energy conversion rates peak closer to the X-line. Still, electrons with gyroradii larger than the current sheet thickness exhibited a non-gyrotropy, and the divergence of the non-gyrotropic electron pressure contributed substantially to the dissipative electric field (Genestreti et al. 2018b). The remainder was supported by the divergence of the gyrotropic pressure (with a small contribution from electron inertia), in particular the gradient of the pressure in the current (M) direction, consistent with some 3-d simulations (e.g., Liu et al. 2013).

MMS studies have thus far characterized few EDRs with strong guide fields. One such example is shown on the right of Fig. 7. Here, the electrons remained mostly magnetized, with the reconnection electric field being in the parallel direction (Eriksson et al. 2016). A relatively negligible work rate associated with the perpendicular electric field was observed, while intense parallel electric fields contributed predominantly to the energy conversion, as is typical for guide field reconnection (Wilder et al. 2018, 2022). While the MMS satellites were in an unfavorable position to quantify the forces balancing that of the electric field, no non-gyrotropy was observed, meaning that either (a) the non-gyrotropic layer was too narrow to be resolved by MMS or (b) the inertial and/or 3-d gyrotropic pressure gradient forces must have dominated. Such 3-d gradients, however, are incompatible with the commonly-assumed translationally-symmetric, quasi-2-d picture of reconnection. This event was observed in a current sheet embedded within strong turbulence. As such, it is not clear whether these signatures are characteristic of reconnection in the strong guide field limit (cf. Nakamura et al. 2025, this collection).

3 Summary and Outlook

We have reviewed theory, modeling, and MMS observational results to summarize the current understanding of how EDR dynamics are determined by the conditions of the background plasma and field, specifically focusing on asymmetries in the density and magnetic field and the presence of a guide magnetic field. The electron dynamics in the EDR are responsible for enabling reconnection. We focused in particular on how (a) energy conversion by the non-ideal electric field, (b) the electron-kinetic forces that balance that of the non-ideal electric field, and (c) the structure of the electron current sheet differs depending on the asymmetry and magnetic shear. These quantities were chosen as they are characteristically different at the dayside magnetopause and magnetotail, the two primary targets of MMS during its prime mission phase. The following general conclusions can be drawn for steady reconnection from works summarized herein:

1. In the symmetric EDR, energy conversion near the X-line is driven primarily by a steady reconnection electric field. In the asymmetric case, electrons are also accelerated across the X-line by a normal electric field.
2. For strongly asymmetric EDRs, the out-of-plane current and energy conversion are increasingly displaced from the X-line toward the high-B, low-n inflow. For increasingly symmetric EDRs and increasingly large guide fields, the current and energy conversion occur nearer the X-line.
3. In the anti-parallel magnetic field limit, the current density and energy conversion are supported by perpendicular fields and flows. For increasingly large guide fields, both become increasingly field aligned.
4. In all cases, the electron pressure divergence is predominantly responsible for balancing the force of the non-ideal electric field. The non gyrotropic portion of the pressure force dominates in the anti-parallel limit. Meandering is relegated to increasingly higher energies for increasingly large guide fields; subsequently, the role of the non-gyrotropic pressure force is reduced for increasingly large guide fields.

The scope of this review is limited to reconnection in the most well understood regimes. Still, many open questions remain (for a comprehensive list, see Nakamura et al. 2025, this collection). Variations in the structure and dynamics of the EDR with a number of other background parameters (e.g., plasma beta, flow shear, etc.) are yet to be fully explored, as detailed in following papers Liu et al. (2025), Norgren et al. (2025), Hwang et al. (2023). In the strong guide field limit, it is not currently understood whether non-gyrotropic pressure forces are commonly responsible for violating the frozen-in condition, as predicted by theories of quasi-2-d reconnection (see Fig. 4 and discussion thereof). Alternatively, 3-d variations in the parallel pressure and/or time-varying electron inertia (Liu et al. 2014) may provide a mechanism for violating the frozen-in condition (see Sect. 2.2.2), though such 3-d pressure gradients must be strongly localized in space and time-varying inertial forces are not indefinitely sustainable. Electron non-gyrotropy is commonly used as a criterion for identifying EDRs (e.g., Webster et al. 2018). Few high-guide-field EDRs have currently been identified. Additional work is required to determine whether different criteria are needed to characterize low-shear reconnection. Understanding the physics that enable reconnection in the strong guide field regime is important, e.g., for understanding the criteria necessary for reconnection onset.

MMS has thus far remained in a tight, electron-scale tetrahedron. Each crossing of an EDR thus provides a snapshot in time of the local electron-kinetic dynamics, which can be associated with the inflow conditions observed by MMS before or after it transits the

EDR. If the fluid-scale inflow is assumed steady and uniform, MMS also provides adequate coverage of the inflow. Major open questions remain:

1. How are reconnection dynamics coupled across intermediate scales, between the electron-inertial and fluid scales?
2. How does reconnection between unsteady / non-uniform plasmas differ from its steady counterpart?

In reality, merging plasmas and their fields are never truly uniform and steady. Instead, turbulent variations are observed across a continuous spectrum of spatial and temporal scales. The efficiency with which electrons and ions are able to respond to these variations differs based on their differing gyro-scales and gyro-periods. While some studies have reported temporal or spatial variability in EDRs (e.g., Genestreti et al. 2018c, 2020, 2022; Cozzani et al. 2019; Zhou et al. 2021; Fargette et al. 2024; Payne et al. 2025), it is unclear whether such variations result from varying inflow conditions, nor whether variability in the EDR impacts the macroscopic jetting, acceleration, or topological change brought about by reconnection. To address these questions, MMS is currently deviating from the electron-scale tetrahedron and operating in a multi-scale configuration, such that the fluid inflow, ion scale, and electron scale physics may be resolved simultaneously. Addressing these questions will represent a major paradigm shift in our understanding of reconnection.

Acknowledgements The authors are grateful for the support and contributions from the participants of the International Space Science Institute (ISSI) workshop, "Magnetic Reconnection: Explosive Energy Conversion in Space Plasmas". We graciously acknowledge the ISSI staff for accommodation and support. Additionally, authors are indebted to the large team of people that have contributed to the success of MMS. This work was supported in part by NASA's MMS FIELDS contract NNG04EB99C.

Funding Information Open access funding provided by Österreichische Akademie der Wissenschaften.

Declarations

Competing Interests The authors declare no competing interests. RM and JLP are Guest Editors of the collection, but were not involved in the peer review process of this article.

References

Barbhuiya MH, Cassak PA, Shay MA, Roytershteyn V, Swisdak M, Caspi A, et al (2022) Scaling of electron heating by magnetization during reconnection and applications to dipolarization fronts and super-hot solar flares. J Geophys Res Space Phys 127:e2022JA030610. https://doi.org/10.1029/2022JA030610

Bessho N, Chen L-J, Hesse M (2016) Electron distribution functions in the diffusion region of asymmetric magnetic reconnection. Geophys Res Lett 43:1828–1836. https://doi.org/10.1002/2016GL067886

Bessho N, Chen L-J, Wang S, Hesse M (2018) Effect of the reconnection electric field on electron distribution functions in the diffusion region of magnetotail reconnection. Geophys Res Lett 45:12,142–12,152. https://doi.org/10.1029/2018GL081216

Birn J, et al (2001) Geospace Environmental Modeling (GEM) magnetic reconnection challenge. J Geophys Res 106(A3):3715–3719. https://doi.org/10.1029/1999JA900449
Burch JL, Moore TE, Torbert RB, Giles BL (2016a) Magnetospheric multiscale overview and science objectives. Space Sci Rev 199(1–4):5–21. https://doi.org/10.1007/s11214-015-0164-9
Burch JL, Torbert RB, Phan TD, Chen L-J, Moore TE, Ergun RE, et al (2016b) Electron-scale measurements of magnetic reconnection in space. Science 352(6290):1176. https://doi.org/10.1126/science.aaf2939
Burch JL, Ergun RE, Cassak PA, Webster JM, Torbert RB, Giles BL, Dorelli JC, Rager AC, Hwang K-J, Phan TD, Genestreti KJ, Allen RC, Chen L-J, Wang S, Gershman D, Le Contel O, Russell CT, Strangeway RJ, Wilder FD, Graham DB, Hesse M, Drake JF, Swisdak M, Price LM, Shay MA, Lindqvist P-A, Pollock CJ, Denton RE, Newman DL (2018a) Localized Oscillatory Energy Conversion in Magnetopause Reconnection. Geophys Res Lett 45(3):1237–1245. https://doi.org/10.1002/2017GL076809
Burch JL, Webster JM, Genestreti KJ, Torbert RB, Giles BL, Fuselier SA, et al (2018b) Wave phenomena and beam-plasma interactions at the magnetopause reconnection region. J Geophys Res Space Phys 123:1118–1133. https://doi.org/10.1002/2017JA024789
Cassak PA, Shay MA (2007) Scaling of asymmetric magnetic reconnection: general theory and collisional simulations. Phys Plasmas 14(10):102114. https://doi.org/10.1063/1.2795630
Cassak PA, Genestreti KJ, Burch JL, Phan TD, Shay MA, Swisdak M, et al (2017) The effect of a guide field on local energy conversion during asymmetric magnetic reconnection: particle-in-cell simulations. J Geophys Res Space Phys 122:11,523–11,542. https://doi.org/10.1004/2017JA024555
Cozzani G, Retinò A, Califano F, Alexandrova A, Le Contel O, Khotyaintsev Y, et al (2019) In situ spacecraft observations of astructured electron diffusion region during magnetopause reconnection. Phys Rev E 99(4):043204. https://doi.org/10.1103/PhysRevE.99.043204
Drake JF, Antiochos SK, Bale SD, et al (2025) Magnetic reconnection in solar flares and the near-sun solar wind. Space Sci Rev https://doi.org/10.1007/s11214-025-01153-x
Dungey JW (1961) Interplanetary magnetic field and the auroral zones. Phys Rev Lett 6(2):47–48. https://doi.org/10.1103/PhysRevLett.6.47
Eastwood JP, Phan TD, Øieroset M, Shay MA (2010) Average properties of the magnetic reconnection ion diffusion region in the Earth's magnetotail: the 2001–2005 cluster observations and comparison with simulations. J Geophys Res 115(A8):A08215. https://doi.org/10.1029/2009JA014962
Eastwood JP, Mistry R, Phan TD, Schwartz SJ, Ergun RE, Drake JF, et al (2018) Guide field reconnection: exhaust structure and heating. Geophys Res Lett 45:4569–4577. https://doi.org/10.1029/2018GL077670
Egedal J, Le A, Daughton W, et al (2016) Spacecraft observations and analytic theory of crescent-shaped electron distributions in asymmetric magnetic reconnection. Phys Rev Lett 117:185101. https://doi.org/10.1103/PhysRevLett.117.185101
Ergun RE, Tucker S, Westfall J, Goodrich KA, Malaspina DM, Summers D, et al (2016) The axial double probe and fields signal processing for the MMS mission. Space Sci Rev 199(1–4):167–188. https://doi.org/10.1007/s11214-014-0115-x
Eriksson S, et al (2016) Magnetospheric multiscale observations of magnetic reconnection associated with Kelvin-Helmholtz waves. Geophys Res Lett 43:5606–5615. https://doi.org/10.1002/2016GL068783
Fargette N, Eastwood JP, Waters CL, Øieroset M, Phan TD, Newman DL, et al (2024) Statistical study of energytransport and conversion in electrondiffusion regions at Earth's dayside magnetopause. J Geophys Res Space Phys 129:e2024JA032897. https://doi.org/10.1029/2024JA032897
Fuselier SA, Lewis WS, Schiff C, et al (2016) Magnetospheric multiscale science mission profile and operations. Space Sci Rev 199:77–103. https://doi.org/10.1007/s11214-014-0087-x
Fuselier SA, Petrinec SM, Reiff PH, et al (2024) Global-scale processes and effects of magnetic reconnection on the geospace environment. Space Sci Rev 220(4):34. https://doi.org/10.1007/s11214-024-01067-0
Genestreti KJ, Burch JL, Cassak PA, Torbert RB, Varsani A, Ergun RE, et al (2017) The effect of a guide field on local energy conversion during asymmetric magnetic reconnection: MMS observations. J Geophys Res Space Phys 122:11,342–11,353. https://doi.org/10.1004/2017JA024247
Genestreti KJ, Cassak PA, Varsani A, Burch JL, Nakamura R, Wang S (2018c) Assessing the time dependence of reconnection with Poynting's theorem: MMS observations. Geophys Res Lett 45(7):2886–2892. https://doi.org/10.1002/2017GL076808
Genestreti KJ, Nakamura TKM, Nakamura R, Denton RE, Torbert RB, Burch JL, et al (2018a) How accurately can we measure the reconnection rate E_M for the MMS diffusion region event of 11 July 2017?. J Geophys Res Space Phys 123:9130–9149. https://doi.org/10.1029/2018JA025711
Genestreti KJ, Varsani A, Burch JL, Cassak PA, Torbert RB, Nakamura R, et al (2018b) MMS observation of asymmetric reconnection supported by 3-D electron pressure divergence. J Geophys Res Space Phys 123:1806–1821. https://doi.org/10.1002/2017JA025019
Genestreti KJ, Liu Y-H, Phan T-D, Denton RE, Torbert RB, Burch JL, et al (2020) Multiscale coupling during magnetopause reconnection: interface between the electron and ion diffusion regions. J Geophys Res Space Phys 125:e2020JA027985. https://doi.org/10.1029/2020JA027985

Genestreti KJ, Li X, Liu Y-H, Burch JL, Torbert RB, Fuselier SA, et al (2022) On the origin of "patchy" energy conversion in electron diffusion regions. Phys Plasmas 29(8):082107. https://doi.org/10.1063/5.0090275

Gershman DJ, Fuselier SA, Cohen IJ, et al (2024) Magnetic reconnection at planetary bodies and astrospheres. Space Sci Rev 220:7. https://doi.org/10.1007/s11214-023-01017-2

Goldman MV, Lapenta G, Newman DL, Markidis S, Che H (2011) Jet deflection by very weak guide fields during magnetic reconnection. Phys Rev Lett 107(13):135001. https://doi.org/10.1103/PhysRevLett.107.135001

Guo F, Liu YH, Zenitani S, et al (2024) Magnetic reconnection and associated particle acceleration in high-energy astrophysics. Space Sci Rev 220:43. https://doi.org/10.1007/s11214-024-01073-2

Hasegawa H, Argall MR, Aunai N, et al (2024) Advanced methods for analyzing in-situ observations of magnetic reconnection. Space Sci Rev 220:68. https://doi.org/10.1007/s11214-024-01095-w

Hesse M, Schindler K, Birn J, Kuznetsova M (1999) The diffusion region in collisionless magnetic reconnection. Phys Plasmas 6(5):1781–1795. https://doi.org/10.1063/1.873436

Hesse M, Neukirch T, Schindler K, et al (2011) The diffusion region in collisionless magnetic reconnection. Space Sci Rev 160:3–23. https://doi.org/10.1007/s11214-010-9740-1

Hesse M, Aunai N, Sibeck D, Birn J (2014) On the electron diffusion region in planar, asymmetric, systems. Geophys Res Lett 41:8673–8680. https://doi.org/10.1002/2014GL061586

Hesse M, Liu Y-H, Chen L-J, Bessho N, Kuznetsova M, Birn J, Burch JL (2016) On the electron diffusion region in asymmetric reconnection with a guide magnetic field. Geophys Res Lett 43:2359–2364. https://doi.org/10.1002/2016GL068373

Hesse M, Liu Y-H, Chen L-J, Bessho N, Wang S, Burch JL, Moretto T, Norgren C, Genestreti KJ, Phan TD, Tenfjord P (2018) The physical foundation of the reconnection electric field. Phys Plasmas 25(3):032901. https://doi.org/10.1063/1.5021461

Heuer SV, Genestreti KJ, Nakamura TKM, Torbert RB, Burch JL, Nakamura R (2022) Calculating the electron diffusion region aspect ratio with magnetic field gradients. Geophys Res Lett 49:e2022GL100652. https://doi.org/10.1029/2022GL100652

Hones EW (1979) Transient phenomena in the magnetotail and their relation to substorms. Space Sci Rev 23:393–410. https://doi.org/10.1007/BF00172247

Hwang KJ, Nakamura R, Eastwood JP, et al (2023) Cross-scale processes of magnetic reconnection. Space Sci Rev 219:71. https://doi.org/10.1007/s11214-023-01010-9

Koide S, Shibata K, Kudoh T (2006) Phys Rev D 74:044005

Kulsrud RM (2001) Magnetic reconnection: Sweet-Parker versus Petschek. Earth Planets Space 53:417–422. https://doi.org/10.1186/BF03353251

Lavraud B, et al (2016) Currents and associated electron scattering and bouncing near the diffusion region at Earth's magnetopause. Geophys Res Lett 43:3042–3050. https://doi.org/10.1002/2016GL068359

Lindqvist P-A, Olsson G, Torbert RB, King B, Granoff M, Rau D, et al (2016) The spin-plane double probe electric field instrument for MMS. Space Sci Rev 199:137–165. https://doi.org/10.1007/s11214-014-0116-9

Liu YH, Daughton W, Karimabadi H, et al (2013) Phys Rev Lett 110(26):265004. https://doi.org/10.1103/PhysRevLett.110.265004

Liu Y-H, Birn J, Daughton W, Hesse M, Schindler K (2014) Onset of reconnection in the near magnetotail: PIC simulations. J Geophys Res Space Phys 119:9773–9789. https://doi.org/10.1002/2014JA020492

Liu Y-H, Hesse M, Guo F, Daughton W, Li H, Cassak PA, Shay MA (2017) Why does steady-state magnetic reconnection have a maximum local rate of order 0.1? Phys Rev Lett 118(8):85101. https://doi.org/10.1103/PhysRevLett.118.085101

Liu YH, Cassak P, Li X, et al (2022) First-principles theory of the rate of magnetic reconnection in magnetospheric and solar plasmas. Commun Phys 5:97. https://doi.org/10.1038/s42005-022-00854-x

Liu YH, Hesse M, Genestreti KJ, et al (2025) Ohm's law, the reconnection rate, and energy conversion in collisionless magnetic reconnection. Space Sci Rev https://doi.org/10.1007/s11214-025-01142-0

Malakit K, Shay MA, Cassak PA, Ruffolo D (2013) New electric field in asymmetric magnetic reconnection. Phys Rev Lett 111(13):135001. https://doi.org/10.1103/PhysRevLett.111.135001

Mozer FS, Bale SD, Phan TD (2002) Evidence of diffusion regions at a subsolar magnetopause crossing. Phys Rev Lett 89:015002. https://doi.org/10.1103/PhysRevLett.89.015002

Nagai T, Fujimoto M, Saito Y, Machida S, Terasawa T, Nakamura R, Yamamoto T, Mukai T, Nishida A, Kokubun S (1998) Structure and dynamics of magnetic reconnection for substorm onsets with geotail observations. J Geophys Res 103(A3):4419–4440. https://doi.org/10.1029/97JA02190

Nagai T, Shinohara I, Fujimoto M, Hoshino M, Saito Y, Machida S, Mukai T (2001) Geotail observations of the Hall current system: evidence of magnetic reconnection in the magnetotail. J Geophys Res 106(A11):25929–25949. https://doi.org/10.1029/2001JA900038

Nakamura R, Baumjohann W, Asano Y, Runov A, Balogh A, Owen CJ, Fazakerley AN, Fujimoto M, Klecker B, Rème H (2006) Dynamics of thin current sheets associated with magnetotail reconnection. J Geophys Res 111:A11206. https://doi.org/10.1029/2006JA011706

Nakamura TKM, Genestreti KJ, Liu Y-H, Nakamura R, Teh W-L, Hasegawa H, et al (2018) Measurement of the magnetic reconnection rate in the Earth's magnetotail. J Geophys Res Space Phys 123:9150–9168. https://doi.org/10.1029/2018JA025713

Nakamura R, Genestreti KJ, Nakamura T, Baumjohann W, Varsani A, Nagai T, et al (2019) Structure of the current sheet in the 11 July 2017 electron diffusion region event. J Geophys Res Space Phys 124:1173–1186. https://doi.org/10.1029/2018JA026028

Nakamura R, Burch JL, Birn J, et al (2025) Outstanding questions and future research on magnetic reconnection. Space Sci Rev 221. https://doi.org/10.1007/s11214-025-01143-z

Ng J, Egedal J, Le A, Daughton W (2012) Phase space structure of the electron diffusion region in reconnection with weak guide fields. Phys Plasmas 19(11):112108. https://doi.org/10.1063/1.4766895

Norgren C, et al (2025) Plasma dynamics in reconnection diffusion regions. Space Sci Rev

Øieroset M, Phan T, Fujimoto M, et al (2001) In situ detection of collisionless reconnection in the Earth's magnetotail. Nature 412:414–417. https://doi.org/10.1038/35086520

Paschmann G, Daly PW (eds) (2000) Analysis methods for multi-spacecraft data. ISSI Scientific Reports, vol SR-001. ISSI, Bern. https://www.issibern.ch/wp-content/uploads/SR-001.pdf

Paschmann G, Sonnerup B, Papamastorakis I, et al (1979) Plasma acceleration at the Earth's magnetopause: evidence for reconnection. Nature 282:243–246. https://doi.org/10.1038/282243a0

Payne DS, Swisdak M, Eastwood JP, et al (2025) In-situ observations of the magnetothermodynamic evolution of electron-only reconnection. Commun Phys 8:36. https://doi.org/10.1038/s42005-025-01948-y

Phan TD, Drake JF, Shay MA, Mozer FS, Eastwood JP (2007) Evidence for an elongated (>60 ion skin depths) electron diffusion region during fast magnetic reconnection. Phys Rev Lett 99(25):255002-1–255002-4. https://doi.org/10.1103/PhysRevLett.99.255002

Phan TD, et al (2016) MMS observations of electron-scale filamentary currents in the reconnection exhaust and near the X line. Geophys Res Lett 43:6060–6069. https://doi.org/10.1002/2016GL069212

Pollock C, Moore T, Jacques A, Burch J, Gliese U, Saito Y, et al (2016) Fast plasma investigation for magnetospheric multiscale. Space Sci Rev 199:331–406. https://doi.org/10.1007/s11214-016-0245-4

Pritchard KR, Burch JL, Fuselier SA, Webster JM, Torbert RB, Argall MR, et al (2019) Energy conversion and electron acceleration in the magnetopause reconnection diffusion region. Geophys Res Lett 46:10,274–10,282. https://doi.org/10.1029/2019GL084636

Pritchett PL, Mozer FS (2009) The magnetic field reconnection site and dissipation region. Phys Plasmas 16(8):080702. https://doi.org/10.1063/1.3206947

Rager AC, Dorelli JC, Gershman DJ, Uritsky V, Avanov LA, Torbert RB, et al (2018) Electron crescent distributions as a manifestation of diamagnetic drift in an electron-scale current sheet: magnetospheric multiscale observations using new 7.5 ms fast plasma investigation moments. Geophys Res Lett 45:578–584. https://doi.org/10.1002/2017GL076260

Runov A, Nakamura R, Baumjohann W, Zhang TL, Volwerk M, Eichelberger H-U, Balogh A (2003) Cluster observation of a bifurcated current sheet. Geophys Res Lett 30:1036. https://doi.org/10.1029/2002GL016136

Schindler K, Biskamp D (1971) Instability of two-dimensional collisionless plasmas with neutral points. Plasma Phys 13:1013

Shay MA, Phan TD, Haggerty CC, Fujimoto M, Drake JF, Malakit K, et al (2016) Kinetic signatures of the region surrounding the X line in asymmetric (magnetopause) reconnection. Geophys Res Lett 43:4145–4154. https://doi.org/10.1002/2016GL069034

Shuster JR, Chen LJ, Hesse M, et al (2015) Spatiotemporal evolution of electron characteristics in the electron diffusion region of magnetic reconnection: implications for acceleration and heating. Geophys Res Lett 42:2586–2593. https://doi.org/10.1002/2015GL063601

Sonnerup BUÖ (1979) Magnetic field reconnection. In: Lanzerotti LT, Kennel CF, Parker EN (eds) Solar System plasma physics, vol 3: Solar system plasma processes. North-Holland, New York, pp 45–108

Sonnerup BU, Cahill LJ Jr (1967) Magnetopause structure and attitude from Explorer 12 observations. J Geophys Res 72(1):171–183. https://doi.org/10.1029/JZ072i001p00171

Speiser TW (1965) Particle trajectories in model current sheets: 1. Analytical solutions. J Geophys Res 70(17):4219–4226. https://doi.org/10.1029/JZ070i017p04219

Stawarz JE, Genestreti KJ (2023) Preface to special topic: plasma physics from the magnetospheric multiscale mission. Phys Plasmas 30(4):040401. https://doi.org/10.1063/5.0148163

Stawarz JE, Muñoz PA, Bessho N, et al (2024) The interplay between collisionless magnetic reconnection and turbulence. Space Sci Rev 220:90. https://doi.org/10.1007/s11214-024-01124-8

Torbert RB, et al (2016b) Estimates of terms in Ohm's law during an encounter with an electron diffusion region. Geophys Res Lett 43:5918–5925. https://doi.org/10.1002/2016GL069553

Torbert RB, Russell CT, Magnes W, et al (2016a) The FIELDS instrument suite on MMS: scientific objectives, measurements, and data products. Space Sci Rev 199:105–135. https://doi.org/10.1007/s11214-014-0109-8

Torbert RB, Burch JL, Phan TD, Hesse M, Argall M, Shuster RJ, et al (2018) Electron-scale dynamics of the diffusion region during symmetric magnetic reconnection in space. Science. https://doi.org/10.1126/science.aat2998

Uzdensky DA, Cerutti B, Begelman MC (2011) Reconnection-powered linear accelerator and gamma-ray flares in the Crab Nebula. Astrophys J Lett 737(2):L40

Webster JM, Burch JL, Reiff PH, Daou AG, Genestreti KJ, Graham DB, et al (2018) Magnetospheric multiscale dayside reconnection electron diffusion region events. J Geophys Res Space Phys 123:4858–4878. https://doi.org/10.1029/2018JA025245

Wilder FD, Ergun RE, Burch JL, Ahmadi N, Eriksson S, Phan TD, et al (2018) The role of the parallel electric field in electron-scale dissipation at reconnecting currents in the magnetosheath. J Geophys Res Space Phys 123:6533–6547. https://doi.org/10.1029/2018JA025529

Wilder FD, Conley M, Ergun RE, Newman DL, Chasapis A, Ahmadi N, et al (2022) Magnetospheric multiscale observations of waves and parallel electric fields in reconnecting current sheets in the turbulent magnetosheath. J Geophys Res Space Phys 127:e2022JA030511. https://doi.org/10.1029/2022JA030511

Zenitani S, Hesse M, Klimas A, Kuznetsova M (2011) New measure of the dissipation region in collisionless magnetic reconnection. Phys Rev Lett 106(19):195003. https://doi.org/10.1103/PhysRevLett.106.195003

Zhou M, Deng XH, Zhong ZH, Pang Y, Tang RX, El-Alaoui M, et al (2019) Observations of an electron diffusion region in symmetric reconnection with weak guide field. Astrophys J 870(1):34. https://doi.org/10.3847/1538-4357/aaf16f

Zhou M, Man H, Yang Y, Zhong Z, Deng X (2021) Measurements of energy dissipation in the electron diffusion region. Geophys Res Lett 48:e2021GL096372. https://doi.org/10.1029/2021GL096372

Space Science Reviews (2025) 221:16
https://doi.org/10.1007/s11214-025-01142-0

Ohm's Law, the Reconnection Rate, and Energy Conversion in Collisionless Magnetic Reconnection

Yi-Hsin Liu[1] · Michael Hesse[2] · Kevin Genestreti[3] · Rumi Nakamura[4,5] · James L. Burch[6] · Paul A. Cassak[7] · Naoki Bessho[8,9] · Jonathan P. Eastwood[10] · Tai Phan[11] · Marc Swisdak[12] · Sergio Toledo-Redondo[13] · Masahiro Hoshino[14] · Cecilia Norgren[15,16] · Hantao Ji[17] · Takuma K.M. Nakamura[4,18]

Received: 4 June 2024 / Accepted: 15 January 2025 / Published online: 10 February 2025

Abstract

Magnetic reconnection is a ubiquitous plasma process that transforms magnetic energy into particle energy during eruptive events throughout the universe. Reconnection not only converts energy during solar flares and geomagnetic substorms that drive space weather near Earth, but it may also play critical roles in the high energy emissions from the magnetospheres of neutron stars and black holes. In this review article, we focus on collisionless plasmas that are most relevant to reconnection in many space and astrophysical plasmas. Guided by first-principles kinetic simulations and spaceborne in-situ observations, we highlight the most recent progress in understanding this fundamental plasma process. We start by discussing the non-ideal electric field in the generalized Ohm's law that breaks the frozen-in flux condition in ideal magnetohydrodynamics and allows magnetic reconnection to occur. We point out that this same reconnection electric field also plays an important role in sustaining the current and pressure in the current sheet and then discuss the determination of its magnitude (i.e., the reconnection rate), based on force balance and energy conservation. This approach to determining the reconnection rate is applied to kinetic current sheets with a wide variety of magnetic geometries, parameters, and background conditions. We also briefly review the key diagnostics and modeling of energy conversion around the reconnection diffusion region, seeking insights from recently developed theories. Finally, future prospects and open questions are discussed.

1 Introduction

Magnetic reconnection is a ubiquitous process that converts magnetic energy into plasma thermal and kinetic energy in laboratory, space, and astrophysical plasmas (Zweibel and Yamada 2009; Yamada et al. 2010). This efficient energy conversion process involves the effective "breaking" and "rejoining" of magnetic field lines (although note that reconnection does not violate Gauss' law, $\nabla \cdot \mathbf{B} = 0$). By altering their connectivity within the so-called "diffusion region" in the microscopic scale, the gray area in Fig. 1, this process imparts energy into outflow plasma jets, the purple arrows. While this picture captures the local process, the resulting change in the magnetic connectivity has far-reaching consequences as

Extended author information available on the last page of the article

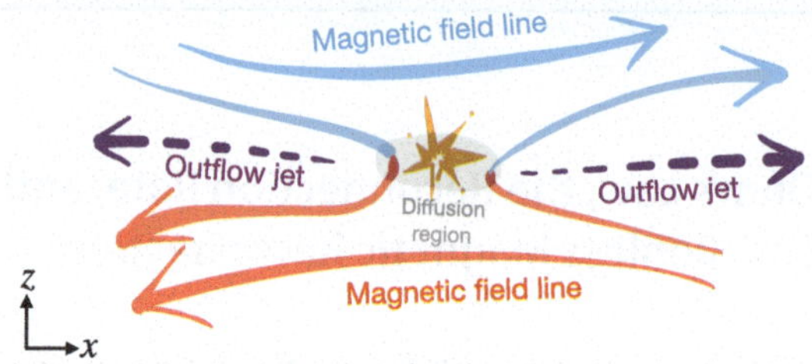

Fig. 1 *Artist's rendition of magnetic reconnection*. The breaking and rejoining of magnetic field lines (red and blue) in the diffusion region (gray) drive plasma outflow jets (purple arrows)

it can lead to energy release at large scales in the surrounding plasma systems, causing solar flares (Carmichael 1964; Sturrock 1966; Hirayama 1974; Kopp and Pneuman 1976; Priest and Forbes 2000), planetary geomagnetic substorms (Dungey 1961), and superflares from other astrophysical objects, for example the Crab nebula (Tavani et al. 2011; Abdo et al. 2011; Cerutti et al. 2014).

In a nutshell, magnetic reconnection is a nonlinear, dynamical process that involves electromagnetism, magnetic field geometry and topology, and complex charged particle motions in a multi-dimensional, multiscale system, where physics occurring at a singular point can lead to tremendous energy release at the macroscale. For these reasons, the study of magnetic reconnection has been a fascinating and challenging subject since it was first formulated in 1953 (Dungey 1953). Its study will continue to thrive with our increasing capability to observe electromagnetic phenomena in the universe (e.g., Bale et al. 2023; Burch and Torbert 2016; Raouafi et al. 2023a; Müller et al. 2020). The development of reconnection theories is guided and constrained by a wealth of data from numerical simulations, in-situ and remote space observations, and laboratory experiments. We do not intend to exhaustively include the many great efforts performed in various communities over the past 70 years in this review paper. Here instead we focus on the progress in the past 20 years on collisionless reconnection, where our understanding has been accelerated by kinetic simulations and in-situ spacecraft observations of NASA's ongoing Magnetospheric Multiscale (MMS) mission (Burch and Torbert 2016), THEMIS/ARTEMIS (Angelopoulos 2008; Sweetser et al. 2011), and Cluster (Escoubet et al. 2001). More exciting results are expected from the Parker Solar Probe (Raouafi et al. 2023a) and Solar Orbiter (Müller et al. 2020) missions, but are not discussed here. It is worth noting that Earth's magnetosphere and the solar wind are the most ideal testing grounds for reconnection physics reachable by human probes with current space technology. Because the size of a single spacecraft is relatively small compared to the electron kinetic scale, and now the cadence of measurement well resolves the dynamic time scale of reconnection therein; see Genestreti et al. (2025, this collection) for the review on current sheets in geospace. A companion review of collisionless reconnection research in the laboratory over the past 20 years, in comparisons with kinetic simulations and space observations, is given by Ji et al. (2023, this collection).

The fundamental questions of reconnection discussed in this review are: (1) what breaks the ideal-magnetohydrodynamic frozen-in flux condition, enabling reconnection to occur on a microscopic/kinetic scale? And what roles does the non-ideal electric field play (Sect. 2)? (2) what determines the rate at which reconnection processes the incoming magnetic flux (Sect. 3)? and (3) how plasmas are energized around the reconnection diffusion region (Sect. 4)? Each topic can be read independently, and we point out connections between different sections. This article serves as a review but also, hopefully, a tutorial for graduate students and early career scientists.

This review focuses on the fluid-type descriptions of reconnection physics within and around the diffusion region but is based on fully kinetic simulations and in-situ space measurements. It is not our intention to discuss all the details of each topic, but to integrate them into a bigger picture. Nevertheless, references that contain the full treatment are provided to interested readers. The discussion of the rich kinetic features and particle distribution functions is beyond the scope of this paper but can be found in Norgren et al. (2025, this collection). For discussions of a broader scope or emphasis on other areas of study, a variety of other papers complement this review (Vasyliunas 1975; Priest and Forbes 2000; Birn and Priest 2007; Zweibel and Yamada 2009; Mozer and Pritchett 2010; Yamada et al. 2010; Gonzalez and Parker 2016; Burch and Torbert 2016; Lee and Lee 2020; Ji et al. 2022; Pontin and Priest 2022; Yamada 2022).

2 Breaking the Frozen-in Flux Condition

Alfvén's frozen-in flux theorem (Alfvén 1942) shows that perfectly conducting fluids, such as those in ideal magnetohydrodynamics (MHD), and embedded magnetic fields are constrained to move together. Mathematically, this occurs when $\mathbf{E} + \mathbf{V} \times \mathbf{B}/c$ vanishes (e.g., Stern 1966; Newcomb 1958).[1] In a hypothetical plasma for which the frozen-in flux theorem is satisfied, the total magnetic flux going through any closed Ampèrian loop in the plasma does not change in time. Note that the magnetic field self-consistently evolves with the moving plasma, which can generate currents that modify the magnetic fields. If the frozen-in condition works everywhere within the system of interest, the connectivity of magnetic field lines within this system cannot change because doing so would change the flux through a closed loop somewhere within the system.

The field line connectivity, nevertheless, can change when some dissipation breaks the frozen-in condition. For instance, the condition breaks down within the diffusion region (DR) in Fig. 1 that is sandwiched by magnetic field lines that point in opposite directions. Within this diffusion region, the inflowing magnetic field lines are "rewired" to form highly curved (blue-red) field lines, which are again frozen to the plasma outside the diffusion region and act like a slingshot, shooting plasma out as outflow jets. Once the plasma is jetted out, the plasma pressure within the diffusion region drops, and plasma flows in from the top and bottom along with the magnetic field for further reconnection. It is thus a self-driven (i.e., spontaneous) non-linear process; once it starts, it does not want to stop as long as more magnetic field is available in the inflow region. In addition, because of Ampère's law, the anti-parallel fields sandwich a current sheet where the DR resides. The singular point inside the DR where field lines reconnect is referred to as the "X-point" because the adjacent reconnected field lines form an X-shape. In 3D, the collection of these X-points extends in the out-of-plane direction to form an "X-line".

2.1 The Generalized Ohm's Law

Magnetic reconnection is the process that changes the field line connectivity in plasmas, and it requires the existence of a reconnection electric field to break the frozen-in flux condition, either at a topological boundary (Vasyliunas 1975), or, more generally, in a localized region parallel to the magnetic field (Hesse and Schindler 1988; Hesse et al. 2005). This requirement is a simple consequence of Maxwell's equations and the need to transport magnetic flux from the inflow to the outflow regions. While it has long been shown that reconnection cannot proceed without the presence of such a reconnection electric field, we only recently are understanding its full physical foundations.

[1] More precisely, any flow with a velocity $\mathbf{V}$ that satisfies $\nabla \times (\mathbf{E} + \mathbf{V} \times \mathbf{B}/c) = 0$ can be regarded as the magnetic flux preserving flow (Vasyliunas 1972).

Beginning with Vasyliunas (1975), it was recognized that the reconnection electric field has to be balanced by one or more terms in the generalized Ohm's law (Vasyliunas 1975; Cai and Lee 1997; Hesse et al. 2011). Writing the electron momentum equation in collisionless plasmas and solving for **E** gives,

$$\mathbf{E}+\frac{\mathbf{V}_e\times\mathbf{B}}{c}=-\frac{\nabla\cdot\mathbf{P}_e}{ne}-\frac{m_e}{e}(\mathbf{V}_e\cdot\nabla)\mathbf{V}_e-\frac{m_e}{e}\frac{\partial}{\partial t}\mathbf{V}_e, \tag{1}$$

where variables $\mathbf{E}$, $\mathbf{B}$, $\mathbf{V}_e$, $\mathbf{P}_e$, n, e, m_e and c are electric field, magnetic field, electron velocity, electron pressure tensor, number density, proton charge, electron mass and the speed of light, respectively, and Gaussian units are used.

Since m_e is small, the last two terms are only appreciable if the electron speed V_e is much larger than the ion speed V_i, so in those terms we can replace $\mathbf{V}_e \simeq -\mathbf{J}/ne$, where $\mathbf{J}$ is the current density. Then, we obtain the generalized Ohm's Law close to that discussed in Vasyliunas (1975),

$$\mathbf{E}+\frac{\mathbf{V}_i\times\mathbf{B}}{c}=\frac{\mathbf{J}\times\mathbf{B}}{nec}-\frac{\nabla\cdot\mathbf{P}_e}{ne}-\frac{m_e}{e^2}\left(\frac{\mathbf{J}}{ne}\right)\cdot\nabla\left(\frac{\mathbf{J}}{n}\right)+\frac{m_e}{e^2}\frac{\partial}{\partial t}\left(\frac{\mathbf{J}}{n}\right). \tag{2}$$

The left-hand side (LHS) of Eq. (2) measures the ion frozen-in condition, which is violated when its value is non-zero. Terms on the right-hand side (RHS) contribute to this violation. In collisionless plasmas, it includes, from left to right, the Hall electric field $[(\mathbf{J}\times\mathbf{B})/nec]$, the divergence of electron pressure term, the spatial derivative of the electron inertia term, and the temporal derivative of the electron inertia term. With collisions, one also needs to include the resistive electric field $\eta\mathbf{J}$, but it is omitted from our treatment.

Here we consider a symmetric, anti-parallel low-β reconnection in a Particle-in-Cell (PIC) simulation. Figure 2 shows the out-of-plane (y) component of the terms in the generalized Ohm's law (Eq. (2)) in a cut through the X-line in the inflow (z) direction. Upstream of the ion diffusion region (IDR) at $|z|>d_i$ (the ion inertial scale $d_i\equiv c/\omega_{pi}$, where $\omega_{pi}=\sqrt{m_i/(4\pi n_i e^2)}$ is the ion plasma frequency), ion convection brings magnetic field in, inducing the motional electric field (in gray). The Hall electric field (in purple) becomes the dominant term supporting the reconnection electric field E_y (in red) between the d_i and the electron inertial scale ($d_e\equiv c/\omega_{pe}$, where $\omega_{pe}=\sqrt{m_e/(4\pi n_e e^2)}$ is the electron plasma frequency). The Hall term arises because of the decoupling of the relatively immobile ions from

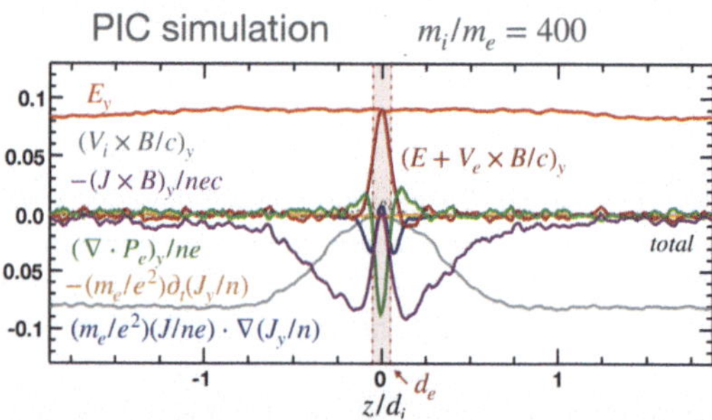

Fig. 2 *The generalized Ohm's law* in symmetric, antiparallel low-β reconnection. The out-of-plane component of terms in the generalized Ohm's law (normalized by $B_{x0}V_{A0}/c$) across the X-line in the inflow (z) direction, based on a particle-in-cell simulation of reconnection. The vertical red transparent band marks the electron diffusion region (EDR), while the ion diffusion region (IDR) expands between $z\in[-d_i,d_i]$. Adapted from Liu et al. (2022), reproduced by permission of Springer Nature

the motion of electrons that remain frozen to the magnetic field (Sonnerup 1979), which becomes significant beneath the ion inertial (d_i)-scale. The divergence of the electron pressure tensor is important within the electron gyro-scale because the off-diagonal component of a species' pressure tensor becomes pronounced only when the gradient scale of the magnetic field is small or comparable to particles' thermal gyro-radius ($\rho_e = m_e c v_{the}/eB$) or bounce lengths (Hesse et al. 2011). The spatial derivative of electron inertia is important within the electron inertial scale. The maximum of the gyro-scale and d_e determines the scale of the electron diffusion region (EDR). The time-derivative electron inertial term is negligible in the steady-state shown here, but it is significant in the initiation stage of reconnection, or in the presence of very fast fluctuations with time scales on the order of the electron plasma period (Vasyliunas 1975).

By inspection of Eq. (2), we see that the Hall term vanishes at the X-line, and so does the spatial-derivative inertia term in the symmetric case where the flow stagnation point ($\mathbf{V}_{e,xz} = 0$) coincides with the X-line. In addition, $\partial/\partial t = 0$ in the steady state. These leave us with the divergence of the electron pressure tensor, $(\nabla \cdot \mathbf{P}_e)_y = \partial_x P_{exy} + \partial_z P_{ezy}$, which, at a quasi-2D reconnection X-line (i.e., $\partial/\partial y = 0$), needs to have off-diagonal pressure components in order to balance the reconnection electric field. These off-diagonal terms around the X-line arise from the non-gyrotropic feature of the electron distributions. Hence, it has been proposed that the electron pressure tensor term should be the main contributor to the reconnection electric field at the reconnection site, at least in 2D symmetric situations (Vasyliunas 1975; Dungey 1988; Lyons and Pridmore-Brown 1990; Cai and Lee 1997; Hesse et al. 1999). The physical origin of the existence of a non-gyrotropic pressure tensor can be traced back to the free acceleration of electrons by the reconnection electric field but only within the unmagnetized EDR (Kulsrud et al. 2005; Hesse et al. 2011).

In an asymmetric configuration (discussed further in Sect. 3.2), the situation is slightly different in that the inertial term in Eq. (2) does not necessarily vanish at the X point. Instead, it is possible that the inertial term contributes part of, or even the majority of the reconnection electric field at this location (Hesse et al. 2014). However, we see from Eq. (2) that non-gyrotropic pressure tensor effects still need to exist at the flow stagnation point (Hesse et al. 2014), which is typically shifted toward the inflow region with a stronger magnetic field (Cassak and Shay 2007, 2008). A simple analysis shows that non-gyrotropic pressure effects are not only expected at the flow stagnation point, but are essential for consistent magnetic flux transport (Hesse et al. 2014). Recent research has further indicated that the reconnection electric field is a consequence of the need to maintain the current density in the electron diffusion region, which would otherwise be reduced by non-gyrotropic electron pressure effects (Hesse et al. 2018). These authors also showed that the thermal interaction of accelerated particles with the adjacent magnetic field, which gives rise to non-gyrotropic pressures and quasi-viscous current reductions, simultaneously leads to electron heating. This electron heating appears to be the key contributor to maintaining pressure balance in the electron diffusion region (see Sect. 2.3 for more discussion).

2.2 Observational Analysis of the Generalized Ohm's Law

Determining which non-ideal terms are responsible for violating the frozen-in flux condition in EDRs was one of the major objectives of NASA's Magnetospheric Multiscale (MMS) mission (Burch et al. 2016a). Note that in the decades preceding MMS observations from many previous satellite missions had confirmed the predominance of the Hall term in the IDR (Nagai et al. 2001; Øieroset et al. 2001; Mozer et al. 2002; Eastwood et al. 2010). The four identical MMS spacecraft are each capable of measuring the three-dimensional electromagnetic field vector (Torbert et al. 2016b) and electron and ion velocity space distribution functions (Pollock et al. 2016) at very high time resolution. The spacecraft orbits are typically maintained such that the fleet flies in a tightly-spaced tetrahedral formation with inter-spacecraft separations that can be on the order of the electron inertial length (Fuselier et al. 2016). During crossings through an EDR, differences in the electron and ion fluid moments are obtained between spacecraft pairs, such that, for the first time, the gradient terms in Eq. (2) can be approximated (Chanteur 1998). For more information on the methods, readers are directed to Hasegawa et al. (2024, this collection) and Paschmann and Daly (1998).

MMS has confirmed that the divergence of the electron pressure tensor dominates other non-ideal terms in EDRs near reconnection X-lines. This finding is consistent with the fact that most reconnection events observed by MMS have small or negligible electron flows in the reconnection plane at the X-line when measured in the co-moving frame of the X-line. Egedal et al. (2019) analyzed a symmetric and nearly-anti-parallel EDR observed by MMS on 11 July 2017, evaluated the electron pressure gradient using MMS data, and compared it with a 2D PIC simulation that used initial conditions based on the observations. Egedal et al. (2019) found that the non-gyrotropic pressure components $\partial_x P_{exy} + \partial_z P_{ezy}$ were predominantly responsible for balancing the reconnection electric field E_y, especially the latter term. Figure 3(a) and (b) show the numerical profile of $(\mathbf{E} + \mathbf{V}_e \times \mathbf{B}/c)_M$ and the pressure gradient $\partial P_{eMN}/\partial N$ with the projected trajectory of the spacecraft, determined by matching the observed magnetic field data to the simulated profile of the current sheet (see more detail in Egedal et al. 2019). Note that the LMN coordinate system is often used for reconnection observations once the quasi-2D reconnection plane is determined. It corresponds to the XYZ coordinate system shown in Fig. 1, used in most theoretical discussions of this review. The inner EDR is marked with a blue color in Fig. 3(c). The profile of $\partial P_{eMN}/\partial N$ is in excellent agreement with both theory and the simulation, being the main contribution to the non-ideal electric field within the inner electron diffusion region (EDR) where the electron frozen-in condition is broken.

An alternative method to calculate the non-gyrotropic pressure term at the inner EDR is to use the theory by Hesse et al. (1999, 2011), where the spatial scale of the electron diffusion is given by the electron orbit excursion in a field reversal, the so-called "bounce widths", λ_L and λ_N in the L and N directions respectively, and expressed as $\lambda_{L,N} = [2m_e T_e/(e^2(\partial_{L,N} B_{N,L})^2)]^{1/4}$. The electric field in the electron diffusion region, i.e., the non-gyrotropic pressure term, can then be expressed as $E_{\mathrm{M,model}} \simeq (1/e)(\partial_L V_{eL})(2m_e T_e)^{1/2}$. This result was confirmed by Nakamura et al. (2019) based on MMS analysis of the same event by determining the spacecraft orbit relative to an X-line as shown in Fig. 3(e). As expected for the inner EDR, the observed E_{M} and $E_{\mathrm{M}'} = (\mathbf{E} + \mathbf{V}_e \times \mathbf{B}/c)_M$ in Fig. 3(f) coincide when the spacecraft was inside the inner diffusion region (bounded by the dashed lines in Fig. 3(e)). The dash-dotted horizontal line is the $E_{\mathrm{M,model}}$ calculated from the velocity gradients obtained from the upper panel of Fig. 3(d). The model shows a good agreement with the observed electric field in the inner EDR, indicating that the theoretical concept of the inner EDR for laminar reconnection presented by Hesse et al. (1999, 2011) is well recovered for this event. This means that the divergence of the non-gyrotopic pressure term obtained with this model is also consistent with the reconnection electric field. The same scheme used by Hesse et al. (1999, 2011) to determine the off-diagonal pressure gradient has been also applied to an EDR event during more turbulent magnetotail (symmetric) reconnection and also obtained good agreement with the observed electric field (Ergun et al. 2022), indicating that the Ohm's law for laminar reconnection can be maintained even in a turbulent environment.

During anti-parallel asymmetric reconnection, the non-ideal electric field is balanced by a combination of the electron inertial and pressure terms, as described at the end of Sect. 2.1

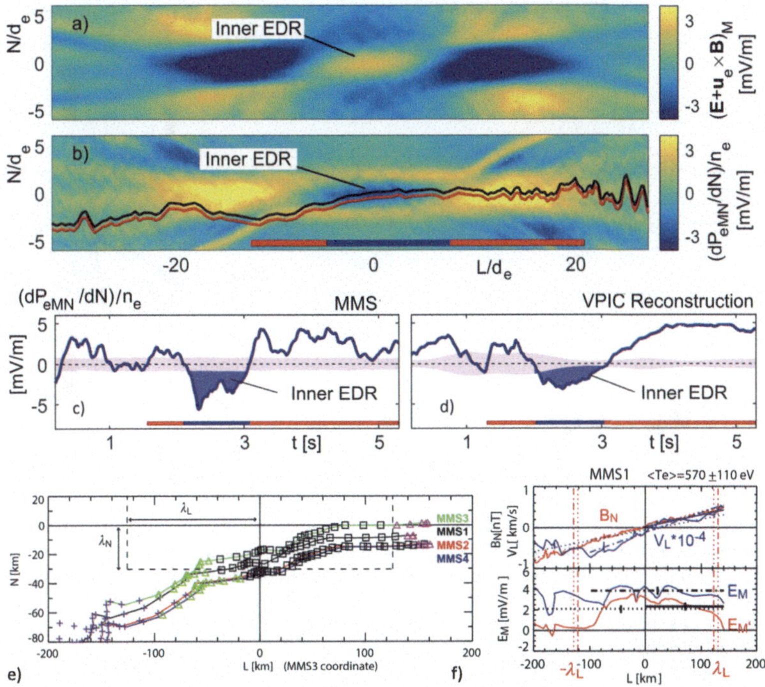

Fig. 3 *Non-gyrotropic pressure gradient in the inner EDR* Simulated 2D profiles of (a) $(\mathbf{E} + \mathbf{V}_e \times \mathbf{B}/c)_M$ and (b) $\partial P_{eMN}/\partial N$. Based on the MMS data, the 1D profile of $\partial P_{eMN}/\partial N$ in (c) was obtained, representing a close match to the simulated profile along the MMS trajectory in (d). Adapted from Egedal et al. (2019). (e) Four MMS spacecraft orbit relative to the inner diffusion region (bounded by dashed lines). (f) B_N and V_{eL} (top panel) and profile of E_M and E'_M around the X-line. The three horizontal lines (i.e., dash-dotted, solid, and dotted) in the bottom panel show the electric field value estimated from Hesse et al. (1999) formula using $\partial_L V_{eL}$ obtained from data points between different intervals indicated by the length of these lines. Adapted from Nakamura et al. (2019)

and confirmed by MMS observations, as shown in Fig. 4. MMS encountered the EDR at around 13:07:02 UT at a negative J_M peak. It is seen that the contributions of the inertial term (Fig. 4d) are generally smaller than the pressure term (Fig. 4c), but at times can be comparable, in particular for the M (green) component only, which is primarily along the reconnection electric field. Overall both electron pressure gradients and electron inertial effects are important, with a ratio of about 4:1. Yet, there are residuals of a few mV/m (30-50% of the $\mathbf{E}'$) during the encounters with the electron stagnation point (Fig. 4f) and it was also found that the error in the gradient approximation was considerable (Torbert et al. 2016a). Rager et al. (2018) analyzed the same event with higher time resolution (7.5 ms) electron data and concluded that Ohm's law could not be fully accurately resolved even with the 7.5 ms data due to time variability on the scale of the energy sweep of the particle instrument and smoothing of spatial structures by the four spacecraft gradient operator. One possibility of the violation of the Generalized Ohm's law has been suggested to be evidence of anomalous resistivity (Torbert et al. 2016a). Yet, the results of the kinetic simulation performed for the event (Torbert et al. 2016a) suggested that its effect is not significant. The small contribution from the anomalous resistivity to the Generalized Ohm's law is supported also by Graham et al. (2022) based on direct estimation of the anomalous resistivity, viscosity, and cross-field electron diffusion (see Eq. (45) in Sect. 3.10.1) associated with lower hy-

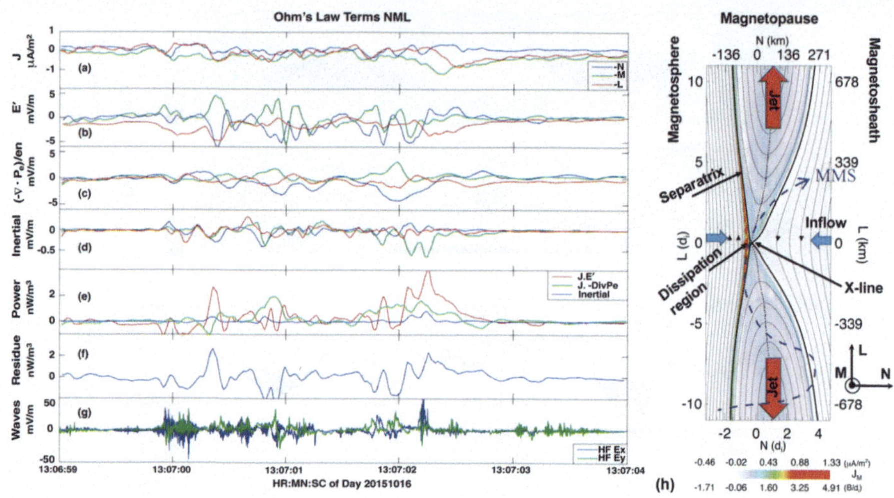

Fig. 4 *The generalized Ohm's law for an asymmetric reconnection event observed by MMS* (a) Three components of the magnetopause current density. (b–d) The comparison of terms in Ohm's law for interval around 13:07:02 UT when the MMS fleet traversed an EDR (e) the total power dissipation from individual terms. (f) The residue, $\mathbf{J}\cdot\{\mathbf{E}'-(-\nabla\cdot\mathbf{P}_e/en_e)-m_e\nabla\cdot[n_e(\mathbf{V}_i\mathbf{V}_i-\mathbf{V}_e\mathbf{V}_e)]/en_e\}$, (g) The full wave amplitude at frequencies up to 4 kHz, (h) The schematic of MMS path through EDR. Adapted from Torbert et al. (2016a) and Burch et al. (2016b)

brid waves during another asymmetric reconnection event measured by MMS. It was shown that the anomalous resistivity is approximately balanced by anomalous viscosity. Hence, although waves do produce an anomalous electron drift and diffusion across the current layer associated with magnetic reconnection, their contribution to the reconnection electric field is considered to be negligible during this observation. More discussions of Ohm's law during the presence of 3D fluctuations will be deferred to Sect. 3.10.1.

The dominant role of the pressure divergence over the inertial term in Ohm's law has also been seen for cases of guide field reconnection in both a symmetric (Wilder et al. 2017) and asymmetric (Genestreti et al. 2018c) current sheet. Genestreti et al. (2018c) also found that both out-of-the-reconnection-plane gradients ∂_M and in-plane $\partial_{L,N}$ in the pressure tensor contribute to energy conversion near the X-point. A finite $\partial_M P_{eMM}\simeq\partial_\parallel P_{e,\parallel}$ near the X-line was also observed in 3D guide field reconnection simulations that have a significant 3D structure (Liu et al. 2013; Stanier et al. 2019).

2.3 The Nature of the Reconnection Electric Field

A question of more than just academic nature is how there is a reconnection electric field at all. This question transcends the simple conclusion from above that there has to be a reconnection electric field for flux to be transferred from inflow to outflow. This existence question was raised by Hesse et al. (2018), who investigated the current and energy balance in the electron diffusion region.

During the initial phase of an evolving symmetric reconnecting current sheet, the time-derivative of the electron inertia term, $(m_e/e)\partial\mathbf{V}_e/\partial t$ in Eq. (1), is the only non-ideal term available to break the frozen-in condition at the X-line. This dominance causes the continuous intensification of the current density ($\mathbf{J}\simeq -en\mathbf{V}_e$) at the X-line, leading to a sharp current density peak around the electron gyro-scale ρ_e, generating a non-gyrotropic particle distribution (Hesse et al. 2011; Aunai et al. 2013; Zenitani and Nagai 2016) that eventually makes $\nabla \cdot \mathbf{P}_e$ the dominant non-ideal term in the quasi-steady ($\partial_t = 0$) phase; note again that $\nabla \cdot \mathbf{P}_e$ is the only term available to support the reconnection electric field at the X-line in the steady state of a symmetric case. This transition of the dominant non-ideal term during this current density intensification is clearly demonstrated in PIC simulations (Liu et al. 2014b).

Focusing on the quasi-steady state, Hesse et al. (2018) investigated the electron momentum equation Eq. (1), rewritten in the form of a current evolution equation:

$$\frac{\partial J_{ey}}{\partial t} = \frac{e^2 n_e}{m_e} E_y + \frac{e^2 n_e}{m_e c} (\mathbf{V}_e \times \mathbf{B})_y + \frac{e}{m_e} \left(\frac{\partial P_{eyz}}{\partial z} + \frac{\partial P_{exy}}{\partial x} \right) - \nabla \cdot (\mathbf{V}_e J_{ey}). \tag{3}$$

The terms on the right-hand side (RHS) of this equation describe, in order, the electric field force (due to the reconnection electric field E_y), conversion of in-plane to out-of-plane current by Lorentz forces, pressure gradient forces, and current convection into or out of the volume of interest.

To evaluate the balance of these terms over a larger domain that contains the singular X-line in the steady state, Hesse et al. (2018) integrated the individual terms of this equation over rectangles of different sizes, centered about the X-line location in a PIC simulation of symmetric magnetic reconnection, as shown in Fig. 5(a). The results of this integration are displayed in Fig. 5(b). The only terms of importance are the non-gyrotropic pressure terms (i.e., $\partial P_{eyz}/\partial z + \partial P_{exy}/\partial x$), which act to reduce the current density, and the electric field term, which acts to increase it. These terms roughly balance each other, keeping the electron current density constant, or varying on very slow time scales to account for the overall system evolution.

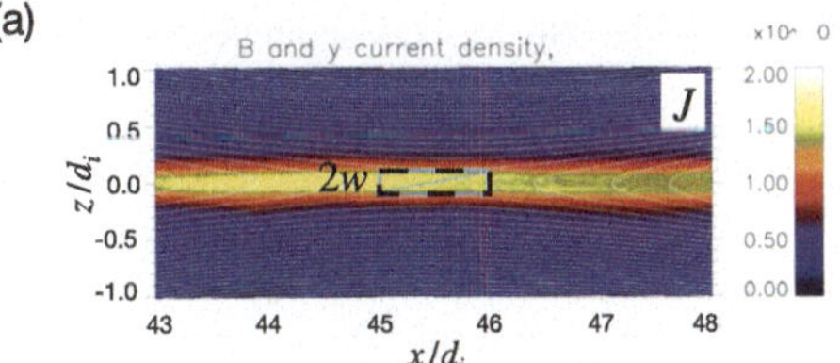

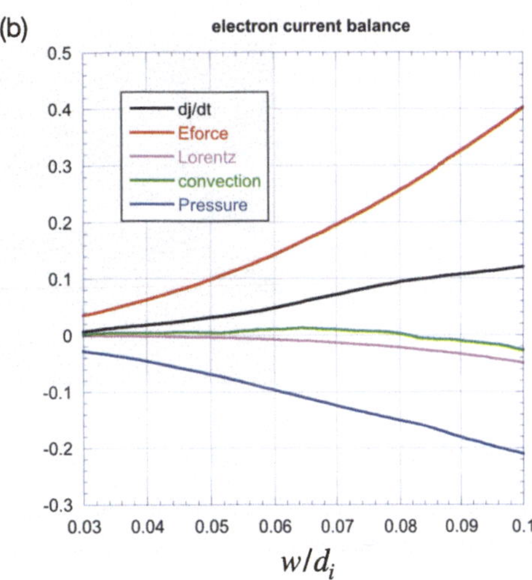

Fig. 5 *Terms that balance the current density in the quasi-steady state*. (a) The largest integration region centered on the X-point is shown as the blue-black rectangle with a thickness of 0.2 d_i and a lateral extent of one d_i. The box size varies from an initially much smaller size while keeping the aspect ratio fixed. (b) Integration of the electron current balance equation as a function of the integration region half-thickness "w". The integrated time derivative of the out-of-place current density is indicated by the black curve, which is also equal to the sum of the four other curves in agreement with Eq. (3). The major contributors are the current increase by the reconnection electric field (red), and the current reduction by pressure effects (blue). The dominance of these two terms holds over the entire range rather than at the X-point alone. Adapted from Hesse et al. (2018)

Hesse et al. (2018) further found that the source of the electron pressure, $P_e \equiv \mathrm{Tr}(\mathbf{P}_e)/3$, within the EDR is essentially exclusively due to the non-gyrotropic effect, i.e., due to complex particle orbits rather than simple, gyrotropic behavior. For this purpose, the equation

$$\frac{\partial P_e}{\partial t} = -\nabla \cdot (\mathbf{V}_e P_e) - \frac{2}{3}\sum_l P_{e,ll}\frac{\partial V_{el}}{\partial x_l} - \frac{1}{3}\sum_{l,i}\frac{\partial Q_{e,lii}}{\partial x_i} - \frac{2}{3}\sum_{l,i\neq l} P_{e,li}\frac{\partial V_{el}}{\partial x_i} \tag{4}$$

was integrated over the varying rectangle, as before, in a way similar to Fig. 5(b). The result (not shown) was that among the terms on the RHS of Eq. (4), the last term provided a positive contribution, whereas negative contributions were provided by the first two terms. The heat flux term $Q_{e,lii}$ is negligible. The dominant non-gyrotropic pressure $P_{e,li}$ contributions here appeared to be the same as the ones acting to reduce the current density in Eq. (3), suggesting that the conversion of the current carrier motion to the plasma pressure plays an important role. This last term plus the second term on the RHS of Eq. (4) is basically the "pressure-strain interaction" (Yang et al. 2017b,a), $-(\mathbf{P}\cdot\nabla)\cdot\mathbf{V}$, that will be further discussed in Sect. 4.3).

In short, the reconnection electric field within the diffusion region converts incoming electromagnetic energy into the current carrier bulk kinetic energy through direct acceleration. The current at the electron gyro-scale is intensified until the current density gradient is strong enough to generate complex, non-gyrotropic particle distribution (i.e., non-zero P_{eyz} and P_{exy}), which funnels the current carrier kinetic energy into the thermal energy through the $-(\mathbf{P}\cdot\nabla)\cdot\mathbf{V}$ term (Eq. (4)). This makes the steady state possible, where $\partial_t J_{ey}$ vanishes and the $\nabla\cdot\mathbf{P}$ becomes strong enough to balance the reconnection electric field in the Ohm's law (Eq. (3) or Eq. (2)). This multifaceted nature of the reconnection electric field is highlighted in Fig. 6.

An alternative viewpoint is also offered in Hesse et al. (2018), which argues that the electric field exists as a consequence of Maxwell's equations, specifically, Ampère's law:

$$\frac{\partial \mathbf{E}}{\partial t} = c\nabla\times\mathbf{B} - 4\pi\mathbf{J}. \tag{5}$$

Imagine that the current density is reduced by "a mechanism" below what is required to balance the $\nabla\times\mathbf{B}$ term. Then Ampère's law, Eq. (5), will immediately signal the need to

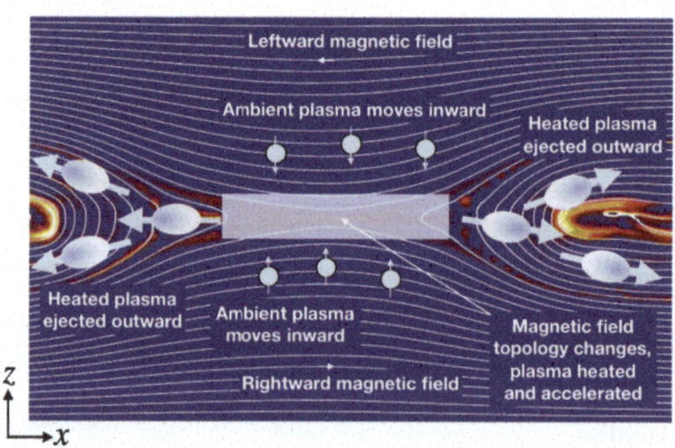

Fig. 6 *The nature of the reconnection electric field.* Shown are in-plane magnetic field lines (white), and the out-of-plane current density contour. In addition to breaking the frozen-in condition and transporting the flux into and out of the diffusion region, the reconnection electric field also sustains the electric current and increases the thermal pressure through the $-(\mathbf{P}\cdot\nabla)\cdot\mathbf{V}$ term within the diffusion region. Reprinted from Hesse and Cassak (2020), with the permission of Wiley

increase the electric field, which accelerates the current carriers and re-establishes balance in the steady state.

In other words, the steady-state reconnection electric field exists because there is a mechanism at work, which attempts to "dissipate" the current density. This conclusion holds irrespective of the dissipation mechanism – for example, classical collisions would have the same effect, as captured in Ohm's law $\mathbf{J} = \sigma \mathbf{E}$ where σ is the conductivity determined by the collisions. In a collisionless plasma, however, the current dissipation is provided by non-gyrotropic pressure effects, which are a manifestation of complex particle orbits, which lead to the scattering of directed motion by the local magnetic geometry. Hesse et al. (2021) extended this investigation to asymmetric systems (defined in Sect. 3.2) and found that the overall conclusions also hold there, even though some of the current reduction was found to be due to convective effects, in addition to the above-discussed non-gyrotropic pressure effects.

3 Collisionless Magnetic Reconnection Rate

In this section, we discuss how one can determine the magnitude of the reconnection electric field E_R (i.e., the E_y at the X-line), which is essentially the reconnection rate that measures how fast reconnection processes the incoming magnetic flux. We will organize the discussions of different regimes using the governing force-balance equation, as it determines the characteristic reconnection outflow speed, being critical to the rate. To avoid a common confusion in the normalization of reconnection rates, we normalize E_R by the "asymptotic" value of reconnecting magnetic field component B_R (or B_{x0}) and the associated proton Alfvén speed in the upstream region $V_A = B_R/\sqrt{4\pi n m_i}$ to define the "normalized reconnection rate" $R \equiv cE_R/B_R V_A$. While we have aimed to unify the notation throughout this review, some differences in subsections are unavoidable in order to strike the balance between simplicity and consistency. New notations, if needed, are defined with respect to the coordinates shown in the relevant figures.

3.1 Standard Symmetric Anti-Parallel Reconnection

We begin with the simplest current sheet, one that has symmetric, antiparallel magnetic fields, as illustrated in Fig. 7. Combining the electron and ion momentum equations, we can derive the MHD force balance equation. In the steady state, it reads

$$\frac{(\mathbf{B} \cdot \nabla)\mathbf{B}}{4\pi} \simeq \frac{\nabla B^2}{8\pi} + \nabla \cdot \mathbf{P} + nm_i(\mathbf{V} \cdot \nabla)\mathbf{V}. \tag{6}$$

This force-balance equation works in most regions, including the ideal MHD region and the ion diffusion region, as long as the electron inertial term is negligible and the quasi-neutral condition holds; that is usually valid in the non-relativistic limit. As we will see, the scaling of reconnection rates in diverse regimes can be more or less captured by the force balance along the inflow and outflow symmetry lines.

3.1.1 Sweet-Parker Scaling

The first quantitative model of the magnetic reconnection rate was derived by Sweet and Parker (Parker 1957; Sweet 1958). From mass conservation $\nabla \cdot (n\mathbf{V}) \simeq 0$ in steady state and

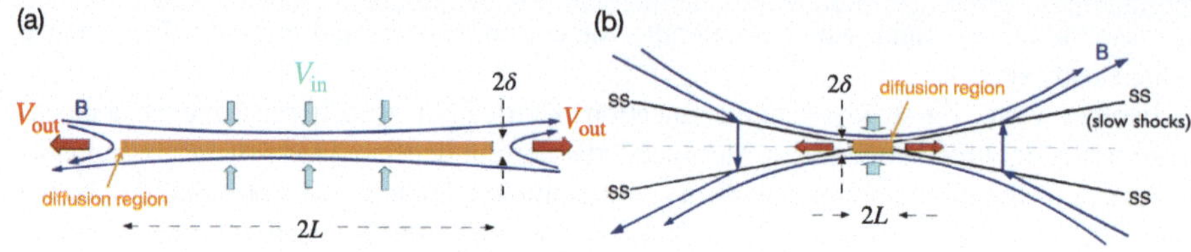

Fig. 7 *Classical reconnection models*. (a) Sweet-Parker solution. Adapted from Sweet (1958), Parker (1957). (b) Petschek solution. Adapted from Petschek (1964)

the incompressible assumption,

$$V_{\text{in}} L \simeq V_{\text{out}} \delta, \tag{7}$$

where δ and L are the half-thickness and half-length of the diffusion region, respectively. From the momentum equation, balancing the magnetic tension and inertia force in Eq. (6), $(\mathbf{B} \cdot \nabla)\mathbf{B}/4\pi \simeq nm_i(\mathbf{V} \cdot \nabla)\mathbf{V}$, one gets

$$V_{\text{out}} \simeq \frac{B_R}{\sqrt{4\pi nm_i}} = V_A. \tag{8}$$

Thus, the outflow speed is the characteristic Alfvén speed based on the upstream magnetic field B_R. It then makes sense to define the *normalized reconnection rate* as

$$R \equiv \frac{V_{\text{in}}}{V_A}, \tag{9}$$

which measures how fast the inflowing plasma transports magnetic flux for processing. Combining Eqs. (7) and (8), one realizes that the normalized reconnection rate is basically the aspect ratio of the diffusion region, i.e., $R \simeq \delta/L$. Note that E_y is uniform in the 2D steady-state per Faraday's law and $E_y = V_{\text{in}} B_R/c$ at the inflow boundary of the diffusion region, thus the definition of the reconnection rate can also be expressed as $R \equiv cE_R/(B_R V_A)$.

After coupling the inflow region to a diffusion region dominated by resistivity (which requires collisions), the full Sweet-Parker solution (omitted here) was derived in 1957. It has a system size long current sheet (Fig. 7(a)), resulting in a low δ/L and thus a reconnection rate that is too low to explain the energy release during solar flares (Parker 1963, 1973). Petschek (1964) proposed that standing slow shocks that bound the exhaust can resolve this challenge by opening out the outflow geometry, and localizing the diffusion region (Fig. 7(b)). However, Petschek reconnection, unlike Sweet-Parker reconnection, is not a solution of resistive-MHD with uniform resistivity; it tends to collapse into the long Sweet-Parker layer in (uniform) resistive-MHD simulations (Sato and Hayashi 1979; Biskamp 1986). Such an elongated reconnection layer can be unstable to the plasmoid instability if collisions are weak enough (Biskamp 1982; Shibata and Tanuma 2001; Bhattacharjee et al. 2009; Loureiro et al. 2007; Pucci and Velli 2014; Comisso et al. 2016), but we will not discuss this resistive-MHD mode further.

3.1.2 $R - S_{\text{lope}}$ Relation and the Maximum Plausible Rate

For collisionless reconnection, the rate in Eq. (9) is not bounded, since δ/L can, in principle, be any value. To fix this problem, one needs to consider force balance along the inflow direction and recognize there is a scale separation between the regions "immediately upstream" and "far (asymptotic) upstream" of the ion diffusion region.

Per geometry, this diffusion region aspect ratio δ/L is also the slope of the separatrix $S_{\rm lope} = \Delta z/\Delta x$, as shown in Fig. 8(a). The $R \simeq \delta/L$ scaling only works when $S_{\rm lope} = \delta/L \ll 1$. In the $S_{\rm lope} \rightarrow 1$ limit (i.e., a localized diffusion region with an open outflow geometry as illustrated in Fig. 8(a)), the upstream magnetic field is indented, which unavoidably induces a magnetic tension force $(\mathbf{B}\cdot\nabla)\mathbf{B}/4\pi$ pointing to the upstream region, as illustrated by the green arrow in Fig. 8(a). In the low-β limit, both the upstream $\nabla\cdot\mathbf{P}$ and $nm_i(\mathbf{V}\cdot\nabla)\mathbf{V}$ terms are negligible; thus, the only term that can counterbalance this tension force is the magnetic pressure gradient force ($-\nabla B^2/8\pi$, black arrow), which requires the reduction of the reconnecting field when it is convected into the diffusion region. Similarly, a finite magnetic pressure gradient force also arises in the outflow direction in the $S_{\rm lope} \rightarrow 1$ limit, as depicted by the black arrow in Fig. 8(b), which slows down the outflow.

Quantitatively, through discretizing the inflow force-balance at point **1** of Fig. 8(c), one can relate the ratio of the magnetic field immediately upstream of the ion diffusion region B_{xi} and the asymptotic value at far upstream B_{x0} to the slope of the reconnection separatrix $S_{\rm lope}$ (Liu et al. 2017) as

$$\frac{B_{xi}}{B_{x0}} \simeq \frac{1-S_{\rm lope}^2}{1+S_{\rm lope}^2}. \tag{10}$$

In the large opening limit ($S_{\rm lope} \rightarrow 1$), the magnetic field B_{xi} that actually reconnects is reduced, as is the rate R.

By analyzing the outflow force balance, including the $nm_i(\mathbf{V}\cdot\nabla)\mathbf{V}$ term at a point within the diffusion region (Fig. 8(b)), one can derive the outflow speed at the outflow edge of the ion diffusion region,

$$V_{{\rm out},i} \simeq V_{Ai}\sqrt{1-S_{\rm lope}^2}, \tag{11}$$

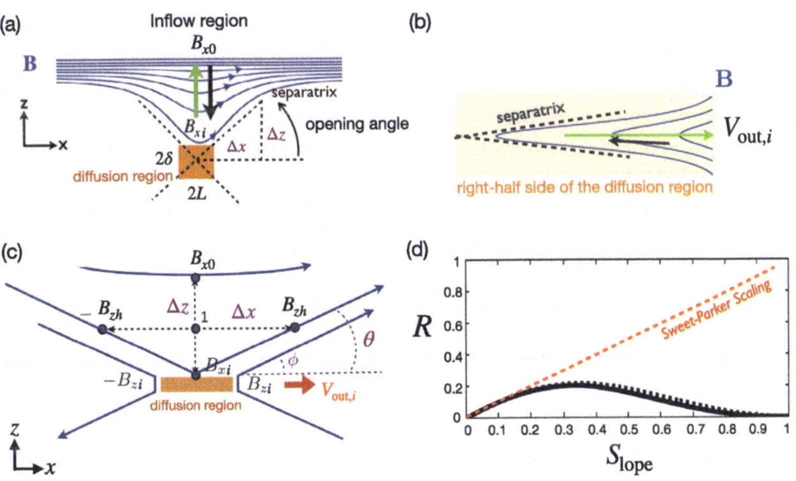

Fig. 8 *General constraints on the maximum plausible rate from mesoscale force-balance.* (a) Field line geometry and force balance upstream of the diffusion region. (b) Field line geometry and force balance along the outflow. (c) The scheme used to analyze the force balance. (d) The predicted reconnection rate R as a function of the separatrix slope $S_{\rm lope}$. Adapted from Liu et al. (2017)

where $V_{Ai} \equiv B_{xi}/\sqrt{4\pi n m_i}$. Equation (11), coupled with Eq. (10), recovers the Alfvén speed $V_{A0} = B_{x0}/\sqrt{4\pi n m_i}$ in the small opening limit, as in the Sweet-Parker analysis. In the large opening ($S_{\rm lope} \rightarrow 1$) limit, the outflow speed is reduced, and so is the rate.

The reconnection rate is $R = cE_y/B_{x0}V_{A0} = (B_{zi}/B_{xi})(B_{xi}/B_{x0})(V_{{\rm out},i}/V_{A0})$. Using Eqs. (10) and (11), and noting that $B_{zi}/B_{xi} \simeq S_{\rm lope}$, we then find the $R - S_{\rm lope}$ relation

$$R = S_{\rm lope}\left(\frac{1 - S_{\rm lope}^2}{1 + S_{\rm lope}^2}\right)^2 \sqrt{1 - S_{\rm lope}^2}, \tag{12}$$

which is shown as the black solid curve in Fig. 8(d). Clearly, these two geometrical constraints along the inflow and outflow bring down the reconnection rate to zero in the $S_{\rm lope} \rightarrow 1$ limit, where the separatrix makes a right angle. The maximum plausible rate is around the value of 0.2. For reference, the Sweet-Parker scaling is shown by the red dashed line, which is unbounded in the large $S_{\rm lope}$ limit. Importantly, the profile of the predicted black curve is relatively flat for a wide range of $S_{\rm lope}$. Thus as long as there is some degree of localization, the predicted rate will be on the order of $\mathcal{O}(0.1)$. Note that this prediction does not depend on the dissipation physics or the thickness of the current sheet. Thus, this value likely also constrains the maximal plausible rate in theorized "turbulent reconnection" where the diffusion region is turbulent and thick, and has large-scale outflow exhausts (Lazarian and Vishniac 1999). On the other hand, this maximum plausible reconnection rate of value $\simeq 0.2$ is clearly demonstrated using a large and spatially localized anomalous resistivity right at the X-line in MHD simulations (Lin et al. 2021; Jiménez et al. 2022).

3.1.3 Localization Mechanism That Leads to Fast Reconnection

Petschek (1964) provided the correct steady-state outflow solution of reconnection that predicts slow shocks and rotational discontinuities farther downstream, but the solution failed to capture the essential *localization mechanism*, that leads to the open geometry in the first place. While Eq. (12) provides the general $R - S_{\rm lope}$ relation, to determine the rate, we still need to identify the (primary) mechanism that localizes the diffusion region, determining the opening geometry captured by $S_{\rm lope}$.

Kinetic simulations beyond the MHD model suggest that antiparallel reconnection with an open outflow geometry occurs when the current sheet thins down to the ion inertial scale (Bhattacharjee 2004; Cassak et al. 2005; Daughton et al. 2009; Jara-Almonte and Ji 2021). When this occurs, the Hall term in the generalized Ohm's law (Vasyliunas 1975; Swisdak et al. 2008) dominates the electric field in the ion diffusion region (IDR), where the ions become demagnetized. The correlation between the Hall effect and fast reconnection was clearly demonstrated in the GEM reconnection challenge study (Birn et al. 2001), as shown in Fig. 9(a). This study showed that simulation models with the Hall term in the generalized Ohm's law (particle-in-cell (PIC), hybrid, and Hall-MHD) realize fast reconnection, while only the uniform resistive-MHD model, which lacks the Hall term, exhibits a slow rate (Parker 1957; Sweet 1958). The value of the fast rate in collisionless plasmas is on the order of 0.1 over a range of electron-ion mass ratios and initial thicknesses, as shown in Fig. 9(b) and (c). For decades, it had been unclear "how" the Hall term localizes the diffusion region, producing an open geometry. The dispersive property of waves arising from the Hall term was proposed as an explanation (Mandt et al. 1994; Shay et al. 1999; Rogers et al. 2001; Drake et al. 2008), but the role of dispersive waves derived from the linear analysis was called into question because reconnection can be fast even in systems that lack dispersive

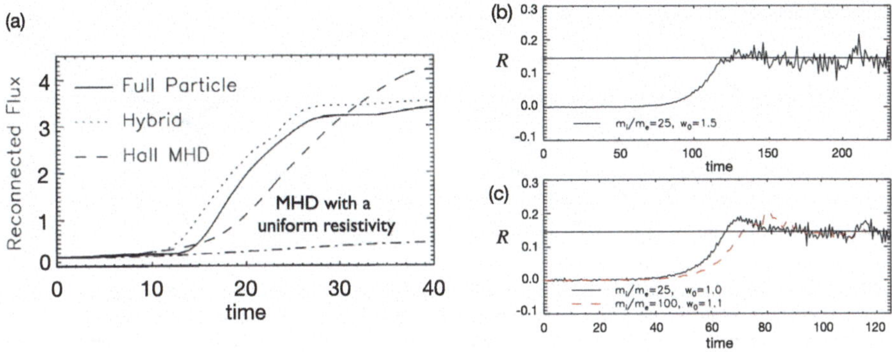

Fig. 9 (a) The GEM reconnection challenge result that compares reconnection rates in four different numerical models. Reprinted from Birn et al. (2001), with the permission of Wiley. Panels (b) and (c) show the normalized reconnection rate in PIC simulations over a range of mass ratio m_i/m_e and initial thickness w_0. Adapted from Shay et al. (2007)

waves (Bessho and Bhattacharjee 2005; Liu et al. 2014b; TenBarge et al. 2014; Stanier et al. 2015a).

While the Hall electromagnetic fields were well-known for being the key feature of the ion diffusion region (Sonnerup 1979), their role in transporting the incoming magnetic energy was less recognized. Figure 10a shows the out-of-plane magnetic field B_y in the nonlinear stage. This out-of-plane quadrupolar Hall magnetic field arises because electrons, the primary current carrier within the IDR (i.e., $\mathbf{J} \simeq -en\mathbf{V}_e$), drag both reconnected and not-yet reconnected magnetic field lines out of the reconnection plane (Mandt et al. 1994; Ren et al. 2005; Drake et al. 2008; Burch et al. 2016b), as illustrated in Fig. 10(c); the laboratory evidence is reviewed in Ji et al. (2023, this collection). Importantly, this Hall quadrupole magnetic field B_y along with the inward-pointing Hall electric field $E_z \simeq V_{ey}B_x/c$, shown in Fig. 10b, constitute a Poynting vector $S_x = -cE_zB_y/4\pi$ in the x-direction. This component diverts the inflowing electromagnetic energy toward the outflow direction. This is shown by the streamlines of $\mathbf{S} = c\mathbf{E} \times \mathbf{B}/4\pi$ in yellow, which bend in the x direction significantly before reaching the outflow symmetry line at $z = 0$.

Since the Hall term dominates the electric field $\mathbf{E} \simeq \mathbf{E}_{\text{Hall}} = \mathbf{J} \times \mathbf{B}/nec$ inside the IDR, then $\nabla \cdot \mathbf{S} = -\mathbf{J} \cdot \mathbf{E} \simeq 0$ per Poynting's theorem in the steady state (n.b., further discus-

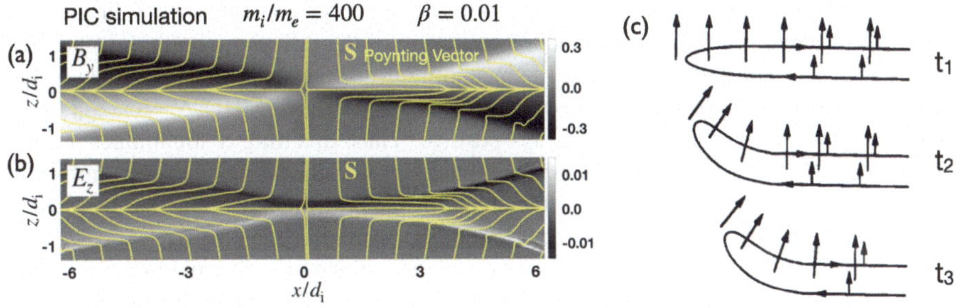

Fig. 10 *Hall electromagnetic fields* (a) The Hall magnetic field B_y and (b) the Hall electric field E_z (normalized by B_{x0}) overlaid with Poynting vector **S** streamlines (yellow). Adapted from Liu et al. (2022), reproduced by permission of Springer Nature. (c) Electrons drag the reconnected field out of the reconnection plane to form the Hall quadrupole magnetic field. Adapted from Mandt et al. (1994)

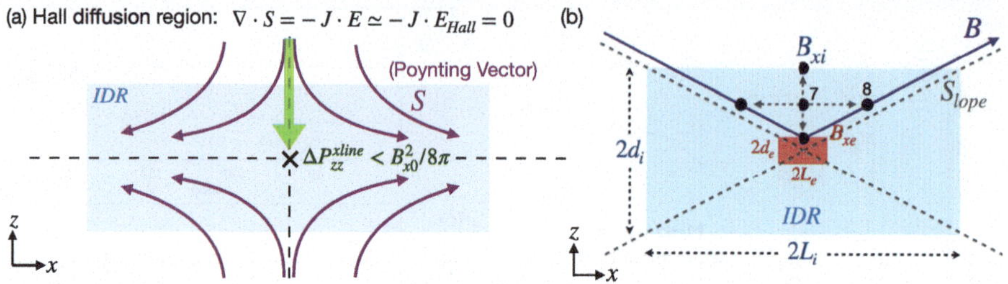

Fig. 11 *Transport patterns of electromagnetic energy in Hall reconnection and the diffusion region structure.* (a) The Hall effect results in this intrinsic, "diverting", Poynting vector **S** streamline pattern in purple, which limits the pressure increase of inflowing plasma (green arrow) that reaches the X-line. (b) The diagram used to derive the slope of the separatrix, $S_{\rm lope}$. The blue (red) box represents the IDR (EDR). The solid blue line depicts an upstream magnetic field **B** line adjacent to the separatrix shown by diagonal dashed lines. Adapted from Liu et al. (2022), reproduced by permission of Springer Nature

sion of Poynting's theorem can be found in Sect. 4.2). When the divergence of a vector field vanishes, like that for magnetic fields (i.e., $\nabla \cdot \mathbf{B} = 0$), Gauss' theorem indicates that the associated flux into a closed volume equals the flux out. The associated flux of **S** can be quantified by the number of streamlines equally spaced at the inflow boundary. These **S** streamlines (purple lines in Fig. 11(a)) into the IDR (blue region) do not end within the region of $\nabla \cdot \mathbf{S} = 0$. Approaching the X-line, the magnetic field strength decreases and eventually vanishes due to the symmetry of this system. The **S** streamlines thus need to get around this singular point and exit at the outflow direction. This results in an intrinsically "diverting" **S** streamline pattern around the X-line, consistent with the presence of $S_x = -cE_z B_y/4\pi$, as discussed in Fig. 10(a) and (b). These streamlines play a role analogous to railroad tracks in guiding the transport of incoming (magnetic) energy through the IDR. From Fig. 11(a), we realize that if $\mathbf{E} = \mathbf{E}_{\rm Hall}$ none of the upstream magnetic energy can be transported to the X-line due to this diverting **S** streamline pattern.

As time proceeds, an energy void centered around the X-line develops. Without energy input, no pressure (either thermal or magnetic) can be built up at the X-line. The upstream magnetic pressure around the energy void will then locally pinch the upstream magnetic field lines. This is a localization mechanism needed for the open outflow geometry and fast reconnection (Liu et al. 2022). Here we point out three more important observations. First, the diverted energy is deposited on the outflow symmetry line (i.e., $z = 0$) downstream of the X-line, which helps establish the pressure balance across the exhaust (in the normal direction), keeping the exhaust open. This difference of energy content at the X-line and its downstream region itself also implies the localization of the diffusion region. Second, this diverting **S** streamline pattern persists even in an (initially) elongated reconnection layer, but such a layer is not sustainable, as noted above. Third, in resistive-MHD, $\nabla \cdot \mathbf{S} = -\mathbf{J} \cdot \mathbf{E} \simeq -\eta J_y^2 < 0$ since the resistivity η is always positive. Thus, diverting **S** streamlines are not required (i.e., $S_x \simeq 0$ is possible). The **S** streamlines can end and distribute energy uniformly on the outflow symmetry line, in favor of maintaining the pressure balance across the X-line. This is why the diffusion region in Sweet-Parker reconnection is not localized.

To quantify the degree of localization, we need to estimate the thermal pressure at the X-line. The key is that $\mathbf{J} \cdot \mathbf{E} \simeq 0$ inside the Hall-dominated IDR, which limits the energy conversion to particles and thus also limits the difference in the zz-component of the pressure tensor between the X-line and the far upstream asymptotic region $\Delta P_{zz}^{\rm xline} \equiv P_{zz}|_{\rm xline} - P_0$ (illustrated as the green arrow in Fig. 11a). Given that magnetic pressure $B^2/8\pi = 0$ at the antiparallel reconnection X-line, as long as $\Delta P_{zz}^{\rm xline} < B_{x0}^2/8\pi$, the inflowing reconnecting field bends toward the X-line to restore the force-balance condition $\nabla(P + B^2/8\pi) = (\mathbf{B} \cdot \nabla)\mathbf{B}/4\pi$ (Liu et al. 2020a). This bending makes the outflow exhausts open out.

These observations can be used to determine the separatrix slope S_{lope}, then using the $R - S_{\text{lope}}$ relation discussed in the previous Sect. 3.1.2; it will provide a first-principles prediction of reconnection rate in collisionless plasmas. Specifically, after recognizing that some limited energy still goes to the ballistically accelerated incoming ions (Wygant et al. 2005; Aunai et al. 2011), one can derive the pressure difference between the d_e- and d_i-scale (Liu et al. 2022),

$$P_{izz}|_{d_i}^{d_e} \simeq \frac{2}{3}\left(\frac{B_{xi}^2 - B_{xe}^2}{8\pi}\right), \tag{13}$$

where B_{xe} is the reconnecting magnetic field at the inflow boundary of the EDR. and the slope of the separatrix associated with the open outflow geometry can then be determined by analyzing the inflow force balance at point **7** of Fig. 11(b),

$$S_{\text{lope}} \simeq \sqrt{\frac{1}{3}\left[\frac{1-(B_{xe}/B_{xi})}{1+(B_{xe}/B_{xi})}\right]}, \tag{14}$$

where

$$\frac{B_{xe}}{B_{xi}} \simeq \left(\frac{m_e}{m_i}\right)^{1/4} \tag{15}$$

is derived through coupling to the EDR. The cross-scale coupling from the mesoscale upstream region down to the IDR, and then the EDR is achieved by recognizing that the magnetic field line tends to straighten itself out (when it is possible). Thus, the separatrix slope S_{lope} is similar in these different regions. For the real proton-to-electron mass ratio $m_i/m_e = 1836$, the total pressure increase along the inflow symmetry line to the X-line was derived to be $\Delta P_{zz}^{xline} \simeq 0.25 B_{x0}^2/8\pi$, and the resulting reconnection rate $R \simeq 0.16$ from Eq. (12), consistent with numerical simulations in Fig. 9, in-situ observations (Genestreti et al. 2018b; Nakamura et al. 2018; Torbert et al. 2018; Nakamura et al. 2019) discussed in Sect. 3.4.1, and other examples discussed in a previous review (Cassak et al. 2017b).

3.2 Asymmetric Reconnection

While "symmetric" magnetic reconnection discussed in the previous subsection (Sect. 3.1) is a reasonable approximation to the energy release process during geomagnetic substorms at Earth's magnetotail (e.g. Angelopoulos et al. 2008; Paschmann et al. 2013), magnetic reconnection at Earth's magnetopause is "asymmetric", as it occurs at the boundary layer between the magnetosphere plasmas and magnetosheath plasmas (e.g. Paschmann et al. 2013, 2005, 1979), where the plasma and magnetic field conditions on two sides of the current sheet can be very different. In this subsection, we will generalize the theoretical modeling into this configuration.

3.2.1 Cassak-Shay Scaling

To predict the rate of asymmetric reconnection in terms of upstream plasma parameters, we make the same simplifying assumptions typically made for such studies: two-dimensionality, steady state, upstream asymptotic magnetic fields are straight and anti-parallel, no bulk flow upstream except for the inflow, and the upstream plasmas are in local thermodynamic equilibrium. The analysis is carried out in the reference frame in which the X-line is stationary. We use subscripts "1" and "2" to denote the two upstream sides of the reconnection site, and the upstream reconnecting magnetic field strengths are B, number densities are n, and temperatures are T. For definiteness, if the magnetic field strength is stronger on one side than the other, we take the stronger magnetic field side to be "2", so that $B_2 \geq B_1$.

First, we note that there must be a pressure balance in the MHD sense across the current sheet in the upstream asymptotic regions

$$P_1 + \frac{B_1^2}{8\pi} = P_2 + \frac{B_2^2}{8\pi}. \tag{16}$$

Here the total plasma pressure is $P = \sum_s^{i,e} n k_B T_s$. In writing Eq. (16), we ignore the ram pressure due to the inflow speed $V_{\rm in}$. This is justifiable *a posteriori* because the resulting normalized reconnection rate R is $\mathcal{O}(0.1)$ and the ratio of the inflow kinetic energy density $(1/2)nm_i V_{\rm in}^2$ to the upstream magnetic pressure $B^2/8\pi$ scales like R^2, so the contribution of the ram pressure due to the inflow is at the 1% level. Pressure balance follows from the momentum equation; if pressure balance were not satisfied, the current sheet would have a net force on it and would accelerate, which would violate the assumption that the system is in a steady state.

The most basic estimate of the asymmetric reconnection rate in terms of upstream parameters is obtained using a generalization of the classical Sweet-Parker analysis (Cassak and Shay 2007). This approach simply relies on conservation laws, which impose that the flux of particles, energy, and magnetic flux coming in the upstream edge of the diffusion region must equal their fluxes leaving at the downstream edge in the steady state. The diffusion region is assumed to be a rectangular box of half-thickness δ in the inflow direction and half-length L in the outflow direction, as sketched in Fig. 12. We first treat the limit in which the process is incompressible (Cassak and Shay 2007).

In the steady-state in two dimensions, Faraday's law implies that the out-of-plane electric field E_y must be uniform. At the upstream edge of the (ion) diffusion region, the ideal-MHD Ohm's law is expected to be valid, so $\mathbf{E} + \mathbf{V} \times \mathbf{B}/c \simeq 0$. An important result follows; defining

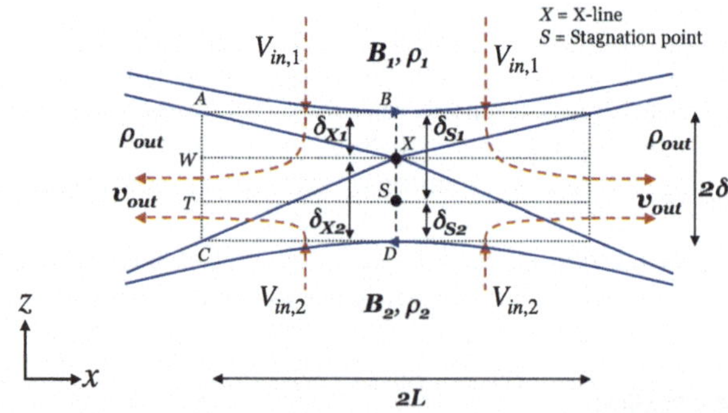

Fig. 12 *Sketch of asymmetric reconnection diffusion region.* Magnetic field lines (blue solid lines) and bulk flow streamlines (red dashed lines) in asymmetric reconnection. The outer gray rectangle denotes the edge of the diffusion region. X denotes the location of the X-line, and S denotes the location of the stagnation point, where the in-plane magnetic field and bulk flow go to zero, respectively. Reprinted from Cassak and Shay (2007), with the permission of AIP Publishing

the inflow speeds as $V_{\text{in},1}$ and $V_{\text{in},2}$, the constancy of E_y implies

$$V_{\text{in},1} B_{x1} \sim V_{\text{in},2} B_{x2}. \tag{17}$$

Since we assume $B_{x2} \geq B_{x1}$, this result implies the stronger magnetic field convects into the diffusion region more slowly. By conservation of particles, the flux of particles entering the diffusion region must equal its flux as it leaves, which is quantified as

$$(n_1 V_{\text{in},1} + n_2 V_{\text{in},2}) L \sim 2 n_{\text{out}} V_{\text{out}} \delta, \tag{18}$$

where n_{out} and V_{out} are the number density and (outflow) bulk speed at the downstream edge of the diffusion region. Similarly, the conservation of energy implies

$$\left(V_{\text{in},1} \frac{B_{x1}^2}{8\pi} + V_{\text{in},2} \frac{B_{x2}^2}{8\pi} \right) L \sim 2 \left(\frac{1}{2} n_{\text{out}} m_i V_{\text{out}}^2 \right) V_{\text{out}} \delta. \tag{19}$$

Finally, it was argued that the outflow number density scales as

$$n_{\text{out}} \sim \frac{n_1 B_{x2} + n_2 B_{x1}}{B_{x1} + B_{x2}}, \tag{20}$$

which follows from the plasmas mixing in proportion to the volume of the flux tubes on either upstream side, since the weaker magnetic field side reconnects more volume than the stronger magnetic field. Putting the results of Eqs. (17) through (20) together give predictions for the outflow speed V_{out} and the reconnection electric field $E_{R,asym}$:

$$V_{\text{out}} \simeq V_{A,\text{asym}} \sim \sqrt{\frac{B_{x1} B_{x2}}{4\pi n_{\text{out}} m_i}}, \tag{21}$$

$$E_{R,\text{asym}} \sim 2 \left(\frac{B_{x1} B_{x2}}{B_{x1} + B_{x2}} \right) \left(\frac{\delta}{L} \right) \frac{V_{\text{out}}}{c}. \tag{22}$$

This gives the desired asymmetric reconnection rate as a function of upstream parameters. For each expression, the result reduces to the standard incompressible Sweet-Parker scaling $V_{\text{out}} \sim V_A$ and $E_R \sim (\delta/L) V_A B_R / c$ in the symmetric limit.

The analysis described thus far did not take compressibility into account, which allows for the heating of the plasma as it passes through the diffusion region. The analysis was extended (Birn et al. 2010) to include these effects. The way to do so involves replacing magnetic energy $B^2/8\pi$ with magnetic enthalpy $B^2/4\pi$ and including the enthalpy flux $[\gamma/(\gamma-1)]P \equiv \kappa P$ (where γ is the ratio of specific heats in the fluid description) in the energy flux balance in Eq. (19). The predicted outflow speed ends up being unchanged from Eq. (21), but the predicted reconnection rate is multiplied by a factor of r given by

$$r = \frac{\kappa (B_{x1} + B_{x2})}{\lambda_1 B_{x2} + \lambda_2 B_{x1}}, \tag{23}$$

where $\lambda_j = (1 + \kappa \beta_j)/(1 + \beta_j)$, and plasma $\beta_j = 8\pi P_j / B_{xj}^2$ for $j = 1, 2$. Taking the incompressible limit with either $\beta_j \to \infty$ or $\gamma \to \infty$ reproduces Eq. (22). A similar scaling analysis was extended to relativistic asymmetric magnetic reconnection (Mbarek et al. 2022).

3.2.2 $R - S_{\rm lope}$ Relation and the Maximum Plausible Rate

Both results from the previous section predict the reconnection rate in terms of asymptotic upstream parameters but have a factor of δ/L. It can be calculated for resistive reconnection analogously to the Sweet-Parker model (Cassak and Shay 2007), but for collisionless reconnection δ/L remained as a free parameter. Empirically from two-fluid simulations, it was found that $\delta/L \sim 0.1$ (Cassak and Shay 2008) for collisionless asymmetric reconnection, just like it does for symmetric reconnection (Shay et al. 1999). However, it is important to better understand why this is the case. The analysis for symmetric reconnection used to show that 0.1 is approximately the maximum reconnection rate allowed (Liu et al. 2017) was extended to asymmetric reconnection (Liu et al. 2018a).

In this model, the reconnecting magnetic fields at the mesoscale bend in towards the reconnection site as they do in symmetric reconnection. Force balance in the inflow direction is analogous to the symmetric reconnection case, where the magnetic curvature force opposes the magnetic pressure force between the X-line and the asymptotic region, reducing the magnetic field at the microscopic scale, as illustrated in Fig. 13(a). For asymmetric reconnection, however, the geometry on the two sides of the current sheet is different, specifically the slopes of the separatrix on the two sides of the current sheet. Analogous to Eq. (10) for the magnetic field strength at the upstream edge of the diffusion region in symmetric reconnection, one gets

$$\frac{B_{xmj}}{B_{xj}} \simeq \frac{1 - S^2_{{\rm lope},j}}{1 + S^2_{{\rm lope},j}}, \tag{24}$$

for the two sides $j = 1, 2$ and the subscript "m" denotes the edges of the "microscopic" ion diffusion region. Here $S_{{\rm lope},j} = \delta_j/L$ is the slope made by the separatrix within the diffusion region, which can be different on each side and $\delta_1 + \delta_2 = \delta$. The reconnected magnetic field B_{zm} at the downstream edge is assumed to be the same for each side.

A brief analysis predicts the outflow speed when taking into account the reduction of the upstream magnetic field and magnetic pressure, giving

$$V_{{\rm out},m} \simeq \sqrt{\frac{B_{xm1}B_{xm2}}{4\pi n_{{\rm out},m} m_i}} \sqrt{1 - 4\frac{B_{xm1}B_{xm2}}{(B_{xm1}+B_{xm2})^2}\left(\frac{\delta}{L}\right)^2}, \tag{25}$$

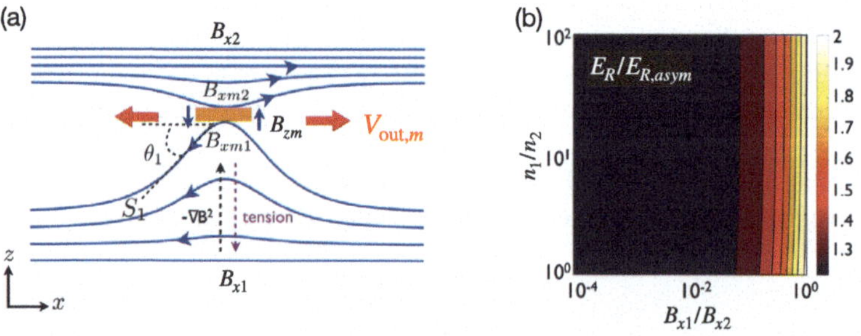

Fig. 13 *Estimation of the maximum plausible reconnection rate in asymmetric reconnection.* (a) The magnetic field geometry of asymmetric reconnection, where the field strength is different on two sides of the diffusion region (orange box). (b) The predicted ratio of the maximum reconnection electric field E_R and $E_{R,asym}$ in Eq. (22) (with an effective $\delta/L = 0.1$) over a wide range of magnetic field and density asymmetries is within a factor of 2. Reprinted from Liu et al. (2018a), with the permission of Wiley

which generalizes Eq. (21) and captures that the outflow speed decreases when the exhausts open out. The subscript "m" again indicates quantities at the edges of the ion diffusion region. With Eqs. (24) and (25), an expression for the normalized reconnection rate R can be obtained that is only a function of $S_{\mathrm{lope},1}$ (or $S_{\mathrm{lope},2}$) and the upstream plasma parameters, as was done for symmetric reconnection in Fig. 8(d). The prediction of the maximum rate shows a similar scaling with magnetic field ratio and density ratio as Eq. (22), as shown in Fig. 13(b). Importantly, this analysis reveals that it is the reduction of the reconnecting magnetic field on the weak field side (i.e., B_{x1} in Fig. 13) that limits the reconnection rate. The slopes of separatrix on two sides are also predicted (Liu et al. 2018a). As of now, there has not been a first-principles calculation of the reconnection rate for collisionless asymmetric reconnection, generalizing the symmetric result in Sect. 3.1.3.

3.2.3 Structure of the Diffusion Region During Asymmetric Reconnection

In addition to the asymmetric conditions modifying the macroscale properties of the reconnection, such as the outflow speed and reconnection rate, they also impact the microscale physics within the diffusion region. One key result is that the X-line (the location at which the magnetic topology changes) and the stagnation point (the location at which the in-plane bulk flow goes to zero) are not in the same location (Hoshino and Nishida 1983; Scholer 1989; La Belle-Hamer et al. 1995; Nakamura and Scholer 2000; Priest et al. 2000; Dorelli et al. 2004; Mirnov et al. 2006; Cassak and Shay 2007). The reason follows from conservation laws (Cassak and Shay 2007). From Eq. (17), the inflow is slower on the high magnetic field side. Counter-intuitively, the rate at which the magnetic energy enters the diffusion region is higher on the high field side: $(V_{\mathrm{in},2}B_{x2}^2/8\pi)/(V_{\mathrm{in},1}B_{x1}^2/8\pi) \sim B_{x2}/B_{x1}$. Since no magnetic flux passes through the X-line, the X-line is displaced in the inflow direction toward the low magnetic field side so that the distance from the X-line to each side, δ_{X1} and δ_{X2} in Fig. 12, has a ratio $\delta_{X1}/\delta_{X2} \simeq B_{x2}/B_{x1}$.

Similarly, the stagnation point has no particle flux across it. The ratio of the incoming particle flux from the 2-side to the 1-side is $n_2 V_{\mathrm{in},2}/n_1 V_{\mathrm{in},1} \simeq n_2 B_{x1}/n_1 B_{x2}$ using Eq. (17). This implies that the stagnation point is offset from the center of the diffusion region toward whichever side has the smaller n/B (Cassak and Shay 2007) as is sketched in Fig. 12. It was further shown that this analysis implies the displacement of the X-line and stagnation point both in the ion diffusion region and the electron diffusion region during collisionless reconnection (Cassak and Shay 2009).

The relative location of the X-line and stagnation point has important implications for the microphysics of reconnection, including the structure of the Hall fields in the ion diffusion region, transport of plasma through the diffusion region, and energizing the plasma. The discussion above treated only asymmetries in the inflow direction. It has been similarly shown that an asymmetry in the outflow direction leads to the X-line and stagnation point being displaced in the outflow direction (Oka et al. 2008; Murphy et al. 2010).

3.3 Guide Field Reconnection

Theories in the previous two subsections (the symmetric case in Sect. 3.1 and the asymmetric case in Sect. 3.2) do not include the effect of an external guide field, B_g, that points out of the reconnection plane. However, in many situations, there is such a magnetic component during reconnection, and we call these cases "guide-field reconnection". For instance, solar wind magnetic fields (interplanetary magnetic field, IMF) in the magnetosheath plasma can touch Earth's magnetopause in all possible orientations, making a wide range of magnetic shear angles with respect to the magnetosphere magnetic fields. The strength of the guide field will depend on this magnetic shear angle and the X-line orientation, as illustrated in Fig. 25(b) in Sect. 3.8.1, and also discussed in Gershman et al. (2024, this collection).

3.3.1 Theory and Simulations

Early studies of guide field reconnection were actually motivated by the Sawtooth crashes in fusion devices (von Goeler et al. 1974; Kadomtsev 1975; Aydemir 1991, 1992; Denton et al. 1987; Yamada et al. 1994; Biskamp and Drake 1994; Beidler and Cassak 2011). Reduced fluid simulations (Kleva et al. 1995) suggest the importance of the ion sound Larmor radius $\rho_s = \sqrt{k_B T_e/m_i}/\Omega_{ci}$, that is, the ion gyro-radius based on the electron temperature. (Some authors use the total temperature $T_e + T_i$ in the definition of the ion sound Larmor radius, e.g., Rogers et al. (2001)). This kinetic spatial scale is a different ion length scale than what appears in anti-parallel reconnection, namely where $\nabla_\parallel P_{e\parallel}$ contributes significantly to $E_\parallel$ in the generalized Ohm's law (Eq. (2)); more complete review on this scale and the relevant diffusion region signature can be found in Ji et al. (2023, this collection). Fluid simulations show that fast reconnection with an open geometry can be realized when ρ_s is much larger than the resistive current sheet thickness of the Sweet-Parker solution and the electron inertial scale d_e (Aydemir 1991; Biskamp and Drake 1994; Cassak et al. 2007). If ρ_s is smaller than these length scales, the current sheet tends to form an elongated Sweet-Parker-type layer but is often prone to secondary island generation, as seen in panels (a)-(d) in Fig. 14 (Drake et al. 2004; Liu et al. 2014b; Stanier et al. 2015b).

In the collisionless limit, there is no dispute that reconnection with a guide field of order B_{x0} or smaller has a similar reconnection rate $R \sim \mathcal{O}(0.1)$ as antiparallel reconnection.

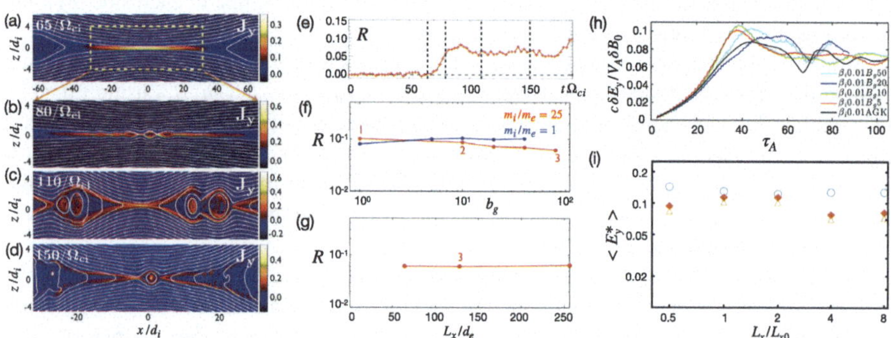

Fig. 14 *Guide field reconnection simulations.* (a)-(d) Evolution of an ion-scale current sheet with the guide field strength $b_g \equiv B_g/B_R = 8$, white contours are flux surfaces. They correspond to the four different times indicated as vertical dashed lines in panel (e), where the time evolution of the normalized reconnection rate is shown. (a) shows the formation of the intensified out-of-plane current density arising from the initial perturbation before the onset of reconnection; (b) formation of secondary magnetic islands within this sheet; (c) coalescence of magnetic islands and their ejection from the X-line; (d) formation of another secondary island. Panels (f) and (g) show the results of runs where the initial current sheets are on the electron scale. Panel (f) shows the rate R as a function of guide field b_g for a fixed system size $L_x/d_e = 128$, while panel (g) shows the rate as a function of system size for fixed guide field $b_g = 80$. Reprinted from Liu et al. (2014b), with the permission of AIP Publishing. Panel (h) shows the time-evolution of reconnection rate with different b_g in gyrokinetic simulations. Adapted from TenBarge et al. (2014), with the permission of AIP Publishing. Panel (i) shows the comparison of reconnection rates between PIC and the two-fluid model in different system sizes. Adapted from Stanier et al. (2015b,a), with the permission of AIP Publishing. They all show a reconnection rate R on the order $\mathcal{O}(0.1)$

However, no consensus has been reached on the large guide field limit. It was suggested that the reconnection rate drops when the guide field weakens the dispersive property of the kinetic Alfvénic wave (KAW) (Rogers et al. 2001; Tharp et al. 2013), that drives the reconnection outflow. However, in several later simulations, including PIC (Liu et al. 2014b), gyrokinetic (TenBarge et al. 2014), and reduced two-fluid models (Huba 2005; Stanier et al. 2015a,b), the reconnection rate $R \equiv cE_R/B_R V_A$, that is normalized to the reconnecting component, appears to be insensitive to a strong guide field that modifies the dispersive nature of the kinetic Alfvén wave (KAW). These results are highlighted in Fig. 14. One should keep in mind that the relevant Alfvén speed $V_A \equiv B_R/\sqrt{4\pi n m_i}$ is based on the reconnecting magnetic field component only, independent of the guide field strength. Even though empirically from simulations, the presence of a guide field may not affect the value of the collisionless reconnection rate significantly in both symmetric and asymmetric reconnection, a first-principles explanation of why it is this case remains missing.

3.4 Reconnection Rate Observation by the MMS Mission

The normalized reconnection rate has been determined from in-situ plasma particle and field measurements in a variety of ways (e.g. Hasegawa et al. 2024, this collection, and references therein), e.g., the normalized reconnection electric field in the diffusion or inflow regions $E_y/B_{x0}V_{Ai0}$, the normal magnetic field component in the exhausts B_z/B_{x0}, the ion inflow speed in the asymptotic inflow region V_{iz0}/V_{Ai0}, the electron inflow speed at the inflow edge of the electron diffusion region (EDR) V_{ez}/V_{Ae}, the magnetic flux transport rate across the separatrices $\partial A_y/\partial t$, the opening angle of the separatrices (see Eq. (12)), the aspect ratio of the EDR $(\partial B_z/\partial x)/(\partial B_x/\partial z)$, etc. Here, all quantities are evaluated in the co-moving frame of the X-line, and subscript "0" denotes quantities evaluated in the asymptotic inflow region. For most observations of reconnection, only a small number of these methods may be applicable, depending on where, relative to the X-line, the spacecraft collected measurements and the types of measurements that were made.

Having a large number of rate measurements is a necessary foundation for determining how the background plasma conditions impact the rate. However, reconnection rate measurements typically are associated with large error bars, making comparative analysis difficult. Errors can arise from the determination of the appropriate coordinate system; for instance, E_y in the EDR is significantly smaller than the normal electric field E_z, meaning that small errors in the coordinate axes can lead to large errors in the rate (e.g. Genestreti et al. 2018b). Additionally, errors may be introduced by the determination of the co-moving frame of the X-line; for instance, this frame velocity is often comparable to the upstream ion inflow speed at the magnetopause or magnetotail current sheets. Lastly, remote quantities that are determined by spacecraft far from the X-line (e.g., quantities determined in the inflow region that are used for normalization) are difficult to associate with measurements near the EDR when reconnection is time-varying and/or occurring in spatially inhomogeneous plasma conditions.

Nevertheless, in many ways, MMS data are ideally suited for determining the reconnection rate. Unlike previous missions, MMS particle measurements are made at a rapid enough cadence to resolve the EDR, the spacecraft measures the full 3-D electric field vector, and the tightly-spaced tetrahedral formation of four spacecraft allows gradients of plasma quantities to be determined accurately. Below, we first review the reconnection rate observations derived by various methods listed above for symmetric antiparallel reconnection cases, followed by the reconnection rate observations for different background conditions, such as the asymmetry across the current sheet and the external guide field strength.

3.4.1 Rate Observations for Symmetric Anti-Parallel Reconnection

The EDR crossing observation shown in Fig. 3 of Sect. 2.2 took place when the average inter-probe separation was approximately 17 km. It is about half of the asymptotic electron inertial length, $d_e \simeq 30$ km. This close spacecraft distance enabled the application of multi-point analysis methods to determine the detailed characteristics of the current sheet for this event in a quantitative way, such as current sheet orientation, structure, and the spacecraft orbits within the EDR. Genestreti et al. (2018b) estimated the reconnection rate $R = E_M / B_{L0} V_{Ai0}$ for this event by using several techniques to find the out-of-plane, M, direction along the reconnection electric field and estimated also the error bars using virtual data from a 2D PIC simulation (Nakamura et al. 2018) performed using the initial conditions from the observation.

Figure 15 shows the reconnection electric field E_M (left axis) and normalized reconnection rate (right axis) estimated using different analysis methods to obtain the LMN coordinate systems, as reviewed in Hasegawa et al. (2024, this collection). The average values for the upstream Alfvén speed and lobe magnetic field, B_{L0} and V_{Ai0}, were used for the normalization: $B_{L0} V_{Ai0} = 18.12$ mV/m. E_M is further corrected to minimize the contamination from the large Hall field (E_N) in the estimation. A similar reconnection rate was obtained from different methods after reasonable adjustments were performed, and the normalized rate ranged between 0.14 and 0.22. The estimated reconnection rate is $E_M = 3.2$ mV/m $\pm\, 0.6$ mV/m, which corresponds to a normalized rate of $R = 0.18 \pm 0.035$. This value well agrees with the normalized reconnection rate $R = E_M / B_{L0} V_{Ai0} = 0.18$ inside the simulated EDR by Nakamura et al. (2018) as shown in Fig. 15c, which was obtained

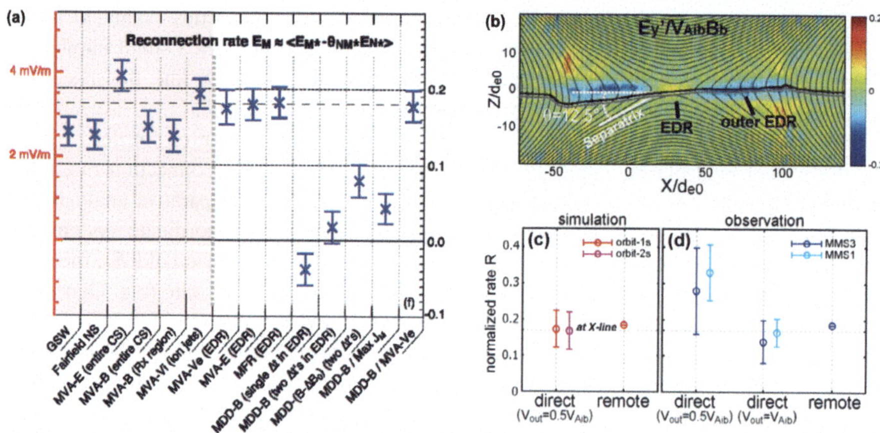

Fig. 15 *Reconnection rate estimation for symmetric anti-parallel reconnection event on 11 July 2017*. (a) The reconnection rate in the X-line frame determined over the period 22:34:03–22:34:04 UT. The error bars mark the standard deviation of the reconnection rate over this period. The reconnection rate determined from near-EDR separatrix method (Nakamura et al. 2018) is marked by the long dashed horizontal line. The data in the red-shaded region are determined using coordinate systems that are not solely based on MMS data from within the EDR. GSW = solar-wind-aberrated geocentric solar magnetospheric; MVA = minimum variance analysis; MDD = maximum directional derivative; MFR = minimization of Faraday residue. Adapted from Genestreti et al. (2018b). (b) Simulated E'_y (or E_R) with the in-plane field lines and the paths of two MMS virtual orbits. The opening angle θ of the separatrix from the simulation is estimated to be 12.5 degree. The normalized reconnection rates R directly obtained from the electric field near the EDR and remotely estimated at the separatrix (c) for two virtual orbits (red and magenta) in the simulation shown in panels (b) and (d) from the MMS3 (blue) and MMS1 (cyan) observations. Adapted from Nakamura et al. (2018)

for the virtual MMS trajectory inside the simulation shown in Fig. 15b. This value is also consistent with the reconnection rate, $R = 0.1$-0.2, approximated using the aspect ratio, which is estimated from the scale-size of the current sheet from the spacecraft motion inside the EDR and the average current density (Torbert et al. 2018).

Nakamura et al. (2018) showed that the observed reconnection rate was consistent with that found in the simulation. Furthermore, the reconnection rate was estimated from the slope of the separatrix using a method introduced by Liu et al. (2017) (see Sect. 3.1.2) giving $R = 0.186$ for the simulation and, from the MMS observations, $R = 0.17$ as shown in (Fig. 15c). Hasegawa et al. (2019) obtained the opening angle of the separatrix field line from the 2D map of the magnetic field and electron streamlines by applying the electron-MHD (EMHD) reconstruction method (see details in Hasegawa et al. 2024, this collection) and obtained a similar reconnection rate, $R = 0.17$. The aspect ratio was also determined directly from the magnetic field gradients by Heuer et al. (2022) for three magnetotail symmetric anti-parallel reconnection events, including the 11 July 2017 event and a similar reconnection rate, $R = 0.1$-0.2, was obtained.

Burch et al. (2022) determined the normalized reconnection rate from the inflow velocities normalized to the electron Alfvén speed (V_{Ae}) at the edge of the EDR, and compared with other methods during another magnetotail (symmetric) reconnection event on 6 July 2017. Figure 16 shows the plasma and field parameters near the EDR region. The vertical lines show the edge of the EDR (i.e., red transparent bands) for each spacecraft. For this event, the spacecraft was northward of the current sheet, and the converging inflow toward the current sheet center can be seen in the negative V_N. The normalized reconnection rates derived from the electron inflow velocity measurements, V_N/V_{AeL}, were 0.11–0.14 using average values of the inflow among 3 MMS spacecraft and 0.15-0.20 when maximum inflow velocity values were used. In comparison, E_M normalized to the lobe inflow quantities $V_{iA}B_L$ indicates reconnection rates of 0.1-0.17. If E_M is normalized to the EDR inflow quantities $V_{eA}B_L$, a lower reconnection rate, around 0.06, was found.

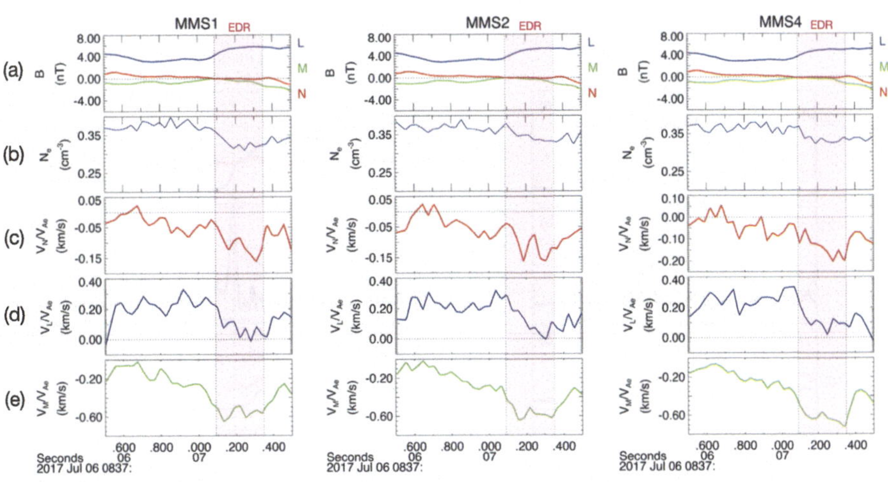

Fig. 16 *Plasma and field parameters near the EDR edge.* The transparent red bands mark the EDR for MMS1, 2, and 4. (a) Magnetic field LMN components. (b) Electron density. (c) V_{eN}/V_{AeL}. (d) V_{eL}/V_{AeL}. (e) V_{eM}/V_{AeL}, where V_{AeL} is the mean electron Alfvén speed with $B = B_L$ for each spacecraft over the first half of each plot (08:37:06.5–08:37:07.0 UT). Adapted from Burch et al. (2022)

3.4.2 Current Sheet Structures and Rate Observations for Asymmetric and/or Guide Field Reconnection

As discussed in the two previous sections, the background asymmetry across the current sheet and/or the existence of the guide fields significantly modifies the structure of the reconnection current sheet, an effect that has been identified in observations. Figure 17 shows two examples of MMS observations from magnetopause reconnection events. The left panels show an event on 16 October 2015 with anti-parallel field geometry, i.e., $B_M/B_L \simeq 0.1$, at the magnetosheath and a large asymmetry, i.e., the magnetosheath to magnetosphere density ratio $n_{sh}/n_{sp} = 16$. The right panels show an event on 8 September 2015 that has a strong guide field $B_M/B_L \simeq 5$ and a smaller density asymmetry, $n_{sh}/n_{sp} = 2.5$.

The 16 October 2015 event (left panels) was first reported by Burch et al. (2016b). The current sheet crossing took place from the magnetosphere (low density) to the magnetosheath (high density). The most pronounced feature of asymmetric reconnection is the deviation between the X-point where the $B_L = 0$ (blue line) and the electron "crescent point" (red line), where the electron velocity distribution function (VDF) has a crescent shape indicating non-gyrotropic distribution in the flow stagnation point; this special point coincides with the peak $\mathbf{J} \cdot \mathbf{E}' = \mathbf{J} \cdot (\mathbf{E} + \mathbf{V}_e \times \mathbf{B}/c)$, that was used to measure the dissipation (Zenitani et al. 2011b).

The 8 September 2015 event (right panels of Fig. 17), first reported by Eriksson et al. (2016b), on the other hand, shows a clear peak in the energy conversion rate around the X-point, $B_L = 0$ due to the dominant parallel components of the current and the electric field. Such features of the strong guide field event (i.e., $B_M/B_L > 0.5$) were obtained also in a statistical study of the energy conversion rate in the magnetosheath reconnection by Wilder et al. (2018). There was no significant non-gyrotropic electron distribution detected at the X-point during the 8 September 2015 event, indicating the effect of small gyroradius relative to the scale size of the current sheet. In modest guide field events, such as the one reported by Chen et al. (2017), energy conversion rate enhancement takes place both at the X-line and the flow stagnation point, and the parallel heating of electrons occurs at both locations. Overall, the separation of the flow stagnation point and the X point is found with

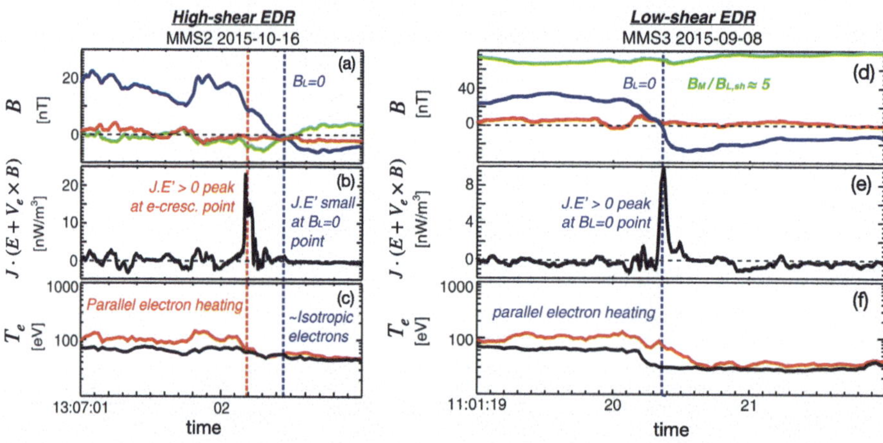

Fig. 17 *Observation of asymmetric reconnection events for antiparallel and guide field geometry.* (a, d) The L(blue), M(green), N(red) components of magnetic fields, (b, e) the local energy conversion rate, $\mathbf{J} \cdot \mathbf{E}'$, (c, f) the parallel (red) and perpendicular (black) electron temperatures. The vertical dashed blue and red lines mark the $B_L = 0$ point and the electron "crescent point". Adapted from Genestreti et al. (2017)

density asymmetry and magnetic field asymmetry (Genestreti et al. 2017), consistent with the prediction in Sect. 3.2.3.

Although the asymmetry, as well as the guide field, significantly modifies the structure of the reconnection current sheet, the observed range of reconnection rates is similar to that in standard anti-parallel symmetric reconnection. Burch et al. (2020) determined the normalized reconnection rate from the electron inflow velocities V_{eN} for four MMS events, including three previously published crossings (Chen et al. 2017; Phan et al. 2018; Pritchard et al. 2019), and obtained values between 0.05 and 0.25. Among these four events, one event was an "electron-only" reconnection event in the magnetosheath (Phan et al. 2018) that will be discussed in the next section.

A survey of asymmetric reconnection rates has been performed by Pritchard et al. (2023), including seven magnetopause events that show values of 0.14 ± 0.09 and seven magnetosheath events that show values of 0.16 ± 0.12. There was no correlation between the normalized reconnection rate and guide field, as has been suggested by simulation (see Sect. 3.3.1). A finite guide field has been also reported for reconnection events in the magnetotail in a current sheet with wave fluctuations (Chen et al. 2019) and varying guide field, 0.14-0.5; the normalized reconnection rate ranges between 0.05 and 0.3. A transient current sheet at the dipolarization front (Hosner et al. 2024) with a guide field 1.8 shows a normalized reconnection rate of 0.16-0.18, which is comparable to that observed during reconnection at the magnetopause and in the magnetosheath.

3.5 Electron-Only Reconnection

3.5.1 Observational Evidence

In turbulent plasma, reconnection has long been suggested to play a role in the dissipation of turbulent energy (e.g., Matthaeus and Lamkin 1986; Servidio et al. 2009). The turbulent magnetosheath region downstream of Earth's quasi-parallel bow shock often contains hundreds of small-scale current sheets in which reconnection could occur (Retinò et al. 2007; Sundkvist et al. 2007; Yordanova et al. 2016; Vörös et al. 2017; Wilder et al. 2018). If standard reconnection were to operate in turbulent current sheets, the ion jets in the extended exhausts would be the easiest reconnection signature to detect. However, the ultra-high time resolution plasma and field measurements of MMS have revealed a lack of ion scale exhausts, although some electron jets were observed. It was suggested that this implies the existence of a new form of reconnection in which ions do not participate, but electrons do. This was dubbed "electron-only" reconnection (Phan et al. 2018; Stawarz et al. 2019, 2022).

In this type of reconnection, the electron outflow jets from the reconnection X-line have speeds comparable to the electron Alfvén speed based on B_L upstream of the electron diffusion region, and the current sheet width is substantially narrower than the ion Larmor radius or ion inertial scale. Importantly, in contrast to the electron diffusion region of standard reconnection, electron-only reconnecting current sheets are not embedded inside ion-scale current sheets (Phan et al. 2018). Figure 18 shows a fortuitous event where pairs of MMS spacecraft simultaneously detected oppositely directed super-ion-Alfvénic electron outflow jets emanating from an X-line (Fig. 18(c), (d)) in an electron-scale current sheet (Fig. 18(b)) (Phan et al. 2018). Strong parallel electric fields (Fig. 18(e)) and enhanced energy conversion (Fig. 18(f)) were present in the current sheet. This current sheet was one of hundreds of electron-scale current sheets in a 10-minute interval downstream of a quasi-parallel shock. Analysis of the statistical properties of this and other magnetosheath intervals measured by MMS reveals that the presence of electron-only reconnection is linked to the correlation length of the turbulence (i.e., the driving scale of the turbulence), with the correlation

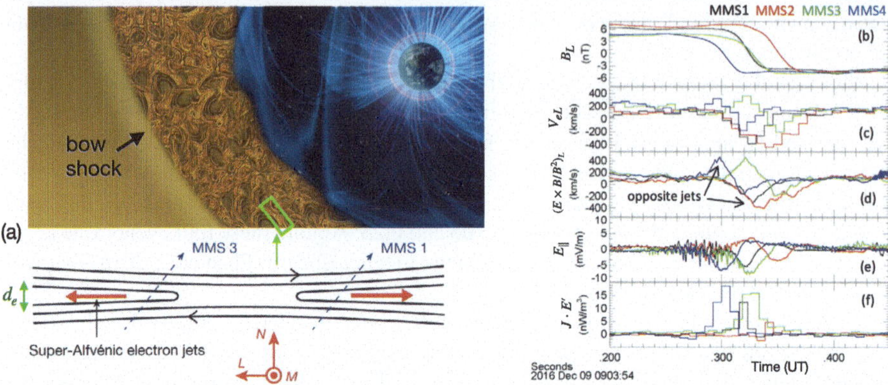

Fig. 18 *Observation of electron-only reconnection.* (a) Schematics showing electron-only reconnection in an electron-scale current sheet embedded in turbulent structures downstream of Earth's quasi-parallel shock, (b) reconnecting magnetic field component (L), (c) electron outflow velocity, (d) $E \times B$ velocity in the outflow direction, (e) parallel electric field, and (f) non-ideal energy conversion $\mathbf{J} \cdot (\mathbf{E} + \mathbf{V}_e \times \mathbf{B}/c)$. MMS 3 and 4 observed positive V_{eL} outflow jets, while MMS 1 and 2 observed negative V_{eL} jets inside the current sheet. Adapted from Phan et al. (2018), reproduced by permission of Springer Nature. The illustration in Panel(a) is credited to NASA's Goddard Space Flight Center

length of the electron-only events being several ion inertial lengths or less (Stawarz et al. 2019, 2022). These observations suggest that electron-only reconnection occurs in small-scale current sheets when there is insufficient space and/or time for the ions to couple to the reconnected magnetic field.

MMS has also detected sites of magnetic reconnection within the bow shock transition layer itself (Gingell et al. 2019; Wang et al. 2019). The predominantly electron-only reconnection events in the shock disentangle the turbulent shock fields and may contribute to the overall shock heating. Along with complementary studies quantifying how current sheets and reconnection in the magnetosheath fit into the energy budget (Schwartz et al. 2021) and are influenced by the properties of the bow shock (Bessho et al. 2019; Gingell et al. 2020; Yordanova et al. 2020; Bessho et al. 2022), MMS has brought together three fundamental areas of plasma physics research – turbulence, shocks, and reconnection. In addition to the bow shock and magnetosheath, MMS has measured electron reconnection in other contexts, most of which involve kinetic-scale turbulent structures. These included foreshock transients (Liu et al. 2020b), electron-scale substructures inside macro-scale magnetic flux ropes (Man et al. 2020), reconnection exhausts (Huang et al. 2021; Norgren et al. 2018), magnetotail dipolarization fronts (Marshall et al. 2020), and the early phase of magnetotail reconnection (Lu et al. 2020). See also the review by Hwang et al. (2023, this collection) for further discussion of many of these phenomena. These findings suggest that electron-only reconnection is prevalent in kinetic-scale current sheets in space plasmas and could play an important role in the dissipation of turbulence energy. In the wake of this MMS discovery, electron-scale reconnection is now also studied in laboratory experiments (Shi et al. 2022; Greess et al. 2022; Chien et al. 2023; Shi et al. 2023).

3.5.2 Theory

A hybrid simulation study of resistive reconnection with no guide field by Mandt et al. (1994) found that ions become decoupled from the magnetic field when the length of the

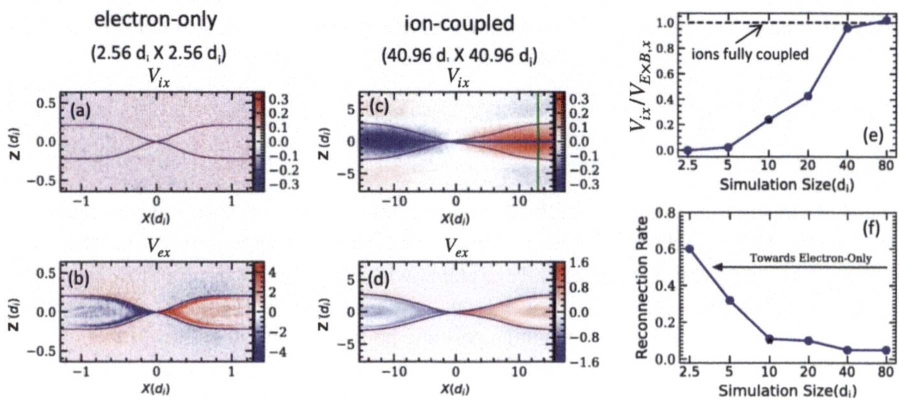

Fig. 19 *Transition from standard ion-coupled reconnection to electron-only reconnection.* (a, b) Ion and electron velocities in the exhaust (x or L) direction for a small 2D PIC simulation. Velocities are normalized to the asymptotic upstream proton Alfvén speed V_A, and $V_{Ae} = 42.8V_A$ in this simulation. (c, d) ion and electron velocities for a larger simulation, (e) ion outflow velocity normalized to the $E \times B$ velocity as a function of simulation domain size, and (f) reconnection rate R (normalized to $B_{x0}V_A$) versus simulation domain size. Adapted from Pyakurel et al. (2019)

current sheet (in the outflow direction) falls below $\sim 10d_i$. Such reconnection without ion-coupling has also been modeled using electron magnetohydrodynamics (EMHD) simulations (Jain and Sharma 2009, 2015). Prompted by the MMS observation (Phan et al. 2018), Pyakurel et al. (2019) investigated the transition from ion-coupled to electron-only reconnection using particle-in-cell (PIC) simulations by varying the simulation domain size systematically. They found that the transition from fully ion-coupled to electron-only reconnection is gradual. This transition is characterized by a gradual increase in the degree to which the ions are frozen-in to the magnetic field as the simulation box size increases, with the ions being fully coupled ($\mathbf{V}_{i\perp} \simeq \mathbf{E} \times \mathbf{B}/B^2$) when the box size reaches $40d_i$ (Fig. 19(e)); note that the boundary conditions are periodic in the outflow direction. On the other hand, ion outflows are weakly coupled to the magnetic field when the domain size is below $20d_i$ and clearly not coupled below $5d_i$. Another study suggested that it is the ion gyro-radius, instead of the inertial scale, that sets the transition to electron-only reconnection (Guan et al. 2023).

Figures 19(a), (b) show an example of the ion and electron velocity along the outflow direction for the smallest simulated domain of $2.56d_i \times 2.56d_i$. The electron outflows are super-ion-Alfvénic outflows (Fig. 19(b)), while there are essentially no ion outflows (Fig. 19(a)). On the other hand, the simulation with a large domain size of $40.96d_i \times 40.96d_i$ exhibits standard ion-coupled reconnection with both ion and electron outflows (Fig. 19(c)-(d)).

Since ions are much more massive than electrons, they remain more or less immobile ($V_i \simeq 0$) compared to electrons within a small spatial and temporal scale, as shown in Fig. 19(a), and the rate of work done on ions ($en\mathbf{E} \cdot \mathbf{V}_i$) thus becomes negligible. The majority of magnetic energy is converted to electron energy. The dynamics are then described by the steady-state electron momentum equation,

$$\frac{(\mathbf{B} \cdot \nabla)\mathbf{B}}{4\pi} - en\mathbf{E} \simeq nm_e(\mathbf{V}_e \cdot \nabla)\mathbf{V}_e, \tag{26}$$

where the magnetic tension works to drive the electron outflow. If we just consider tension and electron inertia, this equation results in electron Alfvénic jets with speed

$$V_{\text{out}} \simeq V_{Ae} \simeq \frac{B_{x0}}{\sqrt{4\pi n m_e}}. \tag{27}$$

The Hall electric field arising from the decoupling between the two species gives a positive enE_x in Eq. (26), which slows down electrons while speeding up ions (i.e., tries to couple electrons and ions again). This partially explains why the peak electron outflow speed V_{ex} is super-ion-Alfvénic but sub-electron-Alfvénic as shown in Fig. 19(b). The back-pressure (not included in Eq. (27)) arising from the periodic boundaries in the outflow direction may also limit the outflow speed, especially within such a small system. For ions, Pyakurel et al. (2019) show that the reduction of the ion outflow speeds as a function of the system's size compares well with the prediction from the "standing wave" approximation, where the Hall effect dominates (Mandt et al. 1994; Rogers et al. 2001; Drake et al. 2008).

Since magnetic flux remains frozen-in to electrons, the magnetic flux transport speed that determines the reconnection rate is now not limited by the ion Alfvén speed but by the faster electron Alfvén speed. The higher flux transport speed (Eq. (27)) will make the normalized rate R (that is normalized to the proton Alfvén speed) higher than $\mathcal{O}(0.1)$, consistent with the simulated rate in the small system size limit, as shown in Fig. 19(f). If the EDR aspect ratio remains on the order of ~ 0.1, a rough estimate of the normalized rate is $R \leq 0.1\sqrt{m_i/m_e} = \sqrt{1836} \times 0.1 = 4.3$, which can only be regarded as the upper bound value because the simulated value appears to be smaller than unity (Fig. 19(f)). An analytical model better than this simple estimation needs to be derived.

Given that electron-only reconnection tends to occur in plasma environments where the magnetic structure correlation length is small (several ion inertial lengths or less) (Stawarz et al. 2022), such structures tend to be highly 3D in nature. Pyakurel et al. (2021) found that the reconnection rate in 3D electron-only reconnection (with a finite X-line) is higher than in 2D. This is because, in addition to reconnection outflows in the standard exhaust direction, there is a differential mass flux out of the diffusion region along the X-line direction, enabling a faster inflow velocity and, thus, a larger reconnection rate. The theoretical findings of higher reconnection rates in 3D electron-only reconnection further suggest that it could play an important role in the dissipation of turbulence energy.

Observationally, Burch et al. (2022) reported a normalized reconnection rate for the Phan et al. (2018) electron-only reconnection event using measurement of the inflow velocity and obtained a value of $0.25 \pm 20\%$. Pritchard et al. (2023) used measurements of the reconnection electric field for this event to determine a very similar normalized reconnection rate of $0.23 \pm 43\%$. These values are at the high end of theoretical prediction, but more measurements are needed to determine whether the reconnection rates are significantly higher than for ion-coupled reconnection.

3.6 Reconnection with Heavy and Cold Ions

Space plasmas often have multiple ion populations, including, for instance, ions heavier than protons, or proton beams that are colder than the background protons. In this subsection, we discuss the effects of multiple ion populations on the magnetic reconnection rate and the extension of the generalized Ohm's law to such plasmas.

3.6.1 Theory

It is of interest to understand how reconnection physics is affected when multiple ion species co-exist in a plasma. Similar to the treatment of a typical two-species (electron-proton) plasma, we can combine the momentum equations of multiple species into a single force balance equation,

$$\frac{(\mathbf{B}\cdot\nabla)\mathbf{B}}{4\pi} \simeq \sum_{s}^{1,2,\ldots} n_{is} m_{is} (\mathbf{V}_{is}\cdot\nabla)\mathbf{V}_{is}, \tag{28}$$

where m_{is} and n_{is} represent the ion mass and density of species "s", respectively. The ion charge q_{is} and density satisfy $\sum_s q_{is} n_{is} \simeq n_e \equiv n$ for quasi-neutrality. Each ion species will have its own diffusion region (Shay and Swisdak 2004). Given a sufficiently large system, all ions will become frozen-in to the reconnection outflow outside the outermost diffusion region. With heavier ions, this will occur at larger spatial scales and longer timescales than that in the typical electron-proton plasma. The increased mass load at outflows can significantly limit the outflow speed; i.e., note that a proton is an ion of the lowest mass. The outflow speed scales as the Alfvén speed based on the effective mass density $\rho_{\text{heavy}} = \sum_s n_{is} m_{is}$

$$V_{\text{out}} \simeq V_{A,\text{heavy}} = \frac{B_R}{\sqrt{4\pi\rho_{\text{heavy}}}}. \tag{29}$$

Thus, based on this full mass load and the $\delta/L \sim \mathcal{O}(0.1)$ assumption, the reconnection rate normalized to $V_{A,heavy}$ is expected to be $\mathcal{O}(0.1)$. If one normalized E_R to the proton Alfvén speed, a lower value is expected. However, if reconnection occurs within a small spatial domain and short time scale, the heavy ions can become decoupled, and the outflowing flux transport speed is not constrained by the Alfvén speed in Eq. (29). The reconnection electric field can thus be higher than the expected $0.1 B_R V_{A,\text{heavy}}$. This situation resembles "electron-only" reconnection, as discussed in Sect. 3.5, where protons are not fully coupled.

3.6.2 Results from Simulations and Observational Evidence

Heavy Ions Several PIC simulations that include three species (electrons, H^+, O^+) have shown the differential behavior of lighter and heavier ions near the x-line of magnetic reconnection, resulting in a multi-layered diffusion region with sizes related to the characteristic length-scales of each species, e.g., Shay and Swisdak (2004), Markidis et al. (2011), Liu et al. (2015c), as illustrated in Fig. 20(a). Observational evidence of the multi-layered nature of the DR in the presence of oxygen has also been shown using Cluster (Escoubet et al. 2001) observations in Earth's magnetotail (Liu et al. 2015c).

Full PIC simulations of magnetic reconnection in Earth's magnetotail that include O^+ have shown that the reduction in reconnection electric field does not really scale with the prediction based on the full mass-load (Shay and Swisdak 2004; Markidis et al. 2011; Tenfjord et al. 2019). Figure 20(b) shows the results of various full PIC simulations of magnetic reconnection with varying amounts of O^+, where the time evolution of the reconnection electric field is plotted. The maximum electric field drops with increasing amounts of oxygen. However, the reduction is consistent with

$$R \simeq \frac{0.1}{1 + n_O/n_p}, \tag{30}$$

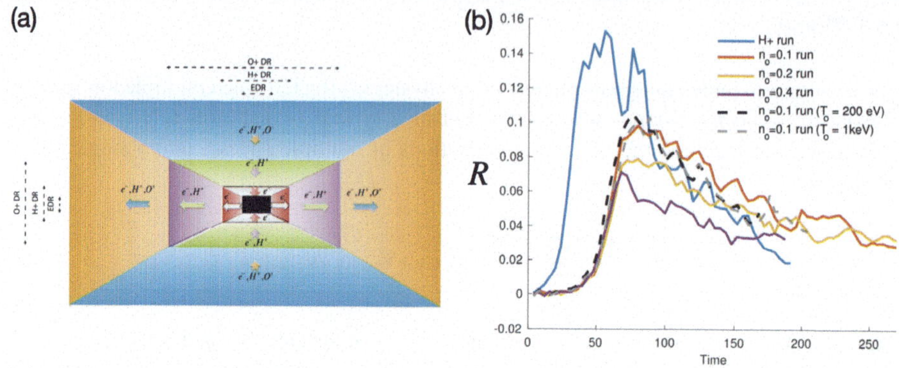

Fig. 20 *Reconnection rate for various amounts of oxygen (i.e., heavy ion) in PIC simulations.* (a) multi-layered diffusion regions. Adapted from Liu et al. (2015c). (b) Evolution of reconnection electric field in full PIC simulations of symmetric magnetic reconnection with different amounts of O^+ (i.e., n_O). Reprinted from Tenfjord et al. (2019), with the permission of Wiley

rather than the reduction expected by the full mass load. i.e., $R \simeq 0.1/[1 + m_O n_O/(m_p n_p)]^{1/2}$. This discrepancy can be explained by the fact that O^+ remains unmagnetized during the typical timescales of the simulations (which are related to the reconnection timescales in the Earth's magnetotail) (Tenfjord et al. 2019; Kolstø et al. 2020). Therefore, O^+ is ballistically accelerated by the non-ideal electric field within its diffusion region, and O^+ acts as an energy sink (reducing the reconnection rate), but the reduction is less severe compared with the prediction based on the full mass load. Another interesting conclusion that can be drawn from Fig. 20(b) is that changing the temperature of the O^+ population does not have an effect on the reconnection electric field. Finally, it is also noted that for larger domains and longer time scales, when all ions are magnetized outside of the diffusion region, the full mass-load scaling of the reconnection rate is expected.

Cold Protons Another common situation in magnetic reconnection in Earth's magnetosphere is having two distinct proton populations: a hot (keV-scale) component coming from the plasma sheet plus a cold (eV-scale) component arising from the Earth's ionosphere. In the following discussion, we refer to "protons" as simply "ions".

Ions demagnetize at length scales below the ion inertial length or the ion gyroradius. The ratio between the two is given by $\rho_i/d_i = \sqrt{\beta_i}$. For high-$\beta$ plasmas, the gyroradius is larger, while for low-β plasmas, the inertial length is larger. At the Earth's magnetopause, the plasma beta is often of the order of 1, and therefore the two length scales are comparable. When hot and cold protons are present, cold protons have the ability to remain magnetized inside narrow structures such as the separatrix or the (hot) ion diffusion region (Toledo-Redondo et al. 2015, 2018; André et al. 2010). Under this situation, a multi-layered diffusion region is also generated due to the different gyroradius of the two proton populations. The multi-layered nature of the DR has been observed both using PIC simulations (Divin et al. 2016; Dargent et al. 2017, 2020) and MMS observations (Toledo-Redondo et al. 2016).

The inclusion of multiple proton populations leads to a mass-loading reduction of the reconnection electric field. However, the normalized reconnection rate ($E_R/V_{A,\text{proton}}B_R$) remains unaffected (Divin et al. 2016; Dargent et al. 2017, 2020; Spinnangr et al. 2021). Figure 21(a)-(b) serves to illustrate the mass-loading effect by adding cold protons to magnetic reconnection at a later time. Figure 21(c) compares the measured reconnection rate (horizontal axis) with the predicted reconnection rate using the full mass-load in the Cassak

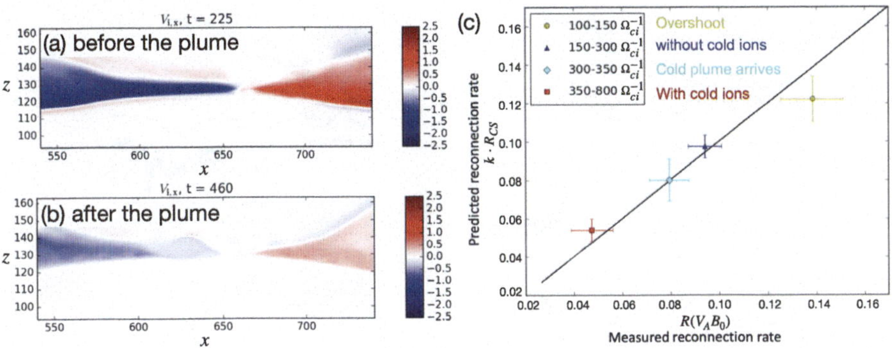

Fig. 21 *Reconnection electric field for various amounts of cold ions in a PIC simulation.* Panel (a) and (b) show the ion outflow speed V_{ix} before and after the arrival of the cold plasma plume during asymmetric reconnection. (c) Predicted versus measured reconnection electric fields at various stages model the impact of a cold, dense plume on Earth's magnetopause reconnection. Except during the overshoot time, the scaling by Cassak and Shay (2007) explains the observed reductions of the reconnection electric field. Reprinted from Dargent et al. (2020), with the permission of Wiley

and Shay (2007) scaling, as discussed in Sect. 3.2. This setup of the PIC simulation mimics the impact of a cold, dense plume on the reconnecting Earth's magnetopause (Dargent et al. 2020). At $t < 150\ \Omega_{ci}^{-1}$, the maximum reconnection rate is reached in the simulation; this time is often referred to as the overshoot time. For $150 < t < 300\ \Omega_{ci}^{-1}$, reconnection proceeds, but the cold, dense plume has not yet reached the reconnection region (dark blue dot). At $t \simeq 300\ \Omega_{ci}^{-1}$, the cold, dense plume reaches the reconnection region, starts mass-loading the reconnecting flux tubes, and reduces the reconnection electric field (cyan dot). For $t > 350\ \Omega_{ci}^{-1}$, more cold ions have entrained reconnection, and the reconnection electric field is even smaller (red dot). Except during the overshoot time (yellow), all measured reconnection electric fields scale well with the predicted asymmetric reconnection electric field (Eq. (22) with the $\delta/L \sim 0.1$ assumption), indicating that the observed reduction corresponds to the effect of mass-load.

When both cold (eV) and hot (keV) proton populations are present in reconnection, Ohm's law can be expressed as (Toledo-Redondo et al. 2015)

$$\mathbf{E} + \frac{n_{ic}}{n}\frac{\mathbf{V}_{ic}\times\mathbf{B}}{c} + \frac{n_{ih}}{n}\frac{\mathbf{V}_{ih}\times\mathbf{B}}{c} = \frac{\mathbf{J}\times\mathbf{B}}{nec} - \frac{\nabla\cdot\mathbf{P}_e}{ne} + \frac{m_e}{e^2}\frac{d}{dt}\left(\frac{\mathbf{J}}{n}\right), \tag{31}$$

where "c" and "h" indicate cold and hot populations, respectively. The electron density $n_e = n \simeq n_{ic} + n_{ih}$, and $\mathbf{J} = en_{ic}\mathbf{V}_{ic} + en_{ih}\mathbf{V}_{ih} - en\mathbf{V}_e$.

Figure 22 shows two independent MMS crossings of the reconnecting dayside magnetopause. The magnetic field rotation can be observed in panels a1 and a2. The magnetic field amplitude at the two crossings, both at the magnetosheath and the magnetosphere, is of the same level. The magnetosheath densities are between 10-20 cm^{-3} on the two crossings, but the density on the magnetosphere for crossing 1 ($\sim$0.5 cm^{-3}) is much smaller than for crossing 2 ($\sim$11 cm^{-3}), due to the presence of a cold ion plume in the latter (see panel d2). The ion velocities (panels c1 and c2) show ion jets consistent with the prediction in Eq. (21) (Cassak and Shay 2007). The normal component (N) of the Ohm's law terms is plotted in panels e1 and e2. For crossing 1, there is no cold ion population ($n_i = n_{ih}$, $\mathbf{V}_i = \mathbf{V}_{ih}$) and the term $\mathbf{E} + \mathbf{V}_i \times \mathbf{B}/c$ is balanced by $\mathbf{J}\times\mathbf{B}/enc$, while for crossing 2 the term $\mathbf{E} + (n_{ih}/n_i)\mathbf{V}_{ih}\times\mathbf{B}/c$

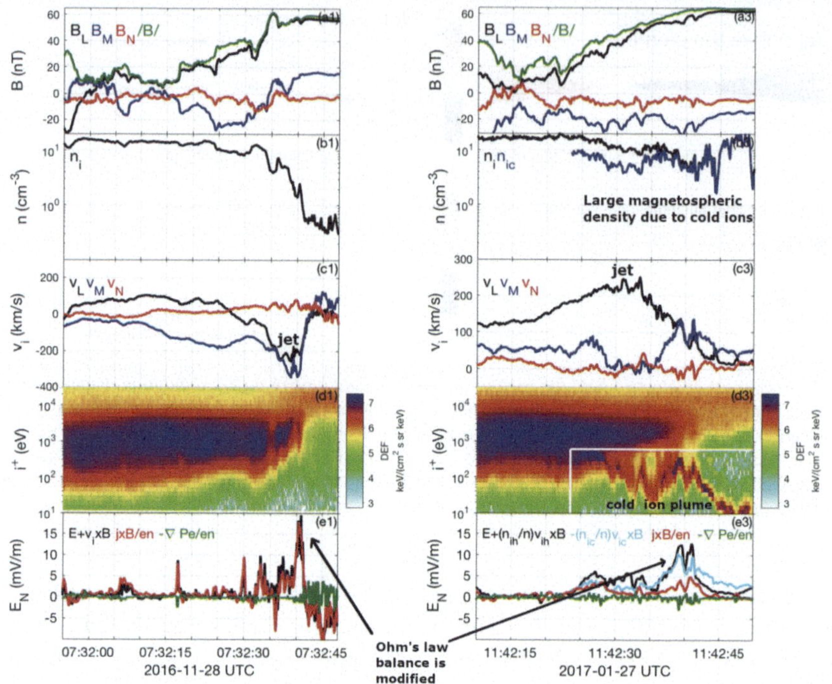

Fig. 22 *Generalized Ohm's law analysis including cold ions at the separatrix of dayside magnetic reconnection.* Comparison of two crossings of the magnetic reconnection separatrices. (a) Magnetic field in LMN coordinates. (b) total ion density and cold ion density. (c) Ion velocity in LMN coordinates. (d) Ion spectrogram. (e) Terms in the generalized Ohm's law (Eq. (31)). Adapted from Toledo-Redondo et al. (2018), with the permission of Wiley

is mostly balanced by $-(n_{ic}/n_i)\mathbf{V}_{ic} \times \mathbf{B}/c$, and $\mathbf{J} \times \mathbf{B}/enc$ contributes only a small fraction. The reason is that for crossing 2, the abundant cold ions remain magnetized inside the separatrix layer and reduce the perpendicular currents (Toledo-Redondo et al. 2018).

3.7 High-β Reconnection

While more magnetic energy is available in the low plasma $\beta \equiv P/(B^2/8\pi) \ll 1$ regime, reconnection also occurs in systems with high $\beta \gg 1$. Such plasmas can be found in the outer heliosphere (β up to 10) (Drake et al. 2010; Schoeffler et al. 2011), at the Galactic center ($\beta \sim 10^1 - 10^2$) (Marrone et al. 2007), or in the hot intracluster medium (ICM) of galaxy clusters ($\beta \sim 10^2 - 10^4$) (Carilli and Taylor 2002; Schekochihin and Cowley 2006).

3.7.1 Theory

In this limit, self-generated pressure anisotropy and/or pressure gradients both upstream (Egedal et al. 2013) and downstream (Bessho and Bhattacharjee 2010; Liu et al. 2011b, 2012; Le et al. 2014; Haggerty et al. 2018) of the diffusion region can affect the force balance that determines the maximum plausible reconnection rate. Thus, unlike in Sect. 3.1.2, we

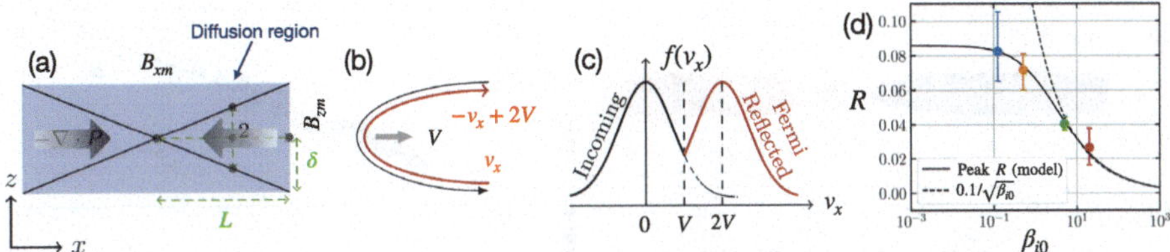

Fig. 23 *Including thermal pressure effect in the reconnection rate model.* (a) Illustrates the back-pressure (grey arrows) within the diffusion region, that opposes the outflow. (b) and (c) show how Fermi reflection is revealed in the particle distribution. (d) Shows that the predicted maximum plausible rate (solid curves) explains well the simulated rates (dots) in cases with different plasma-β. Adapted from Li and Liu (2021), reproduced by permission of the AAS

need to include plasma pressure effects in the mesoscale force balance,

$$\nabla \cdot \left(\varepsilon \frac{\mathbf{BB}}{4\pi} \right) \simeq \frac{\nabla B^2}{8\pi} + \nabla P_\perp + nm_i (\mathbf{V} \cdot \nabla)\mathbf{V} \tag{32}$$

where the pressure anisotropy (i.e., firehose) factor $\varepsilon = 1 - 4\pi(P_\| - P_\perp)/B^2$ and $P_\|$ $(P_\perp)$ refers to the pressure component parallel (perpendicular) to the local magnetic field. Using this force balance, one can follow the framework in Sect. 3.1.2 to derive the general $R - S_{\rm lope}$ relation (Li and Liu 2021).

Specifically, the back-pressure $\nabla P_\perp$ can oppose the outflow, and a pressure anisotropy of $\varepsilon < 1$ can weaken the magnetic tension. It was shown that ion Fermi reflections of ions in the outflow region (illustrated in Fig. 23(b)) play the dominant role in increasing the back-pressure (illustrated in Fig. 23(c)) and reducing the reconnection outflow speed in the high-β limit. The predicted maximum plausible reconnection rate scales as $R_{max} \simeq 0.1/\sqrt{\beta_i}$ in the high upstream ion beta (β_i) limit, comparing well with PIC simulations in Fig. 23(d).

3.7.2 Observational Evidence

The same theory (Li and Liu 2021) also makes an important correction to the outflow speed that can be tested using observation. In the low-β limit,

$$V_{\rm out} \simeq V_{{\rm out},m} \simeq 0.43 V_A. \tag{33}$$

Here subscript "m" denotes the outflow edges of the "microscopic" ion diffusion region. The first equality holds in the small-aspect-ratio limit. This prediction explains why the outflow speed is usually around half of the Alfvénic speed, as is often observed in space. In the high-β limit, the outflow speed is further reduced

$$V_{\rm out} \simeq V_{{\rm out},m} \simeq \frac{\sqrt{\pi}}{4} \frac{\varepsilon_m V_A}{\sqrt{\beta_i}}, \tag{34}$$

where ε_m is the anisotropic parameter at the inflow edge of the IDR. Equation (34) is almost identical to the expression obtained in Haggerty et al. (2018), $V_{\rm out} \simeq (1/3)\varepsilon_m V_A^2/(T_{i\|}/m_i)^{1/2}$, which adapted an empirical factor of $1/3$ in their model. Regardless of the difference in the approach, most importantly, both expressions agree well with 81 kinetic simulations and 14 in situ observations that span a wide range of parameter regimes, as shown in Fig. 24(c)-(d).

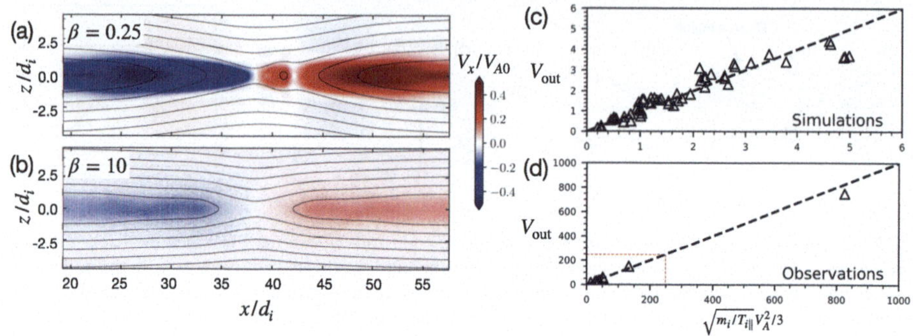

Fig. 24 *The scaling of outflow speed as a function of plasma β.* Ion outflow speeds in (a) low-β and (b) high-β plasmas Adapted from Li and Liu (2021). In panel (c), the asymptotic $E \times B$ outflow velocity (presuming the ion outflow speed for magnetized ions) versus the outflow speed prediction for PIC simulations and for observations (d) in nearly anti-parallel events. Adapted from Haggerty et al. (2018)

3.8 Reconnection Suppression by Diamagnetic Drifts and Sheared Flows

In the usual scenario, the outflow directions from a reconnection X-line ($\pm L$ in LMN coordinates) are equivalent, and hence one expects the outflow jets to be symmetric. However, in certain situations the X-line itself can move in one direction or the other. Not only does this motion break the outflow symmetry, but if the motion is sufficiently fast, it can disrupt an outflow jet. When this happens, reconnection itself can be suppressed. This effect is particularly pronounced in two regimes: asymmetric reconnection in which a pressure gradient extends across the current sheet, such as planetary magnetopauses (see e.g. Gershman et al. 2024, this collection, Phan et al. 2013a, and references therein), and reconnection in the presence of shear flows, such as events at the flanks of Earth's magnetopause (see e.g. Hwang et al. 2023, this collection, and references therein).

3.8.1 Diamagnetic Suppression

In the presence of a magnetic field, any non-parallel pressure gradient produces a diamagnetic drift,

$$\mathbf{V}_s^* = -c\frac{\nabla P_s \times \mathbf{B}}{q_s n B^2}, \tag{35}$$

where $P_s = nk_B T_s$ is the thermal pressure and q_s is the charge of species s. Somewhat famously, $\mathbf{V}_s^*$ is a fluid drift that does not correspond to actual particle motion; nevertheless, it can advect the magnetic field (Coppi 1965; Scott and Hassam 1987). To see this, note that in a system with an invariant direction in $\hat{\mathbf{y}}$ (i.e., $\partial_y = 0$), one can write $\mathbf{B} = \hat{\mathbf{y}} \times \nabla\psi(x, z) + B_y(x, z)\hat{\mathbf{y}}$ where ψ is the magnetic flux function. Taking the cross product of Faraday's law with $\hat{\mathbf{y}}$ yields $\partial_t \nabla\psi - c\nabla E_y = 0$, or $E_y = \partial_t\psi/c$.

Next, dotting the electron momentum equation (Eq. (1)) with $\hat{\mathbf{y}}$ gives

$$E_y = -\frac{1}{c}\hat{\mathbf{y}} \cdot (\mathbf{V}_e \times \mathbf{B}) + E_{y,\text{non-ideal}}. \tag{36}$$

The last term represents the non-ideal electric field that breaks the electron frozen-in condition. For simplicity, we ignore it and substitute for $\mathbf{B}$ to get $E_y = -\mathbf{V}_e \cdot \nabla\psi/c$ that leads to

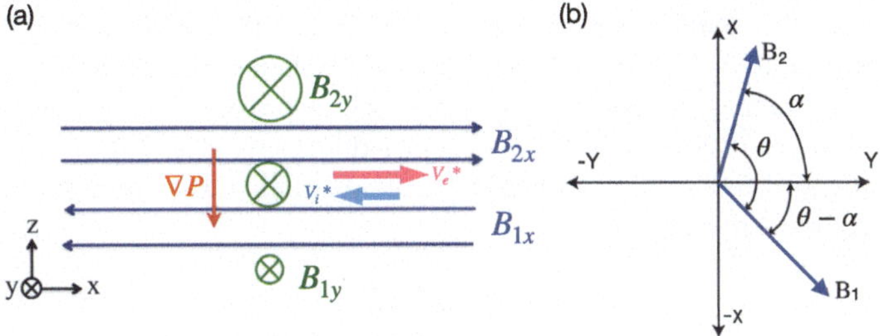

Fig. 25 *Definition of the coordinate system.* (a) The guide field and pressure gradient give rise to diamagnetic drifts during reconnection on the x-z plane. Adapted from Liu and Hesse (2016). (b) The view along the inflow (z) direction. Given the total magnetic shear angle θ, the angle α that tells the orientation of the x-line is, in general, unknown. Adapted from Swisdak and Drake (2007), with the permission of Wiley

an advection equation for the flux (Coppi 1965),

$$\partial_t \psi + \mathbf{V}_e \cdot \nabla \psi = 0. \tag{37}$$

Hence, the electron fluid velocity, which includes a diamagnetic component given by Eq. (35), advects magnetic structures. [Note that if one retains the $E_{y,\text{non-ideal}}$ term that can cause slippage between electrons and magnetic flux within the EDR, the result is the Magnetic Flux Transport (MFT) velocity U_ψ (Liu and Hesse 2016; Liu et al. 2018b; Li et al. 2021a), that is basically the $E \times B$ drift speed based solely on the in-plane magnetic component; also discussed in Hasegawa et al. (2024, this collection).]

Consider then, on a qualitative level, the effect of such a drift on a reconnecting X-line and specifically on the (ion) Alfvénic outflows (in the $\pm x$ direction in the current coordinate system), as shown in Fig. 25(a). A diamagnetically drifting X-line will move in the same direction as one of the two outflows and, for certain parameters, the drift speed can exceed the outflow velocity. This case is roughly analogous to shock propagation in that the X-line's motion is rapid enough that the X-line itself arrives downstream before any newly reconnected field lines. As numerical simulations have shown (Swisdak et al. 2003), the net effect is to choke off and suppress reconnection. The stability criterion is basically,

$$|V_e^* - V_i^*| > V_A, \tag{38}$$

where V_e^* and V_i^* are the electron and ion diamagnetic velocities, respectively. A simple relationship quantifying when such suppression should occur– particularly one that depends only on upstream parameters – would be useful for spacecraft observations. To derive such a condition, begin with a system characterized by magnetic field vectors $\mathbf{B}_1$ and $\mathbf{B}_2$, number densities n_1 and n_2 and pressures P_1 and P_2 on either side of a current layer. The relation $\cos\theta = \mathbf{B}_1 \cdot \mathbf{B}_2 / B_1 B_2$ defines the angle θ between the asymptotic field, with $\theta = 180°$ corresponding to anti-parallel reconnection. The coordinate system, with the unknown angle α, is shown in Fig. 25(b): B_{1x} and B_{2x} are the reconnecting components of the field, and the respective guide fields are B_{1y} and B_{2y}. The X-line points parallel to $\hat{\mathbf{y}}$ while reconnection occurs in the $x - z$ plane.

The stability criterion is obtained in two steps. The first requires determining the direction of the X-line (or, equivalently, the plane in which reconnection occurs) since this choice affects the field components that enter the calculation of V_e^*. Determining this orientation has been the subject of multiple papers (Swisdak and Drake 2007; Schreier et al. 2010; Hesse et al. 2013; Liu et al. 2015b, 2018c) with no clear resolution, but while the exact results differ, there is general agreement that, to a reasonable approximation, the reconnection X-line bisects the angle θ between the two magnetic fields (Hesse et al. 2013; Liu et al. 2018c) (in other words, $\alpha = \theta/2$ in Fig. 25(b)). The resulting outflow velocity from the X-line is given by the hybrid Alfvén speed,

$$V_{A,\text{asym}} = \sqrt{\frac{B_1 + B_2}{4\pi m_i \left(\frac{n_1}{B_1} + \frac{n_2}{B_2}\right)}} \sin\frac{\theta}{2}. \tag{39}$$

This equation agrees with Eq. (21).

Second, we calculate the component of the diamagnetic velocity along the outflow (x-) direction,

$$\langle V_x^* \rangle = -\left\langle \frac{cB_y \partial_z P}{enB^2} \right\rangle, \tag{40}$$

where $\partial_z P$ is the derivative of the total pressure (electron plus ion) in the direction normal to the current layer. The angle bracket $\langle\rangle$ indicates the spatial average across the current sheet of thickness δ.

The stability condition (i.e., $\langle V_x^* \rangle > V_A$) leads to

$$|\beta_1 - \beta_2| > \frac{2\delta}{d_i} \tan\frac{\theta}{2} \tag{41}$$

which was first derived in Swisdak et al. (2010). Here β_1 and β_2 are the plasma betas on two sides.

Simulations suggest that the scale factor δ is $\approx d_i$ when the guide field is small, but $\approx \rho_s$, the sound Larmor radius, is in the opposite limit. However, the approximations made in deriving Eq. (41) mean that the pre-factor on the right-hand side is likely most accurately described as "of order unity" and so exactness is not expected. Immediate consequences of Eq. (41) include: (1) Anti-parallel reconnection ($\theta = \pi$) is never subject to diamagnetic stabilization and (2) Stabilization is likely when the upstream fields nearly align (i.e., for θ small).

Observational Evidence The condition expressed in Eq. (41) has been tested in a variety of locations, e.g., Earth's magnetopause (Phan et al. 2013a), the solar wind (Phan et al. 2010), and the magnetospheres of Mercury (DiBraccio et al. 2013), Jupiter (Desroche et al. 2012), and Saturn (Masters et al. 2012), with reasonable success. Figure 26 shows a representative example from Phan et al. (2013a) examining reconnection at the magnetopause. Both panels show the $\Delta\beta - \theta$ plane. The various curves divide the plane according to Eq. (41) with the upper/leftmost region where diamagnetic suppression should not occur and the lower/rightmost region where it should. The left panel shows current sheet crossings for which reconnection was detected, while the right panel plots crossings for which no reconnection signals were observed. To a large degree, the events are properly segregated by the diamagnetic suppression condition. However, it is important to recognize that non-reconnecting current sheets are, in general, observed even when diamagnetic suppression is not expected to operate, as other factors (e.g., current sheet thickness) may prevent reconnection onset.

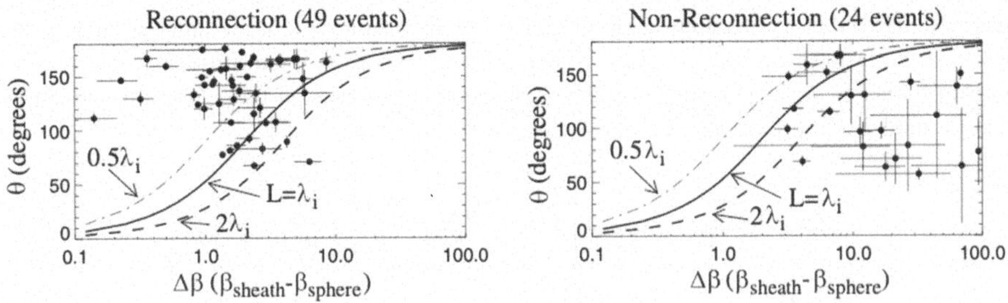

Fig. 26 Results of statistical survey of reconnection (left) and non-reconnection (right) events. Scatter plot of magnetic shear versus $\Delta\beta$ at the magnetopause. Reprinted from Phan et al. (2013a), with the permission of Wiley

3.8.2 Sheared Flow Suppression

Reconnection can also be suppressed by a background in-plane flow shear across the reconnection current sheet, a scenario in which there are different bulk flow speeds on either side of the reconnection site, as illustrated in Fig. 27(a)). The suppression criterion is qualitatively similar to that of the diamagnetic case (Mitchell Jr. and Kan 1978; Chen and Morrison 1990; La Belle-Hamer et al. 1995; Cassak and Otto 2011). The presence of a flow shear V_{shear} opposes the development of reconnection outflows. If the shear flow velocity $V_{\text{shear}} = (V_{x1} - V_{x2})/2$, where V_{x1} and V_{x2} are the bulk flow speed on either side of the reconnection site, is larger than the Alfvén speed,

$$V_{\text{shear}} > V_A, \tag{42}$$

the reconnection outflow can not develop. Such outflow reduction was demonstrated using two-fluid simulations in Fig. 27(b). A similar conclusion is extended to relativistic magnetic reconnection (Peery et al. 2024), but the critical velocity is set by the relativistic Alfvén speed, like that discussed in Sect. 3.11.

Since magnetopause reconnection, where the upstream plasma can have a bulk flow because the magnetosheath plasma moves due to the solar wind, is asymmetric, we will discuss how flow shear impacts asymmetric reconnection. If asymmetric reconnection occurs in a region where there is a bulk flow V_x in the x-direction (along or against the reconnecting

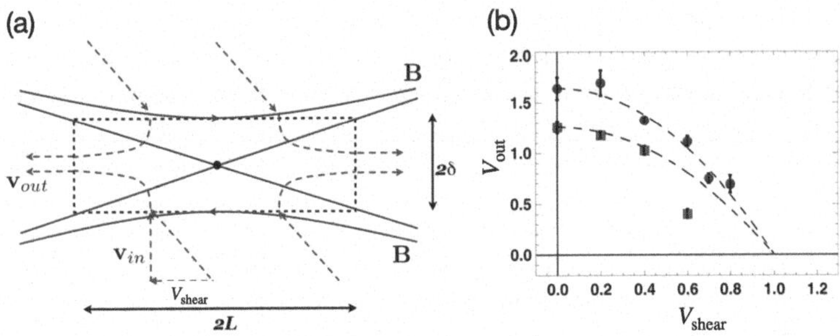

Fig. 27 (a) In-plane shear flow across the reconnection current sheet (reprinted from Cassak and Otto 2011, with the permission of AIP Publishing). (b) The outflow speed as a function of shear flow magnitude V_{shear} during symmetric reconnection. Reprinted from Cassak (2011), with the permission of AIP Publishing

magnetic field) that is different on either side of the diffusion region, reconnection can slow down. If the flow shear $V_{\text{shear}} = (V_{x1} - V_{x2})/2$ is large enough, it can fully suppress reconnection (Doss et al. 2015), just as in the symmetric reconnection case. The reconnection electric field $E_{R,\text{asym,shear}}$ was shown to scale as

$$E_{R,\text{asym,shear}} \sim E_{R,\text{asym}} \left[1 - \frac{V_{\text{shear}}^2}{V_{A,\text{asym}}^2} \frac{4n_1 B_{x2} n_2 B_{x1}}{(n_1 B_{x2} + n_2 B_{x1})^2} \right], \tag{43}$$

where $V_{A,\text{asym}}$ and $E_{R,\text{asym}}$ are defined in Eq. (21)-(22). The critical flow shear at which reconnection is suppressed, when $E_{R,\text{sym,shear}}$ goes to zero, is

$$V_{\text{shear,crit}} \sim V_{A,\text{asym}} \frac{n_1 B_{x2} + n_2 B_{x1}}{2\sqrt{n_1 B_{x2} n_2 B_{x1}}}. \tag{44}$$

In the symmetric limit, the critical flow shear is simply V_A, as has been well known (Mitchell Jr. and Kan 1978; Chen and Morrison 1990; La Belle-Hamer et al. 1995; Cassak and Otto 2011). Interestingly, for asymmetric reconnection, the critical flow shear is faster. This implies that it is more difficult to suppress asymmetric reconnection by flow shear than symmetric reconnection (Doss et al. 2015). It follows that magnetopause reconnection is not expected to be suppressed by flow shear (Doss et al. 2015).

Observational Evidence During radial IMF conditions, the magnetosheath flow and the direction of reconnection jets become roughly aligned at the magnetopause flanks. Toledo-Redondo et al. (2021) took advantage of MMS-Cluster conjunction to investigate the mesoscale of magnetic reconnection along the magnetopause. The MMS fleet was located near the subsolar region, while the Cluster fleet was located in the dusk flank ($X_{GSE} \sim 0$). Based on seven simultaneous crossings (magnetopause current sheet observation within 5 minutes at the two locations), the expected reconnection electric field was $E_{R,\text{asym,shear}}/E_{R,\text{asym}} = 0.71$-$0.98$ in the flank, based on Eq. (43), and thus the effect was negligible for the seven crossings near the subsolar region. While these observations indicate that the shear flow suppression mechanism may have some impact in regulating the global coupling of the magnetosphere to the solar wind during radial IMF conditions, more observations are needed to quantify this impact.

3.9 Reconnection Driven by Converging Flows

While flows that oppose the development of outflows can suppress reconnection, external flows that converge into the current sheet can, in contrast, drive the reconnection process.

To compare the results of driven reconnection in various different types of simulations, a series of studies called the Newton Challenge (Birn et al. 2005; Pritchett 2005; Huba 2006; see Fig. 28) was conducted. Similar to the GEM Reconnection Challenge (Birn et al. 2001), these authors used two-dimensional resistive MHD (with a localized resistivity), Hall MHD, full PIC, and hybrid simulations. Boundary inflows were imposed on both the top and the bottom boundaries with the functional form $V_{\text{in}} = 2a\omega\tanh(\omega t)/\cosh^2(\omega t)$ with $a = 2d_i$ and $\omega = 0.05\Omega_{ci}$, giving the maximum inflow speed $0.08V_A$, where $V_A = B_{x0}/\sqrt{4\pi n_0 m_i}$ with B_{x0} being the asymptotic reconnecting field and n_0 being the initial peak density in the current sheet. These inflows drive the boundary electric field $E_y = B_x V_{\text{in}}/c$ out of the reconnection plane, where the maximum electric field reaches $0.1B_{x0}V_A/c$. Note that the reconnecting component B_x at the boundary increases due to V_{in}, with the maximum value

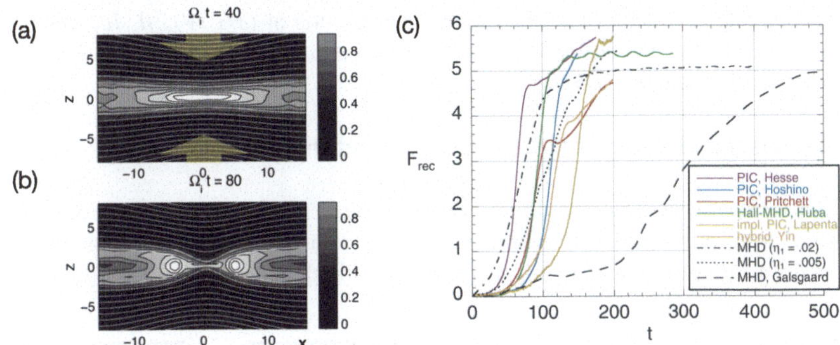

Fig. 28 *Newton Challenge*. (a)-(b) Reconnection driven from the top and bottom inflow boundaries within a PIC simulation. Adapted from Pritchett (2005). (c) The evolution of the reconnected magnetic flux. PIC, Hall, and hybrid simulations show similar reconnection rates (given by the slope of each curve), while resistive MHD simulations show lower reconnection rates. Reprinted from Birn et al. (2005), with the permission of Wiley

around 1.1 to 1.2 times larger than the initial asymptotic value B_{x0}. Such a "pileup" of the upstream magnetic flux may compensate for any local reduction in the reconnection rate due to a weak current sheet dissipation (Dorelli 2019), resolving the debate of whether magnetopause reconnection rates are controlled by the solar wind driving or local reconnection physics (Borovsky et al. 2008; Lopez 2016).

The PIC simulation results of the Newton Challenge by Pritchett (2005) show that the reconnection electric field E_R at the X-line increases with time, reaching a maximum of $\sim 0.12 B_{x0} V_A/c$, slightly larger than the maximum of the boundary E_y, after which E_R decreases to values less than $0.05 B_{x0} V_A/c$. In the later stage, reconnection reaches a quasi-steady state, during which the magnetic field geometry shows an almost equilibrium state with two large magnetic islands. In all of the physical models, the final state reaches a similar equilibrium, even though the reconnection rate (the reconnection electric field) in the resistive MHD simulations in the earlier fast stage is slightly smaller than that in simulations with Hall physics.

Although the final magnetic configurations are the same in all the simulation methods, the "two-stage" evolution of reconnection (the fast phase followed by the slow phase) is observed only in PIC and hybrid simulations, where particle kinetics is included. Pritchett (2005) explains that the following slow phase in the Newton Challenge appears after the outflows reach the periodic boundaries, and at that time, the system does not reach equilibrium with a large magnetic island. In contrast with kinetic simulations, Hall MHD (Huba 2006) and resistive MHD (Birn et al. 2005) simulations do not show the two-step evolution, and the fast reconnection phase continues until the system reaches the equilibrium with a large magnetic island. These results suggest that the kinetics of ions and electrons play a role in the late-stage evolution of driven reconnection.

While the Newton Challenge was primarily a study of 2D-driven reconnection, 3D-driven reconnection was also considered. Sullivan and Rogers (2008) used Hall MHD simulations with external inflows similar to those in the Newton Challenge to compare the results in 2D and 3D simulations. The external inflows in 3D cases are not uniform in the y (electric current) direction but are localized around $y = 0$ in the simulation box. In 2D simulations, they observed the reconnection electric field scaled as predicted, $E_R \sim (\delta/L) V_{Am} B_m/c$, where V_{Am} and B_m are the Alfvén speed and the magnetic field at the edge of the diffusion region, and the aspect ratio of the diffusion region δ/L is 0.1-0.2. In contrast, in 3D runs, the reconnection rate is a factor of 2 larger than the prediction of $(\delta/L)V_{Am}B_m/c$, which is attributed to the fact that the diffusion region is localized in the y direction and not uniformly distributed as in Pritchett (2005).

3.9.1 Reconnection within Vortices, Turbulence and Shocks

MMS has observed evidence of driven reconnection in the flanks of the Earth's magnetopause. During northward IMF reconnection occurs in the high-latitude regions, which transfer the accumulated magnetic flux to the low-latitude magnetopause, forming the so-

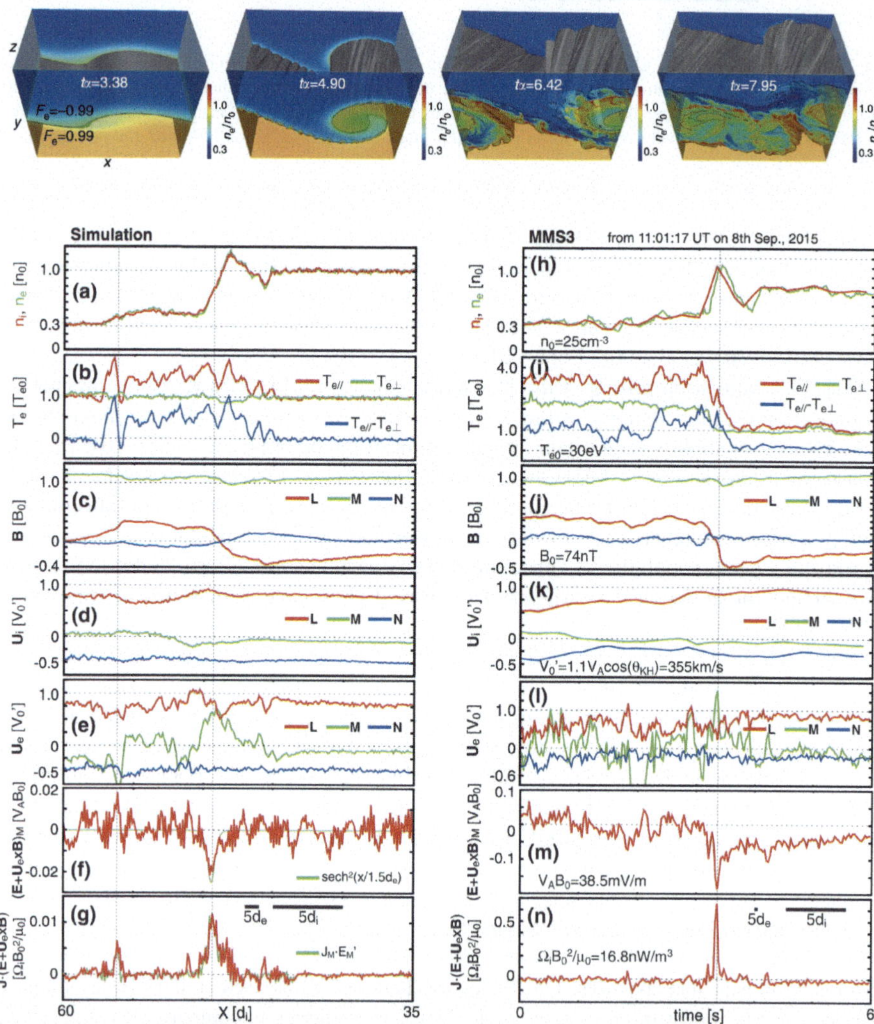

Fig. 29 Top panels show the time evolution of Kelvin-Helmholtz vortex-induced reconnection (VIR) in a 3D PIC simulation (Nakamura et al. 2017a). The rest of the panels show its comparison with MM3 data obtained on September 8, 2015. Especially, panels (f) and (m) compare the non-ideal electric field in the M direction, which shows $0.5V_{AL}B_L$ and $1V_{AL}B_L$, respectively, due to reconnection driven by the vortex flow. Reprinted from Nakamura et al. (2017b), with the permission of Wiley

called "low-latitude boundary layer" (LLBL), in which most of the plasma originates from the magnetosheath. In the flank side of the LLBL, strong velocity shear is unstable to the Kelvin-Helmholtz (KH) instability. The super-Alfvénic shear flows associated with this instability can drive reconnection within these vortices (see e.g. Hwang et al. 2023, this collection, and references therein). MMS detected such KH-driven reconnection (Eriksson et al. 2016a; Li et al. 2016; Nakamura et al. 2017b, 2018; Hwang et al. 2020; Kieokaew et al. 2020; Hwang et al. 2021).

Nakamura et al. (2017b) presented 3D PIC simulations of a KH-driven reconnection event observed by MMS on 8 September 2015, as illustrated in Fig. 29. The right panels indicate that MMS3 crossed a current sheet in the reconnection region from the magnetospheric side to the magnetosheath side, during which it detected a large non-ideal electric field $E'_M \equiv (\mathbf{E} + \mathbf{V}_e \times \mathbf{B}/c)_M = -7$ mV/m; although this value corresponds to $\sim 0.2 V_A B_{x0}$, it is $\sim 1 V_{AL} B_L$ if one normalizes it to the L component of the local magnetic field [2] $B_L \sim 35$ nT $\times$ the Alfvén speed V_{AL} based on B_L and the local density 17 cm^{-3}. This $|E'_M| \sim V_{AL} B_L$ is 10 times larger than a typical standard laminar reconnection value of $0.1 V_{AL} B_L$, perhaps due to the fact that reconnection is driven by the strong flows generated by the KH instability. The observational data are consistent with the 3D PIC simulation in the left panels, where the peak of $|E'_M|$ (panel (f)) is $0.5 V_{AL} B_L$, which is also 5 times larger than the standard reconnection rate.

Driven reconnection can also occur in turbulent environments. The strong flows therein can force reconnection to occur at a variety of rates. Haggerty et al. (2017) performed 2D PIC simulations of turbulent reconnection and observed normalized reconnection rates distributed between 0 to 0.5, suggesting that reconnection rates are not limited to the order of 0.1. Bessho et al. (2020, 2022) demonstrated using 2D PIC simulations that reconnection in turbulence associated with Earth's bow shock is strongly driven by super Alfvénic flows, and normalized reconnection rates in both electron-only reconnection and ion-coupled reconnection can be of the order of unity.

3.10 Turbulent 3D Reconnection

While kinetic-scale reconnection (either "electron-only" or ion-coupled) within turbulent plasmas can be affected by turbulence driving, it has been proposed that the dissipation mechanism of reconnection itself may be affected by turbulence or instabilities within the diffusion region. Turbulent reconnection operates in large, 3D systems, where the additional degree of freedom introduces various types of instabilities (e.g. Daughton et al. 2011), leading to turbulence. Unlike laminar reconnection, it was proposed that turbulence may produce "anomalous resistivity" (e.g., Higashimori et al. 2013) or "anomalous viscosity" (e.g., Che et al. 2011a; Price et al. 2016), modifying the diffusion region physics that breaks the MHD frozen-in condition. Other competing ideas also exist, including coupling to the Goldreich-Sridhar-like turbulence spectrum (Lazarian and Vishniac 1999), field-line super-diffusion (Eyink et al. 2011), and fast field-line separation (Boozer 2012).

Figure 30 shows an example of 3D magnetic reconnection in a large PIC simulation. The entire reconnection layer becomes turbulent because of self-driven instabilities within the current sheet. Unlike in 2D models, an in-plane flux function does not exist that allows a straightforward calculation of the reconnection rate. Instead, Daughton et al. (2014) devised an approach based on the electron mixing across the separatrix in full 3D systems. The

[2] Note that this location, where $B_L = -35$ nT ~ 0.5 B_0, is considered to be close to the edge of the diffusion region.

Fig. 30 Panel (a) shows the field line exponentiation factor σ at two y-locations within a 3D PIC simulation. The solid black lines mark the boundaries of the electron mixing fraction $|\mathcal{F}_e| = 0.99$. Panel (b) shows the reconnection rate computed based on the top (red), bottom (blue), and average (purple) magnetic fluxes and using $|\mathcal{F}_e| = 0.99$. Grey triangles are the inflow rates applied to the bottom region. The black curve is the corresponding 2D reconnection rate measured from the flux function A_y, while the green crosses are the 2D rate obtained from the same electron mixing approach. Reprinted from Daughton et al. (2014), with the permission of AIP Publishing

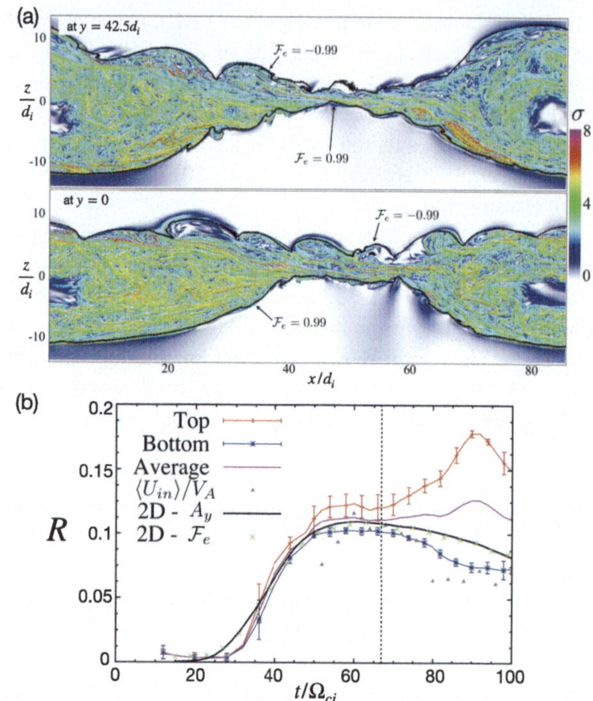

measured reconnection rate is shown in Fig. 30(b), and interestingly, the 3D reconnection rate appears to be similar to its 2D laminar counterpart (also in Le et al. 2018). This 3D rate may still be bounded by the same geometrical constraints discussed in Sect. 3.1.2 if the force balance is taken to work in an average sense. In addition, a broad turbulent reconnection layer is often dominated by a few active diffusion regions at the kinetic scale where the dissipation mechanism may be similar to that in Sect. 2. A thorough investigation is required to validate these assertions. To date, it remains challenging to model turbulent reconnection rate from first principles; more discussion on this topic can be found in Stawarz et al. (2024, this collection), Graham et al. (2025, this collection) and Guo et al. (2024).

3.10.1 Averaged 3D Ohm's Law

Other than those kinetic terms discussed in Sect. 2.1, an alternative view holds that fluctuating electric fields, generated by kinetic instabilities such as Buneman modes or lower-hybrid drift effects, can effectively scatter electrons in the electron diffusion region and consequently lead to effective resistance to the reconnection electric field. The effects of such fluctuations are captured by time- or spatial averaging of the microscopic electron momentum equation (Eq. (1)):

$$
\begin{aligned}
e\langle n_e\rangle\langle\mathbf{E}\rangle = &-\frac{e}{c}\langle n_e\rangle\langle\mathbf{V}_e\rangle\times\langle\mathbf{B}\rangle - \nabla\cdot\langle\mathbf{P}_e\rangle - m_e\nabla\cdot(\langle n_e\mathbf{V}_e\rangle\langle\mathbf{V}_e\rangle) - m_e\frac{\partial\langle n_e\mathbf{V}_e\rangle}{\partial t} \\
&-e\langle\delta n_e\delta\mathbf{E}\rangle - \frac{e}{c}\langle\delta(n_e\mathbf{V}_e)\times\delta\mathbf{B}\rangle + m_e\nabla\cdot\langle\delta(n_e\mathbf{V}_e)\delta\mathbf{V}_e\rangle .
\end{aligned}
\tag{45}
$$

Here, $\langle\rangle$ denotes spatial averages and the fluctuation of a quantity $\mathbf{Q}$ is $\delta\mathbf{Q}\equiv\mathbf{Q}-\langle\mathbf{Q}\rangle$; a similar equation can be obtained for time averaging, with a different form of the time derivative of the electron momentum density (Le et al. 2018). The last three terms on the RHS of Eq. (45) are often referred to as anomalous drag, anomalous momentum transport, and anomalous viscosity, respectively (e.g., Büchner et al. 1998; Che et al. 2011a,b; Price et al. 2016; Le et al. 2018). It should be noted that Eq. (45) does not contain any new or additional information: all information is already included in the microscopic description (Eq. (1)), while Equation (45) is obtained by spatial averaging of this microscopic equation and hence contains less information.

Translationally invariant models demonstrate, without exception, that non-gyrotropic pressure tensor effects dominate at the X-line for symmetric configurations (or, more generally, at the flow stagnation point) (e.g., Pritchett 2001; Schmitz and Grauer 2006). Three-dimensional models of collisionless reconnection can show, however, when averaged, significant contributions of the anomalous terms and the presence of substantial fluctuations at the X point (e.g., Büchner et al. 1998; Che et al. 2011a,b; Fujimoto and Sydora 2012; Price et al. 2016; Muñoz and Büchner 2016). However, some local analyses continued to show the dominance of non-gyrotropic pressure terms (Hesse et al. 2005; Liu et al. 2024), and the magnitude of these anomalous terms are sensitive to how one averages the Ohms' law (Le et al. 2018). A recent, very large simulation demonstrated the near-absence of significant fluctuations at the X-line if effects of periodic boundaries can be excluded (Liu et al. 2018c).

Prior to the Magnetospheric Multiscale mission, these two theories (i.e., anomalous dissipation versus non-gyrotropic electron pressure) were competing, and MMS had, as a key goal, to determine which of these theories was matched by reality. Beginning with the first key observation of an electron diffusion region at the magnetopause (Burch et al. 2016b), observations have shown remarkably quiescent electron diffusion regions, whether they are asymmetric with (Burch and Phan 2016) or without a guide field (Burch et al. 2016b), or whether they are in the tail's plasma sheet (Torbert et al. 2018). While it has been difficult to measure electron pressure tensor effects directly, there has been some indication that these are indeed important (Genestreti et al. 2018c), and a recent observation even shows that the analytic prediction of Hesse et al. (1999, 2011) provides a reasonable match to the observed reconnection electric field (Nakamura et al. 2019). Furthermore, a tailored, translationally invariant, numerical simulation (Nakamura et al. 2018) provides an exceptionally good match between observations and model results. While observations around the outflow region show significant fluctuations and turbulent effects (Ergun et al. 2016, 2018; Burch et al. 2018), there is rapidly increasing evidence that the central electron diffusion region is indeed relatively quiescent and properly described by the quasi-viscous, electron nongyrotropy-based model (Hesse et al. 1999). Therefore, it appears that MMS has accomplished its primary objective: to determine the physics behind the electron diffusion region (Torbert et al. 2018).

3.11 Relativistic Reconnection

In plasmas near compact astrophysical objects, such as neutron stars and black holes, the magnetic field strength is extremely strong (e.g., Uzdensky 2011; Ripperda et al. 2020 and references therein), and the plasma flow speed can become relativistic. Under this condition, assuming an anti-parallel magnetic geometry, the relevant force balance equation becomes

$$\frac{(\mathbf{B}\cdot\nabla)\mathbf{B}}{4\pi} \simeq n' m_i (\mathbf{U}\cdot\nabla)\mathbf{U}, \tag{46}$$

where $\mathbf{U} = \Gamma\mathbf{V}$, $\Gamma \equiv [1-(V/c)^2]^{-1/2}$ is the Lorentz factor, and n' is the plasma proper density. The resulting outflow speed (in the x-direction) is the relativistic Alfvén speed (Liu et al. 2017),

$$V_{\rm out} \simeq V_{Ax} = c\sqrt{\frac{\sigma_R}{1+\sigma_R}}, \tag{47}$$

which can approach the speed of light c when the magnetization parameter $\sigma_R = B_R^2/4\pi n' mc^2 \gg 1$. With an external guide field B_g, the Alfvénic outflow speed becomes

$$V_{\rm out} \simeq V_{Ax} = c\sqrt{\frac{\sigma_R}{1+\sigma_R+\sigma_g}}, \tag{48}$$

where $\sigma_g = B_g^2/4\pi n' mc^2$. This expression can be formally derived after considering the additional momentum carried by the outflowing Poynting vector $S_x = -E_z B_y/4\pi$ (that is not included in Eq. (46), but considered in Peery et al. (2024)), where the motional electric field $E_z = -V_{\rm out} B_g/c$ is associated with the convection of the guide field. It is interesting to note that the guide field can significantly slow down the outflow speed, unlike in the non-relativistic case. To comprehend this fact in another way, we see that with a guide field the total Alfvén speed (i.e., Eq. (47) with σ_R replaced by $\sigma_R + \sigma_g$) is still limited by the speed of light c due to the special relativity, and Eq. (48) is the projection of this total Alfvén velocity along the magnetic field to the outflow direction (Melzani et al. 2014; Liu et al. 2015a), thus its magnitude is expected to be lower than c.

Most theoretical studies of relativistic reconnection rates (Blackman and Field 1994; Lyutikov and Uzdensky 2003; Lyubarsky 2005; Liu et al. 2017; Mbarek et al. 2022; Goodbred and Liu 2022) have been performed in the comoving frame of the X-line, where the X-line stays stationary in the 2D reconnection plane. However, observers at different inertial reference frames will disagree on the magnitude of the reconnection electric field and even the magnetic topology within the diffusion region (Hornig and Schindler 1996), because electric fields and magnetic fields can convert into each other in the Lorentz (frame) transformation.. In the absence of a "special" reference frame, especially in a system that lacks symmetry, a covariant (frame-independent) definition of magnetic reconnection becomes desirable. Scientists have started to address this nontrivial issue (Hornig and Schindler 1996; Asenjo and Comisso 2015; Pegoraro 2016).

Relativistic magnetic reconnection has been proposed to explain the superflares observed in the Crab Nebula and argued to cause fast radio bursts (FRBs) from neutron stars and magnetars (Philippov et al. 2019; Mahlmann et al. 2022). Interested readers are referred to the discussion in Guo et al. (2024, this collection).

4 Energy Conversion within the Diffusion Region

Aside from changing the large-scale magnetic connectivity/topology, perhaps the most important consequence of magnetic reconnection is converting magnetic energy into plasma kinetic energy and thermal energy. In this section, we collect approaches being used to quantify the energetics and energy conversion processes around the diffusion region, with a particular focus on progress enabled by MMS observations as well as recent advances in simulation capabilities. For the discussion of non-thermal particle accelerations during reconnections, a complimentary review can be found in Oka et al. (2023, this collection).

4.1 Energy Conservation and Energy Fluxes

The second moment of the Vlasov equation gives the energy equation, which in the conservative form (Birn and Hesse 2010) reads

$$\frac{\partial u_{\mathrm{total}}}{\partial t} + \nabla \cdot (\mathbf{S} + \mathbf{H} + \mathbf{K} + \mathbf{q}) = 0. \tag{49}$$

Here, $u_{\mathrm{total}} \equiv \sum_s^{i,e} \left(\mathrm{Tr}(\mathbf{P}_s)/2 + nm_s V_s^2/2\right) + (B^2 + E^2)/8\pi$ is the total energy density with $\mathrm{Tr}(\mathbf{P}_s) \equiv \sum_j^{x,y,z} \mathbf{P}_{s,jj}$ being the trace of the pressure tensor, $\mathbf{S} \equiv c\mathbf{E} \times \mathbf{B}/4\pi$ is the Poynting vector, $\mathbf{H} \equiv \sum_s^{i,e} [(1/2)\mathrm{Tr}(\mathbf{P}_s)\mathbf{V}_s + \mathbf{P}_s \cdot \mathbf{V}_s]$ is the enthalpy flux, and $\mathbf{K} \equiv \sum_s^{i,e} (1/2) nm_s V_s^2 \mathbf{V}_s$ is the bulk-flow kinetic energy flux. $\mathbf{q} \equiv \sum_s^{i,e} (m_s/2) \int |\mathbf{v}_s - \mathbf{V}_s|^2 (\mathbf{v}_s - \mathbf{V}_s) f_s d^3 v_s$ is the heat flux, where $\mathbf{v}_s$ is the particle velocity, and f_s is the particle distribution function of species "s".

Figure 31 shows schematically the energy fluxes into and out of a reconnection site that is treated as invariant in the y-direction (out-of-plane). Here, for simplicity, the reconnection process is implicitly taken as being in a steady state ($\partial/\partial t \simeq 0$). We now discuss the nature of the energy fluxes in reconnection before turning to look more closely at the problem of energy conversion.

Based on magnetotail observations in the IDR by the Cluster spacecraft, in anti-parallel, symmetric reconnection the outflowing energy flux is dominated by H_{ix}, taking up ~50% of the total, followed by H_{ex} (~20%) and K_{ix} (~10%); the outflowing S_x (~10%-20%) is comparable to H_{ex} and K_{ix}, and even dominates in certain regions in the IDR, where the Hall term dominates (Eastwood et al. 2013). The relative rankings between different forms of energy flux are qualitatively consistent with other subsequent Cluster observations [even when considering the energetics of O^+ (Typer et al. 2016)], kinetic simulations (e.g., Birn and Hesse 2010; Lapenta et al. 2020) and laboratory experiments (e.g., Yamada et al. 2016; see also Yamada et al. 2018 and Table 2 in Ji et al. 2023, this collection, for a brief summary of progress in this area).

Prior to MMS, it was not possible to access the dynamics of the EDR with sufficient resolution to determine the detailed properties of the energy fluxes. However, recent efforts have now enabled such analysis around the EDR of asymmetric reconnection at the dayside magnetopause (Eastwood et al. 2020), as shown in Fig. 32. The study also confirms previous ion-scale observations, such as smoothly varying ion energy fluxes dominated by the ion enthalpy flux in the exhausts, and demonstrates the influence of the large-scale asymmetries introduced by the magnetopause, finding, for example, the peak of the total ion energy flux to be displaced towards the magnetospheric side.

In the case of the ions, the heat flux was observed to be directed back towards the X-line, a feature also seen in symmetric reconnection simulations (Lu et al. 2018) and can be explained with non-Maxwellian distributions (Hesse et al. 2018). It should, therefore, be

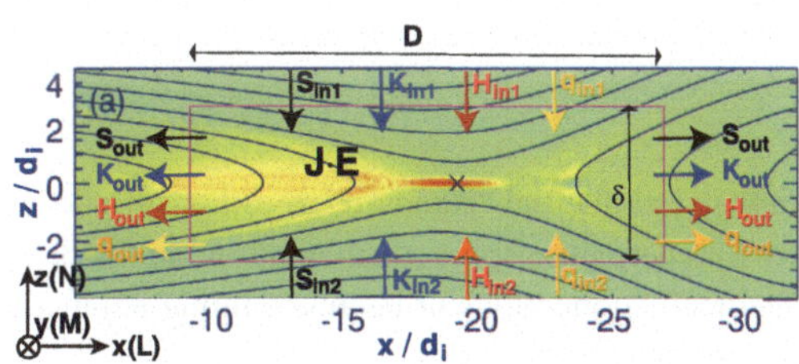

Fig. 31 *Schematic of the energy conversion in the reconnection region*, modified from Eastwood et al. (2013). The color represents $\mathbf{J} \cdot \mathbf{E}$ from a PIC simulation. Reprinted from Lu et al. (2018), with the permission of AIP Publishing

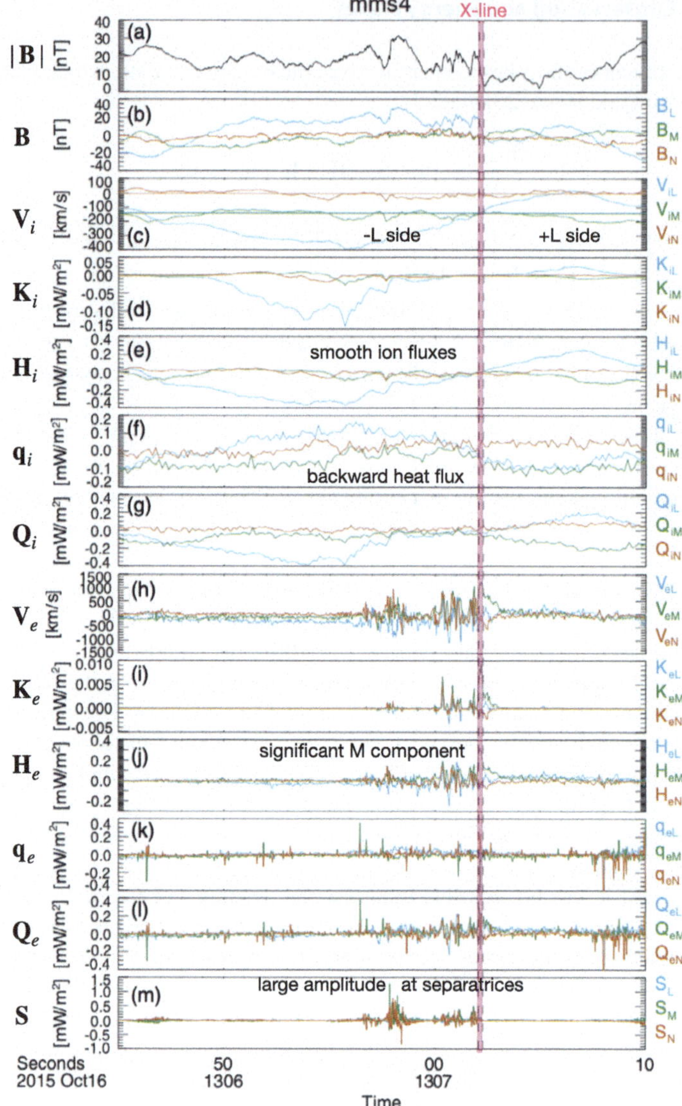

Fig. 32 *MMS observations of energy fluxes around an EDR of dayside magnetopause reconnection*, first reported by Burch et al. (2016b). Panels (d, e, f, g) show the kinetic, enthalpy, heat, and total ion flux. Panels (i, j, k, l) show equivalent electron fluxes. MMS reveals that the ion fluxes are smoothly varying, whereas the electron fluxes are structured and variable. The MMS data shows the existence of a significant out-of-plane electron energy flux at the X-line (marked by vertical pink lines), discussed in more detail in the text. Modified from Eastwood et al. (2020)

emphasized that the "standard" decomposition of energy flux (which is the relevant parameter for energy transport considerations) in the presence of non-Maxwellian distributions or specifically collections of beams/multiple populations should be interpreted with care (see, e.g. Goldman et al. 2020).

In the case of the electrons, the results from MMS are more surprising. At the EDR, it would be expected to observe an enhanced out-of-plane kinetic energy flux because of the enhanced current density at the X-line. However, the small mass of the electrons renders K_e negligible. MMS showed that the combination of electron heating at the EDR together with fast electron motion leads to an out-of-plane electron enthalpy flux density, which is comparable to the ion flux densities in the exhaust (Eastwood et al. 2020). This may have an important impact on the plasma dynamics, particularly in driving electron-scale instabilities out of the plane. The MMS observations raise further questions about the ultimate source and sink of this out-of-plane energy flux at the EDR, and how it varies along the X-line across the magnetopause or in the magnetotail. Answering this requires a more detailed experimental study of both the energy equation (Eq. (49)) as well as the transfer of energy from fields to particles, the latter being controlled by $\mathbf{J} \cdot \mathbf{E}$ as we now discuss.

4.2 Poynting's Theorem and $\mathbf{J} \cdot \mathbf{E}$

To understand the transfer of energy between electromagnetic fields and particles during magnetic reconnection, we can use Poynting's theorem,

$$\frac{\partial u_{EM}}{\partial t} + \nabla \cdot \mathbf{S} = -\mathbf{J} \cdot \mathbf{E}, \tag{50}$$

where $u_{EM} = (B^2 + E^2)/8\pi$ is the energy density of electromagnetic fields and $\mathbf{S} = c(\mathbf{E} \times \mathbf{B})/4\pi$ is the Poynting vector. Since the left-hand side of this equation describes the continuity of the electromagnetic energy, the source term on the right-hand side, $\mathbf{J} \cdot \mathbf{E}$, will measure the energy conversion from electromagnetic energy to plasma energy. A similar equation can be written for the particles, and the sum of these two equations reduces to Eq. (49), i.e., conservation of total energy. The signature of $\mathbf{J} \cdot \mathbf{E}$ in anti-parallel, symmetric reconnection is shown in Fig. 31 based on PIC simulation results (Lu et al. 2018). It is most enhanced within the d_e-scale EDR, but positive values (i.e., energy transfers to the plasma) extend further downstream within the outflow exhaust. $\mathbf{J} \cdot \mathbf{E}$ can also be decomposed according to the electric field components to assist in understanding the energization mechanisms. The reconnection electric field (E_y) is along the reconnection X-line. The electric field in the $x - z$ plane, $\mathbf{E}_{xz}$, is dominated by the Hall electric fields ($\mathbf{E}_{\mathrm{Hall}} = \mathbf{J} \times \mathbf{B}/enc$), which is set up due to the charge separation between the faster-moving electrons and slower ions, with the z component pointing towards the mid-plane and the x component away from the X-line; its effect is to slow down electrons while speeding up ions. It is, therefore, useful to further decompose $\mathbf{J} \cdot \mathbf{E}$ by considering the current density of each species.

Figure 33 shows such decomposition for asymmetric reconnection in PIC simulations, and we will discuss the electron energization first, then ion energization in the next paragraph. Around the EDR, $J_{ey} E_y$ is dominantly positive (Fig. 33(a)), such that electrons gain energy from E_y during the meandering motion. $\mathbf{J}_{e,xz} \cdot \mathbf{E}_{xz}$ is mainly negative, especially further than $\sim 1 d_i$ downstream of the X-line (Fig. 33(b)). Such features also exist for symmetric reconnection, and $\mathbf{J}_{e,xz} \cdot \mathbf{E}_{xz}$ within the EDR has a much smaller amplitude than $J_{ey} E_y$, as, e.g., shown in Payne et al. (2021). Their study further shows that for a well-developed reconnection layer, a region may develop around the end of the EDR with negative $J_{ey} E_y$ (not shown here) and positive $J_{ex} E_x$, as the electron flow turns from the y to the x direction and the electrons become re-magnetized. For asymmetric reconnection, because the stagnation point is on the magnetospheric side of the X-line (seen from the electron flow lines in Fig. 33(d)) and the magnetosheath-pointing E_z extends to the magnetosheath side of the X-line (e.g., Shay et al. 2016; Chen et al. 2016), a region with positive $\mathbf{J}_{e,xz} \cdot \mathbf{E}_{xz}$ exists

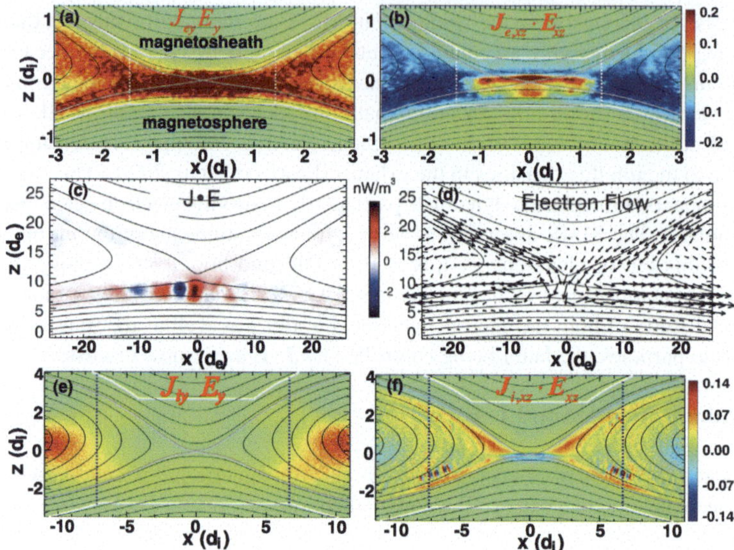

Fig. 33 $\mathbf{J}_s \cdot \mathbf{E}$ *decomposition by the electric field components in PIC simulations* of asymmetric reconnection with zero guide field. Positive $J_{ey}E_y$ dominates the energy conversion to electrons in the EDR (a), and the $\mathbf{E}_{xz}$ does negative work outside of EDR (b). (c) In certain parameter regimes, $\mathbf{J}_e \cdot \mathbf{E}$ can exhibit significant oscillations due to oscillating V_{ez} (d). (e)-(f) energy conversion for ions, where the Hall fields dominate while E_y has a positive contribution in a broad region. Panels (a), (b), (e), (f) are modified from Wang et al. (2018); (c)-(d) are adapted from Swisdak et al. (2017), with the permission of Wiley

near the X-line that contributes additional electron energy gain (Fig. 33(b)). The $\mathbf{J}_{e,xz} \cdot \mathbf{E}_{xz}$ profile exhibits fluctuations, and in certain parameter regimes, the fluctuations can be more significant and dominate the total $\mathbf{J} \cdot \mathbf{E}$ profile (Fig. 5(c), Swisdak et al. 2017). As electrons bounce within the current sheet, they gain a velocity along x by turning around B_z and B_y, so most electrons cannot bounce many times at the same x location to maintain similar densities for populations at positive and negative v_z, leading to non-zero and fluctuating bulk V_{ez} (Fig. 33(d)) and hence oscillating $\mathbf{J} \cdot \mathbf{E}$. Adding a guide field, the amplitude of $\mathbf{J}_{e,xz} \cdot \mathbf{E}_{xz}$ in the central EDR becomes smaller compared to $J_{ey}E_y$ (Cassak et al. 2017a; Wang et al. 2018), as electrons have less freedom to bounce across the current sheet.

Within the IDR (but outside the EDR), the electric field $\mathbf{E} = -\mathbf{V}_e \times \mathbf{B}/c$, thus the rate of the electron energy gain $\mathbf{E} \cdot \mathbf{J}_e$, vanishes. The rate of the ion energy gain is $\mathbf{E} \cdot \mathbf{J}_i = (-\mathbf{V}_i \times \mathbf{B}/c + \mathbf{J} \times \mathbf{B}/nec) \cdot \mathbf{J}_i = \mathbf{E}_{\mathrm{Hall}} \cdot \mathbf{J}_i$. Thus, the Hall electric field dominates the energization of ions overall (Fig. 33(e)-(f)). Since the Hall field is set up due to the ion-electron decoupling, it plays opposite roles in the energization of two species. We may understand the Hall field as a pathway to transfer energies between the two species without energy exchange between fields and particles, as quantified by $\mathbf{E}_{\mathrm{Hall}} \cdot \mathbf{J} = 0$. The Hall electromagnetic fields lead to the diverging Poynting flux streamline patterns around the X-line, which is critical in facilitating fast reconnection (Sect. 3.1.3, Liu et al. 2022). For asymmetric reconnection, $\mathbf{J}_{i,xz} \cdot \mathbf{E}_{xz}$ is negative in a localized region near the X-line (Fig. 33(f)), and coincides with the positive $\mathbf{J}_{e,xz} \cdot \mathbf{E}_{xz}$ in the similar region (Fig. 33(b)). $J_{iy}E_y$ has a smaller net contribution than $\mathbf{J}_{i,xz} \cdot \mathbf{E}_{xz}$ when integrating over the entire diffusion region (Wang et al. 2018). However, $J_{iy}E_y$ dominates close to the X-line and has positive values in a broad region over z due to J_{iy} from the finite Larmor radius effect of meandering ions near the boundary of the ion current layer (Fig. 33(f)).

Turning to observations more specifically, Genestreti et al. (2018a) demonstrated that $\mathbf{J}_e \cdot \mathbf{E} \sim J_{ey} E_y$ in a symmetric reconnection EDR, while $\mathbf{J}_e \cdot \mathbf{E}$ at dayside asymmetric reconnection exhibits significant fluctuations that may be associated with fluctuating upstream conditions (Genestreti et al. 2022) beyond the scope of the simulation discussions here. Genestreti et al. (2018a) and Payne et al. (2020) also used MMS to further evaluate the balance between $\partial u_{EM}/\partial t$ and $-\mathbf{J} \cdot \mathbf{E} - \nabla \cdot \mathbf{S}$ in Poynting's theorem for magnetopause and magnetotail EDRs, respectively. The time-derivative term $\partial u_{EM}/\partial t$ in the X-line frame was calculated based on $du_{EM}/dt = \partial u_{EM}/\partial t + \mathbf{V}_X \cdot \nabla u_{EM}$, where du_{EM}/dt is the temporal evolution in the spacecraft frame, and $\mathbf{V}_X$ is the X-line velocity. The results indicate that $\partial u_{EM}/\partial t$ is close to zero near the X-line, but it has more variations away from the X-line. The 2D PIC simulation exhibits an overall consistent pattern (e.g., Payne et al. 2020), while detailed comparisons suggest that events observed by MMS may be at a locally more unsteady state than what is seen in 2D simulations (Genestreti et al. 2018a). A relevant quantity is $\mathbf{J} \cdot (\mathbf{E} + \mathbf{V}_e \times \mathbf{B}/c)$, that is the energy conversion rate measured in the local bulk electron frame (Zenitani et al. 2011a); this useful quantity is often used to identify EDRs.

4.3 Further Decomposition and $(\mathbf{P} \cdot \nabla) \cdot \mathbf{V}$

We now discuss the evolution of plasma energy, and we treat this by considering two equations that describe the bulk and thermal forms separately. By dotting the momentum equation with $\mathbf{V}_s$, one can write the governing equation of bulk flow kinetic energy $u_{\text{bulk},s} \equiv (1/2)nm_s V_s^2$ in conservative form,

$$\frac{\partial u_{\text{bulk},s}}{\partial t} + \nabla \cdot \mathbf{K}_s = \mathbf{J}_s \cdot \mathbf{E} - \mathbf{V}_s \cdot (\nabla \cdot \mathbf{P}_s). \tag{51}$$

Subtracting Eq. (51) from Eq. (49), the equation for the thermal energy $u_{th} \equiv (1/2)\text{Tr}(\mathbf{P}_s)$ is obtained,

$$\frac{\partial u_{\text{th},s}}{\partial t} + \nabla \cdot \mathbf{H}_s + \nabla \cdot \mathbf{q}_s = \mathbf{V}_s \cdot (\nabla \cdot \mathbf{P}_s). \tag{52}$$

Note that the sum of Eq. (51) and (52) gives the overall particle energy equation that is the direct counterpart of Poynting's theorem. Interestingly, from the source terms on the right-hand side of these two equations, we can tell that the $\mathbf{V}_s \cdot (\nabla \cdot \mathbf{P}_s)$ term re-distributes the energy stored in the bulk and thermal forms.

Figure 34 shows the source terms on the right-hand side of Eqs. (51) and (52). Electrons gain both significant bulk (Fig. 34a) and thermal energies (Fig. 34b) within the EDR. Around

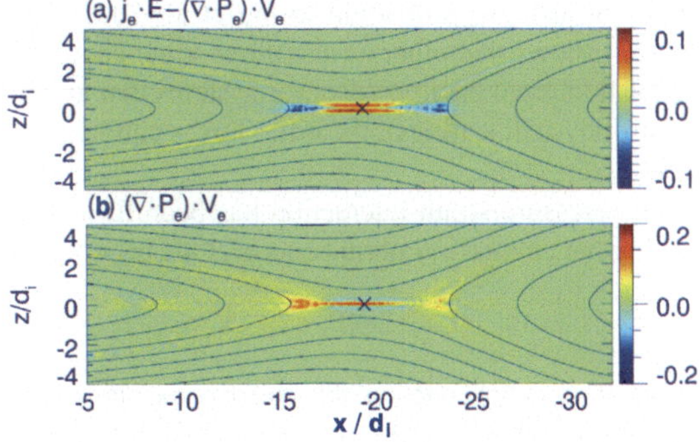

Fig. 34 Profiles of the source terms of the electron bulk energy equation $\mathbf{J}_e \cdot \mathbf{E} - \mathbf{V}_e \cdot (\nabla \cdot \mathbf{P}_e)$ (a) and thermal energy equation $\mathbf{V}_e \cdot (\nabla \cdot \mathbf{P}_e)$ in PIC. Reprinted from Lu et al. (2018), with the permission of AIP Publishing

the end of the EDR, the bulk energy gain is negative, and the thermal energy gain is positive, indicating the conversion from bulk to thermal energies (Lu et al. 2018), which from the kinetic perspective is associated with electron re-magnetization through gyro-turning around the reconnected magnetic field (Shuster et al. 2015; Payne et al. 2021). For ions, comparable bulk and thermal energy gains occur throughout the reconnection region (Lu et al. 2018). The source term for the thermal energy gain can be further decomposed in different forms (e.g., Hesse et al. 2018; Lapenta et al. 2020), and it has been demonstrated, using both simulations (Hesse et al. 2018) and MMS observations (Holmes et al. 2021), that the "quasi-viscous" term associated with the off-diagonal components of the pressure tensor has a dominant contribution in describing electron heating from inflow to outflow regions.

There is another form of the equation that is used to quantify the energy conversion between bulk flow kinetic energy and thermal energy of a species s. A brief calculation shows that $\nabla \cdot \mathbf{H}_s = \nabla \cdot (u_{\mathrm{th},s}\mathbf{V}_s) + \mathbf{V}_s \cdot (\nabla \cdot \mathbf{P}_s) + (\mathbf{P}_s \cdot \nabla) \cdot \mathbf{V}_s$. Substituting this expression into Eq. (52) gives

$$\frac{\partial u_{\mathrm{th},s}}{\partial t} + \nabla \cdot \left(u_{\mathrm{th},s}\mathbf{V}_s + \mathbf{q}_s\right) = -(\mathbf{P}_s \cdot \nabla) \cdot \mathbf{V}_s. \tag{53}$$

Similarly, rearranging Eq. (51) gives

$$\frac{\partial u_{\mathrm{bulk},s}}{\partial t} + \nabla \cdot (\mathbf{K}_s + \mathbf{V}_s \cdot \mathbf{P}_s) = \mathbf{J}_s \cdot \mathbf{E} + (\mathbf{P}_s \cdot \nabla) \cdot \mathbf{V}_s. \tag{54}$$

In this form, it is readily apparent that $-(\mathbf{P}_s \cdot \nabla) \cdot \mathbf{V}_s$ is a source of internal energy. Since the same term appears in the kinetic energy equation with the opposite sign, this term also describes the conversion between bulk kinetic energy and internal energy. This term, with the minus sign, is called "the pressure-strain interaction." A relevant discussion of this term to reconnection electric field can be found in Sect. 2.3, where this term is explicitly related to the non-gyrotropic plasma pressure within the EDR (Hesse et al. 2018)

The pressure-strain interaction has undergone significant study in the MMS era because MMS is uniquely capable of making reliable *in situ* measurements of it. One special property of the pressure-strain interaction is that if one has a closed (infinite or isolated) system, the volume integral over the whole domain V of Eq. (53) reveals (Yang et al. 2017b, 2022),

$$\frac{dU_{\mathrm{th},s}}{dt} = \int_{\mathrm{V}} d^3r[-(\mathbf{P}_s \cdot \nabla) \cdot \mathbf{V}_s], \tag{55}$$

where $U_{\mathrm{th},s} = \int_{\mathrm{V}} u_{\mathrm{th},s} d^3r$ is the total thermal energy in the system. Thus, in a collisionless closed system, the volume-integrated pressure-strain interaction is the *only* source of thermal energy. It is important to emphasize, however, that it is not the only source of thermal energy locally at any given position (Song et al. 2020; Du et al. 2020; Barbhuiya et al. 2024); Eq. (53) shows that other terms (the thermal energy flux and the heat flux) can also change the local internal energy. Also, for systems that are not closed (such as any system in space or astrophysical settings), the other fluxes can lead to a non-zero source or sink for internal energy.

The pressure-strain interaction has been further decomposed to isolate the key physics causing the change in internal energy. One decomposition is to write (Yang et al. 2017b,a)

$$-(\mathbf{P}_s \cdot \nabla) \cdot \mathbf{V}_s = -P_s(\nabla \cdot \mathbf{V}_s) - \mathbf{\Pi}_s : \mathbf{D}_s, \tag{56}$$

where $P_s \equiv (1/3)\mathrm{Tr}(\mathbf{P}_s)$ is the effective (scalar) pressure, $\mathbf{\Pi}_s = \mathbf{P}_s - P_s\mathbf{I}$ is the deviatoric pressure tensor which describes the departure of the pressure tensor from being isotropic, and $D_{s,jk} = (1/2)(\partial V_{s,j}/\partial r_k + \partial V_{s,k}/\partial r_j) - (1/3)\delta_{jk}(\nabla \cdot \mathbf{V}_s)$ is the "traceless strain rate tensor" which describes the incompressible portion of the flow. Thus, the first term on the right of Eq. (56) describes heating or cooling via compression or expansion, and the second term on the right describes incompressible deformation of fluid elements (Del Sarto et al. 2016; Yang et al. 2017b; Del Sarto and Pegoraro 2018). The second term can be further decomposed into incompressible deformation due to normal flow and incompressible deformation due to flow shear (Cassak and Barbhuiya 2022). The latter decomposition can be useful for reconnection studies because it isolates the effect of converging flow and flow shear. The pressure-strain interaction has also been written in magnetic field-aligned coordinates (Cassak et al. 2022), which allows one to determine if the compression, deformation, or shear is parallel or perpendicular to the magnetic field. The pressure-strain interaction and its decompositions have been studied in numerical simulations of magnetic reconnection (Sitnov et al. 2018; Du et al. 2018; Song et al. 2020; Fadanelli et al. 2021; Barbhuiya and Cassak 2022) and turbulence (Parashar et al. 2018; Pezzi et al. 2019; Yang et al. 2019; Hellinger et al. 2022) and in MMS observations (Chasapis et al. 2018; Zhong et al. 2019; Bandyopadhyay et al. 2020, 2021; Zhou et al. 2021; Wang et al. 2021).

4.4 Describing Changes to Internal Moments Beyond Internal Energy

Equation (49) and its subsequent decompositions discussed in Sects. 4.1-4.3 follow from the second moment of the Vlasov equation and contain a complete description of the information about energy conversion associated with the number density (the zeroth moment of the distribution function), bulk flow (the first moment), and the thermal energy (the trace of the second moments). However, the distribution function has an infinite number of moments, and the evolution of the other moments is not described by Eq. (49). For systems close to local thermodynamic equilibrium (LTE), *i.e.*, the distribution function is close to being Maxwellian, the other moments are small and their evolution is typically ignored. Any systems of interest for space and astrophysical environments, however, are far from LTE because they are weakly collisional or essentially collisionless. For such systems, it has been unclear how to quantify changes to the higher-order internal moments beyond density, bulk flow, and temperature. The wealth of particle distribution data from MMS, in particular, is now bringing these questions to the fore.

Recently, an approach to quantify changes associated with higher-order internal moments was suggested (Cassak et al. 2023; Barbhuiya et al. 2024). The key quantity is the so-called relative entropy density $s_{s,\mathrm{rel}}$, given by

$$s_{s,\mathrm{rel}} = -k_B \int f_s \ln\left(\frac{f_s}{f_{sM}}\right) d^3 v_s, \tag{57}$$

where the integral is over all of the velocity space. Here, f_{sM} is the "Maxwellianized" distribution associated with the distribution function f_s, given by a Maxwellian distribution with the number density n_s, the bulk flow $\mathbf{V}_s$ and temperature $T_s = (1/3)\mathrm{Tr}(\mathbf{P}_s)/n_s k_B$ (Grad 1965). This quantity is a measure of how non-Maxwellian a distribution function is, with $s_{s,\mathrm{rel}} = 0$ if f_s is a Maxwellian distribution and it being negative-definite if f_s is anything non-Maxwellian.

Because $s_{s,\mathrm{rel}}$ is a measure of how non-Maxwellian a distribution is, its time derivative describes how rapidly the shape of the distribution is changing to become more or less Maxwellian (Cassak et al. 2023). In particular, if $(d/dt)(s_{s,\mathrm{rel}}/n_s) > 0$, then f_s is becoming more Maxwellian in the comoving (Lagrangian) reference frame, while $(d/dt)(s_{s,\mathrm{rel}}/n_s) <$

0 implies f_s is becoming less Maxwellian. Dividing by n_s to give the relative entropy per particle is done to not include compression, which is described in the energy equation. It was argued (Cassak et al. 2023) that scaling $(d/dt)(s_{s,\mathrm{rel}}/n_s)$ by the temperature gives an effective energy per particle associated with changes to any (and all) of the higher order moments, called the change of relative energy per particle $d\mathcal{E}_{s,\mathrm{rel}}$ and given by

$$\frac{d\mathcal{E}_{s,\mathrm{rel}}}{dt} = T_s \frac{d(s_{s,\mathrm{rel}}/n_s)}{dt}. \tag{58}$$

It is important to note that $\mathcal{E}_{s,\mathrm{rel}}$ is not a form of energy and, therefore, does not appear in the second moment of the Vlasov equation (Eq. (49)), but it does have the same dimensions and therefore is a quantitative measure of the changes to the higher order internal moments of the distribution that can be directly compared to the standard forms of energy.

Understanding the interplay of changes of all of the higher-order internal moments and the lower-order moments is in its infancy. In a single simulation of reconnection using a particle-in-cell code with 25,600 particles per grid cell, it was shown (Cassak et al. 2023) that the relative energy change can locally be important or even dominate the changes of internal. How relative energy and entropy depend on ambient plasma parameters and the time evolution of reconnection remains unknown. Entropy-related quantities have been measured with MMS (Argall et al. 2022), but relative entropy has yet to be measured with MMS.

4.5 Energy Partition between Ions and Electrons

Understanding the energy partition between species is also desirable, particularly for understanding reconnection in settings where data may be incomplete (for example, in remote observations or planetary missions where the experimental payload is not optimized for plasma physics). Most incoming electromagnetic energy is eventually converted to the enthalpy flux at locations away from the X-line, as discussed in Sect. 4.1. In the more general asymmetric reconnection case, the thermal energy gain of each species was modeled as (Wang et al. 2018; Shay et al. 2014)

$$\frac{\Delta U_{\mathrm{th},s}}{U_{\mathrm{in}}} = \frac{\gamma}{\gamma-1}\frac{T_{\mathrm{out},s}-T_{\mathrm{in},s}}{m_i V_{A,\mathrm{asym}}^2}, \tag{59}$$

where $\Delta U_{\mathrm{th},s} \equiv \int \mathbf{J}_s \cdot \mathbf{E} d^3r$ and U_{in} is the input field energy available for conversion. $T_{\mathrm{in},s} = (n_1 T_{1,s} B_2 + n_2 T_{2,s} B_1)/(n_1 B_2 + n_2 B_1)$ represents the inflow temperature, $V_{A,\mathrm{asym}} = (B_1 B_2/(4\pi m_i)(B_1+B_2)/(n_1 B_2 + n_2 B_1))^{1/2}$ is the hybrid Alfvén speed for asymmetric reconnection (Eq. (21)), and $\gamma = 5/3$ is the ratio of specific heats. $T_{\mathrm{out},s} \equiv \langle n V_{x,s} T_s \rangle / \langle n V_{x,s} \rangle$ can be regarded as the outflow temperature averaged over the outflow exhaust with a weighting factor of $n V_{x,s}$. The outflow temperature $T_{\mathrm{out},s}$ was further approximated to be the temperature averaged using n as the weighting factor. The observations suggest that the heating rate of $(T_{\mathrm{out},s} - T_{\mathrm{in},s})/(m_i V_{A,\mathrm{asym}}^2)$ is 1.7% for electrons (Phan et al. 2013b) and 13% for ions (Phan et al. 2014), evaluated using the n-weighted $T_{\mathrm{out},s}$ across exhausts at the far downstream region. PIC simulations show similar results (Shay et al. 2014). A test using PIC indicates that the heating rate based on the n-weighted T_{out} is nearly constant at varying distances from the X-line (Wang et al. 2018). A caveat is that while T_{out} is insensitive to the approximated forms at distances well away from the EDR, the original $n V_x$-weighted T_{out} should be used around the EDR when using Eq. (59). Close to the EDR, the heating rate is only a few percent while the electron enthalpy flux gain is tens of percent of the incoming Poynting flux as the dominant form of energy conversion. The application of Eq. (59) to a

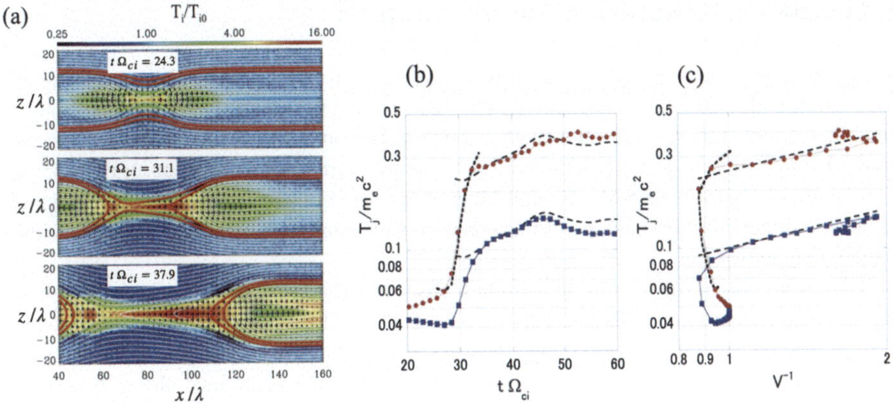

Fig. 35 (a) Time evolution of ion temperature in two-dimensional PIC simulation. (b) Time history of ion (red) and electron (blue) temperatures confined in the magnetic flux tube. The dashed lines are the adiabatic relation with $T_s V^{2/3}$ being a constant, and the dotted line is obtained using the effective Ohmic heating model of $T_i/T_e = (m_i/m_e)^{1/4}$ (Eq. (60)). (c) Relationship between the reciprocal of the flux tube volume (V^{-1}) and the temperatures (T_s) for ions (red) and electrons (blue). The dashed and dotted lines are the same as those in panel (b)). Adapted from Hoshino (2018), reproduced by permission of the AAS

magnetopause reconnection event observed by MMS in between the EDR and IDR boundaries suggests comparable energy partitions between ions and electrons, consistent with the trend predicted by PIC (Wang et al. 2018). We note that the calculation of $nV_{x,s}$-weighted $T_{\text{out},s}$ has significant uncertainties, so quantitative values need to be treated with caution.

Particle energization mechanisms provide insight into understanding the scaling laws of heating. For magnetized ions or electrons within outflow exhausts, the particles can be roughly described as moving along field lines at $\sim V_A$ in the Alfvénic outflow frame. Thus, the superposition of particles from two inflow regions leads to counter-streaming beams in the distribution, so that the effective temperature scales with V_A^2 (e.g., Liu et al. 2011a; Shay et al. 2014). A parallel potential exists in the exhaust, which modulates the beam speeds, and hence modifies the temperature profile and affects the overall $\Delta T_i/\Delta T_e$ (Haggerty et al. 2015). Such modulations of the ion beam speeds have been observed by MMS (Wang et al. 2019).

Inside the diffusion region, the acceleration by E_y (reconnection electric field in Fig. 35) during the meandering motion was considered to be the primary energization mechanism. Hoshino (2018) estimated the ratio of the ion-to-electron temperature enhancement $\Delta T_i/\Delta T_e$ using the effective Ohmic heating rates $E_y J_{ys} V_s$ of the two species, where J_{ys} and V_s are the ion/electron electric current density and the volume of the diffusion region (e.g., Coppi et al. 1966; Coroniti 1985), respectively. This leads to

$$\frac{\Delta \langle T_i \rangle_{\text{flux}}}{\Delta \langle T_e \rangle_{\text{flux}}} \simeq \left(\frac{m_i}{m_e}\right)^{1/4} \left(\frac{T_{i0}}{T_{e0}}\right)^{1/4}, \tag{60}$$

where $\langle T_s \rangle_{\text{flux}}$ is the species temperature averaged over the flux tubes and T_{s0} is the far upstream temperature. This scaling is supported by PIC simulations, as shown in Fig. 35. It is also interesting to note that the averaged temperature follows the adiabatic heating law (i.e., $P/n^{5/3} = \text{const}$) during the contraction of reconnected flux tubes (Fig. 35(c)).

5 Concluding Remarks and Future Prospects

In this tutorial review article, we have presented the basics of collisionless magnetic reconnection and highlighted some recent progress in understanding the generalized Ohm's law, the reconnection rate, and the energy conversion around the diffusion region. We also showed supporting evidence from local kinetic simulations and in-situ spacecraft observations, particularly from NASA's ongoing Magnetospheric Multiscale (MMS) mission, which is capable of performing multi-point measurements on electron-kinetic scale physics within Earth's magnetosphere.

The discussion of theories in this article focuses mostly on 2D models, originating from the classical Sweet-Parker (Parker 1957; Sweet 1958) and Petschek (Petschek 1964) solutions, where the spatial variation scale along the reconnection X-line is assumed to be much longer than the in-plane spatial scale (i.e., near translational invariance along the out-of-plane direction). The difference of our discussion from these classical models is that we treat the collisionless limit as it is relevant to most applications of reconnection in space plasmas. It is interesting to note that an initially three-dimensional (3D), short reconnection X-line within a uniform current sheet is inclined to spread linearly out of the reconnection plane, making the local geometry two-dimensional (Huba and Rudakov 2002; Shay et al. 2003; Karimabadi et al. 2004; Lapenta et al. 2006; Nakamura et al. 2012; Shepherd and Cassak 2012; Li et al. 2020). One should always be aware that Nature works in three-dimensional space, often accompanied by a high degree of complexity, even if insights from reduced dimensions enable one to extract essential physics. Understanding these 2D limits remains indispensable when seeking to single out the inherently 3D effects that only exist in a 3D system.

In the following, we discuss potential topics critical to the further understanding of magnetic reconnection. From the *local perspective*, a complete theory of the rate of collisionless reconnection, similar to that discussed for standard symmetric reconnection in Sect. 3.1.3, is still missing for most regimes discussed in Sect. 3 and deserves further development. The descriptions proposed thus far are primarily based on the moments of Vlasov equations, only minimally considering kinetic effects. Kinetic features not included here could play important roles and are highlighted in Norgren et al. (2025, this collection).

The study of the three-dimensional nature of reconnection X-lines is also important. For instance, how does the reconnection X-line orient itself within an asymmetric current sheet (Sonnerup 1974; Swisdak and Drake 2007; Hesse et al. 2013; Aunai et al. 2016; Liu et al. 2013, 2018c)? How does an X-line spread (Huba and Rudakov 2003; Shay et al. 2003; Lapenta et al. 2006; Nakamura et al. 2012; Shepherd and Cassak 2012; Jain and Büchner 2017; Liu et al. 2019; Li et al. 2020; Arencibia et al. 2023), and what is its minimal length (Shay et al. 2003; Liu et al. 2019; Huang et al. 2020; Pyakurel et al. 2021)? Over a larger spatial scale, the implication of these "local" 3D X-line properties to global solutions (Trattner et al. 2007) and 3D MHD reconnection theories (Priest et al. 2003; Pontin and Priest 2022; Li et al. 2021b), such as fan-spine reconnection, remains unclear. In terms of future observations, the ESA's SMILE mission (Raab et al. 2016), NASA's LEXI telescope (Walsh et al. 2024), and TRACERS (Kletzing 2019) will provide the intriguing possibility of imaging magnetopause dynamics for the first time, providing truly novel experimental data for addressing these questions.

A full 3D system also introduces additional players that are suppressed in two dimensions, including instabilities (Che et al. 2011a; Daughton et al. 2011; Che et al. 2011b; Roytershteyn et al. 2012; Liu et al. 2013), waves (Khotyaintsev et al. 2019; Yoo et al. 2020; Graham et al. 2022; Ng et al. 2023), and turbulence (Ergun et al. 2018; Stawarz et al. 2019), either at MHD or kinetic scales. The impact of these fundamental plasma processes on the reconnection rate and particle energization will continue to be an active direction for research. In particular, while the idea of "anomalous resistivity and transport" is appealing to MHD modeling of magnetic reconnection (Kulsrud 2001; Lin et al. 2021; Jiménez et al. 2022), concrete evidence that links this idea to fast reconnection in collisionless plasmas (Davidson and Gladd 1975; Yoo et al. 2020; Graham et al. 2022; Yoo et al. 2024) remains elusive. Another relevant open question is the existence of turbulent reconnection that has a thick diffusion region (TDR) on the MHD scales (as theorized in, e.g., Lazarian and Vishniac 1999) with well-defined global inflows and outflows. In a separate endeavor, Higashimori et al. (2013) (and more recently Stanish and MacTaggart 2024) demonstrated a nice transition from the Sweet-Parker solution to the Petschek-like solution in their 2D mean-field model, where the turbulence effect can be cranked up to enhance the anomalous resistivity and other non-ideal electric fields in Ohm's law. The enhanced anomalous resistivity also makes the diffusion region thicker, presumably due to the reduced current density (Lin et al. 2021).

It will be interesting to examine whether such a thick diffusion region is sustainable, not collapsing into kinetic scales, in collisionless plasmas. To address this problem, it would be ideal to have open boundaries in kinetic simulations, avoiding the exaggerated turbulence levels and current sheet broadening caused by the recycling of particles and magnetic structures from small periodic boundary conditions (Liu et al. 2018c). More discussion on the simulation approach of turbulent reconnection can be found in Ji et al. (2022). More discussion on the MMS observations of waves and turbulence associated with reconnection can be found in Stawarz et al. (2024, this collection) and Graham et al. (2025, this collection).

While we only discussed collisionless reconnection in this review, reconnection also occurs in collisional (Daughton et al. 2009; Stanier et al. 2019) and partially ionized plasmas (Zweibel 1989; Zweibel et al. 2011; Murphy and Lukin 2015; Ni et al. 2018; Jara-Almonte et al. 2019; Ni et al. 2020). The study of such reconnection could be important to understand the heating in the lower atmosphere of the Sun (e.g., jetlets in Shibata et al. 2007; Raouafi et al. 2023b) or the production of precipitating energetic electrons in ionospheres (e.g., aurora spirals in Huang et al. 2022). The transition from the collisional to collisionless limits could also be critical in understanding the onset problem of reconnection on the Sun, where the initial current sheet can be collisional and as thick as $\sim 10^6 d_i$. The plasmoid instability (Biskamp 1982; Shibata and Tanuma 2001; Bhattacharjee et al. 2009; Loureiro et al. 2007; Pucci and Velli 2014; Comisso et al. 2016) in collisional plasmas may enable a transition into the collisionless regime (Shibata and Tanuma 2001; Daughton et al. 2009; Huang et al. 2017; Stanier et al. 2019; Jara-Almonte and Ji 2021). In contrast, the plasmas in Earth's magnetotail are nearly collisionless, and the onset study of tail reconnection relevant to substorms is concerned more with the stability of a 2D magnetotail geometry (Schindler 1974; Lembege and Pellat 1982; Hesse and Schindler 2001; Pritchett 2005; Sitnov et al. 2009; Liu et al. 2014a; Bessho and Bhattacharjee 2014), where the collisionless tearing instability can be suppressed by the magnetic field normal to the current sheet (because electrons remain magnetized). The question in this context is under what conditions reconnection onset can be triggered in collisionless plasmas, enabling the energy release of geomagnetic substorms. More discussion on the onset problem can be found in Nakamura et al. (2025, this collection).

From the *global perspective*, it is critical to integrate our understanding of the local reconnection physics into the macroscale phenomena of a given system. The multiscale nature of reconnection makes this process interesting but also challenging for both first-principles numerical simulations and analytical theory. While the theoretical framework in Sect. 3.1

had coupled the mesoscale MHD region upstream of the ion diffusion region (IDR) to the electron diffusion region (EDR) in the steady state, it is assumed that the flux-breaking mechanism within the EDR can "passively" match (presumably by thinning) the reconnection electric field dictated by the outer region. A detailed coupling between the EDR particle kinetics (as discussed in Norgren et al. 2025, this collection) and the IDR solution has not yet been established. On the other hand, it remains unclear how one can couple this locally steady-state solution to the global macroscale in general settings, not to mention the difficulty in modeling the full macro-micro coupling in a time-dependent, dynamical system. Important progress may be made through existing state-of-art simulations (e.g., embedded PIC simulations Daldorff et al. 2014; Tóth et al. 2016) and the development of other novel numerical techniques (Shay et al. 2025, this collection). Different macro-micro couplings are summarized in reconnection phase diagrams (Ji and Daughton 2011; Ji et al. 2022) based on the previously mentioned plasmoid instability of long current sheets. During macro-micro coupling, key questions to ask are where and how a current sheet forms in a given global context, when reconnection can be triggered, and how efficiently it works. Such macro-micro coupling, for instance, includes reconnection within Kelvin-Helmholtz vortices (e.g., Nakamura et al. 2022; Blasl et al. 2023), other MHD-scale instabilities (e.g., Kliem and Török 2006; Zuccarello et al. 2014), solar wind-magnetosphere coupling (e.g., Dorelli 2019), solar flares (e.g., Wyper et al. 2017; Dahlin et al. 2022). More discussion can be found in Hwang et al. (2023).

The growing effort in space exploration [e.g., BepiColombo (Heyner et al. 2021) at Mercury, Juno (Bolton et al. 2017) at Jupiter, etc.] provides exciting opportunities to perform comparative studies of planetary magnetospheric reconnection; more discussion can be found in Gershman et al. (2024, this collection) and Fuselier et al. (2024, this collection). Both ground and spaceborne remote sensing/imaginary will further enable our understanding of solar flares, the coronal heating problem, and solar wind drivers; more discussion can be found in Drake et al. (2025, this collection). Meanwhile, terrestrial laboratory experiments [e.g., MRX (Yamada et al. 1997), TS-3/4 (Ono et al. 1993), TREX (Olson et al. 2016), PHASMA (Shi et al. 2022), FLARE (Ji et al. 2018), etc.] provide invaluable studies performed in a controlled, repeatable manner; more discussion can be found in Ji et al. (2023, this collection). Our hope is that what we have learned from magnetic reconnection within our solar system can also be used to understand other astrophysical objects in the Universe, such as the magnetospheres of stars, exoplanets, and the extreme plasmas near compact objects, including black holes and neutron stars; more discussion can be found in Guo et al. (2024, this collection). Going forward, continuous communication across disciplines will be the key to making breakthroughs in understanding this fundamental, exciting plasma process.

Acknowledgements YL thanks Shan Wang for the useful discussion on the section of energy conversion within the diffusion region. YL is supported by NASA MMS Theory and Modeling grant 80NSSC21K2048, NASA grant 80NSSC20K1316, and NSF grant 214230. RN is supported by Austrian Science Fund (FWF): P32175-N27. PAC acknowledges support from NASA Grants 80NSSC24K0172, 80NSSC23K0409, 80NSSC22K0323, and 80NSSC19M0146, NSF Grant PHY-2308669, and DoE Grant DE-SC0020294. JPE acknowledges UKRI/STFC grant ST/W001071/1. STR acknowledges support of MCIN/AEI/PRTR 10.13039/501100011033 (Grant PID2020-471 112805GA-I00) and Seneca Agency from Region of Murcia (Grant 21910/PI/22). HJ acknowledges support by NASA under Grants No. NNH15AB29I and 80HQTR21T0105. CN acknowledges support from the Research Council of Norway under Contract No. 300865, and the Swedish National Space Board under Grant 2022-00121.

Declarations

Competing Interests The authors declare no competing interests.

References

Abdo AA, Ackermann M, Ajello M, et al (2011) Gamma-ray flares from the Crab Nebula. Science 331:739

Alfvén H (1942) Existence of electromagnetic-hydrodynamic waves. Nature 150:405

André M, Vaivads A, Khotyaintsev YV, et al (2010) Magnetic reconnection and cold plasma at the magnetopause. Geophys Res Lett 37(22):L22108

Angelopoulos V (2008) The THEMIS Mission. Space Sci Rev 141(1–4):5–34. https://doi.org/10.1007/s11214-008-9336-1

Angelopoulos V, McFadden JP, Larson D, et al (2008) Tail reconnection triggering substorm onset. Science 321:931–935

Arencibia M, Cassak PA, Shay MA, et al (2023) Three-dimensional magnetic reconnection spreading in current sheets of non-uniform thickness. J Geophys Res Space Phys 128:e2022JA030999

Argall MR, Barbhuiya MH, Cassak PA, et al (2022) Theory, observations, and simulations of kinetic entropy in a magnetotail electron diffusion region. Phys Plasmas 29(2):022902. https://doi.org/10.1063/5.0073248. https://aip.scitation.org/doi/10.1063/5.0073248

Asenjo FA, Comisso L (2015) Generalized magnetofluid connections in relativistic magnetohydrodynamics. Phys Rev Lett 114(11):115003. https://doi.org/10.1103/PhysRevLett.114.115003. arXiv:1502.07461 [physics.plasm-ph]

Aunai N, Belmont G, Smets R (2011) Proton acceleration in antiparallel collisionless magnetic reconnection: kinetic mechanisms behind the fluid dynamics. J Geophys Res 116:A09232

Aunai N, Hesse M, Kuznetsova M (2013) Electron nongyrotropy in the context of collisionless magnetic reconnection. Phys Plasmas 20:092903

Aunai N, Hesse M, Lavraud B, et al (2016) Orientation of the x-line in asymmetric magnetic reconnection. J Plasmas Phys 82:535820401

Aydemir AY (1991) Linear studies of $m = 1$ modes in high-temperature plasmas with a four-field model. Phys Fluids B 3(11):3025

Aydemir A (1992) Nonlinear studies of $m = 1$ modes in high-temperature plasmas. Phys Fluids B 4:3469

Bale SD, Drake JF, McManus MD, et al (2023) Interchange reconnection as the source of the fast solar wind within coronal holes. Nature 618:252

Bandyopadhyay R, Matthaeus WH, Parashar TN, et al (2020) Statistics of kinetic dissipation in the Earth's magnetosheath: MMS observations. Phys Rev Lett 124:255101. https://doi.org/10.1103/PhysRevLett.124.255101

Bandyopadhyay R, Chasapis A, Matthaeus WH, et al (2021) Energy dissipation in turbulent reconnection. Phys Plasmas 28(11):112305. https://doi.org/10.1063/5.0071015. https://aip.scitation.org/doi/10.1063/5.0071015

Barbhuiya MH, Cassak PA (2022) Pressure-strain interaction: III. Particle-in-cell simulations of magnetic reconnection. Phys Plasmas 29:122308

Barbhuiya MH, Cassak PA, Adhikari S, et al (2024) Higher-order nonequilibrium term: effective power density quantifying evolution towards or away from local thermodynamic equilibrium. Phys Rev E 109:015205. https://doi.org/10.1103/PhysRevE.109.015205

Beidler MT, Cassak PA (2011) Model for incomplete reconnection in sawtooth crashes. Phys Rev Lett 107:255002

Bessho N, Bhattacharjee A (2005) Collisionless reconnection in an electron-positron plasma. Phys Rev Lett 95(24):245001. https://doi.org/10.1103/PhysRevLett.95.245001

Bessho N, Bhattacharjee A (2010) Fast magnetic reconnection in low-density electron-positron plasmas. Phys Plasmas 17(10):102104. https://doi.org/10.1063/1.3488963

Bessho N, Bhattacharjee A (2014) Instability of the current sheet in the Earth's magnetotail with normal magnetic field. Phys Plasmas 21(10):102905. https://doi.org/10.1063/1.4899043

Bessho N, Chen LJ, Wang S, et al (2019) Magnetic reconnection in a quasi-parallel shock: two-dimensional local particle-in-cell simulation. Geophys Res Lett 46(16):9352–9361. https://doi.org/10.1029/2019GL083397

Bessho N, Chen LJ, Wang S, et al (2020) Magnetic reconnection and kinetic waves generated in the Earth's quasi-parallel bow shock. Phys Plasmas 27(9):092901. https://doi.org/10.1063/5.0012443
Bessho N, Chen LJ, Stawarz JE, et al (2022) Strong reconnection electric fields in shock-driven turbulence. Phys Plasmas 29(4):042304. https://doi.org/10.1063/5.0077529
Bhattacharjee A (2004) Impulsive magnetic reconnection in the Earth's magnetotail and the solar corona. Annu Rev Astron Astrophys 42:365
Bhattacharjee A, Huang YM, Yang H, et al (2009) Fast reconnection in high-Lundquist-number plasmas due to secondary tearing instabilities. Phys Plasmas 16:112102
Birn J, Hesse M (2010) Energy release and transfer in guide field reconnection. Phys Plasmas 17:012109
Birn J, Priest ER (2007) Reconnection of magnetic fields: magnetohydrodynamics and collisionless theory and observations
Birn J, Drake JF, Shay MA, et al (2001) Geospace environmental modeling (GEM) magnetic reconnection challenge. J Geophys Res 106(A3):3715–3719
Birn J, Galsgaard K, Hesse M, et al (2005) Forced magnetic reconnection. Geophys Res Lett 32:L06105. https://doi.org/10.1029/2004GL022058
Birn J, Borovsky JE, Hesse M, et al (2010) Scaling of asymmetric reconnection in compressible plasmas. Phys Plasmas 17:052108
Biskamp D (1982) Effect of secondary tearing instability on the coalescence of magnetic islands. Phys Lett A 87(7):357–360. https://doi.org/10.1016/0375-9601(82)90844-1
Biskamp D (1986) Magnetic reconnection via current sheets. Phys Fluids 29(5):1520–1531
Biskamp D, Drake JF (1994) Dynamics of the sawtooth collapse in tokamak plasmas. Phys Rev Lett 73:971
Blackman EG, Field GB (1994) Kinematics of relativistic magnetic reconnection. Phys Rev Lett 72:494
Blasl KA, Nakamura TKM, Nakamura R, et al (2023) Electron-scale reconnecting current sheet formed within the lower hybrid wave-active region of Kelvin-Helmholtz waves. Geophys Res Lett 50:e2023GL104309
Bolton SJ, Lunine J, Stevenson D, et al (2017) The Juno Mission. Space Sci Rev 213(1–4):5–37. https://doi.org/10.1007/s11214-017-0429-6
Boozer AH (2012) Separation of magneitc field lines. Phys Plasmas 19:112901
Borovsky JE, Hesse M, Birn J, et al (2008) What determines the reconnection rate at the dayside magnetosphere? J Geophys Res 113:A07210
Büchner J, Kuska JP, Nikutowski B, et al (1998) Three-dimensional reconnection in the Earth's magnetotail: simulations and observations. Geophys Monogr 104:313–326. https://doi.org/10.1029/GM104p0313
Burch JL, Phan TD (2016) Magnetic reconnection at the dayside magnetopause: advances with MMS. Geophys Res Lett 43(16):8327–8338. https://doi.org/10.1002/2016GL069787
Burch JL, Torbert RB (eds) (2016) Magnetospheric multiscale: a mission to investigate the physics of magnetic reconnection. Space Sci Rev 199(1–4)
Burch JL, Moore TE, Torbert RB, et al (2016a) Magnetospheric multiscale overview and science objectives. Space Sci Rev 199(1–4):5–21. https://doi.org/10.1007/s11214-015-0164-9
Burch JL, Torbert RB, Phan T, et al (2016b) Electron-scale measurement of magnetic reconnection in space. Science 352:6290
Burch JL, Ergun RE, Cassak PA, et al (2018) Localized oscillatory energy conversion in magnetopause reconnection. Geophys Res Lett 45(3):1237–1245. https://doi.org/10.1002/2017GL076809. arXiv:1712.05697 [physics.space-ph]
Burch JL, Webster JM, Hesse M, et al (2020) Electron inflow velocities and reconnection rates at Earth's magnetopause and magnetosheath. Geophys Res Lett 47(17):e89082. https://doi.org/10.1029/2020GL089082
Burch JL, Hesse M, Webster JM, et al (2022) The EDR inflow region of a reconnecting current sheet in the geomagnetic tail. Phys Plasmas 29(5):052903. https://doi.org/10.1063/5.0083169
Cai HJ, Lee LC (1997) The generalized Ohm's law in collisionless magnetic reconnection. Phys Plasmas 4(3):509–520
Carilli CL, Taylor GB (2002) Cluster magnetic fields. Annu Rev Astron Astrophys 40:319–348. https://doi.org/10.1146/annurev.astro.40.060401.093852. arXiv:astro-ph/0110655 [astro-ph]
Carmichael H (1964) A process for flares. In: Ness WN (ed) AAS/NASA symposium on the physics of solar flares. NASA, Washington, p 451
Cassak PA (2011) Theory and simulations of the scaling of magnetic reconnection with symmetric shear flow. Phys Plasmas 18:072106
Cassak PA, Barbhuiya MH (2022) Pressure-strain interaction: I. On compression, deformation, viscosity, and their relation to Pi-D. Phys Plasmas 29:122306
Cassak PA, Otto A (2011) Scaling of the magnetic reconnection rate with symmetric shear flow. Phys Plasmas 18:074501. https://doi.org/10.1063/1.3609771

Cassak PA, Shay MA (2007) Scaling of asymmetric magnetic reconnection: general theory and collisional simulations. Phys Plasmas 14:102114
Cassak PA, Shay MA (2008) Scaling of asymmetric Hall magnetic reconnection. Geophys Res Lett 35:L19102
Cassak PA, Shay MA (2009) Structure of the dissipation region in fluid simulations of asymmetric magnetic reconnection. Phys Plasmas 16:055704
Cassak PA, Shay MA, Drake JF (2005) Catastrophe model for fast magnetic reconnection onset. Phys Rev Lett 95:235002
Cassak PA, Drake JF, Shay MA (2007) Catastrophe onset of fast magnetic reconnection with a guide field. Phys Plasmas 14:054502
Cassak PA, Genestreti KJ, Burch JL, et al (2017a) The effect of a guide field on local energy conversion during asymmetric magnetic reconnection: particle-in-cell simulations. J Geophys Res Space Phys 122(11):11523–11542. https://doi.org/10.1002/2017JA024555
Cassak PA, Liu Y-H, Shay MA (2017b) A review of the 0.1 reconnection rate problem. J Plasma Phys 83:715830501
Cassak PA, Barbhuiya MH, Weldon HA (2022) Pressure-strain interaction: II. Decomposition in magnetic field-aligned coordinates. Phys Plasmas 29:122307
Cassak PA, Barbhuiya MH, Liang H, et al (2023) Quantifying energy conversion in higher-order phase space density moments in plasmas. Phys Rev Lett 130(8):085201. https://doi.org/10.1103/PhysRevLett.130.085201
Cerutti B, Werner GR, Uzdensky DA, et al (2014) Gamma-ray flares in the Crab Nebula: a case of relativistic reconnection?. Phys Plasmas 21(5):056501. https://doi.org/10.1063/1.4872024. arXiv:1401.3016 [astro-ph.HE]
Chanteur G (1998) Spatial interpolation for four spacecraft: theory. In: Analysis methods for multi-spacecraft data. ISSI scientific reports series, vol 1, pp 349–370
Chasapis A, Yang Y, Matthaeus WH, et al (2018) Energy conversion and collisionless plasma dissipation channels in the turbulent magnetosheath observed by the magnetospheric multiscale mission. Astrophys J 862(1):32. https://doi.org/10.3847/1538-4357/aac775
Che H, Drake JF, Swisdak M (2011a) A current filamentation mechanism for breaking field magnetic field lines during reconnection. Nature 474:184–187. https://doi.org/10.1038/nature10091
Che H, Goldman MV, Newman DL (2011) Buneman instability in a magnetized current-carrying plasma with velocity shear. Phys Plasmas 18(5):052109. https://doi.org/10.1063/1.3590879. arXiv:1104.5283 [physics.plasm-ph]
Chen XL, Morrison PJ (1990) Resistive tearing instability with equilibrium shear flow. Phys Fluids B 2:495
Chen LJ, Hesse M, Wang S, et al (2016) Electron energization and structure of the diffusion region during asymmetric reconnection. Geophys Res Lett 43:2405
Chen LJ, Hesse M, Wang S, et al (2017) Electron diffusion region during magnetopause reconnection with an intermediate guide field: magnetospheric multiscale observations. J Geophys Res Space Phys 122(5):5235–5246. https://doi.org/10.1002/2017JA024004
Chen LJ, Wang S, Hesse M, et al (2019) Electron diffusion regions in magnetotail reconnection under varying guide fields. Geophys Res Lett 46(12):6230–6238. https://doi.org/10.1029/2019GL082393
Chien A, Gao L, Zhang S, et al (2023) Non-thermal electron acceleration from magnetically driven reconnection in a laboratory plasma. Nat Phys 19(2):254–262. https://doi.org/10.1038/s41567-022-01839-x. arXiv:2201.10052 [physics.plasm-ph]
Comisso L, Lingam M, Huang YM, et al (2016) General theory of the plasmoid instability. Phys Plasmas 23:100702
Coppi B (1965) Current-driven instabilities in configurations with sheared magnetic fields. Phys Fluids 8:2273
Coppi B, Laval G, Pellat R (1966) Dynamics of the geomagnetic tail. Phys Rev Lett 16(26):1207–1210. https://doi.org/10.1103/PhysRevLett.16.1207
Coroniti FV (1985) Explosive tail reconnection: the growth and expansion phases of magnetospheric substorms. J Geophys Res 90(A8):7427–7448. https://doi.org/10.1029/JA090iA08p07427
Dahlin JT, Antiochos SK, Qiu J, et al (2022) Variability of the reconnection guide field in solar flares. Astrophys J 932(2):94. https://doi.org/10.3847/1538-4357/ac6e3d. arXiv:2110.04132 [astro-ph.SR]
Daldorff LKS, Tóth G, Gombosi TI, et al (2014) Two-way coupling of a global Hall magnetohydrodynamics model with a local implicit particle-in-cell model. J Comput Phys 268:236–254. https://doi.org/10.1016/j.jcp.2014.03.009
Dargent J, Aunai N, Lavraud B, et al (2017) Kinetic simulation of asymmetric magnetic reconnection with cold ions. J Geophys Res Space Phys 122(5):5290–5306
Dargent J, Aunai N, Lavraud B, et al (2020) Simulation of plasmaspheric plume impact on dayside magnetic reconnection. Geophys Res Lett 47(4):e2019GL086546

Daughton W, Roytershteyn V, Albright BJ, et al (2009) Transition from collisional to kinetic regimes in large-scale reconnection layers. Phys Rev Lett 103:065004

Daughton W, Roytershteyn V, Karimabadi H, et al (2011) Role of electron physics in the development of turbulent magnetic reconnection in collisionless plasmas. Nat Phys 7:539–542. https://doi.org/10.1038/nphys1965

Daughton W, Nakamura TKM, Karimabadi H, et al (2014) Computing the reconnection rate in turbulent kinetic layers by using electron mixing to identify topology. Phys Plasmas 21:052307

Davidson RC, Gladd NT (1975) Anomalous transport properties associated with the lower-hybrid-drift instability. Phys Fluids 18(10):1327–1335. https://doi.org/10.1063/1.861021

Del Sarto D, Pegoraro F (2018) Shear-induced pressure anisotropization and correlation with fluid vorticity in a low collisionality plasma. Mon Not R Astron Soc 475:181. https://doi.org/10.1093/mnras/stx3083

Del Sarto D, Pegoraro F, Califano F (2016) Pressure anisotropy and small spatial scales induced by velocity shear. Phys Rev E 93(5):053203. https://doi.org/10.1103/PhysRevE.93.053203

Denton RE, Drake JF, Kleva RG (1987) The $m = 1$ convection cell and sawteeth in tokamaks. Phys Fluids 30:1448

Desroche M, Bagenal F, Delamere PA, et al (2012) Conditions at the expanded Jovian magnetopause and implications for the solar wind interaction. J Geophys Res 117:A07202. https://doi.org/10.1029/2012JA017621

DiBraccio GA, Slavin JA, Boardsen SA, et al (2013) MESSENGER observations of magnetopause structure and dynamics at Mercury. J Geophys Res 118:997–1008. https://doi.org/10.1002/jgra.50123

Divin A, Khotyaintsev YV, Vaivads A, et al (2016) Three-scale structure of diffusion region in the presence of cold ions. J Geophys Res Space Phys 121(12):12001–12013

Dorelli JC (2019) Does the solar wind electric field control the reconnection rate at Earth's subsolar magnetopause? J Geophys Res Space Phys 124(4):2668–2681. https://doi.org/10.1029/2018JA025868

Dorelli JC, Hesse M, Kuznetsova MM, et al (2004) A new look at driven magnetic reconnection at the terrestrial subsolar magnetopause. J Geophys Res 109:A12216

Doss C, Komar C, Cassak P, et al (2015) Asymmetric magnetic reconnection with a flow shear and applications to the magnetopause. J Geophys Res Space Phys 120(9):7748–7763

Drake JF, Swisdak M, Hesse M (2004) The structure of the parallel electric field during magnetic reconnection. In: EOS trans. AGU, pp Abstract SM42A–05

Drake JF, Shay MA, Swisdak M (2008) The Hall fields and fast magnetic reconnection. Phys Plasmas 15(4):042306. https://doi.org/10.1063/1.2901194

Drake JF, Opher M, Swisdak M, et al (2010) A magnetic reconnection mechanism for the generation of anomalous cosmic rays. Astrophys J 709:963–974. https://doi.org/10.1088/0004-637X/709/2/963

Drake JF, Antiochos SK, Bale SD, et al (2025) Magnetic reconnection in solar flares and the near-Sun solar wind. Space Sci Rev 221

Du S, Guo F, Zank GP, et al (2018) Plasma energization in colliding magnetic flux ropes. Astrophys J 867:16. https://doi.org/10.3847/1538-4357/aae30e

Du S, Zank GP, Li X, et al (2020) Energy dissipation and entropy in collisionless plasma. Phys Rev E 101:033208. https://doi.org/10.1103/PhysRevE.101.033208

Dungey JW (1953) Conditions for the occurrence of electrical discharges in astrophysical systems. Philos Mag 44:725

Dungey JW (1961) Interplanetary magnetic field and the auroral zones. Phys Rev Lett 6(2):47–48. https://doi.org/10.1103/PhysRevLett.6.47

Dungey JW (1988) Noise-free neutral sheet. In: Proceedings of an international workshop in space plasma, ESA SP-285, p 15

Eastwood JP, Shay MA, Phan TD, et al (2010) Asymmetry of the ion diffusion region Hall electric and magnetic fields during guide field reconnection: observations and comparison with simulations. Phys Rev Lett 104(20):205001. https://doi.org/10.1103/PhysRevLett.104.205001

Eastwood JP, Phan TD, Drake JF, et al (2013) Energy partition in magnetic reconnection in Earth's magnetotail. Phys Rev Lett 110:225001

Eastwood JP, Goldman MV, Phan TD, et al (2020) Energy flux densities near the electron dissipation region in asymmetric magnetopause reconnection. Phys Rev Lett 125:265102

Egedal J, Le A, Daughton W (2013) A review of pressure anisotropy caused by electron trapping in collisionless plasma, and its implications for magnetic reconnection. Phys Plasmas 20:061201

Egedal J, Ng J, Le A, et al (2019) Pressure tensor elements breaking the frozen-in law during reconnection in Earth's magnetotail. Phys Rev Lett 123(22):225101. https://doi.org/10.1103/PhysRevLett.123.225101

Ergun RE, Goodrich KA, Wilder FD, et al (2016) Magnetospheric multiscale satellites observations of parallel electric fields associated with magnetic reconnection. Phys Rev Lett 116:235102

Ergun RE, Goodrich KA, Wilder FD, et al (2018) Magnetic reconnection, turbulence, and particle acceleration: observations in the Earth's magnetotail. Geophys Res Lett 45:3338

Ergun RE, Pathak N, Usanova ME, et al (2022) Observation of magnetic reconnection in a region of strong turbulence. Astrophys J Lett 935(1):L8. https://doi.org/10.3847/2041-8213/ac81d4

Eriksson S, Lavraud B, Wilder FD, et al (2016a) Magnetospheric multiscale observations of magnetic reconnection associated with Kelvin-Helmholtz waves. Geophys Res Lett 43(11):5606–5615. https://doi.org/10.1002/2016GL068783

Eriksson S, Wilder FD, Ergun RE, et al (2016b) Magnetospheric multiscale observations of the electron diffusion region of large guide field magnetic reconnection. Phys Rev Lett 117(1):015001. https://doi.org/10.1103/PhysRevLett.117.015001

Escoubet CP, Fehringer M, Goldstein M (2001) Introduction: the Cluster mission. Ann Geophys 19:1197–1200. https://doi.org/10.5194/angeo-19-1197-2001

Eyink GL, Lazarian A, Vishniac ET (2011) Fast magnetic reconnection and spontaneous stochasticity. Astrophys J 743(1):51. https://doi.org/10.1088/0004-637X/743/1/51. arXiv:1103.1882 [astro-ph.GA]

Fadanelli S, Lavraud B, Califano F, et al (2021) Energy conversions associated with magnetic reconnection. J Geophys Res Space Phys 126(1):A028333. https://doi.org/10.1029/2020JA028333

Fujimoto K, Sydora RD (2012) Plasmoid-induced turbulence in collisionless magnetic reconnection. Phys Rev Lett 109(26):265004. https://doi.org/10.1103/PhysRevLett.109.265004

Fuselier SA, Lewis WS, Schiff C, et al (2016) Magnetospheric multiscale science mission profile and operations. Space Sci Rev 199(1–4):77–103. https://doi.org/10.1007/s11214-014-0087-x

Fuselier SA, Petrinec SM, Reiff PH, et al (2024) Global-scale processes and effects of magnetic reconnection on the geospace environment. Space Sci Rev 220(4):34. https://doi.org/10.1007/s11214-024-01067-0

Genestreti KJ, Burch JL, Cassak PA, et al (2017) The effect of a guide field on local energy conversion during asymmetric magnetic reconnection: MMS observations. J Geophys Res Space Phys 122(11):11342–11353. https://doi.org/10.1002/2017JA024247. arXiv:1706.08404 [physics.space-ph]

Genestreti KJ, Cassak PA, Varsani A, et al (2018a) Assessing the time dependence of reconnection with Poynting's theorem: MMS observations. Geophys Res Lett 45:2886

Genestreti KJ, Nakamura TKM, Nakamura R, et al (2018b) How accurately can we measure the reconnection rate E_M for the MMS diffusion region event of 11 July 2017? J Geophys Res Space Phys 123(11):9130–9149. https://doi.org/10.1029/2018JA025711. arXiv:1808.03603 [physics.space-ph]

Genestreti KJ, Varsani A, Burch JL, et al (2018c) MMS observation of asymmetric reconnection supported by 3-D electron pressure divergence. J Geophys Res Space Phys 123(3):1806–1821. https://doi.org/10.1002/2017JA025019. arXiv:1711.08262 [physics.space-ph]

Genestreti KJ, Li X, Liu YH, et al (2022) On the origin of "patchy" energy conversion in electron diffusion regions. Phys Plasmas 29(8):082107. https://doi.org/10.1063/5.0090275. arXiv:2203.13879 [physics.space-ph]

Genestreti KJ, Nakamura R, Burch JL, et al (2025) Structure of the electron diffusion region during magnetic reconnection. Space Sci Rev 221

Gershman D, Fuselier S, Cohen I, et al (2024) Magnetic reconnection at planetary bodies and astrospheres. Space Sci Rev 220:7. https://doi.org/10.1007/s11214-023-01017-2

Gingell I, Schwartz SJ, Eastwood JP, et al (2019) Observations of magnetic reconnection in the transition region of quasi-parallel shocks. Geophys Res Lett 46(3):1177–1184. https://doi.org/10.1029/2018GL081804. arXiv:1901.01076 [physics.space-ph]

Gingell I, Schwartz SJ, Eastwood JP, et al (2020) Statistics of reconnecting current sheets in the transition region of Earth's bow shock. J Geophys Res Space Phys 125(1):e27119. https://doi.org/10.1029/2019JA027119

Goldman MV, Newman DL, Eastwood JP, et al (2020) Multibeam energy moments of multibeam particle velocity distributions. J Geophys Res Space Phys 125(12):e28340. https://doi.org/10.1029/2020JA028340. arXiv:2005.09113 [physics.plasm-ph]

Gonzalez WD, Parker EN (eds) (2016) Magnetic reconnection: concepts and applications. Astrophysics and Space Science Library, vol 427. Springer, Cham. https://doi.org/10.1007/978-3-319-26432-5

Goodbred M, Liu YH (2022) First-principle theory of the relativistic magnetic reconnection rate in astrophyical pair plasmas. Phys Rev Lett 129:265101. https://doi.org/10.1103/PhysRevLett.129.265101

Grad H (1965) On Boltzmann's H-theorem. J Soc Ind Appl Math 13(1):259–277. http://www.jstor.org/stable/2946404

Graham DB, Khotyaintsev YV, André M, et al (2022) Direct observations of anomalous resistivity and diffusion in collisionless plasma. Nat Commun 13:2954. https://doi.org/10.1038/s41467-022-30561-8

Graham DB, Cozzani G, Khotyaintsev YV, et al (2025) The role of kinetic instabilities and waves in collisionless magnetic reconnection. Space Sci Rev 221. https://doi.org/10.1007/s11214-024-01133-7

Greess S, Egedal J, Stanier A, et al (2022) Kinetic simulations verifying reconnection rates measured in the laboratory, spanning the ion-coupled to near electron-only regimes. Phys Plasmas 29(10):102103. https://doi.org/10.1063/5.0101006. arXiv:2210.04960 [physics.plasm-ph]

Guan Y, Lu Q, Lu S, et al (2023) Reconnection rate and transition from ion-coupled to electron-only reconnection. Astrophys J 958(2):172. https://doi.org/10.3847/1538-4357/ad05b8
Guo F, Liu Y, Zenitani S, et al (2024) Magnetic reconnection and associated particle acceleration in high-energy astrophysics. Space Sci Rev 220:43. https://doi.org/10.1007/s11214-024-01073-2. arXiv:2309.13382
Haggerty C, Shay M, Drake J, et al (2015) The competition of electron and ion heating during magnetic reconnection. Geophys Res Lett 42(22):9657–9665
Haggerty CC, Parashar TN, Matthaeus WH, et al (2017) Exploring the statistics of magnetic reconnection X-points in kinetic particle-in-cell turbulence. Phys Plasmas 24(10):102308. https://doi.org/10.1063/1.5001722. arXiv:1706.04905 [physics.space-ph]
Haggerty C, Shay M, Chasapis A, et al (2018) The reduction of magnetic reconnection outflow jets to sub-Alfvénic speeds. Phys Plasmas 25:102120. https://doi.org/10.1063/1.5050530
Hasegawa H, Denton RE, Nakamura R, et al (2019) Reconstruction of the electron diffusion region of magnetotail reconnection seen by the MMS spacecraft on 11 July 2017. J Geophys Res Space Phys 124(1):122–138. https://doi.org/10.1029/2018JA026051
Hasegawa H, Argall MR, Aunai N, et al (2024) Advanced methods for analyzing in-situ observations of magnetic reconnection. Space Sci Rev 220:68. https://doi.org/10.1007/s11214-024-01095-w
Hellinger P, Montagud-Camps V, Franci L, et al (2022) Ion-scale transition of plasma turbulence: pressure–strain effect. Astrophys J 930:48
Hesse M, Cassak P (2020) Magnetic reconnection in the space science: past, present, and future. J Geophys Res 125:e2018JA025935
Hesse M, Schindler K (1988) A theoretical foundation of general magnetic reconnection. J Geophys Res 93(A6):5559–5567
Hesse M, Schindler K (2001) The onset of magnetic reconnection in the magnetotail. Earth Planets Space 53:645–653
Hesse M, Schindler K, Birn J, et al (1999) The diffusion region in collisionless magnetic reconnection. Phys Plasmas 6:1781–1795
Hesse M, Kuznetsova M, Schindler K, et al (2005) Three-dimensional modeling of electron quasiviscous dissipation in guide-field magnetic reconnection. Phys Plasmas 12(10):100704. https://doi.org/10.1063/1.2114350
Hesse M, Neukirch T, Schindler K, et al (2011) The diffusion region in collisionless magnetic reconnection. Space Sci Rev 160:3–23. https://doi.org/10.1007/s11214-010-9740-1
Hesse M, Aunai N, Zenitani S, et al (2013) Aspects of collisionless magnetic reconnection in asymmetric systems. Phys Plasmas 20:061210
Hesse M, Aunai N, Sibeck D, et al (2014) On the electron diffusion region in planar, asymmetric systems. Geophys Res Lett 41:8673
Hesse M, Liu YH, Chen LJ, et al (2018) The physical foundation of the reconnection electric field. Phys Plasmas 25:032901
Hesse M, Norgren C, Tenfjord P, et al (2021) A new look at the electron diffusion region in asymmetric magnetic reconnection. J Geophys Res Space Phys 126(2):e28456. https://doi.org/10.1029/2020JA028456. arXiv:2007.03379 [physics.space-ph]
Heuer SV, Genestreti KJ, Nakamura TKM, et al (2022) Calculating the electron diffusion region aspect ratio with magnetic field gradients. Geophys Res Lett 49(20):e2022GL100652. https://doi.org/10.1029/2022GL100652
Heyner D, Auster H, Fornacon K, et al (2021) The BepiColombo planetary magnetometer MPO-MAG: what can we learn from the Hermean magnetic field? Space Sci Rev 217:52. https://doi.org/10.1007/s11214-021-00822-x
Higashimori K, Yokoi N, Hoshino M (2013) Explosive turbulent magnetic reconnection. Phys Rev Lett 110(25):255001. https://doi.org/10.1103/PhysRevLett.110.255001. arXiv:1305.6695 [astro-ph.EP]
Hirayama T (1974) Theoretical model of flares and prominences. I: evaporating flare model. Sol Phys 34:323
Holmes JC, Nakamura R, Schmid D, et al (2021) Wave activity in a dynamically evolving reconnection separatrix. J Geophys Res 126:e2020JA028520
Hornig G, Schindler K (1996) Magnetic topology and the problem of its invariant definition. Phys Plasmas 3(3):781–791. https://doi.org/10.1063/1.871778
Hoshino M (2018) Energy partition between ion and electron of collisionless magnetic reconnection. Astrophys J Lett 868:L18
Hoshino M, Nishida A (1983) Numerical simulation of the dayside magnetopause. J Geophys Res 88:6926
Hosner M, Nakamura R, Schmid D, et al (2024) Reconnection inside a dipolarization front of a diverging earthward fast flow. J Geophys Res Space Phys 129(1):e2023JA031976. https://doi.org/10.1029/2023JA031976

Huang YM, Comisso L, Bhattacharjee A (2017) Plasmoid instability in evolving current sheets and onset of fast reconnection. Astrophys J 849(2):75. https://doi.org/10.3847/1538-4357/aa906d. arXiv:1707.01863 [physics.plasm-ph]

Huang K, Liu YH, Lu Q, et al (2020) Scaling of magnetic reconnection with a limited x-line extent. J Geophys Lett 47:e2020GL088147

Huang SY, Xiong QY, Song LF, et al (2021) Electron-only reconnection in an ion-scale current sheet at the magnetopause. Astrophys J 922(1):54. https://doi.org/10.3847/1538-4357/ac2668. arXiv:2109.13051 [physics.plasm-ph]

Huang K, Liu YH, Lu Q, et al (2022) Auroral spiral structure formation through magnetic reconnection in the aurora acceleration region. J Geophys Lett 49:e2022GL100466

Huba J (2005) Hall magnetic reconnection: guide field dependence. Phys Plasmas 12:012322

Huba JD (2006) Forced Hall magnetic reconnection: parametric variation of the "Newton challenge". Phys Plasmas 12:062311. https://doi.org/10.1063/1.2212397

Huba JD, Rudakov LI (2002) Three-dimensional Hall magnetic reconnection. Phys Plasmas 9(11):4435–4438

Huba JD, Rudakov LI (2003) Hall magnetohydrodynamics of neutral layers. Phys Plasmas 10(8):3139–3150

Hwang KJ, Dokgo K, Choi E, et al (2020) Magnetic reconnection inside a flux rope induced by Kelvin-Helmholtz vortices. J Geophys Res Space Phys 125(4):e27665. https://doi.org/10.1029/2019JA027665

Hwang KJ, Dokgo K, Choi E, et al (2021) Bifurcated current sheet observed on the boundary of Kelvin-Helmholtz vortices. Front Astron Space Sci 8:201. https://doi.org/10.3389/fspas.2021.782924

Hwang KJ, Nakamura R, Eastwood JP, et al (2023) Cross-scale processes of magnetic reconnection. Space Sci Rev 219(8):71. https://doi.org/10.1007/s11214-023-01010-9

Jain N, Büchner J (2017) Spreading of electron scale magnetic reconnection with a wave number dependent speed due to the propagation of dispersive waves. Phys Plasmas 24(8):082304. https://doi.org/10.1063/1.4994704

Jain N, Sharma AS (2009) Electron scale structures in collisionless magnetic reconnection. Phys Plasmas 16(5):050704. https://doi.org/10.1063/1.3134045

Jain N, Sharma AS (2015) Evolution of electron current sheets in collisionless magnetic reconnection. Phys Plasmas 22(10):102110. https://doi.org/10.1063/1.4933120

Jara-Almonte J, Ji H (2021) Thermodynamic phase transition in magnetic reconnection. Phys Rev Lett 127(5):055102. https://doi.org/10.1103/PhysRevLett.127.055102

Jara-Almonte J, Ji H, Yoo J, et al (2019) Kinetic simulations of magnetic reconnection in partially ionized plasmas. Phys Rev Lett 122(1):015101. https://doi.org/10.1103/PhysRevLett.122.015101

Ji H, Daughton W (2011) Phase diagram for magnetic reconnection in heliophysical, astrophysical, and laboratory plasmas. Phys Plasmas 18:111207

Ji H, Cutler R, Gettelfinger G, et al (2018) The FLARE device and its first plasma operation. In: APS division of plasma physics meeting abstracts, p CP11.020

Ji H, Daughton W, Jara-Almonte J, et al (2022) Magnetic reconnection in the era of exascale computing and multiscale experiments. Nat Rev Phys 4(4):263–282. https://doi.org/10.1038/s42254-021-00419-x. arXiv:2202.09004 [physics.plasm-ph]

Ji H, Yoo J, Fox W, et al (2023) Laboratory study of collisionless magnetic reconnection. Space Sci Rev 219(8):76. https://doi.org/10.1007/s11214-023-01024-3

Jiménez JPC, Tenfjord P, Hesse M, et al (2022) The role of resistivity on the efficiency of magnetic reconnection in mhd. J Geophys Res 127:e2021JA030134

Kadomtsev BB (1975) Disruptive instability in tokamaks. Sov J Plasma Phys 1:389–391

Karimabadi H, Krauss-Varban D, Huba JD, et al (2004) On magnetic reconnection regimes and associated three-dimensional asymmetries: hybrid, Hall-less hybrid, and Hall-MHD simulations. J Geophys Res Space Phys 109(A9):A09205. https://doi.org/10.1029/2004JA010478

Khotyaintsev YV, Graham DB, Norgren C, et al (2019) Collisionless magnetic reconnection and waves: progress review. Front Astron Space Sci 6:70. https://doi.org/10.3389/fspas.2019.00070

Kieokaew R, Lavraud B, Foullon C, et al (2020) Magnetic reconnection inside a flux transfer event-like structure in magnetopause Kelvin-Helmholtz waves. J Geophys Res Space Phys 125(6):e27527. https://doi.org/10.1029/2019JA027527

Kletzing C (2019) The Tandem Reconnection and Cusp Electrodynamics Reconnaissance Satellites (TRACERS) Mission. In: AGU Fall Meeting abstracts, p A41U–2687

Kleva RG, Drake JF, Waelbroeck FL (1995) Fast reconnection in high temperature plasmas. Phys Plasmas 2(1):23–34

Kliem B, Török T (2006) Torus instability. Phys Rev Lett 96(25):255002. https://doi.org/10.1103/PhysRevLett.96.255002. arXiv:physics/0605217 [physics.plasm-ph]

Kolstø HM, Hesse M, Norgren C, et al (2020) Collisionless magnetic reconnection in an asymmetric oxygen density configuration. Geophys Res Lett 47(1):e2019GL085359

Kopp R, Pneuman G (1976) Magnetic reconnection in the corona and the loop prominence phenomenon. Sol Phys 50(1):85–98

Kulsrud RM (2001) Magnetic reconnection: Sweet-Parker versus Petschek. Earth Planets Space 53:417

Kulsrud R, Ji H, Fox W, et al (2005) An electromagnetic drift instability in the magnetic reconnection experiment and its importance for magnetic reconnection. Phys Plasmas 12:082301. https://doi.org/10.1063/1.1949225

La Belle-Hamer AL, Otto A, Lee LC (1995) Magnetic reconnection in the presence of sheared flow and density asymmetry: applications to the Earth's magnetopause. J Geophys Res 100:11875

Lapenta G, Krauss-Varban D, Karimabadi H, et al (2006) Kinetic simulations of x-line expansion in 3D reconnection. Geophys Res Lett 33:L10102. https://doi.org/10.1029/2005GL025124

Lapenta G, El-Alaoui M, Berchem J, et al (2020) Multiscale mhd-kinetic pic study of energy fluxes caused by reconnection. J Geophys Res 125:e2019JA027276

Lazarian A, Vishniac E (1999) Reconnection in a weakly stochastic field. Astrophys J 517:700

Le A, Egedal J, Ng J, et al (2014) Current sheets and pressure anisotropy in the reconnection exhaust. Phys Plasmas 21(1):012103. https://doi.org/10.1063/1.4861871

Le A, Daughton W, Ohia O, et al (2018) Drift turbulence, particle transport, and anomalous dissipation at the reconnection magnetopause. Phys Plasmas 25:062103. https://doi.org/10.1063/1.5027086

Lee LC, Lee KH (2020) Fluid and kinetic aspects of magnetic reconnection and some related magnetospheric phenomena. Rev Mod Plasma Phys 4(1):9. https://doi.org/10.1007/s41614-020-00045-7

Lembege B, Pellat R (1982) Stability of a thick two-dimensional quasineutral sheet. Phys Fluids 25:1995

Li X, Liu YH (2021) The effect of thermal pressure on collisionless magnetic reconnection rate. Astrophys J 912:152. https://doi.org/10.3847/1538-4357/abf48c

Li W, André M, Khotyaintsev YV, et al (2016) Kinetic evidence of magnetic reconnection due to Kelvin-Helmholtz waves. Geophys Res Lett 43(11):5635–5643. https://doi.org/10.1002/2016GL069192

Li TC, Liu YH, Hesse M, et al (2020) Three-dimensional x-line spreading in asymmetric magnetic reconnection. J Geophys Res 125:e2019JA027094

Li TC, Liu YH, Qi Y (2021a) Identification of active magnetic reconnection using magnetic flux transport in plasma turbulence. Astrophys J Lett 909:L28

Li T, Priest E, Guo R (2021b) Three-dimensional magnetic reconnection in astrophysical plasmas. Proc R Soc Lond Ser A 477(2249):20200949. https://doi.org/10.1098/rspa.2020.0949. arXiv:2104.05174 [astro-ph.SR]

Lin SC, Liu YH, Li X (2021) Fast magnetic reconnection induced by resistivity gradients in 2D magnetohydrodynamics. Phys Plasmas 28:072109

Liu YH, Hesse M (2016) Suppression of collisionless magnetic reconnection in asymmetric current sheets. Phys Plasmas 23:060704

Liu YH, Drake JF, Swisdak M (2011a) The effects of strong temperature anisotropy on the kinetic structure of collisionless slow shocks and reconnection exhausts. I. Particle-in-cell simulations. Phys Plasmas 18:062110. https://doi.org/10.1063/1.3601760

Liu YH, Drake JF, Swisdak M (2011b) The effects of strong temperature anisotropy on the kinetic structure of collisionless slow shocks and reconnection exhausts. II. Theory. Phys Plasmas 18:092102. https://doi.org/10.1063/1.3627147

Liu YH, Drake JF, Swisdak M (2012) The structure of magnetic reconnection exhaust boundary. Phys Plasmas 19:022110. https://doi.org/10.1063/1.3685755

Liu YH, Daughton W, Karimabadi H, et al (2013) Bifurcated structure of the electron diffusion region in three-dimensional magnetic reconnection. Phys Rev Lett 110:265004

Liu YH, Birn J, Daughton W, et al (2014a) Onset of reconnection in the near magnetotail: PIC simulations. J Geophys Res 119:9773

Liu YH, Daughton W, Karimabadi H, et al (2014b) Do dispersive waves play a role in collisionless magnetic reconnection? Phys Plasmas 21:022113

Liu YH, Guo F, Daughton W, et al (2015a) Scaling of magnetic reconnection in relativistic collisionless pair plasmas. Phys Rev Lett 114:095002

Liu YH, Hesse M, Kuznetsova M (2015b) Orientation of x lines in asymmetric magnetic reconnection– mass ratio dependency. J Geophys Res 120:7331

Liu Y, Mouikis C, Kistler L, et al (2015c) The heavy ion diffusion region in magnetic reconnection in the Earth's magnetotail. J Geophys Res Space Phys 120(5):3535–3551

Liu YH, Hesse M, Guo F, et al (2017) Why does steady-state magnetic reconnection have a maximum local rate of order 0.1? Phys Rev Lett 118:085101

Liu YH, Hesse M, Cassak PA, et al (2018a) On the collisionless asymmetric magnetic reconnection rate. Geophys Res Lett 45(8):3311–3318. https://doi.org/10.1002/2017GL076460

Liu YH, Hesse M, Guo F, et al (2018b) Strongly localized magnetic reconnection by the super-Alfvénic shear flow. Phys Plasmas 25:080701. https://doi.org/10.1063/1.5042539

Liu YH, Hesse M, Li TC, et al (2018c) Orientation and stability of asymmetric magnetic reconnection x-line. J Geophys Res 123:4908. https://doi.org/10.1029/2018JA025410

Liu YH, Li TC, Hesse M, et al (2019) Three-dimensional magnetic reconnection with a spatially confined x-line extent: implications for dipolarizing flux bundles and the dawn-dusk asymmetry. J Geophys Res 124:2819. https://doi.org/10.1029/2019JA026539

Liu YH, Lin SC, Hesse M, et al (2020a) The critical role of collisionless plasma energization on the structure of relativistic magnetic reconnection. Astrophys J Lett 892:L13. https://doi.org/10.3847/2041-8213/ab7d3f

Liu TZ, Lu S, Turner DL, et al (2020b) Magnetospheric multiscale (MMS) observations of magnetic reconnection in foreshock transients. J Geophys Res Space Phys 125(4):e27822. https://doi.org/10.1029/2020JA027822

Liu YH, Cassak P, Li X, et al (2022) First-principle theory of the rate of magnetic reconnection in magnetospheric and solar plasmas. Commun Phys 5:97. https://doi.org/10.1038/s42005-022-00854-x

Liu YN, Fujimoto K, Cao JB (2024) Intense magnetic reconnection process embedded in three-dimensional turbulent current sheet. Geophys Res Lett 51(1):e2023GL106466. https://doi.org/10.1029/2023GL106466

Lopez RE (2016) The integrated dayside merging rate is controlled primarily by the solar wind. J Geophys Res Space Phys 121(5):4435–4445. https://doi.org/10.1002/2016JA022556

Loureiro NF, Schekochihin AA, Cowley SC (2007) Instability of current sheets and formation of plasmoid chains. Phys Plasmas 14:100703. https://doi.org/10.1063/1.2783986

Lu S, Pritchett PL, Angelopoulos V, et al (2018) Magnetic reconnection in Earth's magnetotail: energy conversion and its earthward-tailward asymmetry. Phys Plasmas 25:012905

Lu S, Wang R, Lu Q, et al (2020) Magnetotail reconnection onset caused by electron kinetics with a strong external driver. Nat Commun 11:5049

Lyons LR, Pridmore-Brown DC (1990) Force balance near an X line in a collisionless plasma. J Geophys Res 95(A12):20903–20909. https://doi.org/10.1029/JA095iA12p20903

Lyubarsky YE (2005) On the relativistic magnetic reconnection. Mon Not R Astron Soc 358:113–119

Lyutikov M, Uzdensky D (2003) Dynamics of relativistic reconnection. Astrophys J 589:893–901

Mahlmann JF, Philippov AA, Levinson A, et al (2022) Electromagnetic fireworks: fast radio bursts from rapid reconnection in the compressed magnetar wind. Astrophys J Lett 932(2):L20. https://doi.org/10.3847/2041-8213/ac7156

Man HY, Zhou M, Yi YY, et al (2020) Observations of electron-only magnetic reconnection associated with macroscopic magnetic flux ropes. Geophys Res Lett 47(19):e89659. https://doi.org/10.1029/2020GL089659

Mandt ME, Denton RE, Drake JF (1994) Transition to whistler mediated reconnection. Geophys Res Lett 21(1):73–76

Markidis S, Lapenta G, Bettarini L, et al (2011) Kinetic simulations of magnetic reconnection in presence of a background O^+ population. J Geophys Res Space Phys 116(A1):A00K16. https://doi.org/10.1029/2011JA016429

Marrone DP, Moran JM, Zhao JH, et al (2007) An unambiguous detection of Faraday rotation in Sagittarius A*. Astrophys J Lett 654(1):L57–L60. https://doi.org/10.1086/510850. arXiv:astro-ph/0611791 [astro-ph]

Marshall AT, Burch JL, Reiff PH, et al (2020) Asymmetric reconnection within a flux rope-type dipolarization front. J Geophys Res Space Phys 125(1):e27296. https://doi.org/10.1029/2019JA027296

Masters A, Eastwood JP, Swisdak M, et al (2012) The importance of plasma β conditions for magnetic reconnection at Saturn's magnetopause. Geophys Rev Lett 39(8):L08103. https://doi.org/10.1029/2012GL051372

Matthaeus WH, Lamkin SL (1986) Turbulent magnetic reconnection. Phys Fluids 29:2513

Mbarek R, Haggerty C, Sironi L, et al (2022) Relativistic asymmetric magnetic reconnection. Phys Rev Lett 128(14):145101. https://doi.org/10.1103/PhysRevLett.128.145101. arXiv:2109.12125 [physics.plasm-ph]

Melzani M, Walder R, Folini D, et al (2014) Relativistic magnetic reconnection in collisionless ion-electron plasmas explored with particle-in-cell simulations. A & A 570:A111

Mirnov VV, Hegna CC, Prager SC, et al (2006) Two fluid dynamo and edge-resonant $m = 0$ tearing instability in reversed field pinch. In: IAEA FEC conf, China, p TH–P3–18

Mitchell HG Jr, Kan JR (1978) Merging of magnetic fields with field-aligned plasma flow components. J Plasma Phys 20:31

Mozer FS, Pritchett PL (2010) Magnetic field reconnection: a first-principles perspective. Phys Today 63(6):34. https://doi.org/10.1063/1.3455250

Mozer FS, Bale SD, Phan TD (2002) Evidence of diffusion regions at a subsolar magnetopause crossing. Phys Rev Lett 89(1):015002. https://doi.org/10.1103/PhysRevLett.89.015002

Müller D, St. Cyr OC, Zouganelis I, et al (2020) The Solar Orbiter mission. Science overview. Astron Astrophys 642:A1. https://doi.org/10.1051/0004-6361/202038467. arXiv:2009.00861 [astro-ph.SR]
Muñoz PA, Büchner J (2016) Non-Maxwellian electron distribution functions due to self-generated turbulence in collisionless guide-field reconnection. Phys Plasmas 23(10):102103. https://doi.org/10.1063/1.4963773. arXiv:1608.03110 [physics.plasm-ph]
Murphy NA, Lukin VS (2015) Asymmetric magnetic reconnection in weakly ionized chromospheric plasmas. Astrophys J 805(2):134. https://doi.org/10.1088/0004-637X/805/2/134. arXiv:1504.01425 [astro-ph.SR]
Murphy NA, Sovinec CR, Cassak PA (2010) Magnetic reconnection with asymmetry in the outflow direction. J Geophys Res 115:A09206
Nagai T, Shinohara I, Fujimoto M, et al (2001) Geotail observations of the Hall current system: evidence of magnetic reconnection in the magnetotail. J Geophys Res 106(A11):25929–25950. https://doi.org/10.1029/2001JA900038
Nakamura M, Scholer M (2000) Structure of the magnetopause reconnection layer and of flux transfer events: ion kinetic effects. J Geophys Res 105:23179–23191
Nakamura TKM, Nakamura R, Alexandrova A, et al (2012) Hall magnetohydrodynamic effects for three-dimensional magnetic reconnection with finite width along the direction of the current. J Geophys Res 117:A03220
Nakamura TKM, Eriksson S, Hasegawa H, et al (2017a) Mass and energy transfer across the Earth's magnetopause caused by vortex-induced reconnection. J Geophys Res Space Phys 122(11):11505–11522. https://doi.org/10.1002/2017JA024346
Nakamura TKM, Haswgawa H, Daughton W, et al (2017b) Turbulent mass transfer caused by vortex induced reconnection in collisionless magnetospheric plasmas. Nat Commun 8:1582
Nakamura TKM, Genestreti KJ, Liu YH, et al (2018) Measurement of the magnetic reconnection rate in the Earth's magnetotail. J Geophys Res Space Phys 123(11):9150–9168. https://doi.org/10.1029/2018JA025713
Nakamura R, Genestreti KJ, Nakamura T, et al (2019) Structure of the current sheet in the 11 July 2017 electron diffusion region event. J Geophys Res Space Phys 124(2):1173–1186. https://doi.org/10.1029/2018JA026028
Nakamura TKM, Blasl KA, Hasegawa H, et al (2022) Multi-scale evolution of Kelvin-Helmholtz waves at the Earth's magnetopause during southward IMF periods. Phys Plasmas 29(1):012901. https://doi.org/10.1063/5.0067391
Nakamura R, et al (2025) Outlook. Space Sci Rev 221
Newcomb WA (1958) Motion of magnetic lines of force. Ann Phys 3(4):347–385. https://doi.org/10.1016/0003-4916(58)90024-1
Ng J, Yoo J, Chen LJ, et al (2023) 3D simulation of lower-hybrid drift waves in strong guide field asymmetric reconnection in laboratory experiments. Phys Plasmas 30(4):042101. https://doi.org/10.1063/5.0138278
Ni L, Lukin VS, Murphy NA, et al (2018) Magnetic reconnection in strongly magnetized regions of the low solar chromosphere. Astrophys J 852(2):95. https://doi.org/10.3847/1538-4357/aa9edb. arXiv:1712.00582 [astro-ph.SR]
Ni L, Ji H, Murphy NA, et al (2020) Magnetic reconnection in partially ionized plasmas. Proc R Soc Lond Ser A 476(2236):20190867. https://doi.org/10.1098/rspa.2019.0867. arXiv:2003.13233 [physics.plasm-ph]
Norgren C, Graham DB, Khotyaintsev YV, et al (2018) Electron reconnection in the magnetopause current layer. J Geophys Res Space Phys 123(11):9222–9238. https://doi.org/10.1029/2018JA025676
Norgren C, Chen LJ, Graham DB, et al (2025) Electron and ion dynamics in reconnection diffusion regions. Space Sci Rev 221
Øieroset M, Phan TD, Fujimoto M, et al (2001) In situ detection of collisionless reconnection in the Earth's magnetotail. Nature 412(6845):414–417. https://doi.org/10.1038/35086520
Oka M, Fujimoto M, Nakamura TKM, et al (2008) Magnetic reconnection by a self-retreating X line. Phys Rev Lett 101(20):205004
Oka M, Birn J, Egedal J, et al (2023) Particle acceleration by magnetic reconnection in geospace. Space Sci Rev 219(8):75. https://doi.org/10.1007/s11214-023-01011-8. arXiv:2307.01376 [physics.space-ph]
Olson J, Egedal J, Greess S, et al (2016) Experimental demonstration of the collisionless plasmoid instability below the ion kinetic scale during magnetic reconnection. Phys Rev Lett 116(25):255001. https://doi.org/10.1103/PhysRevLett.116.255001
Ono Y, Morita A, Katsurai M, et al (1993) Experimental investigation of three-dimensional magnetic reconnection by use of two colliding spheromaks. Phys Fluids B 5(10):3691–3701. https://doi.org/10.1063/1.860840
Parashar TN, Matthaeus WH, Shay MA (2018) Dependence of kinetic plasma turbulence on plasma β. Astrophys J Lett 864:L21

Parker EN (1957) Sweet's mechanism for merging magnetic fields in conducting fluids. J Geophys Res 62(4):509–520
Parker EN (1963) The solar-flare phenomenon and the theory of reconnection and annihilation of magnetic fields. Astrophys J 8:177
Parker EN (1973) The reconnection rate of magnetic fields. Astrophys J 180:247
Paschmann G, Daly PW (1998) Analysis methods for multi-spacecraft data. ISSI scientific reports series SR-001, ESA/ISSI
Paschmann G, Sonnerup BUÖ, Papamastorakis I,et al (1979) Plasma acceleration at the Earth's magnetopause: evidence for reconnection. Nature 282(5736):243–246
Paschmann G, Schwartz SJ, Escoubet CP, et al (2005) Outer magnetospheric boundaries: cluster results. https://doi.org/10.1007/1-4020-4582-4
Paschmann G, Øieroset M, Phan T (2013) In-situ observations of reconnection in space. Space Sci Rev 178(2–4):385–417. https://doi.org/10.1007/s11214-012-9957-2
Payne DS, Genestreti KJ, Germaschewski K, et al (2020) Energy balance and time dependence of a magnetotail electron diffusion region. J Geophys Res 125:e2020JA028290
Payne DS, Farrugia CJ, Torbert RB, et al (2021) Origin and structure of electromagnetic generator regions at the edge of the electron diffusion region. Phys Plasmas 28:112901
Peery S, Liu YH, Li X (2024) Conditions for relativistic magnetic reconnection under the presence of shear flow and guide field. Astrophys J 964:144
Pegoraro F (2016) Covariant magnetic connection hypersurfaces. J Plasma Phys 82(2):555820201. https://doi.org/10.1017/S0022377816000325. arXiv:1603.03909 [physics.plasm-ph]
Petschek HE (1964) Magnetic field annihilation. In: Proc. AAS-NASA symp. phys. solar flares, pp 425–439
Pezzi O, Yang Y, Valentini F, et al (2019) Energy conversion in turbulent weakly collisional plasmas: Eulerian hybrid Vlasov-Maxwell simulations. Phys Plasmas 26(7):072301. https://doi.org/10.1063/1.5100125. arXiv:1904.07715
Phan TD, Gosling JT, Paschmann G, et al (2010) The dependence of magnetic reconnection on plasma β and magnetic shear: evidence from solar wind observations. Astrophys J Lett 719:L199–L203. https://doi.org/10.1088/2041-8205/719/L199
Phan TD, Paschmann G, Gosling JT, et al (2013a) The dependence of magnetic reconnection on plasma β and magnetic shear: evidence from magnetopause observations. Geophys Res Lett 40(1):11–16. https://doi.org/10.1029/2012GL054528
Phan T, Shay M, Gosling J, et al (2013b) Electron bulk heating in magnetic reconnection at Earth's magnetopause: dependence on the inflow Alfvén speed and magnetic shear. Geophys Res Lett 40(17):4475–4480
Phan T, Drake J, Shay M, et al (2014) Ion bulk heating in magnetic reconnection exhausts at Earth's magnetopause: dependence on the inflow Alfvén speed and magnetic shear angle. Geophys Res Lett 41(20):7002–7010
Phan TD, Eastwood JP, Shay MA, et al (2018) Electron magnetic reconnection without ion coupling in Earth's turbulent magnetosheath. Nature 557(7704):202–206. https://doi.org/10.1038/s41586-018-0091-5
Philippov A, Uzdensky DA, Spitkovsky A, et al (2019) Pulsar radio emission mechanism: radio nanoshots as a low-frequency afterglow of relativistic magnetic reconnection. Astrophys J Lett 876(1):L6. https://doi.org/10.3847/2041-8213/ab1590
Pollock C, Moore T, Jacques A, et al (2016) Fast plasma investigation for magnetospheric multiscale. Space Sci Rev 199(1–4):331–406. https://doi.org/10.1007/s11214-016-0245-4
Pontin DI, Priest ER (2022) Magnetic reconnection: MHD theory and modelling. Living Rev Sol Phys 19(1):1. https://doi.org/10.1007/s41116-022-00032-9
Price L, Swisdak M, Drake JF, et al (2016) The effects of turbulence on three-dimensional magnetic reconnection at the magnetopause. Geophys Res Lett 43:6020
Priest E, Forbes T (2000) Magnetic reconnection. Cambridge University Press, Cambridge
Priest ER, Titov VS, Grundy RE, et al (2000) Exact solutions for reconnective magnetic annihilation. Proc R Soc Lond Ser A 456:1821
Priest ER, Hornig G, Pontin DI (2003) On the nature of three-dimensional magnetic reconnection. J Geophys Res 108(A7):1285. https://doi.org/10.1029/2002JA009812
Pritchard KR, Burch JL, Fuselier SA, et al (2019) Energy conversion and electron acceleration in the magnetopause reconnection diffusion region. Geophys Res Lett 46(10274):10274–10282. https://doi.org/10.1029/2019GL084636
Pritchard KR, Burch JL, Fuselier SA, et al (2023) Reconnection rates at the Earth's magnetopause and in the magnetosheath. J Geophys Res Space Phys 128(9):e2023JA031475. https://doi.org/10.1029/2023JA031475
Pritchett PL (2001) Geospace environment modeling (GEM) magnetic reconnection challenge: simulations with a full particle electromagnetic code. J Geophys Res 106:3783

Pritchett PL (2005) The "Newton challenge": kinetic aspects of forced magnetic reconnection. J Geophys Res 110:A10213. https://doi.org/10.1029/2005JA011228

Pucci F, Velli M (2014) Reconnection of quasi-singular current sheets: the "ideal" tearing mode. Astrophys J Lett 780:L19

Pyakurel PS, Shay MA, Phan TD, et al (2019) Transition from ion-coupled to electron-only reconnection: basic physics and implications for plasma turbulence. Phys Plasmas 26:082307

Pyakurel PS, Shay MA, Drake JF, et al (2021) Faster form of electron magnetic reconnection with a finite length X-line. Phys Rev Lett 127(15):155101. https://doi.org/10.1103/PhysRevLett.127.155101

Raab W, Branduardi-Raymont G, Wang C, et al (2016) SMILE: a joint ESA/CAS mission to investigate the interaction between the solar wind and Earth's magnetosphere. In: den Herder JWA, Takahashi T, Bautz M (eds) Space telescopes and instrumentation 2016: ultraviolet to gamma ray, p 990502. https://doi.org/10.1117/12.2231984

Rager AC, Dorelli JC, Gershman DJ, et al (2018) Electron crescent distributions as a manifestation of diamagnetic drift in an electron-scale current sheet: magnetospheric multiscale observations using new 7.5 ms fast plasma investigation moments. Geophys Res Lett 45(2):578–584. https://doi.org/10.1002/2017GL076260. arXiv:1706.08435 [physics.space-ph]

Raouafi NE, Matteini L, Squire J, et al (2023a) Parker Solar Probe: four years of discoveries at solar cycle minimum. Space Sci Rev 219(1):8. https://doi.org/10.1007/s11214-023-00952-4. arXiv:2301.02727 [astro-ph.SR]

Raouafi NE, Stenborg G, Seaton DB, et al (2023b) Magnetic reconnection as the driver of the solar wind. Astrophys J 945(1):28. https://doi.org/10.3847/1538-4357/acaf6c. arXiv:2301.00903 [astro-ph.SR]

Ren Y, Yamada M, et al (2005) Experiment verification of the Hall effect during magnetic reconnection in a laboratory plasma. Phys Rev Lett 95:055003

Retinò A, Sundkvist D, Vaivads A, et al (2007) In situ evidence of magnetic reconnection in turbulent plasma. Nat Phys 3(4):236–238

Ripperda B, Bacchini F, Philippov AA (2020) Magnetic reconnection and hot spot formation in black hole accretion disks. Astrophys J 900(2):100. https://doi.org/10.3847/1538-4357/ababab. arXiv:2003.04330 [astro-ph.HE]

Rogers BN, Denton RE, Drake JF, et al (2001) Role of dispersive waves in collisionless magnetic reconnection. Phys Rev Lett 87(19):195004

Roytershteyn V, Daughton W, Karimabadi H, et al (2012) Influence of the lower-hybrid drift instability on magnetic reconnection in asymmetric configurations. Phys Rev Lett 108:185001

Sato T, Hayashi T (1979) Externally driven magnetic reconnection and a powerful magnetic energy converter. Phys Fluids 22:1189–1202

Schekochihin AA, Cowley SC (2006) Turbulence, magnetic fields, and plasma physics in clusters of galaxies. Phys Plasmas 13(5):056501. https://doi.org/10.1063/1.2179053. arXiv:astro-ph/0601246

Schindler K (1974) A theory of the substorm mechanism. J Geophys Res 79:2803

Schmitz H, Grauer R (2006) Kinetic Vlasov simulations of collisionless magnetic reconnection. Phys Plasmas 13(9):092309. https://doi.org/10.1063/1.2347101. arXiv:physics/0608175 [physics.plasm-ph]

Schoeffler KM, Drake JF, Swisdak M (2011) The effects of plasma beta and anisotropy instabilities on the dynamics of reconnecting magnetic fields in the heliosheath. Astrophys J 743:70. https://doi.org/10.1088/0004-637X/743/1/70

Scholer M (1989) Undriven magnetic reconnection in an isolated current sheet. J Geophys Res 94(A7):8805–8812

Schreier R, Swisdak M, Drake JF, et al (2010) Three-dimensional simulations of the orientation and structure of reconnection X-lines. Phys Plasmas 17:110704. https://doi.org/10.1063/1.3494218

Schwartz SJ, Kucharek H, Farrugia CJ, et al (2021) Energy conversion within current sheets in the Earth's quasi parallel magnetosheath. Geophys Res Lett 48(4):e91859. https://doi.org/10.1029/2020GL091859

Scott BD, Hassam AB (1987) Analytical theory of nonlinear drift-tearing mode stability. Phys Fluids 30(1):90–101

Servidio S, Matthaeus WH, Shay MA, et al (2009) Magnetic reconnection in two-dimensional magnetohydrodynamic turbulence. Phys Rev Lett 102:115003

Shay MA, Swisdak M (2004) Three-species collisionless reconnection: effect of O^+ on magnetotail reconnection. Phys Rev Lett 93(17):175001

Shay MA, Drake JF, Rogers BN, et al (1999) The scaling of collisionless, magnetic reconnection for large systems. Geophys Res Lett 26(14):2163–2166

Shay MA, Drake JF, Swisdak M, et al (2003) Inherently three dimensional magnetic reconnection: a mechanism for bursty bulk flows? Geophys Res Lett 30(6):1345. https://doi.org/10.1029/2002GL016267

Shay MA, Drake JF, Swisdak M (2007) Two-scale structure of the electron dissipation region during collisionless magnetic reconnection. Phys Rev Lett 99:155002. https://doi.org/10.1103/PhysRevLett.99.155002

Shay MA, Haggerty CC, Phan TD, et al (2014) Electron heating during magnetic reconnection: a simulation scaling study. Phys Plasmas 21:122902

Shay MA, Phan TD, Haggerty CC, et al (2016) Kinetic signatures of the region surrounding the X line in asymmetric (magnetopause) reconnection. Geophys Res Lett 43(9):4145–4154. https://doi.org/10.1002/2016GL069034

Shay M, Adhikari S, Beesho N, et al (2025) Simulation models for exploring magnetic reconnection. Space Sci Rev 221

Shepherd LS, Cassak PA (2012) Guide field dependence of 3-D X-line spreading during collisionless magnetic reconnection. J Geophys Res 117:A10101

Shi P, Srivastav P, Barbhuiya MH, et al (2022) Laboratory observations of electron heating and non-Maxwellian distributions at the kinetic scale during electron-only magnetic reconnection. Phys Rev Lett 128(2):025002. https://doi.org/10.1103/PhysRevLett.128.025002

Shi P, Scime EE, Barbhuiya MH, et al (2023) Using direct laboratory measurements of electron temperature anisotropy to identify the heating mechanism in electron-only guide field magnetic reconnection. Phys Rev Lett 131:155101. https://doi.org/10.1103/PhysRevLett.131.155101

Shibata K, Tanuma S (2001) Plasmoid-induced-reconnection and fractal reconnection. Earth Planets Space 53:473

Shibata K, Nakamura T, Matsumoto T, et al (2007) Chromospheric anemone jets as evidence of ubiquitous reconnection. Science 318(5856):1591. https://doi.org/10.1126/science.1146708. arXiv:0810.3974 [astro-ph]

Shuster JR, Chen LJ, Hesse M, et al (2015) Spatiotemporal evolution of electron characteristics in the electron diffusion region of magnetic reconnection: implications for acceleration and heating. Geophys Res Lett 42:2586

Sitnov MI, Swisdak M, Divin AV (2009) Dipolarization fronts as a signature of transient reconnection in the magnetotail. J Geophys Res 114:A04202

Sitnov MI, Merkin VG, Roytershteyn V, et al (2018) Kinetic dissipation around a dipolarization front. Geophys Res Lett 45(10):4639–4647. https://doi.org/10.1029/2018GL077874

Song L, Zhou M, Yi Y, et al (2020) Force and energy balance of the dipolarization front. J Geophys Res Space Phys 125(9):e2020JA028278. https://doi.org/10.1029/2020JA028278

Sonnerup BUÖ (1974) Magnetopause reconnection rate. J Geophys Res 79(10):1546–1549

Sonnerup BUÖ (1979) Magnetic field reconnection. In: Lanzerotti LJ, Kennel CF, Parker EN (eds) Solar System plasma physics, vol 3. North-Holland, Amsterdam, p 46

Spinnangr SF, Hesse M, Tenfjord P, et al (2021) The micro-macro coupling of mass-loading in symmetric magnetic reconnection with cold ions. Geophys Res Lett 48(13):e2020GL090690

Stanier A, Simakov AN, Chacoń L, et al (2015a) Fast magnetic reconnection with strong guide fields. Phys Plasmas 22:010701

Stanier A, Simakov AN, Chacoń L, et al (2015b) Fluid vs. kinetic magnetic reconnection with strong guide fields. Phys Plasmas 22:101203

Stanier A, Daughton W, Le A, et al (2019) Influence of 3D plasmoid dynamics on the transition from collisional to kinetic reconnection. Phys Plasmas 26:072121

Stanish S, MacTaggart D (2024) On turbulent magnetic reconnection: fast and slow mean steady states. arXiv: 2409.07346 [physics.plasm-ph]

Stawarz JE, Eastwood JP, Phan TD, et al (2019) Properties of the turbulence associated with electron-only magnetic reconnection in Earth's magnetosheath. Astrophys J Lett 877(2):L37. https://doi.org/10.3847/2041-8213/ab21c8

Stawarz JE, Eastwood JP, Phan TD, et al (2022) Turbulence-driven magnetic reconnection and the magnetic correlation length: observations from magnetospheric multiscale in Earth's magnetosheath. Phys Plasmas 29(1):012302. https://doi.org/10.1063/5.0071106

Stawarz JE, Muñoz PA, Bessho N, et al (2024) The interplay between collisionless magnetic reconnection and turbulence. Space Sci Rev 220:90. https://doi.org/10.1007/s11214-024-01124-8

Stern DP (1966) The motion of magnetic field lines. Space Sci Rev 6:147–173. https://doi.org/10.1007/BF00222592

Sturrock PA (1966) Model of the high-energy phase of solar flares. Nature 211:695

Sullivan BP, Rogers BN (2008) The scaling of forced collisionless reconnection. Phys Plasmas 15(10):102106. https://doi.org/10.1063/1.2992136. arXiv:0906.0334 [physics.plasm-ph]

Sundkvist D, Retinò A, Vaivads A, et al (2007) Dissipation in turbulent plasma due to reconnection in thin current sheets. Phys Rev Lett 99(2):025004. https://doi.org/10.1103/PhysRevLett.99.025004

Sweet PA (1958) The neutral point theory of solar flares. In: Lehnet B (ed) IAU symp. in electromagnetic phenomena in cosmical physics. Cambridge University Press, New York, p 123

Sweetser TH, Broschart SB, Angelopoulos V, et al (2011) ARTEMIS Mission design. Space Sci Rev 165(1–4):27–57. https://doi.org/10.1007/s11214-012-9869-1

Swisdak M, Drake JF (2007) Orientaion of the reconnection x-line. Geophys Res Lett 34:L11106

Swisdak M, Rogers BN, Drake JF, et al (2003) Diamagnetic suppression of component magnetic reconnection at the magnetopause. J Geophys Res 108(A5):1218. https://doi.org/10.1029/2002JA009726

Swisdak M, Liu Y-H, Drake JF (2008) Development of a turbulent outflow during electron-positron magnetic reconnection. Astrophys J 680(2):999–1008. https://doi.org/10.1086/588088

Swisdak M, Opher M, Drake JF, et al (2010) The vector direction of the interstellar magnetic field outside the heliosphere. Astrophys J 710(2):1769–1775. https://doi.org/10.1088/0004-637X/710/2/1769

Swisdak M, Drake JF, Price L, et al (2017) Localize and intense energy converson in the diffusion region of asymmetric magnetic reconnection. Geophys Res Lett 45:5260

Tavani M, Bulgarelli A, Vittorini V, et al (2011) Discovery of powerful gamma-ray flares from the Crab Nebula. Science 331(6018):736. https://doi.org/10.1126/science.1200083. arXiv:1101.2311 [astro-ph.HE]

TenBarge JM, Daughton W, Karimabadi H, et al (2014) Collisionless reconnection in the large guide field regime: gyrokinetic versus particle-in-cell simulations. Phys Plasmas 21:020708

Tenfjord P, Hesse M, Norgren C, et al (2019) The impact of oxygen on the reconnection rate. Geophys Res Lett 46(12):6195–6203

Tharp TD, Yamada M, Ji H, et al (2013) Study of the effects of guide field on Hall reconnection. Phys Plasmas 20(5):055705. https://doi.org/10.1063/1.4805244

Toledo-Redondo S, Vaivads A, André M, et al (2015) Modification of the Hall physics in magnetic reconnection due to cold ions at the Earth's magnetopause. Geophys Res Lett 42(15):6146–6154

Toledo-Redondo S, André M, Khotyaintsev YV, et al (2016) Cold ion demagnetization near the x-line of magnetic reconnection. Geophys Res Lett 43(13):6759–6767

Toledo-Redondo S, Dargent J, Aunai N, et al (2018) Perpendicular current reduction caused by cold ions of ionospheric origin in magnetic reconnection at the magnetopause: particle-in-cell simulations and spacecraft observations. Geophys Res Lett 45(19):10033–10042. https://doi.org/10.1029/2018GL079051

Toledo-Redondo S, Hwang KJ, Escoubet CP, et al (2021) Solar wind—magnetosphere coupling during radial interplanetary magnetic field conditions: simultaneous multi-point observations. J Geophys Res Space Phys 126(11):e2021JA029506

Torbert RB, Burch JL, Giles BL, et al (2016a) Estimates of terms in Ohm's law during an encounter with an electron diffusion region. Geophys Res Lett 43(12):5918–5925. https://doi.org/10.1002/2016GL069553

Torbert RB, Russell CT, Magnes W, et al (2016b) The FIELDS instrument suite on MMS: scientific objectives, measurements, and data products. Space Sci Rev 199(1–4):105–135. https://doi.org/10.1007/s11214-014-0109-8

Torbert RB, Burch JL, Phan TD, et al (2018) Electron-scale dynamics of the diffusion region during symmetric magnetic reconnection in space. Science 362(6421):1391–1395. https://doi.org/10.1126/science.aat2998. arXiv:1809.06932 [physics.space-ph]

Tóth G, Jia X, Markidis S, et al (2016) Extended magnetohydrodynamics with embedded particle-in-cell simulation of Ganymede's magnetosphere. J Geophys Res Space Phys 121(2):1273–1293. https://doi.org/10.1002/2015JA021997

Trattner KJ, Mulcock JS, Petrinec SM, et al (2007) Probing the boundary between antiparallel and component reconnection during souwthward interplanetary magneitc field conditions. J Geophys Res 112:A01201

Typer E, Cattell C, Thaller S, et al (2016) Partitioning of integrated energy fluxes in four tail reconnection events observed by cluster. J Geophys Res 121:11798

Uzdensky DA (2011) Magnetic reconnection in extreme astrophysical environments. Space Sci Rev 160(1–4):45–71. https://doi.org/10.1007/s11214-011-9744-5. arXiv:1101.2472 [astro-ph.HE]

Vasyliunas VM (1972) Nonuniqueness of magnetic field line motion. J Geophys Res 77:6271

Vasyliunas VM (1975) Theoretical models of magnetic field line merging, 1. Rev Geophys Space Phys 13(1):303

von Goeler S, Stodiek W, Sauthoff N (1974) Studies of internal disruptions and $m = 1$ oscillations in tokamak discharges with soft-x-ray techniques. Phys Rev Lett 33:1201

Vörös Z, Yordanova E, Varsani A, et al (2017) MMS observation of magnetic reconnection in the turbulent magnetosheath. J Geophys Res Space Phys 122(11):11442–11467. https://doi.org/10.1002/2017JA024535

Walsh BM, Kuntz KD, Busk S, et al (2024) The Lunar Environment Heliophysics X-ray Imager (LEXI) Mission. Space Sci Rev 220:37. https://doi.org/10.1007/s11214-024-01063-4

Wang S, Chen LJ, Bessho N, et al (2018) Energy conversion andpartition in the asymmetrc reconnection diffusion region. J Geophys Res 123:8185

Wang S, Chen LJ, Bessho N, et al (2019) Observational evidence of magnetic reconnection in the terrestrial bow shock transition region. Geophys Res Lett 46(2):562–570. https://doi.org/10.1029/2018GL080944. arXiv:1812.09337 [physics.space-ph]

Wang Y, Bandyopadhyay R, Chhiber R, et al (2021) Statistical survey of collisionless dissipation in the terrestrial magnetosheath. J Geophys Res Space Phys 126(6):e2020JA029000. https://doi.org/10.1029/2020JA029000

Wilder FD, Ergun RE, Eriksson S, et al (2017) Multipoint measurements of the electron jet of symmetric magnetic reconnection with a moderate guide field. Phys Rev Lett 118(26):265101. https://doi.org/10.1103/PhysRevLett.118.265101

Wilder FD, Ergun RE, Burch JL, et al (2018) The role of the parallel electric field in electron-scale dissipation at reconnecting currents inthe magnetosheath. J Geophys Res Space Phys 123(8):6533–6547. https://doi.org/10.1029/2018JA025529

Wygant JR, Cattell CA, Lysak R, et al (2005) Cluster observations of an intense normal component of the electric field at a thin reconnecting current sheet in the tail and its role in shock-like acceleration of the ion fluid into the separatrix region. J Geophys Res 110:A09206

Wyper PF, Antiochos SK, DeVore CR (2017) A universal model for solar eruptions. Nature 544(7651):452–455. https://doi.org/10.1038/nature22050

Yamada M (2022) Magnetic reconnection. A modern synthesis of theory, experiment, and observations. Princeton University Press. https://doi.org/10.1515/9780691232980

Yamada M, Levinton FM, Pomphrey N, et al (1994) Investigation of magnetic reconnection during a sawtooth crash in a high-temperature tokamak plasma. Phys Plasmas 1(10):3269–3276

Yamada M, Ji H, Hsu S, et al (1997) Study of driven magnetic reconnection in a laboratory plasma. Phys Plasmas 4(5):1936–1944. https://doi.org/10.1063/1.872336

Yamada M, Kulsrud R, Ji H (2010) Magnetic reconnection. Rev Mod Phys 82(1):603

Yamada M, Yoo J, Myers CE (2016) Understanding dynamics and energetics of magnetic reconnection in a laboratory plasma: review of recent progress on selected fronts. Phys Plasmas 23:055402

Yamada M, Chen LJ, Yoo J, et al (2018) The two-fluid dynamics and energetics of the asymmetric magnetic reconnection in laboratory and space plasmas. Nat Commun 9:5223

Yang Y, Matthaeus WH, Parashar TN, et al (2017a) Energy transfer, pressure tensor, and heating of kinetic plasma. Phys Plasmas 24(7):072306. https://doi.org/10.1063/1.4990421. arXiv:1705.02054

Yang Y, Matthaeus WH, Parashar TN, et al (2017b) Energy transfer channels and turbulence cascade in Vlasov-Maxwell turbulence. Phys Rev E 95:061201. https://doi.org/10.1103/PhysRevE.95.061201

Yang Y, Wan M, Matthaeus WH, et al (2019) Scale dependence of energy transfer in turbulent plasma. Mon Not R Astron Soc 482(4):4933–4940. https://doi.org/10.1093/mnras/sty2977

Yang Y, Matthaeus WH, Roy S, et al (2022) Pressure–strain interaction as the energy dissipation estimate in collisionless plasma. Astrophys J 929(2):142. https://doi.org/10.3847/1538-4357/ac5d3e

Yoo J, Ji JY, Ambat MV, et al (2020) Lower hybrid drift waves during guide field reconnection. Geophys Res Lett 47(21):e87192. https://doi.org/10.1029/2020GL087192

Yoo J, Ng J, Ji H, et al (2024) Anomalous resistivity and electron heating by lower hybrid drift waves during magnetic reconnection with a guide field. Phys Rev Lett 132(14):145101. https://doi.org/10.1103/PhysRevLett.132.145101

Yordanova E, Vörös Z, Varsani A, et al (2016) Electron scale structures and magnetic reconnection signatures in the turbulent magnetosheath. Geophys Res Lett 43(12):5969–5978. https://doi.org/10.1002/2016GL069191. arXiv:1706.04053 [physics.space-ph]

Yordanova E, Vörös Z, Raptis S, et al (2020) Current sheet statistics in the magnetosheath. Front Astron Space Sci 7:2. https://doi.org/10.3389/fspas.2020.00002

Zenitani S, Nagai T (2016) Particle dynamics in the electron current layer in collisionless magnetic reconnection. Phys Plasmas 23(10):102102. https://doi.org/10.1063/1.4963008. arXiv:1605.07472 [astro-ph.SR]

Zenitani S, Hesse M, Kimas A, et al (2011a) New measure of the dissipation region in collisionless magnetic reconnection. Phys Rev Lett 106:195003

Zenitani S, Hesse M, Klimas A, et al (2011b) The inner structure of collisionless magnetic reconnection: the electron-frame dissipation measure and Hall fields. Phys Plasmas 18(12):122108. https://doi.org/10.1063/1.3662430. arXiv:1110.3103 [astro-ph.SR]

Zhong ZH, Deng XH, Zhou M, et al (2019) Energy conversion and dissipation at dipolarization fronts: a statistical overview. Geophys Res Lett 46(22):12693–12701. https://doi.org/10.1029/2019GL085409

Zhou M, Man H, Yang Y, et al (2021) Measurements of energy dissipation in the electron diffusion region. Geophys Res Lett 48(24):e2021GL096372. https://doi.org/10.1029/2021GL096372

Zuccarello FP, Seaton DB, Mierla M, et al (2014) Observational evidence of torus instability as trigger mechanism for coronal mass ejections: the 2011 August 4 filament eruption. Astrophys J 785(2):88. https://doi.org/10.1088/0004-637X/785/2/88. arXiv:1401.5936 [astro-ph.SR]

Zweibel EG (1989) Magnetic reconnection in partially ionized gases. Astrophys J 340:550. https://doi.org/10.1086/167416

Zweibel EG, Yamada M (2009) Magnetic reconnection in astrophysical and laboratory plasmas. Annu Rev Astron Astrophys 47:291–332. https://doi.org/10.1146/annurev-astro-082708-101726

Zweibel EG, Lawrence E, Yoo J, et al (2011) Magnetic reconnection in partially ionized plasmas. Phys Plasmas 18(11):111211. https://doi.org/10.1063/1.3656960

Authors and Affiliations

Yi-Hsin Liu[1] · Michael Hesse[2] · Kevin Genestreti[3] · Rumi Nakamura[4,5] · James L. Burch[6] · Paul A. Cassak[7] · Naoki Bessho[8,9] · Jonathan P. Eastwood[10] · Tai Phan[11] · Marc Swisdak[12] · Sergio Toledo-Redondo[13] · Masahiro Hoshino[14] · Cecilia Norgren[15,16] · Hantao Ji[17] · Takuma K.M. Nakamura[4,18]

✉ Y.-H. Liu
Yi-Hsin.Liu@dartmouth.edu

✉ J.P. Eastwood
Jonathan.Eastwood@imperial.ac.uk

1 Department of Physics and Astronomy, Dartmouth College, Hanover, NH 03750, USA

2 Ames Research Center, NASA, Moffett Field, CA 94035, USA

3 Southwest Research Institute, Durham, NH 03824, USA

4 Space Research Institute, Austrian Academy of Sciences, Schmiedlstraße 6, 8042 Graz, Austria

5 International Space Science Institute, Bern, Switzerland

6 Southwest Research Institute, San Antonio, TX 78238, USA

7 Department of Physics and Astronomy and Center for KINETIC Plasma Physics, West Virginia University, Morgantown, WV 26506, USA

8 Goddard Space Flight Center, NASA, Greenbelt, MD 20771, USA

9 Department of Astronomy, University of Maryland, College Park, MD 20742, USA

10 Department of Physics, Imperial College London, London, UK

11 Space Science Laboratory, UC Berkeley, Berkeley, CA 94720, USA

12 IREAP, University of Maryland, College Park, MD 20742, USA

13 Department of Electromagnetism and Electronics, University of Murcia, Murcia, Spain

14 Department of Earth and Planetary Science, The University of Tokyo, Tokyo, 113-0033, Japan

15 Swedish Institute of Space Physics, Uppsala, Sweden

16 Department of Physics and Technology, University of Bergen, Bergen, Norway

17 Department of Astrophysical Sciences, Princeton University, Princeton, NJ 08544, USA

18 Krimgen LLC, Hiroshima, 7320828, Japan

Space Science Reviews (2025) 221:73
https://doi.org/10.1007/s11214-025-01197-z

Electron and Ion Dynamics in Reconnection Diffusion Regions

C. Norgren[1,2] · L.-J. Chen[3] · D.B. Graham[1] · N. Bessho[3,4] · J. Egedal[5] · L. Richard[1] · Yu. V. Khotyaintsev[1] · J. Shuster[6] · S. Toledo-Redondo[7] · B. Lavraud[8] · H. Hasegawa[9] · J.P. Eastwood[10] · M. Hesse[11] · Y.-H. Liu[12] · J.C. Holmes[13] · M. Argall[6]

Received: 7 June 2024 / Accepted: 4 July 2025 / Published online: 11 August 2025

Abstract
Magnetic reconnection is a fundamental plasma process responsible for the sometimes explosive release of magnetic energy in space and laboratory plasmas. Inside the diffusion regions of magnetic reconnection, the plasma becomes demagnetized and decouples from the magnetic field, enabling the change in magnetic topology necessary to power the energy release over larger scales. Since it was launched in 2015, the Magnetospheric MultiScale (MMS) mission has significantly advanced the understanding of the particle dynamics key to magnetic reconnection by providing high-resolution, in-situ measurements able to resolve ion and electron kinetic scales, i.e. a fraction of a gyroradius, that have confirmed theoretical predictions, revealed new phenomena, and refined existing models. These breakthroughs are critical for understanding not only space plasmas but also laboratory and astrophysical plasmas where magnetic reconnection occurs. In this work, we review the ion and electron dynamics occurring within the diffusion regions, in the inflow, along the separatrices, and downstream of the diffusion regions, in different reconnection configurations: symmetric, asymmetric, antiparallel, and guide field reconnection.

Keywords Magnetic reconnection · Ion dynamics · Electron dynamics · Kinetic processes · Electron diffusion region · Ion diffusion region · Symmetric reconnection · Asymmetric reconnection · Antiparallel reconnection · Guide field reconnection · Hall electric and magnetic fields

1 Introduction

Magnetic reconnection is a fundamental plasma process responsible for the explosive release of magnetic energy in space and laboratory plasmas (e.g. Ji et al. 2023, this collection, and references therein). While the release of magnetic energy occurs over macroscopic scales (e.g. Oka et al. 2023, this collection) and drives magnetospheric-scale phenomena such as the Dungey cycle (e.g. Fuselier et al. 2024, this collection), it is enabled by the 'breaking' and 'reconnecting' of magnetic field lines inside the electron diffusion region (EDR) – wherein the motions of the magnetic field and the electron fluid decouple (e.g. Liu et al. 2025b, this collection, and references therein). Magnetic reconnection is intrinsically a multi-scale process in which the electron and ion dynamics are intimately coupled. The EDR

Extended author information available on the last page of the article

is small, with a thickness of the order of a couple of electron inertial lengths, which at the dayside magnetopause corresponds to a few kilometers and in the magnetotail current sheet to a few tens of kilometers. Observing it with in situ measurements has, therefore, presented a challenge historically. With the advent of the Magnetospheric MultiScale (MMS) mission, which was specifically designed to unveil the secrets of the EDR, the understanding of the dynamics within these inner reconnection regions has seen major breakthroughs. These breakthroughs were enabled by the high temporal resolution of the plasma measurements together with the electron-scale tetrahedral formation of the four satellites.

For the first time, MMS enabled the observation of electron-scale dynamics within the EDR. Burch et al. (2016) reported the first observations of electron velocity distribution functions (VDFs) associated with the demagnetized electron motion inside the EDR for asymmetric reconnection at the dayside magnetopause. Torbert et al. (2018) reported the first in-depth analysis of a symmetric EDR in the nightside current sheet. The electron acceleration by the reconnection electric field was demonstrated by electron VDFs with multiple crescent populations, presenting evidence of demagnetized meandering trajectories. The MMS mission revealed new regimes and a multitude of dynamic regions where reconnection takes place. One prominent example of a new regime is the electron-only reconnection first observed in the magnetosheath, where the reconnecting current sheet is so small in spatial scale that it does not couple to the ion dynamics (Phan et al. 2018; Yordanova et al. 2016). The regions where MMS first discovered the presence of reconnection include the bowshock and foreshock region, inside primary reconnection exhausts, between interlaced fluxropes, inside fluxropes, between fluxropes and a dipolarization front, and inside Kelvin-Helmholtz vortices (see Hwang et al. 2023, this collection, and references therein). The unprecedented capabilities of MMS to resolve the electron and ion diffusion regions (IDR) have opened new windows to examine how ions are accelerated in the IDR, how ion outflow jets are formed in the close vicinity of the reconnection X line, and how ions may couple with electrons through fields of appropriate scales to cause effective resistivity (Graham et al. 2025, this collection).

The progress in the understanding of magnetic reconnection during the last several decades is due to the efforts of hundreds of scientists, which we can not acknowledge directly. In this review, we will instead focus on the most recent advances, in particular concerning electron dynamics, many of which were made possible with MMS. For prior reviews on the subject, see e.g. Yamada et al. (2010), Yamada (2022), Zhou et al. (2022), Wang et al. (2023b) and other papers in this collection. For a broader perspective on the reconnection process, see the other papers in this collection. To make the content accessible for students and non-experts, we first provide a brief background of some key concepts relevant to grasping the breakthroughs reviewed in this article. Thereafter, we review the state-of-the-art for different reconnection configurations roughly in order of increasing complexity. Throughout most of this review, with the exception of a few reproduced figures, we will adopt LMN coordinates where B_L and B_N are the *reconnecting* and *reconnected* magnetic field components, v_N and v_L are the inflow and outflow velocity components, and the reconnection electric field is contained in E_M.

2 Fundamental Concepts and Particle Dynamics of Magnetic Reconnection

In this section, we introduce some fundamental concepts of magnetic reconnection to provide some intuition with which to interpret the observations. First, we briefly introduce the concepts of diffusion regions for different plasma components, the induction equation and resistivity without classical collisions, and the displacement of a particle from its field line during the diffusion process. Then, we briefly review single-particle motions in three simple 1D current sheets, and in current sheets with a finite magnetic field normal to the current sheet. Thereafter, we give a basic formulation outlining the formation of crescent distributions, i.e. non-gyrotropic finite gyroradius effects which have been shown to appear in reconnection current sheets. At last, we briefly introduce some basics of adiabatic motion.

2.1 Magnetic Diffusion

Magnetic diffusion is an essential concept in the process of magnetic reconnection. It is through magnetic diffusion that the magnetic field can slip through the plasma and then break and reconnect at the X line. The change in topology occurs in a small volume of space, but the resulting change in magnetic topology enables energy conversion over vast scales (e.g. Oka et al. 2023, this collection).

The Diffusion Regions The term *diffusion region* refers to the regions in space where there is significant diffusion of the magnetic field through the plasma or a given plasma population. This occurs when the given plasma population becomes demagnetized, i.e., it can no longer trace a magnetic field line. Due to their different masses and thermal speeds, different plasma populations will become demagnetized at different distances from the magnetic field reversal: heavier particles will become demagnetized before lighter species; species with larger thermal speeds will become demagnetized before species with smaller thermal speeds. This leads to a layered structure where the EDR is embedded in one or several IDRs. For example, when multiple ion species or ion populations of distinct temperatures are present, there can be an inner light-ion diffusion region and an outer heavy-ion diffusion region (Liu et al. 2015), and/or an inner cold-ion diffusion region and an outer warm-ion diffusion region (André et al. 2016).

The Induction Equation Faraday's law, which governs the change in magnetic topology during the reconnection process, requires the existence of an electric field directed normal to the reconnection plane. One of the fundamental questions of magnetic reconnection in collisionless plasmas regards what physical process may sustain such an electric field. An expression for the electric field may come in various forms (see review by Liu et al. 2025b, this collection, and references therein), but follows the standard expression

$$\mathbf{E} + \mathbf{v} \times \mathbf{B} = \mathbf{R} \tag{1}$$

where the so-called non-ideal terms are given by $\mathbf{R}$, and $\mathbf{v}$ represents the bulk velocity of the plasma species or population under consideration, for example, electrons, ions, a given ion species or population, or the plasma mass-centralized velocity $\mathbf{v} = (m_i \mathbf{v}_i + m_e \mathbf{v}_e)/(m_i + m_e)$. By inserting the electric field from Eq. (1) into Faraday's law, we obtain the magnetic induction equation,

$$\frac{\partial \mathbf{B}}{\partial t} = \nabla \times (\mathbf{v} \times \mathbf{B}) - \nabla \times \mathbf{R}. \tag{2}$$

The first term on the right-hand side of Eq. (2) dictates the rate at which magnetic flux is transported with the bulk motion. If $\nabla \times \mathbf{R} = 0$, the displacement of the magnetic field lines exactly corresponds to the displacement of a fluid element during the same time interval.

The second term gives the rate of magnetic diffusion across the plasma. When the plasma approaches the X line, its velocity will not exactly correspond to the displacement of a magnetic field line, i.e. $\mathbf{E}+\mathbf{v}\times\mathbf{B}\neq 0$. In a steady-state situation, the two terms on the right-hand side of Eq. (2) are balanced: the transport of magnetic flux toward the X line with the plasma in the inflow is balanced by the diffusion of the magnetic field across the plasma from the inflow to the outflow. The transition between convective and diffusive electric fields is gradual (see e.g. Liu et al. 2025b, this collection). In the following, we consider a coordinate system where the reconnection magnetic field lies in the LN plane and the reconnection electric field is directed normal to this plane, i.e., along $\hat{e}_M$. When $(\mathbf{E}+\mathbf{v}\times\mathbf{B})_M > 0$, the magnetic flux is transported downstream faster than the plasma, i.e., the magnetic field slips through or diffuses across the plasma. When $(\mathbf{E}+\mathbf{v}\times\mathbf{B})_M < 0$, the particles are instead outrunning, or catching up with, the field. Based on these criteria, the diffusion regions are sometimes separated into the inner and the outer diffusion regions (Karimabadi et al. 2007).

The non-ideal ($\mathbf{R}\neq 0$) electric field must satisfy momentum balance, which is generally described by the generalized Ohm's law (discussed in detail in Liu et al. 2025b, this collection). In the tenuous plasma we consider in this review, the classical resistivity (Spitzer 1962), i.e. $\mathbf{R}=\eta\mathbf{J}$, is negligible (Vasyliunas 1975). However, the expression is instructive in its simplicity – the non-ideal electric field is sustained by some process which regulates the current density. Or, inversely, there is a finite conductivity, $\sigma = 1/\eta$, that permits a finite current $\mathbf{J}=\sigma\mathbf{R}$. Speiser (1970) showed that both the electron inertia, when taking into account a finite dwell time of particles in the diffusion region, and the gyroturning during a Speiser orbit (see Sect. 2.2.2, or Speiser 1965), lead to effective conductivities proportional to the reconnection electric field. The latter mechanism was termed gyro resistivity – instead of the momentum being limited by collisions, it is being removed from the region by the gyroturning. The dwell times, which, for the case of the gyro resistivity, were of the order of the inverse (Larmor) gyrofrequency in the weak (reconnected) magnetic field component normal to the current sheet, replaced the classical collisional time scale. That is, the acceleration by the reconnection electric field, which drives the reconnection current, is limited by the finite dwell time imposed by the magnetic topology of magnetic reconnection, thereby providing a finite effective resistivity. This concept connects the particle dynamics and trajectories of individual particles, which can be considered microscopic properties of the system, to macroscopic properties such as the current or reconnection electric field.

Particle Displacement from Magnetic Field Line The relative motion of the field and the particles is illustrated in Fig. 1 for a group of ions that impinge on the X line in a particle-in-cell simulation of antiparallel symmetric reconnection. The in-plane magnetic flux tubes, where the magnetic field is defined by equicontour lines of the out-of-plane magnetic vector potential $A_M\hat{e}_M$, where $\mathbf{B}=\nabla\times\mathbf{A}$, are illustrated by the colored background. The ions (with a few exceptions) initially dwell on the same (orange) flux tube in the inflow region. When approaching the X line, we can see that the flux tube moves faster than the ions. This delay is increased as the ions meander in the field reversal. The speed the ions gain during the meandering motion makes it possible for them to eventually rejoin their original flux tube downstream (not shown), where they are remagnetized. When they are remagnetized, their gyro-radii are larger due to the acceleration.

The displacement of a particle from a given flux tube can be quantified using the conservation of the canonical momentum $p_M = m_s v_M + q_s A_M$, where A_M is the magnetic vector potential, which defines the magnetic field in the reconnecting plane, $\mathbf{B}_{LN} = \nabla\times(A_M\hat{e}_m)$. For a given particle with constant p_M, any change in v_M is associated with a change in A_M, that is, a displacement from a given magnetic field line: $m_s v_M(t_1) + q_s A_M(t_1) =$

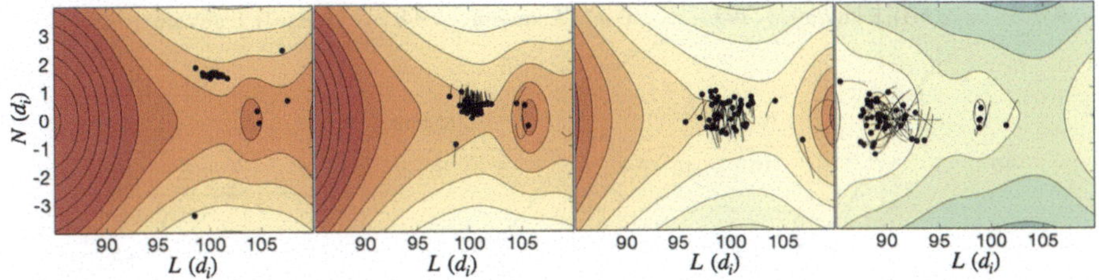

Fig. 1 Illustration of magnetic diffusion with respect to a set of ions inside the IDR. The four frames illustrate the in-plane magnetic field (through equicontour lines of the out-of-plane magnetic vector potential A_M, where $B_L = -\partial_N A_M$ and $B_N = \partial_L A_M$) and an ensemble of ions for four different time steps. As the ions dwell inside the IDR, undergoing meandering motion, the magnetic field slips ahead downstream. During the meandering motion, the ions gain energy, which is partly transferred to bulk motion in the outflow direction through gyroturning. Later on, with the gain of energy and speed, the ions rejoin the magnetic field and become remagnetized (not shown here). For details of the simulation, see Norgren et al. (2021)

$m_s v_M(t_2) + q_s A_M(t_2)$, or $\Delta v_M = -(q_s/m_s)\Delta A_M$, where $\Delta v_M = v_M(t_2) - v_M(t_1)$ and $\Delta A_M = A_M(t_2) - A_M(t_1)$ (e.g. Egedal et al. 2016a). This is easily understood for gyromotion in a uniform magnetic field $B_L\hat{e}_L$: the continuous change in v_M from negative to positive during gyromotion provides information on how much displaced (in M) the particle is on either side of the guiding center magnetic field line. However, it also holds for a more complicated magnetic field topology and particle trajectory, like the Speiser orbit inside the reconnection exhaust, or particle trajectories affected by Hall electric fields (e.g. Egedal et al. 2016a; Lapenta et al. 2017; Zenitani et al. 2017; Norgren et al. 2021).

2.2 Single Particle Motion

In this section, we first briefly cover the fundamentals of single-particle motion in a reversing magnetic field, including the effects of a magnetic guide field and a normal electric field. Thereafter, we give a brief introduction to the particle motion in a reconnection topology, including a reversing normal magnetic field and a reconnection electric field.

2.2.1 Periodic Particle Motion in Reversing Magnetic Field

The motion of a charged particle in an electromagnetic field is described by

$$m\frac{d\mathbf{v}}{dt} = q(\mathbf{E} + \mathbf{v} \times \mathbf{B}), \tag{3}$$

where m is the mass and q is the charge of the particle. In a uniform magnetic field, and in the absence of any electric field, the particle will perform circular gyrations around a given magnetic field line at a frequency $\omega_c = eB/m$ and with a radius of gyration $\rho = v_\perp/\omega_c$, where $\perp$ means perpendicular to the magnetic field. The particle trajectory can be greatly modified if the magnetic field is non-uniform over the spatial or temporal scales of such a gyration.

Of particular interest to magnetic reconnection is the behavior of particles in topologies where the magnetic field reverses direction. In Fig. 2, we show the trajectories of three electrons in a simple reversing Harris current sheet (Harris 1962) with a magnetic field $B_L = B_0 \tanh(N/\delta)$, where the asymptotic magnetic field is $B_0 = 10$ nT and the current sheet half thickness is $\delta = 5$ km. The corresponding magnetic vector potential

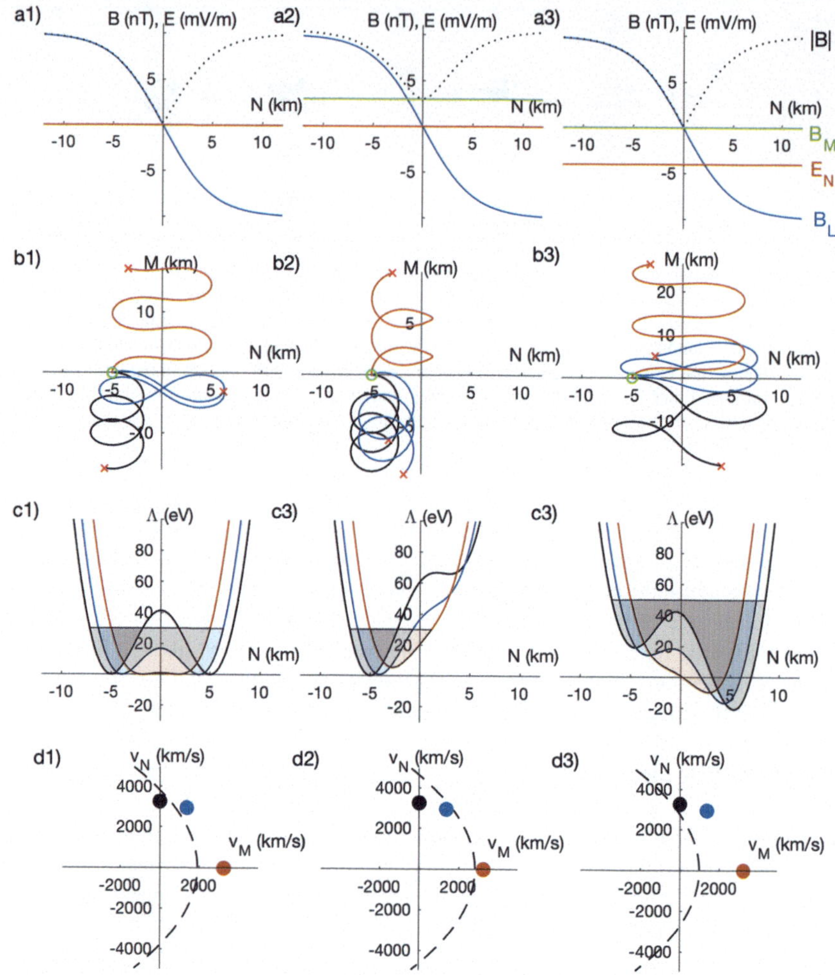

Fig. 2 Electron trajectories for three different current sheet configurations: (1) strictly antiparallel magnetic fields, (2) moderate guide field, and (3) a normal electric field. The magnetic guide field acts as a confining field while the electric field increases the electrons' access to the other side of the current sheet. While the guide field affects ions and electrons similarly, the electric field has opposite effects. (a) The electromagnetic fields. (b) Electron trajectories. (c) Pseudo-potential Λ defined by Eq. (5). (d) Initial electron velocities. The dashed lines show the boundary between crossing and non-crossing trajectories of particles starting from the same x location

is given by $A_M = \delta B_0 \log(\cosh(N/\delta))$. We also show the added effect of a guide field $B_g \hat{e}_M = \hat{e}_M \partial_N A_L = \hat{e}_M \partial_N (B_g N)$, and the added effect of an electric field $E_N = -\partial_N \phi$ directed normal to the current sheet, where ϕ is an electrostatic potential (Fig. 2c). The electrons start at a distance of $N = -5$ km from the center of the current sheet, and all have the same perpendicular speed $v_\perp = 3250$ km/s, corresponding to an energy of 30 eV, but a slightly different initial direction (Fig. 2b). As can be seen, the trajectories strongly depend on the initial direction of the electrons' motion. The electrons that remain on their initial side of the magnetic field reversal are following *non-crossing* orbits. In this case, the motion can be described by the superposition of a circular motion and a *guiding center* drift speed, where the guiding center is the electrons' locations averaged over one gyroperiod. The electrons crossing to the current sheet's other side are following *crossing* orbits. These electrons undergo a meandering motion due to the reversal of the magnetic force $-e(\mathbf{v} \times \mathbf{B})_N = ev_M B_L \hat{e}_N$ when B_L reverses. Crossing orbits are therefore also called *meandering* orbits. With the guide field, the electrons become more confined to their initial side of the current sheet since the magnetic force is stronger and the associated gyroscale is smaller (Fig. 2b2). With the electric field, the electrons from the side of the current sheet towards (from) which the electric field points have easier (more limited) access to the other side (Fig. 2b3). For a given magnetic field and energy, heavier particles perform crossing trajectories over a larger spatial domain than lighter particles.

The periodic motion of a charged particle in a prescribed electromagnetic field can be described using the constants of motion, which are the total energy of the particle $H = \epsilon + q_s\phi = (m_s/2)\left(v_L^2 + v_M^2 + v_N^2\right) + q_s\phi$ and the different components $i = L, M, N$ of the canonical momenta $p_i = m_s v_i + qA_i$, where A_i are the components of the magnetic vector potential. For a one-dimensional reversing magnetic field, $B_L(N) = -\partial_N A_M$, with a uniform guide field component $B_M = \partial_N A_L$, and an electric field $E_N = -\partial_N \phi$, the equation of motion in the direction normal to the current sheet ($\hat{e}_N$) becomes

$$m_s \frac{dv_N}{dt} = -\frac{\partial \Lambda}{\partial N}, \tag{4}$$

where the pseudo-potential Λ is given by

$$\Lambda(N) = \frac{1}{2m_s}\left[(p_L - q_s A_L)^2 + (p_M - q_s A_M)^2 + p_N^2\right] + q_s\phi(N). \tag{5}$$

The pseudo-potential provides a means to examine what parts of physical and velocity space a set of particles can or cannot access (e.g. Egedal et al. 2016a; Lapenta et al. 2017; Zenitani et al. 2017; Drake et al. 2009b). Particles can cross the neutral line $N = 0$ if the sum of their kinetic and potential energies exceeds $\Lambda_0 = \Lambda(N = 0)$, i.e., $\Lambda_0 < H$. The separatrix $\Lambda_0 = H$ in velocity space thus gives a boundary between crossing and non-crossing trajectories (see Fig. 2c and 2d). For an extensive review of the quasi-adiabatic dynamics, discussing classes of charged particle trajectories, including the effects of a B_N, see Zelenyi et al. (2013).

A particle convecting with the magnetic field towards the field reversal will eventually become demagnetized and transition from a non-crossing to a crossing trajectory (e.g. Yoon et al. 2021). If a particle is close to the crossing-non-crossing boundary, a small change in particle velocity, either in magnitude or direction, for example, by scattering off some wave field, may greatly affect the trajectory (see, e.g. Daughton et al. 2004, who discusses this process). In addition, a small magnetic field normal to the current layer (which is likely present in most physical situations) causes the particles to naturally transition between these orbit types in their flow in and out of the layer (as demonstrated for the case of electrons by Egedal 2023).

2.2.2 Speiser Orbits

As already mentioned in Sect. 2.1, the ejection of particles away from the X line limits the dwell time of the particles in the region in which they are demagnetized and can be efficiently accelerated by E_M, thus leading to an effective resistivity since it limits the current that E_M can drive (Speiser 1970). In simple terms, the accelerated particles are continuously replaced by newly arrived, non-accelerated particles. Speiser (1965) analyzed particle motion in a current sheet with a zero or finite magnetic field component along the current sheet normal and an electric field normal to the reconnection plane, $\mathbf{E} = E_M \hat{e}_M$. Initially, magnetized particles drift with the magnetic field toward the field reversal and eventually transition from non-crossing to crossing trajectories, performing meandering motion along $\hat{e}_M$ due to the reversal of B_L. During the meandering motion, the particles are accelerated by E_M, gaining speed $\pm v_M \hat{e}_M$ depending on the particle charge. In a reconnection topology, where there is a finite B_N, the magnetic force $q v_M B_N \hat{e}_L$ turns the particles toward the outflow, ejecting them away from the X line. In the frame where the electric field is zero, the particle energy is conserved, and they move along circular trajectories of constant speed in velocity space (v_L, v_M). This demagnetized but orderly motion is enabled by the large separation between the particle bounce frequency across the current sheet due to B_L reversal, and their slow gyration frequency in B_N.

In a natural reconnection environment, the magnetic field component normal to the current sheet changes with distance from the X line, and as a consequence, so does the frame in which the electric field is zero. This affects the velocity-space contours of constant energy and the magnetization of the particles. The analysis has also been generalized to include a non-constant and reversing $B_N(L)$ and Hall fields. Zenitani and Nagai (2016) showed that some electrons, which experience an outward (away from the current sheet center) force due to the Hall fields, transition to non-crossing Speiser-type orbits in the presence of a Hall electric field. Chen et al. (2011) compared two mechanisms limiting the electron acceleration along $\hat{e}_M$: (1) ejection away from $N = 0$ due to the transition from oscillatory to exponential v_N, and (2) complete turning from $\hat{e}_M$ to $\hat{e}_L$ by B_N. They found that the latter dominated for electrons crossing the central EDR.

Variations of particle orbits occur due to large-scale factors, for example, where the particles enter the current sheet, in particular at what distance from the X line, and potentially on kinetic factors such as the particle gyrophase when they become demagnetized (see e.g. Fig. 2b). Since particles enter the exhaust at different distances from the X line, the particle VDFs inside the exhaust are typically a superposition of several different phases of Speiser-type orbits (e.g. Shuster et al. 2014, 2015; Cheng et al. 2015; Zenitani and Nagai 2016; Nakamura et al. 1998; Wang et al. 2016c).

2.2.3 The Formation of Crescent Velocity Distribution Functions

The meandering motion of particles across the current sheet, as well as sharp boundaries and strong acceleration, can lead to distinctly non-gyrotropic VDFs in the shape of crescents (see e.g. Fig. 4a and 4b), which can be observed with measurements when the instruments can resolve the particle gyroradii. Crescent-type VDFs have been observed both for electrons (Burch et al. 2016; Webster et al. 2018; Chen et al. 2016b) and ions (Nagai et al. 2015; Yamada et al. 2018) in regions of active reconnection. Electron crescents were first predicted to occur on the low-β side of asymmetric reconnection by Hesse et al. (2014) and have since been studied analytically by several authors for a few other types of reconnection configurations. By using the conservation of energy and the conservation of canonical momentum and/or integration of the equations of motion, analytical predictions of the VDFs' shapes are obtained. They have mostly been studied for the case of electrons, but the theory presented here is equally valid for ions with appropriate adjustments for the charge and mass (e.g. Usami and Zenitani 2024). Generally, crescents are formed through finite-gyroradius effects when only particles with sufficiently high speed and right direction can reach a certain observation point (see e.g. Fig. 2). This effect becomes more prominent when plasma of different characteristics interface, or when the plasma is accelerated, e.g. by the Hall fields or by

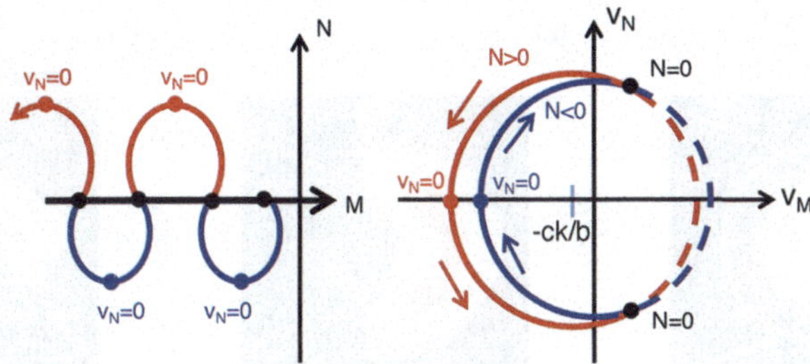

Fig. 3 (Left) Orbit of an electron meandering across the current sheet. In the $N > 0$ region, the electron is accelerated by the Hall electric field $E_N < 0$ and reaches a farther distance from $N = 0$ than in the $N < 0$ region where $E_N = 0$. (Right) The electron trajectory in the velocity plane $v_M v_N$. The energy is conserved in the negative N region since $E_N = 0$, and the blue trajectory is a circle around the origin. In contrast, in the $N > 0$ region, the E_N electric field can be Galilean transformed away by using the $E \times B$ drift (in the plot, it is shown as $-ck/b$) frame of reference. The particle's energy is conserved in that frame, and the red trajectory is a circle around the $E \times B$ drift speed. Figure from Bessho et al. (2016)

the reconnection electric field during meandering motion. The dominating effect differs for different plasma configuration, e.g., the Hall electric fields are dominating for asymmetric conditions while the reconnection electric field dominates for symmetric conditions. Below, we review the analytical formulation.

The Effect of a Normal Electric Field In asymmetric reconnection, the current sheet is often characterized by a strong Hall electric field directed towards the high-β side (e.g., the magnetosheath at the magnetopause). The crescent formation in asymmetric reconnection is mainly due to the acceleration by this Hall E_N field (Fig. 3). The reconnection electric field E_M accelerates meandering electrons, broadening the crescent in the v_M direction (Fig. 4a). Under a guide field, not only v_N but also v_L motion affects the crescent boundary. One guide field effect is widening the crescent boundary along v_N (Fig. 4b). Note that the VDF in the velocity plane $v_L v_M$ has a sharp cutoff (Fig. 4c).

If both the electric field and magnetic field vary linearly with distance across the current sheet, e.g. $E_N = -kN$[1] and $B_L = bN$, the electric drift $v_{E\times B} = -k/b$ is constant. These profiles of B_L and E_N are the first approximations for asymmetric reconnection (neglecting a contribution of the pressure tensor, which provides a small E_N at $N = 0$ as shown in Chen et al. 2016b). These assumptions were used as an example by Bessho et al. (2016), Shay et al. (2016), and Lapenta et al. (2017), although Lapenta et al. (2017) expressed the fields in terms of the corresponding electrostatic and vector potential, i.e., integrated once. The following analysis follows Bessho et al. (2016), but is very similar to the independent analyses by Shay et al. (2016), Lapenta et al. (2017), and Egedal et al. (2016a).

In the $v_{E\times B}$ frame, the particle energy is constant, giving the relation

$$(v_M - v_{E\times B})^2 + v_N^2 = (v_{M0} - v_{E\times B})^2 + v_{N0}^2 > (v_{M0} - v_{E\times B})^2, \tag{6}$$

where the subscript '0' denotes a reference location where both E_N and B_L vanish. The inequality comes from the fact that particles with $v_{N0} = 0$ cannot enter the region $N > 0$, which will be our region of interest.

[1]Note that Bessho et al. (2016) assumed in their analysis that $E_N(N < 0) = 0$ (Fig. 3); however, a linearly varying electric field at $N > 0$ can be transformed away.

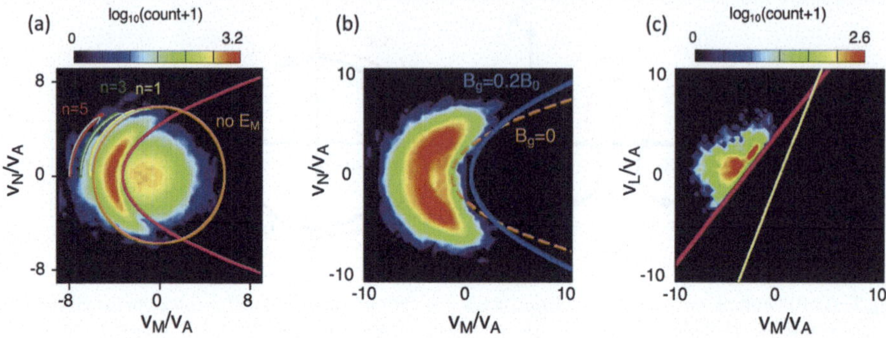

Fig. 4 Electron VDFs from PIC simulations in the EDR with (a) zero and (b)-(c) 20% guide field. (a) VDF at a position slightly away from the X line in the magnetospheric side of the current sheet. The parabolic (magenta) and the circular (orange) curves are predicted by the theory, which takes into account acceleration by the Hall electric field. Yellow, green, and red curves show the effect of the reconnection electric field E_M. Particles are further accelerated as they reflect back to the current sheet more times. The numbers $n = 1, 3$, and 5 represent the number of times the particle is reflected. (b) VDF and the boundary curves comparing zero (orange dashed line) and $B_g = 0.2B_0$ (blue) guide field cases, showing that the effect of the guide field is to widen the crescent along v_N. (c) The cut of the VDF at the plane $v_N = 0$. The cut line predicted by the theory (magenta) lies right against the PIC VDF. The magnetic field direction is shown by the yellow line. Reprinted from Bessho et al. (2017) and Bessho et al. (2019), with permission from AIP Publishing

By integrating the equation of motion in the M direction for the given electric and magnetic fields, we also obtain

$$v_M - v_{E\times B} = v_{M0} - v_{E\times B} + \frac{q}{m}\frac{b}{2}N^2, \tag{7}$$

where we have expressed the motion in the $v_{E\times B}$ frame. The further away in N the particle moves, the larger $v_M - v_{E\times B}$ becomes. Since the energy is conserved in this frame, v_N must decrease with increasing N until it becomes zero and the particle turns around. The decrease in v_N is due to the gyroturning around B_L. By using Eq. (7) to eliminate $v_{M0} - v_{E\times B}$ in Eq. (6), we reach Eq. 6 in Bessho et al. (2016):

$$v_M < (>)\, v_{E\times B} - \frac{v_N^2}{2}\left(\frac{q}{m}\frac{b}{2}N^2\right)^{-1} + \frac{1}{2}\frac{q}{m}\frac{b}{2}N^2, \tag{8}$$

where the inequalities $<$ and $(>)$ hold for negatively and positively charged particles, respectively, and we have written the terms in a way to elucidate the connection to the magnetic vector potential in e.g. Lapenta et al. (2017). The inequality above gives the inner boundary of the crescent population for different locations N. That is, particles that do not satisfy the inequality will not reach the given N location. If we instead substitute $v_M - v_{E\times B}$ in Eq. (6), we get an expression for what particles at $N = 0$ will reach N.

The above analysis did not take into account the reconnection electric field. This was done in a follow-up study by Bessho et al. (2017), where they included an electric field E_M that was relatively weak compared to E_N. They showed that the innermost boundary of the crescent remained unchanged while the outer boundary was modified based on the number of bounces the electron performed before leaving the region (see Fig. 4a).

In the presence of a guide field, the motion in the third direction must also be taken into account. This configuration was analyzed by Bessho et al. (2018) and for the ion case by Usami and Zenitani (2024). In the presence of a guide field, not all of the available energy is used to reach a higher N, but some is instead diverted to v_L. This results in a broader parabola. In the $v_L v_M$ plane, the crescent population is bounded by a straight line. This is illustrated in Fig. 4b-4c.

The Effect of the Reconnection Electric Field In symmetric reconnection, the effect of the reconnection electric field, E_M, becomes more significant than in asymmetric reconnection. The Hall fields are also anti-symmetric across the magnetic field reversal. This configuration was investigated by Bessho et al. (2018). The particles are accelerated along M by E_M:

$$v_M - v_{M0} = \frac{q}{m} E_M t + \frac{q}{m} \frac{b}{2} N^2. \tag{9}$$

Substituting v_M from Eq. (9) into the equation of motion along N gives the equation

$$\frac{d^2 N}{dt^2} = -\frac{q}{m} bN \left(v_{E\times B} + v_{M0} + \frac{q}{m} E_M t \right) \tag{10}$$

where a term of order N^3 has been discarded because only motion close to $N = 0$, or large E_M or long time is considered. Equation (10) is an Airy equation that gives a family of oscillatory solutions. We note that the types of orbits present in the current sheet depend on the initial speed v_{M0} and the $v_{E\times B}$ acquired when entering the Hall region. If these speeds are significant, the beginning of the oscillating orbit will already be dragged out, i.e., more snake-like rather than figure-eight-like. The relation between v_M and v_N becomes hyperbolic instead of parabolic, as shown in Fig. 9a.

Crescent VDFs for Non-meandering Electrons in Asymmetric Reconnection We note that the occurrence of electron crescent distributions does not per se require a meandering trajectory, i.e., the theory is equally valid for a particle that does not observe a field reversal. This was stressed by Egedal et al. (2016a), who showed that the formation of crescent-type VDFs can be a natural consequence of non-meandering electrons along an extended region stretching outside the EDR. Such crescents are observed in asymmetric reconnection near the EDR at the lower-density side of the current layer (e.g. Khotyaintsev et al. 2016). In these cases, the crescents can be explained by a finite-gyroradius effect where electrons from the larger-pressure side (i.e., with either denser or hotter electrons) enter into the lower-pressure side of the boundary. This typically leads to a finite diamagnetic drift, and can be associated with enhanced electron agyrotropy (e.g. Guo et al. 2017).

2.3 Adiabatic Particle Dynamics and Scattering in Curved Magnetic Fields

The adiabatic approximation is useful in describing and predicting particle motion in slowly varying magnetic fields. The term 'slowly' refers to the perspective of the particle itself, which can exhibit both temporal and spatial variations along its trajectory. When this is the case, the particle motion can be well described by a superposition of a guiding center motion and a gyromotion (Northrop 1963). The adiabatic descriptions are independent of field geometry as long as the adiabatic approximation holds. Thus, it is a useful tool for symmetric, asymmetric, antiparallel, and guide field reconnection. The approximation breaks down when the particle gyroradius $\rho = v_\perp/\omega_c$ becomes comparable to the magnetic field curvature radius $R_B = |\hat{b} \cdot \nabla \hat{b}|^{-1}$ (Buechner and Zelenyi 1989). Based on the parameter $\kappa^2 = R_B/\rho$, the trajectory can be classified into adiabatic orbits ($\kappa \gg 1$), chaotic orbits ($\kappa \sim 1$), and quasi-adiabatic orbits ($\kappa \ll 1$). The chaotic orbits lead to non-reversible pitch-angle scattering. For more comprehensive discussions of particle-trajectories in reconnection current sheets and exhaust topologies, see e.g. Zelenyi et al. (2013) and references therein.

Neglecting perpendicular drifts, the average perpendicular energy of a particle is given by $\mathcal{E}_\perp = \mu B$, where $\mu = m{v_\perp}^2/(2B)$ is the magnetic moment. The parallel energy is given by $\mathcal{E}_\parallel = \mathcal{E} - \mu B + q\phi$, where $\phi = -\int E_\parallel ds$ is the potential difference obtained by integrating the parallel electric field along the path of the particle, and $\mathcal{E}$ is the particle energy. By applying Liouville's theorem, i.e., assuming that $df/dt = 0$ along phase-space trajectories defined by constant μB and $\mathcal{E} + q\phi$, it becomes possible to predict how particle VDFs will evolve, or to solve for unknown parameters such as ϕ by comparing VDFs at two different points. Particles that are *trapped* and particles that are *passing* are typically treated separately (e.g. Egedal et al. 2008), and their phase space densities are mapped as follows

$$f(\mathbf{x}, \mathbf{v}) = \begin{cases} f_\infty(0, \mu B_\infty) & \text{trapped} \\ f_\infty(\mathcal{E}_{\parallel,\infty}, \mu B_\infty) & \text{passing} \end{cases} \tag{11}$$

The boundary between trapped and passing particles is given by $\mathcal{E}_{\infty,\parallel} = 0$, which approaches $v_\parallel \approx \sqrt{2q\phi/m}$ for low perpendicular energies and $v_\parallel \approx v_\perp\sqrt{1 - B/B_\infty}$ for large perpendicular energies. Particle trapping occurs when $q\phi > 0$ or $B/B_\infty < 1$ (Ng et al. 2011). That is, particles that have an upstream parallel speed of zero where the magnetic field has its asymptotic value B_∞ and where $E_\parallel = 0$ are, by definition, marginally trapped.

The limit of large perpendicular speed, or low parallel potential, can also be expressed in terms of the particle's pitch angle, where the pitch angle is defined as $\theta = \cos^{-1}(\mathbf{v}\cdot\mathbf{B}/|\mathbf{v}||\mathbf{B}|)$. In this case, under the assumption that the magnetic moment $\mu = mv^2\sin^2\theta$, and total energy $\mathcal{E} = mv^2/2$ of a particle are constant, a change in the magnetic field magnitude from B_1 to B_2 along the trajectory of the electron will lead to a change in pitch angle according to

$$\theta_2 = \sin^{-1}\left(\sin\theta_1\sqrt{\frac{B_2}{B_1}}\right). \tag{12}$$

When the magnetic field magnitude reaches a value such that $\theta_2 = 90^\circ$, the parallel energy is $\mathcal{E}_\parallel = 0$, and the particle will reflect. This is the simple mechanism of a magnetic bottle, which is applicable to both the reconnection inflow and exhaust under certain conditions.

The trapping of particles in specific regions has further implications for the particle dynamics therein. For example, trapping of electrons inside the reconnection exhaust confines them to a region where they may undergo repeated Fermi reflections and associated energization (e.g. Ergun et al. 2020). Particle-in-cell simulations have shown that the parallel electron temperature gradient that develops as a result also plays a role in the energy partition between ions and electrons (Haggerty et al. 2015).

3 Electron Dynamics

The electron dynamics lie at the heart of the magnetic reconnection process. Only when the electrons decouple from the magnetic field can the magnetic topology change. This, in turn, enables the release of magnetic energy over larger scales through the magnetic tension force. While the decoupling of electrons from the magnetic field happens inside the EDR, the chain of events leading up to the change of magnetic topology already starts upstream of the EDR, inside the IDR, where the demagnetized ions strongly influence the motion of electrons.

The first key breakthrough by the MMS mission was reported by Burch et al. (2016), and showed observations of electron VDFs associated with the demagnetized electron motion inside the EDR for asymmetric reconnection at the dayside magnetopause. The measured electron VDFs exhibited outstanding crescent-shaped features and indicated meandering and mixing of energized magnetosheath and magnetosphere electrons in the seemingly laminar reconnecting current sheet. The similarity of the measured VDFs with 2D PIC simulations (e.g. Hesse et al. 2014; Chen et al. 2016a; Bessho et al. 2016; Shay et al. 2016) indicates that the simulations have captured the key physics operating in the EDR, despite the idealized parameters used, such as the mass ratio and electron plasma to cyclotron frequency ratio. Before MMS, the impact of these idealized parameters was unknown, as electron VDFs in the EDR were not resolved by any measurements made in space or the laboratory. MMS demonstrated a stunning picture: the core region of reconnection, which is known as an explosive energy conversion phenomenon, can be quiet and laminar. The second key breakthrough was reported by Torbert et al. (2018), who presented the first in-depth analysis of a symmetric EDR in the nightside current sheet. The electron acceleration by the reconnection electric field was demonstrated by electron VDFs with multiple crescent populations.

The Kinetic Scale of the EDR Prior to the advent of MMS, the decoupling process between fields and electrons could only be studied with numerical simulations and theory. It was predicted that the breaking of the frozen-in condition required the formation of current sheets with thicknesses comparable to electron kinetic scales, $\sim 1d_e$ (e.g. Jara-Almonte et al. 2014). This seems to have been confirmed by MMS observations, which show that the half thickness of the reconnecting current sheet inside the EDR can be as thin as $1d_e$ to $4d_e$ (Nakamura et al. 2019; Tang et al. 2022; Li et al. 2021a; Zhou et al. 2019a; Wang et al. 2020). Whether the variance is due to the distance to the X line or some regulatory factor found in the ambient plasma conditions is unclear. The thinnest current sheets inside the EDR can be both bifurcated and have a central current peak. Whether a bifurcation appears depends strongly on the guide field (where a change in electron dynamics occurred already at guide field levels of 0.1 times the asymptotic field) but also on the inflow temperature, where a hotter plasma tends to obfuscate the current peaks (Swisdak et al. 2005). Thicker current sheets from the outer EDR, or inner IDR, show a tendency to become bifurcated (Eriksson et al. 2016a; Norgren et al. 2018; Øieroset et al. 2021). This is seen in Fig. 5, which shows the magnetic field, B_L, and current, J_M, structure of five current sheets of varying thickness. In particular, both Øieroset et al. (2021) and Eriksson et al. (2016a) reported simultaneous measurements of a non-bifurcated current sheet closer to the X line, while further away it was bifurcated. While current bifurcation of thin current sheets has been connected to the turning-point of meandering electrons (Chen et al. 2011; Shuster et al. 2015; Egedal 2024), it is not clearly demonstrated when the meandering trajectories result in bifurcation and not. Le et al. (2013) predicted that the electron outflows become deflected to the separatrices at low but finite guide fields. In the event reported by Zhou et al. (2019b), a slight bifurcation seems to be correlated with the region of isotropic curvature-scattered electrons. Simulations showed that a bifurcation extending along the separatrices was related to finite-gyroradius effects of exhaust electrons manifesting as a diamagnetic drift (Guo et al. 2017). Overall, the dominant bifurcation mechanisms of current sheets of varying thicknesses are still not fully understood.

Measuring the length of the EDR is more challenging since it requires a fortuitous spacecraft trajectory. Estimates range between $4d_e$ to tens of d_e for the half length of the inner

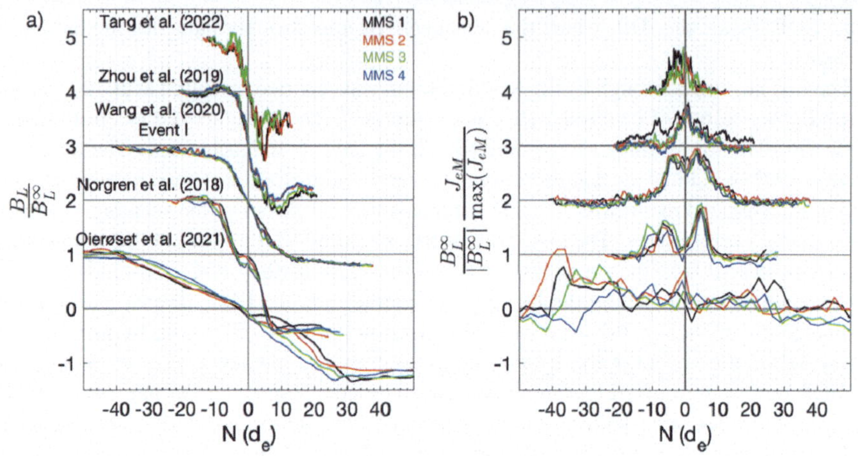

Fig. 5 Structure of five reconnecting current sheets. (a) Magnetic field B_L normalized to the asymptotic magnetic field B_L^∞. (b) Current density J_M normalized to the maximum current density (and multiplied by $B_L^\infty/|B_L^\infty|$ such that $J_M > 0$, for easier comparison). Further away from the X line, where the current sheet becomes thicker, it also tends to bifurcate. Although exceptions also occur, e.g., MMS 2 in the top row shows signs of bifurcation. Note that these events have varying guide fields. Each event is shifted for visibility, and the horizontal solid line shows the respective zero level

EDR (Nakamura et al. 2018; Li et al. 2021a). Simulations predict that the length of the EDR depends on factors such as the guide magnetic field (e.g. Le et al. 2013), the reconnection rate that in turn depends on ambient plasma parameters (e.g. Liu et al. 2025b, this collection, and references therein), and the larger scale magnetic topology that may strongly inhibit or modulate the formation of elongated current sheets (e.g. Phan et al. 2018). Connecting the simulations to the observations would help us understand the governing physical mechanisms behind the larger-scale structure of the EDR.

In the following sections, we will review the electron dynamics within three basic reconnection configurations: symmetric antiparallel, asymmetric antiparallel, and guide field reconnection.

3.1 Symmetric Antiparallel Reconnection

Symmetric reconnection refers to cases where the plasma parameters are roughly similar in both inflow regions. Such conditions are most commonly found in the magnetotail (e.g. Torbert et al. 2018), but also occur in other locations such as in the solar wind (e.g. Gosling 2012, and references therein), the magnetosheath (e.g. Wilder et al. 2017), and inside primary reconnection exhausts (e.g. Norgren et al. 2018).

The first detailed observations of an EDR of low-beta symmetric reconnection was conducted by MMS on 11 July 2017 in the magnetotail, as first reported by Torbert et al. (2018). Studies of this event yielded several significant findings, including the identification of multiple-crescent electron VDFs evidencing the acceleration of meandering electrons by the reconnection electric field (Torbert et al. 2018; Nakamura et al. 2019); the observations of the inner and outer EDRs (Torbert et al. 2018; Nakamura et al. 2019); the determination of the magnitude of the reconnection electric field (Genestreti et al. 2018; Nakamura et al. 2018); and the demonstration that the reconnection electric field was balanced by the gradients of the off-diagonal elements of the electron pressure tensor (Egedal et al. 2019).

In the following sections, we will review some of the recent findings regarding electron dynamics in symmetric magnetic reconnection. We will begin by briefly reviewing the dynamics of the inflow region, as it sets the upstream conditions for the EDR. Thereafter, we will review the dynamics within the EDR and, lastly, the dynamics downstream of the EDR, but still within the IDR.

3.1.1 The Electron Dynamics in the Inflow and at the Separatrices

The electron behavior in the inflow regions of symmetric reconnection is strongly characterized by adiabatic dynamics (introduced in Sect. 2.3), which lead to the development of strong pressure anisotropies, $p_{e\|} > p_{e\perp}$, due to the decreasing magnetic field magnitude (Uzdensky and Kulsrud 2006; Le et al. 2009; Egedal et al. 2013; Le et al. 2010; Egedal et al. 2015). This behavior is well documented in observations, both quantitatively (Wetherton et al. 2019) and qualitatively, i.e., an increase of $T_{e\|}$ and a decrease of $T_{e\perp}$ are observed as the magnetic field magnitude decreases from the far inflow towards the separatrices and X line (Torbert et al. 2018; Norgren et al. 2018; Chen et al. 2019; Yu et al. 2019). The parallel electron temperatures commonly reach values a factor of two larger than the perpendicular ones (e.g. Torbert et al. 2018). Estimates of the parallel potential show that it often exceeds upstream thermal energies (Wetherton et al. 2019; Egedal et al. 2012). Parallel electron energies can reach several keV in the magnetotail (Egedal et al. 2005), and tens of eV in the magnetosheath (e.g. Eriksson et al. 2018). The conservation of the magnetic moment based on the electron temperature has also been demonstrated in observations (Norgren et al. 2018).

The pressure anisotropy that forms in the inflow upstream of the EDR can play an important role in the momentum balance by offsetting the magnetic tension force. Simulations show that at the upstream edge of the EDR, the marginal firehose threshold is often approached and marginally satisfied $p_{e\|} - p_{e\perp} = B^2/\mu_0$ (Egedal et al. 2013; Le et al. 2019). The pressure anisotropy is a large driver of the reconnection current (Egedal et al. 2023; Le et al. 2019), which we will further discuss in Sect. 3.1.2. These upstream effects are strong indications that a holistic approach is crucial for fully understanding how various scales and plasma species interact to enable the explosive release of magnetic energy during magnetic reconnection.

The Separatrix Electron Jet The separatrices of magnetic reconnection are extended boundaries mapping from the topological reconnection X-line, separating the inflow from the outflow. While by this definition they have a vanishing thickness, the word separatrix typically refers to the kinetic boundary of finite width where the electron inflow population meets and overlaps with the outflow population. The separatrices are characterized by electron flows propagating towards the X line. These flows form due to the same basic dynamics as the trapped inflow population, with the difference that they are not trapped but proceed directly to the EDR or outflow (Le et al. 2009). These flows were first observed by Geotail (Nagai et al. 2001) and Wind (Øieroset et al. 2001), who identified them as the outer part of the Hall current loop. When thermal electron streaming alone cannot maintain quasineutrality, strong double layers can develop (e.g. Fujimoto 2014). Egedal et al. (2015) estimated that this occurs roughly when the asymptotic upstream $\beta_e^\infty \lesssim 200 m_e/m_i \sim 0.1$ (assuming that the dominant ion species are protons), which is typically well satisfied in the magnetotail where $\beta_e^\infty \sim \beta_e^{lobe} \ll 0.1$. Estimates of the corresponding acceleration potential ϕ show that the electrons can be accelerated to several keV, which can correspond to several tens of the upstream lobe temperature (Norgren et al. 2020).

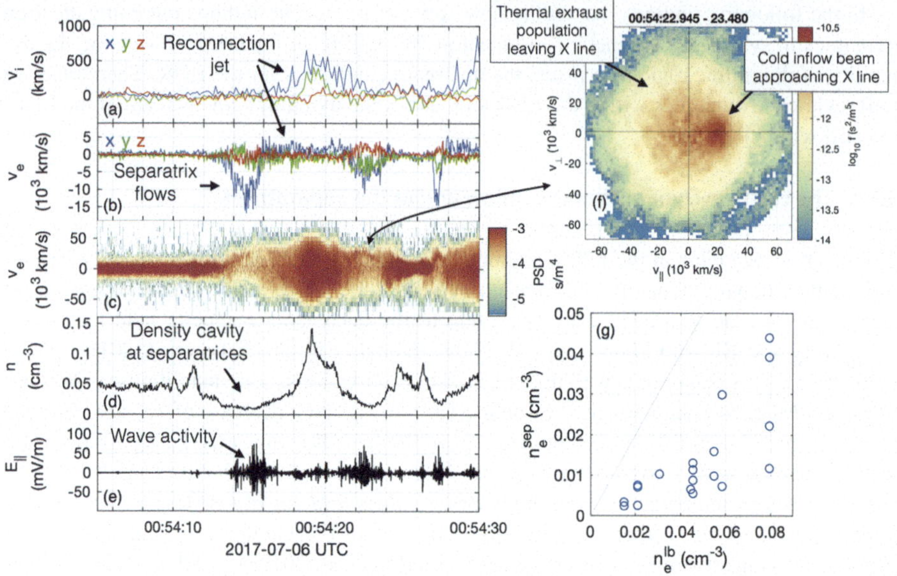

Fig. 6 Three separatrix crossings observed by MMS in the magnetotail ($xyz \sim LMN$). (a) Ion velocity, (b) electron velocity, (c) electron VDF $f_e(v_\parallel)$, (d) density, (e) electric field parallel to the magnetic field, (f) electron VDF $f_e(v_\parallel, v_\perp)$, where $v_\perp = \mathbf{v}_{E\times B}$, (g) scatter plot of densities inside the separatrix density cavity vs. lobe densities for 19 separatrix crossings. The separatrices are characterized by large electron flows toward the X line carried by cold plasma populations accelerated to suprathermal energies that are gradually thermalized in the presence of large-amplitude electrostatic waves. These acceleration channels are associated with density cavities formed as a result of the accelerated electron populations. Figure partly adapted from Norgren et al. (2020)

A crossing of a magnetotail separatrix is shown in Fig. 6. The Earthward reconnection outflow, v_L, best seen in the ion velocities, is bounded by oppositely directed electron flows that are moving toward the X line (e.g. Zhang et al. 2020; Norgren et al. 2020). The acceleration channels, i.e., the region where the electron flow towards the X line is observed, separate the colder lobe plasma from the heated exhaust plasma. The beams at the three crossings are carried by gradually accelerated cold lobe electron populations that overlap with the exhaust at the inner edge of the channel (Fig. 6c and 6f). While the beam carrying the separatrix flows can be suprathermal, the bulk speed is typically subthermal (Liu et al. 2025a). As the accelerated populations must conserve the particle flux, the acceleration channels are associated with density cavities (e.g. Mozer et al. 2002; Liu et al. 2025a) (see also Fig. 6d and 6g). Since the massive ions can not entirely replicate this density drop, the density cavities can also be associated with a locally divergent electric field (e.g. Divin et al. 2012; Yu et al. 2019; Zhou et al. 2019b; Norgren et al. 2020). Fully kinetic particle-in-cell simulations have shown that the width of the active electron separatrices scales with the electron inertial length (Lapenta et al. 2010). The separatrix plasma jets also lead to the formation of plasma waves that play a role in the beam thermalization and scattering (e.g. Norgren et al. 2020; Graham et al. 2025, this collection). The overlapping populations at the separatrices are associated with quasi-viscous heating, which can be interpreted as irreversible scattering by electromagnetic fields (Hesse et al. 2018; Holmes et al. 2021). Flat-top distributions associated with the inflow anisotropy and the thermalized separatrix jets are also observed to extend into the outflow, with energies reaching from a few tens to a few thousand electronvolts (Richard et al. 2025).

3.1.2 The Electron Diffusion Region

Magnetic reconnection affects large volumes of space but is enabled inside the EDR, which measures a few electron inertial lengths in size (see the beginning of Sect. 3). Inside the EDR, the magnetic field disconnects from the electron fluid and changes connectivity (e.g. Liu et al. 2025b, this collection, and references therein). The transfer of magnetic energy to particle energy is mainly dictated by the out-of-plane electric field E_M, i.e., the reconnection electric field.

The electron dynamics within the EDR are affected by the asymmetries of the field and plasma across the reconnecting current sheet and the magnetic guide field. In this section, we start by reviewing the EDR dynamics for a close-to-antiparallel magnetic field, followed by the dynamics in the asymmetric case. Many results obtained for symmetric reconnection also adhere to asymmetric reconnection. The dynamics due to a finite guide field are treated at the end.

The First Detailed View of a Symmetric Electron Diffusion Region On 11 July 2017, MMS encountered a symmetric EDR in the magnetotail first reported by Torbert et al. (2018). Figure 7 shows a brief overview of the event. The observations revealed a large-amplitude electron flow reaching $v_{eM} = -15{,}000$ km/s, which separated two reversing plasma jets reaching maximum amplitudes of roughly $v_{eL} = -10{,}000$ km/s and $v_{eL} = 5000$ km/s, respectively. The out-of-plane electron flow exceeded the local electron Alfvén and thermal speeds. The electron v_{eL} reversal was concurrent with the ion v_{iL} reversal but occurred on a much smaller scale.

The VDFs revealed complex structures both in the plane containing $\mathbf{B}$ and the plane perpendicular to $\mathbf{B}$ (Torbert et al. 2018). The observations in the perpendicular plane, $f(v_{\perp 1}, v_{\perp 2})$ (Fig. 7f), revealed a three- or four-layered structure (four layers are indicated in Fig. 7f) with a cold drifting core and two or three crescent-shaped populations at larger energies. These VDFs are the smoking-gun evidence of the demagnetized electrons being accelerated by the reconnection electric field during their meandering motions across the magnetic field reversal (Torbert et al. 2018; Bessho et al. 2018). In the plane containing the magnetic field and the electron bulk speed, $f(v_{\perp 1}, v_{\parallel})$ (Fig. 7e), the VDFs revealed that the cold drifting core was elongated along the magnetic field and had a fan-like structure at higher speeds. The core, or base, consists of electrons just arriving at the EDR, while the electrons forming the tip of the structure, or crescents, represent electrons that have been accelerated by the reconnection electric field during the meandering motion (e.g. Ng et al. 2011; Egedal et al. 2023).

The electric field in the frame of the electrons, $\mathbf{E}'_M = (\mathbf{E} + \mathbf{v}_e \times \mathbf{B})_M$, was both positive and negative. In an inner region spanning the flow reversal $\mathbf{E}'_M > 0$ (Torbert et al. 2018; Nakamura et al. 2019), showing that MMS crossed the inner diffusion region where the magnetic field is transported downstream faster than the electron bulk flow (Karimabadi et al. 2007). The inner EDR was bordered on both sides by regions with $\mathbf{E}'_M < 0$, indicative of the outer diffusion region. Nakamura et al. (2019) estimated the dimensions of the inner EDR to be $1 \times (4 - 5.5)d_e$, corresponding to an aspect of 0.2-0.18, which is consistent with what is found in particle-in-cell simulations that modeled this event (Nakamura et al. 2018; Egedal et al. 2019).

The reversal of the normal magnetic field B_N was slightly displaced toward the tailward exhaust with respect to the v_{eL} reversal. Based on a reconstruction of the region, Hasegawa

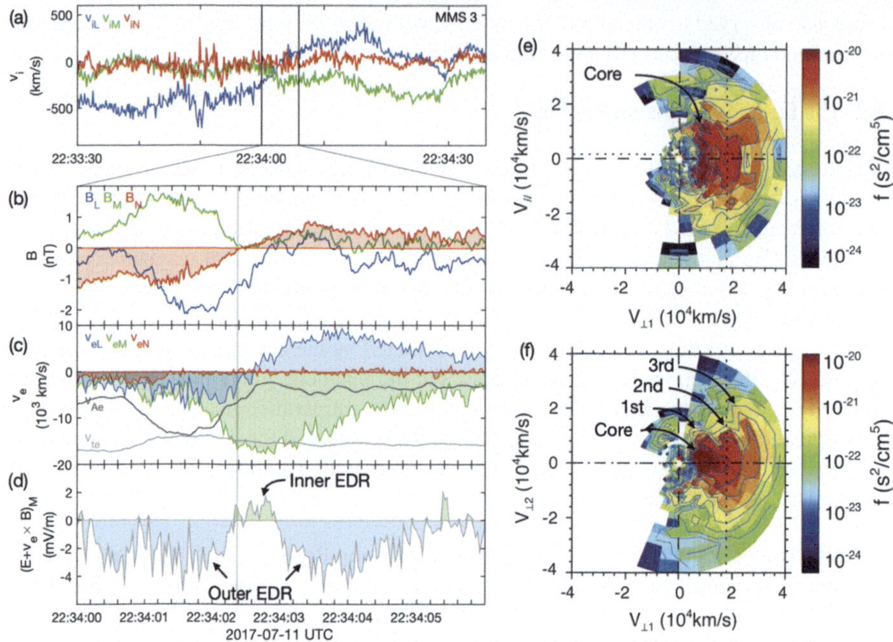

Fig. 7 The electric diffusion region on 11 July 2017. (a) Ion velocity. (d) Magnetic field. (c) Electron velocity and electron Alfvén and thermal speeds. (d) The electric field in the frame of the electrons, $(\mathbf{E} + \mathbf{v}_e \times \mathbf{B})_M$. (e) Electron VDFs in the $v_{e\perp 1}v_{e\parallel}$ plane and the (f) $v_{e\perp 1}v_{e\perp 2}$ plane at the time marked by the vertical shaded line in panels b-d. Here, $\mathbf{v}_{\perp 1} = \mathbf{b} \times (\mathbf{v} \times \mathbf{b})$, $\mathbf{v}_{\perp 2} = \mathbf{v} \times \mathbf{b}$, where $\mathbf{b} = \mathbf{B}/B$. Figure reproduced and adapted from Torbert et al. (2018). Panels (e)-(f) from Torbert et al. (2018), with annotations added. Reprinted with permission from AAAS. The LMN coordinate for panels (a)-(d) are from Nakamura et al. (2019)

et al. (2019) showed that the X line might be displaced with respect to the flow stagnation point by as much as $3d_e$. They suggested it could be due to a tangential flow in the inflow. For a different but similar symmetric EDR event, Hasegawa et al. (2022) showed that magnetic field annihilation occurred in an electron-scale magnetic island, which is consistent with an EDR elongated in $\hat{e}_L$ and with the simulation reported by Nakamura et al. (2021). These results reveal slight departures from the textbook reconnection events that are typically studied in simulations.

Detailed comparisons of the 11 July 2017 event with particle-in-cell simulations showed that MMS initially skimmed the southern separatrix before entering the EDR, where it stayed close to the current sheet midplane, with a brief excursion to the northern side before exiting (Nakamura et al. 2018; Egedal et al. 2019). The similarities between the simulations and MMS observations are striking, strongly indicating that the observed EDR was quasi-2D in nature, without significant effects beyond those found in 2D simulations. The variations observed, such as those in the magnetic field in the tailward exhaust, could be well explained by the relative motion of the spacecraft through an otherwise laminar reconnection region. For more details regarding the reconstruction of the spacecraft trajectories, see Hasegawa et al. (2024, this collection).

While many EDRs in Earth's magnetotail exhibit characteristics consistent with laminar 2D reconnection, Cozzani et al. (2021) reported a strongly perturbed EDR, attributing the perturbations to current sheeting kinking propagating in the out-of-plane direction. Numerical simulations have similarly shown that strong electromagnetic turbulence can develop around the EDR (Fujimoto and Sydora 2023). Although EDRs have been observed amid strong turbulence in the magnetotail, they themselves appear laminar (Ergun et al. 2022; Stawarz et al. 2024, this collection). In MMS observations, EDRs are often identified based on signatures associated with 2D laminar reconnection, raising the question of how prevalent turbulent or perturbed EDRs are in space.

The crossing of the inner EDR enabled direct measurements of the non-ideal contributions to the reconnection electric field: $E_M = -\partial_N p_{eNM}/ne - \partial_L p_{eLM}/ne$ (Egedal et al. 2019). Interestingly, and contrary to some predictions (Ng et al. 2011; Hesse et al. 2001), the contribution from $-\partial_L p_{eML}/ne$ was negative. Egedal et al. (2023) suggested that the negative $-\partial_L p_{eLM}$ was caused by the upstream pressure anisotropy as the cigar distribution rotates towards $\hat{e}_y$ together with the Hall magnetic fields. They theorized that if the upstream pressure anisotropy exceeds the pressure associated with the acceleration of the reconnection electric field, $-\partial_L p_{eLM}$ becomes negative. This is an interesting case in that theoretical predictions are correct in their reasoning (gyro resistivity), but there may be some other factor unaccounted for (upstream pressure anisotropy). The quantitative contributions to $-p_{eLM}$ and $-p_{eMN}$ from various electron populations were explored by Norgren et al. (2025), who confirmed the predictions by Egedal et al. (2023) regarding the effect of the upstream electron anisotropy on $-\partial_L p_{eML}/ne$. Since the upstream pressure anisotropy depends on inflow conditions, different events may also have varying effects. In contrast to $-\partial_L p_{eML}/ne$, Egedal et al. (2019) found that the contribution from $-\partial_N p_{eMN}/ne$ was positive and could balance the reconnection electric field with the offset from the negative contributions.

Recently, Shuster et al. (2019, 2021) used the high spatiotemporal and velocity-space resolution of the thermal plasma instruments aboard MMS to investigate the *gradients* of electron phase space density. This provides a new venue to look into the kinetic origins of, for example, the electron pressure divergence $\nabla \cdot \mathbf{p}_e$ known to support the non-ideal reconnection electric field in the EDR.

The Formation of the Reconnection Current Observations show that the reconnection current, J_M, inside the EDR is essentially exclusively carried by electrons (e.g. Torbert et al. 2018; Zhou et al. 2019a; Huang et al. 2018; Tang et al. 2022). The acceleration of the corresponding electron flow, v_{eM}, can be due to both the reconnection electric field E_M (e.g. Speiser 1965) and the magnetic force $\sim -ev_{eL}B_N$ in the presence of an upstream pressure anisotropy (Ng et al. 2011; Egedal et al. 2023). We will discuss both of these drivers below.

Figure 8 illustrates the trajectory in the MN plane of an electron that moves through the EDR from the inflow to the outflow. The color shows the work done by the reconnection electric field E_M on the electron along the trajectory, i.e., $dW = -eE_M dM$. Initially, the electron has no field-aligned speed v_L (in the inflow $\mathbf{B} = B_L\hat{e}_L$) and simply drifts with the magnetic field toward the field reversal at the speed $v_N = -E_M B_L/B^2$, with zero net energization over the gyro orbit. As the electron approaches the field reversal, it becomes gradually accelerated over several Larmor gyrations by the reconnection electric field during a drift $v_M = E_N B_L/B^2$, where E_N is the Hall electric field (Egedal et al. 2023). The work done is thus $dW = -eE_M dM = -eE_M v_M dt \approx -eE_M E_N B_L/B^2$. The well-defined guiding-center drift is permitted due to the fact that the electron Larmor radius is smaller than the Hall region. The pre-acceleration of the electrons before they reach the current sheet center and commence the meandering part of their trajectory is seen both in simulations (e.g. Ng et al. 2011; Hesse et al. 2014; Shuster et al. 2014) and in observations (e.g. Torbert et al. 2018; Tang et al. 2022). It is identified by a finite drift v_M of the coldest and slowest population, which, presumably due to the lack of heating, has not yet reached the

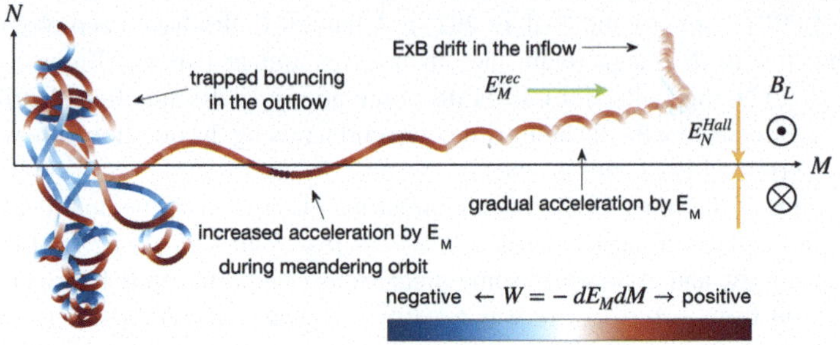

Fig. 8 Characteristic EDR electron trajectory in the LM-plane. The initial drift due to $-E_M B_N/B^2$ transitions into $E_N B_L/B^2$ inside the ion demagnetization layer (where $E_N = \partial_N p_i/ne$). During this $E_N B_L/B^2$ drift, the electron is accelerated by E_M, forming the drifting core. As the particle reaches $N = 0$, it begins to meander, increasing its acceleration and forming the crescents. After crossing the $N = 0$ plane three times, the electron is ejected from the vicinity of the X line and becomes trapped in the exhaust. For details of the simulation, see Norgren et al. (2021)

current sheet midplane. The gradual departure of $v_{E\times B,M}$ from $v_{e\perp,M}$ upstream of the EDR has been reported in the vicinity of an EDR, although any acceleration by E_M can not be affirmed from the observations of that event (Li et al. 2019).

While the particle energization within the EDR is driven by the reconnection electric field, acceleration in the current direction ($\hat{e}_M$) can also be achieved by the Lorentz force $-e(v_N B_L - v_L B_N)\hat{e}_M$ (Ng et al. 2011; Egedal et al. 2023). They showed how it could be, in fact, the main driver of the reconnection current J_M. As discussed in Sect. 3.1.1, a parallel electron pressure anisotropy typically forms in the inflow region upstream of the EDR (Le et al. 2009). Observations show that the anisotropic electron VDF remains field-aligned in the outer EDR (Burch et al. 2023). This is also shown in Fig. 10c-10d. This implies that the Hall magnetic field B_M is an indicator of the angle at which field-aligned upstream electrons are injected into the EDR (Ng et al. 2011). Electrons that arrive at the EDR will thus have distinct field-aligned speeds that, on average, supersede the perpendicular speeds. The initial upstream field-aligned velocity is redirected to $-v_M\hat{e}_M$ by the force $-ev_L\hat{e}_L \times B_N\hat{e}_L$, thus contributing to the reconnecting current $J_M\hat{e}_M$ (Egedal et al. 2023).

In terms of the basic concept of Ohm's law and effective resistivity (Speiser 1970), as discussed in Sect. 2.1, the mechanism whereby the pressure anisotropy drives the current would be associated with an enhanced effective (gyro) conductivity or reduced effective resistivity. This exemplifies how the particle dynamics are intimately related to the macroscopic processes.

Meandering Electron Trajectories Manifested as Crescent Populations Through analytical theory supported by particle-in-cell simulations, Bessho et al. (2014) predicted the gain in speed Δv_M based on the initial speed v_M for meandering electrons inside the EDR and the reconnection electric field E_M. By using the four MMS spacecraft to estimate the current sheet structure, including the reconnection electric field, E_M, the gradient in the normal magnetic field, B_N, and the asymptotic magnetic field strength, B_L, Nakamura et al. (2019) could make a detailed comparison of the observed multi-crescent structure to these analytical predictions. They investigated the difference in speed between the clear three-stepped structure consisting of the core and the first crescent (which blend together), and the two subsequent crescents, and found a quantitative agreement between observations and analytical theory, which provides confirmation of the proposed crescent generation mechanism.

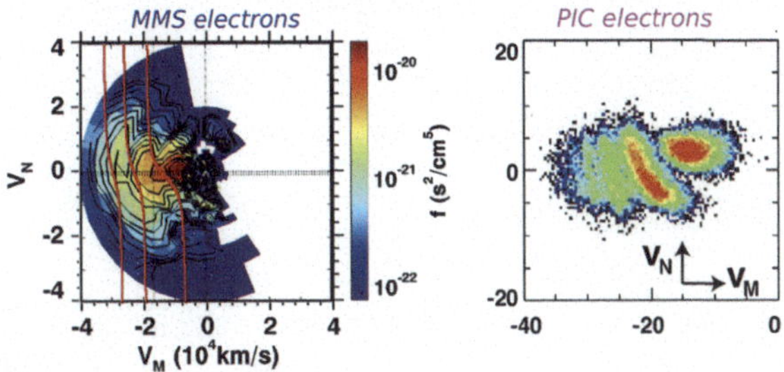

Fig. 9 Multiple electron crescent populations in the $v_M v_N$ plane showing multi-bounce electrons accelerated by the reconnection electric field (Shuster et al. 2015; Torbert et al. 2018). The red lines in the left panel are estimates of the crescent shape based on analytical theory, with the reconnection electric field as a free parameter (Bessho et al. 2018). The observed electron VDF provides an independent way to estimate the reconnection electric field (see also Hasegawa et al. 2024, this collection). The right panel, adapted from (Shuster et al. 2015), shows clear likenesses with the observations

Twisting of crescents due to acceleration by E_M. Bessho et al. (2018) predicted that the acceleration of the electrons throughout the meandering motion would lead to an asymmetry in the crescent-shaped populations: the inbound electrons should have a slightly higher speed than their outbound counterparts due to the significant acceleration during the turning of the meandering trajectory (see Sect. 2.2.3). In Fig. 9a, the analytical predictions of the lower edge in velocity space of the respective crescents are overlaid on the VDF captured during the EDR encountered on 11 July 2017. The spacecraft were located south of the neutral plane, and electrons with $v_N < 0$ are thus moving outward, and vice versa for $v_N > 0$. The red lines show the estimated gain in energy between the outbound and inbound leg of the meandering trajectory. For comparison, Fig. 9b shows the corresponding VDF from a PIC simulation, showing clear similarities. One notable difference includes the azimuthal coverage of the crescents in the $v_M v_N$ plane, or equivalently, the range of v_N. Crescent electron populations have been observed for many symmetric reconnection events (e.g. Ergun et al. 2022; Li et al. 2021a; Burch et al. 2019), whereof only some have displayed clear multiple crescents (e.g. Tang et al. 2022). It is possible that the crescents are not well separated in velocity and thus overlap, forming a more thermal population (e.g. Ergun et al. 2022), or that the electrons are ejected from the meandering region before bouncing more than once, and thus that there is only a single crescent. This is a question that could be explored more in observations. One process that could thermalize and potentially merge distinct crescents is the wave-particle interaction due to high-frequency waves that have been excited by the crescent beam propagating perpendicular to the ambient magnetic field (see e.g. Burch et al., 2019 or Graham et al., 2017b for a magnetopause event). Li et al. (2021a) reported upper-hybrid waves generated by the crescent and discussed how they were potentially connected to the acceleration throughout the meandering electron motion. They demonstrated that the waves were associated with the inbound wing of the crescent population. They hypothesized that the acceleration during one excursion brought the crescent population above the threshold for the beam-plasma instability and that the waves subsequently played a role in dissipating some of that energy. For a more comprehensive discussion of the role of waves in the EDR, we refer to Graham et al. (2025, this collection).

The meandering electron trajectories have been associated with the off-diagonal pressure tensor element p_{eMN} (e.g. Horiuchi and Sato 1997; Bessho et al. 2018). In combination with the convergent inflow that is yet to cross the current sheet midplane, the twisted structure

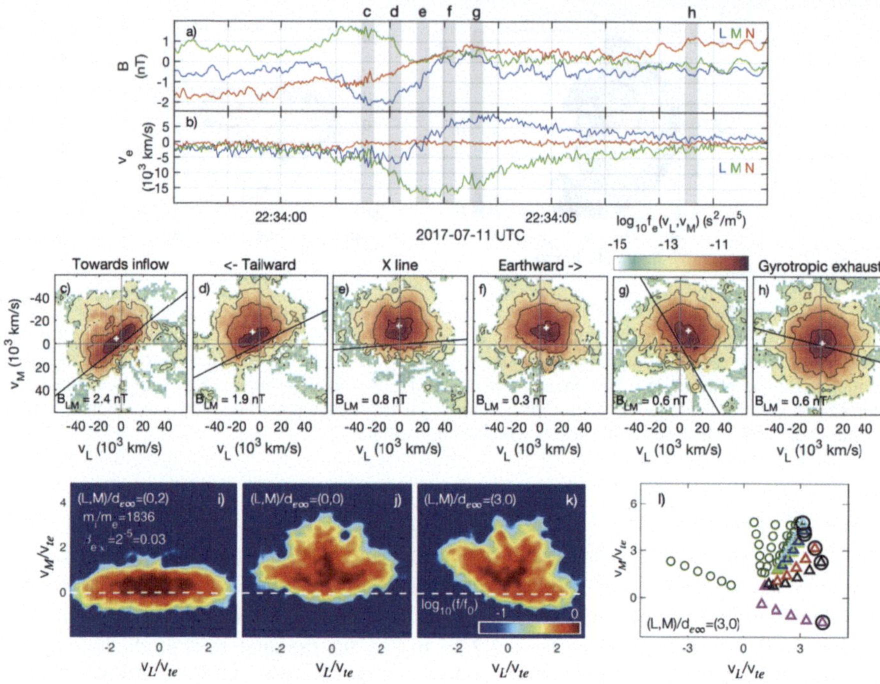

Fig. 10 The formation of the outflow jets (as observed by MMS 3) of a symmetric magnetotail reconnection event first reported by Torbert et al. (2018). (a) Magnetic field. (b) Electron velocity. (c)-(h) Reduced electron VDF $f_e(v_L, v_M)$ for the times shown by shaded areas in panels (a)-(b). The VDFs show how v_{eM} is gyroturned to either direction by B_N. Further downstream (h), the VDF is gyrotropic. The in-plane magnetic field is shown when the average value exceeds the standard deviation by a factor of two. The electron bulk velocity is shown as a white +. (i)-(k) Electron VDFs for a kinetic simulation in the (i) inflow, (j) X line, and (k) slightly downstream of X line. (l) Illustration of the number of $N = 0$ crossings electrons have performed: 0, magenta; 1, black; 2, red; 3, blue; 4, cyan. Panels (a)-(g) are reproduced from Norgren et al. (2025) and panels (i)-(l) are from (Egedal et al. 2023)

leads to a gradient $-\partial_N p_{eMN}$ that contributes positively to E_M (Ng et al. 2011; Egedal et al. 2019).

Formation of the Outflow Jets and Re-Magnetization of Electrons Downstream of the Electron Diffusion Region As the electrons that are accelerated by $-eE_M$ and/or $ev_L B_N$ gain a larger v_M and encounter the normal magnetic field B_N, they are accelerated in the direction away from the X line by $-ev_M B_N$. The spatiotemporal evolution of the electron VDFs inside the EDR has been extensively studied using particle-in-cell simulations (e.g. Ng et al. 2011; Bessho et al. 2014; Shuster et al. 2014, 2015; Egedal et al. 2023). Inside the EDR, the electron VDF is highly non-gyrotropic with a triangular or fan-like structure, where the different layers, or 'fingers', are comprised of electrons having crossed the neutral plane different number of times. The formation of the outflow is illustrated in Fig. 10, where MMS observations from 11 July 2017 are compared with VDFs from a particle-in-cell simulation (Egedal et al. 2023). The base of the triangle comprises electrons that arrive at the location directly from upstream (magenta in Fig. 10l). After each successive crossing of the neutral plane (black, red, blue, cyan), the base is folded inward toward the tip of the triangle. Since the electrons with larger v_M experience a larger force $-ev_M B_N$, the triangular structure rotates in the $v_L v_M$-plane (Fig. 10c-10g), becomes deformed, and eventually gyrotropizes (Shuster et al. 2015). A gyrotropized eVDF slightly downstream of the EDR is shown in Fig. 10h. Egedal et al. (2023) reported simulation results with varying mass ratios and showed that the triangular shape became more blunt at the full electron-proton mass ratio. This was attributed to a decrease in acceleration by the electric field, which they showed scaled with the mass ratio. At the same time, the wings, or base, of the triangles, which originate from the upstream pressure anisotropy, were unchanged. The observed VDFs are similar to the ones from the simulations. Although it is often challenging to unambiguously identify the different layers in observations, sometimes they are clearly present, as shown in Fig. 7f.

During a traversal from one exhaust to the other, the electron bulk velocity is typically seen as a rotation in the $v_L v_M$ plane, where v_M peaks at the reversal of v_L (e.g. Torbert et al. 2018; Li et al. 2021a). For the 11 July 2017 event, it is notable that $v_{eM} > v_{eL}$ for the entirety of the EDR crossing. This seems to be the case for a few other events as well (e.g. Torbert et al. 2018; Ergun et al. 2022; Wang et al. 2020, 2022b; Motoba et al. 2022; Tang et al. 2022; Qi et al. 2024), implicating that the formation of the outflow jets through gyroturning is inherently accompanied by a transformation of the directed plasma flow to thermal motion through some gyrotropization process. Theoretical works suggest this is due to the remagnetization of electrons due to the normal magnetic field component B_N that increases away from the X line (e.g. Hesse et al. 1999; Shuster et al. 2015). However, variations in observations from event to event naturally occur, and the observations may be dependent on the spacecraft trajectories relative to the EDR. Exceptions to the case of $|v_{eM}| > |v_{eL}|$ seem to occur in limited regions primarily when (1) the current sheet is bifurcated, i.e. exhibit a plateau in B_L associated with a decrease in J_M (e.g. Tang et al. 2022), or (2) when the measurements are taken in the off-equatorial region where v_{eL} tends to be more field aligned (e.g. Li et al. 2021a). However, some exceptions do not fall under these categories (Li et al. 2022).

The kinetic structure of the formation of the outflow jets, as seen in the electron VDFs, was shown by Nakamura et al. (2019) for the 11 July 2017 event. This is illustrated in Fig. 10, showing the reduced electron VDFs $f_e(v_L, v_M)$, which is an adaptation of the work by Nakamura et al. (2019), with a few more added time steps. The electrons arriving directly from upstream can be identified as a denser, elongated population aligned with the magnetic field (black line). The electrons undergoing meandering motion form roughly a half-circle. The electron VDF rotates in the $v_L v_M$ plane as the spacecraft first observes the tailward exhaust and thereafter crosses the concurrent v_{eL} reversal and v_{eM} peak into the Earthward exhaust. The rotation of the VDF was also demonstrated by Nakamura et al. (2019) by displaying the time series of phase space density at the different angles in the $v_L v_M$ plane. There, it also becomes evident that the transition between central and outflow jets is not associated with a significant increase in energy. This is consistent with the outflow jet being formed through gyroturning due to the reconnected magnetic field B_N, i.e., there is no significant work done by e.g. E_L. Further downstream, where the electron flow has decreased, the VDF is gyrotropic (Fig. 10h). The VDFs observed by MMS are compared with VDFs from a particle-in-cell simulation by Egedal et al. (2023), which show a clear likeness. The triangular, or fan-like, structures that are clearly visible in the simulations can perhaps be identified in the observations as pointed features along the half-circle, see, e.g., Figure 1K of Torbert et al. (2018) or Figure 1k of Burch et al. (2023). To determine whether these are indeed the same structures requires further analysis.

3.1.3 The Electron Dynamics in the Outflow Region

While the inflow regions are characterized by large temperature anisotropies, the central regions of the reconnection current sheets typically exhibit isotropic temperatures (e.g. Graham et al. 2016a; Torbert et al. 2017; Wang et al. 2020). In the survey of 32 EDRs at the dayside magnetopause by Webster et al. (2018), the minimum temperature ratio $\min(T_{e\parallel}/T_{e\perp})$ observed (their Table 2) had an average value of ~ 0.9. The transition from anisotropic to more isotropic temperatures has been attributed to pitch angle scattering in the strongly curved magnetic field lines when the conservation of the magnetic moment breaks down. Significant pitch angle scattering occurs when the electron gyroradius becomes comparable to the magnetic field curvature radius (Buechner and Zelenyi 1989) (see Sect. 2.3). This behavior has been verified by observations in several cases (Zhang et al. 2016; Zhou et al. 2019a; Eriksson et al. 2020). The energy-dependent scattering mechanism was verified by Zhang et al. (2016), who showed that lower-energy electrons retained their field-aligned anisotropic distribution throughout the reconnection region while higher-energy electrons were pitch-angle scattered to a nearly isotropic distribution.

Downstream of the EDR, inside the exhaust, where the magnetic flux tubes shrink, and the magnetic field strength increases again, simulations and observations show that the electrons again follow adiabatic behavior, bouncing in the magnetic bottle, confined by both the magnetic mirror force and the parallel electric fields that develops there (Egedal et al. 2013; Birn et al. 2013). The magnetic trapping is quantified by studying the change in pitch angle as predicted by the conservation of the magnetic moment (see Sect. 2.3) (see Fig. 15 or Lavraud et al. 2016 for an asymmetric event). Magnetic trapping has been identified as an important mechanism that enhances the electron energization by the motional electric field through repeated Fermi accelerations of (e.g. Wang et al. 2016a; Oka et al. 2023, this collection) or continued betatron acceleration (Wang et al. 2016a). The trapping in magnetic islands has also been identified as a key mechanism that enables the acceleration of electrons to large energies (Drake et al. 2006; Zhong et al. 2020).

3.2 Asymmetric Reconnection

Magnetic reconnection often occurs at plasma interfaces where the densities, temperatures, and magnetic field strengths across the current sheet differ. Here, we focus on the electron dynamics in asymmetric reconnection at Earth's magnetopause, where the upstream magnetospheric and magnetosheath plasmas have distinct properties. The magnetosheath has higher density, lower temperature, typically weaker magnetic field, and higher beta than the magnetosphere. The asymmetry results in electron dynamics distinct from those in symmetric reconnection discussed in Sect. 3.1.2. For example, the conservation of mass and energy requires that the electron flow stagnation point is displaced to the magnetospheric side of the current sheet with respect to the X line (Cassak and Shay 2007; Liu et al. 2025b, this collection). Another distinction concerns the Hall electric field, which is typically stronger on the magnetospheric side, with a positive sign (pointing toward the magnetosheath (e.g. Khotyaintsev et al. 2006)), and plays a major role in the electron dynamics in the diffusion regions (e.g. Pritchett 2008; Tanaka et al. 2008; Bessho et al. 2016; Chen et al. 2016a; Swisdak et al. 2018).

The first key breakthrough of the MMS mission was the measurement of crescent electron VDFs revealing meandering and mixing of magnetosheath and magnetosphere electrons in the reconnecting current sheet at Earth's magnetopause (Burch et al. 2016). The crescent distributions are signatures of energized magnetosheath electrons as they meander across the reconnecting current sheet, carrying the distinct high phase-space density characteristic of the magnetosheath. This first discovery started a tidal wave of subsequent observations and simulations on crescent distributions from the EDR to the downstream regions, reflecting the unique capability of MMS to resolve particle gyroradii at nearly all dynamical boundaries, including the smallest-scale turbulent current sheets in the magnetosheath (Wilder et al. 2017) and magnetotail (Ergun et al. 2022).

Magnetic reconnection in an asymmetric configuration enables the mixing of plasma with different properties. While the two inflow plasmas typically interface with a mixed and heated exhaust, sometimes they are observed as clearly distinct. During magnetic reconnection at the magnetopause, where the plasma asymmetry is very pronounced, the hot magnetospheric population is often easy to identify in the differential energy flux. It is not uncommon to observe a double loss-cone distribution, where hot magnetospheric particles move away from the magnetosphere without returning and colder magnetosheath electrons move away from the magnetosheath without returning (e.g. Chen et al. 2016b; Wang et al. 2022a). This is clear evidence of open magnetic field lines and is often used as a sign of ongoing magnetic reconnection (e.g. Li et al. 2016, who studied a large guide field event). Some observations also show distinct bi-directional beams where both populations are of magnetosheath origin (Tang et al. 2020).

Ergun et al. (2016) showed that electrostatic waves present in the magnetosphere inflow can be generated by the interaction between cold plasmaspheric electrons and warm magnetosheath electrons that mix on newly reconnected field lines. Holmes et al. (2019) similarly showed that electrostatic waves can form when warm magnetosheath plasma interfaces and mixes with hot magnetospheric plasma on the magnetospheric side of the exhaust.

3.2.1 The Inflow and Separatrix Regions

Similar to symmetric reconnection, the inflow regions of asymmetric reconnection are characterized by enhanced parallel electron pressure anisotropy due to adiabatic behavior. However, in contrast to symmetric reconnection, the most intense electron heating occurs on the magnetospheric side of the X line. This behavior was predicted by 2D kinetic simulations of asymmetric reconnection (Egedal et al. 2011; Shay et al. 2016; Chen et al. 2016a). They found that the most intense heating occurred in the magnetospheric inflow region close to the EDR over ion spatial scales. The VDFs were characterized by elongated flat-top VDFs parallel and antiparallel to **B**. The temperature anisotropy was found to be consistent with electron trapping via the parallel potential and decreased magnetic field (e.g., Sect. 2.3).

Significant temperature anisotropies have been found in spacecraft observations of magnetopause reconnection. Using the Cluster spacecraft, Graham et al. (2014) found strong parallel electron heating in the magnetospheric inflow region close to the IDR, consistent with electron trapping. In contrast to 2D simulations (e.g. Shay et al. 2016; Egedal et al. 2011), Graham et al. (2014) concluded that the heated electrons originated from the magnetosheath rather than the magnetosphere, based on the distinct temperatures of the magnetospheric and magnetosheath electrons. Similarly, Tang et al. (2013) showed that the heated electrons had thermal speeds comparable to those in the magnetosheath. With MMS, the electron heating was found to be most extreme in the magnetospheric inflow region (Graham et al. 2016a, 2017a; Wang et al. 2017), reaching $T_{e,\parallel}/T_{e,\perp} \approx 5$. The statistical study of Webster et al. (2018) found a range of peak $T_{e,\parallel}/T_{e,\perp}$ (1.4-4.4) near different EDRs, suggesting the degree of heating may be variable between reconnection events. Graham et al. (2016a) found that the strongest electron heating occurred slightly Earthward of the IDR, i.e. outside the region where the strongest Hall electric fields and cross-field currents are

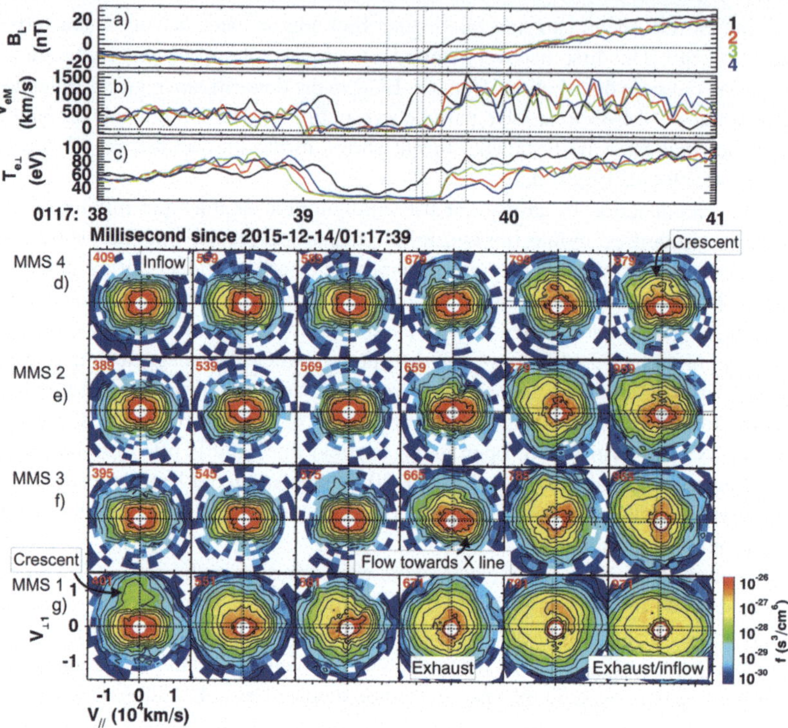

Fig. 11 Kinetic structure of the magnetosheath inflow, separatrix, EDR, and magnetosphere separatrix and inflow during reconnection at the dayside magnetopause. Data are from four MMS spacecraft crossing from one outflow side of the X line to the other. (a) Magnetic field, B_L, showing crossing from the magnetosheath to the magnetosphere side of the current sheet. (b) Electron flow v_{eM}. (c) Perpendicular electron temperature $T_{e\perp}$. (d) Electron VDFs in the $v_\parallel v_{\perp 1}$ plane, where $v_\parallel$ is the electron velocity parallel to the local magnetic field direction $\mathbf{b} = \mathbf{B}/B$, and $v_{\perp 1}$ is the electron velocity along $(\mathbf{b} \times \mathbf{v}_e) \times \mathbf{b}$. The magnetosheath (magnetosphere) inflow is characterized by cold/dense (hot) electron populations elongated along the magnetic field. The EDR is characterized by non-gyrotropic populations within a gyroradius from the current sheet midplane. The two exhausts exhibit larger temperatures. The separatrix (interface between the inflow and the exhaust) is characterized by finite gyroradius effects in the form of crescents (note that the crescents are not seen explicitly in this plane). Figure adapted from Chen et al. (2017). The time steps are slightly changed with respect to the original figure to better highlight the magnetosheath inflow

observed. They also showed that the electron heating occurred over a distance in the normal direction comparable to the ion inertial length, similar to simulations (Egedal et al. 2011; Shay et al. 2016). The electron VDFs were characterized by flat-top VDFs parallel and antiparallel to **B**, which are predicted from the electron trapping model (Egedal et al. 2013; Oka et al. 2023, this collection). Figure 11g (right) shows an electron VDF elongated along **B** typical of heating in the magnetospheric inflow region. I also shows similar elongation, although less pronounced, in the magnetosheath inflow (Fig. 11d). Graham et al. (2014) and Graham et al. (2016a) found that the observed VDFs could be reproduced using the electron trapping model (Egedal et al. 2013) and estimated parallel accelerating potentials of a few hundred Volts. Wang et al. (2017) showed that this parallel heating occurred for a range of dayside reconnection events with different guide field strengths. Some VDFs deviated from the electron trapping model due to an asymmetry between the electron VDF parallel and antiparallel to **B**. These observations also show that the temperature anisotropy decreases as the separatrix or diffusion region is approached and the density increases. The decrease in temperature, which is opposite to what the trapping model predicts, is associated with the appearance of lower-energy magnetosheath electrons. These electrons potentially penetrate to the magnetospheric side due to cross-field diffusion associated with waves in the lower hybrid frequency range (Graham et al. 2022).

The finite Larmor radius ions of magnetosheath origin have been proposed to be responsible for the parallel electron heating observed in the magnetospheric inflow. While this heating is well described by the adiabatic trapping model developed by Egedal et al. (2015), the manner in which the differential ion and electron motions are set up is different. Shay et al. (2016) proposed that due to the smaller gyroradius of electrons, magnetosheath electrons cannot penetrate into the magnetospheric inflow, so magnetospheric electrons must be accelerated into the inflow region along the magnetic field lines to maintain quasi-neutrality. However, in three-dimensional simulations and in observations, electron diffusion can lead to magnetosheath electrons being present where finite Larmor radius ions are observed (Le et al. 2018; Graham et al. 2022). Observations show that the finite Larmor radius ions are strongly correlated with strong parallel electron heating (Khotyaintsev et al. 2016; Graham et al. 2017a, 2019), but it was concluded that the parallel heated electrons consisted of magnetosheath electrons or mixing of magnetospheric and magnetosheath electrons.

This raises the question of whether the observed heated electrons are of magnetosheath origin, magnetospheric origin, or a mixture of both. Graham et al. (2014) argued that the heated electrons were of magnetosheath origin based on the lack of magnetospheric electrons at the energies where electron heating is strongest. Graham et al. (2017a) and Wang et al. (2017) argued that the heated electrons could be a mixture of magnetospheric and magnetosheath electrons. They argued that lower hybrid waves in the diffusion region or separatrices could provide the diffusion required for magnetosheath electrons to be observed in the inflow region. In particular, Wang et al. (2017) argued that the increase in magnetosheath electrons was the cause of the decrease in temperature anisotropy as the diffusion region or separatrix region is approached from the inflow region. They showed that this region was inconsistent with fluid scaling laws for magnetospheric electrons. Similarly, Graham et al. (2017a) and Graham et al. (2019) found that much of the parallel electron heating region was inconsistent with fluid scaling laws for magnetospheric electrons alone and argued that a mixture of magnetospheric and magnetosheath electrons was required.

Recent 3D simulations have shown that a region of mixing between magnetospheric and magnetosheath electrons develops in the magnetospheric inflow region and along the separatrices (Le et al. 2017, 2018). The width of this region was found to be comparable to the ion inertial length, consistent with MMS observations. Moreover, strong parallel electron heating was found over this mixing layer, and the parallel heating was substantially stronger compared with 2D simulations (Le et al. 2017). In both simulations and observations, the mixing of magnetospheric and magnetosheath electrons in the inflow region and along the separatrices was attributed to cross-field diffusion caused by lower hybrid waves (Graham et al. 2017a; Le et al. 2017; Price et al. 2020; Graham et al. 2022). Recently, Tigik et al. (2025) investigated the energy transfer between lower hybrid waves and electrons in the mixing layer. They showed that the waves could contribute to the observed electron heating, although they argued that large-scale electric fields were likely responsible for most of the electron heating. Nevertheless, the relative roles of waves and wave-particle interactions versus larger-scale electric fields in controlling electron heating are not fully understood. The role of waves in the inflow region and IDR is extensively discussed in Graham et al. (2025, this collection).

Observations in the inflow regions on the magnetosheath side of the magnetopause have reported pressure anisotropies in the form of electron VDFs elongated along the local magnetic field (Chen et al. 2017; Lavraud et al. 2016), although these anisotropies are less extreme than those reported in the magnetospheric inflow. These distributions seem to result from the magnetosheath electrons alone. Figure 11d-11g (top left part of the matrix of VDFs) shows the anisotropic electron distributions in the magnetosheath inflow close to the EDR. Here, the kinetic structure of the inflow, separatrix, and EDR is 'imaged' by electron VDFs from the four MMS spacecraft (Chen et al. 2017). The four spacecraft were separated along the normal of the reconnecting current sheet, which permitted the construction of a map of the inflow and diffusion region structure based on the VDFs measured during this transition. As already stated, both inflow regions consist of anisotropic electron VDFs. The outflow in the diffusion region exhibits non-isotropic and non-gyrotropic features (Fig. 11d:679-979, 11e:659-959, 11f:665-965, 11g:401-971). At the interface between the inflow and the exhaust, the VDFs exhibit perpendicular populations related to finite gyroradius effects of exhaust electrons. Toward the exhaust, the electron VDFs become increasingly elongated. While this is hard to discern in the 2D VDFs in Fig. 11, it is clearly seen in the electron pitch-angle spectrogram reported by e.g. Lavraud et al. (2016, for another event, which is also shown in Fig. 15).

In the magnetosheath separatrix, the electrons have been observed to primarily move towards the X line, both in simulations (Pritchett 2008; Chen et al. 2016a) and observations (Graham et al. 2016b; Chen et al. 2016b; Norgren et al. 2016). Additionally, in the magnetosheath separatrices, the energetic magnetospheric electrons can escape along the newly reconnected field lines, resulting in an anisotropic VDF at high energies (Lindstedt et al. 2009; Graham et al. 2016b). These electron VDFs can be used to track magnetosheath field lines that have reconnected with the magnetosphere. Similar to symmetric reconnection, the separatrix can be characterized by a density cavity and a locally divergent electric field (e.g. Norgren et al. 2016; Khotyaintsev et al. 2006), where the latter is set up to maintain quasi-neutrality when the electron density is depleted due to flux conservation.

Chen et al. (2016b) observed D-shaped loss-cone VDFs of magnetosheath electrons that constituted a parallel electron bulk flow towards the X line (see their Figures 2i and 2j). The bulk flow seemed to be due to the loss of the outgoing or returning population rather than any significant acceleration of the inflow population, which had energies very similar to a magnetosheath population observed at an earlier time. In magnetotail reconnection, the separatrix flow has been observed to be carried by suprathermal beams (e.g. Norgren et al. 2020), likely due to lower β_e (Egedal et al. 2015). Closer to the EDR, the magnetosheath electrons formed an elongated and highly anisotropic VDF with energies above and below the inflow energies parallel and perpendicular to the magnetic field, respectively. Wang et al. (2022a) showed the presence of counter-streaming electron populations at the magnetosheath separatrix of another magnetopause reconnection event. The higher-speed population propagated towards the X line while the lower-speed population moving away from the X line had maximum energies more comparable to the magnetosheath inflow.

On the magnetospheric side of the X line, the separatrix regions are characterized by a partial loss due to magnetospheric electrons moving away from the X line, resulting in anisotropic VDFs (Lindstedt et al. 2009; Graham et al. 2016b).

The electron edge on the magnetospheric side of the outflow is the boundary where magnetosheath electrons propagating along newly reconnected field lines are first observed. It is situated close to the outer edge of the exhaust, just inside the separatrix. The distance between the separatrix and the electron edge increases away from the X line due to the time-of-flight effect coupled with the convection of field lines from the inflow to the outflow. Due to the fast speed of the electrons, the electron edge remains close to the separatrix over relatively large distances and is hence the best plasma measure of where the magnetic separatrix is situated. Similarly, magnetospheric electrons propagating away from the X line will be lost when the magnetic field lines connect with the magnetosheath. Closer to the ion outflow region, magnetosheath electrons propagating along the reconnected field lines are observed and can form electron beams propagating away from the X line (Graham et al. 2016b; Khotyaintsev et al. 2016; Wilder et al. 2016, 2017). The resulting electron beams and anisotropic energetic electrons can be unstable to plasma waves (Graham et al. 2025, this collection). Like the inflow region, the magnetospheric separatrices occur over ion or hybrid scales due to plasma mixing associated with lower hybrid waves at the density gradient.

3.2.2 The Electron Diffusion Region

In asymmetric reconnection, the conservation of mass and energy requires that the electron flow stagnation point is displaced to the low-β (magnetospheric) side of the current sheet with respect to the X line (Cassak and Shay 2007; Liu et al. 2025b, this collection). Between the X line and the stagnation point, the bulk of the electrons from the high-β (magnetosheath) thus move in the direction opposite to the convecting magnetic field. In the magnetosphere inflow region, the electrons are gyroturned by the magnetic field B_L, acquiring an increased out-of-plane speed v_M. For negligible guide fields, the reconnection current, $J_M \sim -nv_{eM}$, is the strongest around the stagnation point on the high-β side of the X line (Pritchett 2008; Burch et al. 2016; Genestreti et al. 2017). The electrons that provide this out-of-plane current are associated with strong finite gyroradius effects with VDFs in the shape of crescents (e.g. Hesse et al. 2014; Burch et al. 2016). In contrast to symmetric reconnection, simulations show that the reconnection electric field at the X line is not supported by the divergence of the pressure but instead by the electron inertia $(m_e/e)v_{eN}\partial v_{eM}/\partial N$ (Hesse et al. 2014; Egedal et al. 2018; Liu et al. 2025b, this collection). As v_{eM} is gradually built up during the transit to the low-β side, more v_M-momentum is transported away from the X line than toward it, thus balancing the production of momentum by E_M there. Terms in the generalized Ohm's law have been estimated using MMS measurements. For example, Torbert et al. (2016) found that both pressure and inertial effects could be significant. However, it has not yet been clearly demonstrated how these terms relate to the X line. For a more extensive discussion of these results, see the review by Liu et al. (2025b, this collection).

Due to the pressure asymmetry, a normal electric field, E_N, spans the X line. As shown in Sect. 2.2.1, such a field affects the electron trajectories, introducing an asymmetry, which in this case accelerates them toward, or confines them more to, the low-β side. Shay et al. (2016) showed with kinetic simulations that the E_N field at the X line was balanced by the divergence of the electron pressure tensor due to counter-streaming electrons along $\hat{e}_N$ performing "cusp-like" trajectories (see their Fig. 2p or e.g. the red trajectory in Fig. 2b3). Chen et al. (2016a), Bessho et al. (2016) showed how the electrons of magnetosheath origin underwent meandering trajectories. Chen et al. (2016a) also showed how the distribution was gyroturned toward the outflow, akin to the symmetric case. Swisdak et al. (2018) and Egedal et al. (2018) showed how electrons can also enter the magnetospheric side and oscillate around the magnetospheric separatrix without crossing back to the high-β side (see e.g. Fig. 14). While Egedal et al. (2018) showed a quantitative comparison between simulation and observations (discussed further below), the other results mentioned above are from particle-in-cell simulations. Observational evidence of the characteristics of asymmetric gyroresistivity is still needed.

The first resolved observation of an asymmetric EDR was made by MMS at the magnetopause and reported by Burch et al. (2016). The EDR was identified by the reversal of

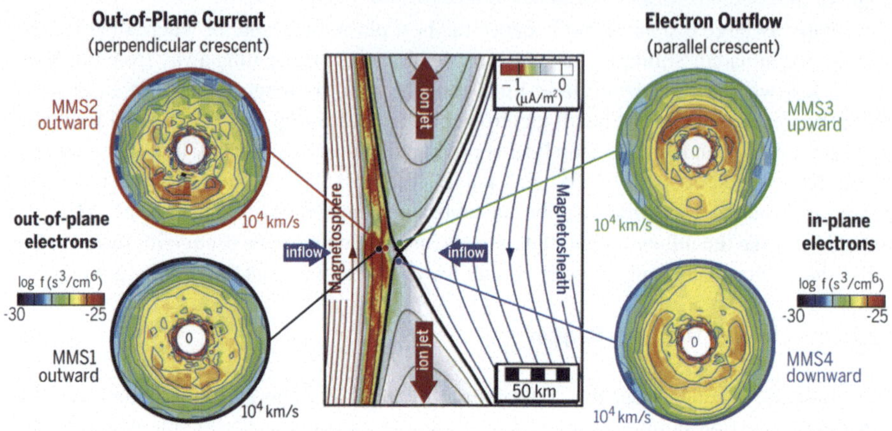

Fig. 12 Electron VDFs observed by MMS inside and in the vicinity of an EDR at the dayside magnetopause. On the magnetosphere side of the current sheet, the observations reveal crescent-shaped electron populations in the plane perpendicular to the magnetic field. Towards the outflows, parallel populations in the shape of double horseshoes are observed (only one plane is shown here). In 3D, these double horseshoes become a dome-like structure. From Burch et al. (2016). Reprinted with permission from AAAS

the electron outflow jets, intense current density in the out-of-plane direction, and nongyrotropic electron velocity space signatures indicative of demagnetized electron motion. Subsequently, several EDRs were identified with the help of the above three signatures (Webster et al. 2018). These signatures have been observed both on the magnetospheric and magnetosheath side of the reconnecting magnetopause, indicating the presence of meandering electrons. In general, to establish that the nongyrotropic populations (such as those marked as "crescent" in the magnetosheath side of the EDR in Fig. 11) are demagnetized meandering electrons, the following are examined (1) the electron gyroradius being larger than the magnetic field gradient scale, (2) comparison of measured distributions in all three velocity planes with PIC distributions at corresponding locations (Chen et al. 2017). For near-X line exhaust electrons, the ratio between the magnetic curvature radius and the local thermal gyroradius will need to be in the demagnetized range (Lavraud et al. 2016; Chen et al. 2017).

The Kinetic Structure of the Asymmetric Electron Diffusion Region Observations of the kinetic structure of the asymmetric EDR by MMS have both confirmed predictions and revealed new features that had not been predicted by numerical simulations. For example, Burch et al. (2016) confirmed the presence of crescent-type populations on the magnetospheric side predicted by Hesse et al. (2014) (Fig. 12). However, Burch et al. (2016) also revealed that the electron outflow jets on the magnetospheric side were formed by double parallel crescent populations (whereof one plane is shown in Fig. 12), which together formed a shell population in the shape of a dome (Fig. 13e), or a partial dome (Egedal et al. 2016b). Several observations from the dayside magnetopause have shown how the electron VDF on the magnetospheric (low-β) side of the current sheet midplane transitions between these two types of distributions, i.e. a perpendicular out-of-plane ($\hat{e}_M$) beam close to the flow reversal (Fig. 13f and 13c), to parallel beams toward the outflow directions ($\hat{e}_L$, Fig. 13e and 13h) (e.g. Khotyaintsev et al. 2016; Burch et al. 2016; Chen et al. 2016b; Webster et al. 2018; Pritchard et al. 2019). The perpendicular beam, carried by an agyrotropic crescent population, shifts to higher energies the further into the magnetospheric inflow regions they reach. This was predicted by simulations (Shay et al. 2016; Bessho et al. 2016;

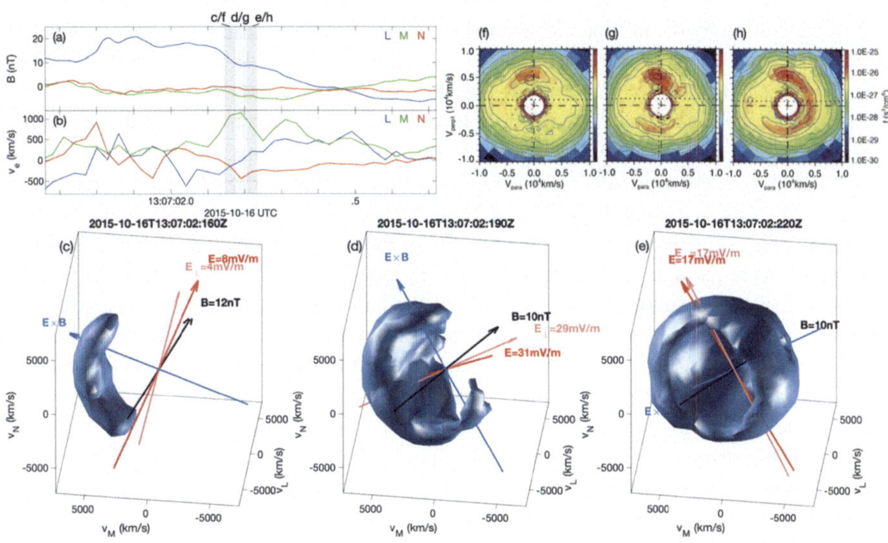

Fig. 13 Evolution of electron VDFs in the vicinity of the X line of asymmetric reconnection. (a) Magnetic field. (b) Electron velocity. (c)-(d) Isocontour surfaces at $f_e(v_L, v_M, v_N) = 6 \times 10^{-27}$ s^3/cm^6 for the times marked in panels (a)-(b) (LMN vectors from Burch et al. 2016). The quivers **B**, **E**, and $\mathbf{E}_\perp$ show the average magnetic and electric field directions observed during the 30 ms interval the VDFs were sampled. We note that the **E** for the second VDF was ~ 0 during half of this time interval and ~ 50 mV/m during the other half. At the largest B, the electron VDF exhibits a crescent well limited in $v_\parallel$. For the second timestep, this thin crescent spread around and towards $\hat{e}_B$. For the third time, the VDF is a gyrotropic shell roughly shaped as a dome, or a mushroom cap. (f)-(h) 2D VDFs for the same timesteps from Burch et al. 2016. Reprinted with permission from AAAS.

Egedal et al. 2016a) and has been confirmed by observations (Pritchard et al. 2019). At the same time as the electrons crescents reached higher energies, they become more confined in the perpendicular plane (Pritchard et al. 2019; Torbert et al. 2017), consistent with theoretical predictions (e.g. Bessho et al. 2016; Shay et al. 2016). Egedal et al. (2016a) noted that the electrons reaching the furthest into the magnetosphere were dominated by their original energy in the magnetosheath.

The parallel beams form gradually as we move away from the stagnation point, and seem to be partially concurrent with the transition to a more gyrotropic distribution (Burch et al. 2016). The transition between the perpendicular crescent and the dome (or mushroom cap) is illustrated in Fig. 13, which reproduces the first three VDFs in Figure 4J in Burch et al. (2016) as isosurfaces. The first VDF (Fig. 13c) is sampled at a location with the highest magnetic field magnitude out of the three. This is consistent with the expectation of a smaller perpendicular coverage further into the magnetospheric inflow. At the second time step (Fig. 13d), the crescent population is observed to spread towards the parallel direction while remaining confined to roughly half the perpendicular plane (i.e., still agyrotropic and carrying a perpendicular bulk electron flow). At the third time step (Fig. 13e), the population has spread all the way to the field-aligned direction and has become gyrotropic. The transition is observed slightly differently on the various spacecraft (Burch et al. 2016; Egedal et al. 2016a), which could be a result of the speed of the transition (i.e., we might catch more snapshots) or depend on where exactly the transition occurs.

The formation of the parallel dome populations (Fig. 13e) has been attributed to a combination of forces (Egedal et al. 2016a). First, the lower energy cut off has been attributed to the acceleration by the normal electric field $E_N > 0$ combined with pitch-angle scattering around the X line. Second, the lower energy cutoff in one of the directions along the magnetic field was attributed to the confining electrostatic potential and the mirror force present on the magnetospheric side of the X line. Third, a lower energy cutoff in the direction perpendicular to the magnetic field was attributed to finite gyroradius effects. If this limit is higher than the overall lower-energy cutoff, also a drop in both of the parallel directions will be observed (see e.g. Figure 3c in Egedal et al. 2016a). Simulations also show direct energization by a parallel electric field in the vicinity of the X line (Shay et al. 2016), which also plays a role in the formation of the beams. As mentioned above, the lower energy cutoff, i.e. the absence of lower energy electrons to fill up the half dome (note that the lowest-energy electrons are attributed to spacecraft effects and not included in Fig. 13), is seen in observations (Burch et al. 2016). Since the cut off also persists when the electron VDF become gyrotropic, it can not be attributed solely to the finite gyroradius effects.

The structure of the outflow jet varies from observation to observation, which might be an effect of the distance to the X line. For example, some observations show clear unidirectional beams with cut-off at higher speeds (Wang et al. 2022a). However, even at higher cut-off speeds, the beam may remain oblique (Khotyaintsev et al. 2016; Burch and Phan 2016). Other events show clear counter-streaming beams with different lower cut-off speeds (Wang et al. 2022a; Burch et al. 2018). Genestreti et al. (2017) illustrated for a range of guide fields how the electron VDFs close to the B_L reversal were characterized by elongation aligned with the magnetic field. They had moderate asymmetries in the parallel and antiparallel directions. At larger magnetic fields B_L, the VDF transitioned into a unidirectional beam. Crescent-type VDFs were observed for moderate-shear events, where the presented events had $B_M/B_L^{sheath} \sim 0.5 - 1$, and B_L^{sheath} is the reconnecting field component on the magnetosheath side of the current sheet.

The electron demagnetization in the vicinity of the X line is illustrated in Fig. 14, which compares particle-in-cell simulations (Fig. 14b) with MMS observations (Fig. 14c) (Egedal et al. 2018). On the high-β side, parallel electron beams converge to the X line but start to deviate from the magnetic field before reaching it (Fig. 14b and 14b I, and see also Fig. 2 of Egedal et al. 2018). While the originally parallel beam continues in the same direction, the magnetic field turns (Fig. 14b-14c II-V). This produced an oblique electron beam satisfying $\mathbf{E} + \mathbf{v}_e \times \mathbf{B} \neq 0$. The deviation of the flow from the magnetic field occurs in a region where the typical electron gyroradius is larger than the magnetic field curvature radius, $\kappa = \sqrt{R_B/\rho_e} > 0.5$.

The electron trajectories in Fig. 14 are seen to oscillate around the magnetosphere separatrix as they move away from the X line and stagnation point (Egedal et al. 2016a). This oscillatory motion is also seen in the electron bulk speed in other simulations (Swisdak et al. 2018), which suggests the EDR can serve as a funnel that is small enough to keep the electrons' phases bunched. The oscillatory electron motion is associated with patchy fluid quantities and energy conversion, such as $(\mathbf{E} + \mathbf{v} \times \mathbf{B})_M$, $E_\parallel$, and charge density (Chen et al. 2016a; Swisdak et al. 2018). These simulations have been connected to (and some motivated by) early MMS observations that saw large, and seemingly patchy, electric fields and strong energy dissipation $\mathbf{E} \cdot \mathbf{J}$ occurred on the low-beta magnetospheric side of the reconnecting magnetopause (e.g. Burch et al. 2016, 2018).

Electron Crescent Distributions on the Magnetosheath Side of the Outflow Electron crescent populations were first predicted to occur on the magnetosphere side of the reconnection site (Hesse et al. 2014). However, spacecraft observations by MMS (Norgren et al. 2016; Graham et al. 2017b; Chen et al. 2017) and simulations (Chen et al. 2017) show that they

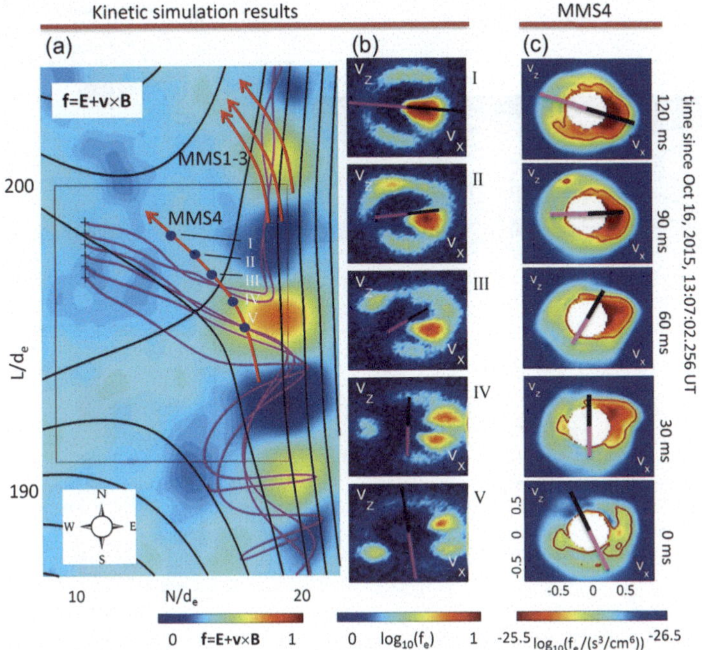

Fig. 14 Demagnetization of electrons in the EDR. (a) Example electron trajectories together with spacecraft trajectories overlaid $(\mathbf{E}+\mathbf{v}_e\times\mathbf{B})_M$. (b) Simulation and (c) observed electron VDFs along the MMS4 trajectory. The parallel electron beams that form along the magnetosheath separatrices deviate from **B** around the X line. When the electrons enter the magnetospheric side, they gyro-oscillate, which leads to oscillatory electric fields $\mathbf{E}'=\mathbf{E}+\mathbf{v}_e\times\mathbf{B}$ and $\mathbf{E}'\cdot\mathbf{J}$. Figure reprinted from Egedal et al. (2018). Copyright (2018) by the American Physical Society

can also occur on the magnetosheath side of the exhaust. Figure 11 illustrates the associated finite gyroradius effects, where the crescents are viewed 'from the side' (Chen et al. 2017). This event (Chen et al. 2017; Graham et al. 2017b) had a low to moderate guide field, exhibiting electron crescent populations on both the magnetosheath and magnetosphere sides of the outflow. There, the electron crescent population, roughly aligned with the local $\mathbf{v}_{E\times B}$ direction, was directed along $\sim\hat{e}_M$. Graham et al. (2017b) demonstrated that the crescent beam gave rise to upper hybrid waves (for a more extensive discussion of their potential role, see Graham et al. 2025, this collection). Norgren et al. (2016) reported another magnetopause crossing with close to zero guide field. An electron crescent population was observed on the magnetosheath side of the exhaust, directed in the outflow direction $\sim\hat{e}_L$, roughly aligned with the local $\mathbf{v}_{E\times B}$ direction. The pressure gradient associated with the crescents was, in this case, directed from the magnetosheath to the exhaust, with the minimum density being associated with the separatrix environment. Li et al. (2020) also reported electron crescent distributions on both sides of the magnetopause. These observations illustrate that the occurrence of finite gyroradius effects resulting in non-gyrotropic electron VDFs is highly dependent on the local environment that develops as reconnection proceeds.

3.2.3 The Outflow Region

Electron Scattering in the Central Current Sheet Up to intermediate guide fields, the electron temperatures at the center of the current sheets are often isotropic (e.g. Graham et al.

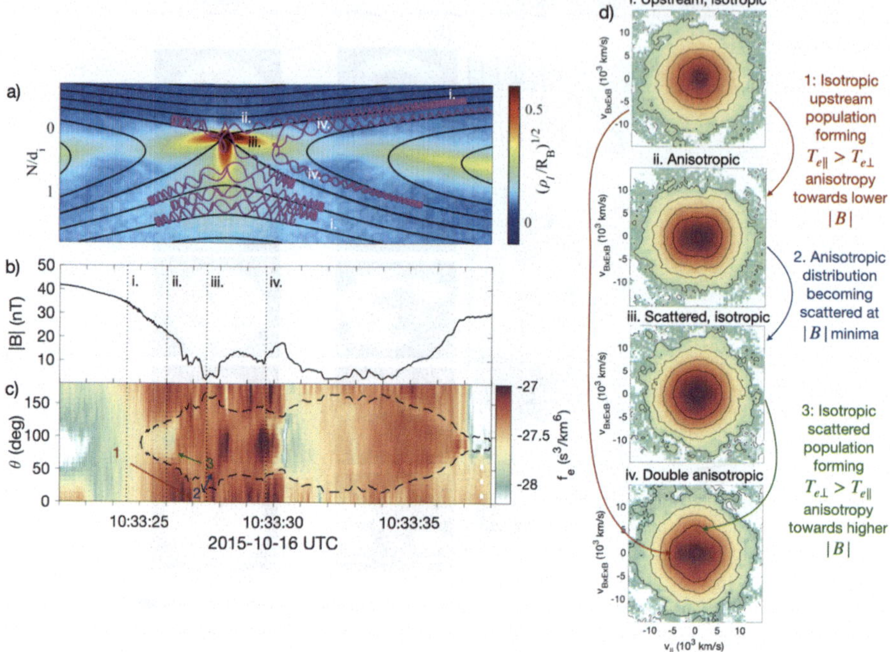

Fig. 15 Particle focusing and scattering in a reconnection current sheet at the magnetopause. (a) Electron trajectory. (b) Magnetic field magnitude during current sheet crossing. (c) Electron pitch angle spectrogram. The dashed line shows the predicted change in pitch angle according to Eq. (12). (d) Reduced electron VDFs at the times marked by dashed vertical lines. Far upstream (i), the VDF is isotropic. When approaching the current sheet center (ii), $|\mathbf{B}|$ decreases and the VDF becomes elongated along $\mathbf{B}$. The energies are elevated above those of the asymptotic VDF, indicating the VDF has also been accelerated by a field-aligned electric field. In the center of the current sheet (iii), the large magnetic curvature scatters the electrons, forming again an isotropic VDF. At the interface between the inflow and exhaust (iv), there is a superposition of two anisotropic populations, one originating upstream in the inflow experiencing a decreasing $|\mathbf{B}|$ and has $T_{e\parallel} > T_{e\perp}$, and one originating from the current sheet center experiencing an increasing $|\mathbf{B}|$ and has $T_{e\parallel} < T_{e\perp}$. Panel (a) reprinted from Egedal et al. (2016a). Copyright (2016) by the American Physical Society. Panels (b)-(c) reconstructed from Lavraud et al. (2016)

2017a; Torbert et al. 2017; Genestreti et al. 2017). Since the inflow VDFs are often highly anisotropic (with $T_{e\parallel} > T_{e\perp}$), this suggests that the electrons are scattered as they enter the regions of highly curved magnetic fields at the current sheet center.

Lavraud et al. (2016) examined the pitch-angle scattering for a reconnection event observed at the dayside magnetopause (see Fig. 15). During this event, MMS passed through the IDR and the outer EDR, and the minimum magnetic curvature radius corresponded to κ values between 1 and 10 for electron energies between 500 eV and 20 eV. Electron VDFs at energies above 500 eV were isotropic throughout the crossing. At subsequently lower energies, the outer region exhibited field-aligned electron VDFs that transitioned into more isotropic VDFs further and further toward the center of the current sheet. That is, lower-energy electrons could remain magnetized for longer than higher-energy electrons when approaching the current sheet center. Lavraud et al. (2016) also illustrated how the current density on the magnetospheric side of the current sheet had a large parallel component.

The change in pitch angle due to constant μ (Eq. (12)) has been observed inside the exhaust in various reconnection events at the dayside magnetopause (e.g. Lavraud et al. 2016;

Zhu et al. 2019; Robertson et al. 2021) but also in events where there, locally, are no signs of active magnetic reconnection (Tang et al. 2019). In either case, they are clear indications of electron trapping in magnetically connected regions. The behavior in the event reported by Lavraud et al. (2016) is shown as a dashed line in Fig. 15c. Both the population inside and outside the line are well bounded by the line. The 'outer' population is relatively isotropic at high $|\mathbf{B}|$ and becomes more field-aligned and anisotropic towards lower $|\mathbf{B}|$. The 'inner' population is very anisotropic at high $|\mathbf{B}|$, and transitions to an isotropic distribution towards low $|\mathbf{B}|$. This can be explained by the two presumed source populations, which are both illustrated by the particle trajectory in Fig. 15a. In the inflow, the source population is likely an isotropic population originating from a region where the magnetic field reaches a certain asymptotic value. As it approaches the current sheet center, the magnetic field magnitude decreases, and the electrons become more field-aligned. Eventually, they become scattered in the curved magnetic field, potentially accelerated, again forming a relatively isotropic VDF. When this population bounces in the outflow and again reaches higher magnetic field amplitudes, it becomes more perpendicular and anisotropic and is potentially further accelerated to form a peak near 90° pitch angles. This is fundamentally similar to the symmetric reconnection configuration (Sect. 3.1.1). This combination of electron populations thus demonstrates the presence of two adjacent magnetic bottles and two different 'isotropic source' populations. Although, as mentioned above, in this reconnection case, the isotropic source of the inner population is the accelerated and scattered outer population. In this manner, one can also deduce a course of events, or partial acceleration history of the particles.

3.3 Guide-Field Reconnection

A magnetic guide field is introduced when the magnetic field on the two sides of the current sheet is not exactly antiparallel, so there is a finite magnetic field at the X-line. This scenario is common at the dayside magnetopause, where the highly variable solar wind dictates the direction of the magnetic field in the magnetosheath. Guide-field reconnection is also prevalent in turbulent regions, such as the magnetosheath (Wilder et al. 2018), the foreshock and bowshock transition regions (Wang et al. 2019b), and the solar wind (Gosling 2012). Additionally, strong guide field magnetic reconnection has been detected at the magnetopause flanks, where the formation of vortices due to the Kelvin-Helmholtz instability leads to the development of thin current sheets that can reconnect (Eriksson et al. 2016a; Li et al. 2016).

The presence of a guide field has two main effects. First, particles become more strongly magnetized (see e.g. Sect. 2.2.1), and second, the reconnection electric field at the X-line acquires a parallel component (e.g., Lapenta et al. 2015). This, in turn, modifies the electron motion at the separatrices and within the current sheet, causing the current system to become distorted compared to the antiparallel case (Lapenta et al. 2010; Pritchett and Coroniti 2004). The increased magnetization inhibits magnetic curvature scattering and the resulting isotropization observed in the antiparallel cases. Studies have shown that even small guide fields can introduce significant effects (Le et al. 2013; Wilder et al. 2018).

3.3.1 Symmetric Guide-Field Reconnection

In symmetric reconnection without a guide field, the electrons stream toward the X line along the magnetic field at all four separatrices. The acceleration is given by the integral of the parallel electric field (Sect. 2.3). With the added magnetic guide field, the reconnection electric field acquires a parallel component, which modifies the parallel acceleration. At

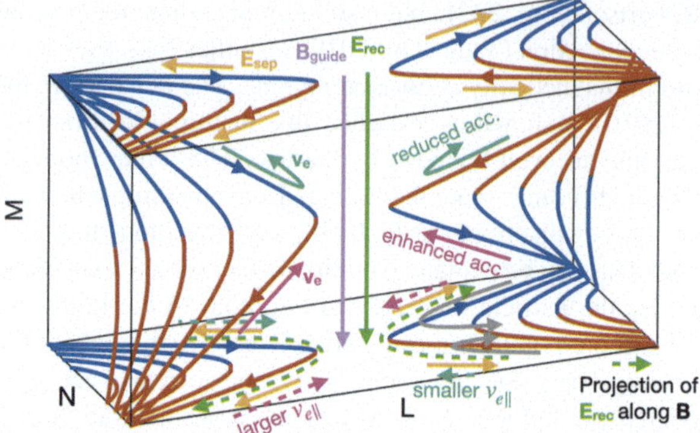

Fig. 16 In the presence of a guide field, the reconnection electric field acquires a parallel component, which either enhances or inhibits the electron acceleration and flows depending on which quadrant the electron enters the region from. Where the projection of the reconnection electric field on **B** (green dashed line) is directed away from the X line, the electron flow will be enhanced, and vice versa. The electrons with enhanced acceleration will be able to penetrate into the other side of the current sheet, which leads to asymmetric convection cells and Hall magnetic and electric fields. Note that the Hall magnetic field would add an asymmetric distortion of the magnetic field line. Similar to Lapenta et al. (2010), the blue and red colors represent the direction of B_L

two opposing separatrices, the parallel component of the reconnection electric adds to the separatrix acceleration, while at the other two, it subtracts from it (Pritchett and Coroniti 2004; Lapenta et al. 2010; Eastwood et al. 2018). This is illustrated in Fig. 16 (note that the effects of the Hall magnetic field, which may further complicate the picture, are not included in this illustration, see e.g. Wetherton et al. 2021), where the projection of the reconnection electric field, E_{rec}, along the field lines is shown as green dashed lines. At the two separatrices where E_{rec} and E_{sep} are in the same direction, the acceleration is enhanced, and vice versa. The electrons arriving from the separatrices with enhanced acceleration will be able to cross into the opposite side of the current sheet, while the electrons arriving from the other two separatrices will be deflected towards the outflow as they approach the current sheet center. The asymmetric electron flows create an asymmetry in the current system, which modifies the Hall magnetic and electric fields (Pritchett and Coroniti 2004; Eastwood et al. 2010; Wang et al. 2012). The guide field also introduces an additional force component, $J_L B_M / n\hat{e}_N$, inside the exhaust that distorts the outflow jet, and thereby the inner Hall current loop as well (e.g. Goldman et al. 2011; Hwang et al. 2021). This leads to modified Hall magnetic and electric fields, which are well documented in observations (e.g. Eastwood et al. 2010). The density cavities and associated electric fields that develop along the separatrices due to the electron streaming also become asymmetric (Lapenta et al. 2010; Eastwood et al. 2018; Fu et al. 2018). Additionally, in the presence of a guide field, the reconnection current, J_M, has a parallel component (e.g. Wang et al. 2020; Li et al. 2019).

Wilder et al. (2018) studied the electron frame dissipation measure $\mathbf{E}' \cdot \mathbf{J} = (\mathbf{E} + \mathbf{v}_e \times \mathbf{B}) \cdot \mathbf{J}$ (Zenitani et al. 2011) for 15 events in the magnetosheath and found a rather sharp transition around a guide field of $B_G / B_0 \sim 0.3$, where B_0 is the asymptotic magnetic field. Below 0.3, the contribution from the perpendicular components $\mathbf{E}'_\perp \cdot \mathbf{J}_\perp$ dominated while above 0.3 the contribution from the parallel component $E'_\parallel J_\parallel$ dominated. Although the dynamics behind this transition are not completely understood, the study shows a clear change in the particle energization from perpendicular to field-aligned.

Le et al. (2013) studied the structure of the current sheet for a range of guide field strengths and delineated four distinct regimes regulated by the marginal firehose condition. The key mechanism providing these breaks is the pressure anisotropy that develops inside the IDR upstream of the EDR. The pressure anisotropy, governed by CGL-like scaling laws (Le et al. 2009), creates a feedback loop that influences the magnetic field strength and electron currents, especially in low guide field scenarios. This feedback maintains electron pressure anisotropy near the marginal firehose condition along elongated current layers. Without the pressure anisotropy, the magnetic tension force, which is the driver of shear Alfvén waves, would normally straighten the magnetic field.

In the literature, there is some inconsistency in the terms used to describe guide-field levels, such as weak, moderate, intermediate, and strong guide fields. In the following paragraphs, we review the literature based on the following classification: weak and moderate ($0.05 \lesssim B_G/B_0 \lesssim 0.2$), intermediate ($0.2 \lesssim B_G/B_0 \lesssim 1$), and strong ($1 < B_G/B_0$) guide fields.

Weak and Moderate Guide Field This guide-field range, $0.05 \lesssim B_G/B_0 \lesssim 0.2$, includes the second and sometimes third regimes (depending on β_e) defined by Le et al. (2013). In their second guide-field regime, a weak guide field, $B_G/B_0 \simeq 0.07 - 0.14$, disrupts the formation of elongated current layers, which instead transitions to a central current region with currents along the separatrices. This behaviour has been seen in simulations (Goldman et al. 2011; Le et al. 2013; Wetherton et al. 2022) and observations (Zhou et al. 2019a; Denton et al. 2020; Schroeder et al. 2022).

Tang et al. (2022) reported observations from the outer EDR in the magnetotail. The current sheet was bifurcated, had a guide field of $B_G/B_0 = 0.1$ in the center of the current sheet, and the thickness was ~ 70 km $\approx 4d_e$. In the center of the current sheet, the pitch-angle distribution was populated at $v_\parallel \lesssim 90°$ and exhibited an oscillatory pattern visible around $\theta \sim 90°$. The 2D reduced VDFs revealed that the flow was oblique ($\sim 45°$ angle between $\hat{e}_B$ and $\hat{e}_E \times \hat{e}_B$) and consisted of two crescent layers seen in the plane perpendicular to the magnetic field. This is strong kinetic evidence of meandering electron trajectories typical for antiparallel reconnection, which indicates that the level of guide field present, $B_G/B_0 \sim 0.1$, was not sufficient to keep the electrons magnetized throughout the crossing.

Another weak guide field event, $B_G/B_0 \sim 0.13 - 0.17$, from the magnetotail was studied by Zhou et al. (2019a), Li et al. (2019), and Cozzani et al. (2021). At the current sheet center, the current was predominantly parallel, the electron temperatures were isotropic, $T_{e\perp} \approx T_{e\parallel}$, and a beam was observed. By investigating the curvature scattering, Zhou et al. (2019a) found that electrons down to 0.2 keV had $\kappa < 3$, suggesting that despite the guide field, the magnetic curvature was sufficient to scatter a significant part of the electrons, accounting for the isotropic VDF at the current sheet center. Slightly away from the current sheet center, clear electron crescent populations were observed in the perpendicular plane.

Chen et al. (2019) reported a magnetotail interval where three crossings of the current sheet over 30 seconds showed three different guide fields. One of the crossings, with a weak guide field of $B_G/B_0 \sim 0.14$, is shown in Fig. 17. Despite the low guide field, magnetized field-aligned electron populations are observed at the center of the current sheet. However, there is also evidence of demagnetized electrons moving perpendicular to the magnetic field. The reduced electron VDF $f_e(v_{\perp 1})$ in Fig. 17c, where $v_{\perp 1} = \hat{b} \times \mathbf{v}_e \times \hat{b}$, shows that the core is also drifting. Close to the center of the current sheet, the antiparallel electrons drop out (Fig. 17f:C-D). At one side, the remaining anisotropic population is oblique, deviating from the magnetic field direction by $\sim 30°$, also evidencing demagnetized behavior (Fig. 17d and 17f:D).

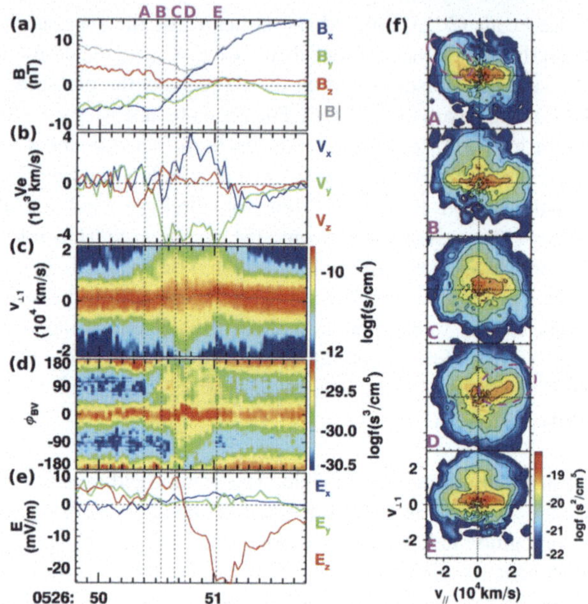

Fig. 17 Observations of an EDR with a weak guide field ($B_g \sim 0.14 B_0$) observed in the magnetotail. (a) Magnetic field. (b) Electron velocity. (c) Reduced electron VDF along the perpendicular electron bulk flow direction, $\mathbf{v}_{\perp 1} = \mathbf{b} \times \mathbf{v}_e \times \mathbf{b}$. (d) Angular VDF in the plane of $\mathbf{B}$ and $\mathbf{v}_{\perp 1}$. (e) Electric field. (f) Reduced electron VDFs at the location marked with dashed vertical lines in the left panels. The electron VDF (C) right next to the B_L reversal consists of an elongated population only at $v_\parallel > 0$, and a perpendicular VDF showing the presence of meandering electrons. Right at the B_L reversal (VDF D), the elongated population is oblique. Both VDFs indicate a departure from fully magnetized motion. Figure adapted from Chen et al. (2019)

Intermediate Guide Field This guide field range $0.2 \lesssim B_G/B_0 \lesssim 1$, includes the third and the fourth regimes defined by Le et al. (2013). Their third regime features extended current layers in a broader exhaust region. Such an event, shown in Fig. 18, was observed by Wilder et al. (2017) and further studied by Wetherton et al. (2022). The magnetic field profile consists of a slowly decreasing magnitude that ends with a sharp reversal at the current sheet center, corresponding to a thin current layer embedded in a broader exhaust. A notable feature is that the highest anisotropy is shifted to one side of the magnetic field reversal, which Wetherton et al. (2022) attributed to asymmetric pitch-angle scattering. The marginal firehose condition, $\mu_0(p_{e\parallel} - p_{e\perp})/B^2 \sim 1$, remains marginally satisfied in the broader current layer, indicating that the pressure anisotropy is at a level to offset the magnetic tension. Note that Wilder et al. (2017) estimated that $B_G/B_0 \sim 0.5$ while Wetherton et al. (2022) compared the observations to a simulation with $B_G/B_0 \sim 0.2$. The observations and the simulation are remarkably similar, indicating that the dominant processes prevail over this range, or that there is an error associated with the guide field estimation from observations.

An intermediate-guide field ($B_G/B_0 \sim 0.3$) crossing reported by Chen et al. (2019), and studied in more detail by Chen et al. (2020), showed how the guide field permitted the lower-hybrid drift wave oscillations to penetrate close to the current sheet center. The wave field led to strong modulations of the electron motion extending all the way into the immediate vicinity of the magnetic field reversal and even modulated the electron crescent population there. The electron dynamics could also be seen in the electron VDF displayed by Chen et al. (2019) and included modulations in the directions perpendicular as well as parallel to the magnetic field. Overall, the electron VDF appeared quite thermal, and no field-aligned populations were observed close to the current sheet center.

At moderate guide fields, the Hall and guide fields can be of similar magnitudes. In such cases, in the quadrants where they are directed opposite to each other, they will act to cancel out each other, while in the quadrants where they are aligned, they will add up. This can create an asymmetry in the magnitude of the magnetic field and its curvature. In the quadrants with decreased magnetic field magnitude, the curvature will become larger, which may lead

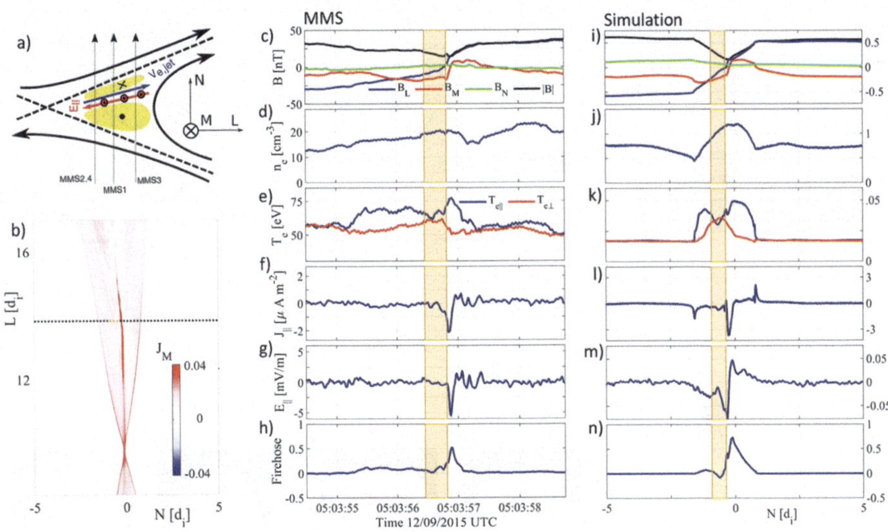

Fig. 18 (a) Illustration of the observed embedded electron jet in an intermediate guide field event. (b) Out-of-plane current density from a VPIC simulation that is compared to the MMS observations. The remaining panels compare data from the MMS event with simulation data along the cut shown in (b), with the highlighted region corresponding to the yellow portion of the cut line. (c) and (i) magnetic field components, (d) and (j) electron density, (e) and (k) electron temperatures, (f) and (l) parallel current density, (g) and (m) parallel electric field, and (h) and (n) firehose threshold, $B^2/\mu_0 - (p_{e\parallel} - p_{e\perp})$, measured across the exhaust. Panel (a) is adapted from Wilder et al. (2017), and the remaining panels from Wetherton et al. (2022)

to enhanced electron scattering (Chen et al. 2017). Norgren et al. (2018) illustrated how the scattering of electrons (out of the loss cone) arriving at the current sheet from the side of enhanced curvature prevented electrons from exiting the current sheet on the opposite side, leading to a dropout of outgoing field-aligned electrons there. Electrons arriving from the other side remained relatively well magnetized and were capable of following the magnetic field throughout the current sheet crossing. Similar observations have been seen in various events with moderate guide field (e.g. Zhou et al. 2019a), although separating the effects of the reconnection electric field from the scattering requires further analysis.

Strong Guide Field This guide field range, $B_G/B_0 \gtrsim 1$, includes the fourth regime defined by Le et al. (2013). The stronger guide fields allow electrons to maintain magnetization throughout the reconnection exhaust and prevent the plasma from reaching the marginal firehose condition, thereby eliminating long current layers. Figure 19 shows observations from two events (Wetherton et al. 2021), with guide fields near ($B_G/B_0 \sim 2$, Fig. 19d-19h) and far downstream ($B_G/B_0 \sim 1$, Fig. 19j-19n) of the X line. The events showed anisotropic electron VDFs consistent with CGL-like models (Le et al. 2009; Egedal et al. 2013). MMS data revealed some differences in electron temperature scaling (c.f. Figures 19g and 19m), attributed to upstream compression effects for the event on 21 January 2016.

Wang et al. (2023a) reported a large guide-field event $B_G/B_0 \sim 1.3$ occurring inside a magnetic island in the magnetotail. The observations showed field-aligned electron acceleration of 2 keV across the current sheet, corresponding to $e\phi/k_B T_e > 1$. On the high-potential side, the electron VDF was isotropic, while on the low-potential side, it was anisotropic with $T_{e\parallel} > T_{e\perp}$. They thus suggested that the parallel electric field could sustain the parallel pressure gradient. Wilder et al. (2017) showed a large guide-field event with $B_G/B_0 \sim 1.3$,

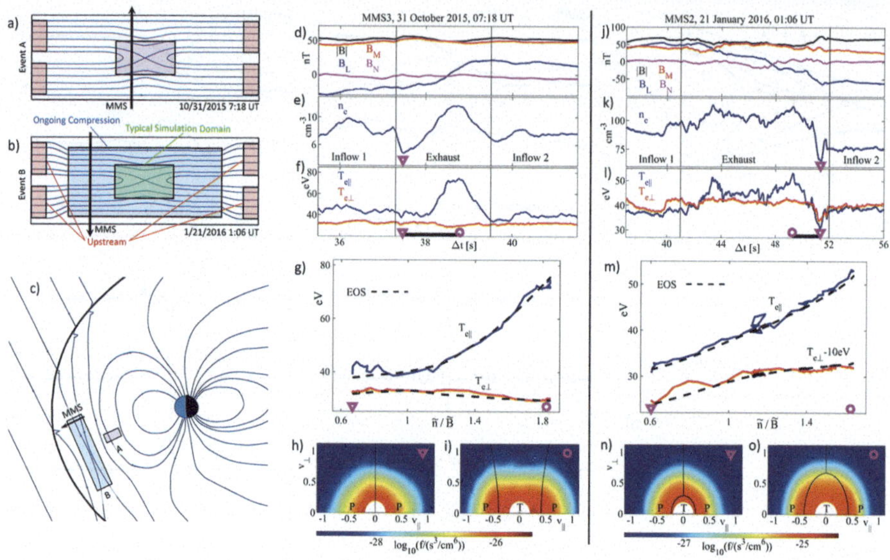

Fig. 19 The evolution of electron VDFs and temperatures in two guide field reconnection events, whereof one is showing evidence of being compressed. (a)-(b) Schematic overview of the events. (c) Locations of events. (d) and (j) magnetic fields; (e) and (k) density, which exhibits a density cavity at one of the two separatrices; (f) and (l) parallel and perpendicular electron temperatures. (g) Evolution of $T_{e\|}$ and $T_{e\perp}$ with $\tilde{n}/\tilde{B}$ in a region between an electron density cavity (marked by the magenta triangles) and peak (marked by the magenta circles), as well as the predictions of the equation of state (EoS) in the dashed lines. (h) and (n) show VDFs measured in the density cavities, while (i) and (o) show the VDFs at the density peaks. The black lines indicate boundaries between trapped and passing populations, where vertical lines at $v_{||} = 0$ separate passing populations coming from either side of the flux tube. Figure adapted from Wetherton et al. (2021). The event on 21 January, 2016, was previously published by Eastwood et al. (2018)

where the perpendicular temperature was more or less constant throughout the crossing, and the parallel temperature was slightly elevated in the vicinity of the current sheet center.

Observations of guide-field reconnection show how the reconnection electric field often develops into a double layer – a localized intensification of the parallel electric field – which is associated with the generation of electrostatic solitary waves (e.g. Eriksson et al. 2016b; Wilder et al. 2018; Øieroset et al. 2021). Although the electric field magnitudes are sometimes comparable to instrument error levels, the repeated observations suggest this is a generic feature for guide field events. Observations indicated that the double layer develops along the separatrix with enhanced acceleration (Øieroset et al. 2021). The waves typically appear on the high-potential side of the integrated electric field (Wilder et al. 2017), i.e., towards the inhibited separatrix, and likely form as the electrons with enhanced acceleration that arrive from the other side of the current sheet mix with plasma from the inhibited side of the current sheet. Wilder et al. (2017) showed that the high-potential side was characterized by a higher parallel temperature seen as an anisotropic electron VDF extended along the direction of the magnetic field. On the low-potential side of the double layer, the electron VDF was colder and more isotropic.

3.3.2 Asymmetric Guide-Field Reconnection

Without a significant guide field, the electrons originating on the high-β side can gain access to the low-β side of the current sheet through the normal electric field and the iner-

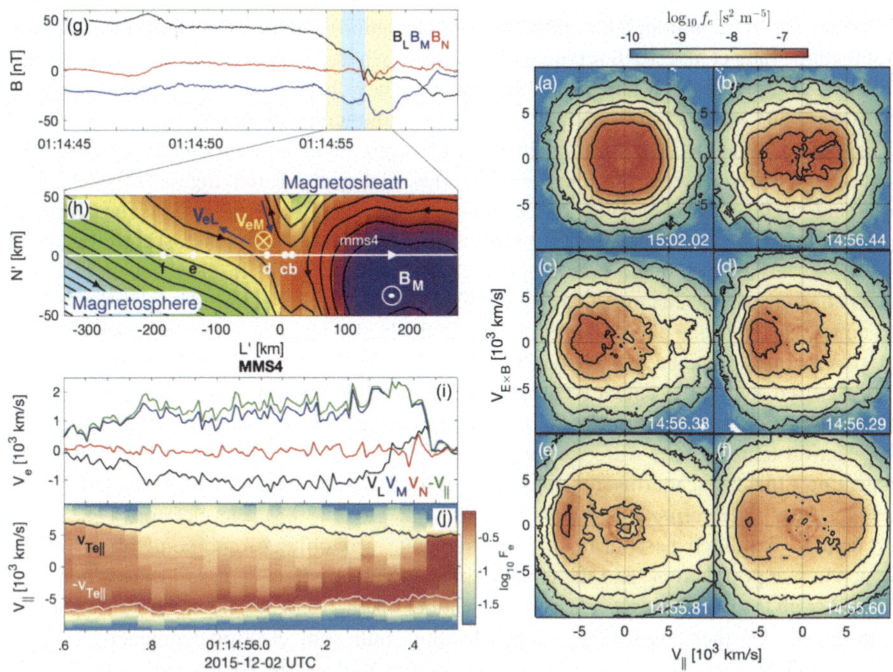

Fig. 20 Electron jet for a finite guide field (50% of the magnetospheric B_L and 100% of the magnetosheath B_L) asymmetric reconnection diffusion region. (a)-(f) shows reduced 2D VDFs $f_e(v_{\|}, v_{E\times B})$, while (g)-(j) provides the context. (a) Magnetosheath inflow. (b) Vicinity of X line. (c)-(e) Parallel reconnection jet. (f) Parallel reconnection jet and trapped electrons in the magnetospheric inflow. (g) Magnetic field. (h) Reconstructed reconnection topology, where the color indicates the amplitude of the magnetic vector potential and the contours represent the in-plane magnetic field. (i) Electron velocity and parallel speed. (j) Reduced electron VDF $f_e(v_{\|})$. Adapted from Khotyaintsev et al. (2020). Copyright (2020) by the American Physical Society

tia acquired on the low-β side. With a finite guide field, the increased magnetization limits this cross-field access (see e.g. Sect. 2.2.1 and Genestreti et al. 2017). The guide field also permits a parallel current to develop at the X line, facilitating the electron acceleration by the reconnection electric field (Genestreti et al. 2017). Therefore, in the presence of a guide field, the electron flow carrying the reconnection current becomes field-aligned (e.g. Khotyaintsev et al. 2020; Eriksson et al. 2016b). Additionally, at the magnetospheric side of the magnetopause reconnection region, the reconnection outflow jets become parallel (e.g. Burch et al. 2016; Khotyaintsev et al. 2016), even for negligible guide fields (Sect. 3.2.2). Figure 20 shows how the parallel electron jet evolves in space from the X line to the magnetospheric side of the magnetic field reversal (Khotyaintsev et al. 2020). Close to the X line, the jet is carried by a clear antiparallel beam that becomes successively accelerated as the spacecraft approaches the magnetosphere side of the X line and simultaneously moves in the downstream direction. During the continued acceleration, the beam gradually thermalizes by electrostatic waves generated by the beam. Further into the magnetosphere, the appearance of electrons also in the direction parallel to **B** (Fig. 20j) indicates the presence of trapped electrons, i.e., electrons that bounce back and forth along the magnetic field through a combination of parallel electric fields and the mirror force (see e.g. Sect. 2.3). Similarly, layered structures are also seen for other events (Wang et al. 2022a) and in simulations (e.g.

Egedal et al. 2016a, although for antiparallel reconnection) and therefore appear to be a general feature. Tang et al. (2020) reported observations where MMS crossed the magnetopause slightly further from the X line, inside the exhaust. They observed distinct populations moving toward and away from the X line. In this case, the temperatures of the beams seemed to indicate that they were both of magnetosheath origin, indicating mixing of plasma of the same origin but with distinct acceleration histories. The electrons moving towards the X line occupied slightly larger energies and a thinner pitch angle range of $\sim 30°$. The electrons moving away from the X line, likely arriving directly from the magnetosheath, occupied slightly lower energies and a broader pitch angle range of $\sim 60°$. In the reduced VDF, the beam character of the electrons leaving the X line was especially clear. The combination of asymmetries across the current sheet and guide fields also introduces asymmetries between the two exhaust (Pritchett and Mozer 2009).

Genestreti et al. (2017) made a comparative study of eleven asymmetric guide field events of varying magnetic shear (guide field strength). At high-shear events ($B_M/B_L^{sheath} < 0.2$), the electron temperatures were isotropic ($T_{e\parallel}/T_{e\perp} \sim 1$) at the center of the current sheet, indicating efficient scattering (e.g. Lavraud et al. 2016) or perpendicular acceleration of the electrons. At intermediate shears ($B_M/B_L^{sheath} \sim 0.2 - 0.3$), the electron VDF at the B_L reversal exhibited parallel beams (whereof one was oblique) and sometimes distinct perpendicular populations, and the temperatures were isotropic $T_{e\parallel}/T_{e\perp} \lesssim 1$. This indicates that although the electron motion has become more magnetized, as evidenced by the parallel beams at $B_L = 0$, there are still agyrotropic features, indicating electron demagnetization. At moderate shears ($B_M/B_L^{sheath} \sim 0.5 - 1$), the electron VDFs at the B_L reversal were strongly anisotropic ($T_{e\parallel}/T_{e\perp} > 1$) without distinct beams. Towards larger B, a parallel beam was present. At low shear ($B_M/B_L^{sheath} \sim 5$), the electron VDF was again anisotropic at the B_L reversal, and no parallel beam was observed at higher B. Although the upstream conditions varied for the studied events, one could conclude that increasing the guide field strength decreases the electron crescent features, consistent with theoretical predictions (Hesse et al. 2014, 2016). The particle dynamics were also connected to regions of energy dissipation. For large shears, $(\mathbf{E} + \mathbf{v_e} \times \mathbf{B}) \cdot \mathbf{J}$, had a single peak in the magnetosphere inflow, associated with agyrotropic VDFs. With decreasing shear, this peak decreased and eventually disappeared for large shears. With decreasing shear, another peak at the B_L reversal appeared, associated with parallel streaming along the guide field direction (see also Genestreti et al. 2025, this collection, for further discussion). This study illustrates the connection between electron dynamics and energy dissipation.

Large Guide-Field Reconnection Inside Kelvin-Helmholtz Vortices The Kelvin-Helmholtz instability evolves at low magnetic shears between the magnetosheath and the flanks of the magnetosphere. The instability results in vortices, which may compress the boundary layer and trigger magnetic reconnection at the trailing edges of the vortices (Eriksson et al. 2016a; Li et al. 2016; Hwang et al. 2023, this collection). Eriksson et al. (2016a) reported the first observations of magnetic reconnection in this environment. They identified evidence of reconnection exhaust at 22 of 42 compressed Kelvin-Helmholtz-related current sheets. The reconnection event was identified by ion and electron outflow jets. The observations also revealed the electron separatrix flows toward the X line. The Hall electric and magnetic structures were asymmetric, as expected from large guide fields and weak pressure asymmetries. For one of the same events and another one, Li et al. (2016) reported kinetic evidence of magnetic reconnection in terms of plasma mixing. They showed that high-energy magnetosphere electrons and lower-energy magnetosheath electrons mixed along open field lines.

4 Ion Dynamics

The unprecedented capabilities of MMS to resolve the electron and ion diffusion regions have opened new windows to examine how ions are accelerated in the IDR (Oka et al. 2023, this collection), how ion outflow jets are formed in the close vicinity of the reconnection X line, and how ions may couple with electrons through fields of appropriate scales to cause effective resistivity (Graham et al. 2025, this collection). Analysis of the ion dynamics in the inner IDR reveals that the dominant ion acceleration is by the in-plane electric field. This same physics operates in laboratory reconnection with asymmetric upstream conditions similar to those at the magnetopause, albeit the physical system size differs by six orders of magnitude (Yamada et al. 2018).

While the ions experience the same electric and magnetic field as the electrons, they are affected differently due to their larger mass and opposite charge (although some behaviors are qualitatively similar to the electrons). As a consequence, ions become demagnetized over much larger spatial scales. The ion dynamics surrounding a reconnection site are generally complex, and the ion VDFs are often far from local thermodynamic equilibrium. VDFs often contain multiple populations (Goldman et al. 2021) and are often both anisotropic and non-gyrotropic (e.g. Nagai et al. 2015; Wang et al. 2016c).

For a more extensive review of the impact and role of ions in magnetic reconnection, for example, regarding their effect on the reconnection rate, see Liu et al. (2025b, this collection, and references therein).

Transition from Electron to Ion Diffusion Region The IDR forms over larger spatial scales than the EDR. In some cases, the ions remain unmagnetized over most of the reconnecting system. This happens in regions where the magnetic topology and associated plasma hinder the formation of a magnetized ion outflow, for example, in the turbulent magnetosheath (Yordanova et al. 2016; Phan et al. 2018) or in larger scale reconnection exhausts (Norgren et al. 2018; Huang et al. 2021). Øieroset et al. (2021) showed, for one event, how the ion jet began to emerge over a distance of $2d_i$ outside of the EDR in conjunction with a broadening and slight bifurcation of the current sheet. Particle-in-cell simulations have shown that the formation of an ion jet is largely dependent on the system size (Pyakurel et al. 2019). The ion outflow jet reached Alfvénic speeds when the system size reached $\sim 80d_i$. The Hall fields, which are a result of the distinct motions of ions and electrons as the ions become demagnetized, appear at roughly the ion gyroradius scale (Stawarz et al. 2021).

4.1 Antiparallel Symmetric Reconnection

The first observations of ion dynamics related to magnetic reconnection in the tail were Earthward field-aligned ion fluxes in the plasma sheet boundary layer. Moebius et al. (1980) observed a highly collimated field-aligned flux, which they concluded was consistent with acceleration by a cross-tail electric field near a steady X line. For another event, Forbes et al. (1981) deduced that the source of acceleration was a tailward retreating X line. The observations of velocity dispersion (Forbes et al. 1981; Andrews et al. 1981; Takahashi and Hones 1988), where the highest-energy ions were observed further out on the lobe side of the boundary layer, were found consistent with a steady acceleration region. The spatial velocity dispersion, which is an effect of plasma convection, was strong evidence of plasma flow across magnetic separatrices. Another important observation was done by Baumjohann et al. (1990), who reported the presence of super-Alfvénic ion jets in the central plasma sheet.

A few basic mechanisms, or combination of forces, govern the ion motion and dynamics inside the reconnecting current sheet: the reconnection electric field, which does work on the ions and accelerates them in the out-of-plane direction; the reconnecting magnetic field that confines the ions around the current sheet midplane through gyroturning and leads to meandering motion across the field reversal; the Hall electric field, which forms a potential well that acts to confine the ions to the current sheet and accelerates them downstream; the gyroturning in the reconnected field component, which leads to the formation of the outflow jets; and the Hall magnetic fields, which aids in the formation of the outflow jets.

The ion motion inside a reconnecting current sheet can roughly be divided into three basic categories depending on the ratio between their bounce frequency in the reversing magnetic field and their gyrofrequency in the normal magnetic field component (Nakamura et al. 1998): Speiser, chaotic, and magnetized motion. In the chaotic region, where the bounce frequency is comparable to the gyro frequency, the ions are efficiently scattered by the curved magnetic field and become gyrotropized (e.g. Nakamura et al. 1998; Richard et al. 2023). In the region where the gyrofrequency is significantly larger than the bounce frequency, the ions remain magnetized for an extended time and can traverse the field reversal like pearls on a string (Nagai et al. 2002; Xu et al. 2019; Norgren et al. 2021). Outside of the IDR, where $\mathbf{E} + \mathbf{v}_i \times \mathbf{B} = 0$, the electric field in the frame of the ion bulk motion is zero, and the energy of the ions is thus conserved. Here, the Speiser motion manifests as a superposition of populations placed on a half circle centered on the bulk speed (which coincides with $\mathbf{v}_{E\times B}$) (Burkhart et al. 1992; Nakamura et al. 1998; Hietala et al. 2015). Inside the IDR, where $\mathbf{E} + \mathbf{v}_i \times \mathbf{B} \neq 0$, the ion energy in the ion bulk frame is not conserved, and the ion motion, which can still be well represented by Speiser motion, can not be described by the conservation of energy alone. The magnetized, Speiser, and chaotic populations often correspond to different phases of the ion motion and can coexist. For example, ions that are just about to enter the current sheet and still retain their 'inflow characteristics' coexist with ions that have already been undergoing meandering motion and have become accelerated and potentially thermalized (e.g. Nagai et al. 2015; Wang et al. 2019a; Norgren et al. 2021).

In the sections below, we will elaborate on these processes. We begin by reviewing the effect of the Hall fields and the reconnection electric field within the IDR. Then, we cover the phase mixing of Speiser orbits in the non-chaotic region and how the ions can be incorporated into, or *picked up* by, the moving exhaust downstream of the IDR. Finally, we cover the heating of ions during reconnection and the dynamics of both heavy ions and cold ions more generally.

4.1.1 The Ion Diffusion Region

Inside the IDR, the ions are strongly affected by both the Hall fields set up by the decoupled motion of ions and electrons and the reconnection electric field. The Hall electric field, $\mathbf{E} = \mathbf{J} \times \mathbf{B}/ne$, has a strong E_N component converging on the current sheet midplane and a weaker E_L component directed away from the X line. It thus forms a potential well that becomes deeper further away from the X line as shown by observations[2] (Wygant et al. 2005), laboratory experiments (e.g. Ji et al. 2023, this collection), and simulations (e.g. Fujimoto 2014). The ratio of magnitudes of E_N and E_L is related to the current sheet aspect ratio. In the IDR, the demagnetized ions become ballistically accelerated by E_N towards the current sheet midplane (e.g. Wygant et al. 2005; Nagai et al. 2015; Divin et al. 2016; Tenfjord et al. 2018). The inward motion leads to counter-streaming populations along v_N, as

[2] Wygant et al. (2005) observed a potential well of $4-6$ keV closer to the X line and $\simeq 8$ keV further downstream.

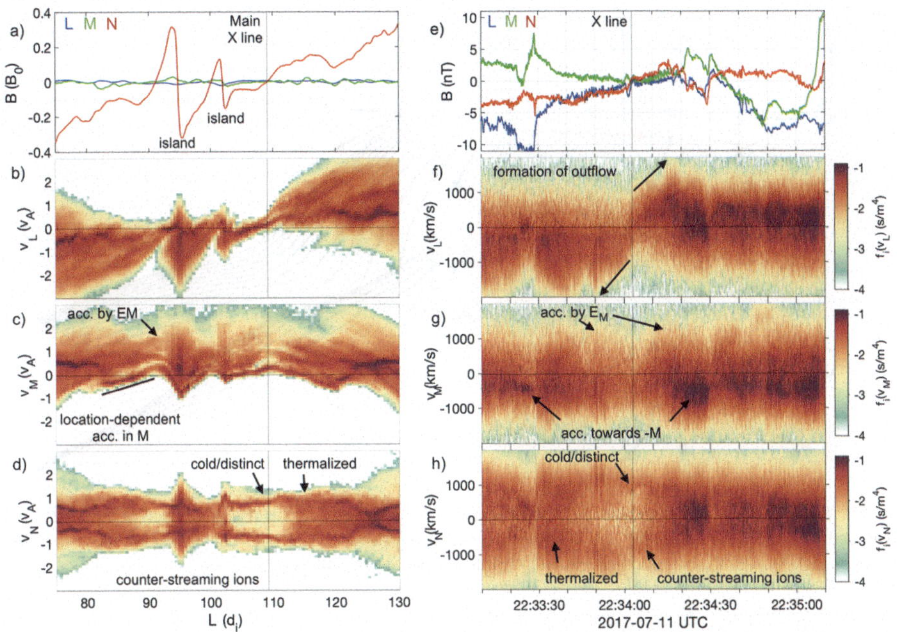

Fig. 21 Reduced ion VDFs $\log_{10} f_i$ around the X line from (a)-(d) a particle-in-cell simulation (at $N = 0$) and (e)-(h) MMS observations from the magnetotail following a trajectory from the tailward to the Earthward exhaust (see e.g. Torbert et al. 2018). (a) and (e) magnetic field. (b) and (f) $f_i(v_L)$, which shows the ion outflow. (c) and (g) $f_i(v_M)$, which shows the ions in the direction of the reconnection current. (d) and (h) $f_i(v_N)$, which shows the ions in the direction normal to the current sheet. The coordinates for the MMS observations are given by $\mathbf{L} = [0.89, -0.45, -0.00]$, $\mathbf{M} = [0.45, 0.89, 0.06]$, $\mathbf{N} = [0.03, 0.05, 1.00]$ (GSE). Simulation data adapted from Norgren et al. (2021)

demonstrated in observations (e.g. Wygant et al. 2005; Nagai et al. 2015; Huang et al. 2012) and simulations (e.g. Arzner and Scholer 2001; Aunai et al. 2011; Divin et al. 2012; Liu et al. 2022). These counter-streaming ions move mainly perpendicular to the local magnetic field, in contrast to counter-streaming ion populations observed in the far exhaust that move parallel to the local magnetic field (e.g., close to dipolarization fronts Nagai et al. 2002; Xu et al. 2019). Figure 21d shows the counter-streaming ion populations from a particle-in-cell simulation initiated with cold inflow ions corresponding to a lobe population (Norgren et al. 2021). The counter-streaming ions are clearly distinct in the vicinity of the X lines and thermalize further downstream and in the two magnetic islands. In Fig. 21h we compare the simulations to observations from the EDR and IDR encounter on 11 July 2017. Also, here, the counter-streaming ions are most distinct in the vicinity of the X line and blend further downstream.

Both observations (Nagai et al. 2015) and simulations (Cheng et al. 2015; Divin et al. 2016) have shown that the relatively cold counter-streaming ions can be moving along $-\hat{e}_M$, i.e., in the general direction opposite to the reconnection electric field, but in the direction of the out-of-plane $\mathbf{E} \times \mathbf{B}$ drift. This has been explained by gyroturning in the reconnection magnetic field B_L. During their motion inward along $\hat{e}_N$, the ions are subject to a magnetic force $-ev_N B_L \hat{e}_M + ev_N B_M \hat{e}_L$, where B_M is the quadrupolar Hall magnetic field. This is illustrated in Fig. 22. The $\hat{e}_M$ component of this force will accelerate the ions toward $-\hat{e}_M$, i.e., in the direction opposite to the reconnection electric field $E_M > 0$ (Cheng et al. 2015;

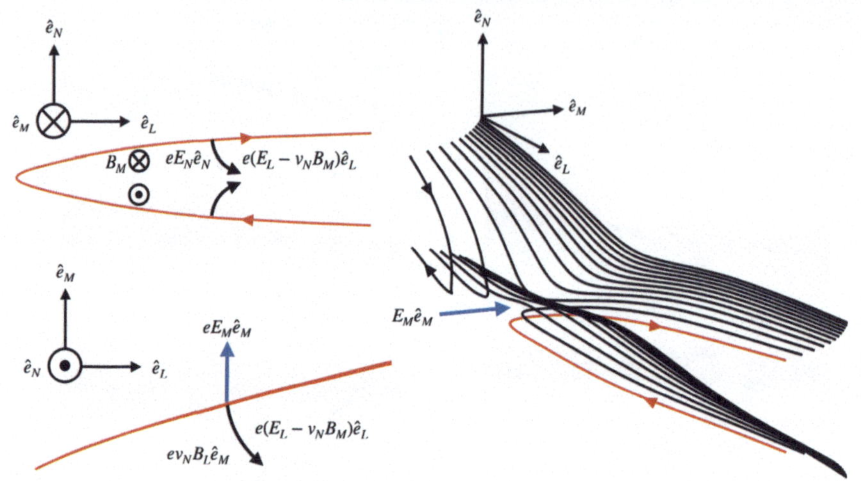

Fig. 22 Illustration of how the Hall field affects how ions are injected into the reconnection exhaust. The demagnetized acceleration by the Hall fields directs the ions toward the outflow. This kickstarts the formation of the outflow before the acceleration by eE_M followed by $ev_M B_N$ takes over

Divin et al. 2016). The total force during the inward motion is $(eE_M + ev_N B_L - ev_L B_N)\hat{e}_M$, and the resulting v_M depends on the depth of the Hall potential well, E_M, and the thermal streaming along $\hat{e}_L$ upstream of the IDR. Considering a steady state, where E_M is constant in the vicinity of the X line, simulation results show that the relative effect of $ev_N B_L$ is larger further away from the X line, where the potential well is deeper, and, consequently, v_N is larger (Divin et al. 2016; Wygant et al. 2005). This is seen in Fig. 21c, where the speed of the population at most negative v_M varies with distance from the X line. In the observations (Fig. 21g), a clear population is also seen at $v_M < 0$, although it seems to have a roughly constant speed of $v_M \sim -500$ km/s throughout the entire interval. While both observations and simulations show a population at $v_M < 0$, and thus are consistent, the differences in spatial variations of its speed remain to be explored. Further out along the separatrices, the ion motion along $-\hat{e}_M$ is not demagnetized, but follows $\mathbf{v}_{E\times B}$ (Alm et al. 2018; Kamaletdinov et al. 2025).

In the outflow direction, both components of the Hall forces $e(E_L - v_N B_M)\hat{e}_L$ will accelerate the ions away from the X line. Drake et al. (2009b) showed that the combination of acceleration by $eE_N\hat{e}_N$ followed by $-ev_N B_M\hat{e}_L$ directed the ions downstream. However, they also stressed that since the Hall field $E_N\hat{e}_N$ arises from a potential, it should not be responsible for the net energization of the ions. That is, the large energy gained while approaching the current sheet midplane will be returned at the corresponding mirror point at the other side of the current sheet. This was also shown by Liu et al. (2015), see Fig. 23, where the work done by the electric fields along a proton trajectory is shown in Fig. 23c. While the initial work is done by E_N, W_{E_N}, the net W_{E_N} is close to zero and is overtaken by W_{E_L} after two bounces across $N = 0$. In this case, the work done by the reconnection electric field W_{E_M} is low due to the relatively small displacement along $\hat{e}_M$ over the presented time interval. Aunai et al. (2011) also stressed the importance of $eE_L\hat{e}_L$ in doing a net work on the ions. By integrating forces along ion fluid streamlines, they showed that $enE_L\hat{e}_L$ dominated. The divergent E_L has been identified in observations, although the authors could not unambiguously conclude its effect (Nagai et al. 2015).

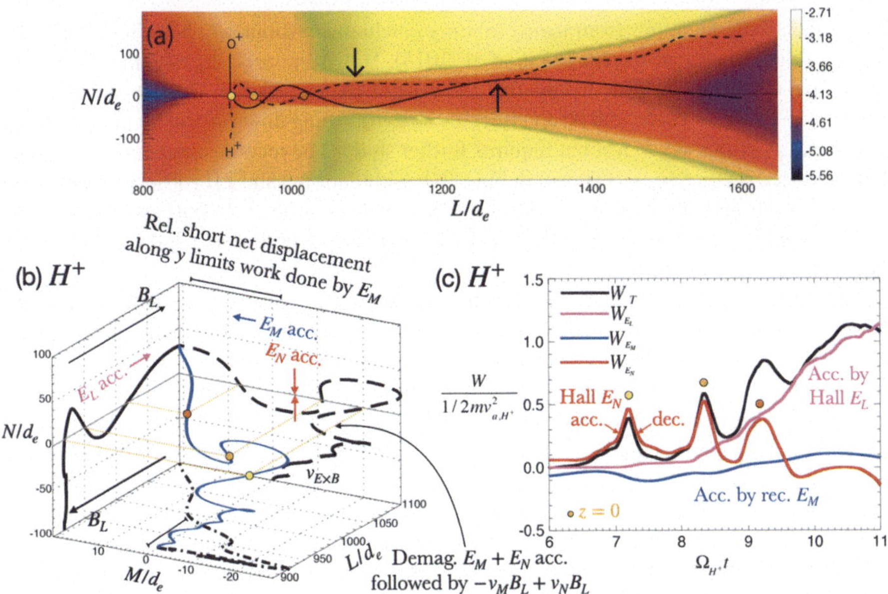

Fig. 23 Work done by the electric fields along a proton trajectory in a 2D particle-in-cell simulation. (a) Electrostatic potential and proton and oxygen ion trajectories. (b) 3-dimensional proton trajectory (blue), with projections on each plane (black). (c) Work done on the proton as a function of time. The yellow/orange circles mark points in (b) space and (c) corresponding times where the proton crosses $N = 0$. Annotated and adapted from Liu et al. (2015)

Note that the effects of the Hall fields inside the IDR is to induce an ion motion similar to that of the magnetized electrons. For instance, the gyroturning caused by $ev_N B_L \hat{e}_M$, which is enabled by the acceleration by the Hall electric field $E_N \hat{e}_N$, accelerates ions towards $-\hat{e}_M$, i.e. in the direction of the electron flow carrying the reconnection current. Additionally, the ions are also partially gyroturned by $-ev_N B_M \hat{e}_L$ (Drake et al. 2009b) or accelerated by $eE_L \hat{e}_L$ (Aunai et al. 2011) away from the X line. The Hall fields thus help to kickstart the formation of the ion outflow, so that they can better 'keep up' with the electrons as they move away from the X line. The question of when the large-scale formation of the outflow jets by $eE_M \hat{e}_M \rightarrow v_M \hat{e}_M \rightarrow ev_M B_N \hat{e}_L$ takes over requires further investigation.

Numerical simulations have demonstrated that the Hall term facilitates rapid reconnection (Birn et al. 2001). Liu et al. (2022) attributed this to a mechanism whereby the Hall-associated Poynting flux is deviated around the X line, creating an energy void, which pinches the current sheet and opens the exhaust, thus leading to faster reconnection rates. Another, potentially connected, reason for the relation between fast rates and the Hall fields could be their role in the initial formation of the ion outflow.

Just like the electrons, the ions are confined to the current sheet midplane by the magnetic force $-ev_M B_L \hat{e}_N$. The reconnection electric field along the current direction accelerates ions during the meandering motion. Simulations show that this leads to a pointed triangle structure in $f_i(v_L, v_M)$ that becomes thinner towards increasing v_M (e.g. Zenitani et al. 2013; Cheng et al. 2015). The base of the triangle consists of ions just arriving from the inflow, while the tip of the triangle consists of ions having undergone acceleration by E_M during their meandering motion. In the work by Zenitani et al. (2013), the base of the triangle is shifted to $v_M < 0$ while in the work by Cheng et al. (2015), the base begins at $v_M \gtrsim 0$. Moving away from the X line toward the outflow, the cyclotron turning around B_N by $ev_M B_N \hat{e}_L$

rotates the accelerated ions from $v_M \hat{e}_M$ to $\pm v_L \hat{e}_L$, which contributes to building up the outflow jet (e.g. Zenitani et al. 2013; Liu et al. 2015). However, only slightly downstream of the X line, Cheng et al. (2015) show that the base of the distribution also has a finite drift towards the outflow. This could be an effect of the gyroturning downstream by the Hall magnetic field, as discussed above, but requires further study. The reconnection outflow, as seen in the reduced VDF $f_i(v_L)$, is shown in Figs. 21b (simulations) and 21f (observations). Both the simulations and observations demonstrate that the formation of the outflow is concurrent with a thermalization process, i.e., the population expands to higher $|v_L|$ at the same time as they cover a larger range of v_L.

Just upstream of the X line, simulations also show the presence of ion crescent populations (Cheng et al. 2015). These are also seen close to the X line in the magnetotail, where they are accompanied by counter-streaming cold ion populations (Nagai et al. 2015; Wygant et al. 2005). This indicates that both the acceleration by the Hall electric field and the reconnection electric field are at work to energize ions there. Interestingly, these counter-streaming ion beams that penetrate the midplane along the inflow direction cause the enhancement of the effective ion pressure $\Delta P_{i,NN}$ at the X-line. However, due to the Hall effect, this $\Delta P_{i,NN}$ is not large enough to balance the upstream asymptotic magnetic pressure, leading to fast reconnection in collisionless plasmas (Liu et al. 2022).

Aunai et al. (2011) showed how $f_i(v_L, v_N)$ consisted of counter-streaming populations in v_N that converged towards larger v_L. By tracing test particles backward in time, they concluded that the ions with higher v_L had undergone more bounces. Similar distributions were observed in the magnetotail inside an IDR (e.g. Zhou et al. 2019b). The observations showed that the two convergent populations also seemed to merge at $v_N = 0$ at the higher-v_L end, thus forming a half circle in the $v_L v_M$ plane.

4.1.2 Phase Mixing and Thermalization Inside the Exhaust

Inside the IDR, the ion distributions will consist of a superposition of different phases of the Speiser motion from ions entering the exhaust at different distances from the X line. Nakamura et al. (1998) showed how the resulting distribution consisted of distinct colder populations distributed along a half circle centered on the velocity of the moving exhaust. Each progressive population along the circle had undergone one more crossing of the current sheet midplane. These half-circle distributions have also been observed in self-consistent particle-in-cell simulations and observations (Hietala et al. 2015).

At intermediate magnetic field curvature, when the bounce time in the B_L reversal is comparable to the gyration time scale around B_N, the ion motion becomes chaotic, and the distribution is thermalized (Nakamura et al. 1998). This is likely to occur somewhere downstream of the IDR, where the magnetic curvature decreases due to the release of the magnetic tension (B_L) and pile-up of the field (B_N) behind the jet front. In addition, the bounce frequency increases with the particle speed (the bounce frequency is roughly $\omega_b \sim \sqrt{eB_0 v/mL}$, for $B_L = B_0 N/L$, see Speiser 1965 or Zenitani and Nagai 2016) which increases during the reconnection process. The thermalization of ion distributions around the current sheet midplane was clearly illustrated by Li et al. (2021b), see Figure 24. They quantified ion VDFs observed inside an ion jet and found counter-streaming distributions in which a colder component typically approached the current sheet midplane while a hotter component was leaving it. In a region centered around $B_L \sim 0$, they found a hot population with or without two counter-streaming cold populations.

With a particle-in-cell simulation, Norgren et al. (2021) illustrated the phase mixing of ion trajectories in the reduced distribution $f_i(L, v_M)$, which can be seen in Fig. 21c. Different phases can be identified as distinct striations. The superposition of various phases occurs

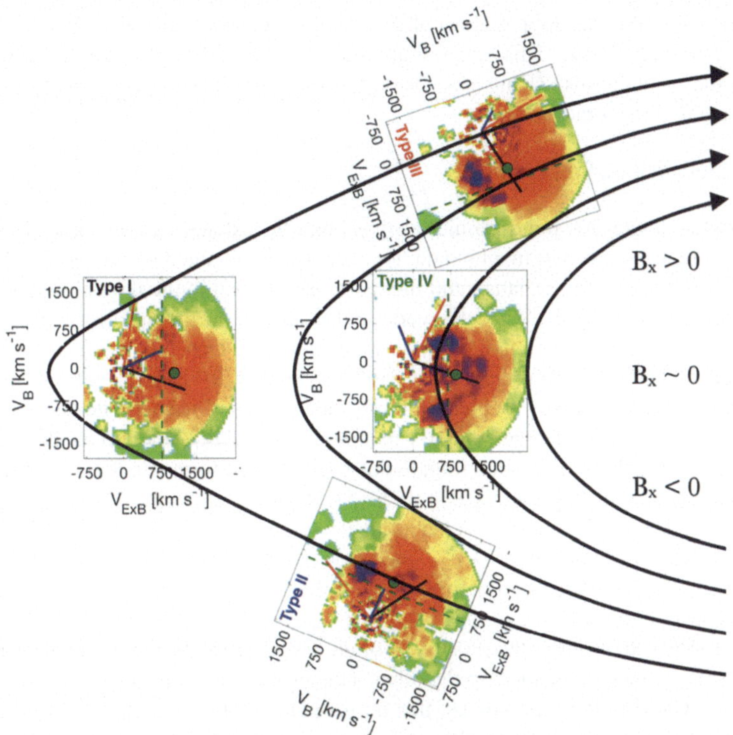

Fig. 24 Characteristic ion VDFs inside a magnetotail reconnection jet. In the off-equatorial region, most often, cold populations are observed approaching the midplane while hotter populations are seen leaving it (Type II and III). Close to the midplane, either a completely thermalized population (Type I) or a three-population VDF with two counter-streaming beams and a more thermal population (Type IV) are most commonly observed. Note that the study did not quantify the distribution of different types of VDFs in the direction of the outflow (i.e., Type IV was not necessarily observed closer to the jet front than Type I). Figure adapted from Li et al. (2021b)

as ions enter the exhaust at different distances from the X line. We note that these striations are not visible in $f_i(v_M)$ from observations in Fig. 21g. This can indicate that additional thermalization effects are in effect in the observations or that the initial temperature of the observations is too high to resolve the individual phases. Norgren et al. (2021) also showed how some thermalization already occurred as the ions crossed the active separatrices characterized by density cavities and parallel electric fields. The thermalization at the separatrices was shown with observations at the dayside magnetopause (Toledo-Redondo et al. 2016b).

Based on a statistical analysis of reconnection jets downstream of the IDR in the magnetotail, Richard et al. (2023) concluded that the isotropization of the ion distribution was due to chaotic and quasiadiabatic ion motion. They showed that the temperature anisotropies inside the jets were well bounded by temperature anisotropy-driven instabilities. However, they found that the instability-generated waves were not able to efficiently isotropize the VDF. Instead, they showed that the pitch-angle scattering due to chaotic motion is much faster and concluded that this is the primary mechanism of ion isotropization in the reconnection outflow.

Further downstream, the hot, kinetic ions can leak across the exhaust boundary to the upstream (Phan et al. 2022; Zhang et al. 2019; Liu et al. 2012), causing the temperature anisotropy within the transition regions of slow shocks and/or rotational discontinuities predicted by Petschek (1964). This temperature anisotropy modifies solution of the Rankine-Hugoniot jump conditions (Zhang et al. 2019; Liu et al. 2012, 2011), coupling the slow and intermediate nonlinear modes.

4.1.3 The Ion Pick-up Process

Neglecting the details, the ion dynamics as they enter the exhaust can be relatively well described by a pick-up process, in which the ions are first accelerated by the motional electric field $E_M \hat{e}_M$ of the moving exhaust, and thereafter gyroturned in the normal magnetic field $B_N \hat{e}_N$ (Burkhart et al. 1992). Since the work on the ions is done by the electric field $E_M \hat{e}_M$, the process is associated with a finite displacement along $\hat{e}_M$. Depending on the scattering by the magnetic field during this process, they can either escape downstream or become trapped in the exhaust and isotropized by phase mixing and continued magnetic field scattering before they eventually leave the region (e.g. Nakamura et al. 1998; Zelenyi et al. 2013; Richard et al. 2023). In the absence of scattering, the distribution can form a half circle in the $v_M v_L$ plane (Nakamura et al. 1998; Hietala et al. 2015). However, at the most basic level, the acceleration can be described and modeled as an elastic reflection in the frame of the moving exhaust, i.e., $\mathbf{v}_{E\times B}\hat{e}_L$. That is, $\mathbf{v}_i^{t2} - \mathbf{v}_{E\times B} = -(\mathbf{v}_i^{t1} - \mathbf{v}_{E\times B})$, such that $\mathbf{v}_i^{t2} = -\mathbf{v}_i^{t1} + 2\mathbf{v}_{E\times B}$, where t_1 and t_2 refers to the times before and after the reflection, respectively. The energy of the ion is constant in the frame of the exhaust ($\mathbf{v}_{E\times B}$), leading to the half-circle distribution mentioned above. In reality, this process is often associated with the crossing of the ions from one side of the current sheet to the other. This process is often called Fermi sling-shot acceleration. Close to the separatrices, this out-of-plane motion is redirected to the downstream direction and becomes field-aligned (e.g. Eastwood et al. 2015). In the guiding center theory, the associated movement along $\hat{e}_y$ is due to the curvature drift.

During their traversal of the current sheet midplane, ions are often thermalized (e.g. Li et al. 2021b), likely due to scattering in the curved magnetic field. The accelerated population is therefore typically much hotter than the non-accelerated one (Cowley and Shull 1983; Li et al. 2021b), although they are typically well separated in $v_\parallel$. Since the ions are initially approximately stationary, and the accelerated ones have a speed $2v_{E\times B} \sim 2v_A$ (assuming the asymptotic outflow speed is roughly the Alfvéen speed), their parallel temperature can be approximated as $T_{i\parallel} \sim m_i v_A^2$ (consider e.g. two drifting populations represented by the delta functions $f_{i1} = n_1\delta(v_\parallel)$ and $f_{i2} = n_2\delta(v_\parallel - 2v_A)$. Drake et al. (2009b) also took into account perpendicular temperatures and found the same parallel temperature but a slightly smaller total temperature $T_i = T_{i\parallel} \sim m_i v_A^2/3$.

The acceleration in the direction normal to the current sheet, which is a part of the sling-shot acceleration, can be due both to eE_N (active inside the IDR) and $-ev_M B_L$ where v_M is acquired through eE_M (active in principle everywhere, but should be dominant outside the IDR). In the case of the latter, first, the magnetic force induces motion towards the current sheet center, and after crossing the magnetic field reversal, it decelerates the particles again (Eastwood et al. 2015). At the current sheet midplane, two populations originating from opposite sides of the current sheet would, therefore, be seen as counter-streaming populations in the direction parallel to the magnetic field, with a finite drift in the perpendicular direction of $v_{E\times B}$ (i.e., co-moving with the exhaust). At the other end of the current sheet, when the ions exit, the speed is often parallel to the magnetic field. The 'parallel acceleration' is due to the turning of the magnetic field (Northrop 1963).

Downstream of the IDR, closer to the jet front in regions of strong B_N, ions can remain relatively magnetized throughout this process. There, the moving reconnection exhaust expands into the downstream population. Depending on the source population, for example,

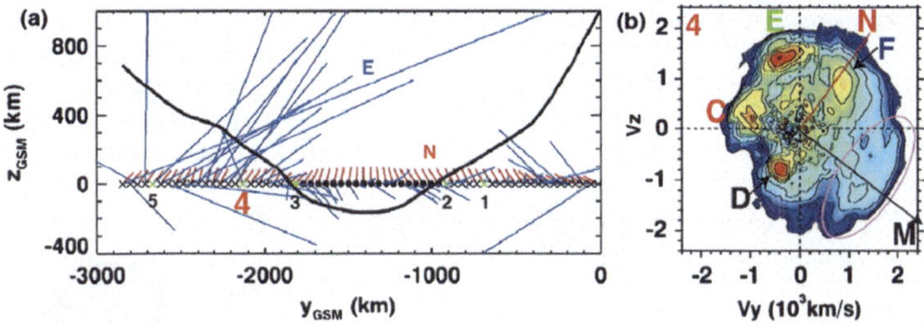

Fig. 25 Ion dynamics in the diffusion region of a flapping current sheet. (a) The structure of the flapping current sheet, where the Hall electric field, is mainly along the local normal direction. (b) Ion distribution from location 4 shows the mixing of multiple populations. Adapted from Wang et al. (2019a)

the hot pre-existing plasma sheet or colder inflow plasma of lobe and/or ionospheric origin, the reflected population is hot (Eastwood et al. 2015) or cold (Xu et al. 2019).

4.1.4 Non-laminar Reconnection

Magnetic reconnection current sheets are often non-laminar, exhibiting turbulence (e.g. Stawarz et al. 2024, this collection, and references therein), the production of islands (e.g. Stawarz et al. 2018), and current sheet corrugation, or flapping motion. The period of the flapping motion can range from a few seconds to a few minutes, with a propagation speed of a few hundred km/s and a wavelength of a few thousand km (corresponding to a few d_i) (e.g. Zhang et al. 2002; Sergeev et al. 2003; Wang et al. 2019a; Wei et al. 2019; Richard et al. 2021). The current sheet flapping is consistent with the kink mode instability (e.g. Karimabadi et al. 2003), and it makes the current sheet corrugated in the yz plane (in GSM coordinates). MMS observations show how ions respond to the flapping current sheets. Around the reconnection diffusion region, ions are still demagnetized, and their decoupling with electrons generates Hall electric fields mostly along the local current sheet normal direction that varies over space (Wang et al. 2019a; Richard et al. 2021). In an examination of an event by Wang et al. (2019a), the VDFs showed complicated structures with multiple populations organized by the local current sheet orientation (Fig. 25). In Fig. 25, one population forms a crescent shape at large speeds along the local M direction of the flapping current sheet, which indicates its meandering motion and possible acceleration by the reconnection electric field. The additional multiple cold beams, when analyzed together with a series of VDFs in different parts of the flapping current sheet, can be understood as becoming accelerated by the Hall electric field along the local normal direction where they enter the current sheet. These beams then move to locations with varying current sheet orientations to mix with one another. Compared to VDFs in a more laminar current sheet, e.g., Fig. 28, the flapping enables the mixture of more ion populations, enhancing the effective heating and thermalization.

4.1.5 The Dynamics of Heavy Ions

Owing to their larger mass, heavy ions typically remain unmagnetized over longer temporal and spatial scales. Cluster observations from the magnetotail showed counter-streaming oxygen populations within a reconnecting current sheet that were accelerated from opposing inflow regions by the Hall electric field converging on the neutral line (Wygant et al. 2005). Liu et al. (2015) presented both particle-in-cell simulations and Cluster observations from

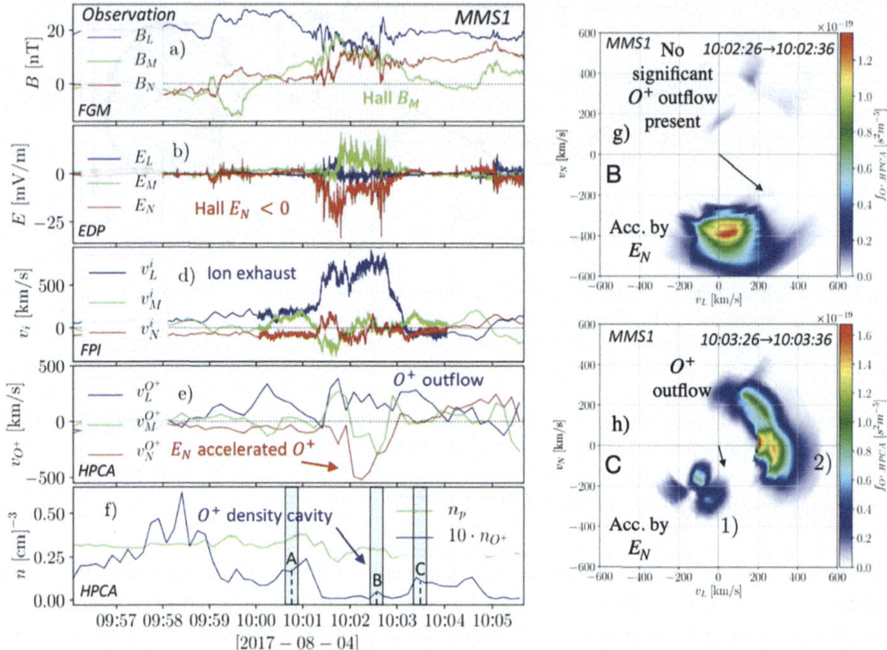

Fig. 26 Observation of ion dynamics inside an oxygen diffusion region. (a) Magnetic field. (b) Electric field. (d) Ion velocity from FPI. (e) Oxygen velocity from HPCA. (f) Proton and oxygen ion density. (g)-(h) Reduced oxygen ion distributions at the two times marked by B and C in panel (f). An oxygen ion flow $v_N^{O^+} < 0$ (e) towards the current sheet center is carried by a single superthermal population (g). Adjacent to this flow, the inward-moving oxygen ions are accompanied by a hotter population moving in the outflow direction (h). Figure adapted from Kolstø et al. (2021)

the magnetotail. The simulation showed how the finite gyroradius effects of the meandering oxygen ions extended further upstream into the inflow than the proton counterpart. The observations showed how the oxygen distribution remained relatively cold in locations where the protons already exhibited phase mixing and thermalization. Similarly, oxygen energization by several keV by the Hall electric field was reported from the distant dayside separatrix region (Lindstedt et al. 2010).

Particle-in-cell simulations have demonstrated that the ballistic acceleration of oxygen ions by the Hall electric field occurs over large scales (Tenfjord et al. 2018), extending even downstream of dipolarization fronts (Liang et al. 2017). According to Tenfjord et al. (2018), the initial acceleration caused by the expanding Hall fields creates a density enhancement front, termed an 'oxygen wave', which expands into the exhaust, followed by a density cavity. These two regions exhibit distinct distributions and were observed by MMS in the magnetotail (Kolstø et al. 2021, see Fig. 26). At the location labeled B in Fig. 26, the O^+ VDF is characterized by a single population moving towards the current sheet midplane, accelerated by the Hall electric field E_N. In contrast, an additional population is present at the locations labeled A (not shown) and C. This second population has a distinct speed in the outflow direction and is more thermalized, which is a characteristic of the ion outflow (e.g. Cowley and Shull 1983).

Early observations of Earthward field-aligned fluxes of protons and helium ions in the plasma sheet boundary layer identified distinct proton and helium ion edges (Moebius et al.

1980). The helium ion fluxes appeared equatorward of the proton fluxes, which was attributed to a quicker ejection of the lighter protons from the X line region. The observations also identified a duskward flux attributed to a density-gradient drift at the boundary layer.

4.2 Asymmetric Reconnection

At the magnetopause, the magnetosheath is characterized by warm (~ 500 eV), dense ($n \sim 10 - 100\,\text{cm}^{-3}$) ions, while the magnetosphere is characterized by hot (several keV), tenuous ($n \sim 1\ \text{cm}^{-3}$) ions, and sometimes (depending on magnetospheric activity) cold ions (~ 1 eV) of varying densities ($n \sim 0.1 - 50\ \text{cm}^{-3}$) (e.g. Toledo-Redondo et al. 2021). When these populations mix and are accelerated during the reconnection process, they form yet other populations with potentially distinct characteristics.

4.2.1 The Effects of the Hall Fields and Finite Larmor Effects

During magnetopause reconnection, the magnetosheath particles typically dominate the thermal pressure to offset the stronger magnetic field pressure on the magnetosphere side of the current sheet. The overall ion pressure gradient leads to a strong Hall electric field E_N located on the magnetospheric side of the current sheet and pointing towards the magnetosheath (e.g. Pritchett 2008; Shay et al. 2016; Graham et al. 2016a). On the magnetosheath side of the current sheet, the Hall electric field is typically smaller and is instead directed toward the magnetospheric side. Similar to symmetric reconnection, the effect of the Hall electric field is to aid the ions in rejoining the electrons in the downstream exhaust. On the magnetospheric side of the X line, the strong E_N pushes most magnetosheath ions back from entering the magnetosphere.

The electric field structure can be characterized by an additional layer due to finite gyroradius effects of ions of magnetosheath origin. These Larmor radius ions in the magnetospheric inflow have been observed in numerous observations of dayside magnetopause reconnection (Phan et al. 2016a; Khotyaintsev et al. 2016; Wang et al. 2016b; Graham et al. 2017a). The observations show that the highest energy magnetosheath ions penetrate further into the magnetospheric inflow region, resulting in a velocity filter effect, where the minimum energy of the observed ions increases with distance from the X line (Khotyaintsev et al. 2016; Graham et al. 2017a). In observations, ion VDFs are characterized by crescent-shaped populations centered around the out-of-plane direction (Wang et al. 2016c; Graham et al. 2017a), consistent with numerical simulations (Shay et al. 2016; Wang et al. 2016b). These crescent populations result in a large out-of-plane bulk velocity $v_{iM} < 0$ (for when $B_L > 0$ on the magnetospheric side) (Wang et al. 2016c; Graham et al. 2017a), which is responsible for an Earthward normal electric field $E_N \approx -(\mathbf{v}_i \times \mathbf{B})_N \sim v_{iM} B_L < 0$ (Malakit et al. 2013). Graham et al. (2017a) and Graham et al. (2019) proposed that these Larmor radius ions could be unstable to lower hybrid wave generation due to the relative cross-field motion between the Larmor radius ions and cold magnetospheric ions (Graham et al. 2017a) or electrons (Graham et al. 2019). In simulations, the finite Larmor radius effect was found to occur close to the X line in the inflow regions, within $\sim 15\,d_i$ from the X line in the L direction (Malakit et al. 2013). Further downstream, ions tend to propagate along field lines, and thus, the observation of finite gyroradius effects might be an indicator of close proximity to the X line (Shay et al. 2016). However, the simulations of Dargent et al. (2017) showed that an $E_N < 0$ extends far along the separatrices when cold magnetospheric ions are present. Genestreti et al. (2018) also observed another layer of a positive (Sunward) electric field deeper within the magnetospheric side of the current layer. This field was associated with finite gyroradius effects of hot magnetospheric ions, as illustrated by Phan et al. (2016b), where the finite gyroradius ions result in a drift $v_{iM} > 0$ such that $E_N = v_{iM} B_L > 0$. Although the finite gyroradius effect of hot magnetospheric ions extends further towards the current sheet center (as seen by the presence of hot ions in the ion differential energy flux in Figure 2a in Genestreti et al. 2018), the finite gyroradius effects of the magnetosheath ions become dominant closer to the current sheet center.

4.2.2 Ion Dynamics in the Vicinity of the X Line

Similar to symmetric reconnection, counter-streaming ion populations are found inside the IDR, even though the Hall electric field strength differs significantly on the two sides of the X line (Wang et al. 2016c). Observations suggest that the Hall electric field, which also has a weaker E_L component, dominates the ion dynamics in the vicinity of the X line (Wang et al. 2016b; Yamada et al. 2018). The analysis indicates that E_N plays a dominant role in reflecting and trapping magnetosheath ions in an inner layer close to the X line. The ion distributions exhibit counter-streaming populations along v_N as supported by both simulations (Wang et al. 2016b) and MMS observations (Yamada et al. 2018; Wang et al. 2016b).

Further on the magnetospheric side, the cyclotron turning around B_L becomes important in reflecting ions. The ion velocities are turned by $ev_N B_L \hat{e}_M$ ($v_N < 0$, $B_L > 0$) from $-\hat{e}_N$ to $-\hat{e}_M$ close to the turn-over location, and additional acceleration occurs along $-v_M$ by the reconnection electric field (which in this topology is along $-\hat{e}_M$), which forms crescent distributions in the $v_M v_N$ plane (Wang et al. 2016c). Although crescent distributions can also occur downstream near the separatrix, ions close to the X line bounce in the current sheet with demagnetized motion, so the nature of the crescent distribution is different from the finite Larmor radius effect of magnetized ions downstream. The observations away from the X line illustrate the build-up of the outflow (Yamada et al. 2018). The distributions exhibit a clear shift of the counter-streaming populations towards $\pm v_L$ (Fig. 27e-27f), which Yamada et al. (2018) attributed to acceleration by E_L. For another event, Wang et al. (2016c) showed how, in the $v_L v_M$ plane, the tip of the distribution exhibited a rotation from v_M to v_L in the shape of a half-circle centered on the local $\mathbf{v}_{E\times B}$, which indicates that ions are also accelerated by the reconnection electric field close to the X line and thereafter cyclotron turn around B_N. Both mechanisms work together to form the outflow jet. Further downstream, at $\sim 80d_i$ from the X line, Wang et al. (2016c) show how finite-gyroradius ions are transitioned into Speiser ions, also forming a half circle in the $v_L v_M$ plane. This is illustrated in Fig. 28c-28d. Fig. 28e shows how the low-v_M cut-off increases away from the midplane, consistent with the finite gyroradius effect due to the cyclotron turning around $B_L > 0$. The Speiser half-circle in Fig. 28c-28d reaches higher energies than the population with $v_L \sim 0$ directly arriving from the magnetosheath. This can indicate additional acceleration by the reconnection electric field.

4.2.3 The Dynamics of Cold and Heavy Ions

Many reconnection environments are characterized by multi-component plasma, both concerning ion species and population temperatures. For example, in addition to protons (H^+), heavy ions (primarily He^+ or O^+) of ionospheric origin often reach the magnetopause, where they become involved in the reconnection process. Furthermore, multiple H^+ populations with different temperatures, e.g. plasma sheet ions (keV) and ionospheric outflows (eV) are commonly observed. For recent reviews on the occurrence and effects of multiple ion populations in magnetic reconnection at the Earth's magnetosphere, see Kronberg et al.

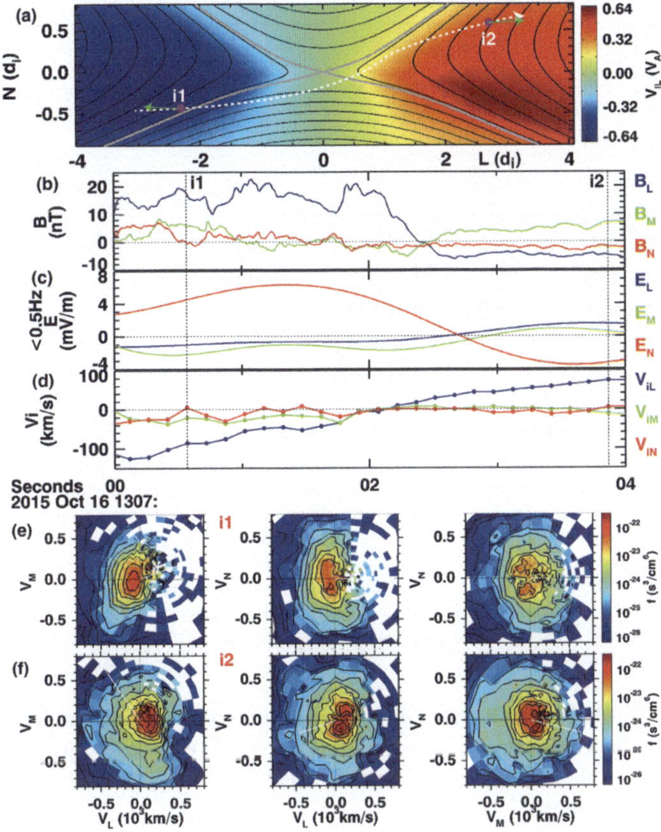

Fig. 27 Ion distributions around the flow reversal of the magnetopause event on 16 Oct 2015 that was first reported by Burch et al. (2016). (a) Simulation of asymmetric reconnection depicting v_{iL} and approximate spacecraft trajectory together with locations of displayed VDFs. The remaining panels are observations. (b) Magnetic field. (c) Low-pass filtered electric field. (d) Ion velocity. (e)-(f) Ion VDFs in all three planes at the locations marked by (e) $i1$ and (f) $i2$ in panels (a)-(d). The X-line ion VDFs exhibit counter-streaming populations along the normal direction $\hat{e}_N$, indicating acceleration by the Hall electric field E_N. A clear motion towards the outflow ($\pm\hat{e}_L$) is also observed, which Yamada et al. (2018) attributed to acceleration by the Hall electric field E_L. Figure based on Yamada et al. (2018)

(2021) and Toledo-Redondo et al. (2021). While heavier ions demagnetize at larger scales than lighter ions, cold ions demagnetize at smaller scales than hot ions. This differential behavior results in multi-layered diffusion regions, as it has been shown using PIC simulations (e.g. Dargent et al. 2017) (see also Divin et al. 2016; Liu et al. 2015; Markidis et al. 2011, who studied a symmetric reconnection configuration but demonstrated the principle of a cold ion diffusion region), and in observations (Lindstedt et al. 2010; André et al. 2016; Toledo-Redondo et al. 2016a; Alm et al. 2018, 2019).

The high-resolution measurements of MMS have allowed, for the first time, the observation of a multi-layered diffusion region in the presence of hot and cold H^+ (Toledo-Redondo et al. 2016a). The observations revealed distinct behavior of the hot and cold ions. Inside a large portion of the hot IDR, of thickness $\sim$140 km, the cold ions remained magnetized, with $\mathbf{v}_i^{cold} \sim \mathbf{v}_{E\times B}$. In a narrow layer of thickness $\sim$20 km, the cold ions were observed to be ballistically accelerated by the normal electric field. The Hall electric field $\mathbf{J}\times\mathbf{B}/ne$ was

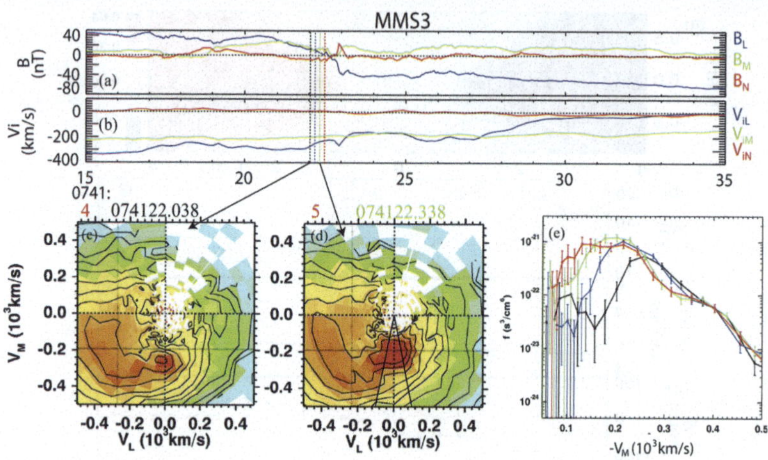

Fig. 28 Ion acceleration during asymmetric reconnection. (a) Magnetic field. (b) Ion velocity. (c)-(d) Reduced ion VDFs $f(v_L, v_N)$ and (e) $f(-v_M)$ for the times marked by like-colored lines in panels (a)-(b). These ion distributions reveal a two-step acceleration process. First, ions are gyroturned by $v_N B_L$ ($v_N < 0$, $B_L > 0$) to $-v_M$. Second, ions are gyroturned by $-v_M B_N$ ($v_M < 0$, $B_N > 0$) to $-v_L$. Although it might be hard to determine the sign of B_N from the time series, the distributions are a clear indication that B_N is directed out of the $v_L v_M$ plane. $f(-v_M)$ illustrates how the minimum ion speed is shifted to higher energies further into the magnetospheric side of the current sheet. Figure adapted from Wang et al. (2016c)

weaker in the hot IDR, which was a direct consequence of the partial cancellation of the Hall current by the cold magnetized ions (Toledo-Redondo et al. 2015; André et al. 2016). The thicknesses of these two regions were comparable to the gyroradius of the respective demagnetized populations .[3]

Heating of Cold Ions at the Dayside Magnetopause A fundamental aspect of magnetic reconnection is the conversion of magnetic energy into particle acceleration and plasma heating. When plasmaspheric plumes reach the reconnecting magnetopause, a significant amount of cold ions enter the reconnection process, where they become accelerated and heated, and mass load the reconnecting flux tubes (e.g. Toledo-Redondo et al. 2016b). MMS enabled direct observations of the energization of these cold populations (Toledo-Redondo et al. 2017; Vines et al. 2017). Vines et al. (2017) distinguished three regions with different energization signatures of the cold ions entraining reconnection. These regions are: (1) The electron edge of the separatrix region, where the magnetospheric ions start mixing with magnetosheath electrons, and a variety of processes leading to energization take place. Parallel heating and formation of ring distributions associated with pick-up acceleration are observed for both H^+, He^+, and O^+ ions. It is suggested that an electron-ion two-stream instability takes place and heats the cold ions (Steinvall et al. 2021, also identified large-amplitude electrostatic waves in the magnetospheric separatrix and outflow strongly correlated to cold ions and proposed they were generated by an ion-acoustic instability due to the cold ions and hot electrons). (2) The region where the first reconnected magnetosheath ions appear and ions of magnetosheath and magnetosphere origin start to mix. Vines et al. (2017) suggest that an ion-ion two-stream instability is formed, resulting in the observed increase in the parallel proton temperature of the cold ions. (3) Inside the exhaust, where perpendicular

[3] $r_{hot} = 115$ km ($T_{hot} = 1$ keV, $B = 40$ nT), and $r_{cold} = 20$ km ($T_{cold} = 30$ eV, $B = 40$ nT).

heating becomes pronounced, particularly for heavy ions. The pick-up process (Drake et al. 2009b) (see Sect. 4.1) is at work inside the exhaust, consistent with nonadiabatic heating and mass-per-charge selection, which can explain the observations. Cold ion parallel beams are also observed inside the exhaust, in particular composed of heavy ions.

Toledo-Redondo et al. (2017) studied four cases of observations of cold ion heating in the separatrix region of magnetic reconnection. They found that significant heating was observed when the gradient length scale of the electric field, $E/\nabla E$, was smaller than the ion pick-up gyroradius, $v_{E\times B}/\omega_{ci}$, which translates to $\nabla E > qB^2/m$. Li et al. (2017) studied an MMS magnetopause encounter with multiple crossings when a plume with cold (few tens of eV) ion densities up to 60 cm^{-3} on the magnetospheric side was entraining reconnection. They observed signatures of the plume ions inside the magnetopause jets and concluded that plume ions that crossed the magnetopause near the ion diffusion region remained cold (tens of eV) inside the exhaust, while cold ions that crossed the magnetopause further downstream (a few hundred ion inertial lengths) were heated to hundreds of eV.

4.3 The Effect of Guide Fields

Eastwood et al. (2018) studied the structure of a guide-field reconnection exhaust $\sim 100d_i$ downstream of the X line. The event showed asymmetries across the current sheet, i.e., a density cavity at one separatrix. The separatrix with the density cavity exhibited a decrease of $T_{e\|}$, but an increase of $T_{i\|}$. In the vicinity of the density cavity, the ion VDFs revealed counter-streaming populations, which also had a finite perpendicular drift. This indicates that the ions are affected similarly to electrons in the presence of a guide field. I.e., due to the opposite acceleration, the ion temperature is elevated while the electron temperature drops. This can be seen in Fig. 19l at the location marked by the magenta triangle.

Drake et al. (2009a) and Knizhnik et al. (2011) studied the effect of the guide field on the ion pick-up process. They showed that the heating of protons becomes less efficient due to the guide field, causing ions to remain adiabatic upon crossing the outflow boundary. On the other hand, they showed that ions with mass-per-charge m/q exceeding a threshold are no longer adiabatic and behave like pick-up ions.

5 Conclusions and Outlook

We have reviewed recent work on particle dynamics key to magnetic reconnection through a combination of high-resolution in-situ measurements, as well as theory and simulations. We highlighted examples of breakthroughs in the structure and dynamics of the diffusion regions, the core of magnetic reconnection.

To understand the macroscopic behavior of a plasma, understanding the underlying particle dynamics is essential. One long-standing question in the field of magnetic reconnection is what term provides the effective resistivity necessary to sustain the reconnection electric field in the generalized Ohm's law (e.g. Liu et al. 2025b, this collection). Another question concerns the nature of the energy conversion during magnetic reconnection (e.g. Oka et al. 2023, this collection; Genestreti et al. 2025, this collection). For example, what trajectories comprise the reconnection current or off-diagonal pressure tensor elements?

In the immediate vicinity of the X line, measurements show that the bulk of the electron motion breaks the frozen-in condition, $E_M + (\mathbf{v}_e \times \mathbf{B})_M \neq 0$, which is key to the reconnection process as it permits the presence of a diffusive electric field. In addition, the associated electron VDFs clearly show that the electrons move in an unmagnetized manner, in agreement with predictions from fully kinetic particle-in-cell simulations and theory. For example, the clear presence of electrons following meandering trajectories with multiple bounces (Torbert et al. 2018) indicates a finite dwell time in the EDR at the same time that the magnetic field would be following a steady convection from the inflow to the outflow. For a symmetric antiparallel reconnection event, the diffusive electric field was quantitatively identified as being based on thermal electron inertia, manifested by the gradients of the off-diagonal electron pressure tensor elements (Egedal et al. 2019). These fluid terms have previously been connected to the meandering electrons in the EDR (e.g. Chen et al. 2011; Ng et al. 2011; Bessho et al. 2017), clearly demonstrating the connection between the particle dynamics and the fluid quantities.

The presence of multiple distinct crescent distributions in some observed electron diffusion regions suggests that wave activity plays a secondary role compared with the laminar electromagnetic fields. This suggests that anomalous resistivity associated with wave-particle interactions may not play a major role in determining the reconnection electric field, and hence the reconnection rate. However, reconnection diffusion regions have been observed with both strong and weak wave activity. Therefore, future research is still needed to determine the precise role of waves in reconnection and to identify the conditions under which wave activity is important.

We also want to highlight the synergetic effect of observations and simulations. Without observational confirmations of many diffusion-associated plasma features prior to MMS, particle-in-cell simulations were valuable assets, but could still only be treated as speculation, regardless of how well-founded they were. Now, when the kinetic structure of the plasma deduced from observations can be compared side-by-side with simulations, the conclusions drawn from simulations have gained more weight. The high level of detail that can often be gained from observations also guides future simulation efforts, further deepening our understanding of magnetic reconnection.

To help the community prepare for future advancements, we list a few outstanding questions for future directions (see also Nakamura et al. 2025, this collection, for a more comprehensive review). How 3D effects modify reconnection is a long-standing open question. The role of out-of-plane instabilities in reconnection remains to be quantified. MMS measurements in the reconnection EDRs have seen cases requiring 3D dynamics such as kinking of the electron current sheet (Cozzani et al. 2021) and obliquely propagating electron vortices (e.g. Chen et al. 2020) or lower hybrid waves (Graham et al. 2025, this collection), even though the 2D laminar picture can successfully describe the electron-scale structure in many observed cases. The contribution of Debye- and sub-electron-gyroradius-scale waves and electrostatic turbulence (Khotyaintsev et al. 2020; Graham et al. 2017b; Li et al. 2021a) to this picture remains to be quantified. On the ion-skin-depth and larger scale, 3D variations along the reconnection current direction have been shown to be needed to account for observations from all four spacecraft (Chen et al. 2016b). With advanced supercomputing capabilities such as exascale computing (Ji et al. 2022), 3D simulations with improved resolution and parameters more close to physical will provide new predictions to be compared with observations of 3D reconnection in complex magnetopause and magnetotail events, as well as within Kelvin-Helmholtz and shock-driven turbulence (Stawarz et al. 2024, this collection).

The findings reported in this review also highlight the importance of cross-scale coupling, which is critical for space plasmas, as well as laboratory and astrophysical plasmas. Open questions in cross-scale coupling have motivated the development of new multi-spacecraft missions such as Plasma Observatory (Retino et al. 2023) to explore plasma energization and HelioSwarm (Klein et al. 2023) to explore turbulence.

Acknowledgements The authors thank the International Space Science Institute (ISSI) for hosting the Magnetic Reconnection: Explosive Energy Conversion in Space Plasmas team. We gratefully acknowledge Shan Wang for her contribution to the section about ion dynamics. We gratefully acknowledge Kevin Genestreti for valuable discussions.

Funding Information Open access funding provided by University of Bergen (incl Haukeland University Hospital). CN acknowledges support from the Research Council of Norway under Contract No. 300865, and the Swedish National Space Agency (SNSA) under Grant 2022-00121. DG acknowledges support from the SNSA under Grant 128/17. STR acknowledges support of Grant PID2023-147331OB-I00 funded by MICIU/AEI/ 10.13039/501100011033 and FEDER, UE, and Seneca Agency from Region of Murcia (Grant 21910/PI/22). LR and YK acknowledge support from SNSA, Grant 139/18 and the Knut and Alice Wallenberg Foundation (Dnr. 2022.0087). JPE acknowledges UKRI/STFC grant ST/W001071/1. MMS data are publicly available through the MMS Science Data Center at https://lasp.colorado.edu/mms/sdc/public.

Declarations

Competing Interests The authors declare no competing interests.

References

Alm L, André M, Vaivads A, et al (2018) Magnetotail Hall physics in the presence of cold ions. Geophys Res Lett 45(20):10941–10950. https://doi.org/10.1029/2018GL079857

Alm L, André M, Graham DB, et al (2019) MMS observations of multiscale Hall physics in the magnetotail. Geophys Res Lett 46(10230):10230–10239. https://doi.org/10.1029/2019GL084137

André M, Li W, Toledo-Redondo S, et al (2016) Magnetic reconnection and modification of the Hall physics due to cold ions at the magnetopause. Geophys Res Lett 43(13):6705–6712. https://doi.org/10.1002/2016GL069665

Andrews MK, Keppler E, Daly PW (1981) Plasma sheet motions inferred from medium-energy ion measurements. J Geophys Res 86(A9):7543–7556. https://doi.org/10.1029/JA086iA09p07543

Arzner K, Scholer M (2001) Kinetic structure of the post plasmoid plasma sheet during magnetotail reconnection. J Geophys Res Space Phys 106(A3):3827–3844. https://doi.org/10.1029/2000JA000179

Aunai N, Belmont G, Smets R (2011) Proton acceleration in antiparallel collisionless magnetic reconnection: kinetic mechanisms behind the fluid dynamics. J Geophys Res Space Phys 116(A9):A09232. https://doi.org/10.1029/2011JA016688

Baumjohann W, Paschmann G, Luehr H (1990) Characteristics of high-speed ion flows in the plasma sheet. J Geophys Res 95(A4):3801–3809. https://doi.org/10.1029/JA095iA04p03801

Bessho N, Chen LJ, Shuster JR, et al (2014) Electron distribution functions in the electron diffusion region of magnetic reconnection: physics behind the fine structures. Geophys Res Lett 41:8688–8695. https://doi.org/10.1002/2014GL062034

Bessho N, Chen LJ, Hesse M (2016) Electron distribution functions in the diffusion region of asymmetric magnetic reconnection. Geophys Res Lett 43(5):1828–1836. https://doi.org/10.1002/2016GL067886

Bessho N, Chen LJ, Hesse M, et al (2017) The effect of reconnection electric field on crescent and U-shaped distribution functions in asymmetric reconnection with no guide field. Phys Plasmas 24(7):072903. https://doi.org/10.1063/1.4989737

Bessho N, Chen LJ, Wang S, et al (2018) Effect of the reconnection electric field on electron distribution functions in the diffusion region of magnetotail reconnection. Geophys Res Lett 45(22):12142–12152. https://doi.org/10.1029/2018GL081216

Bessho N, Chen LJ, Wang S, et al (2019) Effects of the guide field on electron distribution functions in the diffusion region of asymmetric reconnection. Phys Plasmas 26(8):082310. https://doi.org/10.1063/1.5092809

Birn J, Drake JF, Shay MA, et al (2001) Geospace environmental modeling (GEM) magnetic reconnection challenge. J Geophys Res 106(A3):3715–3719

Birn J, Hesse M, Nakamura R, et al (2013) Particle acceleration in dipolarization events. J Geophys Res Space Phys 118(5):1960–1971. https://doi.org/10.1002/jgra.50132

Buechner J, Zelenyi LM (1989) Regular and chaotic charged particle motion in magnetotaillike field reversals. I - basic theory of trapped motion. J Geophys Res 94:11821-11842. https://doi.org/10.1029/JA094iA09p11821

Burch JL, Torbert RB, Phan TD, et al (2016) Electron-scale measurements of magnetic reconnection in space. Science 352:aaf2939. https://doi.org/10.1126/science.aaf2939

Burch JL, Phan TD (2016) Magnetic reconnection at the dayside magnetopause: advances with MMS. Geophys Res Lett 43:8327–8338. https://doi.org/10.1002/2016GL069787

Burch JL, Ergun RE, Cassak PA, et al (2018) Localized oscillatory energy conversion in magnetopause reconnection. Geophys Res Lett 45(3):1237–1245. https://doi.org/10.1002/2017GL076809.

Burch JL, Dokgo K, Hwang KJ, et al (2019) High-frequency wave generation in magnetotail reconnection: linear dispersion analysis. Geophys Res Lett 46(8):4089–4097. https://doi.org/10.1029/2019GL082471

Burch JL, Genestreti KJ, Heuer SV, et al (2023) Electron energy dissipation in a magnetotail reconnection region. Phys Plasmas 30(8):082903. https://doi.org/10.1063/5.0153628

Burkhart GR, Drake JF, Dusenbery PB, et al (1992) A particle model for magnetotail neutral sheet equilibria. J Geophys Res 97(A9):13799–13815. https://doi.org/10.1029/92JA00495

Cassak PA, Shay MA (2007) Scaling of asymmetric magnetic reconnection: general theory and collisional simulations. Phys Plasmas 14(10):102114. https://doi.org/10.1063/1.2795630

Chen LJ, Daughton WS, Lefebvre B, et al (2011) The inversion layer of electric fields and electron phase-space-hole structure during two-dimensional collisionless magnetic reconnection. Phys Plasmas 18(1):012904. https://doi.org/10.1063/1.3529365

Chen LJ, Hesse M, Wang S, et al (2016a) Electron energization and structure of the diffusion region during asymmetric reconnection. Geophys Res Lett 43(6):2405–2412. https://doi.org/10.1002/2016GL068243

Chen LJ, Hesse M, Wang S, et al (2016b) Electron energization and mixing observed by mms in the vicinity of an electron diffusion region during magnetopause reconnection. Geophys Res Lett 43(12):6036–6043. https://doi.org/10.1002/2016GL069215

Chen LJ, Hesse M, Wang S, et al (2017) Electron diffusion region during magnetopause reconnection with an intermediate guide field: magnetospheric multiscale observations. J Geophys Res Space Phys 122:5235–5246. https://doi.org/10.1002/2017JA024004

Chen LJ, Wang S, Hesse M, et al (2019) Electron diffusion regions in magnetotail reconnection under varying guide fields. Geophys Res Lett 46(12):6230–6238. https://doi.org/10.1029/2019GL082393

Chen LJ, Wang S, Le Contel O, et al (2020) Lower-hybrid drift waves driving electron nongyrotropic heating and vortical flows in a magnetic reconnection layer. Phys Res Lett 125(2):025103. https://doi.org/10.1103/PhysRevLett.125.025103

Cheng CZ, Inoue S, Ono Y, et al (2015) Physical processes of driven magnetic reconnection in collisionless plasmas: zero guide field case. Phys Plasmas 22(10):101205. https://doi.org/10.1063/1.4932337

Cowley SW, Shull P (1983) Current sheet acceleration of ions in the geomagnetic tail and the properties of ion bursts observed at the lunar distance. Planet Space Sci 31(2):235–245. https://doi.org/10.1016/0032-0633(83)90058-2

Cozzani G, Khotyaintsev YV, Graham DB, et al (2021) Structure of a perturbed magnetic reconnection electron diffusion region in the Earth's magnetotail. Phys Res Lett 127(21):215101. https://doi.org/10.1103/PhysRevLett.127.215101.

Dargent J, Aunai N, Lavraud B, et al (2017) Kinetic simulation of asymmetric magnetic reconnection with cold ions. J Geophys Res Space Phys 122(5):5290–5306. https://doi.org/10.1002/2016JA023831

Daughton W, Lapenta G, Ricci P (2004) Nonlinear evolution of the lower-hybrid drift instability in a current sheet. Phys Rev Lett 93(10):105004. https://doi.org/10.1103/PhysRevLett.93.105004

Denton RE, Torbert RB, Hasegawa H, et al (2020) Polynomial reconstruction of the reconnection magnetic field observed by multiple spacecraft. J Geophys Res Space Phys 25(2):e2019JA027481. https://doi.org/10.1029/2019JA027481

Divin A, Lapenta G, Markidis S, et al (2012) Numerical simulations of separatrix instabilities in collisionless magnetic reconnection. Phys Plasmas 19(4):042110. https://doi.org/10.1063/1.3698621

Divin A, Khotyaintsev YV, Vaivads A, et al (2016) Three-scale structure of diffusion region in the presence of cold ions. J Geophys Res Space Phys 121(12):12001–12013. https://doi.org/10.1002/2016JA023606

Drake JF, Swisdak M, Che H, et al (2006) Electron acceleration from contracting magnetic islands during reconnection. Nature 443(7111):553–556. https://doi.org/10.1038/nature05116

Drake JF, Cassak PA, Shay MA, et al (2009a) A magnetic reconnection mechanism for ion acceleration and abundance enhancements in impulsive flares. Astrophys J Lett 700(1):L16–L20. https://doi.org/10.1088/0004-637X/700/1/L16

Drake JF, Swisdak M, Phan TD, et al (2009b) Ion heating resulting from pickup in magnetic reconnection exhausts. J Geophys Res Space Phys 114(A5):A05111. https://doi.org/10.1029/2008JA013701

Eastwood JP, Shay MA, Phan TD, et al (2010) Asymmetry of the ion diffusion region Hall electric and magnetic fields during guide field reconnection: observations and comparison with simulations. Phys Res Lett 104(20):205001. https://doi.org/10.1103/PhysRevLett.104.205001

Eastwood JP, Goldman MV, Hietala H, et al (2015) Ion reflection and acceleration near magnetotail dipolarization fronts associated with magnetic reconnection. J Geophys Res Space Phys 120(1):511–525. https://doi.org/10.1002/2014JA020516

Eastwood JP, Mistry R, Phan TD, et al (2018) Guide field reconnection: exhaust structure and heating. Geophys Res Lett 45(10):4569–4577. https://doi.org/10.1029/2018GL077670

Egedal J, Øieroset M, Fox W, et al (2005) In situ discovery of an electrostatic potential, trapping electrons and mediating fast reconnection in the Earth's magnetotail. Phys Rev Lett 94(2):025006. https://doi.org/10.1103/PhysRevLett.94.025006

Egedal J, Fox W, Katz N, et al (2008) Evidence and theory for trapped electrons in guide field magnetotail reconnection. J Geophys Res 113(6):A12207. https://doi.org/10.1029/2008JA013520

Egedal J, Le A, Pritchett PL, et al (2011) Electron dynamics in two-dimensional asymmetric anti-parallel reconnection. Phys Plasmas 18:102901. https://doi.org/10.1063/1.3646316

Egedal J, Daughton W, Le A (2012) Large-scale electron acceleration by parallel electric fields during magnetic reconnection. Nat Phys 8(4):321–324. https://doi.org/10.1038/nphys2249

Egedal J, Le A, Daughton W (2013) A review of pressure anisotropy caused by electron trapping in collisionless plasma, and its implications for magnetic reconnection. Phys Plasmas 20:061201. https://doi.org/10.1063/1.4811092

Egedal J, Daughton W, Le A, et al (2015) Double layer electric fields aiding the production of energetic flat-top distributions and superthermal electrons within magnetic reconnection exhausts. Phys Plasmas 22(10):101208. https://doi.org/10.1063/1.4933055.

Egedal J, Le A, Daughton W, et al (2016a) Spacecraft observations and analytic theory of crescent-shaped electron distributions in asymmetric magnetic reconnection. Phys Rev Lett 117:185101. https://doi.org/10.1103/PhysRevLett.117.185101

Egedal J, Wetherton B, Daughton W, et al (2016b) Processes setting the structure of the electron distribution function within the exhausts of anti-parallel reconnection. Phys Plasmas 23(12):122904. https://doi.org/10.1063/1.4972135

Egedal J, Le A, Daughton W, et al (2018) Spacecraft observations of oblique electron beams breaking the frozen-in law during asymmetric reconnection. Phys Res Lett 120(5):055101. https://doi.org/10.1103/PhysRevLett.120.055101

Egedal J, Ng J, Le A, et al (2019) Pressure tensor elements breaking the frozen-in law during reconnection in Earth's magnetotail. Phys Rev Lett 123:225101. https://doi.org/10.1103/PhysRevLett.123.225101

Egedal J, Gurram H, Greess S, et al (2023) The force balance of electrons during kinetic anti-parallel magnetic reconnection. Phys Plasmas 30(6):062106. https://doi.org/10.1063/5.0130417

Egedal J (2023) On a plasma sheath with a small normal magnetic field separating regions of oppositely directed magnetic field. Phys Plasmas 30(11). https://doi.org/10.1063/5.0170212

Egedal J (2024) The adiabatic 1D kinetic equilibrium of the electron diffusion region during anti-parallel magnetic reconnection. Geophys Res Lett 51(10):e2024GL108895. https://doi.org/10.1029/2024GL108895

Ergun RE, Holmes JC, Goodrich KA, et al (2016) Magnetospheric multiscale observations of large-amplitude, parallel, electrostatic waves associated with magnetic reconnection at the magnetopause. Geophys Res Lett 43(11):5626–5634. https://doi.org/10.1002/2016GL068992

Ergun RE, Ahmadi N, Kromyda L, et al (2020) Particle acceleration in strong turbulence in the Earth's magnetotail. Astrophys J 898(2):153. https://doi.org/10.3847/1538-4357/ab9ab5

Ergun RE, Pathak N, Usanova ME, et al (2022) Observation of magnetic reconnection in a region of strong turbulence. Astrophys J Lett 935(1):L8. https://doi.org/10.3847/2041-8213/ac81d4

Eriksson S, Lavraud B, Wilder FD, et al (2016a) Magnetospheric multiscale observations of magnetic reconnection associated with Kelvin-Helmholtz waves. Geophys Res Lett 43:5606–5615. https://doi.org/10.1002/2016GL068783

Eriksson S, Wilder FD, Ergun RE, et al (2016b) Magnetospheric multiscale observations of the electron diffusion region of large guide field magnetic reconnection. Phys Res Lett 117(1):015001. https://doi.org/10.1103/PhysRevLett.117.015001

Eriksson E, Vaivads A, Graham DB, et al (2018) Electron energization at a reconnecting magnetosheath current sheet. Geophys Res Lett 45(16):8081–8090. https://doi.org/10.1029/2018GL078660

Eriksson E, Vaivads A, Alm L, et al (2020) Electron acceleration in a magnetotail reconnection outflow region using magnetospheric MultiScale data. Geophys Res Lett 47(1):e85080. https://doi.org/10.1029/2019GL085080

Forbes TG, Hones EW, Bame SJ, et al (1981) Evidence for the tailward retreat of a magnetic neutral line in the magnetotail during substorm recovery. Geophys Res Lett 8(3):261–264. https://doi.org/10.1029/GL008i003p00261

Fu S, Huang S, Zhou M, et al (2018) Tripolar electric field structure in guide field magnetic reconnection. Ann Geophys 36(2):373–379. https://doi.org/10.5194/angeo-36-373-2018

Fujimoto K (2014) Wave activities in separatrix regions of magnetic reconnection. Geophys Res Lett 41:2721–2728. https://doi.org/10.1002/2014GL059893

Fujimoto K, Sydora RD (2023) The electron diffusion region dominated by electromagnetic turbulence in the reconnection current layer. Phys Plasmas 30(2):022106. https://doi.org/10.1063/5.0129591

Fuselier SA, Petrinec SM, Reiff PH, et al (2024) Global-scale processes and effects of magnetic reconnection on the geospace environment. Space Sci Rev 220(4):34. https://doi.org/10.1007/s11214-024-01067-0

Genestreti KJ, Nakamura R, Liu YH, et al (2025) Structure of the electron diffusion region during magnetic reconnection. Space Sci Rev 221:59. https://doi.org/10.1007/s11214-025-01188-0

Genestreti KJ, Burch JL, Cassak PA, et al (2017) The effect of a guide field on local energy conversion during asymmetric magnetic reconnection: MMS observations. J Geophys Res Space Phys 122(11):11342–11353. https://doi.org/10.1002/2017JA024247.

Genestreti KJ, Varsani A, Burch JL, et al (2018) MMS observation of asymmetric reconnection supported by 3-D electron pressure divergence. J Geophys Res Space Phys 123(3):1806–1821. https://doi.org/10.1002/2017JA025019.

Goldman MV, Lapenta G, Newman DL, et al (2011) Jet deflection by very weak guide fields during magnetic reconnection. Phys Res Lett 107(13):135001. https://doi.org/10.1103/PhysRevLett.107.135001.

Goldman MV, Newman DL, Eastwood JP, et al (2021) Multi-beam energy moments of measured compound ion velocity distributions. Phys Plasmas 28(10):102305. https://doi.org/10.1063/5.0063431

Gosling JT (2012) Magnetic reconnection in the solar wind. Space Sci Rev 172(1–4):187–200. https://doi.org/10.1007/s11214-011-9747-2

Graham DB, Khotyaintsev YV, Vaivads A, et al (2014) Electron dynamics in the diffusion region of an asymmetric magnetic reconnection. Phys Res Lett 112(21):215004. https://doi.org/10.1103/PhysRevLett.112.215004

Graham DB, Khotyaintsev YV, Norgren C, et al (2016a) Electron currents and heating in the ion diffusion region of asymmetric reconnection. Geophys Res Lett 43:4691–4700. https://doi.org/10.1002/2016GL068613

Graham DB, Vaivads A, Khotyaintsev YV, et al (2016b) Whistler emission in the separatrix regions of asymmetric reconnection. J Geophys Res 121:1934–1954. https://doi.org/10.1002/2015JA021239

Graham DB, Khotyaintsev YV, Norgren C, et al (2017a) Lower hybrid waves in the ion diffusion and magnetospheric inflow regions. J Geophys Res Space Phys 122(1):517–533. https://doi.org/10.1002/2016JA023572

Graham DB, Khotyaintsev YV, Vaivads A, et al (2017b) Instability of agyrotropic electron beams near the electron diffusion region. Phys Rev Lett 119(2):025101. https://doi.org/10.1103/PhysRevLett.119.025101

Graham DB, Khotyaintsev YV, Norgren C, et al (2019) Universality of lower hybrid waves at Earth's magnetopause. J Geophys Res Space Phys 124(11):8727–8760. https://doi.org/10.1029/2019JA027155

Graham DB, Khotyaintsev YV, André M, et al (2022) Direct observations of anomalous resistivity and diffusion in collisionless plasma. Nat Commun 13:2954. https://doi.org/10.1038/s41467-022-30561-8

Graham DB, Cozzani G, Khotyaintsev YV, et al (2025) The role of kinetic instabilities and waves in collisionless magnetic reconnection. Space Sci Rev 221(1):20. https://doi.org/10.1007/s11214-024-01133-7

Guo R, Pu Z, Wei Y (2017) Current structure and flow pattern on the electron separatrix in reconnection region. Geosci Lett 4(1):18. https://doi.org/10.1186/s40562-017-0085-4

Haggerty CC, Shay MA, Drake JF, et al (2015) The competition of electron and ion heating during magnetic reconnection. Geophys Res Lett 42(22):9657–9665. https://doi.org/10.1002/2015GL065961.

Harris EG (1962) On a plasma sheath separating regions of oppositely directed magnetic field. Nuovo Cimento 23(1):115–121. https://doi.org/10.1007/BF02733547

Hasegawa H, Denton RE, Nakamura R, et al (2019) Reconstruction of the electron diffusion region of magnetotail reconnection seen by the MMS spacecraft on 11 July 2017. J Geophys Res Space Phys 124(1):122–138. https://doi.org/10.1029/2018JA026051

Hasegawa H, Denton RE, Nakamura TKM, et al (2022) Magnetic field annihilation in a magnetotail electron diffusion region with electron-scale magnetic island. J Geophys Res Space Phys 127(7):e2022JA030408. https://doi.org/10.1029/2022JA030408

Hasegawa H, Argall MR, Aunai N, et al (2024) Advanced methods for analyzing in-situ observations of magnetic reconnection. Space Sci Rev 220(6):68. https://doi.org/10.1007/s11214-024-01095-w
Hesse M, Schindler K, Birn J, et al (1999) The diffusion region in collisionless magnetic reconnection. Phys Plasmas 6(5, 2):1781–1795. https://doi.org/10.1063/1.873436.
Hesse M, Kuznetsova M, Birn J (2001) Particle-in-cell simulations of three-dimensional collisionless magnetic reconnection. J Geophys Res Space Phys 106 https://doi.org/10.1029/2001JA000075
Hesse M, Aunai N, Sibeck D, et al (2014) On the electron diffusion region in planar, asymmetric, systems. Geophys Res Lett 41:8673–8680. https://doi.org/10.1002/2014GL061586
Hesse M, Liu YH, Chen LJ, et al (2016) On the electron diffusion region in asymmetric reconnection with a guide magnetic field. Geophys Res Lett 43(6):2359–2364. https://doi.org/10.1002/2016GL068373
Hesse M, Norgren C, Tenfjord P, et al (2018) On the role of separatrix instabilities in heating the reconnection outflow region. Phys Plasmas 25(12):122902. https://doi.org/10.1063/1.5054100
Hietala H, Drake JF, Phan TD, et al (2015) Ion temperature anisotropy across a magnetotail reconnection jet. Geophys Res Lett 42(18):7239–7247. https://doi.org/10.1002/2015GL065168
Holmes JC, Ergun RE, Nakamura R, et al (2019) Structure of electron-scale plasma mixing along the dayside reconnection separatrix. J Geophys Res Space Phys 124(11):8788–8803. https://doi.org/10.1029/2019JA026974
Holmes JC, Nakamura R, Schmid D, et al (2021) Wave activity in a dynamically evolving reconnection separatrix. J Geophys Res Space Phys 126(7):e28520. https://doi.org/10.1029/2020JA028520
Horiuchi R, Sato T (1997) Particle simulation study of collisionless driven reconnection in a sheared magnetic field. Phys Plasmas 4(2):277–289. https://doi.org/10.1063/1.872088
Huang SY, Vaivads A, Khotyaintsev YV, et al (2012) Electron acceleration in the reconnection diffusion region: Cluster observations. Geophys Res Lett 39(11):L11103. https://doi.org/10.1029/2012GL051946
Huang SY, Jiang K, Yuan ZG, et al (2018) Observations of the electron jet generated by secondary reconnection in the terrestrial magnetotail. Astrophys J 862(2):144. https://doi.org/10.3847/1538-4357/aacd4c
Huang SY, Xiong QY, Song LF, et al (2021) Electron-only reconnection in an ion-scale current sheet at the magnetopause. Astrophys J 922(1):54. https://doi.org/10.3847/1538-4357/ac2668.
Hwang KJ, Dokgo K, Choi E, et al (2021) Bifurcated current sheet observed on the boundary of Kelvin-Helmholtz vortices. Front Astron Space Sci 8:201. https://doi.org/10.3389/fspas.2021.782924
Hwang KJ, Nakamura R, Eastwood JP, et al (2023) Cross-scale processes of magnetic reconnection. Space Sci Rev 219(8):71. https://doi.org/10.1007/s11214-023-01010-9
Jara-Almonte J, Daughton W, Ji H (2014) Debye scale turbulence within the electron diffusion layer during magnetic reconnection. Phys Plasmas 21(3):032114. https://doi.org/10.1063/1.4867868
Ji H, Daughton W, Jara-Almonte J, et al (2022) Magnetic reconnection in the era of exascale computing and multiscale experiments. Nat Rev Phys 4(4):263–282. https://doi.org/10.1038/s42254-021-00419-x.
Ji H, Yoo J, Fox W, et al (2023) Laboratory study of collisionless magnetic reconnection. Space Sci Rev 219(8):76. https://doi.org/10.1007/s11214-023-01024-3
Kamaletdinov SR, Artemyev AV, Runov A, et al (2025) Ion kinetics in thin current sheets at lunar distances. Geophys Res Lett 52(9):e2024GL114522. https://doi.org/10.1029/2024GL114522
Karimabadi H, Pritchett PL, Daughton W, et al (2003) Ion-ion kink instability in the magnetotail: 2. Three-dimensional full particle and hybrid simulations and comparison with observations. J Geophys Res Space Phys 108(A11):1401. https://doi.org/10.1029/2003JA010109
Karimabadi H, Daughton W, Scudder J (2007) Multi-scale structure of the electron diffusion region. Geophys Res Lett 34(13). https://doi.org/10.1029/2007GL030306
Khotyaintsev YV, Vaivads A, Retinò A, et al (2006) Formation of inner structure of a reconnection separatrix region. Phys Res Lett 97(20):205003. https://doi.org/10.1103/PhysRevLett.97.205003
Khotyaintsev YV, Graham DB, Norgren C, et al (2016) Electron jet of asymmetric reconnection. Geophys Res Lett 43:5571–5580. https://doi.org/10.1002/2016GL069064
Khotyaintsev YV, Graham DB, Steinvall K, et al (2020) Electron heating by Debye-scale turbulence in guide-field reconnection. Phys Res Lett 124(4):045101. https://doi.org/10.1103/PhysRevLett.124.045101.
Klein KG, Spence H, Alexandrova O, et al (2023) HelioSwarm: a multipoint, multiscale mission to characterize turbulence. Space Sci Rev 219(8):74. https://doi.org/10.1007/s11214-023-01019-0.
Knizhnik K, Swisdak M, Drake JF (2011) The acceleration of ions in solar flares during magnetic reconnection. Astrophys J Lett 743(2):L35. https://doi.org/10.1088/2041-8205/743/2/L35.
Kolstø HM, Norgren C, Hesse M, et al (2021) Magnetospheric multiscale observations of an expanding oxygen wave in magnetic reconnection. Geophys Res Lett 48(19):e95065. https://doi.org/10.1029/2021GL095065
Kronberg EA, Grigorenko EE, Ilie R, et al (2021) Impact of ionospheric ions on magnetospheric dynamics. In: Maggiolo R, André N, Hasegawa H, et al (eds) Magnetospheres in the Solar System. Geophysical monograph, vol 259. Wiley / AGU, p 353. https://doi.org/10.1002/9781119815624.ch23

Lapenta G, Markidis S, Divin A, et al (2010) Scales of guide field reconnection at the hydrogen mass ratio. Phys Plasmas 17(8):082106. https://doi.org/10.1063/1.3467503

Lapenta G, Markidis S, Divin A, et al (2015) Separatrices: the crux of reconnection. J Plasma Phys 81(1):325810109. https://doi.org/10.1017/S0022377814000944.

Lapenta G, Berchem J, Zhou M, et al (2017) On the origin of the crescent-shaped distributions observed by mms at the magnetopause. J Geophys Res Space Phys 122(2):2024–2039. https://doi.org/10.1002/2016JA023290

Lavraud B, Zhang YC, Vernisse Y, et al (2016) Currents and associated electron scattering and bouncing near the diffusion region at Earth's magnetopause. Geophys Res Lett 43:3042–3050. https://doi.org/10.1002/2016GL068359

Le A, Egedal J, Daughton W, et al (2009) Equations of State for Collisionless Guide-Field Reconnection. Phys Rev Lett 102(8). https://doi.org/10.1103/PhysRevLett.102.085001

Le A, Egedal J, Daughton W, et al (2010) Magnitude of the Hall fields during magnetic reconnection. Geophys Res Lett 37. https://doi.org/10.1029/2009GL041941

Le A, Egedal J, Ohia O, et al (2013) Regimes of the Electron Diffusion Region in Magnetic Reconnection. Phys Rev Lett 110(13). https://doi.org/10.1103/PhysRevLett.110.135004

Le A, Daughton W, Chen LJ, et al (2017) Enhanced electron mixing and heating in 3-D asymmetric reconnection at the Earth's magnetopause. Geophys Res Lett 44:2096–2104. https://doi.org/10.1002/2017GL072522.

Le A, Daughton W, Ohia O, et al (2018) Drift turbulence, particle transport, and anomalous dissipation at the reconnecting magnetopause. Phys Plasmas 25(6):062103. https://doi.org/10.1063/1.5027086.

Le A, Stanier A, Daughton W, et al (2019) Three-Dimensional Stability of Current Sheets Supported by Electron Pressure Anisotropy. Phys Plasmas 102114. https://doi.org/10.1063/1.5125014

Li W, André M, Khotyaintsev YV, et al (2016) Kinetic evidence of magnetic reconnection due to Kelvin-Helmholtz waves. Geophys Res Lett 43(11):5635–5643. https://doi.org/10.1002/2016GL069192

Li WY, André M, Khotyaintsev YV, et al (2017) Cold ionospheric ions in the magnetic reconnection outflow region. J Geophys Res Space Phys 122(10):10194–10202. https://doi.org/10.1002/2017JA024287

Li X, Wang R, Lu Q, et al (2019) Observation of nongyrotropic electron distribution across the electron diffusion region in the magnetotail reconnection. Geophys Res Lett 46(24):14263–14273. https://doi.org/10.1029/2019GL085014

Li WY, Graham DB, Khotyaintsev YV, et al (2020) Electron Bernstein waves driven by electron crescents near the electron diffusion region. Nat Commun 11:141. https://doi.org/10.1038/s41467-019-13920-w

Li WY, Khotyaintsev YV, Tang BB, et al (2021a) Upper-hybrid waves driven by meandering electrons around magnetic reconnection X line. Geophys Res Lett 48(16):e93164. https://doi.org/10.1029/2021GL093164

Li YX, Li WY, Tang BB, et al (2021b) Quantification of cold-ion beams in a magnetic reconnection jet. Front Astron Space Sci 8:193. https://doi.org/10.3389/fspas.2021.745264

Li X, Wang R, Huang C, et al (2022) Energy conversion and partition in plasma turbulence driven by magnetotail reconnection. Astrophys J 936(1):34. https://doi.org/10.3847/1538-4357/ac84d7

Liang H, Lapenta G, Walker RJ, et al (2017) Oxygen acceleration in magnetotail reconnection. J Geophys Res Space Phys 122(1):618–639. https://doi.org/10.1002/2016JA023060

Lindstedt T, Khotyaintsev YV, Vaivads A, et al (2009) Separatrix regions of magnetic reconnection at the magnetopause. Ann Geophys 27(10):4039–4056. https://doi.org/10.5194/angeo-27-4039-2009

Lindstedt T, Khotyaintsev YV, Vaivads A, et al (2010) Oxygen energization by localized perpendicular electric fields at the cusp boundary. Geophys Res Lett 37(9):L09103. https://doi.org/10.1029/2010GL043117

Liu YH, Drake JF, Swisdak M (2011) The effects of strong temperature anisotropy on the kinetic structure of collisionless slow shocks and reconnection exhausts. II: theory. Phys Plasmas 18:092102. https://doi.org/10.1063/1.3627147

Liu YH, Drake JF, Swisdak M (2012) The structure of magnetic reconnection exhaust boundary. Phys Plasmas 19:022110. https://doi.org/10.1063/1.3685755

Liu YH, Mouikis CG, Kistler LM, et al (2015) The heavy ion diffusion region in magnetic reconnection in the Earth's magnetotail. J Geophys Res Space Phys 120(5):3535–3551. https://doi.org/10.1002/2015JA020982

Liu YH, Cassak P, Li X, et al (2022) First-principle theory of the rate of magnetic reconnection in magnetospheric and solar plasmas. Commun Phys 5:97. https://doi.org/10.1038/s42005-022-00854-x

Liu YH, Hesse M, Genestreti K, et al (2025b) Ohm's law, the reconnection rate, and energy conversion in collisionless magnetic reconnection. Space Sci Rev 221(1):16. https://doi.org/10.1007/s11214-025-01142-0

Liu H, Li W, Tang B, et al (2025a) High-speed electron flows in the Earth magnetotail. AGU Adv 6(2):e2024AV001549. https://doi.org/10.1029/2024AV001549

Malakit K, Shay MA, Cassak PA, et al (2013) New electric field in asymmetric magnetic reconnection. Phys Res Lett 111(13):135001. https://doi.org/10.1103/PhysRevLett.111.135001.

Markidis S, Lapenta G, Bettarini L, et al (2011) Kinetic simulations of magnetic reconnection in presence of a background O^+ population. J Geophys Res Space Phys 116:A00K16. https://doi.org/10.1029/2011JA016429.

Moebius E, Scholer M, Hovestadt D, et al (1980) Observations of a nonthermal ion layer at the plasma sheet boundary during substorm recovery. J Geophys Res 85(A10):5143–5148. https://doi.org/10.1029/JA085iA10p05143

Motoba T, Sitnov MI, Stephens GK, et al (2022) A new perspective on magnetotail electron and ion divergent flows: MMS observations. J Geophys Res Space Phys 127(10):e2022JA030514. https://doi.org/10.1029/2022JA030514

Mozer FS, Bale SD, Phan TD (2002) Evidence of diffusion regions at a subsolar magnetopause crossing. Phys Res Lett 89(1):015002. https://doi.org/10.1103/PhysRevLett.89.015002

Nagai T, Shinohara I, Fujimoto M, et al (2001) Geotail observations of the Hall current system: evidence of magnetic reconnection in the magnetotail. J Geophys Res Space Phys 106:25929–25950. https://doi.org/10.1029/2001JA900038

Nagai T, Nakamura M, Shinohara I, et al (2002) Counterstreaming ions as evidence of magnetic reconnection in the recovery phase of substorms at the kinetic level. Phys Plasmas 9(9):3705–3711. https://doi.org/10.1063/1.1499117

Nagai T, Shinohara I, Zenitani S (2015) Ion acceleration processes in magnetic reconnection: geotail observations in the magnetotail. J Geophys Res Space Phys 120(3):1766–1783. https://doi.org/10.1002/2014JA020737

Nakamura MS, Fujimoto M, Maezawa K (1998) Ion dynamics and resultant velocity space distributions in the course of magnetotail reconnection. J Geophys Res 103(A3):4531–4546. https://doi.org/10.1029/97JA01843

Nakamura TKM, Genestreti KJ, Liu YH, et al (2018) Measurement of the magnetic reconnection rate in the Earth's magnetotail. J Geophys Res Space Phys 123(11):9150–9168. https://doi.org/10.1029/2018JA025713

Nakamura R, Genestreti KJ, Nakamura T, et al (2019) Structure of the current sheet in the 11 July 2017 electron diffusion region event. J Geophys Res Space Phys 124(2):1173–1186. https://doi.org/10.1029/2018JA026028

Nakamura TKM, Hasegawa H, Genestreti KJ, et al (2021) Fast cross-scale energy transfer during turbulent magnetic reconnection. Geophys Res Lett 48(13):e2021GL093524. https://doi.org/10.1029/2021GL093524

Nakamura R, Burch JL, Birn J, et al (2025) Outstanding questions and future research on magnetic reconnection. Space Sci Rev 221(1):17. https://doi.org/10.1007/s11214-025-01143-z

Ng J, Egedal J, Le A, et al (2011) Kinetic structure of the electron diffusion region in antiparallel magnetic reconnection. Phys Res Lett 106(6):065002. https://doi.org/10.1103/PhysRevLett.106.065002

Norgren C, Graham DB, Khotyaintsev YV, et al (2016) Finite gyroradius effects in the electron outflow of asymmetric magnetic reconnection. Geophys Res Lett 43:6724–6733. https://doi.org/10.1002/2016GL069205

Norgren C, Graham DB, Khotyaintsev YV, et al (2018) Electron reconnection in the magnetopause current layer. J Geophys Res Space Phys 123(11):9222–9238. https://doi.org/10.1029/2018JA025676

Norgren C, Hesse M, Graham DB, et al (2020) Electron acceleration and thermalization at magnetotail separatrices. J Geophys Res Space Phys 125(4):e27440. https://doi.org/10.1029/2019JA027440.

Norgren C, Tenfjord P, Hesse M, et al (2021) On the presence and thermalization of cold ions in the exhaust of antiparallel symmetric reconnection. Front Astron Space Sci 8:149. https://doi.org/10.3389/fspas.2021.730061

Norgren C, Hesse M, Phan T, et al (2025) Phase-space contributions to pressure and stress tensors in the electron diffusion region of magnetic reconnection. Phys Plasmas 32(5):052109. https://doi.org/10.1063/5.0246402

Northrop TG (1963) Adiabatic charged-particle motion. Rev Geophys Space Phys 1:283–304. https://doi.org/10.1029/RG001i003p00283

Øieroset M, Phan T, Fujimoto M, et al (2001) In situ detection of collisionless reconnection in the Earth's magnetotail. Nature 412(6845):414–417 https://doi.org/10.1038/35086520

Øieroset M, Phan TD, Ergun R, et al (2021) Spatial evolution of magnetic reconnection diffusion region structures with distance from the X-line. Phys Plasmas 28(12):122901. https://doi.org/10.1063/5.0072182

Oka M, Birn J, Egedal J, et al (2023) Particle acceleration by magnetic reconnection in geospace. Space Sci Rev 219(8):75. https://doi.org/10.1007/s11214-023-01011-8

Petschek HE (1964) Magnetic field annihilation. In: Proc. AAS-NASA symp. Phys. Solar flares, pp 425–439

Phan TD, Eastwood JP, Cassak PA, et al (2016a) MMS observations of electron-scale filamentary currents in the reconnection exhaust and near the X line. Geophys Res Lett 43:6060–6069. https://doi.org/10.1002/2016GL069212

Phan TD, Shay MA, Haggerty CC, et al (2016b) Ion Larmor radius effects near a reconnection X line at the magnetopause: THEMIS observations and simulation comparison. Geophys Res Lett 43(17):8844–8852. https://doi.org/10.1002/2016GL070224

Phan TD, Eastwood JP, Shay MA, et al (2018) Electron magnetic reconnection without ion coupling in Earth's turbulent magnetosheath. Nature 557:202–206. https://doi.org/10.1038/s41586-018-0091-5

Phan TD, Verniero JL, Larson D, et al (2022) Parker Solar Probe observations of solar wind energetic proton beams produced by magnetic reconnection in the near-sun heliospheric current sheet. Geophys Res Lett 49(9):e96986. https://doi.org/10.1029/2021GL096986

Price L, Swisdak M, Drake JF, et al (2020) Turbulence and transport during guide field reconnection at the magnetopause. J Geophys Res Space Phys 125(4). https://doi.org/10.1029/2019JA027498

Pritchard KR, Burch JL, Fuselier SA, et al (2019) Energy conversion and electron acceleration in the magnetopause reconnection diffusion region. Geophys Res Lett 46(10274):10274–10282. https://doi.org/10.1029/2019GL084636

Pritchett P, Coroniti F (2004) Three-dimensional collisionless magnetic reconnection in the presence of a guide field. J Geophys Res 109(A1). https://doi.org/10.1029/2003JA009999

Pritchett PL (2008) Collisionless magnetic reconnection in an asymmetric current sheet. J Geophys Res Space Phys 113:A06210. https://doi.org/10.1029/2007JA012930

Pritchett PL, Mozer FS (2009) Asymmetric magnetic reconnection in the presence of a guide field. J Geophys Res 114. https://doi.org/10.1029/2009JA014343

Pyakurel SP, Shay MA, Phan TD, et al (2019) Transition from ion-coupled to electron-only reconnection: basic physics and implications for plasma turbulence. Phys Plasmas 26(8):082307. https://doi.org/10.1063/1.5090403.

Qi Y, Ergun R, Pathak N, et al (2024) Investigation of a magnetic reconnection event with extraordinarily high particle energization in magnetotail turbulence. Astrophys J Lett 962(2):L39. https://doi.org/10.3847/2041-8213/ad24eb

Retino A, Kepko L, Kucharek H, et al (2023) Unveiling plasma energization and energy transport in the Earth magnetospheric system: the need for future coordinated multiscale observations. A White Paper submitted for the Decadal Survey for Solar and Space Physics. arXiv:2311.09920 [physics.space-ph]

Richard L, Khotyaintsev YV, Graham DB, et al (2021) Observations of short-period ion-scale current sheet flapping. J Geophys Res Space Phys 126(8):e29152. https://doi.org/10.1029/2021JA029152.

Richard L, Khotyaintsev YV, Graham DB, et al (2023) Fast ion isotropization by current sheet scattering in magnetic reconnection jets. Phys Res Lett 131(11):115201. https://doi.org/10.1103/PhysRevLett.131.115201.

Richard L, Khotyaintsev YV, Norgren C, et al (2025) Electron heating by parallel electric fields in magnetotail reconnection. Phys Res Lett 134(21):215201. https://doi.org/10.1103/PhysRevLett.134.215201.

Robertson SL, Eastwood JP, Stawarz JE, et al (2021) Electron trapping in magnetic mirror structures at the edge of magnetopause flux ropes. J Geophys Res Space Phys 126(4):e29182. https://doi.org/10.1029/2021JA029182

Schroeder JM, Egedal J, Cozzani G, et al (2022) 2D reconstruction of magnetotail electron diffusion region measured by MMS. Geophys Res Lett 49(19):e2022GL100384. https://doi.org/10.1029/2022GL100384

Sergeev V, Runov A, Baumjohann W, et al (2003) Current sheet flapping motion and structure observed by Cluster. Geophys Res Lett 30(6). https://doi.org/10.1029/2002GL016500

Shay MA, Phan TD, Haggerty CC, et al (2016) Kinetic signatures of the region surrounding the x line in asymmetric (magnetopause) reconnection. Geophys Res Lett 43(9):4145–4154. https://doi.org/10.1002/2016GL069034

Shuster JR, Chen LJ, Daughton WS, et al (2014) Highly structured electron anisotropy in collisionless reconnection exhausts. Geophys Res Lett 41:5389–5395. https://doi.org/10.1002/2014GL060608

Shuster JR, Chen LJ, Hesse M, et al (2015) Spatiotemporal evolution of electron characteristics in the electron diffusion region of magnetic reconnection: implications for acceleration and heating. Geophys Res Lett 42:2586–2593. https://doi.org/10.1002/2015GL063601

Shuster JR, Gershman DJ, Chen LJ, et al (2019) MMS measurements of the Vlasov equation: probing the electron pressure divergence within thin current sheets. Geophys Res Lett 46(14):7862–7872. https://doi.org/10.1029/2019GL083549

Shuster JR, Gershman DJ, Dorelli JC, et al (2021) Structures in the terms of the Vlasov equation observed at Earth's magnetopause. Nat Phys 17(9):1056–1065. https://doi.org/10.1038/s41567-021-01280-6

Speiser TW (1965) Particle trajectories in model current sheets, 1, analytical solutions. J Geophys Res 70(17):4219–4226. https://doi.org/10.1029/JZ070i017p04219

Speiser TW (1970) Conductivity without collisions or noise. Planet Space Sci 18(4):613–622. https://doi.org/10.1016/0032-0633(70)90136-4

Spitzer L (1962) Physics of fully ionized gases. Interscience, New York

Stawarz JE, Eastwood JP, Genestreti KJ, et al (2018) Intense electric fields and electron-scale substructure within magnetotail flux ropes as revealed by the magnetospheric multiscale mission. Geophys Res Lett 45(17):8783–8792. https://doi.org/10.1029/2018GL079095

Stawarz JE, Matteini L, Parashar TN, et al (2021) Comparative analysis of the various generalized Ohm's law terms in magnetosheath turbulence as observed by magnetospheric multiscale. J Geophys Res Space Phys 126(1):e8447. https://doi.org/10.1029/2020JA028447

Stawarz JE, Muñoz PA, Bessho N, et al (2024) The interplay between collisionless magnetic reconnection and turbulence. Space Sci Rev 220(8):90. https://doi.org/10.1007/s11214-024-01124-8. arXiv:2407.20787 [physics.space-ph]

Steinvall K, Khotyaintsev YV, Graham DB, et al (2021) Large amplitude electrostatic proton plasma frequency waves in the magnetospheric separatrix and outflow regions during magnetic reconnection. Geophys Res Lett 48(5):e90286. https://doi.org/10.1029/2020GL090286

Swisdak M, Drake JF, Shay MA, et al (2005) Transition from antiparallel to component magnetic reconnection. J Geophys Res Space Phys 110(A5):A05210. https://doi.org/10.1029/2004JA010748.

Swisdak M, Drake JF, Price L, et al (2018) Localized and intense energy conversion in the diffusion region of asymmetric magnetic reconnection. Geophys Res Lett 45(11):5260–5267. https://doi.org/10.1029/2017GL076862

Takahashi K, Hones JEW (1988) ISEE 1 and 2 observations of ion distributions at the plasma sheet-tail lobe boundary. J Geophys Res 93(A8):8558–8582. https://doi.org/10.1029/JA093iA08p08558

Tanaka KG, Retino A, Asano Y, et al (2008) Effects on magnetic reconnection of a density asymmetry across the current sheet. Ann Geophys 26(8):2471–2483

Tang X, Cattell C, Dombeck J, et al (2013) THEMIS observations of the magnetopause electron diffusion region: large amplitude waves and heated electrons. Geophys Res Lett 40:2884. https://doi.org/10.1002/grl.50565

Tang BB, Li WY, Graham DB, et al (2019) Crescent-shaped electron distributions at the nonreconnecting magnetopause: magnetospheric multiscale observations. Geophys Res Lett 46(6):3024–3032. https://doi.org/10.1029/2019GL082231

Tang BB, Li WY, Le A, et al (2020) Electron mixing and isotropization in the exhaust of asymmetric magnetic reconnection with a guide field. Geophys Res Lett 47(14):e87159. https://doi.org/10.1029/2020GL087159

Tang BB, Li WY, Khotyaintsev YV, et al (2022) Fine structures of the electron current sheet in magnetotail guide-field reconnection. Geophys Res Lett 49(9):e97573. https://doi.org/10.1029/2021GL097573

Tenfjord P, Hesse M, Norgren C (2018) The formation of an oxygen wave by magnetic reconnection. J Geophys Res Space Phys 123(11):9370–9380. https://doi.org/10.1029/2018JA026026

Tigik SF, Graham DB, Khotyaintsev YV (2025) Electron-scale energy transfer due to lower hybrid waves during asymmetric reconnection. J Geophys Res Space Phys 130(4):e2024JA033503. https://doi.org/10.1029/2024JA033503.

Toledo-Redondo S, Vaivads A, André M, et al (2015) Modification of the Hall physics in magnetic reconnection due to cold ions at the Earth's magnetopause. Geophys Res Lett 42(15):6146–6154. https://doi.org/10.1002/2015GL065129

Toledo-Redondo S, André M, Khotyaintsev YV, et al (2016a) Cold ion demagnetization near the X-line of magnetic reconnection. Geophys Res Lett 43:6759–6767. https://doi.org/10.1002/2016GL069877

Toledo-Redondo S, André M, Vaivads A, et al (2016b) Cold ion heating at the dayside magnetopause during magnetic reconnection. Geophys Res Lett 43(1):58–66. https://doi.org/10.1002/2015GL067187

Toledo-Redondo S, André M, Khotyaintsev YV, et al (2017) Energy budget and mechanisms of cold ion heating in asymmetric magnetic reconnection. J Geophys Res Space Phys 122(9):9396–9413. https://doi.org/10.1002/2017JA024553

Toledo-Redondo S, André M, Aunai N, et al (2021) Impacts of ionospheric ions on magnetic reconnection and Earth's magnetosphere dynamics. Rev Geophys 59(3):e00707. https://doi.org/10.1029/2020RG000707

Torbert RB, Burch JL, Giles BL, et al (2016) Estimates of terms in Ohm's law during an encounter with an electron diffusion region. Geophys Res Lett 43(12):5918–5925. https://doi.org/10.1002/2016GL069553

Torbert RB, Burch JL, Argall MR, et al (2017) Structure and dissipation characteristics of an electron diffusion region observed by MMS during a rapid, normal-incidence magnetopause crossing. J Geophys Res Space Phys 122(12):11901–11916. https://doi.org/10.1002/2017JA024579

Torbert RB, Burch JL, Phan TD, et al (2018) Electron-scale dynamics of the diffusion region during symmetric magnetic reconnection in space. Science 362(6421):1391–1395. https://doi.org/10.1126/science.aat2998.

Usami S, Zenitani S (2024) Three-dimensional crescent-shaped ion velocity distributions created by magnetic reconnection in the presence of a guide field. Phys Plasmas 31(2):022102. https://doi.org/10.1063/5.0171785

Uzdensky DA, Kulsrud RM (2006) Physical origin of the quadrupole out-of-plane magnetic field in Hall-magnetohydrodynamic reconnection. Phys Plasmas 13(6):062305. https://doi.org/10.1063/1.2209627.

Vasyliunas VM (1975) Theoretical models of magnetic-field line merging 1. Rev Geophys 13(1):303–336 https://doi.org/10.1029/RG013i001p00303

Vines SK, Fuselier SA, Trattner KJ, et al (2017) Magnetospheric ion evolution across the low-latitude boundary layer separatrix. J Geophys Res Space Phys 122(10):10247–10262. https://doi.org/10.1002/2017JA024061

Wang R, Nakamura R, Lu Q, et al (2012) Asymmetry in the current sheet and secondary magnetic flux ropes during guide field magnetic reconnection. J Geophys Res Space Phys 117(A7):A07223. https://doi.org/10.1029/2011JA017384

Wang S, Chen LJ, Bessho N, et al (2016a) Electron heating in the exhaust of magnetic reconnection with negligible guide field. J Geophys Res Space Phys 121(3):2104–2130. https://doi.org/10.1002/2015JA021892

Wang S, Chen LJ, Hesse M, et al (2016b) Two-scale ion meandering caused by the polarization electric field during asymmetric reconnection. Geophys Res Lett 43(15):7831–7839. https://doi.org/10.1002/2016GL069842

Wang S, Chen LJ, Hesse M, et al (2016c) Ion demagnetization in the magnetopause current layer observed by MMS. Geophys Res Lett 43(10):4850–4857. https://doi.org/10.1002/2016GL069406

Wang S, Chen LJ, Hesse M, et al (2017) Parallel electron heating in the magnetospheric inflow region. Geophys Res Lett 44:4384–4392. https://doi.org/10.1002/2017GL073404

Wang S, Chen LJ, Bessho N, et al (2019a) Ion behaviors in the reconnection diffusion region of a corrugated magnetotail current sheet. Geophys Res Lett 46(10):5014–5020. https://doi.org/10.1029/2019GL082226

Wang S, Chen LJ, Bessho N, et al (2019b) Observational evidence of magnetic reconnection in the terrestrial bow shock transition region. Geophys Res Lett 46(2):562–570. https://doi.org/10.1029/2018GL080944

Wang R, Lu Q, Lu S, et al (2020) Physical implication of two types of reconnection electron diffusion regions with and without ion-coupling in the magnetotail current sheet. Geophys Res Lett 47(21):e88761. https://doi.org/10.1029/2020GL088761

Wang S, Bessho N, Graham DB, et al (2022a) Whistler waves associated with electron beams in magnetopause reconnection diffusion regions. J Geophys Res Space Phys 127(9):e30882. https://doi.org/10.1029/2022JA030882

Wang S, Chen LJ, Bessho N, et al (2022b) Lower-hybrid wave structures and interactions with electrons observed in magnetotail reconnection diffusion regions. J Geophys Res Space Phys 127(5):e30109. https://doi.org/10.1029/2021JA030109

Wang R, Lu S, Wang S, et al (2023b) Recent progress on magnetic reconnection by in situ measurements. Rev Mod Plasma Phys 7(1):27. https://doi.org/10.1007/s41614-023-00129-0

Wang HW, Tang BB, Li WY, et al (2023a) Electron dynamics in the electron current sheet during strong guide-field reconnection. Geophys Res Lett 50(10):e2023GL103046. https://doi.org/10.1029/2023GL103046

Webster JM, Burch JL, Reiff PH, et al (2018) Magnetospheric multiscale dayside reconnection electron diffusion region events. J Geophys Res Space Phys 123(6):4858–4878. https://doi.org/10.1029/2018JA025245

Wei YY, Huang SY, Rong ZJ, et al (2019) Observations of short-period current sheet flapping events in the Earth's magnetotail. Astrophys J 874(2):L18. https://doi.org/10.3847/2041-8213/ab0f28

Wetherton BA, Egedal J, Le A, et al (2019) Validation of anisotropic electron fluid closure through in situ spacecraft observations of magnetic reconnection. Geophys Res Lett 46(12):6223–6229. https://doi.org/10.1029/2019GL083119

Wetherton BA, Egedal J, Le A, et al (2021) Anisotropic electron fluid closure validated by in situ spacecraft observations in the far exhaust of guide-field reconnection. J Geophys Res Space Phys 126(1):e2020JA028604. https://doi.org/10.1029/2020JA028604

Wetherton BA, Egedal J, Le A, et al (2022) Generation of a strong parallel electric field and embedded electron jet in the exhaust of moderate guide field reconnection. Geophys Res Lett 49(14):e98907. https://doi.org/10.1029/2022GL098907

Wilder FD, Ergun RE, Goodrich KA, et al (2016) Observations of whistler mode waves with nonlinear parallel electric fields near the dayside magnetic reconnection separatrix by the magnetospheric multiscale mission. Geophys Res Lett 43(12):5909–5917. https://doi.org/10.1002/2016GL069473

Wilder FD, Ergun RE, Eriksson S, et al (2017) Multipoint measurements of the electron jet of symmetric magnetic reconnection with a moderate guide field. Phys Rev Lett 118(26):265101. https://doi.org/10.1103/PhysRevLett.118.265101

Wilder FD, Ergun RE, Burch JL, et al (2018) The role of the parallel electric field in electron-scale dissipation at reconnecting currents in the magnetosheath. J Geophys Res Space Phys 123(8):6533–6547. https://doi.org/10.1029/2018JA025529

Wygant JR, Cattell CA, Lysak R, et al (2005) Cluster observations of an intense normal component of the electric field at a thin reconnecting current sheet in the tail and its role in the shock-like acceleration of the ion fluid into the separatrix region. J Geophys Res Space Phys 110:A09206. https://doi.org/10.1029/2004JA010708

Xu Y, Fu HS, Norgren C, et al (2019) Ionospheric cold ions detected by MMS behind dipolarization fronts. Geophys Res Lett 46(14):7883–7892. https://doi.org/10.1029/2019GL083885

Yamada M (2022) Magnetic reconnection. A modern synthesis of theory, experiment, and observations

Yamada M, Kulsrud R, Ji H (2010) Magnetic reconnection. Rev Mod Phys 82(1):603–664. https://doi.org/10.1103/RevModPhys.82.603

Yamada M, Chen LJ, Yoo J, et al (2018) The two-fluid dynamics and energetics of the asymmetric magnetic reconnection in laboratory and space plasmas. Nat Commun 9:5223. https://doi.org/10.1038/s41467-018-07680-2

Yoon YD, Yun GS, Wendel DE, et al (2021) Collisionless relaxation of a disequilibrated current sheet and implications for bifurcated structures. Nat Commun 12:3774. https://doi.org/10.1038/s41467-021-24006-x

Yordanova E, Vörös Z, Varsani A, et al (2016) Electron scale structures and magnetic reconnection signatures in the turbulent magnetosheath. Geophys Res Lett 43(12):5969–5978. https://doi.org/10.1002/2016GL069191.

Yu X, Wang R, Lu Q, et al (2019) Nonideal electric field observed in the separatrix region of a magnetotail reconnection event. Geophys Res Lett 46(19):10744–10753. https://doi.org/10.1029/2019GL082538

Zelenyi LM, Neishtadt AI, Artemyev AV, et al (2013) Quasiadiabatic dynamics of charged particles in a space plasma. Phys Usp 56(4):347–394. https://doi.org/10.3367/UFNe.0183.201304b.0365

Zenitani S, Hesse M, Klimas A, et al (2011) New measure of the dissipation region in collisionless magnetic reconnection. Phys Rev Lett 106(19):195003. https://doi.org/10.1103/PhysRevLett.106.195003.

Zenitani S, Shinohara I, Nagai T, et al (2013) Kinetic aspects of the ion current layer in a reconnection outflow exhaust. Phys Plasmas 20(9):092120. https://doi.org/10.1063/1.4821963.

Zenitani S, Nagai T (2016) Particle dynamics in the electron current layer in collisionless magnetic reconnection. Phys Plasmas 23(10):102102. https://doi.org/10.1063/1.4963008.

Zenitani S, Hasegawa H, Nagai T (2017) Electron dynamics surrounding the X line in asymmetric magnetic reconnection. J Geophys Res Space Phys 122:7396–7413. https://doi.org/10.1002/2017JA023969.

Zhang TL, Baumjohann W, Nakamura R, et al (2002) A wavy twisted neutral sheet observed by Cluster. Geophys Res Lett 29(19):5-1–5-4. https://doi.org/10.1029/2002GL015544

Zhang YC, Shen C, Marchaudon A, et al (2016) First in situ evidence of electron pitch angle scattering due to magnetic field line curvature in the ion diffusion region. J Geophys Res Space Phys 121(5):4103–4110. https://doi.org/10.1002/2016JA022409

Zhang Q, Drake JF, Swisdak M (2019) Particle heating and energy partition in low-β guide field reconnection with kinetic Riemann simulations. Phys Plasmas 26(7):072115. https://doi.org/10.1063/1.5104352.

Zhang M, Wang R, Lu Q, et al (2020) Observation of the tailward electron flows commonly detected at the flow boundary of the earthward ion bursty bulk flows in the magnetotail. Astrophys J 891(2):175 https://doi.org/10.3847/1538-4357/ab72a8

Zhong ZH, Zhou M, Tang RX, et al (2020) Direct evidence for electron acceleration within ion-scale flux rope. Geophys Res Lett 47(1):e85141. https://doi.org/10.1029/2019GL085141

Zhou M, Deng XH, Zhong ZH, et al (2019a) Observations of an electron diffusion region in symmetric reconnection with weak guide field. Astrophys J 870(1):34. https://doi.org/10.3847/1538-4357/aaf16f

Zhou M, Man HY, Zhong ZH, et al (2019b) Sub-ion-scale dynamics of the ion diffusion region in the magnetotail: MMS observations. J Geophys Res Space Phys 124(10):7898–7911. https://doi.org/10.1029/2019JA026817

Zhou M, Zhong Z, Deng X (2022) Kinetic properties of collisionless magnetic reconnection in space plasma: in situ observations. Rev Mod Plasma Phys 6(1):15. https://doi.org/10.1007/s41614-022-00079-z

Zhu C, Zhang H, Fu S, et al (2019) Trapped and accelerated electrons within a magnetic mirror behind a flux rope on the magnetopause. J Geophys Res Space Phys 124(6):3993–4008. https://doi.org/10.1029/2019JA026464

Authors and Affiliations

C. Norgren[1,2] · L.-J. Chen[3] · D.B. Graham[1] · N. Bessho[3,4] · J. Egedal[5] · L. Richard[1] · Yu. V. Khotyaintsev[1] · J. Shuster[6] · S. Toledo-Redondo[7] · B. Lavraud[8] · H. Hasegawa[9] · J.P. Eastwood[10] · M. Hesse[11] · Y.-H. Liu[12] · J.C. Holmes[13] · M. Argall[6]

✉ C. Norgren
cecilia.norgren@irfu.se; cecilia.norgren@uib.no

1 Swedish Institute of Space Physics, Uppsala, Sweden

2 Department of Physics and Technology, University of Bergen, Bergen, Norway

3 NASA Goddard Space Flight Center, Greenbelt, MD, 20771, USA

4 Department of Astronomy, University of Maryland, College Park, MD 20742, USA

5 Department of Physics, University of Wisconsin-Madison, Madison, WI 53706, USA

6 Space Science Center, University of New Hampshire, Durham, NH, 03824, USA

7 Department of Electromagnetism and Electronics, University of Murcia, Murcia, Spain

8 Laboratoire d'Astrophysique de Bordeaux, Univ. Bordeaux, CNRS, Pessac, France

9 Institute of Space and Astronautical Science, JAXA, Sagamihara, Japan

10 Department of Physics, Imperial College London, London, UK

11 Ames Research Center, NASA, Moffett Field, CA 94035, USA

12 Department of Physics and Astronomy, Dartmouth College, Hanover, NH 03750, USA

13 Los Alamos National Laboratory, Los Alamos, NM, USA

III. Waves and Turbulence

Space Science Reviews (2025) 221:20
https://doi.org/10.1007/s11214-024-01133-7

The Role of Kinetic Instabilities and Waves in Collisionless Magnetic Reconnection

D.B. Graham[1] · G. Cozzani[2] · Yu.V. Khotyaintsev[1,3] · V.D. Wilder[4] · J.C. Holmes[5] · T.K.M. Nakamura[6] · J. Büchner[7,8] · K. Dokgo[9] · L. Richard[1] · K. Steinvall[10] · C. Norgren[1,11] · L.-J. Chen[12] · H. Ji[13,14] · J.F. Drake[15] · J.E. Stawarz[16] · S. Eriksson[17]

Received: 6 June 2024 / Accepted: 18 December 2024 / Published online: 14 February 2025

Abstract
Magnetic reconnection converts magnetic field energy into particle energy by breaking and reconnecting magnetic field lines. Magnetic reconnection is a kinetic process that generates a wide variety of kinetic waves via wave-particle interactions. Kinetic waves have been proposed to play an important role in magnetic reconnection in collisionless plasmas by, for example, contributing to anomalous resistivity and diffusion, particle heating, and transfer of energy between different particle populations. These waves range from below the ion cyclotron frequency to above the electron plasma frequency and from ion kinetic scales down to electron Debye length scales. This review aims to describe the progress made in understanding the relationship between magnetic reconnection and kinetic waves. We focus on the waves in different parts of the reconnection region, namely, the diffusion region, separatrices, outflow regions, and jet fronts. Particular emphasis is placed on the recent observations from the Magnetospheric Multiscale (MMS) spacecraft and numerical simulations, which have substantially increased the understanding of the interplay between kinetic waves and reconnection. Some of the ongoing questions related to waves and reconnection are discussed.

Keywords Magnetic reconnection · Waves · Instabilities · Kinetic processes · Methods

1 Introduction

Magnetic reconnection is the process by which magnetic field lines break and reconnect, resulting in explosive energy releases in the form of particle acceleration and heating (Yamada et al. 2010). Magnetic reconnection is a fundamental process in solar and astrophysical plasmas, and occurs in many plasma environments, such as planetary magnetopauses and magnetotails, the solar corona, the solar wind, and accretion disks. In many contexts, such as at Earth's magnetopause and magnetotail, magnetic reconnection is essentially a collisionless process (Coulomb collisions do not play a significant role). As a result, particle distributions can deviate significantly from Maxwellian distributions, resulting in unstable particle distributions and leading to a wide variety of plasma waves and instabilities. Additionally, the strong currents and plasma inhomogeneities associated with magnetic reconnection pro-

Extended author information available on the last page of the article

vide a source of energy for waves. Through wave-particle interactions, waves can modify the local plasma and have important effects on reconnection. For example, waves can heat the plasma, dissipate currents, produce anomalous resistivity and diffusion, and can transfer energy between different particle species. These effects can have important consequences for ongoing reconnection. Likewise, pre-existing waves in the plasma could impact the initialization of reconnection.

The waves associated with magnetic reconnection have been studied extensively using spacecraft observations, laboratory experiments, and numerical simulations. These different approaches have each provided insights into the role waves play in magnetic reconnection. In the case of spacecraft observations, high-resolution measurements of the electric and magnetic fields have enabled waves to be directly characterized and studied, while particle measurements with much lower time resolution than that of the fields have made determining the cause and effects of waves on the plasma challenging. Simulations can model the large-scale reconnection processes along with the development of kinetic waves. However, to achieve this, artificial or idealized plasma conditions often need to be employed.

The relationship between kinetic waves and reconnection has been reviewed several times in the past (Vaivads et al. 2006; Fujimoto et al. 2011; Khotyaintsev et al. 2019). These reviews have shown that important advances have been made using spacecraft observations, such as Cluster, THEMIS (Time History of Events and Macroscale Interactions during Substorms), and MMS (Magnetospheric Multiscale), numerical simulations, and laboratory experiments. Ji et al. (2023) provides a recent review of laboratory experiments on collisionless reconnection, including the associated waves.

The aim of this paper is to review the most recent advances in our understanding of the relation between magnetic reconnection and kinetic plasma waves. We primarily focus on recent in situ observations from the MMS spacecraft and recent numerical simulations. The outline of this paper is as follows: Sect. 2 provides an overview of magnetic reconnection and the associated waves. In Sect. 3, we describe the waves associated with the different regions of reconnection, namely, the diffusion region, separatrices, outflow regions, and jet fronts. In Sect. 4, we focus on larger-scale instabilities, namely, current sheet kinking and tearing mode instabilities, and their interaction with reconnection. In Sect. 5, we discuss the effects of waves on reconnection and ongoing questions. In Sect. 6, we state the conclusions and recent key results.

2 Overview of Waves and Reconnection

In this section, we provide an overview of the waves associated with reconnection. In particular, we describe the types of waves that have been reported and where they are found in relation to the different regions of magnetic reconnection.

Plasmas consist of ions and electrons and are governed by electromagnetic forces acting collectively on the particles. Waves in plasmas are generally defined as oscillations in the electromagnetic fields and particles, which propagate in a plasma and can transport energy without net motion of the plasma. Waves develop in a plasma due to instabilities, which result from free energy that has accumulated in the system. The theory of plasma waves and instabilities has been studied in many papers and textbooks. Comprehensive reviews of the theory of plasma waves and instabilities can be found in Stix (1992), Swanson (1989), Treumann and Baumjohann (1997), Gary (1993).

In magnetic reconnection, and plasmas more generally, there are multiple sources of free energy. These include deformations in the particle distribution functions, inhomogeneities in the plasma, and electric currents. The effect of the growing waves on the plasma is to remove the free energy responsible for the generation of the waves. This can include returning unstable particle distributions to stable distributions, reducing the magnitude of electric currents, and reducing inhomogeneities in the plasma. Additionally, plasma waves can potentially produce anomalous resistivity and diffusion, as well as heat the plasma through wave-particle interactions. Waves can also potentially accelerate particles to non-thermal energies.

In plasmas, waves can be either electrostatic or electromagnetic. Electrostatic waves are characterized by longitudinal fluctuations in the electric field **E**, namely **E** is aligned with the wave vector **k**, due to charge separation and have no associated magnetic field **B** fluctuations. Examples of electrostatic waves include Langmuir waves, ion-acoustic waves, and electron-acoustic waves. Electromagnetic waves are characterized by **E** transverse to the wave vector **k** and have associated magnetic field fluctuations. Examples of electromagnetic waves include magnetic field **B** aligned whistler and Alfvén waves. In general, waves exhibit both electrostatic and electromagnetic components, but are classified as electrostatic or electromagnetic depending on which component of **E** dominates.

In magnetic reconnection regions, a wide variety of waves can develop. Table 1 lists some of the most common waves associated with magnetic reconnection, along with their properties and the sources of instability. The different plasma waves span multiple spatial scales and frequencies. At relatively low frequencies, close to the ion cyclotron frequency f_{ci}, Alfvén or ion-cyclotron waves can develop due to deformations in the ion particle distribution functions. At smaller scales between ion and electron spatial scales, lower hybrid waves can develop. Electron-scale waves include Langmuir and upper hybrid (UH) waves, ion-acoustic waves, and whistler waves.

Figure 1 provides a schematic overview of the different regions associated with magnetic reconnection, namely, the ion and electron diffusion regions, separatrices, outflow regions, and jet fronts. Magnetic reconnection is often characterized by an X-line magnetic field structure, where the magnetic fields associated with inflowing plasmas reconnect. The separatrix regions are narrow boundaries between the inflow and outflow regions. Near the X line are the ion and electron diffusion regions, where ions and electrons, respectively, decouple from the magnetic field (shown in red and blue). In each of these regions, different types of electrostatic and electromagnetic waves commonly develop, some of which are listed in Table 1. Figure 1 represents symmetric reconnection, where the inflowing plasmas have the same properties. Symmetric reconnection typically occurs in Earth's magnetotail. In contrast, reconnection is typically asymmetric at Earth's dayside magnetopause, where reconnection occurs between the hotter tenuous magnetospheric and the cooler, denser magnetosheath plasmas. These different plasma conditions modify the ion and electron distributions observed in the separatrices, outflow regions, and diffusion regions, which can modify the types of instabilities that develop. Additionally, the out-of-plane guide field can vary for reconnection, which can modify the structure and dynamics of the diffusion region. In particular, for a negligible guide field, the currents around the electron diffusion region (EDR) are perpendicular to **B**, while for a strong guide field, strong currents parallel to **B** occur. This, too, can modify the types of instabilities that can develop. In the following section, we discuss the differences between symmetric and asymmetric reconnection, and antiparallel and guide-field reconnection, where applicable.

Table 1 A summary of some of the commonly observed waves associated with magnetic reconnection. The properties of the waves and the sources of free energy

Summary of the most common waves associated with magnetic reconnection		
Waves	Properties and observed characteristics	Source of the waves
Langmuir waves	Electrostatic waves with narrow frequency range near f_{pe}, with $E_{\parallel} \gg E_{\perp}$.	Fast electron beams; bump-on-tail instability.
Upper hybrid waves	Quasi-electrostatic waves with narrow range near f_{uh}, with $E_{\perp} \gg E_{\parallel}$.	Complex electron distributions such as beams, loss cones, and agyrotropic distributions.
Electron holes	Nonlinear localized positive potential structures supported by trapped electrons. Seen as bipolar fluctuations in $E_{\parallel}$.	Nonlinear evolution of streaming instabilities, such as from electron beams and relative drift between electron and ion populations.
Ion-acoustic waves	Electrostatic waves with frequencies near f_{pi}. Typically $E_{\parallel} \gg E_{\perp}$.	Relative drift between electron and ion populations and ion beams. Typically occur when $ZT_e \gg T_i$.
Whistler waves	Typically electromagnetic waves with right-hand circular polarization and frequencies $f_{lh} \lesssim f < f_{ce}$. Identified from ellipticity of **B** fluctuations.	Electron temperature anisotropy $T_{\perp} > T_{\parallel}$, loss cones, electron beams, parallel electron heat flux.
Lower hybrid waves	Typically quasi-electrostatic waves with measurable **B** fluctuations and frequencies near f_{lh}.	Plasma inhomogeneities and cross-field currents, relative drift between electron and ions or different ion populations in the cross-field direction.
Alfvén waves	Magnetic field fluctuations near f_{ci}, often left-hand polarization.	Ion temperature anisotropies $T_{\perp} > T_{\parallel}$, ion beams.

3 Regions of Reconnection

In this section, we review the different types of waves associated with magnetic reconnection. We focus on the different regions associated with reconnection, namely, the diffusion region, separatrix regions, outflow regions, and jet fronts.

3.1 Diffusion Region

A wide variety of waves have been reported in or near reconnection diffusion regions. In the ion diffusion region (IDR), ion meandering and finite Larmor radius effects occur, leading to agyrotropic ion crescent-shaped distributions. Similarly, agyrotropic electron distributions occur in the EDR due to finite Larmor radius effects. In the diffusion region, strong currents and plasma inhomogeneities also develop. All these features can lead to the development of a wide variety of plasma waves. Before MMS, diffusion regions, in particular EDRs, were difficult to identify. Therefore, observations of waves associated with diffusion regions are biased toward MMS observations. Recent research has particularly focused on the nature of lower hybrid drift waves near the diffusion region and UH and electron Bernstein waves in the EDR.

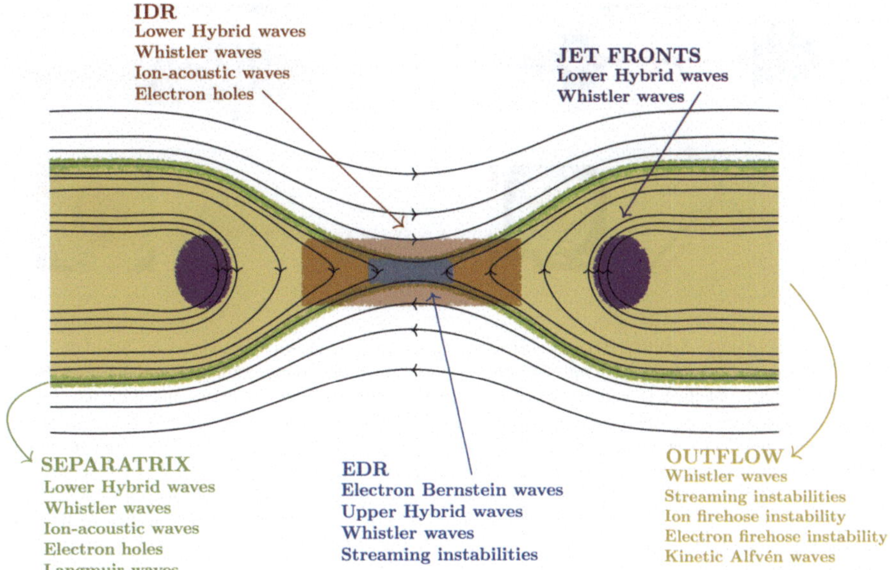

Fig. 1 Schematic of magnetic reconnection showing the different regions, namely, the ion and electron diffusion regions (red and blue), separatrices (green), ion outflow (yellow), and jet front regions (purple). For these regions we list the types of waves expected there based on observations and simulations

Upper Hybrid and Electron Bernstein Waves One of the first results from MMS observations was the verification of agyrotropic crescent-shaped or beam electron distributions in the EDR of symmetric (Torbert et al. 2018) and asymmetric reconnection (Burch et al. 2016). These distributions form due to electron finite Larmor radius effects or electron bouncing near the center of the current sheet (Norgren et al. 2016, 2025, this collection). At Earth's magnetopause these distributions have been observed on both the magnetospheric (Burch et al. 2016) and magnetosheath (Chen et al. 2016) side of the X line. Graham et al. (2017b) found intense UH waves associated with agyrotropic electron beams and crescent-shaped distributions on the magnetosheath side of the X line in the EDR. The source of the waves was a beam-plasma instability between the agyrotropic electron beam and the core population, where the beam component has been demonstrated to be unmagnetized meandering electrons (Chen et al. 2017). UH waves generated by agyrotropic electron distributions in the EDR were subsequently found by Burch et al. (2019) for symmetric reconnection in Earth's magnetotail due to the same beam-plasma instability.

Figures 2a–2c show an example of the UH waves observed at a magnetotail EDR by MMS3 (Burch et al. 2019). In this and the other events, the waves are characterized by $E_\perp \gg E_\parallel$, and the waves have frequency near $f_{pe} \approx f_{uh}$, where f_{pe} and f_{uh} are the electron plasma and upper hybrid frequencies. Like the magnetopause case, the source of instability was shown to be the agyrotropic crescent-shaped distributions. Figure 2c shows an example of the electron distribution observed at the same time as the UH waves. The distribution is characterized by a gyrotropic core population and a higher energy agyrotropic beam or crescent component. A later case study by Li et al. (2021) found that UH waves were generated on both inflow sides of the EDR in Earth's magnetotail. Li et al. (2021) argued that the waves were generated by inbound meandering electrons and that the waves could significantly dissipate energy from the electrons.

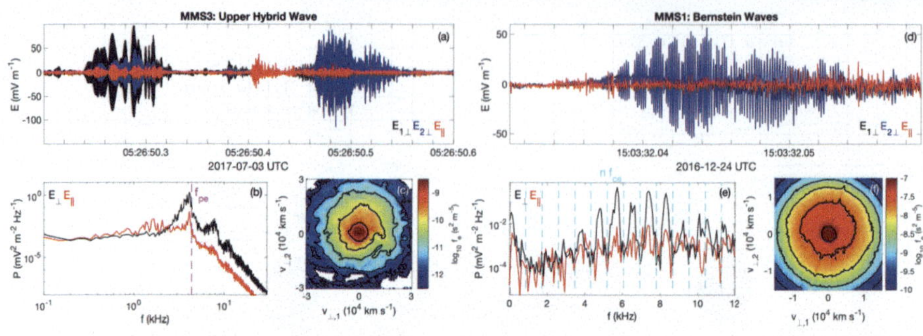

Fig. 2 Example of upper hybrid (UH) waves observed by MMS in magnetotail reconnection, based on Burch et al. (2019) (panels a–c), and an example of electron Bernstein waves in magnetopause reconnection, based on Li et al. (2020) (panels d–f). (a) and (d) Electric field waveform of the UH waves and Bernstein waves, respectively. The waveforms are presented in field-aligned coordinates, where $E_{1\perp}$ and $E_{2\perp}$ (black and blue) are the electric field perpendicular to **B**, while $E_{\parallel}$ (red) is aligned with **B**. (b) and (e) Power spectra of UH and Bernstein waves, respectively. The black and red lines are the powers of $E_{\perp}$ and $E_{\parallel}$, respectively. In panel (b), the magenta dashed line indicates f_{pe}, and in panel (e), the cyan dashed line indicates f_{ce} and its harmonics. (c) and (f) Two-dimensional reduced electron distributions in the plane normal to **B** at the times the waves were observed, where $v_{\perp,2}$ is in the direction of bulk flow perpendicular to **B**, and $v_{\perp,1}$ is orthogonal to **B** and $v_{\perp,2}$

In an EDR at Earth's magnetopause, agyrotropic electron distributions were found by Li et al. (2020) to generate a series of electron Bernstein waves. Similar to the case in Graham et al. (2017a), the waves were observed on the magnetosheath side of the EDR. Figures 2d–2f show these Bernstein waves and the associated electron distribution. The Bernstein waves are characterized by $E_{\perp} \gg E_{\parallel}$ and distinct spectral peaks at frequencies between the harmonics of the electron cyclotron frequency f_{ce}. The associated electron distribution, shown in Fig. 2f, exhibits a gyrotropic core population and an agyrotropic crescent-shaped component. Li et al. (2020) argued that the observed distributions could be unstable to electron Bernstein waves by modeling the agyrotropic crescent distributions as a gyrotropic ring distribution. In each case, the UH or Bernstein waves were associated with small to moderate guide-field reconnection events. To date, electron Bernstein waves have not been observed in magnetotail EDRs.

Crescent distributions are a key element of reconnection, and therefore recent theoretical and numerical results have focused on understanding the generation of the UH and Bernstein waves by agyrotropic distributions and the nonlinear evolution of the UH waves. Dokgo et al. (2020b) derived an electrostatic kinetic dispersion equation for an agyrotropic beam-plasma instability, which can model the growth of UH and electron Bernstein waves. Making use of the dispersion equation and plasma parameters obtained from MMS, they showed that this new model predicted that both types of waves could be excited, depending on the beam properties; when the beam was low density, the frequency of the excited waves was close to f_{pe}, while for denser beams the frequency was downshifted. For faster growth rates $\gamma \gtrsim \Omega_{ce}$ a more continuous beam mode was found, while for lower growth rates discrete Bernstein modes became clearer. By comparing the model with observations from Li et al. (2020) and Burch et al. (2019) they were able to explain the generation of UH and Bernstein waves.

At present, fully kinetic three-dimensional simulations of reconnection, which can resolve these UH and Bernstein waves, are not possible due to the large separation in spatial and temporal scales. Instead, to numerically study these waves, small-scale local homogeneous simulations were performed using initial conditions and electron distributions based on MMS observations. Using these simulations, Dokgo et al. (2019) showed that UH waves were generated by a beam-plasma interaction between the gyrotropic core electrons and agyrotropic beam. Moreover, the UH waves were found to undergo nonlinear processes, namely, the generation of electrostatic harmonic waves and the production of long-wavelength radio waves near f_{pe} and the second harmonic via nonlinear three-wave processes. Evidence of nonlinear electrostatic harmonics has been seen by MMS (Li et al. 2021) (see also Fig. 2b, where spectral peaks are observed at the harmonics of the UH waves). Similar simulations were employed to investigate how UH waves contribute to energy dissipation near the EDR (Dokgo et al. 2020a). Dokgo et al. (2020a) showed that the waves could contribute to dissipation via wave-particle interactions and argued that the profiles of plasma parameters could be modified due to the waves, which could affect the larger-scale energy dissipation processes in the EDR. Overall, these results show that the agyrotropic electron distributions in or near the EDR are often unstable to UH or Bernstein waves.

Lower Hybrid Drift Waves Lower hybrid waves are known to develop at plasma boundaries where the plasma is inhomogeneous. Thus, lower hybrid waves can develop at current sheets, in the IDR of reconnection, and along the separatrices. The waves are typically quasi-electrostatic and develop at short wavelengths with wave number $k\rho_e \sim 1$, where ρ_e is the electron Larmor radius, and develop at the edges of current sheets. However, near the center of the current sheet and close to the EDR, theory and simulations predict that lower hybrid waves become more electromagnetic with longer wavelengths, corresponding to $k\sqrt{\rho_i \rho_e} \sim 1$ (Yoon et al. 2002; Daughton 2003), where ρ_i is the ion Larmor radius. In both cases, the waves tend to propagate in the reconnection out-of-plane direction with wave vectors nearly perpendicular to $\mathbf{B}$. In reconnection, the development of lower hybrid waves is most often attributed to the lower hybrid drift instability (Davidson et al. 1977), where inhomogeneities and cross-field currents drive the instability. Figure 3 shows two examples of three-dimensional (3D) magnetic reconnection where lower hybrid waves develop. The overall structure of the reconnection region is retained, while the lower hybrid waves introduce smaller-scale perturbations in the current density.

Lower hybrid waves are typically identified in spacecraft observations by electric field fluctuations, which exhibit broadband fluctuations around the lower hybrid frequency f_{lh} (Cattell and Mozer 1986). These fluctuations have been reported in the diffusion regions of reconnection at the magnetopause (Bale et al. 2002) and in the magnetotail (Zhou et al. 2009b). In some cases, the properties of the waves, such as wave vector and phase speed,

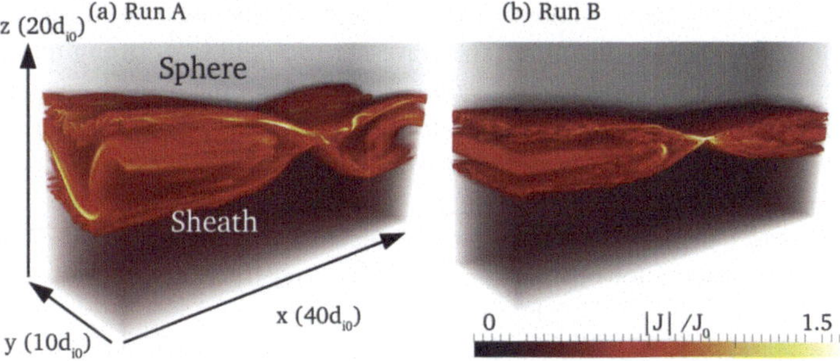

Fig. 3 Volume rendering of current density from 3D PIC asymmetric reconnection simulations with different guide fields B_g. (a) $B_g = 0.099\ B_0$ (weak guide field); (b) $B_g = 0.40\ B_0$ (moderate guide field). The fluctuations at the separatrices are found in both runs and correspond to electrostatic lower hybrid waves. The run with a weak guide field also exhibits a longer-wavelength, electromagnetic lower hybrid mode, which is suppressed in higher guide-field simulations. Adapted from Le et al. (2018)

were determined using multi-spacecraft (Zhou et al. 2009b) and single-spacecraft methods (Norgren et al. 2012). Recent observations by MMS and numerical simulations have significantly improved our understanding of the behavior of the lower hybrid waves and their role in magnetic reconnection, owing to the high temporal resolution of the particle detectors and close tetrahedral configurations of the spacecraft.

At Earth's magnetopause, observations show that lower hybrid waves occur near the EDR and in the IDR, primarily on the low-density magnetospheric side of the X line (Bale et al. 2002; Graham et al. 2014; Khotyaintsev et al. 2016; Graham et al. 2017a, 2019). The electric field fluctuations are characterized by $E_\perp \gg E_\parallel$. The largest amplitude waves were primarily electrostatic, although magnetic field fluctuations associated with the waves occur (Carter et al. 2001; Graham et al. 2017a). As the center of the current sheet is approached, the waves tend to become more electromagnetic (Graham et al. 2019; Yoo et al. 2020). This was attributed to plasma becoming more weakly magnetized near the center of the current sheet (Graham et al. 2019; Yoo et al. 2020). However, close to the center of the current sheet, where the magnetic field is small, the amplitude of lower hybrid waves is significantly reduced (Graham et al. 2019, 2022), which suggests that the waves do not play a significant role in the EDR (Graham et al. 2022).

These results are consistent with recent 3D simulations of asymmetric magnetic reconnection, where intense lower hybrid waves were found in the low-density side of the X line (Roytershteyn et al. 2012; Price et al. 2016, 2017). In some simulations, the power of lower hybrid waves was reduced in the diffusion region compared with the separatrices (Pritchett 2013; Le et al. 2017; Price et al. 2020). However, this has not been clearly seen in observations, where large-amplitude lower hybrid waves have been reported both in the IDR and separatrices.

The presence of cold magnetospheric ions can also be a source of lower hybrid waves. Graham et al. (2017a) found that the relative drift between cold magnetospheric ions and finite Larmor radius magnetosheath ions in the magnetospheric inflow region can generate lower hybrid waves. Similarly, the relative cross-field drift between finite Larmor radius ions and electrons can excite lower hybrid waves via the modified two-stream instability (Graham et al. 2019).

Lower hybrid waves have also been observed close to the EDR during symmetric reconnection in the Earth's magnetotail. Chen et al. (2020) reported MMS observations of lower hybrid waves associated with electron heating and vortical flows at the EDR in an electron-scale current sheet. The observed lower hybrid waves are electron-scale fluctuations ($k_\perp \rho_e \sim 1$, where $k_\perp$ is the wave number perpendicular to the ambient magnetic field) and propagate along the outflow direction. The wave electric field is so strong that the higher-energy electrons are demagnetized and gain energy through the wave potential, while the lower-energy electrons execute the $E \times B$ vortical flow generating a magnetic field comparable to the guide field strength. This event was characterized by a moderate guide field (about 30% of the asymptotic reconnecting magnetic field component), which allows lower hybrid waves with a significant electrostatic component to be present in the center of the current sheet. The observations by Chen et al. (2020) suggest that electron dynamics in the reconnection layer can be modified by the waves.

Cozzani et al. (2021) presented observations of the electromagnetic long-wavelength lower hybrid mode ($k\sqrt{\rho_i \rho_e} \sim 2.7$) in the magnetotail electron-scale current layer with weak guide field (see Figs. 4a–4f). Large-amplitude oscillations of the electron-scale gradients at the separatrix immediately adjacent to the EDR and magnetic-field fluctuations at the center of the current sheet reveal the presence of the electromagnetic drift instability leading to the current sheet kinking. The observed properties of the waves are consistent with the obliquely

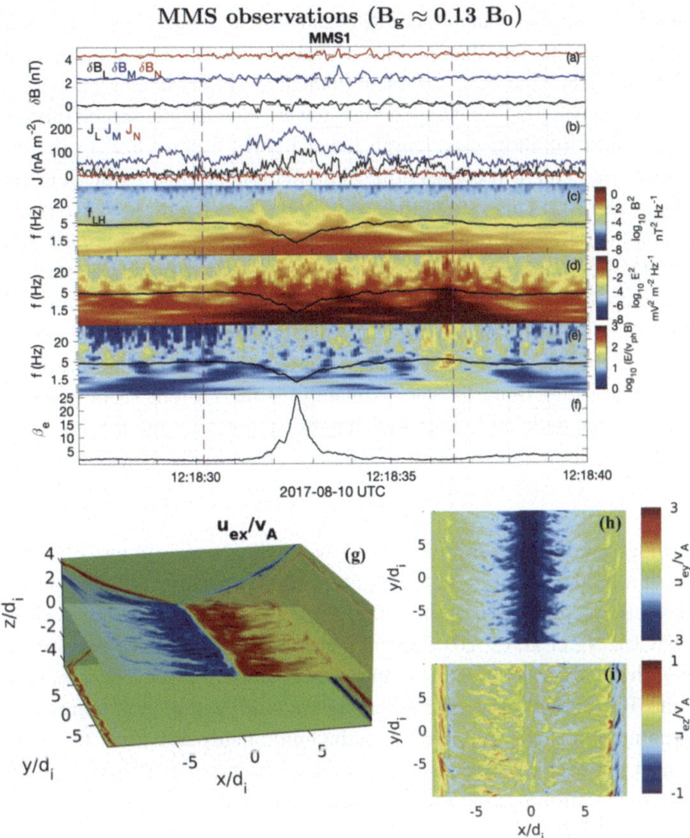

Fig. 4 *Top:* Electromagnetic lower hybrid scale waves observed by MMS during an EDR crossing in the magnetotail. (a) Three components of magnetic field fluctuations $\delta\mathbf{B}$. Offsets of 2.3 nT and 4.3 nT are added to δB_M and δB_N, respectively; (b) current density $\mathbf{J}$ calculated from particle moments. Spectrum of (c) $\mathbf{B}$ wave power; (d) $\mathbf{E}$ wave power; (e) $\log_{10}(E/(B\ v_{ph}))$ quantifying the electromagnetic component of the fluctuations. (f) β_e. The black line indicates f_{lh}. The vertical dashed magenta lines bound the region where the electromagnetic lower hybrid waves are observed. Based on Cozzani et al. (2021). *Bottom:* Electron velocity fluctuations associated with lower hybrid waves near the diffusion region from a 3D PIC simulation of symmetric reconnection with guide field $B_g = 0.3B_0$. (g) 2D slice along the current sheet associated with reconnection. The color shading indicates electron velocities in the x-direction (normal to the current sheet). (h) and (i) the same slices showing electron velocities in the y- and z-directions (out-of-plane and along the reconnecting $\mathbf{B}$), respectively. Adapted from Le et al. (2019)

propagating electromagnetic drift instability (Ji et al. 2004, 2005). The current density fluctuations at approximately the lower hybrid timescale and the broadband feature of the wave fields are comparable to the turbulent electron flow structures of the lower hybrid waves in 3D kinetic simulations (Le et al. 2019) (see Figs. 4g–4i). However, further measurements and 3D simulations are needed to understand the coupling between lower hybrid waves, current sheet instabilities, and the EDR structure.

For zero guide fields, theory and simulations (Daughton 2003) show that the electrostatic short-wavelength mode ($k\rho_e \sim 1$) is confined at the edges of the current sheet, and only the electromagnetic long-wavelength mode ($k\sqrt{\rho_i \rho_e} \sim 1$) is present in the center of the current sheet. In the reconnection context, the zero guide field case has been investigated in detail in a simulation study (Wang et al. 2021) showing a picture consistent with the results from Daughton (2003), namely that the long-wavelength lower hybrid mode develops at the current sheet center, while the short-wavelength mode develops at the separatrices. However, the short-wavelength wave is found to penetrate towards the midplane, and this could explain the MMS observation of short-wavelength lower hybrid waves close to the current sheet midplane during magnetotail reconnection events with negligible guide fields (Wang et al. 2022b). Wang et al. (2021) also show that while long-wavelength lower hybrid waves lead to oscillations of the current density, vortexes do not form in zero guide-field reconnection.

Wang et al. (2022b) investigated lower hybrid waves at MMS magnetotail reconnection diffusion-region crossings characterized by guide field values ranging from 1% to 30% of the upstream magnetic field. The direction of wave propagation and the dominant magnetic field component of the fluctuations vary considerably among the events, although a majority of wave propagation directions lie approximately in the current sheet plane. Both long-wavelength electromagnetic and short-wavelength electrostatic modes are observed in the diffusion regions, including events with negligible guide fields. It was speculated that the short-wavelength mode was excited in separatrix regions and penetrated toward the current sheet mid-plane.

To better understand the generation of the lower hybrid waves in thin electron-scale current layers with a moderate guide field, Ng et al. (2020) performed 2D and 3D PIC simulations of magnetic reconnection with a 30% guide field, mimicking the event reported by Chen et al. (2020). Ng et al. (2020) showed that lower hybrid waves develop immediately downstream of the electron jet reversal in the electron current layer. The lower hybrid waves observed in the 3D simulations propagate along the outflow direction (perpendicular to the ambient magnetic field at the midplane, i.e., the guide field) and affect the flow pattern by creating electron vortexes that are similar to in situ observations (Chen et al. 2020). A 3D kinetic simulation of strong guide-field reconnection similarly showed electrostatic waves developing near the EDR and propagating into the outflow region (Ng et al. 2023). These results demonstrate the importance of the guide field, which allows for the short-wavelength, $k\rho_e \sim 1$, lower hybrid waves to develop at the reconnection midplane, potentially leading to anomalous resistivity and heating, as shown in a recent laboratory experiment (Yoo et al. 2024). These findings open new possibilities in the context of 3D reconnection. For example, as mentioned by Chen et al. (2020), the new dynamics imposed by the lower hybrid waves in the current layer may lead to secondary tearing at the wave scale.

One of the major advances made with MMS is resolving particle distributions at sufficiently high frequency to reveal wave-particle interactions. Thus, MMS, along with numerical simulations, can resolve detailed information on the wave properties. By assuming that electrons remain close to frozen in and that the waves tend to be localized near boundary layers, the electrons form vortex-like motions, which in turn generates parallel magnetic field fluctuations (Norgren et al. 2012). Evidence of this vortex-like motion has been reported in both symmetric and asymmetric reconnection. In Graham et al. (2019), electron moments resolving electron density and velocity fluctuations associated with the waves were used to investigate the properties of the lower hybrid waves. Figures 5a–5b show an example of lower hybrid waves in the IDR at Earth's magnetopause (Khotyaintsev et al. 2016; Graham et al. 2022), where the electron moments can resolve the fluctuations associated with the waves. Figure 5c shows that the electrons remain approximately frozen in to the magnetic field. Density perturbations are also observed in association with the waves (Fig. 5d). Similarly, Ergun et al. (2017) and Ergun et al. (2019) found density and velocity fluctuations associated with these waves and interpreted them as corrugations of the current sheet.

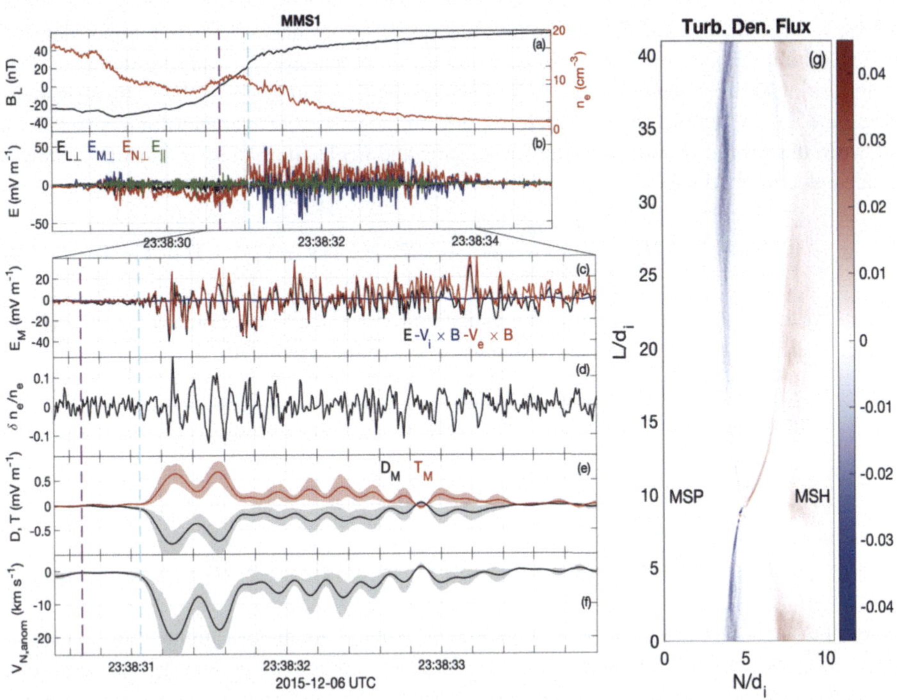

Fig. 5 Example of lower hybrid waves in the inflow region of magnetic reconnection at Earth's magnetopause, based on Graham et al. (2022). (a) Reconnection magnetic field B_L (black) and electron number density n_e (red). (b) Perpendicular and parallel components of the electric field **E**. (c) E_M (black) and **M** components of $-\mathbf{V}_i \times \mathbf{B}$ (blue), and $-\mathbf{V}_e \times \mathbf{B}$ (red). (d) Normalized electron density perturbations $\delta n_e/n_e$. (e) **M** components of **D** (black) and **T** (red). (f) Anomalous electron drift in the direction normal to the magnetopause $V_{N,anom}$. (g) Anomalous density flux of lower hybrid waves in a three-dimensional asymmetric reconnection simulation. Based on Price et al. (2020)

Lower hybrid waves are a prime candidate for anomalous resistivity, as they generate large electric fields and density perturbations (Fig. 5d). Estimates from spacecraft observations have found that the anomalous resistivity is small (Mozer et al. 2011; Graham et al. 2017a). However, these observations relied on the spacecraft potential to infer density fluctuations, which can be unreliable when large electric fields are present. Recently, Graham et al. (2022) used MMS observations to directly evaluate these anomalous terms for quasi-electrostatic lower hybrid waves at Earth's magnetopause by resolving the fluctuations in the electron moments. This was possible using the highest temporal resolution electron moments (Rager et al. 2018). The study focused on the effects of lower hybrid waves on reconnection, in particular anomalous resistivity and diffusion. The anomalous effects can be evaluated by decomposing quantities into quasi-stationary and fluctuating terms, $Q = \langle Q \rangle + \delta Q$, where $\langle \cdots \rangle$ is the ensemble average over space or time and δQ is the fluctuating quantity associated with the waves. This decomposition is then applied to the electron momentum equation

$$m_e \frac{\partial (n_e \mathbf{V}_e)}{\partial t} + m_e \nabla \cdot (n_e \mathbf{V}_e \mathbf{V}_e) + \nabla \cdot \mathbf{P}_e + e n_e (\mathbf{E} + \mathbf{V}_e \times \mathbf{B}) = 0, \tag{1}$$

where e, m_e, n_e, $\mathbf{V}_e$, and $\mathbf{P}_e$ are the unit charge, electron mass, electron number density, electron bulk velocity, and electron pressure tensor, respectively. Applying the above decomposition then averaging yields the momentum equation in terms of the stationary quantities and the anomalous terms, which result from the ensemble average of the correlations between fluctuating quantities. After rearranging and neglecting the time derivative, the averaged electric field can be expressed as:

$$\langle \mathbf{E} \rangle + \langle \mathbf{V}_e \rangle \times \langle \mathbf{B} \rangle = -\frac{\nabla \cdot \langle \mathbf{P}_e \rangle}{\langle n_e \rangle e} - \frac{m_e}{\langle n_e \rangle e} \nabla \cdot (\langle n_e \rangle \langle \mathbf{V}_e \rangle \langle \mathbf{V}_e \rangle) + \mathbf{D} + \mathbf{T} + \mathbf{I}, \tag{2}$$

where we have neglected the time derivative and introduced the anomalous terms **D**, **T**, and **I**. These anomalous terms can be expressed as (Graham et al. 2022)

$$\mathbf{D} = -\frac{\langle \delta n_e \delta \mathbf{E} \rangle}{\langle n_e \rangle}, \tag{3}$$

$$\mathbf{T} = -\frac{\langle n_e \mathbf{V}_e \times \mathbf{B} \rangle}{\langle n_e \rangle} + \langle \mathbf{V}_e \rangle \times \langle \mathbf{B} \rangle, \tag{4}$$

$$\mathbf{I} = -\frac{m_e}{e \langle n_e \rangle} \left[\nabla \cdot (\langle n_e \mathbf{V}_e \mathbf{V}_e \rangle) - \nabla \cdot (\langle n_e \rangle \langle \mathbf{V}_e \rangle \langle \mathbf{V}_e \rangle) \right], \tag{5}$$

where the three terms refer to the anomalous resistivity, anomalous momentum transport, and anomalous inertial terms, respectively. In these expressions n_e and $\mathbf{V}_e$ are treated as separate quantities, while it is common to treat the electron flux $n_e \mathbf{V}_e$ as a single quantity (Le et al. 2018).

Graham et al. (2022) found that **D** and **T** can reach amplitudes comparable to the reconnection electric field ($\sim$1 mV m^{-1}) but mostly cancel each other out, while **I** is negligible. The approximate cancellation of **D** and **T** was found to be due to electron fluctuations remaining close to frozen-in [cf., Fig. 5c] (Graham et al. 2019, 2022). As a result, the contribution to the reconnection electric field and reconnection rate is small. Figure 5e shows the out-of-plane components of **D** and **T**, which approximately cancel each other out. Additionally, the waves do not reach the center of the current sheet, so the anomalous terms were found to be negligible there.

The anomalous terms were also evaluated in three-dimensional simulations of magnetopause reconnection (Le et al. 2017, 2018; Price et al. 2016, 2017, 2020; Pritchett 2013). These results showed that either **D**, **T**, and **I** could provide significant contributions to the reconnection electric field in the diffusion region. The exact contributions depend on the guide-field strength and method of performing the ensemble average (Le et al. 2018). However, a careful examination suggests that the contribution to the anomalous terms mainly results from the large-scale kinking or bending of the current sheet (Le et al. 2018), often with wavelengths on much larger ion scales. Overall, both simulations and observations suggest that the quasi-electrostatic short-wavelength lower hybrid waves do not contribute significantly to the reconnection electric field.

By applying the same procedure to the continuity equation and using Fick's law, an anomalous diffusion coefficient can be defined as

$$D_{\perp} = -\frac{\langle \delta n_e \delta V_{e,N} \rangle}{\nabla \langle n_e \rangle}, \tag{6}$$

where $\langle \delta n_e \delta V_{e,N} \rangle$ is an anomalous flux in the direction normal to the boundary. An anomalous drift speed in the normal direction can be calculated as $V_{N,anom} = \langle \delta n_e \delta V_{e,N} \rangle / \langle n_e \rangle$.

This diffusion coefficient has been calculated from Cluster (Vaivads et al. 2004) and MMS observations (Graham et al. 2022), as well as numerical simulations (Le et al. 2018; Price et al. 2020). From observations, Graham et al. (2022) found that $D_\perp$ was variable at the magnetopause and reached large values of $\sim 10^9\ \mathrm{m^2\,s^{-1}}$, which is sufficient to broaden the magnetopause. Statistically, the largest values of $D_\perp$ tended to occur closer to the EDR rather than along the separatrices. The sign of $D_\perp$ was found to consistently correspond to diffusion from the magnetosheath toward the magnetosphere. Similar electron diffusion was also found in simulations (Le et al. 2018; Price et al. 2020). Although electrons remain close to frozen in, irreversible transport is possible via $\mathbf{E} \times \mathbf{B}$ trapping, in which the vortical motion of electrons associated with large-amplitude lower hybrid waves becomes strong enough for the electron orbits to become chaotic (Kleva and Drake 1984; Price et al. 2020). The net effect of this diffusion is to broaden the boundary layer and facilitate mixing between magnetosheath and magnetospheric electrons. This mixing has been seen in MMS observations (Wang et al. 2017; Graham et al. 2017a, 2022) and kinetic simulations (Le et al. 2017, 2019). In summary, recent observations and simulations suggest anomalous resistivity due to lower hybrid waves is small and does not likely affect the reconnection rate; however, anomalous diffusion is often sufficiently large to broaden the boundaries where the waves occur.

Streaming Instabilities and Electrostatic Waves Near the EDR, strong out-of-plane currents occur. In antiparallel reconnection, these currents are approximately perpendicular to **B**, while for guide-field reconnection these currents become closely aligned with **B**. In addition, the stronger magnetic field due to the guide field at the X line tends to keep electrons magnetized and gyrotropic. The strong currents can become unstable to instabilities, for instance, due to beams, streaming instabilities, and temperature anisotropies. These instabilities can potentially thermalize particle distributions and dissipate the current via wave-particle interactions. Recently, Khotyaintsev et al. (2020) showed that electrostatic Debye-scale waves and turbulence developed in the diffusion region of a guide-field reconnection event at Earth's magnetopause. In the diffusion region, electron distributions composed of a core population and beam were observed. They found that these distributions were unstable to Buneman and beam-mode waves, producing broadband waves with a range of phase speeds. These waves can then thermalize the jet by converting the kinetic energy into thermal energy.

Fully kinetic simulations have been used to investigate streaming instabilities and turbulence in the diffusion region of reconnection. Jara-Almonte et al. (2014) showed using 2D simulations that Debye-scale electrostatic turbulence due to streaming instabilities can develop in the diffusion region for parameters typical of the outer magnetosphere. Specifically, the instabilities developed when there was a significant scale separation between the Debye length and the electron inertial length. Yao et al. (2022a) and Yao et al. (2022b) used the electron distributions from a 3D simulation of reconnection as initial conditions for high-resolution local simulations to study the instability of these distributions. They found that the distributions were unstable to both electrostatic and electromagnetic waves, which propagate both parallel and perpendicular to **B**. Simulations of guide-field reconnection show evidence of the Buneman and beam-driven instabilities in the diffusion region (Che et al. 2010, 2011; Muñoz and Büchner 2018), similar to observations. However, these instabilities occur for strongly magnetized low-beta plasmas. Such parameters are typical of the solar corona but are less common in Earth's magnetosphere. Nevertheless, reconnection with guide fields as strong as the reconnection magnetic field, $B_g \sim B_0$ can occur, particularly at Earth's magnetopause (Øieroset et al. 2016; Zhou et al. 2018). Similarly, strong-guide-field

$B_g > B_0$ reconnection occurs in Earth's magnetosheath (Wilder et al. 2018), although the plasma beta is typically relatively high $\beta \sim 1$.

Whistler Waves Electromagnetic whistler waves are frequently observed in and near the EDR of reconnection. Whistlers observed near the EDR have been attributed to electron temperature anisotropy (Khotyaintsev et al. 2016; Cao et al. 2017). However, it remains unclear if field-aligned whistler waves are generated in or near the EDR because whistlers generated in the separatrices can propagate toward the diffusion region. Recent observations have reported whistlers in the diffusion region at Earth's magnetopause (Zhong et al. 2022). Zhong et al. (2022) found that the local electron distributions were stable to whistler emission and argued that the waves were most likely generated in the separatrices and propagated toward the X line.

In the case of guide-field reconnection, the strong out-of-plane currents and associated electron beams can generate oblique quasi-electrostatic whistler waves via Landau resonance. Khotyaintsev et al. (2020) found evidence for quasi-electrostatic whistler waves generated by electron beams. A recent study by Wang et al. (2022a) of multiple magnetopause crossings near the EDR of guide-field reconnection found that oblique whistlers were generated locally by electron beams, while parallel propagating whistlers were generated by the temperature anisotropy of the background population. Thus, there are potentially multiple sources of whistler waves near the X line.

3.2 Separatrix Regions

The magnetic reconnection separatrices have been shown to exhibit a variety of plasma wave modes. In symmetric magnetotail reconnection, the separatrices are typically characterized by fast, cold electron beams propagating toward the X line. For asymmetric reconnection at Earth's magnetopause, the electron distributions in the separatrices can become more complicated due to magnetospheric and magnetosheath plasmas having distinct densities and temperatures.

Whistler Waves Electromagnetic whistler waves have been reported in the separatrices of both dayside reconnection (Graham et al. 2016b; Le Contel et al. 2016; Wilder et al. 2016) and magnetotail reconnection (Huang et al. 2016; Ren et al. 2019). Field-aligned whistler waves were reported by Graham et al. (2016b) at the magnetopause, who investigated the electron distributions associated with the waves. They found that within the magnetospheric separatrix regions the electron distributions were characterized by an electron beam propagating away from the X line and a loss of energetic magnetospheric electrons propagating away from the X line. Graham et al. (2016b) argued that the loss-cone or temperature anisotropy associated with the partial loss of energetic electrons was the source of the observed whistlers. At the dayside magnetopause, oblique whistler waves associated with electron beams have been found, which coincided with electrostatic solitary waves (ESWs) (Wilder et al. 2016, 2017, 2019). These observations showed the presence of both electron temperature anisotropy ($T_{e\perp} > T_{e\parallel}$) and electron beams at half the electron Alfvén speed, suggesting the presence of competing mechanisms for whistler generation. Wang et al. (2022a) investigated the competition between whistlers driven by beams and electron temperature anisotropy in detail using observations of multiple events. They showed that further away from the X line, whistler waves were more likely to have high ellipticity and a narrow frequency band around half the electron cyclotron frequency. These waves were largely driven by electron anisotropy, with the beams providing additional cyclotron resonance.

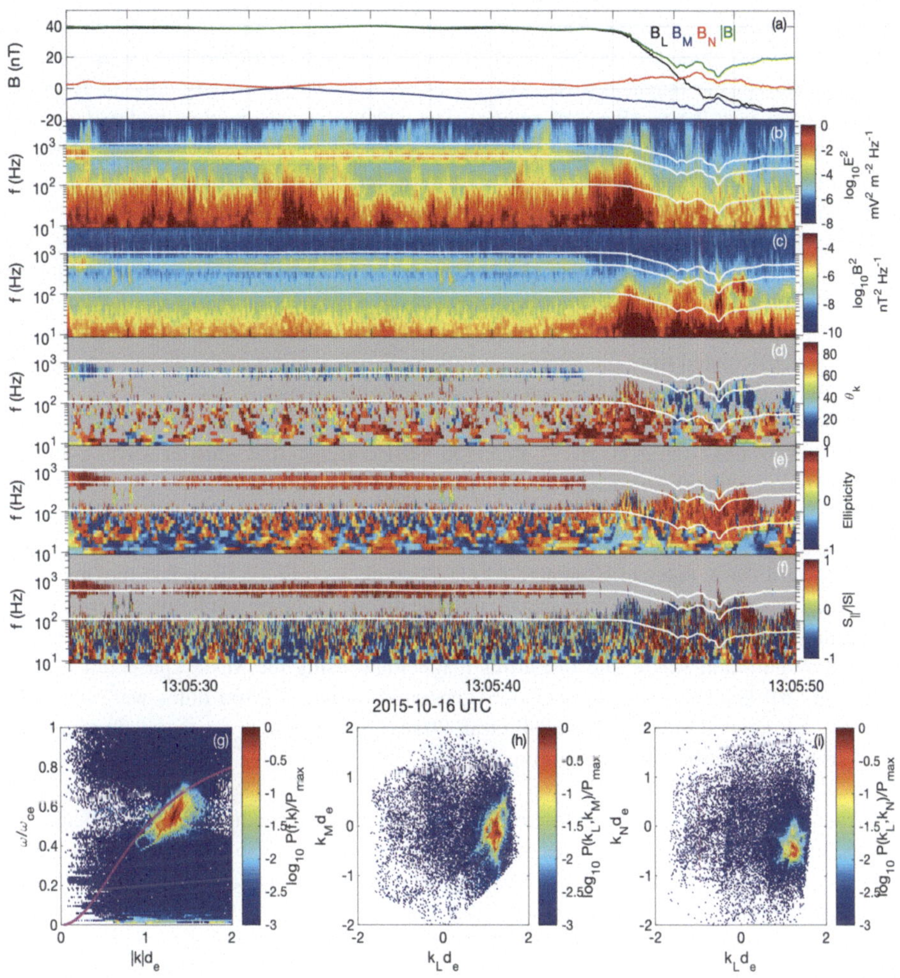

Fig. 6 Example of whistler waves at the magnetopause based on Le Contel et al. (2016) and Yoo et al. (2018). The whistler waves were observed near the separatrix region of asymmetric reconnection. (a) Magnetic field **B** in boundary normal LMN coordinates. (b) and (c) Frequency-time spectrograms of **E** and **B**. (d)–(e) Spectrograms of wave-normal angle θ_k, ellipticity, and ratio of parallel to total Poynting flux $S_{\|}/S$. Data from MMS1 is used in panels (a)–(e). The white lines in the spectrograms indicate $0.1 f_{ce}$, $0.5 f_{ce}$ and f_{ce}. (g) Frequency-wave number spectrum of whistler waves computed using four-spacecraft phase-differencing between 13:05:26.5 and 13:05:27.0 UT. The magenta line is the whistler dispersion relation predicted by cold plasma theory. (h)–(i) Wave power in the k_L–k_M and k_L–k_N planes

Closer to the X line, whistler waves exhibited broadband frequency spectra with varying ellipticity and were more likely to be generated by Landau resonance with the beam. Whistler waves appear to be ubiquitous to magnetic reconnection, and have also been observed during events in the turbulent magnetosheath (Vörös et al. 2019). In guide-field reconnection in the magnetosheath, it was shown that anisotropy-driven whistlers in the separatrix may be damped by parallel electric fields before they reach the EDR (Wilder et al. 2022).

Figure 6 shows an example of whistler waves at Earth's magnetopause observed by MMS1, based on Le Contel et al. (2016) and Yoo et al. (2018). Figure 6a shows **B** in boundary-normal coordinates. The magnetopause crossing occurs around 13:05:46 UT, where **B** reverses direction. In this event, the whistler waves are observed on the magnetospheric side of the magnetopause. The waves are seen in the spectrograms of **E** and **B** at frequencies near $0.5 f_{ce}$ (Figs. 6b and 6c). The whistler waves are identified using polarization analysis of the electromagnetic fields. Figures 6d–6f show spectrograms of wave-normal angle θ_k, ellipticity, and the ratio of parallel to total Poynting flux $S_{\|}/S$. For these waves θ_k tends to vary between 0° and ~40°, meaning the wave propagation varies between field-aligned and oblique. The ellipticity remains close to 1, which is a characteristic signature of whistler waves. Figure 6f shows that $S_{\|}/S$ remains close to 1 corresponding to propagation along **B**, which in this case is toward the reconnection X line (Le Contel et al. 2016). Figures 6g–6i show the dispersion relation for the whistler waves at the beginning of the interval, calculated from the four spacecraft. Such calculations of the dispersion relation are crucial to determining the resonant energies of the electrons.

In the magnetotail, whistler waves have been observed and shown to be associated with fast electron beams (Ren et al. 2019). Ren et al. (2019) argued that whistler waves could be generated via Landau resonance with the inflow electron beam. Similar results were found in kinetic simulations of symmetric reconnection (Fujimoto 2014).

Using a kinetic simulation, Goldman et al. (2014) showed that whistler waves propagating toward the X line can be generated via Cherenkov emission from fast-moving electron holes. These whistlers can propagate into the diffusion region while the electron holes generating them dissipate in the separatrices. Evidence of whistler waves generated via Cherenkov emission was found in Earth's magnetotail using the MMS spacecraft (Steinvall et al. 2019a). However, the whistlers generated by the electron holes were typically found to be damped quickly (Steinvall et al. 2019a), limiting how far they can propagate away from the electron holes generating them.

Electrostatic Solitary Waves and Electron Phase Space Holes In addition to electromagnetic waves, nonlinear electrostatic waves and solitary structures are also prominent at the separatrices. In the magnetotail, cold lobe electrons are accelerated towards the X line forming beams, which are unstable to instabilities that can develop ESWs, which in turn can thermalize the electron population (Fujimoto 2014; Egedal et al. 2015; Norgren et al. 2020). When the electrostatic potential associated with the ESWs is positive, they are often interpreted as the manifestation of electron phase space holes. Direct observation of the particle phase-space structure is challenging due to their short observation times. However, recently Mozer et al. (2018) used MMS Fast Plasma Investigation electron data and a superimposed epoch analysis of a group of ESWs to calculate the electron phase-space density and show that the electron distributions were consistent with electron holes. Norgren et al. (2022) used MMS Electron Drift Instrument data to provide continuous time sampling electron fluxes of a group of ESWs in a limited velocity space interval. The changes in electron fluxes were shown to be consistent with predicted electron distributions across electron holes.

Figure 7 shows an example of electron holes observed by MMS at a magnetotail separatrix crossing (Norgren et al. 2020). In Fig. 7a the cold lobe electrons are accelerated parallel to **B**. In this region, strong parallel electric field fluctuations develop (Fig. 7b). The electric fields are characterized by bipolar fluctuations, which are a characteristic signature of electron holes. Using the four MMS spacecraft, the frequency-wave number power spectrum can be calculated (cf., the Appendix) and is shown in Fig. 7d. Most of the electron holes have speeds comparable to, but slightly below, the electron beam speed, consistent with generation by the beam-plasma instability, while some slower electron holes are observed, consistent with generation via the interaction of the electron beam with ions (Norgren et al.

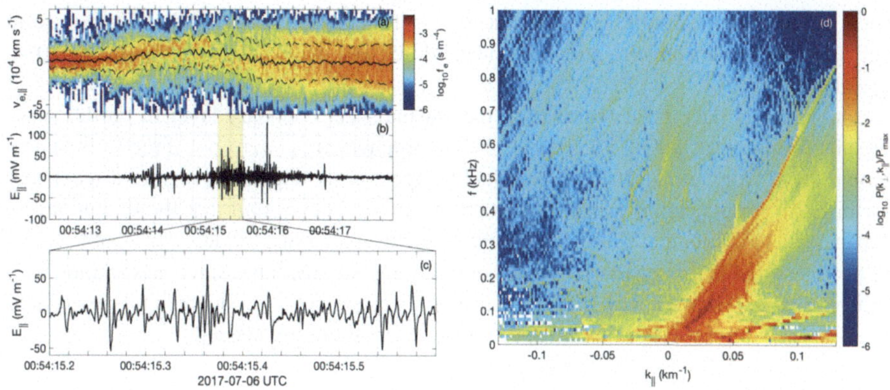

Fig. 7 Example of electron holes observed in the magnetotail separatrix region, based on Norgren et al. (2020), showing the electron distributions in the separatrix region and the associated electric fields. (a) One-dimensional reduced electron distribution along **B**. The solid black line is the parallel electron bulk speed $V_{e,\parallel}$ and the dashed lines indicate $V_{e,\parallel} \pm v_{e,th}$, where $v_{e,th}$ is the electron thermal speed. (b) Parallel electric field $E_{\parallel}$, showing large-amplitude waves. (c) Zoom in of $E_{\parallel}$ where the bipolar signatures of electron holes are seen. (d) Power spectrum of $E_{\parallel}$ as a function of $k_{\parallel}$ and f computed from the four-spacecraft phase-difference method (cf., the Appendix)

2015). These observations are consistent with simulations of symmetric reconnection, which show large amplitude ESWs propagating toward the X line (Egedal et al. 2015).

ESWs have also been reported in the separatrices of asymmetric dayside reconnection. For example, Graham et al. (2015) found that ESWs can develop in both the magnetospheric and magnetosheath separatrices. They had positive potentials and were interpreted as electron holes. At the dayside magnetopause separatrix, ESWs with negative potentials associated with the mixing of cold magnetospheric and warm magnetosheath plasma have been observed (Holmes et al. 2018b).

Ion-Acoustic Waves Ion-acoustic waves have been found to develop in the separatrix region of magnetopause reconnection (Uchino et al. 2017; Steinvall et al. 2021). They are electrostatic waves with frequencies near the ion plasma frequency f_{pi} and are typically characterized by magnetic-field-aligned electric fields. For the ion-acoustic instability to grow, it is often necessary that the ions are cold, $T_i \ll T_e$ (Stringer 1964), which restricts the occurrence of ion-acoustic waves to instances when cold ions are the dominant ion component of one of the reconnecting plasmas. Recent observations by MMS have shown that ion-acoustic waves can develop when cold magnetospheric ions enter the separatrix region, where strong electron-carried currents provide enough free energy to drive the instability (Steinvall et al. 2021). Steinvall et al. (2021) argued that the ion-acoustic instability could lead to current dissipation in the separatrices and cold-ion heating. Similar ion-acoustic waves have also been observed in recent laboratory experiments of magnetic reconnection (Zhang et al. 2023).

Finally, large-amplitude parallel electrostatic waves associated with the mixing of cold magnetospheric and warm magnetosheath plasma have been observed on the magnetospheric side of the dayside magnetopause separatrix (Ergun et al. 2016). These can have amplitudes exceeding 100 mV m^{-1}. A topic of future research will be what additional roles these waves play in magnetic reconnection.

Double Layers Double layers are localized regions with a net potential change. Double layers are observed in the reconnection separatrices and are associated with the fast electron flows. In simulations, double layers accelerate electrons along **B** toward the X line to form beams, which are unstable to streaming instabilities and generate ESWs (Fujimoto 2014; Egedal et al. 2015).

Double layers have been observed in the plasma sheet boundary layer in Earth's magnetotail (Ergun et al. 2009; Yuan et al. 2022), although they were not directly linked to reconnection. Wang et al. (2014) found double layers in the separatrix region of reconnection in Earth's magnetotail using Cluster observations. They found that the double layers propagated away from the X line, and might accelerate electrons. Evidence of double layers has been reported in the separatrix regions of magnetopause reconnection in association with electron mixing (Holmes et al. 2019). Overall, the observations of double layers associated with reconnection are rather limited and further observations are needed.

Lower Hybrid Drift Waves The presence of lower hybrid and related drift waves has also been reported on the separatrices of magnetic reconnection. Indeed, the largest amplitude lower hybrid waves reported in magnetic reconnection simulations tend to occur along the separatrices. In particular, lower hybrid waves develop along the magnetospheric separatrices of dayside reconnection, where density and ion pressure gradients are strong (Graham et al. 2022). Recent observations show that lower hybrid waves can also develop in the magnetosheath separatrices of asymmetric magnetopause reconnection (Tang et al. 2020a). In addition, lower hybrid waves have been reported in the separatrices of reconnection in the magnetosheath (Vörös et al. 2019), and in the magnetotail (Holmes et al. 2021). Magnetic corrugations associated with drift waves can also extend from the separatrix well into the EDR (Wilder et al. 2019; Ergun et al. 2019), and could be associated with patchy reconnection. Like near the diffusion region, the lower hybrid waves can smooth plasma gradients and facilitate plasma mixing.

Langmuir and Upper Hybrid Waves Both Langmuir and UH waves have been observed in or near the separatrices of reconnection at the magnetopause and in the magnetotail. Cluster observations have shown that enhancements in wave power around f_{pe} occur in the separatrices of high-latitude reconnection at Earth's magnetopause (Khotyaintsev et al. 2004; Retinò et al. 2006). In Earth's magnetotail, UH waves were observed in the separatrices close to the diffusion region (Farrell et al. 2002) and were associated with electron beams (Farrell et al. 2003). Similarly, Viberg et al. (2013) observed plasma frequency waves, interpreted as Langmuir waves, in the outer separatrix regions in association with fast electron beams propagating away from the X line. Graham et al. (2023) investigated the statistical occurrence of Langmuir and UH waves in Earth's magnetotail and showed that Langmuir and UH waves were associated with electron beams. A fraction of these beams were consistent with electrons escaping along newly reconnected field lines in the outer separatrix region.

3.3 Ion Outflow

Magnetic reconnection produces non-Maxwellian ion and electron distribution functions that are far from thermal equilibrium. The outflow region of reconnection is particularly rich in non-Maxwellian distributions (Lapenta et al. 2018) leading to the development of kinetic instabilities. The onset of the instabilities produces waves which in turn bring the distribution functions toward isotropization.

Temperature Anisotropy Instabilities A common example of ion distribution functions found in the outflow region are anisotropic distributions, either with the parallel temperature larger than the perpendicular $T_{i,\parallel}/T_{i,\perp} > 1$ or vice versa $T_{i,\parallel}/T_{i,\perp} < 1$ (the parallel and perpendicular direction of $T_{i,\parallel}$ and $T_{i,\perp}$ are based on the direction of the background magnetic field). The ion temperature behavior in the reconnection-related regions has been highlighted by observational and numerical studies (Hietala et al. 2015; Wu et al. 2013). The outflow region, in particular, hosts anisotropic ion distribution functions that could rapidly trigger various instabilities such as the ion firehose (developing if $T_{i,\parallel}/T_{i,\perp} > 1$), mirror, and ion cyclotron instabilities (developing if $T_{i,\parallel}/T_{i,\perp} < 1$) (Gary 1993). As a result, the ion distribution functions are isotropized while low-frequency ion-scale wave modes are generated. This is confirmed by ARTEMIS observations in the deep magnetotail (spacecraft located at $X \sim -45\ R_E$, (Vörös 2011)) and numerical simulations using different models (Finelli et al. 2021) showing that the ion population in the magnetic reconnection outflow regions is shaped by the anisotropy-driven instabilities. ARTEMIS observations in Earthward and tailward reconnection exhausts also showed compressional waves in correspondence with regions unstable to the ion cyclotron, mirror, and firehose instability (Wang et al. 2020). Furthermore, the development of the temperature anisotropy depends on the location with respect to the X line. Wu et al. (2013) used THEMIS observations at different locations along the magnetotail to show that $T_{i,\perp}/T_{i,\parallel}$ is enhanced in correspondence of bursty bulk flows (BBFs). Also, the spread of the ion distribution in the $T_{i,\perp}/T_{i,\parallel}$ versus $\beta_{i,\parallel}$ parameter space is reduced closer to Earth ($\beta_{i,\parallel} = 2\mu_0 n_i T_{i,\parallel}/B^2$, where μ_0 is the vacuum magnetic permeability and n_i is the proton number density).

A large body of work exists about these instabilities constraining ion anisotropy in the solar wind (Gary et al. 1976; Hellinger et al. 2006; Bale et al. 2009; Matteini et al. 2013a). Among these studies, Matteini et al. (2013b) investigated the relationship between the temperature anisotropy-driven instabilities and the onset of reconnection by simulating temperature anisotropic plasma mimicking the solar wind. However, the interplay between reconnection and the ion firehose, mirror and cyclotron instability is still relatively unexplored, especially in near-Earth plasmas. Few recent efforts have been made to investigate the effect of these instabilities on magnetosheath ions using MMS observations (Maruca et al. 2018). Recently, Richard et al. (2023) investigated the anisotropic distributions in the ion outflow regions of magnetotail reconnection. They showed that the observed ion anisotropies could lead to instability, although the effects of the resulting wave-particle interactions were too slow to return the distributions to isotropy. Rather, chaotic and quasi-chaotic motion of the ions in the current sheet was found to more quickly return ion distributions to isotropy.

Ion Firehose Instability The ion firehose instability is a kinetic instability which is driven by temperature anisotropy $T_{i,\parallel}/T_{i,\perp} > 1$ (Gary et al. 1998; Hellinger and Matsumoto 2000). The instability may arise in plasma with sufficiently high temperature anisotropy $T_{i,\parallel}/T_{i,\perp}$ and ion $\beta_{i,\parallel}$. The ion firehose instability has two branches: a low-frequency ($f < f_{ci}$) whistler firehose branch that propagates parallel to the background magnetic field (Gary et al. 1998) and a non-propagating oblique branch, called the Alfvén firehose branch by Hellinger and Matsumoto (2000). The non-propagating ion firehose branch has a linear growth rate comparable to or even higher than the propagating branch. The presence of the ion firehose instability in the ion outflow has been identified by Hietala et al. (2015). Their work presented ARTEMIS spacecraft observations in the magnetotail and PIC simulations showing that the ion firehose threshold is greatly exceeded within patchy regions in the reconnection outflow. Their results show that the driving of the temperature anisotropy by reconnection is stronger and faster than the linear growth of the instability, so the waves cannot efficiently remove the temperature anisotropy. This behavior is explained by ongoing magnetic reconnection constantly refilling the outflow region with unstable plasma. However, as highlighted by Wu et al. (2013), the ion temperature anisotropy highly depends on the location in the outflow with respect to the X line.

Large-scale simulations of reconnection have shown that the current becomes unstable far downstream of the X line, resulting in current sheet disruption and current filamentations forming (Lottermoser et al. 1998; Karimabadi et al. 1999; Arzner and Scholer 2001; Liu et al. 2012; Higashimori and Hoshino 2015). These simulations found that the ion firehose threshold can be satisfied in the outflow region. Using fluid theory, Arzner and Scholer (2001) argued that the instability was similar to the ion firehose instability due to counter-streaming ions, although velocity shears can also contribute to the instability. These results underline the complex interplay between magnetic reconnection and instabilities.

Ion Mirror Instability The interaction between magnetic reconnection and ion mirror instability is still quite unexplored. Laitinen et al. (2010) presented Cluster observations of strong mirror modes in the magnetosheath and strongly time- and space-varying reconnection jets at the magnetopause. Since the timescales of both phenomena were similar (of the order of one minute), these results suggest that mirror mode fluctuations in the magnetosheath can affect magnetopause reconnection, either by imposing modulation of continuous reconnection or by triggering bursts of reconnection. This scenario has been confirmed in hybrid-Vlasov global magnetospheric simulations (Hoilijoki et al. 2017). Hau et al. (2020) provided evidence of the presence of mirror mode waves – identified by observing the anticorrelation between the magnetic field and the density signals – in the vicinity of magnetic reconnection sites at the magnetopause. The two analysed MMS events were located in the magnetosphere dawn flank and the characteristic length associated with the mirror waves was smaller in these events than the size that is typically observed in the outer magnetosheath, at larger distances from the magnetopause. Despite the relatively limited amount of observational studies reporting mirror waves associated with reconnection, theoretical studies suggest that the mirror instability could significantly affect the reconnection process. Chiou and Hau (2003) predict a mixed mirror-tearing instability scenario where the mirror instability could increase the tearing instability growth rate. However, it is important to note that this study investigated the resistive tearing-mode instability. More recently, Winarto and Kunz (2022) investigated the relationship between reconnection and ion mirror instability in a magnetized, collisionless, high-beta plasma using hybrid-PIC simulations. They showed that initially, the mirror modes modify the current sheet by reducing its thickness, and then tearing modes with wavelengths comparable to the mirror mode become unstable, triggering reconnection on smaller scales and at earlier times with respect to the case of the current sheet not perturbed by mirror modes.

Electron Firehose Instability The presence of electron temperature anisotropy $T_{e,\parallel}/T_{e,\perp} > 1$ can provide free energy for the development of the electron firehose instability (Hollweg and Völk 1970). There are two distinct electron firehose wave modes. One electron firehose branch is characterized by parallel propagation with respect to the background magnetic field; the other is non-propagating, and it is predicted to develop for oblique wave-normal angles (Gary and Nishimura 2003). It is now established that the non-propagating mode has a lower threshold and a larger growth rate than the propagating mode. As with its ion counterpart, the electron firehose instability has been investigated quite extensively in the solar wind context (Verscharen et al. 2022, and references therein) since it is invoked as one of the sources of isotropization to explain the quasi-isotropic electron distributions observed in the solar wind.

A few efforts have been devoted to investigating the electron firehose instability in magnetospheric plasmas. Notably, a Cluster statistical study of electron distributions in the magnetosheath shows that the electron distribution function is constrained by the electron firehose instability threshold (Gary et al. 2005). Similar conclusions are presented in a statistical observational study focusing on dipolarization fronts (DFs) in the magnetotail (Zhang et al. 2018). Using MMS, Graham et al. (2021) found near the magnetopause and in the magnetosheath that the parallel temperature anisotropy was constrained by the electron firehose instability threshold, with $T_{e,\parallel}/T_{e,\perp}$ rarely exceeding the threshold. Yet, the role of the electron firehose anisotropy in constraining the temperature anisotropy in the regions associated with reconnection is still largely unknown. Recently, electron firehose fluctuations have been identified in the reconnection outflow in 3D PIC simulations (Le et al. 2019) and in Earth's magnetotail using in situ MMS observations (Cozzani et al. 2023) (see Fig. 8). Figure 8 shows a comparison of electron firehose observations and the associated temperature anisotropies (Cozzani et al. 2023) and a 3D kinetic simulation (Le et al. 2019). In both these cases $T_{e,\parallel}/T_{e,\perp}$ was constrained by the firehose threshold and the observed wave properties were consistent with the oblique electron firehose mode.

Electron Two-Stream Instability The electron two-stream instability (ETSI) is another significant source of isotropization of electron distribution functions in the reconnection outflow region. The development of the ETSI has been reported in the EDR (Jara-Almonte et al. 2014) and at the separatrix (Hesse et al. 2018) in symmetric magnetic reconnection simulation without a guide field and in situ observation of the separatrix region in the magnetotail (Norgren et al. 2020). In these cases, the instability was triggered by the presence of counterstreaming electron beams originating in the two inflow regions and interacting in the EDR or by the mixing of inflow and outflow populations at the thin separatrix boundary (see Sect. 3.2). However, ETSI can develop for asymmetric magnetic reconnection as well. Magnetic reconnection at the magnetopause naturally involves the mixing of multiple electron populations of magnetosheath and magnetospheric origin and typically the guide field is non-zero. Tang et al. (2020b) presented an MMS exhaust crossing during magnetopause reconnection where MMS measures high-frequency electrostatic waves generated by the ETSI in the outflow region. They show unstable electron distributions, which are rapidly isotropized, suggesting that wave-particle interaction is important for electron dynamics in the outflow. Figure 9 shows an example of the electrostatic waves and the associated electron distribution. The electron distribution consisted of a core and a beam, and a fit to the observed distribution yielded linear growth of ETSI.

Kinetic Alfvén Waves Alfvén waves are low-frequency, typically electromagnetic waves that are ubiquitous in magnetized plasmas. Shear Alfvén waves are essentially an MHD wave mode. However, if the perpendicular wavelength $\lambda_\perp$ (with respect to the background magnetic field direction) is comparable to kinetic characteristic spatial scales such as the ion gyroradius ρ_i, kinetic effects become significant and the shear Alfvén waves transition to the Dispersive Alfvén Wave (DAW) modes. In particular, Kinetic Alfvén Waves (KAWs) correspond to the DAW mode that arises in the intermediate-high β regime (Stasiewicz et al. 2000; Chen et al. 2021). KAWs are characterized by $\lambda_\perp \rho_i \sim 1$ and electric fields parallel to the background magnetic field $E_\parallel$ sustained by the electron pressure gradient. The parallel electric fields could have a significant role in enabling particle acceleration and plasma heating. Identification of KAWs is usually based on the transverse spatial scale, which is expected to be comparable to or smaller than the ion gyroradius, and on the ratio between the electric field and magnetic field, which is expected to be close to or larger

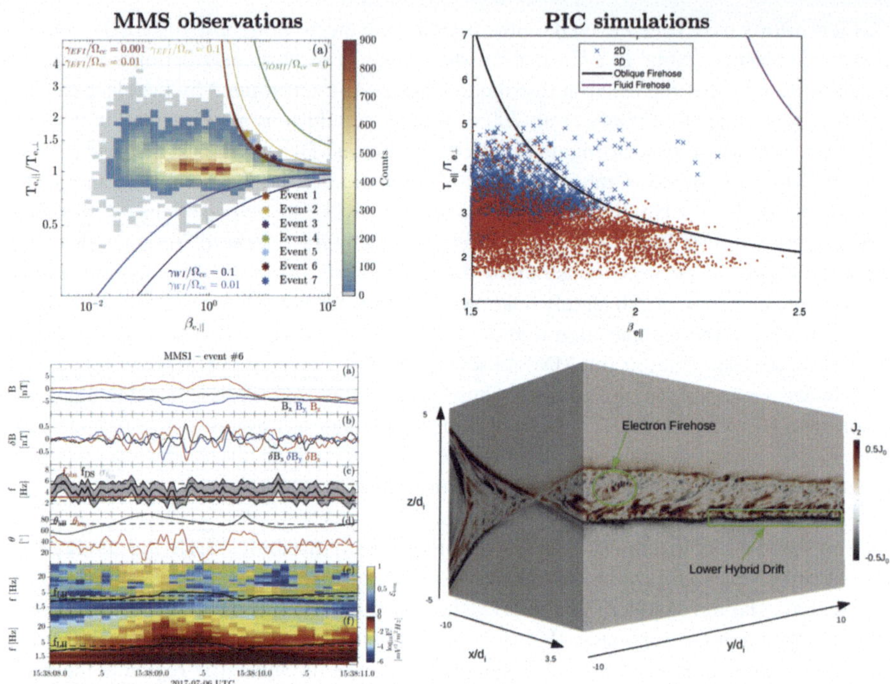

Fig. 8 Electron firehose waves in the reconnection outflows as observed by MMS (left column panels, adapted from Cozzani et al. 2023) and in 3D PIC simulations (right column panels, adapted from Le et al. 2019). *Top left:* (a) Electron distribution in the parameter space $\beta_{e\parallel}$–$T_{e\parallel}/T_{e\perp}$. The electron population is bounded by the electron firehose (EFI) and whistler instability (WI) thresholds. The green curve corresponds to the Weibel ordinary-mode instability (OMI) marginal condition of stability (Ibscher et al. 2012). *Bottom left:* Time series of quantities used to identify the non-propagating electron firehose fluctuations. (a) Magnetic field; (b) Magnetic field fluctuations; (c) Observed frequency f_{obs} (red solid line), Doppler-shift frequency f_{DS} (solid black line) and associated variability range (grey shaded region) with value $\sigma_{f_{DS}}$. (d) Angle between the wave vector direction and background magnetic field direction θ_{kB} and angle between the wave vector direction and electron velocity direction θ_{kv_e}. (e) Spectrogram of $\mathcal{E}_{long} = |\mathbf{E} \cdot \hat{\mathbf{k}}^2|/|\mathbf{E}^2|$. (f) Spectrogram of the electric field power. The black line is the lower hybrid frequency f_{LH}, and the dashed black lines indicate the frequency range of the observed fluctuations ($\Delta f = [2.6, 3.8]$ Hz for this event). *Top right:* distribution in the parameter space $\beta_{e\parallel}$–$T_{e\parallel}/T_{e\perp}$. The magenta curve is the fluid firehose condition $F_e = \mu_0(P_{e\parallel} - P_{e\perp})/B^2 = 1$ and the black curve is the non-propagating electron firehose instability threshold (same as the red curve in the top left panel). The points are taken in the reconnection outflow. Several points of the 2D run lie well above the electron firehose stability threshold, indicating that the firehose instability acts to regulate the electron temperature anisotropy and it is more efficient in doing so in 3D systems. *Bottom right:* Surface plots of the current density J_z. Electron firehose fluctuations are present in the outflow region, while lower hybrid drift modes develop at the separatrices. The electron firehose mode develops in a large number of flux tubes scattered throughout the exhaust at different times depending on the local plasma conditions

than the Alfvén speed. A critical property of KAWs is that they can transport energy over large distances in the form of Poynting flux $\mathbf{S} = \delta\mathbf{E} \times \delta\mathbf{B}/\mu_0$ directed mainly along the ambient magnetic field (see Fig. 10), where $\delta\mathbf{E}$ and $\delta\mathbf{B}$ are the wave electric and magnetic fields. For this reason, they are considered a fundamental process contributing to the mutual magnetosphere-ionosphere coupling, to ion energization and heating, and to electron acceleration via parallel electric fields.

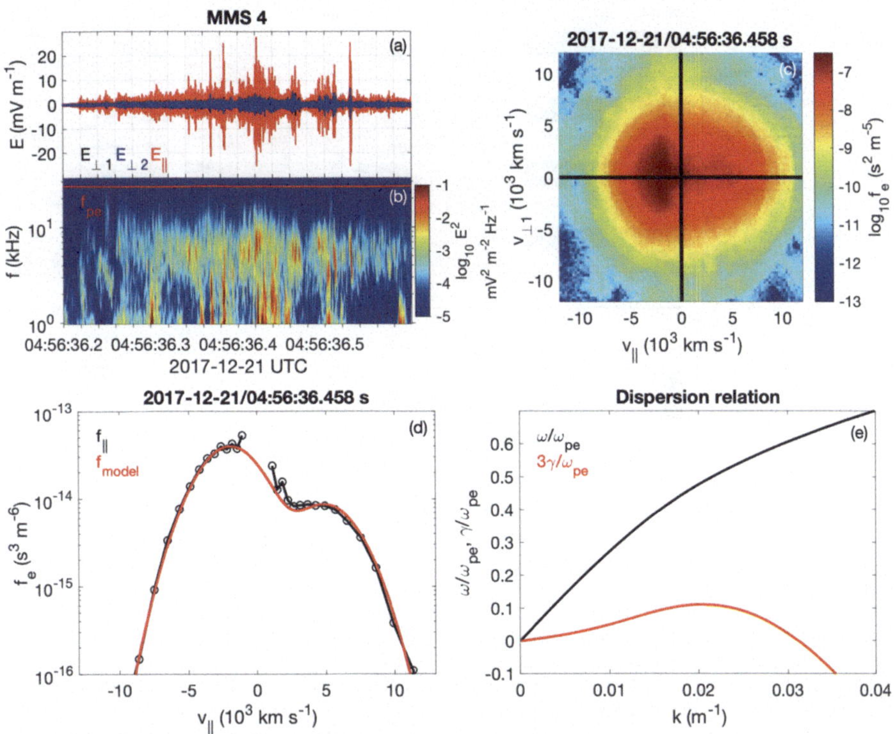

Fig. 9 Electrostatic wave associated with ETSI in the reconnection outflow of dayside magnetic reconnection observed by MMS4. Based on Tang et al. (2020b). (a) High-frequency **E** fluctuations in field-aligned coordinates. (b) Spectrogram of **E**. (c) Two-dimensional reduced electron distribution at the time the waves are observed. (d) One-dimensional cut of the electron distribution parallel and antiparallel to **B** (black) and a two-Maxwellian fit to the data (red). (e) Dispersion relation (black) and growth rate (red) predicted from the fitted distribution

KAWs are intimately related to magnetic reconnection as they are proposed to be associated with the Hall reconnection signatures. In particular, Rogers et al. (2001) suggested that the formation of an open outflow typical of fast magnetic reconnection is associated with the generation of whistler waves and KAWs. Also, Dai (2009) presented an approach in which the Hall fields and currents are incorporated in the Alfvén eigenmode theory. Dai (2009) showed that magnetic and electric fields measured in situ across the IDR (Mozer et al. 2002) can be explained in terms of eigenmode theory as they agree with the $n = 1$ Alfvén eigenmode signatures. KAWs are thought to set the magnetic and electric field Hall structure of the IDR and then propagate away from the reconnection site. Observational evidence of the coexistence of KAWs and magnetic reconnection was presented in several studies (Chaston et al. 2005, 2009; Dai et al. 2011). In particular, Dai et al. (2011) reported observations of KAWs in the magnetotail reconnection outflow region, while Chaston et al. (2009) presented in situ Cluster observations of KAWs in the IDR and in the outflow region of magnetic reconnection. In Chaston et al. (2009) the wave fronts are observed to propagate away from the X line, and the propagation is compatible with KAWs being emitted from a localized source. While the generation mechanism of the KAWs could not be established, this study provides evidence of the presence of KAWs in the exhaust and indicates the X line as the possible generation region. A statistical study by Chaston et al. (2012) considered THEMIS

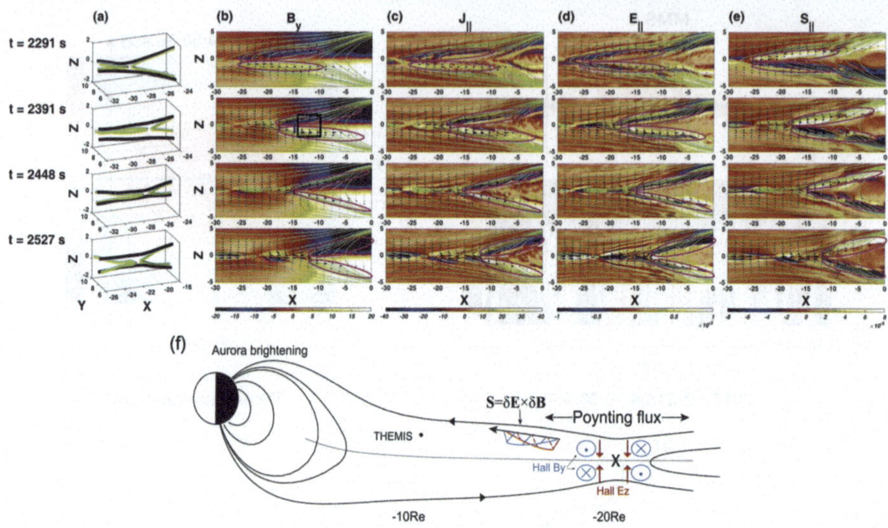

Fig. 10 Kinetic Alfvén Waves associated with magnetotail reconnection. Signatures of KAWs observed in a global 3D hybrid simulation of Earth's magnetosphere, adapted from Cheng et al. (2020) (panels a–e) and a schematic of the magnetic and electric field Hall patterns associated with KAW and magnetic reconnection in the magnetotail, adapted from Duan et al. (2016) (panel f). Time evolution of (a) magnetic field lines at the X line; (b) contours of B_y (GSE); (c) contours of the current density parallel to the background magnetic field $J_{\|}$; (d) contours of the electric field parallel to the background magnetic field $E_{\|}$; (e) contours of the Poynting flux parallel to the background magnetic field $S_{\|}$. The axis X, Y and Z are in units of Earth's radius R_E. The pink ovals show the time evolution of the same wave structure generated by reconnection and its Earthward propagation. (f) Schematic depicting the Earth's magnetotail, the Hall magnetic and electric field structure associated with KAWs at the reconnection site and the KAWs parallel Poynting flux transporting energy towards Earth and likely accounting for aurora brightening

observations of fast flows in Earth's magnetotail and showed that low-frequency electromagnetic waves consistent with KAWs are continuously radiated outward toward the lobe. The fact that KAWs are generated in the IDR and then propagate in the outflow region is further supported by simulation studies (Shay et al. 2011; Liang et al. 2016; Huang et al. 2018; Cheng et al. 2020). A similar conclusion has been drawn from reconnection observed in a laboratory experiment featuring mutually attracting flux ropes in a linear mirror device (Shi et al. 2019). KAWs were identified, and the temporal variation of the KAW's amplitude was correlated with the temporal variation of the reconnection rate, supporting the relationship between the two.

As mentioned above, KAWs are thought to play a primary role in magnetospheric-ionospheric coupling. Notably, the Poynting flux carried by KAWs has been proposed as a possible source for aurora (Angelopoulos et al. 2002; Wygant et al. 2002; Shay et al. 2011), connecting magnetotail and magnetopause reconnection to ionospheric processes. In particular, since KAWs can transport energy at super-Alfvénic velocities, they could account for the observed early onset of aurora with respect to the Alfvénic transit time from the reconnection site (Angelopoulos et al. 2008). Duan et al. (2016) reported THEMIS observations in the near-Earth plasma sheet showing signatures of KAWs during the substorm expansion phase. The KAWs were observed in correspondence with dipolarization, likely due to intermittent reconnection. Several observational studies reporting intense KAWs in correspondence with high-speed flows suggest that the observed waves carry sufficient Poynting flux flowing Earthward that they could account for auroral brightening at low altitudes (Stawarz et al. 2017; Chaston et al. 2012; Angelopoulos et al. 2002). Simulation studies have provided evidence supporting this scenario. Shay et al. (2011) estimated that KAW energy is able to propagate to Earth's ionosphere and create aurora, and similar conclusions were drawn in a simulation study quantifying the attenuation of the Hall quadrupolar magnetic field structure propagation as a KAW during magnetic reconnection (Sharma Pyakurel et al. 2018). These estimations are valuable but constrained by the limited size of the simulation domains. Recently, Gurram et al. (2021) suggested that the main carrier of wave energy away from the reconnection site in magnetotail reconnection was shear Alfvén waves (SAWs) instead of KAWs. By performing fully-kinetic simulations with high mass ratio ($m_i/m_e = 400$) and open boundary conditions, Gurram et al. (2021) found that both KAWs and SAWs are generated and that the damping rate of KAWs is enhanced so that the SAWs are the main wave energy carrier. The wave energy transported by the SAWs is higher by an order of magnitude with respect to previous estimates and could account for the energy deposited in the ionosphere and lead to the aurora. In summary, despite the significant progress in the investigation of KAWs, namely their relation to reconnection and the KAW-related energy transport, more work is needed to fully understand their role in terms of magnetosphere-ionosphere coupling.

High-cadence in situ measurements by MMS enables KAWs to be investigated with an unprecedented level of detail. Zhang et al. (2017) presented MMS observations at the magnetopause confirming that the electric and magnetic field Hall pattern was observed as far as 40 d_i away from the reconnection site, and it is consistent with the KAW mode. The high cadence of MMS particle measurements combined with small inter-spacecraft separation allowed for the calculation of spatial gradients at kinetic scales, enabling the detailed study of wave-particle interactions. Gershman et al. (2017) reported nonlinear wave-particle interactions between electrons and KAWs in the reconnection outflow region at the magnetopause. The trapped population of electrons may contribute to the saturation and damping of the KAWs. Nonlinear wave-particle interactions such as nonlinear Landau damping and stochastic heating play a key role also in the acceleration and heating of ions, as shown in 3D hybrid simulations investigating KAWs associated with magnetic reconnection (Liang et al. 2017). Recently, the set of MMS data presented by Gershman et al. (2017) was analyzed to accurately determine the KAWs perpendicular wavelength $\lambda_\perp$ by using a particle-sounding technique that can only be applied to high-resolution particle instruments. As the required time resolution is higher than one-fifth of the wave period, only MMS observations can be used in this case (Liu et al. 2023). Liu et al. (2023) showed that $\lambda_\perp/\rho_{th,i} \approx 2.4 \pm 0.7$, where $\rho_{th,i} = \sqrt{2m_p T_{i,\perp}/eB_0}$ is the proton thermal gyro-radius (m_p is the proton mass, $T_{i,\perp}$ is the ion temperature perpendicular to the ambient magnetic field B_0). While confirming the ion-scale nature of KAWs, this result imposes a limit to the energy exchange between KAWs and ions. Indeed, the highest perpendicular energy $W_{\perp,m}$ that can be gained by ions via coherent wave-particle interaction is related to the perpendicular wavelength $\sqrt{2m_p W_{\perp,m}/eB_0} \lesssim \lambda_\perp \approx (2.4^2)\rho_{th,i} \approx 5.76\ \rho_{th,i}$.

Apart from the generation in the IDR, several other KAW generation mechanisms have been proposed. KAWs could be generated by mode conversion (Hasegawa and Chen 1976), which is particularly relevant in regions with a significant gradient in Alfvén speed, such as the magnetopause (Shi et al. 2013). Velocity shears could also facilitate the generation of KAWs (Taroyan and Erdélyi 2002; Nykyri et al. 2021). KAWs could also be generated by intermittent reconnection (Cao et al. 2013). Also, KAWs have been proposed as a key constituent of the turbulent reconnection outflows. Huang et al. (2012) show that the dispersion relation computed in the turbulent outflow of magnetic reconnection in the Earth's magnetotail agrees with the Alfvén-whistler wave mode.

3.4 Jet Fronts

Jet fronts are regions where the outflow jets from magnetic reconnection interact with the ambient plasma sheet of the magnetotail. The jet fronts are often ion scale boundaries characterized by sharp increases in the normal to the current sheet component of **B**. In the case of the Earth's magnetotail, this can be interpreted as the field configuration getting closer to a dipole field, and therefore the fronts are often called dipolarization fronts (DFs). There is also a decrease in density as the outflow plasma has a lower density than the plasma sheet. The thickness of the jet front boundaries is typically comparable to the ion kinetic scales, and thus the plasma gradients can be unstable to the generation of various waves. For example, drift waves and interchange instabilities can develop at the fronts, while whistler waves and streaming instabilities develop behind the fronts.

Lower Hybrid Waves Lower hybrid drift waves are among the most commonly observed waves associated with dipolarization fronts. The gradients in **B** and thermal pressure, and the associated cross-field currents, provide the source of instability, similar to the magnetopause. Cluster and THEMIS observations have shown lower hybrid waves developing at dipolarization fronts (Zhou et al. 2009a; Greco et al. 2017; Khotyaintsev et al. 2011; Divin et al. 2015b). Zhou et al. (2009a) found lower hybrid waves in regions of enhanced Hall electric field and suggested that diamagnetic drifts were responsible for the waves. Similarly, Divin et al. (2015b) found that the waves were consistent with generation by the lower hybrid drift instability by comparing Cluster observations with linear kinetic theory (Yoon and Lui 2008). Based on observations, the lower hybrid waves have been proposed to contribute to both ion (Greco et al. 2017) and electron heating (Khotyaintsev et al. 2011).

Recently, the properties of lower hybrid waves have been investigated using MMS. Le Contel et al. (2017) investigated the properties of lower hybrid waves and suggested the waves contribute to electron heating. Pan et al. (2018) investigated a dipolarization front with electron-scale structures across the front. These structures were consistent with a rippling structure along the front due to lower hybrid waves. Figure 11 shows the rippled dipolarization front observed by MMS in Pan et al. (2018). Figure 11a shows the profile of **B** in boundary-normal coordinates. The rippling results in large **J** in the normal direction and large fluctuating $\mathbf{E}_{\perp}$ (both the normal and in-plane components). The electric fields are dominated by the Hall electric field, with a minor contribution from the electron pressure divergence. A simple model of the front ripples assuming pressure balance shows good agreement with the observations (Figs. 11e–11g). Statistical investigation of lower hybrid waves at dipolarization fronts by Hosner et al. (2022) concluded that the waves are a consistent feature of the fronts.

Simulations of magnetotail reconnection show that as reconnection proceeds, the region near the X line is filled with the upstream low-density plasmas, and thus, the density and pressure gradients become smaller than those occurring during reconnection onset in simulations. This can stabilize plasma instabilities, such as LHDI, near the X line (Nakamura et al. 2016). Instead, at this later stage of reconnection, the sharpest gradients develop at the jet fronts, leading to LHDI waves developing there (Vapirev et al. 2013; Lapenta et al. 2014; Divin et al. 2015a; Nakamura et al. 2016). Similarly, simulations show that velocity shear instabilities at DFs can excite waves in the lower hybrid frequency range (Lin et al. 2019). Simulations have also shown that the gradients in the jet front region can also induce the ballooning/interchange instability (BICI), which breaks up the front into ion-scale finger-like structures (Nakamura et al. 2002; TanDokoro and Fujimoto 2005; Guzdar et al. 2010; Lapenta and Bettarini 2011; Pritchett 2015b). 3D kinetic simulations of reconnection

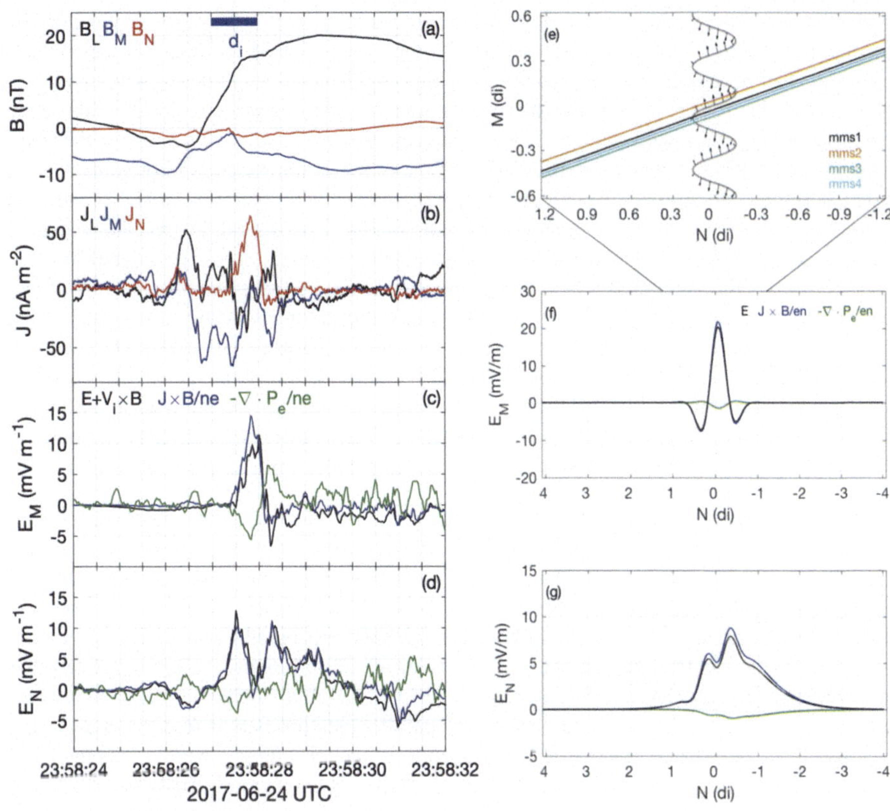

Fig. 11 A rippled dipolarization front observed by MMS in Earth's magnetotail, adapted from Pan et al. (2018). Panels (a)–(d) present MMS observations of the DF, while panels (e)–(g) show a model of the rippled DF. (a) **B**, (b) current density **J** calculated using the Curlometer technique. (c) and (d) Normal **N** and out-of-plane **M** components of $\mathbf{E} + \mathbf{V}_i \times \mathbf{B}$ (black), $\mathbf{J} \times \mathbf{B}/ne$ (blue), and $-\nabla \cdot \mathbf{P}_e/ne$ (green). (e) Model of the perturbed DF where the arrows indicate the direction of **E**. The colored lines indicate the trajectory of MMS across the DF. (f) and (g) Model predictions of the components of **E** in the **M** and **N** directions, respectively

showed that LHDI at jet fronts could nonlinearly couple with the BICI, leading to additional growth of BICI finger-like structures (Nakamura et al. 2016). The nonlinear evolution of LHDI at the front dissipates magnetic energy and diffuses the compressed front layer (Nakamura et al. 2019).

Whistler Waves Pileup of the reconnected magnetic flux (increase of **B**) behind the front drives the electron VDF to perpendicular anisotropy $T_{e\perp} > T_{e\parallel}$ due to conservation of magnetic moment $\mu = m_e v_{e\perp}^2/2B$, known as the betatron effect (Birn et al. 2022). Such anisotropic VDFs eventually become unstable to the generation of field-aligned whistlers due to whistler temperature anisotropy instability (Kennel and Petschek 1966). Generation of whistlers in the flux-pileup region (FPR) behind the front has been observed in numerical simulations (Fujimoto and Sydora 2008) and spacecraft observations (Le Contel et al. 2009; Khotyaintsev et al. 2011). It has been statistically confirmed that whistler generation is a characteristic feature of the FPR, with about half of the observed fronts having associated whistler emission (Viberg et al. 2014; Li et al. 2015). The waves are in cyclotron resonance with suprathermal (>1 keV) electrons, and these electrons with large pitch angles provide the major contribution to the wave growth (Grigorenko et al. 2023). Adiabatic betatron heating is thus followed by the wave generation, which in turn leads to pitch-angle scattering of the resonant electrons reshaping the distribution function (Artemyev et al. 2022), making the overall heating process non-adiabatic. The scattering process is relatively fast: for the magnetotail conditions, the characteristic scale is of the order of one second (Khotyaintsev et al. 2011), and thus, the whistler growth effectively regulates the temperature anisotropy driven by the flux pileup, which has been confirmed statistically (Zhang et al. 2018). The whistlers can also scatter energetic electrons into the loss cone, leading to their precipitation (Tsai et al. 2022). As the generated whistlers will propagate out of the generation region (FPR), they will carry away energy in the form of Poynting flux, corresponding to several percent of the suprathermal electron flux (Khotyaintsev et al. 2011; Zhang et al. 2019). The common occurrence of whistlers at the reconnection jet fronts provides smoking-gun evidence of the active pileup of magnetic flux. Therefore, observation of whistlers in combination with other data can be used to identify such a region in cases where only very limited or no particle data is available, e.g., for observations at Jupiter, Ganymede, Saturn, and other planets.

Electrostatic Waves The complex ion and electron velocity distributions behind the jet front can also drive electrostatic waves, such as electrostatic solitary waves, broadband electrostatic waves, or electron cyclotron waves. In particular, loss-cone and ring-type electron distributions behind the dipolarization front can excite electrostatic electron cyclotron harmonics (ECHs) (Zhou et al. 2009a). Zhou et al. (2009a) attributed these ECHs to the positive slope in the electron perpendicular velocity distribution function. They suggested that these ECHs could accelerate non-resonant electrons through a stochastic mechanism (Farrell et al. 2003). In addition, Fermi-accelerated parallel electron beams (e.g. Fu et al. 2013) may also drive broadband electrostatic waves with frequencies up to several times the electron cyclotron frequency f_{ce} (Hwang et al. 2011, 2014). Hwang et al. (2014) found that the Fermi-accelerated electron beams in the flux tube behind the dipolarization front drive two electrostatic wave modes. The low frequency ($\sim 0.43 f_{ce}$) mode was an oblique ($\theta_{kb} \approx 35°$) whistler mode and the high frequency ($\sim 3.4 f_{ce}$) mode was a parallel ($\theta_{kb} = 0.7°$) electron beam mode. The authors speculated that the former could be responsible for ion heating. Fermi-accelerated ion beam VDFs originating from the pre-existing plasma sheet ahead of the front (Eastwood et al. 2015), and cold ions accelerated across the separatrices (Wygant et al. 2005) can also drive ESWs (Liu et al. 2019; Zhang et al. 2022). Liu et al. (2019) suggested that intense electrostatic solitary waves observed behind the DF are generated by an ion-beam instability driven by the fast cold ion beam. Alternatively, Zhang et al. (2022) suggested that ESWs behind the jet fronts are driven by a drift between the hot ion population in the outflow and the cold ion beam originating from the inner magnetosphere. Lakhina et al. (2021) proposed a mechanism for the generation of these ESW in terms of slow and fast ion-acoustic solitons. Using a multi-fluid description with a hot Maxwellian electron core and two ion beams, Lakhina et al. (2021) found two fast and two slow ion-acoustic modes propagating parallel and antiparallel to the background magnetic field. They found that the predicted ion-acoustic solitons are in good agreement with MMS's observations. Liu et al. (2019) suggested that these ESWs can eventually thermalize the unstable ion beams, heating the ions in the reconnection jet.

4 Large-Scale Waves

In the previous sections, we primarily focused on kinetic-scale microinstabilities. However, larger-scale macroinstabilities associated with current sheets develop and can have important consequences for ongoing reconnection and reconnection onset. We briefly describe current sheet kinking and the tearing mode instabilities. The relation between the Kelvin-Helmholtz instability and reconnection is discussed in Stawarz et al. (2024, this collection).

4.1 Flapping and Kinking of Current Sheets

Flapping motions are large-scale ($\lambda \sim 1\ R_E$) North-South oscillations of the magnetotail current sheet (Speiser and Ness 1967; Lui et al. 1978). In particular, Gao et al. (2018) distinguished two types of flapping motions: the steady flapping, which consists of a global North-South motion of the current sheet, and the sinusoidal wave-like sheet motion (Speiser 1973). From an observational point of view, the flapping motions are seen as large amplitude oscillations of the Earthward/tailward magnetic field with multiple current sheet crossings (Figs. 12a and 12b). The typical period of these oscillations ranges from ~10 minutes (Sergeev et al. 2004) to ~10 seconds (Wei et al. 2019). Multi-spacecraft missions, such as Cluster and MMS, enable the type of flapping motion as well as the propagation velocity to be calculated using multi-spacecraft timing (Vogt et al. 2011). The timing method has been used statistically by Runov et al. (2005) to show that the wave-like flapping motion propagates toward the flank of the magnetosphere along the electric current direction (Fig. 12c). Three-dimensional fully kinetic simulations with open boundaries showed that wave-like flapping motions propagate in the duskward direction with a speed close to the Alfvén speed (Sitnov et al. 2014). The measured propagation velocity of the wave-like flapping motions ranges from $\sim 10\ \mathrm{km\,s^{-1}}$ to $\sim 100\ \mathrm{km\,s^{-1}}$ (Sergeev et al. 2004; Runov et al. 2005).

Various mechanisms have been proposed to excite flapping motions. For instance, Gao et al. (2018) suggested that the steady flapping in the midnight region induces the wave-like flankward propagating flapping motion observed in the flanks. Using MMS, Wang et al. (2019) suggested that solar wind directional change creates a North-South motion of the magnetotail current sheet, which triggers the flapping motions. MMS observed the modulation of the reconnection electric field with the same period as the flapping motions, suggesting that unsteady reconnection may induce the flapping motion (Wei et al. 2019). On the

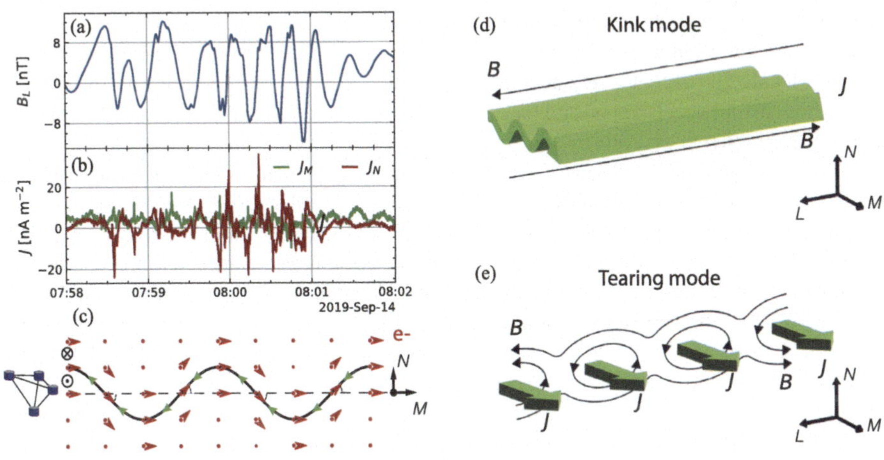

Fig. 12 Illustrations of flapping and tearing modes associated with large-scale current sheets. Panels (a)–(c) present MMS observation of current sheet flapping. Based on Richard et al. (2021). (a) Magnetic field, (b) current density, and (c) schematic of the drift-kink instability-driven current sheet flapping in the ion frame, showing the electron motion (red) and current direction (green). (d) and (e) Illustrations of the kink mode associated with current sheet flapping and the tearing mode

other hand, linear stability analyses have suggested that the wave-like flapping may result from the MHD double gradient instability (Erkaev et al. 2007), the ion-ion kink instability (Karimabadi et al. 2003; Sitnov et al. 2004), and the drift-kink instability (Daughton 1999). In particular, Karimabadi et al. (2003) found that the wave properties observed by Runov et al. (2003) are in very good agreement with the prediction of the ion-ion kink mode. Recently, Richard et al. (2021) found that the wavelength of the wave-like flapping motion is consistent with the prediction of the drift-kink instability, suggesting that this instability can be responsible for the flapping motion.

4.2 Tearing Mode

The tearing mode is particularly interesting because it can generate the conditions required for reconnection onset. The instability develops in current sheets to perturb the magnetic field, potentially leading to the development of X lines and magnetic islands (as illustrated in Fig. 12e). The tearing mode exists in the resistive and collisionless regimes. The resistive tearing mode can be described by the MHD equations in the presence of finite resistivity in the center of the current sheet (Furth et al. 1963). The theory was then extended to collisionless plasma (Coppi et al. 1966) and to plasmas with variable collisionality (Drake and Lee 1977). At Earth's magnetopause and magnetotail, plasmas are essentially collisionless, so other effects must take the place of collisions, such as Landau damping.

Much of the theoretical work has focused on Earth's magnetotail. There, due to the contribution of Earth's dipole magnetic field, the magnetic field configuration has a small but finite component ($B_z/B_0 \sim 0.1$) normal to the current sheet. Therefore, the current sheet is often modeled using a 2D isotropic locally Harris equilibrium (Schindler 1972; Lembege and Pellat 1982). Due to the finite normal component, the electrons remain magnetized, suppressing the electron Landau dissipation and thus the electron tearing mode (Galeev and Zelenyĭ 1976). Assuming the electron damping is unimportant for $T_i/T_e \gg 1$, Schindler (1974) showed that ion Landau dissipation of unmagnetized ions drives an ion tearing instability. However, for many conditions, the growth rate is too low to be important (Sitnov et al. 2019). Additionally, Lembege and Pellat (1982) found that ion tearing is prevented by an electrostatic potential due to the electron density compression associated with the tearing mode. Sitnov and Schindler (2010) showed that the stability criteria may be exceeded in the case of a localized magnetic flux accumulation called a B_z hump. The stability of the non-Harris thin current sheets supported by non-Maxwellian ion VDFs is still debated (Burkhart et al. 1992; Zelenyi et al. 2008).

Kinetic simulations have been used to study the development of tearing instability. The stability of the configuration proposed by Sitnov and Schindler (2010) was tested in multiple two-dimensional PIC simulations (Sitnov et al. 2013; Bessho and Bhattacharjee 2014; Pritchett 2015a). While all studies revealed the development of an instability leading to reconnection, the nature of the driving instability is debated. In particular, Bessho and Bhattacharjee (2014) showed the development of an electron tearing mode for a sufficiently small normal magnetic field. Simulations have generally found growth of an electron tearing mode, where the electron physics drives the instability. Imposing an external driving electric field along the out-of-plane direction compresses the current sheet and reduces the normal field, leading to electron demagnetization and electron tearing (Hesse and Schindler 2001; Liu et al. 2014; Pritchett 2005).

Currently, there is limited observational evidence of tearing mode in space plasmas. A small number of studies have investigated tearing modes within Earth's magnetosphere. For example, Bakrania et al. (2022) used machine learning techniques to identify signatures in electron distributions consistent with tearing modes in Earth's magnetotail. Lu et al.

(2020) argued that a magnetotail reconnection onset event was consistent with electron tearing occurring, while the ion tearing mode was found to be unlikely. Similarly, Genestreti et al. (2023) investigated reconnection onset in the magnetotail and found that onset was consistent with the electron tearing instability. At present, it is unclear to what degree the tearing mode can account for the onset of reconnection in Earth's magnetopause and magnetotail. Some observations have found evidence of the tearing mode in the solar wind (Réville et al. 2020). Analyses of tearing mode remain largely theoretical, and direct observations are currently limited. Further observational work is needed to determine the role of tearing modes in the onset of reconnection.

5 Discussion

The review in Sect. 3 shows that we now understand well the types of waves that can develop in reconnection regions. High-resolution fields measurements from spacecraft enable the types of waves associated with reconnection to be identified and characterized. Spacecraft such as Cluster and MMS have enabled waves to be characterized using multi-spacecraft methods. Similarly, multi-spacecraft methods enable the region where waves develop in relation to reconnection to be identified. Additionally, the observation of background particle distributions and estimates of plasma gradients and electric currents enable the instabilities responsible for the waves to be identified. Similar analyses from kinetic simulations and laboratory experiments enable the location of the waves in relation to reconnection to be identified and the underlying instabilities to be investigated. Thus, the type of waves that develop in different regions, their properties, and the underlying instabilities are generally well understood.

A crucial ongoing question is what effect these waves have on the plasma and how these waves affect reconnection. Generally, the effect of waves will be to return unstable plasmas to stability. Most of the knowledge of the effects of waves on plasmas is based on theoretical work or numerical simulations. Numerical simulations enable wave-particle interactions to be isolated from other plasma processes. Using in situ observations to study the effects of waves on the plasma is generally very difficult. For lower frequency waves up to lower-hybrid frequency, MMS has resolved the fluctuations in the particle distributions and moments associated with the waves (Gershman et al. 2017; Graham et al. 2022). This enables wave-particle interactions to be directly investigated. Additionally, wave-particle interactions for waves at higher frequencies, such as whistlers (Kitamura et al. 2022) and electron holes (Mozer et al. 2018; Norgren et al. 2022), have been investigated. These studies suggest further work can be done to quantify the effects of wave-particle interactions on magnetic reconnection and plasmas more generally with current instrumentation. In particular, further work is needed to determine if higher-frequency waves can contribute to anomalous resistivity and electron diffusion near the EDR.

Another ongoing question is the importance of three-dimensional effects when studying magnetic reconnection. Most simulations and observations have modeled or interpreted reconnection in two dimensions, assuming a uniform out-of-plane direction. Recent simulations and observations of reconnection have focused on three-dimensional effects. One of the primary effects when simulating reconnection in three dimensions is that drift waves in the out-of-plane direction develop. Drift waves can develop close to the X line, resulting in significant perturbations in and around the EDR (Cozzani et al. 2021). Three-dimensional simulations and observations have shown that lower hybrid waves tend to broaden the separatrix regions from electron-scale boundaries to hybrid or ion scale boundaries. In addition, significantly enhanced electron mixing and heating occur in three-dimensional simulations compared with two-dimensional simulations for asymmetric magnetopause reconnection (Le et al. 2017). The cause of the heating due to lower hybrid waves or the associated electron mixing is not fully understood. It has been argued that electron heating can result from Landau damping of lower hybrid waves (Cairns and McMillan 2005; Ren et al. 2022). Observations have shown that these waves have small but finite parallel wave numbers (Graham et al. 2019) so Landau damping of electrons is plausible. Further work is required to determine if the energy transferred to electrons by the waves is sufficient to account for the strong parallel electron heating reported in reconnection observations and simulations.

The recent observations of UH waves in and near the EDR and Langmuir waves in the reconnection separatrices suggest that these regions are possible sources of electron heating, scattering, and diffusion. These processes could modify reconnection at electron scales (Dokgo et al. 2020a). Due to the challenges in modeling these waves in simulations of magnetic reconnection and the current difficulty in directly quantifying these effects with spacecraft observations, further work is needed to determine the role of these waves in reconnection. These observations also suggest that diffusion regions and separatrices may be sources of escaping electromagnetic O- or X-mode waves, which could be observed remotely. Langmuir and UH waves can be converted to O- and X-mode waves near the plasma frequency and its harmonics via linear and nonlinear processes. Recent simulations show that UH waves generated by electron distributions can undergo three-wave decay and coalescence to generate electromagnetic waves at $2f_{pe}$ (Dokgo et al. 2019). Similar results were found for Langmuir and UH waves/electron Bernstein waves by Yao et al. (2022a). However, these simulations used local homogeneous conditions with initial distribution functions based on distributions from observations or simulations of magnetic reconnection. Large-scale three-dimensional kinetic simulations of reconnection, which can resolve Langmuir and UH waves, and the generated electromagnetic waves are not currently feasible. Although electromagnetic waves produced in reconnection regions have been reported on the Sun (Cairns et al. 2018), escaping electromagnetic waves have not yet been reported from in situ observations of magnetic reconnection. Further work is required to better understand magnetic reconnection as a source of escaping electromagnetic waves.

Finally, one of the major ongoing questions is reconnection onset, namely, how is reconnection initiated in thin current sheets and what role do waves and wave-particle interactions play? Tearing modes are often invoked to explain the development of X lines and reconnection onset. However, observational evidence of the tearing mode from spacecraft observations is limited, especially at Earth's magnetopause and in the magnetotail. Identifying signatures of tearing instability in spacecraft observations has proved very challenging and requires further investigation.

6 Conclusion

The role waves play in collisionless magnetic reconnection has been studied extensively using spacecraft observations, numerical simulations, and laboratory experiments. Kinetic plasma waves have been found in the reconnection diffusion region, separatrices, ion outflow regions, and at jet fronts. Significant progress has been made in characterizing the waves developing in these different regions. High-resolution electromagnetic field measurements have enabled the waves to be identified and their properties characterized. Single- and multi-spacecraft observations have been used to determine where these waves occur in relation to magnetic reconnection. Detailed analyses of the particle distribution functions have been used to determine the source of instability generating the waves. As a result, the type of waves that develop and why they occur in relation to magnetic reconnection are generally well understood. Recent major results are that MMS observations have demonstrated for the first time that the agyrotropic electron distributions produced by the EDR are unstable to large-amplitude electrostatic waves, which had not been previously found in simulations. Additionally, the type of instabilities that cause the observed waves have been identified by comparing the wave properties with the observed particle distribution functions. Significant progress has been made in understanding the role of anomalous resistivity in reconnection due to short-wavelength lower hybrid waves, enabled by the direct measurement of fluctuating electron distributions associated with the waves. Observations and simulations have shown that the effects on the reconnection electric field are small, but anomalous diffusion and plasma mixing due to lower hybrid waves can significantly broaden the boundary layer near the diffusion region and separatrices.

Appendix: Wave Measurement Techniques

In this section, we summarize important wave analysis techniques, which have been employed to study plasma associated with magnetic reconnection.

A.1 Multi-Spacecraft Interferometry

Multi-spacecraft methods are most appropriate for waves on scales larger than the probe-to-probe distance and close to the spacecraft separation. Given a collection of spacecraft with known separations, the standard approach assumes a plane wave structure and uses cross-correlation of fields measurements and the resulting time-delays to determine the phase velocity of the wave (Götz Paschmann 1998). For a matrix of vector separations $\mathbf{r}$ between spacecraft indexed α and β, and a corresponding matrix of timings $t_{\alpha\beta}$, the basic equation to solve is given by:

$$\mathbf{r}_{\alpha\beta} \cdot \frac{\hat{\mathbf{n}}}{v_{ph}} = \mathbf{t}_{\alpha\beta}. \tag{A1}$$

Where $\hat{\mathbf{n}}$ is the normal vector in the direction of the plane wave. The system can then be solved for $\hat{\mathbf{n}}/v_{ph}$ using a least squares fitting or other linear algebra technique. In the special case of a tetrahedron, keeping α fixed, there are three baselines, and the system can be solved exactly by inverting $\mathbf{r}_{\alpha\beta}$. The differences in applying this method come from changing how $t_{\alpha\beta}$ are determined, and how the matrix equation is solved.

The most basic approach of cross-correlating $E_{\parallel}$ measurements has been used effectively for solitary waves, which have an isolated pulse signature (Tong et al. 2018; Holmes et al. 2018b,a; Steinvall et al. 2019b; Mozer et al. 2018; Lotekar et al. 2020; Kamaletdinov et al. 2021; Norgren et al. 2022). However, it has also been used for wave packets of lower-hybrid, Buneman, and beam-mode waves (Holmes et al. 2021), and whistler waves (Hull et al. 2020; Zhong et al. 2022).

One can extend the above method to the spectral domain by determining a phase difference (and therefore a time delay from $\Delta t_{\alpha\beta} = \Delta\phi_{\alpha\beta}/\omega$). Averaging over a sliding time window, one can compute the complex-valued cross-spectral density $C(t,\omega)$ as a Fourier transform of the cross-correlation. The phase difference is given by:

$$\Delta\phi = \tan^{-1}(\Im[C(t,\omega)]/\Re[C(t,\omega)]), \tag{A2}$$

where $\Re$ and $\Im$ denote the real and imaginary components of $C(t, \omega)$. Because t, ω and k are directly mapped to one another via the phase velocity and $\mathbf{r}_{\alpha\beta}$, the average power spectrum $P(t, \omega)$ of the fields can be remapped as a function $P(\mathbf{k}, \omega)$. Furthermore, from the full wave-vector $\mathbf{k}$, spectra in terms of the components $k_{\parallel}$ and $k_{\perp}$ can also be compared with theoretical dispersion relations. This method has recently been applied using wavelet rather than Fourier transforms for lower hybrid waves (Graham et al. 2019), whistlers (Zhong et al. 2022), and electron holes (Norgren et al. 2020).

Finally, the k-filtering or wave-telescope technique (see e.g. Capon 1969; Pincon and Lefeuvre 1991; Motschmann et al. 1996; Tjulin et al. 2005) solves the general matrix equation (A3):

$$P(\mathbf{k}, \omega) = \mathrm{Tr}\left[\left(\mathbf{H}^{\dagger}(\mathbf{k})\mathbf{M}^{-1}(\omega)\mathbf{H}(\mathbf{k})\right)^{-1}\right], \tag{A3}$$

where $\mathbf{M}$ is a large matrix of the time-windowed cross-spectral density of the input vector fields and the spacecraft positions, and $\mathbf{H}$ is a "shape function" which is typically a matrix of plane waves describing how waves travel between each spacecraft position. $\mathbf{H}$ could also be chosen to account for three-dimensional structure such as spherical waves or current-sheet shapes. This approach can be further extended by applying various types of adaptive filters or constrained optimization schemes (Narita et al. 2022). The wave-telescope technique plus various extensions has been recently used to identify spectra of mirror mode waves (Narita et al. 2016), kinetic Alfvén waves (Gershman et al. 2018), whistler waves (Yoo et al. 2018), and turbulence in the solar wind from density fluctuations (Roberts et al. 2017).

A.2 Multi-Probe Interferometry

For MMS and many other spacecraft, electric fields are measured using the double probe technique, where the difference in potentials between two probes is used to estimate the component of the electric field between the probes. The probes measure the potential difference V_i between the probe and the spacecraft, given by $V_i = V_{p,i} - V_{SC}$, where $V_{p,i}$ is the probe potential, V_{SC} is the spacecraft potential, and the subscript i refers to the probe number. The probes are typically biased with a current so their potentials remain close to the plasma potential. The electric field is then given by

$$E_{ij} = -\frac{V_j - V_i}{L_{ij}}, \tag{A4}$$

where L_{ij} is the distance between probes i and j. Nominally, the difference is taken from probes on opposite sides of the spacecraft to construct E_{ij}. Each MMS spacecraft consists of six probes, whence the three-dimensional electric field is constructed. Figure A1 shows the probe configuration in the spin plane; the two axial probes (probes 5 and 6) are out of and into the page. In practice, the conducting spacecraft and booms typically lead to a reduction in E_{ij} due to local short-circuiting. Therefore, an effective baseline is used for L_{ij} rather than the physical probe separation (Pedersen et al. 2008).

The fact that the probes are separated in space means that it is possible to use V_j as multi-point measurements, namely multi-probe interferometry. Multi-probe interferometry has been used to estimate the phase speed or velocities by either comparing different V_i directly or reconstructing electric fields from different probe combinations. The time delays Δt (Vasko et al. 2018) or the phase differences (Graham et al. 2016a) between measurements are used to estimate the phase speed. Three different methods have been employed:

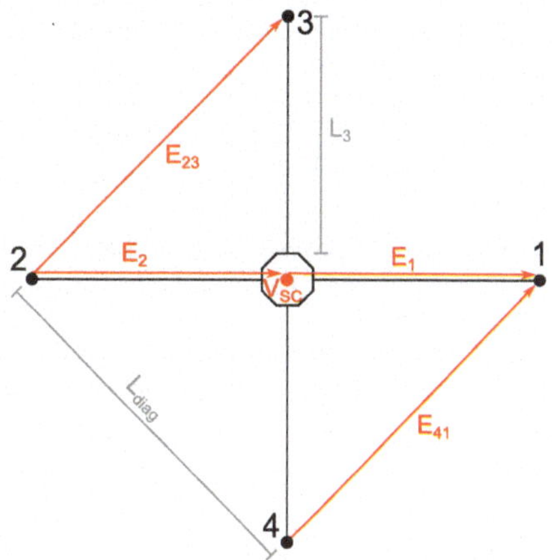

Fig. A1 Schematic of the MMS probe configuration in the spacecraft spin plane. The axial probes are into and out of the page

1. The time delays between opposing probe-to-spacecraft potentials to estimate the phase speeds along the probe displacement. For MMS time delays between V_1 and $-V_2$, V_3 and $-V_4$, and V_5 and $-V_6$. The phase speed and wave vector are calculated using (Vasko et al. 2018):

$$v_{ph} = \left(\frac{\Delta t_{12}^2}{L_1^2} + \frac{\Delta t_{34}^2}{L_3^2} + \frac{\Delta t_{56}^2}{L_5^2} \right)^{-1/2}, k_{ji} = \frac{v_{ph} \Delta t_{ij}}{L_i}, \tag{A5}$$

   where Δt_{ij} is the time delay between probes i and j, L_j is the probe to spacecraft separation and k_{ji} is the wave number along the ji direction. The advantage of this method is that it can in theory provide a three-dimensional wave vector when using the three probe pairs. The disadvantage is that V_j is used, which means that any changes in V_{SC} due to density fluctuations or electric fields affect the results. At high frequencies, one measurement has a δV_{SC} contribution while the other has a $-\delta V_{SC}$ contribution, where δV_{SC} refers to fluctuations in the spacecraft potential.
2. The second method involves constructing two spatially separated electric fields in the same direction. For an electric field along the direction of the 12 probe pair this is achieved using (Khotyaintsev et al. 2011; Graham et al. 2015):

$$E_1 = \frac{V_{SC} - V_1}{L_1}, E_2 = \frac{V_2 - V_{SC}}{L_2}, V_{SC} = \frac{V_3 + V_4}{2}. \tag{A6}$$

   The fields are aligned and correspond to a spatial separation of L_1, as illustrated in Fig. A1. Since the method relies on differences between V_i, the contributions from δV_{SC} should be negligible compared with method 1. However, this method requires the direction of **k** to be known from other methods or needs to be assumed, and is typically only reliable for electrostatic waves with **k** closely aligned with one of the probe pairs, which corresponds to the longest baseline and most reliable estimate of V_{SC}.
3. The third method involves constructing an electric field from adjacent probes around the spacecraft rather than opposing pairs (Vaivads et al. 2004). The fields $E_{41} \parallel E_{23}$ (illustrated in Fig. A1) and $E_{13} \parallel E_{42}$ using equation (A4) and $L_{diag} = \sqrt{2} L_i$ is used as the baseline. Two time delays can be calculated $\Delta t_{41,23}$ and $\Delta t_{42,13}$ to calculate v_{ph} and **k** in

the spacecraft spin plane (Steinvall et al. 2022):

$$v_{ph} = \left(\frac{\Delta t_{41,23}^2}{L_{diag}^2} + \frac{\Delta t_{42,13}^2}{L_{diag}^2} \right)^{-1/2}, \mathbf{k} = \begin{pmatrix} \frac{1}{\sqrt{2}} & -\frac{1}{\sqrt{2}} \\ \frac{1}{\sqrt{2}} & \frac{1}{\sqrt{2}} \end{pmatrix} \begin{pmatrix} \Delta t_{42,13} \\ \Delta t_{41,23} \end{pmatrix} \frac{v_{ph}}{L_{diag}}. \quad \text{(A7)}$$

This method provides v_{ph} and **k** in the spin plane, but this method cannot be applied to the axial probes. Like method (2) the effects of δV_{SC} should be small.

We note that performing multi-probe interferometry is generally difficult and can be prone to error. In particular, in many plasma environments, such as Earth's magnetosphere and magnetotail, the wavelength of the waves is too large compared with the probe separations to measure a time delay. This is particularly a problem for axial probes, which have significantly shorter separations than spin-plane probes. The fact that the axial probes are closer to the spacecraft means that they can be strongly influenced by the spacecraft. Overall, this makes estimating time delays from the axial probes extremely difficult, and results can be quite unreliable. The methods are also restricted by the cadence of the probe measurements (8.192 kHz for MMS in burst mode), typically limiting the frequency of the waves that can be investigated to well below the Nyquist frequency.

Recently, the reliability of these three interferometry methods was investigated by Steinvall et al. (2022), using an analytic model. They showed that method (1) is highly susceptible to spacecraft potential fluctuations and how non-planar the wave is. This is particularly important for electron holes, which are highly localized and often non-planar. The spacecraft potential fluctuation effects were found to be reduced for high-density plasmas. Although it was found that method (2) is generally reliable when wave vector direction is known and closely aligned with one of the probe pairs, method (3) was found to be the preferable method to use to determine the spin-plane component of the phase velocity and wave vector.

Acknowledgements We thank Shan Wang for her contributions to the sections on lower hybrid waves. We thank the International Space Science Institute (ISSI) for hosting the Magnetic Reconnection: Explosive Energy Conversion in Space Plasmas team. DBG acknowledges support from the Swedish National Space Agency (SNSA), Grant 128/17. GC acknowledges support from the Academy of Finland Grant n.345701. CN acknowledges support from the Research Council of Norway under Contract No. 300865, and SNSA, Grant 2022-00121. LR acknowledges support from SNSA, Grant 139/18. JFD acknowledges support from NSF, Grant PHY2109083, and NASA, Grant 80NSSC22K0352. The work was supported by the Knut and Alice Wallenberg foundation (Dnr. 2022.0087). Figures produced for this paper were generated using the irfu-matlab software package (https://github.com/irfu/irfu-matlab). MMS data are publicly available at https://lasp.colorado.edu/mms/sdc/public.

Funding Open access funding provided by Uppsala University.

Declarations

Competing Interests The authors declare no competing interests.

References

Angelopoulos V, Chapman JA, Mozer FS, et al (2002) Plasma sheet electromagnetic power generation and its dissipation along auroral field lines. J Geophys Res Space Phys 107(A8):SMP 14–1–SMP 14–20. https://doi.org/10.1029/2001JA900136

Angelopoulos V, McFadden JP, Larson D, et al (2008) Tail reconnection triggering substorm onset. Science 321(5891):931–935. https://doi.org/10.1126/science.1160495

Artemyev AV, Neishtadt AI, Angelopoulos V (2022) On the role of whistler-mode waves in electron interaction with dipolarizing flux bundles. J Geophys Res Space Phys 127(4):e2022JA030265. https://doi.org/10.1029/2022JA030265

Arzner K, Scholer M (2001) Kinetic structure of the post plasmoid plasma sheet during magnetotail reconnection. J Geophys Res Space Phys 106(A3):3827–3844. https://doi.org/10.1029/2000JA000179

Bakrania MR, Rae IJ, Walsh AP, et al (2022) Direct evidence of magnetic reconnection onset via the tearing instability. Front Astron Space Sci 9:869491. https://doi.org/10.3389/fspas.2022.869491

Bale SD, Mozer FS, Phan T (2002) Observation of lower hybrid drift instability in the diffusion region at a reconnecting magnetopause. Geophys Res Lett 29:2180. https://doi.org/10.1029/2002GL016113

Bale SD, Kasper JC, Howes GG, et al (2009) Magnetic fluctuation power near proton temperature anisotropy instability thresholds in the solar wind. Phys Rev Lett 103:211101. https://doi.org/10.1103/PhysRevLett.103.211101

Bessho N, Bhattacharjee A (2014) Instability of the current sheet in the Earth's magnetotail with normal magnetic field. Phys Plasmas 102:905. https://doi.org/10.1063/1.4899043

Birn J, Hesse M, Runov A (2022) Electron anisotropies in magnetotail dipolarization events. Front Astron Space Sci 9:908730. https://doi.org/10.3389/fspas.2022.908730

Burch JL, Torbert RB, Phan TD, et al (2016) Electron-scale measurements of magnetic reconnection in space. Science 352(6290):aaf2939. https://doi.org/10.1126/science.aaf2939

Burch J, Dokgo K, Hwang K, et al (2019) High-frequency wave generation in magnetotail reconnection: linear dispersion analysis. Geophys Res Lett 46(8):4089–4097. https://doi.org/10.1029/2019GL082471

Burkhart GR, Drake JF, Dusenbery PB, et al (1992) Ion tearing in a magnetotail configuration with an embedded thin current sheet. J Geophys Res Space Phys 97:16749–16756. https://doi.org/10.1029/92JA01523

Cairns IH, McMillan BF (2005) Electron acceleration by lower hybrid waves in magnetic reconnection regions. Phys Plasmas 12:102110. https://doi.org/10.1063/1.2080567

Cairns IH, Lobzin VV, Donea A, et al (2018) Low altitude solar magnetic reconnection, type III solar radio bursts, and X-ray emissions. Sci Rep 8:1676. https://doi.org/10.1038/s41598-018-19195-3

Cao JB, Wei XH, Duan AY, et al (2013) Slow magnetosonic waves detected in reconnection diffusion region in the Earth's magnetotail. J Geophys Res Space Phys 118(4):1659–1666. https://doi.org/10.1002/jgra.50246

Cao D, Fu HS, Cao JB, et al (2017) MMS observations of whistler waves in electron diffusion region. Geophys Res Lett 44(9):3954–3962. https://doi.org/10.1002/2017GL072703

Capon J (1969) High-resolution frequency-wavenumber spectrum analysis. Proc IEEE 57(8):1408–1418. https://doi.org/10.1109/PROC.1969.7278

Carter TA, Ji H, Trintchouk F, et al (2001) Measurement of lower-hybrid drift turbulence in a reconnecting current sheet. Phys Rev Lett 88:015001. https://doi.org/10.1103/PhysRevLett.88.015001

Cattell CA, Mozer FS (1986) Experimental determination of the dominant wave mode in the active near-Earth magnetotail. Geophys Res Lett 13(3):221–224. https://doi.org/10.1029/GL013i003p00221

Chaston CC, Phan TD, Bonnell JW, et al (2005) Drift-kinetic Alfvén waves observed near a reconnection X line in the Earth's magnetopause. Phys Rev Lett 95:065002. https://doi.org/10.1103/PhysRevLett.95.065002

Chaston CC, Johnson JR, Wilber M, et al (2009) Kinetic Alfvén wave turbulence and transport through a reconnection diffusion region. Phys Rev Lett 102:015001. https://doi.org/10.1103/PhysRevLett.102.015001

Chaston CC, Bonnell JW, Clausen L, et al (2012) Energy transport by kinetic-scale electromagnetic waves in fast plasma sheet flows. J Geophys Res Space Phys 117(A9):A09202. https://doi.org/10.1029/2012JA017863

Che H, Drake JF, Swisdak M, et al (2010) Electron holes and heating in the reconnection dissipation region. Geophys Res Lett 37:L11105. https://doi.org/10.1029/2010GL043608

Che H, Drake JF, Swisdak M (2011) A current filamentation mechanism for breaking magnetic field lines during reconnection. Nature 474:184–187. https://doi.org/10.1038/nature10091

Chen LJ, Hesse M, et al (2016) Electron energization and mixing observed by MMS in the vicinity of an electron diffusion region during magnetopause reconnectiond. Geophys Res Lett 43:6036–6043. https://doi.org/10.1002/2016GL069215

Chen LJ, Hesse M, Wang S, et al (2017) Electron diffusion region during magnetopause reconnection with an intermediate guide field: magnetospheric multiscale observations. J Geophys Res Space Phys 122:5235–5246. https://doi.org/10.1002/2017JA024004

Chen LJ, Wang S, Le Contel O, et al (2020) Lower-hybrid drift waves driving electron nongyrotropic heating and vortical flows in a magnetic reconnection layer. Phys Rev Lett 125:025103. https://doi.org/10.1103/PhysRevLett.125.025103

Chen L, Zonca F, Lin Y (2021) Physics of kinetic Alfvén waves: a gyrokinetic theory approach. Rev Mod Plasma Phys 5(1):1

Cheng L, Lin Y, Perez JD, et al (2020) Kinetic Alfvén waves from magnetotail to the ionosphere in global hybrid simulation associated with fast flows. J Geophys Res Space Phys 125(2):e2019JA027062. https://doi.org/10.1029/2019JA027062

Chiou SW, Hau LN (2003) Explosive and oscillatory tearing-mode instability in gyrotropic plasmas. Phys Plasmas 10(10):3813–3816. https://doi.org/10.1063/1.1606682

Coppi B, Laval G, Pellat R (1966) Dynamics of the geomagnetic tail. Phys Rev Lett 16(26):1207–1210. https://doi.org/10.1103/PhysRevLett.16.1207

Cozzani G, Khotyaintsev YV, Graham DB, et al (2021) Structure of a perturbed magnetic reconnection electron diffusion region in the Earth's magnetotail. Phys Rev Lett 127:215101. https://doi.org/10.1103/PhysRevLett.127.215101

Cozzani G, Khotyaintsev YV, Graham DB, et al (2023) Direct observations of electron firehose fluctuations in the magnetic reconnection outflow. J Geophys Res Space Phys 128(5):e2022JA031128. https://doi.org/10.1029/2022JA031128

Dai L (2009) Collisionless magnetic reconnection via Alfvén eigenmodes. Phys Rev Lett 102:245003. https://doi.org/10.1103/PhysRevLett.102.245003

Dai L, Wygant JR, Cattell C, et al (2011) Cluster observations of surface waves in the ion jets from magnetotail reconnection. J Geophys Res Space Phys 116(A12):A12227. https://doi.org/10.1029/2011JA017004

Daughton W (1999) The unstable eigenmodes of a neutral sheet. Phys Plasmas 6(4):1329–1343. https://doi.org/10.1063/1.873374

Daughton W (2003) Electromagnetic properties of the lower-hybrid drift instability in a thin current sheet. Phys Plasmas 10:3103. https://doi.org/10.1063/1.1594724

Davidson RC, Gladd NT, Wu CS, et al (1977) Effects of finite plasma beta on the lower-hybrid-drift instability. Phys Fluids 20:301. https://doi.org/10.1063/1.861867

Divin A, Khotyaintsev YV, Vaivads A, et al (2015a) Evolution of the lower hybrid drift instability at reconnection jet front. J Geophys Res Space Phys 120(4):2675–2690. https://doi.org/10.1002/2014JA020503

Divin A, Khotyaintsev YV, Vaivads A, et al (2015b) Lower hybrid drift instability at a dipolarization front. J Geophys Res Space Phys 120(2):1124–1132. https://doi.org/10.1002/2014JA020528

Dokgo K, Hwang KJ, Burch JL, et al (2019) High-frequency wave generation in magnetotail reconnection: nonlinear harmonics of upper hybrid waves. Geophys Res Lett 46(14):7873–7882. https://doi.org/10.1029/2019GL083361

Dokgo K, Hwang KJ, Burch JL, et al (2020a) The effects of upper-hybrid waves on energy dissipation in the electron diffusion region. Geophys Res Lett 47(19):e2020GL089778. https://doi.org/10.1029/2020GL089778

Dokgo K, Hwang KJ, Burch JL, et al (2020b) High-frequency waves driven by agyrotropic electrons near the electron diffusion region. Geophys Res Lett 47(5):e2020GL087111. https://doi.org/10.1029/2020GL087111

Drake JF, Lee YC (1977) Kinetic theory of tearing instabilities. Phys Fluids 20(8):1341–1353. https://doi.org/10.1063/1.862017

Duan SP, Dai L, Wang C, et al (2016) Evidence of kinetic Alfvén eigenmode in the near-Earth magnetotail during substorm expansion phase. J Geophys Res Space Phys 121(5):4316–4330. https://doi.org/10.1002/2016JA022431

Eastwood JP, Goldman MV, Hietala H, et al (2015) Ion reflection and acceleration near magnetotail dipolarization fronts associated with magnetic reconnection. J Geophys Res Space Phys 120(1):511–525. https://doi.org/10.1002/2014JA020516

Egedal J, Daughton W, Le A, et al (2015) Double layer electric fields aiding the production of energetic flat-top distributions and superthermal electrons within magnetic reconnection exhausts. Phys Plasmas 22:101208. https://doi.org/10.1063/1.4933055

Ergun RE, Andersson L, Tao J, et al (2009) Observations of double layers in Earth's plasma sheet. Phys Rev Lett 102:155002. https://doi.org/10.1103/PhysRevLett.102.155002

Ergun RE, Holmes JC, Goodrich KA, et al (2016) Magnetospheric multiscale observations of large-amplitude, parallel, electrostatic waves associated with magnetic reconnection at the magnetopause. Geophys Res Lett 43:5626–5634. https://doi.org/10.1002/2016GL068992

Ergun RE, Chen LJ, Wilder FD, et al (2017) Drift waves, intense parallel electric fields, and turbulence associated with asymmetric magnetic reconnection at the magnetopause. Geophys Res Lett 44:2978–2986. https://doi.org/10.1002/2016GL072493

Ergun RE, Hoilijoki S, Ahmadi N, et al (2019) Magnetic reconnection in three dimensions: observations of electromagnetic drift waves in the adjacent current sheet. J Geophys Res Space Phys 124(12):10104–10118. https://doi.org/10.1029/2019JA027228

Erkaev NV, Semenov VS, Biernat HK (2007) Magnetic double-gradient instability and flapping waves in a current sheet. Phys Rev Lett 99(23):235003. https://doi.org/10.1103/PhysRevLett.99.235003

Farrell WM, Desch MD, Kaiser ML, et al (2002) The dominance of electron plasma waves near a reconnection X-line region. Geophys Res Lett 29:1902. https://doi.org/10.1029/2002GL014662

Farrell WM, Desch MD, Ogilvie KW, et al (2003) The role of upper hybrid waves in magnetic reconnection. Geophys Res Lett 30:2259. https://doi.org/10.1029/2003GL017549

Finelli F, Cerri SS, Califano F, et al (2021) Bridging hybrid- and full-kinetic models with Landau-fluid electrons - I. 2D magnetic reconnection. Astron Astrophys 653:A156. https://doi.org/10.1051/0004-6361/202140279

Fu HS, Khotyaintsev YV, Vaivads A, et al (2013) Energetic electron acceleration by unsteady magnetic reconnection. Nat Phys 9(7):426–430. https://doi.org/10.1038/nphys2664

Fujimoto K (2014) Wave activities in separatrix regions of magnetic reconnection. Geophys Res Lett 41:2721. https://doi.org/10.1002/2014GL059893

Fujimoto K, Sydora RD (2008) Whistler waves associated with magnetic reconnection. Geophys Res Lett 35:L19112. https://doi.org/10.1029/2008GL035201

Fujimoto M, Shinohara I, Kojima H (2011) Reconnection and waves: a review with a perspective. Space Sci Rev 160:123. https://doi.org/10.1007/s11214-011-9807-7

Furth HP, Killeen J, Rosenbluth MN (1963) Finite-resistivity instabilities of a sheet pinch. Phys Fluids 6(4):459–484. https://doi.org/10.1063/1.1706761

Galeev AA, Zelenyĭ LM (1976) Tearing instability in plasma configurations. Sov Phys JETP 43:1113

Gao JW, Rong ZJ, Cai YH, et al (2018) The distribution of two flapping types of magnetotail current sheet: implication for the flapping mechanism. J Geophys Res Space Phys 123(9):7413–7423. https://doi.org/10.1029/2018JA025695

Gary SP (1993) Theory of space plasma microinstabilities. Cambridge atmospheric and space science series. Cambridge University Press, Cambridge. https://doi.org/10.1017/CBO9780511551512

Gary SP, Nishimura K (2003) Resonant electron firehose instability: particle-in-cell simulations. Phys Plasmas 10(9):3571–3576. https://doi.org/10.1063/1.1590982

Gary SP, Montgomery MD, Feldman WC, et al (1976) Proton temperature anisotropy instabilities in the solar wind. J Geophys Res 81(7):1241–1246. https://doi.org/10.1029/JA081i007p01241

Gary SP, Li H, O'Rourke S, et al (1998) Proton resonant firehose instability: temperature anisotropy and fluctuating field constraints. J Geophys Res Space Phys 103(A7):14567–14574. https://doi.org/10.1029/98JA01174

Gary SP, Lavraud B, Thomsen MF, et al (2005) Electron anisotropy constraint in the magnetosheath: cluster observations. Geophys Res Lett 32(13):L13109. https://doi.org/10.1029/2005GL023234

Genestreti KJ, Farrugia CJ, Lu S, et al (2023) Multi-scale observation of magnetotail reconnection onset: 2. Microscopic dynamics. J Geophys Res Space Phys 128(11):e2023JA031760. https://doi.org/10.1029/2023JA031760

Gershman DJ, F-Viñas A, Dorelli JC, et al (2017) Wave-particle energy exchange directly observed in a kinetic Alfvén-branch wave. Nat Commun 8:14719. https://doi.org/10.1038/ncomms14719

Gershman DJ, F-Viñas A, Dorelli JC, et al (2018) Energy partitioning constraints at kinetic scales in low-β turbulence. Phys Plasmas 25(2):022303. https://doi.org/10.1063/1.5009158

Goldman MV, Newman DL, Lapenta G, et al (2014) Čerenkov emission of quasiparallel whistlers by fast electron phase-space holes during magnetic reconnection. Phys Rev Lett 112:145002. https://doi.org/10.1103/PhysRevLett.112.145002

Götz Paschmann PWD (ed) (1998) Analysis methods for multi-spacecraft data. The International Space Science Institute, Chap. 10

Graham DB, Khotyaintsev YV, Vaivads A, et al (2014) Electron dynamics in the diffusion region of an asymmetric magnetic reconnection. Phys Rev Lett 112:215004. https://doi.org/10.1103/PhysRevLett.112.215004

Graham DB, Khotyaintsev YV, Vaivads A, et al (2015) Electrostatic solitary waves with distinct speeds associated with asymmetric reconnection. Geophys Res Lett 42:215. https://doi.org/10.1002/2014GL062538

Graham DB, Khotyaintsev YV, Vaivads A, et al (2016a) Electrostatic solitary waves and electrostatic waves at the magnetopause. J Geophys Res Space Phys 121:3069–3092. https://doi.org/10.1002/2015JA021527

Graham DB, Vaivads A, Khotyaintsev YV, et al (2016b) Whistler emission in the separatrix regions of asymmetric reconnection. J Geophys Res 121:1934–1954. https://doi.org/10.1002/2015JA021239
Graham DB, Khotyaintsev YV, Norgren C, et al (2017a) Lower hybrid waves in the ion diffusion and magnetospheric inflow regions. J Geophys Res Space Phys. https://doi.org/10.1002/2016JA023572
Graham DB, Khotyaintsev YV, Vaivads A, et al (2017b) Instability of agyrotropic electron beams near the electron diffusion region. Phys Rev Lett 119:025101. https://doi.org/10.1103/PhysRevLett.119.025101
Graham DB, Khotyaintsev YV, Norgren C, et al (2019) Universality of lower hybrid waves at Earth's magnetopause. J Geophys Res Space Phys 124(11):8727–8760. https://doi.org/10.1029/2019JA027155
Graham DB, Khotyaintsev YV, André M, et al (2021) Non-maxwellianity of electron distributions near Earth's magnetopause. J Geophys Res Space Phys 126(10):e2021JA029260. https://doi.org/10.1029/2021JA029260
Graham DB, Khotyaintsev YV, André M, et al (2022) Direct observations of anomalous resistivity and diffusion in collisionless plasma. Nat Commun 13:2954. https://doi.org/10.1038/s41467-022-30561-8
Graham DB, Khotyaintsev YV, André M (2023) Langmuir and upper hybrid waves in Earth's magnetotail. J Geophys Res Space Phys 128(10):e2023JA031900. https://doi.org/10.1029/2023JA031900
Greco A, Artemyev A, Zimbardo G, et al (2017) Role of lower hybrid waves in ion heating at dipolarization fronts. J Geophys Res Space Phys 122(5):5092–5104. https://doi.org/10.1002/2017JA023926
Grigorenko EE, Malykhin AY, Kronberg EA, et al (2023) Quasi-parallel whistler waves and their interaction with resonant electrons during high-velocity bulk flows in the Earth's magnetotail. Astrophys J 943(2):169. https://doi.org/10.3847/1538-4357/acaf52
Gurram H, Egedal J, Daughton W (2021) Shear Alfvén waves driven by magnetic reconnection as an energy source for the aurora borealis. Geophys Res Lett 48(14):e2021GL094201. https://doi.org/10.1029/2021GL094201
Guzdar PN, Hassam AB, Swisdak M, et al (2010) A simple mhd model for the formation of multiple dipolarization fronts. Geophys Res Lett 37(20):L20102. https://doi.org/10.1029/2010GL045017
Hasegawa A, Chen L (1976) Kinetic processes in plasma heating by resonant mode conversion of Alfvén wave. Phys Fluids 19(12):1924–1934. https://doi.org/10.1063/1.861427
Hau LN, Chen GW, Chang CK (2020) Mirror mode waves immersed in magnetic reconnection. Astrophys J Lett 903(1):L12. https://doi.org/10.3847/2041-8213/abbf4a
Hellinger P, Matsumoto H (2000) New kinetic instability: oblique Alfvén fire hose. J Geophys Res Space Phys 105(A5):10519–10526. https://doi.org/10.1029/1999JA000297
Hellinger P, Trávníček P, Kasper JC, et al (2006) Solar wind proton temperature anisotropy: linear theory and wind/swe observations. Geophys Res Lett 33(9):L09101. https://doi.org/10.1029/2006GL025925
Hesse M, Schindler K (2001) The onset of magnetic reconnection in the magnetotail. Earth Planets Space 53:645–653
Hesse M, Norgren C, Tenfjord P, et al (2018) On the role of separatrix instabilities in heating the reconnection outflow region. Phys Plasmas 25(12):122902. https://doi.org/10.1063/1.5054100
Hietala H, Drake JF, Phan TD, et al (2015) Ion temperature anisotropy across a magnetotail reconnection jet. Geophys Res Lett 42(18):7239–7247. https://doi.org/10.1002/2015GL065168
Higashimori K, Hoshino M (2015) Ion beta dependence on the development of Alfvénic fluctuations in reconnection jets. J Geophys Res Space Phys 120(3):1803–1813. https://doi.org/10.1002/2014JA020544
Hoilijoki S, Ganse U, Pfau-Kempf Y, et al (2017) Reconnection rates and X line motion at the magnetopause: global 2D-3V hybrid-Vlasov simulation results. J Geophys Res Space Phys 122(3):2877–2888. https://doi.org/10.1002/2016JA023709
Hollweg JV, Völk HJ (1970) New plasma instabilities in the solar wind. J Geophys Res 75(28):5297–5309. https://doi.org/10.1029/JA075i028p05297
Holmes JC, Ergun RE, Newman DL, et al (2018a) Electron phase-space holes in three dimensions: multi-spacecraft observations by magnetospheric multiscale. J Geophys Res Space Phys 123(12):9963–9978. https://doi.org/10.1029/2018JA025750
Holmes JC, Ergun RE, Newman DL, et al (2018b) Negative potential solitary structures in the magnetosheath with large parallel width. J Geophys Res Space Phys 123(1):132–145. https://doi.org/10.1002/2017JA024890
Holmes JC, Ergun RE, Nakamura R, et al (2019) Structure of electron-scale plasma mixing along the dayside reconnection separatrix. J Geophys Res Space Phys 124(11):8788–8803. https://doi.org/10.1029/2019JA026974
Holmes JC, Nakamura R, Schmid D, et al (2021) Wave activity in a dynamically evolving reconnection separatrix. J Geophys Res Space Phys 126(7):e2020JA028520. https://doi.org/10.1029/2020JA028520
Hosner M, Nakamura R, Nakamura TKM, et al (2022) Statistical investigation of electric field fluctuations around the lower-hybrid frequency range at dipolarization fronts in the near-Earth magnetotail. Phys Plasmas 29(1):012111. https://doi.org/10.1063/5.0067382

Huang SY, Zhou M, Sahraoui F, et al (2012) Observations of turbulence within reconnection jet in the presence of guide field. Geophys Res Lett 39(11):L11104. https://doi.org/10.1029/2012GL052210
Huang SY, Fu HS, Yuan ZG, et al (2016) Two types of whistler waves in the Hall reconnection region. J Geophys Res Space Phys 121(7):6639–6646. https://doi.org/10.1002/2016JA022650
Huang H, Yu Y, Dai L, et al (2018) Kinetic Alfvén waves excited in two-dimensional magnetic reconnection. J Geophys Res Space Phys 123(8):6655–6669. https://doi.org/10.1029/2017JA025071
Hull AJ, Muschietti L, Le Contel O, et al (2020) Mms observations of intense whistler waves within Earth's supercritical bow shock: source mechanism and impact on shock structure and plasma transport. J Geophys Res Space Phys 125(7):e2019JA027290. https://doi.org/10.1029/2019JA027290
Hwang KJ, Goldstein ML, Lee E, et al (2011) Cluster observations of multiple dipolarization fronts. J Geophys Res Space Phys 116:A00I32. https://doi.org/10.1029/2010JA015742
Hwang KJ, Goldstein ML, F-Viñas A, et al (2014) Wave-particle interactions during a dipolarization front event. J Geophys Res Space Phys 119(4):2484–2493. https://doi.org/10.1002/2013JA019259
Ibscher D, Lazar M, Schlickeiser R (2012) On the existence of weibel instability in a magnetized plasma. II. Perpendicular wave propagation: the ordinary mode. Phys Plasmas 19(7):072116. https://doi.org/10.1063/1.4736992
Jara-Almonte J, Daughton W, Ji H (2014) Debye scale turbulence within the electron diffusion layer during magnetic reconnection. Phys Plasmas 21:032114. https://doi.org/10.1063/1.4867868
Ji H, Terry S, Yamada M, et al (2004) Electromagnetic fluctuations during fast reconnection in a laboratory plasma. Phys Rev Lett 92:115001. https://doi.org/10.1103/PhysRevLett.92.115001
Ji H, Kulsrud R, Fox W, et al (2005) An obliquely propagating electromagnetic drift instability in the lower hybrid frequency range. J Geophys Res Space Phys 110(A8):A08212. https://doi.org/10.1029/2005JA011188
Ji H, Yoo J, Fox W, et al (2023) Laboratory study of collisionless magnetic reconnection. Space Sci Rev 219(8):76. https://doi.org/10.1007/s11214-023-01024-3
Kamaletdinov SR, Hutchinson IH, Vasko IY, et al (2021) Spacecraft observations and theoretical understanding of slow electron holes. Phys Rev Lett 127:165101. https://doi.org/10.1103/PhysRevLett.127.165101
Karimabadi H, Krauss-Varban D, Omidi N, et al (1999) Magnetic structure of the reconnection layer and core field generation in plasmoids. J Geophys Res Space Phys 104(A6):12313–12326. https://doi.org/10.1029/1999JA900089
Karimabadi H, Pritchett PL, Daughton W, et al (2003) Ion-ion kink instability in the magnetotail: 2. Three-dimensional full particle and hybrid simulations and comparison with observations. J Geophys Res Space Phys 108(A11):1401. https://doi.org/10.1029/2003JA010109
Kennel CF, Petschek HE (1966) Limit on stably trapped particle fluxes. J Geophys Res 71:1–28. https://doi.org/10.1029/JZ071i001p00001
Khotyaintsev Y, Vaivads A, Ogawa Y, et al (2004) Cluster observations of high-frequency waves in the exterior cusp. Ann Geophys 22:2403–2411. https://doi.org/10.5194/angeo-22-2403-2004
Khotyaintsev YV, Cully CM, Vaivads A, et al (2011) Plasma jet braking: energy dissipation and nonadiabatic electrons. Phys Rev Lett 106:165001. https://doi.org/10.1103/PhysRevLett.106.165001
Khotyaintsev YV, Graham DB, Norgren C, et al (2016) Electron jet of asymmetric reconnection. Geophys Res Lett 43:5571–5580. https://doi.org/10.1002/2016GL069064
Khotyaintsev YV, Graham DB, Norgren C, et al (2019) Collisionless magnetic reconnection and waves: progress review. Front Astron Space Sci 6:70. https://doi.org/10.3389/fspas.2019.00070
Khotyaintsev YV, Graham DB, Steinvall K, et al (2020) Electron heating by debye-scale turbulence in guide-field reconnection. Phys Rev Lett 124:045101. https://doi.org/10.1103/PhysRevLett.124.045101
Kitamura N, Amano T, Omura Y, et al (2022) Direct observations of energy transfer from resonant electrons to whistler-mode waves in magnetosheath of Earth. Nat Commun 13:6259. https://doi.org/10.1038/s41467-022-33604-2
Kleva RG, Drake JF (1984) Stochastic E $\times$ B particle transport. Phys Fluids 27(7):1686–1698. https://doi.org/10.1063/1.864823
Laitinen TV, Khotyaintsev YV, André M, et al (2010) Local influence of the magnetosheath plasma beta fluctuations on magnetopause reconnection. Ann Geophys 28:1053
Lakhina GS, Singh SV, Rubia R (2021) A mechanism for electrostatic solitary waves observed in the reconnection jet region of the Earth's magnetotail. Adv Space Res 68(4):1864–1875. https://doi.org/10.1016/j.asr.2021.04.026
Lapenta G, Bettarini L (2011) Self-consistent seeding of the interchange instability in dipolarization fronts. Geophys Res Lett 38(11):L11102. https://doi.org/10.1029/2011GL047742
Lapenta G, Goldman M, Newman D, et al (2014) Electromagnetic energy conversion in downstream fronts from three dimensional kinetic reconnection. Phys Plasmas 21(5):055702. https://doi.org/10.1063/1.4872028

Lapenta G, Pucci F, Olshevsky V, et al (2018) Nonlinear waves and instabilities leading to secondary reconnection in reconnection outflows. J Plasma Phys 84(1):715840103. https://doi.org/10.1017/S002237781800003X

Le Contel O, Roux A, Jacquey C, et al (2009) Quasi-parallel whistler mode waves observed by themis during near-Earth dipolarizations. Ann Geophys 27(6):2259–2275. https://doi.org/10.5194/angeo-27-2259-2009

Le Contel O, Retinò A, Breuillard H, et al (2016) Whistler mode waves and Hall fields detected by mms during a dayside magnetopause crossing. Geophys Res Lett 43(12):5943–5952. https://doi.org/10.1002/2016GL068968

Le Contel O, Nakamura R, Breuillard H, et al (2017) Lower hybrid drift waves and electromagnetic electron space-phase holes associated with dipolarization fronts and field-aligned currents observed by the magnetospheric multiscale mission during a substorm. J Geophys Res Space Phys 122(12):12236–12257. https://doi.org/10.1002/2017JA024550

Le A, Daughton W, Chen LJ, et al (2017) Enhanced electron mixing and heating in 3-D asymmetric reconnection at the Earth's magnetopause. Geophys Res Lett 44:2096–2104. https://doi.org/10.1002/2017GL072522

Le A, Daughton W, Ohia O, et al (2018) Drift turbulence, particle transport, and anomalous dissipation at the reconnecting magnetopause. Phys Plasmas 25:062103. https://doi.org/10.1063/1.5027086

Le A, Stanier A, Daughton W, et al (2019) Three-dimensional stability of current sheets supported by electron pressure anisotropy. Phys Plasmas 26(10):102114. https://doi.org/10.1063/1.5125014

Lembege B, Pellat R (1982) Stability of a thick two-dimensional quasineutral sheet. Phys Fluids 25(11):1995–2004. https://doi.org/10.1063/1.863677

Li H, Zhou M, Deng X, et al (2015) A statistical study on the whistler waves behind dipolarization fronts. J Geophys Res Space Phys 120(2):1086–1095. https://doi.org/10.1002/2014JA020474

Li WY, Graham DB, Khotyaintsev YV, et al (2020) Electron Bernstein waves driven by electron crescents near the electron diffusion region. Nat Commun 11:141. https://doi.org/10.1038/s41467-019-13920-w

Li WY, Khotyaintsev YV, Tang BB, et al (2021) Upper-hybrid waves driven by meandering electrons around magnetic reconnection x line. Geophys Res Lett 48(16):e2021GL093164. https://doi.org/10.1029/2021GL093164

Liang J, Lin Y, Johnson JR, et al (2016) Kinetic Alfvén waves in three-dimensional magnetic reconnection. J Geophys Res Space Phys 121(7):6526–6548. https://doi.org/10.1002/2016JA022505

Liang J, Lin Y, Johnson JR, et al (2017) Ion acceleration and heating by kinetic Alfvén waves associated with magnetic reconnection. Phys Plasmas 24(10):102110. https://doi.org/10.1063/1.4991978

Lin D, Scales WA, Ganguli G, et al (2019) A new perspective for dipolarization front dynamics: electromagnetic effects of velocity inhomogeneity. J Geophys Res Space Phys 124(9):7533–7542. https://doi.org/10.1029/2019JA026815

Liu YH, Drake JF, Swisdak M (2012) The structure of the magnetic reconnection exhaust boundary. Phys Plasmas 19(2):022110. https://doi.org/10.1063/1.3685755

Liu YH, Birn J, Daughton W, et al (2014) Onset of reconnection in the near magnetotail: pic simulations. J Geophys Res Space Phys 119(12):9773–9789. https://doi.org/10.1002/2014JA020492

Liu CM, Vaivads A, Graham DB, et al (2019) Ion-beam-driven intense electrostatic solitary waves in reconnection jet. Geophys Res Lett 46(22):12702–12710. https://doi.org/10.1029/2019GL085419

Liu ZY, Zong QG, Rankin R, et al (2023) Particle-sounding of the spatial structure of kinetic Alfvén waves. Nat Commun 14(1):2088. https://doi.org/10.1038/s41467-023-37881-3

Lotekar A, Vasko IY, Mozer FS, et al (2020) Multisatellite mms analysis of electron holes in the Earth's magnetotail: origin, properties, velocity gap, and transverse instability. J Geophys Res Space Phys 125(9):e2020JA028066. https://doi.org/10.1029/2020JA028066

Lottermoser RF, Scholer M, Matthews AP (1998) Ion kinetic effects in magnetic reconnection: hybrid simulations. J Geophys Res Space Phys 103(A3):4547–4559. https://doi.org/10.1029/97JA01872

Lu S, Wang R, Lu Q, et al (2020) Magnetotail reconnection onset caused by electron kinetics with a strong external driver. Nat Commun 11:5049. https://doi.org/10.1038/s41467-020-18787-w

Lui ATY, Meng CI, Akasofu SI (1978) Wavy nature of the magnetotail neutral sheet. Geophys Res Lett 5(4):279–282. https://doi.org/10.1029/GL005i004p00279

Maruca BA, Chasapis A, Gary SP, et al (2018) Mms observations of beta-dependent constraints on ion temperature anisotropy in Earth's magnetosheath. Astrophys J 866(1):25. https://doi.org/10.3847/1538-4357/aaddfb

Matteini L, Hellinger P, Goldstein BE, et al (2013a) Signatures of kinetic instabilities in the solar wind. J Geophys Res Space Phys 118(6):2771–2782. https://doi.org/10.1002/jgra.50320

Matteini L, Landi S, Velli M, et al (2013b) Proton temperature anisotropy and magnetic reconnection in the solar wind: effects of kinetic instabilities on current sheet stability. Astrophys J 763(2):142. https://doi.org/10.1088/0004-637x/763/2/142

Motschmann U, Woodward TI, Glassmeier KH, et al (1996) Wavelength and direction filtering by magnetic measurements at satellite arrays: generalized minimum variance analysis. J Geophys Res Space Phys 101(A3):4961–4965. https://doi.org/10.1029/95JA03471

Mozer FS, Bale SD, Phan TD (2002) Evidence of diffusion regions at a subsolar magnetopause crossing. Phys Rev Lett 89:015002

Mozer FS, Wilber M, Drake JF (2011) Wave associated anomalous drag during magnetic field reconnection. Phys Plasmas 18:102902. https://doi.org/10.1063/1.3647508

Mozer FS, Agapitov OV, Giles B, et al (2018) Direct observation of electron distributions inside millisecond duration electron holes. Phys Rev Lett 121:135102. https://doi.org/10.1103/PhysRevLett.121.135102

Muñoz PA, Büchner J (2018) Kinetic turbulence in fast three-dimensional collisionless guide-field magnetic reconnection. Phys Rev E 98:043205. https://doi.org/10.1103/PhysRevE.98.043205

Nakamura MS, Matsumoto H, Fujimoto M (2002) Interchange instability at the leading part of reconnection jets. Geophys Res Lett 29(8):88–1–88–4. https://doi.org/10.1029/2001GL013780

Nakamura TKM, Nakamura R, Baumjohann W, et al (2016) Three-dimensional development of front region of plasma jets generated by magnetic reconnection. Geophys Res Lett 43(16):8356–8364. https://doi.org/10.1002/2016GL070215

Nakamura TKM, Umeda T, Nakamura R, et al (2019) Disturbance of the front region of magnetic reconnection outflow jets due to the lower-hybrid drift instability. Phys Rev Lett 123:235101. https://doi.org/10.1103/PhysRevLett.123.235101

Narita Y, Plaschke F, Nakamura R, et al (2016) Wave telescope technique for mms magnetometer. Geophys Res Lett 43(10):4774–4780. https://doi.org/10.1002/2016GL069035

Narita Y, Glassmeier KH, Motschmann U (2022) The wave telescope technique. J Geophys Res Space Phys 127(2):e2021JA030165. https://doi.org/10.1029/2021JA030165

Ng J, Chen LJ, Le A, et al (2020) Lower-hybrid-drift vortices in the electron-scale magnetic reconnection layer. Geophys Res Lett 47(22):e2020GL090726. https://doi.org/10.1029/2020GL090726

Ng J, Yoo J, Chen LJ, et al (2023) 3D simulation of lower-hybrid drift waves in strong guide field asymmetric reconnection in laboratory experiments. Phys Plasmas 30(4):042101. https://doi.org/10.1063/5.0138278

Norgren C, Vaivads A, Khotyaintsev YV, et al (2012) Lower hybrid drift waves: space observations. Phys Rev Lett 109:055001. https://doi.org/10.1103/PhysRevLett.109.055001

Norgren C, André M, Graham DB, et al (2015) Slow electron holes in multicomponent plasmas. Geophys Res Lett 42.7264. https://doi.org/10.1002/2015GL065390

Norgren C, Graham DB, Khotyaintsev YV, et al (2016) Finite gyroradius effects in the electron outflow of asymmetric magnetic reconnection. Geophys Res Lett 43:6724–6733. https://doi.org/10.1002/2016GL069205

Norgren C, Hesse M, Graham DB, et al (2020) Electron acceleration and thermalization at magnetotail separatrices. J Geophys Res Space Phys 125(4):e2019JA027440. https://doi.org/10.1029/2019JA027440

Norgren C, Graham DB, Argall MR, et al (2022) Millisecond observations of nonlinear wave–electron interaction in electron phase space holes. Phys Plasmas 29(1):012309. https://doi.org/10.1063/5.0073097

Norgren C, Chen LJ, Graham DB, et al (2025) Electron and ion dynamics in reconnection diffusion regions. Space Sci Rev 221

Nykyri K, Ma X, Johnson J (2021) Cross-scale energy transport in space plasmas. In: Maggiolo R et al (eds) Magnetospheres in the Solar System. Wiley, American Geophysical Union, Chap. 7, pp 109–121. https://doi.org/10.1002/9781119815624.ch7

Øieroset M, Phan TD, Haggerty C, et al (2016) MMS observations of large guide field symmetric reconnection between colliding reconnection jets at the center of a magnetic flux rope at the magnetopause. Geophys Res Lett 43:5536–5544. https://doi.org/10.1002/2016GL069166

Pan DX, Khotyaintsev YV, Graham DB, et al (2018) Rippled electron-scale structure of a dipolarization front. Geophys Res Lett 45(22):12116–12124. https://doi.org/10.1029/2018GL080826

Pedersen A, Lybekk B, André M, et al (2008) Electron density estimates derived from spacecraft potential measurements on cluster in tenuous plasma regions. J Geophys Res 113:A07S33

Pincon JL, Lefeuvre F (1991) Local characterization of homogeneous turbulence in a space plasma from simultaneous measurements of field components at several points in space. J Geophys Res Space Phys 96(A2):1789–1802. https://doi.org/10.1029/90JA02183

Price L, Swisdak M, Drake JF, et al (2016) The effects of turbulence on three-dimensional magnetic reconnection at the magnetopause. Geophys Res Lett 43:6020–6027. https://doi.org/10.1002/2016GL069578

Price L, Swisdak M, Drake JF, et al (2017) Turbulence in three-dimensional simulations of magnetopause reconnection. J Geophys Res Space Phys 122:11086–11099. https://doi.org/10.1002/2017JA024227

Price L, Swisdak M, Drake JF, et al (2020) Turbulence and transport during guide field reconnection at the magnetopause. J Geophys Res Space Phys 125(4):e2019JA027498. https://doi.org/10.1029/2019JA027498

Pritchett PL (2005) Externally driven magnetic reconnection in the presence of a normal magnetic field. J Geophys Res Space Phys 110(A5):A05209. https://doi.org/10.1029/2004JA010948
Pritchett PL (2013) The influence of intense electric fields on three-dimensional asymmetric magnetic reconnection. Phys Plasmas 20:061204. https://doi.org/10.1063/1.4811123
Pritchett PL (2015a) Instability of current sheets with a localized accumulation of magnetic flux. Phys Plasmas 22(6):062102. https://doi.org/10.1063/1.4921666
Pritchett PL (2015b) Structure of exhaust jets produced by magnetic reconnection localized in the out-of-plane direction. J Geophys Res Space Phys 120(1):592–608. https://doi.org/10.1002/2014JA020795
Rager AC, Dorelli JC, Gershman DJ, et al (2018) Electron crescent distributions as a manifestation of diamagnetic drift in an electron-scale current sheet: magnetospheric multiscale observations using new 7.5 ms fast plasma investigation moments. Geophys Res Lett 45(2):578–584. https://doi.org/10.1002/2017GL076260
Ren Y, Dai L, Li W, et al (2019) Whistler waves driven by field-aligned streaming electrons in the near-Earth magnetotail reconnection. Geophys Res Lett 46(10):5045–5054. https://doi.org/10.1029/2019GL083283
Ren Y, Dai L, Wang C, et al (2022) Parallel electron heating through Landau resonance with lower hybrid waves at the edge of reconnection ion jets. Astrophys J 928(1):5. https://doi.org/10.3847/1538-4357/ac53fb
Retinò A, Vaivads A, André M, et al (2006) Structure of the separatrix region close to a magnetic reconnection X-line: cluster observations. Geophys Res Lett 33:L06101. https://doi.org/10.1029/2005GL024650
Réville V, Velli M, Rouillard AP, et al (2020) Tearing instability and periodic density perturbations in the slow solar wind. Astrophys J Lett 895(1):L20. https://doi.org/10.3847/2041-8213/ab911d
Richard L, Khotyaintsev YV, Graham DB, et al (2021) Observations of short-period ion-scale current sheet flapping. J Geophys Res Space Phys 126(8):e29152. https://doi.org/10.1029/2021JA029152
Richard L, Khotyaintsev YV, Graham DB, et al (2023) Fast ion isotropization by current sheet scattering in magnetic reconnection jets. Phys Rev Lett 131:115201. https://doi.org/10.1103/PhysRevLett.131.115201
Roberts OW, Narita Y, Li X, et al (2017) Multipoint analysis of compressive fluctuations in the fast and slow solar wind. J Geophys Res Space Phys 122(7):6940–6963. https://doi.org/10.1002/2016JA023552
Rogers BN, Denton RE, Drake JF, et al (2001) Role of dispersive waves in collisionless magnetic reconnection. Phys Rev Lett 87:195004. https://doi.org/10.1103/PhysRevLett.87.195004
Roytershteyn V, Daughton W, Karimabadi H, et al (2012) Influence of the lower-hybrid drift instability on magnetic reconnection in asymmetric configurations. Phys Rev Lett 108:185001. https://doi.org/10.1103/PhysRevLett.108.185001
Runov A, Nakamura R, Baumjohann W, et al (2003) Cluster observation of a bifurcated current sheet. Geophys Res Lett 30(2):1036. https://doi.org/10.1029/2002GL016136
Runov A, Sergeev VA, Baumjohann W, et al (2005) Electric current and magnetic field geometry in flapping magnetotail current sheets. Ann Geophys 23(4):1391–1403. https://doi.org/10.5194/angeo-23-1391-2005
Schindler K (1972) A self-consistent theory of the tail of the magnetosphere. In: McCormac BM (ed) Earth's magnetospheric processes, p 200. https://doi.org/10.1007/978-94-010-2896-7_19
Schindler K (1974) A theory of the substorm mechanism. J Geophys Res 79(19):2803–2810. https://doi.org/10.1029/JA079i019p02803
Sergeev V, Runov A, Baumjohann W, et al (2004) Orientation and propagation of current sheet oscillations. Geophys Res Lett 31(5):L05807. https://doi.org/10.1029/2003GL019346
Sharma Pyakurel P, Shay MA, Haggerty CC, et al (2018) Super-Alfvénic propagation and damping of reconnection onset signatures. J Geophys Res Space Phys 123(1):341–349. https://doi.org/10.1002/2017JA024606
Shay MA, Drake JF, Eastwood JP, et al (2011) Super-Alfvénic propagation of substorm reconnection signatures and Poynting flux. Phys Rev Lett 107:065001. https://doi.org/10.1103/PhysRevLett.107.065001
Shi F, Lin Y, Wang X (2013) Global hybrid simulation of mode conversion at the dayside magnetopause. J Geophys Res Space Phys 118(10):6176–6187. https://doi.org/10.1002/jgra.50587
Shi P, Huang K, Lu Q, et al (2019) Experimental observation of kinetic Alfvén wave generated by magnetic reconnection. Plasma Phys Control Fusion 61(12):125010. https://doi.org/10.1088/1361-6587/ab4f9c
Sitnov MI, Schindler K (2010) Tearing stability of a multiscale magnetotail current sheet. Geophys Res Lett 37(8):L08102. https://doi.org/10.1029/2010GL042961
Sitnov MI, Swisdak M, Drake JF, et al (2004) A model of the bifurcated current sheet: 2. Flapping motions. Geophys Res Lett 31(9):L09805. https://doi.org/10.1029/2004GL019473
Sitnov MI, Buzulukova N, Swisdak M, et al (2013) Spontaneous formation of dipolarization fronts and reconnection onset in the magnetotail. Geophys Res Lett 40(1):22–27. https://doi.org/10.1029/2012GL054701

Sitnov MI, Merkin VG, Swisdak M, et al (2014) Magnetic reconnection, buoyancy, and flapping motions in magnetotail explosions. J Geophys Res Space Phys 119(9):7151–7168. https://doi.org/10.1002/2014JA020205

Sitnov M, Birn J, Ferdousi B, et al (2019) Explosive magnetotail activity. Space Sci Rev 215(4):31. https://doi.org/10.1007/s11214-019-0599-5

Speiser TW (1973) Magnetospheric current sheets. Radio Sci 8(11):973–977. https://doi.org/10.1029/RS008i011p00973

Speiser TW, Ness NF (1967) The neutral sheet in the geomagnetic tail: its motion, equivalent currents, and field line connection through it. J Geophys Res 72(1):131–141. https://doi.org/10.1029/JZ072i001p00131

Stasiewicz K, Bellan P, Chaston C, et al (2000) Small scale Alfvénic structure in the aurora. Space Sci Rev 92(3):423–533. https://doi.org/10.1023/A:1005207202143

Stawarz JE, Eastwood JP, Varsani A, et al (2017) Magnetospheric multiscale analysis of intense field-aligned Poynting flux near the Earth's plasma sheet boundary. Geophys Res Lett 44(14):7106–7113. https://doi.org/10.1002/2017GL073685

Stawarz JE, Muñoz PA, Bessho N, et al (2024) The interplay between collisionless magnetic reconnection and turbulence. Space Sci Rev 220:90. https://doi.org/10.1007/s11214-024-01124-8

Steinvall K, Khotyaintsev YV, Graham DB, et al (2019a) Observations of electromagnetic electron holes and evidence of Cherenkov whistler emission. Phys Rev Lett 123:255101. https://doi.org/10.1103/PhysRevLett.123.255101

Steinvall K, Khotyaintsev YV, Graham DB, et al (2019b) Multispacecraft analysis of electron holes. Geophys Res Lett 46(1):55–63. https://doi.org/10.1029/2018GL080757

Steinvall K, Khotyaintsev YV, Graham DB, et al (2021) Large amplitude electrostatic proton plasma frequency waves in the magnetospheric separatrix and outflow regions during magnetic reconnection. Geophys Res Lett 48(5):e2020GL090286. https://doi.org/10.1029/2020GL090286

Steinvall K, Khotyaintsev YV, Graham DB (2022) On the applicability of single-spacecraft interferometry methods using electric field probes. J Geophys Res Space Phys 127(3):e2021JA030143. https://doi.org/10.1029/2021JA030143

Stix TH (1992) The theory of plasma waves. American Institute of Physics, Melville

Stringer TE (1964) Electrostatic instabilities in current-carrying and counterstreaming plasmas. J Nucl Energy, Part C Plasma Phys Accel Thermonucl Res 6(3):267–279. https://doi.org/10.1088/0368-3281/6/3/305

Swanson DG (1989) Plasma waves. Academic Press, Boston

TanDokoro R, Fujimoto M (2005) Three-dimensional mhd simulation study of the structure at the leading part of a reconnection jet. Geophys Res Lett 32(23):L23102. https://doi.org/10.1029/2005GL024467

Tang BB, Li WY, Graham DB, et al (2020a) Lower hybrid waves at the magnetosheath separatrix region. Geophys Res Lett 47(20):e2020GL089880. https://doi.org/10.1029/2020GL089880

Tang BB, Li WY, Le A, et al (2020b) Electron mixing and isotropization in the exhaust of asymmetric magnetic reconnection with a guide field. Geophys Res Lett 47(14):e2020GL087159. https://doi.org/10.1029/2020GL087159

Taroyan Y, Erdélyi R (2002) Resonant and Kelvin–Helmholtz instabilities on the magnetopause. Phys Plasmas 9(7):3121–3129. https://doi.org/10.1063/1.1481746

Tjulin A, Pincon JL, Sahraoui F, et al (2005) The k-filtering technique applied to wave electric and magnetic field measurements from the cluster satellites. J Geophys Res Space Phys 110(A11):A11224. https://doi.org/10.1029/2005JA011125

Tong Y, Vasko I, Mozer FS, et al (2018) Simultaneous multispacecraft probing of electron phase space holes. Geophys Res Lett 45(21):11513–11519. https://doi.org/10.1029/2018GL079044

Torbert RB, Burch JL, Phan TD, et al (2018) Electron-scale dynamics of the diffusion region during symmetric magnetic reconnection in space. Science 362(6421):1391–1395. https://doi.org/10.1126/science.aat2998

Treumann RA, Baumjohann W (1997) Advanced space plasma physics. Imperial College Press, London

Tsai E, Artemyev A, Zhang XJ, et al (2022) Relativistic electron precipitation driven by nonlinear resonance with whistler-mode waves. J Geophys Res Space Phys 127(5):e30338. https://doi.org/10.1029/2022JA030338

Uchino H, Kurita S, Harada Y, et al (2017) Waves in the innermost open boundary layer formed by dayside magnetopause reconnection. J Geophys Res Space Phys 122(3):3291–3307. https://doi.org/10.1002/2016JA023300

Vaivads A, André M, Buchert SC, et al (2004) Cluster observations of lower hybrid turbulence within thin layers at the magnetopause. Geophys Res Lett 31:L03804. https://doi.org/10.1029/2003GL018142

Vaivads A, Khotyaintsev Y, André M, et al (2006) Plasma waves near reconnection sites. In: LaBelle JW, Treumann RA (eds) Geospace electromagnetic waves and radiation. Lecture Notes in Physics, vol 687. Springer, Berlin, Heidelberg, pp 251–269. https://doi.org/10.1007/3-540-33203-0_10

Vapirev A, Lapenta G, Divin A, et al (2013) Formation of a transient front structure near reconnection point in 3-d pic simulations. J Geophys Res Space Phys 118(4):1435–1449. https://doi.org/10.1002/jgra.50136
Vasko IY, Mozer FS, Krasnoselskikh VV, et al (2018) Solitary waves across supercritical quasi-perpendicular shocks. Geophys Res Lett 45(12):5809–5817. https://doi.org/10.1029/2018GL077835
Verscharen D, Chandran BDG, Boella E, et al (2022) Electron-driven instabilities in the solar wind. Front Astron Space Sci 9:951628. https://doi.org/10.3389/fspas.2022.951628
Viberg H, Khotyaintsev YV, Vaivads A, et al (2013) Mapping HF waves in the reconnection diffusion region. Geophys Res Lett 40:1032. https://doi.org/10.1002/grl.50227
Viberg H, Khotyaintsev YV, Vaivads A, et al (2014) Whistler mode waves at magnetotail dipolarization fronts. J Geophys Res 119:2605. https://doi.org/10.1002/2014JA019892
Vogt J, Haaland S, Paschmann G (2011) Accuracy of multi-point boundary crossing time analysis. Ann Geophys 29(12):2239–2252. https://doi.org/10.5194/angeo-29-2239-2011
Vörös Z (2011) Magnetic reconnection associated fluctuations in the deep magnetotail: Artemis results. Nonlinear Process Geophys 18(6):861–869. https://doi.org/10.5194/npg-18-861-2011
Vörös Z, Yordanova E, Graham DB, et al (2019) Mms observations of whistler and lower hybrid drift waves associated with magnetic reconnection in the turbulent magnetosheath. J Geophys Res Space Phys 124(11):8551–8563. https://doi.org/10.1029/2019JA027028
Wang R, Lu Q, Kotyaintsev YV, et al (2014) Observation of double layer in the separatrix region during magnetic reconnection. Geophys Res Lett 41:4851. https://doi.org/10.1002/2014GL061157
Wang S, Chen LJ, Hesse M, et al (2017) Parallel electron heating in the magnetospheric inflow region. Geophys Res Lett 44:4384–4392. https://doi.org/10.1002/2017GL073404
Wang GQ, Zhang TL, Wu MY, et al (2019) Solar wind directional change triggering flapping motions of the current sheet: MMS observations. Geophys Res Lett 46(1):64–70. https://doi.org/10.1029/2018GL080023
Wang CP, Liu YH, Xing X, et al (2020) An event study of simultaneous Earthward and tailward reconnection exhaust flows in the Earth's midtail. J Geophys Res Space Phys 125(6):e2019JA027406. https://doi.org/10.1029/2019JA027406
Wang S, Chen LJ, Ng J, et al (2021) Lower-hybrid drift waves and their interaction with plasmas in a 3D symmetric reconnection simulation with zero guide field. Phys Plasmas 28(7):072102. https://doi.org/10.1063/5.0054626
Wang S, Bessho N, Graham DB, et al (2022a) Whistler waves associated with electron beams in magnetopause reconnection diffusion regions. J Geophys Res Space Phys 127(9):e2022JA030882. https://doi.org/10.1029/2022JA030882
Wang S, Chen LJ, Bessho N, et al (2022b) Lower-hybrid wave structures and interactions with electrons observed in magnetotail reconnection diffusion regions. J Geophys Res Space Phys 127(5):e2021JA030109. https://doi.org/10.1029/2021JA030109
Wei YY, Huang SY, Rong ZJ, et al (2019) Observations of short-period current sheet flapping events in the Earth's magnetotail. Astrophys J Lett 874(2):L18. https://doi.org/10.3847/2041-8213/ab0f28
Wilder FD, Ergun RE, Goodrich KA, et al (2016) Observations of whistler mode waves with nonlinear parallel electric fields near the dayside magnetic reconnection separatrix by the magnetospheric multiscale mission. Geophys Res Lett 43:5909–5917. https://doi.org/10.1002/2016GL069473
Wilder FD, Ergun RE, Newman DL, et al (2017) The nonlinear behavior of whistler waves at the reconnecting dayside magnetopause as observed by the magnetospheric multiscale mission: a case study. J Geophys Res Space Phys 122(5):5487–5501. https://doi.org/10.1002/2017JA024062
Wilder FD, Ergun RE, Burch JL, et al (2018) The role of the parallel electric field in electron-scale dissipation at reconnecting currents in the magnetosheath. J Geophys Res Space Phys 123(8):6533–6547. https://doi.org/10.1029/2018JA025529
Wilder FD, Ergun RE, Hoilijoki S, et al (2019) A survey of plasma waves appearing near dayside magnetopause electron diffusion region events. J Geophys Res Space Phys 124(10):7837–7849. https://doi.org/10.1029/2019JA027060
Wilder FD, Conley M, Ergun RE, et al (2022) Magnetospheric multiscale observations of waves and parallel electric fields in reconnecting current sheets in the turbulent magnetosheath. J Geophys Res Space Phys 127(9):e2022JA030511. https://doi.org/10.1029/2022JA030511
Winarto HW, Kunz MW (2022) Triggering tearing in a forming current sheet with the mirror instability. J Plasma Phys 88(2):905880210. https://doi.org/10.1017/S0022377822000150
Wu M, Volwerk M, Lu Q, et al (2013) The proton temperature anisotropy associated with bursty bulk flows in the magnetotail. J Geophys Res Space Phys 118(8):4875–4883. https://doi.org/10.1002/jgra.50451
Wygant JR, Keiling A, Cattell CA, et al (2002) Evidence for kinetic Alfvén waves and parallel electron energization at 4–6 re altitudes in the plasma sheet boundary layer. J Geophys Res Space Phys 107(A8):SMP 24–1–SMP 24–15. https://doi.org/10.1029/2001JA900113

Wygant JR, Cattell CA, Lysak R, et al (2005) Cluster observations of an intense normal component of the electric field at a thin reconnecting current sheet in the tail and its role in the shock-like acceleration of the ion fluid into the separatrix region. J Geophys Res Space Phys 110:A09206. https://doi.org/10.1029/2004JA010708
Yamada M, Kulsrud R, Ji H (2010) Magnetic reconnection. Rev Mod Phys 82:603–664. https://doi.org/10.1103/RevModPhys.82.603
Yao X, Muñoz PA, Büchner J (2022b) Non-thermal electron velocity distribution functions due to 3D kinetic magnetic reconnection for solar coronal plasma conditions. Phys Plasmas 29(2):022104. https://doi.org/10.1063/5.0061151
Yao X, Muñoz PA, Büchner J, et al (2022a) Wave emission of nonthermal electron beams generated by magnetic reconnection. Astrophys J 933(2):219. https://doi.org/10.3847/1538-4357/ac7141
Yoo J, Jara-Almonte J, Yerger E, et al (2018) Whistler wave generation by anisotropic tail electrons during asymmetric magnetic reconnection in space and laboratory. Geophys Res Lett 45(16):8054–8061. https://doi.org/10.1029/2018GL079278
Yoo J, Ji JY, Ambat MV, et al (2020) Lower hybrid drift waves during guide field reconnection. Geophys Res Lett 47(21):e2020GL087192. https://doi.org/10.1029/2020GL087192
Yoo J, Ng J, Ji H, et al (2024) Anomalous resistivity and electron heating by lower hybrid drift waves during magnetic reconnection with a guide field. Phys Rev Lett 132:145101. https://doi.org/10.1103/PhysRevLett.132.145101
Yoon PH, Lui ATY (2008) Drift instabilities in current sheets with guide field. Phys Plasmas 15:072101. https://doi.org/10.1063/1.2938386
Yoon PH, Lui ATY, Sitnov MI (2002) Generalized lower-hybrid drift instabilities in current-sheet equilibrium. Phys Plasmas 9(5):1526–1538. https://doi.org/10.1063/1.1466822
Yuan Z, Dong Y, Huang S, et al (2022) Direct observation of acceleration and thermalization of beam electrons caused by double layers in the Earth's plasma sheet. Geophys Res Lett 49(13):e2022GL099483. https://doi.org/10.1029/2022GL099483
Zelenyi L, Artemiev A, Malova H, et al (2008) Marginal stability of thin current sheets in the Earth's magnetotail. J Atmos Sol-Terr Phys 70(2–4):325–333. https://doi.org/10.1016/j.jastp.2007.08.019
Zhang YC, Lavraud B, Dai L, et al (2017) Quantitative analysis of a Hall system in the exhaust of asymmetric magnetic reconnection. J Geophys Res Space Phys 122(5):5277–5289. https://doi.org/10.1002/2016JA023620
Zhang X, Angelopoulos V, Artemyev AV, et al (2018) Whistler and electron firehose instability control of electron distributions in and around dipolarizing flux bundles. Geophys Res Lett 45(18):9380–9389. https://doi.org/10.1029/2018GL079613
Zhang X, Angelopoulos V, Artemyev A, et al (2019) Energy transport by whistler waves around dipolarizing flux bundles. Geophys Res Lett 46(21):11718–11727. https://doi.org/10.1029/2019GL084226
Zhang H, Zhong Z, Tang R, et al (2022) Observations of whistler-mode waves and large-amplitude electrostatic waves associated with a dipolarization front in the bursty bulk flow. Astrophys J 933(1):105. https://doi.org/10.3847/1538-4357/ac739d
Zhang S, Chien A, Gao L, et al (2023) Ion and electron acoustic bursts during anti-parallel magnetic reconnection driven by lasers. Nat Phys 19(6):909–916. https://doi.org/10.1038/s41567-023-01972-1
Zhong ZH, Zhou M, Graham DB, et al (2022) Evidence for whistler waves propagating into the electron diffusion region of collisionless magnetic reconnection. Geophys Res Lett 49(7):e2021GL097387. https://doi.org/10.1029/2021GL097387
Zhou M, Ashour-Abdalla M, Deng X, et al (2009a) Themis observation of multiple dipolarization fronts and associated wave characteristics in the near-Earth magnetotail. Geophys Res Lett 36(20):L20107. https://doi.org/10.1029/2009GL040663
Zhou M, Deng XH, Li SY, et al (2009b) Observation of waves near lower hybrid frequency in the reconnection region with thin current sheet. J Geophys Res 114:A02216. https://doi.org/10.1029/2008JA013427
Zhou M, Berchem J, Walker RJ, et al (2018) Magnetospheric multiscale observations of an ion diffusion region with large guide field at the magnetopause: current system, electron heating, and plasma waves. J Geophys Res Space Phys 123(3):1834–1852. https://doi.org/10.1002/2017JA024517

Authors and Affiliations

D.B. Graham[1] · G. Cozzani[2] · Yu.V. Khotyaintsev[1,3] · V.D. Wilder[4] · J.C. Holmes[5] · T.K.M. Nakamura[6] · J. Büchner[7,8] · K. Dokgo[9] · L. Richard[1] · K. Steinvall[10] ·

C. Norgren[1,11] · L.-J. Chen[12] · H. Ji[13,14] · J.F. Drake[15] · J.E. Stawarz[16] · S. Eriksson[17]

✉ D.B. Graham
dgraham@irfu.se

✉ Y.V. Khotyaintsev
Yuri.Khotyaintsev@rymdfysik.uu.se

1 Swedish Institute of Space Physics, Uppsala, Sweden

2 Department of Physics, University of Helsinki, Helsinki, Finland

3 Department of Physics and Astronomy, Uppsala University, Uppsala, Sweden

4 Department of Physics, University of Texas at Arlington, Arlington, TX, USA

5 T-5, Los Alamos National Laboratory, Los Alamos, NM, USA

6 Space Research Institute, Austrian Academy of Sciences, Graz, Austria

7 Max Planck Institute for Solar System Research, Göttingen, Germany

8 Centre for Astronomy and Astrophysics, Technical University of Berlin, Berlin, Germany

9 Southwest Research Institute, San Antonio, TX, USA

10 Department of Physics, Chalmers University of Technology, Göteborg, Sweden

11 Departement of Physics and Technology, University of Bergen, Bergen, Norway

12 NASA Goddard Space Flight Center, Greenbelt, MD, USA

13 Department of Astrophysical Sciences, Princeton University, Princeton, NJ, USA

14 Princeton Plasma Physics Laboratory, P.O. Box 451, Princeton, NJ, USA

15 Department of Physics, Institute for Physical Science and Technology and Joint Space Science Institute, University of Maryland, College Park, MD, USA

16 Northumbria University, Newcastle upon Tyne, UK

17 Laboratory for Atmospheric and Space Physics, University of Colorado, Boulder, CO, USA

Space Science Reviews (2024) 220:90
https://doi.org/10.1007/s11214-024-01124-8

The Interplay Between Collisionless Magnetic Reconnection and Turbulence

J.E. Stawarz[1] · P.A. Muñoz[2,3] · N. Bessho[4,5] · R. Bandyopadhyay[6] · T.K.M. Nakamura[7,8] · S. Eriksson[9] · D.B. Graham[10] · J. Büchner[2,3] · A. Chasapis[9] · J.F. Drake[11,12] · M.A. Shay[13] · R.E. Ergun[9,14] · H. Hasegawa[15] · Yu.V. Khotyaintsev[10] · M. Swisdak[11] · F.D. Wilder[16]

Received: 8 June 2024 / Accepted: 7 November 2024 / Published online: 25 November 2024

Abstract
Alongside magnetic reconnection, turbulence is another fundamental nonlinear plasma phenomenon that plays a key role in energy transport and conversion in space and astrophysical plasmas. From a numerical, theoretical, and observational point of view there is a long history of exploring the interplay between these two phenomena in space plasma environments; however, recent high-resolution, multi-spacecraft observations have ushered in a new era of understanding this complex topic. The interplay between reconnection and turbulence is both complex and multifaceted, and can be viewed through a number of different interrelated lenses - including turbulence acting to generate current sheets that undergo magnetic reconnection (*turbulence-driven reconnection*), magnetic reconnection driving turbulent dynamics in an environment (*reconnection-driven turbulence*) or acting as an intermediate step in the excitation of turbulence, and the random diffusive/dispersive nature of the magnetic field lines embedded in turbulent fluctuations enabling so-called *stochastic reconnection*. In this paper, we review the current state of knowledge on these different facets of the interplay between turbulence and reconnection in the context of collisionless plasmas, such as those found in many near-Earth astrophysical environments, from a theoretical, numerical, and observational perspective. Particular focus is given to several key regions in Earth's magnetosphere – namely, Earth's magnetosheath, magnetotail, and Kelvin-Helmholtz vortices on the magnetopause flanks – where NASA's *Magnetospheric Multiscale* mission has been providing new insights into the topic.

Keywords Magnetic reconnection · Turbulence · Collisionless plasmas

1 Introduction

Many natural plasmas where magnetic reconnection occurs have wide scale separations between the length scales where energy is injected into the system through dynamical processes and the smaller scales where energy is most effectively dissipated. Such systems are conducive to the excitation of complex and highly-nonlinear fluctuations, known as turbulence, that transfer energy across scales facilitating the dissipation of large-scale free energy.

While the basic physics underpinning individual reconnection sites can be considered using idealized models, magnetic reconnection is fundamentally a nonlinear and multi-scale

Extended author information available on the last page of the article

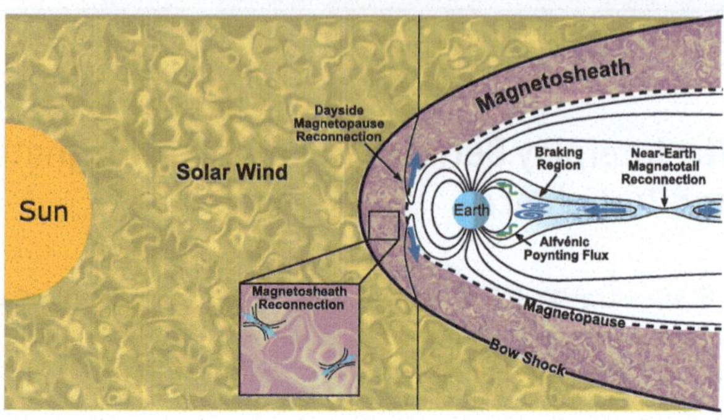

Fig. 1 Diagram illustrating the turbulent regions in near-Earth space along with the locations of the system-scale reconnection events associated with the Dungey cycle in Earth's magnetosphere. Turbulence can both interact with these system-scale reconnection events, as well as generate additional small-scale reconnection events within the turbulent regions

process that both influences and is influenced by the turbulent dynamics in the surrounding plasma. Therefore, a complete picture of the onset, development, and interaction of magnetic reconnection with the surrounding plasma requires coupling it into the turbulent dynamics. The study of the interaction between turbulence and reconnection has a long history; however, recent high-resolution space plasma observations, notably from NASA's *Magnetospheric Multiscale* (*MMS*) mission, have allowed us to observationally examine this interaction in greater detail than ever before.

In this review, we discuss our current understanding of the interaction between magnetic reconnection and turbulence – particularly within the nearly collisionless plasma regime applicable to many space and astrophysical systems – from an observational, numerical, and theoretical perspective. Section 1.1 provides an overview of the varied ways through which turbulence and reconnection can influence each other. Section 1.2 provides an introduction to elements of turbulence theory geared toward those who may be less familiar with the statistical theory of turbulence. Section 2 discusses *turbulence-driven reconnection* in which reconnection occurs at thin current sheets created by the turbulent dynamics, with Earth's magnetosheath highlighted as a key example where recent progress has been made. Section 3 discusses *reconnection-driven turbulence* in which reconnection at a pre-existing current sheet excites turbulence, with Earth's magnetotail highlighted as a key region of recent progress. Section 4 discusses magnetic reconnection as an element in the process of large-scale structures transitioning to a turbulent state, with a focus on the Kelvin-Helmholtz Instability (KHI) on Earth's magnetopause. Section 5 discusses how the stochastic nature of turbulent environments may impact magnetic reconnection. Section 6 summarizes our current state of knowledge and provides an outlook for future areas of research.

1.1 The Interaction Between Magnetic Reconnection and Turbulence

The interaction between magnetic reconnection and turbulence is a complex topic with multiple facets and, as such, it has been examined from a variety of distinct viewpoints in the literature. It is, therefore, important to consider what is meant by the interaction between turbulence and reconnection for a given environment, which may include:

1. *Turbulence-Driven Reconnection* – Turbulent plasmas are well-known to generate many thin current structures embedded within the fluctuations associated with the sheared, twisted, and tangled magnetic field topologies set up by the turbulent dynamics (e.g, Matthaeus et al. 2015, and Sect. 1.2.3), which can be sites where reconnection occurs (as illustrated in Fig. 1 inset). The nature of these current structures is fundamentally linked

to an aspect of turbulence referred to as intermittency, in which the turbulent dynamics have a tendency to generate coherent structures with extreme gradients. Turbulence-driven reconnection can play a role in facilitating both the nonlinear interaction between structures and the dissipation of the turbulent fluctuations. Assessing the importance of these reconnection events in a turbulent medium requires both information about the prevalence of magnetic reconnection at turbulent current sheets and an understanding of the key question of how magnetic reconnection partitions energy as discussed in Liu et al. (2024, this collection).

2. *Reconnection-Driven Turbulence* – Space plasmas also contain regions where reconnection occurs at system-scale current sheets set up by the configuration of the system as a whole, such as the reconnecting current sheets associated with the Dungey cycle at Earth's magnetopause and in the magnetotail (as illustrated in Fig. 1), the heliospheric current sheet, or reconnection associated with the configuration of coronal loops on the Sun. Large-scale reconnection outflows associated with these current sheets can excite turbulent fluctuations in the system through the spontaneous destabilization of the outflows and the interaction of the outflow with the surrounding plasma (Pucci et al. 2017; Ergun et al. 2018; Volwerk et al. 2007; Chaston et al. 2012; Stawarz et al. 2015). In this context, turbulence may impact the reconnection rate ($\mathcal{R}$) through anomalous resistivity/viscosity and can be thought of as a conduit through which the energy that is released by reconnection is re-partitioned across different energy channels in the outflow (although it does not necessarily need to alter the net energy released).
3. *The Transition to Turbulence* – As well as system-scale reconnection driving turbulence, it can also be driven by other processes. Magnetic reconnection can play a key role as a secondary process in the transition of such systems into a fully-developed turbulent state. The degree to which this process can be clearly distinguished from *turbulence-driven reconnection* and *reconnection-driven turbulence* in observations may be more or less clear depending on the system in question; however, a good representative case where this transitory phase is readily accessible to spacecraft observations is the KHI on the flanks of Earth's magnetosphere (e.g., Nykyri and Otto 2001). Numerical simulations of the development of turbulence from other initial configurations, including decaying Alfvénic turbulence (Franci et al. 2017) and systems of multiple current sheets (Gingell et al. 2017), have also highlighted the potential importance of reconnection in transitioning the initially large-scale fluctuations to a fully-developed turbulent state.
4. *Stochastic Reconnection* – Whether reconnection self-generates turbulence or is embedded in a turbulent environment, the stochastic nature of magnetic field lines in a turbulent flow can potentially have a profound impact on the global structure of the reconnection site. For example, the "rough" field line topology associated with the broadband distribution of fluctuations may lead to a broad, patchy region where the frozen-in flux theorem is violated and field line wandering in the turbulent flow may disperse field lines faster than otherwise expected. Since $\mathcal{R}$ is set by the aspect ratio of the diffusion region, these turbulent effects may significantly increase the reconnection rate if they play a dominant role in the dynamics. In this situation, $\mathcal{R}$ would no longer be controlled by microphysical effects, such as resistivity or collisionless processes, but instead by the properties of the turbulent fluctuations, such as the scale at which turbulent energy is injected into the system and fluctuation amplitudes.

While the above scenarios are not necessarily mutually exclusive, they provide a framework for conceptualizing different aspects of the complex multi-faceted interaction between reconnection and turbulence, which we will elaborate on further throughout this review.

1.2 Concepts from Turbulence Theory

While many concepts in turbulence theory can be extended to more complete descriptions of collisionless plasmas, in this section we illustrate some of the basic concepts based on the simplest case of incompressible magnetohydrodynamics (MHD). When considering turbulent dynamics it can be useful to divide the particle and electromagnetic variables into spatially and temporally uniform average parts (denoted by subscript 0) and fluctuations about that mean (denoted by a leading δ), such that for an arbitrary quantity $\mathbf{g}(\mathbf{x},t) = \mathbf{g}_0 + \delta\mathbf{g}(\mathbf{x},t)$. Performing this decomposition, the incompressible MHD equations can be written as

$$\frac{\partial \delta\mathbf{u}}{\partial t} = \overbrace{\frac{(\nabla\times\delta\mathbf{B})\times\mathbf{B}_0}{\mu_0\rho_0} - \frac{\nabla\delta p}{\rho_0}}^{Linear} \overbrace{- \delta\mathbf{u}\cdot\nabla\delta\mathbf{u} + \frac{(\nabla\times\delta\mathbf{B})\times\delta\mathbf{B}}{\mu_0\rho_0}}^{Nonlinear} + \overbrace{\nu\nabla^2\delta\mathbf{u}}^{Dissipative}, \tag{1}$$

$$\frac{\partial \delta\mathbf{B}}{\partial t} = \nabla\times(\delta\mathbf{u}\times\mathbf{B}_0) + \nabla\times(\delta\mathbf{u}\times\delta\mathbf{B}) + \frac{\eta}{\mu_0}\nabla^2\delta\mathbf{B}, \tag{2}$$

$$\nabla^2\delta p = \nabla\cdot\left(\frac{(\nabla\times\delta\mathbf{B})\times\mathbf{B}_0}{\mu_0}\right) - \nabla\cdot\left(\rho_0\delta\mathbf{u}\cdot\nabla\delta\mathbf{u} - \frac{(\nabla\times\delta\mathbf{B})\times\delta\mathbf{B}}{\mu_0}\right), \tag{3}$$

where $\mathbf{u}$ is the single-fluid velocity with $\mathbf{u}_0$ taken to be zero since the equations are Galilean invariant, $\mathbf{B}$ is the (divergence free) magnetic field, p is the particle pressure, $\rho = \rho_0$ is the mass density that we have taken to be spatially and temporally uniform, and μ_0 is the vacuum permeability. We have included collisional viscous and resistive dissipation, which introduce the kinematic viscosity (ν) and the resistivity (η), for illustrative purposes; however, in collissionless plasmas dissipation occurs through kinetic process that are not present in the single-fluid MHD approximation. Equation (3) follows from the incompressibility condition $\nabla\cdot\delta\mathbf{u} = 0$.

By dividing variables into mean and fluctuating parts, three classes of terms are apparent – 1) linear terms, which in isolation (e.g., for sufficiently small amplitude fluctuations, or if the dynamics preserve alignments that minimise the nonlinearities) give rise to linear wave modes, 2) nonlinear terms, which give rise to the turbulent dynamics, and 3) dissipative terms, which remove fluctuation energy from the system. In general, both linear and nonlinear terms contribute to the dynamics in the fully nonlinear system and developing a theoretical description of the turbulence requires understanding the interplay between these dynamics. While under the right circumstances exact nonlinear solutions for, typically isolated, plasma structures may be possible, fully-developed turbulent systems are characterised by dynamics that are sufficiently nonlinear and made up of such a multitude of interacting structures and fluctuations that exact solutions to the equations become intractable.

The relative importance of the nonlinear to dissipative dynamics for a system can be estimated by comparing the typical amplitude of the nonlinear advection term ($-\mathbf{u}\cdot\nabla\mathbf{u}$) to the viscous dissipation term ($\nu\nabla^2\mathbf{u}$) in Eq. (1) such that

$$\frac{|-\delta\mathbf{u}\cdot\nabla\delta\mathbf{u}|}{|\nu\nabla^2\delta\mathbf{u}|} \sim \frac{\mathcal{L}\mathcal{U}}{\nu} \equiv Re, \tag{4}$$

where $\mathcal{L}$ and $\mathcal{U}$ represent a characteristic length scale and velocity for the fluctuations and Re is referred to as the Reynolds number. For MHD systems, two additional analogues to the Reynolds number can be defined based on the nonlinear Lorentz force term in Eq. (1) and the nonlinear magnetic advection term in Eq. (2). For sufficiently large Re, the nonlinear terms significantly dominate over the dissipative terms across a wide range of scales and the system becomes turbulent.

Nearly collisionless plasmas are by definition high-Re in the sense that $\nu \sim \eta \sim 0$; however, some care must be taken because, in the absence of collisions, kinetic phenomena associated with the breakdown of the fluid approximation can introduce dissipative effects at scales larger than those expected from collisions. An alternative way to interpret Re is as a measure of the scale separation between $\mathcal{L}$ and the so-called Kolmogorov microscale, defined as the scale ℓ_{kol} at which dissipative dynamics dimensionally become more important than the nonlinear dynamics, such that $Re = (\mathcal{L}/\ell_{kol})^{4/3}$. It has been suggested that an alternative way to characterize Re for collisionless plasmas is to consider the scale separation between $\mathcal{L}$ and the ion scales (Parashar et al. 2019).

The relative importance of the nonlinear to linear terms can be estimated by the dimensionless parameter

$$\chi = \frac{\tau_L}{\tau_{NL}} \tag{5}$$

where $\tau_L = \omega_L^{-1}$ is the linear timescale given by the inverse of the frequency of the associated wave and τ_{NL} is the timescale associated with the nonlinear dynamics of interest. If $\chi > 1$, nonlinear interactions are faster than linear dynamics (i.e., the propagation of linear waves) and the nonlinear dynamics dominate the behavior, while, if $\chi < 1$, the linear dynamics are faster than the nonlinear interactions. Equation (5) is dimensionally equivalent to comparing the amplitudes of the nonlinear and linear terms in a given equation in analogy to the definition of Re; however, χ is more typically framed in terms of the ratio of timescales as this definition lends itself to comparing the importance of the nonlinear dynamics to the normal modes of the system. Analyses of turbulence in the solar wind, for example, suggest τ_{NL} is faster than τ_L in many cases and, thus, $\chi > 1$ in this environment (Matthaeus et al. 2014) and in Earth's magnetosheath analyses suggest a mixture of intervals where χ is greater than or less than one for the linear and nonlinear terms relevant to the induction equation (Lewis et al. 2023). While in Eqs. (1) – (3), we have split the linear and nonlinear dynamics based on the global $\mathbf{B}_0$, it can also be illustrative to treat $\mathbf{B}_0$ as a scale dependant or locally defined quantity, with the conceptual picture that small-scale fluctuations see their locally averaged field as the background (Chen 2016), although there is some nuance to the interpretation when using a locally defined $\mathbf{B}_0$ due to the stochastically varying reference frame that it introduces (Matthaeus et al. 2012).

1.2.1 Cross-Scale Energy Transfer

Given the complex multiscale nature of turbulent systems, theoretical descriptions are typically formulated in terms of statistical properties of the ensemble of interacting fluctuations - either in physical or spectral space. In physical space, the autocorrelation tensor for an arbitrary variable $\mathbf{g}(\mathbf{x})$ is given by

$$\mathbb{R}^{\mathbf{g}}_{ij}(\mathbf{x}, \boldsymbol{\ell}) = \langle \delta g_i (\mathbf{x} + \boldsymbol{\ell}) \, \delta g_j (\mathbf{x}) \rangle, \tag{6}$$

where $\langle ... \rangle$ represents an ensemble average. The correlation function associated with the total fluctuation energy in $\mathbf{g}(\mathbf{x})$ is given by the trace of $\mathbb{R}^{\mathbf{g}}_{ij}$, such that $R^{\mathbf{g}}(\mathbf{x}, \boldsymbol{\ell}) = \langle \delta \mathbf{g}(\mathbf{x} + \boldsymbol{\ell}) \cdot \delta \mathbf{g}(\mathbf{x}) \rangle$, while the off-diagonal part of the tensor encodes information about the helicity of the fluctuations (Matthaeus and Goldstein 1982). One common measure of the characteristic scale associated with $\delta\mathbf{g}$ is the correlation length ($\lambda_{C,\mathbf{g}}$) defined as

$$\lambda_{C,\mathbf{g}} = \frac{1}{R^{\mathbf{g}}(0)} \int_0^\infty R^{\mathbf{g}}(\ell)\, d\ell, \tag{7}$$

which represents the scale over which structures in $\delta\mathbf{g}$ decorrelate and can have different values along different directions if $R^{\mathbf{g}}(\boldsymbol{\ell})$ is an anisotropic function.

A complementary statistical quantity is the 2nd-order structure function given by

$$S_2^{\mathbf{g}}(\mathbf{x}, \boldsymbol{\ell}) = \langle |\Delta \mathbf{g}(\mathbf{x}, \boldsymbol{\ell})|^2 \rangle = \langle |\mathbf{g}(\mathbf{x} + \boldsymbol{\ell}) - \mathbf{g}(\mathbf{x})|^2 \rangle. \tag{8}$$

A simplifying assumption often, but not always, made in turbulence theory, is that fluctuations are statistically homogeneous and average quantities do not depend on $\mathbf{x}$. In this case there is a relationship between $R^{\mathbf{g}}(\boldsymbol{\ell})$ and $S_2^{\mathbf{g}}(\boldsymbol{\ell})$, such that

$$S_2^{\mathbf{g}}(\boldsymbol{\ell}) = 2R^{\mathbf{g}}(0) - 2R^{\mathbf{g}}(\boldsymbol{\ell}). \tag{9}$$

The statistical evolution of the turbulence can be considered by re-expressing Eqs. (1)-(3) as equations for the evolution of the structure functions or correlation functions associated with the total (bulk kinetic + magnetic) fluctuation energy (i.e., $S_2(\boldsymbol{\ell}) = S_2^{\mathbf{u}}(\boldsymbol{\ell}) + S_2^{\mathbf{B}_A}(\boldsymbol{\ell})$ or $R(\boldsymbol{\ell}) = R^{\mathbf{u}}(\boldsymbol{\ell}) + R^{\mathbf{B}_A}(\boldsymbol{\ell})$ where $\mathbf{B}_A = \mathbf{B}/\sqrt{\mu_0 \rho}$). For $S_2(\boldsymbol{\ell})$ under the assumption of statistical homogeneity of the fluctuations, this gives (Politano and Pouquet 1998a; Adhikari et al. 2023)

$$\frac{\partial S_2(\boldsymbol{\ell})}{\partial t} = \mathcal{P}(\boldsymbol{\ell}) - \frac{\partial}{\partial \boldsymbol{\ell}} \cdot \mathbf{Y}(\boldsymbol{\ell}) + 2D(\boldsymbol{\ell}) - 4\epsilon, \tag{10}$$

where $\mathcal{P}(\boldsymbol{\ell})$ describes the injection of fluctuation energy into the system through some external driving (such driving was not explicitly included in Eqs. (1)-(3) but is included here to aid in the interpretation of Eq. (10)), $\mathbf{Y}(\boldsymbol{\ell})$ is a mixed 3rd-order structure function encoding the effect of the nonlinear terms given by

$$\begin{aligned} \mathbf{Y}(\boldsymbol{\ell}) = \langle\, & \Delta \mathbf{u}(\boldsymbol{\ell})\, |\Delta \mathbf{u}(\boldsymbol{\ell})\,|^2 + \Delta \mathbf{u}(\boldsymbol{\ell})\, |\Delta \mathbf{B}_A(\boldsymbol{\ell})\,|^2 \\ & - 2\Delta \mathbf{B}_A(\boldsymbol{\ell})\, [\Delta \mathbf{u}(\boldsymbol{\ell}) \cdot \Delta \mathbf{B}_A(\boldsymbol{\ell})] \rangle, \end{aligned} \tag{11}$$

$D(\boldsymbol{\ell}) = \nu \frac{\partial^2}{\partial \ell^2} S_2^{\mathbf{u}}(\boldsymbol{\ell}) + \frac{\eta}{\mu_0} \frac{\partial^2}{\partial \ell^2} S_2^{\mathbf{B}_A}(\boldsymbol{\ell})$ describes the impact of dissipation on $S_2(\boldsymbol{\ell})$, and ϵ is the average energy dissipation rate in the system. Equation (10) highlights the closure problem that is one of the core challenges of developing a complete theory of turbulence – the evolution of $S_2(\boldsymbol{\ell})$ (or any statistical quantity) requires knowledge of higher-order structure functions such as $\mathbf{Y}(\boldsymbol{\ell})$, the evolution of which, in turn, require knowledge of progressively higher-order structure functions.

For turbulent systems in a statistically steady state, such that $\partial_t S_2(\boldsymbol{\ell}) = 0$, it is assumed that there will be a significant scale separation between the scales where $\mathcal{P}(\boldsymbol{\ell})$ is significant and the scales where $D(\boldsymbol{\ell})$ is significant. In order for Eq. (10) to be satisfied at intermediate scales, there must then be a range of scales - typically referred to as the *inertial range* - over which the $\mathbf{Y}(\boldsymbol{\ell})$ term dominates and Eq. (10) reduces to

$$\frac{\partial}{\partial \boldsymbol{\ell}} \cdot \mathbf{Y}(\boldsymbol{\ell}) = -4\epsilon. \tag{12}$$

Equation (12) gives an exact relationship between $\mathbf{Y}(\boldsymbol{\ell})$ and ϵ, describing the role of the nonlinear dynamics in transporting energy across scales from the driving to the dissipation scales (Politano and Pouquet 1998b,a) (see also, Kolmogorov (1941b) or Frisch (1995) for a discussion of Eq. (12) for hydrodynamics; and Marino and Sorriso-Valvo (2023) for a complete review of the derivation in plasmas).

Equation (12) can also be extended to more general cases than homogeneous incompressible MHD, for example by including background velocity shears (Wan et al. 2009; Stawarz et al. 2011), compressibility (Banerjee and Galtier 2013), or the Hall effect (Hellinger et al. 2018; Ferrand et al. 2019). The above expression, or variants of it, have been widely applied in both numerical simulations and observations to estimate the energy dissipation rate in turbulent plasmas (e.g. MacBride et al. 2005; Marino et al. 2008; Stawarz et al. 2009; Hadid et al. 2018; Bandyopadhyay et al. 2020b). Applications of Eq. (12) to spacecraft data typically require assumptions about the average geometry of the fluctuations to simplify the divergence with respect to $\boldsymbol{\ell}$, with common examples being isotropy for which $\mathbf{Y}(\boldsymbol{\ell})$ has no angular dependence (Politano and Pouquet 1998a) or hybrid anisotropic geometries with separable variations parallel and perpendicular to the magnetic field (MacBride et al. 2008; Stawarz et al. 2009). However, multispacecraft measurements can be used to explicitly evaluate the divergence (Osman et al. 2011; Pecora et al. 2023) and alternative formulations may also provide ways to estimate ϵ without implicit assumptions about anisotropy (Banerjee and Galtier 2017).

While the above formalism is typically derived and applied in the context of homogeneous turbulent systems, a series of recent studies (Adhikari et al. 2020, 2021, 2023, 2024), have explored the application of Eq. (10) to traditional periodic reconnection simulations (where reconnection is initiated in idealised current sheets at the scale of the periodic box) both with and without guide fields. In these works, it was found that the behavior of the terms in Eq. (10) are qualitatively similar to that found in fully developed turbulent systems, suggesting that the nonlinear dynamics associated with the reconnection process may, in some sense, embody an energy-cascade-like process.

Analogs of Eq. (10) can also be derived in spectral space (e.g., Alexakis et al. 2005; Grete et al. 2017) or using scale-filtering approaches, where a coarse-graining kernel is used instead of structure functions (Aluie 2017; Yang et al. 2017; Manzini et al. 2022). In spectral space, the analogue of $R^{\mathrm{g}}(\boldsymbol{\ell})$ is the energy spectral density, $\mathcal{E}_{\mathbf{g}}(\mathbf{k})$. For statistically homogeneous fluctuations, $\mathcal{E}_{\mathbf{g}}(\mathbf{k})$ is the Fourier transform of $R^{\mathrm{g}}(\boldsymbol{\ell})$ and, given Eq. (9), is also the Fourier transform of $S_2^{\mathbf{g}}(\boldsymbol{\ell})$ for $\mathbf{k} \neq 0$. By manipulating the Fourier transforms of Eqs. (1)-(3) into an expression for the evolution of $\mathcal{E}(\mathbf{k}) = \mathcal{E}_{\mathbf{u}}(\mathbf{k}) + \mathcal{E}_{\mathbf{B}_A}(\mathbf{k})$ or taking the Fourtier transform of Eq. (10) with respect to $\boldsymbol{\ell}$, a spectral representation of Eq. (10) can be obtained (e.g., Pope 2000)

$$\frac{\partial \mathcal{E}(\mathbf{k})}{\partial t} = \mathcal{P}(\mathbf{k}) + \mathcal{T}(\mathbf{k}) - 2D(\mathbf{k}) \tag{13}$$

where $\mathcal{T}(\mathbf{k})$ is the analogue of $-\frac{\partial}{\partial \boldsymbol{\ell}} \cdot \mathbf{Y}(\boldsymbol{\ell})$ – referred to as the transfer function in this context – and $\mathcal{P}(\mathbf{k})$ and $\mathcal{D}(\mathbf{k})$ are the Fourier transforms of $\mathcal{P}(\boldsymbol{\ell})$ and $\mathcal{D}(\boldsymbol{\ell})$, respectively. $\mathcal{T}(\mathbf{k})$ in general takes the form of a complicated set of convolutions associated with the nonlinear terms, representing the net transfer of energy to/from wavevector $\mathbf{k}$ due to the sum of all possible interactions between other wavevectors and can be related to the cross-scale energy flux, $\mathbf{\Pi}(\mathbf{k})$, through a surface in $\mathbf{k}$-space, such that $\mathcal{T}(\mathbf{k}) = -\frac{\partial}{\partial \mathbf{k}} \cdot \mathbf{\Pi}(\mathbf{k})$. Often Eq. (13) is integrated over spherical shells and $\Pi(k)$ is interpreted as the isotropic energy flux; however, other surfaces may be relevant to different types of anisotropy (Yokoyama and Takaoka 2021).

The spectral representation highlights two key assumptions often invoked in turbulence theory - 1) that the nonlinear interactions in fully-developed turbulence are local in k-space, such that the dominant contributions to the convolutions within $\mathcal{T}(\mathbf{k})$ come from wavevectors with similar magnitudes and 2) that the magnitude of $\Pi(k) = \epsilon$ and is constant as a function of k. With these two assumptions, the conceptual picture of the *energy cascade* through the inertial range emerges, where nonlinear interactions incrementally transport energy from scale-to-scale at a constant rate, which in a statistically steady-state, is equal to the average energy dissipation rate. Consequently, ϵ is often referred to as the energy cascade rate in turbulent systems. While the large-scale nonlinear dynamics set the flux of energy to the small scales, for collisionless plasmas understanding which processes are responsible for dissipating that energy remains a key challenge that can have consequences for how dissipated energy is partitioned between particle species and across velocity phase-space and in Sect. 2.3 we discuss the potential role reconnection may play in turbulent dissipation.

1.2.2 Energy Spectra

The shape of $\mathcal{E}(\mathbf{k})$ or $S_2(\boldsymbol{\ell})$ in the inertial range can be estimated by making assumptions about the nature of the dominant nonlinear interactions. The basic ingredients for estimating these scalings amount to 1) taking ϵ to be independent of scale within the inertial rage and 2) estimating the timescale τ_{tr} over which nonlinear interactions transfer energy between scales under a given set of assumptions. Relating $\mathcal{E}(\mathbf{k})$ to τ_{tr} dimensionally gives

$$\mathcal{E}(k) \sim \epsilon \frac{\tau_{tr}}{k}, \tag{14}$$

where, for simplicity, we have assumed the fluctuations are isotropic (although this assumption can be relaxed).[1] In principle, τ_{tr} can be a function of scale, the geometry of the fluctuations, and the physical properties of the medium, and its dependence on these parameters is intrinsically linked to the nature of the underlying nonlinear interactions enabling the cascade, leading to different predictions for $\mathcal{E}(\mathbf{k})$. This link between predictions for the shape of $\mathcal{E}(\mathbf{k})$ and the nature of the nonlinear dynamics is one reason why $\mathcal{E}(\mathbf{k})$ is an important observable quantity for understanding turbulent dynamics, although it does not provide the full picture as discussed in Sect. 1.2.3. Single spacecraft measurements are capable of providing spacecraft-frame frequency spectra; however, assuming the background flow ($\mathbf{U}_0$) is sufficiently fast, comparisons can be made between observed spectra and theoretical predictions for $\mathcal{E}(\mathbf{k})$ by employing the so-called Taylor hypothesis discussed in the Appendix. Table 1 illustrates some example models for $\mathcal{E}(\mathbf{k})$. The first five models are based on MHD, while the final model gives an example for sub-proton-scale dynamics.

Strongly nonlinear models where $Re \gg 1$ and $\chi \gg 1$, as in the *Kolmogorov* model, assume τ_{tr} is governed by the nonlinear timescale (τ_{NL}) associated with the advection terms. Kolmogorov (1941a) originally applied this model to incompressible hydrodynamic turbulence, predicting the well-known $k^{-5/3}$-spectrum. However, since the additional nonlinear advective and Lorentz force terms in the MHD equations are dimensionally equivalent to the advective term in hydrodynamics, an analogous approach can be taken with incompressible MHD, producing an equivalent spectral prediction (Biskamp and Müller 2000).

[1] The units of $\mathcal{E}(\mathbf{k})$ are such that when integrated over all $\mathbf{k}$ it gives the average total fluctuation energy in the system, such that the quantity $\mathcal{E}(\mathbf{k})\, d^3\mathbf{k}$ has units of energy. In some situations it is useful to consider the isotropic (omnidirectional) energy spectrum, defined such that $\mathcal{E}(k)\, dk$ has units of energy, and in plasmas it is often useful to consider the gyrotropically integrated spectrum, defined such that $\mathcal{E}\left(k_{||}, k_{\perp}\right) dk_{||} dk_{\perp}$ has units of energy. Furthermore, in the gyrotropic case, it is sometimes useful to consider the spectra reduced over $k_{||}$ or $k_{\perp}$, such that $\mathcal{E}(k_{\perp})\, dk_{\perp}$ and $\mathcal{E}\left(k_{||}\right) dk_{||}$ have units of energy, respectively.

Table 1 Summary of example models for the turbulent energy spectrum

Model	τ_{tr}	Spectral Prediction	Anisotropy
Kolmogorov[a]	$\tau_{NL} \sim \frac{1}{k\delta u(k)}$	$\mathcal{E}(k) \sim \epsilon^{2/3} k^{-5/3}$	Isotropic
Iroshnikov[b]-Kraichnan[c]	$\frac{\tau_{NL}}{\chi} \sim \frac{V_A}{k[\delta u(k)]^2}$	$\mathcal{E}(k) \sim (\epsilon V_A)^{1/2} k^{-3/2}$	Isotropic
Weak Alfvénic[d]	Derived exactly assuming $\chi \ll 1$	$\mathcal{E}(k_{\|\|}, k_\perp) \sim f(k_{\|\|})(\epsilon V_A)^{1/2} k_\perp^{-2}$	Purely $\perp$
Goldreich-Sridhar[e]	$\chi \sim 1 \rightarrow \frac{1}{k_\perp \delta u_\perp(k_\perp)} \sim \frac{1}{k_{\|\|} V_A}$	$\mathcal{E}(k_\perp) \sim \epsilon^{2/3} k_\perp^{-5/3}$	$k_{\|\|} \sim \epsilon^{1/3} V_A^{-1} k_\perp^{2/3}$
Dynamic Alignment[f]	$\chi \sim 1 \rightarrow \frac{V_A}{k_\perp [\delta u_\perp(k_\perp)]^2} \sim \frac{1}{k_{\|\|} V_A}$	$\mathcal{E}(k_\perp) \sim (\epsilon V_A)^{1/2} k_\perp^{-3/2}$	$k_{\|\|} \sim \epsilon^{1/2} V_A^{-3/2} k_\perp^{1/2}$
Strong Hall MHD[g]	$\tau_{NL} \sim \frac{1}{d_i k^2 \delta B(k)}$	$\mathcal{E}(k) \sim \mathcal{E}_{\mathbf{B}}(k) \sim \epsilon^{2/3} d_i^{-2/3} k^{-7/3}$	Isotropic

[a]Kolmogorov (1941a); [b]Iroshnikov (1964); [c]Kraichnan (1965); [d]Galtier et al. (2000); [e]Goldreich and Sridhar (1995); [f]Boldyrev (2006); [g]Biskamp et al. (1999)

On the other hand, the introduction of a magnetic field, or similarly other effects such as compressibility, collisionless dynamics, etc., can also introduce linear terms describing the effect of waves. For $Re \gg 1$ and $\chi \ll 1$, often referred to as "weak" or "wave" turbulence, nonlinear interactions are strongly mediated by wave-like dynamics. The *Iroshnikov-Kraichnan* model is a simple isotropic model for incompressible MHD in this regime, whereby τ_{tr} is lengthened by a factor of χ^{-1}, since multiple "collisions" between propagating Alfvén wave packets are required for the nonlinearities to fully distort the wave packets and transfer fluctuation energy across scales if the wave propagation is much faster than the nonlinear timescale, resulting in a $k^{-3/2}$-spectrum. However, in the weak turbulence regime, exact analytical progress can also be made (Nazarenko 2011; Galtier 2023). Galtier et al. (2000) applied weak turbulence formalism to derive a *weak Alfvénic* model for incompressible MHD, illustrating that, in the limit of weak turbulence, the cascade is fundamentally anisotropic with the cascade of energy proceeding purely in the direction perpendicular to $\mathbf{B}_0$ and resulting in a spectrum scaling as $k_\perp^{-2}$ and following an arbitrary function of $k_{||}$ set by the driving.

Anisotropic strong turbulence models, such as the *Goldreich-Sridhar* and *Dynamic Alignment* models, often incorporate a constraint known as critical balance, whereby $\chi \sim 1$ such that the linear and nonlinear terms balance scale-by-scale. In these models, τ_{tr} is again related to τ_{NL} associated with nonlinear advection, except now the nonlinear interaction is assumed to be inherently anisotropic with only $k_\perp$ contributing to the nonlinear interaction. Since the dispersion relation for Alfvén waves is also anisotropic, the critical balance constraint then provides a prediction for the anisotropy of the spectrum. While the *Goldreich-Sridhar* model essentially applies analogous phenomenology as the *Kolmogorov* model to predict a $k_\perp^{-5/3}$-spectrum, the *Dynamic Alignment* model argues that geometrical constraints associated with the strength of $\mathbf{B}_0$ reduce the efficiency of nonlinear interactions, producing a $k_\perp^{-3/2}$-spectrum reminiscent of the *Iroshnikov-Kraichnan* model but based on different phenomenological arguments.

The same general framework can be used in systems where other nonlinearities play a dominant role, such as at sub-proton-scales where the single-fluid MHD approximation breaks down. The *Strong Hall MHD* model provides an example of analysing the nonlinear

Hall term in the induction equation under the assumption that the Hall term significantly dominates the dynamics (often referred to as electron-MHD) in a manner similar to the *Kolmogorov* model. In this situation, **B** carries the majority of the fluctuation energy and, because of the presence of an additional derivative in the Hall term, a steeper $k^{-7/3}$-spectrum is obtained. However, other analyses of the sub-proton scale dynamics, such as critical balance models invoking kinetic-scale wave modes like whistler waves (Narita and Gary 2010; Boldyrev et al. 2013; Narita 2016), kinetic Alfvén waves (Boldyrev and Perez 2012), and inertial kinetic Alfvén waves (Chen and Boldyrev 2017), are also commonly employed to explain spacecraft observations.

The models discussed above implicitly assume turbulent fluctuations are space-filling and self-similar. Violations of this assumption, discussed further in Sect. 1.2.3, may also have an impact on $\mathcal{E}(\mathbf{k})$. Such corrections to predicted power laws have been proposed and are often invoked, particularly at sub-proton-scales, to explain steeper power laws of $\sim k^{-2.8}$ or $\sim k^{-3}$ in space plasmas (Boldyrev and Perez 2012). Other complexities may also be present that are not illustrated in the examples provided in Table 1, such as constraints on the alignments between fluctuations in different vector fields that may be imposed by other conserved quantities (e.g., magnetic helicity, cross helicity, or generalized helicity) (Pouquet et al. 2019; Meyrand et al. 2021; Squire et al. 2022). In Sect. 2.3, we discuss further how reconnection, in particular, may alter $\mathcal{E}(\mathbf{k})$.

1.2.3 Intermittency & Current Sheets

The energy spectrum on its own does not provide the full picture of turbulence. Notably, turbulent systems are not simply comprised of uncorrelated randomly superposed normal modes and, in fact, phase correlations between modes at different scales are spontaneously generated by the nonlinear dynamics. These phase correlations manifest as localized *coherent structures* in physical space. Coherent structures can take the form of vorticity sheets and filaments in hydrodynamics (Okamoto et al. 2007) and, with the inclusion of **B**, current sheets and magnetic discontinuities, among other complex structures (Mininni et al. 2006; Greco et al. 2009; Uritsky et al. 2010; Matthaeus et al. 2015). This feature of turbulence is referred to as *intermittency*, so-called because it results in a statistically non-uniform "intermittent" distribution of dissipative structures in the domain. A key aspect of intermittency is that it violates the assumed self-similar and space-filling nature of the nonlinearly interacting fluctuations that are an inherent feature of many of the theoretical models discussed in Sect. 1.2.2. Magnetic shears associated with coherent structures (e.g. current sheets or more generalized current structures) are potential sites where reconnection can occur in turbulent plasmas. The intrinsic link between the intermittent nature of the turbulence and the statistical properties (e.g., structure, prevalence, distribution throughout the domain, etc.) of coherent structures means intermittency is likely an important feature to consider in the case of turbulence-driven reconnection.

Intermittency is typically analyzed by examining higher-order statistics, such as the pth-order structure functions given by

$$S_p^{g_i}(\boldsymbol{\ell}) = \langle [\Delta g_i(\boldsymbol{\ell})]^p \rangle. \tag{15}$$

While $S_2(\boldsymbol{\ell})$ has a direct relationship to $\mathcal{E}(\mathbf{k})$, higher-order statistics encode additional information about cross-scale correlations. The Kolmogorov (1941a) theory of turbulence, which does not include the effect of intermittency, predicts $S_p^{u_\ell}(\boldsymbol{\ell}) \sim \ell^{\zeta_p}$ with $\zeta_p = p/3$ and u_ℓ the component of **u** along $\boldsymbol{\ell}$. Turbulent systems typically exhibit strong deviations from this scaling, with increasingly sub-$p/3$ scalings as p increases, in both hydrodynamic (Anselmet et al. 1984) and plasma systems (Marsch and Tu 1997; Biskamp and Müller 2000). Toy models, such as so-called β, multi-fractal, and random cascade models (see Frisch (1995) for a detailed discussion), which relax assumptions about self-similarity and the statistical homogeneity of dissipation in various ways, demonstrate that intermittency offers an explanation for this behavior.

While some numerical studies indicate intermittency is more intense in MHD than hydrodynamics (Biskamp and Müller 2000), observations from the solar wind suggest, at sub-proton-scales ζ_p is linear with p, as in the non-intermittent case (Kiyani et al. 2009). The behavior at sub-proton-scales may not be universal, however, with observations in Earth's magnetosheath (Chhiber et al. 2018) and numerical simulations (Franci et al. 2015), continuing to show signatures of intermittency well into the sub-proton-scales. A complete understanding of this variation in behaviour, remains an open question and may suggest a dynamically significant variation in the nature or distribution of nonlinear fluctuations at sub-proton-scales across different turbulent environments.

Since $S_p^{g_i}(\boldsymbol{\ell})$ are the statistical moments of the distribution of increments $\Delta g_i(\boldsymbol{\ell})$, characterizing intermittency can be framed as considering how the probability distribution function of increments varies with scale. Typically distributions of $\Delta\mathbf{u}(\boldsymbol{\ell})$ or $\Delta\mathbf{B}(\boldsymbol{\ell})$ exhibit nearly Gaussian shapes for large ℓ and become progressively more heavy-tailed (larger than Gaussian probability of extreme values) for small ℓ (Frisch 1995; Sorriso-Valvo et al. 1999). Given the heavy-tailed nature of the distributions, one common measure of intermittency is the scale-dependant kurtosis given by $\kappa(\boldsymbol{\ell}) = S_4^{g_i}(\boldsymbol{\ell}) / \left[S_2^{g_i}(\boldsymbol{\ell})\right]^2$, which is expected to have a value of 3 at large ℓ where the distribution is Gaussian and then become increasingly larger through the inertial range (Wu et al. 2013). Heuristically, this behavior can be understood from the fact that at large separations ($\gg \lambda_{C,\mathbf{g}}$), quantities at two different points will be uncorrelated and $\Delta\mathbf{g}(\boldsymbol{\ell})$ will be an uncorrelated random variable, while at small separations $\Delta\mathbf{g}(\boldsymbol{\ell})$ will be sensitive to the gradients in these quantities, which are spatially inhomogeneous and intermittent. In fact, detailed examinations of the distributions of the vorticity and current density in turbulent plasmas present a picture where particularly intense vorticity and current structures form at the interfaces between large regions of reduced nonlinear activity, consistent with the heavy-tailed distributions (Servidio et al. 2008; Greco et al. 2009; Pecora et al. 2021, 2023). Furthermore, locating extreme values in $\Delta\mathbf{g}(\boldsymbol{\ell})$ has been proposed as a means of identifying coherent (potentially dissipative) structures, e.g., by examining the so-called partial variance of increments ($PVI = |\Delta\mathbf{g}(\boldsymbol{\ell})|/\sqrt{S_2^{\mathbf{g}}(\boldsymbol{\ell})}$; Greco et al. (2009)) or the integrands of $\mathbf{Y}(\boldsymbol{\ell})$ (i.e., $\Delta\mathbf{u}(\boldsymbol{\ell})|\Delta\mathbf{u}(\boldsymbol{\ell})|^2$, $\Delta\mathbf{u}(\boldsymbol{\ell})|\Delta\mathbf{B}_A(\boldsymbol{\ell})|^2$, and $\Delta\mathbf{B}_A(\boldsymbol{\ell})[\Delta\mathbf{u}(\boldsymbol{\ell})\cdot\Delta\mathbf{B}_A(\boldsymbol{\ell})]$; Sorriso-Valvo et al. (2018a) and Sorriso-Valvo et al. (2018b)).

2 Turbulence-Driven Reconnection

In situ spacecraft observations of turbulent plasmas in near-Earth space, such as the solar wind (Greco et al. 2009), magnetosheath (Gingell et al. 2021; Schwartz et al. 2021), and plasma sheet (Ergun et al. 2018), are filled with current structures and associated magnetic shears that can be sites of reconnection. The extent to which reconnection is self-consistently initiated at these current sheets and the impact that this has on the turbulent dynamics has long been a topic of interest and there is a rich body of literature attempting to explore this issue both observationally and in numerical simulations.

The solar wind is one region of turbulence that has been extensively studied with *in situ* spacecraft observations. Numerous studies have provided evidence for reconnection exhausts at solar wind current sheets (Gosling et al. 2005; Gosling 2007; Osman et al. 2014; Phan et al. 2020; Eriksson et al. 2022, 2024; Fargette et al. 2023). However, solar wind reconnection exhausts are often encountered hundreds or even thousands of ion inertial lengths (d_i) away from the x-line (Mistry et al. 2015b), suggesting that reconnecting current sheets in the solar wind extend over large length-scales. Recent statistical analyses of *Parker Solar Probe* observations further report that observed solar wind reconnection events tend to be found with similar occurrence across radial distance in slow solar wind streams, while they are rare in fast, presumably coronal hole origin, solar wind streams (Eriksson et al. 2024). It is, therefore, challenging to distinguish reconnecting current sheets in the solar wind that may be self-consistently generated by the turbulent dynamics from those associated with the evolution of large-scale solar wind structure (e.g., heliospheric current sheet, stream interaction regions, coronal mass ejections) and it remains an open question as to the extent to which these two populations contribute to the identified reconnection exhausts in the solar wind (Eriksson et al. 2022). Although, sophisticated analyses of magnetic discontinuities embedded within solar wind turbulence suggest a fragmentation of the magnetic discontinuities, with smaller-scale discontinuities embedded within larger-scale discontinuities, which may be indicative of a process such as magnetic reconnection that can disrupt and fragment current sheets within the turbulence (Greco et al. 2016). Recently, however, there has been new progress in observationally identifying and examining turbulence-driven magnetic reconnection in Earth's magnetosheath that has been enabled by high-resolution measurements from *MMS*, which have provided evidence for small-scale reconnection events embedded within the recently excited turbulent fluctuations downstream of Earth's bow shock.

2.1 Observations of Bow Shock & Magnetosheath Reconnection

Earth's bow shock forms at the interface between the solar wind and Earth's magnetosphere, where the super-Alfvénic solar wind suddenly slows down to sub-Alfvénic speeds due to Earth's strong magnetic field and the kinetic energy in the solar wind is converted to magnetic, thermal, and fluctuation energy. In the shock, the electron motion is frozen-in to $\mathbf{B}$, while the ion motion is decoupled from the electron motion and $\mathbf{B}$. Ions can penetrate deep inside the shock transition layer without gyration and, as a result, a shock potential is generated, which reflects some ions to produce ion beams propagating upstream of the shock. The ion-ion beam instability caused by the reflected and solar wind ion populations generates large amplitude electromagnetic waves and the plasma in the shock becomes turbulent. Particles crossing the shock are rapidly heated, forming a region downstream of the shock where $|\mathbf{B}|$, number density (n), and temperature (T) are enhanced compared to the upstream solar wind - referred to as the magnetosheath.

Fluctuations in the magnetosheath have features consistent with turbulent dynamics, including broadband power law spectra as discussed in Sect. 1.2.2 (Sahraoui et al. 2006; Alexandrova et al. 2008; Huang et al. 2014, 2017), higher-order statistics consistent with intermittency as discussed in Sect. 1.2.3 (Yordanova et al. 2008; Chhiber et al. 2018), and evidence of an active cross-scale energy cascade as discussed in Sect. 1.2.1 (Hadid et al. 2018; Bandyopadhyay et al. 2018, 2020b). In contrast to the solar wind, magnetosheath fluctuations typically have a much shorter λ_C that varies from tens to hundreds of d_i (Stawarz et al. 2022). Furthermore, other fluctuation properties - such as the MHD-scale spectral index and turbulent Mach number (ratio of velocity fluctuation amplitude to sound speed) - also vary across the magnetosheath (Huang et al. 2017; Li et al. 2020). These features suggest that the processing of the solar wind plasma by the shock, potentially through processes such as large-amplitude wave-generation associated with the quasi-parallel shock, temperature anisotropy instabilities downstream of the shock, or the interaction of turbulent fluctuations in the solar wind with the shock (Omidi et al. 1994; Bessho et al. 2020; Trotta et al. 2023), drives new fluctuations into the system that interact nonlinearly and evolve into a turbulent state.

Spacecraft observations demonstrate these fluctuations are associated with many current sheets. Using *Cluster* observations, Retinò et al. (2007) found that, among the turbulent current sheets in Earth's magnetosheath, there are current sheets with signatures of reconnection. Retinò et al. (2007) showed that the thickness of one such reconnecting current sheet was $\sim d_i$ and observed an out-of-plane reconnection **E**, quadrupolar Hall **B**, bipolar Hall **E** pointing toward the center of the current sheet, and a positive value of $\boldsymbol{j} \cdot \boldsymbol{E}$, indicating energy exchange from the electromagnetic fields to the particles in accordance with Poynting's theorem. The reconnection inflow inferred from the $\mathbf{E} \times \mathbf{B}$ drift was 0.1 times the Alfvén speed, indicating $\mathcal{R} \sim 0.1$. Notably, due to the small length scale, and associated short time scale over which the event was advected over the spacecraft, all of the signatures of turbulence-driven magnetosheath reconnection identified with *Cluster* were obtained from the electromagnetic field measurements.

Subsequent *MMS* observations also revealed many reconnecting current sheets in the magnetosheath. Yordanova et al. (2016) and Vörös et al. (2016) identified several reconnecting current sheets in Earth's magnetosheath, where reconnection outflows and enhancements of $\boldsymbol{j} \cdot \boldsymbol{E}$ were observed. Vörös et al. (2016) detected electron diffusion region (EDR) signatures by investigating the decoupling of the electron velocity from the $\mathbf{E} \times \mathbf{B}$ drift, the agyrotropy parameter Q (Swisdak 2016), and electron distribution functions. Inside the EDR, $\boldsymbol{j} \cdot \boldsymbol{E}'$, where $\boldsymbol{E}' = \boldsymbol{E} + \boldsymbol{u}_e \times \boldsymbol{B}$, was positive, indicating conversion of magnetic energy to particle kinetic and thermal energy via the non-ideal **E**.

Phan et al. (2018) examined two nearby segments of high-resolution burst data from *MMS* in detail, totalling ~ 21 minutes of magnetosheath observations, revealing that while many current sheets had evidence of reconnection in the form of electron outflows, none had clear evidence of ion outflows. This relative lack of ion outflow observations was unusual because fully accelerated ion outflows should occupy a much larger volume of space than the ion diffusion regions (IDRs) or EDRs, where fully accelerated ion outflows would not be expected. One would, therefore, expect it to be statistically more likely for the spacecraft to encounter ion outflow regions than regions with pure electron outflows. Since electrons appeared to be the only species participating in the reconnection process, this type of reconnection has come to be known as "electron-only" reconnection.

An example of one electron-only reconnection event identified by Phan et al. (2018) is shown in Fig. 2a–j. For the event in Fig. 2a–j the thickness of the current sheet was significantly thinner than the ion scales at ~ 4 electron inertial lengths (d_e). During the current sheet crossing, *MMS3* observed super-Alfvénic electron outflows of ~ 250 km/s in the outflow direction (L direction) relative to an Alfvén speed associated with the reconnecting component of the field of $V_{A,L} \sim 25$ km/s. The other three *MMS* spacecraft also observed super-Alfvénic electron outflows, but in the opposite direction, providing direct evidence for the two oppositely directed outflow jets. The electrons in the current sheet were not frozen-in to **B**, and a large $E_{||}$ was observed, producing a large $\mathbf{j} \cdot \mathbf{E}'$.

It was proposed that the reason ions did not appear to be participating in the reconnection process was because the lengths of the current sheets along the outflow directions were short enough that there was not enough time/space for the reconnected **B** to accelerate the ions before the field fully relaxed. This explanation was supported by idealized numerical

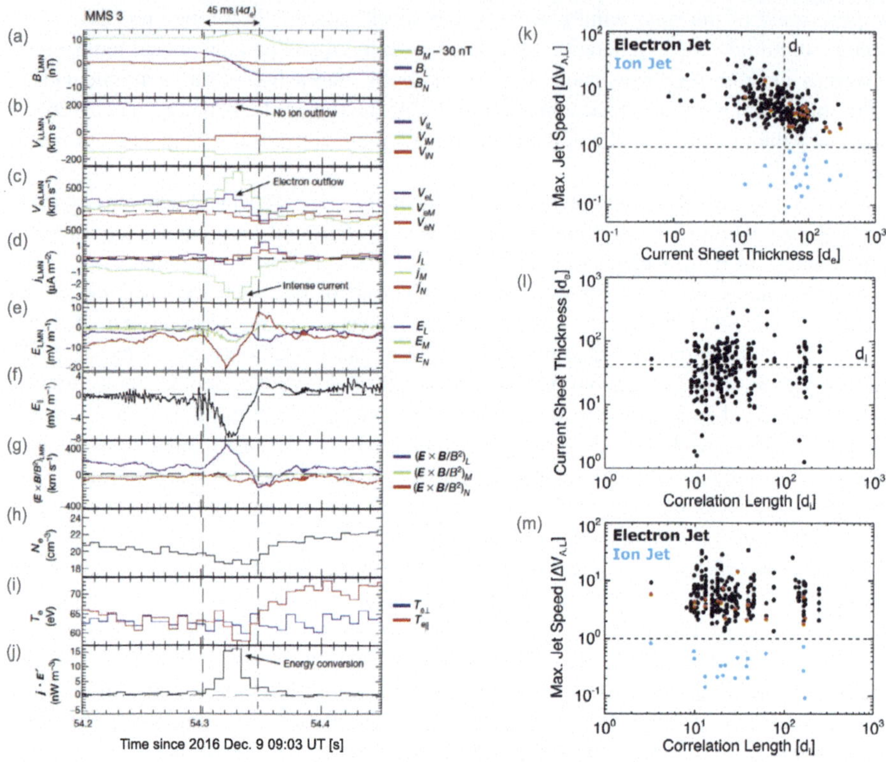

Fig. 2 Example of electron-only reconnection in the magnetosheath showing (a) **B**, (b) ion velocity, (c) electron velocity, (d) **j**, (e) **E**, (f) the component of **E** along **B**, (g) $\mathbf{E} \times \mathbf{B}$-drift velocity, (h) electron number density, (i) electron parallel and perpendicular temperatures, and (j) $\mathbf{j} \cdot \mathbf{E}'$. All vector quantities are in a local current sheet coordinate system based on a hybrid minimum variance analysis (reproduced from Phan et al. 2018). (k)-(m) Statistical study of 256 reconnection events in the turbulent magnetosheath, comparing electron (and ion if present) reconnection jet speeds, current sheet thickness, and $\lambda_{C,\mathbf{B}}$ (reproduced from Stawarz et al. 2022). In panels (k) and (m), ion jet speeds are displayed in blue while the electron jets speeds for the corresponding events are highlighted in red

experiments, where the lengths of reconnecting current sheets were artificially constrained along the outflow direction – resulting in reconnection with unique properties compared to the traditional "ion-coupled" reconnection picture for lengths $\lesssim 10d_i$ (Sharma Pyakurel et al. 2019). These features included the reconnection events only having fast super-Alfvénic electron outflows, having higher reconnection rates than ion-coupled reconnection, and being embedded within thin electron-scale current sheets (in contrast to ion-coupled reconnection, where the electron-scale gradients associated with the EDR are embedded in a broader ion scale current sheet). Similar results were also found in the fluctuations self-consistently generated in shock simulations (Bessho et al. 2019, 2020, 2022, 2023).

Since the Phan et al. (2018) work, examples of both ion-coupled reconnection and electron-only reconnection have been observed in Earth's magnetosheath (Wilder et al. 2018, 2022; Vörös et al. 2019; Stawarz et al. 2019, 2022). Vörös et al. (2019) revealed whistler and lower hybrid waves in the reconnecting current sheets. Wilder et al. (2018, 2022) investigated the energy conversion, demonstrating that reconnection with a small guide field (less than 30% of the reconnecting magnetic field) exhibits $\mathbf{j} \cdot \mathbf{E}'$ primarily associated with $\mathbf{E}_\perp$ and agyrotropic electron distributions. In contrast, large guide fields are associated with $\mathbf{j} \cdot \mathbf{E}'$ generated by $E_{||}$. Bandyopadhyay et al. (2021b) examined the so-called pressure-strain interaction terms, quantifying local energy exchange between the bulk flow and internal energy (defined as the second moment of the distribution function), demonstrating magnetosheath reconnection events are sites of enhanced pressure-strain interaction alongside $\mathbf{j} \cdot \mathbf{E}$. In the small selection of events considered, Bandyopadhyay et al. (2021b) found both positive and negative pressure-strain signatures, suggesting local "heating" and "cooling" (although with a slight preference for positive "heating" signatures), with typically more intense signatures in the electrons compared to the ions.

Stawarz et al. (2022) statistically examined the reconnecting current sheets in the turbulent magnetosheath with *MMS* data. A total of 256 reconnection events (including 18 ion-coupled reconnection events) were investigated to understand the relationship between $\lambda_{C,\mathbf{B}}$ and the type of reconnection present. Stawarz et al. (2022) found thinner current sheets tended to have higher outflow speeds (Fig. 2k), such that when the thickness is $\lesssim d_i$, the electron jet speed becomes super-Alfvénic. These super-Alfvénic flows can reach the order of the electron Alfvén speed, indicating reconnection in the electron-only regime. Stawarz et al. (2022) further showed that intervals with $\lambda_{C,\mathbf{B}} \lesssim 20 d_i$ are associated with thinner reconnecting current sheets and faster super-Alfvénic electron jets on average (Figs. 2l, m). The relationship with $\lambda_{C,\mathbf{B}}$ suggests a scenario where current sheets form at the interface of $\lambda_{C,\mathbf{B}}$-scale magnetic structures – implying $\lambda_{C,\mathbf{B}}$ controls the average length of current sheets along the outflow direction. When $\lambda_{C,\mathbf{B}}$ approaches tens of d_i, it becomes more likely there will be insufficient space for the reconnected field lines to accelerate ion outflows, increasing the prevalence of electron-only reconnection.

In addition to Earth's magnetosheath, the shock transition region and foreshock regions are often turbulent, and many reconnection events have been detected by *MMS* in these regions. Wang et al. (2019) and Gingell et al. (2019) found evidence of electron-only reconnection in the shock transition region. Gingell et al. (2019) observed that T_e rose from 20 eV to 33 eV across the shock ramp and an additional 7 eV increase occurred in the shock transition region, suggesting 35% of the total electron heating across the shock occurs in the transition layer in association with electron-only reconnection. Wang et al. (2019) additionally observed ion-coupled reconnection with both ion and electron jets in the shock transition region. Several studies have further investigated reconnection in the foreshock. Wang et al. (2020) found electron-only reconnection in Short Large-Amplitude Magnetic Structures (SLAMS), suggesting that reconnection occurs due to the compression of the SLAMS. Liu et al. (2020) found electron-only reconnection in shock transients in the ion foreshock. Within the transients, $|\mathbf{B}|$ and n were low and the plasma was turbulent. *MMS* detected high-speed electron jets, enhancements of $\boldsymbol{j} \cdot \boldsymbol{E}'$, and enhancements in T_e during the crossings of current sheets in the transient. Jiang et al. (2021) identified several examples of electron-only reconnection at thin current sheets in the foreshock that were not associated with transient structures.

Gingell et al. (2020) further statistically studied 223 shock crossings by *MMS* - investigating reconnection in the shock transition region of both quasi-parallel and quasi-perpendicular shocks. Gingell et al. (2020) demonstrated reconnection occurs ubiquitously regardless of shock angle, although quasi-parallel shocks and high Alfvén Mach number shocks show slightly higher probability than quasi-perpendicular shocks and low Alfvén Mach number shocks as shown in Fig. 3.

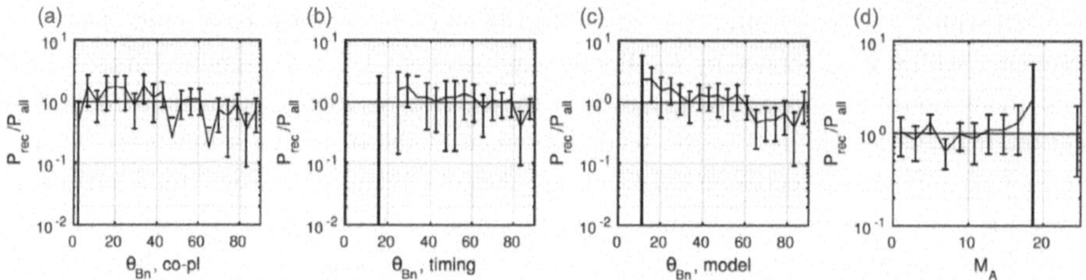

Fig. 3 Probability to detect reconnecting current sheets in the shock-transition-region/magnetosheath as a function of (a)-(c) shock normal angle θ_{Bn} using three different methods and (d) Alfvén Mach number of the upstream shock (reproduced from Gingell et al. 2020)

2.2 Simulations of Turbulence-Driven Reconnection

Over the past several decades, numerous numerical simulations of turbulence have demonstrated the process of turbulence-driven reconnection. While such simulations are able to reveal vital information about the configuration of turbulence-driven reconnection events that are inaccessible with *in situ* observations, the initial identification of reconnection in the simulations presents a challenge, particularly in 3D, which has been a major focus of much of the research. Furthermore, the computational challenge of simultaneously resolving all of the collisionless dynamics alongside the large scale separation characteristic of turbulent plasmas, has prompted the exploration of how different plasma models impact turbulence-driven reconnection (see Shay et al. 2024, this collection, for a detailed discussion of different numerical plasma models).

As early as the 1980s, resistive MHD turbulence simulations demonstrated the presence of coherent structures resembling current sheets, where magnetic reconnection can take place (e.g. Matthaeus and Lamkin 1986; Biskamp and Welter 1989; Politano et al. 1989; Carbone et al. 1990). The characterization and statistical analysis of the distribution and geometry of current sheets formed in those numerical simulations of MHD turbulence, as well as their reconnection rates, have been investigated using diverse methodologies (Dmitruk et al. 2004; Servidio et al. 2009, 2010; Zhdankin et al. 2013, 2014, 2016). Servidio et al. (2009, 2010) statistically analyzed reconnection events in 2D MHD turbulence, quantifying, for example, the distribution of reconnection rates and the aspect ratio at each x-point. In order to find those x-points, they used a topological approach based on classifying the eigenvalues of the Hessian matrix of the vector potential and finding the saddle critical points. A generalization of this method to a 3D geometry for x-points with a guide field, which makes the topological classification more complex, was recently proposed by Wang et al. (2024). Zhdankin et al. (2013) developed an algorithm to identify and characterize the geometrical properties of current sheets in 3D MHD turbulence, and examined how those properties related to the presence of x-points. Zhdankin et al. (2013) quantified the energy dissipation in 3D MHD turbulence simulations, finding relatively uniformly distributed energy dissipation in the current sheets and also that the number of current sheets increases while their thickness decreases as the magnetic Reynolds number increases. The relationship between current sheets and other coherent structures was further numerically explored in the 3D MHD simulations of Zhdankin et al. (2016).

Similar coherent structures have been identified in simulations using other plasma models, such as two fluid and Hall MHD (Donato et al. 2012; Camporeale et al. 2018; Papini et al. 2019) that can resolve d_i and capture the physics of whistler and kinetic Alfvén waves that are thought to play a fundemental role in the sub-proton-scale turbulence in the solar wind and other environments. Donato et al. (2012) compared statistics of reconnection events between Hall-MHD and resistive MHD turbulence simulations. Hall-MHD effects led to faster and broader distributions of reconnection rates compared to resistive MHD. Camporeale et al. (2018) developed a space-filter method to analyze the features of the turbulent cascade due to current sheets in 2D two-fluid simulations, which also included electron inertia. Using this approach, Camporeale et al. (2018) was able to quantify how the cascade was modified due to the presence of coherent structures by means of the cross-scale spectral energy-transfer at localized locations in real space.

In contrast to Hall MHD, hybrid-kinetic models also contain ion kinetic physics, which among other effects can describe temperature-anisotropy instabilities and, in general, any process driven by deviations from an equilibrium Maxwell distribution function. Hybrid-kinetic simulations of turbulence have often been used to investigate and reproduce the ion-scale spectral break and other observed properties of space plasma turbulence (Franci et al. 2015, 2020). Hybrid-kinetic simulations have revealed the process of current sheet formation and how reconnection can play an important role in regulating the turbulent cascade near d_i (Cerri and Califano 2017; Franci et al. 2017; Papini et al. 2019; Hu et al. 2020; Sisti et al. 2021; Manzini et al. 2023). Papini et al. (2019) compared turbulence simulations using Hall-MHD and hybrid-kinetic models, finding good agreement in many turbulent properties between both models. Papini et al. (2019) found similar reconnection rates between both plasma models, and by means of quantifying the energy transfer, that there might exist a reconnection-mediated regime at sub-ion scales. By means of 3D hybrid-kinetic turbulence simulations, Fadanelli et al. (2021) analyzed the energy exchanges near a reconnection site in this scenario. Fadanelli et al. (2021) did not determine the energy budget as usually done, but instead a point-to-point analysis that revealed that the conversion to thermal and kinetic energies is statistically related to the local scale of the system, with the largest conversion rate occuring at scales comparable to a few d_i. Manzini et al. (2023) developed a coarse-graining method to measure the nonlinear cross-scale energy transfer at specific locations. Applying this method to both *MMS* observations in the magnetosheath and hybrid-kinetic turbulence simulations, both cases showed clear indications of a preferential energy transfer to sub-proton scales associated with reconnection. Consolini et al. (2023) quantified the fractal dimension of current sheets in 2D hybrid-kinetic turbulence, finding a relationship between this quantity with the spectral features of $\mathbf{B}$ fluctuations at ion-scales

Other works have developed algorithms to characterize current sheet properties both in 2D and 3D turbulence. In 2D, Hu et al. (2020) developed such an algorithm based on a convolutional neural network, in which first humans detected reconnection events in turbulence based on several physical signatures including a sufficiently strong $\mathbf{j}$, $\mathbf{u}_e$, and $\mathbf{E}'$ in the out-of-plane direction. With this training data, the algorithm automatically identified up to 70% of the reconnection events in turbulence successfully. In 3D, Sisti et al. (2021) developed and compared different methods to extract the three characteristic lengths of the formed current sheets. By comparing with earlier electron-MHD simulations of turbulence by Meyrand and Galtier (2013), Sisti et al. (2021) concluded that the additional physics included in the hybrid-kinetic plasma model changed the shape of the most predominant current sheets from "cigar-like" to "knife-like".

Most hybrid-kinetic simulations have been carried out with a model neglecting electron mass. As a result, reconnection is driven by physical or numerical resistivity, not by collisionless processes as expected. In addition, it has been shown that the current sheet width is mainly limited by the grid size (Azizabadi et al. 2021), so the current sheet thinning process is mainly limited by numerics and not by physical processes. The affect of electron inertia on current sheets formed by turbulence was further investigated in 2D hybrid-kinetic simulations by Jain et al. (2022), which quantified the errors introduced by approximations used in previous hybrid-kinetic codes with electron inertia, concluding that an accurate consideration of electron inertia is important to properly characterize the evolution of electron-scale current sheets in turbulent plasmas. Muñoz et al. (2023) extended this work to 3D, showing how the electron inertia modifies **j**, unexpectedly, even at scales $> d_e$ along the direction parallel to $\mathbf{B}_0$. In addition, Muñoz et al. (2023) emphasized the importance of the electron inertia term in generalized Ohm's law to balance the reconnection electric field in low-β turbulence simulations. Califano et al. (2020) used another hybrid-kinetic plasma model with electron inertia to investigate the presence of electron-only reconnection in turbulence. The presence of electron-only reconnection was dependent on the wavenumber of the injected fluctuations with more electron-only reconnection events when the fluctuations were injected at wavenumbers comparable to the ion scale, while ion-coupled reconnection appeared when the fluctuations were injected at much larger scales, consistent with the observational work of Stawarz et al. (2022). Arró et al. (2020) used a similar code and numerical setup to Califano et al. (2020) in order to determine the influence of electron-only reconnection on the turbulence. However, they did not find significant differences between the turbulent fluctuations and intermittency between the cases with electron-only reconnection or ion-coupled reconnection.

Another approach used to investigate turbulence is gyrokinetics. This is a reduced kinetic model for both electrons and ions that is based on averaging the particle gyromotion and following instead the guiding center of the particles. The model itself is based on an asymptotic expansion of a small parameter associated with small fluctuations of the distribution function or a small value of the wavenumber anisotropy $k_\parallel/k_\perp$, among other quantities. As a consequence, gyroresonances and whistler waves are ruled out of the model, while effects such as electron/ion Landau damping and kinetic Alfvén waves are retained. Turbulence simulations using the gyrokinetic approach have, not only revealed the presence of electron-scale current sheets, but also allowed the quantification of their relative contributions to dissipation (TenBarge and Howes 2013; TenBarge et al. 2013; Howes 2016). For example, it was found that Landau damping plays a fundamental role in dissipating the energy contained in the current sheets. Li et al. (2023) applied a method to identify current sheets based on a measure of the magnetic flux transport from the separatrices to the reconnection x-line in 3D simulations of gyrokinetic turbulence, revealing that the current sheets formed unexpectedly extended x-lines in the turbulent system.

Fully kinetic turbulence simulations have also shown the presence of both ion and electron-scale current sheets and magnetic reconnection (Karimabadi et al. 2013, 2014; Wan et al. 2012, 2015, 2016; Haynes et al. 2014; Haggerty et al. 2017; Vega et al. 2020; Agudelo-Rueda et al. 2021; Franci et al. 2022; Vega et al. 2023). Karimabadi et al. (2013) investigated reconnection in 2D fully kinetic simulations of Kelvin-Helmholtz/shear flow generated turbulence, showing how the current sheets are regions with strong localized electron heating due to the parallel reconnection **E**. This heating was found to be much stronger than the heating due to the damping of waves formed in the same system. Haynes et al. (2014) investigated the relation between reconnection and electron temperature anisotropy via 2D fully kinetic simulations with an implicit PIC code, which allowed them to use a realistic ion-to-electron mass ratio, finding that reconnection sites are associated with strong parallel electron temperature anisotropy, contributing to dissipation. Wan et al. (2016) compared several 2D and 3D fully kinetic and MHD simulations of turbulence in order to determine the relation between intermittency and dissipation in coherent structures, finding the dissipation measure, $\mathbf{j}\cdot\mathbf{E}' - \rho_c\mathbf{u}_e\cdot\mathbf{E}$ where ρ_c is the charge density, scales as $\sim |\mathbf{j}|^2$ in all cases.

Haggerty et al. (2017) analyzed the statistics of reconnection x-points in 2D fully-kinetic turbulence simulations by applying similar methods to those of Servidio et al. (2009) for an MHD model. In contrast to previous MHD simulations, Haggerty et al. (2017) found that the distribution of reconnection electric fields is broader and can reach up to 0.5 of the local Alfvén speed, while keeping an average of 0.1. Vega et al. (2020) focused on electron-only reconnection in 2D fully kinetic simulations, finding that electron-only reconnection occurs at both high and low electron-β with similar reconnection rates. Vega et al. (2023) further analyzed 3D fully kinetic simulations of turbulence using an algorithm based on a medial axis transform from image processing, which was capable of identifying and characterizing arbitrary shaped current structures. The current structures tended to have half-widths of at most one d_e with a length between d_e and d_i. Most energy dissipation took place in current structures occupying $\sim 20\%$ of the total simulation volume and, by identifying large variations in electron flow and characteristic features in the pressure-strain interaction terms, it was estimated that 1% of current sheets were reconnecting - in contrast to the observational work of Stawarz et al. (2022) which found $\sim 10\%$ of intense current sheets underwent reconnection, although this discrepancy may owe to different methodologies for identifying distinct current structures. Using 3D fully-kinetic anisotropic Alfvénic turbulence simulations, Agudelo-Rueda et al. (2021) found that the current sheets generated by the turbulence tended to be less anisotropic than that of the large-scale driving and used several proxies (e.g., $|\mathbf{j}|$, $\mathbf{u}_e$, $\mathbf{u}_i$, and $\mathbf{E}'$) to identify reconnection sites. In a follow-up study, Agudelo-Rueda et al. (2022) further examined the energy transport and dissipation associated with both collionless and effective collision-like terms at the reconnection sites in the simulation. Franci et al. (2022) analyzed reconnection events in 2D fully kinetic simulations of turbulence, identifying several reconnection events with a thickness on the order of d_e. Reminiscent of the work of Califano et al. (2020) and Stawarz et al. (2022), which looked at the statistical prevalence of electron-only reconnection relative to the dynamics of the driving scale, Franci et al. (2022) found that even within a given simulation turbulent current sheets with shorter lengths tended to appear more electron-only-like, while current sheets with longer lengths appeared ion coupled.

While many numerical studies of turbulence-driven reconnection have been performed in idealised periodic boxes of turbulence, a number of works, particularly in recent years, spurred on by the *MMS* results from Earth's turbulent magnetosheath discussed in Sect. 2.1, have begun to examine reconnection events generated by turbulence self-consistently excited in shock simulations. Reconnection driven by shock turbulence has been studied using kinetic simulations of quasi-perpendicular shocks (Matsumoto et al. 2015; Bohdan et al. 2020; Lu et al. 2021; Guo et al. 2023), quasi-parallel shocks (Gingell et al. 2017; Bessho et al. 2019, 2020, 2022, 2023; Lu et al. 2020; Ng et al. 2022, 2024), and across both regimes (Karimabadi et al. 2014; Gingell et al. 2023; Steinvall and Gingell 2024).

2.3 The Role of Reconnection in Turbulent Plasmas

Beyond the existence and identification of turbulence-driven reconnection, it is important to consider how and to what extent reconnection contributes to the turbulent dynamics. There are several avenues through which turbulence-driven reconnection might impact turbulence, with reconnection potentially i) acting as the dominant nonlinear interaction over some range of scales and ii) contributing to the dissipation of the turbulence.

2.3.1 Impact on the Energy Spectrum

As discussed in Sect. 1.2.2, theoretical descriptions of turbulence are built on assumptions about the physical interactions controlling the nonlinear dynamics. Several works have explored how to incorporate reconnection into this framework, with early work basing the analysis on the isotropic weak turbulence formalism of Iroshnikov-Kraichnan (Carbone et al. 1990), while recent works have examined the topic using anisotropic MHD (Loureiro and Boldyrev 2017b; Mallet et al. 2017) and collisionless (Loureiro and Boldyrev 2020; Mallet 2020) turbulence formalisms.

In the theoretical scenarios proposed in these works, turbulent dynamics form current-sheet-like structures with aspect ratios (length relative to thickness) that become progressively more anisotropic at smaller scales in a manner consistent with the spectral anisotropy predicted by an anisotropic turbulence model. The tearing instability is assumed to initiate reconnection at these current sheets once the aspect ratio becomes sufficiently large and alters the turbulence by "disrupting"/"destroying" the elongated current sheets over the timescale of the tearing instability growth, thereby altering the distribution of energy in spectral space. While the tearing instability growth is treated as a linear instability, it is both initiated via the formation of a nonlinear structure and leads to the development of a fully nonlinear perturbation to the current sheet and, in this sense, the linear growth rate is taken to characterise the rate at which the nonlinear reconnection dynamics develop. If a range of scales exists over which this tearing timescale is faster than the dynamical timescales of other nonlinear effects, such as those generating the current sheets, then it is supposed that the tearing timescale will be the relevant timescale to associate with τ_{tr}. This picture relies on reconnection being sufficiently prevalent so as to make a significant impact, the scale-dependant anisotropy of current structures reflecting the spectral anisotropy of the turbulence model, and the linear tearing instability being the correct way to characterize the initiation of reconnection in a turbulent environment.

Based on this scenario, predictions for two key parameters can be derived - the critical scale (a_c) at which reconnection becomes the dominant nonlinear interaction and the power law scaling $\mathcal{E}(\mathbf{k})$. Constraining these parameters requires both a model for the large-scale turbulent dynamics setting up the current sheets and a description of the tearing instability, which depends on whether the system is resistive or collisionless and on the exact profile of the current sheets.

For resistive MHD, several 3D anisotropic turbulence models have been proposed, including the dynamic alignment model of Boldyrev (2006) and intermittency models of Chandran et al. (2015) and Mallet and Schekochihin (2017). Loureiro and Boldyrev (2017b) and Mallet et al. (2017) demonstrated that for these models, at sufficiently large Re, a range of scales exists where the resistive tearing instability is faster than the nonlinear dynamics generating the current sheets, suggesting the presence of a reconnection mediated inertial range. For the case of a hyperbolic tangent current sheet profile, a_c was found to be

$$a_c/\lambda_C \sim \left(V_{A,\lambda_C}\lambda_C/\eta\right)^{-4/7}, \tag{16}$$

where we have identified λ_C with the outer scale of the turbulence and V_{A,λ_C} is the Alfvén speed for fluctuations at the outer scale; and $\mathcal{E}(k_\perp) \sim k_\perp^{-11/5}$ in the reconnection-mediated range. Recent simulations of high-Re MHD turbulence in 2D (Dong et al. 2018) and 3D (Dong et al. 2022) have provided evidence for the presence of reconnection and the expected steepening of the inertial range spectrum at a_c.

Collisionless effects both alter the tearing instability, as well as the nonlinear dynamics of the turbulence. Two scenarios, one where a_c is larger than the ion scales and one where a_c is smaller than the ion scales, can potentially occur. In the former scenario, the analysis proceeds in a similar manner to the resistive case, but with a modified expression for the tearing instability leading to

$$a_c/\lambda_C \sim (d_e/\lambda_C)^{4/9}(\rho_s/\lambda_C)^{4/9} \tag{17}$$

where ρ_s is the ion acoustic scale, and $\mathcal{E}(k_\perp) \sim k_\perp^{-3}$ is obtained for a hyperbolic tangent current sheet profile (Loureiro and Boldyrev 2017a, 2020). The range of spectral indices found for different current sheet profiles is reminiscent of those reported in the so-called transition range of solar wind turbulence, although reconnection is not the only explanation that can produce such slopes (Bowen et al. 2020). The later scenario, where a_c occurs in the kinetic scales, is more challenging due to less well understood anisotropic turbulence models; however, some work has been done on this scenario (Loureiro and Boldyrev 2017a; Boldyrev and Loureiro 2019; Mallet 2020; Loureiro and Boldyrev 2020; Boldyrev and Loureiro 2020), which may be relevant for understanding the impact of electron-only reconnection.

2.3.2 Contribution to Energy Dissipation

Magnetic reconnection can also facilitate the dissipation of turbulence. Simulations (Servidio et al. 2011) and observations (Osman et al. 2012; Chasapis et al. 2017a, 2018) suggest intermittent structures are locations of enhanced temperature and energy conversion. As discussed in Sect. 2.1, missions such as *Cluster* and *MMS* have enabled the identification of thin reconnection events and the direct examination of the local energy conversion associated with them, which may account for nontrivial amounts of energy conversion when compared with estimates of the overall energy budget of the magnetosheath dynamics (Sundkvist et al. 2007; Schwartz et al. 2021). However, other studies, such as that by Hou et al. (2021), which examined integrated $\mathbf{j} \cdot \mathbf{E}'$ at intense PVI structures, while simultaneously identifying PVI structures associated with reconnection, concluded that, while reconnection events may have large energy conversion signatures, their integrated contribution to energy dissipation may be small ($\sim 15\%$ of the dissipation associated with large PVI structures and $\sim 1\%$ of the total integrated $\mathbf{j} \cdot \mathbf{E}'$ in the analysed interval) due to the small size of the diffusion region and limited occurrence rate.

One limitation of local analyses of energy conversion, is that, while the diffusion regions, which contain some of the strongest gradients, occupy a small volume; the entire volume of the reconnection outflows and separatrices, occupy a much larger region of space and can also be energetically important for the particle acceleration and heating. Furthermore, since reconnection involves the inflow of particles from the surrounding environment, in can lead to the acceleration and heating of a larger effective volume of particles than expected from the size of the current sheet alone.

An alternative way to assess the importance of reconnection for turbulent dissipation is to consider the energy budget of the reconnection events generated by the turbulent dynamics. The amount of magnetic energy released by reconnection that is available to each electron-proton pair is given by

$$\mathcal{E}_{rec} = m_i V_{A,inflow}^2, \tag{18}$$

where m_i is the ion mass and $V_{A,inflow}$ is the Alfvén speed associated with the inflowing reconnecting component, B_L, of the magnetic field. Taking into account the potential effect of asymmetry on either side of a reconnecting current sheet, $V_{A,inflow}$ is given by

$$V_{A,inflow} = \sqrt{\frac{|B_{L,1}|\,|B_{L,2}|\left(|B_{L,1}| + |B_{L,2}|\right)}{\mu_0 m_i \left(n_1 |B_{L,2}| + n_2 |B_{L,1}|\right)}}, \tag{19}$$

with subscripts 1 and 2 denoting values on either side of the current sheet. The total rate of energy dissipation associated with reconnection can be quantified by taking into account the fraction of $\mathcal{E}_{rec}$ converted into particle heating (or energetic particle acceleration) and the rate that magnetic flux is reconnected. Denoting the net rate of energy dissipation per unit mass associated with reconnection within a turbulent volume as ϵ_{rec}, which will be equivalent to ϵ if reconnection accounts for all of the dissipation in the volume, gives (Shay et al. 2018; Stawarz et al. 2022)

$$\epsilon_{rec} = \sum_j f_{rec,j} \alpha_j V^2_{A,inflow,j} \left(\frac{V_{A,inflow,j}}{\lambda_j} \mathcal{R}_j \right). \tag{20}$$

In Eq. (20), f_{rec} is the fraction of particles in the turbulent volume processed by a given reconnection event, α is the fraction of $\mathcal{E}_{rec}$ converted into particle heating, λ is the length of the inflow region, and $\mathcal{R}$ is the dimensionless reconnection rate, such that $\mathcal{R} V_{A,inflow}/\lambda$ represents the inverse timescale over which magnetic flux is reconnected. The summation over j represents a sum over each reconnection event in the turbulent region at any given time, such that subscript j denotes a quantity for a given reconnection event.

While Eq. (20) can be straightforwardly computed if all reconnection events within a volume can be characterized, in many cases this is not possible and it can be beneficial to estimate Eq. (20) based on characteristic values for the reconnection events, such that

$$\epsilon_{rec} \sim N_{rec} f_{rec} \alpha V^2_{A,inflow} \left(\frac{V_{A,inflow}}{\lambda} \mathcal{R} \right), \tag{21}$$

where N_{rec} is the number of reconnection events in a turbulent volume. Rough estimates of the above parameters constrained by the reconnection events observed in the systematic survey of turbulence-driven reconnection by Stawarz et al. (2022) obtained dissipation rates of 1×10^4 to 3×10^6 J/kg-s, assuming $\lambda \sim \lambda_{C,\mathbf{B}}$ consistent with reconnection occuring at the interface of correlation length magnetic structures. While there was a large spread, these estimates were in rough agreement with previous independent estimates of ϵ in the magnetosheath obtained from expressions similar to Eq. (12) (Hadid et al. 2018; Bandyopadhyay et al. 2018), suggesting reconnection is a non-trivial contributor to dissipation - potentially alongside other processes.

Shay et al. (2018) derived a related expression for the energy dissipation associated with reconnection, which can be thought of as an extension to Eq. (21) that parameterizes α based on expectations from guide field reconnection. Based on a series of 2D laminar symmetric guide field reconnection simulations Shay et al. (2018) found that ion and electron heating is well parameterized by

$$\Delta T_i = c_i \left(\frac{|B_L|}{|\mathbf{B}|} \right)^2 m_i V^2_{A,inflow} \tag{22}$$

$$\Delta T_e = c_e \left(\frac{|B_L|}{|\mathbf{B}|} \right) m_i V^2_{A,inflow} \tag{23}$$

where $\Delta T_{i,e}$ are increases in ion and electron temperature and $c_{i,e}$ are constants of proportionality associated with ion and electron heating, respectively. While these expressions were empirically derived, Eq. (22) is consistent with theoretical expectations for ion acceleration in contracting magnetic islands (Drake et al. 2009). Based on Eqs. (22) and (23), $\alpha_j = c_i \left(|B_{L,j}|/|\mathbf{B}_j|\right)^2 + c_e \left(|B_{L,j}|/|\mathbf{B}_j|\right)$ in Eq. (20). Using this value of α_j, Eq. (20) is then consistent with Eq. 3 of Shay et al. (2018) divided by the total mass in the turbulent volume, noting that f_{rec} is equivalent to the volume of the magnetic islands that have reconnected in the Shay et al. (2018) formalism divided by the total volume of the region. Due to the different scaling of $\Delta T_{i,e}$, the relative ion to electron heating varies depending on the distribution of local guide field strength at turbulence-driven reconnection sites. Relating the expressions for the ion and electron heating rates to the properties of the turbulent fluctuations in a heuristic manner and comparing with the heating rates obtained from 2.5D PIC simulations of turbulence, Shay et al. (2018) found that, while the individual scalings for ion and electron heating did not agree with the prediction, the ratio did agree well.

The above discussion focuses on quantifying the net enhancement of the internal energy and associated dissipation of the turbulent fluctuations enabled by turbulence-driven reconnection. However, in the collisionless regime, heating and energy conversion can generate complex non-Maxwellian velocity distribution functions for both ions and electrons. With the advent of large kinetic simulations of plasma turbulence and high-resolution *in situ* measurements – particularly from *MMS* – capable of probing the velocity distribution functions at sub-proton-scales, understanding the nature and implications of these non-Maxwellian features for the turbulent dynamics and energy dissipation has become an active area of ongoing research (Servidio et al. 2012; Greco et al. 2012; Servidio et al. 2012, 2017; Sorriso-Valvo et al. 2019; Perri et al. 2020; Agudelo-Rueda et al. 2022). Collisionless magnetic reconnection is well-known to contribute to the generation of non-Maxwellian velocity distributions, including, for example, through the generation of temperature anisotropy (Egedal et al. 2013; Cozzani et al. 2023), agyrotopic crescents (Hesse et al. 2014; Chen et al. 2016; Egedal et al. 2016), and field-aligned beams (Gosling et al. 2005; Phan et al. 2007), suggesting that the action of magnetic reconnection events may be important in considering the role of non-Maxwellian VDFs in the dynamics of turbulent plasmas – although reconnection is not the only process capable of generating such features in a turbulent system.

Without collisions to rapidly scatter and redistribute energy in velocity phase-space, the dynamics are formally reversible and, while the generation of a non-Maxwellian distribution may increase the "temperature" (defined as the second-moment of the distribution), energy conversion can readily move that energy into and out of the internal energy of the particles (Bandyopadhyay et al. 2020b, 2021b; Matthaeus et al. 2020). This energy exchange in and out of the internal energy may couple into the effective nonlinear turbulent dynamics of the system - although scattering through wave-particle interactions or, more generally, through interactions with the ensemble of nonlinear fluctuations may help to convert some fraction of this energy exchange into net energy dissipation in the system. Furthermore, the non-Maxwelian distribution functions can be susceptible to kinetic instabilities that may also excite fluctuations that couple into the nonlinear dynamics (Hellinger et al. 2015; Bandyopadhyay et al. 2022; Opie et al. 2023; Lewis et al. 2024). See also Graham et al. (2024, this collection) for a further discussion of secondary kinetic instabilities and waves that can be excited through non-Maxwellian features generated through the reconnection process. From one point-of-view, the impact of collisionless non-Maxwellian dynamics can be incorporated into estimates of the net dissipation associated with turbulence-driven reconnection such as Eq. (20) via the coefficients α_j, where α_j then represents the fraction of energy released by magnetic reconnection that is "locked into" the internal energy of the distribution via net heating – perhaps through the ensemble of secondary collisionless processes enabled by magnetic reconnection in the reconnection outflow. However, one can be more or less granular in associating the net dissipation that occurs with specific multi-scale plasma processes, and the nuances of collisionless dissipation and non-Maxwellian structure remain an active area of research that will, no doubt, see significant advances in the coming years.

3 Reconnection-Driven Turbulence

The converse of the scenario discussed in Sect. 2, referred to here as reconnection-driven turbulence, in which large-scale reconnection sites set up by system-scale dynamics act to generate turbulent dynamics, has also been an active area of investigation. As discussed in Graham et al. (2024, this collection), reconnection is known to generate a variety of waves in the diffusion regions, along the separatrix, and in the exhausts that are excited by the free energy available in non-Maxwellian distributions or strong gradients. As these waves grow to large amplitudes and nonlinearly interact, they can lead to turbulence. The plasmoid/tearing instability can also destabilize the reconnecting current sheet and spontaneously generate multiple x-lines with plasmoids (or flux ropes in 3D) between them. These plasmoids/flux ropes can interact and merge in the outflows, producing a complex turbulent character to the current layer (Daughton et al. 2011; Oishi et al. 2015; Huang and Bhattacharjee 2016). Turbulent dynamics also can be generated by the interaction of the exhaust with its surroundings, such as through shear instabilities or the exhaust encountering an obstacle.

Given the relatively large amount of data, spanning many correlation lengths, needed to perform typical turbulence analyses, relatively limited observational analyses of the fluctuations within reconnection exhausts have been performed in the solar wind (Miranda et al. 2021; Eastwood et al. 2021; Wang et al. 2023) and at Earth's magnetopause (Ergun et al. 2017), which suggest enhanced turbulent fluctuations within the exhausts. In the solar wind, further evidence suggestive of the secondary processes associated with reconnection-driven turbulence has also been found in the form of recent observations of flux rope merger embedded within reconnection exhausts at the heliospheric current sheet (Phan et al. 2024) and analyses of magnetic discontinuities in the solar wind that suggest a hierarchy of fragmentation with smaller-scale discontinuities embedded within larger-scale discontinuities (Greco et al. 2016). However, Earth's magnetotail has proven to be an ideal environment for the observational examination of reconnection-driven turbulence. This status is partially associated with the central role that system-scale Dungey-Cycle-like reconnection plays in energy transport in the magnetotail and the need to understand the role of turbulence in mediating this transport. Additionally, due to the nature of the system, the effective spacecraft trajectory through a reconnection event in the magnetotail is often such that significant dwell times in the reconnection exhaust are obtained, which is conducive to turbulence analyses (in contrast to solar wind or magnetopause reconnection encounters where significant and sustained components of the background motion normal to the current sheet, combined with relatively narrow outflow thicknesses, mean that rapid transits of the outflow are the norm). Although it may be possible to identify additional regions in near-Earth space, such as the heliospheric current sheet in the inner heliosphere, where extended dwell times are also possible (Phan et al. 2024).

3.1 Observations of Magnetotail Turbulence

As illustrated in Fig. 1, the stretched magnetic field in Earth's magnetotail can be broadly divided into two regions - the northern and southern lobes featuring low n and strong $|\mathbf{B}|$; and the relatively high density plasma sheet surrounding the $\mathbf{B}$ reversal at the center of the magnetotail. The plasma sheet is a dynamic region and early studies examined the fluctuations in the plasma sheet as a whole, demonstrating the region exhibits nonlinear behavior consistent with turbulence albeit with differences relative to classical homogeneous turbulence potentially associated with boundary effects, coupling to Earth's ionosphere, and nonuniform driving (Borovsky et al. 1997; Borovsky and Funsten 2003; Weygand et al. 2005).

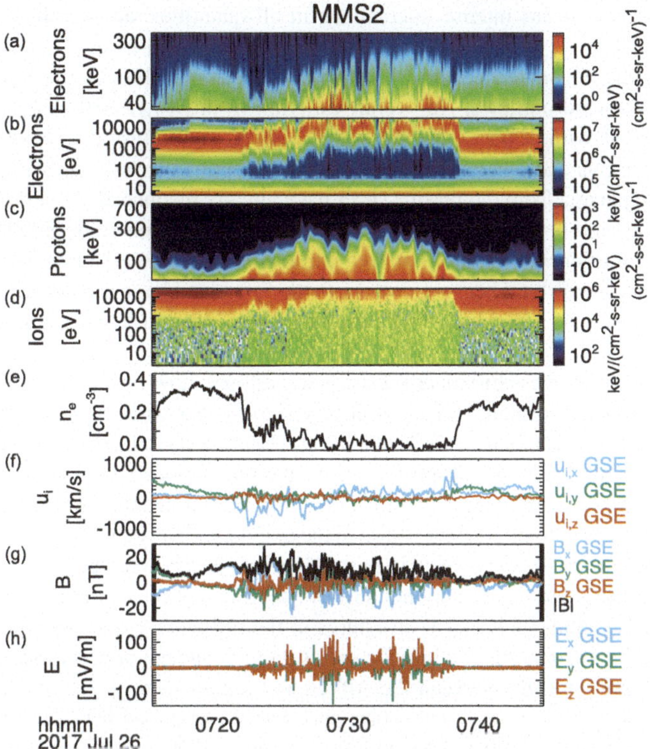

Fig. 4 Example turbulent x-line crossing observed by *MMS*, showing (a) high-energy omnidirectional electron energy fluxes, (b) low-energy omnidirectional electron differential energy fluxes, (c) high-energy omnidirectional proton energy fluxes, (d) low-energy omnidirectional ion differential energy fluxes, (e) electron number density, (f) ion flow velocity, (g) **B**, and (h) **E**. All vector quantities are in Geocentric-Solar-Ecliptic (GSE) coordinates

In the plasma sheet, system-scale reconnection occurs at x-lines near Earth at $\sim 25R_E$ and in the distant tail at $> 60R_E$ from Earth (see Fuselier et al. 2024), driving tailward and Earthward exhausts. These outflows can drive turbulent fluctuations in the plasma sheet and subsequent analyses demonstrated such flows contain some of the clearest and most intense signatures of turbulence in the plasma sheet (Bauer et al. 1995; Vörös et al. 2004, 2006; Stawarz et al. 2015). Reconnection outflows in the magnetotail are typically transient and likely linked with bursty bulk flows (BBFs) in observations. As the flows plow through the surrounding plasma, they relax the stretched $B_{x,GSM}$ component of the field in Geocentric Solar Magnetic (GSM) coordinates and enhance the dipolar $B_{z,GSM}$ component producing so-called dipolarization fronts. On the Earthward side, as the flows impinge on the strong nearly dipolar near-Earth field, the flows are slowed and deflected generating large-scale vorticies (Panov et al. 2010).

Cluster and *MMS* have provided clear examples of extended encounters with reconnection x-lines and adjacent outflows featuring exceptionally intense turbulent fluctuations (Ergun et al. 2018), as well as tailward and Earthward exhausts likely further from the x-line with clear evidence of turbulent dynamics (Vörös et al. 2004; Eastwood et al. 2009; Huang et al. 2012). Figure 4 illustrates a notable example from 26 July 2017 where *MMS* continually observed a near-Earth x-line featuring exceptionally intense fluctuations for 20 minutes. While the event has intense fluctuations in all quantities, an overall variation from $B_{z,GSE} < 0$ to $B_{z,GSE} > 0$ in conjunction with a variation from $u_{x,GSE} < 0$ to $u_{x,GSE} > 0$ occurs during the encounter, consistent with *MMS* traversing through the x-line from the tailward to Earthward outflows. An overall variation from $B_{x,GSE} > 0$ to $B_{x,GSE} < 0$ also occurs, consistent with *MMS* concurrently traveling from southward to northward of the current sheet. In this and other similar events, n is exceptionally low, suggesting field lines carrying dense plasma from the plasma sheet have already reconnected and been evacuated from the region, and now lobe field lines are reconnecting. Ergun et al. (2020a) suggested the large amount of energy imparted to each particle in the reconnection region (~ 3 keV/s per proton/electron pair in this event), owing to the low n and large incoming Poynting flux from the lobes, may be a key reason why such strongly turbulent fluctuations are generated in this type of event.

B and **E** spectra in these turbulent x-lines and outflows show broadband power law scalings across MHD, sub-proton, and electron scales consistent with theory and observations of other turbulent environments (Vörös et al. 2004; Eastwood et al. 2009; Ergun et al. 2018, 2020a; Huang et al. 2012). At sub-proton scales, the fluctuations appear consistent with kinetic Alfvén waves, the dissipation of which may lead to significant proton heating (Chaston et al. 2012, 2014). Due to the variable flows in the magnetotail, even within fast reconnection outflows, some care needs to be taken in the analysis of spectra and other scaling properties, since the Taylor hypothesis is likely not strictly valid. Based on a careful analysis of multipoint timing velocities of **E** and **B** structures in the turbulent x-line shown in Fig. 4g–h, Ergun et al. (2020a) found that, while the direction of propagation was random, the speed of structures were typically between the Alfvén and ion acoustic speeds and within a factor of two of each other. Using the average speed as a conversion between frequency and wavenumber, Ergun et al. (2020a) found good agreement between the location of spectral break points and characteristic plasma length scales, as well as good correspondence between spatial and temporal correlation functions. Bandyopadhyay et al. (2021a) examined Eq. (12) using *MMS* measurements of a reconnection outflow on 16 June 2017, providing evidence of an active cascade of energy to small scales within the reconnection exhaust. By examining higher-order turbulence statistics, several studies have provided evidence of turbulent intermittency within reconnection outflows (Huang et al. 2022, 2023; Jin et al. 2022; Xiong et al. 2024). By dividing an extended reconnection outflow encountered by *MMS* on 28 May 2017 into subintervals, Jin et al. (2022) found evidence of increasingly intermittent behaviour with distance from the x-line, suggesting an evolution of the turbulent dynamics through the outflow region. Based on a combination of numerical simulations and a survey of reconnection outflows encountered by *MMS* in the magnetotail, Xiong et al. (2024) found that, while the intermittency statistics do not show a significant dependence on the out-of-plane guide field, the guide field strength may have an impact on the magnetic topology of the intermittent structures, impacting the energy conversion in the reconnection outflows. Recently, Richard et al. (2024) performed a statistical survey of turbulence in BBFs observed from 15–25R_E downtail of Earth by *MMS*, which confirmed many of the results inferred from case studies of magnetotail reconnection jets. Importantly, it was determined that turbulence quickly develops within a few ion gyroperiods with a driving scale comparable to the size of the jet. An intense energy cascade rate (relative to the solar wind or magnetosheath) was also identified, which extend over an order of magnitude in scale.

While somewhat rare, in several turbulent x-line events, EDRs have been encountered and identified using *MMS* data. Despite the strong fluctuations, the basic properties and structure of the EDR appears broadly consistent with quasi-2D laminar reconnection and the reconnection process continues amidst the turbulent fluctuations for an extended duration

(Ergun et al. 2022a; Qi et al. 2024). Ergun et al. (2022a) found by analysing one event observed on 27 August 2018 that the typical features of reconnection (e.g., a persistent ion jet, **B** profile including a B_L reversal and Hall fields) were all present as expected, but with additional fluctuations on top of them. Off-diagonal terms of the electron stress tensor, which encode off-diagonal contributions to generalised Ohm's law from both the electron pressure and electron inertial effects, were found to both account for the observed reconnection electric field and could be understood from a laminar reconnection perspective. Qi et al. (2024) examined another event, noting that, while large reconnection electric fields may be present, the overall aspect ratio of the event, which provides another estimate of the reconnection rate, gives a value of ~ 0.2 consistent with typical estimates of "fast" reconnection rates.

While turbulence may not significantly alter the electron dynamics at the x-line itself, at least for the few examples that have lent themselves to detailed examination, they may contribute to repartitioning energy in the exhausts. As seen in Fig. 4a and c, intermittent bursts of energetic ions and electrons are interspersed with the fluctuations. Ergun et al. (2018) found that both structures with $\mathbf{j} \cdot \mathbf{E}'$ dominated by the components parallel and perpendicular to **B** contribute to the turbulent dissipation. $\mathbf{j}_\perp \cdot \mathbf{E}'_\perp$ contributed to 80% of the net dissipation and primarily acted at frequencies near the ion cyclotron frequency, while $\mathbf{j}_{||} \cdot \mathbf{E}'_{||}$ primarily acted at higher frequencies and led to the acceleration of some of the most energetic electrons. Further analyses using observations and test particle models demonstrated that magnetic depletions associated with the intense fluctuations enabled particle trapping in regions of energy conversion (Ergun et al. 2020b,a). Notably, perpendicular energization tended to occur making it more difficult for particles to escape, thus leading to further energization and the generation of the non-thermal energetic particles seen in the observations. Bergstedt et al. (2020) performed a statistical analysis of magnetic structures in the same event finding $\mathbf{j}_{||} \cdot \mathbf{E}'_{||}$ dissipation occured at current sheets at the interface of plasmoid-like structures, which may be consistent with the dynamic production and merger of plasmoids by the reconnecting current sheet. Oka et al. (2022) examined particle energization, both in reconnection events like the example in Fig. 4 where lobe field lines reconnect and in events where plasma sheet field lines carrying denser plasma reconnect, demonstrating larger **E** fluctuations are present in lobe reconnection. Interestingly, the stronger **E** fluctuations were linked to a smaller nonthermal energy fraction. This discrepancy with Ergun et al. (2020b) may be linked to differences in the definition of nonthermal particles, with Oka et al. (2022) focusing on the high-energy power law tail to the particle distributions, while Ergun et al. (2020b) focused on a high-energy non-thermal shoulder to the distribution.

Turbulent fluctuations can also be excited through the interaction of the exhaust with the surroundings. Volwerk et al. (2007) found evidence of the Kelvin-Helmholtz instability associated with the velocity shear along the edges of a BBF using a conjunction between *Cluster* and *DoubleStar* and Divin et al. (2015) found evidence of turbulence caused by the LHDI at the dipolarization front. Closer to Earth in the BBF braking region, studies have identified broadband turbulent spectra of kinetic-Alfvén-wave-like fluctuations, the excitation of which is likely partially associated with the flow braking process and may play a role in the conversion of bulk flow kinetic energy into other forms (Chaston et al. 2012; Ergun et al. 2015; Stawarz et al. 2015). Ergun et al. (2015) and Stawarz et al. (2015) found using *THEMIS* data that the BBF braking region was filled with large-amplitude high-frequency $E_{||}$ structures such as double layers and electron phase space holes, potentially excited by field-aligned current instabilities associated with the turbulence. Stawarz et al. (2015) estimated these $E_{||}$ structures could play a significant role in the dissipation of the turbulent fluctuations and heating of the plasma; and subsequent large-scale statistical surveys have supported the picture of the braking region as a region of enhanced solitary wave activity, as well as proton and electron heating (Hansel et al. 2021; Usanova and Ergun 2022; Usanova et al. 2023). Based on a statistical examination of BBF events, Chaston et al. (2014) found an enhanced divergence of field-aligned Poynting flux associated with kinetic Alfvén waves away from the center of the plasma sheet for distances $< 15 R_E$ downtail of Earth, suggesting a fraction of the small-scale fluctuations excited in the region propagate along the field-lines depositing energy in the auroral region. In this scenario, the turbulent dynamics, while not responsible for dissipating the energy radiated from the region, convert bulk flow energy into small-scale fluctuations capable of propagating to the ionosphere. Stawarz et al. (2015) estimated the energy budget in the braking region, finding turbulent energy dissipation, adiabatic heating due to the compression of the magnetic field in the braking region, and radiated Poynting flux were similar in magnitude and the sum of these energy losses were comparable to the energy input into the region due to the bulk flows. At the inner edge of the flow braking region around $\sim 7 R_E$ downtail of Earth, BBFs have been associated with enhanced turbulence and the dissipation of these fluctuations has been implicated in seeding energetic particles in the outer radiation belt (Ergun et al. 2022b).

3.2 Simulations of Reconnection-Driven Turbulence

Numerical simulations have also long been used to examine the turbulence that is self-generated by magnetic reconnection under certain circumstances. Arguably, most of the numerical work on reconnection-driven turbulence has been performed using MHD plasma models, for example in the context of the plasmoid instability or other secondary fluid instabilities; however, the impact of kinetic effects and instabilities more appropriate to the collisionless space plasmas of the near-Earth environment have been historically less well understood. In particular, turbulence near and in the diffusion regions has been a critical area of research due to its potential role in balancing the reconnection electric field, and thereby enabling reconnection, through anomalous resistive and viscous effects that can potentially enhance the reconnection rate. Such anomalous effects arise from the nonlinear contributions to generalized Ohm's law due to fluctuating quantities that are not typically treated in the traditional quasi-laminar reconnection picture. Note that this way in which turbulence affects reconnection is somewhat different from the stochastic reconnection picture, often analysed in the MHD context and discussed further in Sect. 5. In this section, we discuss a selection of the most important findings from the past several years, with a focus on instabilities causing turbulence in kinetic reconnection simulations applied to space plasmas.

In 2D, several works have shown streaming instabilities between different electron and/or ion populations can generate turbulence, particularly at the separatrices that are known to host a variety of waves due to instabilities and other nonlinear mechanisms (Fujimoto 2014; Goldman et al. 2014). Cattell et al. (2005) and Pritchett (2005) performed 2D fully kinetic PIC simulations of magnetic reconnection with a guide field and parameters appropriate to Earth's magnetotail. This configuration produced electron holes and associated turbulence via streaming (Buneman) instabilities at the separatrices, consistent with in-situ observations. Later simulations with more realistic parameters, such as a realistic ion-to-electron mass ratio and larger domain, demonstrated this process occurs for all guide fields and is, therefore, expected under a variety of conditions in Earth's magnetosphere (Lapenta et al. 2011). Muñoz and Büchner (2016) showed using 2D fully kinetic PIC simulations that there is a strong guide field regime in which broadband electrostatic turbulence develops in the separatrices and outflows, with the key finding that the separatrix/outflow turbulence is associated with double-peaked and anisotropic electron distribution functions and wave activity near the lower-hybrid frequency whose strength is correlated with the instantaneous reconnection rate. This turbulence was shown to be associated with anomalous resistivity, which was smaller than the electron inertia or non-gyrotropic electron pressure tensor in generalised Ohm's law (Muñoz et al. 2017). However, since the process was not occurring at the x-line, it was not associated with enabling the reconnection process itself. Using 2D fully-kinetic simulations with a sufficient scale separation between d_e and the Debye length (a parameter that is typically small in PIC simulations of reconnection), Jara-Almonte et al. (2014) showed that Debye-scale turbulence was excited near the x-line as well, through the action of streaming instabilities in the reconnection plane.

3D fully kinetic reconnection simulations tend to be more turbulent than their 2D counterparts due to the presence of the additional degree of freedom enabling more instabilities with wavenumbers along the additional dimension. Drake et al. (2003), Che et al. (2011), and Che (2017) found that in 3D the current layer develops filamentary magnetic structures due to electron-shear flow and Buneman instabilities along the direction parallel to the current sheet using PIC simulations with a relatively strong guide field, small spatial domain, and initially force-free equilibrium. During the non-linear phase of these instabilities, electron holes and turbulence were generated, which produced anomalous resistivity and viscosity. Daughton et al. (2011) carried out 3D PIC simulations but with a smaller guide field, much larger spatial domain, and initialized with a Harris current sheet equilibrium. In these simulations, the current sheet developed flux ropes at electron scales due to the tearing instability, which were not seen in previous simulations because of their smaller spatial domain. These flux ropes underwent complex 3D interactions that lead to continuously self-generated and inhomogeneous turbulence within the electron current layer. Later simulations by Liu et al. (2013) revealed that the turbulence generated under similar conditions to Daughton et al. (2011) does not modify the reconnection rate via anomalous resistivity or viscosity, in contrast to the earlier results by Drake et al. (2003) and Che et al. (2011). The discrepancy was explained as being due to the lack of streaming or electron shear flow instabilities in the higher plasma-β regime appropriate to magnetospheric conditions and enhanced parallel heating that can be developed in simulations with larger spatial domains. Fujimoto and Sydora (2021, 2023) demonstrated that in 3D simulations without a guide field a different scenario occurs, in which electromagnetic turbulence is generated by an electron Kelvin-Helmholtz instability. These fluctuations are capable of producing anomalous viscosity due to electron transport that is capable of breaking the electron frozen-in condition, but not anomalous resistivity since the turbulence mainly effects the electrons, while the ions remain decoupled.

Another source of free energy to drive turbulence is the lower-hybrid drift instability (LHDI) generated by gradients in the density or magnetic field associated with diamagnetic currents. This instability can be particularly important in the case of asymmetric reconnection, such as at Earth's dayside magnetopause, where the strong density gradients across the reconnecting current sheet are conducive to exciting the instability. The LHDI typically generates waves initially at the edge of the current sheet that can spread toward its center. Turbulence associated with the LHDI has been historically considered as a source of anomalous resistivity. However, only in the last decade have 3D fully-kinetic PIC simulations with a large-enough scale separation between electron and ion scales been able to reveal under which conditions this instability can modify the reconnection properties. Roytershteyn et al. (2012) and Pritchett et al. (2012) found that the LHDI can cause enough turbulence near the current sheet center to sustain the reconnection electric field if the plasma-beta is low enough under asymmetric conditions, although most of the LHDI-driven turbulence is confined to the separatrices. Such conditions are unlikely to be met for magnetopause reconnection.

Later similar simulations obtained different conclusions under different conditions. Price et al. (2016, 2017) carried out 3D fully kinetic simulations of an asymmetric magnetopause reconnection event based on observations made by *MMS*. Unlike Roytershteyn et al. (2012), these simulations found LHDI-driven turbulence at both the x-line and sparatricies that was strong enough to balance the reconnection electric field. The discrepancy was attributed to different boundary conditions and a stronger than previously expected density jump across the current sheet in the observed *MMS* event. Although, Le et al. (2017, 2018) modelled the same reconnection event as Price et al. (2016, 2017), concluding that anomalous resistivity was small, but that the turbulence acted to enhance plasma mixing and heating in the event. The simulations by Price et al. (2016) also revealed that crescent-shaped electron distribution functions, an important hallmark of magnetopause reconnection often observed by *MMS*, were not affected by the turbulence developed in the current sheet. Later, Price et al. (2020) explored a similar system but under the presence of a significant guide field, showing that in this case turbulence develops in the diffusion region due to a variant of the LHDI. While the anomalous resistivity produced by the electromagnetic fluctuations at the x-line were small, other anomalous terms were significant, but did not significantly impact the reconnection rate. *MMS* observations of lower hybrid fluctuations associated with reconnection at the magnetopause are often consistent with the properties found in 3D simulations (for a detailed overview of the LHDI associated with reconnection from spacecraft observations and simulations see Graham et al. 2024, this collection) and feature broadband power-law spectra suggestive of turbulent dynamics. However, such fluctuations (both in observations and simulations) are often observed in narrow boundary regions, making it challenging to apply typical turbulence analyses and it remains unclear whether nonlinear processes play a significant or dominant role in the evolution of the waves or if there is significant energy transfer across spatial scales.

As discussed in Sect. 3.1, turbulence is also generated in the reconnection outflows. Pritchett and Coroniti (2010) and Vapirev et al. (2013) showed that in 3D fully kinetic simulations a Rayleigh-Taylor-like interchange instability occurs in association with the density gradients at the dipolarization front, generating a turbulent outflow. Lapenta et al. (2015) investigated the outflows using similar simulations, finding signatures of secondary reconnection events embedded within the complex turbulent fluctuations in the outflows. Subsequent works were able to automatically identify such secondary reconnection events using machine learning methods (Lapenta et al. 2022). Numerical simulations have also demonstrated the ability of turbulence in the outflows to efficiently accelerate electrons into non-thermal power-law tails similar to those observed in the magnetotail observations (Lapenta et al. 2020a).

For extended (long) current sheets, the plasmoid instability is also well-known to occur in numerical simulations. In the context of MHD, theory predicts a long current sheet can generate a chain of secondary magnetic islands with secondary current sheets between them if the Lundqvist number ($S = \mathcal{L}V_A/\eta$; i.e., similar to the magnetic Reynolds number but based on the Alfvén velocity) is large enough relative to the aspect ratio of the current sheet (Shibata and Tanuma 2001; Loureiro et al. 2007). In this regime, the reconnection rate is expected to be independent on resistivity. Each secondary current sheet can also be plasmoid-unstable with possibly leading to a downward cascade of plasmoids in a fractal way until they eventually reach kinetic scales, triggering kinetic reconnection. Turbulence arises due to the interaction between those plasmoids (or flux ropes in 3D). This plasmoid instability has received significant attention and has been mainly analyzed using MHD simulations (see, e.g., Bárta et al. 2011; Huang and Bhattacharjee 2016, and references therein). The transition from a collision-dominated plasmoid instability to kinetic reconnection was studied in 3D PIC simulations including a collision operator (Stanier et al. 2019). But, in general, there has been relatively little work on collisionless kinetic reconection simulations of plasmoids with a focus on the self-generated turbulence. One relevant example of such a work is Fujimoto and Sydora (2012), who simulated 3D reconnection with a long current sheet, finding the plasmoid formation precedes the enhancement of electromagnetic turbulence due to shear flows, associated with anomalous momentum transport. The turbulence is first enhanced around the plasmoid and later, after the plasmoid ejection from the x-line, expands toward the x-line. Fermo et al. (2012) found using 2D simulations that the electron Kelvin-Helmholtz instability can generate plasmoids at small-scales in the kinetic regime. Markidis et al. (2013) simulated a long 3D current sheet, finding evidence for the complex interaction of the resulting plasmoids formed by the tearing instability and bump-on-tail instability together, which generated electron holes and complex electrostatic fluctuations near the plasmoids. Nakamura et al. (2021) used 2D PIC simulations of a current sheet initially seeded with an ensemble of magnetic field perturbations in order to generate multiple x-points. The evolution of the system led to a broadband power-law magnetic energy spectrum with a spectral index of -4 below ion scales. The merging of islands led to a decrease in the reconnection rate and a reduction of the aspect ratio of the electron diffusion region. As a result, magnetic islands/plasmoids can grow within the electron diffusion region, allowing an (inverse) energy transfer to larger scales.

While the above simulation works identified instabilities and the development of seemingly turbulent dynamics during the nonlinear evolution of those instabilities, they generally did not perform the detailed statistical analyses of the fluctuations typical of turbulence theory. Leonardis et al. (2013) analyzed the simulations of Daughton et al. (2011), finding evidence of intermittency both in the increments of $\mathbf{B}$ and in $\mathbf{j} \cdot \mathbf{E}$, indicative of the turbulent nature of the dynamics. Pucci et al. (2017) analyzed the outflows of 3D reconnection simulations with a similar setup and parameters as those of Vapirev et al. (2013) and Lapenta et al. (2015), demonstrating the development of a turbulent cascade and intense dissipation at the boundary between the reconnection outflow and the ambient plasma with turbulence statistics similar to those reported in magnetotail observations. Further work explored the dynamics associated with the collision of two reconnection jets in an O-point/magnetic island geometry, showing the development of broadband non-stationary fluctuations between the ion and electron cyclotron frequencies confined to the interaction region between the jets. Muñoz and Büchner (2018) characterized the magnetic spectra in simulations similar to those of Che et al. (2011), demonstrating the development of broadband kinetic-scale turbulence with typical frequencies between the lower-hybrid and up to the electron-cyclotron frequencies. The magnetic spectra steepened as the instabilities evolved, eventually reaching a power-law of $\sim k^{-2.8}$ (typical of observations of kinetic-scale turbulence) once the reconnection rate reached a normalized value of 0.1. Interestingly, the reconnection rate is enhanced beyond the typical value of 0.1 as the system evolved further, in conjunction with the spectral slope steepening. Zharkova and Xia (2021) performed a similar analysis to Muñoz and Büchner (2018) but for a simulation with a more extended current sheet that led to the formation of more plasmoids, emphasising the role of accelerated particles in generating the unstable beam distributions that excite the turbulence. In contrast to the previous works, Adhikari et al. (2020) analyzed not a turbulent but initially laminar 2D PIC reconnection simulation, analyzing the spectral properties of the resulting fluctuations. In steady state, where the reconnection rate was near 0.1 (in normalized units), the fluctuations featured a double power law spectra following $\sim k^{-5/3}$ for scales larger than d_i (fluid scales) and $\sim k^{-8/3}$ for scales between d_i and d_e (kinetic scales). Similar to the 3D simulations by Muñoz and Büchner (2018), Adhikari et al. (2020) found a correlation between the reconnection rate and the energy spectrum, but mainly for wavenumbers near d_i^{-1}. In addition, Adhikari et al. (2020) determined that, while the initial energy spectrum associated with the current sheet is highly anisotropic, the anisotropy diminishes as reconnection develops. Lapenta et al. (2020b) explored the spectra of the turbulent fluctuations in diverse regions around the main reconnection site using 3D fully kinetic PIC simulations, finding a relationship between the local values of the plasma-β and the coupling between plasma and electromagnetic field fluctuations. In the high-β regions corresponding to the reconnection outflows, plasma and electromagnetic fluctuations are coupled and turbulent, while, in the low-β region corresponding to the reconnection inflow, diffusion region and around the separatrices, the plasma flows appear laminar while the electromagnetic fluctuations appear turbulent. Such results potentially suggest that anomalous resistivity and viscosity primarily play a role in the outflow regions.

4 The Kelvin-Helmholtz Instability & the Role of Reconnection in the Transition to Turbulence

Alongside reconnection, other large-scale structures/instabilities can drive turbulence in a plasma and reconnection can play role as a secondary instability in the initial development of the system into a turbulent state. While this scenario has connections with those discussed in Sects. 2 and 3, it is worth considering some of its unique features. Several numerical studies have looked at the role of reconnection in the destabilization of large-scale initial configurations and the onset of turbulence, noting the apparent non-local energy exchange as reconnection rapidly excites small sub-proton-scale fluctuations (Gingell et al. 2017; Franci et al. 2017; Manzini et al. 2023). One region where the excitation of such a large-scale instability and the subsequent transition to turbulence has been clearly observed in space plasmas is along the strong velocity shear boundary on the flanks of Earth's magnetopause in which the Kelvin-Helmholtz instability (KHI) is well-known to occur. Magnetic and velocity shears, such as those found at the magnetopause, coexist in many boundaries in dynamic plasma environments, such as planetary magnetopauses, the heliopause, solar and stellar flares, and astrophysical jets. While, for sufficiently large magnetic shears, reconnection is expected to be a dominant process, strong velocity shears approaching the Alfvén speed based on the sheared component of **B** are expected to suppress instabilities, such as the tearing instability, that are thought to initiate reconnection (Chen and Morrison 1990; Faganello et al. 2010) and can excite the KHI. Theory and numerical simulations suggest the vortical flow produced by the nonlinear evolution of the KHI can strongly distort and twist **B**, inducing secondary reconnection that may contribute to the evolution of the KHI into a turbulent boundary layer. Recent large-scale simulations further predict that this so-called vortex-induced reconnection (VIR) process can cause mass and energy transfer as efficiently as that caused by regular reconnection induced under large magnetic shears and in-situ observations by various spacecraft have confirmed the evolution of VIR at the magnetopause.

4.1 Vortex-Induced Reconnection

Based on linear MHD theories and 2-D two-fluid simulations, Nakamura et al. (2008) summarized two-types of VIR that can be excited in the 2D vortex plane. Type-I VIR occurs when pre-existing magnetic shear is locally compressed by the non-linear vortex flow (Pu et al. 1990; Knoll and Chacón 2002) as shown in Fig. 5a. Since Type-I VIR reconnects field

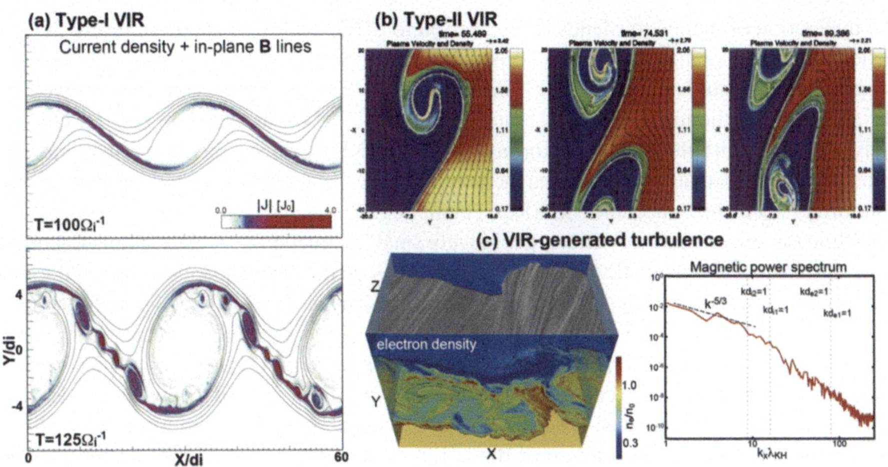

Fig. 5 (a) Current density and in-plane magnetic field lines of a 2D fully kinetic simulation of Type-I VIR, showing formation of multiple magnetic islands in the compressed current layer (reproduced from Nakamura et al. 2013). (b) Plasma density and field lines of a 2-D MHD simulation of Type-II VIR, showing island formation in vortex arms (reproduced from Nykyri and Otto 2001). (c) 3D view of electron density and magnetic power spectrum of a 3D fully kinetic simulation of Type-I VIR, showing generation of turbulence (reproduced from Nakamura et al. 2017b)

lines originally located on different sides of the shear layer, this process can cause rapid plasma mixing across the layer (Nakamura et al. 2011; Nakamura and Daughton 2014). In addition, linear theory predicts Type-I VIR is commonly triggered in boundaries where moderate amplitude magnetic and velocity shears co-exist, as often seen at Earth's magnetopause (Nakamura et al. 2006).

Type-II VIR is driven in highly rolled-up vortices where the wrapped field lines secondarily form thin current sheets (Nykyri and Otto 2001; Nakamura and Fujimoto 2005; Faganello et al. 2008) as shown in Fig. 5b. Nakamura and Fujimoto (2005) performed a parameter study suggesting Type-II VIR is triggered when the velocity shear is significantly larger ($\gtrsim 5\times$) than the Alfvén speed based on the component of **B** along the shear. Since Type-II VIR divides the vortex and forms magnetic islands penetrating through the vortex layer, this process can also cause efficient plasma transport across the boundary (Nykyri and Otto 2001). However, since the magnetic topology change in Type-II VIR occurs within a single wrapped field line, the process of Type-II VIR on its own cannot cause plasma mixing in 2D and additional 3D effects, collisionless cross-field diffusion, and/or the coupling with other types of VIR are necessary to enable cross-field mixing and transport. Based on numerical simulations, both types of VIR are expected to coexist in the KHI vortex layer (Nakamura et al. 2008, 2013; Karimabadi et al. 2013).

In these VIR processes, vortical flows strongly compress the current layers down to electron-scales (Nakamura et al. 2011). When the length of the compressed current layers, which depends on the size of the parent KH vortex, is sufficiently long compared to electron-scales, multiple plasmoids, with initial sizes comparable to election scales, are observed to form in the compressed current layers for both Type-I and Type-II VIR based on numerical simulations (Nakamura et al. 2011, 2013; Karimabadi et al. 2013), as shown in Fig. 5a. Reminiscent of the plasmoid instability discussed in Sect. 3, this tertiary instability contributes to the generation of complex turbulent dynamics within the shear layer through the interaction of magnetic islands from electron to MHD scales. In addition, recent 3D kinetic simulations demonstrated that, Type-I VIR can be triggered and evolve over a broad range of oblique angles, which significantly enhances the rate of plasma mixing, as well as the amplitude of the turbulence (Nakamura et al. 2013, 2017b) as shown in Fig. 5c. Once a fully developed turbulent layer has formed within the KHI vortices, the intermittent current structures and associated complex magnetic topologies created by the turbulent fluctuations can further act as reconnection cites (Rossi et al. 2015).

While the above studies considered relatively small magnetic shears appropriate to northward interplanetary magnetic field (IMF) conditions at the magnetopause, 3D MHD and kinetic simulations have suggested that, when the magnetic shear is large enough, the turbulent evolution of Type-I VIR quickly disturbs and destroys the vortex structure (Ma et al. 2014; Nakamura et al. 2020a). On the other hand, more recent 3D kinetic simulations modeling realistic magnetopause conditions under southward IMF showed that when the density jump across the magnetopause is sufficiently large, which would happen more often under southward IMF, the rapid evolution of the LHDI at the compressed current layers quickly diffuses the layers and suppresses Type-I VIR (Nakamura et al. 2022a), although the substructure produced by the vortex-induced LHDI can itself induce small-scale reconnection (Nakamura et al. 2022b).

Although most of the above assumed the initial equilibrium varied only in the boundary normal direction, the conditions can also vary in different directions in many realistic situations. At Earth's magnetopause, the KHI is unstable at lower latitudes on the magnetopause and stable above and below at higher latitudes (Takagi et al. 2006). Numerical simulations modeling these conditions have demonstrated the KHI vortex motion twists **B** generating additional magnetic shears in the transition region between the low-latitude unstable and high-latitude stable layers, inducing reconnection at mid-latitudes as illustrated in Fig. 6a, b. Contrary to Type I VIR, in this mid-latitude reconnection (ML VIR) magnetic shear is created even if the pre-existing magnetic fields are aligned across the boundary. Furthermore, since the evolution of the KH vortices and thus ML VIR is symmetric with respect to the equatorial plane, reconnection occurs simultaneously in both hemispheres, creating "double-reconnected" field lines topologically connected to the Earth but embedded in the magnetosheath at low-latitudes (Faganello et al. 2012a,b; Borgogno et al. 2015). The creation of such field lines enhances particle transport across the boundary (Faganello et al. 2012b; Borgogno et al. 2015), and can explain the specific entropy increase on the magnetospheric side of the boundary (Johnson and Wing 2009).

In the presence of pre-existing magnetic shear, both Type-I VIR and ML VIR occur, with Type-I VIR dominating at low latitudes. The pre-existing magnetic shear breaks the symmetry in the ML VIR process - gradually enhancing the magnetic shear by differential advection in one hemisphere, while reducing it in the other. The evolution of magnetic topology is quite complex, with a dominant Type-I VIR acting close to the equatorial plane, and ML VIR going on in only one hemisphere, as shown in Fig. 6c, leading to a less efficient, but still important, production of double-reconnected lines (Sisti et al. 2019; Faganello et al. 2022).

4.2 Observations of Vortex-Induced Reconnection

Hasegawa et al. (2004) presented evidence of the coexistence of solar wind and magnetospheric ions on the same field lines, suggesting particle transport across the magnetopause, in association with KH waves using *Cluster* observations. However, magnetic reconnection was initially ruled out as the cause of the plasma transport given the absence of ion

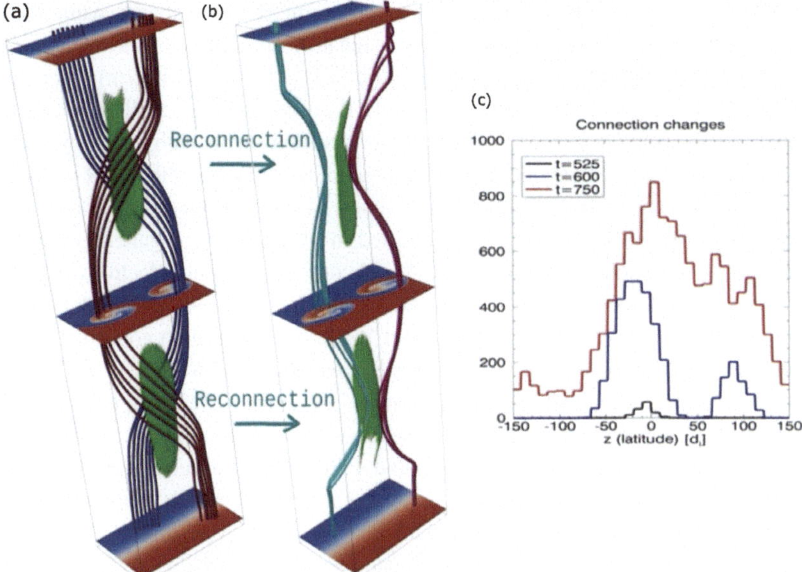

Fig. 6 (a) 3D view of field line deformation due to differential advection in the symmetric case. Blue (magnetospheric) and red (magnetosheath) field lines are initially parallel and twisted by differential advection, creating magnetic shears (green regions). (b) Double-reconnected field lines created by ML VIR, with pale blue lines connecting to magnetospheric (blue) plasma at high-latitudes and magnetosheath (red) plasma at the equator (adapted from Faganello et al. 2012a). (c) Latitude distribution of connection changes (created by reconnection), for different times, in the asymmetric case. At t = 600, the main peak is due to Type-I VIR, while the secondary peak is due to ML VIR (reproduced from Sisti et al. 2019)

reconnection exhaust observations and reconnection-associated D-shaped ion velocity distributions (e.g., Cowley 1982; Fuselier et al. 2014) throughout the KH-active interval. A further analysis of the same event by Hasegawa et al. (2009) provided the first direct indication that magnetic reconnection was involved with the plasma trasport associated with the KHI on the flanks of Earth's magnetopause. High-cadence **B** measurements across the so-called spine-regions between neighboring KH vortices captured several localized current sheets with thicknesses of only a few d_i. One such current sheet showed evidence of bifurcation (Gosling and Szabo 2008) and Alfvénic outflow signatures, inferred from the $\mathbf{E} \times \mathbf{B}$ drift, in agreement with predictions of Type-I VIR (e.g., Pu et al. 1990; Nakamura et al. 2006, 2008), although direct measurement of the outflow jet with the particle measurements was not possible.

MMS provided the high cadence ion measurement capabilities necessary to resolve the ion jets embedded in the KH vortices. Eriksson et al. (2016a) surveyed an extended KH wave-train observed for over an hour at the magnetopause on 8 September 2015, providing the first direct confirmation of ion reconnection exhausts at the quasi-periodic compressed current sheets in the spine region of the KHI associated with Type-I VIR. In total, reconnection exhausts were found at 22 of the 42 currents sheets with equal probability of encountering reconnection outflows in either direction relative to the x-lines. Thicknesses normal to the reconnecting current sheets were $4.4 \pm 1.9 d_i$. In addition to outflow signatures, Hall magnetic field perturbations (Eriksson et al. 2016a) and particle fluxes across the locally open magnetopause (Li et al. 2016; Vernisse et al. 2016) were also identified in these events. Furthermore, in one case, evidence of the EDR was also identified with two of the *MMS* spacecraft (*MMS*1 and *MMS*2) encountering the ion exhaust, while the other two spacecraft did not but instead encountered signatures indicative of the EDR (Eriksson et al. 2016b). In particular, strong parallel electric fields ($E_{||} \sim -15$ mV/m) and enhanced $\mathbf{j} \cdot \mathbf{E}' \sim 8-9$ nW/m^3 were identified. Since the KHI events under generally northward IMF conditions are expected to produce reconnection at current sheets with relatively low magnetic shear, this event provided one of the first direct measurements of the EDR under strong guide field conditions, with an observed guide field $\sim 4\times$ the reconnecting field.

Evidence of ML VIR has also been identified in observations. Faganello et al. (2014) reported a short-duration interval of 100–500 eV counter-streaming electrons in *THEMIS* observations and interpreted them as magnetosheath electrons accelerated along recently closed field lines at two ML VIR regions, and Vernisse et al. (2016) confirmed a similar signature in *MMS* electron measurements in the 8 September 2015 KHI event. Eriksson et al. (2021) also reported several short-duration "bursts" of counter-streaming field-aligned ion beams in the 8 September 2015 KHI event. The ion velocity distributions were typically "D-shaped" in phase-space, as commonly associated with magnetopause reconnection (Cowley 1982). It was concluded that the counter-streaming ion beams, which were encountered in the leading edge of the vortices in contrast with the Type-I VIR events that were encountered at the trailing edge current sheets, were associated with two nearby ML VIR events above and below the spacecraft, in agreement with predictions of double-reconnected field lines in numerical simulations (Sisti et al. 2019).

Many other aspects of plasma dynamics have been explored using *MMS* observations from the 8 September 2015 KHI event, including several detailed reports on the generation of plasma turbulence and plasma wave activity within the vortices (e.g., Stawarz et al. 2016; Wilder et al. 2016; Nakamura et al. 2017b; Sturner et al. 2018; Sorriso-Valvo et al. 2019; Hasegawa et al. 2020; Quijia et al. 2021). Stawarz et al. (2016) showed that electromagnetic fluctuations observed in the KH vortices were characterized by a Kolmogorov-like power spectrum at large MHD scales (Fig. 7b) with signatures of intermittency potentially associated with secondary current sheets formed through the nonlinear KHI development. Hasegawa et al. (2020) further investigated the turbulence suggesting the **B** fluctuations in the KH vortices were convective structures associated with interlinked flux tubes generated through 3D turbulent VIR (Nakamura et al. 2013, 2017a), rather than propagating waves. Further analyses have attempted to probe the evolution of KHI turbulence by examining KHI events encountered at different distances along the magnetopause by *THEMIS* and *Geotail*, although challenges arise from convolving event-to-event variation with the temporal evolution (Di Mare et al. 2019).

While the above studies primarily focused on turbulence generation within KH vortices through the lens of KHI and VIR processes, seed perturbations induced by pre-existing magnetosheath turbulence can also act to enhance the KHI growth, and were suggested as an explanation for the apparently rapid onset of turbulent dynamics in the 8 September 2015 KHI event compared to simulations (Nykyri et al. 2017; Nakamura et al. 2020b). Furthermore, for southward IMF, the KHI growth can not only induce reconnection but also form thin density gradient layers at the edge of the KH waves or vortices, which may become unstable to the LHDI (Blasl et al. 2022). The resulting LHDI waves or turbulence can in turn cause diffusive plasma mixing across the magnetopause (Nakamura et al. 2022b). Thus, there may be an intriguing interplay among turbulence, magnetic reconnection, and MHD and kinetic instabilities, which needs to be further explored.

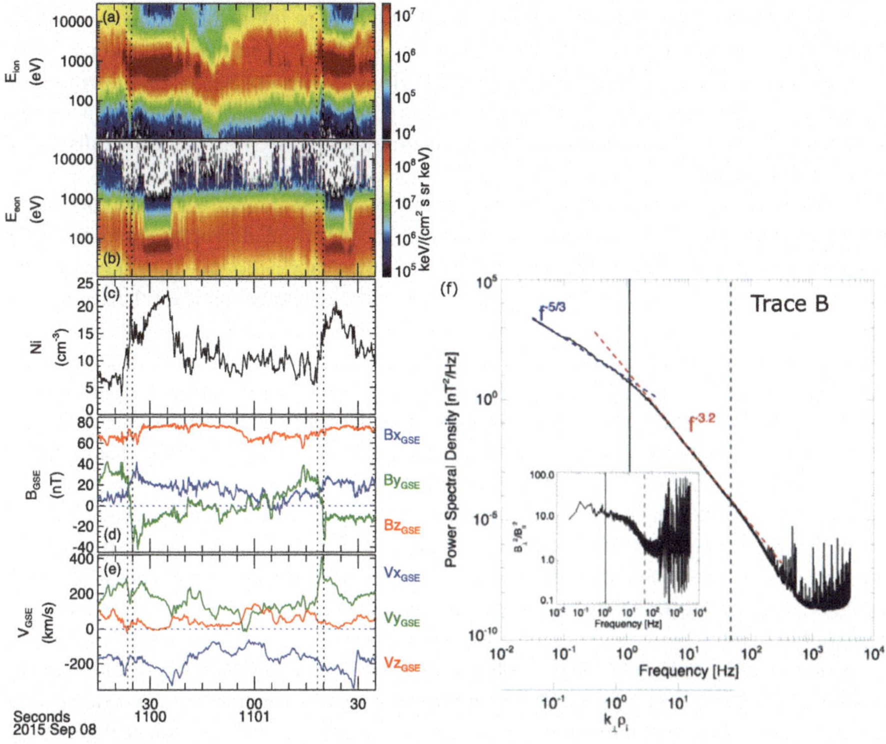

Fig. 7 Example *MMS* observations of two consecutive KH vortices associated with Type-I VIR ion exhausts as highlighted between pairs of vertical dotted lines, showing the (a) omnidirectional ion differential energy flux, (b) omnidirectional electron differential energy flux, (c) ion number density, (d) **B**, and (e) ion bulk flow velocity. Vector quantities are shown in GSE coordinates. (f) Magnetic power spectrum within the KH vortices observed by *MMS* on 8 September 2015, which is characterized by a Kolmogorov-like ($k^{-5/3}$) power law at MHD scales (reproduced from Stawarz et al. 2016)

5 Stochastic Reconnection

A final facet of turbulent reconnection is so-called stochastic reconnection, whereby the wandering of magnetic field lines associated with the turbulent fluctuations plays a dominant role in setting the reconnection rate. The ways in which turbulence may impact magnetic reconnection in this manner were discussed by Matthaeus and Lamkin (1986); however, more complete theoretical treatments were developed by Lazarian and Vishniac (1999) and Eyink et al. (2011). Several reviews discussing the details of this topic have been written over the years (e.g., Lazarian et al. 2012, 2015, 2020) and we will refrain from reiterating a detailed discussion of the process here. However, in this review we provide a brief overview of the conceptual picture and how it interfaces with the other facets of turbulent reconnection discussed in the previous sections.

Stochastic reconnection envisions a "large-scale" magnetic shear, potentially associated with a pre-existing current sheet embedded within a turbulent environment or a variation in the field orientation associated with larger-scale turbulent fluctuations, that is superposed with small-scale turbulent fluctuations introducing a stochastic wandering of the field lines. This field line wandering is akin to Richardson diffusion in hydrodynamic turbulence and, importantly, introduces a rate of diffusion (or perhaps more intuitively dispersion) of the field lines set by the properties of the turbulence, which is independent of the particular non-ideal mechanisms (be that collisional resistivity or collisionless effects) operating in the plasma. In this sense, the stochastic reconnection picture may have applications to both the turbulence-driven reconnection and reconnection-driven turbulence scenarios discussed in the previous sections.

As discussed in Liu et al. (2024, this collection), $\mathcal{R}$ is dictated by the aspect ratio (thickness over length) of the diffusion region. Stochastic reconnection allows the thickness of the diffusion region (often referred to as a reconnection layer in this context) to be set by the stochastic field line wandering, resulting in a much wider reconnection layer than otherwise expected. In this way, stochastic reconnection is capable of producing a reconnection rate that is both fast and independent of microphysical nonideal effects. In terms of generalized Ohm's law (see Liu et al. 2024, this collection), one can view stochastic reconnection as exploring the effect of the nonlinear contribution of the ideal MHD term ($-\delta\mathbf{u} \times \delta\mathbf{B}$) associated with the turbulent fluctuations on the reconnection process (see the discussion of generalized Ohm's law in Eyink 2015; Stawarz et al. 2021; Lewis et al. 2023). Stochastic reconnection is subtly different than facilitating reconnection by invoking anomalous resistivity/viscosity (as discussed in Sect. 3.2) with regard to the order of averaging that one considers. In contrast to anomalous resistivity, where one considers reconnection of an averaged $\mathbf{B}$, stochastic reconnection considers the cumulative stochastic effect of reconnection of the full field lines.

Lazarian and Vishniac (1999) presented a physical picture in which the resulting wider reconnection outflow was supported by a multitude of small-scale local reconnection events enabled by the "rough" perturbed magnetic field line topology of the turbulent fluctuations in the reconnection layer that cumulatively fill the outflow region. For a Goldreich-Sridhar-type spectrum, this picture results in an outflow velocity given by

$$V_{outflow} \approx V_{A,inflow} \min\left[\left(\frac{L_x}{\lambda_{inject}}\right)^{1/2}, \left(\frac{\lambda_{inject}}{L_x}\right)^{1/2}\right] \mathcal{M}^2_{A,\lambda_i}, \tag{24}$$

where L_x is the length of the reconnection layer along the outflow direction, λ_i is the injection scale of the turbulent fluctuations (which might be considered as $\lambda_{C,\mathbf{B}}$), and $\mathcal{M}_{A,\lambda_i} = \delta u_{\lambda_i}/\delta V_{A,\lambda_i}$ is the turbulent Alfvén Mach number for fluctuations at the injection scale. Eyink et al. (2011) and Eyink (2015) supported this picture in an alternative more mathematically rigorous way by showing that for turbulent velocity fields $\mathbf{B}$ is only frozen-in to the flow in a stochastic sense even in the limit of vanishing nonideal effects, obtaining equivalent results to those of Lazarian and Vishniac (1999).

The clearest evidence for stochastic reconnection comes from numerical simulations. Kowal et al. (2009) examined MHD simulations of a reconnecting current layer with different levels of turbulence manually injected into the system, finding $\mathcal{R}$ scaled in a manner similar to the predictions of Lazarian and Vishniac (1999). Further, while in the absence of turbulence $\mathcal{R}$ scaled with resistivity in a manner consistent with Sweet-Parker reconnection, in the presence of turbulence $\mathcal{R}$ was independent of resistivity. The structure of the reconnection layer was broadened in the presence of turbulence and made up of many thinner intense current regions, the cumulative effect of which resulted in the overall $\mathcal{R}$ (Kowal et al. 2009, 2012; Vishniac et al. 2012). Using large MHD turbulence simulations, Eyink et al. (2013) further demonstrated the Richardson-like dispersion behavior of field lines, enabling stochastic breaking of the frozen-in condition.

Clear observational evidence demonstrating stochastic reconnection has not been extensively demonstrated and, in fact, is likely difficult to obtain from the local *in situ* measurements. Lalescu et al. (2015) reported similarities between solar wind reconnection events and turbulent MHD simulations in which the stochastic field line wandering effect was demonstrated to occur - although it remains to be seen whether such qualitative similarities are unique to the proposed scenario. Tangential evidence has also been suggested for the stochastic violation of the frozen-in theorem based on statistical analyses of the Parker spiral (Eyink 2015).

A key aspect of stochastic reconnection is that it requires the existence of an extended MHD-scale inertial range into which the reconnection dynamics of interest are well coupled - that is, the lengths, widths, and thicknesses of the reconnection layers are within the MHD-scale inertial range. In the solar wind, reconnection outflows can be hundreds or even thousands of d_i (Mistry et al. 2017) and the length of the x-line (in the direction orthogonal to the quasi-2D reconnection plane) can be $10^4 d_i$ (Phan et al. 2006, 2009) - both of which are well within the MHD-scale inertial range of solar wind turbulence. The quasi-laminar collisionless reconnection viewpoint would typically treat these large length scales, particularly the width of the observed outflows, as being indicative of traversing the outflow far from the x-line (e.g., Mistry et al. 2015b). Stochastic reconnection, on the other hand, would suggest that these large length scales are driven by the stochastic wandering effect. In the solar wind, there is some evidence for complex distorted structure of reconnection outflow boundaries from multipoint observations (Mistry et al. 2015a), which may be suggestive of the stochastic reconnection picture (although not necessarily conclusive). However, it is generally challenging to tease out direct evidence distinguishing these view points from *in situ* measurements.

This requirement of an extended MHD-scale inertial range may make stochastic reconnection, most relevant to the solar wind, solar corona, and some astrophysical environments with larger dynamical ranges of fluctuations in contrast to the magnetospheric environments such as the magnetosheath, magnetotail, and magnetopause shear layer, which, while present, have more modest MHD-scale inertial ranges. Furthermore, it also does not necessarily preclude the existence of localised kinetic-scale reconnecting structures well described by collisionless reconnection dynamics embedded within the turbulent environment. As discussed in Sects. 2 – 4, clear evidence of such kinetic reconnection dynamics is found within turbulent environments, including electron-only reconnection embedded within the magnetosheath and electron diffusion regions that appear consistent with quasi-2D collisionless reconnection in turbulent magnetotail events. In this light, the further exploration of stochastic reconnection principles in the context of nonlinear collisionless effects, for example the Hall effect or relatedly advection in electron MHD, may be an interesting avenue of research for environments with less extended MHD-scale inertial ranges.

6 Conclusions

As illustrated in this review, the interaction between magnetic reconnection and turbulence is a rich field of study that can be approached from multiple viewpoints, each of which has unique nuances and applications to a variety of plasma environments. In a broad sense, the study of turbulence aims to understand the general nonlinear dynamics arising in a fluid or plasma due to the coupled interaction of a vast array of fluctuations that might consist, for example, of a mixture of waves, current sheets, plasmoids/flux ropes, vortices, and other localized structures. As a fundamental plasma process that can occur in a diverse range of settings, magnetic reconnection events have the potential to be one of these nonlinear structures, which can act to mediate the nonlinear interaction of magnetic structures and facilitate energy dissipation both in systems that are in a fully-developed turbulent state or that are developing into one. In other configurations, where magnetic reconnection is the primary process acting at a system scale current sheet, secondary processes associated with magnetic reconnection can act to excite turbulent fluctuations in the plasma, which mediate the partition of energy released by the reconnection event. Furthermore, the presence of complex turbulent fluctuations can also potentially impact how magnetic reconnection proceeds, for example, by enhancing the reconnection rate through the action of anomalous resistivity/viscosity near the x-line or through stochastic field line wandering.

While these various aspects of turbulent reconnection have been examined in a wide array of theoretical, numerical, and observational studies over the past several decades, recent high-resolution spacecraft measurements, in particular from the *Magnetospheric Multiscale* mission, have ushered in a new era of investigating this complex topic. Analyses enabled by these measurements have provided new insights and observational constraints that have spurred on a range of new theoretical and numerical investigations. Several areas of recent progress that are of particular note and that have been highlighted in this review are:

1. Recent observations of turbulence-driven magnetic reconnection in the transition region and downstream magnetosheath of Earth's bow shock. Systematic surveys of small-scale reconnection events in this environment have allowed statistical examinations of how the turbulent dynamics influence the nature of the magnetic reconnection events and the potential role that magnetic reconnection plays in the turbulent dynamics. Observations in this environment have revealed that so-called electron-only magnetic reconnection can occur when the correlation length of the turbulent magnetic field fluctuations sufficiently limits the extent of the reconnecting current sheets. This was a novel discovery that had not been extensively explored prior to the *MMS* observations, which highlights how the properties of the large-scale turbulent fluctuations can shape the dissipative processes operating in the plasma and may have implications for how the energy dissipated by turbulence is partitioned.
2. Detailed observations of turbulent x-lines and bursty bulk flow braking regions in Earth's magnetotail. These regions have revealed how turbulence facilitates the energization of high-energy non-thermal particles, how turbulence mediates the re-partitioning of energy released by magnetic reconnection both through spontaneously generated turbulence and the interaction with the surrounding environment, and demonstrated that identifiable electron diffusion region signatures, reminiscent of the quasi-laminar picture, can be found even amidst the complex turbulent fluctuations.
3. New observations of the Kelvin-Helmholtz instability on Earth's magnetopause that have both revealed clear evidence of the various forms of vortex-induced magnetic reconnection and the coupled evolution of the instability into a turbulent boundary layer, which can play a key role in the transport of mass, momentum, and energy across the magnetopause and, potentially, other velocity shear boundaries throughout the Universe.

Despite this recent progress a variety of open questions remain. The systematic identification of turbulence-driven reconnection events allows the estimation of the extent to which magnetic reconnection contributes to turbulent dissipation. However, more refined analyses require further constraints on the reconnection rate and particle heating/energisation associated with magnetic reconnection – in particular, in terms of the contrast between electron-only and ion-coupled reconnection. Both these factors could, in principle, be further examined using numerical simulations either of fully turbulent domains or well designed numerical experiments using idealised configurations. Furthermore, understanding the role that non-Maxwellian velocity phase-space structure plays in turbulent plasmas and the potential relationship with turbulence-driven magnetic reconnection remains a key area of active research (spurred on by new theoretical analyses and high-resolution spacecraft observations capable of resolving kinetic-scale phase-space dynamics) that is likely key to understanding the origin and characteristics of irreversible dissipation in turbulent astrophysical plasmas.

Another key area of advancement will be in adapting our understanding of turbulence-driven reconnection to other astrophysical environments. Addressing this point requires further understanding of the configuration of reconnecting current sheets (e.g., whether they are fragmented, "rolled-up", or otherwise deformed) and how this may depend on the Reynolds number/scale separation in the system, as well as how the prevalence of turbulence-driven magnetic reconnection is impacted by the driving, fluctuation characteristics, and ambient plasma conditions. Such factors may have an influence on the number of reconnection events per unit volume in the turbulent domain and the extent to which electron-only reconnection and ion-coupled reconnection are important in different systems. A useful path to pursue in this regard may be in contrasting turbulence-driven reconnection across the different turbulent environments that we have access to in near-Earth space, such as Earth's magnetosheath and the solar wind – particularly given the extreme contrast in effective scale separation between the driving and kinetic scales between these two environments. However, a key component of such studies would need to involve clearly assessing the extent to which reconnection events in the solar wind are related to locally generated turbulent structure rather than large-scale solar wind structure. Upcoming missions, such as NASA's *HelioSwarm* (Klein et al. 2023) which aims to specifically target the multi-scale nature of solar wind turbulence with a large swarm of spacecraft, may aid in assessing the role of turbulence-driven reconnection in the solar wind, but it may also require new technology and mission concepts capable of probing the electron scales in the solar wind to determine the extent to which electron-only reconnection may be relevant at fragmented current sheets in the solar wind (Verscharen et al. 2022). Numerical studies will also likely be a key component for probing how such dynamics may change in more exotic environments that are not directly accessible with spacecraft observations. Given the challenge in identifying 3D magnetic reconnection events in the complex magnetic environments of turbulent plasmas, machine learning or partially machine learning based approaches to systematically identifying turbulence-driven magnetic reconnection events may prove to be a useful tool.

The recent theoretical and numerical work on reconnection-mediated turbulence, which explore the potential role of reconnection in mediating the nonlinear interactions in turbulent plasmas, have also presented intriguing results that still stand to be observationally confirmed. However, further observational tests that go beyond strictly spectral slope based analyses may be necessary to unambiguously confirm the presence of this regime given the similarity of the spectral predictions to commonly observed kinetic-scale turbulent spectra that can be predicted by other means.

In the context of reconnection-driven turbulence, key areas of future research will likely come from developing a further understanding of how the turbulent processes associated with magnetic reconnection couple into and govern the energy transport of the global system. For Earth's magnetotail, such work could involve clear and systematic surveys of the extent to which the turbulent fluctuations accelerate energetic particles, particularly in the braking region where they may act to seed the energetic particles in the radiation belts, exploring the wave radiation associated with confined regions of turbulence that may be generated in reconnection outflows, and systematically quantifying how the partition of energy evolves with both time and distance from the x-line in turbulent reconnection outflows.

As well as having relevance to the magnetotail, such studies may also be relevant to reconnection in solar coronal loops, astrophysical jets, and possibly magnetopause reconnection. The impact of anomalous resistivity/viscosity and stochastic field line wandering on the magnetic reconnection rate, which is particularly useful for its application in parameterising the effects of turbulence in astrophysical environments, still require clear and unambiguous observational confirmation. The potential impacts of turbulent inflow conditions on system-scale reconnection events, such as the impact of turbulent magnetosheath flow on magnetopause reconnection, may also be an interesting and related avenue of research.

Given the recent advances in our understanding of the interplay between turbulence and magnetic reconnection and further upcoming and potential missions such as NASA's *HelioSwarm* (Klein et al. 2023) and ESA's recently proposed *Plasma Observatory* (Retinò et al. 2022) that specifically target turbulence and cross-scale coupling in the solar wind and Earth's magnetosphere, there is a bright future for the further development of our understanding of this complex problem which stands to unlock new insights into the role of magnetic reconnection throughout the Universe.

Appendix: Taylor's Hypothesis

Taylor's "frozen-in flow" hypothesis has been widely used to study the properties of space plasma fluctuations (Taylor 1938). The hypothesis states that the advection of the small-scale fluctuations over a measurement point by the large-scale background flow occurs faster than any significant dynamical evolution of those fluctuations. Therefore, a time series measured at a single point in space can be interpreted as a spatial sample through the system. Using Taylor's hypothesis, $\boldsymbol{\ell}$ is given by $\boldsymbol{\ell} = \mathbf{U}_0\tau$, where τ is a temporal lag. Similarly, the spacecraft-frame frequency (f) spectrum can be interpreted as a k spectrum with the relation $k = 2\pi f/U_0$, although the spectrum computed this way should be interpreted as a reduced spectrum averaged onto the direction of $\mathbf{U}_0$ (Horbury et al. 2012). Taylor's hypothesis is satisfied for MHD scales when $U_0 \gg V_A$ and $U_0 \gg \delta u$. This assumption is typically well satisfied in the solar wind, where background flows are super-Alfvénic; however, in other environments, such as those found in the magnetosphere, the validity is less clear. Furthermore, at sub-proton-scales, the validity may be called into question due to the faster phase speeds at those scales.

Using multi-spacecraft missions, such as *MMS* or *Cluster*, the Taylor hypothesis can be tested at the scale of the multi-spacecraft formation by comparing single-spacecraft statistics (e.g., structure functions or correlation functions) with their multi-point counterparts computed from pairs of spacecraft. For anisotropic turbulence, as expected in the presence of a strong $\mathbf{B}_0$, the correspondence between single- and multi-spacecraft statistics at a given scale may only occur along the direction of $\mathbf{U}_0$; however, the dependence on the angle between the lag and $\mathbf{B}_0$ (θ_ℓ) may provide insight into the validity of the Taylor hypothesis. Stawarz et al. (2019) compared single- and multi-spacecraft estimates of $S_2^{\mathbf{B}}(\boldsymbol{\ell})$ for *MMS* observations of turbulence in the magnetosheath, as shown in Fig. 8. For the super-Alfvénic flows (red triangles and blue sqaures), good agreement is found for both intervals, with no particular dependence on θ, demonstrating the Taylor hypothesis can be valid even down to the small (~ 6 km) separations of the *MMS* formation, which are much smaller than the proton scales, during these intervals. In contrast, the green diamonds display a systematic overestimate of the single-spacecraft estimates relative to the multi-spacecraft estimates for all six spacecraft pairs, indicating that the Taylor hypothesis may not work well during this

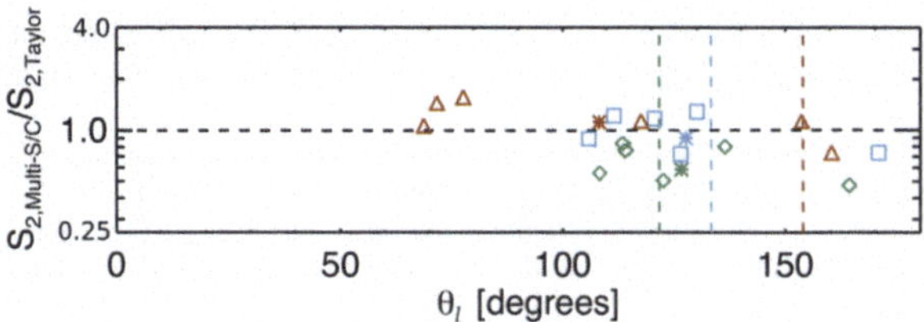

Fig. 8 Ratio of multi-spacecraft to single-spacecraft estimates of $S_2^{\mathbf{B}}(\boldsymbol{\ell})$ as a function of θ_ℓ for several intervals of magnetosheath turbulence measured by *MMS* (blue squares, red triangles, green diamonds). The six points for each interval correspond to the six spacecraft pairs in the formation. Averages of the six points are marked with asterisks, and vertical dashed lines mark the $\mathbf{U}_0$ direction (reproduced from Stawarz et al. 2019)

interval. Stawarz et al. (2022) performed a similar analysis across a range of turbulent intervals in the magnetosheath, demonstrating a clear signature of anisotropy in the turbulent fluctuations as a function of θ_ℓ with increasing strength for $\delta b_{rms}/B_0 < 1$. Good agreement was found between single- and multi-point statistics when θ_ℓ was equivalent to the angle between $\mathbf{B}_0$ and $\mathbf{U}_0$ for intervals where the Taylor hypothesis was valid. Other similar analyses have been performed across the magnetosheath and other regions of the magnetosphere demonstrating the potential validity of the Taylor hypothesis (e.g., Chasapis et al. 2017b; Parashar et al. 2018; Bandyopadhyay et al. 2020a).

Acknowledgements JES is supported by the Royal Society University Research Fellowship URF\R1\201286. R.B. was supported in part by the MMS Early Career Award NASA Grant No. 80NSSC21K1458 and NASA Grant No. 80NSSC21K0739. The authors thank the International Space Science Institute for their support of the "Magnetic Reconnection: Explosive Energy Conversion in Space Plasmas" workshop through which the contributions to this collection were coordinated and prepared.

Declarations

Competing Interests The authors have no conflicts of interest to declare.

References

Adhikari S, Shay MA, Parashar TN, et al (2020) Reconnection from a turbulence perspective. Phys Plasmas 27:042305. https://doi.org/10.1063/1.5128376

Adhikari S, Parashar TN, Shay MA, et al (2021) Energy transfer in reconnection and turbulence. Phys Rev E 104:065206. https://doi.org/10.1103/PhysRevE.104.065206

Adhikari S, Shay MA, Parashar TN, et al (2023) Effect of a guide field on the turbulence like properties of magnetic reconnection. Phys Plasmas 30:082904. https://doi.org/10.1063/5.0150929

Adhikari S, Yang Y, Matthaeus WH, et al (2024) Scale filtering analysis of kinetic reconnection and its associated turbulence. Phys Plasmas 31:020701. https://doi.org/10.1063/5.0185132

Agudelo-Rueda JA, Verscharen D, Wicks RT, et al (2021) Three-dimensional magnetic reconnection in particle-in-cell simulations of anisotropic plasma turbulence. J Plasma Phys 87:905870228. https://doi.org/10.1017/S0022377821000404

Agudelo-Rueda JA, Verscharen D, Wicks RT, et al (2022) Energy transport during 3d small-scale reconnection driven by anisotropic plasma turbulence. Astrophys J 938:4. https://doi.org/10.3847/1538-4357/ac8667

Alexakis A, Mininni PD, Pouquet A (2005) Shell-to-shell energy transfer in magnetohydrodynamics. I. Steady state turbulence. Phys Rev E 72:046301. https://doi.org/10.1103/PhysRevE.72.046301

Alexandrova O, Lacombe C, Mangeney A (2008) Spectra and anisotropy of magnetic fluctuations in the Earth's magnetosheath: cluster observations. Ann Geophys 26:3585–3596. https://doi.org/10.5194/angeo-26-3585-2008

Aluie H (2017) Coarse-grained incompressible magnetohydrodynamics: analyzing the turbulent cascades. New J Phys 19:025008. https://doi.org/10.1088/1367-2630/aa5d2f

Anselmet F, Gagne Y, Hopfinger EJ, Antonia RA (1984) High-order velocity structure functions in turbulent shear flows. J Fluid Mech 140:63–89. https://doi.org/10.1017/S0022112084000513

Arró G, Califano F, Lapenta G (2020) Statistical properties of turbulent fluctuations associated with electron-only magnetic reconnection. Astron Astrophys 642:A45. https://doi.org/10.1051/0004-6361/202038696

Azizabadi AC, Jain N, Büchner J (2021) Identification and characterization of current sheets in collisionless plasma turbulence. Phys Plasmas 28(5):052904. https://doi.org/10.1063/5.0040692

Bandyopadhyay R, Chasapis A, Chhiber R, et al (2018) Incompressive energy transfer in the Earth's magnetosheath: magnetospheric multiscale observations. Astrophys J 866:106. https://doi.org/10.3847/1538-4357/aade04

Bandyopadhyay R, Chasapis A, Gershman DJ, et al (2020a) Observation of an inertial-range energy cascade within a reconnection jet in the Earth's magnetotail. Mon Not R Astron Soc Lett 500:L6–L10. https://doi.org/10.1093/mnrasl/slaa171

Bandyopadhyay R, Sorriso-Valvo L, Aaea C (2020b) In situ observation of Hall magnetohydrodynamic cascade in space plasma. Phys Rev Lett 124:225101. https://doi.org/10.1103/PhysRevLett.124.225101

Bandyopadhyay R, Chasapis A, Gershman DJ, et al (2021a) Observation of an inertial-range energy cascade within a reconnection jet in the Earth's magnetotail. Mon Not R Astron Soc 500:L6–L10. https://doi.org/10.1093/mnrasl/slaa171

Bandyopadhyay R, Chasapis A, Matthaeus WH, et al (2021b) Energy dissipation in turbulent reconnection. Phys Plasmas 28:112305. https://doi.org/10.1063/5.0071015

Bandyopadhyay R, Qudsi RA, Gary SP, et al (2022) Interplay of turbulence and proton-microinstability growth in space plasmas. Phys Plasmas 29:102107. https://doi.org/10.1063/5.0098625

Banerjee S, Galtier S (2013) Exact relation with two-point correlation functions and phenomenological approach for compressible magnetohydrodynamic turbulence. Phys Rev E 87:013019. https://doi.org/10.1103/PhysRevE.87.013019

Banerjee S, Galtier S (2017) An alternative formulation for exact scaling relations in hydrodynamic and magnetohydrodynamic turbulence. J Phys A 50:015501. https://doi.org/10.1088/1751-8113/50/1/015501

Bárta M, Büchner J, Karlický M, Skála J (2011) Spontaneous current-layer fragmentation and cascading reconnection in solar flares. I. Model and analysis. Astrophys J 737:24. https://doi.org/10.1088/0004-637X/737/1/24

Bauer TM, Baumjohann W, Treumann RA, et al (1995) Low-frequency waves in the near-Earth plasma sheet. J Geophys Res 100:9605–9618. https://doi.org/10.1029/95JA00136

Bergstedt K, Ji H, Jara-Almonte J, et al (2020) Statistical properties of magnetic structures and energy dissipation during turbulent reconnection in the Earth's magnetotail. Geophys Res Lett 47:e88540. https://doi.org/10.1029/2020GL088540

Bessho N, Chen LJ, Wang S, et al (2019) Magnetic reconnection in a quasi-parallel shock: two-dimensional local particle-in-cell simulation. Geophys Res Lett 46:9352–9361. https://doi.org/10.1029/2019GL083397

Bessho N, Chen LJ, Wang S, et al (2020) Magnetic reconnection and kinetic waves generated in the Earth's quasi-parallel bow shock. Phys Plasmas 27:092901. https://doi.org/10.1063/5.0012443

Bessho N, Chen LJ, Stawarz JE, et al (2022) Strong reconnection electric fields in shock-driven turbulence. Phys Plasmas 29:042304. https://doi.org/10.1063/5.0077529

Bessho N, Chen LJ, Hesse M, Ng J, Wilson LB, Stawarz JE (2023) Electron acceleration and heating during magnetic reconnection in the Earth's quasi-parallel bow shock. Astrophys J 954:25. https://doi.org/10.3847/1538-4357/ace321

Biskamp D, Müller WC (2000) Scaling properties of three-dimensional isotropic magnetohydrodynamic turbulence. Phys Plasmas 7:4889–4900. https://doi.org/10.1063/1.1322562

Biskamp D, Welter H (1989) Dynamics of decaying two-dimensional magnetohydrodynamic turbulence. Phys Fluids 1:1964–1979. https://doi.org/10.1063/1.859060

Biskamp D, Schwarz E, Zeiler A, et al (1999) Electron magnetohydrodynamic turbulence. Phys Plasmas 6:751–758. https://doi.org/10.1063/1.873312

Blasl KA, Nakamura TKM, Plaschke F, et al (2022) Multi-scale observations of the magnetopause Kelvin-Helmholtz waves during southward IMF. Phys Plasmas 29:012105. https://doi.org/10.1063/5.0067370
Bohdan A, Pohl M, Niemiec J, et al (2020) Kinetic simulations of nonrelativistic perpendicular shocks of young supernova remnants. III. Magnetic reconnection. Astrophys J 893:6. https://doi.org/10.3847/1538-4357/ab7cd6
Boldyrev S (2006) Spectrum of magnetohydrodynamic turbulence. Phys Rev Lett 96(11):115002. https://doi.org/10.1103/PhysRevLett.96.115002
Boldyrev S, Loureiro NF (2019) Role of reconnection in inertial kinetic-Alfvén turbulence. Phys Rev Res 1:012006. https://doi.org/10.1103/PhysRevResearch.1.012006
Boldyrev S, Loureiro NF (2020) Tearing instability in Alfvén and Kinetic-Alfvén turbulence. J Geophys Res 125:e28185. https://doi.org/10.1029/2020JA028185
Boldyrev S, Perez JC (2012) Spectrum of Kinetic-Alfvén turbulence. Astrophys J Lett 758:L44. https://doi.org/10.1088/2041-8205/758/2/L44
Boldyrev S, Horaites K, Xia Q, Perez JC (2013) Toward a theory of astrophysical plasma turbulence at subproton scales. Astrophys J 777:41. https://doi.org/10.1088/0004-637X/777/1/41
Borgogno D, Califano F, Faganello M, Pegoraro F (2015) Double-reconnected magnetic structures driven by Kelvin-Helmholtz vortices at the Earth's magnetosphere. Phys Plasmas 22:032301. https://doi.org/10.1063/1.4913578
Borovsky JE, Funsten HO (2003) MHD turbulence in the Earth's plasma sheet: dynamics, dissipation, and driving. J Geophys Res 108:1284. https://doi.org/10.1029/2002JA009625
Borovsky JE, Elphic RC, Funsten HO, Thomsen MF (1997) The Earth's plasma sheet as a laboratory for flow turbulence in high-β MHD. J Plasma Phys 57:1–34. https://doi.org/10.1017/S0022377896005259
Bowen TA, Mallet A, Bale SD, et al (2020) Constraining ion-scale heating and spectral energy transfer in observations of plasma turbulence. Phys Rev Lett 125:025102. https://doi.org/10.1103/PhysRevLett.125.025102
Califano F, Cerri SS, Faganello M, et al (2020) Electron-only reconnection in plasma turbulence. Front Phys 8. https://doi.org/10.3389/fphy.2020.00317
Camporeale E, Sorriso-Valvo L, Califano F, Retinò A (2018) Coherent structures and spectral energy transfer in turbulent plasma: a space-filter approach. Phys Rev Lett 120:125101. https://doi.org/10.1103/PhysRevLett.120.125101
Carbone V, Veltri P, Mangeney A (1990) Coherent structure formation and magnetic field line reconnection in magnetohydrodynamic turbulence. Phys Fluids 2:1487–1496. https://doi.org/10.1063/1.857598
Cattell C, Dombeck J, Wygant J, et al (2005) Cluster observations of electron holes in association with magnetotail reconnection and comparison to simulations. J Geophys Res 110:A01211. https://doi.org/10.1029/2004JA010519
Cerri SS, Califano F (2017) Reconnection and small-scale fields in 2D-3V hybrid-kinetic driven turbulence simulations. New J Phys 19:025007. https://doi.org/10.1088/1367-2630/aa5c4a
Chandran BDG, Schekochihin AA, Mallet A (2015) Intermittency and alignment in strong RMHD turbulence. Astrophys J 807:39. https://doi.org/10.1088/0004-637X/807/1/39
Chasapis A, Matthaeus WH, Parashar TN, et al (2017a) Electron heating at kinetic scales in magnetosheath turbulence. Astrophys J 836:247. https://doi.org/10.3847/1538-4357/836/2/247
Chasapis A, Matthaeus WH, Parashar TN, et al (2017b) High-resolution statistics of solar wind turbulence at kinetic scales using the magnetospheric multiscale mission. Astrophys J Lett 844:L9. https://doi.org/10.3847/2041-8213/aa7ddd
Chasapis A, Matthaeus WH, Parashar TN, et al (2018) In situ observation of intermittent dissipation at kinetic scales in the Earth's magnetosheath. Astrophys J Lett 856:L19. https://doi.org/10.3847/2041-8213/aaadf8
Chaston CC, Bonnell JW, Clausen L, Angelopoulos V (2012) Energy transport by kinetic-scale electromagnetic waves in fast plasma sheet flows. J Geophys Res 117:A09202. https://doi.org/10.1029/2012JA017863
Chaston CC, Bonnell JW, Salem C (2014) Heating of the plasma sheet by broadband electromagnetic waves. Geophys Res Lett 41:8185–8192. https://doi.org/10.1002/2014GL062116
Che H (2017) How anomalous resistivity accelerates magnetic reconnection. Phys Plasmas 24:082115. https://doi.org/10.1063/1.5000071
Che H, Drake JF, Swisdak M (2011) A current filamentation mechanism for breaking magnetic field lines during reconnection. Nature 474:184–187. https://doi.org/10.1038/nature10091
Chen CHK (2016) Recent progress in astrophysical plasma turbulence from solar wind observations. J Plasma Phys 82:535820602. https://doi.org/10.1017/S0022377816001124
Chen CHK, Boldyrev S (2017) Nature of kinetic scale turbulence in the Earth's magnetosheath. Astrophys J 842:122. https://doi.org/10.3847/1538-4357/aa74e0

Chen XL, Morrison PJ (1990) Resistive tearing instability with equilibrium shear flow. Phys Fluids B 2:495–507. https://doi.org/10.1063/1.859339
Chen LJ, Hesse M, Wang S, et al (2016) Electron energization and structure of the diffusion region during asymmetric reconnection. Geophys Res Lett 43:2405–2412. https://doi.org/10.1002/2016GL068243
Chhiber R, Chasapis A, Bandyopadhyay R, et al (2018) Higher-order turbulence statistics in the Earth's magnetosheath and the solar wind using magnetospheric multiscale observations. J Geophys Res 123:9941–9954. https://doi.org/10.1029/2018JA025768
Consolini G, Alberti T, Benella S, et al (2023) On the fractal pattern of the current structure at ion scales in turbulent space plasmas. Chaos Solitons Fractals 177:114253
Cowley SWH (1982) The causes of convection in the Earth's magnetosphere: a review of developments during the IMS (Paper 2R0568). Rev Geophys Space Phys 20:531. https://doi.org/10.1029/RG020i003p00531
Cozzani G, Khotyaintsev YV, Graham DB André M (2023) Direct observations of electron firehose fluctuations in the magnetic reconnection outflow. J Geophys Res 128:e2022JA031128. https://doi.org/10.1029/2022JA031128
Daughton W, Roytershteyn V, Karimabadi H, et al (2011) Role of electron physics in the development of turbulent magnetic reconnection in collisionless plasmas. Nat Phys 7:539–542. https://doi.org/10.1038/nphys1965
Di Mare F, Sorriso-Valvo L, Retinò A, et al (2019) Evolution of turbulence in the Kelvin–Helmholtz instability in the terrestrial magnetopause. Atmosphere 10:561. https://doi.org/10.3390/atmos10090561
Divin A, Khotyaintsev YV, Vaivads A, et al (2015) Evolution of the lower hybrid drift instability at reconnection jet front. J Geophys Res 120:2675. https://doi.org/10.1002/2014JA020503
Dmitruk P, Matthaeus WH, Seenu N (2004) Test particle energization by current sheets and nonuniform fields in magnetohydrodynamic turbulence. Astrophys J 617:667–679. https://doi.org/10.1086/425301
Donato S, Servidio S, Dmitruk P, et al (2012) Reconnection events in two-dimensional Hall magnetohydrodynamic turbulence. Phys Plasmas 19:092307. https://doi.org/10.1063/1.4754151
Dong C, Wang L, Huang YM, et al (2018) Role of the plasmoid instability in magnetohydrodynamic turbulence. Phys Rev Lett 121:165101. https://doi.org/10.1103/PhysRevLett.121.165101
Dong C, Wang L, Huang YM, et al (2022) Reconnection-driven energy cascade in magnetohydrodynamic turbulence. Sci Adv 8:eabn7627. https://doi.org/10.1126/sciadv.abn7627
Drake JF, Swisdak M, Cattell C, et al (2003) Formation of electron holes and particle energization during magnetic reconnection. Science 299:873–877. https://doi.org/10.1126/science.1080333
Drake JF, Swisdak M, Phan TD, et al (2009) Ion heating resulting from pickup in magnetic reconnection exhausts. J Geophys Res 114:A05111. https://doi.org/10.1029/2008JA013701
Eastwood JP, Phan TD, Bale SD, Tjulin A (2009) Observations of turbulence generated by magnetic reconnection. Phys Rev Lett 102:035001. https://doi.org/10.1103/PhysRevLett.102.035001
Eastwood JP, Stawarz JE, Phan TD, et al (2021) Solar Orbiter observations of an ion-scale flux rope confined to a bifurcated solar wind current sheet. Astron Astrophys 656:A27. https://doi.org/10.1051/0004-6361/202140949
Egedal J, Le A, Daughton W (2013) A review of pressure anisotropy caused by electron trapping in collisionless plasma, and its implications for magnetic reconnection. Phys Plasmas 20:061201. https://doi.org/10.1063/1.4811092
Egedal J, Le A, Daughton W, et al (2016) Spacecraft observations and analytic theory of crescent-shaped electron distributions in asymmetric magnetic reconnection. Phys Rev Lett 117:185101. https://doi.org/10.1103/PhysRevLett.117.185101
Ergun RE, Goodrich KA, Stawarz JE, et al (2015) Large-amplitude electric fields associated with bursty bulk flow braking in the Earth's plasma sheet. J Geophys Res 120:1832–1844. https://doi.org/10.1002/2014JA020165
Ergun RE, Chen LJ, Wilder FD, et al (2017) Drift waves, intense parallel electric fields, and turbulence associated with asymmetric magnetic reconnection at the magnetopause. Geophys Res Lett 44:2978–2986. https://doi.org/10.1002/2016GL072493
Ergun RE, Goodrich KA, Wilder FD, et al (2018) Magnetic reconnection, turbulence, and particle acceleration: observations in the Earth's magnetotail. Geophys Res Lett 45:3338–3347. https://doi.org/10.1002/2018GL076993
Ergun RE, Ahmadi N, Kromyda L, et al (2020a) Observations of particle acceleration in magnetic reconnection-driven turbulence. Astrophys J 898:154. https://doi.org/10.3847/1538-4357/ab9ab6
Ergun RE, Ahmadi N, Kromyda L, et al (2020b) Particle acceleration in strong turbulence in the Earth's magnetotail. Astrophys J 898:153. https://doi.org/10.3847/1538-4357/ab9ab5
Ergun RE, Pathak N, Usanova ME, et al (2022a) Observation of magnetic reconnection in a region of strong turbulence. Astrophys J Lett 935:L8. https://doi.org/10.3847/2041-8213/ac81d4

Ergun RE, Usanova ME, Turner DL, Stawarz JE (2022b) Bursty bulk flow turbulence as a source of energetic particles to the outer radiation belt. Geophys Res Lett 49:e98113. https://doi.org/10.1029/2022GL098113

Eriksson S, Lavraud B, Wilder FD, et al (2016a) Magnetospheric multiscale observations of magnetic reconnection associated with Kelvin-Helmholtz waves. Geophys Res Lett 43:5606–5615. https://doi.org/10.1002/2016GL068783

Eriksson S, Wilder FD, Ergun RE, et al (2016b) Magnetospheric multiscale observations of the electron diffusion region of large guide field magnetic reconnection. Phys Rev Lett 117:015001. https://doi.org/10.1103/PhysRevLett.117.015001

Eriksson S, Ma X, Burch JL, et al (2021) MMS observations of double mid-latitude reconnection ion beams in the early non-linear phase of the Kelvin-Helmholtz instability. Front Astron Space Sci 8:188. https://doi.org/10.3389/fspas.2021.760885

Eriksson S, Swisdak M, Weygand JM, et al (2022) Characteristics of multi-scale current sheets in the solar wind at 1 au associated with magnetic reconnection and the case for a heliospheric current sheet avalanche. Astrophys J 933:181. https://doi.org/10.3847/1538-4357/ac73f6

Eriksson S, Swisdak M, Mallet A, et al (2024) Parker Solar Probe observations of magnetic reconnection exhausts in quiescent plasmas near the sun. Astrophys J 965:76. https://doi.org/10.3847/1538-4357/ad25f0

Eyink GL (2015) Turbulent general magnetic reconnection. Astrophys J 807:137. https://doi.org/10.1088/0004-637X/807/2/137

Eyink GL, Lazarian A, Vishniac ET (2011) Fast magnetic reconnection and spontaneous stochasticity. Astrophys J 743:51. https://doi.org/10.1088/0004-637X/743/1/51

Eyink G, Vishniac E, Lalescu C, et al (2013) Flux-freezing breakdown in high-conductivity magnetohydrodynamic turbulence. Nature 497:466–469. https://doi.org/10.1038/nature12128

Fadanelli S, Lavraud B, Califano F, et al (2021) Energy conversions associated with magnetic reconnection. J Geophys Res 126:e2020JA028333. https://doi.org/10.1029/2020JA028333

Faganello M, Califano F, Pegoraro F (2008) Numerical evidence of undriven, fast reconnection in the solar-wind interaction with Earth's magnetosphere: formation of electromagnetic coherent structures. Phys Rev Lett 101:105001. https://doi.org/10.1103/PhysRevLett.101.105001

Faganello M, Pegoraro F, Califano F, Marradi L (2010) Collisionless magnetic reconnection in the presence of a sheared velocity field. Phys Plasmas 17:062102. https://doi.org/10.1063/1.3430640

Faganello M, Califano F, Pegoraro F, et al (2012a) Magnetic reconnection and Kelvin-Helmholtz instabilities at the Earth's magnetopause. Plasma Phys Control Fusion 54:124037. https://doi.org/10.1088/0741-3335/54/12/124037

Faganello M, Califano F, Pegoraro F, Andreussi T (2012b) Double mid-latitude dynamical reconnection at the magnetopause: an efficient mechanism allowing solar wind to enter the Earth's magnetosphere. Europhys Lett 100:69001. https://doi.org/10.1209/0295-5075/100/69001

Faganello M, Califano F, Pegoraro F, Retinò A (2014) Kelvin-Helmholtz vortices and double mid-latitude reconnection at the Earth's magnetopause: comparison between observations and simulations. Europhys Lett 107:19001. https://doi.org/10.1209/0295-5075/107/19001

Faganello M, Sisti M, Califano F, Lavraud B (2022) Kelvin-Helmholtz instability and induced magnetic reconnection at the Earth's magnetopause: a 3D simulation based on satellite data. Plasma Phys Control Fusion 64:044014. https://doi.org/10.1088/1361-6587/ac43f0

Fargette N, Lavraud B, Rouillard AP, et al (2023) Clustering of magnetic reconnection exhausts in the solar wind: an automated detection study. Astron Astrophys 674:A98. https://doi.org/10.1051/0004-6361/202346043

Fermo RL, Drake JF, Swisdak M (2012) Secondary magnetic islands generated by the Kelvin-Helmholtz instability in a reconnecting current sheet. Phys Rev Lett 108:255005. https://doi.org/10.1103/PhysRevLett.108.255005

Ferrand R, Galtier S, Sahraoui F, et al (2019) On exact laws in incompressible Hall magnetohydrodynamic turbulence. Astrophys J 881:50. https://doi.org/10.3847/1538-4357/ab2be9

Franci L, Verdini A, Matteini L, et al (2015) Solar wind turbulence from MHD to sub-ion scales: high-resolution hybrid simulations. Astrophys J Lett 804:L39. https://doi.org/10.1088/2041-8205/804/2/L39

Franci L, Cerri SS, Califano F, et al (2017) Magnetic reconnection as a driver for a sub-ion-scale cascade in plasma turbulence. Astrophys J Lett 850:L16. https://doi.org/10.3847/2041-8213/aa93fb

Franci L, Stawarz JE, Papini E, et al (2020) Modeling MMS observations at the Earth's magnetopause with hybrid simulations of Alfvénic turbulence. Astrophys J 898:175. https://doi.org/10.3847/1538-4357/ab9a47

Franci L, Papini E, Micera A, et al (2022) Anisotropic electron heating in turbulence-driven magnetic reconnection in the near-sun solar wind. Astrophys J 936:27. https://doi.org/10.3847/1538-4357/ac7da6

Frisch U (1995) Turbulence. The legacy of A.N. Kolmogorov. Cambridge University Press, Cambridge

Fujimoto K (2014) Wave activities in separatrix regions of magnetic reconnection. Geophys Res Lett 41:2721–2728. https://doi.org/10.1002/2014GL059893
Fujimoto K, Sydora RD (2012) Plasmoid-induced turbulence in collisionless magnetic reconnection. Phys Rev Lett 109:265004. https://doi.org/10.1103/PhysRevLett.109.265004
Fujimoto K, Sydora RD (2021) Electromagnetic turbulence in the electron current layer to drive magnetic reconnection. Astrophys J Lett 909:L15. https://doi.org/10.3847/2041-8213/abe877
Fujimoto K, Sydora RD (2023) The electron diffusion region dominated by electromagnetic turbulence in the reconnection current layer. Phys Plasmas 30. https://doi.org/10.1063/5.0129591
Fuselier SA, Petrinec SM, Trattner KJ, Lavraud B (2014) Magnetic field topology for northward IMF reconnection: ion observations. J Geophys Res 119:9051–9071. https://doi.org/10.1002/2014JA020351
Fuselier SA, Petrinec SM, Reiff PH, et al (2024) Global-scale processes and effects of magnetic reconnection on the geospace environment. Space Sci Rev 220:34. https://doi.org/10.1007/s11214-024-01067-0
Galtier S (2023) Physics of wave turbulence. Cambridge University Press, Cambridge. https://doi.org/10.1017/9781009275880
Galtier S, Nazarenko SV, Newell AC, Pouquet A (2000) A weak turbulence theory for incompressible magnetohydrodynamics. J Plasma Phys 63:447–488. https://doi.org/10.1017/S0022377899008284
Gingell I, Sorriso-Valvo L, Burgess D, et al (2017) Three-dimensional simulations of sheared current sheets: transition to turbulence? J Plasma Phys 83:705830104. https://doi.org/10.1017/S0022377817000058
Gingell I, Schwartz SJ, Eastwood JP, et al (2019) Observations of magnetic reconnection in the transition region of quasi-parallel shocks. Geophys Res Lett 46:1177–1184. https://doi.org/10.1029/2018GL081804.
Gingell I, Schwartz SJ, Eastwood JP, et al (2020) Statistics of reconnecting current sheets in the transition region of Earth's bow shock. J Geophys Res 125:e2019JA027119. https://doi.org/10.1029/2019JA027119
Gingell I, Schwartz SJ, Kucharek H, et al (2021) Observing the prevalence of thin current sheets downstream of Earth's bow shock. Phys Plasmas 28:102902. https://doi.org/10.1063/5.0062520
Gingell I, Schwartz SJ, Kucharek H, et al (2023) Hybrid simulations of the decay of reconnected structures downstream of the bow shock. Phys Plasmas 30:012902. https://doi.org/10.1063/5.0129084
Goldman M, Newman DL, Lapenta G, et al (2014) Čerenkov emission of quasiparallel whistlers by fast electron phase-space holes during magnetic reconnection. Phys Rev Lett 112:145002. https://doi.org/10.1103/PhysRevLett.112.145002
Goldreich P, Sridhar S (1995) Toward a theory of interstellar turbulence. II. Strong alfvenic turbulence. Astrophys J 438:763. https://doi.org/10.1086/175121
Gosling JT (2007) Observations of magnetic reconnection in the turbulent high-speed solar wind. Astrophys J Lett 671:L73–L76. https://doi.org/10.1086/524842
Gosling JT, Szabo A (2008) Bifurcated current sheets produced by magnetic reconnection in the solar wind. J Geophys Res 113:A10103. https://doi.org/10.1029/2008JA013473
Gosling JT, Skoug RM, McComas DJ, Smith CW (2005) Direct evidence for magnetic reconnection in the solar wind near 1 AU. J Geophys Res 110:A01107. https://doi.org/10.1029/2004JA010809
Graham DB, Cozzani G, et al (2024) The role of kinetic instabilities and waves in collisionless magnetic reconnection. Space Sci Rev
Greco A, Matthaeus WH, Servidio S, et al (2009) Statistical analysis of discontinuities in solar wind ACE data and comparison with intermittent MHD turbulence. Astrophys J Lett 691:L111–L114. https://doi.org/10.1088/0004-637X/691/2/L111
Greco A, Valentini F, Servidio S, Matthaeus WH (2012) Inhomogeneous kinetic effects related to intermittent magnetic discontinuities. Phys Rev E 86:066405. https://doi.org/10.1103/PhysRevE.86.066405
Greco A, Perri S, Servidio S, et al (2016) The complex structure of magnetic field discontinuities in the turbulent solar wind. Astrophys J Lett 823:L39. https://doi.org/10.3847/2041-8205/823/2/L39
Grete P, O'Shea BW, Beckwith K, et al (2017) Energy transfer in compressible magnetohydrodynamic turbulence. Phys Plasmas 24:092311. https://doi.org/10.1063/1.4990613
Guo A, Lu Q, Lu S, et al (2023) Properties of electron-scale magnetic reconnection at a quasi-perpendicular shock. Astrophys J 955:14. https://doi.org/10.3847/1538-4357/acec48
Hadid LZ, Sahraoui F, Galtier S, Huang SY (2018) Compressible magnetohydrodynamic turbulence in the Earth's magnetosheath: estimation of the energy cascade rate using in situ spacecraft data. Phys Rev Lett 120(5):055102. https://doi.org/10.1103/PhysRevLett.120.055102
Haggerty CC, Parashar TN, Matthaeus WH, et al (2017) Exploring the statistics of magnetic reconnection X-points in kinetic particle-in-cell turbulence. Phys Plasmas 24:102308. https://doi.org/10.1063/1.5001722
Hansel PJ, Wilder FD, Malaspina DM, et al (2021) Mapping MMS observations of solitary waves in Earth's magnetic field. J Geophys Res 126:e29389. https://doi.org/10.1029/2021JA02938910.1002/essoar.10506873.1

Hasegawa H, Fujimoto M, Phan TD, et al (2004) Transport of solar wind into Earth's magnetosphere through rolled-up Kelvin-Helmholtz vortices. Nature 430:755–758. https://doi.org/10.1038/nature02799
Hasegawa H, Retinò A, Vaivads A, et al (2009) Kelvin-Helmholtz waves at the Earth's magnetopause: multiscale development and associated reconnection. J Geophys Res 114:A12207. https://doi.org/10.1029/2009JA014042
Hasegawa H, Nakamura TKM, Gershman DJ, et al (2020) Generation of turbulence in Kelvin-Helmholtz vortices at the Earth's magnetopause: magnetospheric multiscale observations. J Geophys Res 125:e27595. https://doi.org/10.1029/2019JA02759510.1002/essoar.10501080.2
Haynes CT, Burgess D, Camporeale E (2014) Reconnection and electron temperature anisotropy in sub-proton scale plasma turbulence. Astrophys J 783:38. https://doi.org/10.1088/0004-637X/783/1/38
Hellinger P, Matteini L, Landi S, et al (2015) Plasma turbulence and kinetic instabilities at ion scales in the expanding solar wind. Astrophys J Lett 811:L32. https://doi.org/10.1088/2041-8205/811/2/L32
Hellinger P, Verdini A, Landi S, et al (2018) von Kármán-Howarth equation for Hall magnetohydrodynamics: hybrid simulations. Astrophys J Lett 857:L19. https://doi.org/10.3847/2041-8213/aabc06
Hesse M, Aunai N, Sibeck D, Birn J (2014) On the electron diffusion region in planar, asymmetric, systems. Geophys Res Lett 41:8673–8680. https://doi.org/10.1002/2014GL061586
Horbury TS, Wicks RT, Chen CHK (2012) Anisotropy in space plasma turbulence: solar wind observations. Space Sci Rev 172:325–342. https://doi.org/10.1007/s11214-011-9821-9
Hou C, He J, Zhu X, Wang Y (2021) Contribution of magnetic reconnection events to energy dissipation in space plasma turbulence. Astrophys J 908:237. https://doi.org/10.3847/1538-4357/abd6f3
Howes GG (2016) The dynamical generation of current sheets in astrophysical plasma turbulence. Astrophys J Lett 827:L28. https://doi.org/10.3847/2041-8205/827/2/L28
Hu A, Sisti M, Finelli F, et al (2020) Identifying magnetic reconnection in 2d hybrid Vlasov Maxwell simulations with convolutional neural networks. Astrophys J 900:86. https://doi.org/10.3847/1538-4357/aba527
Huang YM, Bhattacharjee A (2016) Turbulent magnetohydrodynamic reconnection mediated by the plasmoid instability. Astrophys J 818:20. https://doi.org/10.3847/0004-637X/818/1/20
Huang SY, Zhou M, Sahraoui F, et al (2012) Observations of turbulence within reconnection jet in the presence of guide field. Geophys Res Lett 39:L11104. https://doi.org/10.1029/2012GL052210
Huang SY, Sahraoui F, Deng XH, et al (2014) Kinetic turbulence in the terrestrial magnetosheath: cluster observations. Astrophys J Lett 789:L28. https://doi.org/10.1088/2041-8205/789/2/L28
Huang SY, Hadid LZ, Sahraoui F, et al (2017) On the existence of the Kolmogorov inertial range in the terrestrial magnetosheath turbulence. Astrophys J Lett 836:L10. https://doi.org/10.3847/2041-8213/836/1/L10
Huang SY, Zhang J, Yuan ZG, et al (2022) Intermittent dissipation at kinetic scales in the turbulent reconnection outflow. Geophys Res Lett 49:e96403. https://doi.org/10.1029/2021GL096403
Huang SY, Zhang J, Xiong QY, et al (2023) Kinetic-scale topological structures associated with energy dissipation in the turbulent reconnection outflow. Astrophys J 958:189. https://doi.org/10.3847/1538-4357/acf847
Iroshnikov PS (1964) Turbulence of a conducting fluid in a strong magnetic field. Sov Astron 7:566
Jain N, Muñoz PA, Tabriz MF, et al (2022) Importance of accurate consideration of the electron inertia in hybrid-kinetic simulations of collisionless plasma turbulence: the 2d limit. Phys Plasmas 29:053902. https://doi.org/10.1063/5.0087103
Jara-Almonte J, Daughton W, Ji H (2014) Debye scale turbulence within the electron diffusion layer during magnetic reconnection. Phys Plasmas 21:032114. https://doi.org/10.1063/1.4867868
Jiang K, Huang SY, Fu HS, et al (2021) Observational evidence of magnetic reconnection in the terrestrial foreshock region. Astrophys J 922:56. https://doi.org/10.3847/1538-4357/ac2500
Jin R, Zhou M, Pang Y, et al (2022) Characteristics of turbulence driven by transient magnetic reconnection in the terrestrial magnetotail. Astrophys J 925:17. https://doi.org/10.3847/1538-4357/ac390c
Johnson JR, Wing S (2009) Northward interplanetary magnetic field plasma sheet entropies. J Geophys Res 114:A00D08. https://doi.org/10.1029/2008JA014017
Karimabadi H, Roytershteyn V, Wan M, et al (2013) Coherent structures, intermittent turbulence, and dissipation in high-temperature plasmas. Phys Plasmas 20:012303. https://doi.org/10.1063/1.4773205
Karimabadi H, Roytershteyn V, Vu HX, et al (2014) The link between shocks, turbulence, and magnetic reconnection in collisionless plasmas. Phys Plasmas 21:062308. https://doi.org/10.1063/1.4882875
Kiyani KH, Chapman SC, Khotyaintsev YV, et al (2009) Global scale-invariant dissipation in collisionless plasma turbulence. Phys Rev Lett 103:075006. https://doi.org/10.1103/PhysRevLett.103.075006
Klein KG, Spence H, Alexandrova O, et al (2023) HelioSwarm: a multipoint, multiscale mission to characterize turbulence. Space Sci Rev 219:74. https://doi.org/10.1007/s11214-023-01019-0
Knoll DA, Chacón L (2002) Magnetic reconnection in the two-dimensional Kelvin-Helmholtz instability. Phys Rev Lett 88:215003. https://doi.org/10.1103/PhysRevLett.88.215003

Kolmogorov A (1941a) The local structure of turbulence in incompressible viscous fluid for very large Reynolds' numbers. Akademiia Nauk SSSR Doklady 30:301–305. [reprinted in (1991) Proc. R. Soc. A, 434:9-13]. https://doi.org/10.1098/rspa.1991.0075

Kolmogorov AN (1941b) Dissipation of energy in locally isotropic turbulence. Akademiia Nauk SSSR Doklady 32:16. [reprinted in (1991) Proc. R. Soc. A, 434:15–17 https://doi.org/10.1098/rspa.1991.0076]

Kowal G, Lazarian A, Vishniac ET, Otmianowska-Mazur K (2009) Numerical tests of fast reconnection in weakly stochastic magnetic fields. Astrophys J 700:63–85. https://doi.org/10.1088/0004-637X/700/1/63

Kowal G, de Gouveia Dal Pino EM, Lazarian A (2012) Particle acceleration in turbulence and weakly stochastic reconnection. Phys Rev Lett 108:241102. https://doi.org/10.1103/PhysRevLett.108.241102

Kraichnan RH (1965) Inertial-range spectrum of hydromagnetic turbulence. Phys Fluids 8(7):1385–1387. https://doi.org/10.1063/1.1761412

Lalescu CC, Shi YK, Eyink GL, et al (2015) Inertial-range reconnection in magnetohydrodynamic turbulence and in the solar wind. Phys Rev Lett 115:025001. https://doi.org/10.1103/PhysRevLett.115.025001

Lapenta G, Markidis S, Divin A, et al (2011) Bipolar electric field signatures of reconnection separatrices for a hydrogen plasma at realistic guide fields. Geophys Res Lett 38:L17104. https://doi.org/10.1029/2011GL048572

Lapenta G, Markidis S, Goldman MV, Newman DL (2015) Secondary reconnection sites in reconnection-generated flux ropes and reconnection fronts. Nat Phys 11:690–695. https://doi.org/10.1038/nphys3406

Lapenta G, Berchem J, Alaoui ME, Walker R (2020a) Turbulent energization of electron power law tails during magnetic reconnection. Phys Rev Lett 125:225101. https://doi.org/10.1103/PhysRevLett.125.225101

Lapenta G, Pucci F, Goldman MV, Newman DL (2020b) Local regimes of turbulence in 3d magnetic reconnection. Astrophys J 888:104. https://doi.org/10.3847/1538-4357/ab5a86

Lapenta G, Goldman M, Newman DL, Eriksson S (2022) Formation and reconnection of electron scale current layers in the turbulent outflows of a primary reconnection site. Astrophys J 940:187. https://doi.org/10.3847/1538-4357/ac98bc

Lazarian A, Vishniac ET (1999) Reconnection in a weakly stochastic field. Astrophys J 517:700–718. https://doi.org/10.1086/307233

Lazarian A, Vlahos L, Kowal G, et al (2012) Turbulence, magnetic reconnection in turbulent fluids and energetic particle acceleration. Space Sci Rev 173:557–622. https://doi.org/10.1007/s11214-012-9936-7

Lazarian A, Eyink G, Vishniac E, Kowal G (2015) Turbulent reconnection and its implications. Philos Trans R Soc A 373:20140144–20140144. https://doi.org/10.1098/rsta.2014.0144

Lazarian A, Eyink GL, Jafari A, et al (2020) 3D turbulent reconnection: theory, tests, and astrophysical implications. Phys Plasmas 27:012305. https://doi.org/10.1063/1.5110603

Le A, Daughton W, Chen LJ, Egedal J (2017) Enhanced electron mixing and heating in 3-D asymmetric reconnection at the Earth's magnetopause. Geophys Res Lett 44:2096–2104. https://doi.org/10.1002/2017GL072522

Le A, Daughton W, Ohia O, et al (2018) Drift turbulence, particle transport, and anomalous dissipation at the reconnecting magnetopause. Phys Plasmas 25:062103. https://doi.org/10.1063/1.5027086

Leonardis E, Chapman SC, Daughton W, et al (2013) Identification of intermittent multifractal turbulence in fully kinetic simulations of magnetic reconnection. Phys Rev Lett 110:205002. https://doi.org/10.1103/PhysRevLett.110.205002

Lewis HC, Stawarz JE, Franci L, et al (2023) Magnetospheric multiscale measurements of turbulent electric fields in Earth's magnetosheath: how do plasma conditions influence the balance of terms in generalized Ohm's law? Phys Plasmas 30:082901. https://doi.org/10.1063/5.0158067

Lewis HC, Stawarz JE, Matteini L, et al (2024) Turbulent Energy Conversion Associated with Kinetic Microinstabilities in Earth's Magnetosheath. e-prints https://doi.org/10.48550/arXiv.2407.20844 [preprint on arXiv:2407.20844]

Li W, André M, Khotyaintsev YV, et al (2016) Kinetic evidence of magnetic reconnection due to Kelvin-Helmholtz waves. Geophys Res Lett 43:5635–5643. https://doi.org/10.1002/2016GL069192

Li H, Jiang W, Wang C, et al (2020) Evolution of the Earth's magnetosheath turbulence: a statistical study based on MMS observations. Astrophys J Lett 898:L43. https://doi.org/10.3847/2041-8213/aba531

Li TC, Liu YH, Qi Y, Zhou M (2023) Extended magnetic reconnection in kinetic plasma turbulence. Phys Rev Lett 131:085201

Liu YH, Daughton W, Karimabadi H, et al (2013) Bifurcated structure of the electron diffusion region in three-dimensional magnetic reconnection. Phys Rev Lett 110:265004. https://doi.org/10.1103/PhysRevLett.110.265004

Liu TZ, Lu S, Turner DL, et al (2020) Magnetospheric multiscale (mms) observations of magnetic reconnection in foreshock transients. J Geophys Res 125:e2020JA027822. https://doi.org/10.1029/2020JA027822

Liu YH, Hessse M, et al (2024) Ohm's law, the reconnection rate, and energy conversion in collisionless magnetic reconnection. Space Sci Rev arXiv:2406.00875

Loureiro NF, Boldyrev S (2017a) Collisionless reconnection in magnetohydrodynamic and kinetic turbulence. Astrophys J 850:182. https://doi.org/10.3847/1538-4357/aa9754

Loureiro NF, Boldyrev S (2017b) Role of magnetic reconnection in magnetohydrodynamic turbulence. Phys Rev Lett 118:245101. https://doi.org/10.1103/PhysRevLett.118.245101

Loureiro NF, Boldyrev S (2020) Nonlinear reconnection in magnetized turbulence. Astrophys J 890:55. https://doi.org/10.3847/1538-4357/ab6a95

Loureiro NF, Schekochihin AA, Cowley SC (2007) Instability of current sheets and formation of plasmoid chains. Phys Plasmas 14:100703. https://doi.org/10.1063/1.2783986

Lu Q, Wang H, Wang X, et al (2020) Turbulence-driven magnetic reconnection in the magnetosheath downstream of a quasi-parallel shock: a three-dimensional global hybrid simulation. Geophys Res Lett 47:e85661. https://doi.org/10.1029/2019GL085661

Lu Q, Yang Z, Wang H, et al (2021) Two-dimensional particle-in-cell simulation of magnetic reconnection in the downstream of a quasi-perpendicular shock. Astrophys J 919:28. https://doi.org/10.3847/1538-4357/ac18c0

Ma X, Otto A, Delamere PA (2014) Interaction of magnetic reconnection and Kelvin-Helmholtz modes for large magnetic shear: 1. Kelvin-Helmholtz trigger. J Geophys Res 119:781–797. https://doi.org/10.1002/2013JA019224

MacBride BT, Forman MA, Smith CW (2005) Turbulence and third moment of fluctuations: Kolmogorov's 4/5 law and its MHD analogues in the solar wind. In: Fleck B, Zurbuchen TH, Lacoste H (eds) Solar wind 11/SOHO 16, connecting sun and heliosphere. ESA Special Publication, vol 592, p 613

MacBride BT, Smith CW, Forman MA (2008) The turbulent cascade at 1 AU: energy transfer and the third-order scaling for MHD. Astrophys J 679:1644–1660. https://doi.org/10.1086/529575

Mallet A (2020) The onset of electron-only reconnection. J Plasma Phys 86:905860301. https://doi.org/10.1017/S0022377819000941

Mallet A, Schekochihin AA (2017) A statistical model of three-dimensional anisotropy and intermittency in strong Alfvénic turbulence. Mon Not R Astron Soc 466:3918–3927. https://doi.org/10.1093/mnras/stw3251

Mallet A, Schekochihin AA, Chandran BDG (2017) Disruption of sheet-like structures in Alfvénic turbulence by magnetic reconnection. Mon Not R Astron Soc 468:4862–4871. https://doi.org/10.1093/mnras/stx670

Manzini D, Sahraoui F, Califano F, Ferrand R (2022) Local energy transfer and dissipation in incompressible Hall magnetohydrodynamic turbulence: the coarse-graining approach. Phys Rev E 106:035202. https://doi.org/10.1103/PhysRevE.106.035202

Manzini D, Sahraoui F, Califano F (2023) Subion-scale turbulence driven by magnetic reconnection. Phys Rev Lett 130:205201. https://doi.org/10.1103/PhysRevLett.130.205201

Marino R, Sorriso-Valvo L (2023) Scaling laws for the energy transfer in space plasma turbulence. Phys Rep 1006:1–144. https://doi.org/10.1016/j.physrep.2022.12.001

Marino R, Sorriso-Valvo L, Carbone V, et al (2008) Heating the solar wind by a magnetohydrodynamic turbulent energy cascade. Astrophys J Lett 677:L71. https://doi.org/10.1086/587957

Markidis S, Henri P, Lapenta G, et al (2013) Kinetic simulations of plasmoid chain dynamics. Phys Plasmas 20:082105. https://doi.org/10.1063/1.4817286

Marsch E, Tu CY (1997) Intermittency, non-Gaussian statistics and fractal scaling of MHD fluctuations in the solar wind. Nonlinear Process Geophys 4:101–124. https://doi.org/10.5194/npg-4-101-1997

Matsumoto Y, Amano T, Kato TN, Hoshino M (2015) Stochastic electron acceleration during spontaneous turbulent reconnection in a strong shock wave. Science 347:974–978. https://doi.org/10.1126/science.1260168

Matthaeus WH, Goldstein ML (1982) Measurement of the rugged invariants of magnetohydrodynamic turbulence in the solar wind. J Geophys Res 87:6011–6028. https://doi.org/10.1029/JA087iA08p06011

Matthaeus WH, Lamkin SL (1986) Turbulent magnetic reconnection. Phys Fluids 29:2513. https://doi.org/10.1063/1.866004

Matthaeus WH, Servidio S, Dmitruk P, et al (2012) Local anisotropy, higher order statistics and turbulence spectra. Astrophys J 750:103. https://doi.org/10.1088/0004-637X/750/2/103

Matthaeus WH, Oughton S, Osman KT, et al (2014) Nonlinear and linear timescales near kinetic scales in solar wind turbulence. Astrophys J 790(2):155. https://doi.org/10.1088/0004-637X/790/2/155

Matthaeus WH, Wan M, Servidio S, et al (2015) Intermittency, nonlinear dynamics and dissipation in the solar wind and astrophysical plasmas. Philos Trans R Soc A 373:20140154–20140154. https://doi.org/10.1098/rsta.2014.0154

Matthaeus WH, Yang Y, Wan M, et al (2020) Pathways to dissipation in weakly collisional plasmas. Astrophys J 891:101. https://doi.org/10.3847/1538-4357/ab6d6a

Meyrand R, Galtier S (2013) Anomalous $k8/3$ spectrum in electron magnetohydrodynamic turbulence. Phys Rev Lett 111:264501. https://doi.org/10.1103/PhysRevLett.111.264501
Meyrand R, Squire J, Schekochihin AA, Dorland W (2021) On the violation of the zeroth law of turbulence in space plasmas. J Plasma Phys 87:535870301. https://doi.org/10.1017/S0022377821000489
Mininni PD, Pouquet AG, Montgomery DC (2006) Small-scale structures in three-dimensional magnetohydrodynamic turbulence. Phys Rev Lett 97:244503. https://doi.org/10.1103/PhysRevLett.97.244503
Miranda RA, Valdivia JA, Chian ACL, Muñoz PR (2021) Complexity of magnetic-field turbulence at reconnection exhausts in the Solar Wind at 1 au. Astrophys J 923:132. https://doi.org/10.3847/1538-4357/ac2dfe
Mistry R, Eastwood JP, Hietala H (2015a) Detection of small-scale folds at a solar wind reconnection exhaust. J Geophys Res 120:30–42. https://doi.org/10.1002/2014JA020465
Mistry R, Eastwood JP, Phan TD, Hietala H (2015b) Development of bifurcated current sheets in solar wind reconnection exhausts. Geophys Res Lett 42:10,513–10,520. https://doi.org/10.1002/2015GL066820
Mistry R, Eastwood JP, Phan TD, Hietala H (2017) Statistical properties of solar wind reconnection exhausts. J Geophys Res 122:5895–5909. https://doi.org/10.1002/2017JA024032
Muñoz PA, Büchner J (2016) Non-Maxwellian electron distribution functions due to self-generated turbulence in collisionless guide-field reconnection. Phys Plasmas 23:102103. https://doi.org/10.1063/1.4963773
Muñoz PA, Büchner J (2018) Kinetic turbulence in fast three-dimensional collisionless guide-field magnetic reconnection. Phys Rev E 98:043205. https://doi.org/10.1103/PhysRevE.98.043205
Muñoz PA, Büchner J, Kilian P (2017) Turbulent transport in 2D collisionless guide field reconnection. Phys Plasmas 24:022104. https://doi.org/10.1063/1.4975086
Muñoz PA, Jain N, Farzalipour Tabriz M, et al (2023) Electron inertia effects in 3D hybrid-kinetic collisionless plasma turbulence. Phys Plasmas 30:092302. https://doi.org/10.1063/5.0148818
Nakamura TKM, Daughton W (2014) Turbulent plasma transport across the Earth's low-latitude boundary layer. Geophys Res Lett 41:8704–8712. https://doi.org/10.1002/2014GL061952
Nakamura TKM, Fujimoto M (2005) Magnetic reconnection within rolled-up MHD-scale Kelvin-Helmholtz vortices: two-fluid simulations including finite electron inertial effects. Geophys Res Lett 32:L21102. https://doi.org/10.1029/2005GL023362
Nakamura TKM, Fujimoto M, Otto A (2006) Magnetic reconnection induced by weak Kelvin-Helmholtz instability and the formation of the low-latitude boundary layer. Geophys Res Lett 33:L14106. https://doi.org/10.1029/2006GL026318
Nakamura TKM, Fujimoto M, Otto A (2008) Structure of an MHD-scale Kelvin-Helmholtz vortex: two-dimensional two-fluid simulations including finite electron inertial effects. J Geophys Res 113:A09204. https://doi.org/10.1029/2007JA012803
Nakamura TKM, Hasegawa H, Shinohara I, Fujimoto M (2011) Evolution of an MHD-scale Kelvin-Helmholtz vortex accompanied by magnetic reconnection: two-dimensional particle simulations. J Geophys Res 116:A03227. https://doi.org/10.1029/2010JA016046
Nakamura TKM, Daughton W, Karimabadi H, Eriksson S (2013) Three-dimensional dynamics of vortex-induced reconnection and comparison with THEMIS observations. J Geophys Res 118:5742–5757. https://doi.org/10.1002/jgra.50547
Nakamura TKM, Eriksson S, Hasegawa H, et al (2017a) Mass and energy transfer across the Earth's magnetopause caused by vortex-induced reconnection. J Geophys Res 122:11,505–11,522. https://doi.org/10.1002/2017JA024346
Nakamura TKM, Hasegawa H, Daughton W, et al (2017b) Turbulent mass transfer caused by vortex-induced reconnection in collisionless magnetospheric plasmas. Nat Commun 8:1582. https://doi.org/10.1038/s41467-017-01579-0
Nakamura TKM, Plaschke F, Hasegawa H, et al (2020a) Decay of Kelvin-Helmholtz vortices at the Earth's magnetopause under pure southward IMF conditions. Geophys Res Lett 47:e87574. https://doi.org/10.1029/2020GL087574
Nakamura TKM, Stawarz JE, Hasegawa H, et al (2020b) Effects of fluctuating magnetic field on the growth of the Kelvin-Helmholtz instability at the Earth's magnetopause. J Geophys Res 125:e27515. https://doi.org/10.1029/2019JA02751510.1002
Nakamura TKM, Hasegawa H, Genestreti KJ, et al (2021) Fast cross-scale energy transfer during turbulent magnetic reconnection. Geophys Res Lett 48. https://doi.org/10.1029/2021GL093524
Nakamura TKM, Blasl KA, Hasegawa H, et al (2022a) Multi-scale evolution of Kelvin-Helmholtz waves at the Earth's magnetopause during southward IMF periods. Phys Plasmas 29:012901. https://doi.org/10.1063/5.0067391
Nakamura TKM, Blasl KA, Liu YH, Peery SA (2022b) Diffusive plasma transport by the magnetopause Kelvin-Helmholtz instability during southward IMF. Front Astron Space Sci 8:247. https://doi.org/10.3389/fspas.2021.809045

Narita Y (2016) Kinetic extension of critical balance to whistler turbulence. Astrophys J 831:83. https://doi.org/10.3847/0004-637X/831/1/83
Narita Y, Gary SP (2010) Inertial-range spectrum of whistler turbulence. Ann Geophys 28:597–601. https://doi.org/10.5194/angeo-28-597-2010
Nazarenko S (2011) Wave turbulence. Springer, Berlin
Ng J, Chen LJ, Bessho N, et al (2022) Electron-scale reconnection in three-dimensional shock turbulence. Geophys Res Lett 49:e99544. https://doi.org/10.1029/2022GL099544
Ng J, Bessho N, Dahlin JT, Chen LJ (2024) Reconnection along a separator in shock turbulence. Astrophys J 962:181. https://doi.org/10.3847/1538-4357/ad2204
Nykyri K, Otto A (2001) Plasma transport at the magnetospheric boundary due to reconnection in Kelvin-Helmholtz vortices. Geophys Res Lett 28:3565–3568. https://doi.org/10.1029/2001GL013239
Nykyri K, Ma X, Dimmock A, et al (2017) Influence of velocity fluctuations on the Kelvin-Helmholtz instability and its associated mass transport. J Geophys Res 122:9489–9512. https://doi.org/10.1002/2017JA024374
Oishi JS, Mac Low MM, Collins DC, Tamura M (2015) Self-generated turbulence in magnetic reconnection. Astrophys J Lett 806:L12. https://doi.org/10.1088/2041-8205/806/1/L12
Oka M, Phan T, Maea Ø (2022) Electron energization and thermal to non-thermal energy partition during Earth's magnetotail reconnection. Phys Plasmas 29:052904. https://doi.org/10.1063/5.0085647
Okamoto N, Yoshimatsu K, Schneider K, et al (2007) Coherent vortices in high resolution direct numerical simulation of homogeneous isotropic turbulence: a wavelet viewpoint. Phys Fluids 19:115109. https://doi.org/10.1063/1.2771661
Omidi N, O'Farrell A, Krauss-Varban D (1994) Sources of magnetosheath waves and turbulence. Adv Space Res 14:45–54. https://doi.org/10.1016/0273-1177(94)90047-7
Opie S, Verscharen D, Chen CHK, et al (2023) The effect of variations in the magnetic field direction from turbulence on kinetic-scale instabilities. Astron Astrophys 672:L4. https://doi.org/10.1051/0004-6361/202345965
Osman KT, Wan M, Matthaeus WH, et al (2011) Anisotropic third-moment estimates of the energy cascade in solar wind turbulence using multispacecraft data. Phys Rev Lett 107:165001. https://doi.org/10.1103/PhysRevLett.107.165001
Osman KT, Matthaeus WH, Wan M, Rappazzo AF (2012) Intermittency and local heating in the Solar Wind. Phys Rev Lett 108:261102. https://doi.org/10.1103/PhysRevLett.108.261102
Osman KT, Matthaeus WH, Gosling JT, et al (2014) Magnetic reconnection and intermittent turbulence in the Solar Wind. Phys Rev Lett 112:215002. https://doi.org/10.1103/PhysRevLett.112.215002
Panov EV, Nakamura R, Baumjohann W, et al (2010) Multiple overshoot and rebound of a bursty bulk flow. Geophys Res Lett 37:L08103. https://doi.org/10.1029/2009GL041971
Papini E, Franci L, Landi S, et al (2019) Can Hall magnetohydrodynamics explain plasma turbulence at sub-ion scales? Astrophys J 870:52. https://doi.org/10.3847/1538-4357/aaf003
Parashar TN, Chasapis A, Bandyopadhyay R, et al (2018) Kinetic range spectral features of cross helicity using the magnetospheric multiscale spacecraft. Phys Rev Lett 121:265101. https://doi.org/10.1103/PhysRevLett.121.265101
Parashar TN, Cuesta M, Matthaeus WH (2019) Reynolds number and intermittency in the expanding Solar Wind: predictions based on Voyager observations. Astrophys J Lett 884:L57. https://doi.org/10.3847/2041-8213/ab4a82
Pecora F, Servidio S, Greco A, Matthaeus WH (2021) Identification of coherent structures in space plasmas: the magnetic helicity-PVI method. Astron Astrophys 650:A20. https://doi.org/10.1051/0004-6361/202039639
Pecora F, Yang Y, Chasapis A, et al (2023) Relaxation of the turbulent magnetosheath. Mon Not R Astron Soc 525:67–72. https://doi.org/10.1093/mnras/stad2232
Perri S, Perrone D, Yordanova E, et al (2020) On the deviation from Maxwellian of the ion velocity distribution functions in the turbulent magnetosheath. J Plasma Phys 86:905860108. https://doi.org/10.1017/S0022377820000021
Phan TD, Gosling JT, Davis MS, et al (2006) A magnetic reconnection X-line extending more than 390 Earth radii in the solar wind. Nature 439:175–178. https://doi.org/10.1038/nature04393
Phan TD, Paschmann G, Twitty C, et al (2007) Evidence for magnetic reconnection initiated in the magnetosheath. Geophys Res Lett 34:L14104. https://doi.org/10.1029/2007GL030343
Phan TD, Gosling JT, Davis MS (2009) Prevalence of extended reconnection X-lines in the solar wind at 1 AU. Geophys Res Lett 36:L09108. https://doi.org/10.1029/2009GL037713
Phan TD, Eastwood JP, Shay MA, et al (2018) Electron magnetic reconnection without ion coupling in Earth's turbulent magnetosheath. Nature 557:202–206. https://doi.org/10.1038/s41586-018-0091-5
Phan TD, Bale SD, Eastwood JP, et al (2020) Parker Solar Probe in situ observations of magnetic reconnection exhausts during encounter 1. Astrophys J Suppl Ser 246:34. https://doi.org/10.3847/1538-4365/ab55ee

Phan TD, Drake JF, Larson D, et al (2024) Multiple subscale magnetic reconnection embedded inside a heliospheric current sheet reconnection exhaust: evidence for flux rope merging. Astrophys J Lett 971:L42. https://doi.org/10.3847/2041-8213/ad6841

Politano H, Pouquet A (1998a) Dynamical length scales for turbulent magnetized flows. Geophys Res Lett 25:273–276. https://doi.org/10.1029/97GL03642

Politano H, Pouquet A (1998b) von Kármán-Howarth equation for magnetohydrodynamics and its consequences on third-order longitudinal structure and correlation functions. Phys Rev E 57:R21–R24. https://doi.org/10.1103/PhysRevE.57.R21

Politano H, Pouquet A, Sulem PL (1989) Inertial ranges and resistive instabilities in two-dimensional magnetohydrodynamic turbulence. Phys Fluids B 1:2330–2339. https://doi.org/10.1063/1.859051

Pope SB (2000) Turbulent flows. Cambridge University Press, Cambridge

Pouquet A, Rosenberg D, Stawarz JE, Marino R (2019) Helicity dynamics, inverse, and bidirectional cascades in fluid and magnetohydrodynamic turbulence: a brief review. Earth Space Sci 6:351–369. https://doi.org/10.1029/2018EA000432

Price L, Swisdak M, Drake JF, et al (2016) The effects of turbulence on three-dimensional magnetic reconnection at the magnetopause. Geophys Res Lett 43:6020–6027. https://doi.org/10.1002/2016GL069578

Price L, Swisdak M, Drake JF, et al (2017) Turbulence in three-dimensional simulations of magnetopause reconnection. J Geophys Res 122:11,086–11,099. https://doi.org/10.1002/2017JA024227

Price L, Swisdak M, Drake JF, Graham DB (2020) Turbulence and transport during guide field reconnection at the magnetopause. J Geophys Res 125:e2019JA027498. https://doi.org/10.1029/2019JA027498

Pritchett PL (2005) Onset and saturation of guide-field magnetic reconnection. Phys Plasmas 12:062301. https://doi.org/10.1063/1.1914309

Pritchett PL, Coroniti FV (2010) A kinetic ballooning/interchange instability in the magnetotail. J Geophys Res 115:A06301. https://doi.org/10.1029/2009JA014752

Pritchett PL, Mozer FS, Wilber M (2012) Intense perpendicular electric fields associated with three-dimensional magnetic reconnection at the subsolar magnetopause. J Geophys Res 117:A06212. https://doi.org/10.1029/2012JA017533

Pu ZY, Yei M, Liu ZX (1990) Generation of vortex-induced tearing mode instability at the magnetopause. J Geophys Res 95:10559–10566. https://doi.org/10.1029/JA095iA07p10559

Pucci F, Servidio S, Sorriso-Valvo L, et al (2017) Properties of turbulence in the reconnection exhaust: numerical simulations compared with observations. Astrophys J 841:60. https://doi.org/10.3847/1538-4357/aa704f

Qi Y, Ergun R, Pathak N, et al (2024) Investigation of a magnetic reconnection event with extraordinarily high particle energization in magnetotail turbulence. Astrophys J Lett 962:L39. https://doi.org/10.3847/2041-8213/ad24eb

Quijia P, Fraternale F, Stawarz JE, et al (2021) Comparing turbulence in a Kelvin-Helmholtz instability region across the terrestrial magnetopause. Mon Not R Astron Soc 503:4815–4827. https://doi.org/10.1093/mnras/stab319

Retinò A, Sundkvist D, Vaivads A, Mozer F (2007) In situ evidence of magnetic reconnection in turbulent plasma. Nat Phys 3:235–238. https://doi.org/10.1038/nphys574

Retinò A, Khotyaintsev Y, Le Contel O, et al (2022) Particle energization in space plasmas: towards a multi-point, multi-scale plasma observatory. Exp Astron 54:427–471. https://doi.org/10.1007/s10686-021-09797-7

Richard L, Sorriso-Valvo L, Yordanova E, et al (2024) Turbulence in magnetic reconnection jets from injection to sub-ion scales. Phys Rev Lett 132:105201. https://doi.org/10.1103/PhysRevLett.132.105201

Rossi C, Califano F, Retinò A, et al (2015) Two-fluid numerical simulations of turbulence inside Kelvin-Helmholtz vortices: intermittency and reconnecting current sheets. Phys Plasmas 22:122303. https://doi.org/10.1063/1.4936795

Roytershteyn V, Daughton W, Karimabadi H, Mozer FS (2012) Influence of the lower-hybrid drift instability on magnetic reconnection in asymmetric configurations. Phys Rev Lett 108:185001. https://doi.org/10.1103/PhysRevLett.108.185001

Sahraoui F, Belmont G, Rezeau L, et al (2006) Anisotropic turbulent spectra in the terrestrial magnetosheath as seen by the cluster spacecraft. Phys Rev Lett 96:075002. https://doi.org/10.1103/PhysRevLett.96.075002

Schwartz SJ, Kucharek H, Farrugia CJ, et al (2021) Energy conversion within current sheets in the Earth's quasi parallel magnetosheath. Geophys Res Lett 48:e91859. https://doi.org/10.1029/2020GL09185910.1002/essoar.10505071.1

Servidio S, Matthaeus WH, Dmitruk P (2008) Depression of nonlinearity in decaying isotropic MHD turbulence. Phys Rev Lett 100(9):095005. https://doi.org/10.1103/PhysRevLett.100.095005

Servidio S, Matthaeus WH, Shay MA, et al (2009) Magnetic reconnection in two-dimensional magnetohydrodynamic turbulence. Phys Rev Lett 102:115003. https://doi.org/10.1103/PhysRevLett.102.115003

Servidio S, Matthaeus WH, Shay MA, et al (2010) Statistics of magnetic reconnection in two-dimensional magnetohydrodynamic turbulence. Phys Plasmas 17:032315. https://doi.org/10.1063/1.3368798
Servidio S, Greco A, Matthaeus WH, et al (2011) Statistical association of discontinuities and reconnection in magnetohydrodynamic turbulence. J Geophys Res 116:A09102. https://doi.org/10.1029/2011JA016569
Servidio S, Valentini F, Califano F, Veltri P (2012) Local kinetic effects in two-dimensional plasma turbulence. Phys Rev Lett 108:045001. https://doi.org/10.1103/PhysRevLett.108.045001
Servidio S, Chasapis A, Matthaeus WH, et al (2017) Magnetospheric multiscale observation of plasma velocity-space cascade: Hermite representation and theory. Phys Rev Lett 119:205101. https://doi.org/10.1103/PhysRevLett.119.205101
Sharma Pyakurel P, Shay MA, Phan TD, et al (2019) Transition from ion-coupled to electron-only reconnection: basic physics and implications for plasma turbulence. Phys Plasmas 26:082307. https://doi.org/10.1063/1.5090403
Shay MA, Haggerty CC, Matthaeus WH, et al (2018) Turbulent heating due to magnetic reconnection. Phys Plasmas 25:012304. https://doi.org/10.1063/1.4993423
Shay MA, Adhikari S, Besho N, et al (2024) Simulation models for exploring magnetic reconnection. Space Sci Rev arXiv:2406.05901
Shibata K, Tanuma S (2001) Plasmoid-induced-reconnection and fractal reconnection. Earth Planets Space 53:473–482. https://doi.org/10.1186/BF03353258
Sisti M, Faganello M, Califano F, Lavraud B (2019) Satellite data-based 3-D simulation of Kelvin-Helmholtz instability and induced magnetic reconnection at the Earth's magnetopause. Geophys Res Lett 46:11,597–11,605. https://doi.org/10.1029/2019GL083282
Sisti M, Fadanelli S, Cerri SS, et al (2021) Characterizing current structures in 3d hybrid-kinetic simulations of plasma turbulence. Astron Astrophys 655:A107. https://doi.org/10.1051/0004-6361/202141902
Sorriso-Valvo L, Carbone V, Veltri P, et al (1999) Intermittency in the solar wind turbulence through probability distribution functions of fluctuations. Geophys Res Lett 26:1801–1804. https://doi.org/10.1029/1999GL900270
Sorriso-Valvo L, Carbone F, Perri S, et al (2018a) On the statistical properties of turbulent energy transfer rate in the inner heliosphere. Sol Phys 293:10. https://doi.org/10.1007/s11207-017-1229-6
Sorriso-Valvo L, Perrone D, Pezzi O, et al (2018b) Local energy transfer rate and kinetic processes: the fate of turbulent energy in two-dimensional hybrid Vlasov-Maxwell numerical simulations. J Plasma Phys 84:725840201. https://doi.org/10.1017/S0022377818000302
Sorriso-Valvo L, Catapano F, Retinò A, et al (2019) Turbulence-driven ion beams in the magnetospheric Kelvin-Helmholtz instability. Phys Rev Lett 122:035102. https://doi.org/10.1103/PhysRevLett.122.035102
Squire J, Meyrand R, Kunz MW, et al (2022) High-frequency heating of the solar wind triggered by low-frequency turbulence. Nat Astron 6:715–723. https://doi.org/10.1038/s41550-022-01624-z
Stanier A, Daughton W, Le A, et al (2019) Influence of 3D plasmoid dynamics on the transition from collisional to kinetic reconnection. Phys Plasmas 26:072121. https://doi.org/10.1063/1.5100737
Stawarz JE, Smith CW, Vasquez BJ, et al (2009) The turbulent cascade and proton heating in the Solar Wind at 1 AU. Astrophys J 697(2):1119–1127. https://doi.org/10.1088/0004-637X/697/2/1119
Stawarz JE, Vasquez BJ, Smith CW, et al (2011) Third moments and the role of anisotropy from velocity shear in the Solar Wind. Astrophys J 736:44. https://doi.org/10.1088/0004-637X/736/1/44
Stawarz JE, Ergun RE, Goodrich KA (2015) Generation of high-frequency electric field activity by turbulence in the Earth's magnetotail. J Geophys Res 120:1845–1866. https://doi.org/10.1002/2014JA020166
Stawarz JE, Eriksson S, Wilder FD, et al (2016) Observations of turbulence in a Kelvin-Helmholtz event on 8 September 2015 by the magnetospheric multiscale mission. J Geophys Res 121(11):11,021–11,034. https://doi.org/10.1002/2016JA023458
Stawarz JE, Eastwood JP, Phan TD, et al (2019) Properties of the turbulence associated with electron-only magnetic reconnection in Earth's magnetosheath. Astrophys J Lett 877:L37. https://doi.org/10.3847/2041-8213/ab21c8
Stawarz JE, Matteini L, Parashar TN, et al (2021) Comparative analysis of the various generalized Ohm's law terms in magnetosheath turbulence as observed by magnetospheric multiscale. J Geophys Res 126:e8447. https://doi.org/10.1029/2020JA02844710.1002/essoar.10503618.1
Stawarz JE, Eastwood JP, Phan TD, et al (2022) Turbulence-driven magnetic reconnection and the magnetic correlation length: observations from magnetospheric multiscale in Earth's magnetosheath. Phys Plasmas 29:012302. https://doi.org/10.1063/5.0071106
Steinvall K, Gingell I (2024) The influence of rotational discontinuities on the formation of reconnected structures at collisionless shocks—hybrid simulations. J Geophys Res 129:e2023JA032101. https://doi.org/10.1029/2023JA032101
Sturner AP, Eriksson S, Nakamura T, et al (2018) On multiple Hall-like electron currents and tripolar guide magnetic field perturbations during Kelvin-Helmholtz waves. J Geophys Res 123:1305–1324. https://doi.org/10.1002/2017JA024155

Sundkvist D, Retinò A, Vaivads A, Bale SD (2007) Dissipation in turbulent plasma due to reconnection in thin current sheets. Phys Rev Lett 99:025004. https://doi.org/10.1103/PhysRevLett.99.025004

Swisdak M (2016) Quantifying gyrotropy in magnetic reconnection. Geophys Res Lett 43:43–49. https://doi.org/10.1002/2015GL066980

Takagi K, Hashimoto C, Hasegawa H, et al (2006) Kelvin-Helmholtz instability in a magnetotail flank-like geometry: three-dimensional MHD simulations. J Geophys Res 111:A08202. https://doi.org/10.1029/2006JA011631

Taylor GI (1938) The spectrum of turbulence. Proc R Soc A 164:476–490. https://doi.org/10.1098/rspa.1938.0032

TenBarge JM, Howes GG (2013) Current sheets and collisionless damping in kinetic plasma turbulence. Astrophys J Lett 771:L27. https://doi.org/10.1088/2041-8205/771/2/L27

TenBarge JM, Howes GG, Dorland W (2013) Collisionless damping at electron scales in solar wind turbulence. Astrophys J 774:139. https://doi.org/10.1088/0004-637X/774/2/139

Trotta D, Pezzi O, Burgess D, et al (2023) Three-dimensional modelling of the shock-turbulence interaction. Mon Not R Astron Soc 525:1856–1866. https://doi.org/10.1093/mnras/stad2384

Uritsky VM, Pouquet A, Rosenberg D, et al (2010) Structures in magnetohydrodynamic turbulence: detection and scaling. Phys Rev E 82:056326. https://doi.org/10.1103/PhysRevE.82.056326

Usanova ME, Ergun RE (2022) Electron energization by high-amplitude turbulent electric fields: a possible source of the outer radiation belt. J Geophys Res 127:e30336. https://doi.org/10.1029/2022JA030336

Usanova ME, Ergun RE, Stawarz JE (2023) Ion energization by turbulent electric fields in fast earthward flows and its implications for the dynamics of the inner magnetosphere. J Geophys Res 128:e2023JA031704. https://doi.org/10.1029/2023JA031704

Vapirev A, Lapenta G, Divin A, et al (2013) Formation of a transient front structure near reconnection point in 3-D PIC simulations. J Geophys Res 118:1435–1449. https://doi.org/10.1002/jgra.50136

Vega C, Roytershteyn V, Delzanno GL, Boldyrev S (2020) Electron-only reconnection in kinetic-Alfven turbulence. Astrophys J Lett 893:L10. https://doi.org/10.3847/2041-8213/ab7eba

Vega C, Roytershteyn V, Delzanno GL, Boldyrev S (2023) Electron-scale current sheets and energy dissipation in 3D kinetic-scale plasma turbulence with low electron beta. Mon Not R Astron Soc 524:1343–1351. https://doi.org/10.1093/mnras/stad1931

Vernisse Y, Lavraud B, Eriksson S, et al (2016) Signatures of complex magnetic topologies from multiple reconnection sites induced by Kelvin-Helmholtz instability. J Geophys Res 121:9926–9939. https://doi.org/10.1002/2016JA023051

Verscharen D, Wicks RT, Alexandrova O, et al (2022) A case for electron-astrophysics. Exp Astron 54:473–519. https://doi.org/10.1007/s10686-021-09761-5

Vishniac ET, Pillsworth S, Eyink G, et al (2012) Reconnection current sheet structure in a turbulent medium. Nonlinear Process Geophys 19:605–610. https://doi.org/10.5194/npg-19-605-2012

Volwerk M, Glassmeier KH, Nakamura R, et al (2007) Flow burst-induced Kelvin-Helmholtz waves in the terrestrial magnetotail. Geophys Res Lett 34:L10102. https://doi.org/10.1029/2007GL029459

Vörös Z, Baumjohann W, Nakamura R, et al (2004) Magnetic turbulence in the plasma sheet. J Geophys Res 109:A11215. https://doi.org/10.1029/2004JA010404

Vörös Z, Baumjohann W, Nakamura R, et al (2006) Bursty bulk flow driven turbulence in the Earth's plasma sheet. Space Sci Rev 122:301–311. https://doi.org/10.1007/s11214-006-6987-7

Vörös Z, Yordanova E, Varsani A, et al (2016) Mms observation of magnetic reconnection in the turbulent magnetosheath. J Geophys Res 122:11,442–11,467. https://doi.org/10.1002/2017JA024535

Vörös Z, Yordanova E, Graham DB, et al (2019) Mms observations of whistler and lower hybrid drift waves associated with magnetic reconnection in the turbulent magnetosheath. J Geophys Res 124:8551–8563. https://doi.org/10.1029/2019JA027028

Wan M, Servidio S, Oughton S, Matthaeus WH (2009) The third-order law for increments in magnetohydrodynamic turbulence with constant shear. Phys Plasmas 16:090703. https://doi.org/10.1063/1.3240333

Wan M, Matthaeus WH, Karimabadi H, et al (2012) Intermittent dissipation at kinetic scales in collisionless plasma turbulence. Phys Rev Lett 109:195001. https://doi.org/10.1103/PhysRevLett.109.195001

Wan M, Matthaeus WH, Roytershteyn V, et al (2015) Intermittent dissipation and heating in 3D kinetic plasma turbulence. Phys Rev Lett 114:175002. https://doi.org/10.1103/PhysRevLett.114.175002

Wan M, Matthaeus WH, Roytershteyn V, et al (2016) Intermittency, coherent structures and dissipation in plasma turbulence. Phys Plasmas 23:042307. https://doi.org/10.1063/1.4945631

Wang S, Chen L, Bessho N, et al (2019) Observational evidence of magnetic reconnection in the terrestrial bow shock transition region. Geophys Res Lett 46:562–570. https://doi.org/10.1029/2018GL080944

Wang S, Chen L, Bessho N, et al (2020) Ion-scale current structures in short large-amplitude magnetic structures. Astrophys J 898:121. https://doi.org/10.3847/1538-4357/ab9b8b

Wang R, Wang S, Lu Q, et al (2023) Direct observation of turbulent magnetic reconnection in the solar wind. Nat Astron 7:18–28. https://doi.org/10.1038/s41550-022-01818-5

Wang Y, Cheng X, Guo Y, et al (2024) A method for determining the locations and configurations of magnetic reconnection within three-dimensional turbulent plasmas. Astron Astrophys 683:A224. https://doi.org/10.1051/0004-6361/202347564

Weygand JM, Kivelson MG, Khurana KK, et al (2005) Plasma sheet turbulence observed by Cluster II. J Geophys Res 110:A01205. https://doi.org/10.1029/2004JA010581

Wilder FD, Ergun RE, Schwartz SJ, et al (2016) Observations of large-amplitude, parallel, electrostatic waves associated with the Kelvin-Helmholtz instability by the magnetospheric multiscale mission. Geophys Res Lett 43:8859–8866. https://doi.org/10.1002/2016GL070404

Wilder FD, Ergun RE, Burch JL, et al (2018) The role of the parallel electric field in electron-scale dissipation at reconnecting currents in the magnetosheath. J Geophys Res 123:6533–6547. https://doi.org/10.1029/2018JA025529

Wilder FD, Conley M, Ergun RE, et al (2022) Magnetospheric multiscale observations of waves and parallel electric fields in reconnecting current sheets in the turbulent magnetosheath. J Geophys Res 127:e2022JA030511. https://doi.org/10.1029/2022JA030511

Wu P, Perri S, Osman K, et al (2013) Intermittent heating in solar wind and kinetic simulations. Astrophys J Lett 763:L30. https://doi.org/10.1088/2041-8205/763/2/L30

Xiong QY, Huang SY, Zhang J, et al (2024) Guide field dependence of energy conversion and magnetic topologies in reconnection turbulent outflow. Geophys Res Lett 51:e2024GL109356. https://doi.org/10.1029/2024GL109356

Yang Y, Matthaeus WH, Parashar TN, et al (2017) Energy transfer, pressure tensor, and heating of kinetic plasma. Phys Plasmas 24:072306. https://doi.org/10.1063/1.4990421

Yokoyama N, Takaoka M (2021) Energy-flux vector in anisotropic turbulence: application to rotating turbulence. J Fluid Mech 908:A17. https://doi.org/10.1017/jfm.2020.860

Yordanova E, Vaivads A, André M, et al (2008) Magnetosheath plasma turbulence and its spatiotemporal evolution as observed by the cluster spacecraft. Phys Rev Lett 100:205003. https://doi.org/10.1103/PhysRevLett.100.205003

Yordanova E, Vörös Z, Varsani A, et al (2016) Electron scale structures and magnetic reconnection signatures in the turbulent magnetosheath. Geophys Res Lett 43:5969–5978. https://doi.org/10.1002/2016GL069191

Zharkova V, Xia Q (2021) Plasma turbulence generated in 3D current sheets with single and multiple X-nullpoints. Front Astron Space Sci 8:178. https://doi.org/10.3389/fspas.2021.665998

Zhdankin V, Uzdensky DA, Perez JC, Boldyrev S (2013) Statistical analysis of current sheets in three-dimensional magnetohydrodynamic turbulence. Astrophys J 771:124. https://doi.org/10.1088/0004-637X/771/2/124

Zhdankin V, Boldyrev S, Perez JC, Tobias SM (2014) Energy dissipation in magnetohydrodynamic turbulence: coherent structures or "nanoflares"? Astrophys J 795:127. https://doi.org/10.1088/0004-637X/795/2/127

Zhdankin V, Boldyrev S, Uzdensky DA (2016) Scalings of intermittent structures in magnetohydrodynamic turbulence. Phys Plasmas 23:055705. https://doi.org/10.1063/1.4944820

Publisher's Note Springer Nature remains neutral with regard to jurisdictional claims in published maps and institutional affiliations.

Authors and Affiliations

J.E. Stawarz[1] · P.A. Muñoz[2,3] · N. Bessho[4,5] · R. Bandyopadhyay[6] · T.K.M. Nakamura[7,8] · S. Eriksson[9] · D.B. Graham[10] · J. Büchner[2,3] · A. Chasapis[9] · J.F. Drake[11,12] · M.A. Shay[13] · R.E. Ergun[9,14] · H. Hasegawa[15] · Yu.V. Khotyaintsev[10] · M. Swisdak[11] · F.D. Wilder[16]

✉ J.E. Stawarz
julia.stawarz@northumbria.ac.uk

1 Department of Mathematics, Physics, and Electrical Engineering, Northumbria University, Ellison Building, Newcastle upon Tyne NE1 8ST, UK

2 Center for Astronomy and Astrophysics, Technical University Berlin, 10623 Berlin, Germany

3 Max Planck Institute for Solar System Research, 37077 Göttingen, Germany

[4] Department of Astronomy, University of Maryland, College Park, MD 20742, USA

[5] NASA Goddard Space Flight Center, Greenbelt, MD 20771, USA

[6] Department of Astrophysical Sciences, Princeton University, Princeton, NJ 08544, USA

[7] Space Research Institute, Austrian Academy of Sciences, 8042 Graz, Austria

[8] Krimgen LLC, Hiroshima, 7320828, Japan

[9] Laboratory for Atmospheric and Space Physics, University of Colorado Boulder, Boulder, CO, USA

[10] Swedish Institute of Space Physics, Uppsala, Sweden

[11] Institute for Research in Electronics and Applied Physics, University of Maryland, College Park, MD 20740, USA

[12] Department of Physics, Institute for Physical Science and Technology and the Joint Space Science Institute, University of Maryland, College Park, MD 20740, USA

[13] Department of Physics and Astronomy, University of Delaware, Newark, DE 19716, USA

[14] Department of Astrophysical and Planetary Sciences, University of Colorado Boulder, Boulder, CO, USA

[15] Institute of Space and Astronautical Science, JAXA, Sagamihara, Japan

[16] University of Texas at Arlington, Arlington, TX, USA

IV. Context and Consequence of Magnetic Reconnection in Geospace

Space Science Reviews (2024) 220:34
https://doi.org/10.1007/s11214-024-01067-0

Global-Scale Processes and Effects of Magnetic Reconnection on the Geospace Environment

S.A. Fuselier[1,2] · S.M. Petrinec[3] · P.H. Reiff[4] · J. Birn[5] · D.N. Baker[6] · I.J. Cohen[7] · R. Nakamura[8] · M.I. Sitnov[7] · G.K. Stephens[7] · J. Hwang[1] · B. Lavraud[9] · T.E. Moore[10] · K.J. Trattner[6] · B.L. Giles[10] · D.J. Gershman[10] · S. Toledo-Redondo[11] · J.P. Eastwood[12]

Received: 17 May 2023 / Accepted: 8 April 2024 / Published online: 19 April 2024

Abstract
Recent multi-point measurements, in particular from the Magnetospheric Multiscale (MMS) spacecraft, have advanced the understanding of micro-scale aspects of magnetic reconnection. In addition, the MMS mission, as part of the Heliospheric System Observatory, combined with recent advances in global magnetospheric modeling, have furthered the understanding of meso- and global-scale structure and consequences of reconnection. Magnetic reconnection at the dayside magnetopause and in the magnetotail are the drivers of the global Dungey cycle, a classical picture of global magnetospheric circulation. Some recent advances in the global structure and consequences of reconnection that are addressed here include a detailed understanding of the location and steadiness of reconnection at the dayside magnetopause, the importance of multiple plasma sources in the global circulation, and reconnection consequences in the magnetotail. These advances notwithstanding, there are important questions about global reconnection that remain. These questions focus on how multiple reconnection and reconnection variability fit into and complicate the Dungey Cycle picture of global magnetospheric circulation.

Keywords Magnetic reconnection · Global magnetospheric circulation · Magnetopause · Magnetospheric cusps · Magnetotail

1 Magnetic Reconnection and the Dungey Cycle

In the absence of magnetic reconnection at the magnetopause, magnetic field lines in the magnetosheath and magnetosphere have no interconnection. Magnetosheath field lines ultimately trace back to the Sun and are considered "solar wind" field lines, with no connection with the Earth's ionosphere. Field lines in the low-latitude magnetosphere trace back to the high-latitude ionosphere in both northern and southern hemispheres and are considered "closed". When the interplanetary magnetic field (IMF) is southward, magnetic reconnection at the magnetopause occurs at relatively low latitudes between these magnetosheath and magnetospheric magnetic field lines. Reconnection at the low-latitude magnetopause "opens" previously closed magnetospheric field lines so that they trace back to the high-latitude ionosphere in one direction along the magnetic field and ultimately to interplanetary

Extended author information available on the last page of the article

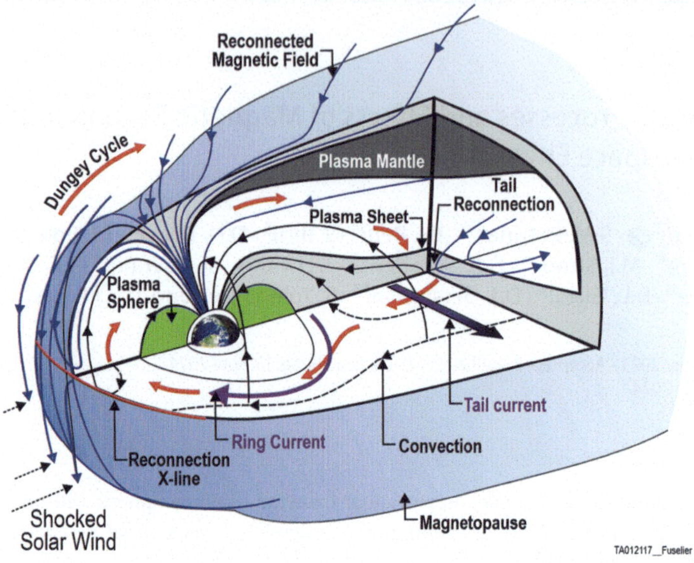

Fig. 1 3-D schematic of the Dungey cycle. For southward IMF and no dipole tilt, reconnection occurs along a long line across the dayside magnetopause. Reconnected field lines convect over the pole and into the magnetotail. Reconnection in the magnetotail closes the previously open field lines. Convection brings these closed field lines from the tail back to the dayside to complete the cycle. This cycle is depicted by the red arrows

space in the opposite direction. This opening of previously closed field lines was depicted schematically for the magnetosphere for the first time by Dungey (1961).

Under the convection of the solar wind, these reconnected field lines are dragged over the poles towards the Earth's magnetotail. Magnetic reconnection in the magnetotail re-connects the previously open field lines, forming newly re-closed field lines. Through the action of magnetospheric convection, these closed field lines convect earthward and finally return to the dayside. A two-dimensional version of this model for an open magnetosphere was first published by Dungey (1962) in a free-hand sketch. However, the model in fact dates back to Dungey's PhD thesis (Dungey 1950). A fascinating and detailed account of the realization that reconnection plays a critical role in magnetospheric dynamics (which was considered quite radical at the dawn of the space age) is found in: Magnetospheric plasma physics: The impact of Jim Dungey's research (2015).

A quasi-three-dimensional schematic depiction of this plasma and field circulation is shown in Fig. 1. With an IMF pointing due southward and no dipole tilt, magnetic field lines in the Earth's magnetosheath reconnect along a long reconnection X-line oriented across the dayside magnetopause through the subsolar region. These reconnected field lines (in blue) are dragged over the poles into the magnetotail. In the magnetotail, the field lines (in black) reconnect and snap sunward. Eventually, they convect around the Earth (a black field line shown in Fig. 1 is convecting sunward and duskward) and return to the subsolar region. The circulation is shown by the red arrows.

In the original model of a magnetosphere driven by reconnection at the magnetopause and in the magnetotail, the reconnection rates at the dayside and in the tail are the same. An important complication in this reconnection-driven circulation model is that the day-side and nightside rates are almost always different. Typically, dayside reconnection occurs continuously (e.g., Russell 1972) with possible variable rate while nightside reconnection is very intermittent (e.g., Russell 1972), especially in the near-Earth magnetotail ~20-30 Earth Radii (R_E) from the Earth. The dayside rate is reflected in the polar cap potential and changes in the dayside and nightside rates are reflected in the size of the polar cap inside the auroral oval.

A revision of the Dungey circulation model that describes the magnetospheric substorm process was introduced by Hones (1977). In this revision, magnetic flux transferred from the dayside builds up in the magnetotail during an interval when there is dayside reconnection and no reconnection in the near-Earth magnetotail. This flux buildup causes the tail to stretch until explosive reconnection in the near-tail re-connects the field lines. During this explosive reconnection, the reconnection rate in the near-tail may exceed that on the dayside. Averaging over a long time, the flux transfer rate from the dayside must match the nightside rate. However, at any given instant in time they can differ: if the dayside rate is larger, the polar cap expands; if the nightside rate is larger, the polar cap shrinks (Cowley and Lockwood 1992). This model is often referred to as the Expanding-Contracting Polar Cap Model (see e.g., Milan et al. 2017)

In this paper, the global-scale processes and effects of magnetic reconnection are reviewed in the context of this revised Dungey circulation model. For Sects. 2-5, the focus is on southward IMF conditions. Section 2 describes global magnetic reconnection at the dayside magnetopause and the initiation of the Dungey cycle. Section 3 describes field line convection from the dayside to the nightside and implications for the Earth's magnetospheric cusps and the polar cap potential. Section 4 describes reconnection in the Earth's magnetotail. Section 5 describes convection of reconnected field lines back to the dayside and the completion of the Dungey cycle. Section 6 describes the changes to the Dungey cycle when the IMF is northward or when the IMF has a dominant B_X component, i.e., when the IMF is nearly radial. Finally, Sect. 7 describes the complications to the Dungey cycle when magnetic reconnection at the dayside is variable in space and/or in time and when nightside reconnection has time and potentially spatial variability outside of the large-scale variability that occurs in the substorm process.

2 Global Magnetic Reconnection at the Dayside Magnetopause

2.1 Evidence for Long "Primary" Reconnection X-Lines

Observations of charged particles within the magnetospheric cusps provide remote sensing evidence for the presence of collisionless magnetic reconnection at the magnetopause. For a cusp-crossing satellite that observes precipitating magnetosheath ions during southward IMF, the extent of the reconnection X-lines is estimated by tracing magnetic field lines from the cusp to the dayside magnetopause. Ion precipitation observations over a range of local times during such a cusp traversal are exploited to determine the large-scale configuration (orientation(s)) of the primary reconnection X-line(s) along the magnetopause. This method yields primary X-lines that are continuous over many R_E (e.g., Trattner et al. (2005), Trattner et al. (2007)). The method was applied to 130 cusp traversals and, because these cusp traversals cover a finite and rather large range of geomagnetic latitudes, these observations provide one of the better estimates of the length of the primary X-line. Continuous X-line lengths of many R_E were observed, with the measured length limited only by the magnetic local time coverage of the cusp traversal. The term "primary" is used here to indicate a long, quasi-stable X-line as opposed to a transient X-line or lines that might be produced in a Flux Transfer Event (FTE) (Fuselier et al. 2019b) (see also Sect. 7).

Another remote sensing method for estimating the reconnection X-line extent is to utilize space-based global auroral imagers (e.g., Fuselier et al. 2002). Doppler-shifted proton auroral emissions result from the precipitation of energetic (>2 keV) protons into the ionosphere. These protons are accelerated in the reconnection process at the magnetopause. Therefore, the precipitation pattern in the ionosphere of these energetic protons provides the foot-points of reconnected magnetic field lines. These foot-points are mapped to the magnetopause using a closed form of the magnetospheric magnetic field to determine the "last closed field line", i.e., the field line that undergoes reconnection to produce the proton precipitation. This mapping method also shows that reconnection X-lines during southward IMF extend over a wide span of local times along the dayside magnetopause, with equivalent extents of $\sim$10 – 25 R_E (Fuselier et al. 2002, 2003; Berchem et al. 2003; Phan et al. 2006; Dunlop et al. 2011). Although this method for estimating the X-line extent has been done for only a few events, these results are in good agreement with the results from the much larger number of cusp traversals discussed above.

For in situ observations at the magnetopause, single-point in situ observations of the presence of magnetopause magnetic reconnection are rather common (typically by observing accelerated ion flows tangent to the magnetopause). However, in situ evidence for long, quasi-continuous reconnection lines is quite rare, because simultaneous in situ observations are required from multiple locations and it is not clear if the reconnection line is continuous between the spacecraft. Such events typically involve multiple, widely spaced spacecraft sampling the magnetopause at approximately the same time during an interval of steady IMF with a southward-directed component, with each observing reconnection signatures (e.g., Peterson et al. 1998; Phan et al. 2000; Dunlop et al. 2011; Toledo-Redondo et al. 2021b). If the spacecraft happen to be positioned such that different spacecraft observe accelerated ion flows in opposing directions (e.g., Phan et al. 2000), then this provides additional information that the extended reconnection line is situated somewhere between the spacecraft. This multi-spacecraft technique for estimating the reconnection X-line length was extended to a number of magnetopause conjunctions by the THEMIS spacecraft (Walsh et al. 2017; Zou et al. 2019, 2020; Atz et al. 2022). While some of the THEMIS conjunctions are consistent with long X-lines, many are not. The discrepancy between these results and all of the other results discussed above may be the use of the Walen test, a very restrictive definition of reconnection at the magnetopause that was employed in the THEMIS conjunctions. The Walen test is a single fluid test that determines if the magnetopause is consistent with an infinite, one-dimensional rotational discontinuity. It fails to take into account the multi-component nature of the plasma, the flow in the magnetosheath, and multiple reconnection at the magnetopause, which is quite common (see, e.g., Vines et al. 2017, and Sect. 7). In essence, this test is sufficient for identifying reconnection at the magnetopause, but it is far from necessary.

2.2 The Maximum Magnetic Shear Model and Long, Continuous Reconnection X-Lines

As described in the previous section, magnetic reconnection readily occurs along extended X-lines over the dayside magnetopause when the IMF is southward, extending from local noon far along the magnetopause flanks. This results in the transport of solar wind flux into the magnetosphere and erosion of the dayside magnetosphere. The location, extent, and configuration of the extended reconnection line along the dayside magnetopause have been the subject of many studies during the past few decades. Some early models predicted long reconnection lines would occur exclusively along anti-parallel merging regions along the magnetopause; extending from low latitudes at the flanks up to the cusp region near local noon (Crooker 1979; Luhmann et al. 1984). Other parameters and models that consider various aspects of collisionless reconnection and include segments of anti-parallel and component reconnection are: a uniform guide field or B_M component (Sonnerup 1974; Gonzalez and Mozer 1974), the angle of bisection between magnetic fields earthward/sunward of the magnetopause (Moore et al. 2002, 2008; Borovsky 2008; Hesse et al. 2013), maxima of the asymmetric reconnection outflow speed (Swisdak and Drake 2007), maxima of the asymmetric Sweet-Parker reconnection rate (Borovsky 2013), maxima of the current density magnitude (Alexeev et al. 1998), and the Maximum Magnetic Shear model (Trattner et al. 2007).

The Maximum Magnetic Shear Model, described in much greater detail in Trattner et al. (2021a), was developed from numerous observations of ion distribution functions and velocity cutoffs within the mid- to high-altitude northern cusp region, systematically traced from the observing spacecraft along magnetic field lines to the magnetopause using a time-of-flight methodology to estimate the locations of magnetopause reconnection sites. The aggregate set of such estimates under varying solar wind conditions and season (dipole tilt angle) then led to the development of this empirical model. Although this is an empirical model, it uses models for the draped magnetosheath magnetic field, the magnetopause boundary, and the internal magnetospheric magnetic field. The magnetosheath and magnetospheric magnetic fields in the models do not interact as would occur, for example, in an MHD simulation. Rather, the magnetosheath and magnetospheric models are used in closed form to determine the shear at the magnetopause boundary (Petrinec and Fuselier 2003; Trattner et al. 2007; Trattner et al. 2021a,b). An example magnetic shear angle color contour plot (2-D projection and 3-D coverage over the magnetopause), along with an extended reconnection line is shown in Fig. 2. This model has been tested extensively (see Trattner et al. 2021a,b).

For the MMS mission, the model was used to predict the number of magnetopause crossings near the reconnection X-line at the dayside magnetopause (Griffiths et al. 2011; Fuselier et al. 2016). After the first year of operations, there were several successful tests conducted to determine the efficacy of these predictions (e.g., Petrinec et al. 2016; Fuselier et al. 2017; Trattner et al. 2017). In particular, Trattner et al. (2017) surveyed the first year of MMS operations and identified 302 instances when the spacecraft were near the X-line and then compared these locations with the predicted location of the X-line from the Maximum Magnetic Shear model. The Maximum Magnetic Shear model, with its long X-lines, was accurate ~85% of the time. For more details on the empirical model and the tests, see the recent review article on the location of the magnetopause X-line (Trattner et al. 2021a).

2.3 IMF Clock Angle and Cone Angle Determine the Location of the Reconnection X-Line

As inferred from an examination of a large set of Polar/TIMAS cusp crossings and low-velocity cutoff mappings to the magnetopause reconnection site using plasma distributions, the location of the extended magnetopause reconnection line at a given date and time (i.e., season and dipole tilt angle) is simply controlled by the IMF clock and cone angles. These IMF parameters dictate how the large-scale magnetic shear angle varies over the entire magnetopause surface. As described in the previous section, regions with the largest magnetic shear angles provide the location where the extended, primary reconnection line is established in the empirical maximum magnetic shear model. This observations-based model had been used to examine multiple MMS orbit scenarios during the mission development phase

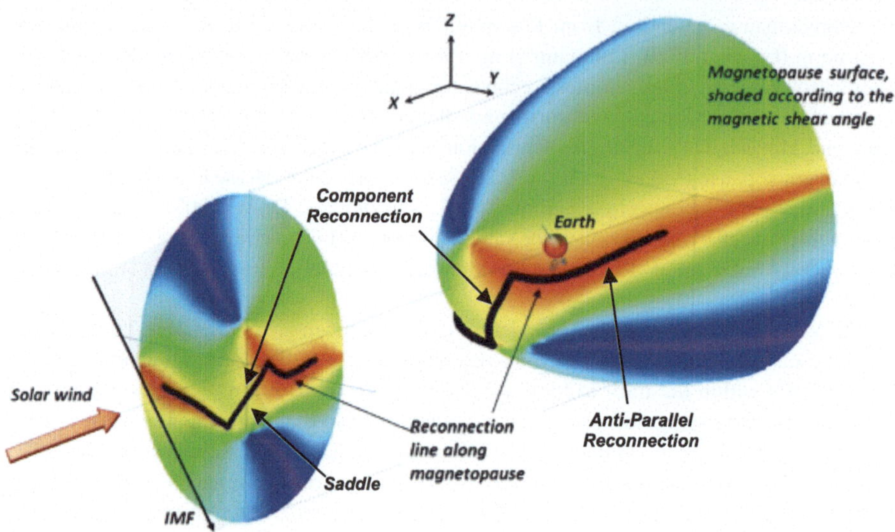

Fig. 2 2-D projection along the Sun-Earth line of the magnetopause surface, and 3-D coverage over the magnetopause, colored according to the magnetic shear angle. A maximum magnetic shear model reconnection X-line across the dayside magnetopause is shown in black. The reconnection line extends from the noon meridian to well along the flanks of the magnetopause

to optimize the likelihood of encountering and directly sampling dayside magnetopause reconnection sites; especially the high-resolution multipoint sampling of the microphysics within the very localized electron diffusion region (Griffiths et al. 2011). This examination proved very beneficial for the prime mission of MMS (Fuselier et al. 2017; Webster et al. 2018). Predicted extended reconnection line locations have subsequently been compared to the MMS observed accelerated ion flow directions. The wide-ranging consistency between the model and observations have provided substantial validation to the conjecture that the location and orientation of the extended reconnection line is controlled primarily by the IMF clock and cone angles at Earth's magnetopause (Petrinec et al. 2016; Trattner et al. 2017), given the boundary condition of the Earth's diopole tilt (Eggington et al. 2020). These single IMF conditions remain true independent of the level of turbulence in the magnetosheath (Petrinec et al. 2022). Obviously, when the clock angle or cone angle rotate, the predicted location of the X-line changes. If the rotation is slow, then the predicted location changes in a way consistent with the Maximum Magnetic Shear model. However, if the rotation is fast, then there is a lag of several minutes as the magnetopause responds to the new IMF orientation (see Trattner et al. 2016)

2.4 Diamagnetic Non-suppression at Earth's Magnetopause

The presence of a pressure gradient (either particle number density or thermal speed) normal to a magnetic field leads to diamagnetic drift. At the magnetopause with magnetic shear across the current layer, an electron pressure gradient normal to a guide field (non-zero component that does not change across the magnetopause) results in an electron diamagnetic drift velocity that is both normal to the guide field and tangent to the magnetopause surface. If component reconnection occurs at the magnetopause along a line representing the guide field, then the diamagnetic drift velocity is along the outflow direction (Cassak and

Fuselier 2016). However, diamagnetic drift may suppress both the onset of reconnection (linear phase) associated with the tearing instability (Coppi et al. 1979; Galeev and Sudan 1984; Zakharov et al. 1993; Rogers and Zakharov 1995), and nonlinear reconnection (Swisdak et al. 2003). Theoretical modeling of this process with the inclusion of some simplifying assumptions leads to a straightforward relation between the jump in the plasma beta across the current layer and the local magnetic shear angle. This relation defines broad regions in parameter space {clock angle (θ), and change in plasma beta across the layer, ($\Delta\beta$) where magnetic reconnection is either 'suppressed' or 'possible'}. For the typical range in the change in plasma beta across the terrestrial magnetopause, reconnection is 'possible' over a large range of magnetic shear angles (Cassak and Fuselier 2016). The lack of diamagnetic suppression for the typical range of plasma conditions is the fundamental reason why the location of reconnection at the Earth's magnetopause is determined by IMF clock and cone angles. There are solar wind conditions for which diamagnetic suppression is expected for parts of the magnetopause surface; however, these conditions are rare in the solar wind at 1 AU and reconnection is typically not suppressed for most of the magnetopause surface the majority of the time.

2.5 Stationarity of Extended Reconnection X-Lines at the Magnetopause

For long intervals of southward IMF, the primary extended reconnection line(s) are also found to be persistent (quasi-stationary) at low- to mid-latitudes (e.g., Frey et al. 2003; Trattner et al. 2021b), extending far along the magnetopause flanks. One example of this behavior was demonstrated with MMS observations by Gomez et al. (2016). During this time, MMS was situated along the dusk flank magnetopause at low latitude and very close to the predicted anti-parallel merging region. Observations of reconnection ion jet reversals tangent to the magnetopause and reversals in heated, streaming electrons switching from parallel to anti-parallel and anti-parallel to parallel to the magnetic field were indicative of a reconnection site passing back and forth over the MMS location. These observations occurred multiple times over a period of $\sim$15 minutes, and small changes in the location of the antiparallel reconnection line coincided with small and slow changes in the IMF clock angle. Surprisingly, the X-line passed back and forth over the MMS location even though the ambient bulk tailward flow in the magnetosheath was approximately equal to the Alfvén speed.

The persistence of a quasi-stationary primary reconnection line from two separate magnetopause encounters during intervals of steady IMF clock angle was also noted in a study by Fuselier et al. (2019b). Ion distribution functions observed near the subsolar magnetopause in these separate encounters showed two magnetosheath ion populations, one entering the magnetosphere and one returning from the ionosphere. These two populations were used to remotely locate and track the reconnection line in the same way that the distance to the reconnection X-line is determined from cusp ion observations (see, Trattner et al. 2021a,b). Analysis of these events showed that the estimated location of the extended reconnection line was consistent with that predicted by the maximum magnetic shear model. Despite the estimated location of the reconnection line appearing along the magnetopause flank where the magnetosheath flow velocity may have been super-Alfvénic, the distance from the sampling spacecraft to the reconnection line was observed to be approximately constant over a span of several minutes.

2.6 Reconnection X-Lines Have a Variable, but Ordered Orientation

A primary extended reconnection line along the low- to mid-latitude dayside magnetopause includes a variety of orientations. These different orientations arise because such extended reconnection X-lines include a mix of reconnection types. These extended X-lines include anti-parallel segments and, especially when a substantial IMF B_y component exists, a component reconnection X-line segment (with non-zero guide field) that is present within a few hours of local noon and intersects the anti-parallel segments along the magnetopause flanks. In general, the intersection between the component and anti-parallel segments of the extended reconnection X-line occurs strictly along the bridge ('saddle') of highest magnetic shear (see Fig. 2). Some deviations from this model are noted during narrow ranges of IMF clock angles ($\sim$120° and $\sim$240°) (Trattner et al. 2018, 2021a) These deviations are a topic of ongoing research.

Varying orientations along an extended reconnection line are associated with a coordinate system oriented by magnetic field variance directions (e.g., the minimum variance analysis (MVA) (Sonnerup and Cahill 1967), minimization of faraday residue (MFR) (Khrabrov and Sonnerup 1998), and the maximum directional derivative of the magnetic field (MDDB) (Shi et al. 2005)). A recent study by Fuselier et al. (2021) showed that when electron diffusion regions were sampled by the MMS spacecraft at the magnetopause along the component reconnection segment of the extended primary reconnection line, the component reconnection line was often oriented along the M-direction (intermediate variance direction). In contrast, when electron diffusion regions were sampled at the anti-parallel reconnection line locations, the large-scale reconnection line was composed of a series of X-lines that are stacked in a stairstep fashion along the L-direction (direction of the reconnecting component of the magnetic field). The X-lines maintain the same orientation over long distances, with component X-lines and some anti-parallel X-lines maintaining orientation over many R_E and other anti-parallel X-lines maintaining orientation over at least 1 R_E.

2.7 Summary of Sect. 2

Reconnection at the dayside magnetopause initiates the Dungey cycle in the magnetosphere. Crucial to this initiation is reconnection at low latitudes, where previously closed magnetic field lines in the magnetosphere are opened. As originally depicted by Dungey (1963), the low latitude reconnection occurs when the IMF is southward. The newly opened field lines convect over the poles into the nightside. At the Earth's magnetopause, low-latitude reconnection occurs along long, quasi-continuous X-lines that extend across the entire dayside from the dawn terminator to the dusk terminator. These primary X-lines are quasi-stationary for quasi-steady IMF conditions. The reconnection rate may vary along these X-lines; however, it is not likely that it goes to zero for any extended period of time. The adjectives that describe reconnection X-lines at the magnetopause are summarized in Table 1.

The location and type of reconnection (component or anti-parallel) of these primary X-lines is well-described by the maximum magnetic shear model and recent studies using MMS have confirmed and extended this model. This empirically developed model uses only the IMF clock angle and cone angle to determine where reconnection occurs on the magnetopause. At the Earth's magnetopause (for southward IMF conditions), the IMF orientation alone controls the reconnection location because, over nearly the full range of solar wind conditions, there is no suppression of reconnection by diamagnetic effects.

3 Field Line Convection

3.1 Reconnection Jets and Convection

Once the magnetic fields of the Earth and solar wind are connected at the dayside X-line, "open" field lines are created. Since the extension of the open field line into the solar wind

Table 1 Adjectives that describe X-lines at the Earth's magnetopause and the definitions of these adjectives

Adjective	Meaning	Comparison to physical parameters	Observed Quantity
Primary	The main X-line at the magnetopause as opposed to secondary X-lines.	N/A	N/A
Long or Extended	Spatial extent of the X-line on the magnetopause.	Much longer than an ion skin depth: $\sim$750 km at the magnetopause	Extending from >1 R_E to >10 R_E or >10 to >100 ion skin depths
Continuous or Quasi-continuous	Unbroken. Observations anywhere along the X-line yield similar meso-scale structure.	Continuous over lengths much longer than an ion skin depth	Quasi-continuous from >1 R_E to >10 R_E or >10 to >100 ion skin depths
Quasi-stationary	Remaining in approximately the same location under quasi-steady IMF conditions.	Remaining approximately in the same location for timescales longer than plasma convection times at the magnetopause: $\sim$1-2 R_E per minute	Quasi-stationary from $\sim$several minutes to >10 minutes.
Quasi-steady	Reconnection rate variations along the X-line can be large, but do not go to zero.	Reconnection rate variability is not from zero to the maximum possible rate.	It is very difficult to quantify reconnection rate variations, but the rate does not go to zero over time periods of minutes to tens of minutes

travels away from the Sun at 300-800 km/s, magnetic tension pulls the field line across the dayside and slings it to the nightside, where it becomes an open tail lobe field line. This $\mathbf{J} \times \mathbf{B}$ force is well modeled by MHD models such as are available in the CCMC.

The ion jets from dayside reconnection partially precipitate into the atmosphere, creating a region called the "cusp", or in some early papers, the "cleft" (Frank 1971; Heikkila and Winningham 1971). Since the particles are convecting tailward at the same time they are precipitating, this leads to an "ion cusp dispersion", with the highest-energy ions precipitating first and thus most equatorward, and the progressively lower energy ions farther poleward (Shelley et al. 1976; Reiff et al. 1977). By examining the details of the particle cutoff energy, the distance to the X-line is determined (Cowley 1982) and the jet distribution reconstructed (Hill and Reiff 1977; Lockwood 1997). Occasionally an electron dispersion is observed between the last closed field line and the beginning of the ion dispersion (Burch et al. 1982). This electron dispersion is the ionospheric footprint of the electron edge of the low-latitude boundary layer at the magnetopause (Gosling et al. 1990).

The convection over the pole of reconnected field lines and reconnection in the tail produces a polar cap convection pattern depicted in Fig. 3 for purely southward IMF. This pattern is driven by the mapping of the solar wind electric field along field lines to the ionosphere as shown in Fig. 3. The dawn-to-dusk electric field (shown as a heavy red arrow) causes dayside-to-nightside $\mathbf{E} \times \mathbf{B}$ flow (shown as dashed lines) and produces the cross-polar cap potential. In principle, the cross-polar cap potential should be related to the length of the X-line or lines at the magnetopause. However, in practice, it is difficult to link the two quantities because of variations of the reconnection rate along the X-lines at the dayside and inefficiencies in plasma transfer on the flanks.

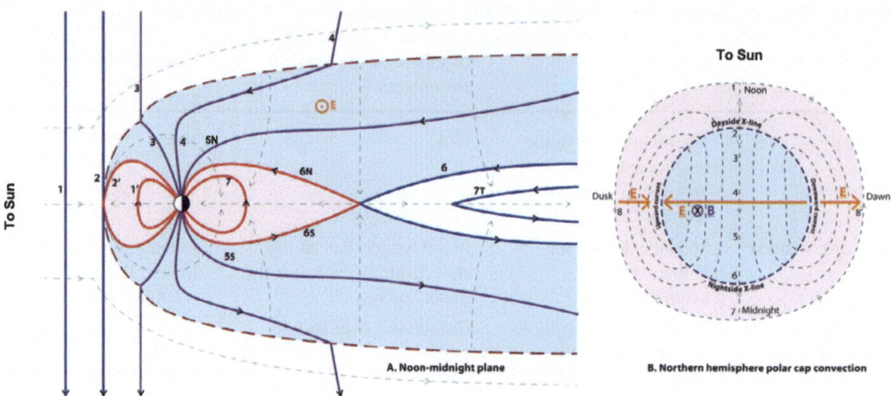

Fig. 3 (left panel) Simplified cross-section of the magnetosphere for due southward IMF. The bow shock and field line draping at the magnetopause were removed to illustrate the convection of a solar wind field line from noon to midnight. Solar wind field lines (blue, 1) reconnect at the magnetopause (2, 2') and form open field lines that convect over the poles (3, 4, 5N, 5S). In the tail, these field lines reconnect (6, 6N, 6S), forming closed field lines (7) earthward of the reconnection site. The closed field lines eventually return to the dayside. (right panel) The magnetopause reconnection, convection, tail reconnection, and the return of the field lines to the dayside produces a 2-cell convection pattern in the ionosphere. The footpoints of the field lines in the left panel are shown in the right panel

Continuing with the polar cap convection patern in Fig. 3, magnetopause reconnection maps to the polar cap boundary near point 2, the field convects over the pole and through the polar cap from point 2 to point 6. Tail reconnection maps to the polar cap boundary near point 6. The polar cap flow continues equatorward and back sunward at lower latitudes. The sunward convecting field lines are closed and ultimately return to the dayside (point 8 and 8' back to point 1 in Fig. 3, with the return field lines not in the noon-midnight meridian cut in the left panel of Fig. 3). Ion outflow from the ionosphere occurs in the cusp, polar cap, and the nightside polar cap boundary.

When the X-line is not parallel to the equatorial plane on the dayside, as it frequently does for a strong non-zero Y-component of the IMF, the ionospheric outflow has a substantial dawnward or duskward flow component – typically reversed in the two hemispheres (Gonzalez and Mozer 1974). When IMF $B_y > 0$, the component reconnection X-line is tilted so that it is above the ecliptic on the duskside. Under these conditions, magnetic field lines convect dawnward and poleward and the cusp flow is dawnward in the northern hemisphere (e.g., Heelis 1984). The reverse is true in the Southern hemisphere or for $B_y < 0$.

The asymmetric flow also extends to the plasma mantle, the extension of the open field lines with cusp-like fluxes to the high latitude magnetotail. The mantle also exhibits a cusp ion dispersion, with the highest energy ions observed farthest from the magnetopause (Rosenbauer et al. 1975). Most of the plasma mantle is lost down the magnetotail (e.g., on field lines 5N and 5S in Fig. 3), leading to significant ionospheric escape (e.g., Schillings et al. 2020). The asymmetry of the convection means that the plasma mantle reaches the equatorial plane at lunar distance only on the favored (higher flow) side (Hardy et al. 1979). Only during times of very high convection is the magnetotail electric field intense enough that the plasma mantle reaches the near-earth neutral sheet. When this mantle flux reaches the neutral sheet during these extreme events, it participates in near-Earth reconnection (Reiff et al. 2016; see next section).

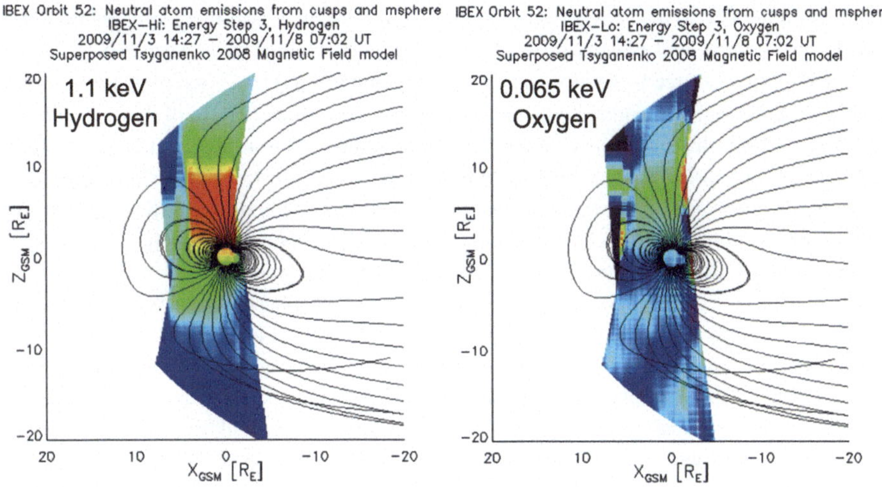

Fig. 4 ENA observations of the magnetosphere from the IBEX mission. The view is from the dusk side of the Earth and magnetic field lines are drawn from a magnetospheric magnetic field model to show the relationship of the ENA fluxes to the Earth's magnetospheric cusps. The left panel shows the precipitation of 1.1 keV hydrogen from the magnetosheath is asymmetric, with more precipitation occurring in the northern hemisphere cusp. The right panel shows that the 0.065 keV O^+ outflow from the ionosphere is also asymmetric with more outflow coming from the northern hemisphere cusp

3.2 Imaging the Cusp – Energetic Neutral Atoms

As solar wind protons enter the magnetosphere through reconnection and are accelerated along field lines with some precipitating in the Earth's magnetospheric cusps, a fraction of these ions charge exchange with the Earth's geocorona and become Energetic Neutral Atoms (ENAs). These ENAs are no longer bound to the Earth's magnetic field lines and propagate away from the Earth in all directions. Although the ENA flux is very low, it has been imaged by ENA cameras on the IMAGE mission (Burch 2000) and on the Interstellar Boundary Explorer (IBEX) mission (McComas et al. 2009). Figure 4 shows images of the Earth's magnetospheric cusps taken by the IBEX ENA cameras (Funsten et al. 2009; Fuselier et al. 2009). The left panel is an image of 1.1 keV Hydrogen, which images the entering magnetosheath population. The right panel is an image of 0.065 keV Oxygen, which images the ionospheric outflow population. The precipitation of magnetosheath ions into the cusp is asymmetric because the location of the reconnection X-line and the entering magnetosheath ion flux depends on dipole tilt (Petrinec et al. 2011). Figure 4 shows that the outflow is also asymmetric, with more O^+ outflow from the northern hemisphere than from the southern hemisphere. The origin of the O^+ outflow could be the cusp/cleft, or just the general auroral oval. The image is averaged over such a long period of time and does not have the spatial resolution to distinguish among these sources.

4 Magnetotail Plasma Sheet

4.1 Overview

As discussed in the previous three sections, magnetopause reconnection plays a crucial role in enabling energy and plasma from the solar wind to enter the magnetosphere. Furthermore, as discussed in Sect. 3, the convection of reconnected field lines over the geomagnetic poles deposits this energy and plasma in the Earth's extended magnetotail. Also in Sect. 3, an additional particle source is the ionosphere, which populates tail lobes and plasma sheet, depending on presumably reconnection-related activity. The ultimate origin of this ionospheric plasma (i.e., the high latitude cusp, auroral oval, and polar cap) is a subject of ongoing study (e.g., Glocer et al. 2020; Kistler 2020; Toledo-Redondo et al. 2021a). Furthermore, the solar wind can also do work on the magnetosphere via compression and waves, thereby causing energy entry that does not directly involve reconnection. However, reconnection remains the dominant driving force in magnetospheric dynamics for southward IMF.

The plasma sheet may become populated directly by ionospheric particles and by plasma that enters through the low-latitude flanks, or indirectly via transport through the lobes, which is part of the Dungey cycle. The latter entry then requires magnetotail reconnection for the transport from the lobes to the plasma sheet. The relative importance of the different sources and entry mechanisms varies considerably depending particularly on distance along the tail and geomagnetic activity (e.g., Wing et al. 2014; Welling et al. 2015; Kistler 2020).

The Dungey cycle includes plasma transport from a nightside reconnection site toward and around the Earth to the dayside. Although the earthward transport from the nightside may be viewed as a quasi-steady process, it has been found to be inconsistent with adiabatic (i.e, entropy and mass conserving) transport (see Sect. 4.3).

The onset of reconnection in these events is apparently preceded by local current sheet thinning. The conditions that determine the location and extent of the thinning regions then are also responsible for the location of the reconnection sites. This section focuses on the global context. The local structure and plasma conditions of thin current sheets (TCSs) are discussed in Hwang et al. (2023, this collection).

Major consequences of reconnection are fast, Alfvénic, outflows. Earthward flow bursts are associated with locally enhanced northward magnetic fields and cross-tail electric fields, which are found to be the dominant mechanism to accelerate ions and electrons to suprathermal energies (see e.g., Phan et al. 2013; Oka et al. 2022; and Oka et al. 2023, this collection). Velocity shear and vorticity at the edges of the flows, particularly in their stopping and diversion region near Earth, are also a mechanism to twist and shear the magnetic field, building up field-aligned currents. These currents provide a connection to auroral streamers and, on larger scale, the substorm current wedge.

4.2 Sources and Entry

As pointed out above, there are two basic sources of plasma sheet particles, the ionosphere and the solar wind. Source differences are relevant for subsequent reconnection in the plasma sheet, as discussed in Chap. 1. Ionospheric ions enter the magnetotail in two ways (e.g., Kistler et al. 2019), either directly onto closed field lines through the auroral region or through the cusp onto open, lobe field lines. In the latter case, tail reconnection is necessary to trap these particles in the plasma sheet, similar to solar wind ions that enter from the lobes. During times of extreme polar cap potential drop, convection brings ionospheric O^+ through the tail lobes to become an accelerated population in the near-Earth plasma sheet (Reiff et al. 2016), as illustrated in Fig. 5.

As described in Sect. 1, for southward IMF solar wind plasma enters the magnetotail and eventually the plasma sheet predominantly by the dayside reconnection process.

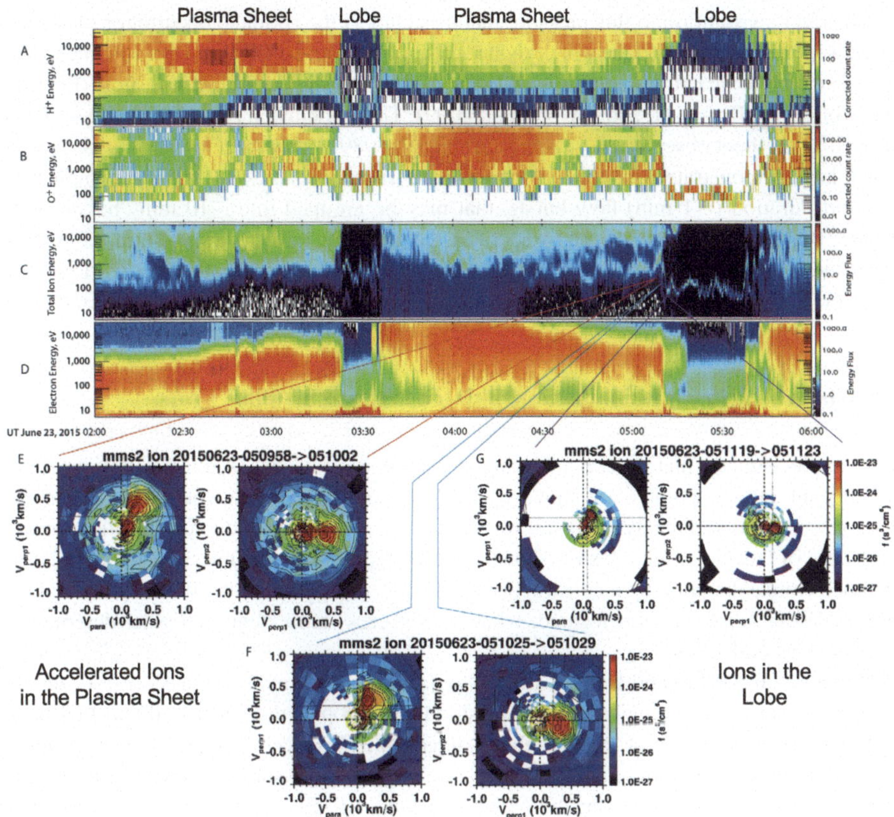

Fig. 5 HPCA (MMS1) and FPI (MMS2) measurements in the magnetotail from 0200 to 0600 on 23 June. (A) Energy spectrograms of the HPCA H^+. Panel B shows an O^+ beam from the ionosphere merging into the plasma sheet at each lobe/plasma sheet transition. FPI total ion and electron spectrograms of differential energy flux are shown respectively in Panels C and D, and clearly shows the transition from lobe conditions (virtually no fluxes except the O^+ beam) to PSBL (where the energized lobe fluxes appear separate from the much more energetic plasma sheet fluxes). The bottom three panels (E, F and G) show three pairs of FPI distribution functions as the spacecraft exited the plasma sheet to the lobe. The left of each pair shows v_{parallel} and v_{perp1} (along $\mathbf{E} \times \mathbf{B}$) components of the particle distribution functions, and the right of each pair shows the two perpendicular velocities, v_{perp1} and v_{perp2}. (From Reiff et al. 2016)

4.3 Transport

As briefly mentioned above, the earthward transport from a distant reconnection site, which is part of the Dungey cycle for southward IMF is visualized as a quasi-steady process within a steady magnetospheric magnetic field. However, Erickson and Wolf (1980) were the first to demonstrate that the average magnetotail configuration is inconsistent with adiabatic (i.e, entropy and mass conserving) transport from the distant to the near tail; such transport would lead to a pressure build-up in the near tail that is not observed and would not be balanced by the observed lobe magnetic pressure. Kivelson and Spence (1988) and Spence et al. (1989) then demonstrated that energy-dependent particle loss from cross-tail drifts may be sufficient only under very weak transport scenarios to reduce the pressure and solve this "pressure inconsistency."

The obvious solution to this pressure inconsistency is the sporadic occurrence of reconnection in the near or mid tail, which reduces the particle and energy content of convecting closed magnetic flux tubes and adds buoyancy effects to the earthward transport (Pontius and Wolf 1990; Birn et al. 2009; Wolf et al. 2009). This conclusion is strongly supported by direct plasma sheet observations (Baumjohann et al. 1989, 1990; Angelopoulos et al. 1992), which have demonstrated that the major earthward transport in the near tail is accomplished by short-duration ($\sim$1 min) flow bursts, that may be grouped into $\sim$10 min "bursty bulk flows" (BBFs), presumably driven by sporadic reconnection in the near- and mid-tail region or by interchange instabilities that allow rarified longitudinal sectors to enter while mass-loaded longitudinal sectors escape down the tail (Wolf et al. 2012; Sorathia et al. 2020).

4.4 Where and How Is Reconnection Initiated in the Near-Tail

The simple Dungey picture contains only one reconnection site (x-line) in the, presumably distant, tail. As pointed out above, a steady convection scenario would be inconsistent with observations and the dynamic tail is more consistent with sporadic reconnection events earthward of that site. Figure 6 shows the tail structure with two reconnection sites (x-lines), obtained from a data mining reconstruction technique, discussed further below (Sitnov et al. 2021). Where are these sites located, what is their cross-tail extent, and what determines their location and extent?

Zwickl et al. (1984) and Daly et al. (1984) used statistical analyses of ISEE-3 observations to show that strong, tailward bulk plasma flow becomes dominant in the distant ($r > 150\ R_E$) plasma sheet. These tailward flows suggest the presence of a typical x-line location inside that distance; that has a tendency to move tailward after substorm onset (Baker et al. 1984). Fast flow observations at the Moon's distance (Kiehas et al. 2018) indicated that frequently the distant neutral line is beyond 60 R_E downtail. Øieroset et al. (2000, 2001) identified a distant reconnection event near 60 R_E. Occasionally, however, the distant x-line location may even move beyond 220 R_E (Schindler et al. 1989).

The location of the reconnection site in the near and mid tail may also vary considerably, being mostly between 20 and 30 R_E downtail and rarely inside of 20 R_E (Nagai et al. 1998; Nagai and Machida 1998). Indeed, for phase 2 of the MMS mission, the spacecraft apogee of 25 R_E (later raised to 29 R_E) was chosen to maximize measurements in the region where near-tail reconnection most often occurs (Burch et al. 2016; Fuselier et al. 2016). It is not well known what determines this location and, prior to onset, the thinning of the current sheet, which enables the onset (McPherron et al. 1987; Pulkkinen et al. 1992; Sergeev et al. 1990; Sanny et al. 1994).

Nagai et al. (2005) used Geotail reconnection events to study the possible solar wind control of the radial distance of the magnetic reconnection site and found it likely that a higher efficiency of the energy input, measured by $V_x B_z$, rather than the total amount of energy input, affects this location. Magnetic reconnection was found to take place closer to the Earth when the energy input was more efficient. This is consistent with indications that reconnection occurs closer to Earth during storm-time substorms within a stressed magnetotail (Nagai 2021; see also Sect. 4.7).

The onset of reconnection is presumably preceded by the thinning of the tail current sheet to ion or sub-ion scale or, more likely, by the formation of a TCS embedded in the wider plasma sheet (for recent reviews, see Sitnov et al. 2019a,b; Runov et al. 2021). Birn and Schindler (2002) and Birn et al. (2004) demonstrated, through magnetostatic theory

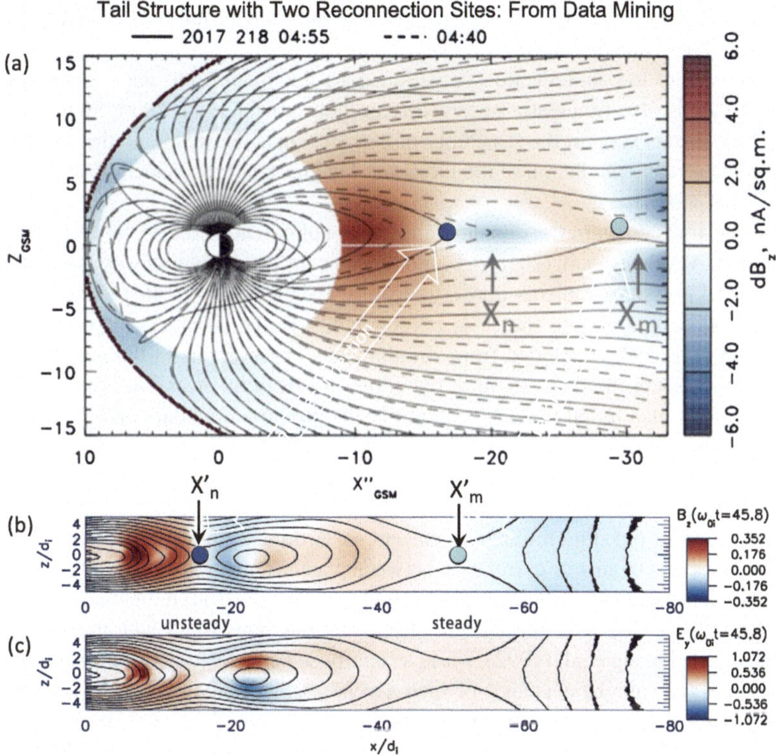

Fig. 6 (a) Regions of unsteady (X_n) and steady (X_m) reconnection in the data-mining reconstruction of the 6 August 2017 substorm shown by the meridional distribution of the normal magnetic field variation $dB_Z = B_Z(t_1)$-$B_Z(t_0)$ with $t_0 = 04$:40 UT and $t_1 = 04$:55 UT; (b) Similar regions of unsteady (X'_n) and steady (X'_m) reconnection found in PIC simulations; (c) the distribution of the electric field Ey for the same PIC run; modified after Sitnov et al. 2021)

and MHD simulations, that high-latitude magnetopause boundary deformations, as expected from magnetic flux addition during the substorm growth phase, could cause the formation of local TCSs embedded in the wider tail plasma/current sheet. An alternative mechanism is based on enhanced equatorial plasma flow toward the dayside, which is part of the Dungey cycle (Hsieh and Otto 2014). This was shown to cause magnetic flux depletion at low latitudes, also leading to local current sheet thinning. Both mechanisms may operate together or at different times and different distances in the tail (Hsieh and Otto 2015; Gordeev et al. 2017).

High-latitude flux addition is consistent with an increase of the lobe field strength during the substorm growth phase, which is documented in many cases (e.g., McPherron 1972; Baker et al. 1996). It is also consistent with observational studies that indicate that substorm onset is more likely when the total open flux is increased (Kamide et al. 1977; Milan et al. 2009; Boakes et al. 2009; Lockwood et al. 2019). Specifically, Boakes et al. found that substorm onset was unlikely when the open flux was below 0.3 GWb, but increased linearly with increasing open flux above that value up to ~0.9 GWb, their observed maximum. This view is consistent with the role of the growth phase in the standard substorm model (Sect. 4.6). The models of TCS formation from high-latitude flux addition indicate a close relationship between thin current sheet formation (and hence substorm onset) and tail flaring (see, also, Lockwood et al. 2019), which may be relevant for determining the location of TCS formation and reconnection onset.

In contrast, the low-latitude flux depletion does not require an increase in lobe field strength, which is also frequently observed (e.g., Petrukovich et al. 2000; Shukhtina et al. 2014). These results support the feasibility of both concepts. However, neither concept permits a direct and easy prediction of the location of the near-tail reconnection site from solar wind or other parameters.

Other questions concern the downtail and cross-tail extent of the TCSs and subsequent reconnection. Artemyev et al. (2016b), found that current sheet thinning was associated with a strong radial pressure gradients $\partial p/\partial r$, indicating a short scale (0.2 R_E) along the current sheet that was comparable to the scale perpendicular to the sheet. At the other extreme, Artemyev et al. (2019) concluded from simultaneous observations in the near and distant tail that current sheet thinning occupies the entire tail from the near-Earth region down to lunar orbit. They also found that convergent plasma pressure gradients $\partial p/\partial y$, measured in equatorial thinning current sheets, indicated a localization of current sheet thinning near midnight. Such a concentration would be consistent with the estimated widths of earthward reconnection outflows of a few R_E (Nakamura et al. 2004)

One of the most puzzling findings of current sheet thinning prior to substorm onset is a collection of reports of an anti-correlation between plasma temperature and density, showing increases in density along with decreases in (both ion and electron) temperature (Artemyev et al. 2016a, 2019, 2021) during current sheet thinning. This trend had already been reported earlier by Huang et al. (1992), while a superposed epoch study by Baumjohann et al. (1991) did not indicate the existence of such a trend. A thermodynamic, e.g., isobaric, scenario that increases plasma density while reducing temperature in a single population (Birn et al. 1994) appears implausible. A more likely interpretation is that the satellites encounter different flux tubes carrying different plasma populations. This seems most plausible near the plasma sheet boundary where thinning brings different plasma populations equatorward combined with a compression. Most likely, the effect cannot be explained by just one plasma population but may involve, for instance, losses of higher-energy particles, which contribute more significantly to the temperature than to the density, or the inclusion of a cold particle population.

Another explanation of the temperature and density correlation is the formation of TCSs due to Speiser orbit effects (Sitnov et al. 2003). In this approach the current sheet thickness scales as the ion gyroradius in the lobe field, which is also close to the ion inertial length when the plasma anisotropy is small (Sitnov and Arnold 2022). The first scaling explains the decrease of the temperature, while the second matches the increase of the plasma density. Moreover, the first scaling is also consistent with the anticorrelation of the TCS thickness with the lobe field strength (e.g., Stephens et al. 2023, Fig. 4).

In principle, global simulations of the solar wind/magnetosphere system might be considered as the ideal tool to investigate the solar wind effect on the initiation of reconnection in the magnetotail. However, x-lines are often found to form too close to the Earth in these simulations (e.g., El-Alaoui et al. 2009; Park 2021). At present, it is not clear whether this is due to the missing non-MHD physics (e.g., Raeder et al. 1996) or to unavoidable numerical dissipation (e.g., Gonzalez and Parker 2016; Raeder 2022). In many of these models x-line formation and reconnection happen despite the absence of an explicit dissipation term in the MHD equations. Despite these deficiencies, large-scale MHD simulations have been highly successful in reproducing major global effects associated with substorms (see, e.g., Birn et al. 1996; Raeder 2003; Merkin et al. 2019), including the location of spacecraft relative to the X-line (Reiff et al. 2016; Torbert et al. 2018; Reiff et al. 2018).

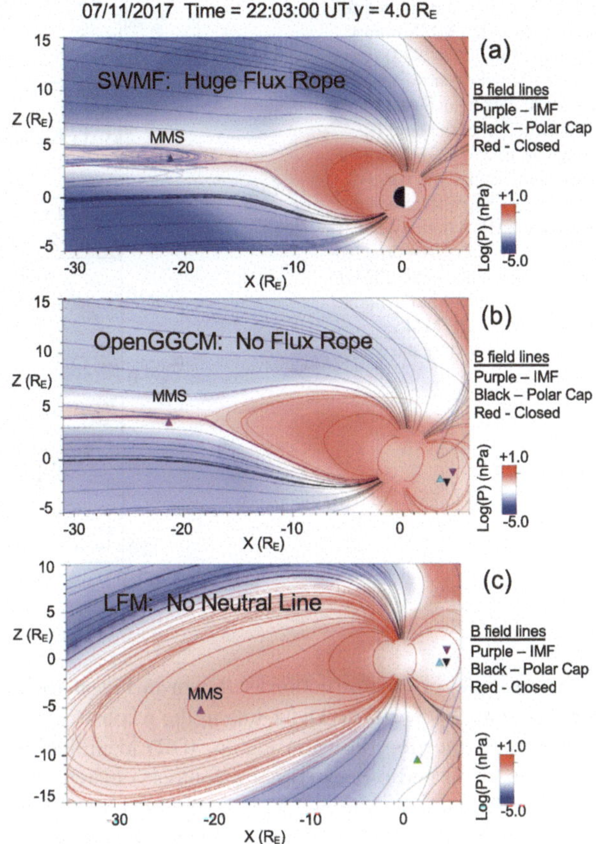

Fig. 7 Simulation results of an event on July 11, 2017, showing magnetic flux contours (solid black lines) and plasma pressure (color) on a logarithmic scale: (a) SWMF simulation in GSM coordinates, (b) OPENGGCM in GSM coordinates, and (c) LFM model in SM coordinates

Figures 7 (a, b, c) show frames from simulations of the July 11, 2017 tail reconnection event observed by MMS. The top figure (a) shows the SWMF simulation, which accurately predicted the MMS crossing the X-line a few minutes prior and predicted the huge flux rope which MMS observed. The middle simulation (b) shows the equivalent OpenGGCM model, which shows an extended neutral sheet but no flux rope; and (c) shows the same time using the LFM model which did not predict a near-earth neutral line at all.

Global simulations yield the evolution of X-lines and reconnection from an evolution driven by a time-dependent solar wind input. An alternative method relies on large statistical data bases of magnetospheric quantities (primarily magnetic fields) measured by satellite missions during varying driving and activity conditions. Reconstruction of the global structure from data is very challenging because of the data paucity with less than a dozen probes available for in-situ observations at any moment. However, modern machine-learning methods, and in particular, the "lazy-learning" approach based on mining multi-decade and multi-mission archives of geomagnetic field data using a nearest neighbor method combined with flexible magnetic field architectures using basis function expansions (Tsyganenko and Sitnov 2007; Sitnov et al. 2008; Stephens et al. 2019; see also Sect. 5 for more detail) have recently provided a breakthrough in this direction (Stephens et al. 2023). A key to the reconstruction has become the recurrent nature of storms and substorms, which allows one to enrich a few points of real observations available at the moment of interest with a super-constellation of up to 50,000 synthetic probes (shown by gray dots in Fig. 8).

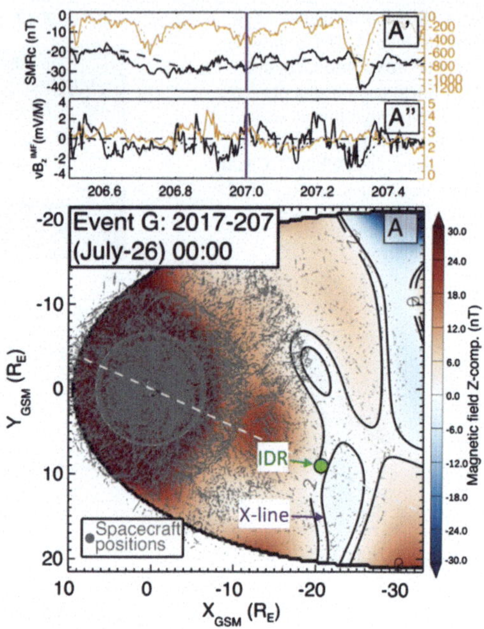

Fig. 8 X-lines reconstructed for event G from the MMS library (Rogers et al. 2019, 2023) (green circle). Gray dots show ~50,000 historical data points used for the reconstruction of the field B_Z (Stephens et al. 2023) assuming zero dipole angle. Top panels show global SuperMag indices SMR and SML (Gjerloev 2012) (SMRc is the pressure-corrected SMR (Tsyganenko et al. 2021)), vB_{zIMF} and solar wind dynamic pressure P_{dyn}. The MMS location is marked by the green circle

An example of reconstruction is shown in Fig. 8 for event G from the MMS library (Rogers et al. 2019, 2023). The x-line in the reconstructed field coincides closely with the reconnection site (ion diffusion region, IDR) identified by MMS. In total, 24 out of 26 reconstructed x- and o-lines ($B_Z = 0$ contours) matched within $\sim 2R_E$ or nearly matched ($B_Z = 2$ nT contours within $\sim 2\ R_E$) the observed MMS IDR locations.

The data mining (DM) method predicts neutral line locations from finding nearest neighbors in parameter space based on five parameters. The parameter V_xB_z directly describes solar wind input, whereas the others, given by the time-averaged Sym-H and its time derivative, as well as the AL index and its time-derivative, characterize storm and substorm dominated states that contain solar wind input only indirectly. The success of the DM method in predicting neutral line locations is surprising in view of the fact that none of these parameters has a direct obvious relationship to the location of the x-lines in the tail. Most recently, these empirical reconstructions have been merged with global MHD models using the explicit resistivity method (Hesse and Birn 1994) where DM "pointed" MHD to the data-derived X-line locations to nudge reconnection there (Arnold et al. 2023).

Through instantaneous fitting of sequences of configurations, the DM approach was also able to demonstrate the stretching and dipolarization during a substorm cycle (Stephens et al. 2019) with the formation of thin ion-scale current sheets embedded within the thicker plasma sheet (Sitnov et al. 2019b; Stephens et al. 2023).

4.5 Consequences of Magnetotail Reconnection: Flow Bursts, Particle Acceleration, Substorm Current Wedge, and Plasmoid Formation

A direct consequence of magnetic reconnection is the conversion of magnetic energy into particle energy, which on fluid scales takes the form of plasma heating and fast bulk flows. On the earthward side of tail reconnection this is most prominent in the form of bursty bulk flows (BBFs, Angelopoulos et al. 1992), consisting of ~10 min fast flow periods of hundreds of km/s with individual peaks of $\sim$1 min duration (For recent reviews, see Sitnov et al. 2019a and Birn et al. 2021). Individual flow bursts tend to be associated with temporary enhancements of the (northward) magnetic field component Bz, which has led to denoting such an event dipolarizating flux bundle" (DFB, Liu et al. 2013a) or "flux pileup region" (FPR, Khotyaintsev et al. 2011). The enhanced Bz in combination with enhanced earthward flow also leads to an enhancement of the (predominantly duskward) electric field, which has led to the concept of "rapid flux transport" (RFT) events (Schödel et al. 2001).

When plasma is transported earthward from a distant reconnection site, particles are expected to become energized adiabatically; consistent with the shortening of magnetic flux tubes and the increase of the magnetic field strength. However, as pointed out above, there is little evidence that this happens in a quasi-steady fashion. Instead, sporadic reconnection events in the near and mid tail play the dominant role in energizing ions and electrons during their transport toward Earth, either in the vicinity of the reconnection site or within its exhaust regions. Numerous investigations based on observations, theory, and simulations have identified the, temporally and spatially localized, electric field of RFT events as the primary mechanism accelerating ions and electrons in the near tail to tens or hundreds of keV, and causing energetic particle "injections", first documented by observations at geosynchronous orbit. This subject is discussed in more detail in Oka et al. (2023, this collection) (see also reviews by Birn et al. 2012, and Fu et al. 2020).

DFBs can also be the source of smaller-scale waves. Specifically, the sharp rise of Bz at the front of a DFB, the "dipolarization front" (DF), may drive lower-hybrid waves, and electron anisotropies in the accelerated population inside the DFB may be the cause of whistler waves. These effects are further discussed in Hwang et al. (2023, this collection).

The earthward flow from a near-tail reconnection site must be stopped closer to Earth and diverted azimuthally eastward and westward, consistent with the Dungey cycle. The flow shear and diversion distort the embedded magnetic field, causing twist and shear that is associated with field-aligned currents, which, in their simplest form flow toward the Earth on the dawn side and away on the dusk side. This paradigm was first developed on the basis of MHD simulations of near-tail reconnection (Birn and Hesse 1991; Scholer and Otto 1991). It is now the most widely accepted view of the build-up of the "substorm current wedge" (SCW), which also includes the closure of the field-aligned currents in the ionosphere through the auroral electrojet (McPherron et al. 1973; Keiling et al. 2009; Birn and Hesse 2014; Kepko et al. 2015).

This picture describes the build-up of the SCW. Once the magnetic field is distorted, the currents can continue to flow, maintained by overall force balance without further need of fast flows, until they are dissipated by ionospheric resistivity. Thus, fast flows typically have a duration of only a few minutes, while the wedge currents may persist for tens of minutes up to 1 hour. Individual flow bursts have been shown to be associated with wedge type currents (Sergeev et al. 1999, 2004; Nakamura et al. 2001a,b) but may collectively contribute as "wedgelets" to the total SCW (Liu et al. 2015; Birn et al. 2019; Merkin et al. 2019).

The major tailside consequence of near-tail reconnection is the severance and tailward ejection of a section of the plasma sheet, denoted as a plasmoid (Hones 1977). Plasmoids consist of loop-like magnetic field lines, which in a generalized 3D picture assume the form of a flux rope (Hughes and Sibeck 1987), some of which may include rather strong axial fields. Since plasmoids are consistent with transient reconnection, they are discussed in more detail in Sect. 7.

The original picture of plasmoids ejected tailward as simple entities is an oversimplification that has been modified by observations and simulations. They gain momentum by the accumulation of accelerated plasma with significant 3D variation (Hesse and Birn 1994).

Their speed may vary both along and across the tail; and some may even be stagnant (Nishida et al. 1986).

4.6 The Near-Earth Neutral Line Substorm Model

The formation of a new, temporary, reconnection site in the near tail is the core element of the neutral line model of substorms (Baker et al. 1996), also referred to as the near-Earth neutral line (NENL) model. Essential features were already formulated by Atkinson (1966) and summarized by Atkinson (1967):

a. The solar wind drags field lines from the region of closed field lines into the tail, either by viscous forces, or by the neutral-point mechanism proposed by Dungey (1958). There is a resulting increase in the tail magnetic field strength and a storing of potential energy.

b. The polar substorm begins when field lines recombine in an implosive fashion at the neutral sheet, in the manner indicated by Petschek (1964). This recombination implies the release of stored potential energy.

c. The recombined flux tubes are added to the night side of the closed region as a giant bulge, causing auroral effects.

d. The flux tubes flow around the closed region toward the day side, causing the magnetic substorm and further auroral effects.

Phase (a) obviously corresponds to the substorm "growth phase," introduced by McPherron (1972) as part of a three-phase phenomenological model, consisting of growth phase, expansion phase and recovery phase (McPherron et al. 1973). Phase (b) represents the onset and phase (c) the expansion phase. Although this was not identified at the early time, phase (d) describes in fact the present view of the build-up of the substorm current wedge (Kepko et al. 2015), presented in Sect. 4.5. Further evidence and a detailed illustration that included a preexisting distant neutral line and the ejection of a plasmoid was then provided by Hones (1977). An updated view and more comprehensive discussions of the features of the NENL model including the context of modeling have been presented by Baker et al. (1996, 2016). The neutral line model represents the background for our understanding of large-scale substorm dynamics, particularly applied to isolated substorms. Some details, however, are still active areas of research, such as the potential role of ballooning/interchange instability prior to the onset of near-tail reconnection or in acceleration and providing cross-tail structure of magnetic flux tubes that are depleted by reconnection and plasmoid ejection (for a recent review, see Sitnov et al. 2019a,b).

4.7 Magnetotail Reconnection in Different Magnetospheric States

Both the solar wind driver and the pre-conditioning of the magnetosphere can modify the location and strength of reconnection in the near-Earth magnetotail (Nagai et al. 2005). The preconditioning, which involves the magnetic field configuration and the plasma sheet parameters, such as temperature, density, and ion composition, results not only from the solar wind history but also from coupling with the ionosphere. The pressure distribution in the magnetotail also affects where reconnection jets get diverted in the inner magnetotail (Dubyagin et al. 2010).

Various scenarios, in addition to isolated substorms, include magnetic storms, steady magnetospheric convection (SMC) events, periodic substorms and sawtooth events, as well as pseudo-breakups. The day/night flux transport and the location of reconnection differ in different magnetospheric states.

The most extreme cases are magnetic storms, occurring under strong solar wind driving conditions, which compress the magnetosphere and increase magnetosphere-ionosphere coupling by enhanced energy deposition, also causing also enhanced ionospheric outflow of cold ions (Kronberg et al. 2021) and enhanced electric fields. Reconnection in storm time substorms can occur much closer to Earth (Miyashita et al. 2005; Angelopoulos et al. 2020) due to its compressed state.

Strong recurrent substorms during geomagnetic storms are called sawtooth events (STEs). They are associated with strong quasi-periodic energetic particle injections in the inner magnetosphere with recurrence times of 2–4 h (Borovsky et al. 1993; Borovsky and Yakymenko 2017). They are also likely related to solar wind Mach number (Lavaud and Borovsky 2008). STEs have a wider azimuthally extended injection region (Clauer et al. 2006; Henderson et al. 2006), a wider night-side ionospheric convection pattern, and a wider local time extent of dipolarization than isolated substorms (Cai et al. 2006a,b). This wider extent indicates also a wider cross-tail extent of the tail reconnection site, which tends to be located much closer to Earth than for isolated substorms (Henderson 2004), consistent with their storm-time occurrence.

In addition to the $\sim$3h period STEs, substorm-like periodic activations with a $\sim$1h period have been reported (e.g., McPherron and Chu 2018; Keiling et al. 2022), associated with dipolarizations, fast plasma flows and particle injections with the same periodicity. They are, however, not associated with storms and can occur over a wide range of AE values.

The mechanisms driving these periodic events are not well understood. For STEs, it might play a role that the nightside magnetosphere is much more stretched prior to the onset (Cai et al. 2006b). The approximate 3h waiting time might reflect an intrinsic oscillation period, assumed under strong driving (Huang et al. 2003; Borovsky and Yakymenko 2017). Global simulations with ionospheric outflow have suggested a possible role of stretching of the magnetotail from increased mass loading by O^+ from the ionosphere, leading to an imbalance between day-night reconnection rates, which may result in a feedback loop in the magnetosphere-ionosphere system (Brambles et al. 2013; Ouellette et al. 2013).

Another possible mechanism driving periodic stretching and release relies on non-MHD effects, which arise from multi-scale modifications to a pure MHD approach, either by adding a physics-based dissipation term (Kuznetsova et al. 2007) or by including a locally embedded PIC simulation into a global MHD approach (Wang et al. 2022). Both approaches found consistency with a wider reconnection site in the tail, but with somewhat shorter sawtooth periods than observed: $\sim$1.5 hrs (Kuznetsova et al. 2007) and 1.5 to 3 hrs (Wang et al. 2022). Wang et al. also reported weaker signatures at geosynchronous orbit and maximum magnetic field stretching near dawn and dusk rather than at local midnight.

These active events are in contrast to intervals called steady magnetospheric convection (SMC). SMCs are extended periods defined by quasi-steady solar wind input under southward IMF Bz, enhanced magnetospheric convection, but an absence of substorms. They are apparently governed by overall balance between dayside and nightside reconnection (e.g., DeJong and Clauer 2005). However, sporadic reconnection appears to happen in the mid tail with intermittent bursty flows (Sergeev et al. 1996), though the build-up of a high-pressure region in the inner magnetosphere prevents high-speed jets from penetrating into the inner magnetosphere (Kissinger et al. 2012).

DeJong et al. (2007) compared auroral features of isolated substorms, sawtooth, and SMC events and found that the polar cap oval measured during individual sawteeth contained, on average, 150% more magnetic flux than the oval measured during isolated substorms or SMC events. However, both isolated substorms and individual sawteeth showed a 30% decrease in polar cap magnetic flux during the dipolarization (expansion) phase, whereas the open polar flux remained steady in SMC events.

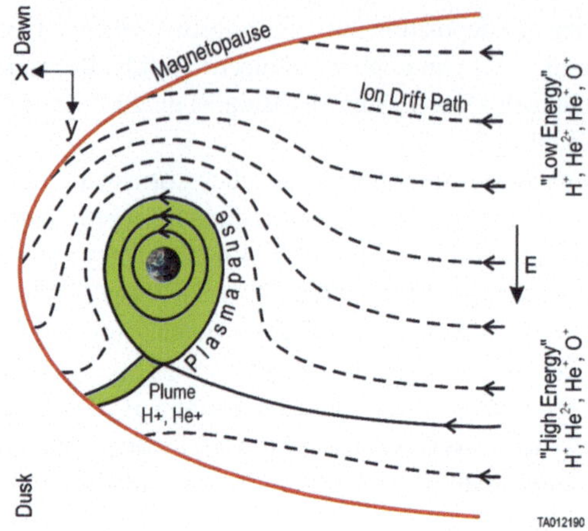

Fig. 9 After Toledo-Redondo et al. 2021 – Schematic of open and closed drift paths in the magnetosphere. Plasma of ionospheric (H^+, He^+, O^+) and solar wind ((H^+, He^{2+}) origin drift to the magnetopause in response to the cross-tail electric field and gradient and curvature drifts. In addition, a plasmaspheric plume (green-shaded extension of the plasmasphere) could intersect the magnetopause. After losses along their drift paths a fraction of ionospheric and solar wind ions and electrons re-encounter the magnetopause and complete the Dungey cycle

Pseudo-breakups are auroral brightenings, much like those at substorm initiation, with similar magnetospheric signatures such as dipolarization fronts, flow bursts, and field-aligned currents (Nakamura et al. 1994). Pseudo-breakups, however, do not develop substorm-like poleward auroral expansions. They may also occur during SMC events (DeJong and Clauer 2005), consistent with small-scale energy releases but the absence of polar expansions. If the poleward expansion of aurora corresponds to tail reconnection proceeding into the lobes, then its absence would indicate that reconnection stops before reaching the lobes. Why reconnection stops before this point is obviously relevant for identifying conditions that stop or quench reconnection.

In summary, the dynamics in the Earth's magnetotail is driven by reconnection. In the original Dungey circulation model, this reconnection is steady and balances the dayside reconnection. However, observations for more than 50 years have shown that reconnection, especially in the near-tail distance of 20 – 30 R_E, is far from steady. Despite this unsteadiness, the Dungey circulation of magnetic flux is largely realized over timescales of hours.

5 Convection Back to the Dayside and the Completion of the Dungey Cycle

With near-tail reconnection, the plasma in the magnetotail is injected Earthward back into the inner magnetosphere. The paths ions take to return to the dayside depend on their energies. Lower energy ions are dominated by corotation and drift eastward while higher energy ions respond to gradient and curvature effects and drift westward to form the ring current (see, e.g., Kistler et al. 1989). Figure 9 shows a schematic of these drift paths that intersect the magnetopause, mostly sunward of the terminator. Fluxes and energies are not constant in this drift process. There are a variety of energy-dependent loss processes, including wave-particle interactions and charge exchange with the Earth's neutral hydrogen geocorona. The high-latitude ionosphere is a source of plasma throughout this convection, providing, for example, the warm plasma cloak (Chappell et al. 2008) and a plasmaspheric plume (e.g., Carpenter et al. 1993), an extension of the plasmasphere (green-shaded region in Fig. 9) that extends to the dayside magnetopause over a limited range of local times during high magnetospheric activity. Thus, a small fraction of the solar wind plasma that entered on the dayside and injected ionospheric plasma from the high latitude ionosphere ultimately returns to the magnetopause and completes the Dungey cycle.

6 The Dungey Cycle for Northward IMF and Near-Radial IMF

6.1 Northward IMF

The previous 5 sections outline the important role of reconnection in the Dungey cycle when the IMF is southward. Reconnection also plays an important role in magnetospheric dynamics when the IMF is northward (see the recent review by Lavraud and Trattner (2021)). Under such conditions, Dungey (1963) originally proposed that reconnection would occur at the magnetopause on magnetospheric field lines poleward of the cusps. In his original, 2-dimensional diagram, reconnection occurs simultaneously poleward of both cusps between magnetosheath and lobe field lines in both hemispheres. This type of reconnection is called dual-lobe reconnection. In 3-dimensions, and with dipole tilt and a finite B_Y component of the IMF, the reconnection is not simultaneous and can also occur on high-latitude, non-lobe field lines in either hemisphere (e.g., Lavraud et al. 2005, 2006; Fuselier et al. 2014; Lavraud et al. 2018). In particular, Lavraud et al. (2005) showed through statistical analysis that dipole tilt determines which hemisphere high-latitude reconnection occurs first, but it does not rule out dual-lobe reconnection. Figure 10 shows non-simultaneous dual-lobe reconnection and the consequences for field line topology and convection (Fuselier et al. 2012). Magnetosheath field lines (dark blue) reconnect first in the southern hemisphere poleward of the cusp. Under the tension force, these open field lines snap sunward and dawnward. The small red circles show where the field lines cross the magnetopause. As the field lines convect dawnward and drape against the magnetopause, they reconnect a second time poleward of the northern cusp, forming closed field lines inside the magnetosphere on the dayside. These field lines are filled with magnetosheath plasma and, because ionospheric outflow from the cusp also occurs along these field lines, they are filled with ionospheric plasma as well (Fuselier et al. 1989, 2019a).

Thus, just as for southward IMF, the Dungey cycle for northward IMF starts with reconnection at the dayside magnetopause. This reconnection produces a boundary layer on the dayside all the time (Fuselier et al. 1995; Øieroset et al. 2008) and has been observed as a quasi-steady process for northward IMF intervals that last for hours (Frey et al. 2003). Unlike southward IMF, the reconnection X-line poleward of the cusps does not extend over the entire dayside magnetopause (Fuselier et al. 2002). Furthermore, this mass-loaded field line convects slowly around the flanks of the magnetosphere rather than relatively rapidly over the poles for southward IMF. Similarly, there is a very slow transfer of magnetic flux to the nightside when the IMF is northward.

The convection of mass-loaded field lines on the flanks of the magnetopause under northward IMF conditions opens the possibility of additional magnetosheath plasma entry into the magnetosphere by the Kelvin-Helmholtz Instability (KHI), (e.g., Otto and Fairfield 2000; Nykyri and Otto 2001; Nykyri et al. 2006, 2021). The wound-up KH vortices are subject to local reconnection. This type of reconnection has been studied extensively for one MMS KHI event (Eriksson et al. 2016; Vernisse et al. 2016; Li et al. 2016). KH modes corrugate

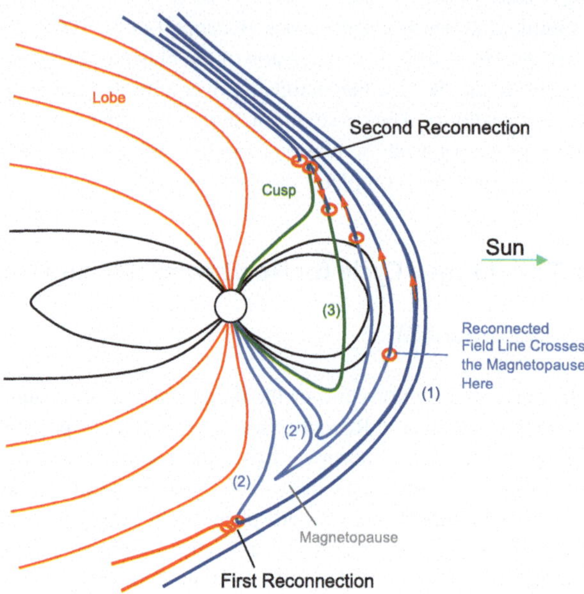

Fig. 10 After Fuselier et al. (2012) – Three-dimensional reconnection scenario. Magnetosheath field lines (1) reconnect poleward of the southern cusp forming field lines 2 and 2′. In the vicinity of the reconnection sites, these field lines convect sunward and dawnward. The points where they cross the magnetopause (small red circles) move northward and dawnward. At some later time, when the field line has convected away from the dayside magnetopause, a second reconnection occurs. This reconnection poleward of the northern hemisphere cusp on the part of the field line that is still in the magnetosheath forms the newly closed field line 3

the low-latitude magnetopause and enable reconnection between plasma sheet and magnetosheath, which traps magnetosheath plasma. These combined KH and reconnection plasma entry mechanisms help populate the plasma sheet on the flanks, with subsequent drift necessary to transport particles toward the center (e.g., Lavraud and Trattner 2021).

When the IMF is northward, high-latitude reconnection and plasma entry on the flanks are manifested in the plasma convection in the cusp and the convection cells in the high latitude ionosphere. The cusp ion dispersion is observed to be "reversed", with higher energy ions at the most poleward latitudes instead of the opposite for southward IMF conditions. Initially, this reversed convection was thought to be a consequence of diffusive entry at the magnetopause; however, it is now well established that the reverse dispersion is a consequence of high latitude reconnection and sunward convection of reconnected field lines (Burch et al. 1980). Imaging of the cusp footprint in the ionosphere, combined with simultaneous in situ observations of reconnection at high latitudes has firmly established this connection between the magnetopause and cusp (Phan et al. 2003). Furthermore, this same imaging has established that the reconnection is quasi-steady and present all the time under northward IMF conditions (Frey et al. 2003).

For northward IMF, magnetosheath particle fluxes from reconnection and Kelvin-Helmholtz entry are observed at low latitudes on the flanks in a relatively wide boundary layer of antisunward flowing plasma. In the ionosphere, the result of this combined reconnection and KH entry is a "four cell" convection pattern. The two cells at high latitude have sunward flow in the central polar cap from lobe reconnection (Burke et al. 1979).

In the magnetotail, there is a distant reconnection site that persists for northward IMF. Cross-tail particle transport typically requires curvature and gradient drifts that exceed earthward $\mathbf{E}\times\mathrm{B}$ drifts, which becomes more effective at higher energies and southward IMF conditions. However, azimuthal flux transport in the magnetotail has been identified also when the IMF is northward and there is significant IMF By. Under these conditions the effects of northward IMF were sufficiently weak so that the Dungey-cycle twin-vortex convection observed for southward IMF was still maintained albeit with high asymmetry (Grocott et al.

2003, 2007). Since these events, apparently driven by reconnection at a distant site, did not exhibit substorm signatures they were called "tail reconnection during IMF-northward, non-substorm intervals" (TRINNIs) (Milan et al. 2005).

While these TRINNI intervals apparently lack near-Earth reconnection, substorms exist also during northward IMF intervals (Lee et al. 2010; Peng et al. 2013), indicating that the onset of near-Earth reconnection does not have to be directly associated with enhanced dayside reconnection and can dissipate previously stored magnetotail energy. However, Peng et al. (2013) found that these substorms occurred mostly soon after southward IMF Bz periods and that intense substorms occurred only during relatively brief northward IMF Bz intervals. These observations suggest that southward IMF is still needed for the strong buildup of flux in the tail and it remains an open question what triggers those events during northward IMF.

In summary, northward IMF also produces a Dungey-like cycle. Magnetic reconnection at the magnetopause initiates this cycle and the persistent, distant tail reconnection site returns plasma sunward. However, the convection from the dayside to the nightside around the flanks of the magnetopause is considerably slower and weaker than the convection over the poles for southward IMF. Magnetic flux buildup in the magnetotail is much weaker and, as long as the northward IMF conditions persist, it is not clear if there is sufficient near-Earth reconnection to transfer plasma back to the dayside to complete the Dungey cycle.

6.2 Near-Radial IMF

IMF-cone angles $<25^\circ$ or $>155^\circ$ occur $\sim 15\%$ of the time at 1 AU (e.g., Pi et al. 2014). During these times of near-radial IMF, the quasi-parallel bow shock is upstream of the subsolar magnetopause. Turbulence generated at and upstream of the bow shock convects downstream and creates significant fluctuations in the plasma and magnetic field magnitude and orientation. Searching for signatures of reconnection at the magnetopause under these conditions is difficult. Furthermore, the magnetosheath draping model used in the maximum magnetic shear model is not valid for these cone angles and the location of the reconnection X-line is uncertain. Nonetheless, signatures of reconnection have been observed under these cone angle conditions, although the steadiness of this reconnection is questionable (e.g., Walsh et al. 2017; Toledo-Redondo et al. 2021b). One promising line of work is the use of large spacecraft datasets to construct a new generation of data-driven field line draping maps, with the potential to improve predictions of magnetic field orientation at the magnetopause (Michotte de Welle et al. 2022).

However, significantly more work is needed to understand how the Dungey cycle is affected by near-radial IMF conditions.

7 Multiple, Transient Reconnection on a Global Scale and Modifications to the Dungey Cycle

Reconnection plays critical roles in the Dungey cycle. Up to this point, there has been no discussion of variability in reconnection at the dayside magnetopause and very little discussion about reconnection variability in the tail. For southward IMF, there are long, primary X-lines at the dayside magnetopause. These primary X-lines are quasi-stationary and quasi-steady (Fuselier et al. 2019b). For quasi-steady reconnection, the reconnection rate does not go to zero over long periods of time as long as the direction of the IMF is also relatively steady. The rate could vary by large amounts within this long time period, which is why the rate is described as quasi-steady.

There is also considerable evidence for transient reconnection at the magnetopause. The original model for flux transfer events (FTEs) is transient in both space and time (Russell and Elphic 1978). FTEs have been modeled also as bulges on the magnetopause between two active reconnection X-lines (Lee and Fu 1986). There is ample evidence of multiple X-lines at the dayside magnetopause (e.g., Hasegawa et al. 2010; Vines et al. 2017; Russell et al. 2017; Trattner et al. 2012; Fuselier et al. 2018, 2022). Some observations are interpreted in terms of a model where there is a quasi-stationary, primary reconnection X-line at the magnetopause and a temporally and/or spatially variable, secondary reconnection X-line separated by up to several R_E (Fuselier et al. 2018, 2022). Thus, there is both quasi-steady and transient reconnection co-existing at the magnetopause for southward IMF.

Multiple X-lines at the magnetopause are manifested in the cusp as overlapping ion dispersions (Fuselier et al. 1997; Lockwood 1995; Onsager et al. 1995; Trattner et al. 1998, 2012). For single reconnection, the precipitating magnetosheath ions disperse in the cusp such that the highest energy ions are observed at the lowest latitude. Overlapping ion dispersions have two (or more) of these dispersions separated in time. The separation provides information on the spacing of the two (or more) reconnection sites at the magnetopause. Recently, observations at the magnetopause and in the cusp have demonstrated the connection between multiple X-lines at the magnetopause and overlapping dispersions in the cusp. An open question about this connection is why multiple X-lines at the magnetopause seem to be common, but overlapping dispersions in the cusp do not appear to be that common (e.g., Trattner et al. 1998).

Multiple X-lines and transient reconnection also occur in the magnetotail. As discussed in Sect. 4.6, the formation of a new, temporary, reconnection site in the near tail is the core element of the neutral line model of substorms. Thus, the magnetotail part of the global Dungey cycle relies on transient reconnection. However, plasma observations both earthward and tailward of the temporary reconnection site indicate temporal and spatial variability on timescales much shorter than the substorm timescale of hours and much smaller than the $\sim$20-40 R_E longitudinal extent of the near-Earth magnetotail.

On the earthward side of the near-Earth reconnection site, plasma flows are intrinsically bursty (BBFs Angelopoulos et al. 1992) and may carry embedded small-scale magnetic island or flux rope structures (e.g., Slavin et al. 2003). Observations in the magnetotail suggest that such near-Earth reconnection sites are activated multiple times at different locations (Nakamura et al. 2011), suggesting that multiple transient reconnection sites are likely key to understanding longer time-scale and larger spatial-scale magnetotail responses related to the Dungey cycle. On the tailward side, large plasmoids typically have multiple-island substructres (e.g., Hones et al. 1984), while scale sizes along the Sun-Earth direction may vary between more than 100 R_E (Hones et al. 1984), several R_E (Moldwin and Hughes 1992; Slavin et al. 2003) and even shorter scales of a few ion inertial lengths (Sun et al. 2019). Two-point measurements at the Moon's distance indicate a cross-tail size of 5-10 R_E. The cross-tail size increases with stronger activity measured by the auroral electrojet, or AE index (Li et al. 2014); however, this cross-tail size is still a fraction of the cross-tail extent of the magnetotail.

The focus of this section has been on the global scale and global consequences of transient reconnection. Transient reconnection produces a wealth of phenomena on the meso- and micro-scale that is considered in much more detail in Hwang et al. (2023, this collection).

8 Conclusions

In summary, this article discusses global-scale structure and consequences of reconnection, conveniently organized around the Dungey Cycle. Magnetic reconnection at the dayside magnetopause and in the magnetotail are clearly the drivers of the global Dungey cycle for southward IMF. The recent advances in the global structure and consequences of reconnection from the MMS mission and modeling include a detailed understanding of the location and steadiness of reconnection at the dayside magnetopause and in the magnetotail and the importance of multiple plasma sources. There are important questions about global reconnection that remain. These questions focus on 1) magnetosheath plasma entry into the magnetosphere and magnetotail dynamics when the IMF is northward and 2) how multiple reconnection and reconnection variability at the magnetopause and in the magnetotail fit into and complicate the classic Dungey Cycle picture of global magnetospheric circulation.

Acknowledgements The authors thank the International Space Science Institute (ISSI) for hosting the Magnetic Reconnection: Explosive Energy Conversion in Space Plasmas team. The MMS mission has been a tremendous success, and the many women and men that helped create this mission share in this success.

Funding Open Access funding provided thanks to the CRUE-CSIC agreement with Springer Nature. Research at Southwest Research Institute is funded by the NASA MMS prime contract NNG04EB99C. Research at LASP is supported by NASA grant NNG04EB99C and 80NSSC20K0688. Research at Lockheed Martin is supported by NASA grant 80NSSC18K1379 and MMS subcontract 499935Q. Research at the University of Murcia is supported by MCIN/AEI/PRTR 10.13039/501100011033 (grants PID2020-112805GA-I00 and TED2021-129357A-I00) and Seneca Agency from Region of Murcia (grant 21910/PI/22). Research at Imperial College London is supported by UKRI(STFC) grant ST/@001071/1.

Declarations

Competing Interests The authors declare that they have no conflicts of interest.

References

Alexeev II, Sibeck DG, Bobrovnikov SY (1998) Concerning the location of magnetopause merging as a function of the magnetopause current strength. J Geophys Res 103(A4):6675–6684. https://doi.org/10.1029/97JA02863

Angelopoulos V, Baumjohann W, Kennel CF, Coroniti FV, Kivelson MG, Pellat R et al (1992) Bursty bulk flows in the inner central plasma sheet. J Geophys Res 97:4027–4039

Angelopoulos V, Artemyev A, Phan TD et al (2020) Near-Earth magnetotail reconnection powers space storms. Nat Phys 16:317–321. https://doi.org/10.1038/s41567-019-0749-4

Arnold H, Sorathia K, Stephens G, Sitnov M, Merkin VG, Birn J (2023) Data mining inspired localized resistivity in global MHD simulations of the magnetosphere. J Geophys Res 128:e2022JA030990. https://doi.org/10.1029/2022JA030990

Artemyev AV, Angelopoulos V, Runov A (2016a) On the radial force balance in the quiet time magnetotail current sheet. J Geophys Res 121:4017–4026. https://doi.org/10.1002/2016JA022480

Artemyev AV, Angelopoulos V, Runov A, Petrokovich AA (2016b) Properties of current sheet thinning at x ~ -10 to -12 R_E. J Geophys Res 121:6718–6731. https://doi.org/10.1002/2016JA022779

Artemyev AV, Angelopoulos V, Runov A, Petrukovich AA (2019) Global view of current sheet thinning: plasma pressure gradients and large-scale currents. J Geophys Res 124:264–278. https://doi.org/10.1029/2018JA026113
Atkinson G (1966) A theory of polar substorms. J Geophys Res 71:5157–5164
Atkinson G (1967) Polar magnetic substorms. J Geophys Res 72:1491–1494
Atz EA, Walsh BM, Broll JM, Zou Y (2022) The spatial extent of magnetopause magnetic reconnection from in situ THEMIS measurements. J Geophys Res 127:e2022JA030894. https://doi.org/10.1029/2022JA030894
Baker DN, Bame SJ, Birn J, Feldman WC, Gosling JT, Hones EW Jr, Zwickl RD, Slavin JA, Smith EJ, Tsurutani BT, Sibeck DG (1984) Direct observations of passages of the distant neutral line (80-140 R_E) following substorm onsets: ISEE-3. Geophys Res Lett 11:1042
Baker DN, Pulkkinen TI, Angelopoulos V, Baumjohann W, McPherron RL (1996) Neutral line model of substorms: past results and present view. J Geophys Res 101(A6):12975–13010. https://doi.org/10.1029/95JA03753
Baker DN et al (2016) A telescopic and microscopic examination of acceleration in the June 2015 geomagnetic storm: magnetospheric multiscale and Van Allen Probes study of substorm particle injection. Geophys Res Lett 43:6051–6059. https://doi.org/10.1002/2016GL069643
Baumjohann W, Paschmann G, Sckopke N, Cattell CA (1989) Average plasma properties in the central plasma sheet. J Geophys Res 94(A6):6597–6606. https://doi.org/10.1029/JA094iA06p06597
Baumjohann W, Paschmann G, Lühr H (1990) Characteristics of high-speed flows in the plasma sheet. J Geophys Res 95:3801–3809
Baumjohann W, Paschmann G, Nagai T, Lühr H (1991) Superposed epoch analysis of the substorm plasma sheet. J Geophys Res 96:11605
Berchem J, Fuselier SA, Petrinec S, Frey HU, Burch JL (2003) Dayside proton aurora: comparisons between global MHD simulations and IMAGE observations. Space Sci Rev 109:313–349. https://doi.org/10.1023/B:SPAC.0000007523.23002.92
Birn J, Hesse M (1991) The substorm current wedge and field-aligned currents in MHD simulations of magnetotail reconnection. J Geophys Res 96:1611
Birn J, Hesse M (2014) The substorm current wedge: Further insights from MHD simulations. J Geophys Res 114. https://doi.org/10.1002/2014JA019863
Birn J, Schindler K (2002) Thin current sheets in the magnetotail and the loss of equilibrium. J Geophys Res 107(A7):SMP18. https://doi.org/10.1029/2001JA0291
Birn J, Schindler K, Janicke L, Hesse M (1994) Magnetotail dynamics under isobaric constraints. J Geophys Res 99:14863
Birn J, Hesse M, Schindler K (1996) MHD simulations of magnetotail dynamics. J Geophys Res 101:12939–12954
Birn J, Dorelli JC, Schindler K, Hesse M (2004) Thin current sheets and loss of equilibrium: three-dimensional theory and simulations. J Geophys Res 109:A02215. https://doi.org/10.1029/2003JA010275
Birn J, Hesse M, Schindler K, Zaharia S (2009) Role of entropy in magnetotail dynamics. J Geophys Res 114. https://doi.org/10.1029/2008JA014015
Birn J, Artemyev AV, Baker DN, Echim M, Hoshino M, Zelenyi LM (2012) Particle acceleration in the magnetotail and aurora. Space Sci Rev 173:49–102. https://doi.org/10.1007/s11214-012-9874-4
Birn J, Liu J, Runov A, Kepko L, Angelopoulos V (2019) On the contribution ofepolarizatig flux bundles to the substorm current wedge and to flux and energy transport. J Geophys Res 124:5408–5420. https://doi.org/10.1029/2019JA026658
Birn J, Runov A, Khotyaintsev Y (2021) Magnetotail processes. In: Maggiolo R, André N, Hasegawa H, Welling DT (eds) Space physics and aeronomy collection volume 2: magnetospheres in the Solar System, 1st edn. Geophysical monograph, vol 259. Am. Geophys. Union, Washington. https://doi.org/10.1002/9781119815624.ch17
Boakes PD, Milan SE, Abel GA, Freeman MP, Chisham G, Hubert B (2009) A statistical study of the open magnetic flux content of the magnetosphere at the time of substorm onset. Geophys Res Lett 36:L04105. https://doi.org/10.1029/2008GL037059
Borovsky JE (2008) The rudiments of a theory of solar wind/magnetosphere coupling derived from first principles. J Geophys Res 113. https://doi.org/10.1029/2007JA012646
Borovsky JE (2013) Physical improvements to the solar wind reconnection control function for the Earth's magnetosphere. J Geophys Res 118:2113–2121. https://doi.org/10.1002/jgra.5011
Borovsky JE, Yakymenko K (2017) Substorm occurrence rates, substorm recurrence times, and solar wind structure. J Geophys Res 122:2973–2998. https://doi.org/10.1002/2016JA023625
Borovsky JE, Nemzek RJ, Belian RD (1993) The occurrence rate of magnetospheric-substorm onsets: random and periodic substorms. J Geophys Res 98:3807–3813. https://doi.org/10.1029/92JA02556

Brambles OJ, Lotko W, Zhang B, Ouellette J, Lyon J, Wiltberger M (2013) The effects of ionospheric outflow on ICME and SIR driven sawtooth events. J Geophys Res 118:6026–6041. https://doi.org/10.1002/jgra.50522

Burch JL (2000) IMAGE mission overview. Space Sci Rev 91:1–14. https://doi.org/10.1023/A:1005245323115

Burch JL, Reiff PH, Spiro RW, Heelis RA, Fields SA (1980) Cusp region particle precipitation and ion convection for northward interplanetary magnetic field. Geophys Res Lett 7:393–396

Burch JL, Reiff PH, Heelis RA, Winningham JD, Hanson WB, Gurgiola C, Menietti JD, Hoffman RA, Barfield JN (1982) Plasma injection and transport in the mid-altitude polar cusp. Geophys Res Lett 9:921–924. https://doi.org/10.1029/GL009i009p00921

Burch JL, Moore TE, Torbert RB et al (2016) Magnetospheric multiscale overview and science objectives. Space Sci Rev 199:5–21. https://doi.org/10.1007/s11214-015-0164-9

Burke WJ, Kelley MC, Sagalyn RC, Smiddy M, Lai ST (1979) Polar cap electric field structures with a northward interplanetary magnetic field. Geophys Res Lett 6:21–24

Cai X, Clauer CR, Ridley AJ (2006a) Statistical analysis of ionospheric potential patterns for isolated substorms and sawtooth events. Ann Geophys 24:1977–1991. https://doi.org/10.5194/angeo-24-1977-2006

Cai X, Henderson MG, Clauer CR (2006b) A statistical study of magneticepolarizationn for sawtooth events and isolated substorms at geosynchronous orbit with GOES data. Ann Geophys 24:3481–3490. https://doi.org/10.5194/angeo-24-3481-2006

Carpenter DL, Giles BL, Chappell CR, Decreau PME, Anderson RR, Persoon AM, Smith AJ, Corcuff Y, Canu P (1993) Plasmasphere dynamics in the duskside bulge region: a new look at an old topic. J Geophys Res 98:19243

Cassak PA, Fuselier SA (2016) Reconnection at Earth's dayside magnetopause. In: Gonzalez W, Parker E (eds) Magnetic reconnection. Astrophysics and space science library, vol 427. Springer, Cham, pp 213–276. https://doi.org/10.1007/978-3-319-26432-5

Chappell CR, Huddleston MM, Moore TE, Giles BL, Delcourt DC (2008) Observations of the warm plasma cloak and an explanation of its formation in the magnetosphere. J Geophys Res 113:A09206. https://doi.org/10.1029/2007JA012945

Clauer CR, Cai X, Welling D, DeJong A, Henderson MG (2006) Characterizing the 18 April 2002 storm-time sawtooth events using ground magnetic data. J Geophys Res 111:A04S90. https://doi.org/10.1029/2005JA011099

Coppi B, Mark JWK, Sugiyama L, Bertin G (1979) Reconnecting modes in collisonless plasmas. Phys Rev Lett 42:1058–1061. https://doi.org/10.1103/PhysRevLett.42.1058

Cowley SWH (1982) The causes of convection in the Earth's magnetosphere: a review of developments during IMS. Rev Geophys 20:531–565

Cowley SWH, Lockwood M (1992) Excitation and decay of solar-wind driven flows in the magnetosphere-ionosphere system. Ann Geophys 10:103–115

Crooker NU (1979) Dayside merging and cusp geometry. J Geophys Res 84:951–959

Daly PW, Sanderson TR, Wenzel K-P (1984) Survey of energetic (E>35 keV) ion anisotropies in the deep geomagnetic tail. J Geophys Res 89:10733–10739. https://doi.org/10.1029/JA089iA12p10733

DeJong AD, Clauer CR (2005) Polar UVI images to study steady magnetospheric convection events: initial results. Geophys Res Lett 32:L24101. https://doi.org/10.1029/2005GL024498

DeJong AD, Cai X, Clauer RC, Spann JF (2007) Aurora and open magnetic flux during isolated substorms, sawteeth, and SMC events. Ann Geophys 25:1865–1876. https://doi.org/10.5194/angeo-25-1865-2007

Dubyagin SV, Sergeev VA, Apatenkov SV, Angelopoulos V, Nakamura R, McFadden J, Larson D, Bonnell J (2010) Pressure and entropy changes in the flow-braking region during magnetic fieldepolarizationn. J Geophys Res 115:A10225. https://doi.org/10.1029/2010JA015625

Dungey JW (1950) Some researches in cosmic magnetism. PhD Thesis, Cambridge University

Dungey JW (1958) Cosmic electrodynamics. Cambridge University Press, Cambridge

Dungey JW (1961) Interplanetary magnetic field and the auroral zones. Phys Rev Lett 6:47–48

Dungey JW (1962) The interplanetary field and auroral theory. J Phys Soc Jpn 17(suppl. a–ii):15–19

Dungey JW (1963) The structure of the ionosphere, or adventures in velocity space. In: DeWitt C, Hiebolt J, Lebeau A (eds) Geophysics: the Earth's environment. Gordon and Breach, New York, pp 526–536

Dunlop MW, Zhang Q-H, Bogdanova YV, Trattner KJ, Pu Z, Hasegawa H, Berchem J, Taylor MGGT, Volwerk M, Eastwood JP, Lavraud B, Shen C, Shi J-K, Wang J, Constantinescu D, Fazakerley AN, Frey H, Sibeck D, Escoubet P, Wild JA, Liu ZX, Carr C (2011) Magnetopause reconnection across wide local time. Ann Geophys 29:1683–1697. https://doi.org/10.5194/angeo-29-1683-2011

Eggington JWB, Eastwood JP, Mejnertsen L, Desai RT, Chittenden JP (2020) Dipole tilt effect on magnetopause reconnection and the steady-state magnetosphere-ionosphere system: global MHD simulations. J Geophys Res Space Phys 125:e2019JA027510. https://doi.org/10.1029/2019JA027510

El-Alaoui M, Ashour-Abdalla M, Walker R, Peroomian V, Richard R, Angelopoulos V, Runov A (2009) Substorm evolution as revealed by themissatellites and a global MHD simulation. J Geophys Res 114:A08221. https://doi.org/10.1029/2009JA014133

Erickson GM, Wolf RA (1980) Is steady convection possible in the Earth's magnetotail? Geophys Res Lett 7:897–900

Eriksson S et al (2016) Magnetospheric Multiscale observations of magnetic reconnection associated with Kelvin-Helmholtz waves. Geophys Res Lett 43. https://doi.org/10.1002/2016GL068783

Frank L (1971) Plasma in the Earth's polar magnetosphere. J Geophys Res 76:5202

Frey HU, Mende SB, Fuselier SA, Immel TJ, Ostgaard N (2003) Proton aurora in the cusp during southward IMF. J Geophys Res 108(A7):1277. https://doi.org/10.1029/2003JA009861

Fu H, Grigorenko EE, Gabrielse C, Liu C, Lu S, Hwang KJ, Zhou X, Wang Z, Chen F (2020) Magnetotailepolarizationn fronts and particle acceleration: a review. Sci China Earth Sci 63:235–256. https://doi.org/10.1007/s11430-019-9551-y

Funsten HO, Allegrini F, Bochsler P, Dunn G, Ellis S, Everett D, Fagan MJ, Fuselier SA, Granoff M, Gruntman M, Guthrie AA, Hanley J, Harper RW, Heirtzler D, Janzen P, Kihara KH, King B, Kucharek H, Manzo MP, Maple M, Mashburn K, McComas DJ, Moebius E, Nolin J, Piazza D, Pope S, Reisenfeld DB, Rodriguez B, Roelof EC, Saul L, Turco S, Walek P, Weidner S, Wurz P, Zaffke S (2009) The Interstellar Boundary Explorer High Energy (IBEX-Hi) neutral atom imager. Space Sci Rev 146(1):75. https://doi.org/10.1007/s11214-009-9495-8

Fuselier SA, Klumpar DM, Peterson WK, Shelley EG (1989) Direct injection of ionospheric O+ into the dayside low latitude boundary layer. Geophys Res Lett 16(10):1121–1124. https://doi.org/10.1029/GL016i010p01121

Fuselier SA, Anderson BJ, Onsager TG (1995) Particle signatures of magnetic topology at the magnetopause: AMPTE/CCE observations. J Geophys Res 100:11805–11822

Fuselier SA, Shelley EG, Peterson WK, Lennartsson OW, Collin HL, Drake JF, Ghielmetti AG, Balsiger H, Burch JL, Johnstone A, Rosenbauer H, Steinberg JT (1997) Bifurcated cusp ion signatures: evidence for quasi-steady re-reconnection? Geophys Res Lett 24:1471–1474

Fuselier SA, Frey HU, Trattner KJ, Mende SB, Burch JL (2002) Cusp aurora dependence on interplanetary magnetic field B_Z. J Geophys Res 107. https://doi.org/10.1029/2001JA900165

Fuselier SA, Mende SB, Moore TE, Frey HU, Petrinec SM, Claflin ES, Collier MR (2003) Cusp dynamics and ionospheric outflow. Space Sci Rev 109:285–312. https://doi.org/10.1023/B:SPAC.0000007522.71147.b3

Fuselier SA, Bochsler P, Chornay D, Clark G, Crew GB, Dunn G, Ellis S, Friedmann T, Funsten HO, Ghielmetti AG, Googins J, Granoff MS, Hamilton JW, Hanley J, Heirtzler D, Hertzberg E, Isaac D, King B, Knauss U, Kucharek H, Kurdirka F, Livi S, Lobell J, Longworth S, Mashburn K, McComas DJ, Möbius E, Moore AS, Moore TE, Nemanich RJ, Nolin J, O'Neal M, Piazza D, Peterson L, Pope SE, Rosmarynowski P, Saul LA, Scheer JA, Scherrer JR, Schlemm C, Schwadron NA, Tillier C, Turco S, Tyler J, Vosbury M, Wieser M, Wurz P, Zaffke S (2009) The IBEX-Lo sensor. Space Sci Rev 146(1):117. https://doi.org/10.1007/s11214-009-9495-8

Fuselier SA, Trattner KJ, Petrinec SM, Lavraud B (2012) Dayside magnetic topology at the Earth's magnetopause for northward IMF. J Geophys Res 117:A08235. https://doi.org/10.1029/2012JA017852

Fuselier SA, Petrinec SM, Trattner KJ, Lavraud B (2014) Magnetic field topology for northward IMF reconnection: Ion observations. J Geophys Res 119. https://doi.org/10.1002/2014JA020351

Fuselier SA, Lewis WS, Schiff C et al (2016) Magnetospheric multiscale science mission profile and operations. Space Sci Rev 199:77–103. https://doi.org/10.1007/s11214-014-0087-x

Fuselier SA et al (2017) Large-scale characteristics of reconnection diffusion regions and associated magnetopause crossings observed by MMS. J Geophys Res 122:5466–5486. https://doi.org/10.1002/2017JA024024

Fuselier SA, Petrinec SM, Trattner KJ, Broll J, Burch JL, Giles BL, Strangeway RM, Russell CT, Lavraud B, Øieroset M, Torbert RB, Farrugia CJ, Vines SK, Gomez RG, Mukherjee J, Cassak PA (2018) Observational evidence of large-scale multiple reconnection at the Earth's dayside magnetopause. J Geophys Res 123:8407–8421. https://doi.org/10.1029/2018JA025681

Fuselier SA, Trattner KJ, Petrinec SM, Denton MH, Toledo-Redondo S, André M et al (2019a) Mass loading the Earth's dayside magnetopause boundary layer and its effect on magnetic reconnection. Geophys Res Lett 46:6204–6213. https://doi.org/10.1029/2019GL082384

Fuselier SA, Trattner KJ, Petrinec SM, Pritchard KR, Burch JL, Cassak PA et al (2019b) Stationarity of the reconnection X-line at Earth's magnetopause for southward IMF. J Geophys Res 124:8524–8534. https://doi.org/10.1029/2019JA027143

Fuselier SA, Webster JM, Trattner KJ, Petrinec SM, Genestreti KJ, Pritchard KR et al (2021) Reconnection x-line orientations at the Earth's magnetopause. J Geophys Res 126. https://doi.org/10.1029/2021JA029789

Fuselier SA, Kletzing CA, Petrinec SM, Trattner KJ, George D, Bounds SR, Sawyer RP, Bonnell JW, Burch JL, Giles BL, Strangeway RJ (2022) Multiple reconnection X-lines at the magnetopause and overlapping cusp ion injections. J Geophys Res 127:e2022JA030354. https://doi.org/10.1029/2022JA030354
Galeev AA, Sudan RN (1984) Basic plasma physics: selected chapters, handbook of plasma physics, vol 1. North-Holland, Amsterdam, p 305
Gjerloev JW (2012) The supermag data processing technique. J Geophys Res 117(A9). https://doi.org/10.1029/2012JA017683
Glocer A, Welling D, Chappell CR, Toth G, Fok MC, Komar C et al (2020) A case study on the origin of near-Earth plasma. J Geophys Res 125:e2020JA028205. https://doi.org/10.1029/2020JA028205
Gomez RG, Vines SK, Fuselier SA, Cassak PA, Strangeway RJ, Petrinec SM, Burch JL, Trattner KJ, Russell CT, Torbert RB, Pollock C, Young DT, Lewis WS, Mukherjee J (2016) Stable reconnection at the dusk flank magnetopause. Geophys Res Lett 43:9374–9382. https://doi.org/10.1002/2016GL069692
Gonzalez WD, Mozer FS (1974) A quantitative model for the potential resulting from reconnection with an arbitrary interplanetary magnetic field. J Geophys Res 79:4186–4194
Gonzalez W, Parker E (2016) Magnetic reconnection: concepts and applications. Astrophysics and space science library, vol 427. Springer, Cham. https://doi.org/10.1007/978-3-319-26432-5
Gordeev E, Sergeev V, Merkin V, Kuznetsova M (2017) On the origin of plasma sheet reconfiguration during the substorm growth phase. Geophys Res Lett 44:8696–8702. https://doi.org/10.1002/2017GL074539
Gosling JT, Thomsen MF, Bame SJ, Onsager TG, Russell CT (1990) Electron edge of the low latitude boundary layer during accelerated flow events. Geophys Res Lett 17:1833–1836
Griffiths ST, Petrinec SM, Trattner KJ, Fuselier SA, Burch JL, Phan TD, Angelopoulos V (2011) A probability assessment of encountering dayside magnetopause diffusion regions. J Geophys Res 116:A02214. https://doi.org/10.1029/2010JA015316
Grocott A, Cowley SWH, Sigwarth JB (2003) Ionospheric flows and magnetic disturbance during extended intervals of northward but by-dominated IMF. Ann Geophys 21:509–538. https://doi.org/10.5194/angeo-21-509-2003
Grocott A, Yeoman TK, Milan SE, Amm O, Frey HU, Juusola L, Nakamura R, Owen CJ, Rème H, Takada T (2007) Multi-scale observations of magnetotail flux transport during IMF-northward non-substorm intervals. Ann Geophys 25:1709–1720
Hardy DA, Reiff PH, Burke WJ (1979) Response of magnetotail plasma at lunar distance to changes in the interplanetary magnetic field, the solar wind plasma, and substorm activity. J Geophys Res 84:1382–1390. https://doi.org/10.1029/gl005i005p00391
Hasegawa H et al (2010) Evidence for a flux transfer event generated by multiple X-line reconnection at the magnetopause. Geophys Res Lett 37:L16101. https://doi.org/10.1029/2010GL044219
Heelis RA (1984) The effects of interplanetary magnetic field orientation on dayside high-latitude ionospheric convection. J Geophys Res 89:2873–2880. https://doi.org/10.1029/JA089iA05p02873
Heikkila WJ, Winningham JD (1971) Penetration of magnetosheath plasma to low altitudes through the dayside magnetospheric cusps. J Geophys Res 76:883–891. https://doi.org/10.1029/JA076i004p00883
Henderson M (2004) The May 2–3, 1986 CDAW-9C interval: a sawtooth event. Geophys Res Lett 31(11):L11804. https://doi.org/10.1029/2004gl019941
Henderson MG, Reeves GD, Skoug R, Thomsen MT, Denton MH, Mende SB, Immel TJ, Brandt PC, Singer HJ (2006) Magnetospheric and auroral activity during the 18 April 2002 sawtooth event. J Geophys Res 111:A01S90. https://doi.org/10.1029/2005JA011111
Hesse M, Birn J (1994) MHD modeling of magnetotail instability for localized resistivity. J Geophys Res 99(A5):8565–8576. https://doi.org/10.1029/94JA00441
Hesse M, Aunai N, Zenitani S, Kuznetsova M, Birn J (2013) Aspects of collisionless magnetic reconnection in asymmetric systems. Phys Plasmas 20(6):061210. https://doi.org/10.1063/1.4811467
Hill TW, Reiff PH (1977) Evidence of magnetospheric cusp proton acceleration by magnetic merging at the dayside magnetopause. J Geophys Res 82:3623–3628. https://doi.org/10.1029/JA082i025p03623
Hones EW Jr (1977) Substorm processes in the magnetotail: comments on "On hot tenuous plasmas, fireballs, and boundary layers in the Earth's magnetotai" by Frank et al. J Geophys Res 82:5633
Hones EW Jr, Birn J, Baker DN et al (1984) Detailed examination of a plasmoid in the distant magnetotail with ISEE 3. Geophys Res Lett 11(10):1046–1049. https://doi.org/10.1029/GL011i010p01046
Hsieh M-S, Otto A (2014) The influence of magnetic flux depletion on the magnetotail and auroral morphology during the substorm growth phase. J Geophys Res 119(5):3430–3443. https://doi.org/10.1002/2013JA019459
Hsieh M-S, Otto A (2015) Thin current sheet formation in response to the loading and the depletion of magnetic flux during the substorm growth phase. J Geophys Res 120(6):4264–4278. https://doi.org/10.1002/2014JA020925
Huang CY, Frank LA, Rostoker G, Fennell J, Mitchell DG (1992) Nonadiabatic heating of the central plasma sheet at substorm onset. J Geophys Res 82:1481–1495

Huang C-S, Reeves GD, Borovsky JE, Skoug RM, Pu ZY, Le G (2003) Periodic magnetospheric substorms and their relationship with solar wind variations. J Geophys Res 108(A6):1255. https://doi.org/10.1029/2002JA009704

Hughes WJ, Sibeck DG (1987) On the 3-dimensional structure of plasmoids. Geophys Res Lett 14(6):636–639. https://doi.org/10.1029/GL014i006p00636

Hwang KJ, Nakamura R, Eastwood JP et al (2023) Cross-scale processes of magnetic reconnection. Space Sci Rev 219:71. https://doi.org/10.1007/s11214-023-01010-9

Kamide Y, Perrault PD, Akasofu S-I, Winningham JD (1977) Dependence of substorm occurrence probability on the interplanetary magnetic field and on the size of the auroral oval. J Geophys Res 82:5521

Keiling A, Angelopoulos V, Runov A, Weygand J, Apatenkov SV, Mende S et al (2009) Substorm current wedge driven by plasma flow vortices: THEMIS observations. J Geophys Res 114. https://doi.org/10.1029/2009JA014114

Keiling A, Ramos C, Vu N, Angelopoulos V, Nosé M (2022) Statistical properties and proposed source mechanism of recurrent substorm activity with one-hour periodicity. J Geophys Res 127:e2021JA030064. https://doi.org/10.1029/2021JA030064

Kepko L, McPherron R, Amm O, Apatenkov S, Baumjohann W, Birn J et al (2015) Substorm current wedge revisited. Space Sci Rev 190:1–46. https://doi.org/10.1007/s11214-014-0124-9

Khotyaintsev YV, Cully CM, Vaivads A, André M, Owen CJ (2011) Plasma jet braking: energy dissipation and nonadiabatic electrons. Phys Rev Lett 106(16):165001. https://doi.org/10.1103/PhysRevLett.106.165001

Khrabrov AV, Sonnerup BUÖ (1998) Orientation and motion of current layers: minimization of the Faraday residue. Geophys Res Lett 25:2373–2376. https://doi.org/10.1029/98GL51784

Kiehas SA, Angelopoulos V, Runov A, Hietala H, Korovinksiy D (2018) Magnetotail fast flow occurrence rate and dawn-dusk asymmetry at XGSM $\sim$60R$_E$. J Geophys Res 123. https://doi.org/10.1002/2017JA024776

Kissinger J, McPherron RL, Hsu T-S, Angelopoulos V (2012) Diversion of plasma due to high pressure in the inner magnetosphere during steady magnetospheric convection. J Geophys Res 117:A05206. https://doi.org/10.1029/2012JA017579

Kistler LM (2020) Ionospheric and solar wind contributions to the storm-time near-Earth plasma sheet. Geophys Res Lett 47:e2020GL090235. https://doi.org/10.1029/2020GL090235

Kistler LM, Ipavich FM, Hamilton DC, Gloeckler G, Wilken B, Kremser G, Stüdemann W (1989) Energy spectra of the major ion species in the ring current during geomagnetic storms. J Geophys Res 94:3579–3599

Kistler LM, Mouikis CG, Asamura K, Yokota S, Kasahara S, Miyoshi Y et al (2019) Cusp and nightside auroral sources of O^+ in the plasma sheet. J Geophys Res 124. https://doi.org/10.1029/2019JA027061

Kivelson MG, Spence HE (1988) On the possibility of quasi-static convection in the quiet magnetotail. Geophys Res Lett 15:1541–1544

Kronberg EA, Daly PW, Grigorenko EE, Smirnov AG, Klecker B, Malykhin AY (2021) Energetic charge particles in the terrestrial magnetosphere: cluster/RAPID results. J Geophys Res 126:E2021JA029273. https://doi.org/10.1029/2021JA029273

Kuznetsova MM, Hesse M, Rasta¨tter L, Taktakishvili A, Toth G, De Zeeuw DL, Ridley A, Gombosi TI (2007) Multiscale modeling of magnetospheric reconnection. J Geophys Res 112:A10210. https://doi.org/10.1029/2007JA012316

Lavaud B, Borovsky JE (2008) Altered solar wind-magnetosphere interaction at low Mach numbers: Coronal mass ejections. J Geophys Res 113. https://doi.org/10.1029/2008JA013192

Lavraud B, Trattner KJ (2021) The polar cusps of the Earth's magnetosphere. In: Moretti R (ed) Magnetospheres in the Solar System, 10th anniversary of the AGU. Geophysical monograph series, vol 2, pp 163–176. https://doi.org/10.1002/9781119815624.ch11

Lavraud B, Thomsen MF, Taylor MGGT, Wang YL, Phan TD, Schwartz SJ et al (2005) Characteristics of the magnetosheath electron boundary layer under northward IMF: implications for high-latitude reconnection. J Geophys Res 110:A06209. https://doi.org/10.1029/2004JA010808

Lavraud B, Thomsen MF, Lefebvre B, Schwartz SJ, Seki K, Phan TD et al (2006) Evidence for newly closed magnetosheath field lines at the dayside magnetopause under northward IMF. J Geophys Res 111(A5):A05211. https://doi.org/10.1029/2005JA011266

Lavraud B, Jacquey C, Achilli T, Fuselier SA, Grigorenko E, Phan TD et al (2018) Concomitant double ion and electron populations in the Earth's magnetopause boundary layers from double reconnection with lobe and closed field lines. J Geophys Res 123:5407–5419. https://doi.org/10.1029/2017JA025152

Lee LC, Fu ZF (1986) Multiple X line reconnection. I. A criterion for the transition from a single X line to multiple X line reconnection. J Geophys Res 91:6807

Lee D-Y, Choi K-C, Ohtani S, Lee JH, Kim KC, Park KS, Kim K-H (2010) Can intense substorms occur under northward IMF conditions? J Geophys Res 115:A01211. https://doi.org/10.1029/2009JA014480

Li S-S, Angelopoulos V, Runov A, Kiehas SA (2014) Azimuthal extent and properties of midtail plasmoids from two-point ARTEMIS observations at the Earth-Moon Lagrange points. J Geophys Res 119:1781–1796. https://doi.org/10.1002/2013JA019292

Li W, André M, Khotyaintsev YV, Vaivads A et al (2016) Kinetic evidence of magnetic reconnection due to Kelvin-Helmholtz waves. Geophys Res Lett 43:5635–5643. https://doi.org/10.1002/2016GL069192

Liu J, Angelopoulos V, Runov A, Zhou X-Z (2013a) On the current sheets surroundingdipolarizing flux bundles in the magnetotail: the case for wedgelets. J Geophys Res 118. https://doi.org/10.1002/jgra.50092

Liu J, Angelopoulos V, Chu X, Zhou XZ, Yue C (2015) Substorm current wedge composition by wedgelets. Geophys Res Lett 42:1669–1676. https://doi.org/10.1002/2015GL063289

Lockwood M (1995) Overlapping cusp ion injections: an explanation invoking magnetopause reconnection. Geophys Res Lett 22:1141

Lockwood M (1997) Energy and pitch-angle dispersions of LLBL/cusp ions seen at middle altitudes: predictions by the open magnetosphere model. Ann Geophys 15:1501–1514. https://doi.org/10.1007/s00585-997-1501-4

Lockwood M, Bentley SN, Owens MJ, Barnard LA et al (2019) The development of a space climatology: 1. Solar wind magnetosphere coupling as a function of timescale and the effect of data gaps. Space Weather 17:133–156. https://doi.org/10.1029/2018SW001856

Luhmann JR, Walker RJ, Russell CT, Crooker NU, Spreiter JR, Stahara SS (1984) Patterns of potential magnetic field merging sites on the dayside magnetopause. J Geophys Res 89:1739–1742

McComas DJ, Allegrini F, Bochsler P, Bzowski M, Collier M, Fahr H, Fichtner H, Frisch P, Funsten HO, Fuselier S, Gloeckler G, Gruntman M, Izmodenov V, Knappenberger P, Lee M, Livi S, Mitchell D, Möbius E, Moore T, Pope S, Reisenfeld D, Roelof E, Scherrer J, Schwadron N, Tyler R, Wieser M, Witte M, Wurz P, Zank G (2009) IBEX – interstellar boundary explorer. Space Sci Rev 146(1):11. https://doi.org/10.1007/s11214-009-9495-8

McPherron RL (1972) Substorm related changes in the geomagnetic tail: the growth phase. Planet Space Sci 20:1521–1539

McPherron RL, Chu X (2018) The midlatitude positive bay index and the statistics of substorm occurrence. J Geophys Res 123:2831–2850. https://doi.org/10.1002/2017JA024766

McPherron RL, Russell CT, Aubry MA (1973) Satellite studies of magnetospheric substorms on August 15 1968:, 9. Phenomenological model for substorms. J Geophys Res 78:3131

McPherron RL, Nishida A, Russell CT (1987) Is near-Earth current sheet thinning the cause of auroral substorm onset? In: Kamide Y, Wolf RA (eds) Quantitative modeling of the magnetosphere-ionosphere coupling processes, p 252. Kyoto Sangyo University, Kyoto

Merkin VG, Panov EV, Sorathia KA, Ukhorskiy AY (2019) Contribution of bursty bulk flows to the globaldipolarization of the magnetotail during an isolated substorm. J Geophys Res 124

Michotte de Welle B, Aunai N, Nguyen G, Lavraud B, Génot V, Jeandet A, Smets R (2022) Global three-dimensional draping of magnetic field lines in Earth's magnetosheath from in-situ spacecraft measurements. J Geophys Res Space Phys 127:e2022JA030996. https://doi.org/10.1029/2022JA030996

Milan SE, Hubert B, Grocott A (2005) Formation and motion of a transpolar arc in response to dayside and nightside reconnection. J Geophys Res 110. https://doi.org/10.1029/2004JA010835

Milan SE, Grocott A, Forsyth C, Imber SM, Boakes PD, Hubert B (2009) A superposed epoch analysis of auroral evolution during substorm growth, onset and recovery: open magnetic flux control of substorm intensity. Ann Geophys 27:659–668. https://doi.org/10.5194/angeo-27-659-2009

Milan SE, Clausen LBN, Coxon JC et al (2017) Overview of solar wind–magnetosphere–ionosphere–atmosphere coupling and the generation of magnetospheric currents. Space Sci Rev 206:547–573. https://doi.org/10.1007/s11214-017-0333-0

Miyashita Y et al (2005) Plasmoids observed in the near-Earth magnetotail at $X \sim -7R_E$. J Geophys Res 110:A12214. https://doi.org/10.1029/2005JA011263

Moldwin MB, Hughes WJ (1992) On the formation and evolution of plasmoids: a survey of ISEE 3 geotail data. J Geophys Res 97(A12):19259–19282. https://doi.org/10.1029/92JA01598

Moore TE, Fok M-C, Chandler MO (2002) The dayside reconnection X line. J Geophys Res 107:1332–1338. https://doi.org/10.1029/2002JA009381

Moore TE, Fok M-C, Delcort DC, Slinker SP, Fedder JA (2008) Plasma plume circulation and impact in an MHD substorm. J Geophys Res 113:A06219. https://doi.org/10.1029/2008JA013050

Nagai T (2021) Magnetic reconnection in the near-Earth magnetotail. In: Maggiolo R, André N, Hasegawa H, Welling DT (eds) Space physics and aeronomy collection volume 2: magnetospheres in the Solar System, 1st edn. Geophysical monograph, vol 259. Am. Geophys. Union, Washington. https://doi.org/10.1002/9781119815624.ch4

Nagai T, Machida S (1998) Magnetic reconnection in the near-Earth magnetotail. In: Nishida A, Baker DN, Cowley SWH (eds) New perspectives on the Earth's magnetotail. Geophysical monograph, vol 105. Am. Geophys. Union, Washington, pp 211–224

Nagai T, Fujimoto M, Saito Y, Machida S, Terasawa T, Nakamura R, Yamamoto T, Mukai T, Nishida A, Kokubun S (1998) Structure and dynamics of magnetic reconnection for substorm onsets with Geotail observations. J Geophys Res 103:4419–4440
Nagai T, Fujimoto M, Nakamura R, Baumjohann W, Ieda A, Shinohara I, Machida S, Saito Y, Mukai T (2005) Solar wind control of the radial distance of the magnetic reconnection site in the magnetotail. J Geophys Res 110. https://doi.org/10.1029/2005JA011207
Nakamura R, Baker DN, Yamamoto T, Belian RD, Bering EA III, Benbrook JR, Theall JR (1994) Particle and field signatures during pseudobreakup and major expansion onset. J Geophys Res 99:207–221
Nakamura R, Baumjohann W, Brittnacher M, Sergeev VA, Kubyshkina M, Mukai T, Liou K (2001a) Flow bursts and auroral activations: onset timing and foot point location. J Geophys Res 106:10777–10789
Nakamura R, Baumjohann W, Schödel R, Brittnacher M, Sergeev VA, Kubyshkina M, Liou K (2001b) Earthward flow bursts, auroral streamers, and small expansions. J Geophys Res 106:10791–10802
Nakamura R, Baumjohann W, Mouikis C, Kistler LM, Runov A, Volwerk M, Asano Y, Vörös Z, Zhang TL, Klecker B, Rème H, Balogh A (2004) Spatial scale of high-speed flows in the plasma sheet observed by Cluster. Geophys Res Lett 31:L09804. https://doi.org/10.1029/2004GL019558
Nakamura R et al (2011) Flux transport,dipolarization, and current sheet evolution during a double-onset substorm. J Geophys Res 116:A00I36. https://doi.org/10.1029/2010JA0156865
Nishida A, Scholer M, Terasawa T, Bame SJ, Gloeckler G, Smith EJ, Zwickl RD (1986) Quasi-stagnant plasmoid in the middle tail: a new preexpansion phase phenomenon. J Geophys Res 91:4245–4255. https://doi.org/10.1029/JA091iA04p04245
Nykyri K, Otto A (2001) Plasma transport at the magnetospheric boundary due to reconnection in Kelvin-Helmholtz vortices. Geophys Res Lett 28:3565–3568. https://doi.org/10.1029/2001GL013239
Nykyri K, Otto A, Lavraud B, Mouikis C, Kistler L, Balogh A, Réme H (2006) Cluster observations of reconnection due to the Kelvin-Helmholtz instability at the dawnside magnetospheric flank. Ann Geophys 24:2619–2643
Nykyri K, Ma X, Johnson J (2021) Cross-scale energy transport in space plasmas: applications to the magnetopause boundary. In: Maggiolo R, André N, Hasegawa H, Welling DT (eds) Space physics and aeronomy collection volume 2: magnetospheres in the Solar System, 1st edn. Geophysical monograph, vol 259. Am. Geophys. Union, Washington. https://doi.org/10.1002/9781119815624.ch6
Øieroset M, Phan TD, Lin RP, Sonnerup BUÖ (2000) J Geophys Res 105(25):247–250
Øieroset M, Phan TD, Fujimoto M, Lin RP, Lepping RP (2001) In situ detection of collisionless reconnection in the Earth's magnetotail. Nature 412:26
Øieroset M, Phan TD, Angelopoulos V, Eastwood JP, McFadden J, Larson D, Carlson CW, Glassmeier K-H, Fujimoto M, Raeder J (2008) THEMIS multi-spacecraft observations of magnetosheath plasma penetration deep into the dayside low-latitude magnetosphere for northward and strong B_Y IMF. Geophys Res Lett 35:L17S11. https://doi.org/10.1029/2008GL033661
Oka M, Phan TD, Øieroset M et al (2022) Electron energization and thermal to non-thermal energy partition during Earth's magnetotail reconnection. Phys Plasmas 29(5):052904. https://doi.org/10.1063/5.0085647
Oka M, Birn J, Egedal J et al (2023) Particle acceleration by magnetic reconnection in geospace. Space Sci Rev 219:75. https://doi.org/10.1007/s11214-023-01011-8
Onsager TG, Chang S-W, Perez JD, Austin JB, Jano LX (1995) Low-altitude observations and modeling of quasi-steady magnetopause reconnection. J Geophys Res 100:11831–11843
Otto A, Fairfield DH (2000) Kelvin-Helmholtz instability at the magnetotail boundary: MHD simulation and comparison with Geotail observations. J Geophys Res 105:21175–21190
Ouellette JE, Brambles OJ, Lyon JG, Lotko W, Rogers BN (2013) Properties of outflow-driven sawtooth substorms. J Geophys Res 118:3223–3232. https://doi.org/10.1002/jgra.50309
Park KS (2021) Global MHD simulation of the weak southward IMF condition for different time resolutions. Front Astron Space Sci 8:758241. https://doi.org/10.3389/fspas.2021.758241
Peng Z, Wang C, Yang YF, Li H, Hu YQ, Du J (2013) Substorms under northward interplanetary magnetic field: statistical study. J Geophys Res 118:364–374. https://doi.org/10.1029/2012JA018065
Peterson WK et al (1998) Simultaneous observations of solar wind plasma entry from FAST and POLAR. Geophys Res Lett 25:2081
Petrinec SM, Fuselier SA (2003) On continuous versus discontinuous neutral lines at the dayside magnetopause for southward interplanetary magnetic field. Geophys Res Lett 30(10):1519. https://doi.org/10.1029/2002GL016565
Petrinec SM, Dayeh MA, Funsten HO, Fuselier SA, Heirtzler D, Janzen P, Kucharek H, McComas DJ, Möbius E, Moore TE, Reisenfeld DB, Schwadron N, Trattner KJ, Wurz P (2011) Neutral atom imaging of the magnetospheric cusps. J Geophys Res 116:A07203. https://doi.org/10.1029/2010JA016357
Petrinec SM, Burch JL, Fuselier SA, Gomez RG, Lewis W, Trattner KJ, Ergun R, Mauk B, Pollock CJ, Schiff C, Strangeway RJ, Russell CT, Phan T-D, Young D (2016) Comparison of Magnetospheric Multiscale

ion jet signatures with predicted reconnection site locations at the magnetopause. Geophys Res Lett 43. https://doi.org/10.1002/2016GL069626

Petrinec SM, Burch JL, Fuselier SA, Trattner KJ, Giles BL, Strangeway RJ (2022) On the occurrence of magnetic reconnection along the terrestrial magnetopause, using Magnetospheric Multiscale (MMS) observations in proximity to the reconnection site. J Geophys Res 127. https://doi.org/10.1029/2021JA029669

Petrukovich AA, Baumjohann W, Nakamura R, Mukai T, Troshichev OA (2000) Small substorms: solar wind input and magnetotail dynamics. J Geophys Res 105:21109–21118. https://doi.org/10.1029/2000JA900057

Petschek HE (1964) Magnetic field annihilation. In: Hess WN (ed) AAS NASA Symp. Phys. Solar Flares. NASA, vol SP–50, pp 425–437

Phan TD, Kistler LM, Klecker B, Haerendel G, Paschmann G, Sonnerup BUÖ, Baumjohann W, Bavassano-Cattaneok MB, Carlson CW, DiLellisk AM, Fornacon K-H, Frank LA, Fujimoto M, Georgescu E, Kokubun S, Moebius E, Mukai T, Oieroset M, Paterson WR, Rème H (2000) Extended magnetic reconnection at the Earth's magnetopause from detection of bi-directional jets. Nature 404:848–850

Phan T, Frey HU, Frey S, Peticolas L, Fuselier S, Carlson C, Reme H, Bosqued J-M, Balogh A, Dunlop M, Kistler L, Mouikis C, Dandouras I, Sauvaud J-A, Mende S, McFadden J, Parks G, Moebius E, Klecker B, Paschmann G, Fujimoto M, Petrinec S, Marcucci MF, Korth A, Lundin R (2003) Simultaneous Cluster and IMAGE observations of cusp reconnection and auroral proton spot for northward IMF. Geophys Res Lett 30(10):1509. https://doi.org/10.1029/2003GL016885

Phan TD, Hasegawa H, Fujimoto M, Oieroset M, Mukai T, Lin RP, Paterson W (2006) Simultaneous geotail and wind observations of reconnection at the subsolar and tail flank magnetopause. Geophys Res Lett 33:9. https://doi.org/10.1029/2006GL025756

Phan TD, Shay MA, Gosling JT et al (2013) Electron bulk heating in magnetic reconnection at Earth's magnetopause: dependence on the inflow Alfvén speed and magnetic shear. Geophys Res Lett 40(17):4475–4480. https://doi.org/10.1002/grl.50917

Pi G, Shue J, Chao J, Němeček Z, Šafránková J, Lin C (2014) A reexamination of long-duration radial imf events. J Geophys Res 119(9):7005–7011. https://doi.org/10.1002/2014ja019993

Pontius D, Wolf RA (1990) Transient flux tubes in the terrestrial magnetosphere. Geophys Res Lett 17:49

Pulkkinen TI, Baker DN, Mitchell DG, McPherron RL, Huang CY, Frank LA (1992) Global and local current sheet thickness estimates during the late growth phase. In: Proceedings of the International Conference on Substorms (ICS-1). European Space Agency, Paris, p 131

Raeder J (2003) Global magnetohydrodynamics – a tutorial. In: Büchner J, Scholer M, Dum CT (eds) Space plasma simulation. Lecture Notes in Physics, vol 615. Springer, Berlin, Heidelberg, pp 212–246. https://doi.org/10.1007/3-540-36530-3_11

Raeder J (2022) Global simulations. In: Maggiolo R, André N, Hasegawa H, Welling DT (eds) Space physics and aeronomy collection volume 2: magnetospheres in the Solar System, 1st edn. Geophysical monograph, vol 259. Am. Geophys Union, Washington. https://doi.org/10.1002/9781119815624

Raeder J, Berchem J, Ashour-Abdalla M (1996) The importance of small scale processes in global MHD simulations: some numerical experiments. In: Chang T, Jasperse JR (eds) The physics of space plasmas, vol 14. MIT Center for Theoretical Geo/Cosmo Plasma Physics, Cambridge, MA, p 403

Reiff PH, Hill TW, Burch JL (1977) Solar-wind plasma injection at the dayside magnetospheric cusp. J Geophys Res 82:479–491. https://doi.org/10.1029/JA082i004p00479

Reiff PH, Daou AG, Sazykin SY et al (2016) Multispacecraft observations and modeling of the 22/23 June 2015 geomagnetic storm. Geophys Res Lett 43:7311–7318. https://doi.org/10.1002/2016GL069154

Reiff PH, Marshall A, Webster J, Sazykin S, Russell CT, Rastaetter L (2018) MMS observations and CCMC modeling of field line stretching at separator lines. Fall AGU e-Lightning poster, https://agu2018fallmeeting-agu.ipostersessions.com/default.aspx?s=B2-10-20-70-BD-2D-A2-4E-35-27-A4-FE-DC-C0-6D-dA. https://doi.org/10.1002/essoar.10502075.1

Rogers B, Zakharov L (1995) Nonlinear ω*-stabilization of the m = 1 mode in tokamaks. Phys Plasmas 2:3420. https://doi.org/10.1063/1.871124

Rogers AJ, Farrugia CJ, Torbert RB (2019) Numerical algorithm for detecting ion diffusion regions in the geomagnetic tail with applications to MMS tail season 1 May to 30 September 2017. J Geophys Res 124:6487–6503. https://doi.org/10.1029/2018JA026429

Rogers AJ, Farrugia CJ, Torbert RB, Rogers TJ (2023) Applying magnetic curvature to MMS data to identify thin current sheets relative to tail reconnection. J Geophys Res 128:e2022JA030577. https://doi.org/10.1029/2022JA030577

Rosenbauer H, Grünwaldt H, Montgomery MD, Paschmann G, Sckopke N (1975) Heos 2 plasma observations in the distant polar magnetosphere: the plasma mantle. J Geophys Res 80:2723–2737. https://doi.org/10.1029/JA080i019p02723

Runov A, Angelopoulos V, Artemyev AV, Weygand JM, Lu S, Li Y, Zhang X-J (2021) Global and local processes of thin current sheet formation during substorm growth phase. J Atmos Sol-Terr Phys 220:105671. https://doi.org/10.1016/j.jastp.2021.105671

Russell CT (1972) The configuration of the magnetosphere. In: Dyer ER (ed) Critical problems in magnetospheric physics. National academy of sciences, p 1

Russell CT, Elphic RC (1978) Initial ISEE magnetometer results: magnetopause observations. Space Sci Rev 22:681–715. https://doi.org/10.1007/BF00212619

Russell CT, Strangeway RJ, Zhao C, Anderson BJ, Baumjohann W et al (2017) Structure, force balance, and topology of Earth's magnetopause. Science 356:960–963

Sanny J, McPherron RL, Russel CT, Baker DN, Pulkkinen TI, Nishida A (1994) Growth-phase thinning of the near-Earth current sheet during CDAW 6 substorm. J Geophys Res 99:5805–5816

Schillings A, Gunell H, Nilsson H, De Spiegeleer A, Ebihara Y, Westerberg LG, Yamauchi M, Slapak R (2020) The fate of O^+ ions observed in the plasma mantle: particle tracing modelling and cluster observations. Ann Geophys 38:645–656. https://doi.org/10.5194/angeo-38-645-2020

Schindler K, Baker DN, Birn J, Hones EW Jr, Slavin JA, Galvin AB (1989) Analysis of an extended period of earthward plasma flow at ~220 RE: CDAW-8. J Geophys Res 94:15177

Schödel R, Baumjohann W, Nakamura R, Sergeev VA, Mukai T (2001) Rapid flux transport in the central plasma sheet. J Geophys Res 106:301–313. https://doi.org/10.1029/2000JA900139

Scholer M, Otto A (1991) Magnetotail reconnection: current diversion and field-aligned currents. Geophys Res Lett 18:7331

Sergeev VA, Tanskanen P, Mursula K, Korth A, Elphic RC (1990) Current sheet thickness in the near-Earth plasma sheet during substorm growth phase. J Geophys Res 95:3819

Sergeev A, Pellinen RJ, Pulkkinen TI (1996) Steady magnetospheric convection: a review of recent results. Space Sci Rev 75:551–604. https://doi.org/10.1007/BF00833344

Sergeev VA, Liou K, Meng CI, Newell PT, Brittnacher M, Parks G, Reeves GD (1999) Development of auroral streamers in association with localized impulsive injections to the inner magnetotail. Geophys Res Lett 26:417–420

Sergeev V, Liou K, Newell PT, Ohtani SI, Hairston MR, Rich F (2004) Auroral streamers: characteristics of associated precipitation, convection and field-aligned currents. Ann Geophys 22:537–548

Shelley EG, Sharp RD, Johnson RG (1976) He^{++} and H^+ flux measurements in the day side cusp: estimates of convection electric field. J Geophys Res 81:2363–2370. https://doi.org/10.1029/JA081i013p02363

Shi QQ, Shen C, Pu ZY, Dunlop MW, Zong Q-G, Zhang H et al (2005) Dimensional analysis of observed structures using multipoint magnetic field measurements: application to cluster. Geophys Res Lett 32:L12105. https://doi.org/10.1029/2005GL022454

Shukhtina MA, Dmitrieva NP, Sergeev VA (2014) On the conditions preceding sudden magnetotail magnetic flux unloading. Geophys Res Lett 41:1093–1099. https://doi.org/10.1002/2014GL059290

Sitnov MI, Arnold H (2022) Equilibrium kinetic theory of weakly anisotropic embedded thin current sheets. J Geophys Res 127:e2022JA030945. https://doi.org/10.1029/2022JA030945

Sitnov MI, Guzdar PN, Swisdak M (2003) A model of the bifurcated current sheet. Geophys Res Lett 30(13):1712. https://doi.org/10.1029/2003GL017218

Sitnov MI, Tsyganenko NA, Ukhorskiy AY, Brandt PC (2008) Dynamical data-based modeling of the storm-time geomagnetic field with enhanced spatial resolution. J Geophys Res 113(A7). https://doi.org/10.1029/2007JA013003

Sitnov M, Birn J, Ferdousi B, Gordeev E, Khotyaintsev Y, Merkin V, Motoba T, Otto A, Panov E, Pritchett P, Pucci F, Raeder J, Runov A, Sergeev V, Velli M, Zhou X (2019a) Explosive magnetotail activity. Space Sci Rev 215:31. https://doi.org/10.1007/s11214-019-0599-5

Sitnov MI, Stephens GK, Tsyganenko NA, Miyashita Y, Merkin VG, Motoba T, Ohtani S, Genestreti KJ (2019b) Signatures of nonideal plasma evolution during sub- storms obtained by mining multimission magnetometer data. J Geophys Res 124(11):8427–8456. https://doi.org/10.1029/2019JA027037

Sitnov M, Stephens G, Motoba T, Swisdak M (2021) Data mining reconstruction of magnetotail reconnection and implications for its first-principal modeling. Front Phys 9. Retrieved from https://www.frontiersin.org/articles/10.3389/fphy.2021.644884. https://doi.org/10.3389/fphy.2021.644884

Slavin JA, Lepping RP, Gjerloev J et al (2003) Geotail observations of magnetic flux ropes in the plasma sheet. J Geophys Res 108(A1):SMP10-1–SMP10-18. https://doi.org/10.1029/2002JA009557

Sonnerup BUÖ (1974) The reconnecting magnetosphere. In: McCormac BM (ed) Magnetospheric physics, proceedings of the advanced summer institute, Sheffield, UK, August 1973. Astrophysics and space science library, vol 44. Reidel, Dordrecht, pp 23–33

Sonnerup BUÖ, Cahill LJ Jr (1967) Magnetopause structure and attitude from explorer 12 observations. J Geophys Res 72:171. https://doi.org/10.1029/JZ072i001p00171

Sorathia KA, Merkin VG et al (2020) Ballooning-Interchange Instability in the Near-Earth Plasma Sheet and Auroral Beads: Global Magnetospheric Modeling at the Limit of the MHD Approximation. Geophys Res Lett 47. https://doi.org/10.1029/2020GL088227

Spence HE, Kivelson MG, Walker RJ, McComas DJ (1989) Magnetospheric plasma pressures in the midnight meridian – observations from 2.5 to 35 RE. J Geophys Res 4:5264–5272

Stephens GK, Sitnov MI, Korth H, Tsyganenko NA, Ohtani S, Gkioulidou M, Ukhorskiy AY (2019) Global empirical picture of magnetospheric substorms inferred from multimission magnetometer data. J Geophys Res 124(2):1085–1110

Stephens GK, Sitnov MI, Weigel RS, Turner DL, Tsyganenko NA, Rogers AJ et al (2023) Global structure of magnetotail reconnection revealed by mining space magnetometer data. J Geophys Res 128:e2022JA031066. https://doi.org/10.1029/2022JA031066

Sun WJ, Slavin JA, Tian AM et al (2019) MMS study of the structure of ion-scale flux ropes in the Earth's cross-tail current sheet. Geophys Res Lett 46(12):6168–6177. https://doi.org/10.1029/2019GL083301

Swisdak M, Drake JF (2007) Orientation of the reconnection X-line. Geophys Res Lett 34:L11106. https://doi.org/10.1029/2007GL029815

Swisdak M, Rogers BN, Drake JF, Shay MA (2003) Diamagnetic suppression of component magnetic reconnection at the magnetopause. J Geophys Res 108(A5):1218. https://doi.org/10.1029/2002JA009726

Toledo-Redondo S, Andre M, Aunai N, Chappell CR, Dargent J, Fuselier SA et al (2021a) Impacts of ionospheric ions on magnetic reconnection and Earth's magnetosphere dynamics. Rev Geophys 59:e2020RG000707. https://doi.org/10.1029/2020RG000707

Toledo-Redondo S, Hwang K-J, Escoubet CP, Lavraud B, Fornieles J, Aunai N et al (2021) Solar wind—magnetosphere coupling during radial interplanetary magnetic field conditions: simultaneous multi-point observations. J Geophys Res Space Phys 126:e2021JA029506. https://doi.org/10.1029/2021JA029506

Toledo-Redondo S, Hwang K-J, Escoubet CP, Lavraud B, Fornieles J, Aunai N et al (2021b) Solar wind—magnetosphere coupling during radial interplanetary magnetic field conditions: simultaneous multi-point observations. J Geophys Res 126:e2021JA029506. https://doi.org/10.1029/2021JA029506

Torbert RB, Burch JL, Phan TD, Hesse M, Argall MR, Shuster JK, Ergun RE, Alm L, Nakamura R et al (2018) Electron-scale dynamics of the diffusion region during symmetric magnetic reconnection in space. Science 362:1391–1395. https://doi.org/10.1126/science.aat2998

Trattner KH, Coates AJ, Fazakerley AN, Johnstone AD, Balsiger H, Burch JL, Fuselier SA, Peterson WK, Rosenbauer H, Shelley EG (1998) Overlapping ion populations in the cusp: polar/TIMAS results. Geophys Res Lett 25:1621–1624

Trattner KJ, Fuselier SA, Petrinec SM, Yeoman TK, Mouikis C, Kucharek H, Reme H (2005) Reconnection sites of spatial cusp structures. J Geophys Res 110:A04207. https://doi.org/10.1029/2004JA010722

Trattner KJ, Mulcock JS, Petrinec SM, Fuselier SA (2007) Probing the boundary between anti-parallel and component reconnection during southwards interplanetary magnetic field conditions. J Geophys Res 112:A08210. https://doi.org/10.1029/2007JA012270

Trattner KJ, Petrinec SM, Fuselier SA, Omidi N, Sibeck DG (2012) Evidence of multiple reconnection lines at the magnetopause from cusp observations. J Geophys Res 117:A01213. https://doi.org/10.1029/2011JA017080

Trattner KJ, Burch JL, Ergun R, Fuselier SA, Gomez RG, Lewis WS, Mauk B, Petrinec SM, Pollock CJ, Phan TD, Vines SK, Wilder FD, Young DT (2016) The response time of the magnetopause reconnection location to changes in the solar wind: MMS case study. Geophys Res Lett 43. https://doi.org/10.1002/2016GL068554

Trattner KJ, Burch JL, Ergun R, Eriksson S, Fuselier SA, Giles BL, Gomez RG, Grimes EW, Lewis WS, Mauk B, Petrinec SM, Russell CT, Strangeway RJ, Trenchi L, Wilder FD (2017) The MMS dayside magnetic reconnection locations during phase 1 and their relation to the predictions of the maximum magnetic shear model. J Geophys Res 122. https://doi.org/10.1002/2017JA024488

Trattner KJ, Burch JL, Cassak PA, Ergun R, Eriksson S, Fuselier SA, Giles BL, Gomez RG, Grimes EW, Petrinec SM, Webster JM, Wilder FD (2018) The transition between anti-parallel and component magnetic reconnection at Earth's dayside magnetopause. J Geophys Res 123. https://doi.org/10.1029/2018JA026081

Trattner KJ, Fuselier SA, Petrinec SM, Burch JL, Ergun R, Grimes EW (2021b) Long and active magnetopause reconnection X-lines during changing IMF conditions. J Geophys Res 126. https://doi.org/10.1029/2020JA028926

Trattner KJ, Petrinec SM, Fuselier SA (2021a) The location of magnetic reconnection at Earth's magnetopause. Space Sci Rev 217(41). https://doi.org/10.1007/s11214-021-00817-8

Tsyganenko NA, Sitnov MI (2007) Magnetospheric configurations from a high- resolution data-based magnetic field model. J Geophys Res 112(A6). https://doi.org/10.1029/2007JA012260

Tsyganenko NA, Andreeva VA, Sitnov MI, Stephens GK, Gjerloev JW, Chu X, Troshichev OA (2021) Reconstructing substorms via historical data mining: is it really feasible? J Geophys Res Space Phys 126:e2021JA029604. https://doi.org/10.1029/2021JA029604

Vernisse Y et al (2016) Signatures of complex magnetic topologies from multiple reconnection sites induced by Kelvin-Helmholtz instability. J Geophys Res 121:9926–9939. https://doi.org/10.1002/2016JA023051

Vines SK, Fuselier SA, Petrinec SM, Trattner KJ, Allen RC (2017) Occurrence frequency and location of magnetic islands at the dayside magnetopause. J Geophys Res 122:4138–4155. https://doi.org/10.1002/2016JA023524

Walsh BM, Komar CM, Pfau-Kempf Y (2017) Spacecraft measurements constraining the spatial extent of a magnetopause reconnection x line. Geophys Res Lett 44(7):3038–3046. https://doi.org/10.1002/2017gl073379

Wang X, Chen Y, Tóth G (2022) Simulation of magnetospheric sawtooth oscillations: the role of kinetic reconnection in the magnetotail. Geophys Res Lett 49:e2022GL099638. https://doi.org/10.1029/2022GL099638

Webster JM, Burch JL, Reiff PH, Daou AG, Genestreti KJ, Graham DB et al (2018) Magnetospheric multiscale dayside reconnection electron diffusion region events. J Geophys Res 123:4858–4878. https://doi.org/10.1029/2018JA025245

Welling DT, André M, Dandouras I, Delcourt D, Fazakerley A, Fontaine D, Foster J et al (2015) The Earth: plasma sources, losses, and transport processes. Space Sci Rev 192(14):145–208. https://doi.org/10.1007/s11214-015-0187-2

Wing S, Johnson JR, Chaston CC et al (2014) Review of solar wind entry into and transport within the plasma sheet. Space Sci Rev 184:33–86. https://doi.org/10.1007/s11214-014-0108-9

Wolf RA, Wan Y, Xing X, Zhang J-C, Sazykin S (2009) Entropy and plasma sheet transport. J Geophys Res 114. https://doi.org/10.1029/2009JA014044

Wolf RA, Chen CX, Toffoletto FR (2012) Thin filament simulations for Earth's plasma sheet: Interchange oscillations. J Geophys Res 117. https://doi.org/10.1029/2011JA016971

Yushkov EV, Petrukovich AA, Artemyev AV, Nakamura R (2021) Thermodynamics of the magnetotail current sheet thinning. J Geophys Res 126:e2020JA028969. https://doi.org/10.1029/2020JA028969

Zakharov L, Rogers B, Migliuolo S (1993) The theory of the early nonlinear stage of m = 1 reconnection in tokamaks*. Phys Fluids B 5:2498. https://doi.org/10.1063/1.860735

Zou Y, Walsh BM, Nishimura Y, Angelopoulos V, Ruohoniemi JM, McWilliams KA, Nishitani N (2019) Local time extent of magnetopause reconnection using space-ground coordination. Ann Geophys 37:215–234. https://doi.org/10.5194/angeo-37-215

Zou Y, Walsh BM, Atz E, Liang H, Ma Q, Angelopoulos V (2020) Azimuthal variation of magnetopause reconnection at scales below an Earth radius. Geophys Res Lett 47:e2019GL086500. https://doi.org/10.1029/2019GL086500

Zwickl RD, Baker DN, Bame SJ, Feldman WC, Gosling JT, Hones EW Jr, McComas DJ, Tsurutani BT, Slavin JA (1984) Evolution of the Earth's distant magnetotail: ISEE 3 electron plasma results. J Geophys Res 89:11007–11012. https://doi.org/10.1029/JA089iA12p11007

Authors and Affiliations

S.A. Fuselier[1,2] · S.M. Petrinec[3] · P.H. Reiff[4] · J. Birn[5] · D.N. Baker[6] · I.J. Cohen[7] · R. Nakamura[8] · M.I. Sitnov[7] · G.K. Stephens[7] · J. Hwang[1] · B. Lavraud[9] · T.E. Moore[10] · K.J. Trattner[6] · B.L. Giles[10] · D.J. Gershman[10] · S. Toledo-Redondo[11] · J.P. Eastwood[12]

✉ S. Toledo-Redondo
sergio.toledo@um.es

[1] Southwest Research Institute, San Antonio, TX, USA

[2] University of Texas at San Antonio, San Antonio, TX, USA

[3] Lockheed Martin Advanced Technology Center, Palo Alto, CA, USA

[4] Rice University, Houston, TX, USA

[5] Space Science Institute, Boulder, CO, USA

[6] University of Colorado, LASP, Boulder, CO, USA

[7] Applied Physics Laboratory, Johns Hopkins University, Laurel, MD, USA

[8] Space Research Institute, Austrian Academy of Sciences, Graz, Austria

[9] Institut de Recherche en Astrophysique et Planétologie, Université de Toulouse, CNRS, UPS, CNES, Toulouse, France

[10] NASA, Goddard Space Flight Center, Greenbelt, MD, USA

[11] Department of Electromagnetism and Electronics, University of Murcia, Murcia, Spain

[12] Department of Physics, Imperial College London, London, UK

Space Science Reviews (2023) 219:71
https://doi.org/10.1007/s11214-023-01010-9

Cross-Scale Processes of Magnetic Reconnection

K.-J. Hwang[1] · R. Nakamura[2] · J.P. Eastwood[3] · S.A. Fuselier[1] · H. Hasegawa[4] · T. Nakamura[5,6] · B. Lavraud[7] · K. Dokgo[1] · D.L. Turner[8] · R.E. Ergun[9] · P.H. Reiff[10]

Received: 13 July 2023 / Accepted: 30 September 2023 / Published online: 30 October 2023

Abstract
Various physical processes in association with magnetic reconnection occur over multiple scales from the microscopic to macroscopic scale lengths. This paper reviews multi-scale and cross-scale aspects of magnetic reconnection revealed in the near-Earth space beyond the general global-scale features and magnetospheric circulation organized by the Dungey Cycle. Significant and novel advancements recently reported, in particular, since the launch of the Magnetospheric Multi-scale mission (MMS), are highlighted being categorized into different locations with different magnetic topologies. These potentially paradigm-shifting findings include shock and foreshock transient driven reconnection, magnetosheath turbulent reconnection, flow shear driven reconnection, multiple X-line structures generated in the dayside/flankside/nightside magnetospheric current sheets, development and evolution of reconnection-driven structures such as flux transfer events, flux ropes, and dipolarization fronts, and their interactions with ambient plasmas. The paper emphasizes key aspects of kinetic processes leading to multi-scale structures and bringing large-scale impacts of magnetic reconnection as discovered in the geospace environment. These key features can be relevant and applicable to understanding other heliospheric and astrophysical systems.

Keywords Magnetic reconnection · Bow shock · Foreshock transient · Magnetosheath · Turbulent reconnection · Flow shear · Flux rope · Flux transfer event · Dipolarization front · Magnetospheric multi-scale mission · Electron-only reconnection · Flow vortex · Electron vortex · Kelvin-Helmholtz instability · Kelvin-Helmholtz wave · Kelvin-Helmholtz vortex · Mid-latitude reconnection · Current sheet flapping · Substorm · Solar wind transport

1 Introduction

This paper addresses recent advancements to understand the cross-scale processes that unify the kinetic physics operating during magnetic reconnection (Genestreti et al., Liu et al., Norgren et al., this collection) and global magnetospheric context of reconnection (Fuselier et al., this collection), as well as reconnection in other regimes encompassing solar and planetary physics (Gershman et al., Drake et al., this collection). Magnetic reconnection is initiated in the electron diffusion region (EDR; microscales) where electrons are demagnetized and electron physics dominates, and then entails dynamics in the ion diffusion region (IDR; mesoscales) where ions are demagnetized and Hall physics due to ion-electron decoupling

Extended author information available on the last page of the article

becomes important. The region where electron jets from the EDR brake and electrons become re-magnetized is called the outer EDR. Across the diffusion region, the energy stored in the magnetic field is converted into particle energy. The separatrix is the narrow line (in 2-D) or surface (in 3-D) that separates inflowing and outflowing plasmas before and after undergoing the reconnection process. Reconnection ultimately propagates its effect to the macroscopic or global (scale sizes of Earth radii, R_E) region where magnetohydro-dynamics (MHD) governs. Even though these sub-structures show distinct physical features, they adjoin each other and interact by exchanging/transporting particles, momentum, and energy. Therefore, the reconnection process is intrinsically multi-scale and cross-scale with different physics predominantly identified at each boundary layer of different scales.

The near-Earth space provides the most accessible laboratory for the multi-scale physics at work during reconnection that can be applied to various heliospheric systems as well as laboratory plasma experiments. In particular, a key advantage of geospace observations is that all relevant (from microscopic to macroscopic) scales are separated to the extents measurable by in situ. E.g., the EDR ranges on the order of electron gyroradius (ρ_e) or electron inertial length (d_e), which is $\sim$5 km in Earth's dayside magnetopause and $\sim$30 km in the nightside magnetotail. On the other hand, e.g., ρ_e in solar corona is of the order of 1 m, much smaller than observable scales. Therefore, comprehensive understanding of the terrestrial plasma system will fill the gaps between theoretical/analytic predictions and in-situ observations of micro-to-macro-scale processes that underly reconnection occurring throughout the heliosphere and the universe.

Two of the most fundamental and long-standing science questions in near-Earth plasma physics are how plasmas are transported and how particles are energized (Oka et al., this collection). The large-scale magnetospheric convection regarding the former has been described as the Dungey Cycle, which is powered by dayside and nightside reconnection (Fuselier et al., this collection). This global and general picture, however, leaves important missing links in cross-scale aspects of reconnection in the terrestrial environment. Frequently-occurring and recently-highlighted ingredients attributed to the multi-scale nature reconnection include, but are not limited to: reconnection upstream of, at, or downstream of the bow shock; the coupled shock-reconnection-turbulence process whose effects convect to the dayside magnetopause and affect dayside reconnection; formation, extent, and orientation of an X-line depending on the background magnetic topologies; development and evolution of multiple X-lines, resultant multi-scale structures, and their interactions with ambient plasmas; and velocity shear effects on reconnection and flow shear-driven reconnection.

This paper focuses on bridging these gaps in cross-scale aspects of magnetic reconnection via understanding the reconnection onset and the structure, evolution, and consequences of reconnection that can be organized in terms of the location in the geospace. For example, reconnection at/around the bow shock is often characterized by turbulent micro-to-meso-scale current sheets. The dayside magnetopause current sheet is typically asymmetric with a significant guide field, whereas the nightside magnetotail current sheet is generally symmetric along the current sheet normal direction with no or little guide field. The tail current sheet is, however, asymmetric along the nominal magnetic field direction, i.e., the x direction in geocentric solar magnetospheric coordinates (GSM). The current sheets formed at higher latitudes than the equatorial dayside magnetopause or developed along the flankside magnetopause are subject to large velocity shear. Although the tearing instability is suppressed by the velocity shear, the flow shear-driven instability can compress an initially thick current sheet, triggering reconnection. These different reconnection geometries and magnetic topologies lead to different extent and orientation of the X-line, which subsequently drives various channels for the evolution and consequence of reconnection including 3-D and/or multi-scale structures. Thus, we structure this paper as follows:

1. Introduction

2. Reconnection in the shock region and magnetosheath

2.1 Introduction
2.2 Reconnection in the shock transition region
2.3 Reconnection in the foreshock
2.4 Interaction of discrete current sheets with the bow shock and associated reconnection onset
2.5 Influence of shock dynamics and reconnection on the downstream magnetosheath

3. Reconnection in the dayside magnetopause

3.1 Introduction
3.2 Multi-scale aspects of dayside reconnection processes
3.3 Extent of reconnection X-lines and 3-D complexities
3.4 Flux ropes and Flux transfer events (FTEs)

4. Reconnection in the flankside magnetopause

4.1 Introduction
4.2 In-plane 2-D reconnection in the presence of a velocity shear
4.3 Mid-latitude reconnection associated 3-D magnetic topologies
4.4 Plasma mixing, transport, and turbulence

5. Reconnection in nightside magnetotail

5.1 Introduction
5.2 Onset conditions for local/global current sheet thinning
5.3 Multi-scale aspects of the nightside reconnection region
5.4 Transient structures in the exhaust region

6. Key aspects with relevance and applications to other plasma systems

The importance of multi-scale perspectives of reconnection has been indicated by recent multi-point in-situ measurements and integration with state-of-the-art numerical simulations. Each section introduces the topic that has drawn wide attention among the magnetospheric community, what is currently known, and what remains elusive about the topic. We highlight what advancements have been achieved by the high time-resolution dataset from the four Magnetospheric Multi-scale (MMS) satellites of inter-spacecraft spacing down to electron scales. We address how these findings can be a key to connect the missing links in the multi-scale reconnection process that is applicable to other heliospheric or astrophysical systems.

2 Reconnection in the Shock Region and Magnetosheath

2.1 Introduction

Shocks, reconnection, and turbulence are often held up as three processes that are fundamental to understanding plasmas, particularly in the space environment. This is, in part, because they all control the transport and repartition of energy. This issue is generally of importance

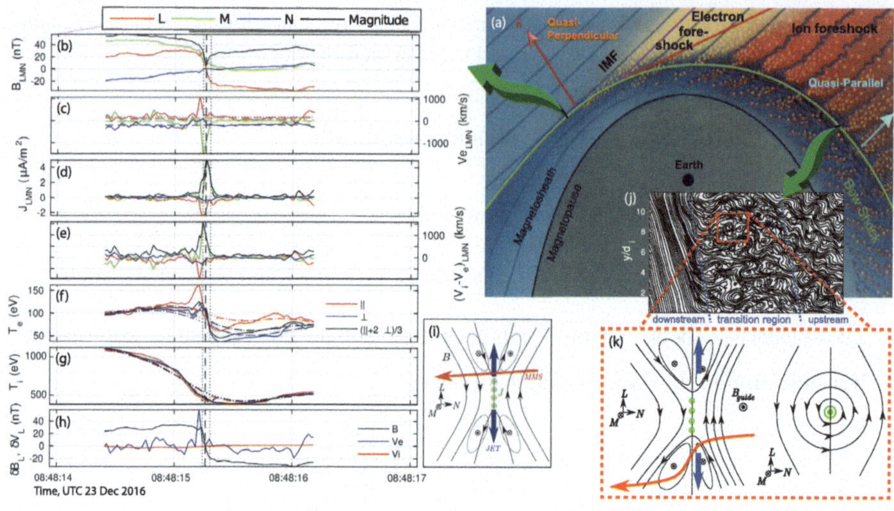

Fig. 1 (a) Schematic of Earth's shock, foreshock, and magnetosheath (adopted from Balogh and Treumann 2013). (b-i) Adopted from Gingell et al. (2020) showing a small-scale reconnection event in the transition region of the quasi-perpendicular shock observed by MMS. The reconnection current carried mostly by electrons (c-e) and electron outflow jet (h; blue profile) indicate that MMS traversed the current sheet undergoing reconnection along the trajectory shown as a red arrow in panel (i). (j, k) Illustration of the quasi-parallel shock structure and the expanded view of the reconnecting current sheet and the magnetic null observed in the turbulent transition region (adopted from Gingell et al. 2019)

in astrophysics where collisionless shocks are believed to be important for creating energetic particles in a variety of circumstances (e.g., Bohdan et al. 2020; Matsumoto et al. 2015).

Prior to the launch of MMS, it was increasingly understood that these three processes may well co-exist in a more tightly-coupled sense (e.g., Karimabadi et al. 2014). The turbulence properties were found to be qualitatively different downstream from the quasi-parallel shock ($\theta_{Bn} < 45°$, where θ_{Bn} is the angle between the shock normal and the interplanetary magnetic field, IMF; Fig. 1a) and quasi-perpendicular shock ($\theta_{Bn} > 45°$). Field-aligned populations backstreaming from the quasi-parallel shock give rise to the formation of the extended foreshock, where the counterstreaming component interacts with the pristine solar wind plasma generating waves, which penetrate the shock, leading to more turbulent magnetosheath (Fig. 1a). Important evidence from Cluster demonstrated the existence of thin ion-scale current sheets in the turbulent magnetosheath exhibiting reconnection signatures in the electric and magnetic field (Retinò et al. 2007; Sundkvist et al. 2007). These studies highlighted the importance of the shock itself as an interplay between turbulence and reconnection.

Notable progress has since stemmed from the analysis of MMS observations of the shock, foreshock, and magnetosheath. A number of studies show that reconnection can operate in all of these regions on a variety of scales, providing a pathway to new understanding of how shocks, turbulence, and reconnection co-exist as well as revealing new questions.

Here we review recent important studies concerning reconnection in the coupled foreshock, bow shock, and magnetosheath system as well as the interplay between turbulence and reconnection. Instead of a comprehensive review of MMS-based magnetosheath reconnection studies detailed in Stawarz et al. (this collection), we focus on multi-scale aspects of the bow shock and its vicinity together with their effect and impact on the dayside magnetopause dynamics to be discussed in Sect. 3.

2.2 Reconnection in the Shock Transition Region

The shock transition region is the front side (ramp) of the shock where the flow speed decreases gradually from the upstream to the shocked downstream region. MMS has provided evidence of magnetic reconnection in the shock transition region for both quasi-perpendicular and quasi-parallel shock configurations. An example for the former is illustrated in the Fig. 1(b-i) event on 23 Dec 2016, where $\theta_{\mathrm{Bn}} \sim 85°$. A coordinate system used in Fig. 1 represents the maximum (L), intermediate (M) and minimum (N) variance directions of the magnetic field over the current sheet crossing. The magnetic field reversal (Fig. 1b, h) is coincident with the electron jet and temperature peak (Fig. 1c, e, h), indicative of electron-engaged reconnection. Another quasi-perpendicular-shock case study from 9 Nov 2016 (Wang et al. 2019), where $\theta_{\mathrm{Bn}} \sim 60°$, identified both the diffusion region and a separate reconnection exhaust. In the diffusion region, Hall electric and magnetic fields were observed as well as enhanced local energy conversion. In the exhaust region observed a few seconds earlier, accelerated ion flow was observed together with evidence of multi-beam mixed populations in a current sheet of $\sim$4 ion inertial lengths (d_i) half-thickness.

At the quasi-parallel shock with $\theta_{\mathrm{Bn}} = 21°$ (Gingell et al. 2019), many small-scale current sheets were observed in a transition region of $\sim$2 minute duration crossing on 26 Jan 2017 as shown in the sketch in Fig. 1j. Reconnection was identified at one such current sheet (Fig. 1k). An unexpected feature was that although the reconnecting current sheet was apparently ion scale in thickness ($\sim 3\ d_i$), only an electron jet was observed. Whilst similar to the MMS observations of electron-only reconnection in the magnetosheath (Phan et al. 2018), it was proposed that the thicker current sheet could be younger than an ion-gyroperiod and therefore in the early stage of its development, consistent with the proximity of the observation to the shock ramp.

Further analysis of both these events, and another event reported in Gingell et al. (2019) used the First Order Taylor Expansion (FOTE) (Fu et al. 2015) to identify the existence of magnetic nulls in the vicinity of MMS (Fig. 1k) when MMS was close to the X-line rather than in the exhaust (Chen et al. 2019). This reconstruction also enabled the reconnection rate to be estimated, finding that it is comparable to observations both in the magnetotail and at the magnetopause.

A follow-up study examining $\sim$900 MMS events of the bow shock observations made in burst mode (Gingell et al. 2020) found that actively reconnecting current sheets were a feature of $\sim$40% of shock crossings and observed in both quasi-parallel and quasi-perpendicular geometries. Given that reconnecting current sheets were encountered for all observed Mach numbers (in addition to θ_{Bn}), it is suggested that reconnection within the shock transition region is a universal process. However, quasi-parallel and high Alfvén Mach number cases exhibited reconnecting current sheets more frequently. In general, the reconnection events were found to typically show only weak ion heating and ion jets with preferential electron coupling. Moreover, the current sheets were reported to be thicker than those observed in the turbulent magnetosheath, and with slower electron jets.

Simulations, as illustrated in Fig. 2, support interpreting and understanding novel MMS observations of reconnection in the shock transition region. The 2-D PIC (particle-in-cell) simulations of a quasi-parallel shock ($\theta_{\mathrm{Bn}} = 25°$) showed that reconnecting current sheets form both in the transition region of the shock and downstream (Bessho et al. 2019), with good agreement to observations made by MMS (Bessho et al. 2022).

Similarly, at quasi-perpendicular shocks it has been shown that cyclical reformation of the shock front leads to the formation of small current sheets where electron-only reconnection may occur (Lu et al. 2021). In general, for high Mach number shocks the shock

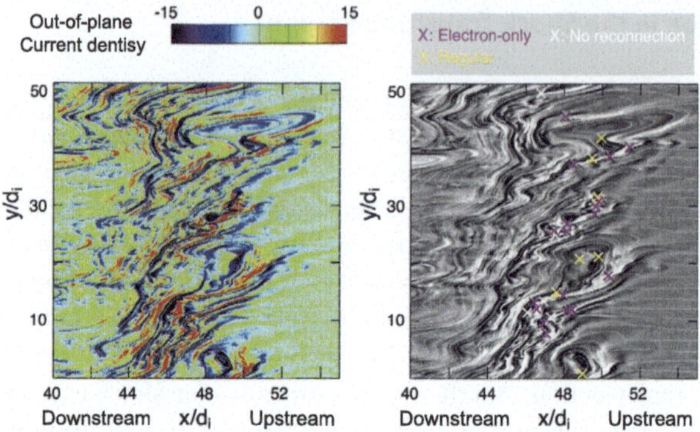

Fig. 2 Abundant current sheets formed in the shock transition and downstream regions of a quasi-parallel shock (adopted from Bessho et al. 2022). In the simulation domain plasmas flow from right (upstream of a shock) to left (downstream). Left panel shows the out-of-plane current density. In the right panel, electron-only and regular (both ion and electron-involving) reconnection sites are marked in magenta and yellow, respectively. Regardless of electron-only or regular reconnection, the reconnection rate ranges between 0.1 and 0.2 (Bessho et al. 2019)

transition tend to become more turbulent with more reconnection sites, exhibiting electron-only reconnection on ion-scale current structures (Bessho et al. 2020). This study further revealed that non-reconnecting current sheets with the magnetic field reversal are also sites for the field-to-particle energy conversion, indicating that this feature is not unique to reconnection events. The energy conversion rate ($\mathbf{J} \cdot \mathbf{E}'$, where $\mathbf{J}$ is the electric current and $\mathbf{E}'$ is the electric field in the electron frame) in non-reconnecting current sheets is, however, smaller than that in reconnecting current sheets.

These data-model analyses provide important context for how shocks, turbulence and reconnection co-exist. In particular, the reconnection rate in the electron-only reconnection current sheets was obtained to be between 0.1 and 0.2, similar to the regular reconnection rate around 0.1 (Bessho et al. 2019). This further poses a new question, whether or not electron-only reconnection is the evolutionary stage before reaching the ion reconnection.

2.3 Reconnection in the Foreshock

The foreshock is defined as the region upstream of the bow shock that is magnetically connected to the shock and contains particles backstreaming from the shock. The combination of backstreaming particles and the inflowing solar wind makes velocity distribution functions subject to a variety of instabilities on a range of spatial and temporal scales, leading to wave generation and particle acceleration/thermalization. The continuously varying solar wind means that the foreshock is not a purely static region but can host a number of transient structures. In particular, the interaction of discontinuities in the solar wind magnetic field with the bow shock can cause changes in particle reflection and localized transient regions of enhanced thermal pressure that may also play a role in particle acceleration processes. These transients are often identified as hot flow anomalies and foreshock bubbles (Archer et al. 2015; Schwartz 1995). Data from MMS has provided new insight and understanding about the role played by reconnection in foreshock dynamics.

In a survey of foreshock transients observed by MMS (Liu et al. 2020), evidence of reconnection was found in five examples of foreshock transients from an examination of 130 events over 2 years of MMS observations. Both anti-parallel and strong guide field reconnection events were reported as shown in Fig. 3. In both cases electron-only reconnection in

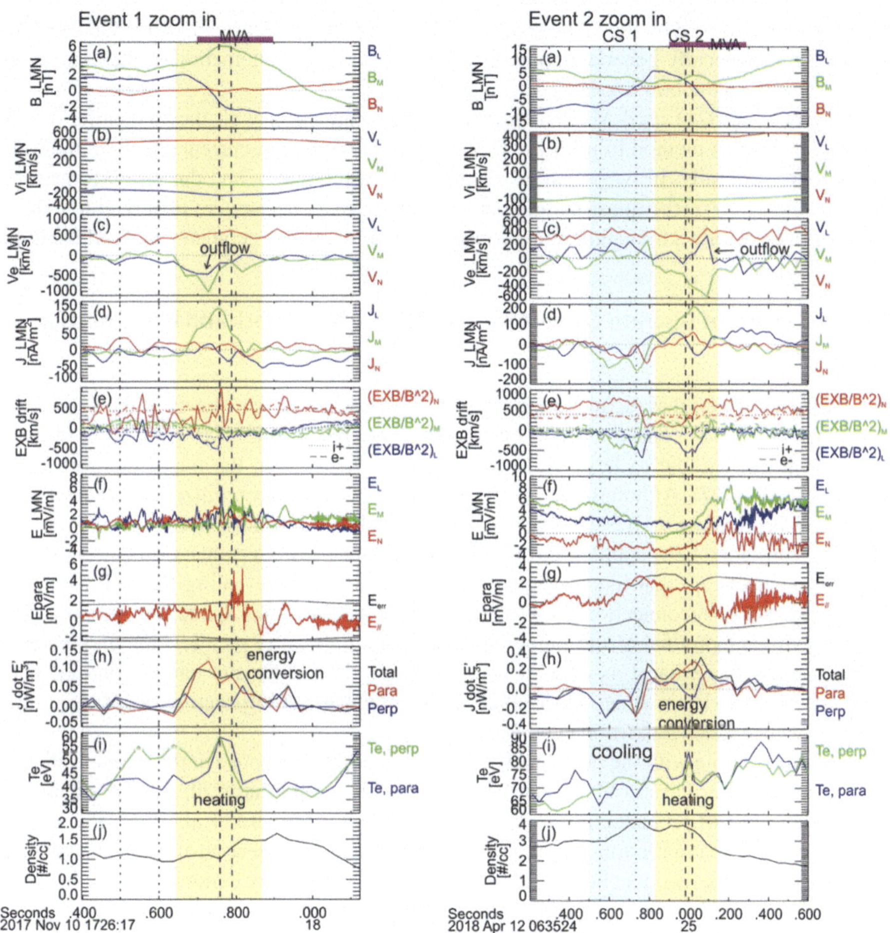

Fig. 3 Adopted from Liu et al. (2020). Two MMS events compare reconnection with a strong guide field (left) and no guide field (right) observed in the turbulent region of a foreshock transient. Both events show electron-only outflows (panel c), indicating electron-only reconnection in a thin ($\sim 1\ d_i$) current sheet. The guide-field reconnection event shows more dominant heating and energy conversion along the parallel direction within the current sheet (bounded by vertical dashed lines)

a thin ($\sim 1\ d_i$) current sheet was identified in the turbulent interior of a foreshock transient. These events (particularly the guide field event) were in many ways similar to observations of electron-only reconnection in the turbulent magnetosheath (Phan et al. 2018). This demonstrates a new role for turbulent reconnection in potentially contributing to the energization of particles in the foreshock, which is relevant for understanding the production of seed populations that undergo subsequent energization during the shock crossing and further downstream the shock.

Magnetic reconnection can be responsible for the formation of flux ropes. Within the foreshock MMS has observed a flux rope inside a hot flow anomaly (Bai et al. 2020). This complements previous (prior to MMS) observations of a flux rope confined inside a hot flow anomaly in the magnetosheath (Hasegawa et al. 2012). Bai et al. (2020) reported a small flux rope (size of 6–8 d_i) encountered at the trailing edge of the hot flow anomaly. A flux rope was not observed by the Wind spacecraft in the upstream solar wind, thus the flux rope was likely to be locally generated, potentially triggered by the interaction of the underlying discontinuity and the bow shock.

Jiang et al. (2021) used MMS observations in conjunction with the first-order Taylor expansion (FOTE) method to infer the existence of reconnecting current sheets in the foreshock more generally. In two separate examples from the dawn- and dusk-side foreshock, but both upstream of the quasi-parallel shock, thin current sheets were observed to exhibit electron-only reconnection signatures, consistent with magnetic geometry derived from the FOTE analysis.

2.4 Interaction of Discrete Current Sheets with the Bow Shock and Reconnection Onset

It is well established that reconnection occurs in large scale current sheets in the solar wind (e.g., Gosling 2012). These current sheets are convected in the solar wind flow through the bow shock, leading to large-scale reconnection observed in the magnetosheath (e.g., Øieroset et al. 2017). Although the MMS instrumentation is not optimized to observe solar wind reconnection per se, it is capable of resolving such current sheets in the magnetosheath. MMS has provided new, detailed information about large-scale reconnection current sheets in the magnetosheath that likely originated in the solar wind, revealing the following features: evidence of reconnection at $\sim 100\ d_i$ downstream from the X-line; asymmetric Hall electric and magnetic fields and inhomogeneous (predominantly parallel) ion and electron heating across the exhaust associated with the guide field; a density cavity confined near one edge of the exhaust where electron cooling and enhanced ion heating occur due to the parallel electric field (Eastwood et al. 2018).

When interacting with the bow shock, or near to the magnetopause in the magnetosheath, a non-reconnecting current sheet can be compressed (Kropotina et al. 2021), which can lead to the onset of reconnection in the magnetosheath (Maynard et al. 2007; Phan et al. 2007). Pre-MMS observations suggested that the interaction of a current sheet with the shock can interrupt reconnection, with it being triggered again by compression at the magnetopause (Phan et al. 2011). During a quasi-perpendicular shock crossing on 20 Dec 2015, a solar wind discontinuity was fortuitously observed by MMS embedded just behind the shock front (Hamrin et al. 2019). Although not in burst mode, MMS provided evidence for Hall fields and particle energization indicating ongoing reconnection. By comparing with multiple other satellites in the solar wind, foreshock, and magnetosheath it was concluded that the observed reconnection was triggered by the compression of the discontinuity at the shock.

Global 3-D hybrid simulations of the dayside shock region have shed further light on this process. A simulation of the interaction of a rotational discontinuity with 180° magnetic shear with the quasi-perpendicular shock suggests that reconnection can be triggered by the shock compression leading to the formation of reconnection jets and flux ropes, initially on ion scales (Guo et al. 2021a). Other simulations focusing on the interaction of rotational discontinuities with the quasi-parallel shock reveal more complex behavior. In particular, the discontinuity can be affected by the enhanced fluctuations in the foreshock, causing more complex structure and local current sheet thinning and reconnection (Guo et al. 2021b).

2.5 Influence of Shock Dynamics and Reconnection on the Downstream Magnetosheath

On the question of how the shock influences the magnetosheath, data from MMS has provided apparently conflicting conclusions. Yordanova et al. (2020) concluded that behind the

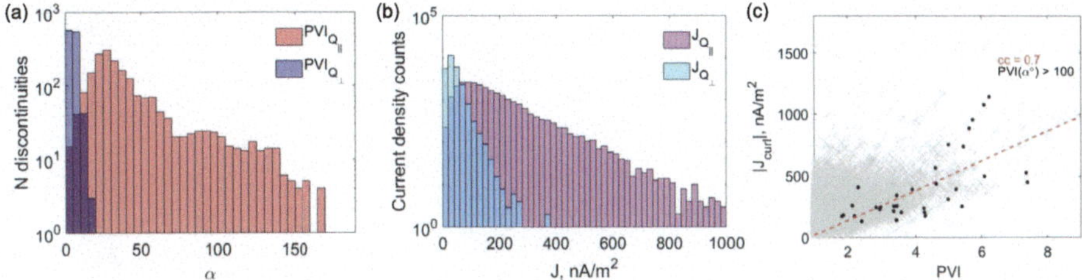

Fig. 4 Adopted from Yordanova et al. (2020). Statistics of the occurrence of discontinuities as a function of the magnetic shear (α) measured from pairs of MMS spacecraft separated by 0.25 s (a). In the quasi-perpendicular magnetosheath, the PVI, a measure of structural coherency, is concentrated mostly at $\alpha < 10°$. The quasi-parallel case shows the distribution over the entire range of α with heavy tail for $\alpha > 90°$. The heavy tail is associated with the presence of currents as confirmed by the histogram of the current density (b). The current density and the PVI (c) for the quasi-parallel magnetosheath shows a good correlation (c) in particular for black dots ($\alpha > 100°$), potentially corresponding to reconnecting current sheets

quasi-parallel shock the magnetosheath exhibits stronger current sheets and discontinuities (Fig. 4) whereas Gingell et al. (2021) examined the prevalence of thin current sheets in the magnetosheath, finding no strong dependence on the local plasma beta or the shock θ_{Bn} and Alfvén Mach number. The two analyses used different approaches. The former employed the Partial Variance of Increments (PVI) approach for detection of coherent structures such as current sheets (Greco et al. 2018) and the latter directly searched for discrete 1-D current sheet structures. Gingell et al. (2021) suggested that the unexpectedly low prevalence in the quasi-parallel magnetosheath could be due to the difficulty of designing automated routines to identify current sheets in a more generally disturbed plasma. In simulations, global 3-D hybrid modelling has shown that foreshock waves can pass through the shock, where they are compressed and can undergo reconnection (Lu et al. 2020). While this type of simulation cannot capture electron-only reconnection, it provides further insight into the differences between the magnetosheath downstream of different bow shock magnetic geometries, which remains an active area of research.

Of particular interest is the role that thin current sheets embedded in the turbulent downstream flow of the shock may play in energization, and therefore the overall energy conversion process that the solar wind experiences as it flows around the magnetosphere. In a case study of a quasi-parallel shock which benefited from a very long acquisition of MMS burst mode data on 21 Dec 2017, thin current sheets were found to occupy 3% of the downstream magnetosheath volume with many of them showing evidence of reconnection (Schwartz et al. 2021). These current sheets were estimated to process 5–11% of the upstream energy flux incident at the bow shock. This indicates that these thin current sheets initiated at the bow shock continue to influence energy processing throughout the downstream magnetosheath to the arrival at the dayside magnetopause. The detailed pathways and impacts of these fluctuations driven by reconnection, shock, and turbulence toward the magnetopause and beyond require further investigation.

3 Reconnection in the Dayside Magnetopause

3.1 Introduction

The dayside magnetopause reconnection current sheet is characterized by local asymmetries that arise from the density gradient across the magnetopause, a velocity shear in the

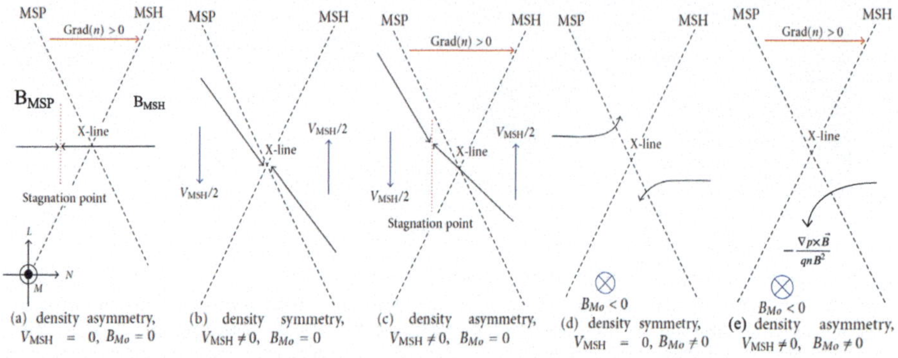

Fig. 5 Three elements driving asymmetries of the dayside magnetopause current sheet are the density gradient (panels a, c, and e), a velocity shear (b, c), and a guide field (d, e). In each panel, the magnetospheric (magnetosheath) region is on the left (right). Effects of each of these and combined effects are illustrated (adopted from Tanaka et al. 2010). The boundary coordinate system represents the maximum (L), intermediate (M) and minimum (N) variance directions of the magnetic field over the current sheet crossing

reconnection plane across the current sheet, and the presence of a guide field (Fig. 5). The density asymmetry leads to a shift in the flow stagnant point toward the low-density magnetospheric side from the X-line due to the high-density magnetosheath plasmas that carry momentum (Fig. 5a). In the presence of a velocity shear (Fig. 5b), which is expected at the high-latitude dayside magnetopause away from the subsolar region, the inflow is slanted from the normal direction. When the two are combined (Fig. 5c), the X-line site is occupied by magnetosheath plasmas, resulting in an L-directional X-line drift. A guide field creates a similar slanted inflow due to the Lorentz force, which occurs near the center of the current sheet (Fig. 5d).

When the density gradient is combined with a guide field (Fig. 5e), the diamagnetic effect controls the drift of an X-line. If the diamagnetic drift is larger than the outflow speed, reconnection ceases to operate (Swisdak et al. 2010, 2003). The presence of a velocity shear (Fig. 5b and 5c) has, in general, a similar effect on the reconnection rate as diamagnetic suppression (Fig. 5e) since the slanted inflow decreases the efficiency of the reconnected field line to drive the outflow (La Belle-Hamer et al. 1995; Cassak and Otto 2011; Doss et al. 2015).

These asymmetries across the dayside magnetopause give rise to significant variations in the multi-scale boundaries of the reconnecting current layer. The density asymmetry (Fig. 5a) leads to the deviation from the quadrupolar Hall fields: strong Hall E_N along the magnetospheric separatrix region; more bipolar than quadrupolar Hall B_M. The effect of a guide field (Fig. 5d) further introduces the quadrupolar density variation and asymmetric quadrupolar Hall fields. (See more details in Genestreti et al., this collection.) The effect of a velocity shear (Fig. 5b), which is most ubiquitous in the flankside magnetopause, is discussed in Sect. 4.1.

The following subsections focus on the multi-scale characteristics of dayside reconnection (Sect. 3.2), the extent and orientation of the X-line over the surface of the dayside magnetopause (Sect. 3.3), and evolution and consequences of dayside reconnection such as flux ropes and flux transfer events (Sect. 3.4).

3.2 Multi-Scale Aspects of Dayside Reconnection Processes

Typical scale lengths of the terrestrial dayside magnetopause current layers are on the order of d_e (the electron inertial length; $\sim$5 km) for EDR, d_i (the ion inertial length; $\sim$200 km) for IDR, and thousands of km to tens of R_E for, e.g., the length of an X-line. The first observation of the EDR at the dayside magnetopause was made by MMS (Burch et al. 2016). The high time-resolution MMS data revealed typical features of the EDR, including electron agyrotropy, i.e., electron crescent distribution in the plane perpendicular to the magnetic field, electron jet reversal, and positive Ohmic energy exchange, as predicted from PIC simulations (Hesse et al. 2014).

Using these common features of the EDR, Webster et al. (2018) reported 32 MMS events of the dayside reconnection EDR. Small separation between the four MMS spacecraft down to electron scales enabled evaluating relative contributions of each term in the generalized Ohm's law. Webster et al. (2018) showed general agreement between the reconnection electric field and the gradient of the electron pressure tensor, while presenting a wide range of electron distribution and energy exchange patterns around the EDR. Torbert et al. (2016) found that the electron inertial term in the generalized Ohm's law is not negligible in the EDR. The EDR was found to be unaffected by turbulence, as seen in a magnetotail reconnection event (Ergun et al. 2022) where the electron physics in a turbulent EDR does not differ from a laminar EDR (Sect. 5.3).

Hwang et al. (2017) reported observations where MMS passed through the edge of the elongated EDR (i.e, the outer EDR) in the dayside reconnection current sheet. Characteristics of the outer EDR included the parallel electron crescent distribution and the out-of-plane electric field exhibiting an opposite polarity to the reconnection electric field. The former (Fig. 6b) is explained by cyclotron turning of the accelerated electrons by the reconnected magnetic field in the outer EDR (Chen et al. 2016). The latter is caused by the electron outflow jet outrunning the moving magnetic field, as evidenced by the observed energy conversion from bulk kinetic energy to field and thermal energy (Zenitani et al. 2011; Chen et al. 2016; Fig. 6a). This indicates the outer EDR as an ingredient in the energy partition during reconnection.

The role of the outer EDR in the cross-scale coupling of reconnection was further investigated by Genestreti et al. (2020). They used the MMS observation of the interface between the EDR and the IDR to assess how electron-scale dynamics affect the ion-scale reconnection rate (Fig. 6c-d). The Hall electron flow in the high-density magnetosheath side leads to the formation of intense pileup of reconnected magnetic flux, consistent with energy conversion from electron kinetic energy to field energy. Thus, electron dynamics affects the opening angle, which is closely related to the reconnection rate.

At the nearby EDR, electron-scale magnetic field fluctuations were observed (Hoilijoki et al. 2021). The higher (lower) reconnection rate can make the opening angle larger (smaller) by increasing (decreasing) the magnetic field pileup. Thus, time variation of the reconnection rate can result in the magnetic field fluctuations. Another generation mechanism for electron-scale magnetic field fluctuations is electron vorticity due to electron Kelvin-Helmholtz instability (Fermo et al. 2012).

The separatrix is an active region where reconnection inflows and outflows are mixed. In the separatrix of dayside reconnection, spiky parallel electric field structures are observed, including double layers and electrostatic solitary waves (Genestreti et al. 2020; Holmes et al. 2019; Hwang et al. 2017). The two-stream instability between inflow and outflow electron plasma, the bump-on-tail instability due to high-speed tail, and the Buneman instability caused by different drifts of ions and electrons can all be important in this region of mixing

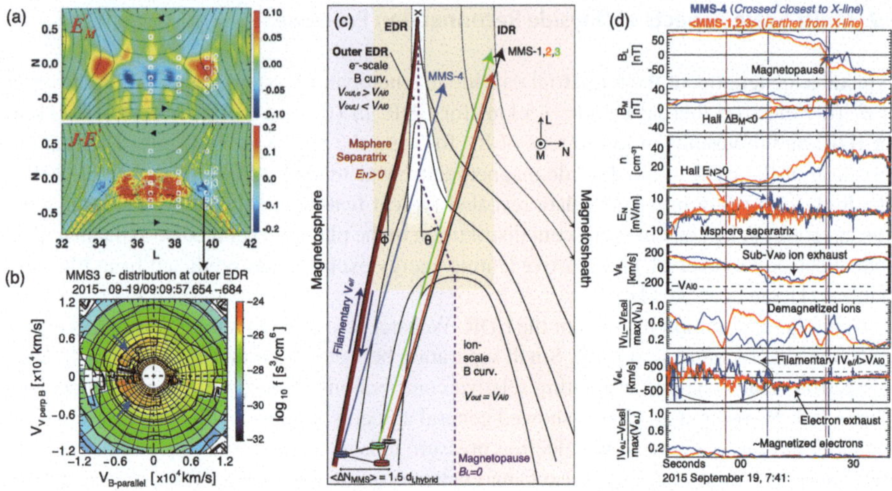

Fig. 6 (a-b) Adapted from Chen et al. (2016) and Hwang et al. (2017) showing the out-of-plane electric field of an opposite polarity to the reconnection electric field, negative Ohmic energy exchange, and parallel electron crescent in the outer EDR. (c-d) Adopted from Genestreti et al. (2020). MMS1-3 passed the IDR, while MMS4 went closer to the EDR through IDR and outer EDR. MMS observed substructures of the IDR including the Hall fields and super-Alfvenic electron flow, sub-Alfvenic ion outflow, and magnetized electron outflow within the exhaust layer. The electron dynamics in the outer EDR can affect the opening angle, thus, the reconnection rate. In this event the opening angles were calculated (magnetosphere side $\phi = 24^\circ \pm 4^\circ$, magnetosheath side $\theta = 6^\circ \pm 4^\circ$)

between magnetospheric and magnetosheath plasmas (Chang et al. 2021; Graham et al. 2017).

The interaction between inflow and outflow can also generate electron vorticity in the separatrix layer (Ahmadi et al. 2022). Such vortices and current loops can induce magnetic field disturbances (Hwang et al. 2019; Stawarz et al. 2018). However, in the magnetospheric separatrix, the net potential of spiky structures is not enough to accelerate the observed electron inflows. Instead, the spiky structures contribute to a rapid electron-time-scale thermalization of plasmas that are transported into the magnetospheric inflow region by ion-scale lower-hybrid drift waves (Holmes et al. 2019). These heating processes engaging multi-scale waves can lead to pre-heating of inflowing plasmas before they enter the EDR.

Beyond this coupling of kinetic reconnection boundaries, the global-scale characteristics of the X-line as well as evolution and consequences of multiple X-lines developed on the dayside magnetopause are discussed in the following Sects. 3.3 and 3.4, respectively. The multi-scale processes and structures occurring in the exhaust region are detailed in Sect. 3.4.

3.3 Extent of Reconnection X-Lines and 3D Complexities

When the IMF is southward, magnetic reconnection at the magnetopause occurs at relatively low latitudes between magnetic field lines in the magnetosheath and field lines in the magnetosphere. This reconnection produces long, continuous, quasi-stationary X-lines that can extend across the entire dayside and possibly along the flanks of the magnetopause. These conclusions about reconnection X-lines were developed from multi-spacecraft observations at the magnetopause (e.g., Dunlop et al. 2011; Phan et al. 2000), global auroral imaging (Fuselier 2002), observations in the magnetospheric cusps (Trattner et al. 2021a, 2007) and, more recently, observations from the MMS mission (Fuselier et al. 2019b; Hasegawa et al. 2016; Trattner et al. 2021b).

Recent MMS observations indicate that X-line length appears to depend on where reconnection is occurring at the magnetopause and whether there is a guide field at the X-line (Fuselier et al. 2021). Together with the outflow velocity data, multi-species ion distributions imply the location of X-line and magnetic topology (Fuselier et al. 2019a; Petrinec et al. 2020). These recent observations also indicate that the orientation of X-lines at the magnetopause is also a complicated function of these two conditions. Figure 7(a) shows the shear angle between the draped magnetosheath and the magnetospheric field lines at the magnetopause. High (low) magnetic-shear regions are shown in red (blue/purple). These magnetic shear angles at the magnetopause are projected onto the Y-Z plane in GSM. Figure 7(a) depicts the magnetic shear conditions when MMS was at the magnetopause on 22 January 2017 and the IMF (brown vector in the lower right) was strongly southward with a duskward component. Under these conditions, the maximum magnetic shear model (Trattner et al. 2007, 2021b) predicts that there is a more-or-less continuous reconnection X-line that stretches from the dawnside along a "ridge" of oppositely-directed (anti-parallel) magnetosheath and magnetospheric field lines. The X-line cuts across the noon meridian ($Y_{GSM} = 0$), forming the component X-line in Fig. 7. This component X-line connects the two anti-parallel ridges on the dawnside and duskside. Anti-parallel reconnection occurs along the ridges, where the shear is 180° and component (or guide field) reconnection occurs along the "component X-line" that cuts across the noon meridian where the shear is $<180°$.

The orientation of the X-line is different for the component X-line and for the anti-parallel X-lines. Fuselier et al. (2021) investigated the X-line orientations for 37 events where the MMS spacecraft were at or very close to the X-line. The X-line is oriented along the M direction in the standard LMN coordinate system at the magnetopause. Figure 7(b) shows that the component X-line is oriented along the M direction. This orientation implies that the component X-line is many R_E long and continuous. Crossings anywhere along this X-line show the same orientation.

For the anti-parallel X-lines on the duskside and dawnside, the orientation and the length of the X-lines are different from the component X-line. The X-lines are oriented perpendicular to the anti-parallel ridge in the region between the "knee" in the ridge and the component X-line as shown in Fig. 7(c). This orientation implies that this part of the anti-parallel X-line is actually composed of short ($\sim 1\ R_E$) X-lines that are stacked as stairsteps along the ridge. Along the anti-parallel ridge from the knee to the beyond the dawn terminator, the X-line is oriented along the ridge.

Thus, the length of the X-line depends on the location at the magnetopause and on the magnetic shear angle where reconnection occurs. While the maximum shear model predicts the extent and orientation of X-lines over the entire magnetopause surface, PIC simulations (Liu et al. 2018b) showed that when a primary X-line misaligns with the optimal orientation due to, e.g., time-varying upstream conditions, secondary X-lines develop to adjust the orientation. Such multiple X-lines, similar to Fig. 7(c), may facilitate the formation of flux ropes and/or secondary islands, which is discussed in Sect. 3.4.

3.4 Flux Ropes and Flux Transfer Events (FTEs)

The flux ropes formed on the surface of the magnetopause, termed flux transfer events (FTEs), are the manifestation of cross-scale or multi-scale aspects of magnetopause reconnection. The observational phenomena of these structures include a bipolar signature in the magnetic field component normal to the nominal magnetopause (B_N), an increase or decrease in the magnetic field strength at (or bounding, in the case of crater FTEs) the center

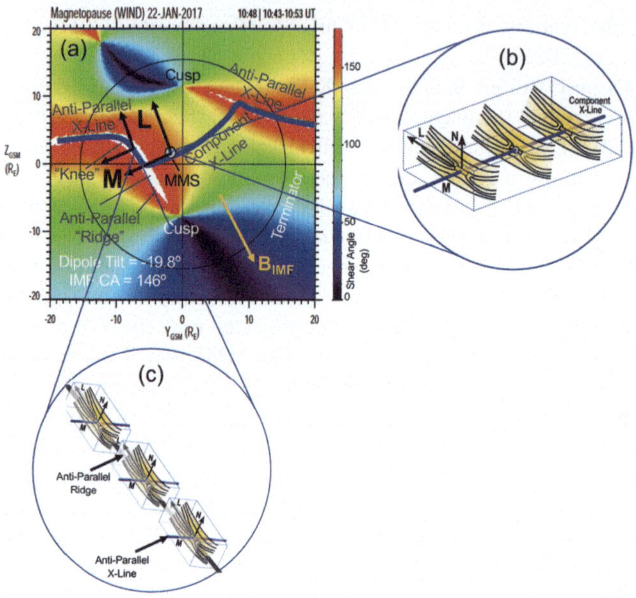

Fig. 7 Shear angle between the magnetosheath and magnetospheric magnetic fields at the magnetopause projected onto the Y-Z_{GSM} plane (a). These shear angles are for 22 January 2017 at 1048 UT when MMS was at the magnetopause and the IMF was strongly southward with a duskward component. The thick blue line traces out a continuous X-line stretching from the dawn flank to the dusk flank magnetopause. The orientations of X-lines are determined by the M direction in the LMN coordinate system. The component X-line is long and continuous and crossings of this X-line show the same orientation of the reconnection structure (b). Part of the anti-parallel ridge is composed of short X-lines that are oriented perpendicular to the ridge (c)

of the B_N reversal, and coexistence of the magnetospheric and magnetosheath plasma inside the FTE.

These characteristics of FTEs are explained by their generation mechanisms invoking 1) transient bursts (spatially and temporally) of dayside reconnection (Russell and Elphic 1978), 2) the turn-on-and-off or temporal modulation of the reconnection rate of single X-line reconnections (Phan et al. 2004; Scholer 1988; Southwood et al. 1988), or 3) multiple X-lines (in 2-D representations) or separator lines (in 3-D representations) (Lee and Fu 1985; Scholer 1995). Different generation mechanisms necessarily give rise to different magnetic topologies or magnetic field connectivities within and around the FTEs.

MMS with its high-resolution measurements enabled us to resolve the detailed substructures of FTEs: multi-layered substructures within a crater FTE (Hwang et al. 2016); reconnecting current sheets between interlinked flux tubes (Kacem et al. 2018; Kieokaew et al. 2020; Øieroset et al. 2019; Hwang et al. 2020b); electron-scale reconnecting current sheets at the leading edge of an FTE or a flux rope (Poh et al. 2019; Zhong et al. 2021); electron- or ion-scale current layers at the interface of two coalescing FTEs (Wang et al. 2017; Zhou et al. 2017); and ion-scale flux ropes growing from an electron-scale current layer (Hasegawa et al. 2023).

Figure 8(b-e) show the observations of a series of the ion-scale FTEs (marked by shaded areas) when MMS crossed between the low-latitude boundary layer (LLBL; green bar at the top of panel b) and the magnetopause current sheet (red bar). Four-spacecraft measurements allow the force analysis across FTEs: the magnetic tension force (Fig. 8c) is relatively

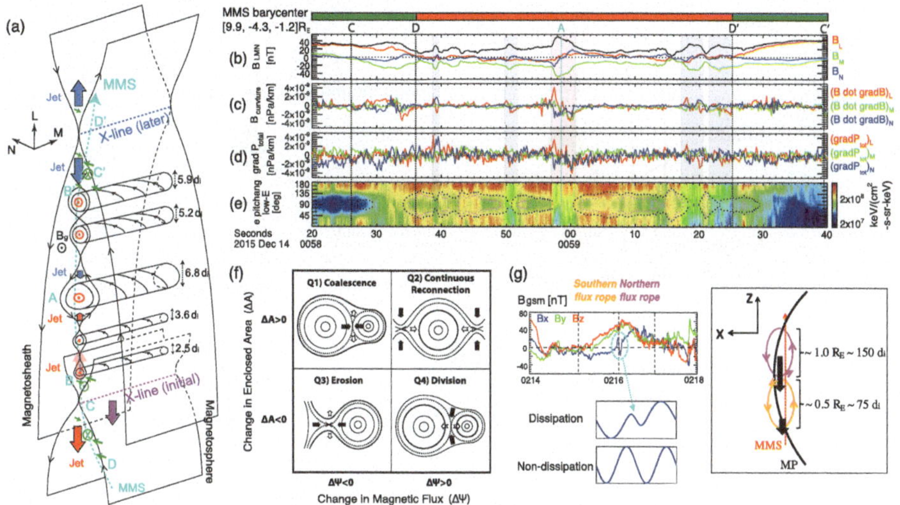

Fig. 8 Adopted from Hwang et al. (2018), Zhou et al. (2017), and Akhavan-Tafti et al. (2020). The observation of a series of ion-scale FTEs generated in the magnetopause current sheet (a-e): across FTEs the magnetic tension force (c) is relatively balanced by the total pressure gradient force (d). The relative motion between the initial and later X-lines (a) facilitates the formation of FTEs via tearing instability. These FTEs can interact via reconnection which dissipates the magnetic field at the interface of the two flux ropes to coalesce (g)

balanced by the total pressure gradient force (Fig. 8d). Throughout the current sheet, the low-energy electrons consist of two populations (Fig. 8e): a field-aligned population accelerated by the electrostatic potential (Egedal et al. 2008) and a trapped (centered on 90° pitch angles) population locally bouncing within the reconnection exhaust (Lavraud et al. 2016). Both indicate ongoing reconnection around the FTEs.

Figure 8(a) illustrates the MMS trajectory (cyan dashed arrow) across these small-scale FTEs. The relative drift between the initial and later X-lines facilitates the formation of multiple secondary islands or FTEs via tearing instability. These FTEs can interact with each other or the ambient magnetic field to either coalesce (Fig. 8g) or erode. Simulation studies by Akhavan-Tafti et al. (2020) predicted the Fig. 8(a-e) FTEs to grow Earth-sized within 10 min, and continuous reconnection at adjacent X-lines between FTEs is the dominant source of magnetic flux and plasma of the resulting large-scale FTEs. Possible interactions between flux ropes and the resulting magnetic geometry are summarized in Fig. 8(f) in terms of the change in the enclosed area inside an FTE and its magnetic flux content (Akhavan-Tafti et al. 2020). These studies indicate that microscale (electron) and mesoscale (ion) processes associated with reconnection occurring around FTEs play a crucial role in the generation, structure, and evolution of FTEs, which, in turn, modulate the structure and evolution of pre-existing or primary reconnection.

Formation and evolution of FTEs are directly linked to magnetic field connectivity with Earth's magnetosphere, affecting solar wind mass, momentum, and energy transfer into the magnetosphere down to the ionosphere. Implications of these FTEs on the large-scale magnetosphere-ionosphere coupling (MIC) processes are obtained from the sequential observations of FTEs, poleward moving auroral forms, and polar cap patches (Hwang et al. 2020b).

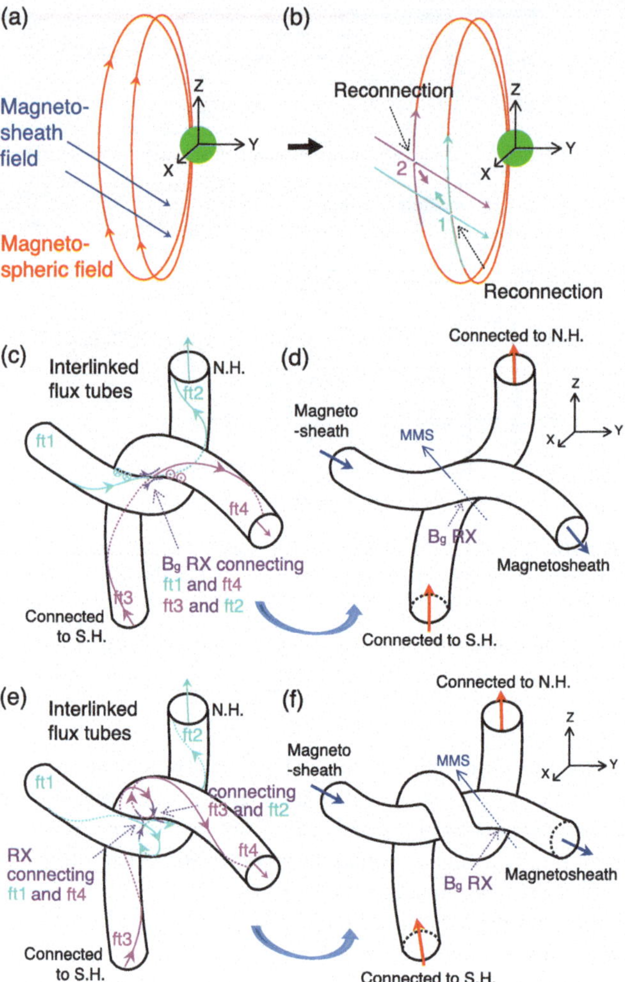

Fig. 9 Adopted from Hwang et al. (2021). Interlinked flux tubes form via multiple reconnection X-lines (a-b-c, a-b-e). Two reconnected flux tubes interact at the center of the interlinked flux tubes via localized reconnection (c, e). Resultant structures keep evolving their magnetic connectivity (d, f). This entanglement can suppress the tailward transition of this type of FTEs

MMS identified the generation and evolution of interlinked flux tubes. Figure 9 shows how the interlinked flux tubes can form and evolve via localized reconnection, which necessarily entails the change of the magnetic connectivity. Figure 9(a, b) illustrate the generation of interlinked flux tubes under the southward IMF with a significant B_y component. This condition leads to two X-lines developed on the dayside magnetopause, and two reconnected flux bundles collide with and entangle each other. A notable difference in the energy-dependent electron pitch-angle distributions before and after the center of the flux rope/FTE indicates such interlinked flux tubes (Hwang et al. 2021; Kacem et al. 2018; Russell and Qi 2020)

At the interface of the two flux tubes (Fig. 9c, e), converging jets together with the magnetic pile up (Øieroset et al. 2016; Øieroset et al. 2019; Maheshwari et al. 2022) lead to the formation of a compressed current sheet giving rise to the onset of reconnection (Fargette et al. 2020). As a result, the two flux tubes change their connectivity either with both ends connected to the magnetosphere or with both ends connected to the magnetosheath (Fig. 9d, f). Furthermore, reconnection at the leading edge of FTEs between the FTE field lines and the magnetospheric (closed), magnetosheath (unconnected), or boundary layer (open or complex topology) field lines will affect whether FTEs significantly contribute to nightside flux transport or not, as will be different for loosely- vs. tightly-interlinked flux tubes (Fig. 9d vs. 9f). These scenarios indicate that the localized physics occurring in FTEs can make macroscopic impacts such as the global circulation of magnetic flux through the magnetosphere and solar wind-magnetosphere coupling, affecting the content and rate of the magnetic flux and solar wind transfer.

Mejnertsen et al. (2021) studied the propagation, evolution, and fate of FTEs generated by multiple reconnection X-lines. They simulated a real solar wind event using the MHD code to find that 1) the extent to which FTEs add magnetic flux to the tail depends on their topology and 2) the resulting FTEs transit around the flankside magnetopause before they eventually dissipate due to non-local (nightside magnetotail) reconnection altering the FTE topology. The FTEs drifting on the flanks of the magnetopause also affect local dynamics occurring there that typically involve velocity-shear driven multi-scale processes, which is the topic of Sect. 4.

4 Reconnection in the Flankside Magnetopause

4.1 Introduction

Figure 5(c) represents the reconnection geometry typically occurring on the flank-side magnetopause. The combined flow shear and density asymmetry have multiple effects on reconnection such as the drift of the X-line (Fig. 5c) and the reconnection efficiency (Cassak and Otto 2011; Doss et al. 2015). These combined effects also drive an additional asymmetry in the reconnection exhaust region (La Belle-Hamer et al. 1995). In the upper (tailward) exhaust of Fig. 5(c), the outflow is in the same direction as the upstream magnetosheath flow. A smaller force is required to drive the outflow. On the other hand, the larger density on the magnetosheath side requires a larger accelerating force to drive the outflow. Thus, the effects of shear flow and density gradient compete on one side but enhance each other on the other side. The tailward exhaust, where the two effects compete, exhibits a broader magnetic field transition region often bounded by bifurcated current sheet. The sunward exhaust, where the two effects enhance each other, becomes a narrow current sheet with the accelerated flow entrained toward the magnetospheric side.

Tanaka et al. (2010), using PIC simulation codes, studied the effects arising from the combination of a guide field as well as a flow shear in density-asymmetric reconnection. They showed that both an initial upstream flow and the Lorentz force acting on inflowing plasmas due to a guide field produce a slanted inflow to the current sheet, giving rise to asymmetries in the quadrupolar exhaust similar to the MHD simulation by La Belle-Hamer et al. (1995). The MMS observation reported by Hwang et al. (2021) indicated the predicted quadrupolar reconnection current layer with asymmetric exhaust patterns. Tanaka et al. (2010) also showed that the X-line motion is controlled either by the ion flow when the shear flow effects dominate or by the electron flow when the guide field effects dominate, as well as the development of asymmetric exhausts.

These studies considered the case where the velocity shear co-exists with the magnetic shear in the reconnection plane. Although this is a valid case for high-latitude dayside reconnection as well, the intrinsic difference of flankside reconnection comes from the fact that Kelvin-Helmholtz instability (KHI) driven by fast magnetosheath shear flows efficiently compresses the magnetopause current sheet, facilitating the onset of reconnection (details in Sect. 4.2).

Furthermore, under mostly southward IMF the current sheet configuration on the flankside magnetopause is typically a largely antiparallel magnetic field with a perpendicular shear flow (Fig. 1 of Ma et al. 2014a). In this case, both reconnection and KHI can simultaneously operate. In the linear stage, the initial perturbations (magnetic shear vs. flow shear) determine the primary process. In the nonlinear stage, the two modes interact with each other: for the case of initially dominant KHI, reconnection is driven and strongly modified by nonlinear KH waves, producing complex flux ropes via patchy reconnection (Ma et al. 2014b, 2014a); for the case of initially dominant reconnection, the onset of reconnection causes a thinned shear flow layer, generating small wavelength KH waves, which in turn modulate the diffusion region and increase the reconnection rate (Ma et al. 2014b).

Reconnection can also occur out of the shear plane due to a 3-D twist of magnetospheric and magnetosheath magnetic fields induced by KH vortices (Sect. 4.3). This section lastly addresses an important question concerning mass transport, i.e., solar wind entry across the flankside magnetopause and plasma mixing in the LLBL induced by KHI-driven reconnection (Sect. 4.4).

4.2 In-Plane 2-D Reconnection in the Presence of a Velocity Shear

When the velocity shear exists in the same 2-D plane as the reconnection plane, the linear growth of the tearing instability is stabilized by the velocity shear (e.g., Chen and Morrison 1990). However, when the shear flow speed V_0 exceeds the Alfvén speed based on the magnetic field component parallel to the velocity shear ($V_0 > V_A$), the KHI becomes unstable overcoming the in-plane magnetic tension (e.g., Chandrasekhar 1961; Miura and Pritchett 1982).

Numerical simulations treating non-linear physics of the KHI demonstrated that under such super-Alfvénic conditions the vortex flow produced by the non-linear growth of the KHI can locally compress the pre-existing magnetic shear layer (current sheet), and secondarily induce reconnection (Knoll and Chacón 2002; Nakamura et al. 2006, 2011; Pu et al. 1990), as shown in Fig. 10(a). This process is termed vortex-induced reconnection (VIR) or the Type-I VIR (Nakamura et al. 2008). Note that when the KHI produces highly rolled-up vortex arms, which can form with a strong velocity shear in a range of $V_0 > 2\text{–}3\ V_A$ (Miura 1984; Nakamura and Fujimoto 2005), the highly-swirled magnetic field lines involved within the arms can newly form thin secondary current sheets and induce reconnection (Faganello et al. 2008; Nakamura and Fujimoto 2005; Nykyri and Otto 2001). This type of VIR is categorized as Type-II VIR (Nakamura et al. 2008).

It is notable that the VIR can be triggered even when the initial shear layer is too thick to cause fast spontaneous reconnection, because the non-linear vortex flow rapidly compresses the thick layer down to electron-scales (Nakamura et al. 2008, 2011) (Fig. 10a). Indeed, evidence of VIR has been reported at the Earth's dayside-to-flank magnetopause (Eriksson et al. 2016; Hasegawa et al. 2009; Hwang et al. 2020a; Nakamura et al. 2013) whose thickness is typically of the order of 10^3 km and larger than ion-scales (e.g., Berchem and Russell 1982). It is also notable that the rate of VIR is basically higher than that of regular reconnection since the super-Alfvénic vortex flow produces a strong inflow towards the

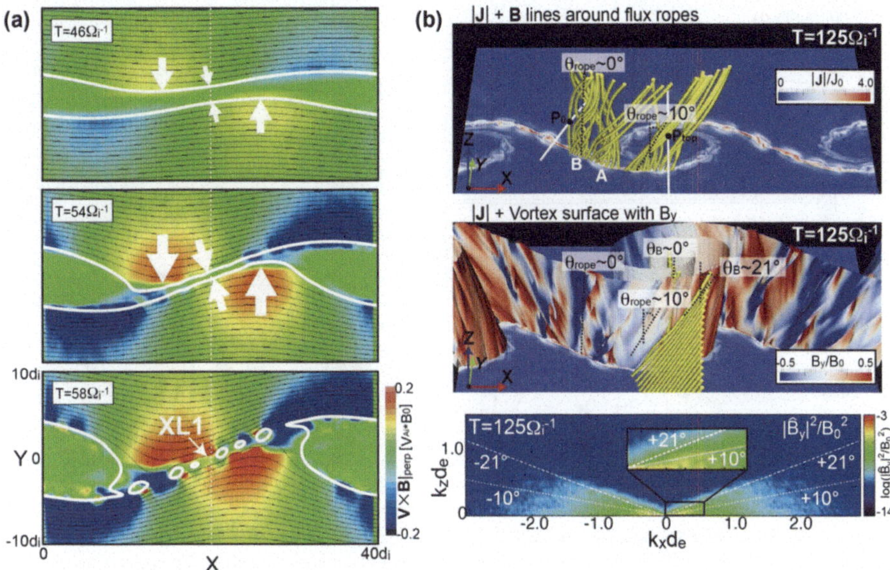

Fig. 10 (a) Time evolution of the perpendicular component of the convection electric field and two selected magnetic field lines in a 2-D fully kinetic simulation, showing the onset of VIR and resulting magnetic island (flux rope) formation in the compressed current sheet (adopted from Nakamura et al. 2011). (b) 3-D views of (top) selected field lines near the VIR-produced transverse/oblique flux ropes and (middle) the vortex surface with the reconnected field component B_y, and (bottom) corresponding magnetic power spectrum (k_x, k_z) in a 3-D fully kinetic simulation, showing the 3-D evolution of VIR and resulting flux ropes over a broad range of oblique angles (adopted from Nakamura et al. 2013)

VIR region (Nakamura et al. 2011). Namely, VIR is a kind of strongly-driven reconnection process, which has been less explored in the reconnection physics. This process results in an efficient plasma mixing and transport across the shear layer along the reconnected field lines (Nakamura et al. 2011).

Recent 3-D kinetic simulations of VIR also demonstrated that the VIR can be triggered at more than one sites in 3-D, with X-line orientations distributed over the entire range of angles between the field direction on one side and that on the other side. As a result, multiple oblique flux ropes are formed along the vortex surface (Fig. 10b), which further enhances the rate of the mass transfer (Nakamura et al. 2013, 2017b).

MMS has provided observational evidence of VIR-generated flux ropes. Figure 11 shows MMS observations of a flux rope locally generated at the KH vortex boundary (Hwang et al. 2020a; Kieokaew et al. 2020). Both outer-leading ('O-L') and outer-trailing ('O-T') edges of the flux rope exhibit reconnection signatures including flow reversals (blue arrows in Figure 11Ab, c). This is consistent with the prediction from the Grad-Shafranov reconstruction of the magnetic field (Fig. 11Ba, b; see details in Hasegawa et al. 2023). The magnetic-field map (black contours) demonstrates an elongation of the flux rope associated with ion flows converging toward the FTE center (colored arrows in Fig. 11Ba). Both plasma flow and density patterns (Fig. 11Ba, b) are consistent with the flux rope generated by VIR (Fig. 10).

At the center of the B_N reversal (marked by vertical dashed black lines, 'C' on the top of Fig. 11A), the rapid B_N change across 'C' with a dip in the magnetic field strength (black profile in Fig. 11Aa; blue in Fig. 11Ad) indicates the existence of local reconnection at the FTE center, as evidenced by ion outflow jets (red arrow in Fig. 11Ab) and out-of-plane

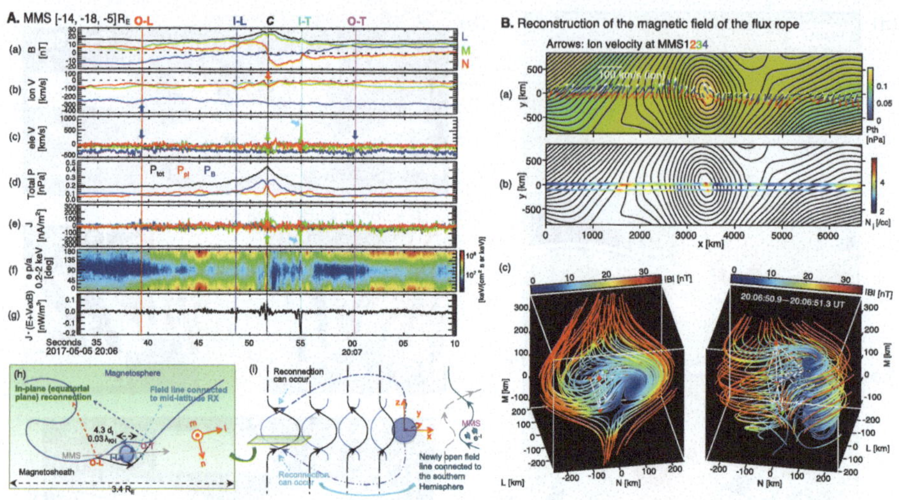

Fig. 11 A. A flankside FTE observed by MMS: (a) the magnetic field, **B** (LMN components and B); (b, c) ion and electron bulk velocities; (d) the plasma (red; P_{pl}) and magnetic (blue; P_B) pressures, and the sum of P_{pl} and P_B (black); (e) the electric current density; (f) the pitch angle distribution for electrons of 0.2–2 keV energy; (g) the energy conversion rate in the electron frame ($\mathbf{J} \cdot \mathbf{E}'$); illustration of VIR-driven flux rope generation (h) and the onset of mid-latitude reconnection (i; Sect. 4.3). Inner-leading (trailing) and outer-leading (trailing) edges are denoted by 'I-L' ('I-T') and 'O-L' ('O-T'), respectively. B. Reconstruction of the magnetic field of the flux rope shown in panels A(a-h) using the Grad-Shafranov (a, b) and SOTE (c) reconstruction methods. Adopted from Hwang et al. (2020a)

current-carrying electron jets (green arrow in Fig. 11Ae), non-zero $\mathbf{J} \cdot \mathbf{E}'$ (Fig. 11Ag), and electron agyrotropy (not shown; Hwang et al. 2020a). Polynomial or second-order Taylor expansion (SOTE) reconstruction methods (Denton et al. 2020; Liu et al. 2019) present an interlinking of two flux tubes (Fig. 11Bc; Sect. 3.4). Indeed, MMS observed notable difference in the electron pitch angle distribution across 'C' (Fig. 11Af) and strong magnetic tension force toward the FTE center (not shown).

The Fig. 11 event demonstrates that reconnection occurs at diverse locations on a variety of scales within the KH vortex, suggesting that KHI and reconnection cooperate on the plasma transport and mixing across the flank magnetopause. 3-D PIC simulations (Nakamura and Daughton 2014; Nakamura et al. 2017a) showed that the rate of the VIR-produced mass transfer across the flank magnetopause, such as the diffusion coefficient ($D_{diff} \sim 10^{10-11}$ m^2/s), could be one-to-two orders of magnitude higher than the previously predicted rate ($D_{diff} \sim 10^9$ m^2/s; Nykyri and Otto 2001) to form the LLBL where the solar wind and the magnetospheric plasmas are mixed. These results indicate that the multi-scale evolution of the VIR may crucially contribute to large-scale mass transfer across the boundary layer where the magnetic and velocity shears co-exist such as the Earth's magnetopause.

4.3 Mid-Latitude Reconnection, Associated 3D Magnetic Topologies

Figure 11A exhibits a distinctive out-of-plane (northward in this event) electron jet at the inner-trailing edge ('I-T') of the flux rope (Fig. 11Ac), carrying most of the current (Fig. 11Ae). The field-aligned jet was not consistent with acceleration by the local electric field, leading to negative $\mathbf{J} \cdot \mathbf{E}'$ (Fig. 11Ag). Observed open field-line topology (illustrated in Fig. 11i) further indicated that the magnetic field lines at 'I-T' might be connected to

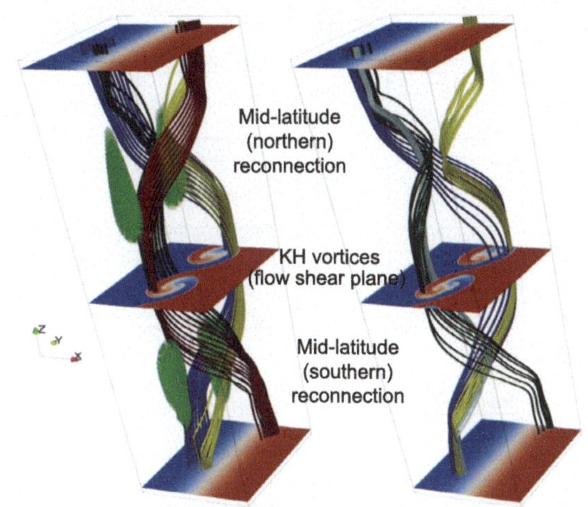

Fig. 12 Adapted from Faganello et al. (2012). Left frame: blue/red colors show tracers of the original side of each field line in the simulations, at the equator and two mid-latitude (X, Y)-planes, light green shows current isosurfaces of large currents, and selected field lines of various colors show the complex magnetic topologies that arise. Right frame: at a later time in the simulation, the green ochre, and grey field lines show pristine magnetospheric, and once-reconnected and double-reconnected field lines, respectively

mid-latitude reconnection between twisted field lines of magnetosheath (black curves in Fig. 11Ai) and magnetosphere (blue) origin in the southern hemisphere.

Thus, in addition to Type-I and Type-II VIR triggered within KH vortices (Sect. 4.2), reconnection can also occur at mid-latitudes in association with the 3-D growth of the KHI. The fact that this type of reconnection is different from VIR occurring in the velocity shear plane is clearly illustrated in Fig. 12 from Faganello et al. (2012).

For the purely northward IMF case as simulated for Fig. 12, Type I and II reconnection would occur in association with the main vortex development in the equatorial plane. Mid-latitude reconnection, by contrast, is triggered away from the equatorial plane as a result of the relative flow shears at the two locations, producing magnetic shears at mid-latitudes that are prone to the triggering of reconnection. The twisting of magnetic field lines as induced by this process also has impacts on coupling to the ionosphere, i.e., the generation of field-aligned currents (e.g., Hwang et al. 2022; Johnson and Wing 2015; Johnson et al. 2021; Petrinec et al. 2022). Mid-latitude reconnection can relax the 3-D twist of the field lines, thus, affecting the large-scale MIC.

Borgogno et al. (2015) studied in further detail the changes in the global magnetic topologies that result from this process in 3-D. They showed in particular that field lines can be reconnected in different ways as the process evolves in the simulation. Magnetic field lines in the simulation domain are found to be reconnected at either one or both reconnection regions at mid-latitudes, with occurrence rates evolving over the time (cf. Faganello et al. 2022; Sisti et al. 2019). Using high-resolution MMS data, Vernisse et al. (2016) showed in-situ signatures of the mid-latitude reconnection process. The observations further suggested that both mid-latitude reconnection (in both hemispheres) and Type-I VIR in the equatorial plane may be occurring at the same time. A direct implication of this possibility is that magnetic field topologies can be even more complex than suggested by Borgogno et al. (2015) since the possible magnetic topologies then become a combination of not only two but three reconnection regions for the same field lines (Type-I or Type-II reconnection in the equatorial plane and the two reconnection regions at mid-latitudes in the northern and southern hemispheres). This is illustrated in Fig. 13, where the various possible combinations are sketched.

Further simulations by Fadanelli et al. (2018), using non-purely northward IMF conditions, showed that in such cases the whole geometry of the KH process is skewed: the distinction between reconnection in the equatorial plane and at mid-latitudes is not clear

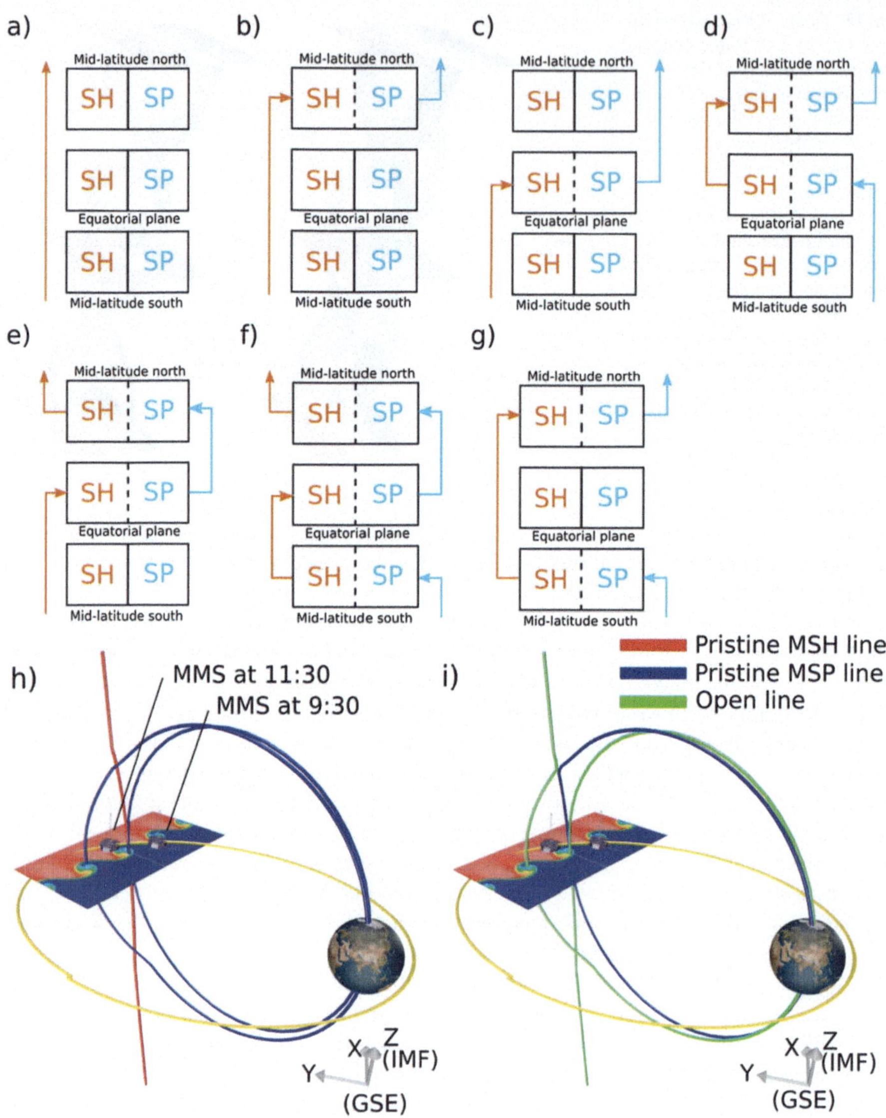

Fig. 13 Adopted from Vernisse et al. (2016). (a-g) Possible topologies of a field line observed on the magnetosheath side of a wavy magnetopause induced by the KHI. SH corresponds to the magnetosheath side and SP to the magnetosphere side. A dashed line within the rectangle between the SH and SP parts means that reconnection is occurring at this particular site. A solid line within the rectangle signals that the magnetopause is closed at this site. (h and i) Example of field line changing topology. Two closed magnetospheric field lines (blue) and one magnetosheath field line (red) before (h) and after (i) reconnection has occurred in the local KH wave and at midlatitude in the northern hemisphere. The resulting open field line is depicted in green

anymore, such that there exists a broad region of possible reconnection from mid-to-low latitudes in one of the hemispheres. The fact that reconnection sites may occur over a broad range of latitudes was later confirmed with MMS observations by Vernisse et al. (2020). Also, similar or even more complex features were found in simulations by Sisti et al. (2019)

or Faganello et al. (2022). For the specific conditions of the MMS KHI event on 8 Sep 2015 (cf. Eriksson et al. 2016; Vernisse et al. 2016), the late nonlinear phase of the KHI simulation (Faganello et al. 2022) showed the development of broad regions of reconnection and vortex pairing at all latitudes. They also found that secondary KHI develops, but only in the northern hemisphere, leading to an enhancement of the occurrence of off-equator reconnection, which is also consistent with observations by Vernisse et al. (2020).

In a symmetric configuration between the northern and southern hemispheres described in Faganello et al. (2012), the double mid-latitude reconnection process with an equal occurrence and rate does not provide a net transport of mass. Under asymmetric conditions, however, transport may occur and be quantified. For such asymmetric conditions Ma et al. (2017) identified doubly-reconnected field lines and estimated the mass transport rate to be on the order of 10^{10} m^2/s, comparable to the VIR-produced mass transport rate (Nakamura and Daughton 2014). The importance of KHI for mass transport into the magnetosphere is now well established, as addressed in a review by Faganello and Califano (2017), which also describes in more details the roles of magnetic reconnection in association with the instability. We further discuss this topic in Sect. 4.4.

4.4 Plasma Mixing, Transport, and Turbulence

Magnetic reconnection induced by the nonlinear development of the KHI at the flank magnetopause has been invoked to have cross-scale or large-scale impacts, including turbulence generation, plasma mixing in KH vortices, and plasma transport across the magnetopause and possibly into the near-Earth portion of the magnetotail. Stawarz et al. (2016) reported a magnetic power spectrum with a power-law index $-5/3$, namely, a traditional Kolmogorov-type turbulence feature at the MHD scale, for the 8 Sep 2015 event of magnetopause KH waves with reconnection jet signatures (Eriksson et al. 2016). Interestingly, a 3-D fully kinetic simulation of this MMS event showed that VIR in 3-D can generate such a turbulent power spectrum even in an early nonlinear stage of the KHI (Nakamura et al. 2017b). Consistently, a study by Hasegawa et al. (2020), in which the same MMS event was compared with a dayside magnetopause crossing event without KHI activity, suggested that magnetic turbulence can be enhanced as a consequence of the KHI growth. Notably, the cross-scale energy transfer rate was shown to be larger on the magnetospheric side of the KH-unstable magnetopause, where the turbulence level was low initially but VIR may grow more vigorously, than on the magnetosheath side where turbulence may be already developed (Quijia et al. 2021). [See Stawarz et al., this collection, for further discussion on an interplay between the KHI, VIR, and turbulence.]

The simulation by Nakamura et al. (2017b) also showed that in a fully developed phase of the KHI and thus of VIR, the rate of particle entry across the magnetopause per one KHI wavelength can be of order 10^{26} s^{-1} and a dense LLBL of ~ 1 R_E thickness can form around the dawn-dusk terminator (Nakamura et al. 2017a). The study thus suggests that VIR significantly contributes to plasma mixing across the flank magnetopause and the LLBL formation. A question from the viewpoint of macroscale impact is then whether LLBL plasmas could be transported beyond the boundary layer into the central or midnight portion of the magnetotail through the KHI or VIR.

Figure 14 shows energy-dispersed ion beams at energies less than 1 keV (thus most likely of magnetosheath origin) observed by MMS in the duskside plasma sheet during a northward IMF period. Nishino et al. (2022) applied both time-of-flight (Lockwood and Smith 1989) and pitch-angle dispersion analyses (Burch et al. 1982) to the observed field-aligned and anti-field-aligned ion beams in order to estimate the source location of these

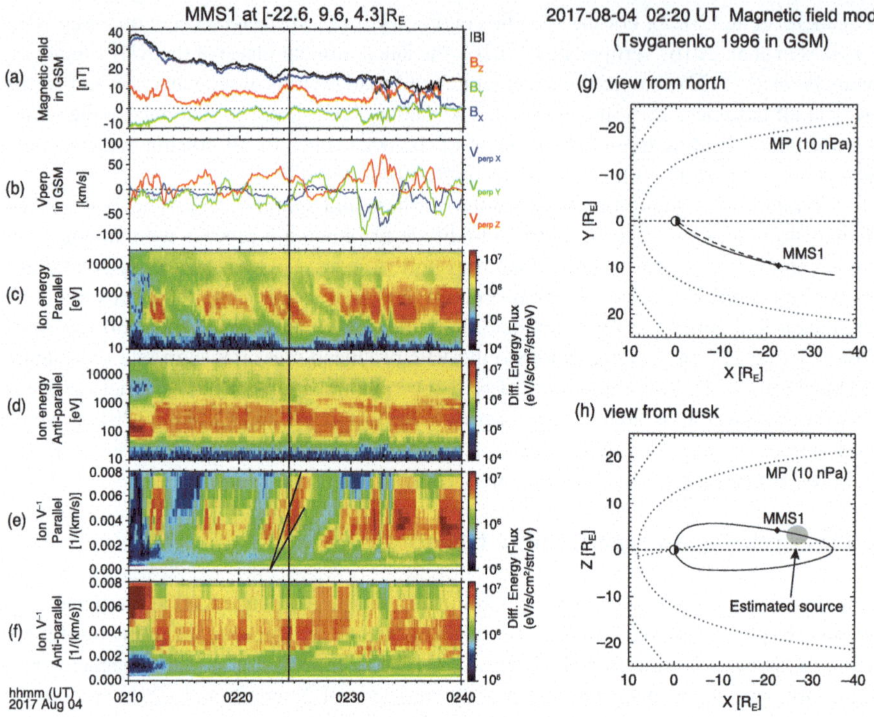

Fig. 14 Adopted from Nishino et al. (2022). MMS observations of repeated energy dispersed magnetic-field-aligned ion beams in the duskside magnetotail during northward interplanetary magnetic field conditions. Panels (c) and (d) show fast and slow dispersions, respectively, consistent with source locations of the beams on the tailward and dusk side of MMS, which may well be a dusk-flank boundary layer where vortex induced reconnection can occur (g, h). Time-of-flight analysis (e) suggests a distance to the source region of $\sim 5\ R_E$

ions. The results show that under the observed solar wind conditions, they can be traced back to a dusk-flank boundary layer at a distance $\sim 5\ R_E$ on the tailward and dusk side of MMS. Interestingly, the field-aligned (earthward traveling) energy-dispersed ions were observed repeatedly with a recurrence time of order a few min (Fig. 14c), which is similar to the KH wave period in the tail flanks (e.g., Hasegawa et al. 2006). This suggests their possible connection to KHI activity; magnetic reconnection induced in a KH-active flank boundary layer may have injected cool ions of magnetosheath origin toward the near-Earth plasma sheet. Their results indicate that, in addition to poleward-of-the-cusp reconnection in both northern and southern hemispheres (e.g., Li et al. 2008), VIR may play a role not only in plasma transport across the magnetopause and mixing in the LLBL but also in large-scale plasma transport, namely, the formation of the cold-dense plasma sheet under northward IMF (Terasawa et al. 1997).

Another aspect from the viewpoint of cross-scale processes is that velocity fluctuations or turbulence in the magnetosheath, which can act as seeds for the KHI and thus VIR, are generally more intense on the dawn than dusk side (e.g., Nykyri et al. 2017). In the presence of such magnetosheath turbulence, larger KH vortices can grow, leading to larger-scale and faster plasma transport (Nakamura et al. 2020; Nykyri et al. 2017). Resulting dawn-dusk asymmetric transport may be responsible, at least partially, for dawn-dusk asymmetries in the plasma sheet density and ion spectra (Wing et al. 2005).

When the plasma and magnetic fluctuations caused by KHI-associated reconnection at the flank magnetopause penetrate toward the midnight region, they provide perturbations in the central tail current sheet that can modify background parameters around the reconnecting current sheet and/or possibly serve triggers of the marginally stable current sheet in the central plasma sheet. Section 5 discusses multi-scale dynamics of reconnection occurring in the magnetotail.

5 Reconnection in Nightside Magnetotail

5.1 Introduction

The background magnetotail current sheet is generally symmetric between the northern and southern hemispheres across the current sheet with no or only small guide field compared to dayside boundaries. Instead, the magnetotail current sheet is asymmetric along the magnetotail axis bounded between the hotter plasma population trapped by Earth's dipole field in the inner magnetosphere at the earthward side and the solar wind plasma on the wake side (tailward side).

Magnetotail reconnection is expected to have, therefore, nearly 2-D geometry locally around the diffusion region in the first place. Corresponding scale sizes for EDR and IDR in the terrestrial magnetotail are on the order of d_e ($d_e < 40$ km) and d_i ($d_i < 1600$ km), respectively. The reconnection X-line is most likely to be localized with an extent of a few R_E in the magnetotail as inferred from the localized cross-tail (dawn-dusk) scales of fast plasma flows, called bursty bulk flows (BBFs; Angelopoulos et al. 1992), about one order smaller than the magnetotail dimension. Hence the evolution and consequence of the reconnection propagating from localized reconnection sites involve 3-D and transient processes, producing complex multi-scale structures.

One of the most important consequences of the energy transport from the reconnection site is a variety of instabilities in the transition region from tail-like to dipolar magnetic fields. In this region the reconnection jets brakes and/or diverts and energy is deposited to the ionosphere (Fuselier et al., this collection) as well as energetic particles are injected toward the inner magnetosphere (Oka et al., this collection).

In the following we highlight recent advances in our understanding of the multi-scale disturbances beyond the diffusion regions (Norgren et al., Genestreti et al., Liu et al., this collection) that are connected to the large-scale magnetotail processes (Fuselier et al., this collection). We focus on the onset (Sect. 5.2), evolution (Sect. 5.3), and consequences (Sect. 5.4) of magnetotail reconnection.

5.2 Onset Conditions for Local/Global Current Sheet Thinning

Formation of the thin current sheet during substorm growth phase (before the onset of reconnection) takes place due to the magnetic flux accumulation in the lobe (Birn and Schindler 2002) and/or enhanced loss of the closed magnetic flux toward the dayside magnetosphere (Hsieh and Otto 2014). Both processes are due to enhanced dayside magnetopause reconnection (Fuselier et al., this collection).

Cluster, THEMIS, and MMS observations in the region between 10–25 R_E, where the near-Earth reconnection takes place, showed that the thin current sheet is embedded in a thicker plasma sheet in stretched magnetic field configuration with a small B_Z (normal) component (Artemyev et al. 2021, 2016; Petrukovich et al. 2007). Such intense ion-scale current sheets are maintained by the stress balance between the enhanced electron anisotropy and agyrotropic ions (Artemyev et al. 2017). Current sheet thinning, therefore, occurs on a macroscopic spatial scale of several to tens of R_E (hundreds of d_i) along the tail axis, while being dominated by kinetic processes that also require the inclusion of multicomponent ion and electron distributions including low-energy plasma from the ionosphere (Runov et al. 2021).

The magnetotail reconnection region, however, is expected to be localized in the dawn-dusk direction based on the few R_E dawn-dusk extent of BBFs (Nakamura et al. 2004). THEMIS observations during a current sheet thinning support this view as the pressure gradient during the thinning was observed to increase both in the dawn-dusk direction and along the tail axis, suggesting that the thinning may take place in a localized manner (Artemyev et al. 2019).

The current sheet configuration can also be modified by the local magnetic flux transport such as dipolarization flux bundles. Nakamura et al. (2021) showed from the conjugate MMS-Cluster event (Fig. 15), where a dipolarization front developed by the fast flow is followed by a thin current sheet configuration at the wake of the flow activity, consistent with the simulation of localized fast flows/BBFs (Birn et al. 2004; Merkin et al. 2019). The localized stretched current sheet facilitates the triggering of reconnection, as predicted in the simulation of a newly formed reconnection at the wake of a dipolarization front (Sitnov et al. 2013). Statistics of the dipolarization fronts and BBFs (e.g., Dubyagin et al. 2011; Richard et al. 2022; Schmid et al. 2011) showed a dawn-dusk asymmetry in the event occurrence, skewed toward the dusk side. This represents an asymmetry in the current sheet along the dawn-dusk direction. Another possible creation of the dawn-dusk inhomogeneity of the current sheet is the flapping waves, although so far the reported flapping modes are rather consequence of reconnection (see Sect. 5.3) than the mechanism of the localized current sheet thinning.

Still there are also observations of a thin current sheet with half-thickness of one ion-gyroradius scale by Cluster (Baumjohann et al. 2007). Wang et al. (2018) reported MMS observations of an even thinner quiet current sheet with a half-thickness of 9 d_e without any reconnection signatures. The current sheet was bifurcated and tilted with the stability criteria of a linear tearing mode satisfied. How the onset condition is fulfilled by what processes for triggering reconnection is yet to be studied by more comprehensive multi-scale observations of the current sheet.

5.3 Multi-Scale Aspects of the Nightside Reconnection Region

Observation of magnetotail reconnection event. Due to the tailward motion of the X-line with a typical speed of a few hundreds km/s obtained from multi-point observations (Alexandrova et al. 2015; Eastwood 2005), at times together with the up-down motion of the current sheet (flapping), an X-line crossing in the magnetotail is often observed as a reversal from tailward to earthward ion flow. Figure 16 shows two examples of X-line crossing observations (A, B) with illustration of the spacecraft motion relative to the X-line (C) and the current sheet flapping (D). While the MMS observations of reconnection EDRs usually lasted only about a few seconds, the large-scale ion flow reversal from tailward to earthward takes several minutes to several tens of minutes as indicated by Fig. 16(A, B). Assuming a constant speed of the X line of few hundred km/s, the overall size of the active current-sheet region can extend up to several tens of R_E scales.

The two examples shown in Fig. 16 are from magnetotail reconnection on 11 Jul 2017 (A) when a 2-D laminar EDR (e.g., Torbert et al. 2018) was detected, and from turbulent

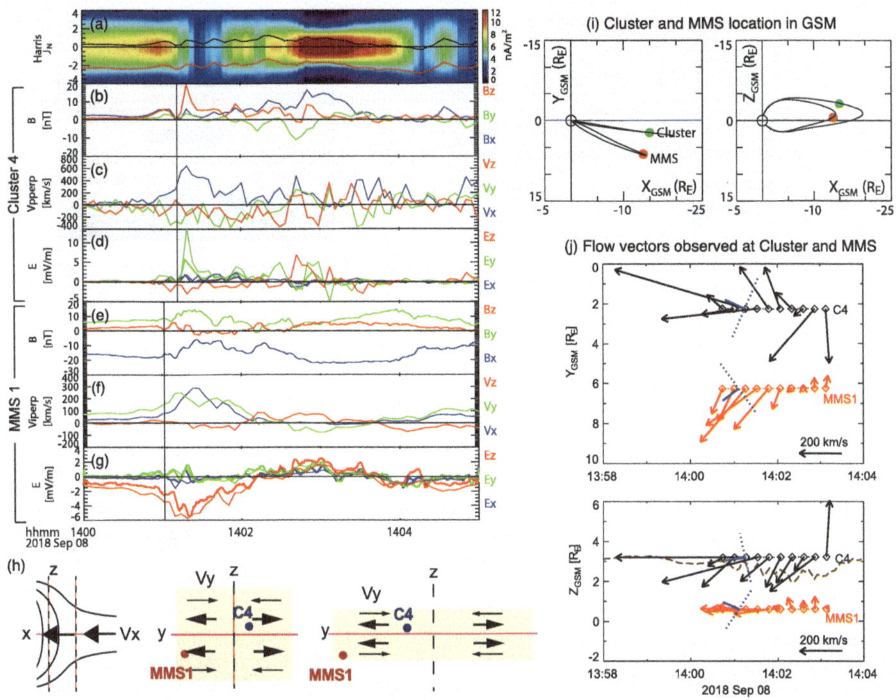

Fig. 15 Adopted from Nakamura et al. (2021). (a) The modeled current density based on conjugate MMS-Cluster observations with location of MMS and Cluster relative to the current sheet center denoted in red and black, respectively. (b-g) The magnetic flux transport associated with a dipolarization front and BBFs, behind which a thin current sheet forms facilitating the onset of reconnection (bottom illustration in panel h). (i) Location of MMS and Cluster in GSM X-Y and X-Z plane with model field lines. (j) Temporal changes of flow vectors from C4 (black) and MMS1 (red) and dipolarization front (blue) in the X-Y plane and in the X-Z plane

reconnection (B) on 26 Jul 2017 (e.g., Ergun et al. 2018) showing significant particle acceleration (for more discussion, see Oka et al. 2023, this collection). The thin current sheet region between the tailward and earthward ion flows contains complex 3-D structures for both events.

2-D and 3-D structuring inside the thin current sheet. Even for the 11 Jul 2017 event when the local EDR contains 2-D laminar reconnection features, multiple flux ropes that are characterized by localized magnetic-field strength enhancements and bipolar B_Z are observed at times indicated by black arrows in Fig. 16A(d). They are ion-scale flux rope with varying axial directions, i.e., 13–55° away from the out-of-reconnection-plane direction and two of them are mostly tilted toward the direction of the reconnecting magnetic field (Teh et al. 2018). The tilted angles agree reasonably well with the predicted angles for secondary tearing modes, suggesting that these ion-scale flux ropes are likely generated by secondary tearing instabilities as those found in 3-D kinetic simulations of turbulent magnetic reconnection (e.g., Daughton et al. 2011). Stawarz et al. (2018) reported an intense, localized electric field within one of the flux ropes associated with an electron-scale vortex. Hence even though the vicinity of EDR can be explained as 2-D reconnection, the thin current sheet contained multiple small-scale 3-D magnetic structures.

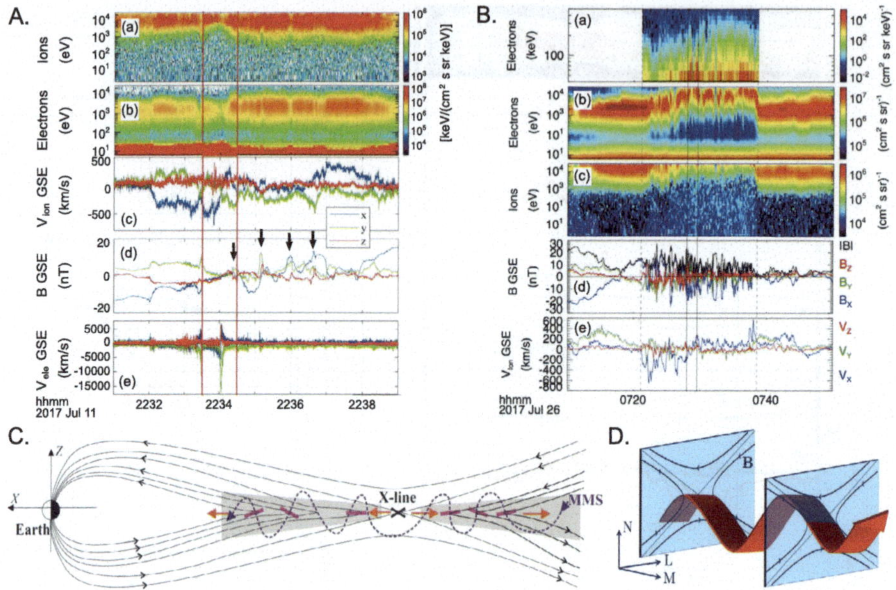

Fig. 16 Adapted from Hasegawa et al. (2019), Ergun et al. (2018), Leonenko et al. (2021), and Cozzani et al. (2021) for panels A, B, C, and D, respectively. Comparison of two MMS observations of a laminar 2-D EDR (A) and a turbulent EDR (B) with illustration of the spacecraft trajectory relative to the X-line (C) and the current sheet flapping (D)

In the prolonged disturbed thin current sheet interval shown in Fig. 16B, on the other hand, turbulent dissipation and particle acceleration are associated with strong magnetic field fluctuations (Figure 16Bd) and large-amplitude electric fields with scale-dependent energy conversion features (Ergun et al. 2018). Turbulent electric fields including significant $E_{||}$ were found to play a central role in accelerating electrons to >100 keV energies. Interesting to note is that even in this turbulent current sheet, the embedded EDR region shows a laminar profile similar to the 11 Jul 2017 event (Ergun et al. 2022). This indicates that the difference in the large-scale consequence such as particle acceleration are rather determined by the ambient parameter of the current sheet.

Effect of flapping in reconnection current sheet and beyond. Flapping kink-like motion of the current sheet is often observed in the magnetotail (Sergeev et al. 2004) associated with the active plasma sheet with fast flows (Sergeev et al. 2006). These oscillations can have frequencies ranging from a fraction of seconds to minutes and can appear on various scales. The current sheet flapping was seen in both reconnection events in Fig. 16 as an oscillating B_X (tail-aligned) component.

Small-scale separation among the four MMS spacecraft enabled the study of flapping in a thinner current sheet than previously reported. Leonenko et al. (2021) reported multiple crossings of an embedded thin current sheet with half-thickness of about few electron gyroradii or less during a prolonged interval involving a flow reversal. Resultant motion of the spacecraft relative to the thin current sheet is given in Fig. 16C. The X-line motion as well as the flapping allows for monitoring spatial/temporal changes of the current sheet. During the same event as reported by Leonenko et al. (2021), Wang et al. (2018) observed a corrugated current sheet with its normal mostly along Y_{GSM} in the flapping current sheet and showed that such current sheet corrugation enhances mixing of demagnetized ions.

The wave-like flapping structure usually propagates along the current direction as illustrated in Fig. 16D. Cozzani et al. (2021) found a flapping oscillation of >1 Hz, near the lower hybrid drift frequency observed inside the EDR. This suggests the importance of current sheet drift instabilities in the electron-scale current sheets. Wei et al. (2019) reported a flapping motion with period of ∼6 seconds within the ion-scale current sheet in the diffusion region that is modulated with the reconnection electric field. They concluded that the flapping motion was likely to be triggered by the periodical unsteady magnetic reconnection. PIC simulation studies (Fujimoto and Sydora 2017) showed that velocity shears of Y-directional electron and ion flows along the current sheet normal (Z) direction in the electron current layer generate the current sheet shear instability. This instability can lead to current sheet flapping with a macroscopic-scale ($\sim R_E$) wavelength (Fujimoto 2016).

Ion-scale current sheet flapping was also observed outside of the diffusion region associated with fast flows at the duskside plasma sheet (Richard et al. 2021). Drift-kink instabilities, caused by the relative drift between ions and electrons, were responsible for the flapping with a 25-second period and a phase velocity comparable to the ion duskward flow. Current sheet flapping, therefore, plays an important role in the evolution and structuring of the thin current sheet along the dawn-dusk direction in the magnetotail both inside and outside the diffusion region.

Evolution of the separatrix region of reconnection. Reconnection drives accelerated plasma beams, which spread along the magnetic field lines leading to energy-dependent dispersed structures called TDS (time-dispersed-structure) originally found in the ion beams (Sauvaud et al. 1999). Injected particles drift also perpendicular to the magnetic field due to $\mathbf{E} \times \mathbf{B}$ velocity, usually toward the plasma sheet center. Since slower particles take longer time to reach the spacecraft, this motion leads a spatial (latitudinal) dispersion due to a velocity filter effect.

The top two panels of Fig. 17A present an example of parallel electrons and ion components showing the high-energy (> a few keV) beams from the reconnection region (highlighted in orange box; Wellenzohn et al. 2021). The energy-dispersed plasma beam in the separatrix region enables remote sensing of the reconnection region, assuming that the injection site along the field line is in the vicinity of the diffusion region. The high-time resolution MMS data enabled, for the first time, applying this method also to the electron data. Figure 17B shows the inverse velocity (1/V) spectra for electrons (upper) and ions (lower), from which the injection time and location were determined (Varsani et al. 2017). Electron and ion TDS inferred the X-line location to be 16–18 R_E downtail, which was rather close to Earth due to the strong storm-time substorm interval. Yet this location was consistent with the estimated values from a conjugate DMSP low-altitude observation of energy dispersed ions and electrons (illustrated in Fig. 17C).

Wellenzohn et al. (2021) also determined the separatrix boundary motion independent from the plasma convection so that the effect of such drift was considered in determining the location of the injection point: the reconnection region was estimated to be located at $X = (-23 \pm 1.9)$ R_E from electrons, comparable to that from ions, $X = (-24.5 \pm 0.7)$ R_E. Interestingly for both events the electron injection time precedes that of the ions for several seconds, which may suggest different acceleration time scales between ions and electrons.

From the speed of the separatrix boundary relative to the convection, the reconnection electric field can be determined (Nakamura et al. 2018) [see more details in Hasegawa et al. 2023]. Wellenzohn et al. (2021) examined different signatures propagated from the reconnection site at the separatrix and used these signatures to determine the reconnection electric field. By obtaining the separatrix motion using the election injection time, start of the electron TDS, and $E_{||}$ waves (Figure 17Ae), the reconnection electric field was estimated to be

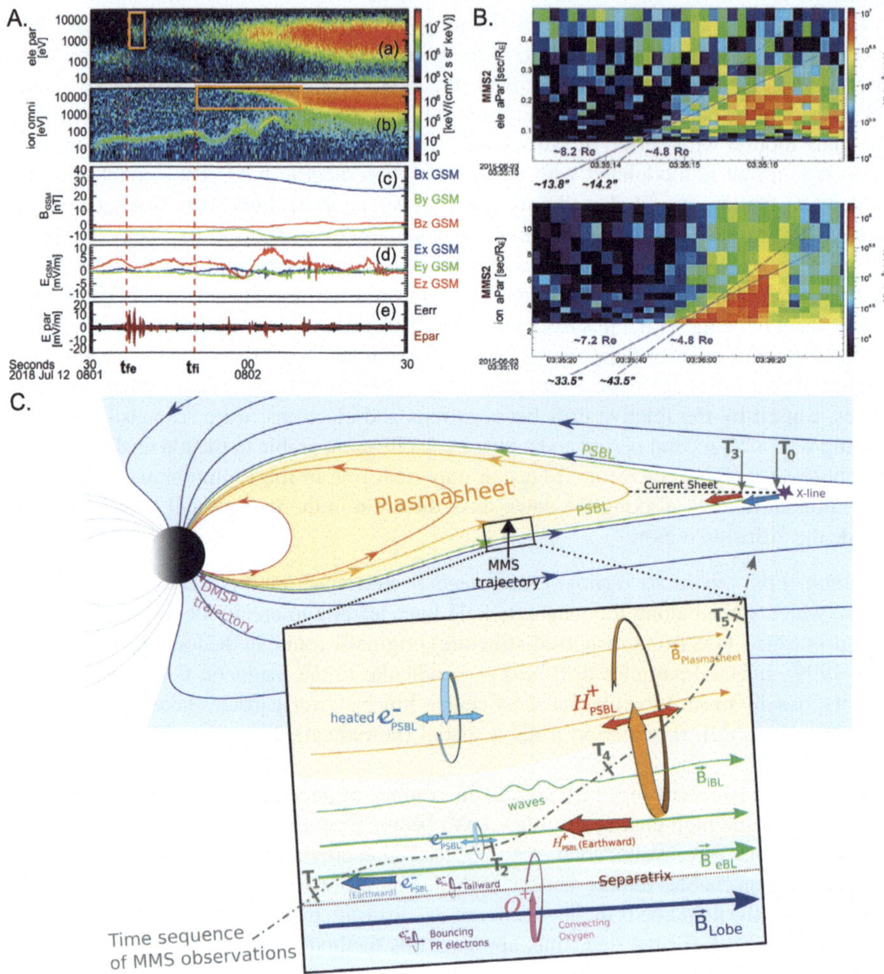

Fig. 17 Adapted from Wellenzohn et al. (2021) and Varsani et al. (2017). A. Energy-dependent time-dispersed structures (TDS) observed for electrons (a) and ions (b) and associated Hall field signatures (c, d) with parallel electric field signatures (e). B. Inverse velocity (1/V) spectra for electrons (upper) and ions (lower) from which the injection time and location can be determined. C. Diagram of TDS and velocity filter effect occurring along/around the PSBL

$E_r = 1.6$–2.4 mV/m. These values are comparable to the observed value of the 11 Jul 2017 EDR event (Fig. 16A), i.e., $E_r = 2$–3 mV/m (e.g., Genestreti et al. 2018).

Sergeev et al. (2021) showed similar patterns to Fig. 17A(b-d), i.e., dispersed ions and associated Hall-field disturbance, and cold ion beams during 9 plasma sheet boundary layer (PSBL) crossings by MMS located at distant ($>100\ d_i$) and mid-distance regions (<30–$50\ d_i$) from the X line. The estimated reconnection electric field from the separatrix motion using ion TDS was $E_r = 1$–8 mV/m. While these remote sensing observations require certain assumptions such as a constant speed of disturbance propagation, these studies confirm the large-scale extent of the region influenced by active reconnection and that these remote signatures can be an alternative method to infer the temporal/spatial changes of magnetotail reconnection.

Wave-particle interaction is an important process in the separatrix region. Instabilities can develop due to the velocity shear created between the inflowing and outflowing beam (see details in Stawarz et al., this collection). High-time resolution measurements from MMS enabled to study the acceleration/heating processes. Norgren et al. (2020) reported an acceleration channel of the cold lobe electrons up to several keV at the separatrix region inside a density cavity and found that this field-aligned beam leads to electrostatic waves that thermalize the cold electron beams. The observed separatrix is adjacent to the ion outflow and hence away from the thin current sheet, being therefore a remote signature of the reconnection.

5.4 Transient Structures in the Exhaust Region

The consequences of magnetotail reconnection are manifested by geomagnetic storms and substorms on global scales (Fuselier et al., this collection), formation of flux ropes via multiple X-lines on micro-, meso-, or macro-scopic scales, and generation of ion-scale dipolarization fronts formed between entrained reconnection jets and the ambient plasma (Oka et al., this collection). The second and third phenomena are frequently observed in the exhaust region of Earth's nightside reconnection. The detailed MMS data allowed resolving down-to-electron-scale structures embedded in such flux ropes and dipolarization fronts as well as interactions between flux ropes or with the ambient geomagnetic field (similar to Fig. 8f).

Electron-scale substructures were suggested to exist within the ion-scale dipolarization front and to play an important role in the energy conversion (Angelopoulos et al. 2013). The MMS measurements have revealed electron-scale density gradients, jets/currents, and electric fields, indicating electron-scale energy conversion, within both the earthward and tailward dipolarization front layer (Liu et al. 2018a; Xu et al. 2021). The dipolarization front is often subject to the ballooning/interchange instability and/or the lower hybrid drift instability (Divin et al. 2015; Hwang et al. 2011; Nakamura et al. 2016; Pan et al. 2018), suggesting the energy conversion via subsequent instabilities. Ongoing reconnection occurring at the leading edge of the dipolarization front was also identified by MMS (Marshall et al. 2020).

Sun et al. (2019) reported two types of ion-scale flux ropes, quasi 1-D and quasi 2-D structures. The former showed higher pressure inside the flux rope and the latter contained larger magnetic flux, possibly representing a later stage of the flux rope evolution. This is consistent with the theoretical expectation that the quasi 2-D structure is in a lower energy state.

Re-reconnection between an earthward propagating flux rope and Earth's dipole magnetic fields was reported by Man et al. (2018). Furthermore, Poh et al. (2019) further investigated the ionospheric response to such interaction using the magnetic field perturbations observed by ground magnetometers. Conjugate ionospheric observations at the ionospheric footprint of the re-reconnection site showed enhanced horizontal currents as observed by AMPERE and increased ionospheric convection by superDARN. This implies the large-scale impacts of the transient structures and their interactions with background fields and pre-existing plasmas occurring in the exhaust region of magnetotail reconnection.

6 Key Aspects with Relevance and Applications to Other Plasma Systems

In this paper we reviewed multi-scale aspects of magnetic reconnection beyond the diffusion region and the global-scale magnetospheric dynamics as described by the Dungey Cycle. Recent high time-resolution in-situ observations and state-of-the-art numerical techniques have significantly advanced our understanding of cross-scale reconnection processes. Table 1 presents the key features of these advancements, focusing on the dominant scales identified for each process. These findings are categorized according to various terrestrial regions, characterized by different magnetic topologies and physical parameters across the current sheet. The results from these studies can be applied to other regimes encompassing solar plasmas, planetary magnetospheres, and other heliospheric and astrophysical systems. We also acknowledge the unsolved problems, which can serve as guidance for future studies on magnetic reconnection in geospace, as well as throughout the heliosphere and the universe.

Acknowledgements The authors thank the International Space Science Institute (ISSI) for hosting the team and relevant meetings. K.-J.H., S.F, and K.D. were supported by NASA's MMS project at SwRI. K.-J.H and K.D. were supported by NASA 80NSSC23K0417 and 80NSSC18K1337. J.E. and H.H were supported by UKRI/STFC ST/W001071/1 and JSPS Grant-in-aid for Scientific Research KAKENHI 21K03504, respectively.

Table 1 Notable findings on multi-scale aspects of reconnection, applications to other systems, and unsolved problems

Location	Key features	Applications to other systems	Unsolved problems
Reconnection at the fore-shock, shock, and down-stream magneto-sheath	*Electron scale*: • Electron-only reconnection with or without a guide field in the foreshock, the shock transition region, and the downstream magnetosheath • The prevalent occurrence of electron-only reconnection, possibly in association with turbulent systems *Ion scale*: • Ion-scale young current sheets exhibiting electron jets only • Cyclical reformation of the shock front leading to the formation of the current sheet • Ion-scale flux ropes in the foreshock *MHD scale*: • 40% of shock crossings contain reconnecting current sheets. • The occurrence rate is strongly or weakly biased toward the quasi-parallel shock at the bow shock and the magnetosheath. • Large-scale discontinuities interacting with the bow shock compress a current sheet, triggering reconnection. • Current sheets initiated at the bow shock continue to influence energy processing downstream	• Planetary magneto-spheres with the bow shock and the shocked solar wind region • The coupled system of the shock, reconnection, and turbulence in the universe	• Whether or not electron-only reconnection is a temporal evolution to ion reconnection • Detailed behavior of the magnetosheath (reconnection and turbulence features) downstream of different bow shock magnetic geometries • Pathway and impacts of fluctuations initiated at the foreshock/shock through the magnetosheath to the magnetopause

Table 1 (*Continued*)

Location	Key features	Applications to other systems	Unsolved problems
Dayside magneto-pause reconnection	*Electron scale*: • Dayside electron diffusion region (EDR) characterized by a perpendicular crescent of electron distribution functions and positive Ohmic energy exchange • Outer EDR characterized by parallel crescents and negative Ohmic energy exchange, with Hall electron flows leading to magnetic pileup affecting the opening angle (reconnection rate). • Electron vortex and phase space holes along the magnetospheric separatrix • EDR detected at the interface of interacting flux ropes or interlinked flux tubes *Ion scale*: • Density asymmetry across the current sheet, leading to the displacement of the flow stagnant point from the X-line, symmetric Hall electric field, and bipolar Hall magnetic field • Multiple ion components implying the location of the X-line and magnetic field topologies • Formation of ion-scale secondary islands/flux ropes/flux transfer events (FTEs) • Coalescence or interaction of secondary islands/flux ropes *MHD scale*: • The extent and orientation of X-lines depend on background magnetic field topologies and a guide field. • Time-varying upstream conditions lead to the formation of secondary X-lines to adjust the orientation. • Generation of interlinked flux tubes that evolve their magnetic connectivity via localized reconnection • Magnetosphere-ionosphere coupling processes caused by unsteady dayside reconnection or FTEs	• Planetary magneto-spheres with the dayside magnetopause across which magnetic shear and density gradient co-exist • Solar/helio-spheric/astro-physical current sheets undergoing asymmetric reconnection with a guide field	• Detailed 3-D topology including the orientation of the X-line(s) with controlling parameters • Mechanisms determining the generation of primary and secondary X-lines and resultant reconnection rates • Evolution and fate of various types of flux ropes/flux transfer events generated on the dayside magnetopause via multiple X-lines

Funding Open access funding provided by Österreichische Akademie der Wissenschaften.

Declarations

Competing Interests All the authors declare that the work was conducted without any commercial or financial relationships that could be construed as potential conflicts of interest.

Table 1 *(Continued)*

Location	Key features	Applications to other systems	Unsolved problems
Flankside magneto-pause reconection	*Electron scale*: • Current sheet thinning to electron scales via a velocity shear leading to Vortex-Induced Reconnection (Type-1 VIR) • Highly swirled field lines via the flow vortex leading to secondary reconnection (Type-II VIR) • In-situ observations of EDR during VIR *Ion scale*: • X-line drift due to the upstream magnetosheath flow • Asymmetric exhaust region due to the combined effects of density gradient and velocity shear across the current sheet • Formation of ion-scale flux ropes along the Kelvin-Helmholtz (KH) vortex boundary *MHD scale*: • In-situ evidence for mid-latitude reconnection that occurs due to a 3-D twist of magnetospheric and magnetosheath fields induced by KH vortices • The latitudinal extent of both in-plane (VIR) and mid-latitude reconnection that becomes wider with nonlinear KH instability evolution • VIR-driven turbulence forming a mixing boundary layer and large-scale mass transport beyond the boundary layer toward the central magnetosphere	• Planetary magneto-spheres with the flankside magnetopause across which density gradient and velocity shear co-exist • Solar/helio-spheric/astro-physical current sheets undergoing asymmetric reconnection with a velocity shear	• The combined effects of density asymmetry, velocity shear, and a guide field (parametric study) • In-plane (VIR) and out-of-plane (mid-latitude) reconnection occurrence with their combined effect on the solar wind transport • The propagation of flankside VIR-driven fluctuations and their impact on the central plasma sheet behavior
Magnetotail reconnection	*Electron scale*: • Nightside EDR characterized by multiple (perpendicular) crescents, positive Ohmic energy exchange, and large electron vorticity • Laminar EDR detected during both 2-D laminar and turbulent reconnection • EDR detected at the interface of interacting flux ropes or at the leading edge of a flux rope (re-reconnection) • Electron-scale structures embedded within the dipolarization front *Ion scale*: • Existence of multiple ion scales in association with cold and/or heavy ion components • Ion-scale flapping current sheet enhancing the mixing of demagnetized ions • Transient structures shown in the exhaust region, such as ion-scale flux ropes and both earthward and tailward dipolarization fronts	• Planetary magneto-spheres with the nightside current sheet along which plasma and magnetic field gradients co-exist • Solar/helio-spheric/astro-physical current sheets undergoing symmetric reconnection	• Onset mechanism of the triggering of reconnection in a marginally stable magnetotail current sheet • Local or ambient parameters determining the consequences of reconnection (e.g., reconnection rate) • Causality between reconnection and current sheet flapping

Table 1 (*Continued*)

Location	Key features	Applications to other systems	Unsolved problems
	MHD scale: • Dawn-dusk extent of the magnetotail X-line implied by BBFs • Current sheet thinning after an earthward-moving dipolarization front, triggering reconnection • Current sheet flapping with low to high frequency (similar to lower-hybrid frequency) • Remote sensing of tail reconnection location using both electron and ion dispersion signatures • Re-reconnection influencing the high-latitude ionospheric currents		

References

Ahmadi N, Eriksson S, Newman D, Andersson L, Ergun RE et al (2022) J Geophys Res Space Phys 127(8):1–12. https://doi.org/10.1029/2022JA030702

Akhavan-Tafti M, Palmroth M, Slavin JA, Battarbee M, Ganse U et al (2020) J Geophys Res Space Phys 125(7):e2019JA027410. https://doi.org/10.1029/2019JA027410

Alexandrova A, Nakamura R, Semenov VS, Nakamura TKM (2015) Geophys Res Lett 42(12):4685–4693. https://doi.org/10.1002/2015GL064421

Angelopoulos V, Baumjohann W, Kennel CF, Coroniti FV, Kivelson MG et al (1992) J Geophys Res 97(A4):4027. https://doi.org/10.1029/91JA02701

Angelopoulos V, Runov A, Zhou XZ, Turner DL, Kiehas SA et al (2013) Science 341(6153):1478–1482. https://doi.org/10.1126/science.1236992

Archer MO, Turner DL, Eastwood JP, Schwartz SJ, Horbury TS (2015) Planet Space Sci 106:56–66. https://doi.org/10.1016/j.pss.2014.11.026

Artemyev AV, Angelopoulos V, Runov A, Petrokovich AA (2016) J Geophys Res Space Phys. https://doi.org/10.1002/2016JA022779

Artemyev AV, Angelopoulos V, Liu J, Runov A (2017) Geophys Res Lett. https://doi.org/10.1002/2016GL072011

Artemyev AV, Angelopoulos V, Runov A, Petrukovich AA (2019) J Geophys Res Space Phys 124(1):264–278. https://doi.org/10.1029/2018JA026113

Artemyev A, Lu S, El-Alaoui M, Lin Y, Angelopoulos V et al (2021) Geophys Res Lett 48(6):1–9. https://doi.org/10.1029/2020GL092153

Bai SC, Shi Q, Liu TZ, Zhang H, Yue C et al (2020) Geophys Res Lett 47(5):1–9. https://doi.org/10.1029/2019GL085933

Balogh A, Treumann RA (2013) Physics of collisionless shocks. ISSI scientific report series, vol 12. Springer, New York. https://doi.org/10.1007/978-1-4614-6099-2

Baumjohann W, Roux A, Le Contel O, Nakamura R, Birn J et al (2007) Ann Geophys 25(6):1365–1389. https://doi.org/10.5194/angeo-25-1365-2007

Berchem J, Russell CT (1982) J Geophys Res 87(A4):2108. https://doi.org/10.1029/JA087iA04p02108

Bessho N, Chen L-J, Wang S, Hesse M, Wilson LB (2019) Geophys Res Lett 46(16):9352–9361. https://doi.org/10.1029/2019GL083397

Bessho N, Chen L-J, Wang S, Hesse M, Wilson LB et al (2020) Phys Plasmas 27(9):092901. https://doi.org/10.1063/5.0012443

Bessho N, Chen L-J, Stawarz JE, Wang S, Hesse M et al (2022) Phys Plasmas 29(4):042304. https://doi.org/10.1063/5.0077529

Birn J, Schindler K (2002) J Geophys Res 107(A7):1117. https://doi.org/10.1029/2001JA000291
Birn J, Raeder J, Wang YL, Wolf RA, Hesse M (2004) Ann Geophys. https://doi.org/10.5194/angeo-22-1773-2004
Bohdan A, Pohl M, Niemiec J, Vafin S, Matsumoto Y et al (2020) Astrophys J 893(1):6. https://doi.org/10.3847/1538-4357/ab7cd6
Borgogno D, Califano F, Faganello M, Pegoraro F (2015) Phys Plasmas. https://doi.org/10.1063/1.4913578
Burch JL, Reiff PH, Heelis RA, Winningham JD, Hanson WB et al (1982) Geophys Res Lett 9(9):921–924. https://doi.org/10.1029/GL009i009p00921
Burch JL, Torbert RB, Phan TD, Chen LJ, Moore TE et al (2016) Science 352(6290):aaf2939. https://doi.org/10.1126/science.aaf2939
Cassak PA, Otto A (2011) Phys Plasmas 18(7):074501. https://doi.org/10.1063/1.3609771
Chandrasekhar S (1961) Oxford University Press, London
Chang C, Huang K, Lu Q, Sang L, Lu S et al (2021) J Geophys Res Space Phys 126(7):1–11. https://doi.org/10.1029/2021JA029290
Chen XL, Morrison PJ (1990) Phys Fluids, B Plasma Phys 2(3):495–507. https://doi.org/10.1063/1.859339
Chen LJ, Hesse M, Wang S, Bessho N, Daughton W (2016) Geophys Res Lett 43(6):2405–2412. https://doi.org/10.1002/2016GL068243
Chen ZZ, Fu HS, Wang Z, Liu CM, Xu Y (2019) Geophys Res Lett 46(17–18):10209–10218. https://doi.org/10.1029/2019GL084360
Cozzani G, Khotyaintsev YV, Graham DB, Egedal J, André M et al (2021) Phys Rev Lett 127(21):1–8. https://doi.org/10.1103/PhysRevLett.127.215101
Daughton W, Roytershteyn V, Karimabadi H, Yin L, Albright BJ et al (2011) Nat Phys 7(7):539–542. https://doi.org/10.1038/nphys1965
Denton RE, Torbert RB, Hasegawa H, Dors I, Genestreti KJ et al (2020) J Geophys Res Space Phys 125(2):1–28. https://doi.org/10.1029/2019ja027481
Divin A, Khotyaintsev YV, Vaivads A, André M (2015) J Geophys Res Space Phys 120(2):1124–1132. https://doi.org/10.1002/2014JA020528
Doss CE, Komar CM, Cassak PA, Wilder FD, Eriksson S et al (2015) J Geophys Res Space Phys 120(9):7748–7763. https://doi.org/10.1002/2015JA021489
Dubyagin S, Sergeev V, Apatenkov S, Angelopoulos V, Runov A et al (2011) Geophys Res Lett 38(8):L08102. https://doi.org/10.1029/2011GL047016
Dunlop MW, Zhang QH, Bogdanova YV, Trattner KJ, Pu Z et al (2011) Ann Geophys 29(9):1693–1697. https://doi.org/10.5194/angeo-29-1683-2011
Eastwood JP (2005) Geophys Res Lett 32(11):L11105. https://doi.org/10.1029/2005GL022509
Eastwood JP, Mistry R, Phan TD, Schwartz SJ, Ergun RE et al (2018) Geophys Res Lett. https://doi.org/10.1029/2018GL077670
Egedal J, Fox W, Katz N, Porkolab M, Øieroset M et al (2008) J Geophys Res Space Phys 113:A12207. https://doi.org/10.1029/2008JA013520
Ergun RE, Goodrich KA, Wilder FD, Ahmadi N, Holmes JC et al (2018) Geophys Res Lett. https://doi.org/10.1002/2018GL076993
Ergun RE, Pathak N, Usanova ME, Qi Y, Vo T et al (2022) Astrophys J Lett 935(1):L8. https://doi.org/10.3847/2041-8213/ac81d4
Eriksson S, Lavraud B, Wilder FD, Stawarz JE, Giles BL et al (2016) Geophys Res Lett 43(11):5606–5615. https://doi.org/10.1002/2016GL068783
Fadanelli S, Faganello M, Califano F, Cerri SS, Pegoraro F et al (2018) J Geophys Res Space Phys 123(11):9340–9356. https://doi.org/10.1029/2018JA025626
Faganello M, Califano F (2017) J Plasma Phys 83(6):535830601. https://doi.org/10.1017/S0022377817000770
Faganello M, Califano F, Pegoraro F (2008) Phys Rev Lett 101(10):105001. https://doi.org/10.1103/PhysRevLett.101.105001
Faganello M, Califano F, Pegoraro F, Andreussi T (2012) Europhys Lett 100(6):69001. https://doi.org/10.1209/0295-5075/100/69001
Faganello M, Sisti M, Califano F, Lavraud B (2022) Plasma Phys Control Fusion 64(4):044014. https://doi.org/10.1088/1361-6587/ac43f0
Fargette N, Lavraud B, Øieroset M, Phan TD, Toledo-Redondo S et al (2020) Geophys Res Lett 47(6):1–9. https://doi.org/10.1029/2019GL086726
Fermo R, Drake J, Swisdak M (2012) Phys Rev Lett 108(25):255005. https://doi.org/10.1103/PhysRevLett.108.255005
Fu HS, Vaivads A, Khotyaintsev YV, Olshevsky V, André M et al (2015) J Geophys Res Space Phys. https://doi.org/10.1002/2015JA021082
Fujimoto K (2016) Geophys Res Lett 43(20):10557–10564. https://doi.org/10.1002/2016GL070810

Fujimoto K, Sydora RD (2017) J Geophys Res Space Phys 122(5):5418–5430. https://doi.org/10.1002/2017JA024079

Fuselier SA (2002) J Geophys Res 107(A7):1111. https://doi.org/10.1029/2001JA900165

Fuselier SA, Trattner KJ, Petrinec SM, Denton MH, Toledo-Redondo S et al (2019a) Geophys Res Lett 46(12):6204–6213. https://doi.org/10.1029/2019GL082384

Fuselier SA, Trattner KJ, Petrinec SM, Pritchard KR, Burch JL et al (2019b) J Geophys Res Space Phys 124(11):8524–8534. https://doi.org/10.1029/2019JA027143

Fuselier SA, Webster JM, Trattner KJ, Petrinec SM, Genestreti KJ et al (2021) J Geophys Res Space Phys 126(12):1–16. https://doi.org/10.1029/2021JA029789

Genestreti KJ, Nakamura TKM, Nakamura R, Denton RE, Torbert RB et al (2018) J Geophys Res Space Phys 123(11):9130–9149. https://doi.org/10.1029/2018JA025711

Genestreti KJ, Liu Y-H, Phan T-D, Denton RE, Torbert RB et al (2020) J Geophys Res Space Phys 125(10):1–16. https://doi.org/10.1029/2020JA027985

Gingell I, Schwartz SJ, Eastwood JP, Burch JL, Ergun RE et al (2019) Geophys Res Lett 46(3):1177–1184. https://doi.org/10.1029/2018GL081804

Gingell I, Schwartz SJ, Eastwood JP, Stawarz JE, Burch JL et al (2020) J Geophys Res Space Phys 125(1):e2019JA027119. https://doi.org/10.1029/2019JA027119

Gingell I, Schwartz SJ, Kucharek H, Farrugia CJ, Trattner KJ (2021) Phys Plasmas 28(10):102902. https://doi.org/10.1063/5.0062520

Gosling JT (2012) Space Sci Rev 172(1–4):187–200. https://doi.org/10.1007/s11214-011-9747-2

Graham DB, Khotyaintsev YV, Norgren C, Vaivads A, André M et al (2017) J Geophys Res Space Phys 122(1):517–533. https://doi.org/10.1002/2016JA023572

Greco A, Matthaeus WH, Perri S, Osman KT, Servidio S et al (2018) Space Sci Rev 214(1):1. https://doi.org/10.1007/s11214-017-0435-8

Guo Z, Lin Y, Wang X (2021a) J Geophys Res Space Phys 126(4):1–20. https://doi.org/10.1029/2020JA028853

Guo Z, Lin Y, Wang X, Du A (2021b) J Geophys Res Space Phys 126(12):e2021JA029979. https://doi.org/10.1029/2021JA029979

Hamrin M, Gunell H, Goncharov O, De Spiegeleer A, Fuselier S et al (2019) J Geophys Res Space Phys 124(11):8507–8523. https://doi.org/10.1029/2019JA027006

Hasegawa H, Fujimoto M, Takagi K, Saito Y, Mukai T et al (2006) J Geophys Res 111:A09203. https://doi.org/10.1029/2006JA011728

Hasegawa H, Retinò A, Vaivads A, Khotyaintsev Y, André M et al (2009) J Geophys Res Space Phys 114(12):1–20. https://doi.org/10.1029/2009JA014042

Hasegawa H, Zhang H, Lin Y, Sonnerup BU, Schwartz SJ et al (2012) J Geophys Res Space Phys 117(9):1–11. https://doi.org/10.1029/2012JA017920

Hasegawa H, Kitamura N, Saito Y, Nagai T, Shinohara I et al (2016) Geophys Res Lett 43(10):4755–4762. https://doi.org/10.1002/2016GL069225

Hasegawa H, Denton RE, Nakamura R, Genestreti KJ, Nakamura TKM et al (2019) J Geophys Res Space Phys 124(1):122–138. https://doi.org/10.1029/2018JA026051

Hasegawa H, Nakamura TKM, Gershman DJ, Nariyuki Y, Viñas AF et al (2020) J Geophys Res Space Phys 125(3):1–19. https://doi.org/10.1029/2019JA027595

Hasegawa H, Denton RE, Dokgo K, Hwang K-J, Nakamura TKM et al (2023) J Geophys Res Space Phys. https://doi.org/10.1029/2022ja031092

Hesse M, Aunai N, Sibeck D, Birn J (2014) Geophys Res Lett 41(24):8673–8680. https://doi.org/10.1002/2014GL061586

Hoilijoki S, Pucci F, Ergun RE, Schwartz SJ, Wilder FD et al (2021) J Geophys Res Space Phys 126(5):1–13. https://doi.org/10.1029/2020JA029046

Holmes JC, Ergun RE, Nakamura R, Roberts O, Wilder FD et al (2019) J Geophys Res Space Phys 124(11):8788–8803. https://doi.org/10.1029/2019JA026974

Hsieh MS, Otto A (2014) J Geophys Res Space Phys 119(5):3430–3443. https://doi.org/10.1002/2013JA019459

Hwang KJ, Goldstein ML, Lee E, Pickett JS (2011) J Geophys Res Space Phys 116(4):1–16. https://doi.org/10.1029/2010JA015742

Hwang KJ, Sibeck DG, Giles BL, Pollock CJ, Gershman D et al (2016) Geophys Res Lett 43(18):9434–9443. https://doi.org/10.1002/2016GL070934

Hwang KJ, Sibeck DG, Choi E, Chen LJ, Ergun RE et al (2017) Geophys Res Lett 44(5):2049–2059. https://doi.org/10.1002/2017GL072830

Hwang KJ, Sibeck DG, Burch JL, Choi E, Fear RC et al (2018) J Geophys Res Space Phys 123(10):8473–8488. https://doi.org/10.1029/2018JA025611

Hwang KJ, Choi E, Dokgo K, Burch JL, Sibeck DG et al (2019) Geophys Res Lett 46(12):6287–6296. https://doi.org/10.1029/2019GL082710
Hwang K-J, Dokgo K, Choi E, Burch JL, Sibeck DG et al (2020a) J Geophys Res Space Phys 125(4):e2019JA027665. https://doi.org/10.1029/2019JA027665
Hwang KJ, Nishimura Y, Coster AJ, Gillies RG, Fear RC et al (2020b) J Geophys Res Space Phys 125(6):1–15. https://doi.org/10.1029/2019JA027674
Hwang K-J, Burch JL, Russell CT, Choi E, Dokgo K et al (2021) Astrophys J 914(1):26. https://doi.org/10.3847/1538-4357/abf8b1
Hwang K-J, Weygand JM, Sibeck DG, Burch JL, Goldstein ML et al (2022) Front Astron Space Sci 9:1–15. https://doi.org/10.3389/fspas.2022.895514
Jiang K, Huang SY, Fu HS, Yuan ZG, Deng XH et al (2021) Astrophys J 922(1):56. https://doi.org/10.3847/1538-4357/ac2500
Johnson JR, Wing S (2015) J Geophys Res Space Phys 120(5):3987–4008. https://doi.org/10.1002/2014JA020312
Johnson JR, Wing S, Delamere P, Petrinec S, Kavosi S (2021) J Geophys Res Space Phys 126(2):1–15. https://doi.org/10.1029/2020JA028583
Kacem I, Jacquey C, Génot V, Lavraud B, Vernisse Y et al (2018) J Geophys Res Space Phys 123(3):1779–1793. https://doi.org/10.1002/2017JA024537
Karimabadi H, Roytershteyn V, Vu HX, Omelchenko YA, Scudder J et al (2014) Phys Plasmas 21(6):062308. https://doi.org/10.1063/1.4882875
Kieokaew R, Lavraud B, Foullon C, Toledo-Redondo S, Fargette N et al (2020) J Geophys Res Space Phys 125(6):1–18. https://doi.org/10.1029/2019JA027527
Knoll DA, Chacón L (2002) Phys Rev Lett. https://doi.org/10.1103/PhysRevLett.88.215003
Kropotina JA, Webster L, Artemyev AV, Bykov AM, Vainchtein DL et al (2021) Astrophys J 913(2):142. https://doi.org/10.3847/1538-4357/abf6c7
La Belle-Hamer AL, Otto A, Lee LC (1995) J Geophys Res 100(A7):11875–11889. https://doi.org/10.1029/94JA00969
Lavraud B, Zhang YC, Vernisse Y, Gershman DJ, Dorelli J et al (2016) Geophys Res Lett 43(7):3042–3050. https://doi.org/10.1002/2016GL068359
Lee LC, Fu ZF (1985) Geophys Res Lett 12(2):105–108. https://doi.org/10.1029/GL012i002p00105
Leonenko MV, Grigorenko EE, Zelenyi LM, Malova HV, Malykhin AY et al (2021) J Geophys Res Space Phys 126(11):e2021JA029641. https://doi.org/10.1029/2021JA029641
Li W, Raeder J, Thomsen MF, Lavraud B (2008) J Geophys Res Space Phys 113(A4):A04204. https://doi.org/10.1029/2007JA012604
Liu CM, Fu HS, Xu Y, Khotyaintsev YV, Burch JL et al (2018a) Geophys Res Lett. https://doi.org/10.1029/2018GL077928
Liu YH, Hesse M, Li TC, Kuznetsova M, Le A (2018b) J Geophys Res Space Phys 123(6):4908–4920. https://doi.org/10.1029/2018JA025410
Liu YY, Fu HS, Olshevsky V, Pontin DI, Liu CM et al (2019) Astrophys J Suppl Ser 244(2):31. https://doi.org/10.3847/1538-4365/ab391a
Liu TZ, Lu S, Turner DL, Gingell I, Angelopoulos V et al (2020) J Geophys Res Space Phys 125(4):1–11. https://doi.org/10.1029/2020JA027822
Lockwood M, Smith MF (1989) Geophys Res Lett 16(8):879–882. https://doi.org/10.1029/GL016i008p00879
Lu Q, Wang H, Wang X, Lu S, Wang R et al (2020) Geophys Res Lett 47(1):1–6. https://doi.org/10.1029/2019GL085661
Lu Q, Yang Z, Wang H, Wang R, Huang K et al (2021) Astrophys J 919(1):28. https://doi.org/10.3847/1538-4357/ac18c0
Ma X, Otto A, Delamere PA (2014a) J Geophys Res Space Phys 119(2):808–820. https://doi.org/10.1002/2013JA019225
Ma X, Otto A, Delamere PA (2014b) J Geophys Res Space Phys 119(2):781–797. https://doi.org/10.1002/2013JA019224
Ma X, Delamere P, Otto A, Burkholder B (2017) J Geophys Res Space Phys 122(10):10,382–10,395. https://doi.org/10.1002/2017JA024394
Maheshwari K, Phan TD, Øieroset M, Fargette N, Lavraud B et al (2022) Astrophys J 940(2):177. https://doi.org/10.3847/1538-4357/ac9405
Man HY, Zhou M, Deng XH, Fu HS, Zhong ZH et al (2018) Geophys Res Lett 45(17):8729–8737. https://doi.org/10.1029/2018GL079778
Marshall AT, Burch JL, Reiff PH, Webster JM, Torbert RB et al (2020) J Geophys Res Space Phys 125(1):1–9. https://doi.org/10.1029/2019JA027296

Matsumoto Y, Amano T, Kato TN, Hoshino M (2015) Science 347(6225):974–978. https://doi.org/10.1126/science.1260168
Maynard NC, Burke WJ, Ober DM, Farrugia CJ, Kucharek H et al (2007) J Geophys Res Space Phys 112(A12):A12219. https://doi.org/10.1029/2007JA012293
Mejnertsen L, Eastwood JP, Chittenden JP (2021) Front Astron Space Sci 8:1–15. https://doi.org/10.3389/fspas.2021.758312
Merkin VG, Panov EV, Sorathia KA, Ukhorskiy AY (2019) J Geophys Res Space Phys 124(11):8647–8668. https://doi.org/10.1029/2019JA026872
Miura A (1984) J Geophys Res 89(A2):801. https://doi.org/10.1029/JA089iA02p00801
Miura A, Pritchett PL (1982) J Geophys Res. https://doi.org/10.1029/JA087iA09p07431
Nakamura TKM, Daughton W (2014) Geophys Res Lett 41(24):8704–8712. https://doi.org/10.1002/2014GL061952
Nakamura TKM, Fujimoto M (2005) Geophys Res Lett 32:L21102. https://doi.org/10.1029/2005GL023362
Nakamura R, Baumjohann W, Mouikis C, Kistler LM, Runov A et al (2004) Geophys Res Lett 31(9):L09804. https://doi.org/10.1029/2004GL019558
Nakamura TKM, Fujimoto M, Otto A (2006) Geophys Res Lett. https://doi.org/10.1029/2006GL026318
Nakamura TKM, Fujimoto M, Otto A (2008) J Geophys Res Space Phys. https://doi.org/10.1029/2007JA012803
Nakamura TKM, Hasegawa H, Shinohara I, Fujimoto M (2011) J Geophys Res Space Phys. https://doi.org/10.1029/2010JA016046
Nakamura TKM, Daughton W, Karimabadi H, Eriksson S (2013) J Geophys Res Space Phys 118(9):5742–5757. https://doi.org/10.1002/jgra.50547
Nakamura R, Sergeev VA, Baumjohann W, Plaschke F, Magnes W et al (2016) Geophys Res Lett 43(10):4841–4849. https://doi.org/10.1002/2016GL068768
Nakamura TKM, Eriksson S, Hasegawa H, Zenitani S, Li WY et al (2017a) J Geophys Res Space Phys 122(11):11505–11522. https://doi.org/10.1002/2017JA024346
Nakamura TKM, Hasegawa H, Daughton W, Eriksson S, Li WY et al (2017b) Nat Commun. https://doi.org/10.1038/s41467-017-01579-0
Nakamura TKM, Nakamura R, Varsani A, Genestreti KJ, Baumjohann W et al (2018) Geophys Res Lett 45(9):3829–3837. https://doi.org/10.1029/2018GL078340
Nakamura TKM, Stawarz JE, Hasegawa H, Narita Y, Franci L et al (2020) J Geophys Res Space Phys 125(3):1–18. https://doi.org/10.1029/2019JA027515
Nakamura R, Baumjohann W, Nakamura TKM, Panov EV, Schmid D et al (2021) J Geophys Res Space Phys 126(10):1–19. https://doi.org/10.1029/2021JA029518
Nishino MN, Hasegawa H, Saito Y, Kitamura N, Miyashita Y et al (2022) J Geophys Res Space Phys 127(1):1–16. https://doi.org/10.1029/2021JA029747
Norgren C, Hesse M, Graham DB, Khotyaintsev YV, Tenfjord P et al (2020) J Geophys Res Space Phys 125(4):e2019JA027440. https://doi.org/10.1029/2019JA027440
Nykyri K, Otto A (2001) Geophys Res Lett 28(18):3565–3568. https://doi.org/10.1029/2001GL013239
Nykyri K, Ma X, Dimmock A, Foullon C, Otto A et al (2017) J Geophys Res Space Phys 122(9):9489–9512. https://doi.org/10.1002/2017JA024374
Øieroset M, Phan TD, Haggerty C, Shay MA, Eastwood JP et al (2016) Geophys Res Lett 43(11):5536–5544. https://doi.org/10.1002/2016GL069166
Øieroset M, Phan TD, Shay MA, Haggerty CC, Fujimoto M et al (2017) Geophys Res Lett 44(15):7598–7606. https://doi.org/10.1002/2017GL074196
Øieroset M, Phan TD, Drake JF, Eastwood JP, Fuselier SA et al (2019) Geophys Res Lett 46(4):1937–1946. https://doi.org/10.1029/2018GL080994
Pan D, Khotyaintsev YV, Graham DB, Vaivads A, Zhou X et al (2018) Geophys Res Lett 45(22):12116–12124. https://doi.org/10.1029/2018GL080826
Petrinec SM, Burch JL, Chandler M, Farrugia CJ, Fuselier SA et al (2020) J Geophys Res Space Phys 125(7):e2020JA027778. https://doi.org/10.1029/2020JA027778
Petrinec SM, Burch JL, Fuselier SA, Trattner KJ, Giles BL et al (2022) J Geophys Res Space Phys 127(6):e2021JA029669. https://doi.org/10.1029/2021JA029669
Petrukovich AA, Baumjohann W, Nakamura R, Runov A, Balogh A et al (2007) J Geophys Res Space Phys 112(A10):A10213. https://doi.org/10.1029/2007JA012349
Phan TD, Kistler LM, Klecker B, Haerendel G, Paschmann G et al (2000) Nature 404(6780):848–850. https://doi.org/10.1038/35009050
Phan TD, Dunlop MW, Paschmann G, Klecker B, Bosqued JM et al (2004) Ann Geophys 22(7):2355–2367. https://doi.org/10.5194/angeo-22-2355-2004
Phan TD, Paschmann G, Twitty C, Mozer FS, Gosling JT et al (2007) Geophys Res Lett 34(14):L14104. https://doi.org/10.1029/2007GL030343

Phan TD, Love TE, Gosling JT, Paschmann G, Eastwood JP et al (2011) Geophys Res Lett 38:L17101. https://doi.org/10.1029/2011GL048586

Phan TD, Eastwood JP, Shay MA, Drake JF, Sonnerup BUO et al (2018) Nature 557(7704):202–206. https://doi.org/10.1038/s41586-018-0091-5

Poh G, Slavin JA, Lu S, Le G, Ozturk DS et al (2019) J Geophys Res Space Phys 124(9):7477–7493. https://doi.org/10.1029/2018JA026451

Pu ZY, Yei M, Liu ZX (1990) J Geophys Res 95(A7):10559–10566. https://doi.org/10.1029/JA095iA07p10559

Quijia P, Fraternale F, Stawarz JE, Vásconez CL, Perri S et al (2021) Mon Not R Astron Soc 503(4):4815–4827. https://doi.org/10.1093/mnras/stab319

Retinò A, Sundkvist D, Vaivads A, Mozer F, André M et al (2007) Nat Phys 3(4):235–238. https://doi.org/10.1038/nphys574

Richard L, Khotyaintsev YV, Graham DB, Sitnov MI, Le Contel O et al (2021) J Geophys Res Space Phys 126(8):e2021JA029152. https://doi.org/10.1029/2021JA029152

Richard L, Khotyaintsev YV, Graham DB, Russell CT (2022) Geophys Res Lett 49(22):e2022GL101693. https://doi.org/10.1029/2022GL101693

Runov A, Grandin M, Palmroth M, Battarbee M, Ganse U et al (2021) Ann Geophys 39(4):599–612. https://doi.org/10.5194/angeo-39-599-2021

Russell CT, Elphic RC (1978) Space Sci Rev 22(6):681–715. https://doi.org/10.1007/BF00212619

Russell CT, Qi Y (2020) Geophys Res Lett 47(15):1–7. https://doi.org/10.1029/2020GL087620

Sauvaud J-A, Popescu D, Delcourt DC, Parks GK, Brittnacher M et al (1999) J Geophys Res Space Phys 104(A12):28565–28586. https://doi.org/10.1029/1999JA900293

Schmid D, Volwerk M, Nakamura R, Baumjohann W, Heyn M (2011) Ann Geophys. https://doi.org/10.5194/angeo-29-1537-2011

Scholer M (1988) Geophys Res Lett. https://doi.org/10.1029/GL015i004p00291

Scholer M (1995) In: Geophysical monograph series, vol 90, pp 235–245. https://doi.org/10.1029/GM090p0235

Schwartz S (1995) Adv Space Res 15(8–9):107–116. https://doi.org/10.1016/0273-1177(94)00092-F

Schwartz SJ, Kucharek H, Farrugia CJ, Trattner K, Gingell I et al (2021) Geophys Res Lett 48(4):e2020GL091859. https://doi.org/10.1029/2020GL091859

Sergeev V, Runov A, Baumjohann W, Nakamura R, Zhang TL et al (2004) Geophys Res Lett. https://doi.org/10.1029/2003GL019346

Sergeev VA, Sormakov DA, Apatenkov SV, Baumjohann W, Nakamura R et al (2006) Ann Geophys 24(7):2015–2024. https://doi.org/10.5194/angeo-24-2015-2006

Sergeev VA, Apatenkov SV, Nakamura R, Plaschke F, Baumjohann W et al (2021) J Geophys Res Space Phys 126(2):e2020JA028694. https://doi.org/10.1029/2020JA028694

Sisti M, Faganello M, Califano F, Lavraud B (2019) Geophys Res Lett 46(21):11597–11605. https://doi.org/10.1029/2019GL083282

Sitnov MI, Buzulukova N, Swisdak M, Merkin VG, Moore TE (2013) Geophys Res Lett 40(1):22–27. https://doi.org/10.1029/2012GL054701

Southwood DJ, Farrugia CJ, Saunders MA (1988) Planet Space Sci 36(5):503–508. https://doi.org/10.1016/0032-0633(88)90109-2

Stawarz JE, Eriksson S, Wilder FD, Ergun RE, Schwartz SJ et al (2016) J Geophys Res Space Phys 121(11):11021–11034. https://doi.org/10.1002/2016JA023458

Stawarz JE, Eastwood JP, Genestreti KJ, Nakamura R, Ergun RE et al (2018) Geophys Res Lett 45(17):8783–8792. https://doi.org/10.1029/2018GL079095

Sun WJ, Slavin JA, Tian AM, Bai SC, Poh GK et al (2019) Geophys Res Lett 46(12):6168–6177. https://doi.org/10.1029/2019GL083301

Sundkvist D, Retinò A, Vaivads A, Bale SD (2007) Phys Rev Lett 99(2):025004. https://doi.org/10.1103/PhysRevLett.99.025004

Swisdak M, Rogers BN, Drake JF, Shay MA (2003) J Geophys Res 108:1218. https://doi.org/10.1029/2002JA009726

Swisdak M, Opher M, Drake JF, Alouani Bibi F (2010) Astrophys J 710(2):1769–1775. https://doi.org/10.1088/0004-637X/710/2/1769

Tanaka KG, Fujimoto M, Shinohara I (2010) Int J Geophys. https://doi.org/10.1155/2010/202583

Teh WL, Nakamura T, Nakamura R, Umeda T (2018) J Geophys Res Space Phys 123(10):8122–8130. https://doi.org/10.1029/2018JA025775

Terasawa T, Fujimoto M, Mukai T, Shinohara I, Saito Y et al (1997) Geophys Res Lett 24(8):935–938. https://doi.org/10.1029/96GL04018

Torbert RB, Burch JL, Giles BL, Gershman D, Pollock CJ et al (2016) Geophys Res Lett 43(12):5918–5925. https://doi.org/10.1002/2016GL069553

Torbert RB, Burch JL, Phan TD, Hesse M, Argall MR et al (2018) Science 362(6421):1391–1395. https://doi.org/10.1126/science.aat2998

Trattner KJ, Mulcock JS, Petrinec SM, Fuselier SA (2007) J Geophys Res Space Phys 112(A8):A08210. https://doi.org/10.1029/2007JA012270

Trattner KJ, Fuselier SA, Petrinec SM, Burch JL, Ergun R et al (2021a) J Geophys Res Space Phys 126(4):e2020JA028926. https://doi.org/10.1029/2020JA028926

Trattner KJ, Petrinec SM, Fuselier SA (2021b) Space Sci Rev 217(3):41. https://doi.org/10.1007/s11214-021-00817-8

Varsani A, Nakamura R, Sergeev VA, Baumjohann W, Owen CJ et al (2017) J Geophys Res Space Phys 122(11):10891–10909. https://doi.org/10.1002/2017JA024547

Vernisse Y, Lavraud B, Eriksson S, Gershman DJ, Dorelli J et al (2016) J Geophys Res Space Phys 121(10):9926–9939. https://doi.org/10.1002/2016JA023051

Vernisse Y, Lavraud B, Faganello M, Fadanelli S, Sisti M et al (2020) J Geophys Res Space Phys 125(5):e2019JA027333. https://doi.org/10.1029/2019JA027333

Wang R, Lu Q, Nakamura R, Baumjohann W, Russell CT et al (2017) J Geophys Res Space Phys 122(10):10436–10447. https://doi.org/10.1002/2017JA024482

Wang R, Lu Q, Nakamura R, Baumjohann W, Huang C et al (2018) Geophys Res Lett 45(10):4542–4549. https://doi.org/10.1002/2017GL076330

Wang S, Chen L, Bessho N, Hesse M, Wilson LB et al (2019) Geophys Res Lett 46(2):562–570. https://doi.org/10.1029/2018GL080944

Webster JM, Burch JL, Reiff PH, Daou AG, Genestreti KJ et al (2018) J Geophys Res Space Phys 123(6):4858–4878. https://doi.org/10.1029/2018JA025245

Wei YY, Huang SY, Rong ZJ, Yuan ZG, Jiang K et al (2019) Astrophys J 874(2):L18. https://doi.org/10.3847/2041-8213/ab0f28

Wellenzohn S, Nakamura R, Nakamura TKM, Varsani A, Sergeev VA et al (2021) J Geophys Res Space Phys 126(1):e2020JA028917. https://doi.org/10.1029/2020JA028917

Wing S, Johnson JR, Newell PT, Meng C-I (2005) J Geophys Res Space Phys 110(A8):A08205. https://doi.org/10.1029/2005JA011086

Xu Y, Fu H, Cao J, Liu C, Norgren C et al (2021) Geophys Res Lett 48(6):e2020GL092232. https://doi.org/10.1029/2020GL092232

Yordanova E, Vörös Z, Raptis S, Karlsson T (2020) Front Astron Space Sci 7:1–11. https://doi.org/10.3389/fspas.2020.00002

Zenitani S, Hesse M, Klimas A, Kuznetsova M (2011) Phys Rev Lett 106:195003. https://doi.org/10.1103/PhysRevLett.106.195003

Zhong ZH, Zhou M, Deng XH, Song LJ, Graham DB et al (2021) Geophys Res Lett 48(1):2020GL090946. https://doi.org/10.1029/2020GL090946

Zhou M, Berchem J, Walker RJ, El-Alaoui M, Deng X et al (2017) Phys Rev Lett. https://doi.org/10.1103/PhysRevLett.119.055101

Authors and Affiliations

K.-J. Hwang[1] · R. Nakamura[2] · J.P. Eastwood[3] · S.A. Fuselier[1] · H. Hasegawa[4] · T. Nakamura[5,6] · B. Lavraud[7] · K. Dokgo[1] · D.L. Turner[8] · R.E. Ergun[9] · P.H. Reiff[10]

✉ K.-J. Hwang
joo.hwang@swri.org

✉ R. Nakamura
rumi.nakamura@aeaw.ac.at

1 Southwest Research Institute, San Antonio, TX, USA

2 Space Research Institute, Austrian Academy of Sciences ÖAW, Graz, Austria

3 Imperial College, London, UK

[4] Institute of Space and Astronautical Science, Japan Aerospace Exploration Agency, Sagamihara, Japan

[5] Austrian Academy of Sciences ÖAW, Space Research Institute, Vienna, Austria

[6] Krimgen LLC, 732-0828, Hiroshima, Japan

[7] Institut de Recherche en Astrophysique et Planétologie, CNRS, UPS, CNES, Université de Toulouse, Toulouse, France

[8] The Johns Hopkins Applied Physics Laboratory, Laurel, MD, USA

[9] Laboratory for Atmospheric and Space Physics, University of Colorado at Boulder, Boulder, CO, USA

[10] Rice University, Houston, TX, USA

Space Science Reviews (2023) 219:75
https://doi.org/10.1007/s11214-023-01011-8

Particle Acceleration by Magnetic Reconnection in Geospace

Mitsuo Oka[1] · Joachim Birn[2,3] · Jan Egedal[4] · Fan Guo[3] · Robert E. Ergun[5,6] · Drew L. Turner[7] · Yuri Khotyaintsev[8] · Kyoung-Joo Hwang[9] · Ian J. Cohen[7] · James F. Drake[10]

Received: 4 May 2023 / Accepted: 5 October 2023 / Published online: 7 November 2023

Abstract
Particles are accelerated to very high, non-thermal energies during explosive energy-release phenomena in space, solar, and astrophysical plasma environments. While it has been established that magnetic reconnection plays an important role in the dynamics of Earth's magnetosphere, it remains unclear how magnetic reconnection can further explain particle acceleration to non-thermal energies. Here we review recent progress in our understanding of particle acceleration by magnetic reconnection in Earth's magnetosphere. With improved resolutions, recent spacecraft missions have enabled detailed studies of particle acceleration at various structures such as the diffusion region, separatrix, jets, magnetic islands (flux ropes), and dipolarization front. With the guiding-center approximation of particle motion, many studies have discussed the relative importance of the parallel electric field as well as the Fermi and betatron effects. However, in order to fully understand the particle acceleration mechanism and further compare with particle acceleration in solar and astrophysical plasma environments, there is a need for further investigation of, for example, energy partition and the precise role of turbulence.

Keywords Particle acceleration · Magnetic reconnection · Magnetosphere · Magnetospheric MultiScale

1 Introduction

1.1 Motivation and Structure

Particles are accelerated to very high, non-thermal energies during explosive energy-release phenomena in space, solar, and astrophysical plasma environments. Unlike remote-sensing measurements of distant astrophysical objects that are often difficult to resolve spatially, *in-situ* measurements of Earth's magnetosphere provide unique opportunities to directly study particle acceleration and its spatial and temporal variations down to the kinetic scale. In fact, through decades of study, it is now established that magnetic reconnection — a plasma process that converts magnetic energy into particle energy — plays an important role in the dynamics of the energy-release process in the magnetotail (e.g. Zweibel and Yamada 2009; Ji and Daughton 2011; Hwang et al. 2023; Fuselier et al. 2023, and references therein). However, it remains unclear how magnetic reconnection can further explain particle acceleration

Extended author information available on the last page of the article

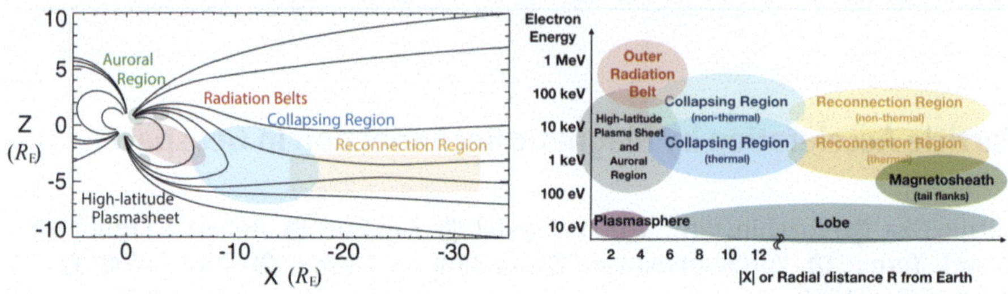

Fig. 1 Schematic illustrations of Earth's magnetotail, demonstrating key regions and the typical electron energy in those regions. Left: Magnetic field lines of a semi-empirical model, with the key regions highlighted in color. The Geocentric Solar Ecliptic (GSE) coordinate is used with the unit of Earth's radii $R_E \sim 6371$ km. Right: The typical electron energy of key regions as a function of distance from Earth. Adapted from Oka et al. (2018)

to non-thermal energies (typically $\gtrsim 10$ keV) during explosive energy-release phenomena in Earth's magnetosphere, although significant progress has been made in the past decades with spacecraft missions such as *Geotail, WIND, Cluster, THEMIS/ARTEMIS*, and *MMS*, combined with theories and simulations.

Thus, the main purpose of this paper is to review the most recent advances in our understanding of particle acceleration by magnetic reconnection in geospace which includes the magnetotail and the dayside magnetosphere. Many observational reports of particle acceleration come from the magnetotail probably because the environmental parameter $m_i V_A^2$ can be much larger in the magnetotail, where m_i is the ion mass and V_A is the Alfvén speed and therefore the energization, both heating and acceleration to non-thermal energies, becomes significant (e.g. Phan et al. 2013; Shay et al. 2014; Oka et al. 2022).

It should be noted that the term 'particle acceleration' typically refers to the process of energizing particles to non-thermal energies and does not include the meaning of heating, an increase of the plasma temperature. Therefore, a discussion of particle acceleration usually involves a power-law form of energy spectrum. However, in some cases, the term 'acceleration' is used in its literal sense, as shown in the equation of motion, $ma = F$ where m, a, and F represent the particle mass, acceleration, and force, respectively. This usage does not differentiate between thermal and non-thermal components. For example, Fermi acceleration in the guiding-center approximation (which will be discussed in the following section) applies to both thermal and non-thermal particles. In this paper, we have attempted to use the phrase 'acceleration to non-thermal energies' when the discussion pertains to the non-thermal component. Also, we sometimes used the term 'energization' when we do not differentiate thermal and non-thermal components.

There are already relevant review articles on particle acceleration in geospace that focus on theories (e.g. Birn et al. 2012; Li et al. 2021) and specific topics such as power-law index (Oka et al. 2018) and dipolarization front (Fu et al. 2020). However, this paper will provide a more general overview of observations and simulations of particle acceleration to non-thermal energies both near the 'reconnection region' (highlighted in yellow in Fig. 1), which is referred to as X-line in simplified (e.g., two-dimensional or north-south symmetric) geometry, and at large scale where the intrinsic dipole field of the magnetosphere becomes important (i.e., the 'collapsing region' as highlighted in blue in Fig. 1). For an up-to-date overview of the relevant context of magnetic reconnection at global scales and its associated cross-scale aspects, readers are referred to Fuselier et al. (2023) and Hwang et al. (2023) in this collection, respectively.

The paper is structured as follows:

1.2 Key Theories

1.2.1 Particle Acceleration Mechanisms in the Guiding-Center Approximation

Observations (e.g. Baumjohann et al. 2007) and simulations have shown that the thickness of a current sheet ought to be less than a typical ion gyroradius or inertial length to enable reconnection or other activity. Thus, ions are expected to be non-adiabatic near the reconnection sites. In contrast, electrons may show adiabatic behavior much closer to an X-line, such that a guiding center approach seems more reasonable.

In the guiding-center approximation, the main acceleration mechanisms are Fermi acceleration, betatron acceleration, and the direct acceleration by the parallel electric field (e.g. Northrop 1963; Birn et al. 2012; Dahlin 2020; Li et al. 2021). Fermi acceleration occurs when a particle encounters a dynamically evolving, curved magnetic field. The betatron acceleration describes the process where the increasing magnetic field leads to the energy gain in the perpendicular direction due to the conservation of the first adiabatic invariant, whereas during the direct acceleration particles stream along the magnetic field and gain energy if a significant parallel electric field exists. Figure 2a shows several patterns where these acceleration mechanisms may happen. The main energy gain of a single particle under the guiding-center limit is:

$$\frac{d\varepsilon}{dt} = q\mathbf{E}_{\|} \cdot \mathbf{v}_{\|} + \frac{\mu}{\gamma}\left(\frac{\partial B}{\partial t} + \mathbf{u_E} \cdot \nabla B\right) + \gamma m_e v_{\|}^2(\mathbf{u_E} \cdot \boldsymbol{\kappa}) \tag{1}$$

Here, q is the particle charge, m_e is the electron mass, $\mathbf{v}_{\|}$ and $\mathbf{v}_{\perp}$ are the parallel and perpendicular components of the particle velocity, respectively, μ is the magnetic moment, γ is the Lorentz factor, and $\mathbf{u_E} = \mathbf{E} \times \mathbf{B}/B^2$ is the electric drift velocity, $\boldsymbol{\kappa}$ is the curvature of magnetic field lines. The first term on the right is the parallel electric field acceleration, the second term corresponds to the betatron acceleration, and the third term is associated

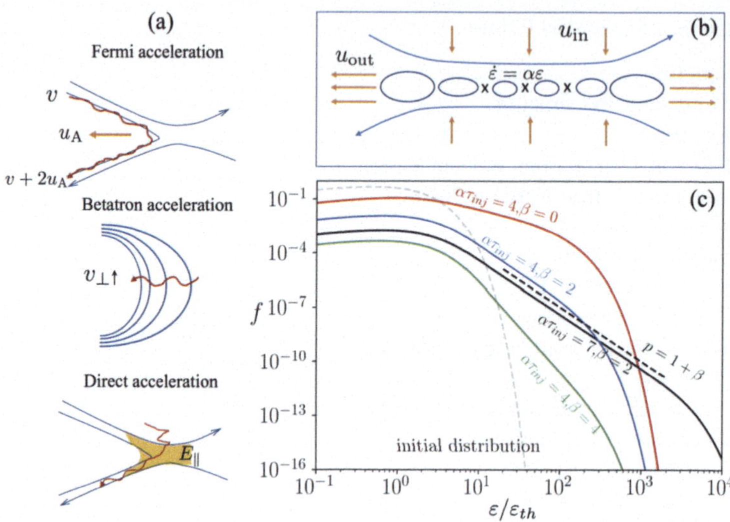

Fig. 2 a) Illustration of the main particle acceleration patterns, b) a cartoon for particle acceleration in the reconnection layer and development of power-law distribution, c) solutions to Equation (4) that show the energy spectra for a few different $\alpha\tau_{inj}$ and escape parameter β

with the Fermi acceleration, corresponding to the curvature drift acceleration. In Fig. 2a, the Fermi acceleration is assumed to be driven by the curved magnetic field that drifts at the Alfvén speed u_A.

It is to be noted that Fermi acceleration is traditionally viewed as bouncing between converging magnetic mirror points (1st order Fermi acceleration of type A) (Northrop 1963) when the energization can be inferred from conservation of the second adiabatic invariant. However, parallel, i.e., Fermi acceleration may also result from an encounter with a strongly curved, moving magnetic field structure, akin to a slingshot effect (1st order Fermi acceleration of type B) (Northrop 1963). Thus, Fermi 'reflections' may occur when a particle encounters a sudden change in magnetic topology and/or field strength, resulting in a sudden change in pitch angle and potentially a reflection. If the sudden change in topology is associated with a moving magnetic structure, then the particle gains energy corresponding to the speed of the structure in the particle's rest frame (i.e., the rate of energy gain is nearly proportional to the particle's energy). For electrons, a single encounter produces only a relatively small energy gain, because the speed of the structure is small compared to the electron thermal speed. However, multiple encounters can add up to substantial energy gains. Such Fermi reflections can occur in collapsing closed field regions (Fig. 1) (Birn et al. 2004) or in contracting or merging islands (Drake et al. 2006; Zank et al. 2014).

It is also useful to note that betatron and first-order Fermi acceleration can be viewed as $\mathbf{E} \times \mathbf{B}$ drift toward increasing magnetic field strength or in the direction of a magnetic field curvature vector, respectively (Eq. (1)), but equivalently also as grad B drift or curvature drift, respectively, in the direction of an electric field (opposite for electrons) (e.g. Birn et al. 2013). The former indicates that the relative role of the two mechanisms depend on the magnetic field geometry, while the latter indicates a pitch angle dependence (as the curvature drift speed depends on the parallel particle energy and grad B drift depends on the perpendicular one).

One can statistically evaluate the importance of the mechanisms by ensemble averaging particle motions, where the collective perpendicular particle current density for each species s is:

$$\mathbf{J}_{s\perp} = p_{s\parallel}\frac{\mathbf{B}\times(\mathbf{B}\cdot\nabla)\mathbf{B}}{B^4} + p_{s\perp}\frac{\mathbf{B}\times\nabla\mathbf{B}}{B^3} - [\nabla\times\frac{p_{s\perp}\mathbf{B}}{B^2}] + \rho_s\mathbf{u}_E - n_s m_s\frac{d\mathbf{v}_s}{dt}\times\frac{\mathbf{B}}{B^2} \tag{2}$$

where $p_{s\parallel}$ and $p_{s\perp}$ are parallel and perpendicular pressures to the local magnetic field, respectively, ρ_s is the charge density, n_s is particle number density, m_s is particle mass, $\mathbf{v_s}$ is the species flow velocity, and $d/dt \equiv \partial_t + \mathbf{v_s}\cdot\nabla$. The terms on the right shows the current due to curvature drift, grad B drift, the perpendicular magnetization, electric drift, and the polarization drift. The total energization can be shown with $\mathbf{J}\cdot\mathbf{E}$. Another equivalent expression for the $\mathbf{J}\cdot\mathbf{E}$, after a rearrangement, is

$$\mathbf{J}_{s\perp}\cdot\mathbf{E}_\perp = \nabla\cdot(p_{s\perp}\mathbf{u}_E) - p_s\nabla\cdot\mathbf{u}_E - (p_{s\parallel} - p_{s\perp})\mathbf{b}_i\mathbf{b}_j\sigma_{ij} \tag{3}$$

where $\sigma_{ij} = 0.5(\partial_i\mathbf{u}_{Ej} + \partial_j\mathbf{u}_{Ei} - (2\nabla\cdot\mathbf{u}_E\delta_{ij}/3))$ is the shear tensor of $\mathbf{u_E}$ flow, $p_s \equiv (p_{s\parallel} + 2p_{s\perp})/3$ is the effective scalar pressure, and we have ignored the effect of the polarization drift (Li et al. 2017). This expression shows the role of fluid compression and velocity shear in the energy gain (Li et al. 2018). These can be connected with the recent work of pressure-strain terms for gaining insight in turbulent plasmas (e.g., Yang et al. 2017; Du et al. 2018; Li et al. 2019a).

Over the past decade, particle-in-cell simulations and test-particle simulations have been widely used to evaluate these acceleration mechanisms. Several particle kinetic simulations (e.g. Guo et al. 2014; Dahlin et al. 2014; Li et al. 2017; Arnold et al. 2021), modeling the formation of multiple magnetic islands or flux ropes and their merging, indicated an overall dominance of Fermi acceleration over betatron acceleration. In contrast, a test-particle simulation of electron drifts in a collapsing magnetic arcade with strong guide field indicated a dominance of betatron acceleration (Birn et al. 2017a). This confirms that the relative role of the two mechanisms depends on the field geometry, for instance, betatron acceleration may be expected to be important particularly inside of reconnection jet fronts and collapsing magnetic traps. In addition, parallel electric fields are shown to contribute to particle energization and modify the distribution functions (e.g. Egedal et al. 2009). As the guide field increases, the Fermi acceleration becomes less efficient, but acceleration by the parallel electric field is not very sensitive to the guide field (Dahlin et al. 2016; Li et al. 2018).

1.2.2 Formation of Nonthermal Power-Law Energy Spectra in Reconnection Acceleration

Power-law energy spectra are a main feature of nonthermal acceleration and are of great interest to reconnection studies (e.g. Li et al. 2019b; Zhang et al. 2021; Arnold et al. 2021; Nakanotani et al. 2022). There have been debates and some confusion about the formation of nonthermal power-law energy spectra during particle acceleration in magnetic reconnection; therefore this journal is worth clarifying. As shown by Fig. 2b, we illustrate a simple case where the main acceleration term is a Fermi-like relation $\dot{\varepsilon} = \alpha\varepsilon$ (α is the acceleration rate) in an energy continuity equation (Guo et al. 2014, 2015), which represents the case when the first-order Fermi acceleration dominates the acceleration process:

$$\frac{\partial f}{\partial t} + \frac{\partial}{\partial\varepsilon}(\dot{\varepsilon}f) = \frac{f_{inj}}{\tau_{inj}} - \frac{f}{\tau_{esc}} \tag{4}$$

As reconnection proceeds, the ambient plasma is continuously injected into the reconnection layer through an inflow speed u_{in}. τ_{inj} is the timescale for the injection of low-energy particles f_{inj}, and τ_{esc} is the escape timescale. For illustration purposes, we assume that the upstream distribution is a Maxwellian distribution $f_{inj} = (2N_{inj}/\sqrt{\pi})\sqrt{\varepsilon_0}exp(-\varepsilon_0)$, where $\varepsilon_0 = \varepsilon/\varepsilon_{th}$ is energy normalized by the thermal energy. With these assumptions, the solution to Eq. (4) can be written as

$$f(\varepsilon, t) = \frac{2N_{inj}}{\sqrt{\pi}(\alpha\tau_{inj}\varepsilon_0^{1+\beta})}[\Gamma_{3/2+\beta}(\varepsilon_0 e^{-\alpha t}) - \Gamma(3/2+\beta)(\varepsilon_0)], \tag{5}$$

where $\beta = 1/(\alpha\tau_{esc})$ and $\Gamma_s(x)$ is the upper incomplete Gamma function. Figure 2c illustrates this simple solution for a few different escape time and $\alpha\tau_{inj}$. As reconnection proceeds, new particles are injected and accelerated in the reconnection, and a power-law distribution can form when $\alpha\tau_{inj}$ is large. Note that the derivation also shows that power-law distribution can still form even for the case with no escape term as shown in Fig. 2 (red curve). However, if the population of particles is initially in the current sheet, it can be shown the distribution remains a Maxwellian (Guo et al. 2020). It is often argued that some loss mechanism is needed to form a power-law distribution, but the simple analytical solution does not support it. Here the main physics for forming a power-law is due to the continuous injection and Fermi acceleration. Meanwhile, it is still important to understand the escape term, as it can strongly change the shape of the distribution. Other acceleration can, in principle, form a power-law, and the steady-state solution has the spectral index

$$p = 1 + \frac{1}{\alpha\tau_{esc}} + \frac{\partial \ln \alpha}{\partial \ln \varepsilon}. \tag{6}$$

This equation includes the case where the acceleration rate has an energy dependence. When the product of $\alpha\tau_{esc}$ does not depend on energy and α has a power-law dependence on energy ε over a certain energy range, p is a constant across this range (power-law energy spectra).

In the context of magnetic reconnection, magnetic islands (or flux ropes in 3D) can play an important role in particle acceleration (e.g. Drake et al. 2006, 2013; Oka et al. 2010b; Hoshino 2012; Guo et al. 2014; Zank et al. 2014; le Roux et al. 2015, 2018). Using a more formal, particle transport equation that captures the essential physics of particle acceleration in a multi-island region, the possibility of compressible flux-rope contraction and merging in a turbulent media was considered in Zank et al. (2014), le Roux et al. (2015, 2018). It was shown theoretically that both curvature drift and betatron acceleration, due to an increasing flux-rope magnetic field strength, contribute to kinetic energy gain, and the particle acceleration is a first-order Fermi acceleration process when the particle distribution is isotropic or nearly isotropic.

1.2.3 Beyond Guiding-Center Approximation

Although numerical simulations have shown that the guiding-center approximation can well describe the acceleration of particles in the reconnection region, the particle dynamics in the reconnection region can be more complicated. Close to the X-line, it is well known that the gyrotropic approximation is not valid. Although the X-line may not be the region with the strongest acceleration, they may support local acceleration that is of interest to *in situ* observations. The X-line region with a weak guide field can support chaotic orbits of particles (Zenitani and Nagai 2016). During particle motion, waves and turbulence can modify the particle distribution. However, as long as the particle distribution is gyrotropic, Eq. (2) can still statistically describe the acceleration, even if significant pitch-angle scattering occurs (Hazeltine and Meiss 2003; Egedal et al. 2013). Therefore, some caution is needed when interpreting the results of the analysis. In addition, waves and turbulence may lead to stochastic heating and acceleration (e.g. Zank et al. 2015; Ergun et al. 2020a). The guiding-center description does not distinguish electrons and ions, meaning multiple species can be accelerated (e.g. Zhang et al. 2021, 2022). However, in the context of the magnetosphere, ions may not be well described by the guiding center approximation, as their gyroradii can be fairly large compared to characteristic scales at which the fields evolve. In particular, close to the center of the magnetotail current sheet, the gyroradii can approach the curvature radius of the magnetic field, and the ions experience strong scattering when crossing the current sheet (e.g. Richard et al. 2023a). Furthermore, instabilities, waves, and turbulence that are generated during magnetic reconnection may lead to more efficient acceleration of particles (e.g. Dahlin et al. 2017; Li et al. 2019b; Zhang et al. 2021; Johnson et al. 2022).

1.3 Example Observations and Challenges

The possible limitation of the guiding-center theory and the importance of turbulence may be glimpsed in recent examples of magnetotail reconnection. Figure 3 shows two cases of particle acceleration during magnetotail reconnection, obtained by *Magnetospheric Multi-Scale (MMS)*. The 2017 July 11 event (left column) is a case with less-enhanced heating and turbulence and has been studied by many authors (e.g. Torbert et al. 2018; Genestreti et al. 2018; Nakamura et al. 2018; Egedal et al. 2019; Nakamura et al. 2019; Jiang et al. 2019; Sitnov et al. 2019; Burch et al. 2019; Hasegawa et al. 2019; Torbert et al. 2020; Hwang et al. 2019; Cohen et al. 2021; Turner et al. 2021a; Oka et al. 2022). On the other hand, the 2017 July 26 event (right column) is a case with significantly enhanced heating and turbulence and has also been studied intensively (e.g. Ergun et al. 2018, 2020b,a; Cohen et al. 2021; Oka et al. 2022).

A puzzle is that, some properties of magnetic reconnection (e.g., heating and turbulence) appear differently in these two cases, and yet particles (both ions and electrons) are accelerated to non-thermal energies in both cases. For electrons, the non-thermal, power-law tail may even be softer in the significantly heated and turbulent case (e.g. Zhou et al. 2016; Oka et al. 2022) but it remains unclear how the observed power-law index can be explained. For ions, the energy spectrum could be more complicated. While the bulk flow component could peak around 1 keV, the higher-energy end of the spectrum may be influenced by the physical size of the energization region, as often argued in the shock physics (e.g. Blandford and Eichler 1987). In the Earth's magnetotail, the gyroradii of ions with energies greater than $\sim$100 keV may exceed several ion skin depths. In any case, the similarities and differences of these two cases lead to questions such as 'What is the precise condition of particle acceleration?', 'What is the precise role of turbulence?', 'How particle energies are partitioned between thermal and non-thermal energies?', and ultimately 'How are particles heated and accelerated to non-thermal energies?'. By reviewing recent progress in more detail in this paper, we hope to clarify what we know so far, what ideas have been discussed, and what we need to work on in the near future.

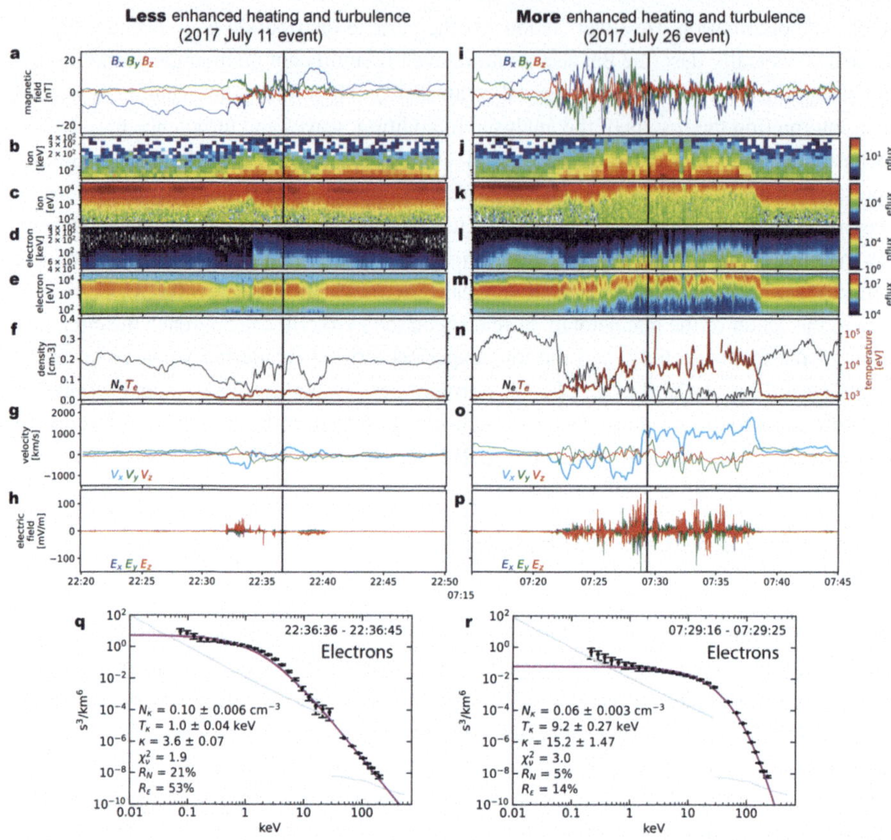

Fig. 3 Example observations demonstrating particle acceleration during magnetic reconnection with less enhanced and much enhanced heating/turbulence obtained on 2017 July 11 (e.g. Torbert et al. 2018) and 2017 July 26 (e.g. Ergun et al. 2018), respectively. The bottom panels show the typical energy spectrum of electrons for each event obtained at the time indicated by the vertical dashed lines in the upper panels (Oka et al. 2022). The best-fit kappa distribution model is displayed in magenta and the obtained parameters are annotated. Note the significant heating but soft (steep) power-law tail for the case of 2017 July 26

2 Particle Acceleration Near the X-Line

2.1 Active Vs. Quiet

Early magnetotail studies showed that, in the plasma sheet, the energy spectra become non-thermal above $\sim$10 keV for ions and $\sim$1 keV for electrons and are often represented by the kappa distribution (e.g., Christon et al. 1988, 1989, 1991, and references therein). The power-law index Γ is often in the range of $\Gamma \gtrsim 4$, as shown in Fig. 4a. Øieroset et al. (2002) showed that these energetic particles may be the result of energization occurring during reconnection in the tail. They found that the hardest particle spectrum (i.e., the most energetic particles) was observed near the center of an ion diffusion region traversed by the Wind spacecraft in the deep (60 R_E) terrestrial magnetotail. This work was followed up by Cohen et al. (2021), who explored whether a similar result would be found for a set of six electron diffusion regions discovered by MMS. Comparing the results to a statistical dataset of 133

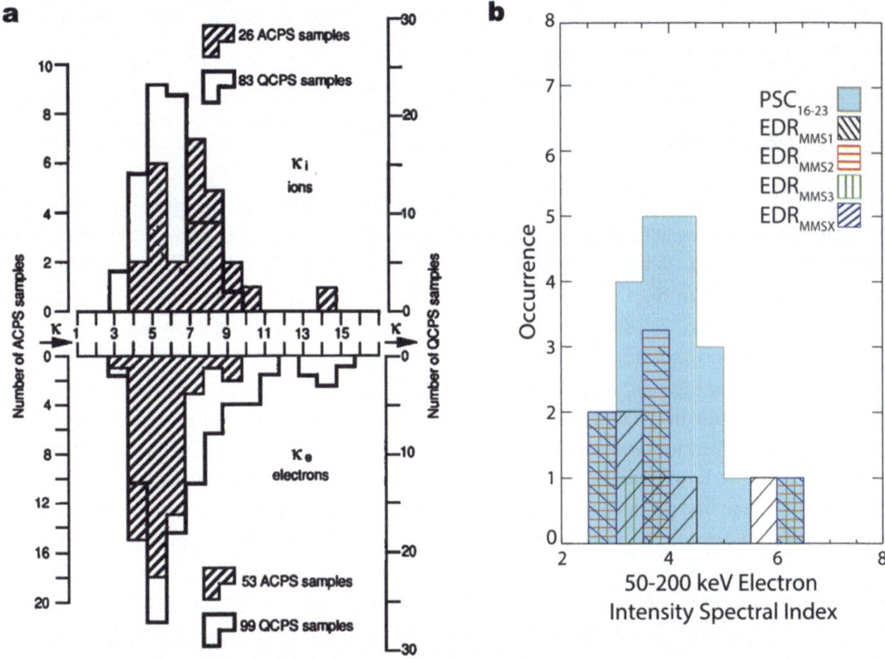

Fig. 4 Statistical analysis of the electron power-law index in the ISEE-era (left; Christon et al. 1991) and the MMS-era (right; Cohen et al. 2021). These analyses demonstrate that the non-thermal tail remains significant even during the quiet-time plasma sheet

quiet-time (i.e., AE* $<$ 300 nT and no fast ($|v_{\text{ion avg}}| \geq$ 100 km/s) flows) plasma sheet crossings (PSC), the authors found that the electron diffusion region (EDR) events did in fact have harder spectra (i.e., more energetic particles) (Fig. 4b). The result suggests that these energetic electrons are coming from a local source associated with active reconnection (e.g., Ergun et al. 2020b). In fact, an observational study reported significant heating within the EDR, followed by an appearance of the non-thermal tail in the immediate downstream of the EDR (e.g. Oka et al. 2016). MMS observations also reported a significantly enhanced flux of energetic electrons within the EDR, although the non-thermal, power-law tail was soft with the power-law index of $\sim$8 as measured in the phase space density (Li et al. 2022). Interestingly, Turner et al. (2021a) reported coherent gyrophase bunching of $>$ 50 keV electrons in the immediate downstream of the EDR and argued that it can be caused by the first-order Fermi acceleration Type B off of the outflowing exhaust structure, evidencing electron acceleration at the reconnection site and possibly also in the outflowing exhaust jets of the active reconnection.

Despite the possible importance of magnetic reconnection, it has also been reported that the non-thermal component is significant even during periods of low geomagnetic activity (AE $<$ 100 nT) (Christon et al. 1989; Cohen et al. 2021; Oka et al. 2022). This is also illustrated by the overlap of the PSC and EDR histograms in Fig. 4. Cohen et al. (2021) argued that such energetic particles may be sourced by remote down-tail reconnection sites or processes not directly related to reconnection at all. This is consistent with an earlier report of significant non-thermal tail during low geomagnetic activity (Christon et al. 1989). Similarly, Oka et al. (2022) examined the spatial variation across the reconnection region and reported that the non-thermal power-law tail can exist even outside the reconnection region (Hall region) where there is no significant plasma flows and turbulence. Therefore, the relationship between the production of energetic electrons and the geomagnetic activity remains unclear, let alone the importance of the EDR.

2.2 Fermi Vs Betatron

As reviewed in Sect. 1.2.1, the main acceleration mechanisms in the guiding-center approximation are Fermi acceleration, betatron acceleration, and the direct acceleration by the parallel electric field. While the parallel electric field might be important for heating (as separately reviewed in Sect. 2.3) or for acceleration to non-thermal energies in some cases (e.g. Zhou et al. 2016, 2018a), many studies argue that Fermi and betatron acceleration are predominantly important during magnetic reconnection.

In observational studies, a pitch angle anisotropy has been the key feature for diagnosing Fermi and betatron acceleration (e.g. Smets et al. 1999). Particles experiencing Fermi and betatron acceleration tend to exhibit parallel and perpendicular anisotropy, respectively. However, with the launch of MMS in 2015, electron data with the time resolution of $\sim$100 times higher than its predecessors became available. Such data sets, combined with the multi-spacecraft approach which is necessary to estimate the magnetic field curvature, have enabled us to evaluate each term in Eq. (1), providing a more direct diagnostics of the acceleration mechanism, i.e., Fermi acceleration, betatron acceleration, and the direct acceleration by the parallel electric field. Significant progress has been made with such analysis and will be reviewed below. It is to be noted that most of the discussion in this subsection is focused on electron acceleration, although there have been some studies of ion acceleration by simulations (e.g. Birn et al. 2015b; Ukhorskiy et al. 2017) and observation (e.g. Wang et al. 2019).

2.2.1 Outflows Near the X-Line

Early studies argued that, in the outflow region immediately downstream of the X-line, magnetic field magnitude increases and that electrons are accelerated by the gradient B and/or curvature drift (Hoshino et al. 2001; Imada et al. 2005, 2007). However, it was also argued that, above a few keV, the $\tilde{\kappa}$ value of electrons can approach ~ 1 where $\tilde{\kappa}^2$ is the ratio of the magnetic field curvature and the particle gyro-radius. In such a condition, a non-adiabatic behavior or scattering becomes important.

Wu et al. (2015) proposed that electrons are first pre-energized at the X-line, accelerated non-adiabatically in the pileup region in the immediate downstream region, and then further accelerated adiabatically in association with burst bulk flows (BBFs) in the outflow region. While energetic electron events tend to be rare tailward of the X-line in the tail, Chen et al. (2019) reported three cases of outflow jets on the tailward side and argued based on anisotropy that electrons were accelerated adiabatically by both Fermi and betatron effects. The observations were made on the tailward (or 'unconfined') side where the effect of the intrinsic, dipole magnetic field can be neglected, but the outflow speeds were increasing in time (or 'growing') leading to the compression or strengthening of the magnetic field.

More recently, *MMS* has enabled to study the acceleration mechanism with the guiding-center approximation described by Eq. (1). A case study of tailward outflows reported that the dominant mechanism, both on average and the peak values, was Fermi acceleration with a peak power density of about +200 pW/m^3 (Eriksson et al. 2020). During the most intense Fermi acceleration, the magnetic field curvature was comparable to the electron gyroradius (i.e., $\kappa \sim 1$), suggesting electrons were being scattered efficiently. Figure 5 shows the

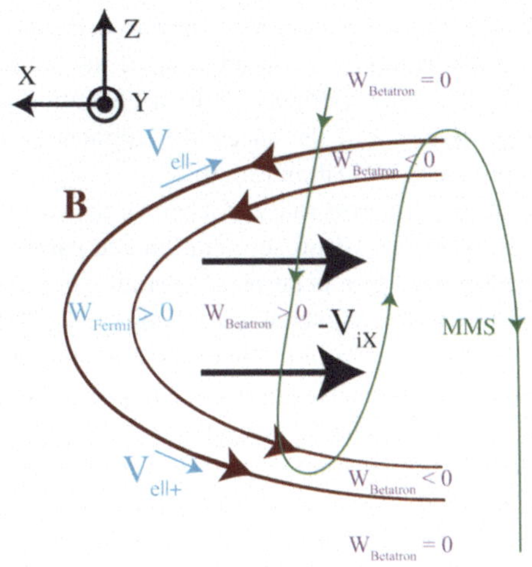

Fig. 5 A schematic illustration of the outflow region near the X-line, demonstrating expected Fermi and betatron acceleration signatures (Eriksson et al. 2020). W_{Fermi} and W_{betatron} represents the power-density of electron acceleration due to Fermi and betatron acceleration, respectively

schematic illustration of their interpretation. In the current sheet center, the power-density of electron acceleration due to Fermi acceleration, W_{Fermi}, and betatron acceleration, W_{betatron}, are positive because the magnetic field magnitude increases with the increasing distance from the X-line. At the edges of the current sheet, however, incoming electrons experience decreasing magnetic field and hence negative values of W_{betatron}. Interestingly, some of these findings (such as moderately non-adiabatic behaviors, energy loss at the edges, etc.) are consistent with earlier simulation results (Hoshino et al. 2001).

2.2.2 Flux Ropes

A magnetic flux rope is one of the key structures associated with magnetic reconnection. It is often referred to as a magnetic island especially in 2D theoretical pictures (e.g. Birn et al. 2012; Zank et al. 2014, and references therein). A distinction is typically made based on the absence or presence of a magnetic field component along the center of the island or rope structure. Many observations indicate that electrons are accelerated to non-thermal energies within the flux ropes both in the magnetotail (e.g. Chen et al. 2008, 2009; Retinò et al. 2008; Wang et al. 2010a,b; Huang et al. 2012; Sun et al. 2022; Wang et al. 2023) and in the magnetopause (e.g. Oieroset et al. 2011). Also, multi-island coalescence may be a key process for the energy conversion during reconnection and associated acceleration of particles (e.g. Oka et al. 2010b; le Roux et al. 2015; Teh et al. 2023, and references therein).

The standard theory for electron acceleration in flux ropes is the contracting island mechanism, whereby particles receive a Fermi-type energization kick at each end of an actively contracting magnetic island (e.g. Drake et al. 2006; Zank et al. 2014; Arnold et al. 2021) (but see also e.g. Stern 1979; Kliem 1994). The process requires an escape process in order to explain the observed spectral indices of energetic particles.

Observations also indicate the importance of Fermi acceleration in addition to betatron acceleration (e.g. Huang et al. 2012; Zhong et al. 2020; Jiang et al. 2021; Sun et al. 2022). Zhong et al. (2020) studied electron acceleration within ion-scale flux ropes by evaluating the equation for adiabatic electrons with the guiding center approximation (GCA, Eq. (1) in Sect. 1.2.1). Their analysis indicated that the lower energy (<10 keV), field-aligned electrons experienced predominantly Fermi acceleration in a contracting flux rope, while the higher energy (>10 keV) electrons with perpendicular anisotropy gained energy mainly from betatron acceleration. They argued that the dominance of betatron acceleration at high energies could be a consequence of the 3D nature of the flux rope. The field-aligned electrons that can experience Fermi acceleration would quickly escape along the axis of the flux rope. Because of the successful application of the GCA theory, the study was positively commented by Dahlin (2020).

In another case study of a pair of tailward traveling flux ropes, Sun et al. (2022) reported that, while the Fermi acceleration and parallel potential are strong near the X-line between the flux rope pair, betatron acceleration is strong on flux rope boundaries. For electron acceleration at the magnetopause, Wang et al. (2023) reported an interaction of two filamentary currents (FCs) within a flux rope and argued that the electrons were mainly accelerated by the betatron mechanism in the compressed region caused by the FC interaction.

However, electron acceleration in flux ropes might not always be adiabatic (e.g. Oka et al. 2010a; Fujimoto and Cao 2021; Sun et al. 2022; Wang et al. 2023), and the parallel electric fields might become important, particularly for small-scale, secondary flux ropes that form at and around the primary X-line with intensified current (e.g. Wang et al. 2017; Zhou et al. 2018a; Jiang et al. 2021). There can also be intense wave activities, turbulence, and current filaments inside flux ropes (e.g. Fu et al. 2017; Huang et al. 2019; Jiang et al. 2021; Sun et al. 2022; Wang et al. 2023) that can lead to stochastic acceleration. Recent 3D simulations have demonstrated that such turbulence and associated induced electric field can result in strong heating of electrons (Fujimoto and Cao 2021).

2.3 Parallel Electric Field

In the earlier years of magnetic reconnection studies, it was proposed that electrons are accelerated directly in the reconnection electric field along the magnetic X-lines (e.g. Litvinenko 1996, and references therein). However, recent studies of magnetic reconnection have revealed a new adiabatic picture in which the parallel electric field plays an important role, as reviewed in this subsection. Here, it is worth noting that, while the rate of energy gain is roughly proportional to the particle's energy for the cases of Fermi and betatron acceleration, the rate of energy gain scales only with the particle speed v for the case of direct acceleration by parallel electric field (Sect. 1.2). Nevertheless, the acceleration by parallel electric field can boost thermal particles by orders of magnitude in energy and hereby provide a preenergized seed populations subject to further Fermi and betatron energization (Egedal et al. 2015).

For many plasma physics problems, it is important to understand how rapidly thermal (and super-thermal) electrons travel along the magnetic lines. As an example, we may consider the July 11, 2017, reconnection event recorded at about $20 R_E$ into the Earth's magnetotail with a typical electron temperature of 1 keV (See Fig. 3, left column, in Sect. 1.3). It follows that the electron thermal speed ($v_{te} \simeq 20 \cdot 10^6$m/s) is about 400 times faster than the expected reconnection inflow speed $v_{in} \simeq v_A/10 \simeq 50 \cdot 10^3$m/s. Thus, during the course of a fluid element (say, initially $1 d_i \simeq 1 \cdot 10^6$m upstream of the reconnection site) traversing the reconnection region, a typical electron will travel a distance of about $(v_{te}/v_{in}) d_i \simeq 80 R_E$ (larger than the distance from Earth to the moon). This means that electrons, once energized, would escape instantly from the energization site and would not exhibit a localized, enhanced flux at and around the energization site, if there were no confinement or trapping. In reality, however, energetic electrons are observed in the localized region of magnetic reconnection (See, for example, Fig. 3 and other studies such as Øieroset et al. (2002) and Oka et al. (2022)). Therefore, we need a model for electron confinement or trapping to explain the observations.

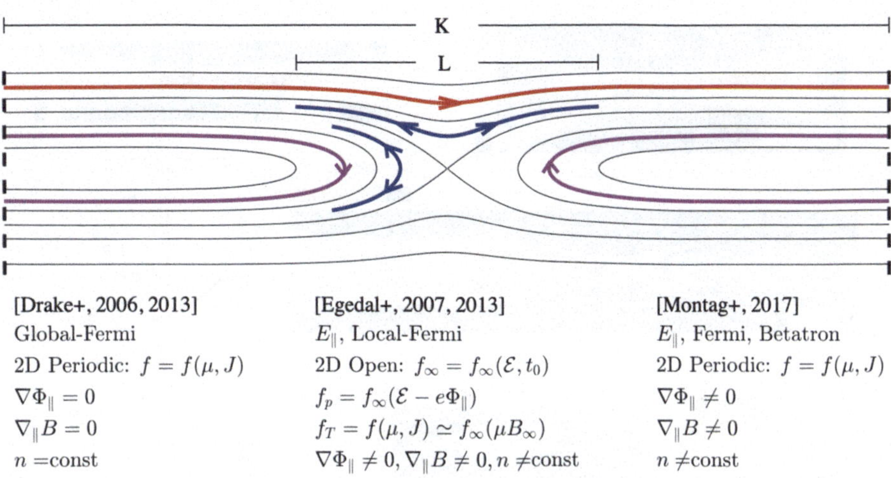

Fig. 6 Examples of trapped and passing electron orbits in 2D magnetic configurations, as well as key assumptions applied in adiabatic models for electron energization during magnetic reconnection. Adapted from Montag et al. (2017)

Due to the fast streaming of the electrons along the magnetic field lines, their parallel action, $J = \oint v_{\|} dl$, is typically a well conserved adiabatic invariant. As illustrated in Fig. 6, this J-invariance has been explored in a range of theoretical models for electron heating. The model of Drake et al. (2006, 2013) considers a 2D periodic and incomprehensible system and in essence applies Jeans' theorem (Jeans 1915) that the gross evolution of the electrons is governed by a double adiabatic assumption, $f = f(J, \mu)$ where μ is the magnetic moment, augmented with phenomenological pitch angle scattering. This Fermi heating model is only concerned with the large-scale energization of the electrons and ignores any variation in f along magnetic field lines.

Meanwhile, Egedal et al. (2008, 2013) assumes that the reconnection region is embedded in a large open system, where the plasma in the ambient regions provides fixed sources of electrons, f_∞. Electrons may here be characterized as either passing or trapped. The passing electrons instantaneously travel along the field lines with their total energy conserved, $U = \mathcal{E} - e\Phi_{\|}$, where $\Phi_{\|} = \int_x^\infty E_{\|} dl$ is the acceleration potential (Egedal et al. 2009). Meanwhile, the trapped electrons again follow Jeans' theorem, $f_T = f_\infty(J, \mu)$. Furthermore, with the imposed boundary conditions it can be shown that $f_T \simeq f_\infty(\mu B_\infty)$, and a relatively simple analytical form is obtained:

$$f = f_\infty(\mathcal{E} - e\Phi_{\|}) \text{ (passing)}, \quad f = f_\infty(\mu B_\infty) \text{ (trapped)}. \tag{7}$$

These types of distributions are common in measurements within reconnection regions and have been observed by multiple spacecraft missions including Wind, Cluster, THEMIS and MMS (Egedal et al. 2005, 2010; Oka et al. 2016; Eriksson et al. 2018; Wetherton et al. 2019, 2021; Wang et al. 2021). The global model by Drake et al. and the local model in Eq. (7) can be obtained as two separate limits of the more general framework recently developed by Montag et al. (2017). The model by Drake et al. follows directly by imposing the conditions of $n =$ constant and $\nabla_{\|} B = 0$, while Eq. (7) is recovered in the limit $K \gg L$, where as illustrated in Fig. 6, K is the size of the periodic domain and L is the typical length scale for electron trapping in the reconnection region.

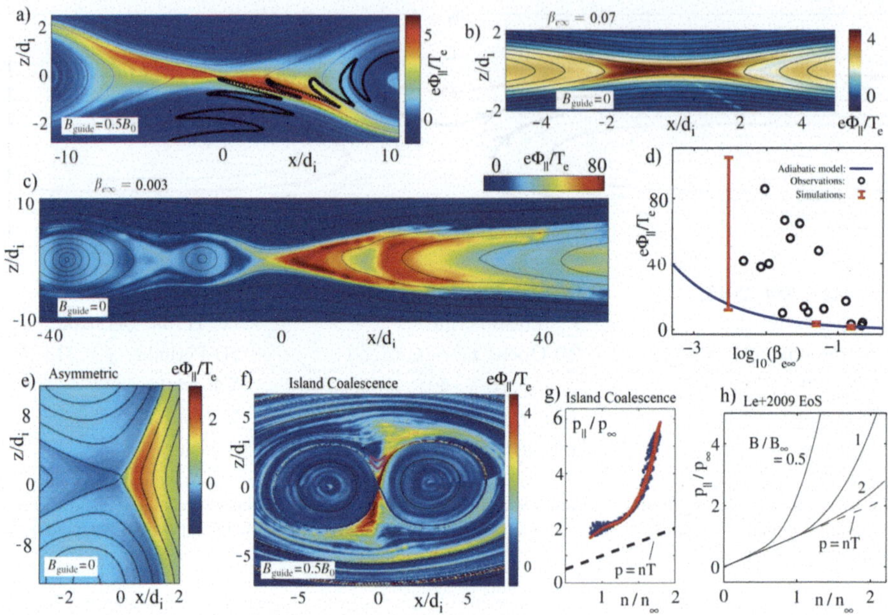

Fig. 7 a-c,e,f) Example profiles of $\Phi_\|$ observed in kinetic simulation under various reconnection scenarios. d) The blue line is the adiabatic prediction for $\Phi_\|$, while Cluster observations and kinetic simulations show strongly enhanced values of $\Phi_\|$ for low values of $\beta_{e\infty}$. g) $p_{e\|}$ as a function of n observed in island coalescence demonstrating parallel heating beyond the level predicted by the Lê2009 EoS (in h)). Collected and adapted from Le et al. (2012, 2016) and Egedal et al. (2013)

Observations suggest that the bulk electron heating for a range of reconnection scenarios is largely governed by Eq. (7). The trapped electrons have negligible heat-exchange with the ambient plasma, and when the majority of the thermal electrons are trapped the pressure components along and perpendicular to the magnetic field follow the CGL (Chew et al. 1956) scaling laws $p_\| \propto n^3/B^2$ and $p_\perp \propto nB$. This is also the asymptotic limit (at large n/B) of the equation of state derived directly from Eq. (7) by Le et al. (2009) (hereafter referred to as Lê2009 EoS).

Consistent with $E_\| \simeq -\nabla p_\|/(en)$, Egedal et al. (2013) found that $e\Phi_\|/T_{e\infty} \propto n^2/B^2$. This dependency is much stronger than the typical Boltzmann scaling of $e\Phi_\|/T_{e\infty} \propto \log(n/n_0)$ and within reconnection regions $\Phi_\|$ typically becomes large and is responsible for trapping and heating the majority of thermal electrons. Figure 7 shows profiles of $\Phi_\|$ recorded in a range of numerical simulations. A value of $\beta_{e\infty} = nT_{e\infty}/(B_\infty^2/2\mu_0) \simeq 0.1$ is often applicable to reconnection within Earth's magnetosphere, yielding the profiles of $\Phi_\|$ displayed in Fig. 7(a, b). Meanwhile, on occasions in the Earth's magnetotail when lobe plasma reaches a reconnection region, the normalized pressure can drop dramatically with $\beta_{e\infty} \ll 0.1$ (see Fig. 7(d)). From the principle of quasi-neutrality, it can be shown that the required parallel streaming of electrons then exceeds their thermal speed. The dynamics then enter a non-adiabatic regime with enhanced values of $e\Phi_\|/T_{e\infty} \gg 10$ over much extended spatial regions (see Fig. 7(c) as well as Egedal et al. (2012, 2015)), likely relevant to recent MMS observations (Ergun et al. 2022a). In Fig. 7(e), for asymmetric reconnection the largest values of $\Phi_\|$ and $p_\|/p_\perp$ are observed in the low-$\beta_{e\infty}$ inflow (Egedal et al. 2011; Burch et al. 2016).

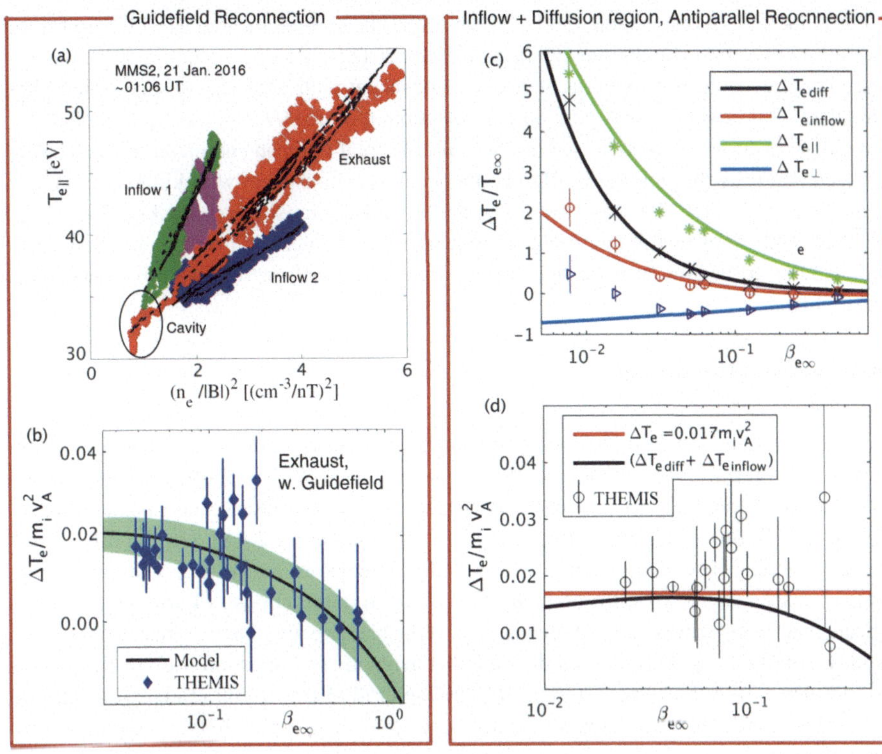

Fig. 8 a) MMS observations of $T_{e\parallel} \propto (n_e/B)^2$ for the event studied in Eastwood et al. (2018), Wetherton et al. (2021). b) Green band indicates the electron energization in fluid simulations (Ohia et al. 2015) for guide-field reconnection applying the Lê2009 EoS, in agreement with observations by THEMIS (Phan et al. 2013). c) Analytical predictions (based on the Lê2009 EoS) for the electron heating within the inflow and EDR of anti-parallel reconnection, validated by kinetic simulations results (Le et al. 2016). d) The black line show the total electron heating (sum of red and black lines in c)), compared to results from THEMIS (Phan et al. 2013)

During island coalescence (in Fig. 7(f) with a guide magnetic field), the effect of $\Phi_\parallel$ is also noticeable, and for this case the $p_\parallel$ values in Fig. 7(g) are enhanced by Fermi acceleration of the contracting island (Drake et al. 2006, 2013) above the levels predicted by Lê2009 EoS outlined in Fig. 7(h). The red line in Fig. 7(g), represents the predictions by the Lê2009 EoS. It corresponds to the curve of $B/B_\infty = 1$ in Fig. 7(h) but enhanced by a factor of 2 due to an inclusion of Fermi heating in the model. Again, these two effects are both captured by the formalism in Montag et al. (2017).

The Lê2009 EoS has been verified directly by MMS during exhaust crossings of guide field reconnection, both close to ($\sim 10d_i$, Montag et al. 2017) and far ($\sim 100d_i$, Wetherton et al. 2021) from the X-line. For example, the data in Fig. 8(a) is from the event far from the X-line first studied in Eastwood et al. (2018), where $T_{e\parallel}$ measured in the two inflows (green and blue) as well as in the reconnection exhaust (red) is observed to follow the aforementioned CGL limit of the Lê2009 EOS where $T_{e\parallel} \propto (n/B)^2$. Slightly asymmetric inflow conditions set different values of proportionality, and the exhaust comprised of a mixture of the two populations falls in the middle. The black lines represent the Lê2009 EoS predictions (which also accurately account for the $T_{e\perp}$ observations, not shown here).

For guide-field reconnection, the Lê2009 EoS has been implemented as a closure for the electrons in two-fluid simulations (Ohia et al. 2012, 2015) and as shown in Fig. 8(b), the predicted heating levels as a function of $\beta_{e\infty}$ are consistent with THEMIS observations (Phan et al. 2013). Likewise, for anti-parallel reconnection, the Lê2009 EoS has been applied (Le et al. 2016) to derive theoretical scaling laws for the total electron energization as electrons approach and pass through the EDR. The theory is also consistent with kinetic simulation results as well as THEMIS observations in the reconnection exhausts (see Fig. 8(c, d)). In Fig. 8(d), compared to the empirical scaling by the red line, the black theoretical curve predicts reduced heating at large $\beta_{e\infty}$. Both curves fall mostly within the error bars of the measurements.

2.4 Waves and Turbulence

Magnetic reconnection in pre-existing turbulence is often referred to as 'turbulent reconnection' (e.g. Lazarian and Vishniac 1999; Lazarian et al. 2015). Earth's magnetosheath is such an environment, where turbulence appears to drive (smaller-scale) magnetic reconnection (e.g. Retinò et al. 2007; Ergun et al. 2016; Phan et al. 2018). On the other hand, magnetic reconnection can generate waves and turbulence in return (e.g. Daughton et al. 2011; Leonardis et al. 2013; Ergun et al. 2016). In-situ observations in Earth's magnetotail indicate that strong waves and turbulence exist in the reconnecting plasma sheet even though the upstream, lobe region is quiet, indicating that magnetic reconnection itself excites waves and turbulence (e.g. Eastwood et al. 2009; Osman et al. 2015; Ergun et al. 2018; Richard et al. 2023b) (See also Cattell and Mozer 1986; Hoshino et al. 1994). In this section, we provide a brief review on particle energization associated with waves and turbulence observed near the X-line, including outflow jets. Similar waves and turbulence are also found at large scales near the flow-braking region, as reviewed in Sect. 3.

2.4.1 Ion Acceleration and Turbulence During Magnetic Reconnection

As suggested by MMS observations (e.g. Ergun et al. 2018), the physical process of ion and electron acceleration can differ. Because ions have larger scale sizes (skin depth and gyroradii), they are the first in line to absorb the magnetic energy. In a region of turbulence, E spectra (Fig. 9) have high enough energy density to explain the high ion energization rates though cyclotron-resonance (Chang et al. 1986; Ergun et al. 2020a,b). However, cyclotron resonance alone does not explain an accelerated tail or other details in the ion distributions (Fig. 9). Instead, a stochastic process needs to be considered, and it requires waves and turbulence that span a wide frequency range.

In principle, ions can undergo Speiser-like orbits at the neutral sheet during magnetic reconnection. Nevertheless, when strong turbulence coexists within the neutral sheet, unmagnetized ions, which do not necessary follow Speiser-like orbits, are more likely to gain energy from large impulses in the turbulent electric fields. It is worth noting that ions with initially high energies not only absorb more powerful, larger-scale electromagnetic energy, but also have a higher probability to be unmagnetized and pass through the neutral sheet. As a result, energization favors ions with initially higher energies and an accelerated tail in the ion distributions could emerge. The kinetic process of ion energization in turbulence is an active, ongoing study in which MMS observations have given good insight (e.g. Richard et al. 2023b).

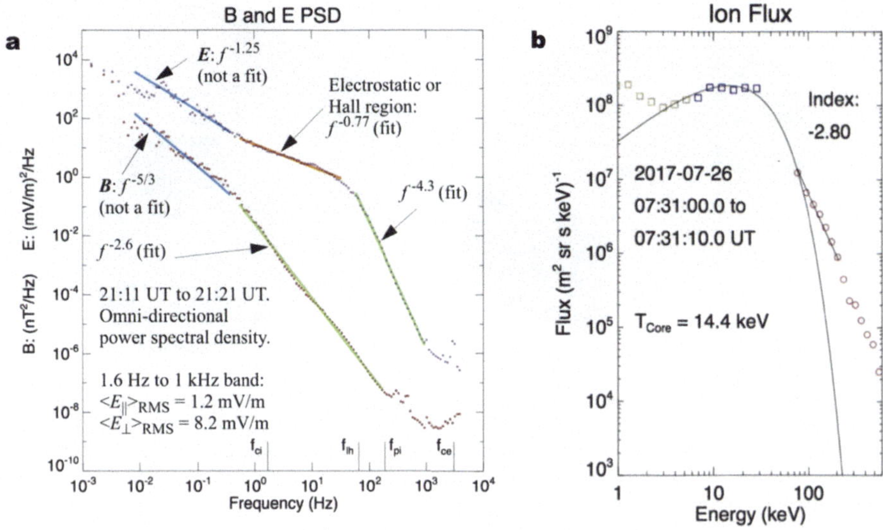

Fig. 9 (a) An example of B and E spectra in the magnetotail. The B spectrum has classic properties of turbulence, with a Kolmogorov-like inertial region ($-5/3$ index) and a sharp break in the region of ion dissipation. The E spectra has a shallower index in the inertial region and has an electrostatic build-up at higher frequencies, before a sharp drop. The electrostatic energy density is linked to electron acceleration. Adapted from Ergun et al. (2022a). (b) An example of energized and accelerated ions as measured in a region of strong turbulence. The core of the distribution is heated from $\sim$4 keV (outside of the turbulent region) to $\sim$16 keV. A high-energy tail has ions greater than 100 keV. Adapted from Ergun et al. (2020b)

2.4.2 Electron Acceleration and Turbulence During Magnetic Reconnection

Recent observations (e.g. Ergun et al. 2018; Li et al. 2022; Oka et al. 2022) and simulations (e.g. Lapenta et al. 2020; Zhang et al. 2021) have provided convincing evidence that turbulence plays a significant role in accelerating electrons to non-thermal energies in the magnetotail (See Sect. 4.2 for further discussion). The observations are so detailed that the specific process of interaction between the turbulence electric field and electrons can be discussed, as summarized below (Ergun et al. 2018, 2020a,b, 2022a).

Perpendicular electron energization requires circumvention of the first adiabatic invariant ($\mu = p_{\perp}^2/2\gamma m_0 B$). Contrary to the case with ions, there is little power at or above the electron cyclotron frequency (Fig. 9) and $E_{\|}$ is small (written on plot) which suggests that electron energization should be negligible. It is found, however, that energization can occur if the correlation length scale (d_{corr}) in the E turbulence is sufficiently small (Usanova and Ergun 2022; Ergun et al. 2022a). If an electron's parallel velocity is high enough that $d_{corr}/v_{\|} < 1/f_{ce}$, it experiences changes in E in less than $1/f_{ce}$ in its frame and therefore can be energized perpendicular to **B**. Furthermore, if an electron's gyroradius is such that $\rho_e \geq d_{corr}$, it can experience enhanced parallel energization, perpendicular energization, and pitch-angle scattering.

Figure 10 illustrates the underlying process of electron acceleration by turbulent and electrostatic **E**. As it gyrates, a low-energy electron (2 keV in the figure) experiences a nearly constant E whereas a higher-energy electron (20 keV in the figure) transits regions of changing E during its gyration. Even though **E** is primarily electrostatic, the particle does not necessarily return to the same location in the perpendicular plane or in the same

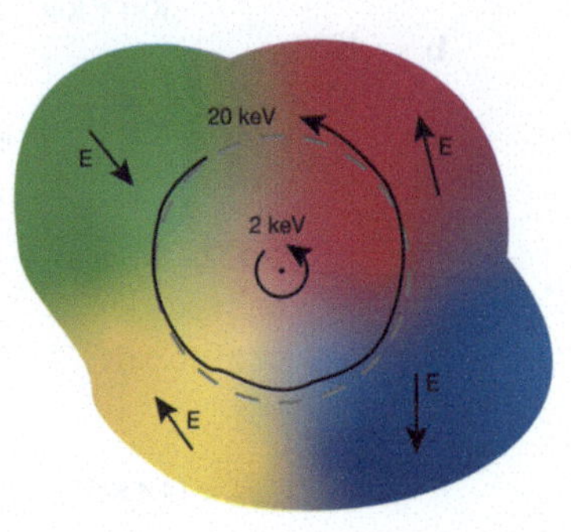

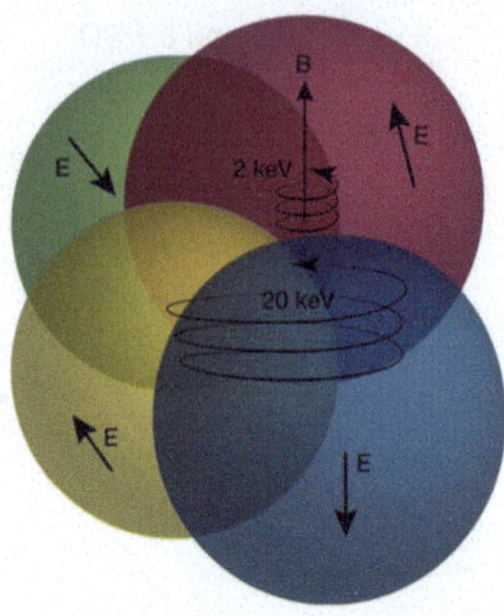

Fig. 10 A drawing of electron orbits in an uncorrelated, electrostatic **E**, illustrating how turbulent acceleration favors higher-energy electrons. (Left) A view of the orbital plane. The higher-energy (20 keV) electron's orbit transits several uncorrelated regions of E (including $E_{||}$) as it gyrates and, therefore, can gain or lose energy. A lower-energy electron (2 keV) sees very little change in E over an orbit. (b) A 3D view of an electron's helical path along **B**. Adapted from Ergun et al. (2022b)

location along B and therefore can experience energy change. A finite $\nabla \times \mathbf{E}$ can enhance acceleration.

The velocity dependence is such that, once again, electrons with initially higher energies are favorably energized, which results in acceleration. Interestingly, the electron energization process can be greatly enhanced by trapping in magnetic depletion (Ergun et al. 2020a,b). Electrons can transit a turbulent region in the magnetotail in a matter of seconds, which greatly limits its energization. If trapped, the electron experiences energization for a significantly longer time, leading to much higher energization. This kinetic picture of ion an electron acceleration suggests that further study is needed.

2.4.3 Electron Energization Associated with Waves

It is instructive to discuss more specifics of what constitutes turbulence. Previous observations have shown that waves are excited over a broad range of frequency during magnetic reconnection and that they can be identified as lower hybrid waves, Langmuir waves, electrostatic solitary waves, and whistler waves (e.g. Khotyaintsev et al. 2019, and references therein). Perpendicular anisotropies in the region behind a dipolarization front (Sect. 3) could act as a source of whistler waves (e.g. Le Contel et al. 2009; Khotyaintsev et al. 2011; Viberg et al. 2014; Breuillard et al. 2016) or electron-cyclotron waves (Zhou et al. 2009).

Many studies have shown that a specific type of waves can play an important role in particle heating. For example, Debye-scale electrostatic waves and structures have been detected and discussed in the context of electron heating (or energization below ~ 1 keV) near the X-line both at the magnetopause (Mozer et al. 2016; Khotyaintsev et al. 2020) and the magnetotail (Norgren et al. 2020). Also, an association between whistler waves and intense bursts of energetic (10s to a few 100 keV) electrons near the reconnection separatrix has been reported in the context of magnetopause reconnection (Jaynes et al. 2016; Fu et al. 2019a), followed by a statistical study (Chepuri et al. 2022). Fu et al. (2019a) analyzed the energy spectrum carefully and showed that such energetic electrons are not contaminated by the magnetospheric population and yet indeed non-thermal.

3 Particle Acceleration at Large Scales

3.1 Overview

While magnetic reconnection ultimately occurs at 'microscopic', electron-kinetic scales within a plasma, reconnection results in macroscopic to global scale reconfiguration of the magnetic field topology and dynamics within a plasma. In the inner magnetotail, this involves inductive electric fields that are responsible for particle acceleration far removed from the actual reconnection site itself. The earthward exhaust region is characterized by transient or more persistent increases in the northward magnetic field, called dipolarizations. Transient events are typically associated with rapid flow bursts, which come to rest and/or get diverted azimuthally in a 'flow-braking region' near or inside of about 10 R_E distance downtail. This is not a fixed distance, however. The fact that dispersionless energetic particle flux increases at tens to hundreds of keV (denoted 'injections') are frequently observed at geosynchronous orbit (e.g. Lezniak and Winckler 1970; Baker et al. 1979), or even inside, is an indication that impulsive electric fields can often penetrate more deeply than the fast flows.

The transient dipolarization events and their associated flows are related to motional electric fields, which may exceed the electric field defining the rate of reconnection. These electric field enhancements are sometimes referred to as 'rapid flux transport' (RFT) events (e.g. Schödel et al. 2001). The dipolarization events typically include sharp increases of the northward magnetic field B_z, called 'dipolarization fronts' (DFs; e.g. Nakamura et al. 2002; Runov et al. 2009; Sitnov et al. 2009), followed by an interval of increased B_z, denoted 'dipolarizing flux bundle' (DFB; Liu et al. 2013) or 'Flux Pileup Region' (FPR; Khotyaintsev et al. 2011). Further details on the terminology and properties of particle acceleration are given in recent reviews (Sitnov et al. 2019; Fu et al. 2020; Birn et al. 2021b).

Transient DFs typically separate a colder denser population in the pre-existing plasma sheet from the hotter, more tenuous population in the DFB, presumably ejected out from the reconnecting X-line (e.g. Runov et al. 2011, 2015). Similar structures are detected for tailward flows as well (with $B_z < 0$), and thus a more generalized term 'reconnection front' is also used to combine both earthward and tailward cases (e.g. Angelopoulos et al. 2013).

Dipolarizations in the flow-braking region tend to show more persistent increases in B_z (e.g. Runov et al. 2015) as well as low or decreasing earthward flow speeds, which may include tailward bounces and oscillations (e.g. Panov et al. 2010; Liu et al. 2017b). They are commonly accompanied by strong electric fields, which may exceed the motional electric field of the transient events by one or more orders of magnitude up to about 100 mV/m (e.g. Ergun et al. 2015, 2022b). In contrast to the RFT electric fields, which are typically duskward, the high-frequency fields also include significant field-aligned components.

Numerous investigations have confirmed that the inductive electric fields associated with dipolarization events are the eminent cause of energetic particle flux increases, including injections observed at geosynchronous orbit. Their properties are briefly reviewed in Sects. 3.2 – 3.4. The effects of the fluctuating strong electric fields in the flow-braking region are not as well documented. They presumably arise from the turbulence associated with the flow-braking and diversion of the earthward flow and may provide a mechanism for particle energization, separate from, or in addition to, the effects of the transient fields, and contribute a source population for the outer radiation belt (Ergun et al. 2022a), as well as a mechanism for energy dissipation (e.g. Stawarz et al. 2015; Ergun et al. 2018).

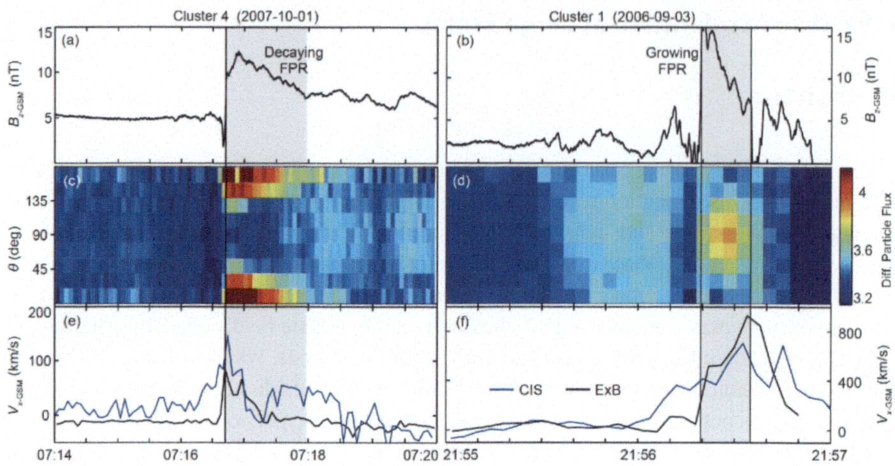

Fig. 11 Two different observations of dipolarization fronts, demonstrating features consistent with predominantly Fermi (left column) and betatron (right column) acceleration (Adapted from Fu et al. 2011). When the outflow speed is decreasing (Panel (e)), the Flux Pileup Region (FPR) is considered decaying (Panel (a)) and the energetic (>40 keV) electrons exhibit parallel anisotropy (Panel (c)). In contrast, when the outflow speed is increasing (Panel (f)), the FPR is considered growing (Panel (b)) and the energetic electrons exhibit perpendicular anisotropy (Panel (d))

3.2 Anisotropies in Dipolarization Events

Many observations indicate that particles can be accelerated to non-thermal energies at and around the transient dipolarization events (e.g. Apatenkov et al. 2007; Runov et al. 2009; Fu et al. 2011, 2013; Ashour-Abdalla et al. 2015; Liu et al. 2017b), with anisotropies of the energetic particle distributions providing major clues of the underlying mechanism. Figure 11 shows two example observations by Cluster reported by Fu et al. (2011). One event was obtained when the bulk flow speed was decreasing and thus the main magnetic structure (denoted FPR, in this case) was considered decaying (left column). The other event was obtained when the bulk flow speed was increasing, and thus the FPR was considered growing (right column). The energetic (> 40 keV) electrons showed parallel anisotropy (indicating Fermi acceleration) and perpendicular anisotropy (indicating betatron acceleration) in the decaying and growing cases, respectively. Based on a statistical analysis of pitch-angle anisotropy, Wu et al. (2013) consistently argued that, because outflow jets have higher speeds in the mid-tail region ($X \lesssim -15 R_E$), there could be more efficient compression of the local magnetic field, leading to more frequent formation of the perpendicular anisotropy by betatron acceleration in the mid-tail region.

It is important to distinguish full particle acceleration, which involves the history of a particle motion, from the local acceleration rate. Estimating the latter, several investigations concluded that locally betatron acceleration was dominant at the dipolarization front (DF) proper (e.g. Xu et al. 2018; Fu et al. 2019b; Ma et al. 2020) and that the Fermi acceleration would be more effective at a larger spatial scale. Using MMS data, Ma et al. (2020) showed that betatron acceleration rate dominates at many dipolarization fronts in the magnetotail in the $X < -10$ R_E range. Such a conclusion is consistent with the earlier, global-scale picture in which electrons are expected to experience predominantly Fermi acceleration in the stretched magnetic field in the magnetotail but undergo betatron acceleration as the magnetic field increases (e.g. Smets et al. 1999).

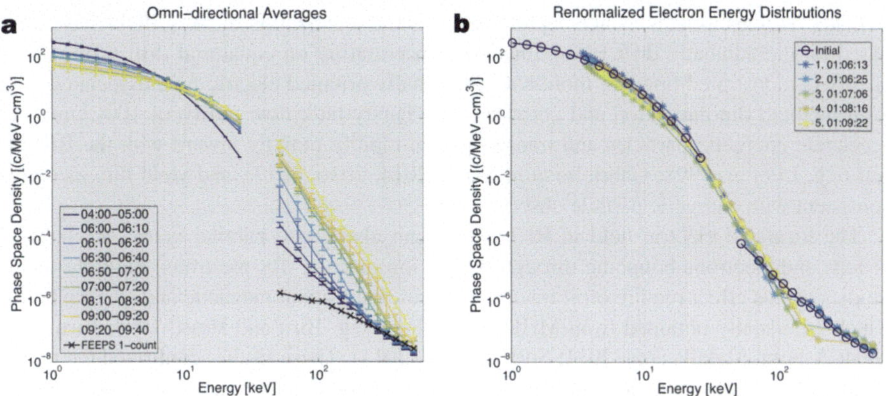

Fig. 12 Electron energy spectra obtained by MMS, demonstrating the importance of the betatron process. (a) Electron energy spectra observed during a series of dipolarization events. The times are listed in minutes and seconds (mm:ss); all of them have the same hour, 01:00 UT. (b) The same energy spectra but renormalized using the simple model that assumes betatron acceleration. Adapted from Turner et al. (2016)

Turner et al. (2016) also used observations from NASA's MMS mission to demonstrate how electron acceleration associated with a dipolarization structures and BBFs in the magnetotail were energy-dependent but consistent with betatron acceleration (Fig. 12). Malykhin et al. (2018) examined 13 dipolarization events using Cluster data and concluded that the electron acceleration up to 90 keV was consistent with betatron acceleration. Vaivads et al. (2021) used Cluster data in the magnetic structures (flux rope and dipolarization) of an Earthward reconnection jet, and found that in the dipolarization structure, electron acceleration was generally consistent with betatron acceleration, while within the flux rope, electron acceleration was more consistent with Fermi acceleration. While most conclusions are from single point measurements, Nakamura et al. (2021) used a constellation between MMS and Cluster satellites to infer consistency with adiabatic acceleration of electrons trapped within a dipolarization structure.

Electron anisotropies may vary not only with distance from the Earth or from the reconnection site but also with respect to the distance from the neutral sheet ($B_x \sim 0$). Runov et al. (2013) reported pancake type distributions (90° peaked) near the neutral sheet and mostly cigar type (0° and 180° peaked) distributions away from the neutral sheet, consistent with a predominance of betatron acceleration of ~90° particles close to the neutral sheet and Fermi acceleration for field-aligned electrons reaching higher latitudes.

Ions can also be accelerated in association with BBFs and dipolarization events, as studied by recent MMS observations (e.g. Bingham et al. 2020; Richard et al. 2023a) and simulations (e.g. Parkhomenko et al. 2019; Birn et al. 2015b). Because the ion gyro-radii are relatively large, they do not conserve the adiabatic moment, except in some average sense, and often behave non-adiabatically. Ion acceleration is further discussed in Sect. 3.3.

3.3 Acceleration Mechanisms

3.3.1 Electrons

The spatially and temporally localized cross-tail electric field associated with earthward propagating dipolarization fronts can result in trapping, earthward transport, and rapid acceleration of energetic particles, leading to the betatron effect from drift toward increasing

B-fields. Various models, which capture the essential localization of the E-field, have been based on the adiabatic drift approximation, concentrating on equatorial drift orbits. They clearly demonstrated how the motional, azimuthally oriented electric field associated with a magnetotail dipolarization and corresponding bursty bulk flow (BBF) of 100s km/s can accelerate energetic particles and transport them rapidly radially inward with the BBF itself (e.g. Li et al. 1998; Gabrielse et al. 2012, 2014, 2016, 2017), and yield flux increases consistent with energetic particle observations.

The localized electric field in RFT events can also cause parallel Fermi acceleration of ions and electrons bouncing through this region once or (for electrons) multiple times. Studies of this effect require orbit tracing in three-dimensional magnetic and electric fields, which are usually obtained from MHD simulations (e.g. Birn and Hesse 1994; Birn et al. 2004; Ashour-Abdalla et al. 2011; Sorathia et al. 2017). These studies confirmed the mechanism of temporal magnetic trapping within the magnetic field structures of DFBs, not only for electrons but also for ions (Birn et al. 2015b; Ukhorskiy et al. 2017, 2018), and showed the rapid acceleration via betatron (perpendicular to the B-field) and/or Fermi (parallel to B) effects (Sect. 1.2.1). They demonstrated not only parallel and perpendicular anisotropies of energetic electron distributions, but also so-called 'rolling pin' distributions (Liu et al. 2017a) with peaks at 0°, 90°, and 180° pitch angles (Runov et al. 2013; Birn et al. 2014, 2022), depending on energy, time and location.

Here, it is worth emphasizing again that even electron motion is not necessarily always adiabatic, especially at and around the X-line, in strongly curved low-B fields, or in regions of strong waves and turbulence, e.g. near the reconnection site. In such cases, the parallel electric field carried by whistler waves (Sect. 2.4) or kinetic Alfvén waves (e.g. Guo et al. 2017) might be important in addition to Fermi and betatron acceleration.

The fate of DFBs and associated energetic particles has also been investigated within the Rice-Convection-Model (RCM-E; Toffoletto et al. 2003), covering the energy-dependent drift of depleted magnetic flux tubes (also denoted 'bubbles') within a quasi-static inner magnetosphere model (Yang et al. 2013, 2015). In this regard, accelerated electrons were demonstrated to be important as a likely seed population of the Earth's radiation belt. Sorathia et al. (2018) conducted test-particle simulations of electrons in high-resolution, dynamic MHD fields to show how energetic electron injections from the magnetotail likely contribute a significant and possibly even dominant source of outer radiation belt electrons in the 100s of keV range in the inner magnetosphere. Turner et al. (2021b) conducted a phase space density analysis using a combination of Van Allen Probes in the outer radiation belt and MMS in the magnetotail plasma sheet to demonstrate also that relativistic electron acceleration in the plasma sheet can result in sufficient intensities to serve as a direct source for outer radiation belt electrons. Here, it is worth emphasizing that the intensities in the magnetotail can get up to radiation belt levels yet the residence time of those electrons in the tail is only a few minutes, in contrast to the several days residence times in the outer radiation belt.

3.3.2 Ions

Acceleration of ions in dipolarization events can be similar to that of electrons. Details are summarized in recent reviews by Sitnov et al. (2019) and Birn et al. (2021b) with references therein. Simulations by Birn et al. (2015a) showed how the acceleration of protons in the central plasma sheet (CPS) is generally consistent with the betatron effect (with an average conservation of the first adiabatic invariant in the presence of an increase in magnetic field strength). This is consistent with conclusions of Ukhorskiy et al. (2017, 2018), which were based on test particle tracing in high-resolution global MHD simulations. The simulations also demonstrated parallel acceleration (similar to Fermi acceleration of type B) by single (or, in rarer cases, multiple) encounters of a dipolarization front. In contrast to electrons, a single encounter of, or reflection at, a dipolarization front may result in observable, albeit moderate-energy proton beams or precursor populations preceding a DF (e.g. Zhou et al. 2010, 2011; Birn et al. 2015b). Presumably, a similar process can also happen at reconnection fronts on the tailward/anti-earthward side of magnetotail reconnection. Due to the mass dependence of the gyroradius that characterizes the encounter or reflection, this energization is even more effective for heavier ions, such as oxygen. As the energy gain essentially results from picking up the speed of the moving structure, it was also likened to a 'pick-up' process (Delcourt and Sauvaud 1994; Eastwood et al. 2015; Bingham et al. 2021; Birn et al. 2021a).

The simulations have yielded characteristics of ion distributions, dominated by protons, that are consistent with observations right after passage of a DF. At low distance from the neutral sheet, in the central plasma sheet, distributions show perpendicular anisotropy (Runov et al. 2015, 2017; Birn et al. 2017c; Zhou et al. 2018b), consistent with the betatron effect, which may be accompanied by lower intensity, lower energy field-aligned counter-streaming beams. At larger distance from the neutral sheet, close to the plasma sheet boundary, the distributions consist of crescent-shaped earthward field-aligned beams (e.g. Zhou et al. 2012). At the distance slightly away from the plasma sheet boundary layer (PSBL) and closer to the neutral sheet, such crescent-shaped earthward beam can be accompanied by tailward beams, which apparently result from mirroring closer to Earth. In such a region, sometimes multiple earthward and tailward beams are observed, which may be considered the counterparts of the field-aligned electron populations, however, involving only few bounces (Birn et al. 2017b). It is noteworthy that crescent-shaped earthward ion beams (including their tailward streaming counterparts) can also result from reconnection deeper in the tail (Andrews et al. 1981; Forbes et al. 1981; Williams 1981).

At higher energies, or for heavier ions, the gyroradius becomes comparable to, or larger than the size of the dipolarizing acceleration region, and the ions may encounter this region exhibiting Speiser-type orbits or even traverse the acceleration region of the enhanced electric field in a demagnetized fashion (Birn et al. 2021a; Richard et al. 2023a). The energy gain is essentially given by

$$\Delta W = q \int E_y dy \tag{8}$$

where q and E_y is the particle charge and the enhanced electric field (Birn et al. 2021a). This provides an upper limit to the possible acceleration of a given species, which is higher for multiply-charged ions. In agreement with that conclusion, the E/q dependence of particle fluxes in MMS observations of energetic particle events associated with fast flows (Bingham et al. 2020, 2021) indicated that He^{++} and O^{6+} of solar wind origin dominated the particle fluxes at highest energies (> 400 keV). The non-adiabatic acceleration effects also lead to non-gyrotropic, phase bunched, velocity distributions of heavy ions (Delcourt et al. 1997; Birn et al. 2021a).

3.4 Sources and Seeding

Observations do not give direct information about the sources of the accelerated particles. The fact that transient DFs typically separate a hotter, more tenuous population inside a DFB from the colder, denser plasma ahead of it indicates that the pre-DF population is not

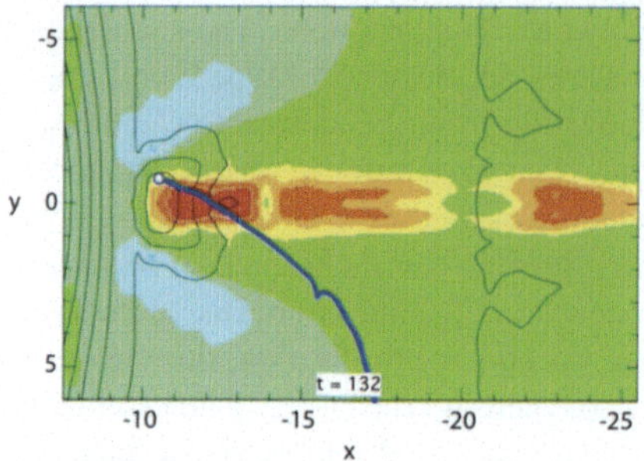

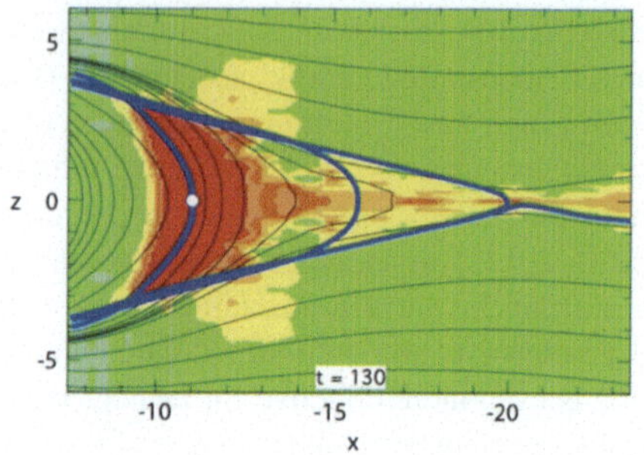

Fig. 13 Simulation results demonstrating two main sources of accelerated electrons: (a) near-equatorial drift from the dusk flank which leads to the energization by the localized motional electric fields and (b) a bounce orbit that originates from the reconnection region and leads to Fermi and betatron acceleration. Adapted from Birn et al. (2014)

the source of the energized population inside the DFB. More definite conclusions about the sources come from modeling, particularly from particle tracing in fields modeling the inward propagation of DFBs. Through backward tracing in dynamic MHD fields, Birn et al. (2012, 2014) demonstrated how particles are seeded onto the reconnected field lines inside a DFB, thus gaining access to the acceleration processes, via two mechanisms: i) local cross-tail particle drifts in the plasma sheet configuration, and ii) direct entry enabled by remote reconnection of field lines (Fig. 13). The entry mechanisms are energy-dependent: at low energies, charged particles are closely tied to the field lines that undergo reconnection before participating in the inner tail collapse, whereas at higher energies, cross-tail drifts or even non-adiabatic cross-tail motions become more important and particles can enter the acceleration region from the flanks earthward of the reconnection site.

Turner et al. (2016) examined MMS observations of a series of dipolarizations associated with magnetotail reconnection and found that for electrons with energy >10 keV, extending into the relativistic range, the observed acceleration was largely consistent with betatron acceleration, and one important consequence of those observational results was that the source of electrons in the ambient, background plasma sheet must have been relatively uniform over a large portion of the magnetotail surrounding the MMS spacecraft. The upper energy limitation of electron flux increases found by Turner et al. (2016) and earlier by Birn et al. (1997) is consistent with the change of particle motion and source regions at high energies mentioned above.

Using again MHD/test particle simulations, Birn et al. (2022) further demonstrated energy and space dependence of source regions of accelerated electrons. Consistent with earlier conclusions, they explained the drop in fluxes observed at energies <10 keV (consistent also with the results of Turner et al. (2016)) as being related to the drop in density from the seed populations in the plasma sheet boundary layer (PSBL) and lobes, despite the fact that these particles were also adiabatically accelerated.

Figure 14, modified after Fig. 4 of Birn et al. (2022), illustrates some important conclusions from modeling electron pitch angle distributions (PADs) right after the passage of a DF. The MHD configuration is indicated in the top panel (a).

1. Panels b and e demonstrate characteristic anisotropies of cigar-type (field-aligned) and pancake-type (perpendicular) away from, and close to, the neutral sheet, respectively; the two locations are indicated by the crosses in panel a. This result agrees with observations by Runov et al. (2013).

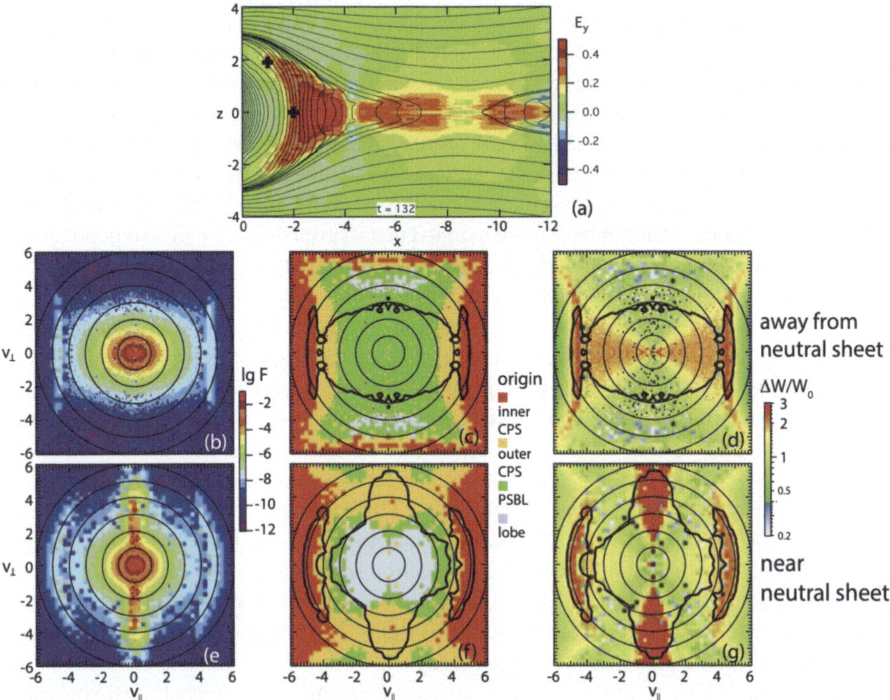

Fig. 14 Electron pitch angle distributions (PADs) obtained from a modeling approach (modified after Birn et al. 2022). The top panel shows a snapshot of the cross-tail electric field of the underlying MHD simulation, obtained right after the arrival of a dipolarization front; crosses indicate two locations where PADs were obtained (panels b-g). Panels b and e show the PADs at the two locations. Panels c and f indicate the origins of the particles contributing to the PADs, and panels d and g show the relative energy gain along each particle trajectory. Based on the chosen parameters, the velocity unit corresponds to $(m_p/m_e)^{1/2}$ 1000 km/s = 42,850 km/s, or an energy of 5.2 keV

2. Panels c and f show the origins of the particles contributing to these PADs, demonstrating that they are composed of different sources: At the highest energies particles originate from the inner CPS, as illustrated in Fig. 13a, whereas at lower energies the outer CPS, the PSBL, and the lobes contribute, as shown in Fig. 13b.
3. Panels d and g show the relative energy gain along the phase space trajectory, represented by the ratio between the final energy and the energy at the source location. These panels illustrate the effects of 'heating', increasing particle energies by a similar factor over a wide energy range, versus the acceleration of particles in a limited energy (and pitch angle) range. The distribution away from the neutral sheet in panel d shows the Fermi 'heating' at pitch angles around 0 and 180 degrees, whereas the distribution near the neutral sheet in panel g shows the betatron heating near 90° pitch angles. Both panels show an accelerated field-aligned population at $v \sim 4$ - 5 (corresponding to ~ 80 - 130 keV for the chosen units), although in panels b, d this is distinct from the 'heated' population mainly by the source locations in the inner CPS, where densities are higher.
4. The three peaks near 0°, 90°, and 180° in panels e and g also illustrate the formation of the 'rolling-pin' distribution, documented observationally (e.g. Runov et al. 2013; Liu et al. 2017a). It is a combination of dominantly parallel ('cigar'-shaped) and perpendicular ('pancake'-shaped) distributions.

The model particle tracing provides information on the immediate source regions, such as plasma sheet vs. lobes, which cannot easily be inferred from observations. Ultimately, particles originate from two sources, the solar wind and the ionosphere. The distinction between the two source regions was made traditionally on the basis of ion composition experiments, with H^+ indicating solar wind origin, while the presence of O^+ indicated an ionospheric source (e.g. Shelley et al. 1972). This view has been extended and modified significantly. On one hand, detailed test particle tracing studies in global MHD models of storm time magnetosphere evolution have demonstrated that ionospheric H^+ can also populate the plasma sheet and provide a seed population (Glocer et al. 2020). On the other hand, detailed studies of the energy/charge dependence of enhanced energetic particle fluxes showed that the contribution of heavy energetic ions to enhanced fluxes is not a fixed percentage but rather depends on energy and charge status, with O^+ (of ionospheric origin) dominating at lower energies of tens of keV, while multiply charged oxygen, particularly O^{6+} of solar wind origin, was found to dominate at energies of hundreds of keV (Cohen et al. 2017; Bingham et al. 2020, 2021).

Reconnection on the dayside might also contribute to seeding of energetic particles in the near-Earth space environment. Fennell et al. (2016) reported on 'microinjections' of relativistic electrons observed by MMS; microinjections are frequent and rapid, energy-dispersed to dispersionless enhancements of electron intensities observed around $\sim 10R_E$ geocentric distance along the tailward-flanks of the magnetosphere. By tracing dispersed particle signatures back to their dispersionless origins, Fennell et al. (2016) demonstrated that dispersed microinjection observations along the dusk-side of the magnetosphere map back to near the subsolar and early afternoon magnetopause. Kavosi et al. (2018) showed that the observed periodicity of microinjection electrons is consistent with a combination of Kelvin-Helmholtz (KH) waves and flux transfer events (FTEs) along the dayside magnetopause. Those results indicate that microinjected electrons might result from bursts of reconnection associated with KH instability and FTEs along the dayside magnetopause. Conversely, the drops in fluxes around microinjection electrons might also be the signature of losses of energetic electrons through the magnetopause and to the magnetosheath; however that electron loss process too is only enabled via reconnection resulting in magnetic connectivity across the magnetopause (e.g. Kim and Lee 2014; Mauk et al. 2016). Tracing test particles in a dynamically evolving MHD model has reproduced even the salient features of losses (including detailed variations both in space and time and the depth of penetration and persistence of particles in the magnetosheath) for different species in agreement with MMS observations (e.g. Sorathia et al. 2017).

3.5 Diamagnetic Cavities

An interesting topic that drew some attention in the recent decade is the diamagnetic cavities that form at high magnetic latitudes in the cusp region as a consequence of large-scale, dayside magnetopause reconnection (e.g. Lavraud et al. 2002, 2005). This region has a substantially reduced magnetic field magnitude and is filled with dense, sheath-like plasma with high-energy (> 30 keV) electrons and ions (including heavy ions). The high-energy particles exhibit perpendicular anisotropy (e.g. Nykyri et al. 2019), and test-particle simulations suggest that those high-energy particles are produced locally via betatron and/or Fermi mechanisms while being trapped in the magnetic bottle like configuration associated with the cavity (Nykyri et al. 2012, 2019; Sorathia et al. 2019; Burkholder et al. 2021). The relatively large size of the diamagnetic cavities, i.e., 3-5 R_E in width (Nykyri et al. 2019) indicates that they can be a major source of plasma (electrons, protons and oxygen ions) into Earth's magnetosphere as well as providing a high-energy particle source (Nykyri et al. 2021).

4 Outstanding Problems

There remain unsolved problems in the topic of particle acceleration by magnetic reconnection in geospace. Here, we describe two topics, energy partition and the precise role of turbulence. These problems are very relevant to particle acceleration in solar flares.

4.1 Energy Partition

For solar flares, it has been reported that non-thermal electrons alone carry up to 50% of the released magnetic energy (e.g. Lin and Hudson 1976; Aschwanden et al. 2017). In fact, more detailed studies argue that thermal electrons can indeed carry much less energy than non-thermal electrons, even in coronal sources (e.g. Krucker et al. 2010; Krucker and Battaglia 2014; Fleishman et al. 2022). This is in stark contrast to the case of Earth's magnetotail (in particular the reconnection region) where non-thermal electrons appear to carry only a minuscule fraction of released energy (e.g. Øieroset et al. 2002).

While the plasma parameters in the magnetotail differ greatly from those in the solar atmosphere, it is still instructive to know how energy is partitioned between thermal and non-thermal components in the magnetotail. A caveat is that the typical particle energy spectrum in the magnetotail does not exhibit a clear spectral break, and it is difficult to separate those components at a certain energy E_c (e.g. Christon et al. 1988, 1989, 1991; Øieroset et al. 2002; Oka et al. 2018). Fortunately, the energy spectrum is often well approximated by the kappa distribution and the non-thermal fraction of particle energy (and also density) can be calculated analytically without introducing a sharp boundary at E_c (Oka et al. 2013, 2015).

Based on the kappa distribution model, it was shown that, for the above-the-looptop (ALT) hard X-ray coronal sources in solar flares, the fraction of non-thermal electron energies was at most $\sim$50%, indicating equipartition between thermal and non-thermal components (Oka et al. 2013, 2015). Similar values of non-thermal fraction were obtained by self-consistent particle simulations of magnetic reconnection (e.g. Arnold et al. 2021; Zhang et al. 2021), as well as *in situ* observations of electron energy spectra during magnetotail reconnection (Oka et al. 2022).

A puzzle is that, even when electrons are significantly heated (for example, the event of 2017 July 26, Fig. 3 right), the non-thermal tail does not necessarily become harder (Runov et al. 2015; Zhou et al. 2016; Oka et al. 2022). This is counter-intuitive because the non-thermal tail is often expected to be enhanced as the temperature increases. When electrons are not significantly heated (for example, the event of 2017 July 11, Fig. 3 left), the non-thermal tail becomes harder (softer) as the spacecraft approaches toward (moves away from) the X-line. The two distinct types of reconnection events, i.e., less heated and much heated events, can be interpreted by the concept of 'plasma sheet reconnection' and 'lobe plasma reconnection', respectively (e.g. Oka et al. 2022, and references therein). However, it remains unclear, at least from the observational point of view, what controls the energy partition between thermal and non-thermal components of electrons. One caveat that has to be considered in the magnetotail events is that the particle distribution observed prior to an event is generally not (or not identical to) the source of the population observed afterward, as discussed in Sect. 3.4.

For ions, the energy partition between thermal and non-thermal components is much less studied in the magnetotail, although ions do form a clear power-law tail in the magnetotail (e.g. Christon et al. 1988, 1989, 1991; Øieroset et al. 2002; Ergun et al. 2020b). Recent particle simulations of magnetic reconnection have shown that ions and electrons form a very similar power-law tail, but non-thermal protons gain $\sim 2\times$ more energy than non-thermal electrons (Zhang et al. 2021). It was argued that the primary mechanism of acceleration is Fermi acceleration and that the strong field-line chaos associated with the flux-rope kink instability allows particles to be transported out of flux ropes for further acceleration. It is to be noted that energetic ions in the magnetotail especially in the dipolarization region can have multiple sources and thus the process of energy partition might be a little more complex. Birn et al. (2015b) argued that an enhanced flux of energetic ions can result from not only acceleration of thermal ions in the reconnection region but also a drift entry of pre-energized ions from the magnetotail flanks. Observational validation of these scenarios of ion acceleration and associated partition of energy is left for future work.

4.2 The Precise Role of Turbulence

Many theoretical and simulation studies have shown that the guiding-center approximation is effective in explaining particle acceleration during magnetic reconnection and that particle acceleration is achieved by a Fermi-type mechanism involving curvature drift (Sect. 1.2). However, turbulence may also play a significant role (Sect. 2.4), although its importance and specific role in particle acceleration are not fully understood, at least from an observational standpoint. For example, Ergun et al. (2020a) argued theoretically that the turbulence with high-frequency electric fields in a magnetic depletion region can energize electrons up to non-thermal energy. Zhang et al. (2021) have demonstrated that the flux-rope kink instability leads to strong field-line chaos, allowing particles to be transported out of flux ropes for further acceleration by other flux ropes. Also, Fujimoto and Cao (2021) have shown that the turbulence-induced electric field at the core of flux ropes can scatter electrons, resulting in heating rather than acceleration to non-thermal energy. The turbulence in these theoretical models has different roles, and such roles have not been fully explored in observational studies.

It is to be noted again that an enhanced turbulence may not necessarily lead to an enhanced non-thermal tail in the reconnection region (Sects. 1.3 and 4.1), although turbulence appears correlated with enhancements of non-thermal tail in the flow braking region (e.g. Ergun et al. 2022b, and references therein). Also, hard electron spectra have been found even in a quiet-time plasma sheet (Sect. 2.1), raising a question whether turbulence can confine electrons. After all, what makes reconnection more turbulent and how important turbulence is for the process of particle acceleration remain unsolved.

5 Summary and Conclusion

In the past decade, a key theme of particle acceleration studies was whether the guiding-center approximation can describe particle acceleration and which of the key mechanisms, i.e., Fermi acceleration, betatron acceleration, and the direct acceleration by parallel electric field, is more dominant. The MMS mission has enabled the evaluation of each term and supported the earlier idea that both Fermi and betatron acceleration are important in many cases of electron acceleration during reconnection. In the collapsing region where the intrinsic dipole field becomes more important, the betatron acceleration dominates in the central plasma sheet. While some populations originate from the flank of the magnetotail without much increase in energy, other populations experience energization at localized dipolarization while being transported earthward from the reconnection region. In addition to the Fermi and betatron acceleration, a parallel potential develops near the reconnection X-line and traps incoming electrons, resulting in a significant energization. The electric field associated with turbulence can also accelerate electrons but such process might invalidate the assumption of adiabatic particle motion. Ions are more likely to behave non-adiabatically even near Earth, away from the reconnection region. Outstanding problems remain regarding, for example, energy partition between thermal and non-thermal components and the precise role of turbulence in the particle acceleration process. Solving these problems might be helpful for understanding the particle acceleration mechanism in other plasma environments, such as the solar corona.

Acknowledgements We thank Seiji Zenitani for proofreading the manuscript prior to submission. This work was initiated and partly carried out with support from the International Space Science Institute (ISSI) in the framework of a workshop entitled 'Magnetic Reconnection: Explosive Energy Conversion in Space Plasmas', led by Rumi Nakamura and James L. Burch.

Funding MO was supported by NASA grants 80NSSC18K1002, 80NSSC18K1373, and 80NSSC22K0520 at UC Berkeley. JB acknowledges support from NASA grants 80NSSC18K1452 and 80NSSK0834, and NSF Grant 1602655. FG acknowledges supports in part from NASA grant 80HQTR20T0073, 80HQTR21T0087 and 80HQTR21T0104, as well as the LDRD program at Los Alamos National Laboratory. YK acknowledges support from the Swedish National Space Agency.

Declarations

Competing Interests The authors have no conflict of interest to declare that is relevant to the content of this article.

References

Andrews MK, Daly PW, Keppler E (1981) Geophys Res Lett 8(9):987–990. https://doi.org/10.1029/gl008i009p00987

Angelopoulos V, Runov A, Zhou XZ et al (2013) Science 341(6153):1478–1482. https://doi.org/10.1126/science.1236992

Apatenkov SV, Sergeev VA, Kubyshkina MV et al (2007) Ann Geophys 25(3):801–814. https://doi.org/10.5194/angeo-25-801-2007

Arnold H, Drake JF, Swisdak M et al (2021) Phys Rev Lett 126(13):135101. https://doi.org/10.1103/PhysRevLett.126.135101

Aschwanden MJ, Caspi A, Cohen CMS et al (2017) Astrophys J 836(17). https://doi.org/10.3847/1538-4357/836/1/17

Ashour-Abdalla M, El-Alaoui M, Goldstein ML et al (2011) Nat Phys 7(4):360–365. https://doi.org/10.1038/Nphys1903

Ashour-Abdalla M, Lapenta G, Walker RJ et al (2015) J Geophys Res 120(6):4784–4799. https://doi.org/10.1002/2014ja020316

Baker DN, Belian RD, Higbie PR et al (1979) J Geophys Res 84(A12):7138. https://doi.org/10.1029/ja084ia12p07138
Baumjohann W, Roux A, Le Contel O et al (2007) Ann Geophys 25(6):1365–1389. https://doi.org/10.5194/angeo-25-1365-2007
Bingham ST, Cohen IJ, Mauk BH et al (2020) J Geophys Res 125(10). https://doi.org/10.1029/2020ja028144
Bingham ST, Nikoukar R, Cohen IJ et al (2021) Geophys Res Lett 48(4). https://doi.org/10.1029/2020gl091697
Birn J, Hesse M (1994) J Geophys Res 99(A1):109. https://doi.org/10.1029/93ja02284
Birn J, Thomsen MF, Borovsky JE et al (1997) J Geophys Res 102(A2):2309–2324. https://doi.org/10.1029/96ja02870
Birn J, Thomsen MF, Hesse M (2004) Phys Plasmas 11:1825–1833. https://doi.org/10.1063/1.1704641
Birn J, Artemyev AV, Baker DN et al (2012) Space Sci Rev 173(1–4):49–102. https://doi.org/10.1007/s11214-012-9874-4
Birn J, Hesse M, Nakamura R et al (2013) J Geophys Res 118:1960–1971. https://doi.org/10.1002/jgra.50132
Birn J, Runov A, Hesse M (2014) J Geophys Res 119:3604. https://doi.org/10.1002/2013ja019738
Birn J, Hesse M, Runov A et al (2015a) J Geophys Res 120(9):7522–7535. https://doi.org/10.1002/2015ja021573
Birn J, Runov A, Hesse M (2015b) J Geophys Res 120(9):7698–7717. https://doi.org/10.1002/2015ja021372
Birn J, Battaglia M, Fletcher L et al (2017a) Astrophys J 848(2). https://doi.org/10.3847/1538-4357/aa8ad4
Birn J, Chandler M, Moore T et al (2017b) J Geophys Res 122(8):8026–8036. https://doi.org/10.1002/2017ja024231
Birn J, Runov A, Zhou XZ (2017c) J Geophys Res 122(8):8014–8025. https://doi.org/10.1002/2017ja024230
Birn J, Hesse M, Bingham ST et al (2021a) J Geophys Res 126(7). https://doi.org/10.1029/2021ja029184
Birn J, Runov A, Khotyaintsev Y (2021b). In: Maggiolo R et al (eds) Magnetospheres in the Solar System. AGU, pp 245–275. https://doi.org/10.1002/9781119815624.ch17
Birn J, Hesse M, Runov A (2022) Front Astron Space Sci 9. https://doi.org/10.3389/fspas.2022.908730
Blandford R, Eichler D (1987) Phys Rep 154(1). https://doi.org/10.1016/0370-1573(87)90134-7
Breuillard H, Le Contel O, Retino A et al (2016) Geophys Res Lett 43(14):7279–7286. https://doi.org/10.1002/2016gl069188
Burch JL, Torbert RB, Phan TD et al (2016) Science 352(6290):aaf2939. https://doi.org/10.1126/science.aaf2939
Burch JL, Dokgo K, Hwang KJ et al (2019) Geophys Res Lett 46:4089. https://doi.org/10.1029/2019gl082471
Burkholder BL, Nykyri K, Ma X et al (2021) J Geophys Res 126(12). https://doi.org/10.1029/2021ja029738
Cattell CA, Mozer FS (1986) Geophys Res Lett 13(3):221–224. https://doi.org/10.1029/gl013i003p00221
Chang T, Crew GB, Hershkowitz N et al (1986) Geophys Res Lett 13(7):636–639. https://doi.org/10.1029/gl013i007p00636
Chen LJ, Bhattacharjee A, Puhl-Quinn PA et al (2008) Nat Phys 4(1):19–23. https://doi.org/10.1038/nphys777
Chen LJ, Bessho N, Lefebvre B et al (2009) Phys Plasmas 16(5). https://doi.org/10.1063/1.3112744
Chen G, Fu HS, Zhang Y et al (2019) Astrophys J 881:L8. https://doi.org/10.3847/2041-8213/ab3041
Chepuri SNF, Jaynes AN, Baker DN et al (2022) Front Astron Space Sci 9. https://doi.org/10.3389/fspas.2022.926660
Chew GF, Goldberger ML, Low FE (1956) Proc R Soc Lond Ser A, Math Phys Sci 236(1204):112–118. https://doi.org/10.1098/rspa.1956.0116
Christon SP, Mitchell DG, Williams DJ et al (1988) J Geophys Res 93:2562. https://doi.org/10.1029/JA093iA04p02562
Christon SP, Williams DJ, Mitchell DG et al (1989) J Geophys Res 94:13409. https://doi.org/10.1029/JA094iA10p13409
Christon SP, Williams DJ, Mitchell DG et al (1991) J Geophys Res 96(1). https://doi.org/10.1029/90ja01633
Cohen IJ, Mitchell DG, Kistler LM et al (2017) J Geophys Res 122(9):9282–9293. https://doi.org/10.1002/2017ja024351
Cohen IJ, Turner DL, Mauk BH et al (2021) Geophys Res Lett 48:e90087. https://doi.org/10.1029/2020gl090087
Dahlin JT (2020) Geophys Res Lett 47(11). https://doi.org/10.1029/2020gl087918
Dahlin JT, Drake JF, Swisdak M (2014) Phys Plasmas 21(9):092304. https://doi.org/10.1063/1.4894484
Dahlin JT, Drake JF, Swisdak M (2016) Phys Plasmas 23:120704. https://doi.org/10.1063/1.4972082
Dahlin JT, Drake JF, Swisdak M (2017) Phys Plasmas 24:092110. https://doi.org/10.1063/1.4986211
Daughton W, Roytershteyn V, Karimabadi H et al (2011) Nat Phys 7(7):539–542. https://doi.org/10.1038/nphys1965
Delcourt DC, Sauvaud JA (1994) J Geophys Res 99(A1):97. https://doi.org/10.1029/93ja01895

Delcourt DC, Sauvaud JA, Moore TE (1997) J Geophys Res 102(A11):24313–24324. https://doi.org/10.1029/97ja02039

Drake JF, Swisdak M, Che H et al (2006) Nature 443:553–556. https://doi.org/10.1038/nature05116

Drake JF, Swisdak M, Fermo R (2013) Astrophys J 763(1):L5. https://doi.org/10.1088/2041-8205/763/1/l5

Du S, Guo F, Zank GP et al (2018) Astrophys J 867(1). https://doi.org/10.3847/1538-4357/aae30e

Eastwood JP, Phan TD, Bale SD et al (2009) Phys Rev Lett 102(3):035001. https://doi.org/10.1103/PhysRevLett.102.035001

Eastwood JP, Goldman MV, Hietala H et al (2015) J Geophys Res 120(1):511–525. https://doi.org/10.1002/2014ja020516

Eastwood JP, Mistry R, Phan TD et al (2018) Geophys Res Lett 45(10):4569–4577. https://doi.org/10.1029/2018gl077670

Egedal J, Oieroset M, Fox W et al (2005) Phys Rev Lett 94(2):025006. https://doi.org/10.1103/PhysRevLett.94.025006

Egedal J, Fox W, Katz N et al (2008) J Geophys Res 113:A12207. https://doi.org/10.1029/2008ja013520

Egedal J, Daughton W, Drake JF et al (2009) Phys Plasmas 16(5):050701. https://doi.org/10.1063/1.3130732

Egedal J, Lê A, Katz N et al (2010) J Geophys Res 115:A03214. https://doi.org/10.1029/2009ja014650

Egedal J, Le A, Pritchett PL et al (2011) Electron dynamics in two-dimensional asymmetric anti-parallel reconnection. https://doi.org/10.1063/1.3646316

Egedal J, Daughton W, Le A (2012) Nat Phys 8(4):321–324. https://doi.org/10.1038/nphys2249

Egedal J, Le A, Daughton W (2013) Phys Plasmas 20(6):061201. https://doi.org/10.1063/1.4811092

Egedal J, Daughton W, Le A et al (2015) Phys Plasmas 22(10):101208. https://doi.org/10.1063/1.4933055

Egedal J, Ng J, Le A et al (2019) Phys Rev Lett 123(22). https://doi.org/10.1103/physrevlett.123.225101

Ergun RE, Goodrich KA, Stawarz JE et al (2015) J Geophys Res 120(3):1832–1844. https://doi.org/10.1002/2014ja020165

Ergun RE, Holmes JC, Goodrich KA et al (2016) Geophys Res Lett 43(11):5626–5634. https://doi.org/10.1002/2016gl068992

Ergun RE, Goodrich KA, Wilder FD et al (2018) Geophys Res Lett 45(8):3338–3347. https://doi.org/10.1002/2018gl076993

Ergun RE, Ahmadi N, Kromyda L et al (2020a) Astrophys J 898:153. https://doi.org/10.3847/1538-4357/ab9ab5

Ergun RE, Ahmadi N, Kromyda L et al (2020b) Astrophys J 898:154. https://doi.org/10.3847/1538-4357/ab9ab6

Ergun RE, Pathak N, Usanova ME et al (2022a) Astrophys J Lett 935(1):L8. https://doi.org/10.3847/2041-8213/ac81d4

Ergun RE, Usanova ME, Turner DL et al (2022b) Geophys Res Lett 49(11). https://doi.org/10.1029/2022gl098113

Eriksson E, Vaivads A, Graham DB et al (2018) Geophys Res Lett 45(16):8081–8090. https://doi.org/10.1029/2018gl078660

Eriksson E, Vaivads A, Alm L et al (2020) Geophys Res Lett 47:e85080. https://doi.org/10.1029/2019gl085080

Fennell JF, Turner DL, Lemon CL et al (2016) Geophys Res Lett 43(12):6078–6086. https://doi.org/10.1002/2016gl069207

Fleishman GD, Nita GM, Chen B et al (2022) Nature 606(7915):674–677. https://doi.org/10.1038/s41586-022-04728-8

Forbes TG, Hones EW, Bame SJ et al (1981) Geophys Res Lett 8(3):261–264. https://doi.org/10.1029/gl008i003p00261

Fu HS, Khotyaintsev YV, André M et al (2011) Geophys Res Lett 38:L16104. https://doi.org/10.1029/2011gl048528

Fu HS, Khotyaintsev YV, Vaivads A et al (2013) Nat Phys 9(7):426–430. https://doi.org/10.1038/nphys2664

Fu HS, Vaivads A, Khotyaintsev YV et al (2017) Geophys Res Lett 44(1):37–43. https://doi.org/10.1002/2016gl071787

Fu HS, Peng FZ, Liu CM et al (2019a) Geophys Res Lett 46(11):5645–5652. https://doi.org/10.1029/2019gl083032

Fu HS, Xu Y, Vaivads A et al (2019b) Astrophys J 870(2). https://doi.org/10.3847/2041-8213/aafa75

Fu H, Grigorenko EE, Gabrielse C et al (2020) Sci China Earth Sci 63(2):235–256. https://doi.org/10.1007/s11430-019-9551-y

Fujimoto K, Cao J (2021) Geophys Res Lett 48(19). https://doi.org/10.1029/2021gl094431

Fuselier S, Petrinec SM, Reiff PH et al (2023) Space Sci Rev 219

Gabrielse C, Angelopoulos V, Runov A et al (2012) J Geophys Res 117(A10):n/a–n/a. https://doi.org/10.1029/2012ja017873

Gabrielse C, Angelopoulos V, Runov A et al (2014) J Geophys Res 119(4):2512–2535. https://doi.org/10.1002/2013ja019638
Gabrielse C, Harris C, Angelopoulos V et al (2016) J Geophys Res 121(10):9560–9585. https://doi.org/10.1002/2016ja023061
Gabrielse C, Angelopoulos V, Harris C et al (2017) J Geophys Res 122(5):5059–5076. https://doi.org/10.1002/2017ja023981
Genestreti KJ, Nakamura TKM, Nakamura R et al (2018) J Geophys Res 123:9130. https://doi.org/10.1029/2018ja025711
Glocer A, Welling D, Chappell CR et al (2020) J Geophys Res 125(11). https://doi.org/10.1029/2020ja028205
Guo F, Li H, Daughton W et al (2014) Phys Rev Lett 113:155005. https://doi.org/10.1103/PhysRevLett.113.155005
Guo F, Liu YH, Daughton W et al (2015) Astrophys J 806(2). https://doi.org/10.1088/0004-637x/806/2/167
Guo Z, Wu M, Du A (2017) Astrophys Space Sci 362(7). https://doi.org/10.1007/s10509-017-3093-0
Guo F, Liu YH, Li X et al (2020) Phys Plasmas 27(8):080501. https://doi.org/10.1063/5.0012094
Hasegawa H, Denton RE, Nakamura R et al (2019) J Geophys Res 124(1):122–138. https://doi.org/10.1029/2018ja026051
Hazeltine RD, Meiss JD (2003) Plasma confinement. Dover, Mineola
Hoshino M (2012) Phys Rev Lett 108:135003. https://doi.org/10.1103/PhysRevLett.108.135003
Hoshino M, Nishida A, Yamamoto T et al (1994) Geophys Res Lett 21(25):2935–2938. https://doi.org/10.1029/94gl02094
Hoshino M, Mukai T, Terasawa T et al (2001) J Geophys Res 106(A11):25979–25997. https://doi.org/10.1029/2001ja900052
Huang SY, Vaivads A, Khotyaintsev YV et al (2012) Geophys Res Lett 39(/a–n/a). https://doi.org/10.1029/2012gl051946
Huang SY, Jiang K, Yuan ZG et al (2019) Geophys Res Lett 46(2):580–589. https://doi.org/10.1029/2018gl081099
Hwang KJ, Choi E, Dokgo K et al (2019) Geophys Res Lett 46(12):6287–6296. https://doi.org/10.1029/2019GL082710
Hwang KJ, Nakamura R, Eastwood JP et al (2023) Space Sci Rev 219
Imada S, Hoshino M, Mukai T (2005) Geophys Res Lett 32(9):2–5. https://doi.org/10.1029/2005gl022594
Imada S, Nakamura R, Daly PW et al (2007) J Geophys Res 112(A3):A03202. https://doi.org/10.1029/2006ja011847
Jaynes AN, Turner DL, Wilder FD et al (2016) Geophys Res Lett 43(14):7356–7363. https://doi.org/10.1002/2016gl069206
Jeans JH (1915) Mon Not R Astron Soc 76(2):70–84. https://doi.org/10.1093/mnras/76.2.70
Ji H, Daughton W (2011) Phys Plasmas 18:111207. https://doi.org/10.1063/1.3647505
Jiang K, Huang SY, Yuan ZG et al (2019) Astrophys J Lett 881(2). https://doi.org/10.3847/2041-8213/ab36b9
Jiang K, Huang SY, Yuan ZG et al (2021) Geophys Res Lett 48(11). https://doi.org/10.1029/2021gl093458
Johnson G, Kilian P, Guo F et al (2022) Astrophys J 933(1). https://doi.org/10.3847/1538-4357/ac7143
Kavosi S, Spence HE, Fennell JF et al (2018) J Geophys Res 123(7):5364–5378. https://doi.org/10.1029/2018ja025244
Khotyaintsev YV, Cully CM, Vaivads A et al (2011) Phys Rev Lett 106(16):165001. https://doi.org/10.1103/PhysRevLett.106.165001
Khotyaintsev YV, Graham DB, Norgren C et al (2019) Front Astron Space Sci 6. https://doi.org/10.3389/fspas.2019.00070
Khotyaintsev YV, Graham DB, Steinvall K et al (2020) Phys Rev Lett 124(4). https://doi.org/10.1103/physrevlett.124.045101
Kim KC, Lee DY (2014) J Geophys Res 119(7):5495–5508. https://doi.org/10.1002/2014ja019880
Kliem B (1994) Astrophys J Suppl Ser 90:719. https://doi.org/10.1086/191896
Krucker S, Battaglia M (2014) Astrophys J 780(1):107. https://doi.org/10.1088/0004-637x/780/1/107
Krucker S, Hudson HS, Glesener L et al (2010) Astrophys J 714:1108–1119. https://doi.org/10.1088/0004-637X/714/2/1108
Lapenta G, Berchem J, Alaoui ME et al (2020) Phys Rev Lett 125(22). https://doi.org/10.1103/physrevlett.125.225101
Lavraud B, Dunlop MW, Phan TD et al (2002) Geophys Res Lett 29(20):56-1–56-4. https://doi.org/10.1029/2002gl015464
Lavraud B, Rème H, Dunlop MW et al (2005) Surv Geophys 26:135–175. https://doi.org/10.1007/s10712-005-1875-3
Lazarian A, Vishniac ET (1999) Astrophys J 517(2):700–718. https://doi.org/10.1086/307233

Lazarian A, Eyink G, Vishniac E et al (2015) Philos Trans R Soc A, Math Phys Eng Sci 373(2041):20140144. https://doi.org/10.1098/rsta.2014.0144
Le Contel O, Roux A, Jacquey C et al (2009) Ann Geophys 27(6):2259–2275. https://doi.org/10.5194/angeo-27-2259-2009
le Roux JA, Zank GP, Webb GM et al (2015) Astrophys J 801(2). https://doi.org/10.1088/0004-637x/801/2/112
le Roux JA, Zank GP, Khabarova OV (2018) Astrophys J 864(2):158. https://doi.org/10.3847/1538-4357/aad8b3
Le A, Egedal J, Daughton W et al (2009) Phys Rev Lett 102(8):085001. https://doi.org/10.1103/PhysRevLett.102.085001
Le A, Karimabadi H, Egedal J et al (2012) Phys Plasmas 19(7):072120. https://doi.org/10.1063/1.4739244
Le A, Egedal J, Daughton W (2016) Phys Plasmas 23(10). https://doi.org/10.1063/1.4964768
Leonardis E, Chapman SC, Daughton W et al (2013) Phys Rev Lett 110(20). https://doi.org/10.1103/physrevlett.110.205002
Lezniak TW, Winckler JR (1970) J Geophys Res 75(34):7075–7098. https://doi.org/10.1029/ja075i034p07075
Li X, Baker DN, Temerin M et al (1998) Geophys Res Lett 25(14):2561–2564. https://doi.org/10.1029/98gl00036
Li XC, Guo F, Li H et al (2017) Astrophys J 843(1):21. https://doi.org/10.3847/1538-4357/aa745e
Li X, Guo F, Li H et al (2018) Astrophys J 855(2). https://doi.org/10.3847/1538-4357/aaacd5
Li X, Guo F, Li H (2019a) Astrophys J 879(1). https://doi.org/10.3847/1538-4357/ab223b
Li X, Guo F, Li H et al (2019b) Astrophys J 884(2):118. https://doi.org/10.3847/1538-4357/ab4268
Li X, Guo F, Liu YH (2021) Phys Plasmas 28(5):052905. https://doi.org/10.1063/5.0047644
Li X, Wang R, Lu Q et al (2022) Nat Commun 13(1). https://doi.org/10.1038/s41467-022-31025-9
Lin RP, Hudson HS (1976) Sol Phys 50:153. https://doi.org/10.1007/bf00206199
Litvinenko YE (1996) Astrophys J 462(2):997–1004. https://doi.org/10.1086/177213
Liu J, Angelopoulos V, Runov A et al (2013) J Geophys Res 118(5):2000–2020. https://doi.org/10.1002/jgra.50092
Liu CM, Fu HS, Xu Y et al (2017a) Geophys Res Lett 44(13):6492–6499. https://doi.org/10.1002/2017gl074029
Liu CM, Fu HS, Xu Y et al (2017b) J Geophys Res 122(1):594–604. https://doi.org/10.1002/2016ja023437
Ma WQ, Zhou M, Zhong ZH et al (2020) Astrophys J 903(2):84. https://doi.org/10.3847/1538-4357/abb8cc
Malykhin AY, Grigorenko EE, Kronberg EA et al (2018) Geomagn Aeron 58:744–752. https://doi.org/10.1134/s0016793218060099
Mauk BH, Cohen IJ, Westlake JH et al (2016) Geophys Res Lett 43(9):4081–4088. https://doi.org/10.1002/2016gl068856
Montag P, Egedal J, Lichko E et al (2017) Phys Plasmas 24(6):062906. https://doi.org/10.1063/1.4985302
Mozer FS, Agapitov OA, Artemyev A et al (2016) Phys Rev Lett 116(14):145101. https://doi.org/10.1103/PhysRevLett.116.145101
Nakamura R, Baumjohann W, Klecker B et al (2002) Geophys Res Lett 29(20):1942. https://doi.org/10.1029/2002gl015763
Nakamura TKM, Genestreti KJ, Liu YH et al (2018) J Geophys Res 123:9150. https://doi.org/10.1029/2018ja025713
Nakamura R, Genestreti KJ, Nakamura T et al (2019) J Geophys Res 124:1173. https://doi.org/10.1029/2018ja026028
Nakamura R, Baumjohann W, Nakamura TKM et al (2021) J Geophys Res 126(10). https://doi.org/10.1029/2021ja029518
Nakanotani M, Zank GP, Zhao L (2022) Front Astron Space Sci 9. https://doi.org/10.3389/fspas.2022.954040
Norgren C, Hesse M, Graham DB et al (2020) J Geophys Res 125:e27440. https://doi.org/10.1029/2019ja027440
Northrop TG (1963) The adiabatic motion of charged particles. Interscience, New York
Nykyri K, Otto A, Adamson E et al (2012) J Atmos Sol-Terr Phys 87(88):70–81. https://doi.org/10.1016/j.jastp.2011.08.012
Nykyri K, Chu C, Ma X et al (2019) J Geophys Res 124(1):197–210. https://doi.org/10.1029/2018ja026131
Nykyri K, Johnson J, Kronberg E et al (2021) Geophys Res Lett 48(9). https://doi.org/10.1029/2021gl092466
Ohia O, Egedal J, Lukin VS et al (2012) Phys Rev Lett 109(11):115004. https://doi.org/10.1103/PhysRevLett.109.115004
Ohia O, Egedal J, Lukin VS et al (2015) Geophys Res Lett 42(24):10549–10556. https://doi.org/10.1002/2015gl067117
Øieroset M, Lin RP, Phan TD et al (2002) Phys Rev Lett 89:195001. https://doi.org/10.1103/PhysRevLett.89.195001

Oieroset M, Phan TD, Eastwood JP et al (2011) Phys Rev Lett 165:007. https://doi.org/10.1103/PhysRevLett.107.165007
Oka M, Fujimoto M, Shinohara I et al (2010a) J Geophys Res 115:A08223. https://doi.org/10.1029/2010ja015392
Oka M, Phan TD, Krucker S et al (2010b) Astrophys J 714(1):915–926. https://doi.org/10.1088/0004-637x/714/1/915
Oka M, Ishikawa S, Saint-Hilaire P et al (2013) Astrophys J 764(6). https://doi.org/10.1088/0004-637x/764/1/6
Oka M, Krucker S, Hudson HS et al (2015) Astrophys J 799:129. https://doi.org/10.1088/0004-637x/799/2/129
Oka M, Phan TD, Øieroset M et al (2016) J Geophys Res 121:1955. https://doi.org/10.1002/2015ja022040
Oka M, Birn J, Battaglia M et al (2018) Space Sci Rev 214:82. https://doi.org/10.1007/s11214-018-0515-4
Oka M, Phan TD, Øieroset M et al (2022) Phys Plasmas 29(5):052904. https://doi.org/10.1063/5.0085647
Osman KT, Kiyani KH, Matthaeus WH et al (2015) Astrophys J 815(2). https://doi.org/10.1088/2041-8205/815/2/l24
Panov EV, Nakamura R, Baumjohann W et al (2010) Geophys Res Lett 37:L08103. https://doi.org/10.1029/2009GL041971
Parkhomenko EI, Malova HV, Grigorenko EE et al (2019) Phys Plasmas 26(4):042901. https://doi.org/10.1063/1.5082715
Phan TD, Shay MA, Gosling JT et al (2013) Geophys Res Lett 40(17):4475–4480. https://doi.org/10.1002/grl.50917
Phan TD, Eastwood JP, Shay MA et al (2018) Nature 557:202. https://doi.org/10.1038/s41586-018-0091-5
Retinò A, Sundkvist D, Vaivads A et al (2007) Nat Phys 3:235–238. https://doi.org/10.1038/nphys574
Retinò A, Nakamura R, Vaivads A et al (2008) J Geophys Res 113(A12):n/a–n/a. https://doi.org/10.1029/2008ja013511
Richard L, Khotyaintsev YV, Graham DB et al (2023) Phys Rev Lett 131:115201. https://doi.org/10.1103/physrevlett.131.115201
Richard L, Sorriso-Valvo L, Yordanova E et al (2023). arXiv e-prints arxiv:2303.08693
Runov A, Angelopoulos V, Sitnov MI et al (2009) Geophys Res Lett 36:L14106. https://doi.org/10.1029/2009gl038980
Runov A, Angelopoulos V, Zhou XZ et al (2011) J Geophys Res 116:A05216. https://doi.org/10.1029/2010ja016316
Runov A, Angelopoulos V, Gabrielse C et al (2013) J Geophys Res 118(2):744–755. https://doi.org/10.1002/jgra.50121
Runov A, Angelopoulos V, Gabrielse C et al (2015) J Geophys Res 120(6):4369–4383. https://doi.org/10.1002/2015ja021166
Runov A, Angelopoulos V, Artemyev A et al (2017) J Geophys Res 122(6):5965–5978. https://doi.org/10.1002/2017ja024010
Schödel R, Baumjohann W, Nakamura R et al (2001) J Geophys Res 106(A1):301–313. https://doi.org/10.1029/2000ja900139
Shay MA, Haggerty CC, Phan TD et al (2014) Phys Plasmas 21(12):122902. https://doi.org/10.1063/1.4904203
Shelley EG, Johnson RG, Sharp RD (1972) J Geophys Res 77(31):6104–6110. https://doi.org/10.1029/ja077i031p06104
Sitnov MI, Swisdak M, Divin AV (2009) J Geophys Res 114:A04202. https://doi.org/10.1029/2008ja013980
Sitnov M, Birn J, Ferdousi B et al (2019) Space Sci Rev 215(4). https://doi.org/10.1007/s11214-019-0599-5
Smets R, Delcourt D, Sauvaud JA et al (1999) J Geophys Res 104(A7):14571–14581. https://doi.org/10.1029/1998ja900162
Sorathia KA, Merkin VG, Ukhorskiy AY et al (2017) J Geophys Res 122(9):9329–9343. https://doi.org/10.1002/2017ja024268
Sorathia KA, Ukhorskiy AY, Merkin VG et al (2018) J Geophys Res 123(7):5590–5609. https://doi.org/10.1029/2018ja025506
Sorathia KA, Merkin VG, Ukhorskiy AY et al (2019) J Geophys Res 124(7):5461–5481. https://doi.org/10.1029/2019ja026728
Stawarz JE, Ergun RE, Goodrich KA (2015) J Geophys Res 120(3):1845–1866. https://doi.org/10.1002/2014ja020166
Stern DP (1979) J Geophys Res 84(A1):63–71. https://doi.org/10.1029/ja084ia01p00063
Sun W, Turner DL, Zhang Q et al (2022) J Geophys Res 127(12). https://doi.org/10.1029/2022ja030721
Teh WL, Nakamura TKM, Zenitani S et al (2023) Astrophys J 947(1). https://doi.org/10.3847/1538-4357/acc2bf

Toffoletto F, Sazykin S, Spiro R et al (2003) Space Sci Rev 107(1/2):175–196. https://doi.org/10.1023/a:1025532008047
Torbert RB, Burch JL, Phan TD et al (2018) Science 362:1391–1395. https://doi.org/10.1126/science.aat2998
Torbert RB, Dors I, Argall MR et al (2020) Geophys Res Lett 47(3). https://doi.org/10.1029/2019gl085542
Turner DL, Fennell JF, Blake JB et al (2016) Geophys Res Lett 43(15):7785–7794. https://doi.org/10.1002/2016gl069691
Turner DL, Cohen IJ, Bingham ST et al (2021a) Geophys Res Lett 48(2). https://doi.org/10.1029/2020gl090089
Turner DL, Cohen IJ, Michael A et al (2021b) Geophys Res Lett 48(21). https://doi.org/10.1029/2021gl095495
Ukhorskiy AY, Sitnov MI, Merkin VG et al (2017) J Geophys Res 122(3):3040–3054. https://doi.org/10.1002/2016ja023304
Ukhorskiy AY, Sorathia KA, Merkin VG et al (2018) J Geophys Res 123(7):5580–5589. https://doi.org/10.1029/2018ja025370
Usanova ME, Ergun RE (2022) J Geophys Res 127(7). https://doi.org/10.1029/2022ja030336
Vaivads A, Khotyaintsev YV, Retinò A et al (2021) J Geophys Res 126(8). https://doi.org/10.1029/2021ja029545
Viberg H, Khotyaintsev YV, Vaivads A et al (2014) J Geophys Res 119(4):2605–2611. https://doi.org/10.1002/2014ja019892
Wang R, Lu Q, Du A et al (2010a) Phys Rev Lett 104:175003. https://doi.org/10.1103/PhysRevLett.104.175003
Wang R, Lu Q, Li X et al (2010b) J Geophys Res 115:A11201. https://doi.org/10.1029/2010JA015473
Wang H, Lu Q, Huang C et al (2017) Phys Plasmas 24(5):052113. https://doi.org/10.1063/1.4982813
Wang S, Chen L, Bessho N et al (2019) Geophys Res Lett 46(10):5014–5020. https://doi.org/10.1029/2019gl082226
Wang S, Wang R, Lu Q et al (2021) Geophys Res Lett 48(23). https://doi.org/10.1029/2021gl094879
Wang S, Wang R, Lu Q et al (2023) Geophys Res Lett 50(11). https://doi.org/10.1029/2023gl103203
Wetherton BA, Egedal J, Lê A et al (2019) Geophys Res Lett 46(12):6223–6229. https://doi.org/10.1029/2019gl083119
Wetherton BA, Egedal J, Le A et al (2021) J Geophys Res 126(1). https://doi.org/10.1029/2020ja028604
Williams DJ (1981) J Geophys Res 86(A7):5507. https://doi.org/10.1029/ja086ia07p05507
Wu M, Lu Q, Volwerk M et al (2013) J Geophys Res 118(8):4804–4810. https://doi.org/10.1002/jgra.50456
Wu MY, Huang C, Lu QM et al (2015) J Geophys Res 120(8):6320–6331. https://doi.org/10.1002/2015ja021165
Xu Y, Fu HS, Liu CM et al (2018) Astrophys J 853(1):11. https://doi.org/10.3847/1538-4357/aa9f2f
Yang J, Wolf RA, Toffoletto FR et al (2013) Geophys Res Lett 40(23):6017–6022. https://doi.org/10.1002/2013gl058253
Yang J, Toffoletto FR, Wolf RA et al (2015) J Geophys Res 120(9):7416–7432. https://doi.org/10.1002/2015ja021398
Yang Y, Matthaeus WH, Parashar TN et al (2017) Phys Plasmas 24(7):072306. https://doi.org/10.1063/1.4990421
Zank GP, le Roux JA, Webb GM et al (2014) Astrophys J 797(1):28. https://doi.org/10.1088/0004-637x/797/1/28
Zank GP, Hunana P, Mostafavi P et al (2015) Astrophys J 814(2):137. https://doi.org/10.1088/0004-637x/814/2/137
Zenitani S, Nagai T (2016) Phys Plasmas 23. https://doi.org/10.1063/1.4963008
Zhang Q, Guo F, Daughton W et al (2021) Phys Rev Lett 127(18). https://doi.org/10.1103/physrevlett.127.185101
Zhang Q, Guo F, Daughton W et al (2022). arXiv e-prints https://doi.org/10.48550/arxiv.2210.04113
Zhong ZH, Zhou M, Tang RX et al (2020) Geophys Res Lett 47(1). https://doi.org/10.1029/2019gl085141
Zhou M, Ashour-Abdalla M, Deng X et al (2009) Geophys Res Lett 36(20). https://doi.org/10.1029/2009gl040663
Zhou XZ, Angelopoulos V, Sergeev VA et al (2010) J Geophys Res 115:A00I03. https://doi.org/10.1029/2010ja015481
Zhou XZ, Angelopoulos V, Sergeev VA et al (2011) J Geophys Res 116:A03222. https://doi.org/10.1029/2010ja016165
Zhou XZ, Angelopoulos V, Runov A et al (2012) J Geophys Res 117:A10216. https://doi.org/10.1029/2012ja018171
Zhou M, Li T, Deng X et al (2016) J Geophys Res 121(4):3108–3119. https://doi.org/10.1002/2015ja022085
Zhou M, El-Alaoui M, Lapenta G et al (2018a) J Geophys Res 123(10):8087–8108. https://doi.org/10.1029/2018ja025502

Zhou X, Runov A, Angelopoulos V et al (2018b) J Geophys Res 123(1):429–442. https://doi.org/10.1002/2017ja024901

Zweibel EG, Yamada M (2009) Annu Rev Astron Astrophys 47(1):291–332. https://doi.org/10.1146/annurev-astro-082708-101726

Publisher's Note Springer Nature remains neutral with regard to jurisdictional claims in published maps and institutional affiliations.

Authors and Affiliations

Mitsuo Oka[1] · Joachim Birn[2,3] · Jan Egedal[4] · Fan Guo[3] · Robert E. Ergun[5,6] · Drew L. Turner[7] · Yuri Khotyaintsev[8] · Kyoung-Joo Hwang[9] · Ian J. Cohen[7] · James F. Drake[10]

✉ M. Oka
moka@berkeley.edu

1 Space Sciences Laboratory, University of California Berkeley, 7 Gauss Way, Berkeley, 94720, CA, USA

2 Center for Space Plasma Physics, Space Science Institute, 4765 Walnut Street, Boulder, 80301, CO, USA

3 Los Alamos National Laboratory, Los Alamos, 87545, NM, USA

4 Department of Physics, University of Wisconsin-Madison, 1150 University Avenue, Madison, 53706, WI, USA

5 Laboratory for Atmospheric and Space Physics, University of Colorado, 1234 Innovation Drive, Boulder, 80303, CO, USA

6 Department of Astrophysical and Planetary Sciences, University of Colorado, 2000 Colorado Avenue, Boulder, 80309, CO, USA

7 The Johns Hopkins Applied Physics Laboratory, 11100 Johns Hopkins Road, Laurel, 20723, MD, USA

8 Swedish Institute of Space Physics, Uppsala, 75121, Sweden

9 Southwest Research Institute, 6220 Culebra Road, San Antonio, 78238, TX, USA

10 Department of Physics, The Institute for Physical Science and Technology and The Joint Space Science Institute, University of Maryland, College Park, 20742, MD, USA

V. Magnetic Reconnection beyond Geospace

Space Science Reviews (2025) 221:27
https://doi.org/10.1007/s11214-025-01153-x

Magnetic Reconnection in Solar Flares and the Near-Sun Solar Wind

J.F. Drake[1] · S.K. Antiochos[2] · S.D. Bale[3] · Bin Chen[4] · C.M.S. Cohen[5] · J.T. Dahlin[6] · Lindsay Glesener[7] · F. Guo[8] · M. Hoshino[9] · Shinsuke Imada[9] · M. Oka[10] · T.D. Phan[10] · Katherine K. Reeves[11] · M. Swisdak[2]

Received: 29 April 2024 / Accepted: 26 February 2025 / Published online: 12 March 2025

Abstract
An overview is presented of our current understanding and open questions related to magnetic reconnection in solar flares and the near-sun (within around $20R_s$) solar wind. The solar-flare-related topics include the mechanisms that facilitate fast energy release and that control flare onset, electron energization, ion energization and abundance enhancement, electron and ion transport, and flare-driven heating. Recent observations and models suggesting that interchange reconnection of multipolar magnetic fields within coronal holes could provide the energy required to drive the fast solar wind are also discussed. Recent *in situ* observations that reconnection in the heliospheric current sheet close to the sun drives energetic ions are also presented. The implications of *in situ* observations of reconnection in the Earth space environment for understanding flares are highlighted. Finally, the impact of emerging computational and observational tools for understanding flare dynamics are discussed.

1 Introduction

The explosive release of magnetic energy and the associated production of energetic particles at the Sun, the Earth space environment and the solar wind constitutes an open problem of great scientific importance with broad implications for understanding impulsive phenomena throughout the universe. Solar flare energy release is often linked to the generation of high-speed flows and magnetic flux that drives space weather in the near-Earth environment. The problem is scientifically complex and, in spite of progress, has remained open since the first observations of flares at the Sun (Carrington 1859).

The analysis of early observations of solar flares quickly established that the source of energy was the coronal magnetic field (Rust 1976), a realization that led to the formulation of the concept of magnetic reconnection (Parker 1957; Sweet 1958). The absence of *in situ* spacecraft measurements has made the understanding of flare dynamics a major challenge. Nevertheless, remote observations of flares over a broad range of photon energies, as well as measurements by satellites of energetic particles escaping from the solar atmosphere, have established key characteristics of the reconnection associated with flares. Magnetic energy released in flares is partitioned between thermal and non-thermal particles (both electrons

Extended author information available on the last page of the article

and ions) (Lin and Hudson 1971; Emslie et al. 2012; Aschwanden et al. 2016). Electrons are characterized by hot thermal distributions with temperatures above a keV that transition to powerlaw distributions with energies that can extend up to several MeVs (Lin et al. 2003; Vilmer 2012). Energetic ions from flares gain energies that can extend beyond a GeV in very energetic gamma-ray flares (Lin et al. 2005). However, the ion spectra below around an MeV are not well documented. The mechanisms for the production of energetic electrons and ions in flares and the factors that control the spectral characteristics are emerging from models (Drake et al. 2006a; Dahlin et al. 2014; Guo et al. 2014; Dahlin et al. 2017; Li et al. 2019; Zhang et al. 2021), but not fully understood.

Because magnetic reconnection is a generic physical process much of the progress in understanding energy conversion through reconnection has resulted from the complementary information that comes from observations both at the Sun and the Earth space environment. On the one hand, the solar flare observations clearly established that electrons gained a significant fraction of the released energy in the form of a hot thermal component and a non-thermal powerlaw tail. These observations fostered a search for comparable spectra in magnetotail events (Øieroset et al. 2002). The observations in the Earth space environment revealed many of the fundamental characteristics of reconnection, including the scaling of the reconnection outflow velocity with the upstream Alfvén speed V_A (Paschmann et al. 1986; Phan et al. 2006; Haggerty et al. 2018), the scaling of the electron (Phan et al. 2013b; Øieroset et al. 2023) and ion (Drake et al. 2009; Phan et al. 2014) heating with the available magnetic energy per particle $B^2/4\pi n = m_p V_A^2$, and the rate of magnetic reconnection (Torbert et al. 2018; Burch et al. 2020).

The conditions under which magnetic reconnection produces a significant population of energetic electrons and ions is not fully understood. However, a key control parameter is likely the ratio of the available magnetic energy (incoming Poynting flux $B^2/4\pi$) to the ambient particle thermal energy ($3nT/2$),

$$\sigma_T = B_r^2/6\pi nT, \tag{1}$$

with B_r the component of the magnetic field that is undergoing reconnection and nT the plasma pressure. In the solar atmosphere the parameter $\sigma_T \sim 100$ so the available energy for particle acceleration is substantial. Thus, the large magnetic energy release per particle underlies the efficient production of energetic particles in flares. Opportunities for *in situ* exploration of reconnection in the high σ_T regime include the solar wind close to the Sun (Desai et al. 2022; Phan et al. 2022) and the Earth's magnetotail (Øieroset et al. 2002; Ergun et al. 2020) where energetic spectra driven by reconnection have been documented. In a recent development the data from the Parker Solar Probe suggests that interchange reconnection between open and closed magnetic flux within coronal holes may power the fast solar wind. Further the data suggests that the wind has an energetic component with protons and alpha particles energies extending to 100 keV/nucleon (Bale et al. 2021, 2023).

Solar flare energy release also has significant implications for understanding the threat of space weather to humanity's technological infrastructure. While it is the high-speed flows and magnetic flux associated with Coronal Mass Ejections (CMEs) that play the primary role in disrupting the terrestrial space environment, flare onset in the corona often occurs together with CME ejection. The intense flare UV/EUV emission can heat up the upper atmosphere leading to large increases in satellite drag. A major scientific goal is to understand how the buildup of magnetic energy in the corona triggers energy release, which is critical to understanding both CME launch and flare initiation. Thus, resolving the flare onset problem will also constitute a major advance for space weather prediction.

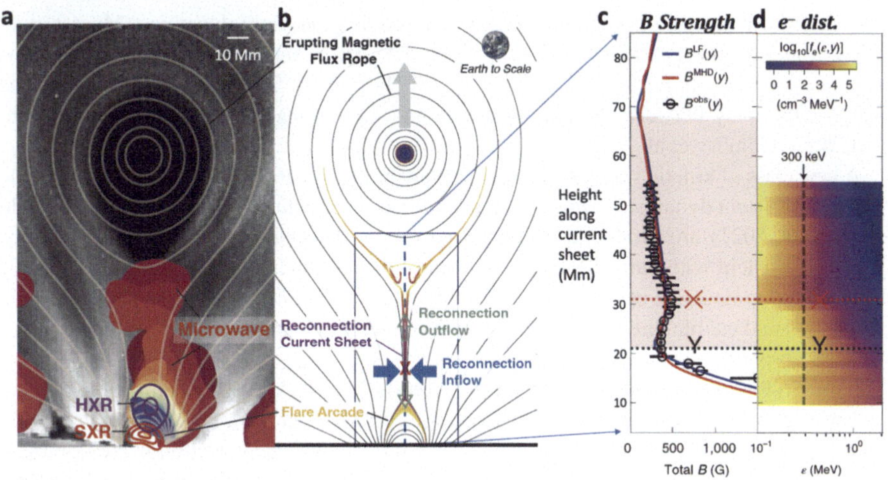

Fig. 1 Measurements of magnetic field and high-energy electrons along a reconnection current sheet Chen et al. (2020). (a) Microwave (red to blue for increasing frequency), soft x-ray (SXR: red) and hard X-ray (HXR; purple) sources associated with energetic electrons in the current sheet region. (b) A matching MHD model of the flare geometry. (c) Spatial variation of the magnetic field strength along the current sheet from the microwave data (black), which matches the model predictions (red/blue). (d) Spatial–energy distribution of energetic electrons along the current sheet. (adapted from Chen et al. 2020)

There is now strong evidence that magnetic reconnection drives flare events in magnetized environments across the universe (Abdo et al. 2011) See discussions by Oka et al. (2023) and Guo et al. (2024) in this collection. The magnetic energy release per particle can be extreme, with values that exceed the particle rest energy. As in solar flares, particle spectra take the form of powerlaws but the mechanisms that control these spectra are under debate (Sironi and Spitkovsky 2014; Guo et al. 2014). Thus, the understanding of energy conversion and energetic particle production in the heliosphere has advanced the knowledge of energetic particle production broadly across the universe.

Several observational platforms have recently come online to make critical new measurements of flare and reconnection dynamics. At the same time, theory and modeling efforts have made inroads in establishing the dominant mechanisms for particle acceleration that have led to new computational models for direct simulation of energetic particle production in macro-scale systems.

Microwave (MW) emission from gyro-synchrotron emission measured by the Expanded Owens Valley Solar Array (EOVSA) is yielding space-time measurements of flare-driven energetic electrons and coronal magnetic field strengths (Gary et al. 2018; Chen et al. 2020) (see Fig. 1). The Daniel K. Inouye Solar Telescope (DKIST) (Tritschler et al. 2016) promises further advances in the measurement of the magnetic geometry of flares. Measurements of high-energy flare emissions from Solar Orbiter (SO/STIX) (Krucker et al. 2020) are supplementing those from Fermi (Omodei et al. 2018), NuSTAR (Grefenstette et al. 2016; Glesener et al. 2017), and the Reuven Ramaty High Energy Solar Spectroscopic Imager (RHESSI) (Lin et al. 2002). The Parker Solar Probe (PSP) (Bale et al. 2016; Kasper et al. 2016; McComas et al. 2016) will reach $\sim 10R_\odot$ on its closest approach to the Sun and is already measuring energetic particles streaming outward from solar reconnection events (Bale et al. 2021, 2023) and *in situ* particle acceleration close to the corona (see Fig. 6) (Phan et al. 2022; Desai et al. 2022). The differing viewing angles of flares by SO, PSP and satellites at 1 AU offer a unique opportunity to gain new insights into flare dynamics (Musset et al. 2021). Imaging and spectroscopic data from Hinode, IRIS, and SDO are tracking the thermal energy release in flares (Simões et al. 2015; Harra et al. 2020) as well as probing the complex interplay between the chromosphere and the corona during these events (Polito et al. 2015; Gömöry et al. 2016).

A new class of kinetic simulation codes that are built with an MHD backbone but include the self-consistent dynamics of particles that undergo acceleration (Drake et al. 2019; Arnold et al. 2019, 2021) are producing extended powerlaw distributions of energetic electrons (Fig. 3) consistent with flare observations (see also discussion in Sect. 5). Such models have the potential to describe energy release and particle acceleration in realistic flare magnetic geometries and enable direct comparisons with observations in macroscale systems such as the solar corona.

In the following sections we present brief overviews of flare physics, some of the recent measurements of reconnection at the heliospheric current sheet from Parker Solar Probe and observational and analysis data that suggest that interchange reconnection is the dominant driver of the fast solar wind. Reconnection in each of these examples drives an energetic particle component. We finally discuss new computation models and observations that have the potential to significantly advance our understanding of particle energization in flares and the HCS close to the sun.

2 Key Scientific Questions in Solar Flare Energy Release

The physics of solar flare energy release is discussed in the framework of the six key science questions outlined below. Also discussed is the development of computational and observational analysis tools. We emphasize that the issues that underlie these science questions are cross-linked so that the boundaries are porous and progress on any specific topic has implications for other topics. A canonical example is the correlation between energetic electrons and ions in flares (Shih et al. 2009).

2.1 What Mechanisms Facilitate the Fast Release of Magnetic Energy in Impulsive Flares?

The mechanism responsible for the fast release of magnetic energy in impulsive flares has remained a central scientific question since the idea of magnetic reconnection was first proposed. That electric fields during the impulsive phase greatly exceed the Dreicer runaway field led to reconnection models based on collisionless dynamics. Key results from the modeling effort are: (1) the rate of magnetic energy dissipation is insensitive to any kinetic scale (Shay et al. 1999; Birn et al. 2001; Shay et al. 2007; Daughton et al. 2014) so that volumetric rates of energy dissipation must scale like

$$k(B_r^2/4\pi)(C_{Ar}/L) \tag{2}$$

with B_r the reconnecting magnetic field, C_{Ar} the corresponding Alfvén speed, L the size of the region of energy release and k a constant of order 0.1 (Birn et al. 2001; Shay et al. 2007; Liu et al. 2022) and (2) that reconnection is intrinsically turbulent in 3D systems, with multiple reconnection sites and chaotic magnetic fields (Daughton et al. 2011, 2014; Dahlin et al. 2015; Huang and Bhattacharjee 2016; Zhang et al. 2021). The latter suggests that the historic separation between laminar models of reconnection and particle acceleration and turbulence models (Miller et al. 1997; Petrosian and Liu 2004) has ended. The scaling of the rate of energy release has been confirmed in the magnetosphere (Nakamura et al. 2018; Burch et al. 2020) but not in flare data.

Estimates of the rate of reconnection in flares can be obtained by measuring the rate at which flare ribbons spread from the polarity inversion lines in on-disc flares. The rate at which the newly brightened ribbons sweep over the integrated radial magnetic field at the solar surface yields the rate at which flux is reconnected in the coronal current sheet (Hesse et al. 2005; Qiu et al. 2017). However, connecting this flux change rate with the inflow reconnection velocity given in Eq. (2) is challenging because of uncertainties in the flare geometry and parameters at the coronal reconnection site. Direct measurements of inflow rates in limb flares using EUV measurements is challenging because the cadence of measurement is not sufficiently high (Chen et al. 2020). While there is evidence from flare observations of non-thermal broadening from turbulent flows (Harra et al. 2013; Kontar et al. 2017), the ubiquity of turbulence must be tested with observations and through MHD modeling in realistic flare geometries.

Thus, an overview of questions related to the rate of flare energy release are:

- What is the rate of magnetic energy release in flares?
- Is magnetic energy release in flares inherently turbulent and, if so, what are the characteristics of this turbulence?
- What causes energy release to vary in space and time?

2.2 What Controls the Onset of Flare Energy Release?

Flares, CMEs, and coronal jets result from the explosive release of energy stored in the highly stressed magnetic field of a filament channel (Forbes 2000; Sterling et al. 2015; Wyper et al. 2017), but the onset mechanism for the eruption remains one of the outstanding problems in heliophysics. This onset mechanism is also likely common to giant stellar and magnetar flares (Davenport 2016; Kaspi and Beloborodov 2017). Thus, solving the flare onset problem would be a far-reaching science advance as well as being essential for developing a predictive capability for the most destructive space weather.

Theories for flare onset fall into two distinct classes (Schmieder et al. 2015): the torus/kink instability and its variations (Kliem and Török 2006); and magnetic reconnection as in the breakout (Antiochos et al. 1999) and tether-cutting (Jiang et al. 2021) models. Soon after onset, however, all models produce the standard eruptive morphology (Fig. 3): a twisted CME flux rope escaping at high speed and fast reconnection in the flare current sheet below the flux rope. As a result, it has not yet been possible to establish definitively from observations whether Alfvénic motions cause onset and drive reconnection or vice versa. Even in simulations it is far from straightforward to separate the two mechanisms.

Beyond the uncertainty in the mechanism for large-scale forcing, the dynamics of reconnection in the flare current layer itself will span an enormous range of scales from the macro- to the kinetic scale. How this complex dynamics couples to and impacts energy release at the largest scales remains uncertain.

Thus, the flare onset problem can be characterized by the following two questions:

- What role do ideal instabilities versus reconnection play in flare onset?
- How do the dynamics of reconnecting current layers at kinetic scales couple to energy release at the macro-scale?

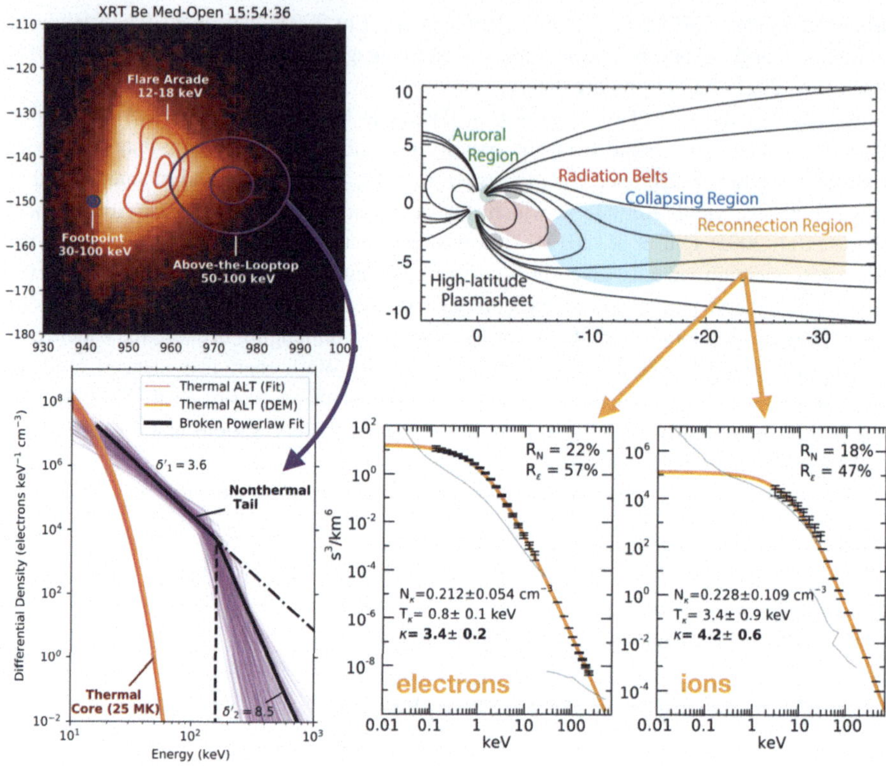

Fig. 2 Particle spectra from a solar flare (left) and a magnetotail reconnection event (right). The flare electrons are from the 'above-the-looptop' source (the purple contour in the upper left image) (Chen et al. 2021). For the magnetotail both ions and electrons are measured *in situ* (lower panels) in the reconnection region (orange in the upper right panel) (Oka et al. 2022). (adapted from Chen et al. 2021 and Oka et al. 2022)

2.3 Why and How do Flares Transfer a Large Fraction of Released Energy into the Energetic Electrons

Observations suggest that the magnetic energy released in flares is partitioned into non-thermal and thermal electrons and ions (Lin and Hudson 1971; Emslie et al. 2012) with recent analysis suggesting that the energy in non-thermal electrons exceeds that in thermal electrons in large flares (Warmuth and Mann 2016). X-ray spectra from RHESSI reveal a hot thermal component at low energy (corresponding to temperatures of 10–30 MK) that roll into powerlaw (or broken powerlaw) spectra that can extend up to several MeV (see Fig. 2) (Lin et al. 2003; Holman et al. 2003; Vilmer 2012). Observations from NuSTAR have established that non-thermal electrons are produced even in some very small flares (Grefenstette et al. 2016; Glesener et al. 2017). The deployment of EOVSA enables direct space-time observations of the energy release domain in the corona (see Fig. 1) (Gary et al. 2018; Chen et al. 2020; Fleishman et al. 2020).

The basic equations describing electron acceleration are easiest to understand in the guiding center limit. Since the Larmor radius of even very energetic electrons is small, these are the relevant equations for reconnection in macro-scale systems. The rate of gain of energy ε

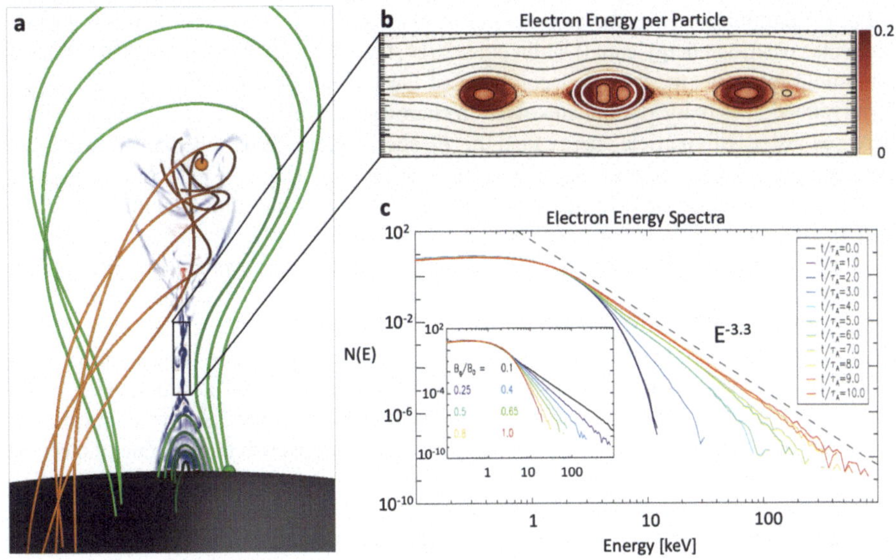

Fig. 3 Global 3D MHD simulation of explosive flare energy release from the Adaptively Resolved MHD Solver (ARMS) code DeVore (1991), Antiochos et al. (1999) (a). Multi-island reconnection (b) and electron spectra (c) from the *kglobal* model (see further details in Sect. 5) where the Alfvén speed is assumed to be 1000 km s^{-1}. (adapted from Dahlin et al. 2019 and Arnold et al. 2021)

of a single electron is given by Dahlin et al. (2014)

$$\frac{d\varepsilon}{dt} = qE_{\parallel}v_{\parallel} + \frac{\mu}{\gamma}\left(\frac{\partial B}{\partial t} + \mathbf{u}_E \cdot \nabla B\right) + \gamma m_e v_{\parallel}^2 \mathbf{u}_E \cdot \boldsymbol{\kappa} \tag{3}$$

where $\mathbf{u}_E = c\mathbf{E} \times \mathbf{B}/B^2$, $\mu = m_e\gamma^2 v_{\perp}^2/2B$ is the magnetic moment and $\boldsymbol{\kappa} = \mathbf{b} \cdot \nabla\mathbf{b}$ is the magnetic curvature. The first term in Eq. (3) is the acceleration by the parallel electric field. The second term corresponds to perpendicular heating due to the conservation of μ. The third term drives parallel acceleration and arises from the first-order Fermi mechanism described in Kliem (1994), Drake et al. (2006a). Freshly reconnected field lines downstream from a reconnecting X-line accelerate as a result of the tension force ($\sim B^2\boldsymbol{\kappa}$). Particles that reflect from this moving field line receive a Fermi "kick" and thereby gain energy. A particle trapped in a magnetic island whose ends are contracting due to these tension forces will repeatedly gain energy as it reflects from the ends of the island. This acceleration can be equivalently described as the conservation of the second adiabatic invariant

$$\oint v_{\parallel} dl. \tag{4}$$

The length reduction of a magnetic field line as an island contracts and releases magnetic energy requires that $v_{\parallel}$ of a particle circulating in that island increases. This can take place as magnetic islands grow in size during reconnection (Drake et al. 2006b) or as a result of the merger of magnetic islands in multi-island reconnection (Oka et al. 2010; Drake et al. 2010).

PIC modeling efforts have led to significant progress in understanding the mechanisms for energetic electron production during reconnection and specifically which of the terms in Eq. (3) dominate. When the ambient out-of-plane guide field is weak, Fermi reflection dominates electron energy gain (Dahlin et al. 2014; Guo et al. 2014; Li et al. 2019). A large guide field increases the radius of curvature of the reconnected field line and suppresses Fermi reflection so that $E_{\|}$ dominates electron energy gain. Further, because Fermi reflection increases faster with particle energy than the $E_{\|}$ energy gain, Fermi reflection increasingly dominates as particles gain energy. The conservation of μ causes a reduction in particle energy since the magnetic field strength decreases during reconnection (Dahlin et al. 2014; Li et al. 2019).

In 3D computational models turbulent reconnection develops spontaneously as x-lines with various tilt angles grow in systems with non-zero guide fields (Daughton et al. 2011, 2014; Dahlin et al. 2015, 2017) or when flux rope kink instability grows (Zhang et al. 2021). That reconnection is intrinsically turbulent suggests that the historical separation between models of flare particle acceleration based on *ad hoc* models of MHD turbulence (Miller et al. 1997; Petrosian 2012) and direct simulations of particle energy gain in reconnection is now outdated. A surprise from the 3D reconnection simulations was that the rate of electron energy gain increases in systems with turbulent reconnection (Dahlin et al. 2015; Li et al. 2019; Zhang et al. 2021). Instead of remaining trapped in fully contracted magnetic islands as in 2D systems, energetic electrons wander chaotically and continue to undergo Fermi reflection in localized regions where $\boldsymbol{\kappa} \cdot \mathbf{u_E}$ is positive (see Eq. (3)).

A failure of the PIC modeling effort has been its inability to produce the extended powerlaw spectra that are ubiquitous in the observations (Li et al. 2019; Zhang et al. 2021). A major advance has been the success of the *kglobal* model, which facilitates the exploration of electron acceleration in macro-systems (Drake et al. 2019; Arnold et al. 2019) and produces extended powerlaw distributions of energetic electrons (Fig. 3) (Arnold et al. 2021). The PIC approach appears to fail because of inadequate separation between kinetic and macro-scales which causes electrons to become demagnetized in unrealistically small magnetic structures as they gain energy. Demagnetization clearly impacts efficient energy gain from Fermi reflection but the link between demagnetization and the limited range of powerlaws obtained in PIC models need further exploration.

The *kglobal* simulations have established that the ambient guide magnetic field is the key control factor in the spectral index of non-thermal electrons in flares. This prediction has received some support in recent EOVSA observations (Fleishman et al. 2022) and recent global MHD simulations of flare dynamics that suggest that the soft to hard transition in flares might be a consequence of the weakening of the guide field in time (Dahlin et al. 2022). The analysis of flare ribbons with simultaneous measurements of hard x-rays with RHESSI observations also suggests that the efficient production of non-thermal electrons is delayed until the ambient guide field is reduced (Qiu et al. 2017; Qiu and Cheng 2022; Qiu et al. 2023). The role of the guide field in the production of energetic electrons needs to be tested with further observations.

The theoretical models of particle acceleration from reconnection suggest that the efficient production of energetic electrons in flares is a consequence of Fermi reflection in merging flux ropes in coronal current sheets (Li et al. 2019; Arnold et al. 2021; Zhang et al. 2021). Combined measurements from EOVSA and RHESSI support the source region of the energetic electrons being in the region "above-the-loop-top" (Chen et al. 2021). On the other hand, the distribution of electrons in the February 10, 2017, event from EOVSA data suggests that the number density of energetic electrons peaks at the low edge of the coronal current sheet (see Fig. 1) (Chen et al. 2020). Further, the non-thermal electrons in this same event appear to be localized in a crescent-shaped region at the base of the current sheet in a region of depleted magnetic field. Thus, whether these non-thermal electrons are produced in the overlying current sheet and convect down to the loop top or are produced locally remains uncertain. Clearly, the spatial localization of the non-thermal electrons in flares provides important information on the mechanism of electron energy gain.

Thus, the following key questions summarize the major issues on energetic electron production in flares:

- Where is the primary site of electron energization?
- Why does the energetic electron spectrum take the form of a powerlaw and what controls the spectral index and its typical soft-hard-soft variation in time?
- What controls the partition of energy into thermal and non-thermal electrons?

2.4 What Mechanism Drives Energization of Ions and the Abundance Enhancements of Some Ion Species During Impulsive Flares?

Observations suggest that ions receive a significant fraction of the magnetic energy released in flares (Lin and Hudson 1971; Emslie et al. 2004, 2005, 2012). Ion energies can reach in excess of a GeV in large flares and spectral indices for protons with energy above 2.5 MeV range from 3.5-4.0 (Ramaty et al. 1996; Lin et al. 2003; Emslie et al. 2004, 2012). Such measurements, however, are limited to large flares where γ-ray emission is strong since these measurements only yield information on ions whose energy exceeds around an MeV. Additional evidence for energetic ion production in impulsive flares has come from extensive *in situ* measurements at 1 AU from Wind/ACE/STEREO. Protons, ^{4}He, and other low-mass ion species exhibit powerlaw distributions that extend down to at least 0.1 MeV/amu (Reames et al. 1997; Mason et al. 2000), well below the proton energies that can be measured with γ-rays. This result suggests that the energy content of ions in flares might be greater than that inferred from γ-ray fluxes. The *in situ* data have also revealed large abundance enhancements, compared with coronal values, of ^{3}He as well as energetic high mass-to-charge ions (Mason 2007). Historically, the explanation of the abundance enhancement of high M/Q ions has been based on acceleration in MHD turbulence that is presumed to develop during flares (Miller et al. 1997; Liu et al. 2006). The large enhancements in fluxes of energetic ^{3}He have been attributed to electron (Fisk 1978; Roth and Temerin 1997) or ion (Fitzmaurice et al. 2024) beam driven Cyclotron waves.

The general equation for ion energy gain during reconnection is the same as for the electrons given in Eq. (3) in the guiding center limit. This equation, however, does not describe acceleration through the cyclotron resonance that is expected to be important for ^{3}He and possibly other high M/Q species. A recent development from PIC simulations is the emergence of simultaneous powerlaws (though with limited energy range) in both electrons and ions with a prediction for the low energy limit of the ion powerlaw that scales as $m_i C_{Ar}^2$ (Zhang et al. 2021). For typical flare parameters this energy is of order 10-20 keV, which is well below that which can be directly measured in flares and below what is typically measured in flare ejecta in the solar wind. These simulations further suggests that the energy content of energetic ions might be much greater than expected. The *klobal* model has recently been extended to include particle ions and is now revealing extended powerlaw distributions of both electrons and protons during reconnection (Yin et al. 2024). An important result is that the protons gain more energy than electrons. Thus, establishing the full range of energetic ions produced in flares is critically important and is likely to be possible with *in situ* measurements of flare ejecta in the solar wind by the Solar Orbiter and the Parker Solar Probe (Bale et al. 2016; McComas et al. 2016; Kasper et al. 2016). Low density reconnection events in the Earth's magnetotail also reveal distinct powerlaw spectra in both protons and electrons as well as the ambient electric and magnetic fields responsible for particle energization (Øieroset et al. 2002; Ergun et al. 2020). Energetic particles reach energies of several hundred keV. These events also reveal that both species develop a hot thermal core of 5-10 keV. Thus, the energetic magnetotail events reveal particle energization that is comparable to that in flares and are an important resource for the exploration of the physics of reconnection-driven particle acceleration.

The following key questions summarize the issues rlated to energetic ion production in flares:

- What is the dominant ion acceleration mechanism during flare energy release in the corona?
- Why do the energetic ion spectra often take the form of a powerlaw and what controls the upper and lower energy limits of the powerlaw?
- Why do high-M/Q ions and ^{3}He in coronal flares exhibit abundance enhancements?

2.5 What Mechanisms Control Energetic Particle Transport in Flares?

Understanding energetic particle production in flares requires a full accounting of particle transport, especially for electrons (Petrosian 2012). The time required for magnetic energy release in flares is linked to the Alfvén transit time across the system, typically $\sim 10L/C_{Ar}$, with L the characteristic scale size of the energy release domain with the factor of 10 a reflection of the rate of reconnection being sub-Alfvénic. In contrast, the transit time of relativistic electrons across the domain scales like L/c, which is much shorter than the energy release time. Observations suggest the existence of a flare electron confinement mechanism (Aschwanden et al. 1996; Simões and Kontar 2013; Fleishman et al. 2022). Magnetic mirroring as well as pitch angle-scattering can limit transport (Schlickeiser 1989; Kontar et al. 2014). PSP observations have revealed the scattering of strahl electrons by oblique whistler waves in the solar wind (Cattell and Vo 2021). Modeling with PIC simulations also suggests the potential role of such waves in energetic electron scattering (Karimabadi et al. 1992; Roberg-Clark et al. 2016, 2018; Vasko et al. 2019; Roberg-Clark et al. 2019; Ma et al. 2023).

Ion escape from the energy release domain in flares can also be significant for very energetic particles. Proton scattering by Alfvén waves plays a key role in theories of particle acceleration at shocks (Blandford and Ostriker 1978; Lee 1983; Schlickeiser 1989) and ion beams can drive Alfvén waves (Gary et al. 1984; Winske and Leroy 1984). However, there is minimal observational evidence of ion transport within the flare site. RHESSI observations reveal that γ-ray emission can be displaced from that of hard X-ray footpoint emission (Lin et al. 2003), but it is unclear if this is a result of transport or the acceleration mechanism.

The following questions summarize the major issues related to energetic particle transport in flares:

- Are electron and ion pitch-angle scattering and associated transport suppression required to explain observations in flares?
- What are the mechanisms for scattering non-thermal electrons and ions?

2.6 How Does Reconnection Heat Plasma in Flares and Small Events (Nanoflares) That May Be Responsible for Heating the Corona?

The X-ray spectra from RHESSI reveal a hot thermal component at low energy (corresponding to temperatures of $10-30$ MK) that roll into powerlaw spectra at higher energy

(Fig. 2). Some of this hot thermal component arises from energetic electrons that ionize chromospheric neutrals that then fill coronal loops (the Neupert effect). However, the presence of the hot thermal component in coronal hard X-ray sources (Lin et al. 2003; Krucker et al. 2010; Oka et al. 2013) and other "hot onset" events (Hudson et al. 2021) suggests that magnetic reconnection drives both the hot thermal and energetic electrons. The corresponding temperature and energy content of flare ions in the range of tens of keV is not known. Direct flare measurements are limited to energies above ~ MeV while *in situ* measurements of ions are typically above 0.1 MeV/nuc (Reames et al. 1997; Mason et al. 2000). Measurements of electron and ion temperatures during *in situ* reconnection at 1AU yield temperature increments that scale with the magnetic energy released per particle

$$W_0 = B_r^2/4\pi n = m_i C_{Ar}^2, \tag{5}$$

with ions scaling like $0.13 W_0$ and electrons scaling like $0.017 W_0$ (Drake et al. 2009; Phan et al. 2013a, 2014; Øieroset et al. 2023). However, this partitioning is not likely to be universal (Shay et al. 2014; Haggerty et al. 2015). For typical flare parameters ($B \sim 50G$, $n \sim 5 \times 10^9/cm^3$) with half of the available energy going to bulk flow and one quarter each into ions and electrons, hot temperatures are around 30MK, which is in the range of hot thermal electrons in flares (Lin et al. 2003; Krucker et al. 2010; Oka et al. 2013). Nanoflares are potential drivers of coronal heating (Klimchuk 2006, 2015; Chhabra et al. 2021) and are basically just very small flares where the guide field dominates the reconnecting component of the magnetic field (Klimchuk 2015), which can have a major impact on the development of the reconnection (Leake et al. 2020).

The following are key questions on electron and ion bulk heating in flares:

- What controls the temperature of the hot thermal electrons and ions in flares?
- What fraction of the released magnetic energy goes to hot thermal electrons and ions?
- How is the energy associated with ion bulk flow, including the turbulent flows associated with the non-thermal broadening of spectral lines, ultimately dissipated?

3 Interchange Reconnection in the Low Corona as the Driver of the Fast Solar Wind

Winds of hot plasma and embedded magnetic field are produced by stars and other objects throughout the universe. The existence of a wind produced by the Sun was first proposed by Eugene Parker (Parker 1958, 1965) and was motivated by the observation that the tails of comets pointed away from the Sun. The existence of the wind was confirmed in observations by the Mariner II satellite (Neugebauer and Snyder 1962). The original Parker theory was based on an isothermal model in which the expansion force of the plasma pressure was sufficient to overcome the gravitational attraction of the Sun. Subsequent models included temperature equations with coronal heating profiles that were motivated by Alfvén wave turbulence propagating upwards from the solar surface. The "furnace" model presumed that these waves were injected into the corona from magnetic reconnection in the chromosphere (Axford and McKenzie 1992; Axford et al. 1999). Observations from Hinode/SOT of the transverse motion of spicules suggested that the dynamic convection zone was a source of these waves (De Pontieu et al. 2007). On the other hand, the energy deposition rate of these waves in the low corona of around $10^5 ergs/cm^2 s$ was somewhat below the $5-10 \times 10^5 ergs/cm^2 s$ required to drive the wind (Axford et al. 1999). In a more recent model the injection of MHD turbulence from magnetic reconnection was invoked to drive sufficient

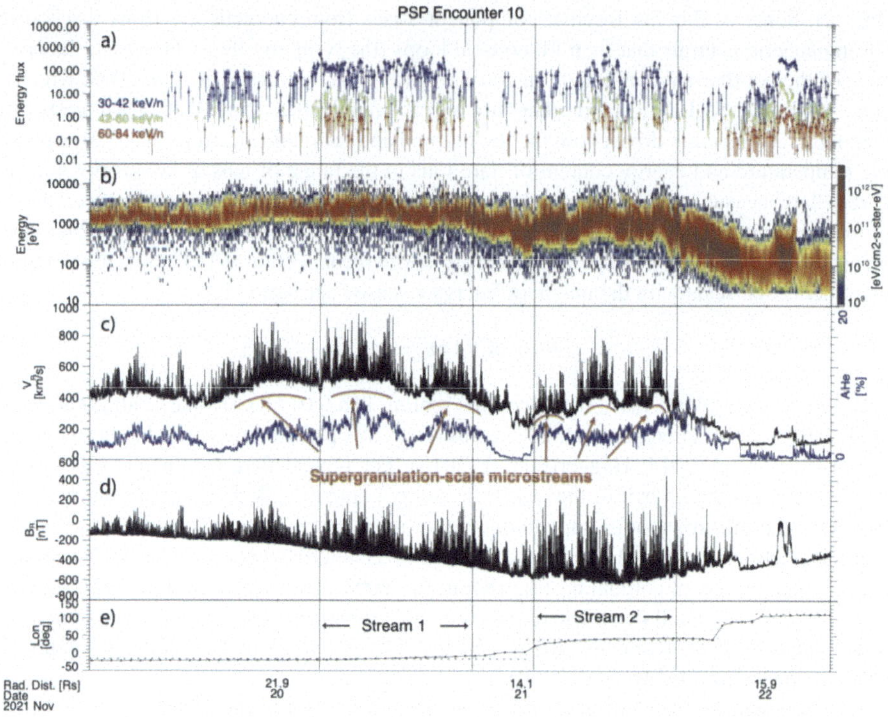

Fig. 4 Parker Solar Probe (PSP) observations of the solar wind during Encounter 10 at around 13 solar radii. The units in (a) are $\#/(\mathrm{cm}^2 - \mathrm{st} - \mathrm{s} - \mathrm{keV} - n^{-1})$

coronal heating to drive the wind (Zank et al. 2021). A significant caveat to the often-invoked reconnection drive mechanism for the turbulence required to heat the corona and drive the wind is that *in situ* measurements of reconnection outflow exhausts do not reveal that a significant fraction of the energy is carried in the MHD Poynting flux. Indeed, magnetotail observations suggest that this MHD Poynting flux is an order of magnitude below the energy carried in the ion enthalpy flux and bulk flow kinetic energy (Eastwood et al. 2013).

Recent observations of the structure of the fast solar wind from the Parker Solar Probe mission close to the Sun are for the first time revealing that the young solar wind is highly structured and this structure is providing clues on the wind drive mechanism. First, the radial wind speed is bursty and produces local reversals in the radial magnetic field, "switchbacks" (Bale et al. 2019). Further, this bursty wind is modulated on spatial scales that are linked to the corresponding spatial structure of the supergranulation network magnetic fields at the base of coronal holes (see Fig. 4 from encounter 10 near perihelion at around 13 solar radii) (Bale et al. 2021, 2023). The peaks in the radial flow speed in Fig. 4(c) are correlated with increases in the proton energy (in (b)) tails up to around 85 keV (in (a)), enhanced abundance of alpha particles (in (c)), and bursty reversals of the radial magnetic field (in (d)). The data from this time interval emerges from two corona holes as shown in (e). The modulation of the wind on scales that match that of the network magnetic fields suggests that interchange reconnection between closed loops of magnetic flux and open field lines is the source of energy that drives the wind and the associated modulations (Bale et al. 2021, 2023). These observational data have been used to infer the basic characteristics of interchange reconnec-

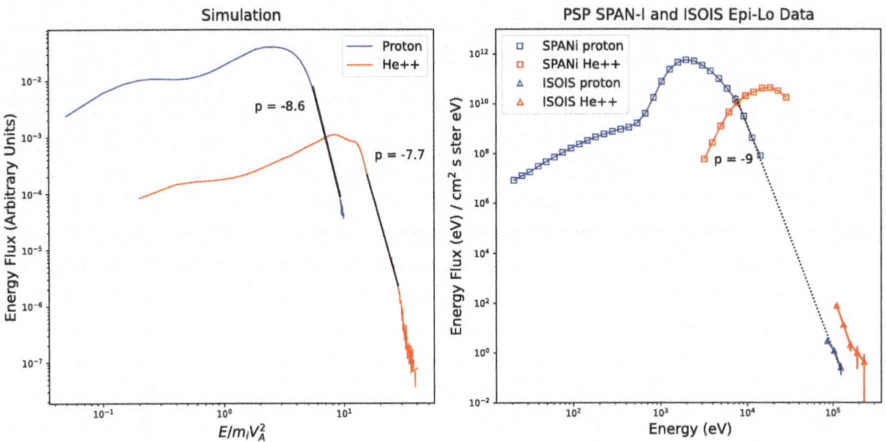

Fig. 5 Energy fluxes from the exhaust from a 2D PIC simulation of interchange reconnection in (a) and the corresponding fluxes from PSP SPANi and ISOIS Epi-Lo during Encounter 10 at around 13 solar radii

tion: magnetic fields of around 4.5G, plasma densities of around $10^9/\text{cm}^3$, Alfvén speeds of between 300 and 400 km/s, reconnection inflow speeds of around 3 km/s (deeply in the collisionless regime) with energy release rates of around 5×10^5 ergs/cm²s. This rate of energy release is sufficient to power the wind (McKenzie et al. 1995; Axford et al. 1999).

A major discovery of PSP observations of the young solar wind is that the wind is an "energetic" wind in that there is a high energy component to the wind as revealed in the data in Fig. 4(a)-(b). In Fig. 5 we show the spectra of the energy flux of protons and alpha particles from a 2D PIC simulation of interchange reconnection (in (a)) and from data from SPANi and ISOIS Epi-Lo during the interval 4:00:00-19:00:00 on November 20, 2021, from Fig. 4 (in (b)) (Bale et al. 2023). All of the spectra roll over into soft powerlaws at high energy with the most energetic ions having energies greater than 100 keV. The simulations have a free parameter, $W_0 = m_i C_{Ar}^2$. Matching the proton energy at the low energy limit of the powerlaw in the simulation with that from the data yields $m_i C_{Ar}^2 \sim 1.6$ keV at the coronal reconnection site. The spectral index in the simulations is controlled by the ambient guide magnetic field, which is $0.55B_r$ for the data shown. The surprisingly good agreement between the spectra from the simulation and from the observations strongly supports the conclusion that interchange reconnection is the driver of the energetic wind documented in the PSP observations.

4 Energetic Ion Spectra Resulting from Reconnection at the Near Coronal Heliospheric Current Sheet

The heliospheric current sheet (HCS) separates the oppositely directed magnetic fields in the Northern and Southern latitudes of the heliospheric magnetic field. Satellite *in situ* measurements in the 1AU solar wind only occasionally reveal evidence for reconnection even though the magnetic shear across the HCS is close to 180 degrees. Identified HCS reconnection exhausts reveal bifurcated current sheets in which the magnetic field within the exhaust sharply drops to a small value with corresponding increases in the exhaust velocity, the plasma density and the ion temperature (Gosling et al. 2005). The change in the ambient electron temperature is typically not measurable. Thus, the exhausts are of the general form predicted in Petschek's model (Petschek 1964) where a pair of switch-off slow-mode shocks bound the exhaust. The increase in the ion temperature typically follows the scaling law $\Delta T \sim 0.13 W_0$ (see Sect. 2.4) and is in the range of 10-20 eV. The measured dropout of the strahl electrons in reconnection exhausts with velocity increments in the anti-sunward direction is consistent with reconnected field lines being disconnected from the Sun (Gosling et al. 2006a). The observations of bi-directional Strahl electrons within reconnection exhausts with velocity increments in the sunward direction are consistent with magnetic fields on both sides of the exhaust maintaining their connection to the Sun (Gosling et al. 2006b). None of the solar wind reconnection events measured at 1AU revealed evidence that local reconnection produced an energetic particle spectrum (Gosling 2007). On the other hand, the ratio of the available magnetic to the particle energy σ (see Eq. (1)) is of order unity so there may be insufficient magnetic energy available for strong particle energization.

Encounters of the PSP with the HCS close to the Sun (in the range of 30-100 R_s) have revealed that, in contrast with measurements at 1AU, the HCS is almost always undergoing reconnection (Phan et al. 2021). This result is a surprise since the observations also reveal that the measured widths of the exhausts are typically thousands of d_i with $d_i = c/\omega_{pi}$ the ion inertial scale. The expectation was that reconnection would only be triggered in current sheets that have widths that are of order of d_i. The basic characteristics of HCS reconnection exhausts close to the Sun are similar to those at 1AU: bifurcated current sheets with strahl dropouts as expected depending on the orientation of the outflow with respect to the Sun.

PSP observations of HCS reconnection events inside around $20R_s$ are for the first time revealing energetic ions. The strongest evidence that reconnection at the HCS was a driver of an energetic ion spectrum was from an event on April 29, 2021, during Encounter 08, that is shown in Fig. 6 (Phan et al. 2022). The basic characteristics of this event were similar to those further from the Sun. In this event a spectrum of energetic ions up to around 40 keV measured by IS⊙IS/EPI-Lo spanned the entire exhaust and extended into the ambient solar wind on either side (Fig. 6(a)). The escape of energetic particles upstream as seen in this data is expected if the conventional Petschek picture of the exhaust boundary as a rotational discontinuity is correct. In this model all the magnetic flux downstream of the reconnection separatrix threads through the exhaust with the consequence that energetic particles within the exhaust can escape upstream. The field lines that thread the exhaust but lie outside of the exhaust are labeled as the separatrix region in the schematic at the bottom of Fig. 6. In this event there is also clear evidence that the highest energy protons that make up the core of the exhaust also escape upstream (see the region to the left of the exhaust in Fig. 6(b) and the beam-like features in the particle distributions in Fig. 6(k-l)). These ions exhibit the usual time-of-flight signature in which the highest energies escape further upstream than the lower energies. The dropout of the strahl electrons (Fig. 6(d)) confirms that the magnetic fields just prior to the entry of PSP into the exhaust are disconnected from the Sun and are therefore separatrix field lines. The peak energy of ions of around 40 keV greatly exceeds the characteristic counter-streaming Alfvénic velocities associated with a single ion encounter with a reconnected field line (Hoshino et al. 1998; Gosling et al. 2005). This suggests that reconnection in this event must involve multiple x-line dynamics somewhere upstream of the PSP crossing location. The core heating of ions in the exhaust that is evident in the velocity distributions (see Fig. 6(m-n)) is further evidence that a mechanism for strong ion energization occurs during reconnection at the HCS. A surprise in these data is that the core energetic ions only leak upstream on the entry side of the exhaust and not on the exit side of the exhaust. This suggests that the exit side of the exhaust is a tangential rather than a rotational discontinuity. On the other hand, the most energetic ions were measured

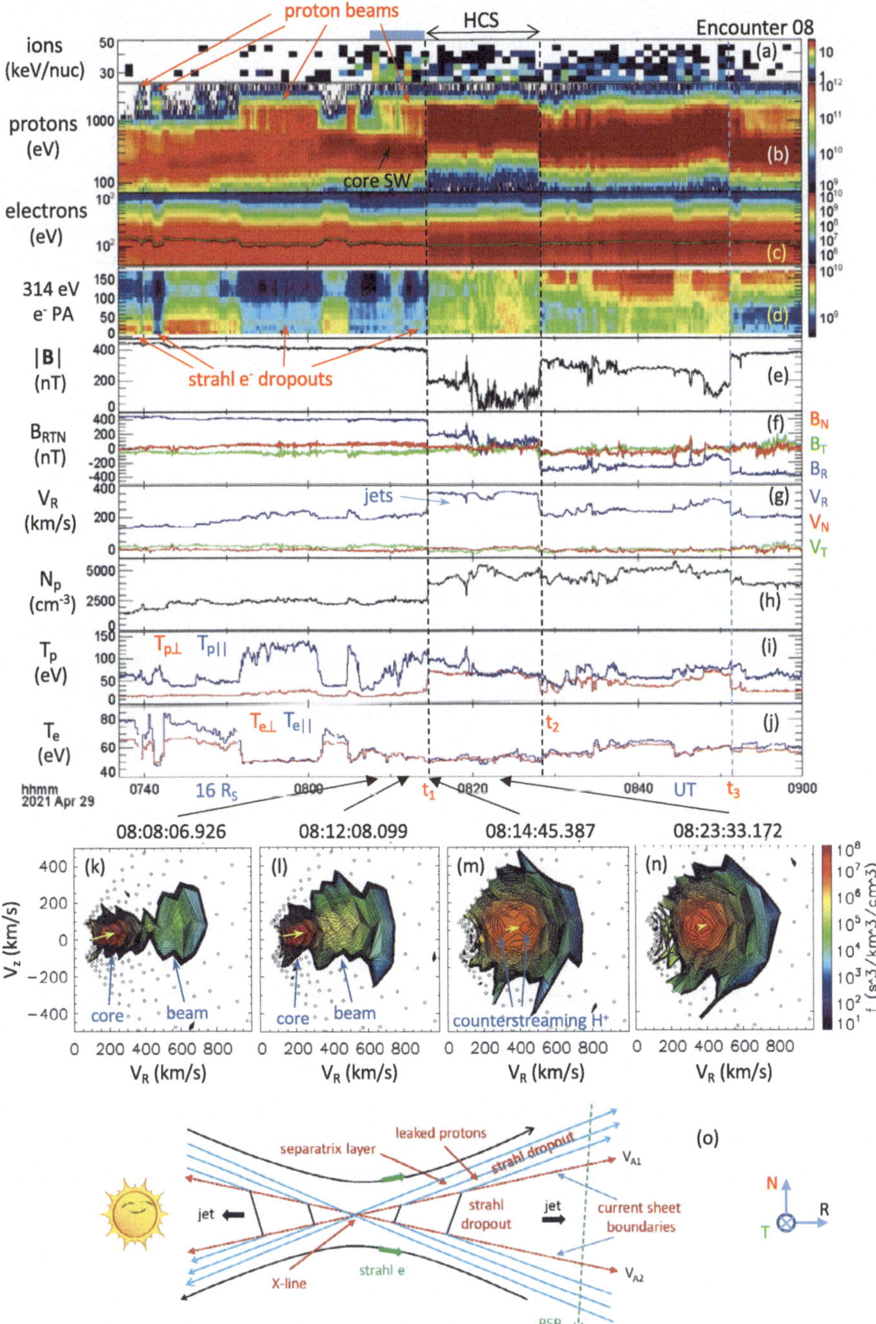

Fig. 6 Parker Solar Probe (PSP) observations of magnetic reconnection at the heliospheric current sheet at $R \sim 16R_{\odot}$, revealing the generation of energetic ions. The IS⊙IS/EPI-Lo data were obtained from double coincidence time-of-flight measurements, which contain background counts

outside of the exhaust on the exit side (Phan et al. 2022), suggesting that the boundary was a rotational discontinuity. These results indicate that major uncertainties in the structure of these exhausts remain.

Significant uncertainties about the conditions under which reconnection at the HCS drives a spectrum of energetic particles remain. The data in Fig. 6(c) do not reveal significant electron energization, perhaps because heated electrons escape from the exhaust. In a January 17, 2021, encounter with a HCS reconnection event a spectrum of energetic ions was also measured but only just outside and not inside of the exhaust, suggesting that the energetic ions were not locally produced (Desai et al. 2022; Phan et al. 2022). On the other hand, there was no evidence of time delay signatures, which would be expected for a remote source.

5 New Macro-Scale Kinetic Modeling and Observational Tools

New computational and observational analysis tools will play a major role in further advancing our understanding of energetic reconnection events throughout the heliosphere. A major challenge in the effort to understand the energization of electrons and ions in flares has been the inability to explore the kinetic dynamics of particles in very large systems. Because particle-in-cell (PIC) models have to resolve kinetic scales, the upper limit on the computational domains that they describe is $\sim 10^3$ m in the corona, compared with the characteristic size of large flares, $\sim 10^4$ km. The consequence is that although PIC simulations have been successful in identifying some of the dominant acceleration mechanisms of electrons and ions (Drake et al. 2006b; Dahlin et al. 2014; Guo et al. 2014; Li et al. 2019), their success in producing the extended powerlaws seen in flare observations has been limited. The largest PIC simulations have revealed electron powerlaws that extend only a single decade in energy (Li et al. 2019) while observations reveal powerlaws that extend across many decades (Lin et al. 2003; Vilmer 2012). The production of powerlaw distributions of ions has been an even greater challenge (Zhang et al. 2021, 2024).

However, PIC modeling revealed that Fermi reflection during the growth and merger of large-scale flux ropes dominates particle acceleration (Dahlin et al. 2016; Guo et al. 2014; Li et al. 2019), a result that led to a new computational model, *kglobal*, that combined MHD fluid and particle descriptions while eliminating the kinetic scales that constrained the macro-scale modeling of flares (Drake et al. 2019; Arnold et al. 2019). A major accomplishment was the first exploration of magnetic reconnection-driven electron acceleration in macro-systems which produced electron powerlaw distributions extending nearly three decades in energy and revealed that the ambient guide field is the dominant control factor of the powerlaw index (Arnold et al. 2021). The success of the *kglobal* model motivates two major extensions: to include particle ions so that the partitioning of energy between the two species can be explored; and to incorporate the *kglobal* particle algorithm into a global simulation code such as the Adaptively Refined MHD Solver (ARMS) flare simulation code (DeVore 1991; Antiochos et al. 1999; Karpen et al. 2012; Dahlin et al. 2019). The resulting model would be able to simulate macroscale flares with energetic electrons and produce synthetic photon spectra for comparison with observations.

Hard x-ray measurements from RHESSI and NuSTAR along with EOVSA MW observations have already played a critical role in exploring the evolution of energetic particle spectra (primarily electron) in flares and the STIX instrument on the Solar Orbiter will provide additional data during the upcoming solar cycle. However, past and present hard x-ray observations suffer from insufficient dynamic range, with the consequence that emission from chromospheric footpoints blinds x-ray emission in the lower density corona where magnetic energy is released. The deployment of a FOXSI-like instrument with much greater dynamic range would greatly enhance the exploration of electron spectra in the source region of flares (Christe et al. 2016). Further development of MW analysis tools by adding polarization measurements would facilitate direct measurements of the coronal magnetic field. Simulations have suggested the importance of the guide field in mediating particle acceleration and, while EOVSA has for the first time measured the dynamics of total B in the flaring region, measuring the evolving out-of-plane component is essential for comparison with theory.

Finally, an instructive counterpart to particle energization in solar flares arises in magnetotail reconnection events. Despite the stark difference in parameters from the coronal case ($n \sim 0.1\ \mathrm{cm}^{-3}$, $B \sim 10$ nT), the magnetic energy released per particle $W_0 = m_i C_{Ar}^2$ is comparable to the coronal case. Direct observations of magnetotail events show hot thermal components in the range of 5-10 keV and powerlaw distributions extending up to nearly 500 keV (Ergun et al. 2020). The ability to explore the full range of particle energy spectra with *in situ* measurements combined with direct measurements of the electromagnetic fields make these events a unique natural laboratory for exploring flare-relevant particle acceleration.

We expect that major progress in answering the questions posed above and understanding magnetic reconnection, in general, will be made by combining the remote-sensing macro-scale observations with in situ micro-scale measurements, just like *kglobal* has delivered important advances in theoretical understanding by combining the global-scale MHD with micro-scale PIC models.

Funding Open Access funding provided by The University of Tokyo. We acknowledge support from the FIELDS team of the Parker Solar Probe (NASA Contract No. NNN06AA01C), the NASA Drive Science Center on Solar Flare Energy Release (SolFER) under Grant 80NSSC20K0627, NASA Grants 80NSSC20K1277 and 80NSSC22K0352 and NSF Grant PHY2109083.

Declarations

Competing Interests The authors have no competing interests to declare that are relevant to the content of this article.

References

Abdo AA, et al (2011) Science 331:739. https://doi.org/10.1126/science.1199705

Antiochos SK, DeVore CR, Klimchuk JA (1999) Astrophys J 510:485. https://doi.org/10.1086/306563

Arnold H, Drake JF, Swisdak M, Dahlin J (2019) Phys Plasmas 26:102903. https://doi.org/10.1063/1.5120373

Arnold H, Drake JF, Swisdak M, et al (2021) Phys Rev Lett 126:135101. https://doi.org/10.1103/PhysRevLett.126.135101

Aschwanden MJ, Hudson H, Kosugi T, Schwartz RA (1996) Astrophys J 464:985. https://doi.org/10.1086/177386

Aschwanden MJ, Holman G, O'Flannagain A, et al (2016) Astrophys J 832:27. https://doi.org/10.3847/0004-637X/832/1/27
Axford WI, McKenzie JF (1992) In: Marsch E, Schwenn R (eds) Solar Wind Seven, Pergamon Press, Oxford, pp 1–5. https://doi.org/10.1016/B978-0-08-042049-3.50004-1
Axford WI, McKenzie JF, Sukhorukova GV, et al (1999) Space Sci Rev 87:25. https://doi.org/10.1023/A:1005197529250
Bale SD, Goetz K, Harvey PR, et al (2016) Space Sci Rev 204:49. https://doi.org/10.1007/s11214-016-0244-5
Bale SD, Badman ST, Bonnell JW, et al (2019) Nature 576:237. https://doi.org/10.1038/s41586-019-1818-7
Bale SD, Horbury TS, Velli M, et al (2021). Astrophys J 923:174. https://doi.org/10.3847/1538-4357/ac2d8c
Bale SD, Drake JF, McManus MD, et al (2023) Nature 618:252. https://doi.org/10.1038/s41586-023-05955-3
Birn J, Drake JF, Shay MA, et al (2001) J Geophys Res 106:3715
Blandford RD, Ostriker JP (1978) Astrophys J Lett 221:L29. https://doi.org/10.1086/182658
Burch JL, Webster JM, Hesse M, et al (2020) Geophys Res Lett 47:e89082. https://doi.org/10.1029/2020GL089082
Carrington RC (1859) Mon Not R Astron Soc 20:13. https://doi.org/10.1093/mnras/20.1.13
Cattell C, Vo T (2021) Astrophys J Lett 914:L33. https://doi.org/10.3847/2041-8213/ac08a1
Chen B, Shen C, Gary DE, et al (2020) Nat Astron 4:1140–1147. https://doi.org/10.1038/s41550-020-1147-7
Chen B, Battaglia M, Krucker S, Reeves KK, Glesener L (2021) Astrophys J Lett 908:L55. https://doi.org/10.3847/2041-8213/abe471
Chhabra S, Klimchuk JA, Gary DE (2021). Astrophys J 922:128. https://doi.org/10.3847/1538-4357/ac2364
Christe S, Glesener L, Buitrago-Casas C, et al (2016) J Astron Instrum 5:1640005. https://doi.org/10.1142/S2251171716400055
Dahlin JT, Drake JF, Swisdak M (2014) Phys Plasmas 21:092304. https://doi.org/10.1063/1.4894484
Dahlin JT, Drake JF, Swisdak M (2015) Phys Plasmas 22:100704. https://doi.org/10.1063/1.4933212
Dahlin JT, Drake JF, Swisdak M (2016) Phys Plasmas 23:120704. https://doi.org/10.1063/1.4972082
Dahlin JT, Drake JF, Swisdak M (2017) Phys Plasmas 24:092110. https://doi.org/10.1063/1.4986211
Dahlin JT, Antiochos SK, DeVore CR (2019) Astrophys J 879:96. https://doi.org/10.3847/1538-4357/ab262a
Dahlin JT, Antiochos SK, Qiu J, DeVore CR (2022) Astrophys J 932:94. https://doi.org/10.3847/1538-4357/ac6e3d
Daughton W, Roytershteyn V, Yin L, Karimabadi H, et al (2011) Nat Phys 7:539
Daughton W, Nakamura TKM, Karimabadi H, Roytershteyn V, Loring B (2014) Phys Plasmas 21:052307. https://doi.org/10.1063/1.4875730
Davenport JRA (2016) Astrophys J 829:23. https://doi.org/10.3847/0004-637X/829/1/23
De Pontieu B, McIntosh SW, Carlsson M, et al (2007) Science 318:1574. https://doi.org/10.1126/science.1151747
Desai MI, Mitchell DG, McComas DJ, et al (2022) Astrophys J 927:62. https://doi.org/10.3847/1538-4357/ac4961
DeVore CR (1991) J Comput Phys 92:142. https://doi.org/10.1016/0021-9991(91)90295-V
Drake JF, Swisdak M, Che H, Shay MA (2006a) Nature 443:553–556. https://doi.org/10.1038/nature05116
Drake JF, Swisdak M, Schoeffler KM, Rogers BN, Kobayashi S (2006b) Geophys Res Lett 33:13105
Drake JF, Swisdak M, Phan TD, et al (2009) J Geophys Res 114:A05111
Drake JF, Opher M, Swisdak M, Chamoun JN (2010) Astrophys J 709:963
Drake JF, Arnold H, Swisdak M, Dahlin JT (2019) Phys Plasmas 26:012901
Eastwood JP, Phan TD, Drake JF, et al (2013) Phys Rev Lett 110:225001. https://doi.org/10.1103/PhysRevLett.110.225001
Emslie AG, Kucharek H, Dennis BR, et al (2004) J Geophys Res Space Phys 109:10104. https://doi.org/10.1029/2004JA010571
Emslie AG, Dennis BR, Holman GD, Hudson HS (2005) J Geophys Res 110:A11103
Emslie AG, Dennis BR, Shih AY, et al (2012) Astrophys J 759:71. https://doi.org/10.1088/0004-637X/759/1/71
Ergun RE, Ahmadi N, Kromyda L, et al (2020) Astrophys J 898:154. https://doi.org/10.3847/1538-4357/ab9ab6
Fisk LA (1978) Astrophys J 224:1048
Fitzmaurice A, Drake JF, Swisdak M (2024). Astrophys J 964:97. https://doi.org/10.3847/1538-4357/ad217f
Fleishman GD, Gary DE, Chen B, et al (2020) Science 367:278. https://doi.org/10.1126/science.aax6874
Fleishman GD, Nita GM, Chen B, Yu S, Gary DE (2022) Nature 606:674. https://doi.org/10.1038/s41586-022-04728-8
Forbes TG (2000) J Geophys Res 105:23153. https://doi.org/10.1029/2000JA000005
Gary SP, Foosland DW, Smith CW, Lee MA, Goldstein ML (1984) Phys Fluids 27:1852. https://doi.org/10.1063/1.864797

Gary DE, Chen B, Dennis BR, et al (2018) Astrophys J 863:83. https://doi.org/10.3847/1538-4357/aad0ef
Glesener L, Krucker S, Hannah IG, et al (2017) Astrophys J 845:122. https://doi.org/10.3847/1538-4357/aa80e9
Gömöry P, Veronig AM, Su Y, Temmer M, Thalmann JK (2016) Astron Astrophys 588:A6. https://doi.org/10.1051/0004-6361/201527403
Gosling JT (2007) Astrophys J Lett 671:L73
Gosling JT, Skoug RM, Haggerty DK, McComas DJ (2005) Geophys Res Lett 32:L14113
Gosling JT, McComas DJ, Skoug RM, Smith CW (2006a) Geophys Res Lett 33:L17102. https://doi.org/10.1029/2006GL027188
Gosling JT, McComas DJ, Skoug RM, Smith CW (2006b) Geophys Res Lett 33:L17102. https://doi.org/10.1029/2006GL027188
Grefenstette BW, Glesener L, Krucker S, et al (2016) Astrophys J 826:20. https://doi.org/10.3847/0004-637X/826/1/20
Guo F, Li H, Daughton W, Liu Y-H (2014) Phys Rev Lett 113:155005. https://doi.org/10.1103/PhysRevLett.113.155005
Guo F, Liu Y-H, Zenitani S, Hoshino M (2024). Space Sci Rev 220:43. https://doi.org/10.1007/s11214-024-01073-2
Haggerty CC, Shay MA, Drake JF, Phan TD, McHugh CT (2015) Geophys Res Lett 42:9657. https://doi.org/10.1002/2015GL065961
Haggerty CC, Shay MA, Chasapis A, et al (2018) Phys Plasmas 25:102120. https://doi.org/10.1063/1.5050530
Harra LK, Matthews S, Culhane JL, et al (2013) Astrophys J 774:122. https://doi.org/10.1088/0004-637X/774/2/122
Harra L, Matthews S, Long D, et al (2020) Sol Phys 295:34. https://doi.org/10.1007/s11207-020-01602-6
Hesse M, Forbes TG, Birn J (2005) Astrophys J 631:1227. https://doi.org/10.1086/432677
Holman GD, Sui L, Schwartz RA, Emslie AG (2003) Astrophys J 595:L97
Hoshino M, Mukai T, Yamamoto T (1998) J Geophys Res 103:4509
Huang Y-M, Bhattacharjee A (2016) Astrophys J 818:20. https://doi.org/10.3847/0004-637X/818/1/20
Hudson HS, Simões PJA, Fletcher L, Hayes LA, Hannah IG (2021) Mon Not R Astron Soc 501:1273. https://doi.org/10.1093/mnras/staa3664
Jiang C, Feng X, Liu R, et al (2021) Nat Astron 5:1126–1138. https://doi.org/10.1038/s41550-021-01414-z
Karimabadi H, Krauss-Varban D, Terasawa T (1992) J Geophys Res 97:13. https://doi.org/10.1029/92JA00997
Karpen JT, Antiochos SK, DeVore CR (2012) Astrophys J 760:81. https://doi.org/10.1088/0004-637X/760/1/81
Kasper JC, Abiad R, Austin G, et al (2016) Space Sci Rev 204:131. https://doi.org/10.1007/s11214-015-0206-3
Kaspi VM, Beloborodov AM (2017) Annu Rev Astron Astrophys 55:261. https://doi.org/10.1146/annurev-astro-081915-023329
Kliem B (1994) Astrophys J 90:719
Kliem B, Török T (2006) Phys Rev Lett 96:255002. https://doi.org/10.1103/PhysRevLett.96.255002
Klimchuk JA (2006) Sol Phys 234:41. https://doi.org/10.1007/s11207-006-0055-z
Klimchuk JA (2015) Philos Trans R Soc Lond Ser A 373:20140256. https://doi.org/10.1098/rsta.2014.0256
Kontar EP, Bian NH, Emslie AG, Vilmer N (2014) Astrophys J 780:176. https://doi.org/10.1088/0004-637X/780/2/176
Kontar EP, Perez JE, Harra LK, et al (2017) Phys Rev Lett 118:155101. https://doi.org/10.1103/PhysRevLett.118.155101
Krucker S, Hudson HS, White SM, et al (2010) Astrophys J 714:1108
Krucker S, Hurford GJ, Grimm O, et al (2020) Astron Astrophys 642:A15. https://doi.org/10.1051/0004-6361/201937362
Leake JE, Daldorff LKS, Klimchuk JA (2020) Astrophys J 891:62. https://doi.org/10.3847/1538-4357/ab7193
Lee WW (1983) Phys Fluids 26:556–562. https://doi.org/10.1063/1.864140
Li X, Guo F, Li H, Stanier A, Kilian P (2019) Astrophys J 884:118. https://doi.org/10.3847/1538-4357/ab4268
Lin RP, Hudson HS (1971) Sol Phys 17:412–435. https://doi.org/10.1007/BF00150045
Lin RP, Dennis BR, Hurford GJ, et al (2002) Sol Phys 210:3. https://doi.org/10.1023/A:1022428818870
Lin RP, Krucker S, Hurford GJ, et al (2003) Astrophys J 595:L69. https://doi.org/10.1086/378932
Lin J, et al (2005) Astrophys J 622:1251. https://doi.org/10.1086/428110
Liu S, Petrosian V, Mason GM (2006) Astrophys J 636:462
Liu Y-H, Cassak P, Li X, et al (2022) Commun Phys 5:97. https://doi.org/10.1038/s42005-022-00854-x

Ma H, Drake JF, Swisdak M (2023) Astrophys J 954:21. https://doi.org/10.3847/1538-4357/ace59e
Mason GM (2007) Space Sci Rev 130:231–242. https://doi.org/10.1007/s11214-007-9156-8
Mason GM, Dwyer JR, Mazur JE (2000) Astrophys J Lett 545:L157. https://doi.org/10.1086/317886
McComas DJ, Alexander N, Angold N, et al (2016) Space Sci Rev 204:187. https://doi.org/10.1007/s11214-014-0059-1
McKenzie JF, Banaszkiewicz M, Axford WI (1995) Astron Astrophys 303:L45
Miller JA, Cargill PJ, Emslie AG, et al (1997) J Geophys Res 102:14631
Musset S, Maksimovic M, Kontar E, et al (2021). Astron Astrophys 656:A34. https://doi.org/10.1051/0004-6361/202140998
Nakamura TKM, Genestreti KJ, Liu YH, et al (2018) J Geophys Res Space Phys 123:9150. https://doi.org/10.1029/2018JA025713
Neugebauer M, Snyder CW (1962) Science 138:1095. https://doi.org/10.1126/science.138.3545.1095.a
Øieroset M, Lin RP, Phan TD, Larson DE, Bale SD (2002) Phys Rev Lett 89:195001
Øieroset M, Phan TD, Oka M, et al (2023) Astrophys J 954:118. https://doi.org/10.3847/1538-4357/acdf44
Oka M, Phan TD, Krucker S, Fujimoto M, Shinohara I (2010) Astrophys J 714:915
Oka M, Ishikawa S, Saint-Hilaire P, Krucker S, Lin RP (2013) Astrophys J 764:6. https://doi.org/10.1088/0004-637X/764/1/6
Oka M, Phan T, Øieroset M, et al (2022) Phys Plasmas 29:052904. https://doi.org/10.1063/5.0085647
Oka M, Birn J, Egedal J, et al (2023) Space Sci Rev 219:75. https://doi.org/10.1007/s11214-023-01011-8
Omodei N, Pesce-Rollins M, Longo F, Allafort A, Krucker S (2018) Astrophys J Lett 865:L7. https://doi.org/10.3847/2041-8213/aae077
Parker EN (1957) J Geophys Res 62:509–520. https://doi.org/10.1029/JZ062i004p00509
Parker EN (1958) Astrophys J 128:664. https://doi.org/10.1086/146579
Parker EN (1965) Planet Space Sci 13:9–49. https://doi.org/10.1016/0032-0633(65)90131-5
Paschmann G, Papamastorakis I, Baumjohann W, et al (1986) J Geophys Res 91:11099. https://doi.org/10.1029/JA091iA10p11099
Petrosian V (2012) Space Sci Rev 173:535. https://doi.org/10.1007/s11214-012-9900-6
Petrosian V, Liu S (2004) Astrophys J 610:550
Petschek HE (1964) In: Ness WN (ed) The physics of solar flares. NASA, Washington, p 425–439
Phan TD, Gosling JT, Davis MS, et al (2006) Nature 439:175
Phan TD, Paschmann G, Gosling JT, et al (2013a) Geophys Res Lett 40:11. https://doi.org/10.1029/2012GL054528
Phan TD, Shay MA, Gosling JT, et al (2013b) Geophys Res Lett 40:4475. https://doi.org/10.1002/grl.50917
Phan TD, Drake JF, Shay MA, et al (2014) Geophys Res Lett 41:7002. https://doi.org/10.1002/2014GL061547
Phan TD, Lavraud B, Halekas JS, et al (2021) Astron Astrophys 650:A13. https://doi.org/10.1051/0004-6361/202039863
Phan TD, Verniero JL, Larson D, et al (2022) Geophys Res Lett 49:e96986. https://doi.org/10.1029/2021GL096986
Polito V, Reeves KK, Del Zanna G, Golub L, Mason HE (2015) Astrophys J 803:84. https://doi.org/10.1088/0004-637X/803/2/84
Qiu J, Cheng J (2022) Sol Phys 297:80. https://doi.org/10.1007/s11207-022-02003-7
Qiu J, Longcope DW, Cassak PA, Priest ER (2017) Astrophys J 838:17. https://doi.org/10.3847/1538-4357/aa6341
Qiu J, Alaoui M, Antiochos SK, et al (2023) Astrophys J 955:34. https://doi.org/10.3847/1538-4357/acebeb
Ramaty R, Mandzhavidze N, Kozlovsky B (1996) AIP Conf Proc 374:172–183. https://doi.org/10.1063/1.50953
Reames DV, Barbier LM, Von Rosenvinge TT, et al (1997) Astrophys J 483:515. https://doi.org/10.1086/304229
Roberg-Clark GT, Drake JF, Reynolds CS, Swisdak M (2016) Astrophys J Lett 830:L9. https://doi.org/10.3847/2041-8205/830/1/L9
Roberg-Clark GT, Drake JF, Reynolds CS, Swisdak M (2018) Phys Rev Lett 120:035101. https://doi.org/10.1103/PhysRevLett.120.035101
Roberg-Clark GT, Agapitov O, Drake JF, Swisdak M (2019) Astrophys J 887:190. https://doi.org/10.3847/1538-4357/ab5114
Roth I, Temerin M (1997) Astrophys J 477:940
Rust DM (1976) Sol Phys 47:21. https://doi.org/10.1007/BF00152243
Schlickeiser R (1989) Astrophys J 336:243. https://doi.org/10.1086/167009
Schmieder B, Aulanier G, Vršnak B (2015) Sol Phys 290:3457. https://doi.org/10.1007/s11207-015-0712-1
Shay MA, Drake JF, Rogers BN, Denton RE (1999) Geophys Res Lett 26:2163
Shay MA, Drake JF, Swisdak M (2007) Phys Rev Lett 99:155002

Shay MA, Haggerty CC, Phan TD, et al (2014) Phys Plasmas 21:122902. https://doi.org/10.1063/1.4904203
Shih AY, Lin RP, Smith DM (2009) Astrophys J 698:L152
Simões PJA, Kontar EP (2013) A & A 551:A135. https://doi.org/10.1051/0004-6361/201220304
Simões PJA, Graham DR, Fletcher L (2015) Astron Astrophys 577:A68. https://doi.org/10.1051/0004-6361/201424795
Sironi L, Spitkovsky A (2014) Astrophys J 783:L21. https://doi.org/10.1088/2041-8205/783/1/L21
Sterling AC, Moore RL, Falconer DA, Adams M (2015) Nature 523:437. https://doi.org/10.1038/nature14556
Sweet PA (1958) In: Lehnert B (ed) Electromagnetic phenomena in cosmical physics. Cambridge University Press, Cambridge, p 123
Torbert RB, Burch JL, Phan TD, et al (2018) Science 362:1391. https://doi.org/10.1126/science.aat2998
Tritschler A, Rimmele TR, Berukoff S, et al (2016) Astron Nachr 337:1064. https://doi.org/10.1002/asna.201612434
Vasko IY, Krasnoselskikh V, Tong Y, et al (2019) Astrophys J Lett 871:L29. https://doi.org/10.3847/2041-8213/ab01bd
Vilmer N (2012) Philos Trans R Soc Lond Ser A 370:3241. https://doi.org/10.1098/rsta.2012.0104
Warmuth A, Mann G (2016) Astron Astrophys 588:A116. https://doi.org/10.1051/0004-6361/201527475
Winske D, Leroy MM (1984) J Geophys Res 89:2673. https://doi.org/10.1029/JA089iA05p02673
Wyper PF, Antiochos SK, DeVore CR (2017) Nature 544:452. https://doi.org/10.1038/nature22050
Yin Z, Drake JF, Swisdak M, et al (2024) Astrophys J 974:74. https://doi.org/10.3847/1538-4357/ad7131
Zank GP, Zhao LL, Adhikari L, et al (2021) Phys Plasmas 28:080501. https://doi.org/10.1063/5.0055692
Zhang Q, Guo F, Daughton W, Li X, Li H (2021) Phys Rev Lett 127:185101. https://doi.org/10.1103/PhysRevLett.127.185101
Zhang Q, Guo F, Daughton W, et al (2024) Phys Rev Lett 132:115201. https://doi.org/10.1103/PhysRevLett.132.115201

Authors and Affiliations

J.F. Drake[1] · S.K. Antiochos[2] · S.D. Bale[3] · Bin Chen[4] · C.M.S. Cohen[5] · J.T. Dahlin[6] · Lindsay Glesener[7] · F. Guo[8] · M. Hoshino[9] · Shinsuke Imada[9] · M. Oka[10] · T.D. Phan[10] · Katherine K. Reeves[11] · M. Swisdak[2]

✉ J.F. Drake
drake@umd.edu

✉ S. Imada
imada@eps.s.u-tokyo.ac.jp

1 Department of Physics, Institute for Physical Science and Technology and the Joint Space Science Institute, University of Maryland, College Park, MD 20742, USA

2 Institute for Research in Electronics and Applied Physics, University of Maryland, College Park, MD 20742, USA

3 Physics Department and Space Sciences Laboratory, University of California, Berkeley, CA 94720, USA

4 Center for Solar-Terrestrial Research, New Jersey Institute of Technology, 323 M L King Jr Blvd, Newark, NJ 07102-1982, USA

5 California Institute of Technology, Pasadena, CA 91125, USA

6 Heliophysics Science Division, NASA Goddard Space Flight Center, Greenbelt, MD 20771, USA

7 University of Minnesota, Minneapolis, MN, USA

8 Los Alamos National Laboratory, Los Alamos, NM 87545, USA

[9] Department of Earth and Planetary Science, The University of Tokyo, Tokyo 113-0033, Japan

[10] Space Sciences Laboratory, University of California, Berkeley, CA 94720, USA

[11] Center for Astrophysics, Harvard & Smithsonian, Cambridge, MA 02138, USA

Space Science Reviews (2024) 220:7
https://doi.org/10.1007/s11214-023-01017-2

Magnetic Reconnection at Planetary Bodies and Astrospheres

Daniel J. Gershman[1] · Stephen A. Fuselier[2] · Ian J. Cohen[3] · Drew L. Turner[3] · Yi-Hsin Liu[4] · Li-Jen Chen[1] · Tai D. Phan[5] · Julia E. Stawarz[6] · Gina A. DiBraccio[1] · Adam Masters[7] · Robert W. Ebert[2] · Weijie Sun[8] · Yuki Harada[9] · Marc Swisdak[10]

Received: 5 May 2023 / Accepted: 12 October 2023 / Published online: 19 January 2024

Abstract
Magnetic reconnection is a fundamental mechanism for the transport of mass and energy in planetary magnetospheres and astrospheres. While the process of reconnection is itself ubiquitous across a multitude of systems, the techniques used for its analysis can vary across scientific disciplines. Here we frame the latest understanding of reconnection theory by missions such as NASA's Magnetospheric Multiscale (MMS) mission for use throughout the solar system and beyond. We discuss how reconnection can couple magnetized obstacles to both sub- and super-magnetosonic upstream flows. In addition, we address the need to model sheath plasmas and field-line draping around an obstacle to accurately parameterize the possibility for reconnection to occur. We conclude with a discussion of how reconnection energy conversion rates scale throughout the solar system. The results presented are not only applicable to within our solar system but also to astrospheres and exoplanets, such as the first recently detected exoplanet magnetosphere of HAT-11-1b.

Keywords Reconnection · Planetary magnetospheres · Astrospheres · Magnetosheath · Magnetized obstacle · Plasma beta

✉ D.J. Gershman
daniel.j.gershman@nasa.gov

✉ J.E. Stawarz
julia.stawarz@northumbria.ac.uk

1 NASA Goddard Space Flight Center, Greenbelt MD, USA

2 Southwest Research Institute, San Antonio TX, USA

3 Johns Hopkins Applied Physics Laboratory, Laurel MD, USA

4 Dartmouth College, Hanover, NH, USA

5 University of California Berkeley, Berkeley, CA, USA

6 Northumbria University, Newcastle upon Tyne, UK

7 Imperial College London, London, UK

8 University of Michigan, Ann Arbor, MI, USA

9 Kyoto University, Kyoto, Japan

10 University of Maryland College Park, College Park, MD, USA

1 Introduction

Magnetic reconnection has been observed as a ubiquitous process in the space environment of magnetized obstacles embedded in collisionless plasmas. The most comprehensive studies of reconnection have taken place at Earth, where not only are there multi-point and multi-scale in-situ observations that elucidate the physics of reconnection itself, but a ground network of magnetometers and radars that can track global dynamics and provide geomagnetic indices to correlate with upstream conditions. Other articles in this collection have provided comprehensive reviews of the latest reconnection theory and observations developed as part of MMS-era exploration of Earth's magnetosphere (Genestreti et al., Liu et al., Norgren et al., Graham et al., Stawarz et al., Fuselier et al., Hwang et al., Oka et al., this collection). Here we shift our focus beyond Earth and onto a diverse set of planetary magnetospheres and astrospheres that each provide unique laboratories for the testing and scaling of reconnection theory.

The magnetized bodies within and including our heliosphere, have been accessible to *in situ* exploration. Such missions beyond Earth are often limited to single-point measurements and constrained in the plasma instrumentation and telemetry rates available, especially when compared to multi-spacecraft formations like THEMIS, Cluster, or MMS at Earth. There are also no corresponding ground-based observations or upstream monitors to provide global activity indices for each planet. The instruments deployed on extraterrestrial missions are nonetheless highly capable and have been used to obtain tremendous insight into the dynamics of planetary magnetospheres and our heliosphere. Overall, there have been sufficient observations to characterize the effective size of an obstacle to the upstream flow and obtain critical information about plasma properties and magnetic field configurations needed to parametrize the role of reconnection at a given system.

This paper is not intended to provide an in-depth study of different planetary magnetospheres and the relative importance of magnetic reconnection in each system. There are a number of comprehensive studies available (Bagenal 2013; Kivelson 2007, and Kivelson and Bagenal 2014) that are highly relevant and will be leveraged throughout this study. Instead, here we attempt to scale the discoveries from recent missions like MMS about the microphysics of reconnection to the global dynamics of a planetary magnetosphere or astrosphere. We also attempt to link the terminology and approaches used by the planetary magnetosphere community to those of the reconnection theory community to maximize the application of MMS-era findings to ongoing, previous, and future studies of reconnection across and beyond the solar system.

We begin with a general discussion of flow around magnetized obstacles in Sect. 2 and relevant considerations for parameterizing the reconnection process at the relevant magnetic boundary (e.g., magnetopause, heliopause). In Sect. 3, we then discuss the typical analyses undertaken at planetary magnetospheres in terms of reconnection theory. Section 4 then provides an overview of reconnection signatures observed throughout the solar system. Finally, in Sect. 5 we discuss the scaling of the reconnection energy available to a magnetized obstacle across the solar system and to exoplanets and astrospheres.

2 Flow Around Magnetized Obstacles

Plasma flows must divert around any embedded magnetized obstacle. The magnetosonic Mach number of this upstream plasma, defined as the ratio of the flow speed (V_U) to the magnetosonic speed ($\sqrt{(V_A^2 + V_S^2)}$) dictates the relevant physical regime for the boundary interaction. Here, V_A is the plasma Alfvén speed ($B/\sqrt{(\mu_o \rho)}$) and V_S is the plasma sound speed ($\sqrt{(\gamma P/\rho)}$), where ρ is the mass density, γ is the ratio of specific heats, often taken as 5/3, μ_o is the permeability of vacuum, B is the magnetic field magnitude, and P is the thermal pressure. Upstream flow regimes can vary from sub-magnetosonic ($M_{MS} < 1$), largely in the case of moons embedded in planetary magnetospheres (e.g., Ganymede and Triton), to marginally-magnetosonic ($M_{MS} \sim 1$) in the case of our heliosphere embedded in the Local Interstellar Medium (LISM) flow, to super-magnetosonic ($M_{MS} > 1$) like planetary bodies embedded in the solar or stellar winds. A summary of typical values of upstream conditions at magnetized bodies reported in and including the heliosphere, and at HAT-P-11b, a Neptune-like exoplanet recently reported to have a magnetosphere (Ben-Jaffel et al. 2022), are included in Table 1.

As shown in Fig. 1, in the $M_{MS} \ll 1$ case, upstream plasmas interact directly with a magnetized obstacle. In this scenario, standing Alfvén waves generated at the magnetopause boundary propagate away from the obstacle and into the flow, forming a set of so-called "Alfvén wings" whose angles depends on upstream M_A (Neubauer 1980). These wings carry Poynting fluxes that can be relevant for describing interactions with their host star or planet (Fischer and Saur 2022 and references therein). The magnetic field lines in Alfvén wings are analogous to the open-field lobes in an intrinsically magnetized planetary magnetosphere, and the effective obstacle to the upstream flow is highly flared, blunted, and quasi-cylindrical (Kivelson and Jia 2013). In the most common scenario in our solar system, the focus of most of this paper, the upstream super-magnetosonic solar wind is significantly heated and compressed at a planetary bow shock, forming a magnetosheath where the flow divertsaround the obstacle. In the $M_{MS} \lesssim 1$ case (marginally sub-magnetosonic), a bow wave forms, where the upstream plasma is slightly heated and compressed over a long upstream distance before it reaches the magnetized body (e.g., heliopause (Zank et al. 2012)). In all regimes, the upstream magnetic field drapes over the embedded obstacle.

There is also a class of unmagnetized planetary bodies and moons (e.g., Mars, Venus, Titan) that exhibit magnetospheric-like behavior. In these situations, the upper atmosphere and ionosphere become the obstacle to the impinging solar wind and provide a highly conductive layer that results in the formation of a so-called 'induced magnetosphere.' Here the induced magnetopause boundary (IMB) acts as the obstacle around which the upstream fields drape and interact (Luhmann et al. 2004; Ness et al. 1982). Some obstacles (e.g., Mars) have remanant crustal magnetic fields that represent a more complex planetary obstacle to the solar wind while providing a pathway for solar wind access (Wang et al. 2021; Harada et al. 2018; Fang et al. 2018). Mars, in particular forms a 'hybrid magnetosphere' (Axford 1991), sharing properties of both intrinsic and induced magnetospheres.

The relevant upstream conditions of flow-embedded obstacles within and including our own heliosphere have been measured by in situ spacecraft and range from sub-magnetosonic to super-magnetosonic regimes. However, unexplored exoplanetary magnetospheres and astrospheres spread throughout the universe also extend over this range of sampled regimes, with many exhibiting extreme values well beyond what we have observed to date (Khodachenko et al. 2013; Belenkaya et al. 2015). Some stars will act similar to our own (i.e., G-type), providing a super-sonic flow of stellar wind plasma throughout their astrospheres and upstream of planetary obstacles (Belenkaya et al. 2022). Others, such as M-type stars, have extremely strong magnetic fields that result in an Alfvén point, i.e., the astrocentric distance at which the flow speed exceeds the Alfvén speed, that extends far out into their respective astrospheres. This extension results in sub-magnetosonic flow upstream of closely orbiting exoplanets (Vidotto et al. 2014; Garraffo et al. 2016). However, regardless of the flow regime, the upstream fields and flows must drape and divert, respectively, around an em-

Table 1 List of upstream parameters for planetary bodies, magnetized moons, the heliosphere, and reported exoplanet magnetospheres

	n^{e} (cm^{-3})	<m/q> (amu/e)	B^{e} (nT)	T^{d} (K)	$V_U{}^{e}$ (km/s)	V_A (km/s)	V_S (km/s)	M_A	M_S	M_{MS}
Mercury	46	1	30	200,000	420	96	52	4.4	8.0	3.8
Venus	14	1	11	150,000	420	61	45	6.9	9.2	5.5
Earth	7.0	1	6.0	100,000	420	49	37	8.5	11	6.8
Mars	3.0	1	3.3	70,000	420	42	31	10	14	8.1
Jupiter	0.26	1	0.83	20,000	420	36	17	12	25	11
Saturn	0.076	1	0.44	15,000	420	35	14	12	29	11
Uranus	0.019	1	0.22	10,000	420	35	12	12	36	11
Neptune	0.0078	1	0.14	9000	420	35	11	12	38	11
Ganymede[a]	2-4	14	75-110	70,000,000	140	233-467	260	0.3-0.6	0.5	0.2-0.4
Triton[b]	0.11	1	5.1-8.2	754,000	43	120-200	50	0.2-0.3	0.8-0.9	0.2-0.3
Heliosphere[c]	0.041	1	0.29	30,000	25.82	31.7	20.3	0.81	1.27	0.69
HAT-P-11b[f]	3.3×10^3	1	439	$1.3\text{-}1.5 \times 10^6$	500-600	167	185-200	3.0-3.6	2.7-3.0	2.0-2.3

a – from Jia et al. (2008), b – from Liuzzo et al. (2021), c – from Schwadron and McComas (2021), Richardson et al. (2022), d – temperatures from Smith et al. (2001), e – using Parker model (Parker 1958, 1963; Burlaga et al. 2002) and Gershman and DiBraccio (2020) averages at 1 AU, f – from Ben Jaffel et al. (2022).

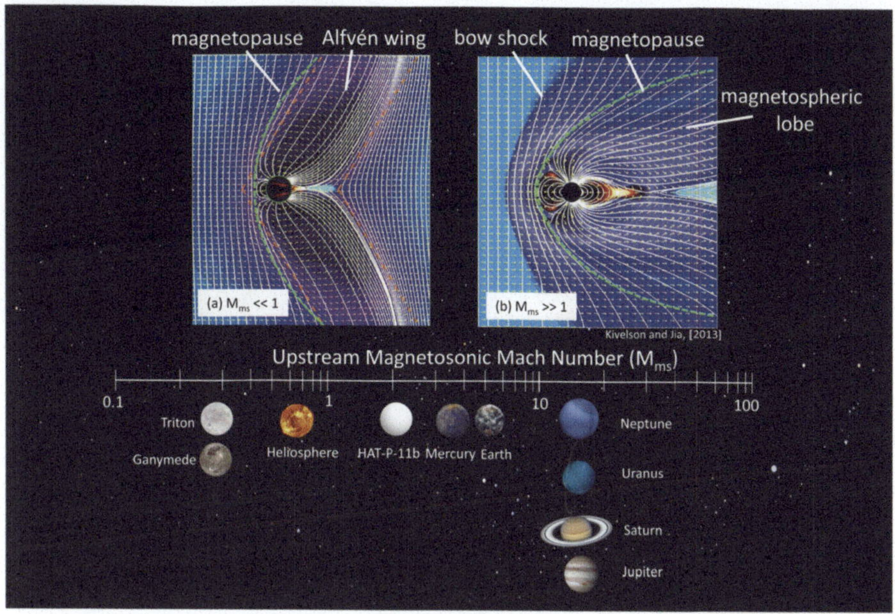

Fig. 1 (After Kivelson and Jia 2013.) Illustration of magnetized bodies embedded in (a) sub-magnetosonic upstream flows, and (b) super-magnetosonic. The magnetosphere in (a) has no bow-shock, with a blunted magnetopause boundary encompassing Alfvén wings that form due to standing Alfvén waves generated at the interface. The magnetosphere in (b) has a well-defined bow shock and bullet-shape magnetopause boundary, with heated and compressed plasmas flowing around the obstacle in a magnetosheath

bedded obstacle. It is these modified plasmas at the magnetopause, astropause, or IMB that interact directly with the planetary or stellar obstacle. Magnetic reconnection is a key process to consider at all these interfaces as it results in the transport of significant amounts of mass, energy, and magnetic flux. Because the physics of magnetic reconnection depends on local plasma properties, it is critical to understand how near-magnetopause/astropause plasmas may be modified from their upstream values (Borovsky et al. 2008; Borovsky 2021).

2.1 The Role of the Magnetosheath

For systems embedded in super-magnetosonic flow, within a planetary magnetosheath, magnetic fields drape over a magnetopause boundary, and the shocked plasmas are heated and compressed compared to their upstream values. Given the complexity of this regime, modeling of the magnetosheath becomes critical to be able to connect upstream flow properties with the likelihood for magnetic reconnection to occur at the magnetopause. For complex geometries and more self-consistent physics, magnetohydrodynamic (MHD) simulations of the flow-embedded obstacle may be required. Here we provide an overview of some common tools used for non-computationally intensive analytical modeling of magnetosheath properties used to predict near-magnetopause plasma parameters relevant for reconnection. These models are most effective when considering a quasi-perpendicular bow shock geometry, as additional acceleration mechanisms and magnetosheath jets may be formed in quasi-parallel geometries that can significantly modify magnetosheath properties (Fuselier et al. 1994; Dimmock et al. 2015; Archer and Horbury 2013; Hietala and Plaschke 2013; Plaschke et al. 2013; Karlsson et al. 2021) and are not as readily modeled.

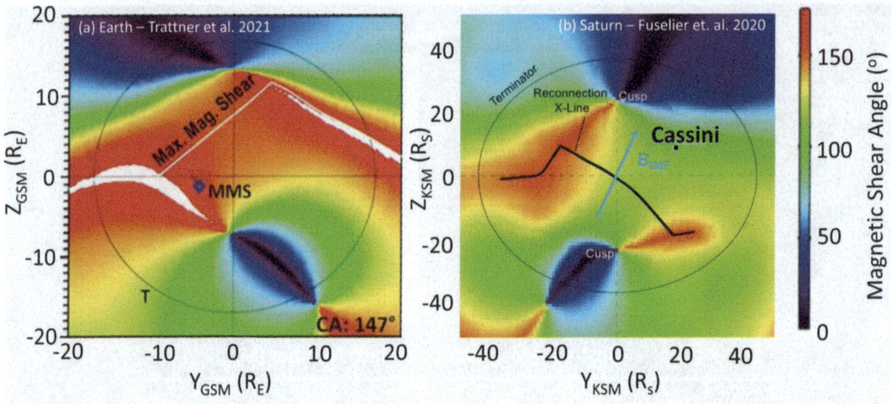

Fig. 2 Examples of the maximum magnetic shear model applied to (a) Earth (adapted from Trattner et al. 2021) and (b) Saturn (adapted from Fuselier et al. 2020b), to contextualize observations of reconnection signatures at MMS and Cassini, respectively

Using analytical magnetosheath models at Earth, the maximum magnetic shear model was developed to identify the most likely location of the reconnection X-line along the magnetopause (Trattner et al. 2007). This model and its application to MMS data are discussed in detail by (Trattner et al. 2012, 2021 and Fuselier et al., this collection). As shown in Fig. 2, maximum shear models have also recently been applied to Saturn (Fuselier et al. 2014, 2020b) and Mars (Bowers et al. 2023). The successful comparisons of the maximum magnetic shear model to spacecraft observations indicate that there is typically a single dominant X-line, or at least active localized reconnection extended generally along the expected dominant X-line at a magnetopause for an intrinsic magnetosphere. This paradigm will be shown in Sect. 5 to be critical for scaling the energy available from magnetopause reconnection to other planetary bodies.

2.1.1 Early Magnetosheath Models and Plasma Depletion Layers

For planetary magnetospheres embedded in the super-magnetosonic solar wind, Spreiter et al. (1966) developed the first early models of magnetosheath flow using gas dynamics and the frozen-in flow condition. Across the bow shock, the Rankine-Hugoniot shock jump conditions were used to determine the downstream conditions at the shock. Then, using hydrodynamic flow and a fixed, impenetrable magnetopause boundary at a given standoff distance from the obstacle, the density, velocity, and temperature of magnetosheath plasmas were determined. This model was used not only at Earth under varying conditions, but to consider flow around various planetary magnetospheres (Spreiter and Alksne 1970). Upstream parameters, magnetopause standoff distance (i.e., effective obstacle size), and shock jump strength/heating were the tunable parameters.

Including electromagnetic forces in models of the magnetosheath is essential to model the near magnetopause plasmas. Zwan and Wolf (1976) and then later Southwood and Kivelson (1995) included such effects in their modeling of magnetic field evolution in the magnetosheath, using the Spreiter et al. (1966) models as an initial condition for the magnetosheath plasmas. Here, as shown in Fig. 3, a plasma depletion layer (PDL) was identified, where magnetic field piles-up against the boundary, squeezing plasma around the nose, resulting in a reduction of density from the magnetosheath to the magnetopause, with a more pre-

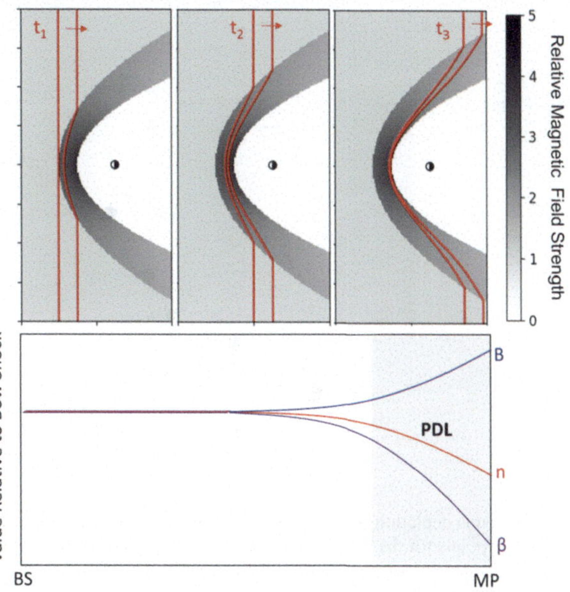

Fig. 3 (After Zwan and Wolf (1976).) Illustration of magnetic flux-pile-up against a magnetized planetary obstacle embedded in super-magnetosonic flow for three successive time intervals with $t_1 < t_2 < t_3$. For a given magnetic flux tube in the upstream flow, indicated with red lines, the magnetic field strength increases and flux tube area decreases in the subsolar region as it drapes against the magnetopause. Plasma within these draped regions is squeezed out along the flux tube around the obstacle between the bow-shock (BS) and magnetopause (MP), resulting in decreased plasma density (i.e., a plasma depletion layer (PDL)) and further increased magnetic field strength, both contributing to a smaller near-magnetopause plasma β

cipitous drop along the subsolar magnetopause. Although its effects are large-scale, plasma depletion itself is a kinetic process, with depletion first taking place through the loss of parallel particles along the field, followed by instability growth and scattering (Anderson and Fuselier 1993).

PDLs have been observed at almost every magnetized obstacle in the solar system and represent a reduction in both near-magnetopause plasma β (the ratio of thermal pressure to magnetic pressure) and a reduction in density that impacts reconnection parameters such as the Alfvén speed, the stability of the high-latitude reconnection at Earth in cases of northward interplanetary magnetic field (IMF) (e.g., Fuselier et al. 2000) and, as will be discussed, the conditions for diamagnetic suppression of reconnection onset at the dayside magnetopause.

As shown in Fig. 4, the plasma depletion process is readily scalable to different objects via a characteristic distance that is required to achieve a certain reduction in plasma β. This depletion length scale is approximately 5-10% of the obstacle size (Gershman et al. 2013; Cairns and Fuselier 2017). Here we add recent observations of plasma depletion layers at the heliopause (Cairns and Fuselier 2017) and at Neptune (Jasinski et al. 2022) to the initial analysis by Gershman et al. (2013). Plasma β at a planetary magnetopause can reduce by up to an order of magnitude from its downstream value at the subsolar region, with more modest modifications elsewhere around the obstacle. Plasma depletion is most pronounced in planetary obstacles that are embedded in low upstream M_A plasma (Gershman et al. 2013), where there is a tremendous amount of incident magnetic flux available to drape over the obstacle, and the scale of the magnetosheath is relatively larger compared to the size of the obstacle ($\sim 1/M_A^2$), allowing for longer depletion times (Zwan and Wolf 1976). Through their modification of the near-magnetopause β, PDLs have been shown to impact magnetic reconnection rates (Anderson et al. 1997; Dorelli et al. 2004).

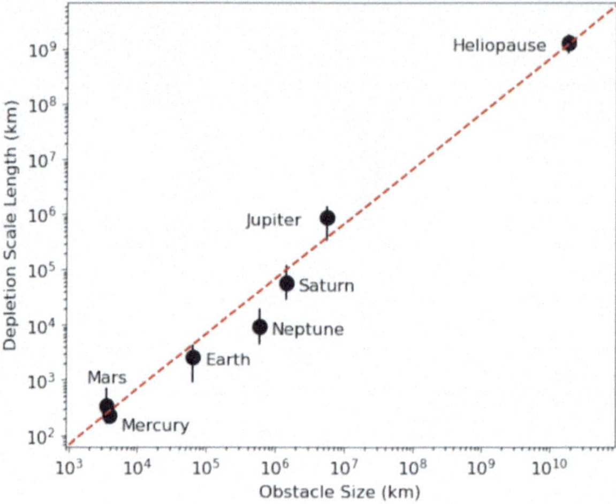

Fig. 4 Plasma depletion scaling throughout the solar system, adapted from Gershman et al. (2013), with updated data points for the heliopause (Cairns and Fuselier 2017) using data from the study and Neptune (Jasinski et al. 2022), assuming ~30 min PDL transit length corresponding to ~2.2 R_N size and a $\beta_{MP}/\beta_{BS} \sim 0.3$. A best fit line of slope 0.073 ± 0.043 is shown, where the mean and standard deviation were calculated from the ratio of depletion scale length to obstacle size. Factors of 2 error bars were added for each point when variations were not provided in the corresponding table by Gershman et al. (2013)

2.1.2 Scaling Magnetosheath and Magnetic Field Models

To model the evolution of magnetosheath plasma around any obstacle, Kobel and Flückiger (1994) developed an analytical model that uses potential fields and paraboloids of revolution for the shape of the bow shock and magnetopause to solve for the magnetic field everywhere in the magnetosheath using only standoff distances and IMF direction as inputs. In addition, Petrinec and Russell (1997) provide flow vectors, mass density, and plasma pressure at any point on the magnetopause surface in the hydrodynamic case. These models can be combined to develop models for plasma and magnetic flux transport along the magnetopause (Petrinec et al. 2003; Cooling et al. 2001). Such expressions do not include the additional reduction in density and increase in magnetic field associated with PDLs, though such effects can be incorporated in studies of near-magnetopause plasma properties (Masters 2014, 2015a; Petrinec et al. 2003). The hydrodynamic assumptions in the Petrinec and Russell (1997) model also results in unphysical super-Alfvénic flows along the flanks of the magnetopause (Petrinec et al. 2003). These, and other magnetosheath models (such as those extracted from MHD simulations) have been used extensively to predict regions of high flow and magnetic shear across magnetopause boundaries throughout the solar system (Fuselier et al. 2014, 2020b; Masters 2014, 2015a; Desroche et al. 2013), providing estimates of under what conditions reconnection may be possible.

The Kobel-Fluckiger (K-F) equations have been converted into GSE-like coordinates (x, y, z) at Earth by Petrinec et al. (2003),

$$B_x^{dis} = \frac{\frac{x'}{r'} - 1}{(r' - x')^2} \left(\frac{r' - x'}{2} B_x^{IMF} - y' B_y^{IMF} - z' B_z^{IMF} \right), \tag{2.1}$$

$$B_y^{dis} = \frac{1}{r'(r' - x')} \left(\frac{y'}{2} B_x^{IMF} + \left(r' - \frac{y'^2}{r' - x'} \right) B_y^{IMF} - \frac{y'z'}{(r' - x')^2} B_z^{IMF} \right), \tag{2.2}$$

$$B_z^{dis} = \frac{1}{r'(r'-x')}\left(\frac{z'}{2}B_x^{IMF} - \frac{y'z'}{(r'-x')^2}B_y^{IMF} + \left(r' - \frac{y'^2}{r'-x'}\right)B_z^{IMF}\right), \tag{2.3}$$

$$\vec{B} = \left(1 + \frac{v_{mp}^2}{v_{bs}^2 - v_{mp}^2}\right)\vec{B}_{IMF} + \left(\frac{v_{mp}^2 v_{bs}^2}{v_{bs}^2 - v_{mp}^2}\right)\vec{B}_{dis} \tag{2.4}$$

with $x' = -x + \mathrm{R_{MP}}/2$, $y' = -y$, and $z' = -z'$ with $r' = \sqrt{}(x'^2 + y'^2 + z'^2)$. $\mathrm{v_{bs}} = \sqrt{}(2\mathrm{R_{BS}} - \mathrm{R_{MP}})$, $\mathrm{v_{MP}} = \sqrt{}\mathrm{R_{MP}}$ and 'dis' referring to a magnetic disturbance field following nomenclature by Petrinec et al. (2003). These equations are valid between the magnetopause and magnetosheath. Petrinec et al. (2003) also provide analytical equations for the flow speed at the magnetopause, modified for the presence of a **JxB** force.

A simple representation of many planetary dynamos is the offset-tilted-dipole (OTD) model, where the planetary field is modeled as a single magnetic dipole with an origin that can be offset from the center of the planetary body with a tilt angle with respect to the rotation axis (Ness et al. 1976). While these models result in significant errors in estimating the field close to the planet (Stanley and Bloxham 2006; Bagenal 2013; Soderlund and Stanley 2020), they can be quite effective for modeling the near-magnetopause fields.

The equations for a dipole field in a cartesian coordinate system $(x_\mathrm{d}, y_\mathrm{d}, z_\mathrm{d})$ where the $\mathrm{Z_d}$ axis is aligned with the dipole moment **M** are,

$$\vec{B}^{dip} = \left(\frac{3Mx_d z_d}{r_d^5}, \frac{3My_d z_d}{r_d^5}, \frac{M\left(3z_d - r_d^2\right)}{r_d^5}\right). \tag{2.5}$$

Here, to obtain the magnetic field vector in GSE-like coordinates (x, y, z), a vector (x, y, z) is transformed into the frame of an offset dipole with origin $(x, y, z) = (x_\mathrm{do}, y_\mathrm{do}, z_\mathrm{do})$. The resultant dipole magnetic field vector is then transformed into the original coordinate system.

Solving for the external magnetic field source required to obtain a given magnetopause is non-trivial and can be computationally intense. However, Masters (2014, 2015a) provide a scheme to estimate the magnetic field vector at the magnetopause using only a model for the internal field and the magnetopause shape. They first take the planetary field vector at the magnetopause $\mathbf{B_{mp}}$ boundary and then zero out the component normal to the boundary, i.e., $\mathbf{B_{mp}} = \mathbf{B_{mp}} - \mathbf{B_{mp}} \cdot \mathbf{n}$, where for the K-F magnetopause, the outward normal vector in GSE coordinates is defined as:

$$\hat{n}_{mp} = \left(\frac{1 - \frac{x'}{r'}}{\sqrt{2\left(1 - \frac{x'}{r'}\right)}}, \frac{-y'}{r'\sqrt{2\left(1 - \frac{x'}{r'}\right)}}, \frac{-z'}{r'\sqrt{2\left(1 - \frac{x'}{r'}\right)}}\right). \tag{2.6}$$

This approach provides a computationally simple and straightforward way to estimate magnetic shear across the magnetopause. The magnetic field magnitude directly inside the magnetopause can also be adjusted to ensure there is pressure balance across the magnetopause (Masters 2014, 2015a). The magnetic pressure inside the magnetopause can be estimated from the upstream dynamic pressure following Spreiter and Alksne (1970),

$$\frac{B_{ms}^2(\psi)}{2\mu_o} \approx \rho_u v_u^2 \kappa \cos^2\psi, \tag{2.7}$$

where $\kappa \approx 0.881$ and ψ is the angle between the magnetopause normal vector and the solar wind flow.

The magnetopause standoff distance can also be estimated from the upstream pressure and surface magnetic field as (Bagenal 2013),

$$R_{MP} \approx \xi \left(\frac{B_o^2}{2\mu_o \rho_u v_u^2} \right)^{1/6}, \tag{2.8}$$

where $\xi = 1.4$ and B_o corresponding to the surface magnetic field strength of a planetary dipole. The specific κ, ξ, and 1/6 dependence assume low β magnetospheric environments, such that giant planets with significant magnetodisc structures exhibit different scalings and the stand-off distance using equation (2.8) is underestimated (Jackman et al. 2019; Arridge et al. 2006).

There is also subtlety in determination of the bow shock standoff distance with respect to the magnetopause, including the magnetopause shape, angle between upstream magnetic field and upstream flow, and Mach numbers (Cairns and Lyon 1996). However, a simple, yet effective model is provided by Farris et al. (1991); Formisano et al. (1971), and Song (2001) that uses the fast magnetosonic Mach number,

$$\frac{R_{BS}}{R_{MP}} \approx 1 + 1.1 \frac{(\gamma - 1) M_{ms}^2 + 2}{(\gamma + 1) M_{ms}^2}. \tag{2.9}$$

For $\gamma = 5/3$ and the values of M_{ms} in Table 1, reasonable estimates of bow shock standoff distances are produced (e.g., a thicker magnetosheath at Mercury where $M_{ms} \sim 3.8$ compared to the outer planets with $M_{ms} \sim 11$).

Equations (2.1)-(2.9) provide a generalized framework for modeling draping and estimating magnetic shear for a magnetized object embedded in supermagnetosonic flow. Examples of draping and magnetic shear plots calculated with this approach are shown in Fig. 5 for several planetary bodies. These plots are then combined with tools like the maximum magnetic shear model (Trattner et al. 2007, 2021) to estimate reconnection X-line formation and propagation along a magnetopause. At Earth, the large amount of data available has enabled more accurate empirical models of IMF draping (Michotte de Welle et al. 2022). Because the Earth's magnetosheath geometry is similar to other planetary magnetospheres, such a model may be adaptable to other systems.

Magnetospheres and astrospheres embedded in sub-Alfvénic flow appear as near cylindrical obstacles (Czechowski and Grygorczuk 2017; Neubauer 1999; Kaweeyanun et al. 2020). Some analytical works have been developed to study Alfvén-wing properties (Neubauer 1980; Simon 2015), though the equivalent of K-F—like and Petrinec-like magnetosheath models for the near-magnetopause plasmas in these systems has not been reported to our knowledge. Kaweeyanun et al. (2020) provide some empirical parameterization of upstream conditions of Ganymede as a cylindrical obstacle with compressions varying as the cosine of the angle between the nose of the obstacle and magnetopause normal.. This approach could be readily applied to other magnetized obstacles embedded in sub-Alfvénic flow.

Finally, at induced magnetospheres, the upstream fields drape around the magnetic pile-up boundary, but also wrap around the obstacle, resulting in a magnetotail-like configuration, but instead of being connected to a polar cap of a planetary field, they are connected to the upstream field. For these obstacles embedded in super magnetosonic flows like the solar wind, the K-F flow vectors may be appropriate using reasonable boundaries for the bow shock and magnetic pile-up boundary. On the dayside, a K-F model may also be relevant to describe the draping over the MPB, though to our knowledge has not yet been applied to Mars or Venus. On the flanks, models of potential flow around a conducting sphere may be more appropriate (Romanelli et al. 2014; Luhmann et al. 2004).

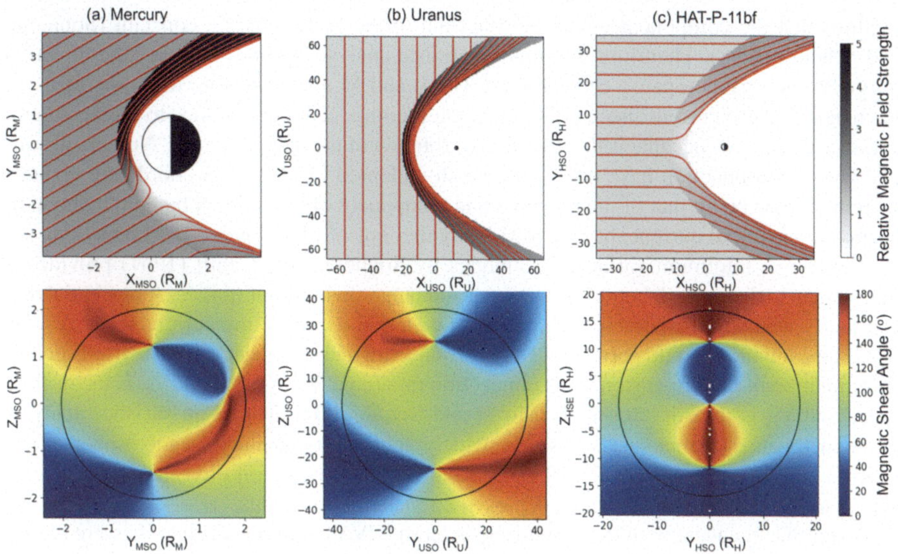

Fig. 5 Example calculation of magnetic field draping over the XY plane and magnetic shear across a planetary magnetopause projected on the YZ plane for (a) Mercury ($-0.2\ R_M$ z-offset in dipole, no tilt), (b) Uranus (effective 39.2^o tilt in XZ plane following Masters (2014), comparable with their Fig. 4(e), and (c) exoplanet HAT-P-11bf (no offset or tilt). Upstream conditions are taken from Table 1 and used to derive magnetopause and bow shock standoff distances. The IMF was taken in the X-Y plane with $Bx/By = 2$ for Mercury, $Bx = 0$ for Uranus, and $By = 0$ for HAT-P-11bf (Ben-Jaffel et al. 2022). The black circle of radius $R_{MP}\sqrt{2}$ corresponds to the $x = 0$ plane crossing of the magnetopause surface

3 Reconnection at Magnetized Obstacles and the Local Plasma Environment

Here we discuss reconnection in the context of theoretical framework often applied to data from Cluster, THEMIS, and MMS and those often employed by those studying planetary magnetospheres and astrospheres.

3.1 L-M-N Coordinates and Minimum Variance Analysis

Defining the relevant coordinate system in which to study reconnection is critical for being able to estimate reconnection rates, to study the transport of reconnected structures such as flux ropes, and to be able to compare observations with numerical models. The reconnection community at large has adopted a boundary-normal 'L-M-N' coordinate system first introduced by Russell and Elphic (1979) to study flux transfer events propagating along Earth's dayside magnetopause. The N-direction was defined as the boundary normal to the magnetopause, the L-direction was along the projection of the solar-magnetospheric Z-direction perpendicular to the magnetopause normal, and the M-direction completed the right-hand coordinate system.

Over time, this concept of global LMN coordinate system was merged with concept of a locally defined coordinate system from minimum variance analysis of magnetic field data across a boundary (Sonnerup and Cahill 1967). Here, the minimum variance direction, i.e., that with the smallest eigenvalue, corresponds to the boundary normal N direction. Across a one-dimensional structure, the condition that the divergence of **B** must be equal to zero requires that the component of the field normal to the layer must be constant (Sonnerup and Scheible 1998). Therefore, minimum variance analysis naturally derives the direction normal to the magnetopause current sheet The L and M axes therefore lie in the plane of the magnetopause current sheet. Nominally, the L component corresponds to the maximum variance direction since that is the direction of rotation of the magnetic field. Numerical simulations of reconnection have nearly universally adopted local L-M-N coordinate systems with the L-axis being that of the reconnecting component of the field, N being the direction normal to the initial current sheet, and M being the 'out-of-plane' guide-field direction.

The use of minimum variance techniques by Sonnerup and Cahill (1967) provided a mechanism to distinguish between rotational and tangential discontinuities. A tangential discontinuity would correspond to a $B_N/B \sim 0$, i.e., a non-reconnecting current sheet. A rotational discontinuity would correspond to a reconnecting current sheet, with the B_N/B giving an estimation of the dimensionless reconnection rate (e.g., V_{in}/V_A, where V_{in} is the in-flow speed and V_A is the Alfven speed) (Mozer and Retinò 2007). Minimum variance analysis of magnetopause current sheets has been widely applied to planetary magnetospheres (DiBraccio et al. 2013; Slavin et al. 2014) due to their reliance only on fluxgate magnetometer data, which are nearly ubiquitously deployed on planetary missions.

A significant caveat with these single-spacecraft MVA techniques is that they are highly trajectory dependent or dependent on the duration of the time interval used to derive the coordinate systems. Care must be taken when trying to connect minimum variance-derived coordinate systems and true L-M-N coordinates for a broad set of observations. Often, the ratio of eigenvalues of the minimum variance direction to that of the intermediate variance direction is used to provide a proxy for the quality of the coordinate system definition. Typically, a factor of ~ 10 is considered as a rough rule of thumb for a quality normal-vector direction determination, though even large ratios may not guarantee accurate results (Sonnerup and Scheible 1998). As an example, 2D PIC simulations of non-reconnecting and reconnecting current sheets (see simulation setup in Chen et al. 2016 of the Burch et al. 2016 reconnection event) are shown in Fig. 6. For spacecraft trajectories along the N direction in the vicinity of the current sheet, we perform minimum variance analysis on the magnetic field vectors and compare the derived minimum variance (B_N) direction with the N-axis (i.e., the nominal normal direction), and the eigenvalues for the minimum to intermediate variance direction. The B_N/B values are very small for the non-reconnecting case (Fig. 6a), as expected, and are significant (i.e., ~ 0.6) for the reconnecting case (Fig. 6b). The determination of the N-direction is accurate to within $<5^o$ for the non-reconnecting case, and accurate only to within $<20^o$ for the reconnecting case, despite large ($\sim$10-50) ratios of the eigenvalues. The presence of significant B_M and B_N variations in the magnetic field in the vicinity of reconnection diffusion regions significantly complicates this type of analysis.

Data from MMS have provided a unique opportunity to assess the accuracy of MVA techniques based on magnetometer data alone with other methods of coordinate system and boundary normal determination. Denton et al. (2018) compared LMN directions derived from both single- and multi-spacecraft techniques, and found that minimum variance analysis of reconnecting current sheets is most reliable in terms of deriving the L-direction, i.e. the maximum variance direction. The differentiation between the M- and N- directions were less clear. Genestreti et al. (2018) also found that different techniques resulted in up to $\sim 35^o$ variation in the directions of the L-M-N coordinate systems. Overall, a combination of techniques was found to be most effective, though this approach is not necessarily possible at most planetary environments. If minimum variance is the only option, its coordinate system should be critically examined and compared to model expectations and other observations that may be available (Denton et al. 2018).

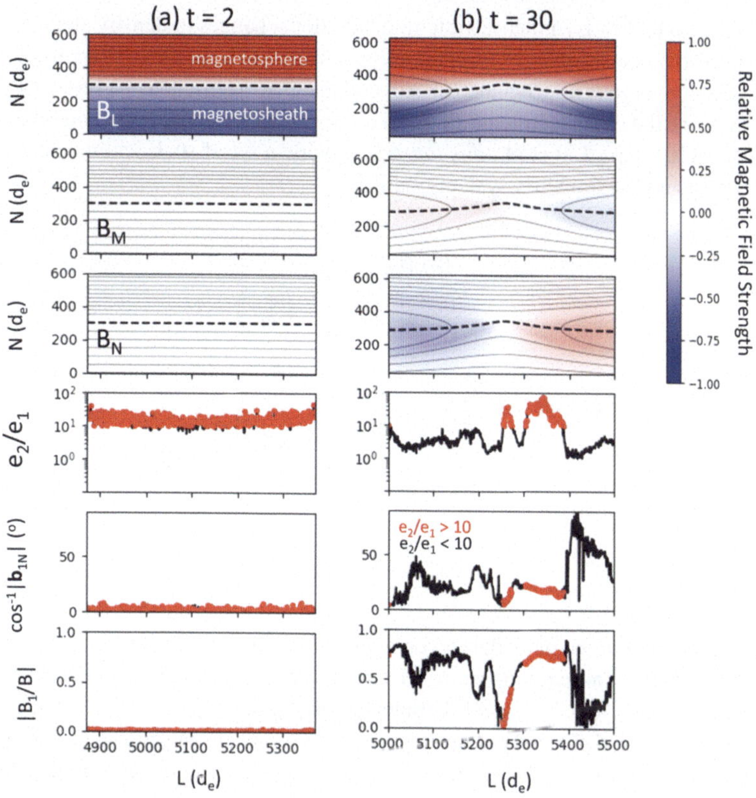

Fig. 6 Two time periods from an example 2-D PIC simulation of a reconnecting current sheet that show (a) before the onset of fast reconnection and (b) after a fast reconnection X-line has formed. For each, an MVA analysis was performed on data $\pm 50 d_e$ from the current sheet at each L. The ratio of intermediate to minimum eigenvalues (e_2/e_1) is shown, with red data points corresponding to $e_2/e_1 > 10$. The angle between the minimum variance direction and N-axis shows how accurately MVA determined the current normal coordinate system. The ratio of the minimum variance component to the total field providing a proxy for the reconnection rate. For the non-reconnecting current sheet, MVA analysis provides a well-resolved coordinate system ($2^{\circ} \pm 2^{\circ}$) with statistically zero reconnection rate 0.007 ± 0.005. For the reconnecting current sheet, MVA-based determination of the coordinate system becomes more challenging, but for well-separated eigenvalues can provide a coordinate system within $17 \pm 4^{\circ}$ and provide a finite reconnection rate of 0.6 ± 0.2

3.2 Guide Field, Symmetry, Shear Angle

As discussed, reconnection studies tend to organize observations using a local L-M-N coordinate system. In such a framework, the L-component of the magnetic field represents the reconnecting component and the M-component represents the so-called 'guide-field.' The relative strengths of the L-component on either side of a reconnecting current sheet defines the symmetry of the system. This framework, used for studying spacecraft observations and in numerical simulations of reconnection, is discussed in detail in Genestreti et al., this collection.

Across a given interface we define $\mathbf{B}_2$ as the vector on the side with the larger field strength (e.g., magnetosphere) and $\mathbf{B}_1$ as the vector on the side with the smaller field strength (e.g., the magnetosheath). A symmetric system corresponds to $B_1 \sim B_2$, while $B_2 \gg B_1$

corresponds to a highly asymmetric system. In addition, changes in density and temperature across the magnetopause also result in an asymmetric system. In the asymmetric case, it is not immediately obvious how magnetosheath and magnetospheric plasma properties should be used to define the reconnection outflow Alfvén speed. However, Cassak and Shay (2007), Cassak et al. (2017a,b), Liu et al., *this journal*, provide a set of such scalings that apply to both symmetric and asymmetric systems, namely,

$$B_{asym,L} = \frac{2B_{1L}B_{2L}}{B_{1L}+B_{2L}} = B_{2L}\frac{2\left(\frac{B_{1L}}{B_{2L}}\right)}{\left(\frac{B_{1L}}{B_{2L}}\right)+1}, \tag{3.1}$$

$$\rho_{asym,L} = \frac{\rho_1 B_{2L}+\rho_2 B_{1L}}{B_{1L}+B_{2L}} = \frac{\rho_1+\rho_2\left(\frac{B_{1L}}{B_{2L}}\right)}{\left(\frac{B_{1L}}{B_{2L}}\right)+1}, \tag{3.2}$$

and

$$V_{A,asym,L} = \frac{B_{asym,L}}{\sqrt{\mu_o \rho_{asym,L}}}. \tag{3.3}$$

Here, ρ_1 and ρ_2 refer to the mass density on either side of the interface (1 = magnetosheath, 2 = magnetosphere). These relationships have been applied to component reconnection processes by applying them only to the L-component of magnetic field.

Often in the study of planetary magnetospheres, the magnetic shear angle (θ) and change in plasma β across a current sheet is used to describe an interface rather than terms such as 'guide-field' and 'symmetry.' These seemingly disparate approaches can be linked together through a model of component reconnection, such as that of Swisdak and Drake (2007) and Hesse et al. (2013), where the X-line approximately bisects the angle between the magnetic fields on either side of the boundary. An alternative formulation was presented by Sonnerup (1974), where an X-line would orient itself in a way where a constant guide-field would form across the reconnection region. Simulations have indicated that the bisection model may be more appropriate (Liu et al. 2015, 2018), though Wang et al. (2015) demonstrated that these different approaches often yield coordinate systems within a few degrees of one another. A key difference in these models is that in the Sonnerup (1974) formulation, it is possible to define a set of vectors across a boundary for which no constant guide field can be determined and reconnection cannot occur. In this scenario, the components of the two magnetic field vectors perpendicular to the predicated X-line direction are parallel rather than anti-parallel. The Sonnerup (1974) therefore predicts a form of 'geometric suppression' (very high-guide-field and low shear angle) that is not present in the bisection model. By allowing the guide-field strength to vary across the interface, geometric suppression of reconnection does not occur (Swisdak and Drake 2007).

Here, as illustrated in Fig. 7, we first define the N-direction as normal to a plane containing $\mathbf{B}_1$ and $\mathbf{B}_2$. Following the Swisdak and Drake (2007) model, An X-line along the M-direction is defined that forms an angle $\approx\theta/2$ from $\mathbf{B}_1$. In this formulation, the guide field (M-direction) can be different on either side of the interface (Fig. 7b), namely,

$$B_{2M} = B_2 \cos\left(\frac{\theta}{2}\right) \tag{3.4}$$

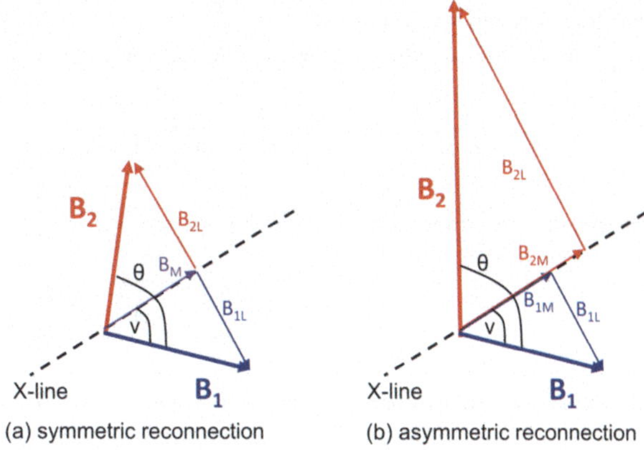

Fig. 7 (After Sonnerup 1974.) Illustration of L-M-N coordinate system defined by two vectors on either side of a plasma interface. Here vectors are shown in the L-M plane. The N direction is normal to the plane of the figure. Examples of symmetric and asymmetric reconnection with a guide field are shown in (a) and (b), respectively. Using the bisection model, $\nu \approx \theta/2$ (Swisdak and Drake 2007; Hesse et al. 2013)

and

$$B_{1M} = B_1 \cos\left(\frac{\theta}{2}\right) \tag{3.5}$$

with reconnecting (L-direction) components,

$$B_{1L} = B_1 \sin\left(\frac{\theta}{2}\right) \tag{3.6}$$

and

$$B_{2L} = B_2 \sin\left(\frac{\theta}{2}\right). \tag{3.7}$$

The ratio of the reconnecting components is therefore equal to the ratio of magnitudes, i.e.,

$$\frac{B_{1L}}{B_{2L}} = \frac{B_1}{B_2}. \tag{3.8}$$

With equations (3.4)-(3.8), the guide fields can be normalized by the reconnecting component, resulting in,

$$\frac{B_{2M}}{B_{2L}} = \frac{B_{1M}}{B_{1L}} = \cot\left(\frac{\theta}{2}\right). \tag{3.9}$$

In this normalization, the guide field can be much larger than 1. In addition, both $B_{\text{asym},L}/B_{\text{asym}}$ and $v_{\text{asym},L}/v_{\text{asym}}$ vary as $\sin(\theta/2)$.

As will be discussed, the change in plasma β across an interface ($\Delta\beta \equiv \beta_1 - \beta_2$) is an important parameter for examining the possibility of reconnection to occur. Across a stagnant asymmetric interface, thermal pressure and magnetic pressure are expected to approximately balance each other (i.e., neglecting dynamic pressure for flow lines that move around the magnetopause, and neglecting magnetic tension forces associated with the curvature of the field). Since the plasma flows are largely tangential to the magnetopause boundary, their dynamic pressure does not significantly contribute to the normal pressure.

$$n_1 k_B T_1 + \frac{B_1^2}{2\mu_o} \approx n_2 k_B T_2 + \frac{B_2^2}{2\mu_o}. \tag{3.10}$$

Dividing both sides of equation (3.10) $\frac{B_2^2}{2\mu_o}$ and rearranging provides a relationship between the plasma betas on either side of the interface and ratios of the magnetic fields,

$$\beta_1 + 1 \approx \left(\frac{B_2}{B_1}\right)^2 (\beta_2 + 1), \tag{3.11}$$

which can be rearranged in terms of $\Delta\beta$ as,

$$\Delta\beta \approx \beta_2 \left(\left(\frac{B_2}{B_1}\right)^2 - 1\right) + \left(\frac{B_2}{B_1}\right)^2 - 1. \tag{3.12}$$

For a low-β magnetosphere, i.e., $\beta_2 \ll 1$, equation (3.12) can be approximated as,

$$\Delta\beta \approx \left(\frac{B_2}{B_1}\right)^2 - 1, \tag{3.13}$$

providing a relationship between the change in plasma β and ratio of magnetic field strengths across an interface. Equations (3.13), (3.4), and (3.9) can therefore be used to 'translate' between $(\Delta\beta, \boldsymbol{\theta})$ to a guide field strength and system symmetry.

Furthermore, if $\beta_1 \gg \beta_2$, then $\Delta\beta \approx \beta_1$ and equation (3.13) can be written as,

$$\beta_1 \approx \left(\frac{B_2}{B_1}\right)^2 - 1. \tag{3.14}$$

If both magnetosheath and magnetospheric plasmas are low β, then equations (3.9)-(3.11) reduces to the symmetric case of $B_1/B_2 \approx 1$, with $\Delta\beta = 0$. Even if the magnetic fields are symmetric, the mass densities may nonetheless be asymmetric across the interface if the temperatures correspondingly vary to allow for pressure balance (Cassak and Shay 2007). Here, the densities and temperatures are either side of the interface will both impact the calculation of plasma β, and the densities impact the asymmetric mass density in equation (3.2).

Finally, for the typical case at a magnetopause interface of $\beta_2 \ll 1$ and $\rho_1 > \rho_2$, we can approximate the asymmetric mass density from equation (3.2) only as a function of the magnetosheath mass density and the ratio of magnetic field strengths across the boundary, i.e.,

$$\rho_{asym} \approx \frac{\rho_1}{\left(\frac{B_1}{B_2}\right) + 1}. \tag{3.15}$$

3.3 Suppressed Reconnection Onset

There are different conditions at the interface of plasmas that can inhibit the ability of reconnection to occur. This type of binary analysis has been useful to study planetary systems and parametrize the role of reconnection. Here we discuss diamagnetic suppression, flow-shear-based suppression, and spatial suppression.

The diamagnetic suppression of component reconnection at interfaces with large density asymmetries was studied in detail by Swisdak et al. 2003, 2010 and summarized in Liu et al., *this journal*. Here, through particle-in-cell simulations, it was found that if the drift speed of the X-line from was larger than that of the reconnection outflow speed, reconnection is suppressed. This condition led to the criterion of

$$\Delta\beta > \frac{2\Delta}{d_i}\tan\left(\frac{\theta}{2}\right). \tag{3.16}$$

where Δ corresponds to the thickness of the current sheet, often taken to be within the range $\Delta = 0.5d_i$ to $2d_i$, and d_i is the ion inertial length. The ion inertial length is the scale at which ions decouple from electrons and reconnection can take place, i.e.,

$$d_i \equiv \left(\frac{m}{q}\right)\frac{1}{\sqrt{\rho_{asym}\mu_o}}. \tag{3.17}$$

For a given plasma composition, the ion inertial length scales only with the number density, a fact that will impact scaling reconnection across different systems.

Kobayashi et al. (2014) and Liu and Hesse (2016) studied the diamagnetic suppression of asymmetric reconnection for the case of both strong density and temperature gradients that lead to a change in plasma β across an interface. They found that diamagnetic drift generated by a strong density asymmetry led to suppression as expected. The drift of the X-line is slowed and overtaken by the faster moving ion flow. However, strong diamagnetic drifts associated with only a gradient in temperature were not as effective in suppressing reconnection. Kobayashi et al. (2014), in particular, found that temperature-based gradients led to the generation of other instabilities that destabilized the boundary to reconnection. Nevertheless, as shown in Fig. 8, equation (3.16) has been highly successful at parameterizing reconnection at Earth (Phan et al. 2013a), planetary magnetospheres (DiBraccio et al. 2013; Masters et al. 2012; Montgomery et al. 2022; Fuselier et al. 2014; Jasinski et al. 2021; Sun et al. 2020a), and at the heliopause (Fuselier and Cairns 2017; Fuselier et al. 2020a), enabling delineation between groups of non-reconnecting and reconnecting current sheets.

Figure 9 provides a summary of the relationship between $\Delta\beta$, θ, and B_M/B_{Lasym}. We add curves corresponding to the criteria for diamagnetic suppression. Strong-guide-field reconnection occurs for low shear angles. The Swisdak et al. (2003, 2010) criterion is most effective when applied locally to a specific plasma interface. Average scalings of Gershman and DiBraccio 2020; Masters 2018 throughout the solar system and models of reconnection at Jupiter (Desroche et al. 2012) predict largely suppressed reconnection at the subsolar and dawnside magnetopauses, respectively, yet observations of reconnection signatures (Ebert et al. 2017; Montgomery et al. 2022; Jasinski et al. 2021) are nonetheless reported. The near ubiquitous presence of boundary layers along planetary magnetopauses (Sonnerup and Lotko 1990; Anderson et al. 2011; Masters et al. 2011; Gershman et al. 2016) tend to systematically reduce the change in plasma beta across interfaces and enable reconnection where it may have otherwise been predicted to be suppressed. In addition, as discussed in Gershman and DiBraccio (2020), increased solar activity also systematically lowers the upstream Alfvénic Mach number, which tends to decrease the plasma β at planetary magnetopause boundaries. Weaker shock compression factors and increased plasma depletion also contribute to a reduced $\Delta\beta$. When diamagnetic suppression criteria are fulfilled and signatures of reconnection are observed in the particle data, it may indicate that reconnection occurred elsewhere along the boundary (Montgomery et al. 2022; Fuselier et al. 2020b).

Various reconnection suppression mechanisms are illustrated in Fig. 10. In addition to diamagnetic suppression, flow-based suppression by Cassak and Otto (2011) and Doss et al. (2015) was predicted to occur when the differential shear speed across the interface exceeded

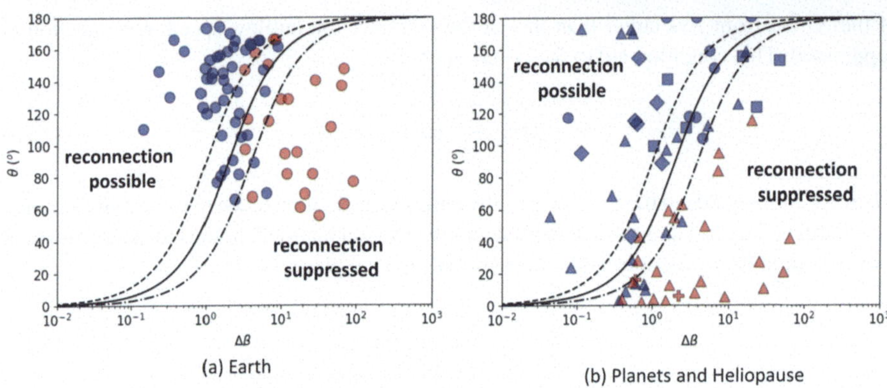

Fig. 8 Reported reconnecting (blue data points) and non-reconnecting (red data points) current sheets at (a) Earth (Phan et al. 2013a,b (●)) and (b) planetary magnetopauses (Mercury from DiBraccio et al. 2013 (♦), Saturn from Jasinski et al. 2016 and Fuselier et al. 2014 (■) and Jupiter from Montgomery et al. 2022 (▲)) and at the heliopause (Fuselier et al. 2020a (+)). The Swisdak et al. (2020) criterion provides a clear grouping between the two sets of events. Data was digitized from the referenced publications. For the DiBraccio et al. (2013) data points, only the high $|B_N/B_{MP}| > 0.25$ data points were taken. For Montgomery et al. (2022) the reconnection events were taken with 2 or 3 reconnection signatures, and the non-reconnecting events were taken as those with 0 reconnection signatures. Uncertainties on shear angles range from 5-20^0 and β up to a factor of 2, though were not necessarily reported for all studies

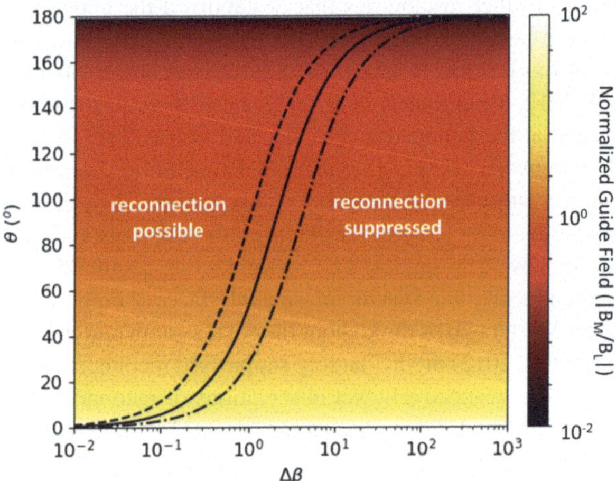

Fig. 9 Normalized guide field ($|B_M/B_L|$) as a function of magnetic shear angle and change in plasma β across an interface, assuming $\Delta\beta \approx \left(\frac{B_2}{B_1}\right)^2 - 1$ and following the bisection model for the X-line orientation. Smaller magnetic shear angles correspond to larger relative guide fields. The solid black, dashed black, and dash-dotted black lines correspond to the criterion of diamagnetic suppression from Swisdak et al. (2010) for $L = 1d_i$, $L = 2d_i$, and $L = 0.5d_i$, respectively. Below these lines, diamagnetic drift of the X-line at the interface of two plasmas exceeds that of the reconnection outflow speed and reconnection is suppressed

the asymmetric Alfvén speed by a factor,

$$V_{shear,crit} \approx V_{A,asym}\left(\frac{\rho_1 B_1 + \rho_2 B_2}{2\sqrt{\rho_1 B_1 \rho_2 B_2}}\right). \tag{3.18}$$

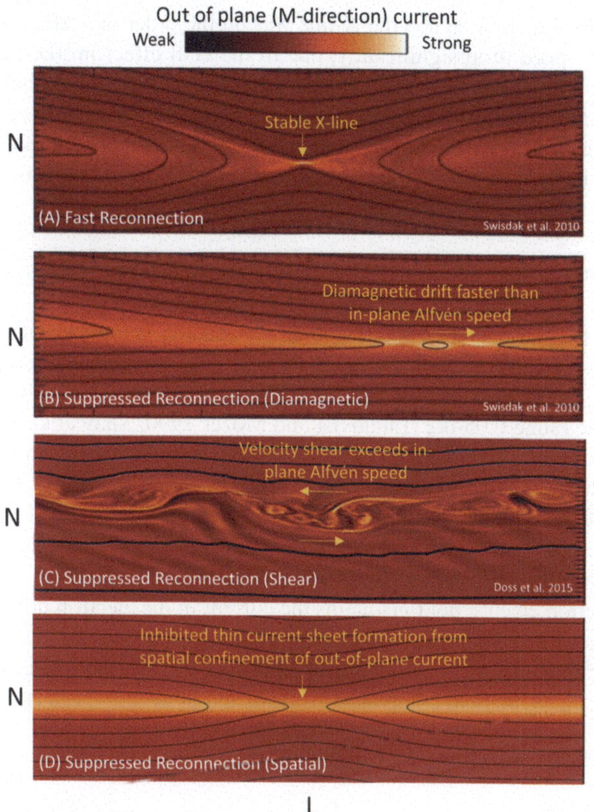

Fig. 10 Suppression mechanisms of reconnection illustrated with out-of-plane current density and magnetic field lines in the L-N plane. Dimensions and current densities are for reference only and not necessarily quantitatively comparable across the panels. Typical fast reconnection is shown in (a), adapted from Swisdak et al. (2010). Diamagnetic suppression is shown in (b), also adapted from Swisdak et al. (2010), where sufficiently large density gradients are introduced across the interface to generate a diamagnetic drift faster than the in-plane Alfvén speed such that the reconnection rate significantly reduces. Shear-based suppression is shown in (c), adapted from Doss et al. (2015), where relative shear flows across the interface significantly exceed the in-plane Alfvén speed and the interface becomes Kelvin-Helmholtz unstable. Spatial suppression, as recently studied by Liu et al. (2019), is shown in (d), where thin current sheets are not able to form. Figure (d) current densities and fields were derived using initial condition from Wilson et al. (2016)

Flow-shear-based suppression was investigated at the magnetopauses of Jupiter and Saturn (Sawyer et al. 2019; Desroche et al. 2012, 2013), where strong internal corotation-based flows may create increased shears at the magnetopause, in particular along the dawn flank. However, such super-Alfvénic shear flows have not been reported at either planet. Quite the contrary, reconnection was observed at Saturn in a region where flow-shear-based suppression was thought to be present (Sawyer et al. 2019). Flow-shear-based suppression at Jupiter and Saturn is based on extension of corotation flows out to the magnetopause, and observations of these flows have been difficult. Furthermore, as discussed in Sect. 2.1.2, hydrodynamic models of magnetosheath (e.g., Petrinec and Russell 1997) tend to overestimate the flow speed that may lead to overestimates of flow-shear-based suppression.

Finally, Liu et al. (2019) study the concept of 'spatial suppression' of magnetic reconnection. Such suppression was observed in 3-D particle-in-cell simulations of X-lines spatially confined in the M-direction. When the X-line was on the order of $<10d_i$, the reconnection rate and outflow speed drop significantly, due to the Hall effect in 3D. Per the author of that study, this minimum X-line extent may explain the smallest azimuthal scales of dipolarization flux bundles at Earth, and the cause of a dawn-dusk asymmetry of reconnection in Mercury's magnetotail As will be discussed in Sect. 5, spatial suppression of reconnection may also serve to reduce the overall efficiency of magnetopause reconnection at smaller planetary magnetospheres.

3.4 Scaling the Reconnection Rate and Energy Partitioning

The dimensionless reconnection rate is commonly studied and evaluated in reconnection simulations andstudies of planetary magnetospheres. The dimensionless reconnection rate (α), namely the ratio of inflow velocity to outflow velocity (V_{in}/V_A) has been found at Earth to be in the range 0.05-0.2 (Lindqvist and Mozer 1990; Shay et al. 1999; Phan et al. 2007; Mozer and Retinò 2007; Cassak et al. 2017b; Hesse et al. 2018; Genestreti et al. 2018; Burch et al. 2020; Sun et al. 2020b; Burch et al. 2022; Li and Liu 2021). Similar ranges of numbers have been derived at planetary magnetospheres (DiBraccio et al. 2013; Gershman et al. 2016). The use of rates on the order of $\alpha \sim 0.1$ is therefore sufficient to parameterize reconnection dynamics.

The reconnection electric field maintains the current in the electron diffusion region and regulates the energy conversion from the inflow to the outflow (Hesse et al. 2018). This field is defined as the dimensionless reconnection rate times the local asymmetric Alfven speed and asymmetric magnetic field component (Cassak and Shay 2007; Liu et al. 2018),

$$E_R \approx \alpha V_{A,asym,L} B_{asym,L} \tag{3.19}$$

For a given magnetized body, the relative strength of the reconnection electric field compared to other processes can be critical to evaluate to understand the potential role reconnection could have in driving dynamics. The outflow speed of reconnection can be compared against the corotating speed inside the magnetopause, giving a measure of what dominates a planetary magnetosphere (Kivelson 2007).

Of critical important at the diverse set of magnetized bodies in our solar system is how reconnection varies with the presence of multi-species plasmas or cold ion populations. Toledo-Redondo et al. (2017, 2018, 2021) and (Norgren et al., this collection) summarize MMS-era findings of both simulations and spacecraft data with regards to the role of cold protons and O^+ in reconnection rate. In the presence of multi-species, multi-temperature plasma, a multi-scale ion diffusion region (IDR) forms that adds complexity to the reconnection site. This multi-scale IDR can result in different energization and scattering processes for different species, or different populations (Dargent et al. 2023). Despite this complexity at the micro-scales, at a macro-scale reconnection rate (i.e., the use of ~0.1) appears largely unaffected by the presence of multi-species plasmas. However, in order to accurately calculate the local outflow speed, i.e., the Alfvén speed, one must utilize the correct mass density that accounts for the relevant plasma composition. Mass-loading effects, therefore, can become important for reducing the overall rate of transport of magnetic flux (i.e., $V_{A,asym,L} B_{asym,L}$) in a system (Norgren et al., this collection).

As discussed by Liu et al., this collection) and Phan et al. (2013b, 2014), approximately 50% of the available magnetic energy per particle, $m_i V_{AL}^2$ (Shay et al. 2014), goes into the outflow jet, where V_{AL} is the inflow Alfven speed based on the reconnecting magnetic field component. Studies by Phan et al. (2013a,b, 2014) and Drake et al. (2009) have found 13%

of the inflowing magnetic energy (or Poynting flux) per particle goes into increase of ion bulk temperature and 2% goes into electron bulk temperature increase. In terms of enthalpy flux, 33% of magnetic energy per particle is converted into ion enthalpy flux, while 4% goes into electron enthalpy flux. Toledo-Redondo et al. and Dargent et al. (2023) reported that up to 25% of the ion bulk heating can go into a cold ion population. These findings should be generally applicable to reconnecting systems throughout the universe and have already scaled to the solar wind (Phan et al. 2022), to the heliopause (Cairns and Fuselier 2018), to black holes (Chael et al. 2018), and to reconnection at sub-Alfven flows (Kaweeyanun et al. 2020) at Ganymede.

4 Reconnection Signatures in Planetary Magnetospheres

Magnetic reconnection has been observed at practically every explored magnetized body in the solar system. At planetary systems, where the plasma observations may be more limited, reconnection signatures are typically limited to the 'magnetopause' and 'magnetotail' regions. However, the high resolution and multipoint measurements provided by MMS has demonstrated that reconnection is ubiquitous in turbulent plasmas such as the magnetosheath (Yordanova et al. 2016; Vörös et al. 2017; Eriksson et al. 2018; Wilder et al. 2018; Phan et al. 2018; Stawarz et al. 2022) Kelvin-Helmholtz along the flanks of planetary magnetopauses is often framed as a competing process with dayside reconnection for the dominant source of solar wind mass and energy transport into a magnetosphere (e.g., Desroche et al. 2013; Masters 2018). However, reconnection has also been observed within Kelvin-Helmholtz vortices (Eriksson et al. 2016; Li et al. 2016; Vernisse et al. 2016) and may in fact lead to increased transport of mass and energy than may be predicted by MHD descriptions of the KHI instability (Nakamura et al. 2017, 2022).

Near-magnetopause plasma β throughout the solar system vary across a wide dynamic range of ~0.05-100. Lower β values are typically observed in the inner heliosphere at Mercury, where the upstream M_A is low, though these nonetheless vary over several orders of magnitude (DiBraccio et al. 2013; Gershman et al. 2013; Slavin et al. 2014; Sun et al. 2022). In the outer solar system (Masters et al. 2012) and at the heliopause (Fuselier et al. 2020a), the near-magnetopause β tend to be larger, in the 1-100 range, which tends to limit the shear angles under which reconnection ispossible to nearly anti-parallel. Magnetic cloud Interplanetary Coronal Mass Ejections (ICMEs) tend to systematically reduce M_A throughout the heliosphere, leading to the potential for lower magnetopause plasma β during times of increased solar activity (Farrugia et al. 1997; Lavraud and Borovsky 2008; Gershman and DiBraccio 2020).

Signatures of active reconnection along a magnetopause-like boundary are (a) heated electrons (Montgomery et al. 2022; Fuselier et al. 2020b), (b) $|B_n/B_{mp}| \gg 0.1$ (Sonnerup 1974; DiBraccio et al. 2013), (c) plasma jets and Hall-field signatures (Ebert et al. 2022; Harada et al. 2018; Cravens et al. 2020; Wang et al. 2021; Collinson et al. 2018), (d) flux ropes and traveling compression regions (Slavin et al. 1993, 2012; Jasinski et al. 2016, 2022; Imber et al. 2014; Russell 1995; Romanelli et al. 2022), and thick boundary layers (Masters et al. 2011; Fuselier et al. 2020b; Gershman et al. 2017). The majority of these signatures have been observed in $M_{MS} > 1$ systems, but several been reported for the $M_{MS} < 1$ Ganymede environment (Collinson et al. 2018; Ebert et al. 2022). As an example of similar signatures observed in different systems, reconnecting current sheets at Mars (Harada et al. 2018) and Jupiter (Ebert et al. 2017) are shown in Fig. 11 utilizing data from MAVEN and Juno, respectively. In both events, an ion jet was observed in the 'L' direction, indicative of

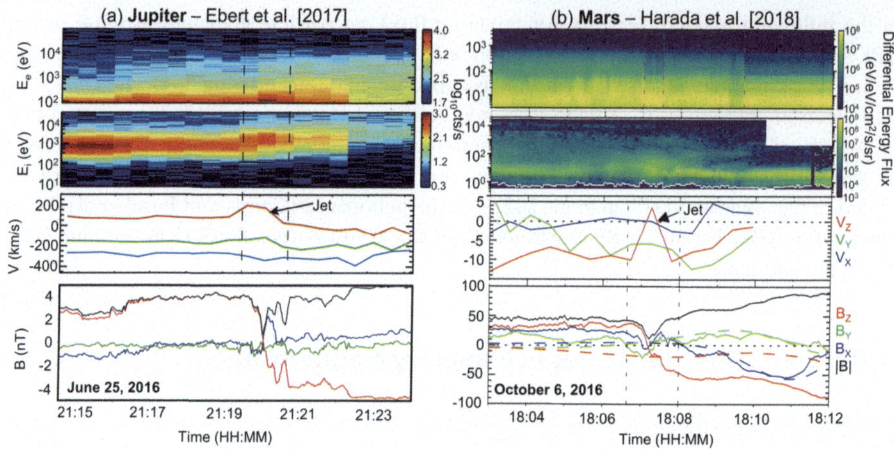

Fig. 11 Reconnecting current sheets at (a) Jupiter (adapted from Ebert et al. 2017) and (b) Mars (adapted from Harada et al. 2018). Magnetic field data are shown in planetocentric coordinates, Mars-Solar-Orbit (MSO) and Jupiter-Solar-Orbit (JSO), respectively. In both of these examples, the (X, Y, Z) coordinates from planetocentric coordinates correspond to the (M, N, L) coordinates of the current sheet. An L-direction ion jet is shown for each in the vicinity of the current sheet crossing

a reconnection outflow. Despite significant differences in the Martian and Jovian systems, these events appear remarkably similar.

Magnetic flux ropes formed along the dayside magnetopause, typically referred to as 'flux-transfer-events' (FTEs) have been observed at nearly every planetary magnetopause (Russell and Elphic 1979; Russell 1995). FTEs have been attributed with multiple possible formation mechanisms such as time varying reconnection (Southwood et al. 1988) and multiple X-line reconnection (Lee and Fu 1986; Raeder 2006). The core field strength and cadence of these structures varies depending on the system, as shown in Fig. 12 with a series of rapid, small-scale set of 'FTE showers' that can account for the majority of the transported flux at Mercury (Slavin et al. 2012; Imber et al. 2014; Sun et al. 2020a,b), to large-scale FTEs that are a signature of reconnection, but may not account for significant flux transport at the Giant Planets (Jasinski et al. 2016). When the magnetic shear at the magnetopause is low, MMS data has shown that the helicity sign of FTEs is correlated with the direction of the IMF, suggestive of guide-field-based-ordering of FTE structure. This correlation was not as strong in the presence of high magnetic shear (Dahani et al. 2022).

In the magnetotails of planets, reconnection is typically more symmetric, with $\Delta\beta \approx 0$ across the tail current sheet. This plasma symmetry enables reconnection to occur for almost any shear angle if the current sheet is sufficiently thin. In planetary magnetotails, minimum variance analysis of magnetic field data is commonly used to identify the presence of reconnection such as dipolarization fronts (Vogt et al. 2020; Sundberg et al. 2012; Dewey et al. 2018; Jackman et al. 2015), flux ropes and plasmoids (Jackman et al. 2011; DiBraccio et al. 2015; Hara et al. 2022; DiBraccio and Gershman 2019; Zhang et al. 2012), and TCRs (Slavin et al. 2009, 2012; Vogt et al. 2014; Jackman et al. 2014). The loading and unloading of the magnetotail also varies from system-to-system, which leads to varying lobe pressures and corresponding reconnection outflow speeds. At Mercury, significantly sheared tail lobes from the interaction between the planet and the solar wind can lead to significant guide fields in the magnetotail, resulting in flux-rope structures with large core fields, significant increases in magnetic pressure, and dawn-dusk asymmetries (DiBraccio et al. 2015;

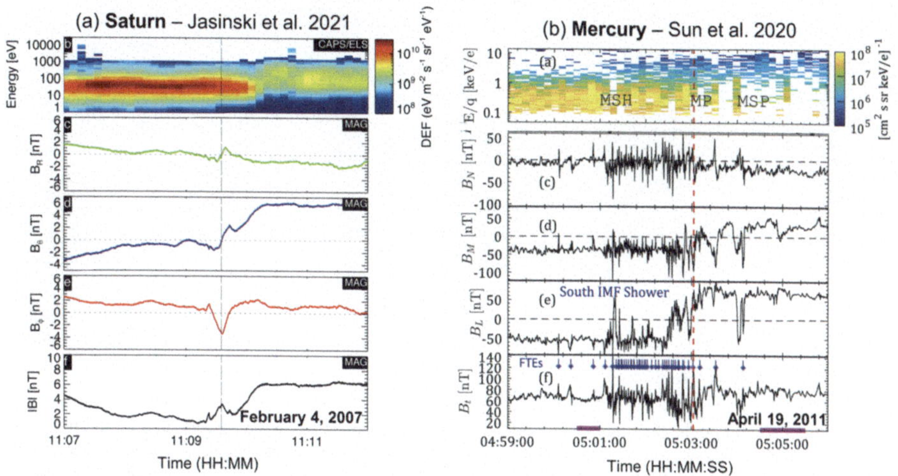

Fig. 12 Example FTE observations at (a) Saturn (adapted from Jasinski et al. 2021) and (b) Mercury (adapted from Sun et al. 2020a). For FTEs, an enhanced magnetic field magnitude is observed within each structure along with a rotation in the field direction. At Saturn, the FTE is ~1 min in duration. At Mercury, there are dozens of ~1-10 s duration events each spaced by ~1-10 s observed along the magnetopause, forming a so-called 'FTE shower'

Sun et al. 2016; Poh et al. 2017) At the Giant Planets, reconnection in the magnetotail is nearly anti-parallel with no guide field, and loop-like plasmoids exhibit signatures of so-called 'O-line' reconnection, where there is no core field and instead there is a depression in magnetic pressure and enhancement in plasma pressure (Vogt et al. 2014; Jackman et al. 2011; DiBraccio and Gershman 2019). At a hybrid magnetosphere such as Mars, the twisted tail associated with changing solar wind conditions (Luhmann et al. 2015; DiBraccio et al. 2018, 2022) and complex draping geometry results in variable core field strength (Briggs et al. 2011; DiBraccio et al. 2015; Hara et al. 2017). A comparison of flux ropes observed in the magnetotails of Mercury, and Saturn are shown in Fig. 13, illustrating similar signatures in the magnetic field data, but with significantly different spatiotemporal scales and core field strengths.

The differences in temporal and spatial scales of reconnection structures in planetary magnetospheres each lead to a varying role and relative importance of reconnection in driving magnetospheric dynamics (Russell 2000; Bagenal 2013). The assessment of the contribution of magnetic reconnection at a planetary magnetopause or magnetotail is a strong function of the specific properties of that magnetospheric system. However, the above techniques combined with suitable in situ plasma and fields data have been demonstrated to be highly successful at identifying the presence of active reconnection and can therefore serve as a reliable starting point for more system-level analyses.

5 Scaling Reconnection Across and Beyond the Solar System

As a final consideration, we evaluate the total amount of magnetic energy a magnetized body can extract from its upstream environment. Determination of this energy requires evaluation of the energy conversion rate of magnetic reconnection. The reconnection energy conversion rate (Mozer and Hull 2010; Goodbred et al. 2021) has been less discussed in the planetary

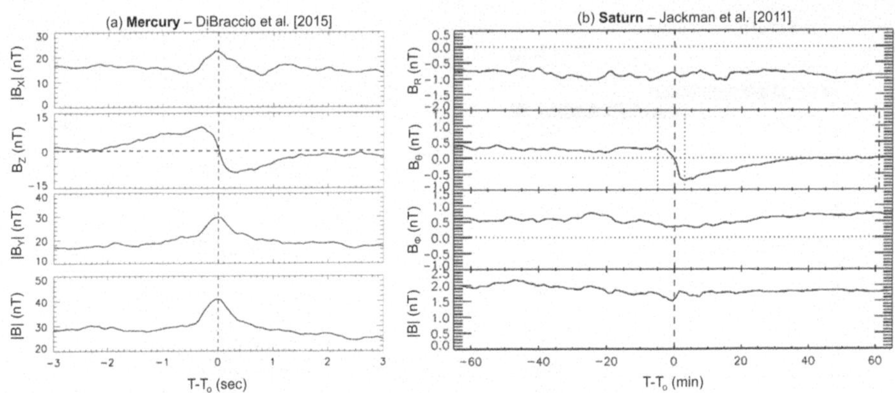

Fig. 13 Superposed epoch analysis of tailward traveling flux ropes at (a) Mercury (adapted from DiBraccio et al. 2015) and (b) Saturn (adapted from Jackman et al. 2011) organized in planetocentric coordinates. In each, bipolar signatures in the B_Z or B_θ component shows a Northward-Southward orientation, indicating tailward flow of the structure. In the Mercury flux rope, the structure duration is ~1 s and it contains a strong core field as evidenced by the significant enhancement of the magnetic field magnitude (X-line reconnection). In the Saturn case, the structure scale is ~5 min, with little-to-no core field (O-line reconnection)

magnetospheres literature with the primary focus being the dimensionless rate of reconnection (α) (Kennel and Coroniti 1977; Holzer and Slavin 1978; Nichols et al. 2006; Kivelson 2007; Slavin et al. 2009, 2010; DiBraccio et al. 2013; Gershman et al. 2016; Zhong et al. 2018; Arridge 2020) or the reconnection electric field (E_R) (Nichols et al. 2006; DiBraccio et al. 2013; Slavin et al. 2009; Masters 2014, 2015a, 2018; Newell et al. 2007; Milan et al. 2012).

The reconnection electric field is in units of energy per length (e.g., mV/m). Integrated that electric field along an X-line results in the "reconnection voltage" or "reconnection potential", i.e., the effective potential difference between opposite ends of the X-line (Masters 2015b). These voltages are typically 10 s of kV. Reconnection voltages are typically used to assess how much energy can be input into a magnetosphere via dayside reconnection as well as quantify the rate of production of open magnetic flux (Nichols et al. 2006; Badman et al. 2014; Zhang et al. 2021). While analyses of reconnection voltage produce instantaneous estimates of how much energization particles can experience (units of energy), they do not typically provide a measurement of the reconnection power, i.e., the rate at which the energy input to a magnetosphere occurs which requires units of energy over time (Tenfjord and Østgaard 2013).

The available magnetic power for an obstacle can be written as the upstream Poynting flux across the area of the obstacle in SI units following Koskinen and Tanskanen (2002) as

$$P_u = \frac{4\pi}{\mu_o} v_u B_u^2 \left(\pi R_{MP}^2\right), \tag{5.1}$$

where P_u scales with the effective area of the magnetopause to the upstream obstacle.

Many coupling functions have been developed for Earth starting with Perreault and Akasofu (1978), Akasofu (1981), and Vasyliunas et al. (1982). These provide scaling for the amount of energy and then are empirically compared with estimates of the total energy in the magnetosphere. After converting to SI units (Koskinen and Tanskanen 2002), the

Perrault-Akasofu parameter takes the form of

$$P_{MP} = \frac{4\pi}{\mu_o} v_u B_u^2 \left(l_o^2\right) \sin\left(\frac{\theta}{2}\right)^4 . \tag{5.2}$$

Here, θ represents the shear angle between the fields across the magnetopause, and l_o is a characteristic scale length, empirically determined for Earth to be ~7 R_E or ~0.7 R_{MP} (Perreault and Akasofu 1978). We define the coupling efficiency as the ratio P_{MP}/P_u. Evaluating this efficiency for $\theta = 180^o$ gives ~0.16. This relatively high efficiency compared to other estimates (~1%) of solar-wind-magnetospheric coupling is due to our use of magnetic energy in P_u instead of solar wind kinetic energy (Tenfjord and Østgaard 2013)

The Perrault-Akasofu parameter has been scaled directly to other planets in several studies (Desch and Kaiser 1984; Desch and Rucker 1985; Ip et al. 2004; Zarka 2007). It is important to note that equation (5.2) implicitly assumes that reconnection can occur uniformly over an obstacle and therefore its energy conversion scales as R_{MP}^2. This assumption can result in large systematic errors in predictions of reconnection-generated energies at non-Earth magnetospheres, as remarked by Baker and Bargatze (1985). Newell et al. (2007) and Tenfjord and Østgaard (2013) investigated the correlation between multiple geomagnetic indices at Earth and different coupling functions (including Perrault-Akasofu). These correlations, although highly effective at Earth for parameterizing the geomagnetic response to changing upstream conditions, are tuned to terrestrial dynamics, limiting their scaling to other magnetospheres.

To scale how much energy can be extracted from an upstream flow by a magnetized obstacle, we directly consider reconnection along a primary magnetopause X-line. Goodbred et al. (2021) recently investigated the scaling of the reconnection energy conversion rate using simulations of symmetric reconnection. They found that the energy conversion rate (converted to SI units here following Koskinen and Tanskanen 2002) scales as,

$$P_{MP} = \frac{8\pi}{\mu_o} \alpha \xi V_A B_A^2 L_x L_y , \tag{5.3}$$

where α is the reconnection rate (~0.05-0.2), ξ is a factor on the order of unity, L_X is the length of the Petschek exhaust region in the presence of secondary tearing (~60 di) and Ly is the length of the reconnection X-line, taken as ~1.5 R_{MP} (Trattner et al. 2021). Here, we apply this equation and use the asymmetric Alfvén speed, mass density and magnetic field expressions from Cassak and Shay (2007) and assume that Ly scales with R_{MP}. The coupling efficiency of reconnection therefore scales as:

$$\frac{P_{MP}}{P_U} \propto \left(\frac{V_{A,asym}}{v_u}\right) \left(\frac{B_{asym}^2}{B_u^2}\right) \left(\frac{d_i}{R_{MP}}\right) . \tag{5.4}$$

The key distinction between a magnetopause-based scaling and that of the Perrault-Akasofu approach is that the energy conversion rate scales as $d_i R_{MP}$ (i.e., $L_x L_y$) instead of R_{MP}^2 (obstacle size) which produces a fundamentally different scaling across the solar system. As an example, in Fig. 14 we compare the energy conversion surfaces superimposed on example magnetopause magnetic shear plots for Earth, Jupiter, and Uranus. For simplicity, we consider cases where the planetary dipole axis is in the X-Z plane and the IMF is oriented to generate anti-parallel reconnection at the subsolar magnetopause (southward for Earth and Uranus and northward for Jupiter). The left panels on Fig. 15 show surface areas equal to $(0.7\ R_{MP})^2$ (corresponding to a disc with radius ~0.4 R_{MP}) while the right panels show

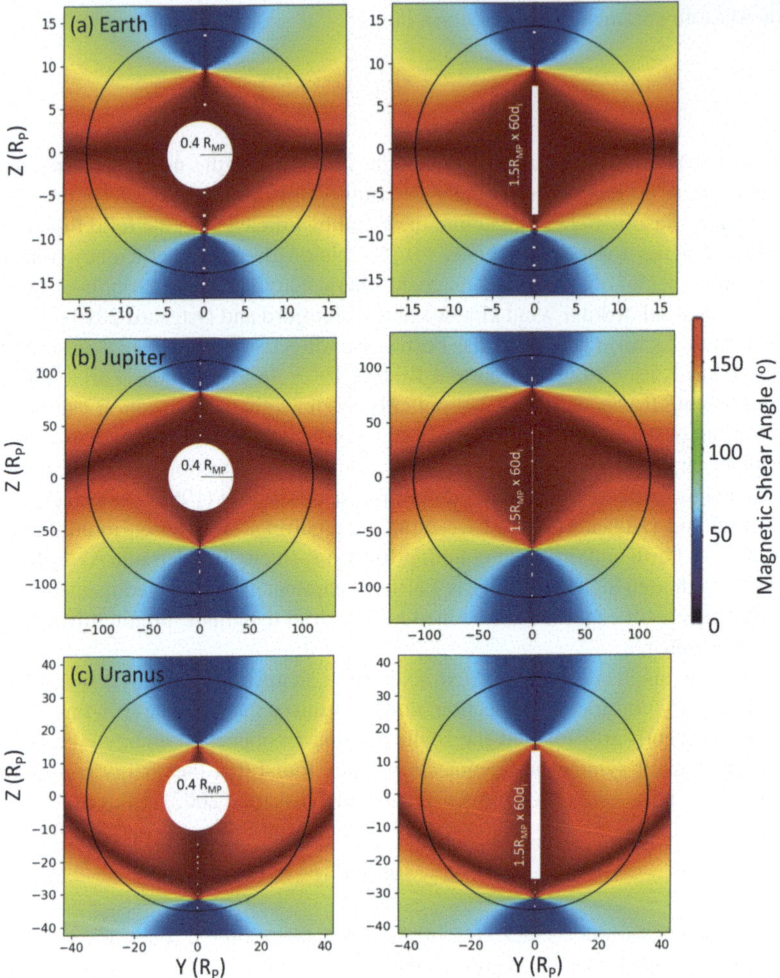

Fig. 14 Maximum energy conversion surfaces (shown in white) for magnetopause reconnection using the (left) Perrault-Akasofu parameter (i.e., scales with obstacle area) and the (right) Goodbred et al. (2021) approach (i.e., scales with X-line length and ion inertial length) for (a) Earth ($60d_i \sim 0.5\ R_E$), (b) Jupiter ($60d_i \sim 0.02\ R_J$), and (c) Uranus ($60d_i \sim 2.5\ R_U$). Surfaces are overlaid on a magnetic shear plot of the surface of magnetopause projected into the YZ-plane with southward IMF for Earth and Uranus and northward IMF for Jupiter (i.e., $\theta = 180^o$). The Perrault-Akasofu scaling results in the same conversion efficiency of upstream energy at the magnetopause. The Goodbred et al. scaling results in more and less efficient conversion of upstream magnetic energy at Uranus and Jupiter respectively

example X-line with length $\sim 1.5\ R_{MP}$ and thickness $60d_i$. When compared to Earth, the Perrault-Akasofu scaling over- and under-estimates the total interaction area at Jupiter and Uranus, respectively.

Use of the Goodbred et al. (2021) scaling to parameterize energy-conversion at the magnetopause is enabled by MMS-era confirmations that reconnection at Earth's magnetopause is typically localized along a primary X-line (Trattner et al. 2021; Fuselier et al., *this journal*). This approach is only valid for $L_X \ll L_Y$, as otherwise the exhaust size approaches

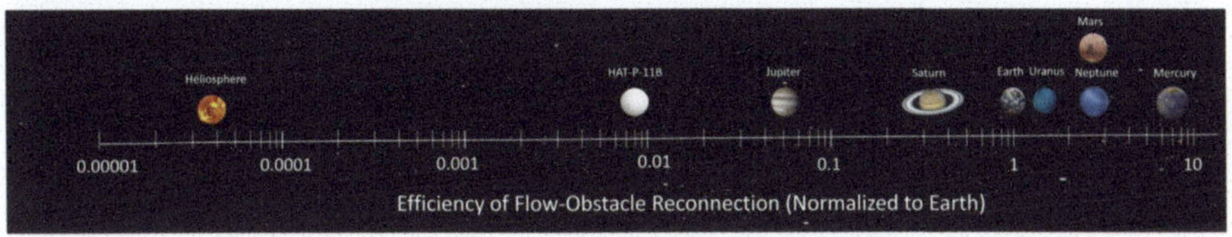

Fig. 15 Efficiency of magnetic reconnection at planetary magnetopauses and the heliopause at extracting magnetic energy from their respective upstream flows. All values are normalized to Earth per Table 2

the size of the X-line and the geometry defined by Goodbred et al. is no longer directly applicable.

Estimates of reconnection parameters and coupling efficiency at different magnetized bodies are provided in Table 2. Magnetopause standoff distances were calculated from equation (2.8), and planetary size and distances from the Sun were taken from Bagenal (2013). The size of the heliosphere was taken as 120 AU converted to solar radii. In order to derive ρ_{asym}, $\mathrm{B}_{\mathrm{asym}}$, and $\mathrm{V}_{A,\mathrm{asym}}$, we need to obtain estimates of the plasma conditions at the subsolar magnetopause.

For planets in our solar system and HAT-B, which all have upstream $\mathrm{M}_{\mathrm{MS}} \gg 1$, we first take the relationships,

$$\left(\frac{B_2}{B_1}\right) \approx \sqrt{1 + d_F\left(2M_A^2\left(\frac{\eta-1}{\eta^3}\right) - 1\right)}, \tag{5.5}$$

and

$$\rho_1 \approx d_F \eta \rho_u, \tag{5.6}$$

which can be derived using the scalings provided by Gershman and DiBraccio (2020) taking their equations 2 and 6 and solving for $\mathrm{B}_2/\mathrm{B}_1$ in equation (3.14) of this study. Here we use a shock compression factor (η) of 4 and PDL depletion factor $\mathrm{d}_{\mathrm{F}} = 0.85$ (Masters 2018)) We then combine these equations with the relationship between the upstream solar wind pressure and magnetospheric field from equation (2.7) using $\psi = 0$. These values for B_1, B_2, and ρ_1 are used to calculate asymmetric reconnection parameters using the Cassak and Shay (2007) formulas assuming $\rho_1 < \rho_2$ and $\beta_1 > \beta_2$ and no guide field, i.e., equations (3.1)-(3.3).

For magnetized moons embedded in sub-magnetosonic flows, we take $\rho_1 = \rho_u$, and $\mathrm{B}_1 = \mathrm{B}_{\mathrm{u}}$ and used the above equations with the relationship (3.14) to derive reconnection parameters $\mathrm{B}_{\mathrm{asym}}$, $\mathrm{V}_{\mathrm{asym}}$, and ρ_{asym}.

For induced magnetospheres, the IMB altitude is determined by a pressure balance between the thermal pressure in the magnetosheath and a combination of the magnetic pressure from piled-up flux, crustal fields in the case of Mars, plus thermal pressure from the ionosphere (e.g., Li et al. 2020). In these environments, there can be significant concentrations of cold, heavy ions such that $\rho_2 \gg \rho_1$. Equations (3.2) therefore can be simplified as be combined to estimate $\rho_{\mathrm{asym}} \approx \frac{\rho_2\left(\frac{B_1}{B_2}\right)}{\left(\frac{B_1}{B_2}\right)+1}$. For Mars we assume the dominant species is O_2^+ and take $\mathrm{n}_2 \sim 50/\mathrm{cc}$ though we note that there is significant variability about both the composition and density (Harada et al. 2018; Ma et al. 2015; Chen et al. 2022; Wang et al. 2021; Nagy et al. 2004; Matsunaga et al. 2017). We then apply equation (5.5) with $\mathrm{M}_{\mathrm{A}} = 10$ to find $\mathrm{B}_1/\mathrm{B}_2 = 0.35$ and $\rho_{\mathrm{asym}} = 415\ \mathrm{amu\,cm}^{-3}$.

Table 2 Reconnection Parameters at Magnetized Bodies in the Heliosphere

	$\langle R_{MP} \rangle$ (body radii)	$\langle R_{sun} \rangle$ (AU)	R (km)	ρ_{asym} (amu/cm^3)	B_{asym} (nT)	$V_{A,asym}$ (km/s)	d_i (km)	$V_{A,asym}/V_U$	B_{asym}/B_U	d_i/R_{MP}	Relative P_{MP}/P_u (Earth-normalized)	Absolute Peak P_{MP}/P_U
Mercury	1.6	0.39	2440	88	151	351	24	0.84	5.1	0.0062	7.1	1.1*
Earth	10	1.0	6371	17	39	209	55	0.50	6.6	8.7×10^{-4}	1.0	0.16
Mars	1.3	1.5	3390	415	23	25	357	0.06	6.8	0.081	2.7	0.43
Jupiter	78	5.2	69,911	0.7	5	158	276	0.38	7.2	5.1×10^{-6}	0.053	8.5×10^{-3}
Saturn	25	9.6	58,232	0.2	3	156	508	0.37	7.2	3.5×10^{-4}	0.36	0.058
Uranus	25	19.2	25,362	0.05	1	155	1018	0.37	7.2	0.0016	1.7	0.27
Neptune	24	30.0	24,622	0.02	1	155	1593	0.37	7.2	0.0027	2.8	0.45
Ganymede	2	N/A	2634	24.6	108	477	643	3.4	1.2	0.12	–	–
Triton	2	N/A	1353	0.056	6.7	620	962	14	1.0	0.36	–	–
Heliosphere	25,804	N/A	696,340	0.017	0.45	75	1742	2.9	1.6	9.6×10^{-8}	3.6×10^{-5}	5.8×10^{-6}
HAT-P-11B	12	N/A	27,800	6600	1913	513	2.8	0.9	4.4	8.4×10^{-6}	0.0079	1.3×10^{-3}

*$P_{MP}/P_U > 1$ indicates likely very efficient but may indicate systems scales where model is not fully valid.

For the heliopause, we take averaged values on either side of the heliopause observed by Voyager 1 and 2 reported by Fuselier et al. (2020a,b) and calculate quantities directly, including the ratio B_1/B_2. The presence of a heliosheath inside of the heliopause boundary results in a higher plasma β than is present in the VLISM such that the approximations made above are no longer valid. A heliosheath is likely common among astrospheres with supermagnetosonic stellar winds where a termination shock would result in compressed and heated plasmas inside a heliopause. By taking the same approximations as above, we find B_{asym} (and $V_{A,asym}$) underestimated by a factor of $\sim \sqrt{\beta_2+1}$ (arising from the approximation in equation (3.11) going to equation (3.14)) and the efficiency underestimated by a factor of $\sim (\beta_2+1)^{3/2}$. These uncertainties should be taken into account when attempting to scale these results to an astrosphere with less well-constrained plasma properties.

The factors from equation (5.4) (i.e., $V_{A,asym}/Vu$, B_{asym}/B_U, and d_i/R_{MP}) are included in the table to understand what parameters drive an increased or reduced energy conversion rate between different planetary bodies. The conversion efficiencies P_{MP}/P_U in Table 2 normalized to Earth are determined by evaluating equation (5.3) for each planet and dividing by the corresponding value for Earth. These values provide a relative measure of the efficiency of an obstacle at extracting magnetic energy from their upstream flow and are visualized in Fig. 15.

The peak efficiencies derived here assume that reconnection is allowed to take place and has no guide field. The suppression conditions for reconnection are still applicable in this context and may inhibit reconnection at a given magnetopause boundary. The absolute efficiency at a given time will also be a function of the magnetic shear angle across the magnetopause due to the effect of component reconnection, i.e., the use of $V_{Aasym,L}$ and $B_{asym,L}$ instead of $V_{A,asym}$ and B_{asym} in equation (5.3).

Absolute conversion energy conversion rates can be estimated by comparing the Goodbred et al. scaling and Perrault-Akasofu scalings at Earth, where we expect $P_{MP}/P_U \approx 0.16$. For this case, we find,

$$2\alpha\xi\left(\frac{L_x}{60d_i}\right)\left(\frac{L_y}{R_{MP}}\right)\approx 0.6. \tag{5.7}$$

Equation (5.7) can be readily satisfied for reasonable parameters $L_y \sim 1.5\ R_{MP}$, $\xi \sim 1$, and $\alpha \sim 0.2$, and $L_X \sim 60d_i$. For these properties, the energy conversion rate in SI units can be estimated as,

$$P_{MP} \approx 3.6\times 10^8 V_{A,asymL} B_{asymL}^2 d_i R_{MP}. \tag{5.8}$$

From Sect. 3.2, there is an effective factor of $\sim\sin^3(\theta/2)$ variation of the reconnection magnetic energy density that should apply here due to the $V_{A,asymL}$ $B_{asymL}{}^2$ factor. A 90° shear across the magnetopause reduces the energy extracted from anti-parallel reconnection by a factor of ~3. This $\sin^3(\theta/2)$ variation in the energy conversion rate is in reasonable agreement with the $\sin^{8/3}(\theta/2)$ correlation by Newell et al. (2007) found at Earth.

The peak absolute efficiency of Mercury is close to ~1 due to its small obstacle size and upstream Mach number, and the peak efficiency of Jupiter is ~0.01 due to its extremely large obstacle size. Neptune and Uranus, are surprisingly more efficient at extracting magnetic energy from the solar wind than Earth. This efficiency is driven by significantly larger ratios of ion inertial length to magnetopause at the Ice Giants. We note that HAT-B, which is Neptune-sized, has a significantly lower relative reconnection efficiency than Neptune due to its very low B_{asym}/B_U ratio.

Table 3 Peak ($\theta = 180^o$) reconnection-generated power at the magnetopause at each planet for typical upstream conditions. Powers are reduced by a factor of ~ 3 for $\theta = 90^o$

a – Clarke (2013), b – Herbert and Sandel (1994), Herbert (2009), c – Lamy (2020), d – Kivelson and Bagenal (2014), e – Haewsantati et al. (2021).

	Typical (GW)	Auroral Input Power (GW)
Mercury	204	–
Earth	308	1-100 [a]
Jupiter	2277	10,000-100,000 [a]
Saturn	320	100-1000 [a]
Uranus	69	30-70 [b]
Neptune	41	~0.2-0.8 [c, d]

Comparable efficiencies were not calculated for Ganymede and Triton because $60d_i \gg R_{MP}$, indicating that the Goodbred et al. (2021) scaling is likely not applicable and spatial suppression effects (see Liu et al. 2019 and Sect. 3) may become relevant. It is possible that for systems with significant Alfvén wings, the effective obstacle size to the upstream flow is much larger, leading to a larger L_Y. In addition, the ratio $60d_i/R_{MP}$ at Mercury is ~ 0.4 such that its calculated reconnection efficiency may be somewhat overestimated. These smaller bodies may be capable of extracting upstream magnetic energy over nearly their entire dayside magnetopause surface.

Table 3 provides estimates of maximum available power from peak reconnection efficiency at each magnetized planet assuming 0.16 peak efficiency at Earth. For comparison, we include estimated total input powers for planetary aurora, which provide a proxy for estimating the energy driving a magnetosphere. Input powers are coarse estimates that are derived by taking by the total observed auroral emission and dividing by an efficiency, assumed to be $\sim 10\%$. The uncertainties in these values are quite large, though it can be highly instructive to compare their orders of magnitude with the available energy conversion rate from magnetic reconnection. From equation (2.8), significant variations in standoff distances observed at Saturn and Jupiter will also result in a different R_{MP} and therefore variations in energy conversion rates, though not by significant enough factors to put them on par with Earth.

At Earth and Neptune, the amount of energy from magnetopause reconnection is more than sufficient to power the auroras, even when considering a factor of ~ 3 reduction due to a more typical shear angle of $\theta = 90^o$. At Jupiter, the magnetopause reconnection power is far less than that required, though it is well known that the dominant energy source for the auroras at the gas giants are internally generated within the magnetosphere (Bagenal 2013). However, at Jupiter, for example, polar bright spots (Haewsantati et al. 2021) that may map to the magnetopause require only $\sim$100-1000 GW of input power, such that it is possible that magnetopause reconnection at Jupiter can power some amount of auroral emission. Interestingly, estimates of auroral input power at Uranus and Saturn are comparable to those expected from magnetopause reconnection, though not necessarily when a factor of ~ 3 reduction is included. For Saturn, modulations in auroral brightness can be attributed to variations in the upstream solar wind (Clarke et al. 2009), though these variations are typically associated with changes in dynamic pressure and the compression of Saturn's magnetosphere rather than magnetopause reconnection (Bagenal 2013), and internal flow shears have been suggested to be the source of the main aurora (Cowley et al. 2004; Bader et al. 2019). For Uranus, there may also be significant internal effects associated with icy moons that lead to additional particle energization within the magnetosphere as suggested by Cheng (1984) and Eviatar and Richardson (1986).

Finally, the above equations and scalings should apply to exoplanets and astrospheres throughout the universe, both for $M_{MS} \ll 1$ and $M_{MS} \gg 1$ systems. The key challenge of modeling reconnection at exoplanets and astrospheres is to develop reasonable estimates for the parameters B_{asym}, ρ_{asym}, and the relevant upstream values. The capability to provide these estimates, both from the observational perspective and modeling tools are growing despite numerous challenges (Boro Saikia et al. 2020; See et al. 2020; Thomas et al. 2021). Using Zeeman-Doppler imaging data, coarse estimates of equatorial magnetic field strength (Semel 1989; Donati and Brown 1997; Carter et al. 1996; Hackman et al. 2016) have been taken as input to coronal heating models (Vidotto 2017; van der Holst et al. 2014; Cohen et al. 2014; Sakaue and Shibata 2021) and generate estimates of stellar winds and upstream parameters. As discussed, Ben-Jaffel et al. reported the first detection of an exoplanet magnetosphere by identifying an extended tail of carbon ions around the planet and use PIC simulations with a model of the local stellar wind to constrain the internal magnetic field strength of the planet. Likewise, many astrospheres have a visible bow shock that can be used to constrain MHD models (Baalmann et al. 2021 and reference therein). As studies of exoplanets and astrospheres continue, and remote sensing techniques are further refined, it may yet be possible to infer reconnection parameters, providing key constraints on the dynamics of an exoplanet magnetosphere.

6 Concluding Remarks

It is unlikely that there will be an MMS-like constellation flown at a different planetary body in the foreseeable future. Fortunately, the observations and studies of MMS data as well as the theoretical studies and modeling tools developed and refined over the past several years are invaluable for understanding the fundamental plasma physics associated with magnetic reconnection and for revealing how to scale results to other magnetized bodies. The presence of small-scale structures like electron crescents may not be directly observable at other planets, but are almost certainly present, and the macro effects of reconnection, cross-scale coupling, energy transfer, and in-flow/out-flow parameters are all directly relevant in the study of planetary magnetospheres. MMS observations and other spacecraft observations at Earth serve as a crucial test of theoretical models and frameworks that can then be scaled elsewhere. Combined with this theoretical framework, plasma measurements at each body can then be used to estimate the role reconnection can play in the dynamics at a given body. While parametric scaling may assist in the development of an initial high-level understanding of a given system, local conditions play a major role in dynamics. Underexplored magnetospheres in the solar system therefore require in situ plasma measurements to accurately quantify solar-wind-magnetospheric coupling.

Funding This research was supported by the National Aeronautics and Space Administration (NASA) Magnetospheric Multiscale Mission (MMS) in association with NASA contract NNG04EB99C. YL is grateful for the support from NASA grants 80NSSC21K2048 and 80NSSC20K131.

Declarations

Competing Interests The majority of this research was supported by the National Aeronautics and Space Administration (NASA) Magnetospheric Multiscale Mission (MMS) in association with NASA contract NNG04EB99C. In addition, YL is supported from NASA grants 80NSSC21K2048 and 80NSSC20K1316, and J.E.S. is supported by the Royal Society Research Fellowship URF\R1\201286. The authors have no competing interests to declare that are relevant to the content of this article.

References

Akasofu SI (1981) Energy coupling between the solar wind and the magnetosphere. Space Sci Rev 28:121–190. https://doi.org/10.1007/BF00218810

Anderson BJ, Fuselier SA (1993) Magnetic pulsations from 0.1 to 4.0 Hz and associated plasma properties in the Earth's subsolar magnetosheath and plasma depletion layer. J Geophys Res 98:1461–1479

Anderson BJ, Phan TD, Fuselier SA (1997) Relationships between plasma depletion and subsolar reconnection. J Geophys Res 102(A5):9531–9542. https://doi.org/10.1029/97JA00173

Anderson BJ, Slavin JA, Korth H, Boardsen SA, Zurbuchen TH, Raines JM, Gloeckler G, McNutt RL, Solomon SC (2011) The dayside magnetospheric boundary layer at Mercury. Planet Space Sci 59(15):2037–2050. https://doi.org/10.1016/j.pss.2011.01.010

Archer MO, Horbury TS (2013) Magnetosheath dynamic pressure enhancements: occurrence and typical properties. Ann Geophys 31:319–331. https://doi.org/10.5194/angeo-31-319-2013

Arridge C (2020) Solar wind: interaction with planets. Oxf Res Encycl Phys: 1–34. https://doi.org/10.1093/acrefore/9780190871994.013.15

Arridge CS, Achilleos N, Dougherty MK, Khurana KK, Russell CT (2006) Modeling the size and shape of Saturn's magnetopause with variable dynamic pressure. J Geophys Res 111(A11227):1–13. https://doi.org/10.1029/2005JA011574

Axford W (1991) A commentary on our present understanding of the Martian magnetosphere. Planet Space Sci 39(1–2):167–173. https://doi.org/10.1016/0032-0633(91)90139-2

Baalmann LR, Scherer K, Kleimann J, Fichtner H, Bomans DJ, Weis K (2021) Simulating observable structures due to a perturbed interstellar medium in front of astrospheric bow shocks in 3D MHD. Astron Astrophys 650(A36):1–13. https://doi.org/10.1051/0004-6361/202039836

Bader A, Badman SV, Cowley SWH, Yao ZH, Ray LC, Kinrade J et al (2019) The dynamics of Saturn's main aurorae. Geophys Res Lett 46:10283–10294. https://doi.org/10.1029/2019GL084620

Badman SV, Jackman CM, Nichols JD, Clarke JT, Gérard JC (2014) Open flux in Saturn's magnetosphere. Icarus 231:137–145. https://doi.org/10.1016/j.icarus.2013.12.004

Bagenal F (2013) Planetary magnetospheres. In: Oswalt TD, French LM, Kalas P (eds) Planets, stars and stellar systems, vol 3: Solar and stellar planetary systems. Springer, Dordrecht, pp 251–307. https://doi.org/10.1007/978-94-007-5606-9_6

Baker DN, Bargatze LF (1985) Proper solar wind power esimtation and planetary radiometric efficiencies. Nature 314:455–456

Belenkaya ES, Khodachenko ML, Alexeev II (2015) Alfvén radius: a key parameter for astrophysical magnetospheres. In: Lammer H, Khodachenko M (eds) Characterizing stellar and exoplanetary environments. Astrophys space sci lib, vol 411. Springer, Cham. https://doi.org/10.1007/978-3-319-09749-7_12

Belenkaya ES, Alexeev II, Blokhina MS (2022) Modeling of magnetospheres of terrestrial exoplanets in the habitable zone around G-type stars. Universe 8(231):1–8. https://doi.org/10.3390/universe8040231

Ben-Jaffel L, Ballester GE, Muñoz AG et al (2022) Signatures of strong magnetization and a metal-poor atmosphere for a Neptune-sized exoplanet. Nat Astron 6:141–153. https://doi.org/10.1038/s41550-021-01505-x

Boro Saikia S, Jin M, Johnstone CP, Lüftinger T, Güdel M, Airapetian VS, Kislyakova KG, Folsom CP (2020) The solar wind from a stellar perspective. Astron Astrophys 635(A178):1–21. https://doi.org/10.1051/0004-6361/201937107

Borovsky JE (2021) Is our understanding of solar-wind/magnetosphere coupling satisfactory? Front Astron Space Sci 8:1–7. https://doi.org/10.3389/fspas.2021.634073

Borovsky JE, Hesse M, Birn J, Kuznetsova MM (2008) What determines the reconnection rate at the dayside magnetosphere? J Geophys Res 113(A07210):1–17. https://doi.org/10.1029/2007JA012645

Bowers CF, DiBraccio GA, Slavin JA, Gruesbeck JR, Weber T, Xu S et al (2023) Exploring the solar wind-planetary interaction at Mars: implication for magnetic reconnection. J Geophys Res Space Phys 128:e2022JA030989. https://doi.org/10.1029/2022JA030989

Briggs J, Brain D, Cartwright M, Eastwood J, Halekas J (2011) A statistical study of flux ropes in the Martian magnetosphere. Planet Space Sci 59(13):1498–1505. https://doi.org/10.1016/j.pss.2011.06.010

Burch JL, Torbert RB, Phan TD, Chen LJ, Moore TE, Ergun RE, Eastwood JP, Gershman DJ, Cassak PA, Argall MR, Wang S, Hesse M, Pollock CJ, Giles BL, Nakamura R, Mauk BH, Fuselier SA, Russell CT, Strangeway RJ et al (2016) Electron-scale measurements of magnetic reconnection in space. Science 352(6290):aaf2939. https://doi.org/10.1126/science.aaf2939

Burch JL, Webster JM, Hesse M, Genestreti KJ, Denton RE, Phan TD et al (2020) Electron inflow velocities and reconnection rates at Earth's magnetopause and magnetosheath. Geophys Res Lett 47:e2020GL089082. https://doi.org/10.1029/2020GL089082

Burch JL, Hesse M, Webster JM, Genestreti KJ, Torbert RB, Denton RE, Ergun RE, Giles BL, Gershman DJ, Russell CT, Wang S, Chen LJ, Dokgo K, Hwang KJ, Pollock CJ (2022) The EDR inflow region of a reconnecting current sheet in the geomagnetic tail. Phys Plasmas 29(5):052903. https://doi.org/10.1063/5.0083169

Burlaga LF, Ness NF, Wang YM, Sheeley NR (2002) Heliospheric magnetic field strength and polarity from 1 to 81 AU during the ascending phase of solar cycle 23. J Geophys Res 107(A11):1410. https://doi.org/10.1029/2001JA009217

Cairns IH, Fuselier SA (2017) The plasma depletion layer beyond the heliopause: evidence, implications, and predictions for Voyager 2 and new horizons. Astrophys J 834(2):197. https://doi.org/10.3847/1538-4357/834/2/197

Cairns IH, Fuselier SA (2018) Electron and ion heating due to magnetic reconnection at the heliopause. J Phys Conf Ser 1100:012004. https://doi.org/10.1088/1742-6596/1100/1/012004

Cairns IH, Lyon JG (1996) Magnetic field orientation effects on the standoff distance of Earth's bow shock. Geophys Res Lett 23(21):2883–2886. https://doi.org/10.1029/96gl02755

Carter B, Brown S, Donati JF, Rees D, Semel M (1996) Zeeman Doppler imaging of stars with the AAT. Publ Astron Soc Aust 13(2):150–155. https://doi.org/10.1017/s1323358000020701

Cassak PA, Otto A (2011) Scaling of the magnetic reconnection rate with symmetric shear flow. Phys Plasmas 18(7):074501. https://doi.org/10.1063/1.3609771

Cassak PA, Shay MA (2007) Scaling of asymmetric magnetic reconnection: general theory and collisional simulations. Phys Plasmas 14(10):102114. https://doi.org/10.1063/1.2795630

Cassak PA, Genestreti KJ, Burch JL, Phan TD, Shay MA, Swisdak M et al (2017a) The effect of a guide field on local energy conversion during asymmetric magnetic reconnection: particle-in-cell simulations. J Geophys Res Space Phys 122:11523–11542. https://doi.org/10.1002/2017JA024555

Cassak PA, Liu YH, Shay M (2017b) A review of the 0.1 reconnection rate problem. J Plasma Phys 83(5):1–17. https://doi.org/10.1017/s0022377817000666

Chael A, Rowan M, Narayan R, Johnson M, Sironi L (2018) The role of electron heating physics in images and variability of the Galactic Centre black hole Sagittarius A*. Mon Not R Astron Soc 478(4):5209–5229. https://doi.org/10.1093/mnras/sty1261

Chen L-J, Hesse M, Wang S, Bessho N, Daughton W (2016) Electron energization and structure of the diffusion region during asymmetric reconnection. Geophys Res Lett 43:2405–2412. https://doi.org/10.1002/2016GL068243

Chen YQ et al (2022) ApJ 927(171). https://doi.org/10.3847/1538-4357/ac497d

Cheng AF (1984) Magnetosphere, rings, and moons of Uranus. In: NASA conference publication, vol 2330, pp 541–556

Clarke JT (2013) Auroral processes on Jupiter and Saturn. In: Auroral phenomenology and magnetospheric processes: Earth and other planets. Geophys monograph series, vol 197, pp 113–122. https://doi.org/10.1029/2011gm001199

Clarke JT et al (2009) Response of Jupiter's and Saturn's auroral activity to the solar wind. J Geophys Res 114:A05210. https://doi.org/10.1029/2008JA013694

Cohen O, Drake JJ, Glocer A, Garraffo C, Poppenhaeger K, Bell JM, Ridley AJ, Gombosi TI (2014) Magnetospheric structure and atmospheric Joule heating of habitable planets orbitinG M-dwarf stars. Astrophys J 790(1):57. https://doi.org/10.1088/0004-637x/790/1/57

Collinson G, Paterson WR, Bard C, Dorelli J, Glocer A, Sarantos M, Wilson R (2018) New results from Galileo's first flyby of Ganymede: reconnection-driven flows at the low-latitude magnetopause boundary, crossing the cusp, and icy ionospheric escape. Geophys Res Lett 45:3382–3392. https://doi.org/10.1002/2017GL075487

Cooling BMA, Owen CJ, Schwartz SJ (2001) Role of the magnetosheath flow in determining the motion of open flux tubes. J Geophys Res 106(A9):18763–18775. https://doi.org/10.1029/2000JA000455

Cowley SWH, Bunce EJ, O'Rourke JM (2004) A simple quantitative model of plasma flows and currents in Saturn's polar ionosphere. J Geophys Res 109:A05212. https://doi.org/10.1029/2003JA010375

Cravens TE, Fowler CM, Brain D, Rahmati A, Xu S, Ledvina SA et al (2020) Magnetic reconnection in the ionosphere of Mars: the role of collisions. J Geophys Res Space Phys 125:e2020JA028036. https://doi.org/10.1029/2020JA028036

Czechowski A, Grygorczuk J (2017) Heliosphere in a strong interstellar magnetic field. J Phys Conf Ser 900:012004. https://doi.org/10.1088/1742-6596/900/1/012004
Dahani S, Kieokaew R, Génot V, Lavraud B, Chen Y, Michotte de Welle B et al (2022) The helicity sign of flux transfer event flux ropes and its relationship to the guide field and Hall physics in magnetic reconnection at the magnetopause. J Geophys Res Space Phys 127:e2022JA030686. https://doi.org/10.1029/2022JA030686
Dargent J, Toledo-Redondo S, Divin A, Innocenti ME (2023) Energy conversion by magnetic reconnection in multiple ion temperature plasmas. Geophys Res Lett 50:e2023GL103324. https://doi.org/10.1029/2023GL103324
Denton RE, Sonnerup BUÖ, Russell CT, Hasegawa H, Phan TD, Strangeway RJ et al (2018) Determining L-M-N current sheet coordinates at the magnetopause from magnetospheric multiscale data. J Geophys Res Space Phys 123:2274–2295. https://doi.org/10.1002/2017JA024619
Desch MD, Kaiser ML (1984) Predictions for Uranus from a radiometric Bode's law. Nature 310(5980):755–757. https://doi.org/10.1038/310755a0
Desch M, Rucker H (1985) Saturn radio emission and the solar wind: Voyager-2 studies. Adv Space Res 5(4):333–336. https://doi.org/10.1016/0273-1177(85)90159-0
Desroche M, Bagenal F, Delamere PA, Erkaev N (2012) Conditions at the expanded Jovian magnetopause and implications for the solar wind interaction. J Geophys Res 117(A07202):1–18. https://doi.org/10.1029/2012JA017621
Desroche M, Bagenal F, Delamere PA, Erkaev N (2013) Conditions at the magnetopause of Saturn and implications for the solar wind interaction. J Geophys Res Space Phys 118:3087–3095. https://doi.org/10.1002/jgra.50294
Dewey RM, Raines JM, Sun W, Slavin JA, Poh G (2018) MESSENGER observations of fast plasma flows in Mercury's magnetotail. Geophys Res Lett 45:10110–10118. https://doi.org/10.1029/2018GL079056
DiBraccio GA, Gershman DJ (2019) Voyager 2 constraints on plasmoid-based transport at Uranus. Geophys Res Lett 46:10710–10718. https://doi.org/10.1029/2019GL083909
DiBraccio GA, Slavin JA, Boardsen SA, Anderson BJ, Korth H, Zurbuchen TH, Raines JM, Baker DN, McNutt RL, Solomon SC (2013) MESSENGER observations of magnetopause structure and dynamics at Mercury. J Geophys Res Space Phys 118:997–1008. https://doi.org/10.1002/jgra.50123
DiBraccio GA, Slavin JA, Imber SM, Gershman DJ, Raines JM, Jackman CM, Boardsen SA, Anderson BJ, Korth H, Zurbuchen TH, McNutt RL, Solomon SC (2015) MESSENGER observations of flux ropes in Mercury's magnetotail. Planet Space Sci 115:77–89. https://doi.org/10.1016/j.pss.2014.12.016
DiBraccio GA, Luhmann JG, Curry SM, Espley JR, Xu S, Mitchell DL et al (2018) The twisted configuration of the Martian magnetotail: MAVEN observations. Geophys Res Lett 45:4559–4568. https://doi.org/10.1029/2018GL077251
DiBraccio GA, Romanelli N, Bowers CF, Gruesbeck JR, Halekas JS, Ruhunusiri S et al (2022) A statistical investigation of factors influencing the magnetotail twist at Mars. Geophys Res Lett 49:e2022GL098007. https://doi.org/10.1029/2022GL098007
Dimmock AP, Osmane A, Pulkkinen TI, Nykyri K (2015) A statistical study of the dawn-dusk asymmetry of ion temperature anisotropy and mirror mode occurrence in the terrestrial dayside magnetosheath using THEMIS data. J Geophys Res Space Phys 120:5489–5503. https://doi.org/10.1002/2015JA021192
Donati JF, Brown SF (1997) Zeeman-Doppler imaging of active stars. V. Sensitivity of maximum entropy magnetic maps to field orientation. Astron Astrophys 326:1135–1142
Dorelli JC, Hesse M, Kuznetsova MM, Rastaetter L, Raeder J (2004) A new look at driven magnetic reconnection at the terrestrial subsolar magnetopause. J Geophys Res 109:A12216. https://doi.org/10.1029/2004JA010458
Doss CE, Komar CM, Cassak PA, Wilder FD, Eriksson S, Drake JF (2015) Asymmetric magnetic reconnection with a flow shear and applications to the magnetopause. J Geophys Res Space Phys 120:7748–7763. https://doi.org/10.1002/2015JA021489
Drake JF, Swisdak M, Phan TD, Cassak PA, Shay MA, Lepri ST, Lin RP, Quataert E, Zurbuchen TH (2009) Ion heating resulting from pickup in magnetic reconnection exhausts. J Geophys Res 114:A05111. https://doi.org/10.1029/2008JA013701
Ebert RW et al (2017) Accelerated flows at Jupiter's magnetopause: evidence for magnetic reconnection along the dawn flank. Geophys Res Lett 44:4401–4409. https://doi.org/10.1002/2016GL072187
Ebert RW, Fuselier SA, Allegrini F, Bagenal F, Bolton SJ, Clark G et al (2022) Evidence for magnetic reconnection at Ganymede's upstream magnetopause during the PJ34 Juno flyby. Geophys Res Lett 49:e2022GL099775. https://doi.org/10.1029/2022GL099775
Eriksson S et al (2016) Magnetospheric multiscale observations of magnetic reconnection associated with Kelvin-Helmholtz waves. Geophys Res Lett 43:5606–5615. https://doi.org/10.1002/2016GL068783
Eriksson E, Vaivads A, Graham DB, Divin A, Khotyaintsev YV, Yordanova E et al (2018) Electron energization at a reconnecting magnetosheath current sheet. Geophys Res Lett 45:8081–8090. https://doi.org/10.1029/2018GL078660

Eviatar A, Richardson JD (1986) Predicted satellite plasma tori in the magnetosphere of Uranus. Astrophys J 300:L99–L102. https://doi.org/10.1086/184611

Fang X, Ma Y, Luhmann J, Dong Y, Brain D, Hurley D et al (2018) The morphology of the solar wind magnetic field draping on the dayside of Mars and its variability. Geophys Res Lett 45:3356–3365. https://doi.org/10.1002/2018GL077230

Farris MH, Petrinec SM, Russell CT (1991) The thickness of the magnetosheath: constraints on the polytropic index. Geophys Res Lett 18(10):1821–1824. https://doi.org/10.1029/91gl02090

Farrugia CJ, Erkaev NV, Biernat HK, Lawrence GR, Elphic RC (1997) Plasma depletion layer model for low Alfvén Mach number: comparison with ISEE observations. J Geophys Res 102(A6):11315–11324. https://doi.org/10.1029/97JA00410

Fischer C, Saur J (2022) Star–planet interaction. Astron Astrophys 668(A10):1–20. https://doi.org/10.1051/0004-6361/202243346

Formisano V, Hedgecock P, Moreno G, Sear J, Bollea D (1971) Observations of Earth's bow shock for low Mach numbers. Planet Space Sci 19(11):1519–1531. https://doi.org/10.1016/0032-0633(71)90011-0

Fuselier SA, Cairns IH (2017) Reconnection at the heliopause: predictions for Voyager 2. J Phys Conf Ser 900:012007. https://doi.org/10.1088/1742-6596/900/1/012007

Fuselier SA, Anderson BJ, Gary SP, Denton RE (1994) Inverse correlations between the ion temperature anisotropy and plasma beta in the Earth's quasi-parallel magnetosheath. J Geophys Res 99(A8):14931–14936. https://doi.org/10.1029/94JA00865

Fuselier SA, Trattner KJ, Petrinec SM (2000) Cusp observations of high- and low-latitude reconnection for northward interplanetary magnetic field. J Geophys Res 105(A1):253–266. https://doi.org/10.1029/1999JA900422

Fuselier SA, Frahm R, Lewis WS, Masters A, Mukherjee J, Petrinec SM, Sillanpaa IJ (2014) The location of magnetic reconnection at Saturn's magnetopause: a comparison with Earth. J Geophys Res Space Phys 119:2563–2578. https://doi.org/10.1002/2013JA019684

Fuselier SA, Petrinec SM, Bobra MG, Cairns IH (2020a) Reconnection at the heliopause: comparing the Voyager 1 and 2 heliopause crossings. J Phys Conf Ser 1620:012004. https://doi.org/10.1088/1742-6596/1620/1/012004

Fuselier SA, Petrinec SM, Sawyer RP, Mukherjee J, Masters A (2020b) Suppression of magnetic reconnection at Saturn's low-latitude magnetopause. J Geophys Res Space Phys 125:1–16. https://doi.org/10.1029/2020JA027895

Garraffo C, Drake JJ, Cohen O (2016) The missing magnetic morphology term in stellar rotation evolution. Astron Astrophys 595(A110):1–7. https://doi.org/10.1051/0004-6361/201628367

Genestreti KJ, Nakamura TKM, Nakamura R, Denton RE, Torbert RB, Burch JL et al (2018) How accurately can we measure the reconnection rate EM for the MMS diffusion region event of 11 July 2017? J Geophys Res Space Phys 123:9130–9149. https://doi.org/10.1029/2018JA025711

Gershman DJ, DiBraccio GA (2020) Solar cycle dependence of solar wind coupling with giant planet magnetospheres. Geophys Res Lett 47:e2020GL089315. https://doi.org/10.1029/2020GL089315

Gershman DJ, Slavin JA, Raines JM, Zurbuchen TH, Anderson BJ, Korth H, Baker DN, Solomon SC (2013) Magnetic flux pileup and plasma depletion in Mercury's subsolar magnetosheath. J Geophys Res Space Phys 118:7181–7199. https://doi.org/10.1002/2013JA019244

Gershman DJ, Dorelli JC, DiBraccio GA, Raines JM, Slavin JA, Poh G, Zurbuchen TH (2016) Ion-scale structure in Mercury's magnetopause reconnection diffusion region. Geophys Res Lett 43:5935–5942. https://doi.org/10.1002/2016GL069163

Gershman DJ et al (2017) Juno observations of large-scale compressions of Jupiter's dawnside magnetopause. Geophys Res Lett 44:7559–7568. https://doi.org/10.1002/2017GL073132

Goodbred M, Liu YH, Chen B, Li X (2021) The relation between the energy conversion rate and reconnection rate in Petschek-type reconnection—implications for solar flares. Phys Plasmas 28(8):082103. https://doi.org/10.1063/5.0050557

Hackman T, Lehtinen J, Rosén L, Kochukhov O, Käpylä MJ (2016) Zeeman-Doppler imaging of active young solar-type stars. Astron Astrophys 587(A28):1–7. https://doi.org/10.1051/0004-6361/201527320

Haewsantati K, Bonfond B, Wannawichian S, Gladstone GR, Hue V, Versteeg MH et al (2021) Morphology of Jupiter's polar auroral bright spot emissions via Juno-UVS observations. J Geophys Res Space Phys 126:e2020JA028586. https://doi.org/10.1029/2020JA028586

Hara T et al (2017) On the origins of magnetic flux ropes in near-Mars magnetotail current sheets. Geophys Res Lett 44:7653–7662. https://doi.org/10.1002/2017GL073754

Hara T, Huang Z, Mitchell DL, DiBraccio GA, Brain DA, Harada Y, Luhmann JG (2022) A comparative study of magnetic flux ropes in the nightside induced magnetosphere of Mars and Venus. J Geophys Res Space Phys 127:e2021JA029867. https://doi.org/10.1029/2021JA029867

Harada Y, Halekas JS, DiBraccio GA, Xu S, Espley J, McFadden JP et al (2018) Magnetic reconnection on dayside crustal magnetic fields at Mars: MAVEN observations. Geophys Res Lett 45:4550–4558. https://doi.org/10.1002/2018GL077281

Herbert F (2009) Aurora and magnetic field of Uranus. J Geophys Res 114:A11206. https://doi.org/10.1029/2009JA014394
Herbert F, Sandel BR (1994) The Uranian aurora and its relationship to the magnetosphere. J Geophys Res 99(A3):4143–4160. https://doi.org/10.1029/93JA02673
Hesse M, Aunai N, Zenitani S, Kuznetsova M, Birn J (2013) Aspects of collisionless magnetic reconnection in asymmetric systems. Phys Plasmas 20(6):061210. https://doi.org/10.1063/1.4811467
Hesse M, Liu YH, Chen LJ, Bessho N, Wang S, Burch JL, Moretto T, Norgren C, Genestreti KJ, Phan TD, Tenfjord P (2018) The physical foundation of the reconnection electric fields. Phys Plasmas 25(3):032901. https://doi.org/10.1063/1.5021461
Hietala H, Plaschke F (2013) On the generation of magnetosheath high-speed jets by bow shock ripples. J Geophys Res Space Phys 118:7237–7245. https://doi.org/10.1002/2013JA019172
Holzer RE, Slavin JA (1978) Magnetic flux transfer associated with expansions and contractions of the dayside magnetosphere. J Geophys Res 83(A8):3831–3839. https://doi.org/10.1029/JA083iA08p03831
Imber SM, Slavin JA, Boardsen SA, Anderson BJ, Korth H, McNutt RL, Solomon SC (2014) MESSENGER observations of large dayside flux transfer events: do they drive Mercury's substorm cycle? J Geophys Res Space Phys 119:5613–5623. https://doi.org/10.1002/2014JA019884
Ip WH, Kopp A, Hu JH (2004) On the star-magnetosphere interaction of close-in exoplanets. Astrophys J 602(1):L53–L56. https://doi.org/10.1086/382274
Jackman CM, Slavin JA, Cowley SWH (2011) Cassini observations of plasmoid structure and dynamics: implications for the role of magnetic reconnection in magnetospheric circulation at Saturn. J Geophys Res 116:A10212. https://doi.org/10.1029/2011JA016682
Jackman CM et al (2014) Saturn's dynamic magnetotail: a comprehensive magnetic field and plasma survey of plasmoids and traveling compression regions and their role in global magnetospheric dynamics. J Geophys Res Space Phys 119:5465–5494. https://doi.org/10.1002/2013JA019388
Jackman CM, Thomsen MF, Mitchell DG, Sergis N, Arridge CS, Felici M, Badman SV, Paranicas C, Jia X, Hospodarksy GB, Andriopoulou M, Khurana KK, Smith AW, Dougherty MK (2015) Field dipolarization in Saturn's magnetotail with planetward ion flows and energetic particle flow bursts: evidence of quasi-steady reconnection. J Geophys Res Space Phys 120:3603–3617. https://doi.org/10.1002/2015JA020995
Jackman CM, Thomsen MF, Dougherty MK (2019) Survey of Saturn's magnetopause and bow shock positions over the entire Cassini mission: boundary statistical properties and exploration of associated upstream conditions. J Geophys Res Space Phys 124:8865–8883. https://doi.org/10.1029/2019JA026628
Jasinski JM, Slavin JA, Arridge CS, Poh G, Jia X, Sergis N, Coates AJ, Jones GH, Waite JH (2016) Flux transfer event observation at Saturn's dayside magnetopause by the Cassini spacecraft. Geophys Res Lett 43:6713–6723. https://doi.org/10.1002/2016GL069260
Jasinski JM, Akhavan-Tafti M, Sun W, Slavin JA, Coates AJ, Fuselier SA et al (2021) Flux transfer events at a reconnection-suppressed magnetopause: Cassini observations at Saturn. J Geophys Res Space Phys 126:e2020JA028786. https://doi.org/10.1029/2020JA028786
Jasinski JM, Murphy N, Jia X, Slavin JA (2022) Neptune's pole-on magnetosphere: dayside reconnection observations by Voyager 2. Planet Sci J 3(4):76. https://doi.org/10.3847/psj/ac5967
Jia X, Walker RJ, Kivelson MG, Khurana KK, Linker JA (2008) Three-dimensional MHD simulations of Ganymede's magnetosphere. J Geophys Res Space Phys 113:A06212. https://doi.org/10.1029/2007ja012748
Karlsson T, Raptis S, Trollvik H, Nilsson H (2021) Classifying the magnetosheath behind the quasi-parallel and quasi-perpendicular bow shock by local measurements. J Geophys Res Space Phys 126:e2021JA029269. https://doi.org/10.1029/2021JA029269
Kaweeyanun N, Masters A, Jia X (2020) Favorable conditions for magnetic reconnection at Ganymede's upstream magnetopause. Geophys Res Lett 47:e2019GL086228. https://doi.org/10.1029/2019GL086228
Kennel CF, Coroniti FV (1977) Possible origins of time variability in Jupiter's outer magnetosphere, 2. Variations in solar wind magnetic field. Geophys Res Lett 4(6):215–218. https://doi.org/10.1029/GL004i006p00215
Khodachenko ML, Sasunov Y, Arkhypov OV, Alexeev II, Belenkaya ES, Lammer H, Kislayakova KG, Odert P, Leitzinger M, Güdel M (2013) Stellar CME activity and its possible influence on exoplanets' environments: importance of magnetospheric protection. Proc Int Astron Union 8(S300):335–346. https://doi.org/10.1017/S1743921313011174
Kivelson MG (2007) Planetary magnetospheres. In: Kamide Y, Chain A (eds) Handbook of the solar-terrestrial environment. Springer, Berlin, pp 469–492. https://doi.org/10.1007/978-3-540-46315-3_19
Kivelson MG, Bagenal F (2014) Planetary magnetospheres. In: Spohn T, Breuer D, Johnson TV (eds) Encyclopedia of the solar system. Elsevier, Amsterdam, pp 137–157. https://doi.org/10.1016/B978-0-12-415845-0.00007-4

Kivelson MG, Jia X (2013) An MHD model of Ganymede's mini-magnetosphere suggests that the heliosphere forms in a sub-Alfvénic flow. J Geophys Res Space Phys 118:6839–6846. https://doi.org/10.1002/2013JA019130

Kobayashi S, Rogers BN, Numata R (2014) Gyrokinetic simulations of collisionless reconnection in turbulent non-uniform plasmas. Phys Plasmas 21(4):040704. https://doi.org/10.1063/1.4873703

Kobel E, Flückiger EO (1994) A model of the steady state magnetic field in the magnetosheath. J Geophys Res 99(A12):23617–23622. https://doi.org/10.1029/94JA01778

Koskinen HEJ, Tanskanen E (2002) Magnetospheric energy budget and the epsilon parameter. J Geophys Res 107(A11):1415. https://doi.org/10.1029/2002JA009283

Lamy L (2020) Auroral emissions from Uranus and Neptune. Philos Trans R Soc A, Math Phys Eng Sci 378(2187):20190481. https://doi.org/10.1098/rsta.2019.0481

Lavraud B, Borovsky JE (2008) Altered solar wind-magnetosphere interaction at low Mach numbers: coronal mass ejections. J Geophys Res 113:A00B08. https://doi.org/10.1029/2008JA013192

Lee LC, Fu ZF (1986) Multiple x-line reconnection: 1. A criterion for the transition from a single x-line to a multiple x-line reconnection. J Geophys Res 91:6807–6815

Li X, Liu Y-H (2021) The effect of thermal pressure on collisionless magnetic reconnection rate. Astrophys J 912(2):152. https://doi.org/10.3847/1538-4357/abf48c

Li W et al (2016) Kinetic evidence of magnetic reconnection due to Kelvin-Helmholtz waves. Geophys Res Lett 43:5635–5643. https://doi.org/10.1002/2016GL069192

Li S, Lu H, Cui J, Yu Y, Mazelle C, Li Y, Cao J (2020) Effects of a dipole-like crustal field on solar wind interaction with Mars. Earth Planet Phys 4:23–31. https://doi.org/10.26464/epp2020005

Lindqvist P-A, Mozer FS (1990) The average tangential electric field at the noon magnetopause. J Geophys Res 95:17137–17144

Liu YH, Hesse M (2016) Suppression of collisionless magnetic reconnection in asymmetric current sheets. Phys Plasmas 23(6):060704. https://doi.org/10.1063/1.4954818

Liu Y-H, Hesse M, Kuznetsova M (2015) Orientation of X lines in asymmetric magnetic reconnection—mass ratio dependency. J Geophys Res Space Phys 120:7331–7341. https://doi.org/10.1002/2015JA021324

Liu Y-H, Hesse M, Li TC, Kuznetsova M, Le A (2018) Orientation and stability of asymmetric magnetic reconnection x line. J Geophys Res Space Phys 123:4908–4920. https://doi.org/10.1029/2018JA025410

Liu YH, Li TC, Hesse M, Sun WJ, Liu J, Burch J et al (2019) Three-dimensional magnetic reconnection with a spatially confined X-line extent: implications for dipolarizing flux bundles and the dawn-dusk asymmetry. J Geophys Res Space Phys 124:2819–2830. https://doi.org/10.1029/2019JA026539

Liuzzo L, Paty C, Cochrane C, Nordheim T, Luspay-Kuti A, Castillo-Rogez J et al (2021) Triton's variable interaction with Neptune's magnetospheric plasma. J Geophys Res Space Phys 126(11):1–27. https://doi.org/10.1029/2021JA029740

Luhmann J, Ledvina S, Russell C (2004) Induced magnetospheres. Adv Space Res 33(11):1905–1912. https://doi.org/10.1016/j.asr.2003.03.031

Luhmann JG, Dong C, Ma Y, Curry SM, Mitchell D, Espley J, Connerney J, Halekas J, Brain DA, Jakosky BM et al (2015) Implications of MAVEN Mars near-wake measurements and models. Geophys Res Lett 42:9087–9094. https://doi.org/10.1002/2015GL066122

Ma YJ, Russell CT, Fang X, Dong Y, Nagy AF, Toth G, Halekas JS, Connerney JEP, Espley JR, Mahaffy PR et al (2015) MHD model results of solar wind interaction with Mars and comparison with MAVEN plasma observations. Geophys Res Lett 42:9113–9120. https://doi.org/10.1002/2015GL065218

Masters A (2014) Magnetic reconnection at Uranus' magnetopause. J Geophys Res Space Phys 119:5520–5538. https://doi.org/10.1002/2014JA020077

Masters A (2015a) Magnetic reconnection at Neptune's magnetopause. J Geophys Res Space Phys 120:479–493. https://doi.org/10.1002/2014JA020744

Masters A (2015b) The dayside reconnection voltage applied to Saturn's magnetosphere. Geophys Res Lett 42:2577–2585. https://doi.org/10.1002/2015GL063361

Masters A (2018) A more viscous-like solar wind interaction with all the giant planets. Geophys Res Lett 45:7320–7329. https://doi.org/10.1029/2018GL078416

Masters A, Mitchell DG, Coates AJ, Dougherty MK (2011) Saturn's low-latitude boundary layer: 1. Properties and variability. J Geophys Res 116:A06210. https://doi.org/10.1029/2010JA016421

Masters A, Eastwood JP, Swisdak M, Thomsen MF, Russell CT, Sergis N, Crary FJ, Dougherty MK, Coates AJ, Krimigis SM (2012) The importance of plasma β conditions for magnetic reconnection at Saturn's magnetopause. Geophys Res Lett 39:L08103. https://doi.org/10.1029/2012GL051372

Matsunaga K, Seki K, Brain DA, Hara T, Masunaga K, Mcfadden JP et al (2017) Statistical study of relations between the induced magnetosphere, ion composition, and pressure balance boundaries around Mars based on MAVEN observations. J Geophys Res Space Phys 122:9723–9737. https://doi.org/10.1002/2017JA024217

Michotte de Welle B, Aunai N, Nguyen G, Lavraud B, Génot V, Jeandet A, Smets R (2022) Global three-dimensional draping of magnetic field lines in Earth's magnetosheath from in-situ spacecraft measurements. J Geophys Res Space Phys 127:e2022JA030996. https://doi.org/10.1029/2022JA030996
Milan SE, Gosling JS, Hubert B (2012) Relationship between interplanetary parameters and the magnetopause reconnection rate quantified from observations of the expanding polar cap. J Geophys Res 117:A03226. https://doi.org/10.1029/2011JA017082
Montgomery J, Ebert RW, Clark G, Fuselier SA, Allegrini F, Bagenal F et al (2022) Investigating the occurrence of magnetic reconnection at Jupiter's dawn magnetopause during the Juno era. Geophys Res Lett 49:e2022GL099141. https://doi.org/10.1029/2022GL099141
Mozer FS, Hull A (2010) Scaling the energy conversion rate from magnetic field reconnection to different bodies. Phys Plasmas 17(10):102906. https://doi.org/10.1063/1.3504224
Mozer FS, Retinò A (2007) Quantitative estimates of magnetic field reconnection properties from electric and magnetic field measurements. J Geophys Res 112:A10206. https://doi.org/10.1029/2007JA012406
Nagy AF et al (2004) The plasma environment of Mars. In: Winterhalter D, Acuña M, Zakharov A (eds) Mars' magnetism and its interaction with the solar wind. Space sciences series of ISSI, vol 18. Springer, Dordrecht. https://doi.org/10.1007/978-0-306-48604-3_2
Nakamura TKM, Hasegawa H, Daughton W et al (2017) Turbulent mass transfer caused by vortex induced reconnection in collisionless magnetospheric plasmas. Nat Commun 8:1582. https://doi.org/10.1038/s41467-017-01579-0
Nakamura TKM, Blasl KA, Hasegawa H, Umeda T, Liu YH, Peery SA, Plaschke F, Nakamura R, Holmes JC, Stawarz JE, Nystrom WD (2022) Multi-scale evolution of Kelvin–Helmholtz waves at the Earth's magnetopause during southward IMF periods. Phys Plasmas 29(1):012901. https://doi.org/10.1063/5.0067391
Ness N, Behannon K, Lepping R, Whang Y (1976) Observations of Mercury's magnetic field. Icarus 28(4):479–488. https://doi.org/10.1016/0019-1035(76)90121-4
Ness NF, Acuna MH, Behannon KW, Neubauer FM (1982) The induced magnetosphere of Titan. J Geophys Res 87(A3):1369–1381. https://doi.org/10.1029/JA087iA03p01369
Ness NF, Acuña MH, Behannon KW, Burlaga LF, Connerney JEP, Lepping RP, Neubauer FM (1986) Magnetic fields at Uranus. Science 233(4759):85–89. https://doi.org/10.1126/science.233.4759.85
Neubauer F (1980) Nonlinear standing Alfvén wave current system at Io: theory. J Geophys Res 85(A3):1171–1178. https://doi.org/10.1029/JA085iA03p01171
Neubauer FM (1999) Alfvén wings and electromagnetic induction in the interiors: Europa and Callisto. J Geophys Res 104(A12):28671–28684. https://doi.org/10.1029/1999JA900217
Newell PT, Sotirelis T, Liou K, Meng CI, Rich FJ (2007) A nearly universal solar wind-magnetosphere coupling function inferred from 10 magnetospheric state variables. J Geophys Res 112:A01206. https://doi.org/10.1029/2006JA012015
Nichols JD, Cowley SWH, McComas DJ (2006) Magnetopause reconnection rate estimates for Jupiter's magnetosphere based on interplanetary measurements at ∼5 AU. Ann Geophys 24:393–406. https://doi.org/10.5194/angeo-24-393-2006
Parker EN (1958) Dynamics of the interplanetary gas and magnetic fields. Astrophys J 128:664–675. https://doi.org/10.1086/146579
Parker EN (1963) Interplanetary dynamical processes. Interscience, New York
Perreault P, Akasofu SI (1978) A study of geomagnetic storms. Geophys J Int 54(3):547–573. https://doi.org/10.1111/j.1365-246x.1978.tb05494.x
Petrinec SM, Russell CT (1997) Hydrodynamic and mhd equations across the bow shock and along the surfaces of planetary obstacles. Space Sci Rev 79(3–4):757–791. https://doi.org/10.1023/A:1004938724300
Petrinec SM, Trattner KJ, Fuselier SA (2003) Steady reconnection during intervals of northward IMF: implications for magnetosheath properties. J Geophys Res 108(A12):1458. https://doi.org/10.1029/2003JA009979
Phan TD, Drake JF, Shay MA, Mozer FS, Eastwood JP (2007) Evidence for an elongated (>60 ion skin depths) electron diffusion region during fast magnetic reconnection. Phys Rev Lett 99:255002
Phan TD, Paschmann G, Gosling JT, Oieroset M, Fujimoto M, Drake JF, Angelopoulos V (2013a) The dependence of magnetic reconnection on plasma β and magnetic shear: evidence from magnetopause observations. Geophys Res Lett 40:11–16. https://doi.org/10.1029/2012GL054528
Phan TD, Shay MA, Gosling JT, Fujimoto M, Drake JF, Paschmann G, Oieroset M, Eastwood JP, Angelopoulos V (2013b) Electron bulk heating in magnetic reconnection at Earth's magnetopause: dependence on the inflow Alfvén speed and magnetic shear. Geophys Res Lett 40:4475–4480. https://doi.org/10.1002/grl.50917
Phan TD, Drake JF, Shay MA, Gosling JT, Paschmann G, Eastwood JP, Oieroset M, Fujimoto M, Angelopoulos V (2014) Ion bulk heating in magnetic reconnection exhausts at Earth's magnetopause: dependence

on the inflow Alfvén speed and magnetic shear angle. Geophys Res Lett 41:7002–7010. https://doi.org/10.1002/2014GL061547

Phan TD, Eastwood JP, Shay MA et al (2018) Electron magnetic reconnection without ion coupling in Earth's turbulent magnetosheath. Nature 557:202–206. https://doi.org/10.1038/s41586-018-0091-5

Phan TD, Verniero JL, Larson D, Lavraud B, Drake JF, Øieroset M et al (2022) Parker Solar Probe observations of solar wind energetic proton beams produced by magnetic reconnection in the near-Sun heliospheric current sheet. Geophys Res Lett 49:e2021GL096986. https://doi.org/10.1029/2021GL096986

Plaschke F, Hietala H, Angelopoulos V (2013) Anti-sunward high-speed jets in the subsolar magnetosheath. Ann Geophys 31:1877–1889. https://doi.org/10.5194/angeo-31-1877-2013

Poh G, Slavin JA, Jia X, Raines JM, Imber SM, Sun WJ, Gershman DJ, DiBraccio GA, Genestreti KJ, Smith AW (2017) Coupling between Mercury and its nightside magnetosphere: cross-tail current sheet asymmetry and substorm current wedge formation. J Geophys Res Space Phys 122:8419–8433. https://doi.org/10.1002/2017JA024266

Raeder J (2006) Flux transfer events: 1. Generation mechanism for strong southward IMF. Ann Geophys 24:381–392

Richardson JD, Burlaga LF, Elliott H et al (2022) Observations of the outer heliosphere, heliosheath, and interstellar medium. Space Sci Rev 218(4):35. https://doi.org/10.1007/s11214-022-00899-y

Romanelli N, Gómez D, Bertucci C, Delva M (2014) Steady-state magnetohydrodynamic flow around an unmagnetized conducting sphere. Astrophys J 789(1):43. https://doi.org/10.1088/0004-637x/789/1/43

Romanelli N, DiBraccio GA, Modolo R, Connerney JEP, Ebert RW, Martos YM et al (2022) Juno magnetometer observations at Ganymede: comparisons with a global hybrid simulation and indications of magnetopause reconnection. Geophys Res Lett 49:e2022GL099545. https://doi.org/10.1029/2022GL099545

Russell C (1995) A study of flux transfer events at different planets. Adv Space Res 16(4):159–163. https://doi.org/10.1016/0273-1177(95)00224-3

Russell C (2000) Reconnection in planetary magnetospheres. Adv Space Res 26(3):393–404. https://doi.org/10.1016/s0273-1177(99)01077-7

Russell CT, Elphic RC (1979) ISEE observations of flux transfer events at the dayside magnetopause. Geophys Res Lett 6(1):33–36. https://doi.org/10.1029/GL006i001p00033

Sakaue T, Shibata K (2021) An M dwarf's chromosphere, corona, and wind connection via nonlinear Alfvén waves. Astrophys J 919(1):29. https://doi.org/10.3847/1538-4357/ac0e34

Sawyer RP, Fuselier SA, Mukherjee J, Petrinec SM (2019) An investigation of flow shear and diamagnetic drift effects on magnetic reconnection at Saturn's dawnside magnetopause. J Geophys Res Space Phys 124:8457–8473. https://doi.org/10.1029/2019JA026696

Schwadron NA, McComas DJ (2021) Between local interstellar magnetic and dynamic pressure balance of heliospheric boundaries measured with the IBEX ribbon—a new paradigm. Astrophys J 914(2):129. https://doi.org/10.3847/1538-4357/abfe6b

See V, Lehmann L, Matt SP, Finley AJ (2020) How much do underestimated field strengths from Zeeman–Doppler imaging affect spin-down torque estimates? Astrophys J 894(1):69. https://doi.org/10.3847/1538-4357/ab7918

Semel M (1989) Zeeman-Doppler imaging of active stars. I – basic principles. Astron Astrophys 225(2):456–466

Shay MA, Drake JF, Rogers BN, Denton RE (1999) The scaling of collisionless, magnetic reconnection for large systems. Geophys Res Lett 26(14):2163–2166. https://doi.org/10.1029/1999gl900481

Shay MA, Haggerty CC, Phan TD, Drake JF, Cassak PA, Wu P, Oieroset M, Swisdak M, Malakit K (2014) Electron heating during magnetic reconnection: a simulation scaling study. Phys Plasmas 21(12):122902

Simon S (2015) An analytical model of sub-Alfvénic moon-plasma interactions with application to the hemisphere coupling effect. J Geophys Res Space Phys 120:7209–7227. https://doi.org/10.1002/2015JA021529

Slavin JA, Smith MF, Mazur EL, Baker DN, Hones EW, Iyemori T, Greenstadt EW (1993) ISEE 3 observations of traveling compression regions in the Earth's magnetotail. J Geophys Res 98(A9):15425–15446. https://doi.org/10.1029/93JA01467

Slavin JA, Acuña MH, Anderson BJ, Baker DN, Benna M, Boardsen SA, Gloeckler G, Gold RE, Ho GC, Korth H, Krimigis SM, McNutt RL, Raines JM, Sarantos M, Schriver D, Solomon SC, Trávníček P, Zurbuchen TH (2009) MESSENGER observations of magnetic reconnection in Mercury's magnetosphere. Science 324(5927):606–610. https://doi.org/10.1126/science.1172011

Slavin JA et al (2010) MESSENGER observations of large flux transfer events at Mercury. Geophys Res Lett 37:L02105. https://doi.org/10.1029/2009GL041485

Slavin JA et al (2012) MESSENGER observations of a flux-transfer-event shower at Mercury. J Geophys Res 117:A00M06. https://doi.org/10.1029/2012JA017926

Slavin JA et al (2014) MESSENGER observations of Mercury's dayside magnetosphere under extreme solar wind conditions. J Geophys Res Space Phys 119:8087–8116. https://doi.org/10.1002/2014JA020319

Smith CW, Matthaeus WH, Zank GP, Ness NF, Oughton S, Richardson JD (2001) Heating of the low-latitude solar wind by dissipation of turbulent magnetic fluctuations. J Geophys Res Space Phys 106(A5):8253–8272. https://doi.org/10.1029/2000ja000366

Soderlund KM, Stanley S (2020) The underexplored frontier of ice giant dynamos. Philos Trans R Soc A, Math Phys Eng Sci 378(2187):20190479

Song P (2001) Model predictions of magnetosheath conditions. In: Song P, Singer HJ, Siscoe GL (eds) Space weather. Geophysical monograph, vol 125. Am Geophys Union, pp 249–255. https://doi.org/10.1029/GM125p0249

Sonnerup BUÖ (1974) Magnetopause reconnection rate. J Geophys Res 79(10):1546–1549. https://doi.org/10.1029/JA079i010p01546

Sonnerup BUÖ, Cahill LJ (1967) Magnetopause structure and attitude from Explorer 12 observations. J Geophys Res 72(1):171–183. https://doi.org/10.1029/JZ072i001p00171

Sonnerup BUÖ, Lotko W (1990) The magnetopause boundary layer. In: Defense Technical Information Center (AD-A229 061). Geophysics Laboratory Air Force Systems Command United States Air Force. Retrieved April 18, 2023. https://apps.dtic.mil/sti/pdfs/ADA229061.pdf

Sonnerup BUÖ, Scheible M: (1998) Minimum and maximum variance analysis. ISSI Scientific Reports Series 1:185–220

Southwood DJ, Kivelson MG (1995) Magnetosheath flow near the subsolar magnetopause: Zwan-Wolf and Southwood-Kivelson theories reconciled. Geophys Res Lett 22(23):3275–3278. https://doi.org/10.1029/95gl03131

Southwood DJ, Farrugia CJ, Saunders MA (1988) What are flux transfer events? Planet Space Sci 36:503–508

Spreiter JR, Alksne AY (1970) Solar-wind flow past objects in the solar system. Annu Rev Fluid Mech 2(1):313–354. https://doi.org/10.1146/annurev.fl.02.010170.001525

Spreiter JR, Summers AL, Alksne AY (1966) Hydromagnetic flow around the magnetosphere. Planet Space Sci 14(3):223–253. https://doi.org/10.1016/0032-0633(66)90124-3

Stanley S, Bloxham J (2006) Numerical dynamo models of Uranus' and Neptune's magnetic fields. Icarus 184(2):556–572. https://doi.org/10.1016/j.icarus.2006.05.005

Stawarz JE, Eastwood JP, Phan TD, Gingell IL, Pyakurel PS, Shay MA et al (2022) Turbulence-driven magnetic reconnection and the magnetic correlation length: observations from magnetospheric multiscale in Earth's magnetosheath. Phys Plasmas 29(1):012302. https://doi.org/10.1063/5.0071106

Sun WJ, Fu SY, Slavin JA, Raines JM, Zong QG, Poh GK, Zurbuchen TH (2016) Spatial distribution of Mercury's flux ropes and reconnection fronts: MESSENGER observations. J Geophys Res Space Phys 121:7590–7607. https://doi.org/10.1002/2016JA022787

Sun WJ, Slavin JA, Dewey RM, Chen Y, DiBraccio GA, Raines JM et al (2020b) MESSENGER observations of Mercury's nightside magnetosphere under extreme solar wind conditions: reconnection-generated structures and steady convection. J Geophys Res Space Phys 125:e2019JA027490. https://doi.org/10.1029/2019JA027490

Sun WJ, Slavin JA, Smith AW, Dewey RM, Poh GK, Jia X et al (2020a) Flux transfer event showers at Mercury: dependence on plasma β and magnetic shear and their contribution to the Dungey cycle. Geophys Res Lett 47:e2020GL089784. https://doi.org/10.1029/2020GL089784

Sun W, Slavin JA, Nakamura R, Heyner D, Trattner KJ, Mieth JZD, Zhao J, Zong QG, Aizawa S, Andre N, Saito Y (2022) Dayside magnetopause reconnection and flux transfer events under radial interplanetary magnetic field (IMF): BepiColombo Earth-flyby observations. Ann Geophys 40:217–229. https://doi.org/10.5194/angeo-40-217-2022

Sundberg T et al (2012) MESSENGER observations of dipolarization events in Mercury's magnetotail. J Geophys Res 117:A00M03. https://doi.org/10.1029/2012JA017756

Swisdak M, Drake JF (2007) Orientation of the reconnection X-line. Geophys Res Lett 34:L11106. https://doi.org/10.1029/2007GL029815

Swisdak M, Rogers BN, Drake JF, Shay MA (2003) Diamagnetic suppression of component magnetic reconnection at the magnetopause. J Geophys Res 108(A5):1218. https://doi.org/10.1029/2002JA009726

Swisdak M, Opher M, Drake JF, Alouani Bibi F (2010) The vector direction of the interstellar magnetic field outside the heliosphere. Astrophys J, 710(2):1769–1775. https://doi.org/10.1088/0004-637x/710/2/1769

Tenfjord P, Østgaard N (2013) Energy transfer and flow in the solar wind-magnetosphere-ionosphere system: a new coupling function. J Geophys Res Space Phys 118:5659–5672. https://doi.org/10.1002/jgra.50545

Thomas AEL, Chaplin WJ, Basu S, Rendle B, Davies G, Miglio A (2021) Impact of magnetic activity on inferred stellar properties of main-sequence Sun-like stars. Mon Not R Astron Soc 502(4):5808–5820. https://doi.org/10.1093/mnras/stab354

Toledo-Redondo S et al (2017) Energy budget and mechanisms of cold ion heating in asymmetric magnetic reconnection. J Geophys Res Space Phys 122:9396–9413. https://doi.org/10.1002/2017JA024553
Toledo-Redondo S, Dargent J, Aunai N, Lavraud B, André M, Li W et al (2018) Perpendicular current reduction caused by cold ions of ionospheric origin in magnetic reconnection at the magnetopause: particle-in-cell simulations and spacecraft observations. Geophys Res Lett 45:10,033–10,042. https://doi.org/10.1029/2018GL079051
Toledo-Redondo S, André M, Aunai N, Chappell CR, Dargent J, Fuselier SA et al (2021) Impacts of ionospheric ions on magnetic reconnection and Earth's magnetosphere dynamics. Rev Geophys 59:e2020RG000707. https://doi.org/10.1029/2020RG000707
Trattner KJ, Mulcock JS, Petrinec SM, Fuselier SA (2007) Probing the boundary between antiparallel and component reconnection during southward interplanetary magnetic field conditions. J Geophys Res 112:A08210. https://doi.org/10.1029/2007JA012270
Trattner KJ, Petrinec SM, Fuselier SA, Phan TD (2012) The location of reconnection at the magnetopause: testing the maximum magnetic shear model with THEMIS observations. J Geophys Res 117:A01201. https://doi.org/10.1029/2011JA016959
Trattner KJ, Petrinec SM, Fuselier SA (2021) The location of magnetic reconnection at Earth's magnetopause. Space Sci Rev 217:41. https://doi.org/10.1007/s11214-021-00817-8
van der Holst B, Sokolov IV, Meng X, Jin M, Manchester IV WB, Tóth G, Gombosi TI (2014) Alfvén wave solar model (AWSoM): coronal heating. Astrophys J 782(2):81. https://doi.org/10.1088/0004-637x/782/2/81
Vasyliunas VM, Kan JR, Siscoe GL, Akasofu SI (1982) Scaling relations governing magnetospheric energy transfer. Planet Space Sci 30(4):359–365. https://doi.org/10.1016/0032-0633(82)90041-1
Vernisse Y et al (2016) Signatures of complex magnetic topologies from multiple reconnection sites induced by Kelvin-Helmholtz instability. J Geophys Res Space Phys 121:9926–9939. https://doi.org/10.1002/2016JA023051
Vidotto A (2017) Stellar coronal and wind models: impact on exoplanets. In: Deeg H, Belmonte J (eds) Handbook of exoplanets. Springer, Cham, pp 1–20. https://doi.org/10.1007/978-3-319-30648-3_26-1
Vidotto AA, Gregory SG, Jardine M, Donati JF, Petit P, Morin J, Folsom CP, Bouvier J, Cameron AC, Hussain G, Marsden S, Waite IA, Fares R, Jeffers S, do Nascimento JD (2014) Stellar magnetism: empirical trends with age and rotation. Mon Not R Astron Soc 441(3):2361–2374. https://doi.org/10.1093/mnras/stu728
Vogt MF, Jackman CM, Slavin JA, Bunce EJ, Cowley SWH, Kivelson MG, Khurana KK (2014) Structure and statistical properties of plasmoids in Jupiter's magnetotail. J Geophys Res Space Phys 119:821–843. https://doi.org/10.1002/2013JA019393
Vogt MF, Connerney JEP, DiBraccio GA, Wilson RJ, Thomsen MF, Ebert RW et al (2020) Magnetotail reconnection at Jupiter: a survey of Juno magnetic field observations. J Geophys Res Space Phys 125:e2019JA027486. https://doi.org/10.1029/2019JA027486
Vörös Z, Yordanova E, Varsani A, Genestreti KJ, Khotyaintsev YV, Li W et al (2017) MMS observation of magnetic reconnection in the turbulent magnetosheath. J Geophys Res Space Phys 122:11442–11467. https://doi.org/10.1002/2017JA024535
Wang S, Kistler LM, Mouikis CG, Petrinec SM (2015) Dependence of the dayside magnetopause reconnection rate on local conditions. J Geophys Res Space Phys 120:6386–6408. https://doi.org/10.1002/2015JA021524
Wang J, Yu J, Xu X, Cui J, Cao J, Ye Y et al (2021) MAVEN observations of magnetic reconnection at Martian induced magnetopause. Geophys Res Lett 48:e2021GL095426. https://doi.org/10.1029/2021GL095426
Wilder FD, Ergun RE, Burch JL, Ahmadi N, Eriksson S, Phan TD et al (2018) The role of the parallel electric field in electron-scale dissipation at reconnecting currents in the magnetosheath. J Geophys Res Space Phys 123:6533–6547. https://doi.org/10.1029/2018JA025529
Wilson F, Neukirch T, Hesse M, Harrison MG, Stark CR (2016) Particle-in-cell simulations of collisionless magnetic reconnection with a non-uniform guide field. Phys Plasmas 23(3):032302. https://doi.org/10.1063/1.4942939
Yordanova E et al (2016) Electron scale structures and magnetic reconnection signatures in the turbulent magnetosheath. Geophys Res Lett 43:5969–5978. https://doi.org/10.1002/2016GL069191
Zank GP, Heerikhuisen J, Wood BE, Pogorelov NV, Zirnstein E, McComas DJ (2012) Heliospheric structure: the bow wave and the hydrogen wall. Astrophys J 763(1):20. https://doi.org/10.1088/0004-637x/763/1/20
Zarka P (2007) Plasma interactions of exoplanets with their parent star and associated radio emissions. Planet Space Sci 55(5):598–617. https://doi.org/10.1016/j.pss.2006.05.045
Zhang TL et al (2012) Giant flux ropes observed in the magnetized ionosphere at Venus. Geophys Res Lett 39:L23103. https://doi.org/10.1029/2012GL054236

Zhang B, Delamere PA, Yao Z, Bonfond B, Lin D, Sorathia KA, Brambles OJ, Lotko W, Garretson JS, Merkin VG, Grodent D, Dunn WR, Lyon JG (2021) How Jupiter's unusual magnetospheric topology structures its aurora. Sci Adv 7(15):1–6. https://doi.org/10.1126/sciadv.abd1204

Zhong J, Wei Y, Pu ZY, Wang XG, Wan WX, Slavin JA, Cao X, Raines JM, Zhang H, Xiao CJ, Du AM, Wang RS, Dewey RM, Chai LH, Rong ZJ, Li Y (2018) MESSENGER observations of rapid and impulsive magnetic reconnection in Mercury's magnetotail. Astrophys J 860(2):L20. https://doi.org/10.3847/2041-8213/aaca92

Zwan BJ, Wolf RA (1976) Depletion of solar wind plasma near a planetary boundary. J Geophys Res 81(10):1636–1648. https://doi.org/10.1029/JA081i010p01636

Space Science Reviews (2024) 220:43
https://doi.org/10.1007/s11214-024-01073-2

Magnetic Reconnection and Associated Particle Acceleration in High-Energy Astrophysics

Fan Guo[1] · Yi-Hsin Liu[2] · Seiji Zenitani[3] · Masahiro Hoshino[4]

Received: 23 September 2023 / Accepted: 29 April 2024 / Published online: 4 June 2024

Abstract
Magnetic reconnection occurs ubiquitously in the universe and is often invoked to explain fast energy release and particle acceleration in high-energy astrophysics. The study of relativistic magnetic reconnection in the magnetically dominated regime has surged over the past two decades, revealing the physics of fast magnetic reconnection and nonthermal particle acceleration. Here we review these recent progresses, including the magnetohydrodynamic and collisionless reconnection dynamics as well as particle energization. The insights in astrophysical reconnection strongly connect to the development of magnetic reconnection in other areas, and further communication is greatly desired. We also provide a summary and discussion of key physics processes and frontier problems, toward a better understanding of the roles of magnetic reconnection in high-energy astrophysics.

Keywords Magnetic reconnection · Particle acceleration · High-energy astrophysics

1 Introduction

Magnetic reconnection is a ubiquitous process that occurs in many space, solar, astrophysical, and laboratory systems. It was initially proposed to explain the fast energy release and acceleration of particles in space and astrophysical systems (e.g., Parker 1957; Sweet 1958; Petschek 1964). During reconnection, magnetic topology changes lead to the rapid release

✉ F. Guo
guofan@lanl.gov

✉ S. Zenitani
seiji.zenitani@oeaw.ac.at

Y.-H. Liu
yi-hsin.liu@dartmouth.edu

M. Hoshino
hoshino@eps.s.u-tokyo.ac.jp

1 Los Alamos National Laboratory, Los Alamos, NM 87545, USA

2 Department of Physics and Astronomy, Dartmouth College, Hanover, NH 03755, USA

3 Space Research Institute, Austrian Academy of Sciences, Schmiedlstraße 6, 8042 Graz, Austria

4 Department of Earth and Planetary Science, The University of Tokyo, Tokyo, 113-0033, Japan

of magnetic energy in highly conducting plasmas that cannot be explained by magnetic diffusion. Magnetic reconnection is now widely considered as a pivotal process for explosive energy release, high-energy particle acceleration and radiation in the universe (Uzdensky 2011; Hoshino and Lyubarsky 2012; Arons 2012; Blandford et al. 2017; Guo et al. 2020; Ji et al. 2022).

In high-energy astrophysics, magnetic reconnection can occur in pulsar wind nebulae (PWNe) and pulsar magnetosphere, relativistic jets of active galactic nuclei (AGNs) and gamma-ray bursts, accretion disks and coronae surrounding massive compact objects, as well as strong magnetic field regions in magnetars, etc. (see Sect. 1.1 for a more extensive discussion). The reconnection region is expected to be much larger than the kinetic scale, so a magnetohydrodynamic (MHD) description is necessary. However, many regimes of high-Lundquist-number magnetic reconnection show multiple X-lines and flux ropes (islands in 2D) developing as the secondary tearing instability is active in a current layer. Kinetic processes are important in a collisionless system (Daughton and Karimabadi 2007; Guo et al. 2015; Sironi et al. 2016) or when a hierarchy of collisional plasmoids (Biskamp 1986; Shibata and Tanuma 2001a; Loureiro et al. 2007; Bhattacharjee et al. 2009; Uzdensky et al. 2010) develop kinetic-scale current layers that may trigger collisionless reconnection (Daughton et al. 2009; Ji and Daughton 2011; Stanier et al. 2019). Magnetic reconnection has been proposed as a mechanism to explain a broad range of high-energy astrophysical phenomena and radiation signatures. This includes high-energy radiation flares and their fast variability and polarized emission signature (Zhang et al. 2018, 2020, 2021a, 2022; Zhang and Giannios 2021; Petropoulou et al. 2016), as well as emissions from the accretion flows recently observed by the Event Horizon Telescope (EHT) (Ripperda et al. 2020) and fast radio bursts (FRBs) (Philippov et al. 2019; Mahlmann et al. 2022).

In many high-energy astrophysical systems, it is often estimated that magnetic reconnection, if occurs, will proceed in a magnetically dominated environment. The parameter for measuring the dominance is the so-called magnetization parameter:[1]

$$\sigma = B^2/(4\pi w) \tag{1}$$

with the enthalpy density $w = nmc^2 + [\Gamma_a/(\Gamma_a - 1)]P$. Here B is the magnitude of the magnetic field, n is the proper plasma density, c is the speed of light, Γ_a is the ratio of specific heats, and m and P are the rest mass and proper pressure of the plasma particles under study, which can be the electron-positron pairs or electron-proton pairs, or a mixture of different species. It is expected that in many situations, σ can be much larger than unity and the Alfvén speed $v_A/c = \sqrt{\sigma/(\sigma+1)}$ is close to the speed of light. In terms of the energy budget, $\sigma_r = B_r^2/(4\pi w)$ is the ratio between the potential free energy carried by the Poynting flux to the particle energy density flux into the reconnection region (B_r represents the reconnecting field component). Analytical theories and MHD simulations have been developed to understand relativistic magnetic reconnection (Sect. 2). Particle-in-cell (PIC) simulations have greatly enhanced our understanding of relativistic magnetic reconnection in the high-σ regime. Magnetic reconnection has been found to support a fast reconnection rate $R \sim 0.1 - 0.3$, indicating fast energy release (Lyubarsky 2005; Liu et al. 2015, 2017, 2020; Werner et al. 2018; Goodbred and Liu 2022). It has been shown to efficiently convert a sizeable fraction of the magnetic energy, leading to a strong particle energization. These reconnection layers are shown to strongly accelerate particles to high energy, leading to

[1] In the nonrelatvistic case, it is more adequate to use plasma β or $\sigma_T = B^2/(6\pi nT)$ to measure the magnetic field dominance. See Drake et al. (2024, this collection).

power-law energy spectra $f = dN/dE = f_0 E^{-p}$ with spectral index approaching $p \sim 1$ for large σ (Zenitani and Hoshino 2001; Sironi and Spitkovsky 2014; Guo et al. 2014, 2015; Werner et al. 2016). The key physics and observational implications of these results are being actively studied. We discuss collisionless reconnection physics, and plasma heating and particle acceleration in relativistic magnetic reconnection in Sects. 3 and 4, respectively. Note that this magnetically-dominated regime, collisionless shocks may be inefficient in dissipating magnetically dominated flows and accelerating energetic particles (e.g., Guo et al. 2015; Sironi et al. 2015), magnetic reconnection is the primary candidate for dissipating and converting magnetic energy into relativistic particles and subsequent radiation.

While the initial studies focused on electron-positron plasmas, recent studies have extended into electron-proton plasmas. For high-σ regime ($\sigma \sim \sigma_i \gg 1$), the behavior of reconnection is similar to the pair plasma case, and both electrons and protons are efficiently accelerated (Guo et al. 2016b; Zhang et al. 2018). Recent studies have also studied the so-called trans-relativistic regime, where $\sigma_i < 1$ but $\sigma_e \gg 1$, which can lead to strong electron energization (Werner et al. 2018; Ball et al. 2018; Kilian et al. 2020). The transrelativistic regime is also a bridge for connecting the highly relativistic regime ($p \gtrsim 1.5$) with the nonrelativistic reconnection studies ($p \sim 4$) (Dahlin et al. 2014; Li et al. 2021, 2019; Zhang et al. 2021) (See Oka et al. 2023 and Drake et al. 2024 in this collection). Traditionally, relativistic magnetic reconnection in the collisionless regime (usually pair plasma) has been less of a focus compared to the nonrelativistic cases. The connection between astrophysical reconnection and reconnection in other fields of research is strong, and communications should be continuously encouraged.

In this paper, we review the recent progress in understanding magnetic reconnection in the relativistic magnetically dominated regime. We discuss the astrophysical systems that host magnetic reconnection and how magnetic reconnection may explain high-energy astrophysics observations in Sect. 1.1. We introduce a list of outstanding problems in relativistic magnetic reconnection in Sect. 1.2. Section 2 discusses MHD models of relativistic magnetic reconnection. In Sect. 3, we discuss the reconnection structure and rate, as well as the generalized Ohm's law in a collisionless plasma. Section 4 discusses the heating and acceleration due to relativistic magnetic reconnection. We will discuss basic acceleration mechanisms, the main features of nonthermal particle energy spectra, and the physics that determines the spectra, such as the low-energy injection and energy partition, power-law formation, and high-energy rollover. Section 5 provides a final remark and possible future directions.

1.1 Where and How Magnetic Reconnection May Happen in Astrophysical Systems?

Relativistic outflows such as pulsar winds and relativistic jets are launched with energy carried largely in the form of Poynting flux (Coroniti 1990; Li et al. 2006; Spruit 2010). However, the magnetic energy in the flows must be eventually converted into energies in thermal and nonthermal particles to power the observed emission signatures. There is ample observational evidence indicating that astrophysical systems with strong magnetic field dissipation support efficient particle acceleration and high-energy radiation (Abdo et al. 2011; Tavani et al. 2011; Abeysekara et al. 2017; Zhang et al. 2015; Ackermann et al. 2016). Magnetic reconnection is a leading mechanism that explains this underlying process.

Figure 1 summarizes several systems and environments where magnetic reconnection can be found. In pulsar wind nebulae (PWNe), the antiparallel component of the pulsar dipole field gives rise to a current sheet. The fast rotation of the obliquely oriented dipole leads to

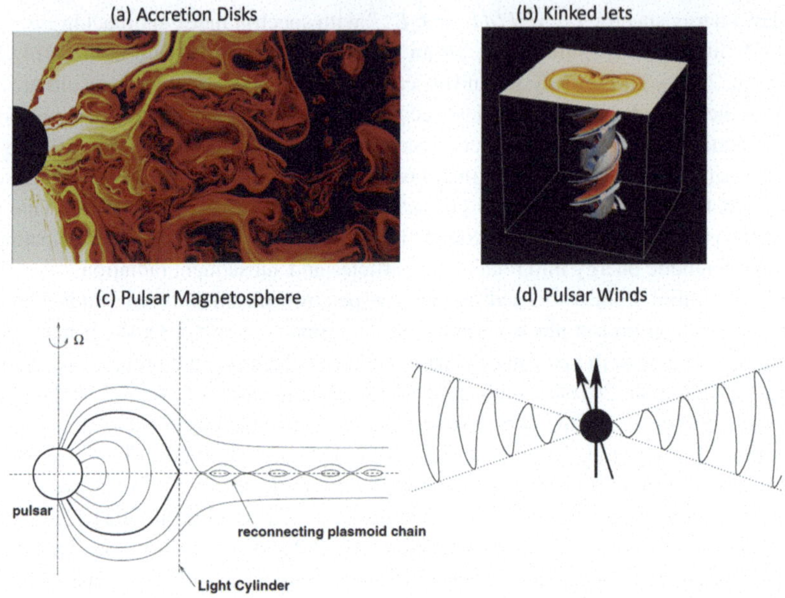

Fig. 1 Several examples of high-energy astrophysical systems that host magnetic reconnection: a) Accretion disks of black holes (Figure created by Bart Ripperda using global general relativistic MHD simulations (Ripperda et al. 2020)), b) Kinked relativistic jets (Zhang et al. 2018), c) pulsar magnetosphere (Uzdensky and Spitkovsky 2014), and d) pulsar wind nebulae (Kirk and Skjæraasen 2003). b), c), and d) are reproduced by permission of the AAS

the so-called striped wind, where magnetic field directions reverse alternatively and can support magnetic reconnection in the equatorial region. Magnetic reconnection may occur starting from the pulsar magnetosphere (Arons 2012; Uzdensky and Spitkovsky 2014; Philippov and Kramer 2022) to the pulsar wind (Lyubarsky and Kirk 2001; Kirk and Skjæraasen 2003), and in the downstream of the termination shock driven by the shock compression (Sironi and Spitkovsky 2011; Lu et al. 2021). A similar situation can happen in relativistic jets, so-called striped jets (Zhang and Giannios 2021). For the pulsar wind and pulsar magnetosphere, it is known that close to the neutron star, σ is very high. In fact, how the magnetic field energy is dissipated before pulsar winds reach the termination shock is one of the unsolved science questions in the physics of PWNe ("σ problem"). Magnetic reconnection in the wind and/or at the termination shock is the key process for solving the problem (Coroniti 1990; Kirk and Skjæraasen 2003; Sironi and Spitkovsky 2011; Lu et al. 2021; Zrake 2016). In the more extreme limit, the magnetic field in magnetars can be strong enough to trigger QED effects (Uzdensky 2011). Current sheets may develop during neutron star coalescence as their magnetospheres interact, leading to magnetic reconnection (Palenzuela et al. 2013). Magnetic reconnection in highly magnetized regimes of pulsar magnetosphere has been proposed to drive fast radio bursts (e.g., Philippov et al. 2019; Mahlmann et al. 2022).

Close to black holes, magnetic reconnection has also been proposed to explain emissions from accretion disks and relativistic jets. Magnetic reconnection can happen in the accretion disks and their coronae (Hoshino 2015; Ball et al. 2018; Ripperda et al. 2020, 2022; Nathanail et al. 2022; Lin et al. 2023; Hakobyan et al. 2023). As kink instability in jets evolves nonlinearly (Zhang et al. 2017; Bodo et al. 2021), it can develop field-line reversals and support magnetic reconnection.[2] In gamma-ray burst models, collisions of relativistic outflows with different magnetic field orientations are discussed (Zhang and Yan 2011). Magnetic reconnection may provide the efficient energy dissipation needed in explaining gamma-ray bursts (Zhang and Yan 2011; McKinney and Uzdensky 2012). Magnetic reconnection may explain polarized radiation signatures (Zhang et al. 2018). Recent general relativistic (GR) MHD simulations show that magnetic reconnection can happen in the accretion disk. Especially, there exists a low-β corona that may have the right condition for strong particle acceleration and powering the nonthermal emission (Ripperda et al. 2020; Yang et al. 2022).

A number of radiation scenarios have been developed to explain high-energy emissions from astrophysical objects. One of the strong motivations to consider magnetic reconnection is the Crab flare, where particles are likely to be accelerated to 10^{15} eV, and it is difficult for a shock model to explain it and avoid synchrotron cooling. Uzdensky and colleagues have proposed that extreme acceleration can happen in the reconnection region with a weak magnetic field ($E > B$) and exceed the "burnoff limit" (Uzdensky et al. 2011; Cerutti et al. 2013). Whether this can be realized is an active field of research. In particular, whether the global observable effect or only beaming at kinetic scale can be observed is under investigation (Mehlhaff et al. 2020). Reconnection has also been used to explain the high energy emission and fast radiation variability in blazars from black holes as the reconnection outflows can Lorentz boost the radiation (Giannios et al. 2009; Agarwal et al. 2023). However, kinetic simulations observe only small regions/structures close to the upper limit of the Alfvén Lorentz factor, and therefore it is still uncertain if the "minijet model" can actually work (Guo et al. 2015; Sironi et al. 2016). In addition, the condition for obtaining these minijets requires a very weak guide field (Liu et al. 2017). How to achieve these conditions is still unclear and requires further study. Magnetic reconnection may also provide an explanation for time-dependent emissions (Petropoulou et al. 2016) and polarized emissions in blazars (Zhang et al. 2018, 2021). However, it is unclear if these radiation features are still preserved in the 3D reconnection, where the flux ropes are shown to be highly dynamic and can easily be disrupted (Guo et al. 2021, 2016a).

It is also interesting to mention that magnetic reconnection can happen in the foreshock region for high Mach number shocks (Matsumoto et al. 2015). Several numerical simulations have indicated that a number of current sheets can be generated by the ion Weibel instability during the interaction of incoming and reflected ions at the foreshock region (Kato and Takabe 2008; Spitkovsky 2008; Burgess et al. 2016), and it is suggested that magnetic reconnection may play an important role on electron heating and acceleration as the so-called shock injection process into the first-order Fermi acceleration (Bohdan et al. 2020; Amano et al. 2022).

1.2 Key Physics Issues

Before getting into the detailed discussion, we close this section by discussing a list of frontier physical problems that are currently undergoing active studies in relativistic magnetic reconnection.

The *Rate Problem and Reconnection Dynamics* is one of the long-standing problems in magnetic reconnection and is concerned with how fast magnetic reconnection proceeds. Previous analytical studies have made various predictions on the rate and reconnection dynamics in the relativistic regime. While the Sweet-Parker-like model cannot support fast

[2] We note that the kink instability itself can also support magnetic energy conversion and particle acceleration (Alves et al. 2018).

reconnection, recent MHD simulations have shown relativistic Petschek reconnection and plasmoid-dominated reconnection with the rate insensitive to the Lundquist number. Kinetic simulations and analysis have shown the rate can be up to $R \sim 0.3$. Figuring out how fast magnetic energy is dissipated can explain astrophysical magnetic energy release. Learning the reconnecting electric field helps us to understand the upper limit of particle acceleration that magnetic reconnection can explain. In Sects. 2 and 3, we will discuss recent studies of relativistic reconnection dynamics via fluid approach and fully kinetic approach, respectively.

The *Particle Heating and Acceleration Problem* is to understand how magnetic energy is converted and partitioned into thermal and nonthermal particles, and how a population of particles are accelerated to high energy. An important goal of the particle energization problem is to build a complete understanding of the physics processes involved and predict the resulting distributions of energetic particles during relativistic magnetic reconnection. A major achievement over the past decade is the robust evidence that magnetic reconnection is a source of nonthermal particles, producing clear power-law particle energy distribution. Competing theories have been proposed based on various reconnection models, but there is still no general consensus on the origin of nonthermal distributions observed during magnetic reconnection. In Sect. 4, we summarize recent progress on particle acceleration in relativistic reconnection.

3D Reconnection and Effects of Turbulence: The main issue here is to understand how reconnection properties and associated particle acceleration change compared to 2D reconnection. A closely related problem is the role of turbulence, either externally driven or self-generated, during the reconnection process. Recently, 3D kinetic simulations and high-Lundquist-number MHD simulations (with and without preexisting turbulence) have been carried out. However, it is still unclear if and how reconnection physics is strongly influenced by turbulence. Meanwhile, there seems to be promising evidence suggesting that particle acceleration in 3D becomes substantially different. We will discuss the 3D effects and role of turbulence in the following sections as we discuss individual topics.

The Onset Problem: Most studies have been focusing on pre-existing current sheets. However, the current sheet needs to form in the first place, which is a dynamic and sometimes prolonged process (e.g., Uzdensky and Loureiro 2016; Tenerani et al. 2016; Huang et al. 2017; Comisso et al. 2017; Lyutikov et al. 2017). In addition, how magnetic energy is accumulated and stored prior to the onset is unknown. Thus, it is important to include the formation of current sheets and their effect. In addition, learning how reconnection onsets helps us understand the conditions that lead to explosive energy release in astrophysics.

Multiscale Problem and Reconnection in Global Models: Reconnection in most astrophysical problems must involves both MHD scales and kinetic scales. While MHD models offer a basic description of large-scale magnetic reconnection, studies have shown kinetic effects are essential. Developing a self-consistent treatment that includes multi-scale effects is of central importance. In addition, developing global particle acceleration models is essential for describing particle acceleration in realistic systems and connecting with observations.

The *Radiation Problem* is to study the effect of radiation cooling (on reconnection structure), pair production, radiation pressure, etc. In addition, there is strong interest in modeling the radiation signature to explain observations. This is not the focus of this review, but we refer interested readers to the recent papers by Jaroschek and Hoshino (2009) and Uzdensky (2011).

2 Fluid Simulations of Relativistic Reconnection

2.1 Early Theories

Blackman and Field (1994) presented the first theoretical models of steady relativistic reconnection. By constructing relativistic Sweet–Parker and Petschek models, they pointed out that plasma density in the outflow region increases due to Lorentz contraction. Since this, in turn, requires relativistically fast plasma inflow, they claimed that the reconnection rate can approach unity, $\sim \mathcal{O}(1)$. Lyutikov and Uzdensky (2003) further examined the relativistic Sweet–Parker model and suggested that reconnection outflow speed may exceed inflow Alfven speed. Later, Lyubarsky (2005) developed theoretical models of relativistic Sweet–Parker and Petschek reconnection, taking compressibility into account. In the relativistic regime, he pointed out that the internal energy density in the outflow region is high enough to increase the effective plasma inertia. For this reason, he claimed that the reconnection outflow is only mildly relativistic. In addition, considering the full momentum balance in Petschek's model, he predicted that the opening angle of slow shocks will be smaller. This leads Lyubarsky to argue that the reconnection rate remains on the order of $\mathcal{O}(0.1)$, because it is difficult to transport energy through a narrow outflow exhaust.

2.2 Basic Equations

To validate the theories, several numerical models have been developed over the past decades. Roughly speaking, two kinds of numerical models have been used, relativistic resistive magnetohydrodynamics (RRMHD) and relativistic multifluid dynamics.

The RRMHD model was developed by Watanabe and Yokoyama (2006) and by Komissarov (2007). Combining relativistic fluid equations, Ohm's law, and Maxwell's equations, they have organized the following set of RRMHD equations,

$$\partial_t(\Gamma\rho) + \nabla\cdot(\rho\mathbf{U}) = 0, \tag{2}$$

$$\partial_t(\Gamma w\mathbf{U} + \mathbf{E}\times\mathbf{B}) + \nabla\cdot\Big((P + \frac{B^2+E^2}{2})\mathbb{I} + w\mathbf{U}\mathbf{U} - \mathbf{B}\mathbf{B} - \mathbf{E}\mathbf{E}\Big) = 0, \tag{3}$$

$$\partial_t(\Gamma^2 w - P + \frac{B^2+E^2}{2}) + \nabla\cdot(\Gamma w\mathbf{U} + \mathbf{E}\times\mathbf{B}) = 0, \tag{4}$$

$$\partial_t\mathbf{B} + \nabla\times\mathbf{E} = 0, \qquad \partial_t\mathbf{E} - \nabla\times\mathbf{B} = -\mathbf{J}, \tag{5}$$

$$\partial_t\bar{\rho}_c + \nabla\cdot\mathbf{J} = 0, \tag{6}$$

$$\Gamma\Big(\mathbf{E} + \mathbf{V}\times\mathbf{B} - (\mathbf{E}\cdot\mathbf{V})\mathbf{V}\Big) = \eta(\mathbf{J} - \bar{\rho}_c\mathbf{V}) \tag{7}$$

Here, we have used Lorentz–Heaviside units with $c = 1$. In the equations, $\Gamma \equiv 1/\sqrt{1-(V/c)^2}$ is the Lorentz factor, ρ is the proper density, $\mathbf{U} = \Gamma\mathbf{V}$ is the 4-vector, w is the relativistic enthalpy, P is the proper pressure, $\mathbb{I}$ is the identity matrix, and $\bar{\rho}$ is the observer-frame charge density. The internal energy in the relativistic enthalpy is often approximated by a simple equation of state with the adiabatic index $\Gamma_a = 4/3$. In such a case, the enthalpy is given by $w = \rho c^2 + [\Gamma_a/(\Gamma_a - 1)]P$.

This form of the Ohm's law may not be familiar to all the readers (Eq. (7); Komissarov (2007)). In relativity, we consider the covariant electric field $e^\mu \equiv [\Gamma(\mathbf{E}\cdot\mathbf{V}), \Gamma(\mathbf{E}+\mathbf{V}\times\mathbf{B})]$ and the electric current $J^\mu \equiv (\bar{\rho}_c, \mathbf{J})$ in the four-vector form. The latter is further split into the conduction current j^μ and the convection current $\rho_c U^\mu$, a projection of the motion of the rest-frame charge ρ_c. The Ohm's law relates the electric field and the conduction current, i.e., $e^\mu = \eta j^\mu = \eta(J^\mu - \rho_c U^\mu)$. In the three-vector form, this gives

$$\Gamma\Big(\mathbf{E}\cdot\mathbf{V}\Big) = \eta(\bar{\rho}_{\mathbf{c}} - \rho_c\Gamma) \tag{8}$$

$$\Gamma\Big(\mathbf{E}+\mathbf{V}\times\mathbf{B}\Big) = \eta(\mathbf{J} - \rho_c\Gamma\mathbf{V}) \tag{9}$$

The Ohm's law (Eq. (7)) is obtained by subtracting Eq. (8) $\times\mathbf{V}$ from Eq. (9). The $\mathbf{E}\cdot\mathbf{V}$ term is purposely added to make the Ohm's law simulation-friendly. Watanabe and Yokoyama (2006) have used a different but equivalent form.

Practically, Eqs. (5) are often replaced by the following equations (Komissarov 2007).

$$\partial_t\mathbf{B} + \nabla\times\mathbf{E} + \nabla\Phi = 0, \qquad \partial_t\mathbf{E} - \nabla\times\mathbf{B} + \nabla\Psi = -\mathbf{J}, \tag{10}$$

$$\partial_t\Phi + \nabla\cdot\mathbf{B} = -\kappa\Phi, \qquad \partial_t\Psi + \nabla\cdot\mathbf{E} = \rho_c - \kappa\Psi, \tag{11}$$

Here, Φ, Ψ, κ are virtual potentials and the decay coefficient in order to reduce numerical errors in $\nabla\cdot\mathbf{B}$ and $\nabla\cdot\mathbf{E}$ by the so-called hyperbolic divergence cleaning method (Munz et al. 2000; Dedner et al. 2002). In the RRMHD equations, the electric field is known to be very stiff, when the resistivity η is low. To deal with these issues, various schemes have been developed such as the operator splitting (Komissarov 2007), implicit schemes (Palenzuela et al. 2009; Dumbser and Zanotti 2009; Mignone et al. 2019), and a method of characteristics (Takamoto and Inoue 2011). Other extensions include the Galerkin method (Dumbser and Zanotti 2009) and generalized equations of state (EoSs) (Mizuno 2013).

Another approach is to use relativistic multifluid equations. Zenitani et al. (2009a,b) have proposed the following relativistic multifluid equation systems. For electron-positron two-fluid plasma, this equation system is also known as relativistic two-fluid electrodynamics. They consist of relativistic fluid equations and Maxwell equations.

$$\partial_t(\Gamma_p n_p) = -\nabla\cdot(n_p\mathbf{U}_p), \tag{12}$$

$$\partial_t\Big(\Gamma_p w_p \mathbf{U}_p\Big) = -\nabla\cdot\Big(w_p\mathbf{U}_p\mathbf{U}_p + P_p\mathbb{I}\Big) + \Gamma_p n_p q_p(\mathbf{E}+\mathbf{V}_p\times\mathbf{B}) - \tau n_p n_e(\mathbf{U}_p - \mathbf{U}_e), \tag{13}$$

$$\partial_t\Big(\Gamma_p^2 w_p - P_p\Big) = -\nabla\cdot(\Gamma_p w_p\mathbf{U}_p) + \Gamma_p n_p q_p(\mathbf{V}_p\cdot\mathbf{E}) - \tau n_p n_e(\Gamma_p - \Gamma_e), \tag{14}$$

$$\partial_t(\Gamma_e n_e) = -\nabla\cdot(n_e\mathbf{U}_e), \tag{15}$$

$$\partial_t\Big(\Gamma_e w_e \mathbf{U}_e\Big) = -\nabla\cdot\Big(w_e\mathbf{U}_e\mathbf{U}_e + P_e\mathbb{I}\Big) + \Gamma_e n_e q_e(\mathbf{E}+\mathbf{V}_e\times\mathbf{B}) - \tau n_p n_e(\mathbf{U}_e - \mathbf{U}_p), \tag{16}$$

$$\partial_t\Big(\Gamma_e^2 w_e - P_e\Big) = -\nabla\cdot(\Gamma_e w_e\mathbf{U}_e) + \Gamma_e n_e q_e(\mathbf{V}_e\cdot\mathbf{E}) - \tau n_p n_e(\Gamma_e - \Gamma_p), \tag{17}$$

$$\partial_t\mathbf{B} = -\nabla\times\mathbf{E}, \qquad \partial_t\mathbf{E} = \nabla\times\mathbf{B} - 4\pi(q_p n_p\mathbf{U}_p + q_e n_e\mathbf{U}_e). \tag{18}$$

In these equations, the subscript p indicates positron properties (and e for electrons), n is the proper number density, and τ is a friction coefficient. For simplicity, the rest mass m and the light speed c are set to 1. Note that interspecies friction terms and the τ parameter are added to the right-hand sides of the momentum and energy equations (Zenitani et al. 2009b).

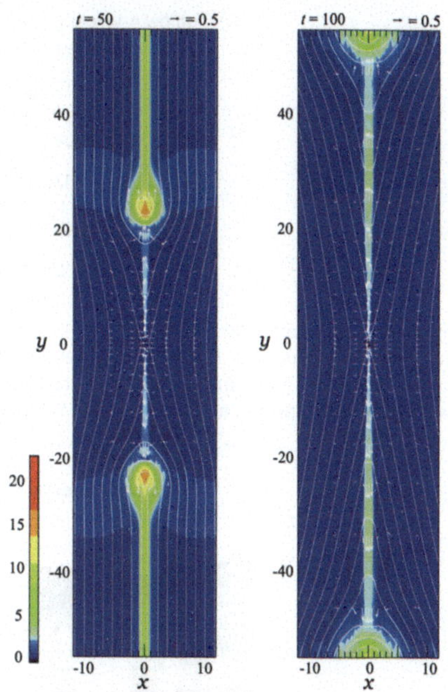

Fig. 2 Evolution of plasma density from RRMHD simulation with $\sigma = 4.0$ and localized resistivity in the reconnection plane. The solid lines and the arrows show magnetic field lines and velocity vectors. Reproduced with permission from Watanabe and Yokoyama (2006), copyright by AAS

The friction terms control the collision between the two species. Therefore, they (and fluid inertial terms) act as a resistivity. By setting τ to small, the friction terms are essentially unused in the inflow region, but we sometimes increase τ near the X-line when we desire a localized resistivity. Similarly, divergence cleaning potentials are often used to improve the numerical accuracy (Eqs. (10) and (11)). Numerical schemes to better solve these equations have been actively developed over the years (Barkov et al. 2014; Balsara et al. 2016; Amano 2016).

2.3 Relativistic Petschek Reconnection

Using RRMHD equations, Watanabe and Yokoyama (2006) have pioneered the MHD-scale evolution of relativistic magnetic reconnection. They have assumed a spatially localized resistivity $\eta = \eta(x, y)$ in the relativistic Ohm's law (Eq. (7)), and then they have obtained a well-developed picture of relativistic Petschek reconnection. As shown in Fig. 2, a narrow reconnection jet extends from the central reconnection point. The reconnection jet is surrounded by a pair of slow shocks, similar to nonrelativistic Petschek reconnection (Petschek 1964). It appears that the outflow channel is much narrower than in the nonrelativistic case. These features were further examined by subsequent studies by two-fluid (Zenitani et al. 2009a,b) and the RRMHD simulations (Zenitani et al. 2010; Zanotti and Dumbser 2011).

It has been found that the typical outflow speed is approximated by the upstream Alfvén speed (Zenitani et al. 2010),

$$V_{\text{out}} \approx c_{\text{A,up}} = c\sqrt{\frac{\sigma}{1+\sigma}}, \tag{19}$$

where $\sigma \equiv B^2/w$ (in the Lorentz–Heaviside units) is the magnetization parameter in the upstream region. This relation in the 4-velocity form is indicated by the white squares and the

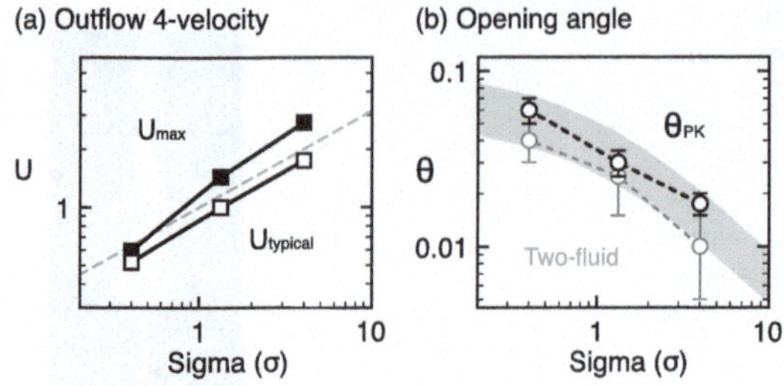

Fig. 3 (a) Maximum and typical outflow 4-velocities U as a function of the inflow σ. The dashed line indicates an Alfvén speed in Eq. (19). (b) Opening angles of the Petschek slow shocks in RRMHD (black line) and in two-fluid (gray) simulations. The shadow indicates a predicted scaling of $\propto (1+\sigma)^{-1}$. Reproduced with permission from Zenitani et al. (2010), copyright by AAS

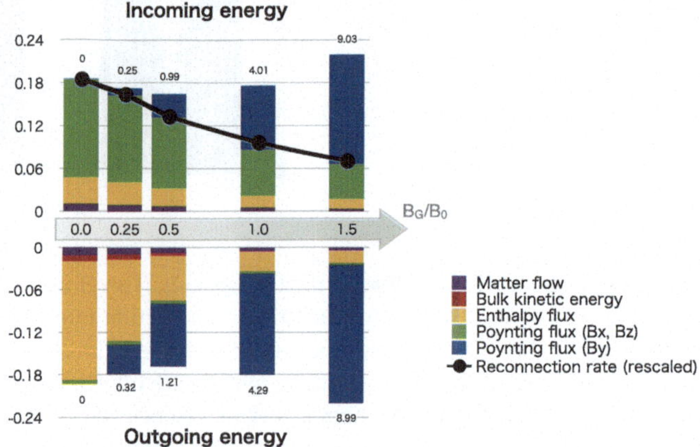

Fig. 4 The incoming and outgoing energy fluxes around the reconnection region. Note that we respect the coordinate system in the original articles. In this case, the antiparallel magnetic field component is along the x direction, the inflow is along z, and the guide field is along y. The guide field Poynting flux (B_y), the rest part of the Poynting flux (B_x, B_z), the plasma enthalpy flux, the bulk kinetic energy, and the matter flow are presented. The black curve indicates a rescaled reconnection rate. Reproduced with permission from Zenitani et al. (2009b), copyright by AAS

dashed line in Fig. 3(a). Also, the opening angle of the Petschek outflow becomes narrower and narrower as the system becomes relativistic, as confirmed in Fig. 3(b). These results are in excellent agreement with theoretical predictions by Lyubarsky (2005). Numerical simulations have revealed that the reconnection rate is $\mathcal{R} \sim \mathcal{O}(0.1)$ or even faster. Theoretically, such a fast reconnection was questioned before because the outflow channel may be too narrow to eject a sufficient amount of energy from the reconnection region (Lyubarsky 2005). Based on the numerical results in the relativistic two-fluid model (Zenitani et al. 2009b), we explain this logical gap in the following way. Figure 4 presents the composition of the incoming and outgoing energy flow during the quasi-steady stage of reconnection. In the antiparallel (leftmost) case with $B_{\rm G} = 0$, it has been found that the energy is mostly carried away in the form of the relativistic enthalpy flux, $\sim \sum_{i=p,e} 4\Gamma^2 P\mathbf{V}$, which was often overlooked by the earlier theories. In other words, since the enthalpy flux can carry a huge amount of energy per unit rest mass, it allows fast reconnection even through the narrow outflow channel.

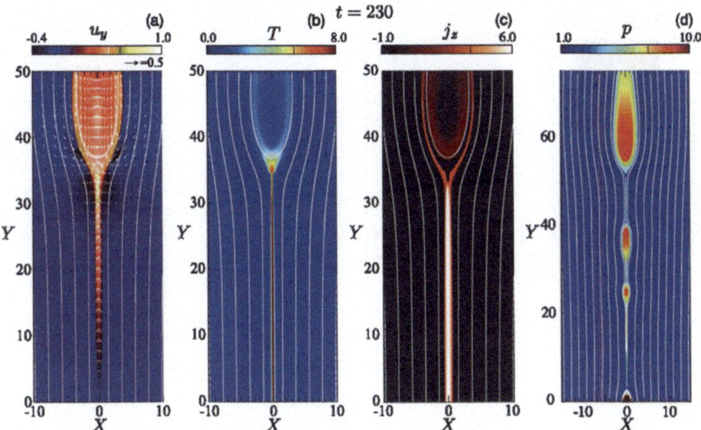

Fig. 5 Sweet–Parker results from RRMHD simulations with a uniform resistivity in the reconnection plane. (a) The outflow component of the four-velocity (U_y), (b) the plasma temperature ($T = P/\rho$), and (c) the out-of-plane electric current density (J_z) are presented. The Lundquist number is $S \sim 10^{3.5}$. (d) The plasma pressure (P) with $S \sim 10^4$. The solid lines show the magnetic field lines. Reproduced with permission from Takahashi et al. (2011), copyright by AAS

2.4 Relativistic Sweet-Parker Reconnection

Takahashi et al. (2011) have studied basic properties of Sweet–Parker reconnection. They employed a uniform resistivity in Ohm's law (Eq. (7)), and then they examined an RRMHD evolution of Sweet–Parker reconnection. A laminar Sweet–Parker current sheet was successfully reproduced in Figs. 5(a)–(c). By increasing the Lundquist number S ($\propto \eta^{-1}$) from 2×10^3 to 2×10^4, they have confirmed that the reconnection rate scales like $\propto S^{-1/2}$, as predicted by the relativistic Sweet–Parker theory (Lyubarsky 2005). It was also reported that the outflow speed is sub-Alfvénic, because of the larger inertia by the relativistic enthalpy.

Similar to the nonrelativistic MHD reconnection, when the Lundquist number exceeds $S \gtrsim \mathcal{O}(10^4)$, the Sweet–Parker current sheet becomes turbulent because of the repeated formation of plasmoids (Biskamp 1986; Loureiro et al. 2007; Bhattacharjee et al. 2009; Uzdensky et al. 2010). An early signature of the plasmoid-dominated regime can be seen in Fig. 5(d), but an RRMHD version of plasmoid-dominated turbulent reconnection has been studied by Takamoto (2013). Figure 6(a) shows a representative result for $S \sim 10^{5.5}$. One can see plasmoids of various sizes. Figure 6(b) presents the S-dependence of the reconnection rate. The magenta line indicates the rate by the relativistic Sweet–Parker theory, which was numerically verified by Takahashi et al. (2011). For higher-S regime of $S \gtrsim \mathcal{O}(10^4)$, the reconnection system becomes plasmoid-dominated. As a result, the reconnection rate deviates from the Sweet–Parker rate, and it remains fast $\sim \mathcal{O}(0.01)$ regardless of S.

2.5 Dependence to the Resistivity Model

Similar to the nonrelativistic case, the system evolution is sensitive to the effective resistivity model. Relativistic Petschek reconnection develops under a spatially-localized resistivity in the RRMHD and relativistic two-fluid models (Watanabe and Yokoyama 2006; Zenitani et al. 2009a, 2010; Zanotti and Dumbser 2011). In the RRMHD model, Sweet-Parker (Takahashi et al. 2011) and plasmoid-dominated reconnections (Takamoto 2013) develop under

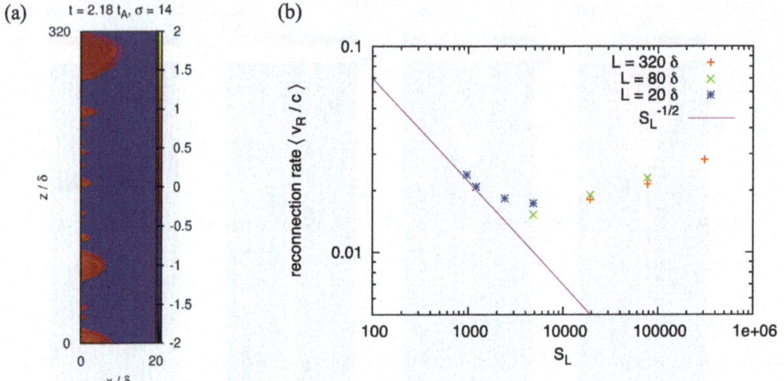

Fig. 6 (a) Plasma temperature T/mc^2 from an RRMHD simulation with $S \sim 10^{5.5}$ and $\sigma = 14$. The figure shows half of the reconnection region, and the reconnecting magnetic field is along the z direction. (b) Time-averaged reconnection rate in the $\sigma = 14$ runs, as a function of the Lundquist number S. The L parameter indicates the length of the simulation box, in unit of the initial current-sheet thickness δ. Reproduced with permission from Takamoto (2013), copyright by AAS

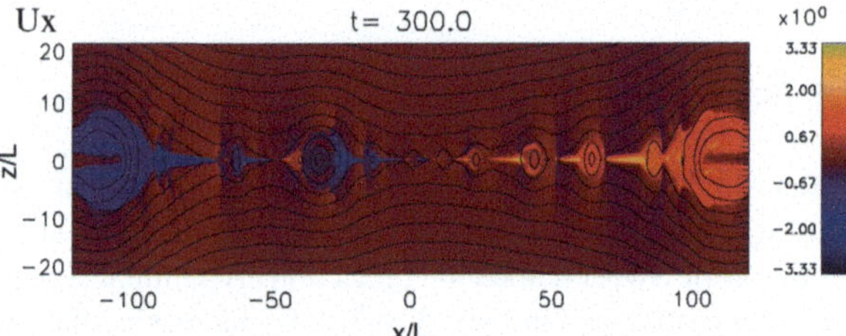

Fig. 7 The x-component of the plasma 4-velocity $U_X = (\Gamma V)_x$ (outflow) from a relativistic two-fluid simulation with uniform resistivity. The black lines show magnetic field lines. Reproduced with permission from Zenitani et al. (2009a), copyright by AAS

a spatially uniform resistivity. Interestingly, when we employ a uniform friction parameter (τ) in the relativistic two-fluid model, multiple plasmoids appear in the reconnecting current sheet, as shown in Fig. 7. This differs from the RRMHD results because the dissipation mechanism is no longer the same. In the relativistic two-fluid model, the fluid inertia terms also play a role similar to resistivity. Finally, under a current-dependent resistivity, repeated formation of plasmoids is observed in the RRMHD model (Zenitani et al. 2010). For practical applications of the RRMHD model, an accurate form of parameter-dependent resistivity needs to be developed.

2.6 Shocks in the Reconnection System

We discuss another feature of relativistic magnetic reconnection in the fluid regime — the system is often dominated by shocks. We remind the readers that the relativistic sound speed is slower than $c/\sqrt{3}$. In contrast, the outflow speed can be faster, approaching the speed of light c, as the magnetization σ increases in the upstream region (Eq. (19)). Comparing these relations, we find that the outflow jet always becomes supersonic when $\sigma > 1/2$. In such a supersonic regime, the outflow jets and the outflow-driven plasmoids generate shocks.

In single-reconnection systems, Zenitani et al. (2010) have reported various shocks around the plasmoid ahead of the reconnection jet, as indicated in Figs. 8(a) and (b). These shocks are essentially attributed to supersonic or transonic reconnection jets, but they need

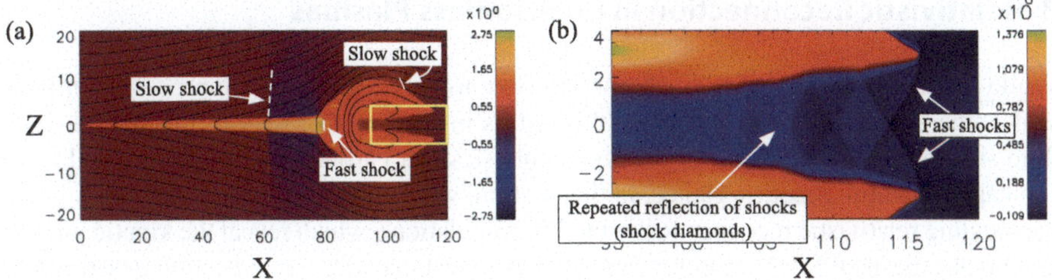

Fig. 8 (a) A spatial profile of the x-component of the plasma 4-velocity $U_x = (\Gamma V)_x$ in the outflow direction from an RRMHD simulation with a localized resistivity. The black lines show magnetic field lines. (b) A profile of U_x in the front side of the plasmoid, $x \in [95, 120]$ and $z \in [-4, 4]$ (corresponding to the yellow box in panel (a)). Modified with permission from Zenitani et al. (2010), copyright by AAS

to be studied in further detail. Even though Takamoto (2013) has explored other important aspects, no one has explored shocks in plasmoid-dominated systems. We expect that the plasmoid-dominated turbulent reconnection is also shock-dominated in high-σ regimes, as potential signatures can be seen in vertical discontinuities in Fig. 7. To numerically deal with these shocks, we need to use shock-capturing simulation codes. Most of the recent codes are capable of shocks by using HLL-type upwind schemes (Komissarov 2007; Palenzuela et al. 2009; Mizuno 2013; Mignone et al. 2019).

2.7 Discussion

These simulations in Sect. 2 have revealed fluid-scale properties of relativistic magnetic reconnection. Relativistic MHD reconnections are qualitatively similar to nonrelativistic MHD reconnections – relativistic Petschek (Watanabe and Yokoyama 2006; Zenitani et al. 2009a, 2010), relativistic Sweet–Parker (Takahashi et al. 2011), and relativistic plasmoid-mediated reconnections (Takamoto 2013) are reported. The resistivity model (uniform vs localized) and inertial effects determine the system evolution, as discussed in Sect. 2.5. In these regimes, the system can be dominated by shocks. At this point, the number of RRMHD reconnection studies is still limited. Many issues, such as the influence of environmental parameters, remain unsolved. For example, dependence on magnetization parameters: $\sigma \equiv B^2/w$ or $\sigma_m \equiv B^2/\rho$, the effects of guide-field, flow-shear, and asymmetry need to be explored.

Beyond the special relativity, several groups have been actively developing advanced Runge-Kutta codes for general relativistic resistive MHD (GR-RMHD) (Bucciantini and Del Zanna 2013; Dionysopoulou et al. 2013; Ripperda et al. 2019). Inda-Koide et al. (2019) have also developed an HLL-type GR-RMHD code to study magnetic reconnection close to the black hole. These GR-RMHD codes are successfully used to study the black hole and its accretion disk systems, however, the influence of GR effects on the local reconnection physics remains unclear. Combining radiative transfer equations with the RRMHD equations, Takahashi and Ohsuga (2013) have developed a relativistic radiative resistive MHD (RRRMHD) model. The numerical results of RRRMHD reconnections look similar to the conventional RRMHD results, however, the number of works is quite limited. There is a strong demand for further studies, to understand GR and/or radiation effects on RRMHD reconnection.

3 Relativistic Reconnection in Collisionless Plasmas

While MHD simulations provide decent descriptions of the reconnection process over large scales, the diffusion region is likely collisionless in many active regions of interest. Two-fluid simulations describe part of the kinetic physics, but Particle-in-Cell (PIC) simulations can capture the full kinetics. Hence, in this section, we report the up-to-date progress in understanding relativistic reconnection using PIC simulations, which reveal the kinetic physics that breaks the ideal MHD condition and the key to fast magnetic reconnection in astrophysical collisionless plasmas.

3.1 Relativistic Generalized Ohm's Law

During magnetic reconnection, magnetic flux is transported across the X-line. This requires the violation of the ideal condition $\mathbf{E} + \mathbf{V} \times \mathbf{B}/c = 0$, where $\mathbf{V}$ is the bulk plasma velocity. Understanding the physical mechanism that breaks the ideal MHD condition is one of the most important topics in reconnection physics, and the generalized Ohm's law is critical in determining such a mechanism, as also discussed in the non-relativistic limit (Liu et al. 2024, this collection).

The extension of the generalized Ohm's law to the relativistic regime can be nontrivial, and different formalism was derived from the electron momentum equation (Hesse and Zenitani 2007; Zenitani 2018). Here we discuss the latest form derived in Zenitani (2018). For simplicity, the speed of light is set to be $c = 1$. We begin with the stress-energy tensor

$$T^{\alpha\beta} = \int f(u) u^\alpha u^\beta \frac{d^3u}{\gamma}. \tag{20}$$

where $u^\alpha = (\gamma, \gamma\mathbf{v})$ is the particle four-velocity and particle Lorentz factor $\gamma = 1/\sqrt{1-(v/c)^2}$. A Greek index (e.g., α, β) runs from 0 to 3 to account for four-dimensional spacetime. Using a four-velocity of the bulk flow $U^\alpha = (\Gamma, \Gamma\mathbf{V})$ where the fluid Lorentz factor $\Gamma = 1/\sqrt{1-(V/c)^2}$, the metric tensor $g^{\alpha\beta} = \mathrm{diag}(-1, 1, 1, 1)$, and the projection operator $\Delta^{\alpha\beta} = g^{\alpha\beta} + U^\alpha U^\beta$, the stress-energy tensor can be decomposed into (Eckart 1940):

$$T^{\alpha\beta} = \mathcal{E} U^\alpha U^\beta + q^\alpha U^\beta + q^\beta U^\alpha + P^{\alpha\beta} \tag{21}$$

Here, $\mathcal{E} \equiv T^{\alpha\beta} U_\alpha U_\beta$ is the invariant energy density, $q^\alpha \equiv -\Delta^\alpha_\beta T^{\beta\gamma} U_\gamma$ is the energy flux (heat flow), and $P^{\alpha\beta} \equiv \Delta^\alpha_\gamma \Delta^\beta_\delta T^{\gamma\delta}$ is the pressure tensor. The momentum part of Eq. (21) is reduced to a familiar combination of the dynamic pressure and the pressure tensor, $T^{ij} = mnV^iV^j + P^{ij}$, in the non-relativistic limit, where a Roman index (e.g., i, j) runs from 1 to 3 to account for the three-dimensional space.

Note that the choice of U^α in Eq. (21) is arbitrary, and several choices can be considered (See Zenitani (2018) for more detail). Among them, here we employ the average plasma velocity $\mathbf{V}$ that carries the electric charge, and it thus satisfies the relation of $J_{\beta,(e)} = -en' U_{\beta,(e)}$, as usual. Here e is the unit charge, n' is the proper number density, and $J_{\beta,(e)}$ is the electric current carried by electrons. We use the prime to denote the proper quantities in Sect. 3.

We then use the energy-momentum equation of electrons,

$$\partial_\beta T^{\alpha\beta}_{(e)} = F^{\alpha\beta} J_{\beta,(e)} = -en' F^{\alpha\beta} U_{\beta,(e)} \tag{22}$$

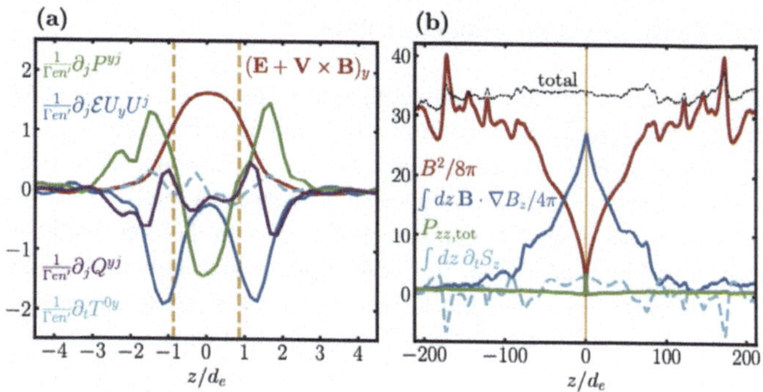

Fig. 9 (a) The relativistic Ohm's law (Eq. (23)) in the (y-) direction out of the reconnection plane. Each term is color-coded. (b) The force balance along the inflow (z-) direction. The yellow vertical dashed lines in (a) and (b) mark the predicted diffusion region edges. Reproduced with permission from Goodbred and Liu (2022), copyright by APS

where $F^{\alpha\beta}$ the electromagnetic tensor. Subscript (e) indicates electron fluid properties. However, for brevity, we omit this subscript hereafter in this subsection.

From the spatial parts (i.e., $\beta = 1, 2, 3$) of Eqs. (21) and (22), we obtain the electron Ohm's law,

$$\mathbf{E} = -\mathbf{V}\times\mathbf{B} - \frac{1}{\Gamma e n'}\Big[\partial_t T^{i0} + \partial_j(\mathcal{E}U^iU^j + Q^{ij} + P^{ij})\Big]. \tag{23}$$

where $\mathbf{V}$ is the bulk velocity and $Q^{\alpha\beta} \equiv q^\alpha U^\beta + q^\beta U^\alpha$ is the heat-flow part of the stress-energy tensor. The relativistic effects appear in Γ, $\mathcal{E}$ ($> n'mc^2$), Q, and their time derivatives.

Figure 9 displays the out-of-plane (y-) component of the electron Ohm's law across the X-line in relativistic magnetic reconnection. One key feature is the dominance of the inertial-like term $\partial_j(\mathcal{E}U_yU^j)$ in blue at the edge of the diffusion region where $(\mathbf{E}+\mathbf{V}\times\mathbf{B})_y$ in red is finite. This fact can be used to show that the diffusion region thickness is on the electron inertial scale (Goodbred and Liu 2022), as marked by the yellow dashed vertical lines in Fig. 9(a). This scale will be used to derive the first-principles reconnection rate in Sect. 3.2.2. The purple curve indicates the "heat-flow inertial" term, which corresponds to the momentum transport by the energetic electrons that carry the heat flow in the rest frame of electrons (Zenitani 2018). Relations between the kinetic physics and the terms in Eq. (23) are largely unknown and require more study, because these equations were formulated relatively recently.

The Eckart decomposition is also useful for evaluating the energy balance. The $0i$ components of the plasma stress-energy tensor in Eq. (21) can be further decomposed into the matter flow, the bulk kinetic energy flux, the enthalpy flux, and the heat flux, as respectively shown in the right-hand side of the following equation,

$$T^{0i} = n'U^i + (\Gamma - 1)n'U^i + \Big[(\mathcal{E} - n')\Gamma U^i + P^{0i}\Big] + Q^{0i}. \tag{24}$$

Zenitani (2018) reported that most incoming electromagnetic energy is converted into the relativistic enthalpy flux in the downstream region during relativistic magnetic reconnection, in agreement with RRMHD discussion in Sect. 2.3.

3.2 Relativistic Collisionless Reconnection Rate

We divide the reconnection rate problem in collisionless pair plasmas into two separate subsections, with one modeling the maximum plausible rate and another one modeling the key localization mechanism that leads to fast reconnection. Combining these two subsections, one can derive the reconnection rate as a function of magnetization from the first principles. Interested readers are encouraged to compare Sect. 3.2 with the non-relativistic reconnection rate discussed in Liu et al. (2024, this collection), which lays out the same approach to tackle the rate problem in kinetic current sheets of a wide variety of magnetic geometry, parameters and background conditions for heliophysics applications.

3.2.1 R-S_{lope} Relation and the Maximum Plausible Rate

As pointed out in earlier sections, in strongly magnetized plasmas, the plasma flow speed can be relativistic. During anti-parallel reconnection, the relevant force balance can be described by

$$\frac{(\mathbf{B}\cdot\nabla)\mathbf{B}}{4\pi} \simeq \frac{\nabla B^2}{8\pi} + \nabla\cdot\mathbf{P} + n'm_i(\mathbf{U}\cdot\nabla)\mathbf{U}, \tag{25}$$

where $\mathbf{U} = \Gamma\mathbf{V}$, $\Gamma \equiv [1-(V/c)^2]^{-1/2}$ is the Lorentz factor of the bulk flow, and primed quantities are the proper quantities. Balancing the magnetic tension with the plasma inertia along the outflow direction, the resulting outflow speed is the relativistic Alfvén speed (Blackman and Field 1994; Lyutikov and Uzdensky 2003; Lyubarsky 2005; Zenitani et al. 2010; Comisso and Asenjo 2014; Liu et al. 2015)

$$V_{A0} = c\sqrt{\frac{\sigma_{x0}}{1+\sigma_{x0}}}. \tag{26}$$

which is limited by the speed of light when the cold magnetization parameter $\sigma_{x0} = B_{x0}^2/4\pi n'mc^2 \gg 1$. Here B_{x0} is the asymptotic value of the reconnecting magnetic field component.

However, when the outflow geometry opens out, as shown in Fig. 10 (a), one needs to consider the difference in quantity magnitudes at the boundary of the microscopic diffusion region (denoted by subscript "m") and the mesoscale at the asymptotic region (denoted by subscript "0"). Recognizing this scale separation is the key to obtaining an upper-bounded reconnection rate. By analyzing the force balance along the inflow (e.g., Fig. 9(b)) and retaining the magnetic pressure ($\nabla B^2/8\pi$), Liu et al. (2017) derived

$$\frac{B_{xm}}{B_{x0}} \approx \frac{1-S_{lope}^2}{1+S_{lope}^2}. \tag{27}$$

where S_{lope} is the slope of the separatrix that separates the reconnecting and reconnected field lines along the outflow exhaust boundary. Figure 10(a) illustrates what these quantities mean. Similarly, analyzing the force balance along the outflow, retaining the magnetic pressure ($\nabla B^2/8\pi$), one can derive the outflow speed at the downstream boundary of the diffusion region

$$V_{out,m} \simeq c\sqrt{\frac{(1-S_{lope}^2)\sigma_{xm}}{1+(1-S_{lope}^2)\sigma_{xm}}}, \tag{28}$$

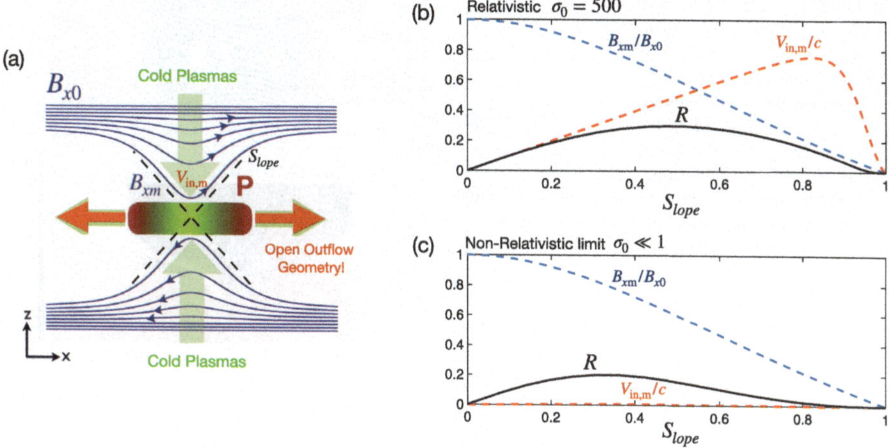

Fig. 10 *The $R - S_{lope}$ relation.* (a) The geometry and notation. (b) The predicted reconnection rate R, microscale inflow speed $V_{in,m}/c \simeq E_y/B_{xm}$, and field reduction B_{xm}/B_{x0} as a function of separatrix S_{lope} in the relativistic regime; (c) The predictions in the non-relativistic limit [Modified from Liu et al. (2017)]

which is smaller than V_{A0} in Eq. (26). Finally, the normalized reconnection rate

$$R \equiv \frac{cE_y}{B_{x0}V_{A0}} = \left(\frac{B_{zm}}{B_{xm}}\right)\left(\frac{B_{xm}}{B_{x0}}\right)\left(\frac{V_{out,m}}{V_{A0}}\right), \tag{29}$$

can be cast into a function of the separatrix S_{lope} after plugging in Eqs. (26)–(28), and realizing that $B_{zm}/B_{xm} \simeq S_{lope}$ due to the geometry.

As indicated by the blue curve in Fig. 10(b), when the separatrix slope increases, the B_{xm}/B_{x0} ratio decreases. Simulation (Liu et al. 2017) suggests that in the high-σ limit, the B_{xm}/B_{x0} ratio can be significantly lower than that in the non-relativistic limit, which leads to a much higher microscopic inflow speed $V_{in,m}$ at the upstream boundary of the diffusion region, as predicted by the red curve in Fig. 10(b). In spite of this relativistic inflow speed ($\sim 0.8c$), a value around 0.3 still upper bounds the relativistic reconnection rate R. In comparison, the maximum plausible rate in the non-relativistic limit (Fig. 10(c)) is around 0.2, and the microscopic inflow speed is much lower than the speed of light. More discussion on the non-relativistic electron-proton plasmas can be found in Liu et al. (2024, this collection).

3.2.2 Localization Mechanism That Leads to Fast Reconnection

In electron-positron (pair) plasmas, the Hall effect critical to facilitate fast reconnection in the electron-proton plasma (Sonnerup 1979; Mandt et al. 1994; Shay et al. 1999; Rogers et al. 2001; Drake et al. 2008; Liu et al. 2014, 2022) is absent, but the pressure depletion at the X-line appears to be significant, and it has an important consequence. This depletion is evident in Fig. 11, which shows the positron pressure P_{izz} contour in Fig. 11(a) and its cut along the outflow symmetry line in Fig. 11(b). The initial positron (also electron) pressure that can balance the asymptotic magnetic pressure is marked by the red dashed horizontal line of value $100m_ec^2$, while the P_{izz} cut at this nonlinear stage indicates a much lower value $\sim \mathcal{O}(m_ec^2)$ at the X-line (Note those peaks are secondary plasmoids that will be discussed later). This depleted pressure can cause the implosion of upstream plasmas into the X-line,

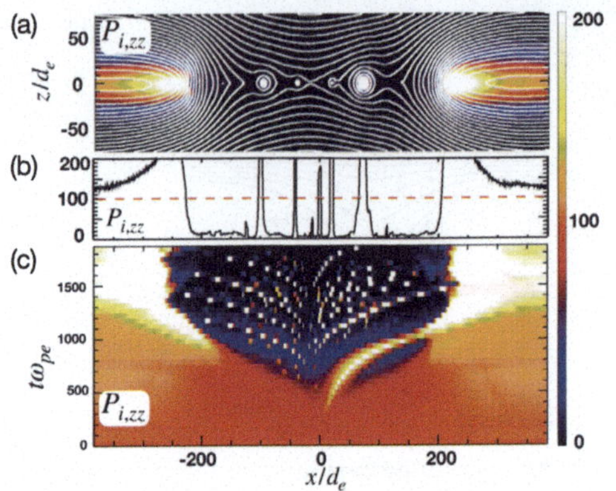

Fig. 11 *The pressure depletion and bursty nature of relativistic reconnection* with $\sigma_{x0} \simeq 89$. (a) The P_{izz} contour. (b) The P_{izz} cut along the outflow symmetry line (i.e., $z = 0$). The red dashed horizontal line marks the initial plasma pressure. (c) The time stack plot of this cut. Reproduced with permission from Liu et al. (2020), copyright by AAS

providing the localization mechanism needed for fast reconnection in ultra-relativistic astrophysical plasmas. This idea is also illustrated by the cartoon in Fig. 10(a).

Goodbred and Liu (2022) managed to show analytically that the relativistic Lorentz factor associated with the large electric current density is responsible for this drastic pressure depletion. By considering the energy conservation along the narrow inflow channel toward the X-line, they derived the upper bound value of the X-line pressure,

$$P_{zz}|_{xline} \leq 2n'_{xline}mc^2 \left[\frac{\langle \gamma(v_z) \rangle_{xline}}{\Gamma_y} - \frac{\Gamma_y}{\langle \gamma(v_z) \rangle_{xline}} \right]. \tag{30}$$

This expression clarifies the factors limiting the X-line thermal pressure. Here $\langle \gamma(v_z) \rangle_{xline}$ measures the v_z-averaged available energy at the X-line, and γ is the Lorentz factor of a particle. On the other hand, $\Gamma_y \equiv [1 - (V_y/c)^2]^{-1/2}$ only measures the bulk flow velocity of the current carriers that drift in the y-direction, whose magnitude is determined by the relativistic inertial scale that breaks the ideal MHD condition (Sect. 3.1). If all available energy is used to drive the current, then $\Gamma_y \simeq \langle \gamma(v_z) \rangle_{xline}$, and $P_{zz}|_{xline}$ becomes very small. Conversely, if only a small fraction of the total energy is needed to drive the current, then $\Gamma_y \ll \langle \gamma(v_z) \rangle_{xline}$ and $P_{zz}|_{xline}$ can become significant. The predicted X-line pressure normalized to the asymptotic magnetic pressure as a function of σ_{x0} is shown in Fig. 12(a) as the solid lines, which scales as $\simeq 2(2/\sigma_{x0})^{1/2}$; It predicts a more severe pressure depletion in the large σ_{x0} limit, capturing the decreasing trend of simulated X-line pressure shown by symbols.

Meanwhile, using the force balance along the inflow symmetry line within the diffusion region (ignoring the inflow inertia in Eq. (25)), one can relate the separatrix slope to the pressure difference between the X-line and upstream region, that is $P_{zz}|_{xline}$ (Eq. (30)) in the cold upstream limit,

$$S_{lope}^2 \approx 1 - \frac{8\pi P_{zz}|_{xline}}{B_{xm}^2}. \tag{31}$$

Since the upstream magnetic field line adjacent to the separatrix tends to straight out (when possible, due to the magnetic tension), we can couple this diffusion region solution with the larger upstream solution at the mesoscale (discussed in Sect. 3.2.1) to get the reconnection rate. Namely, by plugging Eq. (30) into (31) to determine the S_{lope}, then we can fully determine the reconnection rate from the R-S_{lope} relation in Fig. 10(b). The prediction

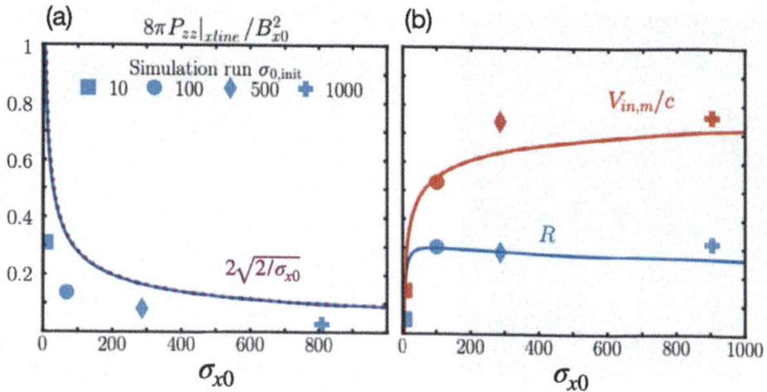

Fig. 12 *First-principles theory*. (a) The predicted upper-bound value of the X-line pressure (blue curve) as a function of σ_{x0} that can be approximated as $2\sqrt{2/\sigma_{x0}}$ (purple dashed curve). The measured values in PIC simulations are shown as symbols. (b) The predicted reconnection rate (blue curve) and the microscale inflow speed $V_{in,m}$ (red curve) as a function of σ_{x0}. The measured values in PIC simulations are shown as symbols with the same color coding. Reproduced with permission from Goodbred and Liu (2022), copyright by APS

of R is shown in Fig. 12(b) as the solid blue curve, which agrees well with the simulated reconnection rates shown as the blue symbols.

3.3 Bursty Nature of Relativistic Reconnection

Most existing literature has focused on the application of plasmoid formation in astrophysics systems and its implication for particle acceleration. As to the origin of these plasmoids in PIC simulations, does it really resemble the "high-Lindquist number plasmoid instability" derived in the uniform resistivity MHD model (Bhattacharjee et al. 2009; Loureiro et al. 2007; Shibata and Tanuma 2001b)? The physics of the tearing instability can be different in collisionless plasmas. Hoshino (2020) shows that the collisionless tearing instability can actually be stabilized by the relativistic drift of current carriers, which was not captured in the resistive MHD model. Instead, in collisionless pair plasmas, the secondary plasmoids may be generated because the pressure within the reconnection exhausts is also depleted in the nonlinear stage, as shown in Fig. 11(b); thus, plasmoids are violently produced from the collapse of the pressure-imbalanced current sheet. The formation of plasmoids within the primary exhausts helps balance the force, but only temporarily because they will be expelled out by outflows of the primary X-line. The evolution is settled into such a repetitive, dynamical balance, as shown in the time-stack plot in Fig. 11(c). These provide an alternative explanation to the bursty nature of relativistic reconnection in the antiparallel limit. It is interesting to note that a similar bursty nature was also reported in two-fluid simulation (Zenitani et al. 2009a) in Fig. 7 of Sect. 2.5. Liu et al. (2020) further demonstrated that the generation of secondary plasmoids can be suppressed in the presence of external guide fields, which is also not expected in the resistive-MHD model. As a potential application in plasma astrophysics, the mergers of secondary plasmoids are proposed to be a plausible generation mechanism of Fast Radiation Bursts (FRB) from neutron star magnetospheres (Philippov et al. 2019; Mahlmann et al. 2022).

3.4 3D Relativistic Turbulent Reconnection

The discussion thus far in this section focuses on the features in 2D kinetic simulations. It is interesting to explore the differences and similarities in a 3D system where the current sheet coexists with background turbulence. Figure 13 shows an example of relativistic turbulent

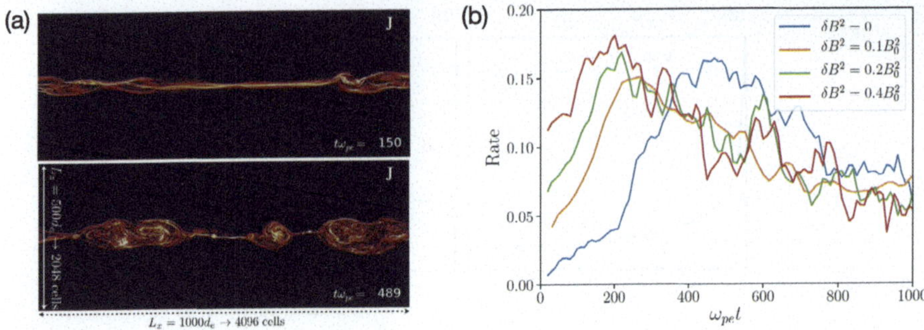

Fig. 13 *Relativistic turbulent 3D reconnection in electron-positron (pair) plasmas and its reconnection rates.* Panel (a) shows 2D cuts of current density during turbulent reconnection in the reconnection region at two different times; Panel (b) shows the evolution of the reconnection rate in cases with different initial turbulent fluctuation levels. Note that the rates are bounded by the predicted maximum plausible value ($\simeq 0.3$) shown in Fig. 10(b) even with a strong fluctuation. Reproduced with permission from Guo et al. (2021), copyright by AAS

reconnection in a 3D electron-positron plasma PIC simulation (Guo et al. 2021, also see Fig. 19). Figure 13(a) shows the current density at two different times. The reconnection layer becomes highly structured because of the turbulent fluctuations initially imposed and later the self-driven turbulence arising from secondary oblique tearing instability (Daughton et al. 2011) and flux-rope kink instability (Zhang et al. 2021, 2024). Figure 13(b) shows the evolution of the global reconnection rates in runs with different turbulence fluctuation levels. Interestingly, no matter how strong the imposed turbulence fluctuation is, the reconnection rate is still well-bounded by the value of 0.3, as predicted by the maximum plausible rate in Sect. 3.2.1. Intriguingly, a similar result was demonstrated even in nonrelativistic resistive MHD simulations (Yang et al. 2020) that constantly drive turbulence within the simulation domain. Since the turbulent reconnection in these simulations still develops large-scale, coherent inflows and outflows, we anticipate that the force balance and the global geometrical constraint still apply on average. Analytically, one can show that the prediction in Sect. 3.2.1 will work, as long as the outflow speed is on the order of ion Alfvén speed (Liu et al. 2017), no matter how thick (Lin et al. 2021) and complex (Liu et al. 2018) the diffusion region is. On the other hand, even though a thick turbulent diffusion region was theorized (Lazarian and Vishniac 1999), the reconnection process in 3D PIC simulations, in fact, is still dominated by a few active diffusion regions in the kinetic scale, as also seen in Fig. 13(a). As long as the current sheet is in kinetic scale, the localization mechanism based on the fast drifting current carriers and pressure depletion, discussed in Sect. 3.2.2, should also work; and this will lead to fast reconnection. Nevertheless, a full resolution to this complex setting remains largely unknown, and there are several other competing ideas on turbulent reconnection (Lazarian and Vishniac 1999; Eyink et al. 2011; Boozer 2012; Higashimori et al. 2013); this topic continues to be an active research area of great interest.

4 Plasma Heating and Particle Acceleration

Since magnetic reconnection in the magnetically dominated regime is likely associated with strong energy release, the heating and nonthermal particle acceleration during reconnection are of strong interest. The past two decades have witnessed unprecedented progress in our

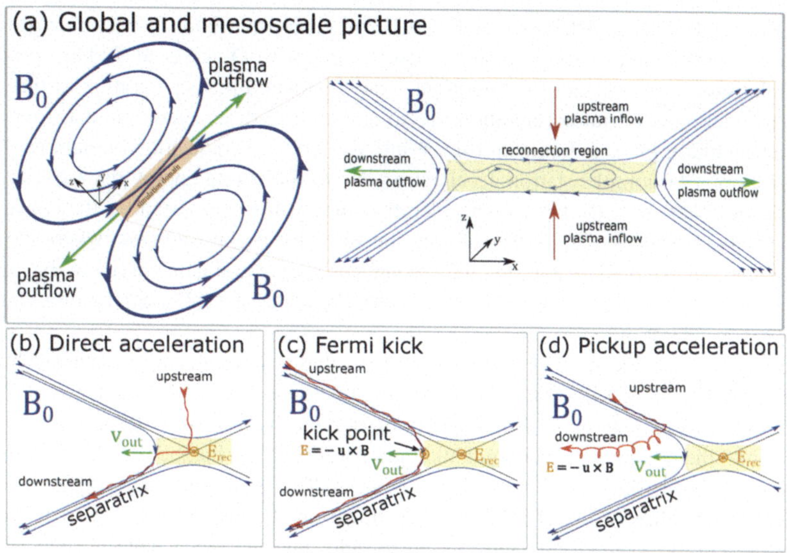

Fig. 14 Sketches of global and mesoscale reconnection configurations and several particle energization mechanisms. (a) The surrounding astrophysical context of the reconnection region (highlighted in orange). (b) Direct acceleration from the reconnection electric field near an X-line. (c) Fermi acceleration in the exhaust region (d) Acceleration by the pickup process, where the particle becomes unmagnetized as it crosses the exhaust boundary. In panels (b-d), B_0 is the reconnecting magnetic field, $E_{\rm rec}$ is the reconnection electric field, and $V_{\rm out} \simeq V_{Ax0}$ is the reconnection outflow speed, approximately equal to the in-plane Alfvén speed. Reproduced from French et al. (2023), copyright by the author(s)

understanding of nonthermal particle acceleration in relativistic magnetic reconnection. An important discovery is that relativistic magnetic reconnection supports strong particle acceleration and development of power-law energy distribution (Zenitani and Hoshino 2001; Sironi and Spitkovsky 2014; Guo et al. 2014, 2015; Werner et al. 2016). These discoveries are first found in the highly relativistic pair plasma and later in the mildly relativistic regime. This development also motivates the new studies in the nonrelativistic low-β regime (Li et al. 2019; Zhang et al. 2021, 2024; Arnold et al. 2021).

4.1 Basic Acceleration Mechanisms

The basic acceleration mechanisms during magnetic reconnection can be broadly categorized into two types. The ones that accelerate particles via non-ideal electric fields and the ones via the motional electric fields $\mathbf{E} = -\mathbf{V} \times \mathbf{B}/c$. PIC simulations have uncovered several basic acceleration mechanisms such as Fermi-type acceleration, acceleration at X-line regions, and betatron acceleration, etc. In addition, analytical theories have been proposed and built to understand the acceleration processes and the resulting energy spectra.

Figure 14 shows a broad reconnection region where several acceleration mechanisms may occur. At X-lines, the ideal Ohm's law is broken, and a strong non-ideal electric field exists $E \sim R V_{A0} B_0$, where R is the reconnection rate. In relativistic magnetic reconnection without a guide field, X-line acceleration is often approximated by regions where the electric field is stronger than the reconnecting magnetic field $E > B$ (Zenitani and Hoshino 2001; Sironi and Spitkovsky 2014). However, significant acceleration has also been found in the broader region of the reconnection layer (Zenitani and Hoshino 2007; Guo et al.

2019). Guo et al. (2014, 2015) proposed that Fermi acceleration due to curvature drift motions in contacting and merging magnetic lands (similar to Drake et al. (2006), proposed in nonrelativistic reconnection) is important. They compared the acceleration of the parallel electric field and Fermi acceleration due to curvature drift acceleration, and found that Fermi acceleration plays a dominant role. While the whole reconnection domain may have a negative contribution of betatron acceleration due to the strong energy release (decaying magnetic field $\partial B/\partial t < 0$), betatron acceleration may still be important in the converging islands (Hakobyan et al. 2021). In addition, the so-called pickup process, where particles become unmagnetized when entering the reconnection layer and gain energy in the outflow (Drake et al. 2009; Sironi and Beloborodov 2020; French et al. 2023), can be important for low-energy acceleration.

There have been recent efforts evaluating the relative importance of each mechanism (Guo et al. 2014, 2015; Kilian et al. 2020; Sironi 2022; Guo et al. 2023; French et al. 2023). One possible way to distinguish different acceleration mechanisms is to decompose the electric field into the components perpendicular and parallel to the magnetic field ($E_\perp$ and $E_\parallel$) (Dahlin et al. 2014; Ball et al. 2019; Kilian et al. 2020), or the motional electric field and non-ideal electric field (Guo et al. 2019).[3] As mentioned above, the $E > B$ regions may better represent the X-line for a vanishing guide field. The analyses generally show that the acceleration associated with the motional electric field / perpendicular electric field (associated with Fermi/betatron or pickup process) dominates for a weak guide field and weakens for a stronger guide field (French et al. 2023), likely because of the compressibility of the layer (Li et al. 2018a). The contribution of the perpendicular electric field also increases with the domain size. It is important to recognize that, during magnetic reconnection, most energy conversion is through a large-scale process by the magnetic tension release, rather than at small kinetic scales. Since the nonthermal particles take a large fraction of the released energy in relativistic reconnection, they must somehow 'tap' the motional electric field. The mechanisms like Fermi acceleration would correspond to the motional electric field and, therefore, must be responsible for the main part of the nonthermal spectra.

Fermi acceleration can be more generalized in reconnection systems (Hoshino 2012; Lemoine 2019). de Gouveia dal Pino and Lazarian (2005) and Drury (2012) proposed a model considering the net compression in a reconnection layer and the effect of escape. Instead of Fermi acceleration by magnetic islands, here, particles are accelerated in the electric field induced by the reconnection inflow. Future studies are needed to show which Fermi acceleration is the most dominant one.

4.2 Nonthermal Acceleration Uncovered by PIC Simulations

Early PIC simulations show that the energy spectrum near the X-line resembles a power-law distribution (Zenitani and Hoshino 2001), whereas the energy distribution over the broader reconnection layer is much softer (Zenitani and Hoshino 2007). Recent large-scale PIC simulations of relativistic magnetic reconnection show a power-law energy spectrum $f(\gamma - 1) \propto \mathcal{E}^{-p}$ when integrated over the whole reconnection region with various spectral indices p as hard as $p \sim 1$ (Sironi and Spitkovsky 2014; Guo et al. 2014, 2015; Melzani et al. 2014; Werner et al. 2016). The nonthermal spectra have been also discussed for electron-proton plasmas, but with a reduced σ as $\sigma \sim \sigma_i \sim \sigma_e/(m_i/m_e)$ (Werner et al. 2018; Ball

[3] As discussed in Lemoine (2019), any perpendicular electric field satisfying $\mathbf{E} \cdot \mathbf{B} = 0$ and $E^2 - B^2 < 0$ may support a generalized Fermi acceleration. The speed of the collision center is not necessarily the speed of the MHD flow.

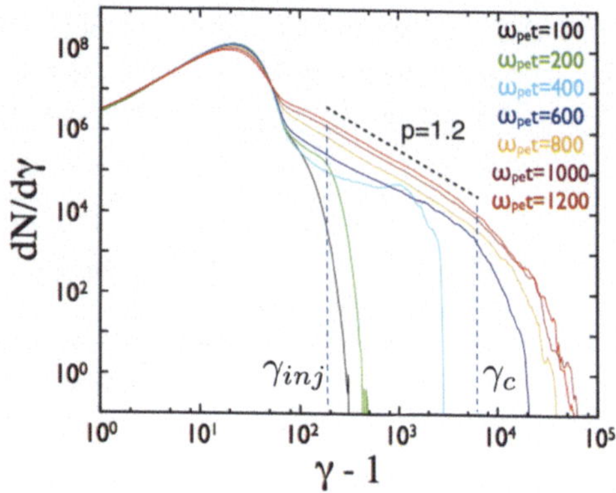

Fig. 15 The resulting energy spectrum in relativistic pair plasma reconnection from a sample PIC simulation starting from a force-free current sheet with $\sigma = 3200$ at different time steps generated from the simulation. Adapted from Guo et al. (2020)

et al. 2018; Kilian et al. 2020; Li et al. 2023). In this regime, the energy spectra became softer ($p \gtrsim 2$). This trend naturally connects to the results from nonrelativistic reconnection, where the power-law spectra are much softer ($p \gtrsim 3-4$) (Li et al. 2019; Arnold et al. 2021; Zhang et al. 2021, 2024).

As shown in Fig. 15, the nonthermal signatures found in PIC simulations can be understood in several aspects. A remarkably clear nonthermal power-law distribution is observed starting from a Lorentz factor a fraction of σ to high energy. We term this lower-energy bound as the injection energy γ_{inj}, above which particles are accelerated into the nonthermal energies. This transition energy is also important for understanding the partition between thermal and nonthermal distribution (Hoshino 2023; French et al. 2023). As the guide field becomes stronger, the energy spectra are softer, and the high-energy cutoff is suppressed (French et al. 2023). Finally, the energy spectra roll over at high energy, and the cutoff energy γ_c scales with the simulation domain and time (Petropoulou and Sironi 2018; French et al. 2023).

4.3 Physics That Determine the Acceleration Results

The strong nonthermal features in the particle energy spectra indicate several key processes, including how particle energization transits from thermal to nonthermal energies ("injection"), how power-law energy spectra develop ("power-law formation"), and the high-energy extension of the power-law ("roll over"). Below, we review the progress in understanding how they determine the energy spectra.

4.3.1 Formation of Nonthermal Power-Law Energy Spectra

Can magnetic reconnection support a clear nonthermal power-law spectrum, and if so, by what mechanism and under what condition? This has been a major theoretical issue in the particle energization during magnetic reconnection (e.g., Drake et al. 2013; Guo et al. 2014).

Zenitani and Hoshino (2001) proposed a simple power-law model. Around the reconnection site, a particle is directly accelerated by the reconnection electric field $E_{\rm rec}$ in the out-of-plane direction, perpendicular to the reconnection plane. Then its energy gain is approximated by

$$\frac{d\mathcal{E}}{dt} = eE_{\rm rec}c. \tag{32}$$

They also estimate the loss rate of particles. Since particles travel through (relativistic) Speiser motion, their typical time scale $\tau(\mathcal{E})$ in the reconnection site can be approximated by a quarter-gyration by a typical reconnected magnetic field $\bar{B}_z$. Then the particle loss rate is estimated by

$$\frac{1}{N}\frac{dN}{dt} = -\frac{4}{2\pi}\left(\frac{e\bar{B}_z}{\gamma mc}\right) = -\frac{2ce\bar{B}_z}{\pi\mathcal{E}} \tag{33}$$

A key point is that the loss rate is energy-dependent. Because of larger inertia γm, higher-energy particles are less likely to escape from the acceleration site. Combining Eqs. (32) and (33), we see that the particle number density follows the power-law distribution,

$$N \propto \mathcal{E}^{-(2\bar{B}_z)/(\pi E_{\rm rec})} \tag{34}$$

Since $\bar{B}_z/B_0 \sim \mathcal{O}(0.1)$ and $E_{\rm rec}/B_0 \sim \mathcal{O}(0.1)$, we see that the power-law index is on order of $\mathcal{O}(1)$. Similar models have been further developed by Uzdensky (2022) and Zhang et al. (2023), including magnetic flux ropes/islands as an escape region.

Recently, several new works discussed the formation of power-law distributions in a broad reconnection layer. Sironi and Spitkovsky (2014) have proposed that the power-law form is established as the particles accelerated at the X-lines with $E > B$. They argue that this process is essential for the power-law formation and determines the spectra index of the energy spectra. However, since Fermi/betatron acceleration is the dominant acceleration in the broad reconnection region, it is unclear how X-line acceleration can solely determine the formation of the power law. Guo et al. (2014, 2015) proposed that the power-law distributions are produced by a Fermi-like process and continuous injection. In general, one can evaluate a Fokker-Planck-like equation for a reconnection layer (Guo et al. 2020; Li et al. 2021, 2023)

$$\partial_t f + \partial_{\mathcal{E}}(\alpha_{\rm acc}\mathcal{E} f) = \partial^2_{\mathcal{E}}(D_{\mathcal{E}\mathcal{E}} f) - \alpha_{\rm esc} f + \frac{f_{\rm inj}}{\tau_{\rm inj}}, \tag{35}$$

where $\mathcal{E} = (\gamma - 1)mc^2$ is the kinetic energy, $\alpha_{\rm acc}$ is the acceleration rate, $D_{\mathcal{E}\mathcal{E}} = D_0\mathcal{E}^2$ is the energy diffusion coefficient, $\alpha_{\rm esc} \equiv \tau^{-1}_{\rm esc}$ is the escape rate, $f_{\rm inj}$ is the injected particle distribution, and $\tau_{\rm inj}$ is particle injection time scale. $\alpha_{\rm acc} = (\partial_t\mathcal{E} + \partial_{\mathcal{E}} D_{\mathcal{E}\mathcal{E}})\mathcal{E}^{-1}$ describes a combination of the first-order Fermi processes and the accompanying first-order term associated with second-order Fermi mechanisms.

As reconnection proceeds, the ambient plasma continuously flows into the reconnection layer through an inflow speed V_{in} and undergoes a selective injection process. For a simple case with $D_{\mathcal{E}\mathcal{E}} = 0$, and α_{acc} and α_{esc} independent of energy, the solution of Eq. (35) naturally recovers the classical solution $p = 1 + 1/(\alpha_{\rm acc}\tau_{esc})$. A power-law distribution can form from an upstream thermal distribution when $\alpha_{\rm acc}\tau$ is large (τ is the duration of acceleration) (Guo et al. 2014, 2015). In the limit that $\alpha_{\rm acc}$ is large, the spectral index approaches $p = 1$, consistent with existing PIC simulations. It was usually argued that an escape mechanism is necessary for forming a power-law distribution. This statement is incorrect or at least misleading. The power-law distribution can still form even for the case with no escape term (Guo et al. 2014, 2015). The main physics for forming a power-law is due to the continuous injection and Fermi acceleration. However, it is still important to understand the escape term, as it determines the shape of the distribution. Other acceleration with different α_{acc} can, in principle, form a power-law as long as the combined spectral index p does not depend on energy,

$$p = 1 + \frac{1}{\alpha_{\rm acc}\tau_{esc}} + \frac{\partial \ln \alpha_{\rm acc}}{\partial \ln \mathcal{E}} = \text{const} \tag{36}$$

Hakobyan et al. (2021) has developed an analytical model based on betatron acceleration to explain the high-energy nonthermal part of the spectra. Li et al. (2023) measured the acceleration and escape rates in PIC simulations, and the numerical solution of Eq. (35) achieved a nice agreement with the PIC simulation results.

4.3.2 Energy Partition of Thermal and Nonthermal Particles in Reconnection

Magnetic reconnection is known to be the most important mechanism not only for plasma thermalization up to the equivalent temperature of the Alfvén velocity, but also for accelerating nonthermal particles whose energies exceed their thermal energies (e.g., Birn and Priest 2007; Zweibel and Yamada 2009; Hoshino and Lyubarsky 2012; Uzdensky 2016; Blandford et al. 2017). It was known that a non-negligible fraction of nonthermal particles is generated during reconnection through satellite observations of the Earth's magnetosphere and the solar atmosphere (e.g., Øieroset et al. 2002; Lin et al. 2003) and PIC simulation studies for non-relativistic plasmas (e.g., Hoshino et al. 2001; Drake et al. 2006; Pritchett 2006; Oka et al. 2010). Furthermore, as magnetic reconnection is a ubiquitous process in the plasma universe, reconnection is believed to occur in many high-energy astrophysical phenomena such as in pulsar magnetosphere, accretion disks, and magnetar (e.g., Remillard and McClintock 2006; Done et al. 2007; Madejski and Sikora 2016; Kirk 2004; Lyutikov and Uzdensky 2003). PIC simulation studies also revealed that relativistic reconnection whose Alfvén velocity is close to the speed of light can effectively generate nonthermal particles with a harder power-law energy spectrum than that generated in non-relativistic plasmas (e.g. Zenitani and Hoshino 2001, 2005a,b; Jaroschek et al. 2004; Jaroschek and Hoshino 2009; Cerutti et al. 2012a,b, 2013; Liu et al. 2011; Sironi and Spitkovsky 2011, 2014; Guo et al. 2014). Since then, regardless of whether the plasma is nonrelativistic or relativistic, magnetic reconnection has gained attention as a mechanism of nonthermal particle acceleration in various space and astrophysical sources. However, the energy partitioning of thermal and nonthermal plasmas during magnetic reconnection is not understood. The energy partition plays an important role in the dynamical evolution of magnetic reconnection.

Recently, the energy partitioning has been quantitatively investigated by using PIC simulations for a pair plasma (Hoshino 2022, 2023). The downstream heated plasma by reconnection has been found to be well modeled by a composed distribution function consisting of a Maxwell distribution function $N_M(\gamma)$ and a kappa distribution function $N_\kappa(\gamma)$, where $N_M(\gamma) = n_M \gamma \sqrt{\gamma^2 - 1} \exp(-(\gamma - 1)/(T_M/mc^2))$, and $N_\kappa(\gamma) = n_\kappa \gamma \sqrt{\gamma^2 - 1} \left(1 + (\gamma - 1)/(\kappa T_\kappa/mc^2)\right)^{-(1+\kappa)} f_{cut}(\gamma)$, where $f_{cut}(\gamma)$ represents the high energy cutoff function given by $\exp(-(\gamma - \gamma_{cut})/\gamma_{cut})$ for $\gamma > \gamma_{cut}$.[4] Based on the model fitting, Fig. 16 shows (a) the average thermal temperature of the Maxwellian and kappa distributions, which is described as $T_{th} = (n_M T_M + n_\kappa T_\kappa)/(n_M + n_\kappa)$, (b) κ index, and (c) the fraction of the nonthermal energy density $\mathcal{E}_{\rm ene}$ as a function of the plasma temperature (T_0/mc^2) and guide field (B_G/B_0). The nonthermal fraction $\mathcal{E}_{\rm ene}$ is defined as $\mathcal{E}_{\rm ene} = \int_1^\infty (\gamma - 1)(N_\kappa(\gamma) - N_\kappa^{\rm M}(\gamma))d\gamma / \int_1^\infty (\gamma - 1)N_{\rm M+\kappa}(\gamma)d\gamma$, where $N_\kappa^{\rm M}(\gamma)$ represents

[4] It is interesting to note that the kappa distribution function is widely observed in space plasma environments (Vasyliunas 1968; Livadiotis and McComas 2013), and the formation of the kappa distribution function may be related to the idea of non-extensive statistics (Tsallis 1988).

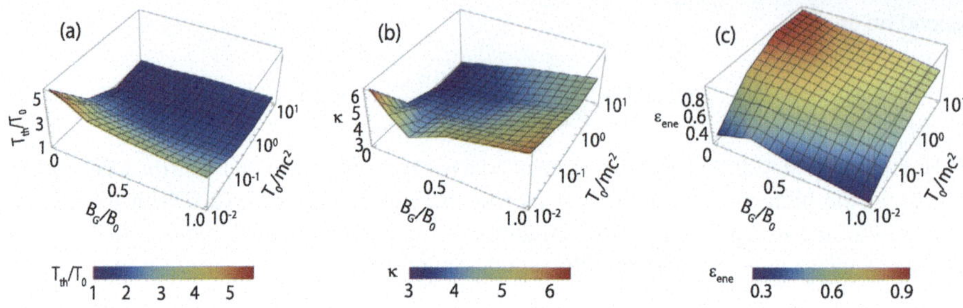

Fig. 16 Model fitting results as functions of the initial plasma temperatures $T/mc^2 = 10^{-2} \sim 10^1$ and the guide magnetic fields of $B_G/B_0 = 0 \sim 1$. Three panels show (a) the average temperatures normalized by the initial background temperature T_0, (b) κ index, and (c) the efficiency of nonthermal particles against thermal plasmas as a function of the initial plasma temperature T/mc^2 and guide magnetic field B_G/B_0. Reproduced with permission from Hoshino (2023), copyright by the AAS

the portion of the Maxwellian distribution function in the κ distribution function. For anti-parallel magnetic field topology without a guide magnetic field, it was found that while the nonthermal energy density in relativistic reconnection can occupy more than 90% of the total kinetic plasma energy density, most dissipated magnetic field energy can be converted into thermal plasma heating in nonrelativistic reconnection. For magnetic reconnection with a guide field, it is found that the fraction of nonthermal particles $\mathcal{E}_{\text{ene}}$ basically decreases with the increase of the guide field. However, for non-relativistic reconnection with $T/mc^2 \ll 1$ and a weak guide field, the nonthermal fraction $\mathcal{E}_{\text{ene}}$ is found to increase.

In the relativistic regime, a recent study by French et al. (2023) determines injection energy beyond which the particle energy distribution is well described by a power-law spectrum. They define the acceleration efficiency by the "number" and "energy", where the contribution of downstream nonthermal particles over the total downstream population is calculated. Consistently, they find that a stronger guide field suppresses the acceleration efficiency, and the efficiency may saturate to a final asymptotic value for a sufficiently large domain (see Fig. 17).

So far, these studies of energy partitioning have mainly been done in a two-dimensional system, but it is important to study a three-dimensional effect where turbulent magnetic reconnection can occur (Daughton et al. 2011).

4.3.3 Injection Problem

There have been several studies on the injection problem with energization up until the injection energy γ_{inj}. While Ball et al. (2019) focused on the work done by the parallel electric field ($W_{\parallel}$), Kilian et al. (2020) studied the roles of both parallel and perpendicular electric fields. They showed that both parallel and perpendicular electric fields play a role, and perpendicular electric fields become more important for particle injection, especially when the simulation domain becomes large. While acknowledging the importance of $W_{\perp}$ during the development of the power-law distribution, Sironi (2022) suggests that $E > B$ regions are important for injecting particles by accelerating particles, and they claimed all injected particles need to cross the X-lines. This apparent correlation is further studied by Guo et al. (2023), and importantly, shown to have no significant contributions to the injection process. Guo et al. (2023) showed that $E > B$ regions contribute very little to injection ($\sim 10\%\gamma_{inj}$) as they only host particles for a short time, insufficient for boosting the particles to injection. French et al. (2023) studied three different injection mechanisms (parallel

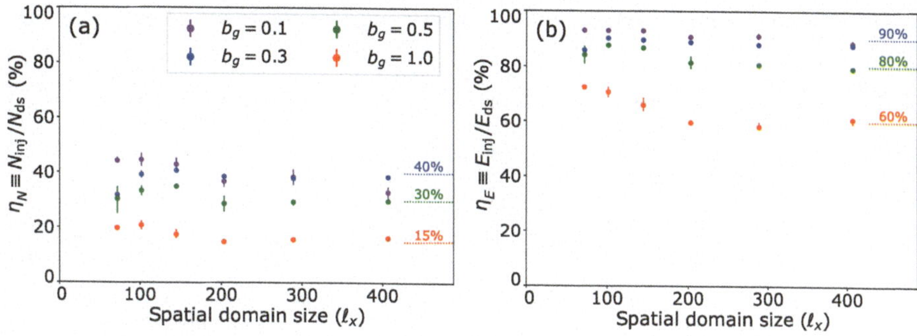

Fig. 17 Acceleration efficiencies from PIC simulations adapted from French et al. (2023). (a) The number efficiency (the ratio between the injected particle number to the downstream particle number) $\eta_N \equiv N_{\mathrm{inj}}/N_{\mathrm{ds}}$. (b) The energy efficiency $\eta_E \equiv E_{\mathrm{inj}}/E_{\mathrm{ds}}$ for the ratio between the sum of injected particle energy to the energy of the whole downstream population. Reproduced with permission from French et al. (2023), copyright by the author(s)

electric field, Fermi reflection, and pickup process) and attempted to quantify their relative importance. Figure 18 shows that the Fermi and pickup processes, related to the electric field perpendicular to the magnetic field, govern the injection for weak guide fields and larger domains. Meanwhile, parallel electric fields are important for injection in the strong guide-field regime. Totorica et al. (2023) isolated the energy gain during injection from the nonideal field. They reached a different conclusion, because they focused on high-energy particles. To reach general consensus and to deepen our understanding, more research is needed on this important topic.

4.3.4 High-Energy Roll-over

Early results with a weak guide field show that the spectral index can be as hard as $p \sim 1$ for high σ, meaning the energy contained in such a spectrum is dominated by high-energy particles. The cutoff energy is therefore limited by the amount of dissipated magnetic energy $\gamma_c \sim [2\delta\sigma(2-p)]^{1/(2-p)}$, where δ is the fraction of the dissipated magnetic energy channeled into each species (Guo et al. 2016b). For $p \sim 1$ this gives a high-energy roll-over at a few times σ (Werner et al. 2016). It was suggested that as reconnection proceeds, the acceleration may evolve into softer spectra ($p \sim 2$), and the high-energy acceleration can proceed for a long time (Petropoulou and Sironi 2018; French et al. 2023), although the σ dependence still exists. Meanwhile, a stronger guide field or escape effects would make a difference. A significant guide field leads to a softer spectrum and lowers the maximum energy (French et al. 2023; Li et al. 2023). As reconnection proceeds, the escape process needs to be considered to correctly understand high-energy roll-over.

4.4 The Roles of 3D Reconnection and Turbulence

Earlier discussions on 3D relativistic reconnection primarily focused on the drift kink instability (Zenitani and Hoshino 2007; Liu et al. 2011). However, the flux rope kink instability can also grow and drive turbulence (Zhang et al. 2021; Guo et al. 2021). Both of these can be suppressed by a strong guide field (Zenitani and Hoshino 2008). In the strong guide field regime, the oblique tearing instability is likely dominant (Guo et al. 2021). As instabilities evolve, it is well established that 3D reconnection can spontaneously generate magnetic

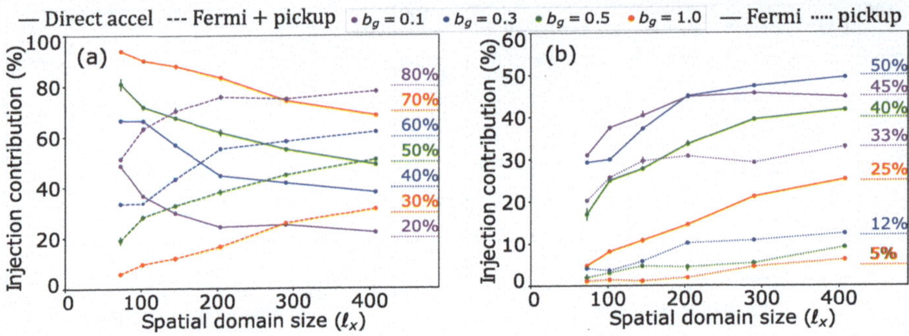

Fig. 18 Cumulative percentage of injected electrons that are injected by each particle acceleration mechanism. (a): Decomposition between particles injected by $W_\parallel$ (solid) and $W_\perp$ (dashed). (b): Decomposition between Fermi-injected particles (solid) and pickup-injected particles (dotted). Reproduced with permission from French et al. (2023), copyright by the author(s)

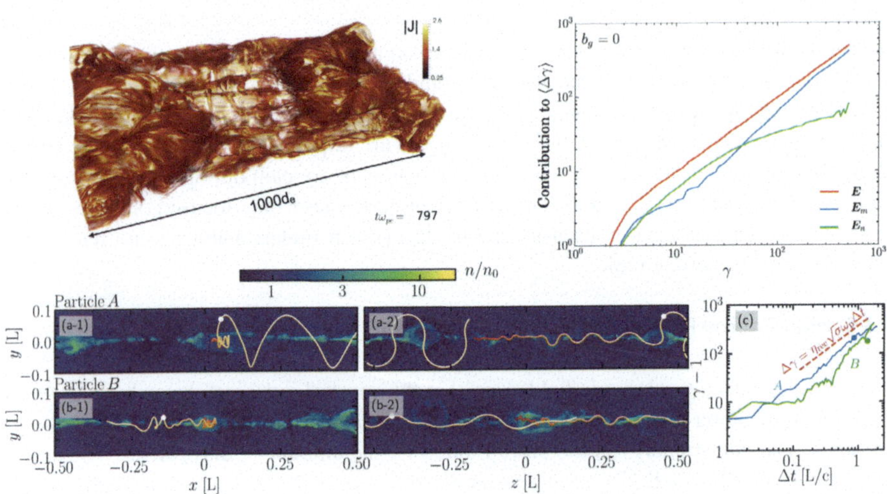

Fig. 19 Several recent studies on 3D relativistic magnetic reconnection and associated particle acceleration. Upper panels: The reconnection layer becomes significantly turbulent as the flux rope kink instability grows. The acceleration to high energy is dominated by the motional electric field ($\mathbf{E}_m$), and the non-ideal electric field ($\mathbf{E}_n$) plays a subdominant role. Lower panels: 3D relativistic reconnection supports additional acceleration pattern, as particles can meander across the reconnection and gain energy in the upstream electric field. Reproduced with permission from [top] Guo et al. (2021) and [bottom] from Zhang et al. (2021b), copyright by AAS

turbulence, as shown in Fig. 19. Recent 3D studies have shown that 3D effects can be important for efficient acceleration in reconnection (Dahlin et al. 2017; Li et al. 2019; Zhang et al. 2021, 2022). In 2D magnetic field configuration, particles are trapped in magnetic islands due to restricted particle motion across magnetic field lines (Jokipii et al. 1993; Jones et al. 1998; Giacalone et al. 1994; Johnson et al. 2022), and high-energy particle acceleration can be prohibited. Chaotic field lines and turbulence due to the 3D evolution of the oblique tearing instability (e.g., Daughton et al. 2011; Liu et al. 2013) and flux rope kink instability (e.g., Guo et al. 2021; Zhang et al. 2021) make particles leave the flux rope and can lead to efficient transport of particles in the reconnection region, which is found to be important for further acceleration in the reconnection region (Dahlin et al. 2017; Li et al. 2019). Bottom panels of Fig. 19 show that particles can escape from flux ropes and get acceleration in the reconnection inflow regions via Speiser-like orbits (Zhang et al. 2022).

4.5 The Problem of Scale Separation and Macroscopic Approach

PIC simulations have developed to the point that it is very useful to study the elementary processes of relativistic magnetic reconnection and particle acceleration (see Shay et al. 2024, this collection), and they have been used to explore global physics in heliophysics applications. However, one should remember that the scale separation in astrophysical problems is far beyond the reach of PIC simulations. For example, the ratio between the system size and the plasma skin depth can be $\sim 10^{13} - 10^{17}$ for pulsar wind nebulae and extragalactic jets. The conclusions made by PIC simulations need to be extrapolated to a large scale, and it is unknown if all the conclusions can still hold on a large scale. The solution to this serious issue is to develop a large-scale model that contains basic acceleration physics learned from PIC simulations.

At large scales where the domain of the reconnection layer becomes much larger than the gyromotion scale, one can often use the guiding-center approximation, which includes Fermi/betatron acceleration identified as primary acceleration mechanisms (Guo et al. 2014; Dahlin et al. 2014; Li et al. 2017). An alternative solution is to use energetic particle transport theory. Li et al. (2018a) have shown that the acceleration in the reconnection layer can be described as compression and shear, which are the main acceleration physics included in the energetic particle transport theory (Parker 1957, 1965; Zank 2014; le Roux et al. 2015). This approach has been used to study large-scale reconnection acceleration (Li et al. 2018b, 2022). In addition, Fermi acceleration can be studied by following the momentum changes of particles through a sequence of local frames where the local electric field vanishes (Lemoine 2019). Monte Carlo simulations (e.g., Seo et al. 2022) including this description may alleviate the complicity of the particle transport equations when the plasma flow becomes relativistic (Webb 1989).

PIC simulations show that nonthermal particles can take a significant energy of the system, and therefore it may be important to include the feedback of energetic particles. Drake et al. (2019) presented a set of equations where the guiding-center particles feedback in the MHD equations so the total energy of the system for the fluid and 'hot' particles is conserved. The feedback is through the pressure tensor of the energetic particles in the fluid equations. Arnold et al. (2021) successfully showed magnetic islands in the reconnection layer accelerate particles via Fermi acceleration and lead to the development of power-law energy spectra. This description has not included the effect of particle scattering in turbulence, which is expected to be important in 3D turbulent reconnection. Seo et al. (2024) developed a new computational method for including the backreaction of energetic particles as a pressure term, while the distribution of the energetic particles is evolved through Parker's transport equation (Parker 1965). An alternative (equivalent) way to include the feedback is through the electric current carried by energetic particles (Bai et al. 2015; Sun and Bai 2023).

5 Final Remarks

This review summarizes recent progress in relativistic magnetic reconnection, focusing on its fluid description and simulations, collisionless reconnection physics, and particle energization. We remark on several directions that may become frontier problems and need further attention.

One of the most important topics is 3D magnetic reconnection. Although many progress has been made in understanding 2D and 3D reconnection, the 3D evolution is still an open issue. At this point, it is unclear whether magnetic reconnection always proceeds through a thin layer on the kinetic or a broader resistive scale. It is widely believed that the coupling between magnetic reconnection/tearing instability and current-driven instabilities, such as drift-kink and flux-rope kink instabilities, play an important role in the formation of turbulent current sheets. When 3D turbulence broadens the kinetic layer, its subsequent evolution can be sensitive to the current sheet and ambient plasma parameters. The 3D system evolution inevitably influences the way particles are accelerated, and therefore we need to clarify whether particle acceleration in 3D is fundamentally different from that in 2D. These aspects have not been settled, and future progress is still very much needed to gain further insights.

As will be discussed in another article (see Nakamura et al. 2024, this collection), it is important to understand the onset mechanism of magnetic reconnection. Kinetic simulations typically start from a thin and long Harris sheet in relativistic reconnection problems, but the current sheet is highly unstable to kinetic instabilities which grow in a few gyro-periods. Then, it is not clear whether the Harris sheet is the best initial condition. In order to justify many earlier works, or in order to discover the best initial condition, we demand more research on the formation and stability of the "initial" current sheet.

In the future, in order to study the magnetosphere or even larger systems, we need to simultaneously resolve kinetic scales of magnetic reconnection near the central object and the large-scale evolution of the entire system. Numerical simulation across such many scales will be challenging, in particular in astrophysical settings. To deal with the cross-scale problem, several attempts have been made to interconnect small-scale kinetic simulations and large-scale MHD simulations (Sugiyama and Kusano 2007; Usami et al. 2013; Daldorff et al. 2014) for nonrelativistic problems, however, the number of these MHD-PIC models is very limited. Unfortunately, no similar attempts have been reported in relativistic astrophysics, partly because it is difficult to translate kinetic quantities into fluid quantities as discussed in Sect. 3. A lot more work is necessary to develop relativistic MHD-PIC models. As we have discussed in Sect. 4.5, it is important to consider nonthermal particle acceleration in MHD models, studying and predicting energetic particles and their signatures in large-scale systems.

Beyond the description discussed in this review article, incorporating the radiation (Jaroschek and Hoshino 2009; Cerutti et al. 2013; Zhang et al. 2018; Werner et al. 2019; Sironi and Beloborodov 2020; Schoeffler et al. 2023) and other QED processes (Schoeffler et al. 2019), such as pair production and annihilation, during reconnection is also important for one to delve into the rich physics in such extreme astrophysical plasmas. On the other hand, this endeavor could also help extract observable signatures for distant observers, constraining our understanding of relativistic magnetic reconnection.

With recent breakthroughs, relativistic magnetic reconnection has become an important topic in reconnection studies and a key process for understanding astrophysical energy release, particle acceleration, and high-energy radiation. Although difficult, it is highly anticipated that transformative progress will be achieved in the near future in key areas (Sect. 1.2) of relativistic magnetic reconnection, reaching a more complete physics understanding, achieving a more advanced modeling capability, and better connecting with astrophysical observations.

Acknowledgements F. G. acknowledges the support from Los Alamos National Laboratory through the LDRD program, DOE Office of Science, and NASA programs through 80HQTR20T0073, 80HQTR21T0087 and 80HQTR21T0104, and the ATP program. The work by Y. L. is funded by the National Science Foundation grant PHY-1902867 through the NSF/DOE Partnership in Basic Plasma Science and Engineering and

NASA 80NSSC21K2048. The work by S. Z. was funded by Japan Society for the Promotion of Science (JSPS) KAKENHI, Grant No. 21K03627. The work by M. H. was supported by JSPS KAKENHI, Grant Nos. 19H01949 and 20K20908.

Funding Open access funding provided by Österreichische Akademie der Wissenschaften.

Declarations

Competing Interests The authors have no conflict of interest to declare that is relevant to the content of this article.

References

Abdo AA, Ackermann M, Ajello M et al (2011) Science 331(6018):739. https://doi.org/10.1126/science.1199705

Abeysekara AU, Albert A, Alfaro R et al (2017) Science 358(6365):911–914. https://doi.org/10.1126/science.aan4880

Ackermann M, Anantua R, Asano K et al (2016) Astrophys J Lett 824(2):L20. https://doi.org/10.3847/2041-8205/824/2/L20

Agarwal S, Banerjee B, Shukla A et al (2023) Mon Not R Astron Soc 521(1):L53–L58. https://doi.org/10.1093/mnrasl/slad023

Alves EP, Zrake J, Fiuza F (2018) Phys Rev Lett 121(24):245101. https://doi.org/10.1103/PhysRevLett.121.245101

Amano T (2016) Astrophys J 831(1):100. https://doi.org/10.3847/0004-637X/831/1/100

Amano T, Matsumoto Y, Bohdan A et al (2022) Rev Mod Plasma Phys 6(1):29. https://doi.org/10.1007/s41614-022-00093-1

Arnold H, Drake JF, Swisdak M et al (2021) Phys Rev Lett 126(13):135101. https://doi.org/10.1103/PhysRevLett.126.135101

Arons J (2012) Space Sci Rev 173(1–4):341–367. https://doi.org/10.1007/s11214-012-9885-1

Bai XN, Caprioli D, Sironi L et al (2015) Astrophys J 809(1):55. https://doi.org/10.1088/0004-637X/809/1/55

Ball D, Sironi L, Özel F (2018) Astrophys J 862(1):80. https://doi.org/10.3847/1538-4357/aac820

Ball D, Sironi L, Özel F (2019) Astrophys J 884(1):57. https://doi.org/10.3847/1538-4357/ab3f2e

Balsara DS, Amano T, Garain S et al (2016) J Comput Phys 318:169–200. https://doi.org/10.1016/j.jcp.2016.05.006

Barkov M, Komissarov SS, Korolev V et al (2014) Mon Not R Astron Soc 438(1):704–716. https://doi.org/10.1093/mnras/stt2247

Bhattacharjee A, Huang YM, Yang H et al (2009) Phys Plasmas 16(11):112102. https://doi.org/10.1063/1.3264103

Birn J, Priest ER (2007) Reconnection of magnetic fields: magnetohydrodynamics and collisionless theory and observations. Cambridge University Press, Cambridge. https://doi.org/10.1017/CBO9780511536151

Biskamp D (1986) Phys Fluids 29(5):1520–1531. https://doi.org/10.1063/1.865670

Blackman EG, Field GB (1994) Phys Rev Lett 72(4):494–497. https://doi.org/10.1103/PhysRevLett.72.494

Blandford R, Yuan Y, Hoshino M et al (2017) Space Sci Rev 207(1–4):291–317. https://doi.org/10.1007/s11214-017-0376-2

Bodo G, Tavecchio F, Sironi L (2021) Mon Not R Astron Soc 501(2):2836–2847. https://doi.org/10.1093/mnras/staa3620

Bohdan A, Pohl M, Niemiec J et al (2020) Astrophys J 904(1):12. https://doi.org/10.3847/1538-4357/abbc19

Boozer AH (2012) Phys Plasmas 19(11):112901. https://doi.org/10.1063/1.4765352
Bucciantini N, Del Zanna L (2013) Mon Not R Astron Soc 428(1):71–85. https://doi.org/10.1093/mnras/sts005
Burgess D, Hellinger P, Gingell I et al (2016) J Plasma Phys 82(4):905820401. https://doi.org/10.1017/S0022377816000660
Cerutti B, Uzdensky DA, Begelman MC (2012a) Astrophys J 746(2):148. https://doi.org/10.1088/0004-637X/746/2/148
Cerutti B, Werner GR, Uzdensky DA et al (2012b) Astrophys J Lett 754(2):L33. https://doi.org/10.1088/2041-8205/754/2/L33
Cerutti B, Werner GR, Uzdensky DA et al (2013) Astrophys J 770(2):147. https://doi.org/10.1088/0004-637X/770/2/147
Comisso L, Asenjo FA (2014) Phys Rev Lett 113(4):045001. https://doi.org/10.1103/PhysRevLett.113.045001
Comisso L, Lingam M, Huang YM et al (2017) Astrophys J 850(2):142. https://doi.org/10.3847/1538-4357/aa9789
Coroniti FV (1990) Astrophys J 349:538. https://doi.org/10.1086/168340
Dahlin JT, Drake JF, Swisdak M (2014) Phys Plasmas 21(9):092304. https://doi.org/10.1063/1.4894484
Dahlin JT, Drake JF, Swisdak M (2017) Phys Plasmas 24(9):092110. https://doi.org/10.1063/1.4986211
Daldorff LKS, Tóth G, Gombosi TI et al (2014) J Comput Phys 268:236–254. https://doi.org/10.1016/j.jcp.2014.03.009
Daughton W, Karimabadi H (2007) Phys Plasmas 14(7):072303. https://doi.org/10.1063/1.2749494
Daughton W, Roytershteyn V, Albright BJ et al (2009) Phys Rev Lett 103(6):065004. https://doi.org/10.1103/PhysRevLett.103.065004
Daughton W, Roytershteyn V, Karimabadi H et al (2011) Nat Phys 7(7):539–542. https://doi.org/10.1038/nphys1965
de Gouveia dal Pino EM, Lazarian A (2005) Astron Astrophys 441(3):845–853. https://doi.org/10.1051/0004-6361:20042590
Dedner A, Kemm F, Kröner D et al (2002) J Comput Phys 175(2):645–673. https://doi.org/10.1006/jcph.2001.6961
Dionysopoulou K, Alic D, Palenzuela C et al (2013) Phys Rev D 88(4):044020. https://doi.org/10.1103/PhysRevD.88.044020
Done C, Gierliński M, Kubota A (2007) Astron Astrophys Rev 15(1):1–66. https://doi.org/10.1007/s00159-007-0006-1
Drake JF, Swisdak M, Che H et al (2006) Nature 443(7111):553–556. https://doi.org/10.1038/nature05116
Drake JF, Shay MA, Swisdak M (2008) Phys Plasmas 15(042):306
Drake JF, Cassak PA, Shay MA et al (2009) Astrophys J Lett 700(1):L16–L20. https://doi.org/10.1088/0004-637X/700/1/L16
Drake JF, Swisdak M, Fermo R (2013) Astrophys J Lett 763(1):L5. https://doi.org/10.1088/2041-8205/763/1/L5
Drake JF, Arnold H, Swisdak M et al (2019) Phys Plasmas 26(1):012901. https://doi.org/10.1063/1.5058140
Drake JF, Antiochos SK, Bale SD et al (2024) Magnetic reconnection in solar flares and the near-Sun solar wind. Space Sci Rev 220
Drury LO (2012) Mon Not R Astron Soc 422(3):2474–2476. https://doi.org/10.1111/j.1365-2966.2012.20804.x
Dumbser M, Zanotti O (2009) J Comput Phys 228(18):6991–7006. https://doi.org/10.1016/j.jcp.2009.06.009
Eckart C (1940) Phys Rev 58(10):919–924. https://doi.org/10.1103/PhysRev.58.919
Eyink GL, Lazarian A, Vishniac ET (2011) Astrophys J 743(1):51. https://doi.org/10.1088/0004-637X/743/1/51
French O, Guo F, Zhang Q et al (2023) Astrophys J 948(1):19. https://doi.org/10.3847/1538-4357/acb7dd
Giacalone J, Jokipii JR, Kota J (1994) J Geophys Res 99(A10):19351–19358. https://doi.org/10.1029/94JA01213
Giannios D, Uzdensky DA, Begelman MC (2009) Mon Not R Astron Soc 395(1):L29–L33. https://doi.org/10.1111/j.1745-3933.2009.00635.x
Goodbred M, Liu YH (2022) Phys Rev Lett 129(26):265101. https://doi.org/10.1103/PhysRevLett.129.265101
Guo F, Li H, Daughton W et al (2014) Phys Rev Lett 113(15):155005. https://doi.org/10.1103/PhysRevLett.113.155005
Guo F, Liu YH, Daughton W et al (2015) Astrophys J 806(2):167. https://doi.org/10.1088/0004-637X/806/2/167
Guo F, Li H, Daughton W et al (2016a) Phys Plasmas 23(5):055708. https://doi.org/10.1063/1.4948284
Guo F, Li X, Li H et al (2016b) Astrophys J Lett 818(1):L9. https://doi.org/10.3847/2041-8205/818/1/L9

Guo F, Li X, Daughton W et al (2019) Astrophys J Lett 879(2):L23. https://doi.org/10.3847/2041-8213/ab2a15
Guo F, Liu YH, Li X et al (2020) Phys Plasmas 27(8):080501. https://doi.org/10.1063/5.0012094
Guo F, Li X, Daughton W et al (2021) Astrophys J 919(2):111. https://doi.org/10.3847/1538-4357/ac0918
Guo F, Li X, French O et al (2023) Phys Rev Lett 130(18):189501. https://doi.org/10.1103/PhysRevLett.130.189501
Hakobyan H, Petropoulou M, Spitkovsky A et al (2021) Astrophys J 912(1):48. https://doi.org/10.3847/1538-4357/abedac
Hakobyan H, Ripperda B, Philippov AA (2023) Astrophys J Lett 943(2):L29. https://doi.org/10.3847/2041-8213/acb264
Hesse M, Zenitani S (2007) Phys Plasmas 14(11):112102. https://doi.org/10.1063/1.2801482
Higashimori K, Yokoi N, Hoshino M (2013) Phys Rev Lett 110(25):255001. https://doi.org/10.1103/PhysRevLett.110.255001
Hoshino M (2012) Phys Rev Lett 108(13):135003. https://doi.org/10.1103/PhysRevLett.108.135003
Hoshino M (2015) Phys Rev Lett 114(6):061101. https://doi.org/10.1103/PhysRevLett.114.061101
Hoshino M (2020) Astrophys J 900(1):66. https://doi.org/10.3847/1538-4357/aba59d
Hoshino M (2022) Phys Plasmas 29(4):042902. https://doi.org/10.1063/5.0086316
Hoshino M (2023) Astrophys J 946(2):77. https://doi.org/10.3847/1538-4357/acbfb5
Hoshino M, Lyubarsky Y (2012) Space Sci Rev 173(1–4):521–533. https://doi.org/10.1007/s11214-012-9931-z
Hoshino M, Mukai T, Terasawa T et al (2001) J Geophys Res 106(A11):25979–25998. https://doi.org/10.1029/2001JA900052
Huang YM, Comisso L, Bhattacharjee A (2017) Astrophys J 849(2):75. https://doi.org/10.3847/1538-4357/aa906d
Inda-Koide M, Koide S, Morino R (2019) Astrophys J 883(1):69. https://doi.org/10.3847/1538-4357/ab345f
Jaroschek CH, Hoshino M (2009) Phys Rev Lett 103(7):075002. https://doi.org/10.1103/PhysRevLett.103.075002
Jaroschek CH, Treumann RA, Lesch H et al (2004) Phys Plasmas 11(1):1151–1163. https://doi.org/10.1063/1.1644814
Ji H, Daughton W (2011) Phys Plasmas 18:111207. https://doi.org/10.1063/1.3647505
Ji H, Daughton W, Jara-Almonte J et al (2022) Nat Rev Phys 4(4):263–282. https://doi.org/10.1038/s42254-021-00419-x
Johnson G, Kilian P, Guo F et al (2022) Astrophys J 933(1):73. https://doi.org/10.3847/1538-4357/ac7143
Jokipii JR, Kota J, Giacalone J (1993) Geophys Res Lett 20(17):1759–1761. https://doi.org/10.1029/93GL01973
Jones FC, Jokipii JR, Baring MG (1998) Astrophys J 509(1):238–243. https://doi.org/10.1086/306480
Kato TN, Takabe H (2008) Astrophys J Lett 681(2):L93. https://doi.org/10.1086/590387
Kilian P, Li X, Guo F et al (2020) Astrophys J 899(2):151. https://doi.org/10.3847/1538-4357/aba1e9
Kirk JG (2004) Phys Rev Lett 92(18):181101. https://doi.org/10.1103/PhysRevLett.92.181101
Kirk JG, Skjæraasen O (2003) Astrophys J 591(1):366–379. https://doi.org/10.1086/375215
Komissarov SS (2007) Mon Not R Astron Soc 382(3):995–1004. https://doi.org/10.1111/j.1365-2966.2007.12448.x
Lazarian A, Vishniac ET (1999) Astrophys J 517(2):700–718. https://doi.org/10.1086/307233
le Roux JA, Zank GP, Webb GM et al (2015) Astrophys J 801(2):112. https://doi.org/10.1088/0004-637X/801/2/112
Lemoine M (2019) Phys Rev D 99(8):083006. https://doi.org/10.1103/PhysRevD.99.083006
Li H, Lapenta G, Finn JM et al (2006) Astrophys J 643(1):92–100. https://doi.org/10.1086/501499
Li X, Guo F, Li H et al (2017) Astrophys J 843(1):21. https://doi.org/10.3847/1538-4357/aa745e
Li X, Guo F, Li H et al (2018a) Astrophys J 855(2):80. https://doi.org/10.3847/1538-4357/aaacd5
Li X, Guo F, Li H et al (2018b) Astrophys J 866(1):4. https://doi.org/10.3847/1538-4357/aae07b
Li X, Guo F, Li H et al (2019) Astrophys J 884(2):118. https://doi.org/10.3847/1538-4357/ab4268
Li X, Guo F, Liu YH (2021) Phys Plasmas 28(5):052905. https://doi.org/10.1063/5.0047644
Li X, Guo F, Chen B et al (2022) Astrophys J 932(2):92. https://doi.org/10.3847/1538-4357/ac6efe
Li X, Guo F, Liu YH et al (2023) Astrophys J Lett 954(2):L37. https://doi.org/10.3847/2041-8213/acf135
Lin RP, Krucker S, Hurford GJ et al (2003) Astrophys J Lett 595(2):L69–L76. https://doi.org/10.1086/378932
Lin SC, Liu YH, Li X (2021) Phys Plasmas 28(7):072109. https://doi.org/10.1063/5.0052317
Lin X, Li YP, Yuan F (2023) Mon Not R Astron Soc 520(1):1271–1284. https://doi.org/10.1093/mnras/stad176
Liu W, Li H, Yin L et al (2011) Phys Plasmas 18(5):052 https://doi.org/10.1063/1.3589304
Liu YH, Daughton W, Karimabadi H et al (2013) Phys Rev Lett 110(26):265004. https://doi.org/10.1103/PhysRevLett.110.265004

Liu YH, Daughton W, Karimabadi H et al (2014) Phys Plasmas 21(022):113
Liu YH, Guo F, Daughton W et al (2015) Phys Rev Lett 114(9):095002. https://doi.org/10.1103/PhysRevLett.114.095002
Liu YH, Hesse M, Guo F et al (2017) Phys Rev Lett 118(8):085101. https://doi.org/10.1103/PhysRevLett.118.085101
Liu YH, Hesse M, Guo F et al (2018) Phys Plasmas 25(8):080701. https://doi.org/10.1063/1.5042539
Liu YH, Lin SC, Hesse M et al (2020) Astrophys J Lett 892(1):L13. https://doi.org/10.3847/2041-8213/ab7d3f
Liu YH, Cassak P, Li X et al (2022) Commun Phys 5(1):97. https://doi.org/10.1038/s42005-022-00854-x
Liu YH, Hesse M, Genestreti K et al (2024) Ohm's law, the reconnection rate, and energy conversion in collisionless magnetic reconnection. Space Sci Rev 220. arXiv:2406.00875
Livadiotis G, McComas DJ (2013) Space Sci Rev 175(1–4):183–214. https://doi.org/10.1007/s11214-013-9982-9
Loureiro NF, Schekochihin AA, Cowley SC (2007) Phys Plasmas 14(10):100
Lu Y, Guo F, Kilian P et al (2021) Astrophys J 908(2):147. https://doi.org/10.3847/1538-4357/abd406
Lyubarsky YE (2005) Mon Not R Astron Soc 358(1):113–119. https://doi.org/10.1111/j.1365-2966.2005.08767.x
Lyubarsky Y, Kirk JG (2001) Astrophys J 547(1):437–448. https://doi.org/10.1086/318354
Lyutikov M, Uzdensky D (2003) Astrophys J 589(2):893–901. https://doi.org/10.1086/374808
Lyutikov M, Sironi L, Komissarov SS et al (2017) J Plasma Phys 83(6):635830601. https://doi.org/10.1017/S0022377817000629
Madejski GG, Sikora M (2016) Annu Rev Astron Astrophys 54:725–760. https://doi.org/10.1146/annurev-astro-081913-040044
Mahlmann JF, Philippov AA, Levinson A et al (2022) Astrophys J Lett 932(2):L20. https://doi.org/10.3847/2041-8213/ac7156
Mandt ME Denton RE Drake JF (1994) Geophys Res Lett 21(1):73–76
Matsumoto Y, Amano T, Kato TN et al (2015) Science 347(6225):974–978. https://doi.org/10.1126/science.1260168
McKinney JC, Uzdensky DA (2012) Mon Not R Astron Soc 419(1):573–607. https://doi.org/10.1111/j.1365-2966.2011.19721.x
Mehlhaff JM, Werner GR, Uzdensky DA et al (2020) Mon Not R Astron Soc 498(1):799–820. https://doi.org/10.1093/mnras/staa2346
Melzani M, Walder R, Folini D et al (2014) Astron Astrophys 570:A112. https://doi.org/10.1051/0004-6361/201424193
Mignone A, Mattia G, Bodo G et al (2019) Mon Not R Astron Soc 486(3):4252–4274. https://doi.org/10.1093/mnras/stz1015
Mizuno Y (2013) Astrophys J Suppl 205(1):7. https://doi.org/10.1088/0067-0049/205/1/7
Munz CD, Omnes P, Schneider R et al (2000) J Comput Phys 161(2):484–511. https://doi.org/10.1006/jcph.2000.6507
Nakamura R et al (2024) Space Sci Rev 220
Nathanail A, Mpisketzis V, Porth O et al (2022) Mon Not R Astron Soc 513(3):4267–4277. https://doi.org/10.1093/mnras/stac1118
Øieroset M, Lin RP, Phan TD et al (2002) Phys Rev Lett 89(19):195001. https://doi.org/10.1103/PhysRevLett.89.195001
Oka M, Phan TD, Krucker S et al (2010) Astrophys J 714(1):915–926. https://doi.org/10.1088/0004-637X/714/1/915
Oka M, Birn J, Egedal J et al (2023). https://doi.org/10.48550/arXiv.2307.01376. arXiv e-prints arXiv:2307.01376
Palenzuela C, Lehner L, Reula O et al (2009) Mon Not R Astron Soc 394(4):1727–1740. https://doi.org/10.1111/j.1365-2966.2009.14454.x
Palenzuela C, Lehner L, Ponce M et al (2013) Phys Rev Lett 111(6):061105. https://doi.org/10.1103/PhysRevLett.111.061105
Parker EN (1957) J Geophys Res 62(4):509–520. https://doi.org/10.1029/JZ062i004p00509
Parker EN (1965) Planet Space Sci 13(1):9–49. https://doi.org/10.1016/0032-0633(65)90131-5
Petropoulou M, Sironi L (2018) Mon Not R Astron Soc 481(4):5687–5701. https://doi.org/10.1093/mnras/sty2702
Petropoulou M, Giannios D, Sironi L (2016) Mon Not R Astron Soc 462(3):3325–3343. https://doi.org/10.1093/mnras/stw1832
Petschek HE (1964) Magnetic field annihilation. In: Proceedings of the AAS-NASA symposium. NASA special publication, vol 50. p 425

Philippov A, Kramer M (2022) Annu Rev Astron Astrophys 60:495–558. https://doi.org/10.1146/annurev-astro-052920-112338

Philippov A, Uzdensky DA, Spitkovsky A et al (2019) Astrophys J Lett 876(1):L6. https://doi.org/10.3847/2041-8213/ab1590

Pritchett PL (2006) J Geophys Res Space Phys 111(A10):A10212. https://doi.org/10.1029/2006JA011793

Remillard RA, McClintock JE (2006) Annu Rev Astron Astrophys 44(1):49–92. https://doi.org/10.1146/annurev.astro.44.051905.092532

Ripperda B, Bacchini F, Porth O et al (2019) Astrophys J Suppl 244(1):10. https://doi.org/10.3847/1538-4365/ab3922

Ripperda B, Bacchini F, Philippov AA (2020) Astrophys J 900(2):100. https://doi.org/10.3847/1538-4357/ababab

Ripperda B, Liska M, Chatterjee K et al (2022) Astrophys J Lett 924(2):L32. https://doi.org/10.3847/2041-8213/ac46a1

Rogers BN, Denton RE, Drake JF et al (2001) Phys Rev Lett 195:004

Schoeffler KM, Grismayer T, Uzdensky D et al (2019) Astrophys J 870(1):49. https://doi.org/10.3847/1538-4357/aaf1b9

Schoeffler KM, Grismayer T, Uzdensky D et al (2023) Mon Not R Astron Soc 523(3):3812–3839. https://doi.org/10.1093/mnras/stad1588

Seo J, Ryu D, Kang H (2022) Astrophys J 944:199. https://doi.org/10.3847/1538-4357/acb3ba

Seo J, Guo F, Li X et al (2024). ArXiv e-prints, arXiv:2404.12276 [astro-ph.HE]

Shay M et al (2024) Space Sci Rev 220

Shay MA, Drake JF, Rogers BN et al (1999) Geophys Res Lett 26(14):2163–2166

Shibata K, Tanuma S (2001a) Earth Planets Space 53:473–482. https://doi.org/10.1186/BF03353258

Shibata K, Tanuma S (2001b) Earth Planets Space 53:473–482. https://doi.org/10.1186/BF03353258

Sironi L (2022) Phys Rev Lett 128(14):145102. https://doi.org/10.1103/PhysRevLett.128.145102

Sironi L, Beloborodov AM (2020) Astrophys J 899(1):52. https://doi.org/10.3847/1538-4357/aba622

Sironi L, Spitkovsky A (2011) Astrophys J 741(1):39. https://doi.org/10.1088/0004-637X/741/1/39

Sironi L, Spitkovsky A (2014) Astrophys J Lett 783(1):L21. https://doi.org/10.1088/2041-8205/783/1/L21

Sironi L, Petropoulou M, Giannios D (2015) Mon Not R Astron Soc 450(1):183–191. https://doi.org/10.1093/mnras/stv641

Sironi L, Giannios D, Petropoulou M (2016) Mon Not R Astron Soc 462(1):48–74. https://doi.org/10.1093/mnras/stw1620

Sonnerup BUO (1979) Magnetic field reconnection. In: Lanzerotti L (ed) Solar System plasma physics. North-Holland, Amsterdam, pp 46

Spitkovsky A (2008) Astrophys J Lett 673(1):L39. https://doi.org/10.1086/527374

Spruit HC (2010) Theory of magnetically powered jets. In: Belloni T (ed) Lecture notes in physics, vol 794. Springer, Berlin, p 233. https://doi.org/10.1007/978-3-540-76937-8_9

Stanier A, Daughton W, Le A et al (2019) Phys Plasmas 26(7):072121. https://doi.org/10.1063/1.5100737

Sugiyama T, Kusano K (2007) J Comput Phys 227(2):1340–1352. https://doi.org/10.1016/j.jcp.2007.09.011

Sun X, Bai XN (2023) Mon Not R Astron Soc 523(3):3328–3347. https://doi.org/10.1093/mnras/stad1548

Sweet PA (1958) Il Nuovo Cimento 8(S2):188–196. https://doi.org/10.1007/BF02962520.

Takahashi HR, Ohsuga K (2013) Astrophys J 772(2):127. https://doi.org/10.1088/0004-637X/772/2/127

Takahashi HR, Kudoh T, Masada Y et al (2011) Astrophys J Lett 739(2):L53. https://doi.org/10.1088/2041-8205/739/2/L53

Takamoto M (2013) Astrophys J 775(1):50. https://doi.org/10.1088/0004-637X/775/1/50

Takamoto M, Inoue T (2011) Astrophys J 735(2):113. https://doi.org/10.1088/0004-637X/735/2/113

Tavani M, Bulgarelli A, Vittorini V et al (2011) Science 331(6018):736. https://doi.org/10.1126/science.1200083

Tenerani A, Velli M, Pucci F et al (2016) J Plasma Phys 82(5):535820501. https://doi.org/10.1017/S002237781600088X

Totorica SR, Zenitani S, Matsukiyo S et al (2023) Astrophys J Lett 952(1):L1. https://doi.org/10.3847/2041-8213/acdb60

Tsallis C (1988) J Stat Phys 52(1–2):479–487. https://doi.org/10.1007/BF01016429

Usami S, Horiuchi R, Ohtani H et al (2013) Phys Plasmas 20(6):061208. https://doi.org/10.1063/1.4811121

Uzdensky DA (2011) Space Sci Rev 160(1–4):45–71. https://doi.org/10.1007/s11214-011-9744-5

Uzdensky DA (2016) Radiative magnetic reconnection in astrophysics. In: Gonzalez W, Parker E (eds) Magnetic reconnection: concepts and applications. Springer, Cham, pp 473. https://doi.org/10.1007/978-3-319-26432-5_12

Uzdensky DA (2022) J Plasma Phys 88(1):905880114. https://doi.org/10.1017/S0022377822000046

Uzdensky DA, Loureiro NF (2016) Phys Rev Lett 116(10):105003. https://doi.org/10.1103/PhysRevLett.116.105003

Uzdensky DA, Spitkovsky A (2014) Astrophys J 780(1):3. https://doi.org/10.1088/0004-637X/780/1/3
Uzdensky DA, Loureiro NF, Schekochihin AA (2010) Phys Rev Lett 105(23):235002. https://doi.org/10.1103/PhysRevLett.105.235002
Uzdensky DA, Cerutti B, Begelman MC (2011) Astrophys J Lett 737(2):L40. https://doi.org/10.1088/2041-8205/737/2/L40
Vasyliunas VM (1968) J Geophys Res 73(9):2839–2884. https://doi.org/10.1029/JA073i009p02839
Watanabe N, Yokoyama T (2006) Astrophys J Lett 647(2):L123–L126. https://doi.org/10.1086/507520
Webb GM (1989) Astrophys J 340:1112. https://doi.org/10.1086/167462
Werner GR, Uzdensky DA, Cerutti B et al (2016) Astrophys J Lett 816(1):L8. https://doi.org/10.3847/2041-8205/816/1/L8
Werner GR, Uzdensky DA, Begelman MC et al (2018) Mon Not R Astron Soc 473(4):4840–4861. https://doi.org/10.1093/mnras/stx2530
Werner GR, Philippov AA, Uzdensky DA (2019) Mon Not R Astron Soc 482(1):L60–L64. https://doi.org/10.1093/mnrasl/sly157
Yang L, Li H, Guo F et al (2020) Astrophys J Lett 901(2):L22. https://doi.org/10.3847/2041-8213/abb76b
Yang H, Yuan F, Li H et al (2022) Science Advances 10(12):eadn3544. https://doi.org/10.1126/sciadv.adn3544
Zank GP (2014) Transport processes in space physics and astrophysics. Lecture Notes in Physics, vol 877. Springer, Berlin. https://doi.org/10.1007/978-1-4614-8480-6
Zanotti O, Dumbser M (2011) Mon Not R Astron Soc 418(2):1004–1011. https://doi.org/10.1111/j.1365-2966.2011.19551.x
Zenitani S (2018) Plasma Phys Control Fusion 60(1):014028. https://doi.org/10.1088/1361-6587/aa8f17
Zenitani S, Hoshino M (2001) Astrophys J Lett 562(1):L63–L66. https://doi.org/10.1086/337972
Zenitani S, Hoshino M (2005a) Astrophys J Lett 618(2):L111–L114. https://doi.org/10.1086/427873
Zenitani S, Hoshino M (2005b) Phys Rev Lett 95(9):095001. https://doi.org/10.1103/PhysRevLett.95.095001
Zenitani S, Hoshino M (2007) Astrophys J 670(1):702–726. https://doi.org/10.1086/522226
Zenitani S, Hoshino M (2008) Astrophys J 677(1):530–544. https://doi.org/10.1086/528708
Zenitani S, Hesse M, Klimas A (2009a) Astrophys J 696(2):1385–1401. https://doi.org/10.1088/0004-637X/696/2/1385
Zenitani S, Hesse M, Klimas A (2009b) Astrophys J 705(1):907–913. https://doi.org/10.1088/0004-637X/705/1/907
Zenitani S, Hesse M, Klimas A (2010) Astrophys J Lett 716(2):L214–L218. https://doi.org/10.1088/2041-8205/716/2/L214
Zhang H, Giannios D (2021) Mon Not R Astron Soc 502(1):1145–1157. https://doi.org/10.1093/mnras/stab008
Zhang B, Yan H (2011) Astrophys J 726(2):90. https://doi.org/10.1088/0004-637X/726/2/90
Zhang H, Chen X, Böttcher M et al (2015) Astrophys J 804(1):58. https://doi.org/10.1088/0004-637X/804/1/58
Zhang H, Li H, Guo F et al (2017) Astrophys J 835(2):125. https://doi.org/10.3847/1538-4357/835/2/125
Zhang H, Li X, Guo F et al (2018) Astrophys J Lett 862(2):L25. https://doi.org/10.3847/2041-8213/aad54f
Zhang H, Li X, Giannios D et al (2020) Astrophys J 901(2):149. https://doi.org/10.3847/1538-4357/abb1b0
Zhang Q, Guo F, Daughton W et al (2021) Phys Rev Lett 127(18):185101
Zhang H, Li X, Giannios D et al (2021a) Astrophys J 912(2):129. https://doi.org/10.3847/1538-4357/abf2be
Zhang H, Sironi L, Giannios D (2021b) Astrophys J 922(2):261. https://doi.org/10.3847/1538-4357/ac2e08
Zhang H, Li X, Giannios D et al (2022) Astrophys J 924(2):90. https://doi.org/10.3847/1538-4357/ac3669
Zhang H, Sironi L, Giannios D et al (2023) Astrophys J Lett 956:L36. https://doi.org/10.3847/2041-8213/acfe7c
Zhang Q, Guo F, Daughton W et al (2024) Phys Rev Lett 132(11):115201. https://doi.org/10.1103/PhysRevLett.132.115201
Zrake J (2016) Astrophys J 823(1):39. https://doi.org/10.3847/0004-637X/823/1/39
Zweibel EG, Yamada M (2009) Annu Rev Astron Astrophys 47(1):291–332. https://doi.org/10.1146/annurev-astro-082708-101726

Space Science Reviews (2023) 219:76
https://doi.org/10.1007/s11214-023-01024-3

Laboratory Study of Collisionless Magnetic Reconnection

H. Ji[1,2] · J. Yoo[2] · W. Fox[2] · M. Yamada[2] · M. Argall[3] · J. Egedal[4] · Y.-H. Liu[5] · R. Wilder[6] · S. Eriksson[7] · W. Daughton[8] · K. Bergstedt[1] · S. Bose[2] · J. Burch[9] · R. Torbert[3] · J. Ng[10,11,2] · L.-J. Chen[11]

Received: 15 June 2023 / Accepted: 3 November 2023 / Published online: 15 November 2023

Abstract
A concise review is given on the past two decades' results from laboratory experiments on collisionless magnetic reconnection in direct relation with space measurements, especially by the Magnetospheric Multiscale (MMS) mission. Highlights include spatial structures of electromagnetic fields in ion and electron diffusion regions as a function of upstream symmetry and guide field strength, energy conversion and partitioning from magnetic field to ions and electrons including particle acceleration, electrostatic and electromagnetic kinetic plasma waves with various wavelengths, and plasmoid-mediated multiscale reconnection. Combined with the progress in theoretical, numerical, and observational studies, the physics foundation of fast reconnection in collisionless plasmas has been largely established, at least within the parameter ranges and spatial scales that were studied. Immediate and long-term future opportunities based on multiscale experiments and space missions supported by exascale computation are discussed, including dissipation by kinetic plasma waves, particle heating and acceleration, and multiscale physics across fluid and kinetic scales.

Keywords Magnetic reconnection · Laboratory experiment · Magnetospheric MultiScale

1 Introduction

The history of laboratory studies of magnetic reconnection goes back to 1960s (e.g. Bratenahl and Yeates 1970), not long after the development of early theoretical models (Sweet 1958; Parker 1957; Dungey 1961; Petschek 1964). As briefly reviewed by Yamada et al. (2010), these early experiments were motivated by solar flares, and were carried out in a collision-dominated MHD regime at low Lundquist numbers ($S < 10$). The subsequent landmark experiments performed by Stenzel and Gekelman (1979) were also at low Lundquist numbers ($S < 10$), but in the electron-only reconnection regime where ions are unmagnetized even with a strong guide field. While these experiments provided insights into the rich physics of magnetic reconnection in relatively collisional regimes, they are not directly relevant to collisionless reconnection in space, which is the focus of this book; therefore, they are not included in this short review paper except in a few relevant places.

Modern reconnection experiments started with merging magnetized plasmas (Yamada et al. 1990; Ono et al. 1993; Brown 1999) using technologies developed during nuclear

Extended author information available on the last page of the article

Table 1 A non-exhaustive list of relevant experiments on collisionless reconnection

Facility / location	Main features / topics	Main or relevant references
Linear device / UCLA	electron-only / waves, non-thermal electrons, plasmoids	Stenzel and Gekelman (1979), Gekelman and Stenzel (1984, 1985), Stenzel et al. (1986)
TS-3/4 / U. Tokyo	toroidal plasma merging / heating, plasmoids	Yamada et al. (1990), Ono et al. (1993, 2011)
MRX / Princeton	axisymmetric current sheet, toroidal plasma merging / reconnection rate, structure, heating, waves, 3D, plasmoids	Yamada et al. (1997, 2006, 2014, 2018), Ji et al. (1998, 2004, 2005, 2008), Hsu et al. (2000), Carter et al. (2001), Ren et al. (2005, 2008), Kulsrud et al. (2005), Tharp et al. (2012), Lawrence et al. (2013), Dorfman et al. (2013, 2014), Yoo et al. (2013, 2014b, 2018, 2023), Jara-Almonte et al. (2016), Fox et al. (2017, 2018), Bose et al. (2023)
SSX / Swarthmore	toroidal plasma merging / heating	Brown (1999), Brown et al. (2002, 2006)
VTF / MIT	axisymmetric current sheet, strong guide field / structure, heating, waves, onset	Egedal et al. (2000, 2003), Egedal and Fasoli (2001), Stark et al. (2005), Katz et al. (2010), Fox et al. (2008, 2010, 2012)
RSX / Los Alamos National Lab	linear plasma merging / onset, 3D	Intrator et al. (2009)
RWX / U. Wisconsin	liner geometry / onset	Bergerson et al. (2006)
TREX / U. Wisconsin	axisymmetric current sheet / structure, plasmoids	Olson et al. (2016, 2021), Greess et al. (2021)
MAGPIE / Imperial	Z-pinch / heating, plasmoids	Hare et al. (2017)
PHASMA / West Virginia U.	linear plasma merging, electron-only / heating	Shi et al. (2022)
Capacitor coil powered by laser / U. Rochester	current sheet / electron acceleration, waves	Chien et al. (2023), Zhang et al. (2023)
FLARE / Princeton	axisymmetric current sheet / multiscale	Ji et al. (2018, 2022)

fusion research. These were followed by driven reconnection experiments in an axisymmetric geometry: Magnetic Reconnection Experiment or MRX (Yamada et al. 1997), Versatile Toroidal Facility or VTF (Egedal et al. 2000), and Terrestrial Reconnection Experiment or TREX (Olson et al. 2016); and in a linear geometry: Rotating Wall Experiment (RWX) (Bergerson et al. 2006), Reconnection Scaling Experiment (RSX) (Intrator et al. 2009), and the more recent Phase Space Mapping experiment (PHASMA) (Shi et al. 2022). There exist also experiments using Z-pinches (Hare et al. 2017) and lasers (e.g. Chien et al. 2023) in relevant conditions. A non-exhaustive list of relevant experiments to this paper are listed in Table 1, including the upcoming Facility for Laboratory Reconnection Experiments or FLARE (Ji et al. 2018, 2022). Many of these experiments were able to reach higher Lundquist number, up to $S \sim 10^3$, with magnetized ions. As a result, plasma conditions local to the reconnecting current sheets in these experiments were nearly collisionless, motivating quantitative comparisons with *in-situ* measurements by spacecraft in near-Earth space as well as predictions by Particle-In-Cell (PIC) kinetic simulations.

The topics on magnetic reconnection for such comparative research include kinetic structures of diffusion regions, energy conversion from magnetic field to plasma, various plasma wave activities, and multiscale reconnection via plasmoid instability of reconnecting current sheets. This paper concisely reviews results from these comparative research activities and highlights several recent achievements, especially in relation to the Magnetospheric Multiscale (MMS) mission. Summary of magnetic reconnection research in a broader scope can be found in review papers by Zweibel and Yamada (2009) and Yamada et al. (2010), as well as in more recent reviews (Yamada 2022; Ji et al. 2022). The latter review paper especially focuses on the future development of magnetic reconnection research by emphasizing its multiscale nature.

The rest of this review is organized into the following sections: kinetic structures of reconnecting diffusion regions in Sect. 2 including both the ion and electron diffusion regions (IDR and EDR), reconnection energetics in Sect. 3, plasma waves in Sect. 4, plasmoids during reconnection in Sect. 5, and the future outlook in Sect. 6.

2 Kinetic Structures of Diffusion Regions

It is interesting that detailed studies of magnetic reconnection based on *in-situ* measurements in modern experiments (e.g. Yamada et al. 1997) and in space (e.g. Fujimoto et al. 1996; Øieroset et al. 1997) began nearly contemporaneously with detecting kinetic structures of diffusion regions near the X-line, as the research focus was the origin of fast reconnection in collisionless plasmas. The origin of kinetic structures which support the reconnection electric field in collisionless plasmas can be understood via the generalized Ohm's law,

$$\boldsymbol{E} + \boldsymbol{V} \times \boldsymbol{B} = \eta_s \boldsymbol{j} + \frac{\boldsymbol{j} \times \boldsymbol{B}}{en} - \frac{\nabla p_e}{en} - \frac{\nabla \cdot \boldsymbol{\Pi}_e}{en} - \frac{m_e}{e}\frac{d\boldsymbol{V}_e}{dt}, \tag{1}$$

where $\boldsymbol{E}$, $\boldsymbol{V}$, $\boldsymbol{B}$, and $\boldsymbol{j}$ are electric field, velocity, magnetic field, and current density, respectively, and η_s is the Spitzer resistivity. n, $\boldsymbol{V}_e$, m_e, and e are the electron density, fluid velocity, mass, and charge, respectively. The full electron pressure tensor is expressed as a sum of a diagonal isotropic pressure tensor and a stress tensor which includes an off-diagonal pressure tensor: $\boldsymbol{P}_e \equiv p_e \boldsymbol{I} + \boldsymbol{\Pi}_e$ where $\boldsymbol{I}$ is the unit tensor. The RHS of Eq. (1) represents non-ideal-MHD electric field in diffusion regions where $\boldsymbol{V} \times \boldsymbol{B}$ diminishes while $\boldsymbol{E}$ remains large for fast reconnection. Each of these non-ideal-MHD terms is associated with a spatial structure in steady state on the corresponding scale in electromagnetic field or electron quantities.

In collisional MHD plasmas, the only non-ideal electric field is due to collisional resistivity, $\eta_s \boldsymbol{j}$, while ions and electrons are closely coupled to behave as a single fluid, moving at the MHD fluid velocity $\boldsymbol{V}$. In contrast, collisional resistivity is negligible in collisionless plasmas where a non-ideal-MHD electric field must come from other terms on the RHS of Eq. (1). In such plasmas, ions and electrons decouple from each other as they approach the current sheet. Ions get demagnetized in a larger ion diffusion region (IDR) while electrons get demagnetized closer to the X-line in a smaller electron diffusion region (EDR), see Fig. 1. In general, the second and third terms on the RHS of Eq. (1), $\boldsymbol{j} \times \boldsymbol{B}/en - \nabla p_e/en$, are responsible for non-ideal-MHD electric field in the IDR depending on the guide field strength, while the last two terms are responsible for non-ideal electric field in the EDR. Below we review the laboratory studies of kinetic structures in both the IDR and the EDR, in comparison with space measurements and numerical simulations, with or without a guide field, as well as with and without symmetries between the two upstream reconnection regions.

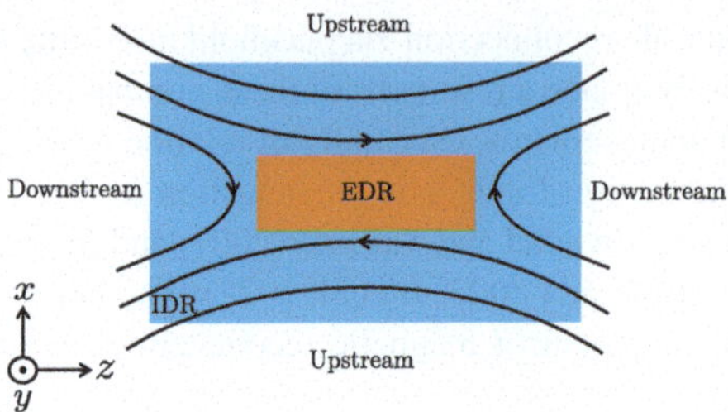

Fig. 1 Schematics of magnetic reconnection geometry and coordinates. Plasma flows in from upstream with oppositely directed magnetic field components towards the ion diffusion region (IDR), where ions become demagnetized, before reaching the electron diffusion region (EDR), where electrons become demagnetized and reconnection occurs. The reconnected plasma flows downstream. Unless stated explicitly, the reconnection plane (z, x) is defined so that z is along the reconnecting field component and x is the direction across the current sheet. y completes a right-handed coordinate system. In most of the laboratory experiments summarized here, x is along the radial direction R

2.1 IDR Structures Without a Guide Field

When the guide field is negligible, the reconnection electric field E_y is perpendicular to the magnetic field, which is mostly within the reconnection plane of (z, x). A natural candidate to balance the required non-ideal-MHD electric field perpendicular to the local magnetic field is the second term on the RHS of Eq. (1), $\boldsymbol{j} \times \boldsymbol{B}/en$, which is often called the Hall term. The Hall term originates from the differences in the in-plane ion and electron motions as expected in the IDR. Since such motions preserve symmetry between both upstreams and also both downstreams (unless distant asymmetries are imposed; see below), a quadrupolar structure in the out-of-the-plane (Hall) magnetic field component B_y on the scale of the ion skin depth has been predicted theoretically (Sonnerup 1979; Terasawa 1983) and numerically (Birn et al. 2001, and references therein).

In addition to the inductive reconnection electric field in the out-of-the-plane direction, E_y, there may exist an in-plane electric field $\boldsymbol{E}_{\text{in-plane}}$. At the outer scales (regions outside of the IDR) where ideal MHD applies, the RHS of Eq. (1) vanishes, resulting in $\boldsymbol{E}_{\text{in-plane}} = -(\boldsymbol{V} \times \boldsymbol{B})_{\text{in-plane}}$. Without a guide field, $E_{\text{in-plane}} = -V_y B$, which vanishes unless there exists a significant out-of-the-plane flow, V_y.

However, a significant $\boldsymbol{E}_{\text{in-plane}}$, called the Hall electric field, arises even without an ion flow V_y in the IDR. This is because in the IDR, but outside of the EDR, only ion dynamics are dissipative (see Sect. 3.2 on the effects on energy dissipation) and electron dynamics are ideal. Therefore, $\boldsymbol{E} \approx -\boldsymbol{V}_e \times \boldsymbol{B}$ and $E_{\text{in-plane}} \approx -V_{ey} B \approx j_y B/en$. It also follows that $\boldsymbol{E} \cdot \boldsymbol{B} \approx 0$, and without a guide field, $\boldsymbol{E}_{\text{in-plane}} \cdot \boldsymbol{B} \approx 0$. In other words, $\boldsymbol{E}_{\text{in-plane}}$ is perpendicular to the local magnetic field everywhere, which by symmetry must have a quadrupolar structure around X-line, consistent with numerical predictions (e.g. Shay et al. 1998). By the virtue of Faraday's Law in quasi-steady state ($\partial B_y/\partial t \approx 0$), $\boldsymbol{E}_{\text{in-plane}}$ is curl-free and can be well represented by an electrostatic potential, $\boldsymbol{E}_{\text{in-plane}} \approx -\nabla\phi$. Therefore ϕ must have a saddle-type quadrupolar structure determined by the significant out-of-the-plane j_y in the IDR.

The presence of both ϕ and B_y in the IDR enables fast reconnection by diverting a significant amount of incoming magnetic energy directly downstream in the outflow direction via the Poynting vector $E_x B_y/\mu_0$, without having to pass through the X-line. Note here that the electric field normal to the current sheet E_x is part of $\boldsymbol{E}_{\text{in-plane}}$ and peaks along the separatrix with a width on electron scales in the EDR and extending to the ion scales further downstream (Chen et al. 2008). Over time, the depleted total pressure at the X-line pulls in more upstream magnetic pressure leading to the open-outflow geometry necessary for fast reconnection (Liu et al. 2022). This magnetic structure of the open-outflow geometry is consistent with the earlier physics explanation of fast reconnection based on whistler dynamics which involves only electrons in the IDR (e.g. Shay and Drake 1998). The prediction of both Hall magnetic and electric fields motivated an intensive search for such field structures as first evidence of fast collisionless reconnection.

2.1.1 Symmetric Anti-Parallel Reconnection

A textbook example measurement of the Hall magnetic and electric structures was by the Polar spacecraft (Mozer et al. 2002), where a bipolar signature for both B_y and E_x was detected as the spacecraft traversed across a current sheet in one of the outflows of a rare event of symmetric, anti-parallel reconnection in Earth's magnetopause. Later, with the multiple spacecraft of Cluster, 2D structures of Hall magnetic and electric fields were mapped statistically around the X-line in Earth's magnetotail (Eastwood et al. 2010).

Aiming to go beyond the 1D measurements by spacecraft, an effort was made in laboratory experiments to directly capture instantaneous 2D quadrupolar structures in B_y during anti-parallel reconnection. Figure 2(a) and (b) show the first such measurements from Magnetic Reconnection eXperiment or MRX (Ren et al. 2005) and Swarthmore Spheromak eXperiment or SSX (Brown et al. 2006), respectively. Furthermore, quantitative comparisons were made between MRX and 2D PIC simulations using corresponding parameters, showing excellent agreements on ion scales (Ji et al. 2008), see Fig. 2(c). Since ions control the overall reconnection rate in collisionless reconnection (Biskamp et al. 1995; Hesse et al. 1999), the convergence on the ion-scale kinetic structures between numerical prediction, laboratory experiment and space measurement essentially validated the concept of collisionless fast reconnection. In addition, since collisionality can be actively controlled in the laboratory, continuous transition has been demonstrated from slow Sweet-Parker collisional reconnection (Ji et al. 1998) without a significant B_y structure to fast collisionless reconnection with a significant B_y structure (Yamada et al. 2006). The Hall electric potential ϕ was also simultaneously measured by multiple spacecraft in the magnetotail on the ion scale at downstream (Wygant et al. 2005), and on the electron scale across the current sheet (Chen et al. 2008). The structure is consistent with the 2D measurements in MRX where half of the saddle-type quadrupolar potential structure is shown, see Fig. 3.

2.1.2 Asymmetric Anti-Parallel Reconnection

Magnetic reconnection in nature often occurs with significant differences in the density, temperature, and magnetic field strength across the current sheet. A best example of this asymmetric reconnection is reconnection at the magnetopause (Mozer and Pritchett 2011), where the density ratio across the current sheet ranges from 10–100 and a magnetic field strength ratio of 2–4. The asymmetry is expected to significantly alter the structure of the diffusion regions as well as scaling of the reconnection process (e.g. Cassak and Shay 2007).

In the laboratory, reconnection with a strong density asymmetry across the current sheet has been extensively studied and compared to space observations at the subsolar magnetopause (Yoo et al. 2014b, 2017; Yamada et al. 2018). The ratio of the two upstream densities ranges from 5 to 10. It has been shown that strong density asymmetry alters the electric and magnetic field structures in the diffusion regions. In the IDR, the uniform reconnection electric field E_y is approximately balanced by the Hall term $\boldsymbol{j}_{\text{in-plane}} \times \boldsymbol{B}/en$ on both upstreams. The asymmetry in density has to be compensated by an asymmetry in $\boldsymbol{j}_{\text{in-plane}}$ since the in-plane magnetic field components are similar, while the pressure balance is maintained by temperature asymmetry. The much larger $\boldsymbol{j}_{\text{in-plane}}$ significantly enlarges B_y on the

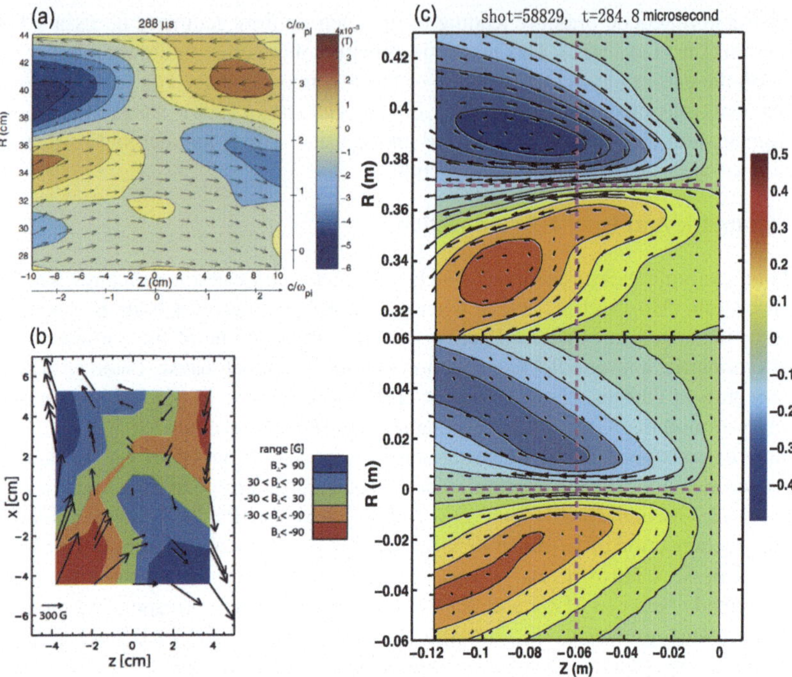

Fig. 2 Measured instantaneous quadrupolar structure of the out-of-the-plane magnetic field component during anti-parallel collisionless reconnection. (a) data from Magnetic Reconnection eXperiment or MRX (Ren et al. 2005); (b) data from Swarthmore Spheromak eXperiment or SSX (Brown et al. 2006); (c) comparison between MRX data (top panel) and 2D PIC simulation using corresponding parameters (bottom panel) in one half of the reconnection plane showing excellent agreements on ion scales (Ji et al. 2008). Arrows indicate electron flow velocity

higher density side so that the quadrupolar structure becomes almost bipolar, as shown in Fig. 4(a) and (b) (Yoo et al. 2014b). In contrast, the in-plane electric field is much larger on the low density side since $\boldsymbol{E}_{\text{in-plane}} \approx j_y B/en$ where j_y and B are similar between the two upstreams. As a result, the in-plane bipolar electrostatic field becomes almost unipolar (Yoo et al. 2017). All these features agree with space observations (e.g. Mozer and Pritchett 2011; Burch et al. 2016). Figure 5 shows excellent agreements between MRX and example MMS measurements at Earth's magnetopause on profiles of magnetic field components, density, ion outflow, and in-plane electric field.

Strong density asymmetry also causes a shift of the electron and ion inflow stagnation points (Yoo et al. 2014b; Yamada et al. 2018). The ion inflow stagnation point is the location where the in-plane ion flow velocity vanishes. As shown in Fig. 4(c) and (d), the ion inflow stagnation point is shifted to the low-density side by about 3 cm ($\sim$0.5 d_i; d_i is the ion skin depth) for the asymmetric case, while it is very close to the X-point for the symmetric case.

The electron inflow stagnation point is also shifted to the low-density side, as shown in the Fig. 6. The stagnation point denoted by the black dot is shifted by about 1 cm, which is about 0.15 d_i. These shifts are caused by the imbalance in the electron and ion inflows due to the density asymmetry. This overshooting of electrons from the magnetosheath (high-density) side is consistent with the well-known crescent-shape electron distribution function near the stagnation point (Hesse et al. 2014), which is observed by MMS (Burch et al. 2016).

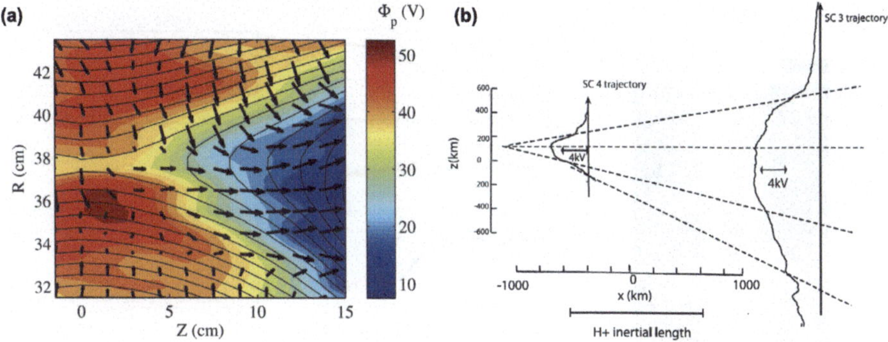

Fig. 3 (a) Measured 2D Hall electric potential and ion in-plane flow in MRX during anti-parallel reconnection where half of the saddle-type quadrupolar structure is shown. Adapted from Yamada et al. (2015). (b) Measured Hall electric potential by two Cluster spacecraft during a magnetotail reconnection event, consistent with the expectation that the potential is deeper and wider further from the X-line. Here X is along the reconnecting field direction while Z is the direction across current sheet. Adapted from Wygant et al. (2005)

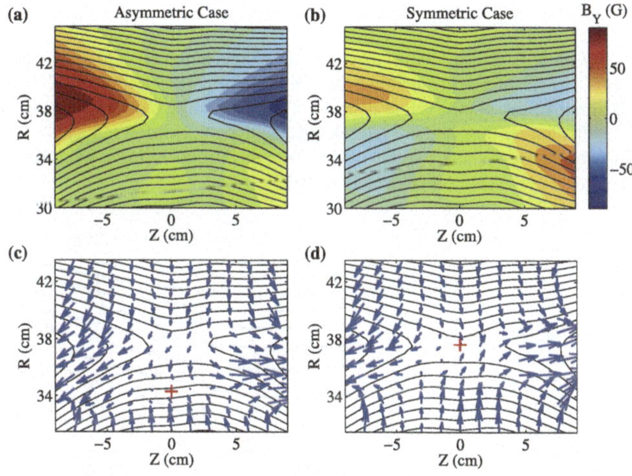

Fig. 4 2-D profiles of the out-of-plane magnetic field (B_y) with contours of the poloidal flux for asymmetric (a) and symmetric (b) cases. Compared to the symmetric case, the quadrupole magnetic field component is enhanced on the high-density side ($R > 37.5$ cm) and suppressed on the low-density side ($R < 37.5$ cm). Black lines indicate contours of the poloidal magnetic flux, which represent magnetic field lines. In-plane ion flow vector profiles for asymmetric (c) and symmetric (d) cases. For the asymmetric case, the ion inflow stagnation point is shifted to the low-density side. The upstream density ratio (n_1/n_2) for the asymmetric case is about 6, while it is about 1.2 for the symmetric case. Figure from Yoo et al. (2014b)

The TREX experiment also explored asymmetric anti-parallel reconnection with the plasma density at large radii inflow being suppressed by a factor of about 4. Numerically, the TREX configuration was implemented in the cylindrical version of the VPIC code (Bowers et al. 2009), where properly scaled current sources increasing over time were added at the drive coil locations. Initial density and magnetic field profiles were set at the simulation based on experimental data. As shown in Fig. 7, magnetic field and current structures similar to those of MRX are observed, and reproduced with remarkable agreement through matching numerical simulations (Olson et al. 2021; Greess et al. 2021).

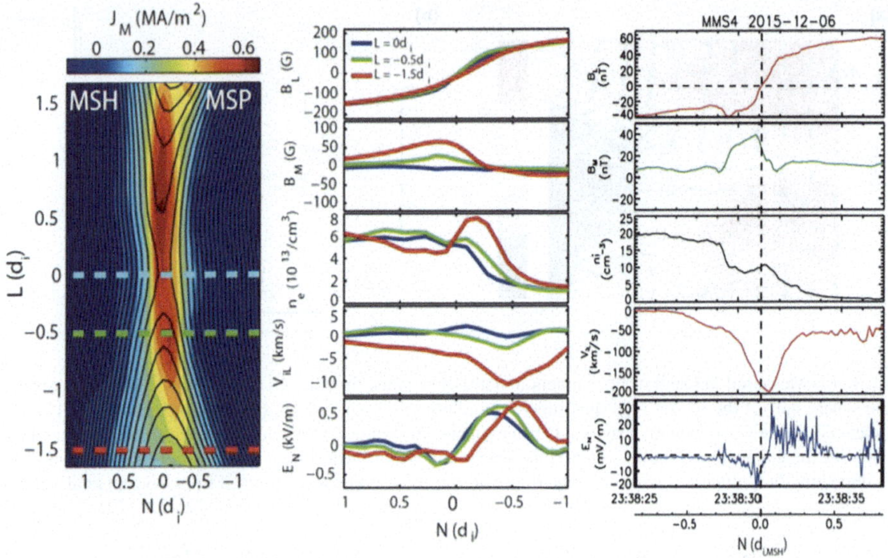

Fig. 5 Comparisons of various profiles across asymmetric reconnection current sheets between MRX and MMS. Here the LMN coordinates correspond to the ZYX coordinates. (left panel) 2-D profiles of reconnecting field lines and out-of-the-plane current density in MRX. (middle panel) Cross-current-sheet profiles of magnetic field, density, ion outflow and in-plane electric field at three different locations marked in the left panel. (right panel) Cross-current-sheet profiles of the same quantities during a magnetopause asymmetric reconnection event observed by MMS on December 6, 2015

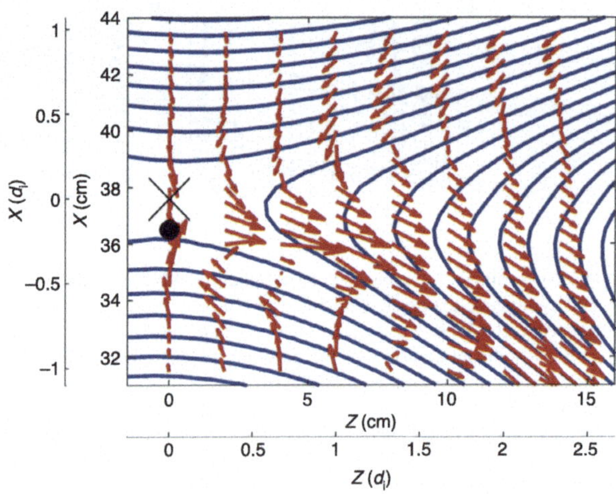

Fig. 6 Electron dynamics observed during asymmetric reconnection in MRX. In the reconnection plane, electrons flow together with reconnecting field lines. The X marker at $(R, Z) = (37.6, 0)$ is the X-line and the black circle denotes the stagnation point of in-plane electron flow. Figure from Yamada et al. (2018)

2.2 IDR Structures with a Guide Field

Anti-parallel reconnection is a rather special magnetic geometry in nature, whereas reconnection occurs often with a finite guide field B_g. With the addition of B_g, the reconnecting field lines meet at an angle less than 180°, and a sufficiently strong guide field modifies the reconnection process by magnetizing the electrons and ions in the layer. The characteris-

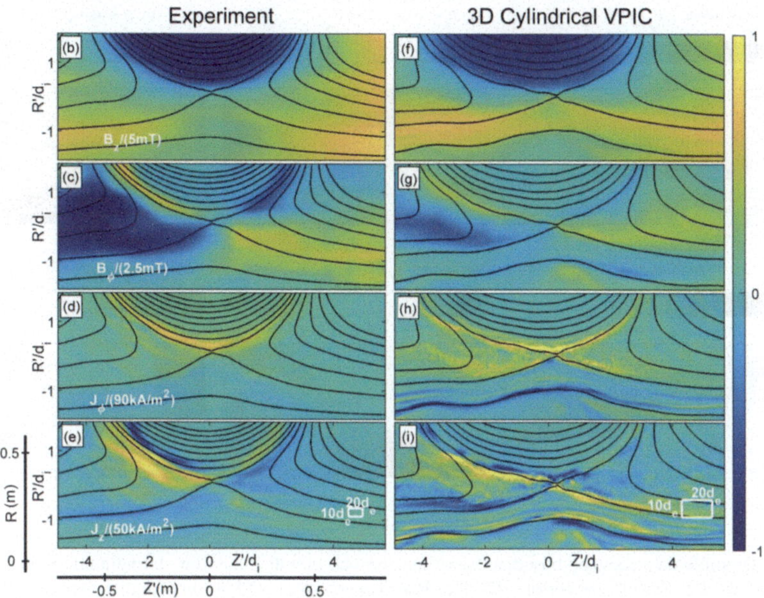

Fig. 7 (Panel b-e) Magnetic field and current components recorded in TREX during reconnection. (Panel f-i) Matched 3D kinetic simulation results reproducing the experimental results. After Greess et al. (2021)

tic kinetic scale across the collisionless current sheet transitions from ion skin depth to ion sound Larmor radius (ρ_s) as B_g increases.

A finite B_g also introduces an in-plane electric field structure at the outer ideal scales even without a significant V_y. This is because in this case $E_{\text{in-plane}} = V_{\text{in-plane}} B_g$ where $V_{\text{in-plane}}$ is the in-plane flow due to reconnection. This $E_{\text{in-plane}}$ is required to satisfy the ideal MHD condition $\boldsymbol{E} \cdot \boldsymbol{B} = E_y B_g + \boldsymbol{E}_{\text{in-plane}} \cdot \boldsymbol{B} = 0$ as the reconnection electric field E_y now has a parallel component which can extend over a large area. At upstream where the *reconnecting* component B_z dominates over the *reconnected* component B_x, $E_z \approx -E_y(B_y/B_z)$ can even dominate the reconnection electric field E_y under strong-guide field conditions. Correspondingly, in the downstream where B_z is small, $E_x \approx -E_y(B_y/B_x)$. As before, under quasi-steady conditions ($\partial B_y/\partial t \approx 0$) the in-plane electric field is well represented by a quadrupolar potential structure, $\boldsymbol{E}_{\text{in-plane}} = -\nabla\phi$. This potential structure, in turn, drives $\boldsymbol{E} \times \boldsymbol{B}$ drift for both electrons and ions to support the required in-plane, incompressible reconnection flow $\boldsymbol{V}_{\text{in-plane}}$. This quadrupolar potential structure on the outer scales was observed in the VTF (Egedal and Fasoli 2001; Egedal et al. 2003) with a strong guide field and shown to balance the global reconnection electric field in the upstream, as well as interact with global MHD modes that drive reconnection (Katz et al. 2010). However, this quadrupolar potential structure on the outer ideal scales has not been reported by space measurements.

This quadrupolar potential structure persists from the outer ideal scales to the IDR with a characteristic scale of ρ_s during guide field reconnection. When approaching ρ_s scale, in addition to the *incompressible* $\boldsymbol{u}_E = \boldsymbol{E} \times \boldsymbol{B}/B^2$ drift, the in-plane ion polarization drift, $\boldsymbol{u}_p = (m_i/eB^2)(\boldsymbol{u}_E \cdot \nabla)\boldsymbol{E}_{\text{in-plane}}$, becomes increasingly important. Here m_i is ion mass. This cross-field ion polarization drift is *compressible*, and it can generate density variation with electrons moving along the field line to satisfy quasineutrality (Kleva et al. 1995). Combined with the continuity equation, $(\boldsymbol{u}_E \cdot \nabla)n + n\nabla \cdot \boldsymbol{u}_p = 0$, the predicted density variation

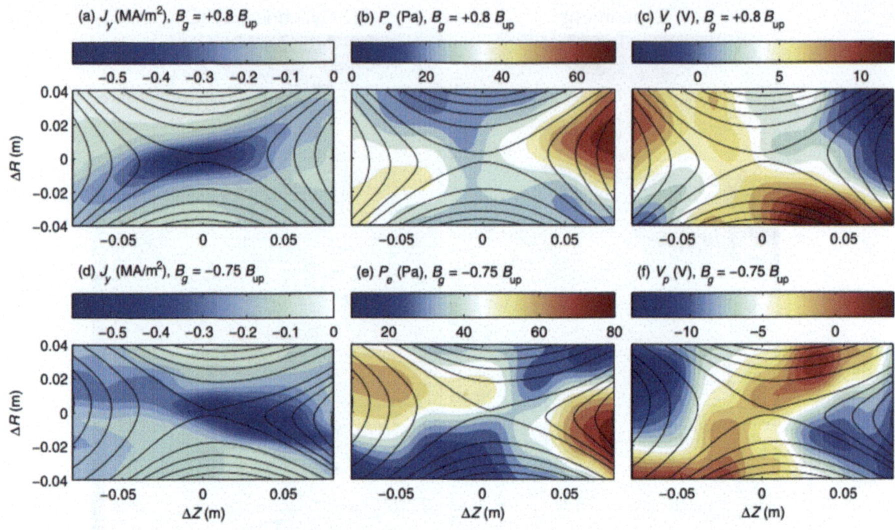

Fig. 8 2-D profile data showing observations of quadrupolar pressure variation during guide field magnetic reconnection. (a,d) Plasma current profile; (b,e) Plasma pressure; (c,f) Plasma potential. Between (a-c) and (d-f) the sign of the guide field was reversed, leading to a change in the orientation of the quadrupolar profiles. After Fox et al. (2017)

obeys $\ln(n/n_0) = (m_i/eB^2)\nabla^2\phi$ with a quadrupolar structure. This density structure develops large electron pressure variations along the field lines until the third term on the RHS of Eq. (1) becomes important so that

$$E_\parallel = -\frac{\nabla_\parallel p_e}{en} \approx -\rho_s^2 \nabla_\parallel \nabla^2 \phi \tag{2}$$

to reach a steady state. Since we also have $E_\parallel = -\nabla_\parallel \phi$, Eq. (2) implies that the spatial scale of ϕ variation is on the order of ρ_s, the characteristic scale of the IDR with a guide field. The quadrupolar density structure has been directly measured on MRX during guide field reconnection as shown in Fig. 8. Such a structure was originally predicted from two-fluid extended MHD simulations (Aydemir 1992; Kleva et al. 1995). Øieroset et al. (2016) have measured a plasma density variation consistent with such a quadrupolar structure during a current sheet crossing by MMS. The correspondence was observed in a symmetric guide-field reconnection event, and inferred through comparison with simulations. The crossing of the current sheet was sufficiently downstream that only a bipolar variation (half a quadrupole) was observed.

2.3 EDR Structures

The last two terms in Eq. (1) are responsible in collisionless plasmas for magnetic field dissipation within the electron diffusion region or EDR, where electrons are demagnetized typically on the order of electron skin depth (d_e) or gyro radius (ρ_e). The EDR is the location where magnetic field lines are finally reconnected from upstream to downstream. In particular, the importance of off-diagonal terms in the electron pressure tensor in the EDR has been predicted theoretically (Vasyliunas 1975; Lyons and Pridmore-Brown 1990), demonstrated

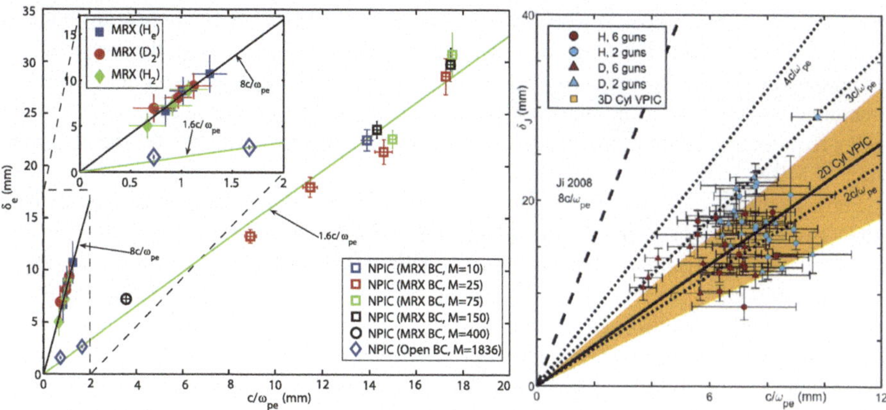

Fig. 9 (a) Measured half width of the EDR on MRX compared with 2D PIC simulations in Cartesian geometry (Ji et al. 2008) (b) measured half width of the EDR on TREX compared with 2D (solid line) and 3D (orange region) PIC simulations in cylinderical geometry (Greess et al. 2021)

numerically (Cai and Lee 1997; Hesse et al. 1999; Pritchett 2001), and explained physically (Kulsrud et al. 2005). Unmagnetized electrons with an in-plane thermal speed v_x or v_z are subject to free acceleration by the reconnection electric field E_y, generating a large off-diagonal pressure P_{xy} or P_{zy}, respectively, during their transit time in EDR. This manifests as spatial derivatives in the y component of $\nabla \cdot \boldsymbol{\Pi}_e$ in Eq. (1). The competing alternative to this dissipation mechanism is the so-called anomalous resistivity based on 3D kinetic instabilities (Papadopoulos 1977, and references therein), which has been used numerically to reproduce the Petschek solution of fast reconnection (Ugai and Tsuda 1977; Sato and Hayashi 1979) since the early phase of reconnection research. There has been evidence from the MMS measurements for the laminar off-diagonal pressure tensor effect (Torbert et al. 2018; Egedal et al. 2018, 2019) and also for the possible importance of anomalous resistivity or 3D effects (Torbert et al. 2016; Ergun et al. 2017; Cozzani et al. 2021).

The EDR has been also identified in anti-parallel reconnection on MRX (Ren et al. 2008) as outgoing electron jets between two quadrants in the B_y structure shown in Fig. 2(c). The importance of the off-diagonal pressure tensor in the EDR is closely related to the magnitude and width of such electron jets (Hesse et al. 1999). Compared with 2D PIC simulations in Cartesian geometry, however, the electron jet speed is much slower and the layer half width is 3-5 times thicker (Ji et al. 2008), as shown in Fig. 9(a). This discrepancy persisted even after incorporating finite collisions (Roytershteyn et al. 2010) and 3D effects via Lower Hybrid Drift Waves (LHDW, see later) (Roytershteyn et al. 2013) in the simulations when averaged over the y direction. In contrast, the EDR has been recently studied on TREX and their measured half width agrees well with the predictions by 2D PIC simulations in cylindrical geometry (Greess et al. 2021), shown in Fig. 9(b). 3D effects via LHDW can distort the EDR in the out-of-the-plane direction, weakly broadening the numerical directions of the EDR width [orange region in Fig. 9(b)], but the off-diagonal pressure tensor effect remains dominant at each location.

In addition to the differences in simulation geometries, there are several possibilities to resolve these different results. First, the anti-parallel reconnection in this comparison was driven symmetrically on MRX (Fig. 2) but asymmetrically on TREX (Fig. 7). It is unclear whether symmetry plays a role in determining EDR thickness. Second, the colder ion temperature $T_i \ll T_e$ at TREX may favor triggering LHDW which can distort the EDR (Roytershteyn et al. 2012), compared with MRX where $T_i \sim T_e$. Third, there are also differences in measuring the EDR: the "jogging" method in which the EDR is rapidly swept over a 1D probe array in TREX may have higher effective spatial resolutions, but requires that the structures remain in the same shape as confirmed experimentally (Olson et al. 2021), while such a requirement is not needed but the spatial resolution is less effective for the 2D probe array on MRX.

Furthermore, if there is sufficient scale separation between the electron skin depth (d_e) and Debye length (λ_D) during anti-parallel reconnection, $d_e/\lambda_D = c/v_{\mathrm{th,e}} > 30$, the counter-streaming electron beams in the unmagnetized EDR are unstable to streaming instabilities (Jara-Almonte et al. 2014), possibly leading to efficient dissipation broadening the EDR. Interestingly, this condition is equivalent to $T_e < 570$ eV which is generally satisfied in space, solar and laboratory plasmas, except in Earth's magnetotail and also in the typical PIC simulations where laminar anti-parallel reconnection is dominated by electron pressure tensor effects (e.g. Torbert et al. 2018; Egedal et al. 2019). For guide field reconnection, this condition should be revised to $\rho_e/\lambda_D = \omega_{pe}/\omega_{ce} = (\sqrt{\beta_e/2})d_e/\lambda_D > 30$ implying the importance of electron beta β_e. Obviously, further research is needed to resolve these differences in order to understand better when and how 2D laminar or 3D anomalous effects dominate the dissipation in the EDR.

3 Energy Conversion and Partitioning

3.1 Magnetic Energy Dissipation at the X-Point

The primary consequence of magnetic reconnection is the impulsive dissipation of excessive free energy in the magnetic field to plasma charged particles. The energy dissipation near the X-point (inside the EDR) is dominated by electron dynamics, as the electron current is much stronger than the ion current in the EDR. The rate of the energy conversion from magnetic to plasma kinetic energy per unit volume can be quantified by $\boldsymbol{j} \cdot \boldsymbol{E}$. In the EDR this is not much different from the often-used dissipation measure at the electron rest frame $\boldsymbol{j} \cdot \boldsymbol{E}'$, where $\boldsymbol{E}' = \boldsymbol{E} + \boldsymbol{V}_e \times \boldsymbol{B}$ (Zenitani et al. 2011), especially near the X-point where electrons are unmagnetized without significant flow. Thus, we will only discuss the quantity of $\boldsymbol{j} \cdot \boldsymbol{E}$ here for simplicity.

During anti-parallel reconnection, magnetic energy dissipation near the X-point is dominated by the perpendicular component of $\boldsymbol{j}_e \cdot \boldsymbol{E}$, $\boldsymbol{j}_{e\perp} \cdot \boldsymbol{E}_\perp$, in both symmetric (Yamada et al. 2014, 2016) and asymmetric cases (Yoo et al. 2017; Yamada et al. 2018). Figure 10 shows a clear dominance of $\boldsymbol{j}_{e\perp} \cdot \boldsymbol{E}_\perp$ (panel b) over $j_{e\parallel}E_\parallel$ (panel a) near the X-point at $(R, Z) = (37.5, 0)$ cm during symmetric, anti-parallel reconnection in MRX. This agrees well with space, where $\boldsymbol{j}_{e\perp} \cdot \boldsymbol{E}_\perp$ is strongest near the stagnation point (Burch et al. 2016; Yamada et al. 2018). Furthermore, the perpendicular electric field near the X-point is dominated by the out-of-the-plane reconnection electric field, which can directly accelerate electrons (Zenitani and Hoshino 2001) as shown during a magnetotail reconnection event measured by MMS (Torbert et al. 2018), and also recently during anti-parallel reconnection driven by lasers (Chien et al. 2023) where an accelerated electron beam was detected.

If there is a significant guide field, however, the energy conversion is dominated by the parallel component, $j_{e\parallel}E_\parallel$ (Fox et al. 2018; Pucci et al. 2018; Bose et al. 2023), consistent with MMS observation (Wilder et al. 2018). This difference is mainly related to the fact that the energy conversion inside the EDR is mostly through the out-of-plane reconnection electric field. Without a guide field, the reconnection electric field is mostly perpendicular to the

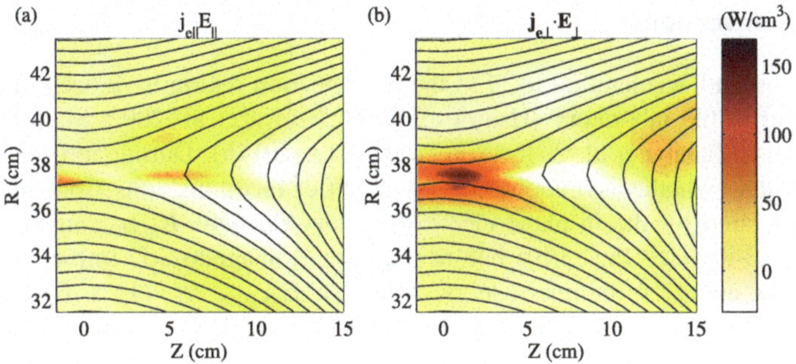

Fig. 10 Comparison of two compositions of energy deposition rate measured in MRX for symmetric, anti-parallel magnetic reconnection; (a) $j_{e\|} E_{\|}$ and (b) $\boldsymbol{j}_{e\perp} \cdot \boldsymbol{E}_{\perp}$. Figure from Yamada et al. (2016)

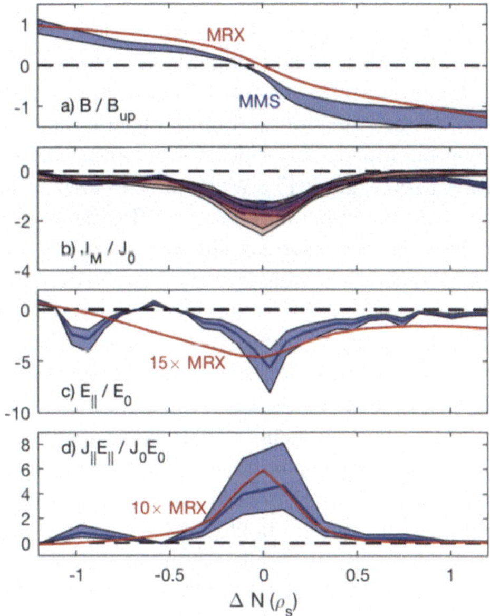

Fig. 11 Scaled comparison of MRX (red curves and bands) and MMS (blue bands) data from the event of Eriksson et al. (2016), for cuts across the current sheet of (a) the reconnecting magnetic field, (b) out-of-the plane current density, (c) parallel electric field, and (d) the parallel component of energy dissipation rate. From Fox et al. (2018)

magnetic field, while it becomes mostly parallel to the magnetic field with a sizable guide field. Figure 11 shows direct and scaled comparisons between MRX data with a guide field of about 0.6 times the reconnecting field (Fox et al. 2017) and MMS data with a guide field of about 3.5 times the reconnecting field (Eriksson et al. 2016). When normalized properly, the profiles of the magnetic field and current density agree with each other within error bars. A similar conclusion was obtained when compared with another MMS event with lower guide field (Wilder et al. 2018). In both cases $\boldsymbol{j} \cdot \boldsymbol{E}$ in the current sheet is dominated by $j_{\|} E_{\|}$, consistent with numerical predictions (Pucci et al. 2018). The peak values of the parallel electric field, however, are larger by an order of magnitude in MMS than in MRX. This highlights the importance in our further understanding energy conversion by reconnection (Ergun et al. 2016a), including questions on where these intense parallel electric fields come from and what effects they have on plasma heating and acceleration.

3.2 Energy Conversion

Particle heating and acceleration local to the reconnection region have been directly measured in detail in the laboratory (Hsu et al. 2000; Brown et al. 2002; Stark et al. 2005; Ono et al. 2011; Tanabe et al. 2015; Yoo et al. 2013, 2014b). During anti-parallel reconnection in MRX, whether symmetric or asymmetric, incoming ions from upstream are directly accelerated by the in-plane electrostatic electric field $\boldsymbol{E}_{\text{in-plane}}$ in the IDR (Yoo et al. 2013, 2014b) (see Fig. 3(b)) before they are "remagnetized" further downstream, converting flow energy to thermal energy. Although $\boldsymbol{E}_{\text{in-plane}} \approx -(\boldsymbol{V}_e \times \boldsymbol{B})_{\text{in-plane}}$ is non-dissipative for electrons within the IDR (but outside the EDR), it can energize ions via $\boldsymbol{j}_i \cdot \boldsymbol{E}_{\text{in-plane}} \approx en\boldsymbol{V}_i \cdot (\boldsymbol{V}_e \times \boldsymbol{B})_{\text{in-plane}}$ (Liu et al. 2022). This has been confirmed experimentally and numerically (Yoo et al. 2014a; Yamada et al. 2018).

During strong guide field reconnection in VTF, ion heating was observed and interpreted (Stark et al. 2005) as magnetic moment conservation being broken due to strong motional variation of the in-plane electric field (Egedal et al. 2003), $(\boldsymbol{v} \cdot \nabla)\boldsymbol{E}_{\text{in-plane}}$. A key dimensionless parameter $e\nabla^2\phi/m_i B^2 \gtrsim 1$ was identified to demagnetize and energize ions (Stark et al. 2005). Ions are heated downstream of magnetic reconnection during plasma merging with a significant guide field (Ono et al. 2011).

Electron heating is mostly localized to the EDR near the X-line during symmetric anti-parallel reconnection as implied by the large value of $\boldsymbol{j} \cdot \boldsymbol{E}$ there (Yoo et al. 2014a) or along the low-density side of separatrices during asymmetric anti-parallel reconnection on MRX (Yoo et al. 2017). While parallel electric field is expected to explain a large fraction of the electron temperature increase (Egedal et al. 2013; Yoo et al. 2017), other mechanisms, such as various wave activities (see below), are not excluded (Ji et al. 2004; Zhang et al. 2023). Electron heating is also measured during guide field reconnection in the electron-only region (Shi et al. 2022) and in the electron-ion region on MRX (Bose et al. 2023). Strong electron heating was observed within the current sheet during plasma merging (Tanabe et al. 2015). These results are in general agreement with MMS results on significant electron energization within the EDR (Eastwood et al. 2020).

Direct measurements of particle acceleration local to the reconnection region are generally difficult in the laboratory, despite many acceleration mechanisms having been proposed and studied intensively numerically (Ji et al. 2022). They include direct acceleration by the reconnection electric field (Zenitani and Hoshino 2001), the parallel electric field (Egedal et al. 2013), Fermi acceleration (Drake et al. 2006), and betatron acceleration (Hoshino et al. 2001). Accelerated electrons along the magnetic field were measured by an energy analyzer (Gekelman and Stenzel 1985) during reconnection, although in a different region. On VTF where reconnection is driven dynamically with a strong guide field, the population of energized tail electrons along the field line were seen to increase by several times, doubling an effective temperature from ~ 20 eV to up to 40 eV (Fox et al. 2010, 2012). Electron jets at the electron Alfvén speed have been directly detected by Thomson scattering diagnostics during guide field electron-only reconnection (Shi et al. 2022). More recently, non-thermal electrons with energies of $\sim 100T_e$ due to the reconnection electric field of anti-parallel reconnection at low-β driven by lasers were directly detected with an angular dependence consistent with simulation (Chien et al. 2023). The later supports an astrophysical conjecture to accelerate electrons by reconnection to high energies beyond the synchrotron burnoff limit (Cerutti et al. 2013).

3.3 Energy Partitioning

One of the advantages of laboratory experiments over space measurements is that 2D profiles of key plasma and field parameters can be obtained by repeating measurements over a

Table 2 Summary of the energy inventory studied in the laboratory for three cases and their counterparts based on PIC simulations for two cases (Yamada et al. 2014; Yoo et al. 2017; Yamada et al. 2018; Bose et al. 2023). Typical errors for these numbers are about 10–20%. The guide field was about 0.7 times the reconnecting field for the guide field reconnection case. One study of space data for a symmetric antiparallel case in Earth's magnetotail (Eastwood et al. 2013) is also listed despite the large uncertainties in determining incoming magnetic energy and size of the volume (Yamada et al. 2015)

Case	Incoming (MW)	Outgoing	Electron	Ion
Symmetric, antiparallel, lab	1 (1.9 ± 0.2)	0.45	0.20	0.35
Symmetric, antiparallel, PIC	1	0.42	0.22	0.34
Symmetric, antiparallel, space	1	0.1-0.3	0.18	0.39
Asymmetric, antiparallel, lab	1 (1.4 ± 0.2)	0.44	0.25	0.31
Asymmetric, antiparallel, PIC	1	0.43	0.25	0.32
Symmetric, guide field, lab	1 (1.5 ± 0.2)	0.65	0.15	0.29

similar set of discharges. These 2D profiles can be used for a quantitative study of energy conversion and partitioning inside the IDR on MRX (Yamada et al. 2014; Yoo et al. 2017; Bose et al. 2023), where the method of the energy inventory analysis has been explained in detail. The incoming magnetic energy, for example, can be obtained by integrating the corresponding Poynting flux ($E_y B_z/\mu_0$) at the boundary surface. The electron (ion) energy gain can be obtained by integrating $\boldsymbol{j}_e \cdot \boldsymbol{E}$ ($\boldsymbol{j}_i \cdot \boldsymbol{E}$) over the entire volume of the analysis.

Table 2 summarizes the energy partitioning for three cases in the lab, two cases in numerical simulations, and one case from space measurements. In all cases, the ion energy gain exceeds that of electrons. Compared to antiparallel reconnection, the total energy conversion is less effective for the case with a guide field at a strength comparable to the reconnecting field component. In all cases, both electron and ion energy gain is dominated by an increase in the thermal energy; the flow energy increase is negligible especially for electrons. These results are in general agreement with space observations (Eastwood et al. 2013) which is also listed in the table for comparison, though they carry large uncertainties due to limited available data. Nonetheless, the fact that all these numbers roughly agree with each other suggests that energy conversion and partitioning in locations near the X-line during collisionless reconnection are reasonably quantified.

4 Plasma Waves

While magnetic reconnection converts magnetic energy to plasma energy, various free energy sources for waves and instabilities are available especially in or near the diffusion regions and separatrices, such as spatial inhomogeneity, relative drift between ions and electrons (or electric current), or kinetic structures in particles' velocity distribution functions. This section reviews relevant studies of plasma waves generated in the vicinity of diffusion regions of collisionless reconnection in the laboratory in comparison with space measurements.

4.1 Whistler Waves

One of these types of waves is whistler waves, which can be generated by either electron beams or temperature anisotropy as summarized by Khotyaintsev et al. (2019). During

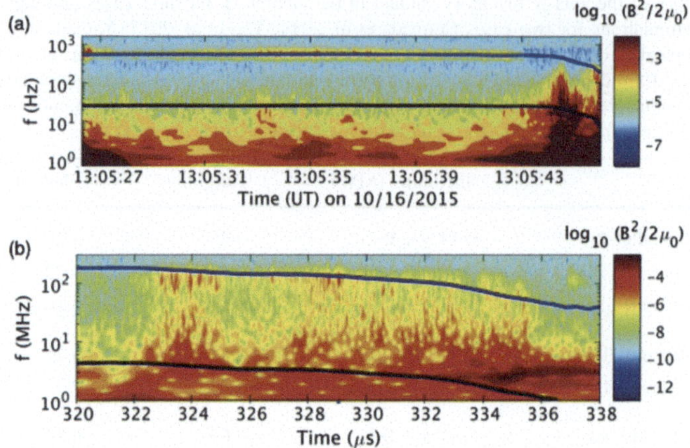

Fig. 12 Comparison of the whistler wave activity during asymmetric reconnection observed in space (a) and MRX (b). Blue lines indicate half of the local electron cyclotron frequency (f_{ce}), while black lines indicate the local lower hybrid frequency (f_{LH}). Near the separatrix on the low-density side, whistler waves near 0.5 f_{ce} are observed. After Yoo et al. (2018)

asymmetric reconnection, the separatrix region on the low-density (magnetospheric) side is unstable to lower hybrid drift waves (LHDW) (Krall and Liewer 1971, see below) due to the large density gradient across the magnetic field. This instability enhances the electron transport and heating near the separatrix region (Le et al. 2017). In this region, electrons with a high parallel velocity can be quickly transported to the exhaust region along the turbulent field lines due to LHDW, leaving behind a population of electrons with temperature anisotropy due to a tail with higher perpendicular energy. This temperature anisotropy generates whistler waves around $0.5 f_{ce}$ near the separatrix on the low-density side (Yoo et al. 2018, 2019).

Figure 12 shows this anisotropy-driven whistler wave observed by MMS (a) and in MRX (b). The color contour shows the energy in fluctuations in the magnetic field. Clear whistler wave activity around the half of the local electron cyclotron frequency ($0.5 f_{ce}$), which is indicated by blue solid lines, is observed in both space and laboratory. In both cases, the measurement location was initially just outside of the separatrix region and moved to the exhaust region around 13:05:43 for the panel (a) and 334 μs for the panel (b). Broad fluctuations mostly below the local lower hybrid frequency (f_{LH}, denoted by black lines) also exist in both measurements. Note that LHDW-driven fluctuations are strongest just before the measurement location enters into the exhaust region. It should be also noted that the whistler wave activity disappears in the exhaust region.

It is worth mentioning that whistler waves were also observed in an earlier reconnection experiment (Gekelman and Stenzel 1984). These waves propagate obliquely with respect to the magnetic field and their amplitudes correlate with the reconnection current. Both of these characteristics are consistent with the observation of electromagnetic LHDW on MRX (Ji et al. 2004), which are explained by a local two-fluid theory (Ji et al. 2005). LHDW will be discussed below in Sect. 4.3.

4.2 Electrostatic Waves

A variety of electrostatic high-frequency waves have also been observed in the laboratory during reconnection events. Above f_{LH}, these waves have multiple names, including R-waves [after the $R = 0$ branch in the Clemmow-Mullaly-Allis (CMA) diagram (Stix 1992)], electrostatic whistlers, or Trivelpiece-Gould modes [from early laboratory contexts (Trivelpiece and Gould 1959)]. These waves extend from $\sim f_{LH}$ to $\min(f_{pe}, f_{ce})$. Under most laboratory as well as space conditions, $f_{ce} < f_{pe}$, so the waves exist up to f_{ce}. For the waves to be electrostatic $kd_e > 1$ must be satisfied, where k is the wavenumber. The electrostatic branch has the dispersion relation $\omega = \omega_{ce}k_{\|}/k$, which allows a broadband collection of waves with parallel phase velocities $\omega/k_{\|}$ resonant with super-thermal electron populations. At longer wavelength, when $kd_e < 1$, these waves transition to the classical electromagnetic whistlers ($\omega = \omega_{ce}d_e^2 k_{\|}k$). At lower frequencies $f \sim f_{\rm LH}$, the waves increasingly interact with the ions. In those cases, the perpendicular group velocity of the waves becomes very small, so that wave packets can stay localized to regions with energized electrons for efficient growth. Theory predicts that there are multiple sources of free energy which can drive the waves, including beam resonance (inverse Landau damping); gyro-resonance driven by $T_{\|} > T_{\perp}$; or gradients in density, temperature, or in fast electron components (Fox et al. 2010). Most interestingly, the waves driven by gradients lead to maximum growth in the lower-hybrid range frequencies ($f \sim f_{LH}$), and are related to quasi-electrostatic lower-hybrid drift waves (see below).

Gekelman and Stenzel (1985) also reported the detection of these waves and suggested that they are generated by the measured energetic electron tail in the 3D velocity space, either by anisotropy mechanisms or inverse Landau damping. High-frequency electrostatic waves were also detected on VTF when guide field reconnection was strongly driven (Fox et al. 2010). This was consistent with a picture where the reconnection events would drive energetic electrons, which in turn would drive waves. The parallel phase speed was observed to be resonant with superthermal electrons, $\omega/k_{\|} > v_{te}$. The spectrum typically consisted of a broad spectrum from near $f_{\rm LH}$ and extending to a very clear cutoff at f_{ce} (Fox et al. 2010).

Given strong beam components, electrostatic waves can often be driven to very large amplitude, which can lead to the formation of non-linear wave structures. One such mechanism is that the waves can grow to large amplitudes and trap resonant electrons. This leads to so-called "electron phase-space hole" structures, also called Bernstein-Greene-Kruskal (BGK) solitary structures (Bernstein et al. 1957), or electrostatic solitary waves (ESW). The latter have been observed in many places in space including during reconnection events in the magnetopause (Matsumoto et al. 2003) and magnetotail (Cattell et al. 2005), as was summarized recently by Khotyaintsev et al. (2019). These electron phase space holes were directly observed on VTF (Fox et al. 2008, 2012) and indicate that the strong electric fields in the reconnection region pull-out strong beam components of the electron population, exciting these hole structures. Electron holes have also been directly generated in electron-beam experiments (Lefebvre et al. 2010). Figure 13 shows observations of electron hole phenomena during the strong wave turbulence during VTF reconnection events. The structures are positive potential ($\phi > 0$) which is consistent with electron trapping. More recently, ESW or electron space holes have been observed during guide field reconnection within the diffusion region (Khotyaintsev et al. 2020) and in the separatrix (Ahmadi et al. 2022) in the magnetopause where they may play an important role in electron heating.

There is a renewed interest in the ion acoustic wave (IAW) (Papadopoulos 1977, and references therein), which is an unmagnetized short-wavelength electrostatic wave. The IAW can be driven unstable by relative drift between ions and electrons or equivalently electric current, which is expected to be intense around the X-line. Anomalous resistivity based on IAW-like waves has been used to numerically generate Petschek solution fast reconnection since Ugai and Tsuda (1977), Sato and Hayashi (1979). Despite a pioneering laboratory detection during relatively collisional reconnection (Gekelman and Stenzel 1984), however,

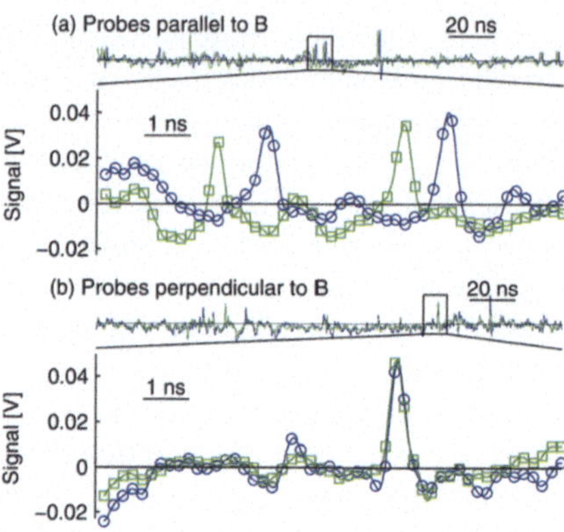

Fig. 13 Observation of phase-space-hole electrostatic structures driven during magnetic reconnection events. a) Propagation between two closely-spaced probes parallel to the magnetic field, b) simultaneous observation on two probes oriented perpendicular to the magnetic field. The time delays combined with known probe separate give the typical size and velocity of the electron holes, which is superthermal compared to the electron temperature. From Fox et al. (2012)

the importance of IAWs for reconnection has been quickly dismissed due to the widely observed high ion temperature $T_i \sim ZT_e$, which is known to stabilize IAW via strong ion Landau damping. However, in a very recent laboratory experiment using lasers (Zhang et al. 2023), strong IAW bursts and the associated electron acoustic wave (EAW) bursts were detected by collective Thomson scattering in the exhaust of anti-parallel reconnection where $T_i \ll ZT_e$ due to high $Z(\sim 18)$ of ions. These IAW and EAW burst were successfully reproduced by PIC simulations showing that strong IAWs generate a double layer, which induces electron two-stream instabilities leading to EAW bursts and electron heating as observed experimentally. These new experimental results are consistent with recent space observations (Uchino et al. 2017; Steinvall et al. 2021) which detected IAWs during reconnection when sufficient cold ions were present, and may be relevant to the outstanding questions on large parallel electric fields measured by MMS (Ergun et al. 2016b). These new results also raised a legitimate question on whether the high ion temperature is a universal observation and thus whether IAW should be dismissed as an anomalous dissipation mechanism in collisionless plasmas. In fact, recent detection of monochromatic IAWs and associated electron heating in solar wind when ions are cold (Mozer et al. 2022) speaks for the needs to revisit this topic, as direct measurements of ion temperature are rare for solar and astrophysical plasmas in general.

4.3 Lower Hybrid Drift Waves and Current Sheet Kinking

Lower hybrid drift waves (LHDWs) have been a candidate for anomalous resistivity and transport in the diffusion region due to their ability to interact with both electrons and ions. The free energy source of LHDWs is the current perpendicular to the magnetic field (Davidson and Gladd 1975). Depending on the local plasma and field parameters, LHDWs may be either quasi-electrostatic (ES-LHDW) (Carter et al. 2001; Hu et al. 2021) or electromagnetic (EM-LHDW) (Ji et al. 2004; Yoo et al. 2014b). With a similar electron temperature and perpendicular current, plasma beta (β) is the key parameter to determine the type of waves; for low β (typically below unity), the ES-LHDW mode propagating nearly perpendicular to the local magnetic field is unstable, while the EM-LHDW mode propagating obliquely to the magnetic field is excited when β is high (Yoo et al. 2020).

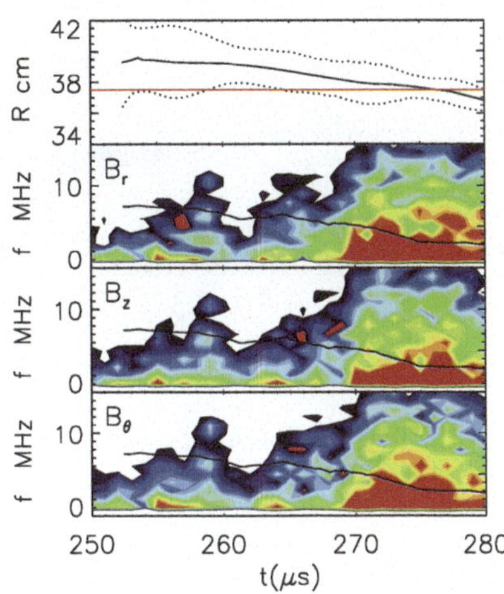

Fig. 14 Detection of electromagnetic lower-hybrid drift waves in the current sheet center during anti-parallel reconnection on MRX. Wave powers are color coded (red high and white low) in spectrograms where lower hybrid frequency is indicated by the black line using upstream reconnecting field. Top panel shows the location of the probe (red) and the current sheet (center as black solid line and edges as dashed lines). When the current sheet center moves close to the probe, high-frequency magnetic fluctuations are detected. Figure from Ji et al. (2004)

During anti-parallel reconnection, plasma β varies rapidly in the current sheet. At the current sheet edge where β is low, the ES-LHDW mode has been observed (Carter et al. 2001; Yoo et al. 2020) consistent with theoretical expectation (Daughton 2003) and space observation by Polar spacecraft (Bale et al. 2002). The obliquely propagating EM-LHDW mode has been observed in the current sheet center where plasma β is high and electric current is large (Ji et al. 2004; Yoo et al. 2014b), as well as in the immediate downstream (Ren 2007). An example is shown in Fig. 14 from MRX where large-amplitude electromagnetic waves were detected when the current sheet center moved close to the probe during anti-parallel reconnection (Ji et al. 2004), consistent with numerical simulations (Daughton et al. 2004). Both ES-LHDWs and obliquely propagating EM-LHDWs have also been observed by Cluster spacecraft in a thin current sheet in magnetotail (Zhou et al. 2009) and recently by MMS in magnetopause (Ergun et al. 2017). More recent measurements on MRX show that the EM-LHDW becomes increasingly organized with larger amplitude with guide field (von Stechow et al. 2018). For more measurements of LHDWs in and around diffusion regions in space with varying influence on anomalous resistivity and viscosity, see recent reviews by Khotyaintsev et al. (2019) and Graham et al. (2023).

Many of the observed wave characteristics of EM-LHDWs, such as propagation direction and polarization, have been qualitatively explained by a local two-fluid theory (Ji et al. 2005) as an instability caused by reactive coupling between the backward propagating whistler wave and the forward propagating sound wave when the relative drifts between electrons and ions are large. The wave amplitude has been observed to correlate positively with fast reconnection (Ji et al. 2004), consistent with quasilinear theory on their possible importance for anomalous resisitivity (Kulsrud et al. 2005). The waves have also been reproduced in 3D PIC simulations performed in MRX geometry in a Cartesian coordinate, but they failed to explain the observed broadened width of the EDR (Roytershteyn et al. 2013). Possible solutions to this discrepancy include differences in the simulation geometry and parameters, as well as measurement resolutions as discussed in Sect. 2.3. It is noted that the current sheet kinking that was observed on TREX and associated simulations (Greess et al. 2021) and in space (e.g. Ergun et al. 2019) could result in broadened current sheets due to limited spatial and/or time resolutions.

With a sizable guide field, however, ES-LHDWs can be unstable inside the IDR and EDR, affecting electron and reconnection dynamics. For example, following a multi-spacecraft

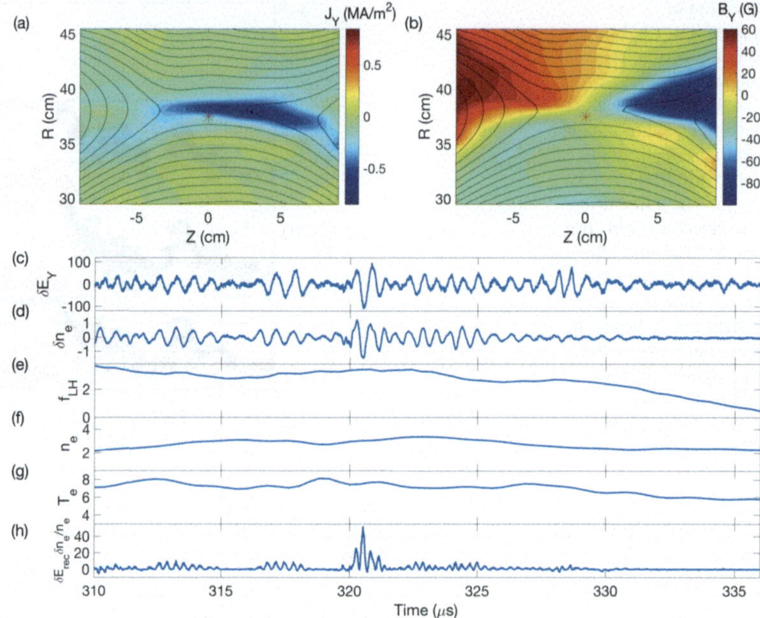

Fig. 15 Measured ES-LHDWs. (a,b) Out-of-plane current or magnetic field component (color) with the poloidal flux contours (black lines) representing the magnetic field lines at 326 μs. The red asterisk indicates the location of the probe. The upper side ($R > 37.5$ cm) has a higher density. (c) Time series of $\delta E_{\rm rec}$ in V/m. Wave activity near the lower hybrid frequency ($f_{\rm LH} \sim 2$ MHz) is detected while the probe stays near the reconnection site. The amplitude of the fluctuation is comparable to the mean reconnection electric field ($\langle E_{\rm rec} \rangle \sim 100$ V/m). (d) Time series of $\delta n_{\rm e}$ in 10^{13} cm^{-3} during the quasi-steady reconnection period. Time series of $f_{\rm LH}$ (e), averaged density ($\langle n_{\rm e} \rangle$) in 10^{13} cm^{-3} (f), and electron temperature ($T_{\rm e}$) in eV (g) are shown. A sharp decrease of $f_{\rm LH}$ is observed with the approach of the X-point to the probe. Time series of $\delta E_{\rm rec} \delta n_{\rm e} / \langle n_{\rm e} \rangle$ are shown in (h). Positive correlation between $\delta E_{\rm rec}$ and $\delta n_{\rm e}$ indicates that the wave is capable of generating anomalous resistivity. Figure from Hu et al. (2021)

analysis using Cluster (Norgren et al. 2012), a recent observation (Chen et al. 2020) using MMS showed that strong ES-LHDWs produce non-gyrotropic electron heating and vortical flows inside the EDR of reconnection with a guide field. These electron vortices have been successfully reproduced by corresponding 3D PIC simulations (Ng et al. 2020) and suggest that further reconnection may occur inside the LHDW vortex tubes as dissipation at smaller scales. Other space observations of guide field reconnection show that ES-LHDWs are capable of generating anomalous resistivity between electrons and ions (Yoo et al. 2020; Graham et al. 2022).

Recently, ES-LHDW measurements were revisited on MRX combined with the simultaneous measurements of electron density at the same location (Hu et al. 2021). Figure 15 shows measurements of ES-LHDWs at the edge of the current sheet during anti-parallel reconnection. Panels (a) and (b) show the 2D profile of the out-of-plane current density and magnetic field, respectively. The black lines are contours of the poloidal magnetic flux, representing magnetic field lines. The red asterisk is the location of the probe that measures high-frequency fluctuations in the reconnection electric field (panel c) and electron density (panel d) (Hu et al. 2021). Due to the positive correlation between two fluctuating quantities, the quantity of $\delta E_y \delta n_{\rm e} / \langle n_{\rm e} \rangle$, which is anomalous resistivity along the out-of-plane direction (Che et al. 2011), becomes positive. These measurements of ES-LHDWs have been further extended on MRX to cases with a sizable guide field demonstrating significant anomalous resistivity and electron heating (Yoo et al. 2023). The initial corresponding 3D simulation show that ES-LHDWs propagating along the outflow are triggered by the difference between electron and ion outflows in regions of low β_e (Ng et al. 2023), consistent with the MRX experiment results.

5 Multiscale Reconnection

The physics of collisionless magnetic reconnection has been studied mostly in locations nearby the local X-line as discussed in the previous sections, such as the IDR and EDR as well as separatrices. If measured in the unit of ion kinetic scales, their distances from the local X-line are not too far. However, the collisionless plasmas in space and astrophysics where reconnection is believed to occur are vastly larger - their normalized sizes have been surveyed (Ji and Daughton 2011) ranging from $\sim 10^3$ for Earth's magnetosphere to $\sim 10^{14}$ for extragalactic jets. In these large plasmas, magnetic reconnection inevitably occurs in the multiple X-line regimes as illustrated in the reconnection phase diagram (Ji and Daughton 2011, 2022).

While there has been abundant evidence for collisionless multiple X-line reconnection in Earth's magnetopause as Flux Transfer Events (FTEs) (Russell and Elphic 1979) and in the magnetotail as plasmoids (Baker et al. 1984), there have been only relatively few laboratory works in this area with (Stenzel et al. 1986; Ono et al. 2011) or without a guide field (Dorfman et al. 2013; Olson et al. 2016; Jara-Almonte et al. 2016; Hare et al. 2017). When plasmoids form and are subsequently ejected from the current sheet, reconnection tends to proceed in an impulsive and intermittent fashion (Ono et al. 2011; Dorfman et al. 2013; Jara-Almonte et al. 2016), qualitatively consistent with space observations of the non-steadiness of multiscale reconnection (e.g. Chen et al. 2008, 2012; Ergun et al. 2018).

Quantifying non-steady reconnection with multiple X-lines or "turbulent" reconnection is non-trivial. There have been several studies that quantified size distributions of plasmoids, or magnetic structure in general, during multiscale reconnection, as shown in Fig. 16. Two are from the laboratory (Dorfman et al. 2014; Olson et al. 2016), two from Earth's magnetopause (Fermo et al. 2011; Akhavan-Tafti et al. 2018), one from Earth's magnetotail (Bergstedt et al. 2020), and one from solar observation (Guo et al. 2013). Other than the last study, the others are on plasmoids on kinetic scales, but all of them are more consistent with an exponential distribution rather than a power-law distribution. It is not surprising to have an exponential distribution on kinetic scales as they are dissipative scales in collisionless plasmas, but it would be a surprise if the exponential distributions also apply to fluid scales, over which the self similar power laws should apply at least in the inertial range. We note that there are interesting statistical *in-situ* studies of heliospheric current sheets (e.g. Eriksson et al. 2022) and flux ropes (Janvier et al. 2014) on a larger scale in the solar wind. The upcoming multiscale experiments, numerical simulations and observatories should shed more light onto these important questions (Ji et al. 2022).

6 Future Prospects

A concise review was given on the recent highlights from controlled laboratory studies of collisionless magnetic reconnection on a variety of topics including ion and electron kinetic

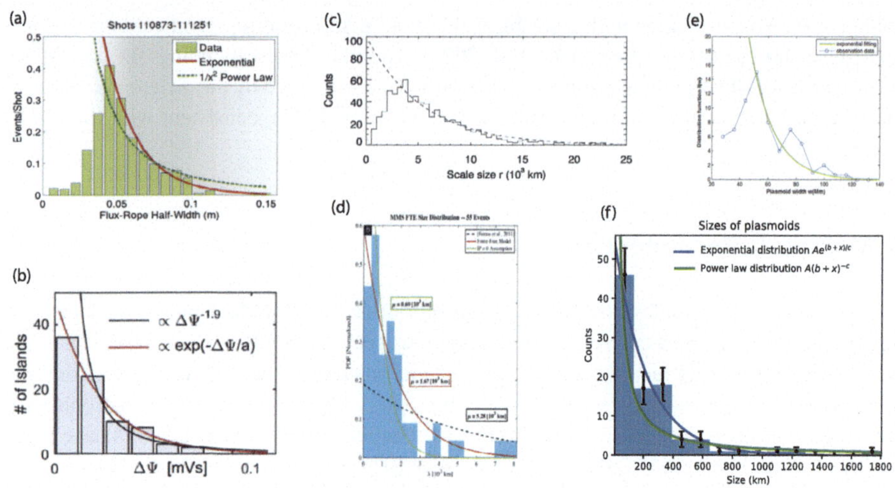

Fig. 16 Plasmoid size distributions (a) Dorfman et al. (2014) and (b) Olson et al. (2016) from the lab; (c) Fermo et al. (2011) and (d) Akhavan-Tafti et al. (2018) from the space observation; (e) Guo et al. (2013) from the solar observation (reproduced by permission of the AAS); and (f) Bergstedt et al. (2020) from the space observation. All of them are more consistent with an exponential distribution rather than a power-law distribution

structures in electromagnetic fields, energy conversion and partitioning, various electromagnetic and electrostatic kinetic plasma waves, as well as plasmoid-mediated multiscale reconnection. While unresolved issues still remain, many of these highlighted results compare well with numerical predictions and space observations, especially by the MMS mission. Thus, it is not an overstatement that the physics foundation of fast reconnection in collisionless plasmas has been largely established, at least within the parameter ranges and spatial scales that were studied.

Nonetheless, there still exist outstanding questions on single X-line collisionless reconnection. The first question is about what dissipates magnetic fields within the EDR when 2D laminar pictures do not apply. We still have cases in the laboratory where the reconnection electric field or the thickness of the EDR is not fully accounted for (Ji et al. 2008; Roytershteyn et al. 2013), while in space we also have cases where 2D laminar reconnection pictures do not tell the whole story (e.g. Cozzani et al. 2021). Does anomalous resistivity exist in its conventional forms, as hinted by electrostatic LHDWs observed during guide field reconnection (Yoo et al. 2023) or by IAWs observed recently during anti-parallel reconnection at low ion temperature (Zhang et al. 2023)? Alternatively, do anomalous effects manifest as kinking of otherwise laminar 2D reconnecting current sheets (Greess et al. 2021) or is anomalous resisitivity cancelled by anomalous viscosity leaving no wave dissipative effects in the EDR (Graham et al. 2022)? Further research using well-controlled experiments with adequate diagnostics, supported by matching numerical simulations, is needed to settle this long standing question.

Another outstanding question is about how magnetic energy is dissipated to a combination of flow, thermal and non-thermal energies of electrons and ions, as a function of field geometry, symmetry, and upstream plasma β. Substantial progress has been made on this subject with laboratory experiments, numerical simulations, and space observation, as summarized in Table 2 in terms of energy partitioning, but there remain a number of unanswered questions, especially on particle acceleration. Recent progress in directly detecting electrons accelerated by the reconnection electric field (Chien et al. 2023) and non-thermal electrons by Thomson scattering (Shi et al. 2022) is an encouraging sign that more results are coming. The predicted scaling of electron heating and acceleration by the parallel electric field with regard to upstream β (Le et al. 2016) is in agreement with certain spacecraft observations (Oka et al. 2023), but its laboratory study sensitively depends on plasma collisionality (Le et al. 2015). High Lundquist number regimes offered by the upgraded TREX (Olson et al. 2016) and the upcoming Facility for Laboratory Reconnection Experiments or FLARE (Ji et al. 2018, 2022) will allow first laboratory accesses to the collisionless regimes required to study this important issue of collisionless reconnection.

Looking further into the future, laboratory access to multiscale regimes of magnetic reconnection is an important step as guided by the reconnection phase diagram (Ji and Daughton 2011, 2022). In addition to high Lundquist numbers, space and astrophysical plasmas have large normalized plasma system sizes, significantly expanding the parameter space over which global fluid scales and local kinetic scales are coupled. The solar corona is an excellent example where the typical mean-free path of thermal particles is much longer than any kinetic scales so that locally physics is collisionless or kinetic, while the mean-free path is much shorter than system sizes so that globally physics is collisional or fluid-like. How does multiscale physics across fluid and kinetic scales operate self-consistently in this regime to generate solar flares as observed, in terms of their impulsive onset and energetic consequences on thermal heating and particle acceleration? Answering multiscale physics questions like this requires going far beyond what has been traditionally done in reconnection research in which the detailed dynamics are studied around local X-lines based on either fluid or kinetic physics.

Statistical properties of multiscale physics need to be quantified in order to identify self-similar behavior across scales. In the case of plasmoid-mediated multiscale reconnection, despite theoretical advances in predicting power-law scaling of plasmoid sizes (e.g. Uzdensky et al. 2010; Huang and Bhattacharjee 2012; Pucci and Velli 2014; Comisso et al. 2016; Majeski et al. 2021), no power-laws have been found from the laboratory or space data thus far. This may be due to the fact that data used are close to dissipative kinetic scales, and thus the accessibility of data on fluid scales is critical. To simultaneously study fluid and kinetic physics, especially under realistic conditions in 3D, effectively exploiting new exascale computing capabilities is crucial (Ji et al. 2022), as highlighted by a recent example in modeling Earth's magnetotail (Palmroth et al. 2023). In addition, exascale computers will permit fully kinetic simulations to more closely match important dimensionless parameters, such as the ion to electron mass ratio (m_i/m_e) and the ratio of the electron skin depth to the electron Debye length (d_e/λ_D), both of which influence the spectrum and nature of instabilities present with reconnection layers (Jara-Almonte et al. 2014) (see Sect. 2.3). We anticipate that exascale computing will permit larger system sizes (S, L/d_i) and permit 3D global kinetic modeling of laboratory experiments. Furthermore, to process a huge amount of existing and new observational, numerical, and laboratory data for statistical studies, there exist promising opportunities to use novel techniques based on data science such as machine learning (e.g. Bergstedt and Ji 2023).

One of the direct consequences of multiscale collisionless reconnection is its ability to accelerate particles into power-law distributions which are often observed during reconnection events. There has been a recent surge of theoretical and numerical work on this subject including reconnection under extreme conditions in astrophysics using kinetic models (e.g. Dahlin 2020; Li et al. 2021; Guo et al. 2020, and references therein) and MHD models (Arnold et al. 2021; Majeski and Ji 2023); however, there have been no laboratory counterparts on this subject. It is imperative to develop new platforms (e.g. Chien et al. 2023) for such studies as well as new diagnostics (e.g. Fox et al. 2010; Shi et al. 2022) to detect accelerated non-thermal particles in laboratory experiments, including upcoming multiscale experiments such as FLARE (Ji et al. 2018, 2022). A concerted effort from exascale modeling, data science, as well as from the scheduled or proposed multiscale space missions such as HelioSwarm (Klein et al. 2023) and Plasma Observatory (Retinò et al. 2022) is critical to address these important questions.

Funding This work is supported in part by the NASA MMS mission. H.J. acknowledges support of this work by NASA under Grants No. NNH15AB29I and 80HQTR21T0105, and by the U.S. Department of Energy, Office of Fusion Energy Sciences under Contract No. DE-AC0209CH11466.

Declarations

Competing Interests The authors declare no competing interests.

References

Ahmadi N, Eriksson S, Newman D et al (2022) Observations of electron vorticity and phase space holes in the magnetopause reconnection separatrix. J Geophys Res Space Phys 127(8):e30702. https://doi.org/10.1029/2022JA030702

Akhavan-Tafti M, Slavin JA, Le G et al (2018) Mms examination of ftes at the Earth's subsolar magnetopause. J Geophys Res Space Phys 123(2):1224–1241. https://doi.org/10.1002/2017JA024681

Arnold H, Drake JF, Swisdak M et al (2021) Electron acceleration during macroscale magnetic reconnection. Phys Rev Lett 126:135101. https://doi.org/10.1103/PhysRevLett.126.135101

Aydemir AY (1992) Nonlinear studies of $m = 1$ modes in high-temperature plasmas. Phys Fluids B 4:3469–3472. https://doi.org/10.1063/1.860355

Baker DN, Bame SJ, Birn J et al (1984) Direct observations of passages of the distant neutral line (80-140 R_E) following substorm pnsets: ISEE-3. Geophys Res Lett 11(10):1042–1045. https://doi.org/10.1029/GL011i010p01042

Bale S, Mozer F, Phan T (2002) Observation of lower hybrid drift instability in the diffusion region at a reconnecting magnetopause. Geophys Res Lett 29:2180. https://doi.org/10.1029/2002GL016113

Bergerson W, Forest C, Fiksel G et al (2006) Onset and saturation of the kink instability in a current-carrying line-tied plasma. Phys Rev Lett 96:015004. https://doi.org/10.1103/PhysRevLett.96.015004

Bergstedt K, Ji H (2023) A novel method to train classification models for structure detection in in-situ spacecraft data. Submitted. https://doi.org/10.22541/essoar.168167402.23523807/v1

Bergstedt K, Ji H, Jara-Almonte J et al (2020) Statistical properties of magnetic structures and energy dissipation during turbulent reconnection in the Earth's magnetotail. Geophys Res Lett 47(19):e88540. https://doi.org/10.1029/2020GL088540

Bernstein IB, Greene JM, Kruskal MD (1957) Exact nonlinear plasma oscillations. Phys Rev 108(3):546–550. https://doi.org/10.1103/PhysRev.108.546

Birn J, Drake J, Shay M et al (2001) Geomagnetic environmental modeling (GEM) magnetic reconnection challenge. J Geophys Res 106(A3):3715. https://doi.org/10.1029/1999JA900449

Biskamp D, Schwarz E, Drake J (1995) Ion-controlled collisionless magnetic reconnection. Phys Rev Lett 75:3850. https://doi.org/10.1103/PhysRevLett.75.3850

Bose S, Fox W, Ji H et al (2023) Conversion of magnetic energy to plasma kinetic energy during guide field magnetic reconnection in the laboratory. Submitted

Bowers K, Albright B, Yin L et al (2009) Advances in petascale kinetic simulations with VPIC and Roadrunner. J Phys Conf Ser 180:012055. https://doi.org/10.1088/1742-6596/180/1/012055

Bratenahl A, Yeates CM (1970) Experimental study of magnetic flux transfer at the hyperbolic neutral point. Phys Fluids 13:2696–2709. https://doi.org/10.1063/1.1692853
Brown M (1999) Experimental studies of magnetic reconnection. Phys Plasmas 6:1717. https://doi.org/10.1063/1.873430
Brown M, Cothran C, Landreman M et al (2002) Experimental observation of energetic ions accelerated by three-dimensional magnetic reconnection in a laboratory plasma. Astrophys J 577:L63. https://doi.org/10.1086/344145
Brown MR, Cothran CD, Fung J (2006) Two fluid effects on three-dimensional reconnection in the swarthmore spheromak experiment with comparisons to space data. Phys Plasmas 13(5):056503. https://doi.org/10.1063/1.2180729
Burch JL, Torbert RB, Phan TD et al (2016) Electron-scale measurements of magnetic reconnection in space. Science 352:aaf2939. https://doi.org/10.1126/science.aaf2939
Cai HJ, Lee LC (1997) The generalized Ohm's law in collisionless magnetic reconnection. Phys Plasmas 4:509. https://doi.org/10.1063/1.872178
Carter T, Ji H, Trintchouk F et al (2001) Measurement of lower-hybrid drift turbulence in a reconnecting current sheet. Phys Rev Lett 88:015001. https://doi.org/10.1103/PhysRevLett.88.015001
Cassak PA, Shay MA (2007) Scaling of asymmetric magnetic reconnection: general theory and collisional simulations. Phys Plasmas 14(10):102114. https://doi.org/10.1063/1.2795630
Cattell C, Dombeck J, Wygant J et al (2005) Cluster observations of electron holes in association with magnetotail reconnection and comparison to simulations. J Geophys Res 110:A01211. https://doi.org/10.1029/2004JA010519
Cerutti B, Werner GR, Uzdensky DA et al (2013) Simulations of particle acceleration beyond the classical synchrotron burnoff limit in magnetic reconnection: an explanation of the crab flares. Astrophys J 770(2):147. https://doi.org/10.1088/0004-637X/770/2/147
Che H, Drake JF, Swisdak M (2011) A current filamentation mechanism for breaking magnetic field lines during reconnection. Nature 474(7350):184–187. https://doi.org/10.1038/nature10091
Chen LJ, Bessho N, Lefebvre B et al (2008) Evidence of an extended electron current sheet and its neighboring magnetic island during magnetotail reconnection. J Geophys Res 113:A12213. https://doi.org/10.1029/2008JA013385
Chen LJ, Daughton W, Bhattacharjee A et al (2012) In-plane electric fields in magnetic islands during collisionless magnetic reconnection. Phys Plasmas 19(11):112902. https://doi.org/10.1063/1.4767645
Chen LJ, Wang S, Le Contel O et al (2020) Lower-hybrid drift waves driving electron nongyrotropic heating and vortical flows in a magnetic reconnection layer. Phys Rev Lett 125(2):025103. https://doi.org/10.1103/PhysRevLett.125.025103
Chien A, Gao L, Zhang S et al (2023) Non-thermal electron acceleration from magnetically driven reconnection in a laboratory plasma. Nat Phys 19:254–262. https://doi.org/10.1038/s41567-022-01839-x
Comisso L, Lingam M, Huang YM et al (2016) General theory of the plasmoid instability. Phys Plasmas 23(10):100702. https://doi.org/10.1063/1.4964481
Cozzani G, Khotyaintsev YV, Graham DB et al (2021) Structure of a perturbed magnetic reconnection electron diffusion region in the Earth's magnetotail. Phys Rev Lett 127(21):215101. https://doi.org/10.1103/PhysRevLett.127.215101
Dahlin JT (2020) Prospectus on electron acceleration via magnetic reconnection. Phys Plasmas 27(10):601. https://doi.org/10.1063/5.0019338
Daughton W (2003) Electromagnetic properties of the lower-hybrid drift instability in a thin current sheet. Phys Plasmas 10:3103. https://doi.org/10.1063/1.1594724
Daughton W, Lapenta G, Ricci P (2004) Nonlinear evolution of the lower-hybrid drift instability in a current sheet. Phys Rev Lett 93(10):105004. https://doi.org/10.1103/PhysRevLett.93.105004
Davidson R, Gladd N (1975) Anomalous transport properties associated with the lower-hybrid drift instability. Phys Fluids 18:1327. https://doi.org/10.1063/1.861021
Dorfman S, Ji H, Yamada M et al (2013) Three-dimensional, impulsive magnetic reconnection in a laboratory plasma. Geophys Res Lett 40:233–238. https://doi.org/10.1029/2012GL054574.
Dorfman S, Ji H, Yamada M et al (2014) Experimental observation of 3-D, impulsive reconnection events in a laboratory plasma. Phys Plasmas 21(1):012109. https://doi.org/10.1063/1.4862039
Drake JF, Swisdak M, Che H et al (2006) Electron acceleration from contracting magnetic islands during reconnection. Nature 443:553–556. https://doi.org/10.1038/nature05116
Dungey J (1961) Interplanetary magnetic field and the auroral zones. Phys Rev Lett 6(2):47. https://doi.org/10.1103/PhysRevLett.6.47
Eastwood JP, Phan TD, Øieroset M et al (2010) Average properties of the magnetic reconnection ion diffusion region in the Earth's magnetotail: the 2001-2005 cluster observations and comparison with simulations. J Geophys Res Space Phys 115:A08215. https://doi.org/10.1029/2009JA014962

Eastwood JP, Phan TD, Drake JF et al (2013) Energy partition in magnetic reconnection in Earth's magnetotail. Phys Rev Lett 110(22):225001. https://doi.org/10.1103/PhysRevLett.110.225001

Eastwood JP, Goldman MV, Phan TD et al (2020) Energy flux densities near the electron dissipation region in asymmetric magnetopause reconnection. Phys Rev Lett 125(26):265102. https://doi.org/10.1103/PhysRevLett.125.265102

Egedal J, Fasoli A (2001) Single-particle dynamics in collisionless magnetic reconnection. Phys Rev Lett 86(22):5047. https://doi.org/10.1103/PhysRevLett.86.5047

Egedal J, Fasoli A, Porkolab M et al (2000) Plasma generation and confinement in a toroidal magnetic cusp. Rev Sci Instrum 71:3351–3361. https://doi.org/10.1063/1.1287340

Egedal J, Fasoli A, Nazemi J (2003) Dynamical plasma response during driven magnetic reconnection. Phys Rev Lett 90:135003. https://doi.org/10.1103/PhysRevLett.90.135003

Egedal J, Le A, Daughton W (2013) A review of pressure anisotropy caused by electron trapping in collisionless plasma, and its implications for magnetic reconnection. Phys Plasmas 20(6):061201. https://doi.org/10.1063/1.4811092

Egedal J, Le A, Daughton W et al (2018) Spacecraft observations of oblique electron beams breaking the frozen-in law during asymmetric reconnection. Phys Rev Lett 120:055101. https://doi.org/10.1103/PhysRevLett.120.055101

Egedal J, Ng J, Le A et al (2019) Pressure tensor elements breaking the frozen-in law during reconnection in Earth's magnetotail. Phys Rev Lett 123:225101. https://doi.org/10.1103/PhysRevLett.123.225101

Ergun RE, Goodrich KA, Wilder FD et al (2016a) Magnetospheric multiscale satellites observations of parallel electric fields associated with magnetic reconnection. Phys Rev Lett 116(2):235102. https://doi.org/10.1103/PhysRevLett.116.235102

Ergun RE, Holmes JC, Goodrich KA et al (2016b) Magnetospheric multiscale observations of large-amplitude, parallel, electrostatic waves associated with magnetic reconnection at the magnetopause. Geophys Res Lett 43:5626–5634. https://doi.org/10.1002/2016GL068992

Ergun RE, Chen LJ, Wilder FD et al (2017) Drift waves, intense parallel electric fields, and turbulence associated with asymmetric magnetic reconnection at the magnetopause. Geophys Res Lett 44(7):2978–2986. https://doi.org/10.1002/2016GL072493

Ergun RE, Goodrich KA, Wilder FD et al (2018) Magnetic reconnection, turbulence, and particle acceleration: observations in the Earth's magnetotail. Geophys Res Lett 45(8):3338–3347. https://doi.org/10.1002/2018GL076993

Ergun RE, Hoilijoki S, Ahmadi N et al (2019) Magnetic reconnection in three dimensions: modeling and analysis of electromagnetic drift waves in the adjacent current sheet. J Geophys Res Space Phys 124:10085–10103. https://doi.org/10.1029/2019JA027275

Eriksson S, Wilder FD, Ergun RE et al (2016) Magnetospheric multiscale observations of the electron diffusion region of large guide field magnetic reconnection. Phys Rev Lett 117(1):015001. https://doi.org/10.1103/PhysRevLett.117.015001

Eriksson S, Swisdak M, Weygand JM et al (2022) Characteristics of multi-scale current sheets in the solar wind at 1 au associated with magnetic reconnection and the case for a heliospheric current sheet avalanche. Astrophys J 933:181. https://doi.org/10.3847/1538-4357/ac73f6

Fermo RL, Drake JF, Swisdak M et al (2011) Comparison of a statistical model for magnetic islands in large current layers with Hall MHD simulations and cluster FTE observations. J Geophys Res 116:A09226. https://doi.org/10.1029/2010JA016271

Fox W, Porkolab M, Egedal J et al (2008) Laboratory observation of electron phase-space holes during magnetic reconnection. Phys Rev Lett 101:255003. https://doi.org/10.1103/PhysRevLett.101.255003

Fox W, Porkolab M, Egedal J et al (2010) Laboratory observations of electron energization and associated lower-hybrid and Trivelpiece-Gould wave turbulence during magnetic reconnection. Phys Plasmas 17:072303. https://doi.org/10.1063/1.3435216

Fox W, Porkolab M, Egedal J et al (2012) Observations of electron phase-space holes driven during magnetic reconnection in a laboratory plasma. Phys Plasmas 19:032118. https://doi.org/10.1063/1.3692224

Fox W, Sciortino F, von Stechow A et al (2017) Experimental verification of the role of electron pressure in fast magnetic reconnection with a guide field. Phys Rev Lett 118:125002. https://doi.org/10.1103/PhysRevLett.118.125002

Fox W, Wilder F, Eriksson S et al (2018) Energy conversion by parallel electric fields during guide field reconnection in scaled laboratory and space experiments. Geophys Res Lett 45(23):12677–12684. https://doi.org/10.1029/2018GL079883

Fujimoto M, Nakamura MS, Nagai T et al (1996) New kinetic evidence for the near-Earth reconnection. Geophys Res Lett 23(18):2533–2536. https://doi.org/10.1029/96GL02429

Gekelman W, Stenzel R (1984) Magnetic field line reconnection experiments: 6. Magnetic turbulence. J Geophys Res Space Phys 89(A5):2715–2733. https://doi.org/10.1029/JA089iA05p02715

Gekelman W, Stenzel R (1985) Measurement and instability analysis of three-dimensional anisotropic electron distribution functions. Phys Rev Lett 54(22):2414. https://doi.org/10.1103/PhysRevLett.54.2414
Graham DB, Khotyaintsev YV, André M et al (2022) Direct observations of anomalous resistivity and diffusion in collisionless plasma. Nat Commun 13:2954. https://doi.org/10.1038/s41467-022-30561-8
Graham D, Khotyaintsev Y, Cozzani G et al (2023) The role of kinetic instabilities and waves in collisionless magnetic reconnection. In preparation
Greess S, Egedal J, Stanier A et al (2021) Laboratory verification of electron-scale reconnection regions modulated by a three-dimensional instability. J Geophys Res Space Phys 126(7):e29316. https://doi.org/10.1029/2021JA029316
Guo LJ, Bhattacharjee A, Huang YM (2013) Distribution of plasmoids in post-coronal mass ejection current sheets. Astrophys J Lett 771:L14. https://doi.org/10.1088/2041-8205/771/1/L14
Guo F, Liu YH, Li X et al (2020) Recent progress on particle acceleration and reconnection physics during magnetic reconnection in the magnetically-dominated relativistic regime. Phys Plasmas 27(8):080501. https://doi.org/10.1063/5.0012094
Hare JD, Suttle L, Lebedev SV et al (2017) Anomalous heating and plasmoid formation in a driven magnetic reconnection experiment. Phys Rev Lett 118(8):085001. https://doi.org/10.1103/PhysRevLett.118.085001
Hesse M, Schindler K, Birn J et al (1999) The diffusion region in collisionless magnetic reconnection. Phys Plasmas 6:1781. https://doi.org/10.1063/1.873436
Hesse M, Aunai N, Sibeck D et al (2014) On the electron diffusion region in planar, asymmetric, systems. Geophys Res Lett 41(24):8673–8680. https://doi.org/10.1002/2014GL061586
Hoshino M, Mukai T, Terasawa T et al (2001) Suprathermal electron acceleration in magnetic reconnection. J Geophys Res 106(A11):25979–25998. https://doi.org/10.1029/2001JA900052
Hsu S, Fiksel G, Carter T et al (2000) Local measurement of nonclassical ion heating during magnetic reconnection. Phys Rev Lett 84:3859. https://doi.org/10.1103/PhysRevLett.84.3859
Hu Y, Yoo J, Ji H et al (2021) Probe measurements of electric field and electron density fluctuations at megahertz frequencies using in-shaft miniature circuits. Rev Sci Instrum 92:033534. https://doi.org/10.1063/5.0035135
Huang YM, Bhattacharjee A (2012) Distribution of plasmoids in high-lundquist-number magnetic reconnection. Phys Rev Lett 109(26):265002. https://doi.org/10.1103/PhysRevLett.109.265002
Intrator TP, Sun X, Lapenta G et al (2009) Experimental onset threshold and magnetic pressure pile-up for 3D reconnection. Nat Phys 5:521–526. https://doi.org/10.1038/NPHYS1300
Janvier M, Démoulin P, Dasso S (2014) In situ properties of small and large flux ropes in the solar wind. J Geophys Res Space Phys 119(9):7088–7107. https://doi.org/10.1002/2014JA020218
Jara-Almonte J, Daughton W, Ji H (2014) Debye scale turbulence within the electron diffusion layer during magnetic reconnection. Phys Plasmas 21(3):032114. https://doi.org/10.1063/1.4867868
Jara-Almonte J, Ji H, Yamada M et al (2016) Laboratory observation of resistive electron tearing in a two-fluid reconnecting current sheet. Phys Rev Lett 117(9):095001. https://doi.org/10.1103/PhysRevLett.117.095001
Ji H, Daughton W (2011) Phase diagram for magnetic reconnection in heliophysical, astrophysical, and laboratory plasmas. Phys Plasmas 18(11):111207. https://doi.org/10.1063/1.3647505
Ji H, Daughton W (2022) Preface for frontiers of magnetic reconnection research in heliophysical, astrophysical, and laboratory plasmas. Phys Plasmas 29(7):070401. https://doi.org/10.1063/5.0104925
Ji H, Yamada M, Hsu S et al (1998) Experimental test of the Sweet-Parker model of magnetic reconnection. Phys Rev Lett 80:3256. https://doi.org/10.1103/PhysRevLett.80.3256
Ji H, Terry S, Yamada M et al (2004) Electromagnetic fluctuation during fast reconnection in a laboratory plasma. Phys Rev Lett 92:115001. https://doi.org/10.1103/PhysRevLett.92.115001
Ji H, Kulsrud R, Fox W et al (2005) An obliquely propagating electromagnetic drift instability in the lower hybrid frequency range. J Geophys Res 110:A08212. https://doi.org/10.1029/2005JA011188
Ji H, Ren Y, Yamada M et al (2008) New insights into dissipation in the electron layer during magnetic reconnection. Geophys Res Lett 35:L13106. https://doi.org/10.1029/2008GL034538
Ji H, Cutler R, Gettelfinger G et al (2018) The FLARE device and its first plasma operation. In: APS meeting abstracts. p CP11.020, http://meetings.aps.org/link/BAPS.2018.DPP.CP11.20
Ji H, Daughton W, Jara-Almonte J et al (2022) Magnetic reconnection in the era of exascale computing and multiscale experiments. Nat Rev Phys 4:263–282. https://doi.org/10.1038/s42254-021-00419-x
Katz N, Egedal J, Fox W et al (2010) Laboratory observation of localized onset of magnetic reconnection. Phys Rev Lett 104(25):255004. https://doi.org/10.1103/PhysRevLett.104.255004
Khotyaintsev YV, Graham DB, Norgren C et al (2019) Collisionless magnetic reconnection and waves: progress review. Front Astron Space Sci 6:70. https://doi.org/10.3389/fspas.2019.00070
Khotyaintsev YV, Graham DB, Steinvall K et al (2020) Electron heating by debye-scale turbulence in guide-field reconnection. Phys Rev Lett 124:045101. https://doi.org/10.1103/PhysRevLett.124.045101

Klein KG, Spence H, Alexandrova O et al (2023) HelioSwarm: a multipoint, multiscale mission to characterize turbulence. Space Sci Rev 219:74. https://doi.org/10.1007/s11214-023-01019-0. ArXiv e-prints arXiv:2306.06537

Kleva R, Drake J, Waelbroeck F (1995) Fast reconnection in high temperature plasmas. Phys Plasmas 2(23):23–34. https://doi.org/10.1063/1.871095

Krall N, Liewer P (1971) Low-frequency instabilities in magnetic pulses. Phys Rev A 4(5):2094. https://doi.org/10.1103/PhysRevA.4.2094

Kulsrud R, Ji H, Fox W et al (2005) An electromagnetic drift instability in the magnetic reconnection experiment and its importance for magnetic reconnection. Phys Plasmas 12:082301. https://doi.org/10.1063/1.1949225

Lawrence E, Ji H, Yamada M et al (2013) Laboratory study of hall reconnection in partially ionized plasmas. Phys Rev Lett 110:015001. https://doi.org/10.1103/PhysRevLett.110.015001

Le A, Egedal J, Daughton W et al (2015) Transition in electron physics of magnetic reconnection in weakly collisional plasma. J Plasma Phys 81(1):305810108. https://doi.org/10.1017/S0022377814000907

Le A, Egedal J, Daughton W (2016) Two-stage bulk electron heating in the diffusion region of anti-parallel symmetric reconnection. Phys Plasmas 23(10):102109. https://doi.org/10.1063/1.4964768

Le A, Daughton W, Chen LJ et al (2017) Enhanced electron mixing and heating in 3-D asymmetric reconnection at the Earth's magnetopause. Geophys Res Lett 44:2096–2104. https://doi.org/10.1002/2017GL072522.

Lefebvre B, Chen LJ, Gekelman W et al (2010) Laboratory measurements of electrostatic solitary structures generated by beam injection. Phys Rev Lett 105:115001. https://doi.org/10.1103/PhysRevLett.105.115001

Li X, Guo F, Liu YH (2021) The acceleration of charged particles and formation of power-law energy spectra in nonrelativistic magnetic reconnection. Phys Plasmas 28:052905 https://doi.org/10.1063/5.0047644

Liu YH, Cassak P, Li X et al (2022) First-principles theory of the rate of magnetic reconnection in magnetospheric and solar plasmas. Commun Phys 5:97. https://doi.org/10.1038/s42005-022-00854-x

Lyons LR, Pridmore-Brown DC (1990) Force balance near an X line in a collisionless plasma. J Geophys Res 95:20903. https://doi.org/10.1029/JA095iA12p20903

Majeski S, Ji H (2023) Super-Fermi acceleration in multiscale MHD reconnection. Phys Plasmas 30(4):042106. https://doi.org/10.1063/5.0139276

Majeski S, Ji H, Jara-Almonte J et al (2021) Guide field effects on the distribution of plasmoids in multiple scale reconnection. Phys Plasmas 28(9):092106. https://doi.org/10.1063/5.0059017

Matsumoto H, Deng XH, Kojima H et al (2003) Observation of electrostatic solitary waves associated with reconnection on the dayside magnetopause boundary. Geophys Res Lett 30(6):1326. https://doi.org/10.1029/2002GL016319

Mozer FS, Pritchett PL (2011) Electron physics of asymmetric magnetic field reconnection. Space Sci Rev 158(1):119–143. https://doi.org/10.1007/s11214-010-9681-8

Mozer FS, Bale S, Phan TD (2002) Evidence of diffusion regions at a subsolar magnetopause crossing. Phys Rev Lett 89:015002. https://doi.org/10.1103/PhysRevLett.89.015002

Mozer FS, Bale SD, Cattell CA et al (2022) Core electron heating by triggered ion acoustic waves in the solar wind. Astrophys J Lett 927:L15. https://doi.org/10.3847/2041-8213/ac5520

Ng J, Chen LJ, Le A et al (2020) Lower-hybrid-drift vortices in the electron-scale magnetic reconnection layer. Geophys Res Lett 47:e2020GL090726. https://doi.org/10.1029/2020GL090726

Ng J, Yoo J, Chen LJ et al (2023) 3d simulation of lower-hybrid drift waves in strong guide field asymmetric reconnection in laboratory experiments. Phys Plasmas 30:042101. https://doi.org/10.1063/5.0138278

Norgren C, Vaivads A, Khotyaintsev YV et al (2012) Lower hybrid drift waves: space observations. Phys Rev Lett 109:055001. https://doi.org/10.1103/PhysRevLett.109.055001

Øieroset M, Sandholt PE, Lühr H et al (1997) Auroral and geomagnetic events at cusp/mantle latitudes in the prenoon sector during positive IMF B_y conditions: signatures of pulsed magnetopause reconnection. J Geophys Res 102(A4):7191–7206. https://doi.org/10.1029/96JA03716

Øieroset M, Phan TD, Haggerty C et al (2016) MMS observations of large guide field symmetric reconnection between colliding reconnection jets at the center of a magnetic flux rope at the magnetopause. Geophys Res Lett 43(1):5536–5544. https://doi.org/10.1002/2016GL069166

Oka M, Birn J, Egedal J et al (2023). Particle acceleration by magnetic reconnection in geospace. Space Sci Rev 219:75. https://doi.org/10.1007/s11214-023-01011-8

Olson J, Egedal J, Greess S et al (2016) Experimental demonstration of the collisionless plasmoid instability below the ion kinetic scale during magnetic reconnection. Phys Rev Lett 116(25):255001. https://doi.org/10.1103/PhysRevLett.116.255001

Olson J, Egedal J, Clark M et al (2021) Regulation of the normalized rate of driven magnetic reconnection through shocked flux pileup. J Plasma Phys 87(3):175870301. https://doi.org/10.1017/S0022377821000659

Ono Y, Morita A, Katsurai M et al (1993) Experimental investigation of three-dimensional magnetic reconnection by use of two colliding spheromaks. Phys Fluids B 5:3691. https://doi.org/10.1063/1.860840
Ono Y, Tanabe H, Hayashi Y et al (2011) Ion and electron heating characteristics of magnetic reconnection in a two flux loop merging experiment. Phys Rev Lett 107(18):185001. https://doi.org/10.1103/PhysRevLett.107.185001
Palmroth M, Pulkkinen TI, Ganse U et al (2023) Magnetotail plasma eruptions driven by magnetic reconnection and kinetic instabilities. Nat Geosci 16:570–576. https://doi.org/10.1038/s41561-023-01206-2
Papadopoulos K (1977) A review of anomalous resistivity for the ionosphere. Rev Geophys Space Phys 15:113. https://doi.org/10.1029/RG015i001p00113
Parker E (1957) Sweet's mechanism for merging magnetic fields in conducting fluids. J Geophys Res 62:509. https://doi.org/10.1029/JZ062i004p00509
Petschek H (1964) Magnetic field annihilation. NASA Spec Publ 50:425
Pritchett PL (2001) Geospace environment modeling magnetic reconnection challenge: simulations with a full particle electromagnetic code. J Geophys Res 106:3783. https://doi.org/10.1029/1999JA001006
Pucci F, Velli M (2014) Reconnection of quasi-singular current sheets: the "ideal" tearing mode. Astrophys J 780(2):L19. https://doi.org/10.1088/2041-8205/780/2/L19
Pucci F, Usami S, Ji H et al (2018) Energy transfer and electron energization in collisionless magnetic reconnection for different guide-field intensities. Phys Plasmas 25(12):122111. https://doi.org/10.1063/1.5050992
Ren Y (2007) Studies of non-MHD effects during magnetic reconnection in a laboratory plasma. PhD thesis, Princeton University
Ren Y, Yamada M, Gerhardt S et al (2005) Experimental verification of the Hall effect during magnetic reconnection in a laboratory plasma. Phys Rev Lett 95(5):055003. https://doi.org/10.1103/PhysRevLett.95.055003
Ren Y, Yamada M, Ji H et al (2008) Identification of the electron diffusion region during magnetic reconnection in a laboratory plasma. Phys Rev Lett 101:085003. https://doi.org/10.1103/PhysRevLett.101.085003
Retinò A, Khotyaintsev Y, Le Contel O et al (2022) Particle energization in space plasmas: towards a multi-point, multi-scale plasma observatory. Exp Astron 54:427–471. https://doi.org/10.1007/s10686-021-09797-7
Roytershteyn V, Daughton W, Dorfman S et al (2010) Driven reconnection near the dreicer limit. Phys Plasmas 17:055706. https://doi.org/10.1063/1.3399787
Roytershteyn V, Daughton W, Karimabadi H et al (2012) Influence of the lower-hybrid drift instability on magnetic reconnection in asymmetric configurations. Phys Rev Lett 108:185001. https://doi.org/10.1103/PhysRevLett.108.185001
Roytershteyn V, Dorfman S, Daughton W et al (2013) Electromagnetic instability of thin reconnection layers: comparison of 3d simulations with mrx observations. Phys Plasmas 20:061212. https://doi.org/10.1063/1.4811371
Russell CT, Elphic RC (1979) ISEE observations of flux transfer events at the dayside magnetopause. Geophys Res Lett 6:33–36. https://doi.org/10.1029/GL006i001p00033
Sato T, Hayashi T (1979) Externally driven magnetic reconnection and a powerful magnetic energy converter. Phys Fluids 22:1189. https://doi.org/10.1063/1.862721
Shay MA, Drake JF (1998) The role of electron dissipation on the rate of collisionless magnetic reconnection. Geophys Res Lett 25:3759–3762. https://doi.org/10.1029/1998GL900036.
Shay M, Drake J, Denton R et al (1998) Structure of the dissipation region during collisionless magnetic reconnection. J Geophys Res 103:9165. https://doi.org/10.1029/97JA03528
Shi P, Srivastav P, Barbhuiya MH et al (2022) Laboratory observations of electron heating and non-Maxwellian distributions at the kinetic scale during electron-only magnetic reconnection. Phys Rev Lett 128:025002. https://doi.org/10.1103/PhysRevLett.128.025002
Sonnerup BUÖ (1979) Magnetic field reconnection. In: Lanzerotti L, Kennel C, Parker E (eds) Solar system plasma physics. Cambridge University Press, New York. p 45
Stark A, Fox W, Egedal J et al (2005) Laser-induced fluorescence measurement of the ion-energy-distribution function in a collisionless reconnection experiment. Phys Rev Lett 95:235005. https://doi.org/10.1103/PhysRevLett.95.235005
Steinvall K, Khotyaintsev YV, Graham DB et al (2021) Large amplitude electrostatic proton plasma frequency waves in the magnetospheric separatrix and outflow regions during magnetic reconnection. Geophys Res Lett 48(5):e90286. https://doi.org/10.1029/2020GL090286
Stenzel R, Gekelman W (1979) Experiments on magnetic field line reconnection. Phys Rev Lett 42:1055. https://doi.org/10.1103/PhysRevLett.42.1055
Stenzel RL, Gekelman W, Urrutia JM (1986) Lessons from laboratory experiments on reconnection. Adv Space Res 6(1):135–147. https://doi.org/10.1016/0273-1177(86)90025-6

Stix T (1992) Waves in plasmas. American Institute of Physics, New York
Sweet P (1958) The neutral point theory of solar flares. In: Lehnert B (ed) Electromagnetic phenomena in cosmical physics. Cambridge University Press, New York, p 123
Tanabe H, Yamada T, Watanabe T et al (2015) Electron and ion heating characteristics during magnetic reconnection in the mast spherical tokamak. Phys Rev Lett 115:215004. https://doi.org/10.1103/PhysRevLett.115.215004
Terasawa T (1983) Hall current effect on tearing mode instability. Geophys Res Lett 10:475. https://doi.org/10.1029/GL010i006p00475
Tharp T, Yamada M, Ji H et al (2012) Quantitative study of guide-field effects on Hall reconnection in a laboratory plasma. Phys Rev Lett 109:169002
Torbert RB, Burch JL, Giles BL et al (2016) Estimates of terms in Ohm's law during an encounter with an electron diffusion region. Geophys Res Lett 43:5918–5925. https://doi.org/10.1002/2016GL069553
Torbert RB, Burch JL, Phan TD et al (2018) Electron-scale dynamics of the diffusion region during symmetric magnetic reconnection in space. Science 362(6421):1391–1395. https://doi.org/10.1126/science.aat2998
Trivelpiece AW, Gould RW (1959) Space charge waves in cylindrical plasma columns. J Appl Phys 30(11):1784–1793. https://doi.org/10.1063/1.1735056
Uchino H, Kurita S, Harada Y et al (2017) Waves in the innermost open boundary layer formed by dayside magnetopause reconnection. J Geophys Res Space Phys 122:3291–3307. https://doi.org/10.1002/2016JA023300
Ugai M, Tsuda T (1977) Magnetic field-line reconnection by localized enhancement of reconnection. 1. Evolution in a compressible mhd fluid. J Plasma Phys 17:337. https://doi.org/10.1017/S0022377800020663
Uzdensky DA, Loureiro NF, Schekochihin A (2010) Fast magnetic reconnection in the plasmoid-dominated regime. Phys Rev Lett 105:235002. https://doi.org/10.1103/PhysRevLett.105.235002
Vasyliunas V (1975) Theoretical models of field line merging, I. Rev Geophys Space Phys 13:303. https://doi.org/10.1029/RG013i001p00303
von Stechow A, Fox W, Jara-Almonte J et al (2018) Electromagnetic fluctuations during guide field reconnection in a laboratory plasma. Phys Plasmas 25:052120. https://doi.org/10.1063/1.5025827
Wilder FD, Ergun RE, Burch JL et al (2018) The role of the parallel electric field in electron-scale dissipation at reconnecting currents in the magnetosheath. J Geophys Res Space Phys 123:6533–6547. https://doi.org/10.1029/2018JA025529
Wygant J, Cattell C, Lysak R et al (2005) Cluster observations of an intense normal component of the electric field at a thin reconnecting current sheet in the tail and its role in the shock-like acceleration of the ion fluid into the separatrix region. J Geophys Res 110:A09206. https://doi.org/10.1029/2004JA010708
Yamada M (2022) Magnetic reconnection: a modern synthesis of theory, experiment, and observations. Princeton University Press, Princeton
Yamada M, Ono Y, Hayakawa A et al (1990) Magnetic reconnection of plasma toroids with co- and counter-helicity. Phys Rev Lett 65:721. https://doi.org/10.1103/PhysRevLett.65.721
Yamada M, Ji H, Hsu S et al (1997) Study of driven magnetic reconnection in a laboratory plasma. Phys Plasmas 4:1936 https://doi.org/10.1063/1.872336
Yamada M, Ren Y, Ji H et al (2006) Experimental study of two-fluid effects on magnetic reconnection in a laboratory plasma with variable collisionality. Phys Plasmas 13:052119. https://doi.org/10.1063/1.2203950
Yamada M, Kulsrud R, Ji H (2010) Magnetic reconnection. Rev Mod Phys 82:603. https://doi.org/10.1103/RevModPhys.82.603
Yamada M, Yoo J, Jara-Almonte J et al (2014) Conversion of magnetic energy in the magnetic reconnection layer of a laboratory plasma. Nat Commun 5:4774. https://doi.org/10.1038/ncomms5774
Yamada M, Yoo J, Jara-Almonte J et al (2015) Study of energy conversion and partitioning in the magnetic reconnection layer of a laboratory plasmas. Phys Plasmas 22(5):056501. https://doi.org/10.1063/1.4920960
Yamada M, Yoo J, Myers CE (2016) Understanding the dynamics and energetics of magnetic reconnection in a laboratory plasma: review of recent progress on selected fronts. Phys Plasmas 23(5):055402. https://doi.org/10.1063/1.4948721
Yamada M, Chen LJ, Yoo J et al (2018) The two-fluid dynamics and energetics of the asymmetric magnetic reconnection in laboratory and space plasmas. Nat Commun 8:5223. https://doi.org/10.1038/s41467-018-07680-2
Yoo J, Yamada M, Ji H et al (2013) Observation of ion acceleration and heating during collisionless magnetic reconnection in a laboratory plasma. Phys Rev Lett 110:215007. https://doi.org/10.1103/PhysRevLett.110.215007
Yoo J, Yamada M, Ji H et al (2014a) Bulk ion acceleration and particle heating during magnetic reconnection in a laboratory plasma. Phys Plasmas 21(5):055706. https://doi.org/10.1063/1.4874331

Yoo J, Yamada M, Ji H et al (2014b) Laboratory study of magnetic reconnection with a density asymmetry across the current sheet. Phys Rev Lett 113(9):095002. https://doi.org/10.1103/PhysRevLett.113.095002
Yoo J, Na B, Jara-Almonte J et al (2017) Electron heating and energy inventory during asymmetric reconnection in a laboratory plasma. J Geophys Res 122:9264–9281. https://doi.org/10.1002/2017JA024152
Yoo J, Jara-Almonte J, Yerger E et al (2018) Whistler wave generation by anisotropic tail electrons during asymmetric magnetic reconnection in space and laboratory. Geophys Res Lett 45(16):8054–8061. https://doi.org/10.1029/2018GL079278
Yoo J, Wang S, Yerger E et al (2019) Whistler wave generation by electron temperature anisotropy during magnetic reconnection at the magnetopause. Phys Plasmas 26(5):052902. https://doi.org/10.1063/1.5094636
Yoo J, Ji JY, Ambat MV et al (2020) Lower hybrid drift waves during guide field reconnection. Geophys Res Lett 47(21):e87192. https://doi.org/10.1029/2020GL087192
Yoo J, Ng J, Ji H et al (2023) Anomalous resistivity and electron heating by lower hybrid drift waves during magnetic reconnection with a guide field. Submitted
Zenitani S, Hoshino M (2001) The generation of nonthermal particles in the relativistic magnetic reconnection of pair plasmas. Astrophys J Lett 562:L63–L66. https://doi.org/10.1086/337972
Zenitani S, Hesse M, Klimas A et al (2011) New measure of the dissipation region in collisionless magnetic reconnection. Phys Rev Lett 106:195003. https://doi.org/10.1103/PhysRevLett.106.195003
Zhang S, Chien A, Gao L et al (2023) Ion and electron acoustic bursts during anti-parallel reconnection driven by lasers. Nat Phys 19:909–916. https://doi.org/10.1038/s41567-023-01972-1
Zhou M, Deng XH, Li SY et al (2009) Observation of waves near lower hybrid frequency in the reconnection region with thin current sheet. J Geophys Res 114:2216. https://doi.org/10.1029/2008JA013427
Zweibel E, Yamada M (2009) Magnetic reconnection in astrophysical and laboratory plasmas. Annu Rev Astron Astrophys 47(1):291. https://doi.org/10.1146/annurev-astro-082708-101726

Authors and Affiliations

H. Ji[1,2] · J. Yoo[2] · W. Fox[2] · M. Yamada[2] · M. Argall[3] · J. Egedal[4] · Y.-H. Liu[5] · R. Wilder[6] · S. Eriksson[7] · W. Daughton[8] · K. Bergstedt[1] · S. Bose[2] · J. Burch[9] · R. Torbert[3] · J. Ng[10,11,2] · L.-J. Chen[11]

✉ H. Ji
hji@pppl.gov

1 Department of Astrophysical Sciences, Princeton University, 4 Ivy Lane, Princeton, 08544, New Jersey, USA

2 Princeton Plasma Physics Laboratory, P.O. Box 451, Princeton, 08543, New Jersey, USA

3 Institute for the Study of Earth, Oceans, and Space, University of New Hampshire, 8 College Road, Durham, 03824, New Hampshire, USA

4 Department of Physics, University of Wisconsin - Madison, 1150 University Avenue, Madison, 53706, Wisconsin, USA

5 Department of Physics and Astronomy, Dartmouth College, 17 Fayerweather Hill Road, Hanover, 03755, New Hampshire, USA

6 Department of Physics, University of Texas at Arlington, 701 S. Nedderman Drive, Arlington, 76019, Texas, USA

7 Laboratory for Atmospheric and Space Physics, University of Colorado at Boulder, 1234 Innovation Drive, Boulder, 80303, Colorado, USA

8 Los Alamos National Laboratory, P.O. Box 1663, Los Alamos, 87545, New Mexico, USA

[9] Southwest Research Institute, 6220 Culebra Road, San Antonio, 78238, Texas, USA

[10] Department of Astronomy, University of Maryland, 4296 Stadium Drive, College Park, 20742, Maryland, USA

[11] Goddard Space Flight Center, Mail Code 130, Greenbelt, 20771, Maryland, USA

VI. Data Analysis Methods and Simulation Models

Space Science Reviews (2024) 220:68
https://doi.org/10.1007/s11214-024-01095-w

Advanced Methods for Analyzing in-Situ Observations of Magnetic Reconnection

H. Hasegawa[1] · M.R. Argall[2] · N. Aunai[3] · R. Bandyopadhyay[4] · N. Bessho[5,6] · I.J. Cohen[7] · R.E. Denton[8] · J.C. Dorelli[6] · J. Egedal[9] · S.A. Fuselier[10,11] · P. Garnier[12] · V. Génot[12] · D.B. Graham[13] · K.J. Hwang[10] · Y.V. Khotyaintsev[13,14] · D.B. Korovinskiy[15] · B. Lavraud[12,16] · Q. Lenouvel[12] · T.C. Li[8] · Y.-H. Liu[8] · B. Michotte de Welle[3] · T.K.M. Nakamura[15,17] · D.S. Payne[18] · S.M. Petrinec[19] · Y. Qi[20] · A.C. Rager[6] · P.H. Reiff[21] · J.M. Schroeder[9] · J.R. Shuster[2] · M.I. Sitnov[7] · G.K. Stephens[7] · M. Swisdak[18] · A.M. Tian[22] · R.B. Torbert[23,24] · K.J. Trattner[20] · S. Zenitani[15,25]

Received: 25 May 2023 / Accepted: 19 July 2024 / Published online: 2 September 2024

Abstract
There is ample evidence for magnetic reconnection in the solar system, but it is a nontrivial task to visualize, to determine the proper approaches and frames to study, and in turn to elucidate the physical processes at work in reconnection regions from in-situ measurements of plasma particles and electromagnetic fields. Here an overview is given of a variety of single- and multi-spacecraft data analysis techniques that are key to revealing the context of in-situ observations of magnetic reconnection in space and for detecting and analyzing the diffusion regions where ions and/or electrons are demagnetized. We focus on recent advances in the era of the Magnetospheric Multiscale mission, which has made electron-scale, multi-point measurements of magnetic reconnection in and around Earth's magnetosphere.

Keywords Magnetic reconnection · Data analysis techniques · In-situ measurements · Magnetosphere · Electron diffusion region

1 Introduction

Magnetic reconnection occurring in geospace is in the collisionless regime, so that the reconnection and surrounding regions have multi-scale structures: magnetohydrodynamic (MHD) regions where both the ion and electron fluids satisfy the frozen-in condition, ion diffusion regions (IDRs) where ions are demagnetized but electrons remain magnetized, and electron diffusion regions (EDRs) where both ions and electrons are demagnetized and magnetic topology changes (e.g., Daughton et al. 2006; Paschmann et al. 2013). Since its launch in 2015, the Magnetospheric Multiscale (MMS) mission (Burch et al. 2016b) has been making electron- or sub-ion-scale (unprecedented high spatial- and temporal-resolution) measurements of these regions in and around Earth's magnetosphere, especially in the magnetotail and at the magnetopause (Fig. 1), to elucidate the microphysics of magnetic reconnection. A number of novel techniques for analyzing electromagnetic field and plasma data taken in

Extended author information available on the last page of the article

and around the reconnection regions have been developed in preparation for and during the MMS mission.

The present review provides an overview of updated data analysis methods for in-situ observations of magnetic reconnection in space. This includes a range of prior applications of the methods, so that it can be used by the community and early career researchers to decide whether some of the methods is appropriate for their research. Thorough reviews of various single- and multi-spacecraft methods for analyzing MHD- and ion-scale aspects of reconnection and other space plasma processes were given by Paschmann and Daly (1998, 2008) in the era of the Cluster mission (e.g., Escoubet et al. 1997; Paschmann et al. 2005). Magnetic reconnection also involves inherently multi-dimensional structures and often occurs in highly nonuniform environments, as in the case at the magnetopause with substantial jumps across the current sheet in the plasma density, temperature, and magnetic field intensity (Fig. 1). Thus, analysis methods previously reviewed and those assuming a uniform or weakly nonuniform background, such as wave analysis techniques (e.g., Narita 2017), are not covered in this review, except for some essential ones.

The analysis of a magnetic reconnection event may proceed as follows: (1) identification of electric current sheets or localized plasma bulk flows where reconnection may occur, (2) revealing the large-scale and local context of the reconnection event, based on upstream solar wind and geomagnetic field conditions, and the geometry (Fig. 1) and structures of the reconnecting current sheet, and (3) detection and analysis of microscopic regions key to the reconnection process, such as the diffusion and energy conversion regions. For single or a few event analysis, step (1) can be done by identifying rapid magnetic field rotations, current density enhancements, and/or Alfvénic plasma velocity changes, which are intermittently seen in time series data. On the other hand, steps (2) and (3) require in-depth data analysis, empirical modeling, and/or numerical simulation performed specifically for the event of interest; they are the main topic of this review.

The rest of the paper is organized as follows. Section 2 presents a variety of methods for both large-scale and local contexts, including estimation of the coordinate system and frame velocity of the current sheet, and reconstruction of two- or three-dimensional plasma and electromagnetic field structures around the diffusion regions. Section 3 focuses on methods to identify and analyze the diffusion regions, including estimation of the reconnection rate and electric field (Fig. 1). Section 4 gives a brief summary and outlook. In Appendix A, an overview is given of methods for the purpose of mission operations and automated identification of plasma regions and current sheets in and around the magnetosphere. Appendix B briefly explains how higher time resolution plasma moments are computed from the MMS data. In Appendix C we provide, as a quick user guide, tables (Tables 1-7) that summarize for each of the methods (1) required input data, (2) output, (3) fundamental theory, concept, or technique(s) that underlies the method, (4) model or underlying assumption(s), (5) relevant references, etc.

For a general overview of magnetic reconnection as a plasma physical process and primary scientific results from the MMS mission, many of which were obtained by use of either of the methods discussed in this review, see other articles in this collection (e.g., Norgren et al. 2024, Fuselier et al. 2024, Hwang et al. 2023, and Oka et al. 2023). We also note that many of the methods summarized in this review can be used for the analysis of in-situ measurements in other regions of the solar system, and some may be applicable to reconnection observed in laboratory and solar (i.e., remotely sensed) plasmas. In particular, the method for estimating the reconnection rate, discussed in Sect. 3.3.3, can be applied to imaging observations as well, and the concept underlying the one reviewed in Sect. 3.3.2 is common to a method for analyzing reconnection observed during solar flares (Qiu et al. 2007).

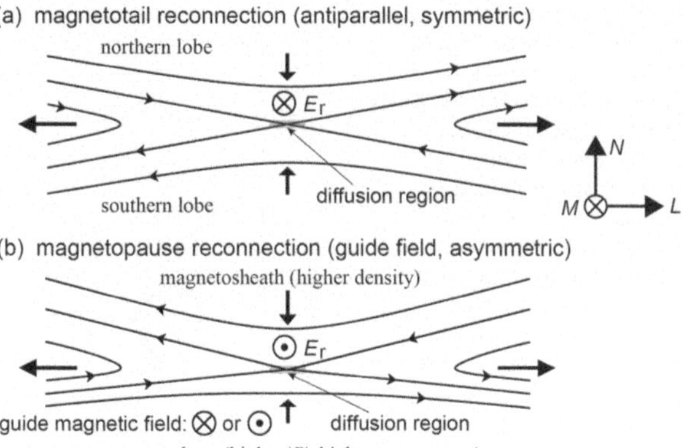

Fig. 1 Typical geometry (magnetic field and plasma inflow and outflow pattern) of (a) magnetotail (approximately antiparallel and symmetric) reconnection and (b) magnetopause (normally guide-field and asymmetric) reconnection. Note that the reconnection electric field E_r in LMN coordinates typically has a different polarity for the two cases

2 Methods for Context

Reconnection regions on kinetic scales (of order 1-500 km) are much smaller than the size of the geospace or the magnetosphere (of order 10^5 km) and are localized in space and often in time. It is thus important to understand a large-scale context and boundary conditions of spacecraft observations of reconnection-related phenomena and the field geometry and structures around the observing spacecraft. This section gives an overview of methods for revealing such contexts.

2.1 Large-Scale Context

2.1.1 Maximum Magnetic Shear Model

The maximum magnetic shear model can predict the location on an empirical model magnetopause where the magnetic shear across the magnetopause current sheet is large or maximized, which is a plausible location of magnetopause reconnection, for interplanetary magnetic field (IMF) and geomagnetic dipole tilt conditions given as input (Trattner et al. 2021 and references therein). It was originally developed as a product of a polar cusp study using data from the NASA Polar satellite. The study determined the dayside magnetopause reconnection location (Trattner et al. 2007) for southward IMF conditions by using time-of-flight characteristics of cusp ions and the low-velocity cutoff method originally developed by Onsager et al. (1990, 1991) for the magnetotail reconnection location.

Figure 2 shows the general geometry of the low-velocity cutoff method that is used to estimate the dayside magnetopause reconnection location from cusp observations. Shown are the geomagnetic field lines (green), the reconnection location at the magnetopause (X), the satellite position in the cusp (⊖), the ionospheric magnetic mirror point on the cusp field line (M), and the orbit path of a satellite passing through the cusp (red curve). A cusp-traversing satellite simultaneously observes slower magnetosheath ions that arrive from the site of magnetopause reconnection (incident ion beam) and faster magnetosheath ions that reached the ionospheric mirror point and returned to the high-altitude cusp-traversing satellite (mirrored ion beam).

The color inlay of Fig. 2, centered along the cusp field line, shows an H^+ velocity distribution acquired by the TIMAS (Toroidal Imaging Mass Angle Spectrometer) instrument (Shelley et al. 1995) on board Polar in the cusp on 20 October 1997 from 14:05:59 to 14:06:11 UT. The H^+ distribution is presented in magnetic field-aligned coordinates after

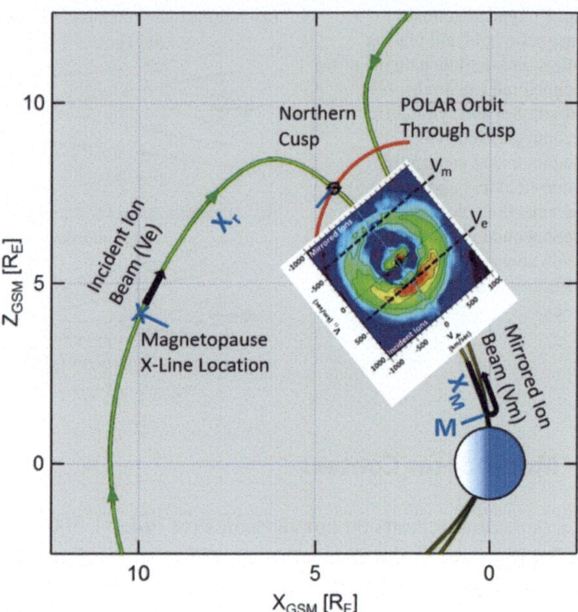

Fig. 2 Schematic of the northern cusp region with the Polar satellite simultaneously observing incident ions on newly opened magnetic field lines which originate at the magnetopause reconnection location and mirrored ions returning from the ionosphere. The color distribution function shows the cutoff velocities V_e and V_m of the incident and mirrored ion beams, respectively

removing the effect of the H^+ bulk flow transverse to the magnetic field, and shows the incident magnetosheath ions injected at the location of magnetopause reconnection in addition to the mirrored ions that returned from the ionospheric mirror points.

The distance X_r along the field line between a satellite in the cusp and the magnetopause reconnection site can be computed by

$$X_r/X_m = 2V_e/(V_m - V_e), \quad (1)$$

derived from equating the flight times of the incident and mirrored ion beams. Here V_e and V_m are the cutoff velocities of the incident and mirrored beams, respectively, and X_m is the distance between the satellite and the mirror point (Fig. 2). To determine the cutoff velocities, the peaks of the ion beams are fit with Gaussian distributions. The cutoff velocities are defined at the low-speed side of the peaks where the ion flux is 1/e of the peak flux (e.g., Fuselier et al. 2000; Trattner et al. 2007, 2005). The low velocity cutoffs are marked with black dashed lines in the color inlay of Fig. 2.

To determine X_m, the geomagnetic field line at the satellite position in the cusp is traced down to the ionospheric mirror point by using the T96 model (Tsyganenko 1995). The model field line is also used to trace the calculated distance X_r back to the reconnection location on the magnetopause. These end points of the field line traces mark the location of dayside magnetopause reconnection where the magnetosheath plasma enters the magnetosphere (e.g., Fuselier et al. 2000; Trattner et al. 2007, 2012, 2021).

Figure 3 (top panel) shows the distance to the reconnection site derived from Eq. (1) versus the Polar/TIMAS observation time during the cusp crossing on 11 April 1996. The distance to the reconnection site ranges from about 6 to 12 R_E, which is most likely caused by changes in the satellite local time position. The uncertainties in the distance calculation are determined by those in measuring the low-velocity cutoff velocities. It is defined as 1/2 the difference between the velocity at the peak and the low-velocity cutoff (Fuselier et al. 2000; Trattner et al. 2007).

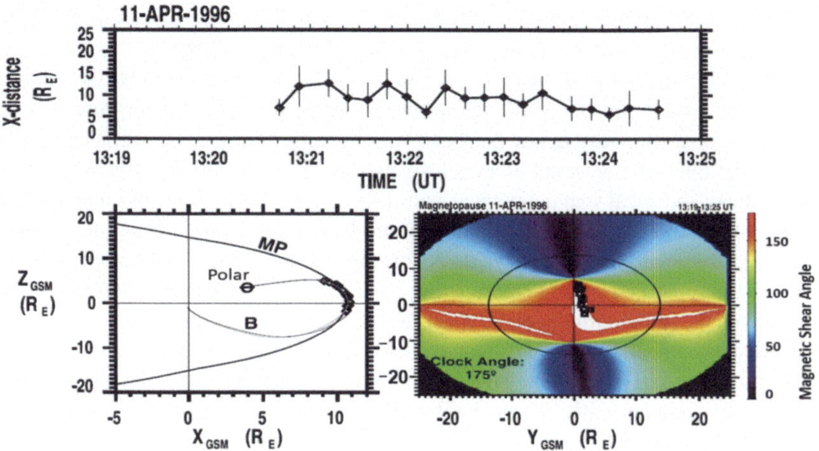

Fig. 3 The distance to the magnetopause reconnection site from the cusp position of the Polar satellite on 11 April 1996 (top panel). The location of the magnetopause reconnection site as seen from dawn (bottom left panel). The magnetopause magnetic shear angle with the reconnection site location (black squares) as seen from the Sun for the 11 April 1996 Polar cusp crossing (bottom right panel)

The lower left panel of Fig. 3 shows the magnetopause shape as viewed from dawn. Starting at the cusp location of the Polar satellite (⊖), the distance to the reconnection site is traced along the T96 model geomagnetic field lines to the magnetopause where their end points are marked with black diamonds.

The lower right panel of Fig. 3 shows the plot of magnetic shear angle on the magnetopause for the Polar cusp crossings. The shear angles are estimated by using the T96 model (internal field) together with the Kobel and Flückiger (1994) magnetosheath magnetic field draping model (external field), merged at the dayside ellipsoidal magnetopause shape of the Sibeck et al. (1991) model (e.g., Trattner et al. 2007, 2021). In the magnetic shear angle plot, the red areas represent the magnetopause antiparallel reconnection region with magnetic shear angles >160°. The white areas in the shear angle plots represent regions where the model magnetic fields are within 3° of being exactly antiparallel. The black circle represents the terminator plane at the magnetopause with the black squares showing the plasma entry points at the magnetopause, the end points of the cusp field line traces.

Because of the southward IMF conditions (IMF clock angle of 175°), the antiparallel reconnection region (red) covers most of the dayside magnetopause with the white regions for the highest magnetic shear shifted to the southern hemisphere due to the tilt of Earth's magnetic dipole. An exception is the dusk region close to local noon where the field-line trace points are also located.

The maximum magnetic shear model has been tested and validated for IMF conditions with $|B_x|/B < 0.7$ (e.g., Trattner et al. 2017). The model predicts long continuous X-lines that extend over the dayside magnetopause (e.g., Fuselier et al. 2002; Phan et al. 2006; Trattner et al. 2007; Dunlop et al. 2011; Trattner et al. 2021). For dominant IMF B_y conditions, the model merges a component reconnection tilted X-line near the subsolar magnetopause with the two branches of the antiparallel reconnection regions, starting at the cusps and continuing towards the magnetotail along the flanks. The model was expanded to northward IMF conditions using observations by Trenchi et al. (2008, 2009) and confirming the existence of a dayside X-line down to an IMF clock angle of 50° (see also Gosling et al. 1990;

Trattner et al. 2017). It highlighted the importance of antiparallel reconnection in constraining the location of the component reconnection line (Trattner et al. 2018).

The maximum magnetic shear model shows anomalies for dominant IMF B_x conditions ($|B_x|/B > 0.7$) which are the result of the limitations of the IMF draping models used to determine the magnetopause magnetic shear. As shown by Michotte de Welle et al. (2022), using a global three-dimensional and exclusively data driven model for the magnetopause magnetic shear, the local magnetic shear can differ significantly from the magnetic shear determined from the currently used numerical models (Sect. 2.1.4), causing the anomalies in predicting the location of the dayside X-line. In addition, large magnetopause surveys (Trattner et al. 2007, 2017, 2021), comparing observed X-line locations with the predicted locations from the maximum magnetic shear model, also showed anomalies for events at the spring and fall equinoxes, specifically for events with IMF clock angles around 120 and 240 degrees, respectively. The fact that the equinox anomalies occur for specific narrow parameter ranges points to a currently unknown effect influencing the location of the magnetopause X-line under these conditions.

2.1.2 Event-Specific Global MHD Modeling

Global MHD models for simulating the solar wind-magnetosphere interaction can be used to provide an important large-scale context and connectivity information to assist in interpreting electron-scale observations of MMS. Runs on demand are readily requested through the Community Coordinated Modeling Center (CCMC) (Table 1 in Appendix C). Reiff and coauthors have now run the "SWMF" (Space Weather Modeling Framework) model (BATS-R-US with "Rice Convection Model" (RCM)) (Tóth et al. 2005; see also Graham et al. 2024) for twelve instances where MMS observed crescent-shaped electron velocity distributions (Sects. 3.1.6 and 3.3.1), both in the dayside magnetopause region and in the tail (Reiff et al. 2017; Marshall et al. 2020, 2022). In each case, the SWMF model placed an X-line (or its neighboring separatrix sheet) within 1 R_E and 2 minutes of the time of the MMS encounter.

MHD models that include RCM (e.g., SWMF) appear to do a better job in predicting the location of the reconnection sites in the tail. For example, the model predicted that MMS should be at the lobe-plasma sheet boundary layer interface near the X-line for an event on 23 June 2015 (Figure 5A,B in Reiff et al. 2016). It also predicted that the X-line for that event would be patchy across the tail (Figure 5B in Reiff et al. 2016). For a 11 July 2017 substorm event, the model not only accurately predicted the near-earth neutral line location but also predicted a huge plasmoid which was observed by MMS (Torbert et al. 2018; Reiff et al. 2018). See Fig. 7 of Fuselier et al. (2024) for comparison images for that event of CCMC versus other global MHD models such as GGCM (Geospace General Circulation Model) (Raeder et al. 2017) and LFM (Lyon-Fedder-Mobarry model) (Lyon et al. 2004).

CCMC models have also been helpful in determining the time for field line reconfiguration, stretching and distortion for dayside events (Reiff et al. 2018). In a recent study, the location of the dayside X-line on 24 December 2016 moved dramatically as a result of a change in the Y-component of the IMF, with the model predicting not only reconnection at MMS but also connection of the Geotail spacecraft in the magnetosheath to the open field line on which MMS was situated (Fig. 4). The Geotail data showed O+ fluxes just a few minutes after the model predicted a connection to the northern polar cap and a close conjunction with MMS.

In another study, an X-line that appears locally quite two-dimensional shows a dramatic difference in connection to the northern and southern ionospheres by field lines quite close on either side of the X-line, and electron fluxes correspondingly show a dramatic change in pitch angle (Marshall et al. 2022).

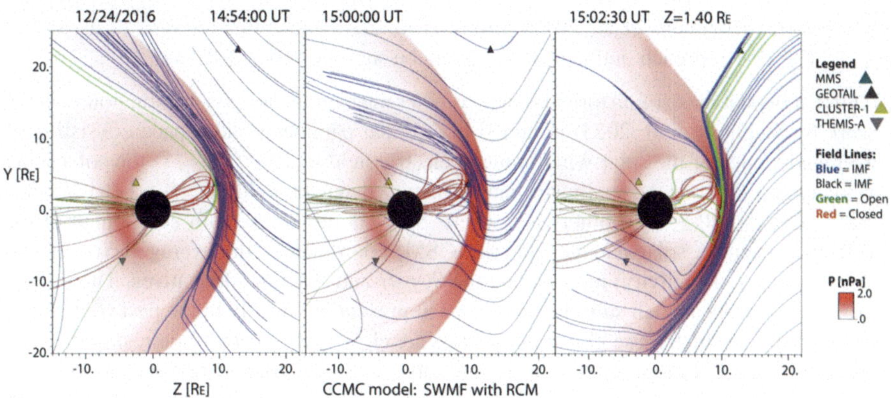

Fig. 4 Evolution of the magnetic field line topology and spacecraft locations for an MMS dayside magnetopause reconnection event on 24 December 2016. Field lines are traced from MMS and from Geotail, and from other start locations. At time 14:54 UT, MMS was predicted by the SWMF model to be near the X-line, and on open field lines (green) connected to the southern cusp. Then the IMF B_y changed sign, and at 15:20:30 UT, both MMS and Geotail were predicted to be on open field lines (green) connected to the northern polar cap, and their mapped field lines passed less than a half R_E (R_E: Earth radius) apart at the magnetopause. About a minute after the predicted connection, Geotail started observing O+ presumably from the magnetosphere, evidence of that connection

MHD models have also been run with embedded Particle-In-Cell (PIC) simulations, to overcome the inherent limitations of the MHD in reproducing kinetic-scale physics, e.g., MHD-EPIC (Chen et al. 2020). A recent CCMC workshop had dozens of presentations on linking CCMC models to solar, interplanetary, PIC and ionosphere/atmosphere models, many using open source modules [https://ccmc.gsfc.nasa.gov/ccmc-workshops/ccmc-2022-workshop/].

2.1.3 Data-Mining Approach to Reconstruction of the Global Reconnection Structure

The major problem in the global empirical reconstruction of the magnetosphere is data paucity: At any moment the huge volume of the magnetosphere ($\gtrsim 10^5\ R_E^3$) is usually probed by less than a dozen spacecraft (e.g., Sitnov et al. 2020). In the past 15 years, it has been understood that this sparse data problem can be resolved or at least substantially mitigated due to the recurrent nature of the main space weather actors, storms and substorms. The storm occurrence depends on their intensity, as well as the strength and phase of the solar cycle. For medium intensity storms the recurrence period is about two weeks (Reyes et al. 2021). The recurrence time of periodic substorms is 2–4 h, while other substorm types have longer recurrence times depending on solar wind conditions (Borovsky and Yakymenko 2017). As a result, the historical records of spaceborne magnetometer observations can be organized using a multi-dimensional state-space, formed from the global storm and substorm activity indices and the solar wind input parameter. This allows the magnetic field for the event of interest to be reconstructed from its nearest neighbors in this state-space and not only from observations during the event. A specific data-mining (DM) technique leveraging this repeatability, the k-Nearest Neighbor (kNN) classifier (Wettschereck et al. 1997; Sitnov et al. 2008), combined with flexible and extensible magnetic field architectures (Tsyganenko and Sitnov 2007; Stephens et al. 2019), helped organize multi-decade archives of spaceborne magnetometer data to reconstruct storms (Tsyganenko and Sitnov 2007; Sitnov et al. 2008) and substorms (Stephens et al. 2019; Sitnov et al. 2019; hereafter referred to as SST19 model). The DM approach outlined in Fig. 5 can be summarized as follows:

(a) First, a big database of historical magnetometer measurements (8.6 million points in the work by Stephens et al. 2023) is mined in a global parameter state-space consisting of averaged values of the solar wind induced electric field $u_{sw}B_s^{IMF}$ (u_{sw} is the solar wind velocity and B_s^{IMF} is the southward interplanetary magnetic field: $B_s^{IMF} = -B_z^{IMF}$ when $B_z^{IMF} < 0$ and $B_s^{IMF} = 0$ otherwise, where B_z^{IMF} is the north-south component of the IMF in geocentric solar magnetospheric (GSM) coordinates), the averaged *Sym-H* and AL (geomagnetic activity) indices, and the *Sym-H* and AL time derivatives. The mining procedure selects a small subset of moments at present but mostly in the past (red circles in Fig. 5a), for which these global parameters are close to the event of interest in the state space (blue circle in Fig. 5a). Events in this subset are called the nearest neighbors.
(b) The resulting subset of the magnetic field database (gray dots in Fig. 5b), which is much larger than the handful of actual satellites available at that moment, is used to fit the free parameters of a very flexible magnetic field architecture ($\sim 10^3$ free parameters) and to reveal details of the magnetosphere such as the formation of new X-lines (at the earthward part of the $B_z = 0$ isocontour in Fig. 5b).
(c) The obtained empirical model allows one to reconstruct a detailed 3D magnetic field structure, as is shown in Fig. 5c for the 11 July 2017 MMS EDR event (Torbert et al. 2018).

The magnetospheric state shown in Fig. 5a is characterized using geomagnetic indices and solar wind conditions. It can be described by a 5-D state-space vector, $\mathbf{G}(t) = (G_1,\ldots,G_5)$, formed from the geomagnetic storm index (*Sym-H*), substorm index (AL), their time derivatives, and the solar wind electric field parameter ($u_{sw}B_s^{IMF}$). Most recently (Stephens et al. 2023), the *Sym-H* and AL indices have been replaced by the *SMR* and *SML* indices provided by the SuperMag project (Gjerloev 2012). The global binning parameters $G_{1--5}(t)$ are normalized by their standard deviations, smoothed over storm or substorm scales, and sampled at a 5-min cadence, as is detailed in Stephens and Sitnov (2021). Including the time derivatives of these activity indices allows the DM procedure to differentiate between storm and substorm phases as well to capture memory effects of the magnetosphere as a dynamic system (Sitnov et al. 2001). The space magnetometer archive contains data from 22 satellites (including four MMS probes) spanning the years 1995–2020 resulting in 8,649,672 magnetic field measurements after being averaged over 5 or 15 min time windows (Stephens et al. 2023).

Every query moment in time $t = t_q$ corresponds to a particular point in the 5-D state-space, $G_{(q)} = G(t_q)$. Its k_{NN} nearest neighbors (NNs) will be other points, $G_{(i)}$, in close proximity to it: $R_i = |G_{(i)} - G_{(q)}| < R_{NN}$ (in the Euclidean metric). The specific choice of k_{NN} (and hence R_{NN}), is determined by a balance between over- and under-fitting. Stephens and Sitnov (2021) found the optimal number to be $k_{NN} = 32{,}000$, corresponding to $\sim 1\%$ of the total database ($\sim 10^7$ sampling cases). The resulting set is composed of a very small number ($\sim$1–10) of real (available at the moment of interest) and a much larger number ($\sim 10^5$) of virtual (from other events in the database) satellites.

The large number of NNs provided by such synthetic satellite observations enables the use of new magnetic field architectures (Tsyganenko and Sitnov 2007; Stephens et al. 2019), which differ from classical empirical models with custom-tailored modules (e.g., Tsyganenko and Sitnov 2005) by utilizing regular basis function expansions for the major magnetospheric current systems. In particular, the equatorial current system, which was previously described by ring and tail current modules, is now described by two expansions representing arbitrary current distributions of thick and thin current sheets with different thicknesses.

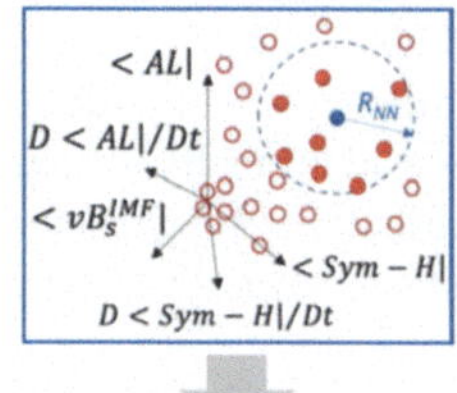

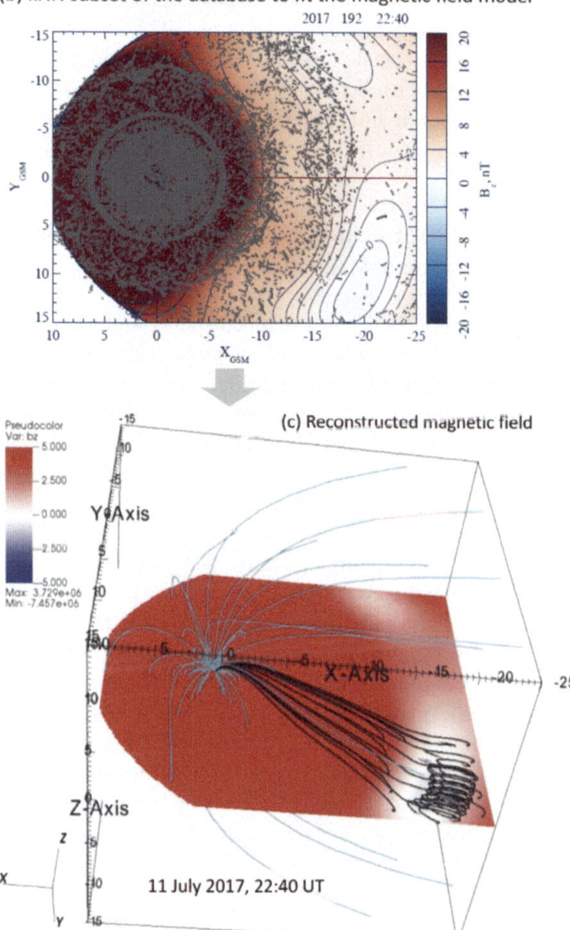

Fig. 5 The kNN DM method outline (Sitnov et al. 2021): (**a**) selecting nearest neighbors for the event of interest (blue circle) in the 5D global parameter state-space; (**b**) finding the corresponding subset in the magnetic field database (gray dots overplotted on the color-coded equatorial B_Z distribution) and using it to fit the magnetic field model and to yield 2D magnetic field distributions (here the equatorial slice) as well as (**c**) 3D magnetic field distributions. The example shown here is for the 11 July 2017 MMS EDR event Torbert et al. 2018) with the color-coded equatorial B_Z (in nT, saturated at 5 nT for better visualization) and a few sample field lines. Panels (**a**) and (**b**) are adapted from Sitnov et al. (2019)

This architecture accounts for the multiscale structure of the tail current sheet with an ion-scale thin current sheet (TCS), with a thickness D_{TCS}, forming inside a much thicker current sheet, with a thickness $D \gg D_{TCS}$, during the substorm growth phase and then decaying during the expansion phase (e.g., Sergeev et al. 2011). The independence of the current sheet expansions is provided by the constraint $D_{TCS} < D_0 < D$, where D_0 is the ad hoc parameter $\sim 1 R_E$. The proper reconstruction of substorms also requires a flexible description of the field-aligned currents, which is provided in the SST19 model using a set of distorted conical modules (Tsyganenko 1991) distributed in latitude and local time, as is discussed in more detail in Sitnov et al. (2017).

To improve the reconstructions, while fitting the magnetic field model with the NN subset, the spacecraft data were additionally weighted: in the real space, to mitigate the inhomogeneity of their radial distribution (Tsyganenko and Sitnov 2007), and in the state-space, to reduce the uncertainty and bias toward weaker activity regions (Sitnov et al. 2020; Stephens et al. 2020).

The SST19 model successfully describes the TCS buildup during the substorm growth phase and its decay during the expansion phase accompanied by the formation of the substorm current wedge (McPherron et al. 1973). It also identifies X-lines in the tail (Sitnov et al. 2019), which match in-situ MMS observations (Stephens et al. 2023), as is described in more detail in Fuselier et al. (2024, this collection). The model has been extensively validated using both in-situ observations (Sitnov et al. 2019; Stephens et al. 2019, 2020, 2023) and uncertainty quantification using DM binning statistics (Sitnov et al. 2019; Stephens et al. 2023).

2.1.4 Global 3D Structure of the Magnetosheath Using in Situ Measurements: Application to Magnetic Field Draping

The dynamics of the Earth's magnetosphere and its coupling to the solar wind importantly depends on how the solar wind interacts at the bow shock and, in particular, on how the plasma is decelerated, heated and deflected there and on how the interplanetary magnetic field drapes around the magnetospheric obstacle in the magnetosheath. The specific structure of the draping, in particular, plays a major role for the reconnection of magnetic field lines at the magnetopause. Magnetic field draping was thus the focus of a study by Michotte de Welle et al. (2022) that permits, based on large-scale statistics of in situ spacecraft measurements, to reconstruct the global 3D structure of the magnetosheath.

Magnetic field draping is a fairly well understood concept, resulting from the frozen-in condition ruling the evolution of magnetized plasmas on large scales. However, our knowledge of the 3D global draping structure in the Earth's magnetosheath is very limited and is mostly described by analytical and numerical models. Michotte de Welle et al. have recently succeeded in reconstructing the 3D structure of the magnetic draping over the whole dayside of the magnetosphere, using only in situ observations, and as a function of the IMF orientation. Two decades of data from Cluster, Double Star, THEMIS and MMS missions have been used for that purpose. The measurements made in the magnetosheath were extracted automatically using a Gradient Boosting Classifier trained to classify magnetosphere, magnetosheath and solar wind data points (Nguyen et al. 2022a). About 50 million measurements were extracted and then associated with a causal solar wind and IMF conditions from OMNI data using a solar wind propagation method (Safrankova et al. 2002). The position of each data point relative to the bow shock and the magnetopause at the time of the measurement is then estimated using a Gradient Boosting Regression model of the boundaries, parameterized with solar wind and IMF conditions. All points are then repositioned between a standard bow shock and magnetopause boundary, determined for average solar wind conditions, and rotated into the solar wind interplanetary (SWI) magnetic field coordinate system (Zhang et al. 2019) in which the upstream IMF direction is parallel to the XY plane. This last step is crucial to ensure each point falls in the right sector of the magnetosheath (quasi-parallel or quasi-perpendicular bow shock sides) with respect to its causal IMF. Magnetic field lines are then integrated with a standard ordinary differential equation integrator, using at each step the weighted average of the k-Nearest Neighbor magnetic field measurements close to the current iteration step position (with k=45000).

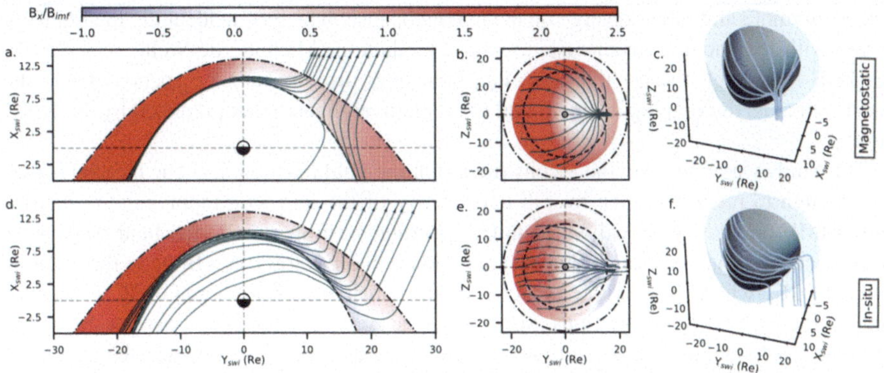

Fig. 6 From left to right: representation of the magnetic field lines in the XY (left), YZ (middle) planes and in 3D (right) as predicted by the KF94 magnetostatic model (top panels) or reconstructed from in situ data (bottom panels). On the four leftmost panels, the color codes the value of the B_X component of the magnetic field. Coordinates are from the SWI system

The bottom three panels of Fig. 6 show the obtained draping in the XY and YZ planes of the SWI coordinate system and in 3D on the rightmost panel. As a comparison, the top three panels show, for the same points of view, the draping obtained with the magnetostatic model of Kobel and Flückiger (1994) (referred to as the KF94 model). In this configuration, the represented data is the subset of all measurements for which the associated IMF cone angle falls between 20° and 30° from the Sun-Earth axis. The figure reveals that the observed draping is fundamentally different from the modeled one. In the modeled draping, the magnetic field appears to diverge as it approaches the magnetopause in the region downstream of the quasi-parallel bow shock. This is the result of the only two constraints imposed by the model to the magnetic field. Indeed, on the one hand, the magnetic field in the quasi-parallel (positive Y) region is mostly conserved as it crosses the bow shock. On the other hand, the field must be tangential to the magnetopause. While these two constraints also apply in reality (if one neglects magnetopause reconnection as a first approximation), magnetic flux is also bound to the plasma as it flows upstream of the bow shock and circumvents the magnetopause in the magnetosheath. In other words, fluid elements connected to a magnetic flux tube entering the quasi-parallel region must remain connected to those that entered earlier on the quasi-perpendicular bow shock (negative Y) side. The considerable slowing down of the flow in the subsolar region forces all field lines entering in the quasi-parallel magnetosheath to head to the dayside where the flux piles up, rather than diverge partly to the nightside as the magnetostatic model predicts. This large-scale kink in magnetic field lines is thus associated with a macroscopic current sheet at mid-depth of the quasi-parallel side of the magnetosheath, an effect that is not seen with the vacuum magnetostatic model. The main consequence is that for this range of IMF cone angles, a large part of the magnetopause on the quasi-parallel region sees a magnetic shear that is vastly different from that predicted using the KF94 model, potentially adding difficulties to the maximum shear angle reconnection model in that cone angle regime.

Interestingly, when the IMF becomes quasi-radial (B_x dominant), in practice when the cone angle is less than about 12°, magnetically connected solar wind elements are so far apart along the Sun-Earth line that by the time fluid elements arrive on the quasi-parallel side of the bow shock, connected fluid elements entered in the subsolar region have long ago re-accelerated and joined the nightside of the system. As a result, field lines are not kinked anymore and rather diverge on the magnetopause, which coincidentally qualitatively agrees with the vacuum magnetostatic model prediction (Michotte de Welle et al. 2022). For IMF cone angles larger than 45°, i.e., field lines arriving rather perpendicular to the Sun-Earth axis, the draping is also found to qualitatively match that predicted by the KF94 model.

This study was limited to the reconstruction of the field line draping. The method is precise enough to reconstruct the overall dependency of the magnetic topological properties on the IMF orientation. Local and detailed quantitative properties of the field such as its divergence-free character cannot be ensured, although on average $|\nabla \cdot \mathbf{B}|$ is on the order of $0.01 B_{\mathrm{IMF}}/\mathrm{R_E}$, where B_{IMF} is the upstream field intensity and Earth's radius $\mathrm{R_E}$ is comparable to the scale of field variations. Similar analysis can be made to reconstruct the global distribution of any physical quantities, providing the capability to reconstruct the global 3D structure of the solar wind–dayside magnetosphere interaction globally. The amount of data now available and modern statistical learning methods will prove useful to understand how physical parameters distribute on and around critical regions such as the magnetopause for various upstream conditions, which is a topic of on-going work.

2.2 Coordinate System, Frame Velocity, and Spacecraft Trajectory Estimation

Current sheets where reconnection may occur are never strictly stationary, their local normal direction can be highly variable in space and time, and the X-line, possibly embedded in those current sheets, may be moving, depending on the external conditions and instabilities excited in the current sheets. It is thus indispensable, for each current sheet crossing or reconnection event, to be able to obtain a proper coordinate system and frame velocity of the current sheet structure or reconnection regions. In this section, we briefly review various methods to estimate the characteristic orientations and motion of the structures from in-situ measurements.

2.2.1 Dimensionality and Coordinate Systems

Here we review methods for estimating the dimensionality and coordinate systems of magnetic or plasma structures in space from in-situ data. Since an overview was given by Sonnerup et al. (2006a) and Shi et al. (2019) on various single- and multi-spacecraft analysis methods for estimating the orientation and motion of plasma discontinuities (one-dimensional (1D) structures, such as planar current sheets), we focus only on recent developments.

Minimum Directional Derivative In Minimum Directional Derivative (MDD) analysis, a multi-spacecraft method applicable to four-spacecraft measurements at any instant of the magnetic (or any vector) field, one takes the gradient of the magnetic field vector, multiplies the resulting matrix by its transpose, and then solves for the eigenvectors of the resulting matrix, finding time-dependent maximum, intermediate, and minimum gradient eigenvalues, λ_{max}, λ_{int}, and λ_{min}, which represent the squared gradient in the respective time-dependent directions, $\hat{\mathbf{e}}_n$, $\hat{\mathbf{e}}_l$, and $\hat{\mathbf{e}}_m$, respectively (Shi et al. 2005, 2019) (alternatively, the eigenvalues can be defined as the square root of these quantities, proportional to the gradient in the respective directions). If $\lambda_{\mathrm{max}} \gg \lambda_{\mathrm{int}}, \lambda_{\mathrm{min}}$, the system is roughly one dimensional with variation mainly in the maximum gradient ($\hat{\mathbf{e}}_n$) direction. If $\lambda_{\mathrm{max}}, \lambda_{\mathrm{int}} \gg \lambda_{\mathrm{min}}$, the system is roughly two-dimensional (2D) with variation mainly in the maximum and intermediate gradient directions ($\hat{\mathbf{e}}_n$ and $\hat{\mathbf{e}}_l$, respectively). If all eigenvalues are comparable, the system is 3D with variation in all three directions, $\hat{\mathbf{e}}_n$, $\hat{\mathbf{e}}_l$, and $\hat{\mathbf{e}}_m$ (Shi et al. 2019). Rezeau et al. (2018)

introduced dimensionality parameters that are useful for determining the dimensionality of the system, $D_{1D} = (\lambda_{max} - \lambda_{int})/ \lambda_{max}$, $D_{2D} = (\lambda_{int} - \lambda_{min})/ \lambda_{max}$, and $D_{3D} = \lambda_{min}/ \lambda_{max}$; D_{1D}, D_{2D}, and D_{3D} quantify the degree to which the system is 1D, 2D, or 3D, respectively.

In practice, when studying magnetic reconnection events, λ_{max} is often significantly greater than the other two eigenvalues in the vicinity of the current sheet (D_{1D} close to unity). However, the system can be somewhat two-dimensional if $\lambda_{int} \gg \lambda_{min}$ so that the variation in the minimum gradient direction can be neglected relative to that in the other two directions, yielding a system that can be analyzed as quasi-2D. (Unfortunately, the minimum MDD eigenvalue direction is not always the M direction as defined below (Denton et al. 2016, 2018).) The coordinate system usually used to describe magnetic reconnection (the so-called LMN coordinate system) has the L direction in the direction of the reconnecting magnetic field and the N direction normal to the current sheet; the M direction completes the triad. Because λ_{max} is often very large, the MDD maximum gradient direction, $\hat{\mathbf{e}}_{N'}$, found from the time dependent $\hat{\mathbf{e}}_n$ direction, is often the most accurately determined direction in the system, and can usually be used to define the normal direction across the current sheet. Then in order to define the reconnection coordinate system, it remains to find one more direction.

Hybrid Methods Denton et al. (2016, 2018), studying the 16 October 2015 magnetopause reconnection event of Burch et al. (2016a), determined the L direction as the maximum variance direction of Minimum Variance Analysis (MVA) (Sonnerup and Cahill 1967; Sonnerup and Scheible 1998) of the magnetic field. This is reasonable seeing as the reconnection magnetic field reverses across the current sheet, leading to large variance. The M direction can be taken to be the direction of the cross product between $\hat{\mathbf{e}}_{N'}$ defined by MDD and $\hat{\mathbf{e}}_{L'}$ defined by MVA, but if $\hat{\mathbf{e}}_{N'}$ and $\hat{\mathbf{e}}_{L'}$ are not exactly orthogonal, a choice must be made to determine the N and L directions. For instance, one could take $\hat{\mathbf{e}}_N = \hat{\mathbf{e}}_{N'}$, and find L from $\hat{\mathbf{e}}_L = \hat{\mathbf{e}}_M \times \hat{\mathbf{e}}_N$, which is what Denton et al. (2016) did. Denton et al. (2018) proposed a hybrid method weighting the influence of $\hat{\mathbf{e}}_{N'}$ and $\hat{\mathbf{e}}_{L'}$ based on the ratio of the maximum MDD eigenvalue to the maximum MVA eigenvalue.

Genestreti et al. (2018) found that Denton et al.'s (2018) method did not work well for the 11 July 2017 magnetotail reconnection event studied by Torbert et al. (2018). Instead, Genestreti et al. used the maximum variance direction of MVAVe (Minimum Variance Analysis of the electron bulk velocity $\mathbf{u}_e$) to determine the L direction. Large variance in the velocity moments along the L direction are expected since the reconnection outflow will be along that direction. (Another possibility is to use MVAE, using the variance of the electric field.) Heuer et al. (2022) recently proposed a hybrid system similar to that of Denton et al. (2018), except that $\hat{\mathbf{e}}_{L'}$ is determined from MVAB (MVA using the magnetic field) only when the spacecraft have a significant velocity component across the current sheet in the frame of the magnetic structure. If the velocity of the spacecraft relative to the magnetic structure is mostly in the L direction, they recommend using MVAVe to determine $\hat{\mathbf{e}}_{L'}$.

Magnetic Configuration Analysis Among the analysis methods that are enabled by four-spacecraft measurements are those that allow the determination of the geometrical properties of the magnetic field. Following the main ideas of the magnetic MDD (Shi et al. 2005) and magnetic rotational analysis procedure (Shen et al. 2007), Fadanelli et al. (2019) derived a new method named the "magnetic configuration analysis" (MCA). The method in effect determines the main axes of the magnetic field rotation rate in space, in a normalized fashion, and permits the categorization of magnetic field geometries in terms of planarity and elongation properties, for instance. MCA is thus designed to estimate the spatial scales on which the magnetic field varies locally and to determine the actual magnetic field shape and dimensionality from multi-spacecraft data. Case studies using MMS data showed that the method is capable of determining, for example, the planar and cigar shapes of structures such as current sheets and small flux ropes, respectively. An interesting property of such a method is that the determination is made very locally, at the scale of the inter-spacecraft separation, which is much smaller than that of the current sheet or flux rope itself.

Fadanelli et al. (2019) also statistically applied the MCA method to magnetic field observations in different near-Earth regions (magnetosphere, magnetosheath, and solar wind). The findings show that the magnetic field structure is typically elongated at small scales (cigar and blade shapes), is less frequently planar (pancake shapes generally associated with current sheets), but rarely shows an isotropic variance in the magnetic field rotation rate. The occurrence frequency of the type of magnetic geometries observed and, most importantly, their scale lengths, strongly depend on the region sampled and plasma β. Interestingly, the most invariant direction is statistically aligned with the electric current, suggesting that electromagnetic forces are fundamental in determining the magnetic field configuration at small scales.

2.2.2 Velocity of the Magnetic Structure

In addition to determining the coordinate system, it is beneficial to determine the velocity of the magnetic structure in order to determine a reference frame in which the magnetic structure is approximately time stationary (what Shi et al. 2019 call the "proper reference frame"). In homogeneous regions, such a velocity can be the $\mathrm{E} \times \mathrm{B}$ velocity or the ion velocity perpendicular to the background magnetic field for MHD-scale structures, and in IDRs the perpendicular components of the electron velocity can be used. A related approach, applicable to inhomogeneous regions, is deHoffmann-Teller (HT) analysis, which finds a frame with minimum electric field, and hence the frame in which ion or electron flows are roughly aligned with the spatially varying magnetic field (De Hoffmann and Teller 1950; Khrabrov and Sonnerup 1998). However, these approaches are unreliable in the EDR.

Four spacecraft timing analysis (Dunlop and Woodward 1998), which assumes that the spatial structure varies in only one direction, can yield the velocity component along that direction, namely, the velocity normal to the plane along which spatial gradient is negligible. However, results may vary depending on the input quantity used. Minimum Faraday residue analysis (MFR) is another approach to get the normal velocity of MHD discontinuities from single-spacecraft data (Terasawa et al. 1996; Khrabrov and Sonnerup 1998).

Shi et al. (2006) introduced the Spatio-Temporal Difference (STD) method, which solves for the structure velocity from the convection equation for steady magnetic structures ($\partial \mathbf{B}/\partial t = 0$)

$$\frac{d\mathbf{B}}{dt} = (\mathbf{V}_{\mathrm{sc}} \cdot \nabla)\,\mathbf{B} = -\,(\mathbf{V}_{\mathrm{str}} \cdot \nabla)\,\mathbf{B}, \tag{2}$$

using instantaneous values of the magnetic gradient at one time, and a centered time step around that time for the total time derivative $d\mathbf{B}/dt$ observed by the spacecraft. Here $\mathbf{V}_{\mathrm{sc}}$ is the spacecraft velocity relative to the magnetic structure, and $\mathbf{V}_{\mathrm{str}}$ is the structure velocity relative to the spacecraft. Using Eq. (2) assumes that the velocity is constant on the spatial scale of the four spacecraft and on the time scale for motion across that spatial scale. However, in most cases, only one or at most two velocity components can be determined, because when the gradient is very small, the velocity component in that direction is unreliable (Shi et al. 2019; Denton et al. 2021). Here we examine the 16 October 2015 magnetopause reconnection event of Burch et al. (2016a) and introduce a modification of STD to get as much information as possible from STD.

Figure 7a shows the MDD eigenvalues normalized to $(0.1\ \mathrm{nT}/d_{sc})^2$, where 0.1 nT is the maximum calibration error of the MMS magnetometers and d_{sc} is the average spacecraft spacing; $(0.1\ \mathrm{nT}/d_{sc})^2$ represents a reasonable minimum value required for accuracy of the squared gradient. In Fig. 7a, the maximum and intermediate eigenvalues are always well above this value, suggesting that they can be satisfactorily determined. However, where the minimum eigenvalue becomes significantly smaller than this value, the STD velocity component in that direction has unrealistically large values. Where all three normalized eigenvalues are significantly above unity, however, as sometimes happens for this event (like after $t = 2.5$ s in Fig. 7a), we may be able to determine a three-dimensional velocity. For the calculations leading to Fig. 7, we required that an eigenvalue be at least $20(0.1\ \mathrm{nT}/d_{sc})^2$ in order to include the STD velocity component associated with that eigenvalue. (A large value was required to yield consistent velocities.) Otherwise, that component was set equal to zero. Figure 7b-d show the velocity components calculated in the time-dependent MDD intermediate, minimum, and maximum gradient directions, $\hat{\mathbf{e}}_L$, $\hat{\mathbf{e}}_M$, and $\hat{\mathbf{e}}_N$, respectively. In Fig. 7c, the dotted green curve is the instantaneous STD velocity component for the MDD local minimum gradient ($\hat{\mathbf{e}}_m$) direction. Black circles mark the data points where all three velocity components are determined, and at data points without black circles, the dotted green curves drop to zero.

Figure 7f–h show the resulting velocity in the fixed LMN coordinate system that was found using the hybrid method of Denton et al. (2018). The N direction agrees well with the instantaneous MDD $\hat{\mathbf{e}}_n$ direction, but the L direction is found from the direction of maximum variance of the magnetic field. Note that whereas the l and m components of the velocity reverse at $t = 2.3$ s in Figs. 7b and 7c, the velocity components in the fixed L and M directions are well behaved in Figs. 7f and 7g. Thus in Fig. 7, we have calculated a three-dimensional structure velocity at the times indicated by the black circles. The velocity is less reliable at the other times because of the omission of the minimum gradient component.

Denton et al. (2021) used polynomial reconstruction (Sect. 2.3.1) to track the motion of the X-line, and the resulting velocity was in rough agreement with the STD velocity in the LN plane for times for which the X line was less than two d_{sc} from the centroid of the MMS spacecraft. Another approach is to match MMS observations to simulation data in order to find a velocity though the simulation fields (Shuster et al. 2017; Nakamura et al. 2018b; Egedal et al. 2019; Schroeder et al. 2022; Sect. 2.2.3).

2.2.3 Spacecraft Trajectory Estimation

MMS observations of a magnetotail EDR can be directly compared to 2D kinetic PIC simulation, for determination of spacecraft trajectories in the event specific LN-plane. As will be described in detail in Sect. 2.3.3, the trajectory of the MMS constellation further allows for reconstruction of the MMS data in a 2D format, which can in turn be compared against the simulation data. Note that these methods rely on spacecraft observations with sufficient features to well constrain the trajectory. The MMS event considered here (10 August 2017) features strong electron pressure anisotropy followed by large electric field gradients that indicate a path that closely follows a separatrix layer. For events that do not exhibit such features, one may find difficulty in accurate determination of the spacecraft trajectory.

Spacecraft trajectory optimization The spacecraft paths are found through a χ^2-optimization procedure, in which a penalty function made up of a sum of squared deviations of spacecraft measurements from corresponding simulation quantities is minimized, similar to that laid out in Egedal et al. (2019). However, some adjustments are made to suit the MMS event at hand. Similarly to the previous method, PIC simulation units are converted

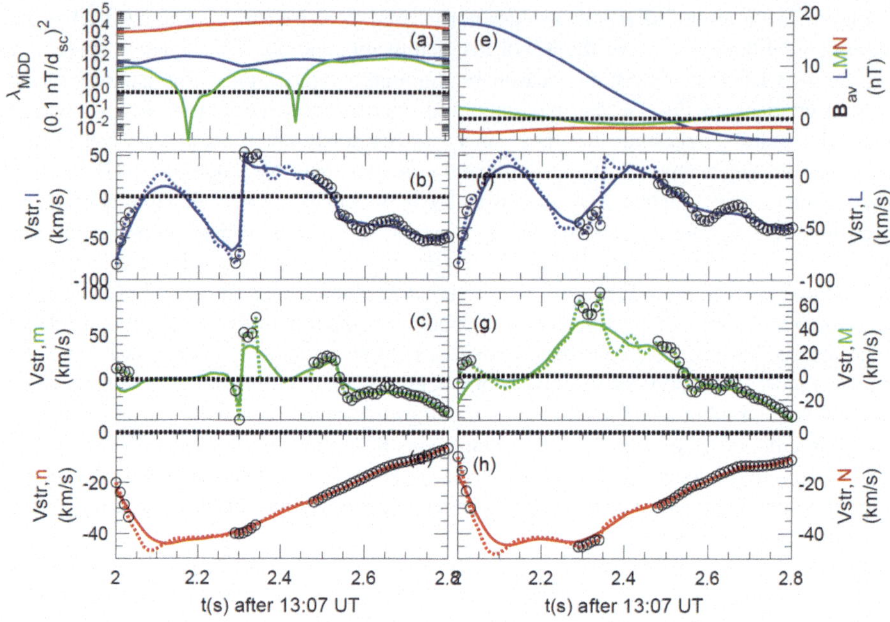

Fig. 7 STD analysis for the 16 Oct 2015 magnetopause reconnection event. (**a**) MDD eigenvalues, (**b–d**) STD structure velocity components in the local MDD gradient directions, $\hat{\mathbf{e}}_l$, $\hat{\mathbf{e}}_m$, and $\hat{\mathbf{e}}_n$ respectively, (**e**) magnetic field averaged over the four MMS spacecraft, (**f–h**) STD structure velocity in the fixed LMN coordinates. The dotted curves are the instantaneous velocities, and the solid curves are smoothed over a time scale of 0.5 s

to physical MMS units using two parameters, the ratio of PIC to MMS densities and the ratio of PIC to MMS temperatures. Once simulation units are converted, a direct numerical optimization scheme to fit the spacecraft path is tractable.

To ensure a well-fit magnetic field profile the path is optimized in such a way that it is constrained to be on PIC simulation contours that match the B_L values measured by MMS1. This amounts to optimizing the MMS1 position along a unique simulation B_L contour at each time point, reducing a 2D problem to a 1D problem and largely simplifying the numerical method. The choice of MMS1 in optimizing the path is arbitrary; one may choose any other spacecraft or use mean magnetic field value at the centroid and can achieve near identical results. The event-specific LMN axes (Sect. 2.2.1) combined with the converted simulation units allows for determination of the spacecraft positions relative to MMS1 in the simulation LN-plane.

At each time point, a penalty function $h(r)$ is evaluated, where h is a sum of weighted χ^2-differences between spacecraft measurements and corresponding simulation data parameterized by r, the distance along the given B_L contour that matches MMS1 data. The signals included in the penalty function are all components of the electromagnetic fields (except for $B_{L,MMS1}$), electron and ion flow velocities, parallel and perpendicular electron pressures, and the ratio of parallel to perpendicular electron pressures. An additional contribution to the penalty function, $g(r)$, penalizes solutions whose positions r are too far away from that at the previous time step r_{previous} and ensures a continuous trajectory. This contribution takes the exact form $g(r) = (r - r_{\text{previous}})^2/\sigma_g^2$, where σ_g is an adjustable weight to enforce a smooth trajectory.

The optimization problem is solved by stepping through time and taking the MMS1 position to be the r-value corresponding to the minimum value of the penalty function. For further details of this method, see Supporting Information of Schroeder et al. (2022).

2.3 Methods for Reconstructing 2D/3D Structures

2.3.1 Field Reconstruction Using Quadratic Expansion

To understand the context of reconnection events, it is desirable to have a reconstruction of the magnetic field in the vicinity of the spacecraft. Without an explicit reconstruction, researchers map the location of the spacecraft by comparing the time series of **B** to the nominal diffusion region picture seen in many 2D simulations (e.g., see Torbert et al. 2018), often using the LMN coordinate system as determined in different ways and described in Sect. 2.2. MMS provides new measurements that allow reconstructions that depend only on the data and the vanishing divergence of the magnetic field. This is made possible because of: 1) the very high fidelity of the current density measurements using only particle data (Pollock et al. 2016; Phan et al. 2016); and 2) the very high accuracy of the magnetometers (Russell et al. 2016), assisted with independent measurements of the field magnitude by the Electron Drift Instrument (EDI) (Torbert et al. 2016b). Using a "modified" curlometer (see Dunlop et al. 1988 for the original method), which employs both temporal and spatial variations of **B** to estimate the current density, Torbert et al. (2017) showed that the particle data matched the magnetic variations at the highest cadence available on MMS within an EDR, where the current density is far from uniform.

If the current density **j** from the particle measurements at four spacecraft locations is assumed correct, then one can extend the linear curlometer approximation (Dunlop et al. 1988) to the second order and reconstruct the magnetic topology in the vicinity of the MMS tetrahedron to sense the locations of X-lines and the four spacecraft within the diffusion region. Torbert et al. (2020) implemented such a reconstruction, using a 24-parameter Taylor expansion around the barycenter of the tetrahedron. Given that there are 24 knowns (3 components times 4 spacecraft measurements of **B** and **j**), this gives a solution for the field that exactly matches the data, with the divergence of **B** zero everywhere. However, such an exact solution requires the addition of at least one cubic term in the expansion because of the constraint that the divergence of the current density, which in the expansion is computed from the curl of **B**, must be zero (neglecting the displacement current for a nonrelativistic system). The measured current values almost never have this property, for the primary reason that they are taken at separated spatial locations where the current may be highly varying. Torbert et al. (2020) assumed that the second derivative in the MDD minimum gradient direction was zero, and arrived at an exact fit for spacecraft measurements of **B** and **j**by using a superposition of cubic terms weighted by the inverse of the coefficient required for each term.

Given that MMS can measure the electron distribution and compute the current density every 30 ms (and sometimes every 7.5 ms (Rager et al. 2018; see also Appendix A)), a reconstruction can be computed for every such time step. As an example, such a reconstruction is given in Fig. 8, for a time when the MMS constellation approached an EDR at the magnetopause on 16 October 2015, as reported by Burch et al. (2016a). The field is computed in a 3D cubic lattice, and the field lines are traced in this lattice. The field lines are then projected into the shown LN plane. The field topology is insensitive to the actual weighting of the 18 solutions using different cubic terms. Using such a reconstruction with synthetic data from simulation as input, Torbert et al. (2020) showed that these reconstructions are very

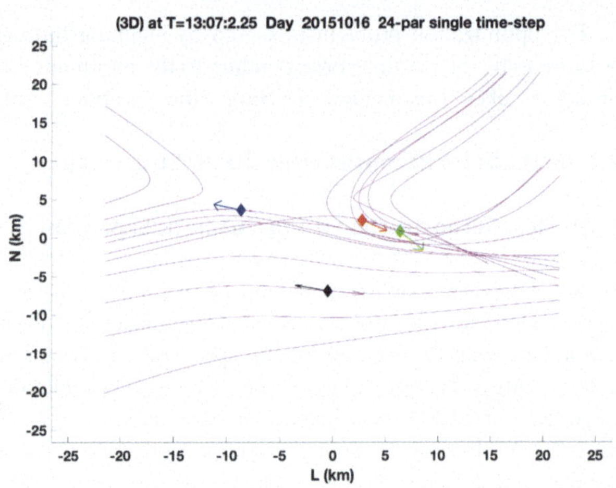

Fig. 8 A reconstruction of the magnetic field lines as MMS approached an electron diffusion region at 13:07:02.25 UT on 16 October 2015. The four MMS spacecraft locations are colored diamonds: (black, red, green, blue) are the standard colors for (MMS1, 2, 3, 4) respectively. The colored arrows show the projection of the electron flow velocity into this LN plane. The purple arrows at each spacecraft show the direction of the field and the lengths the relative magnitude of the magnetic field at that location

representative of the simulation data within a volume whose linear extent is about twice that of the spacecraft tetrahedron.

Denton et al. (2020) implemented a modification of this technique, using only quadratic terms, based on scaling arguments for the various terms and the concern that the exact solutions may lead to over-fitting of the data and show spurious X-lines when far from the tetrahedron. The number of terms in the expansion is reduced using estimates of their relative scaling, and the coefficients are then determined by a least-squares fitting procedure with an assumed weighting between the **B** and **j** values, depending on their accuracies. Although the data cannot exactly match the model for reasons described above, these reconstructions appear to give better results without false X-lines when sensing the presence of X-lines out further from the tetrahedron.

Figure 9 shows the reconstructed magnetic field close to the MMS spacecraft using the reduced quadratic model of Denton et al. (2020) at the same time as that plotted in Fig. 8. Figure 9 shows some interesting features, the sheared field with an X-line close to MMS4, the field line of MMS2 approaching the X-line even closer, and the tilt of the magnetic island structure toward more positive L at more positive M, suggesting that the invariant direction has an L component. Some features are possibly unrealistic. For instance, the flux rope in the island might well be larger than Fig. 9 suggests. Also, some features of the reconstruction are sensitive to details of the reconstruction procedure, such as the amount of smoothing and adjustment of the electron density (scaling of the electron density from the particle instruments to better agree on average with the current density from the curl of the magnetic field, as described by Denton et al. 2020), so the exact field line structure is not known. However, the reconstruction well shows the positions of the MMS spacecraft relative to the X line.

Another approach to the over-fitting problem is to use the data at multiple times and assume that the magnetic topology has not changed over this time interval. The assumption is that the spacecraft are moving through a semi-stationary structure. After all, this is the essence of what researchers have done in the past to draw cartoons of the reconnection regions from time series data over much longer intervals. In producing a data-based reconstruction, the velocity of the spacecraft relative to the structure is required. This can be estimated from time-of-flight analysis or STD, or found from the best fit to the data. The best fit method requires an iterative procedure. For example, for the encounter of the EDR seen in Fig. 8, Burch et al. (2016a) estimated that over an interval of about 0.2 s, the spacecraft were

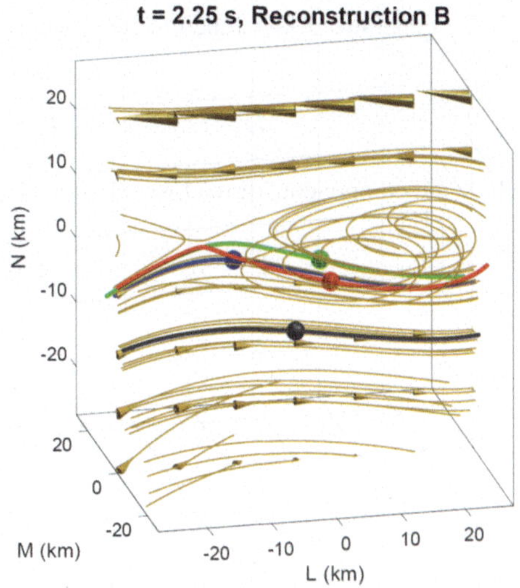

Fig. 9 Reduced quadratic reconstruction for 16 October 2015, 13:07:02.25 UT. The black, red, green, and blue spheres and curves show the positions and magnetic field lines passing through MMS1, 2, 3, and 4, respectively. The gold curves are other magnetic field lines with the cones indicating the direction and magnitude

moving through the structure with $V_{sc,N} = 45$ km/s. For a multiple time-step reconstruction, a least-squares fit is required for the 72 data elements (24 at 3 times around 13:07:02.25 UT) of the 24-parameter expansion. The iterative solution produced a different velocity ($V_{sc,N}$ = 21 km/s, while hardly moving in the L direction with $V_{sc,L} = 1 \pm 5$ km/s). The reconstruction at an earlier time of ∼13:07:02.05 UT showed $V_{sc,N}$ = ∼55 km/s, closer to that determined by Burch et al. (2016a). Research is ongoing into the accuracy of the velocity determined in this way, but the multiple time-step solution appears to be more stable than the single one.

A reduced quadratic reconstruction using the method of Denton et al. (2022) with multiple input times (as described above) yields a result similar to that in Fig. 9, except that the field line passing through MMS2 wraps around the magnetic island (not shown). Differences in the path of this field line are not surprising considering that the field line passing through MMS2 comes very close to the X-line in Fig. 9. A similar result is found with a complete quadratic reconstruction using the Denton et al. (2022) method (not shown).

2.3.2 3D Empirical Reconstruction Using Stochastic Optimization Method

Zhu et al. (2022) developed a new model for empirical reconstruction of the 3D magnetic field and current density field using a stochastic optimization method called simultaneous perturbation stochastic approximation (SPSA) (e.g., Spall 1998; Zhu and Spall 2002; Spall 2003). The model employs an empirical approach by fitting the prescribed analytic functions for the magnetic field to the point-wise measurements from a constellation of spacecraft using physical constraints derived from a set of Maxwell equations. The fitness of the reconstruction is defined by a general loss function (G), which consists of both the differences between the model and in-situ measurements and the model deviations from linear or nonlinear physical constraints. While most applications of SPSA utilize loss functions that include only the differences between the modeled and measured quantities (e.g., Chin 1999; Spall 2003), the new model characterizes the physical robustness of the reconstructed fields. The SPSA approach also has an additional feature that the algorithm includes the effects of random measurement errors. Zhu et al. (2022) demonstrated this new model using MMS measurements of the magnetic field and current density ($\hat{\mathbf{B}}, \hat{\mathbf{j}}$) for the 11 July 2017 magnetotail EDR event (Torbert et al. 2018), which was previously explored by a least-squares method (e.g., Denton et al. 2020; Torbert et al. 2020) introduced in Sect. 2.3.1.

The generalized loss function (G) used in this new empirical reconstruction model has the form of

$$G = G_O + w_A \varepsilon_A G_A + w_B \varepsilon_B G_B + w_C \varepsilon_C G_C, \tag{3}$$

where the components of the loss function (G_O, G_A, G_B, G_C) are defined as

$$G_O = \frac{1}{12} \sum_{a=1}^{4} \sum_{i=1}^{3} \left[B_i(\mathbf{r}_a) - \hat{B}_{a,i} \right]^2, \tag{4}$$

$$G_A = \frac{1}{12} \sum_{a=1}^{4} \sum_{i=1}^{3} \left[j_i(\mathbf{r}_a) - \hat{j}_{a,i} \right]^2, \tag{5}$$

$$G_B = \frac{1}{9} \left[\delta^2(\mathbf{r}_0) + \sum_{a=1}^{4} \delta^2(\mathbf{r}_a) + \sum_{a=1}^{4} \delta^2(\mathbf{r}_{Fa}) \right] \text{ or } G_B^* = \frac{1}{5} \left[\delta^2(\mathbf{r}_0) + \sum_{a=1}^{4} \delta^2(\mathbf{r}_a) \right], \tag{6}$$

and

$$G_C = \frac{1}{4} \sum_{a=1}^{4} \left[\mu_0 \mathbf{j}(\mathbf{r}_{Fa}) \cdot (\Delta\mathbf{r}_{\beta\gamma} \times \Delta\mathbf{r}_{\beta\delta}) - (\overline{B}_{\beta\gamma} \cdot \Delta\mathbf{r}_{\beta\gamma} + \overline{B}_{\gamma\delta} \cdot \Delta\mathbf{r}_{\gamma\delta} + \overline{B}_{\delta\beta} \cdot \Delta\mathbf{r}_{\delta\beta}) \right]^2. \tag{7}$$

Here, $\Delta\mathbf{r}_{\beta\gamma} = (\mathbf{r}_\gamma - \mathbf{r}_\beta)$ is the edge vector connecting the vertices $\mathbf{r}_\beta$ and $\mathbf{r}_\gamma$ and $\overline{\mathbf{B}}_{\beta\gamma} = \left(\hat{\mathbf{B}}_\beta + \hat{\mathbf{B}}_\gamma\right)$ is the mean magnetic field on the edge $\Delta\mathbf{r}_{\beta\gamma}$ calculated using the measured $\hat{\mathbf{B}}$ field by applying a linear approximation between the two spacecraft observations along that edge. G_O and G_A each comprises twelve terms and quantifies the model-measurement difference at each vertex of the tetrahedron ($\mathbf{r}_a$). G_B comprises nine physical constraints and requires minimization of $\delta^2(\mathbf{r}) = (\nabla \cdot \mathbf{B})^2$ at nine spatial points across the tetrahedron (i.e., the barycenter $\mathbf{r}_0$, each of the four vertices $\mathbf{r}_a$, and the center of each of the four faces $\mathbf{r}_{Fa}$). The face centers can be disregarded by replacing G_B with G_B^*. G_C comprises four approximate physical constraints derived from applying Stokes' theorem to Ampere's law $\left(\mu_0 \iint_S \tilde{\mathbf{j}} \cdot d\mathbf{S} = \oint_C \hat{\mathbf{B}} \cdot dl\right)$ on each of the four faces of the tetrahedron; the current density components normal to the tetrahedron faces ($\tilde{\mathbf{j}}$) are derived from the curlometer method with the measured $\hat{\mathbf{B}}$. G_C results from minimizing the difference between $\tilde{\mathbf{j}}$ and $\mathbf{j}$ projecting onto the normal of each of the four tetrahedron faces. Specification of the weighting factors (w_A, w_B, w_C) in G determines which loss function components are included in the reconstruction. The scaling parameters ($\varepsilon_A, \varepsilon_B, \varepsilon_C$) are dependent on the spatial separations of the four spacecraft and are defined so that the different components of the loss function are of the same order of magnitude. The applied SPSA approach also determines the model parameters that minimize a dimensionless loss function that includes a random perturbation that captures the effects of measurement errors.

Zhu et al. (2022) validated the empirical reconstruction by introducing indices (γ_B, γ_j) defined as the normalized magnitude of the differences between the measured ($\hat{\mathbf{B}}, \hat{\mathbf{j}}$) and modeled ($\mathbf{B}, \mathbf{j}$) fields. These sets of indices (γ_B, γ_j), shown in Fig. 10, provide a qualitative measure of the accuracy to the reconstructed fields. Additionally, a model quality indicator, Q_{model}, is introduced – based on the quality indicator Q_{curl}, which is a measure of the ratio

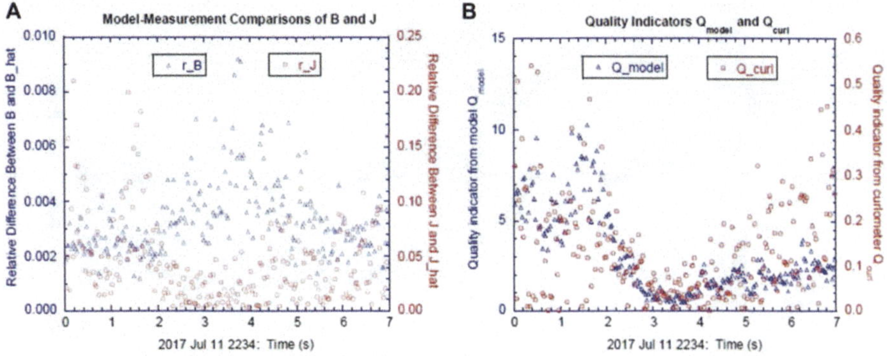

Fig. 10 (**a**) Relative differences (γ_B, γ_j) and (**b**) quality indicators ($Q_{\text{model}}, Q_{\text{curl}}$) for a sensitivity run with weighting factors w_B and w_C both set to 0. The very small relative difference values ($\ll 1$) highlights that the empirical model results in a very good fit between the modeled and measured fields at the prescribed spatial points

$\left|\nabla \cdot \hat{\mathbf{B}}\right| / \left|\nabla \times \hat{\mathbf{B}}\right|$, introduced by Dunlop et al. (1988) – to provide a quantitative assessment of the robustness of the modeled field in terms of the physical property of $\nabla \cdot \mathbf{B} = 0$. These indices respectively represent the two sets of constraints applied to the model-measurement differences and the deviations of the model considered when designing the applied generalized loss function. Zhu et al. (2022) examined the error sources in the reconstructed fields previously noted by studies applying the curlometer method and found that these curlometer-calculated errors in the current density primarily arose from the application of the linear approximation to what is in reality a nonlinear configuration of the 3D magnetic fields.

2.3.3 2D Reconstruction of Reconnection Events Assisted by Simulation

For some spacecraft events a 2D reconstruction can shed light on physics of interest or validate models. For an MMS magnetotail reconnection event (on 10 August 2017) shown in this section, reconstruction allowed for revealing whether the time series of data is consistent with a laminar 2D reconnection geometry or if 3D dynamics are required to explain the observations. Here we present an interpolative method that assumes steady-state reconnection. For the given event, it allows for reconstruction of a physical area extending about $40d_e \times 10d_e$ (where $d_e = c/\omega_{pe}$ is the electron inertial length) around the x-line, an area much larger than that allowed in methods that rely on Taylor expansion (Sect. 2.3.1) or electron magnetohydrodynamics equations (Sect. 2.3.4).

For a given event, once one optimizes spacecraft trajectories parameterized in 2D space and time, using the method as introduced in Sect. 2.2.3, the MMS signals can be used to construct 2D field maps, as shown in Fig. 11. In the event considered here, the trajectories closely follow the magnetic separatrix of the reconnection geometry (shown as the nearly horizontal thick black curve in Fig. 11). There it can be assumed that electron-scale gradients are mostly perpendicular to the magnetic separatrix because electrons thermally stream along the field lines. Therefore, a grid is defined to have cells elongated approximately parallel with the separatrix in order to capture variation in signals across the topological boundary. To achieve this, the spacecraft trajectories are rotated clockwise by an angle $\theta = 17^\circ$ to a coordinate system in which the magnetic separatrix followed by the spacecraft becomes nearly

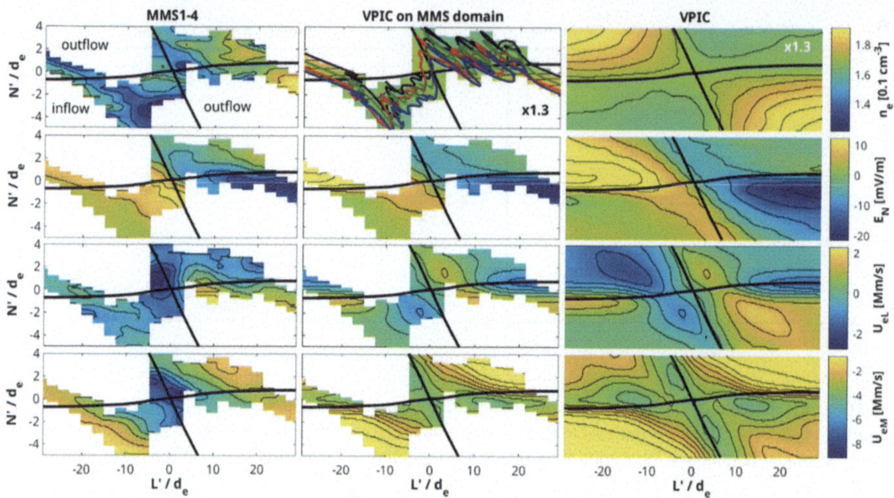

Fig. 11 The left column shows fields measured by MMS constructed into a 2D map based on the spacecraft trajectory. The middle column shows PIC simulation data in the same region of the reconnection geometry for comparison, while the right column shows the simulation data plotted over the entire domain. Adapted from Schroeder et al. (2022)

horizontal. These rotated coordinates are defined as (L', N'); the L'-coordinate approximately represents the distance along the separatrix, and the N'-coordinate approximately represents the distance perpendicular to the separatrix.

Raw MMS data are then distributed spatially according to the optimized spacecraft trajectories. Data are placed into spatial bins with lengths $\Delta L' \simeq 2d_e$ and $\Delta N' \simeq 0.08d_e$, such that the aspect ratio of each cell is approximately 25. All bins through which neither of the trajectories passes are left as empty cells (NaN values). Thus, each spatial grid cell contains data values from all times when either of the MMS spacecraft paths falls within its area. The final value for each cell is calculated by a simple average of all data values contained in that cell.

The particle-in-cell data inherently fills out the entire 2D simulation domain (right panels of Fig. 11). However, to allow for direct comparison between the measurement maps and simulation, the simulation data are binned and averaged on the same grid as the spacecraft data (middle column of Fig. 11). We note that $\nabla \cdot \mathbf{B} = 0$ is not guaranteed by this method, because the spacecraft paths are found by the method discussed in Sect. 2.2.3, so that the measured and thus reconstructed field values do not strictly agree with the simulation values.

2.3.4 Grad-Shafranov and Electron Magnetohydrodynamics Reconstruction

Fundamental equations, such as a set of MHD equations, can be used to reconstruct steady, 2D structures around the path of an observing spacecraft from in-situ measurements of the electromagnetic field and plasma. In standard numerical simulations, a set of equations governing the system is solved as an initial value problem for studying temporal evolution of the system or physical quantities. This can be done by setting the initial and boundary conditions at every part of the simulation domain. On the other hand, the reconstruction as explained below solves a time-independent form of the governing equation(s) as a spatial initial value problem to get 2D field maps of physical quantities. This is possible by setting, based on the measurements, the initial conditions at points along the spacecraft path and solving the equation(s) for spatial development of the corresponding quantities. The first such method, Grad-Shafranov (GS) reconstruction technique, was introduced by Sonnerup and Guo (1996), and was further extended to include MHD (Sonnerup and Teh 2008) and Hall-MHD effects (Sonnerup and Teh 2009). An overview and reviews of these earlier types of reconstruction were given by Sonnerup et al. (2006b, 2008) and Hasegawa (2012). Here we describe more recent developments of the reconstruction techniques along the same line.

For GS reconstruction schemes, in which 2D and steady structures are assumed, one needs to find a proper or comoving frame of reference in which the structure looks approximately time-independent, and a reconstruction plane that is perpendicular to the invariant axis ($\hat{\mathbf{z}}$) along which the structure has negligible spatial gradients. The 2D maps of plasma and magnetic fields are recovered on that plane. From single spacecraft observations, the velocity of the proper frame ($\mathbf{V}_{\text{str}}$) can be obtained by the HT analysis (Khrabrov and Sonnerup 1998; Sect. 2.2.2). If observations from four spacecraft are available, the STD method (Shi et al. 2006; Sect. 2.2.2) can also be used to determine the velocity of the structure.

The invariant axis ($\hat{\mathbf{z}}$) can be determined by rotating one of the eigenvectors from Minimum Variance Analysis (MVA) (Sonnerup and Scheible 1998), taken as a trial invariant axis, by some angle until measured data points in the parameter plane of a field line invariant (such as the axial component of the magnetic field B_z and the transverse pressure $P_t = p + B_z^2/(2\mu_0)$ where p is the plasma pressure) versus partial vector potential A (out-of-plane component of the vector potential) are approximately expressed by a single curve, namely, an exponential or polynomial function: $B_z = B_z(A)$ and $P_t = P_t(A)$ (e.g., Hu and Sonnerup 2002). By ingesting multi-spacecraft data, the optimal axis could also be found in such a way that the correlation coefficient between the reconstructed magnetic fields based on one spacecraft data and the measured magnetic fields from other spacecraft reaches the maximum value (Hasegawa et al. 2004). Using magnetic field data from four-point measurements, the minimum gradient direction from the MDD analysis can be taken as the invariant axis (Shi et al. 2005; Sect. 2.2.1). In some cases, the results of MDD and MVA can be combined to provide a reconstruction coordinate system, in which not only the invariant axis (parallel to $\hat{\mathbf{e}}_M$) but also the L and N axes are properly defined (Denton et al. 2016, 2018; Hasegawa et al. 2017; Tian et al. 2020; Sect. 2.2.1). The x axis is defined as being antiparallel to the projection of the structure velocity $\mathbf{V}_{\text{str}}$ onto the plane perpendicular to the invariant axis ($\hat{\mathbf{z}}$), thus representing the spacecraft path in the reconstruction (xy) plane. A right-handed orthogonal system is formed by $\hat{\mathbf{y}} = \hat{\mathbf{z}} \times \hat{\mathbf{x}}$.

Grad-Shafranov Reconstruction with pressure anisotropy effects In collisionless plasma there can be pressure anisotropy, that is $p_\perp \neq p_\parallel$, where $p_\perp$ and $p_\parallel$ are the thermal pressures perpendicular and parallel to the magnetic field, respectively. Taking the effects of pressure anisotropy and field aligned flow into account, Sonnerup et al. (2006b) derived a new GS equation by considering the double-polytropic energy laws (Hau et al. 1993) $\mathrm{d}\left\{p_\perp/(\rho B^{\gamma_\perp - 1})\right\}/dt = 0$ and $\mathrm{d}\left(p_\parallel B^{\gamma_\parallel - 1}/\rho^{\gamma_\parallel}\right)/dt = 0$, where ρ is the mass density, B is the magnetic field strength, $\gamma_\perp$ and $\gamma_\parallel$ are polytropic exponents, which can be inferred from observations. Different values of $\gamma_\perp$ and $\gamma_\parallel$ represent different thermodynamic conditions (e.g., Hau et al. 2020). Chen and Hau (2018) developed a GS code for anisotropic and field-aligned flow for the first time and benchmarked it with an analytical model. The application of this code to a magnetopause crossing event showed that the recovered magnetic islands inside the magnetopause had larger widths than that from the GS reconstruction for isotropic plasma.

There are also some space plasma structures with anisotropic pressure in quasi-static equilibrium, such as mirror-mode structures and magnetospheric ultra-low frequency compressional waves (drift mirror-mode wave). Tian et al. (2020) reconstructed the magnetic field structure of the ultra-low frequency compressional wave by the GS method including the pressure anisotropy effect. They call it the reduced GS-like method, because the corresponding GS-like equation,

$$\nabla \cdot [(1-\alpha)\nabla A] = \mu_0 \rho \left[T_\perp \frac{dS_\perp}{dA} + T_\parallel \frac{dS_\parallel}{dA} - \frac{dH}{dA} \right] - B_z \frac{dC_z}{dA}, \tag{8}$$

can be derived by removing terms containing the bulk velocity in the equations given by Sonnerup et al. (2006b), which contain both the anisotropy and field-aligned flow effects. Here, $S_\perp = c_{v\perp} \cdot \ln(p_\perp/\rho B^{\gamma_\perp - 1})$ and $S_\parallel = c_{v\parallel} \cdot \ln(p_\parallel B^{\gamma_\parallel - 1}/\rho^{\gamma_\parallel})$ are the perpendicular and parallel pseudo entropies, respectively, $H = [p_\perp/\{(\gamma_\perp - 1)\rho\}] + [\gamma_\parallel p_\parallel/\{(\gamma_\parallel - 1)\rho\}]$ is the total enthalpy, $C_z = (1-\alpha)B_z$, $\alpha = (p_\parallel - p_\perp)\mu_0/B^2$ is the pressure anisotropy factor, μ_0 is the vacuum permeability, B_z is the magnetic field component along the invariant axis, $c_{v\perp} = R/(\gamma_\perp - 1)$ and $c_{v\parallel} = R/(\gamma_\parallel - 1)$ with the ordinary gas constant $R = c_{p\parallel} - c_{v\parallel}$. All of $S_\perp$, $S_\parallel$, H and C_z are field line invariants and are functions of A only. Reduced auxiliary equations,

$$\mathbf{M}\mathbf{X}^T = \mathbf{Y}^T, \tag{9}$$

are used to spatially advance quantities α, ρ, $p_\perp$, $p_\parallel$, B_z, B^2 and $\partial^2 A/\partial^2 y$ in y, along with spatial integration of A and B_x. Here, the superscript T denotes the matrix transpose, and $\mathbf{M}$ is a 7×7 matrix expressed as follows:

$$\mathbf{M} = \begin{bmatrix} \frac{c_{v\perp}}{p_\perp} & 0 & -\frac{c_{v\perp}}{\rho} & 0 & -\frac{R}{2B^2} & 0 & 0 \\ 0 & \frac{c_{v\parallel}}{p_\parallel} & -\frac{c_{p\parallel}}{\rho} & 0 & \frac{R}{2B^2} & 0 & 0 \\ \frac{c_{v\perp}}{R\rho} & \frac{c_{p\parallel}}{R\rho} & -\frac{(c_{v\perp}p_\perp + c_{p\parallel}p_\parallel)}{R\rho^2} & 0 & 0 & 0 & 0 \\ 0 & 0 & 0 & 1-\alpha & 0 & -B_z & 0 \\ -1 & 1 & 0 & 0 & -\alpha & -B^2 & 0 \\ 0 & 0 & 0 & B_z & -0.5 & 0 & B_x \\ 0 & 0 & 0 & 0 & 0 & \frac{B_x}{\alpha - 1} & 1 \end{bmatrix}, \tag{10}$$

$$\mathbf{X} = [\frac{\partial p_\perp}{\partial y}, \frac{\partial p_\parallel}{\partial y}, \frac{\partial \rho}{\partial y}, \frac{\partial B_z}{\partial y}, \frac{\partial B^2}{\partial y}, \frac{\partial \alpha}{\partial y}, \frac{\partial^2 A}{\partial y^2}], \tag{11}$$

$$\mathbf{Y} = [B_x \frac{dS_\perp}{dA}, B_x \frac{dS_\parallel}{dA}, B_x \frac{dH}{dA}, B_x \frac{dC_z}{dA}, B_y \frac{\partial B_x}{\partial x}, \frac{Q}{1-\alpha}], \tag{12}$$

where Q in $\mathbf{Y}$ is RHS$-\partial[(1-\alpha)\partial A/\partial x]/\partial x$, and RHS is the right-hand side quantity in Eq. (8). One difficulty in this method exists in determining the proper polytropic exponents $\gamma_\perp$ and $\gamma_\parallel$. Hau et al. (2020) inferred these parameters by using the measured magnetosheath data, and recovered the 2D topology of a mirror-mode structure observed in the magnetosheath.

Aiming at reconstruction of anisotropic plasma structures, Teh (2019) developed another simple extended GS equation. He did not use the double polytropic energy laws, but assumed that parameters α, $p_\perp$, and $p_\parallel$ are functions of the magnetic field strength B only to derive a relatively simple GS-like equation. This assumption might not be valid. Nevertheless, basic features of magnetic mirror-mode and flux rope with pressure anisotropy were revealed by this extended GS solver (Teh 2019; Teh and Zenitani 2020).

Electron Magnetohydrodynamics Reconstruction The electron magnetohydrodynamics (EMHD) reconstruction is a single-spacecraft method for the reconstruction of steady, 2D electromagnetic fields and electron streamlines in regions where ions are fully demagnetized and thus electron dynamics dominates. It was developed to recover the field geometry in and around the EDR of magnetic reconnection, where electrons are demagnetized. The original version (Sonnerup et al. 2016) is based on an inertia-less and time-independent form of the electron MHD equation (Kingsep et al. 1990 and references therein) and assumes uniform electron density (electron incompressibility) and temperature. A recent version incorporates electron inertia effects in the streamline reconstruction, and the effects of nonuniform density and temperature and a guide magnetic field component B_z in the reconnection region (Hasegawa et al. 2021). The density and temperature are, however, assumed to be preserved along the magnetic field lines, because such assumptions are roughly satisfied around the EDR of symmetric, antiparallel reconnection (Korovinskiy et al. 2020). These conditions are not well satisfied for guide-field or asymmetric reconnection; therefore, further model developments are needed. Nonetheless, under the above assumptions, the magnetic field can be reconstructed by use of a Grad-Shafranov-type equation

$$\nabla^2 A = -\mu_0 j_z (A) = \mu_0 e n_e (A) u_{ez} (A) . \tag{13}$$

Reconstruction of the EDR requires some kind of dissipation term and, for antiparallel reconnection, makes use of a term corresponding to the component of the divergence of the electron pressure tensor $\mathbf{P}$ in the direction $\hat{\mathbf{z}}$ of reconnection electric field (or X-line) (see review by Hesse et al. 2011)

$$(\nabla \cdot \mathbf{P}) \cdot \hat{\mathbf{z}} = n_e \sqrt{2 m_e k_B T_e} \frac{\partial u_{eL}}{\partial L} . \tag{14}$$

In the case of guide-field reconnection, see Hasegawa et al. (2021) for some recipes. The reconstruction is performed in the rest frame of magnetic field structures, as introduced in the second paragraph of Sect. 2.3.4.

Figure 12 shows 2D maps of the magnetic field and electron streamlines from the EMHD reconstruction with electron inertia effects for the magnetotail EDR event on 11 July 2017, first reported by Torbert et al. (2018). Magnetic field, electric field, and electron moment data taken by the MMS3 spacecraft, which made the closest approach to the X-line, are used to set the initial conditions on the x axis. The final frame velocity was determined by a multi-spacecraft method (Hasegawa et al. 2017), in which the correlation coefficient is maximized between the components of the magnetic fields and electron velocities measured by the three spacecraft (MMS1, MMS2, and MMS4) not used as input and those predicted from the maps along the spacecraft paths. The final z axis was optimized by a method based on the y component of a time-independent and 2D form of Ampère's law, $-\partial B_z/\partial x = \mu_0 j_y$; when the z axis is properly chosen, this relation approximately holds for the particle current density and magnetic field data taken along the spacecraft path (x axis) (Hasegawa et al. 2019).

Figure 12a shows a clear X-type magnetic field geometry, as seen in simulations. We note that the information on the separatrix opening angle gained from the reconstructed field map, as seen in Fig. 12a, can be used to estimate the reconnection electric field by a method explained in Sect. 3.3.3. Figure 12b essentially shows expected patterns of electron inflow and outflow and corresponding Hall magnetic fields B_z. Interestingly, the electron stagnation point is displaced in the earthward (outflow) direction by $\sim 3 d_e$ ($d_e \sim 27$ km/s) from the X-point, a new feature revealed by the reconstruction. The method has also been successfully applied to an EDR of magnetopause reconnection (Hasegawa et al. 2017) and an ion-scale magnetic flux rope in the magnetopause current sheet (Hasegawa et al. 2023).

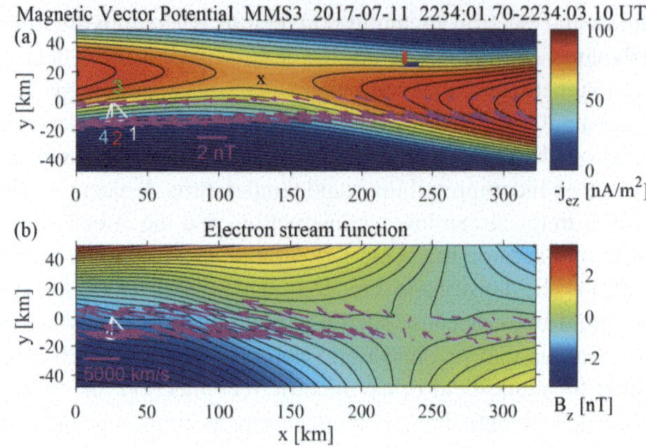

Fig. 12 2D maps of the magnetic field (**a**) and electron streamlines (**b**) from the EMHD reconstruction applied to data from the magnetotail EDR encounter by MMS3 on 11 July 2017. The arrows show the projection onto the reconstruction plane of the measured magnetic fields (**a**) and electron velocities in the structure-rest frame (**b**). The axial current density (**a**) and axial magnetic field component (**b**) are shown in color. The blue and red bars in Fig. 12 are the projection of the GSM x and z axes, respectively. Adapted from Hasegawa et al. (2021)

MMS-Tailored Electron Magnetohydrodynamics Reconstruction An alternative approach to the EMHD GS reconstruction of EDR is represented by the MMS-tailored model, introduced by Korovinskiy et al. (2021). Adopting an assumption of steady-state two-dimensional magnetoplasma configuration and assuming additionally the uniform number density n, the problem is reduced to calculation of two quantities: the magnetic potential A of the in-plane magnetic field and the out-of-plane magnetic field component B_z. The out-of-plane component of Ampère's law makes the magnetic potential to obey the equation,

$$\nabla^2 A = \mu_0 e n u_{ez}, \tag{15}$$

where the right-hand side is represented, in general, by a function of two variables (contrary to Eq. (13)). Then, with neglected ion current and $\partial/\partial z = 0$, the in-plane components of Ampère's law reveal that the quantity $-B_z/(\mu_0 e)$ serves as a stream function for the in-plane electron flow $n\mathbf{u}_{e\perp} = (nu_{ex}, nu_{ey})$: $\mu_0 e n u_{ex} = -\partial B_z/\partial y$ and $\mu_0 e n u_{ey} = +\partial B_z/\partial x$. With uniform number density, this yields the equation for B_z,

$$\nabla^2 B_z = \mu_0 e n \left(\partial u_{ey}/\partial x - \partial u_{ex}/\partial y\right) = Q. \tag{16}$$

Here, the model function Q represents the contribution of the electron inertia and anisotropy (see Eq. 14–16 of Korovinskiy et al. 2021).

Since the Jacobian of a variable transform $(x, y) \rightarrow (A, B_z)$ may turn to zero or infinity at a manifold of Lebesgue measure zero only (see Eq. 9 of Korovinskiy et al. 2021), the right-hand sides of Eqs. (15,16) can be considered as the functions of (A, B_z). Assuming that Cartesian coordinates correspond to the local co-moving LMN coordinate system (e.g., Denton et al. 2018), where the x axis coincides with $\hat{\mathbf{e}}_L$ and the y axis coincides with $\hat{\mathbf{e}}_N$, one can note that the stretched configuration of EDR, dictated by a low reconnection rate $\varepsilon \sim 0.1$ (e.g., Liu et al. 2017), brings the ratio $\partial/\partial x \ll \partial/\partial y$. The analogous scaling ratio $\partial/\partial B_z \ll \partial/\partial A$ (see, e.g., Eqs. 12 and 13 of Korovinskiy et al. 2021) is valid in the variable space (A, B_z). Omitting the minor dependence on B_z, one can consider the zeroth-order reconstruction model, where the model functions depend on A only. This way Eq. (15) turns into the Grad-Shafranov Eq. (13) and in Eq. (16) we have $Q = Q(A)$. Notably, under this equality the mathematical self-consistency of the solution demands an extra term $\sim B_z$ on the right-hand side of Eq. (15); however, the contribution of this term is found to be negligible (see Eq. 21 and Fig. 7 of Korovinskiy et al. 2021), so it can be omitted.

The model functions $u_{ez}(A)$ and $Q(A)$ are evaluated from the boundary conditions. Particularly, the latter is calculated by using the data of the four MMS probes. Thus, the reconstruction model development is completed and the only problem left is the solution of the ill-posed problem, stated by Eq. (15,16), with the boundary conditions specified along the probe trajectory. In an approach discussed by Korovinskiy et al. (2020), the problem regularization was performed by utilizing the so-called boundary layer approximation (BLA) (Schlichting 1979). Namely, the second-order small terms $\partial^2/\partial x^2 \sim \varepsilon^2\partial^2/\partial y^2$ are omitted, reducing the problem to the system of ordinary differential equations of the second order. To benefit from the simplicity of this method, the local LMN coordinate system must be accurately determined. Besides that, this method is less universal, as it is not applicable to EDR crossings in the direction normal to the current sheet.

The described model, named 'Model 2' in Korovinskiy et al. (2021), was tested by reconstruction of the MMS event on 11 July 2017 (Torbert et al. 2018). The advantages of this model are the following. First, it does not depend on the out-of-plane electric field E_z, which is assumed to be constant, while in reality it may be considerably oscillating (see Fig. 2a in Korovinskiy et al. 2021). The nonuniformity of E_z brings appreciable uncertainty to the reconstruction results obtained by utilizing Eq. (14), representing the approximation derived by Hesse et al. (1999) for the electron pressure anisotropy (see 'Model 3' of Korovinskiy et al. 2021). Since this approximation is not used, reconstruction errors, which can appear due to its possible inaccuracy (see Fig. 2 in Korovinskiy et al. (2020), and the corresponding discussion), are also eliminated. Second, with BLA the model allows evaluation of the small terms B_y and u_{ey}, but these quantities do not affect other computations, particularly, the computation of B_z and u_{ex}. This property brings advantage when the translational symmetry of the configuration is corrupted, since in this case $\partial B_x/\partial z \neq 0$, and hence the formula $\mu_0 enu_{ey} = \partial B_z/\partial x$, used in the two-dimensional models, fails.

Thus, the major advantages of the discussed MMS-tailored technique consist in its comparative simplicity and increased accuracy due to the reduced sensitivity to violations of the ideal theoretical conditions. This is achieved at the expense of the lost universality, since the model requires current sheet crossing in the direction with a nonzero angle with respect to $\hat{\mathbf{e}}_N$ and multi-spacecraft data for evaluating the model function Q. The major disadvantage – the assumption of uniform number density, which limits the model applicability to the internal EDR – can be relaxed by substituting $n = n(A)$ in Eqs. (12–16) of Korovinskiy et al. (2021). Another limitation is related to the adopted assumption of a not-too-small guide field value. For a case of small or zero guide field, the corresponding (rather straightforward) modifications are required; in particular, the symmetry considerations demand $Q = Q(A, B_z)$, while the zeroth-order model equation $j_{ez} = j_{ez}(A)$ stays unchanged. The extended compressible model and discussion of the ways to further improve the EMHD GS reconstruction technique can be found in Korovinskiy et al. (2023).

2.3.5 3D Field Reconstruction Using Modified Radial Basis Functions

Another method for the reconstruction of 3D local magnetic field structures is based on the use of toroidal and poloidal magnetic potentials that are expressed as linear combinations of radial basis functions (e.g., Buhmann 2003). In this approach, the magnetic field is represented by

$$\mathbf{B} = \nabla \times \left(\frac{\psi_1}{r} \mathbf{r} \right) + \nabla \times \nabla \times (\psi_2 \mathbf{r}), \tag{17}$$

where ψ_1 and ψ_2 are the toroidal and poloidal potentials, respectively. Instead of using orthogonal basis functions, the potentials are individually expanded into a series $\psi_j = \sum_i \alpha_{ji} \chi_i$ of a modified form of radial basis functions (Andreeva and Tsyganenko 2016; Tsyganenko and Andreeva 2016)

$$\chi_i (\mathbf{r}; D, \mathbf{L}) = \left\{ \left(\frac{x - R_{i,x}}{L_x} \right)^2 + \left(\frac{y - R_{i,y}}{L_y} \right)^2 + \left(\frac{z - R_{i,z}}{L_z} \right)^2 + D^2 \right\}^{1/2}. \tag{18}$$

Here, the vectors $\mathbf{R}_i$ are coordinates of meshwork grid nodes, $\mathbf{L} = \{L_x, L_y, L_z\}$ a characteristic length set equal to the node separation along each axis, and D an adjustable regularization parameter. An advantage of the magnetic field thus defined is its divergence-free nature (Stern 1976). The expansion coefficients $\{\alpha_i\}$ can be determined by optimally fitting to magnetic field data taken by multiple spacecraft, such as Cluster and MMS, for a time interval.

Chen et al. (2019) tested this technique for 2D and 3D model magnetic fields, and applied it to magnetic field data from the Cluster mission. The structure velocity relative to the spacecraft was treated as a hyper parameter, and was tuned by minimizing the average magnitude of an error vector field $\Delta\mathbf{B} \equiv \mathbf{B}_{\text{data}} - \mathbf{B}_{\text{rec}}$, where $\mathbf{B}_{\text{data}}$ and $\mathbf{B}_{\text{rec}}$ are the measured and reconstructed magnetic field vectors, respectively. Their study suggests that the method can be used to identify and investigate the properties of characteristic structures in 3D magnetic reconnection, including magnetic nulls (Sect. 3.1.7) and separator lines.

3 Detection and Analysis of in-Situ Observations of the Diffusion Regions

This section mostly focuses on the methods for detecting and analyzing in-situ observations of the electron diffusion region (EDR). See Phan et al. (2005) for a brief review of in-situ observations and analysis of ion diffusion regions (IDRs) with effects of the Hall term in the generalized Ohm's law (e.g., Liu et al. 2024).

3.1 Diffusion Region Identification

This section reviews several measures that can be used to identify the diffusion region or nearby regions. We stress that none of the following measures are uniquely non-zero or significant only in the diffusion region, and that none of them by themselves necessarily identify "dissipation" in the sense of an irreversible process, although some are conventionally called a dissipation measure or have been used to identify the region where a process leading to entropy increase may occur. Therefore, one should use these measures in the context of other measurements of the reconnection process to identify a candidate diffusion region; the candidate may or may not turn out to be an actual diffusion region after full quantitative analysis to make sure whether other reconnection and diffusion region signatures are observed.

3.1.1 Plasma-Frame Dissipation Measure

Since magnetic reconnection converts electromagnetic field energy to plasma kinetic energy, we expect the energy conversion rate $\mathbf{j} \cdot \mathbf{E}$ or its variants ($\mathbf{j}_i \cdot \mathbf{E}$ and $\mathbf{j}_e \cdot \mathbf{E}$) to be significant in magnetic reconnection. These quantities are important, but may not be useful to identify the EDR because they can be nonzero even in ideal regions (see Eq. (30) below).

One problem is that $\mathbf{j} \cdot \mathbf{E}$ is measured in the stationary or observer's (laboratory) frame; if the EDR is moving, it might be better to evaluate $\mathbf{j} \cdot \mathbf{E}$ in a particular frame. Starting from this, we consider the energy conversion in a moving frame that travels at a velocity $\mathbf{V}_r$. In this reference frame, the electric current density and the electric field are given by

$$\mathbf{j}'_r = \mathbf{j}'(\mathbf{V}_r) = \mathbf{j} - \rho_c \mathbf{V}_r, \tag{19}$$

$$\mathbf{E}'_r = \mathbf{E}'(\mathbf{V}_r) = \mathbf{E} + \mathbf{V}_r \times \mathbf{B}, \tag{20}$$

where ρ_c is the charge density and the primed quantities denotes those in the moving frame. The energy conversion rate in the $\mathbf{V}_r$-moving frame yields

$$D_r = D(\mathbf{V}_r) \equiv \mathbf{j}' \cdot \mathbf{E}' = \mathbf{j} \cdot (\mathbf{E} + \mathbf{V}_r \times \mathbf{B}) - \rho_c (\mathbf{V}_r \cdot \mathbf{E}). \tag{21}$$

Here, the reference velocity is arbitrary, so we call Eq. (21) the frame-independent dissipation measure. Employing the electron fluid velocity as the reference velocity, we define the electron-frame dissipation measure $D_e = D(\mathbf{u}_e)$, and employing the ion fluid velocity, we obtain the ion-frame dissipation measure $D_i = D(\mathbf{u}_i)$.

Importantly, these measures are Galilean invariants, in other words, frame-independent. By using the Lorentz factor $\gamma_r = [1 - (V_r/c)^2]^{1/2}$, we obtain a Lorentz invariant form (Zenitani et al. 2011a)

$$D_r = \gamma_r [\mathbf{j} \cdot (\mathbf{E} + \mathbf{V}_r \times \mathbf{B}) - \rho_c (\mathbf{V}_r \cdot \mathbf{E})]. \tag{22}$$

It is seen that as long as we choose a unique reference velocity $\mathbf{V}_r$, Eq. (22) always gives the same result regardless of the observer's velocity or direction.

Let us discuss properties of the dissipation measures. We focus on the nonrelativistic limit of $\gamma_r \to 1$ for simplicity. In a plasma, the charge density and the electric current density are

$$\rho_c = \sum_s q_s n_s, \qquad \mathbf{j} = \sum_s q_s n_s \mathbf{u}_s, \tag{23}$$

where s denotes the plasma species. We consider a charge-weighted sum of Eq. (20)

$$\sum_s q_s n_s \mathbf{E}'(\mathbf{u}_s) = \rho_c \mathbf{E} + \mathbf{j} \times \mathbf{B}. \tag{24}$$

Applying $\mathbf{j}\cdot$ to both sides, and using Eq. (23), we obtain

$$\mathbf{j} \cdot \sum_s q_s n_s \mathbf{E}'(\mathbf{u}_s) = \rho_c \mathbf{j} \cdot \mathbf{E} = \rho_c \left(\sum_s q_s n_s \mathbf{u}_s \right) \cdot \mathbf{E}, \tag{25}$$

$$\sum_s q_s n_s \left(\mathbf{j} \cdot \mathbf{E}'_s - \rho_c \mathbf{u}_s \cdot \mathbf{E} \right) = 0. \tag{26}$$

This provides a useful relation

$$\sum_{s} q_s n_s D_s = 0. \tag{27}$$

Next we consider the relevance to resistive MHD. We define the MHD quantities,

$$\rho_{\text{mhd}} \equiv \sum_{s} m_s n_s, \qquad \mathbf{u}_{\text{mhd}} \equiv \frac{\sum_s m_s n_s \mathbf{u}_s}{\sum_s m_s n_s} = \frac{\sum_s m_s n_s \mathbf{u}_s}{\rho_{\text{mhd}}}. \tag{28}$$

Since Eq. (21) only uses linear operators, we obtain

$$D_{\text{mhd}} = D\left(\mathbf{u}_{\text{mhd}}\right) = D\left(\frac{\sum_s m_s n_s \mathbf{u}_s}{\rho_{\text{mhd}}}\right) = \frac{\sum_s m_s n_s D\left(\mathbf{u}_s\right)}{\rho_{\text{mhd}}} = \frac{\sum_s m_s n_s D_s}{\rho_{\text{mhd}}}. \tag{29}$$

From the Ohm's law, we obtain the energy conversion rate

$$\mathbf{E} + \mathbf{u}_{\text{mhd}} \times \mathbf{B} = \eta \mathbf{j}, \qquad \mathbf{j} \cdot \mathbf{E} = (\mathbf{j} \times \mathbf{B}) \cdot \mathbf{u}_{\text{mhd}} + \eta j^2. \tag{30}$$

Rearranging the MHD-frame dissipation $D_{\text{mhd}} = D\left(\mathbf{u}_{\text{mhd}}\right)$, we immediately obtain from Eq. (21)

$$\mathbf{j} \cdot \mathbf{E} = (\mathbf{j} \times \mathbf{B} + \rho_c \mathbf{E}) \cdot \mathbf{u}_{\text{mhd}} + D_{\text{mhd}}. \tag{31}$$

In a quasineutral ion-electron plasma, from Eqs. (27) and (29), we obtain

$$D_{\text{mhd}} \approx D_i \approx D_e. \tag{32}$$

From Eqs. (30) and (31) and considering that ρ_c is negligible in MHD, we see that the D_{mhd} term ($\approx D_e$) plays the same role as the nonideal energy conversion rate ηj^2.

PIC simulations have revealed that $D_e > 0$ marks a compact physically-significant region surrounding the X-line (Zenitani et al. 2011a). Although the resolution was limited and it was only partially evaluated, Zenitani et al. (2012) reported $D_e > 0$ during a magnetotail reconnection event observed by the Geotail spacecraft (Nagai et al. 2011). Recent observations by MMS unambiguously reported $D_e > 0$ near the EDR (Burch et al. 2016a; Phan et al. 2018; Torbert et al. 2018). Based on these results, it is fair to say that $D_e > 0$ is an important signature of the EDR.

We raise unsolved issues here. First, there is often a weakly negative region of $D_e < 0$ in the downstream side of the EDR, where the electrons outrun the $\mathbf{E} \times \mathbf{B}$ velocity (Karimabadi et al. 2007; Nakamura et al. 2018b; Pritchett 2001; Shay et al. 2007). We often see $D_e < 0$ at the jet termination region where the reconnected magnetic field is compressed (Payne et al. 2021), while $D_e \lesssim 0$ is also seen inside the elongated electron jet (Zenitani et al. 2011b). By clarifying the underlying mechanisms, we may be able to predict the negative amplitude of D_e. Second, one can split the electron-frame measure into its perpendicular and parallel contributions, $D_e = \mathbf{j}' \cdot \mathbf{E}' \approx \mathbf{j}_\perp \cdot \mathbf{E}'_\perp + j_\| E_\|$. Inside the EDR, we expect that electron meandering motion provides $\mathbf{j}_\perp \cdot \mathbf{E}'_\perp > 0$ in antiparallel reconnection, and that the electron parallel motion leads to $j_\| E_\| > 0$ in guide-field reconnection. Wilder et al. (2018) evaluated the energy conversion rate $\mathbf{j} \cdot \mathbf{E}' = \mathbf{j}_\perp \cdot \mathbf{E}'_\perp + j_\| E_\|$ in the diffusion regions for multiple reconnection events seen by MMS. They organized the results as a function of the guide-field amplitude B_g/B_0, and reported that the perpendicular contribution is dominant in the antiparallel cases ($B_g/B_0 \leq 0.3$) while the parallel contribution is dominant in the guide-field cases ($B_g/B_0 \geq 0.3$) (Fig. 10 in Wilder et al. 2018). However, the critical guide field between the two regimes has not been addressed before, theoretically or numerically. Further research is thus necessary to make a quantitative discussion. Third, strictly speaking, the term "dissipation" is ambiguously used in this section, because these measures do not always lead to an irreversible energy conversion in a collisionless plasma. Connection between the dissipation measures and a true irreversible dissipation process needs to be clarified. This might be better understood from the viewpoint of the entropy production in a kinetic plasma (Liang et al. 2019; Sect. 3.2.4).

3.1.2 Agyrotropy

Due to their small Larmor radii and short gyroperiods, electrons are closely tied to the magnetic field, much more so than heavier ions. As a result, they efficiently probe the field's structure so that locations of topological interest, such as reconnection X-lines and magnetic separatrices, should leave signatures in electron distribution functions. Vasyliunas (1975) identified one such signature by noting that for reconnection to occur, the electron pressure tensor must be non-gyrotropic at the X-point. Spacecraft measurements of non-gyrotropic electron distributions can hence be used as a measure of the proximity to the X-point.

Other authors (Scudder and Daughton 2008; Aunai et al. 2013) have proposed quantifications of agyrotropy, but the version discussed here follows the presentation in Swisdak (2016). A pressure tensor in a field-aligned coordinate system can always be put in the form

$$\mathbf{P} = \begin{pmatrix} P_{\parallel} & P_{12} & P_{13} \\ P_{12} & P_{\perp} & P_{23} \\ P_{13} & P_{23} & P_{\perp} \end{pmatrix}, \tag{33}$$

where $P_{\parallel}$ and $P_{\perp}$ represent the pressure parallel and perpendicular to the field, respectively.

In the gyrotropic case the off-diagonal components vanish; measures of agyrotropy attempt to quantify the size of these components relative to the diagonal ones. Pressure tensors are positive semidefinite (i.e., have non-negative eigenvalues) and thus satisfy the inequalities

$$P_{12}^2 \leq P_{\parallel} P_{\perp}, \qquad P_{13}^2 \leq P_{\parallel} P_{\perp}, \qquad P_{23}^2 \leq P_{\perp}^2, \tag{34}$$

which leads to a natural definition of a gyrotropy parameter

$$Q = \frac{P_{12}^2 + P_{13}^2 + P_{23}^2}{P_{\perp}^2 + 2P_{\perp}P_{\parallel}}. \tag{35}$$

For gyrotropic distributions $Q = 0$, while maximally agyrotropy occurs when $Q = 1$. By using certain rotationally invariant quantities it is possible to calculate Q while in any coordinate system (i.e., the pressure tensor needs not be in the form of Eq. (33)):

$$Q = 1 - \frac{4I_2}{(I_1 - P_{\parallel})(I_1 + 3P_{\parallel})}. \tag{36}$$

The necessary factors are the trace, $I_1 = P_{xx} + P_{yy} + P_{zz}$, the sum of principal minors, $I_2 = P_{xx}P_{yy} + P_{xx}P_{zz} + P_{yy}P_{zz} - (P_{xy}^2 + P_{xz}^2 + P_{yz}^2)$, and the parallel pressure $P_{\parallel} = \hat{\mathbf{b}} \cdot \mathbf{P} \cdot$

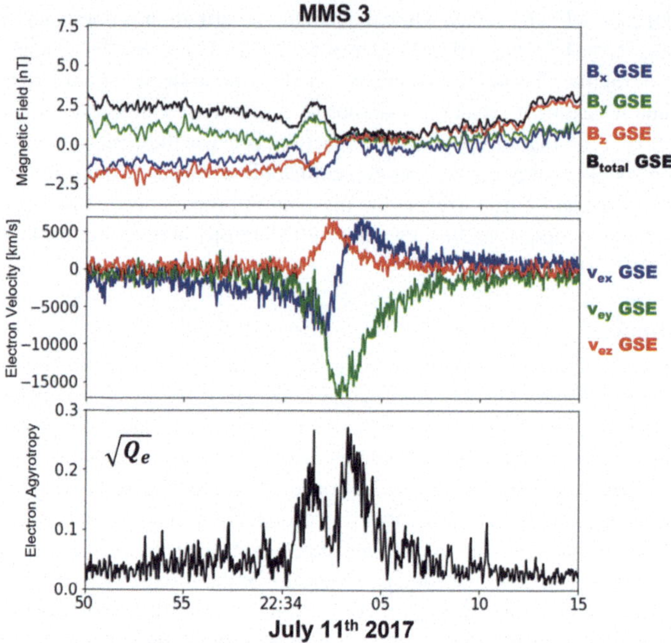

Fig. 13 Magnetospheric Multiscale (MMS) spacecraft 3 observations of an electron diffusion region at 2233:50-2234:15 UT on 11 July 2017. The top panel shows the three components and the strength of the magnetic field in geocentric solar ecliptic (GSE) coordinates (B_x, blue; B_y, green; B_z, red; and $|\mathbf{B}|$, black). The second panel shows the components of the electron velocity in the same color scheme. The final panel shows $\sqrt{Q_e}$

$\hat{\mathbf{b}} = b_x^2 P_{xx} + b_y^2 P_{yy} + b_z^2 P_{zz} + 2\left(b_x b_y P_{xy} + b_x b_z P_{xz} + b_y b_z P_{yz}\right)$. Here $\hat{\mathbf{b}}$ is the unit vector aligned with the magnetic field.

Figure 13 shows MMS data from a well known crossing of an EDR (Torbert et al. 2018) in the magnetotail with a small guide field (B_y component). The top two panels, which show the components of the magnetic field and the electron velocity, respectively, exhibit the expected signatures of an EDR crossing: a null in B_x associated with a minimum in B and a divergence or reversal in u_{ex}. Within the EDR $\sqrt{Q_e}$ rises sharply from its background value of ≈ 0.05 to a peak of ≈ 0.25. The bifurcated structure in $\sqrt{Q}$ has been seen in PIC simulations of antiparallel reconnection (Swisdak 2016). Calculations of Q_i (not shown) exhibit a much broader peak, as expected, since ions decouple from the magnetic field, and hence can acquire non-gyrotropic distribution functions on the much larger scales of the IDR.

We note that non-gyrotropic electron velocity distributions are not a unique signature of magnetic reconnection or the EDR; they have been observed at a non-reconnecting magnetopause current sheet (Tang et al. 2019).

3.1.3 Pressure-Strain Interaction in Reconnection Diffusion Region

Energy conversion in magnetic reconnection is a topic of fundamental importance. Several energy conversion measures, such as the Zenitani measure (Zenitani et al. 2011a; Sect. 3.1.1), have been used to evaluate energy conversion at the reconnection site. Recent studies have revealed the role of pressure-strain interaction in the conversion of bulk kinetic energy to the random (or thermal, or internal) component in collisionless plasmas (Cerri 2016; Yang et al. 2017; Fadanelli et al. 2020). Here, we show the evaluation of pressure-strain interaction near reconnection X-lines and discuss related interpretations (Bandyopadhyay et al. 2021).

The equation of the thermal or random energy density E_α^{th}, computed from the moments of the Vlasov equations, is given by

$$\partial E_\alpha^{\text{th}}/\partial t + \nabla \cdot \left(E_\alpha^{\text{th}} \mathbf{u}_\alpha + \mathbf{h}_\alpha\right) = -(\mathbf{P}_\alpha \cdot \nabla) \cdot \mathbf{u}_\alpha, \tag{37}$$

where α indicates a specific charged species, $\mathbf{u}_\alpha$ is the fluid velocity, $\mathbf{P}_\alpha$ is the pressure tensor, $E_\alpha^{\text{th}} = 3p^{(\alpha)}/2 = P_{ii}^{(\alpha)}/2$ with $P_{ii}^{(\alpha)}$ as the trace of the pressure tensor, and $\mathbf{h}_\alpha$ is the heat flux.

When integrated over a closed domain, the terms on the left-hand side within the divergence operator average to zero. Therefore, these may be interpreted as transport terms that do not contribute to net change in the form of energy but simply move energy from one location to another spatially. However, if we are concerned with quantifying conversion between different kinds of energy instead of transport, we see that the quantity on the right-hand side is responsible for the conversion of bulk kinetic energy to or from the thermal or random energy. We note that $-(\mathbf{P}_\alpha \cdot \nabla) \cdot \mathbf{u}_\alpha$ is not single-signed, and therefore it does not quantify irreversible conversion of energy. The net energy can convert in or out of the random component, depending on the sign of the pressure-strain.

The pressure-strain can be decomposed into compressive and incompressive components as

$$-(\mathbf{P}_\alpha \cdot \nabla) \cdot \mathbf{u}_\alpha = -p^{(\alpha)}\theta^{(\alpha)} - \Pi_{ij}^{(\alpha)} D_{ij}^{(\alpha)}, \tag{38}$$

where $-p^{(\alpha)}\theta^{(\alpha)}$ represents the energy conversion due to compressive motion, and $\theta^{(\alpha)} = \nabla \cdot \mathbf{u}_\alpha$ is the dilatation. Therefore, the remaining term $-\Pi_{ij}^{(\alpha)} D_{ij}^{(\alpha)}$ corresponds to incompressive energy conversion with the deviatoric pressure tensor $\Pi_{ij}^{(\alpha)} = P_{ij}^{(\alpha)} - p^{(\alpha)}\delta_{ij}$ and the traceless strain rate tensor $D_{ij}^{(\alpha)} = (1/2)\left(\nabla_i u_j^{(\alpha)} + \nabla_j u_i^{(\alpha)}\right) - (1/3)\theta^{(\alpha)}\delta_{ij}$ (Yang et al. 2017; Cassak and Barbhuiya 2022 and references therein).

In recent times, two major advances have facilitated the study of the pressure-strain interaction. First, PIC simulations have become sufficiently accurate to evaluate the pressure tensor, and supercomputers have become adequately powerful to perform plasma simulations with higher number of particles and larger systems. Secondly, an evaluation of $-(\mathbf{P}_\alpha \cdot \nabla) \cdot \mathbf{u}_\alpha$ requires accurate evaluation of the full pressure tensor as well as spatial derivatives of the fluid velocity down to kinetic scales. This was not possible observationally before the Magnetospheric Multiscale (MMS) Mission. The MMS mission, consisting of 4 spacecraft separated by a small distance, with high cadence instruments, provides the first and the only opportunity yet to study pressure-strain interactions using in-situ data.

Figure 14 shows an example of MMS observation of the reconnection diffusion region in the magnetopause current sheet, presented by Burch et al. (2016a). The top panel shows the magnetic field measurement in GSE coordinates. The next panel plots the electromagnetic energy conversion rate, as measured by the Zenitani measure (narrow black), and the energy conversion rate to internal energy, quantified by the total (ion+electron) pressure-strain rate (broad blue). We use the extension of the multi-spacecraft curlometer method (Dunlop et al. 1988; Paschmann and Daly 1998), along with the averaged pressure tensor from all 4 MMS spacecraft to measure $-(\mathbf{P}_\alpha \cdot \nabla) \cdot \mathbf{u}_\alpha$ for ions and electrons. Both conversion rates show an elevated signal in the diffusion region with similar magnitude, but the value of $-(\mathbf{P}_\alpha \cdot \nabla) \cdot \mathbf{u}_\alpha$ is smaller than $\mathbf{j} \cdot \mathbf{E}'$. This observation indicates that only part of the magnetic energy is being converted to random energy. The third panel shows that the electrons are responsible for the majority of the conversion. Finally, the bottom two panels show that for both electrons and ions, the compressive heating rate is stronger than the incompressive part.

These results show that the pressure-strain interaction can be used as an independent diagnostic of plasma energization in reconnection regions. Bandyopadhyay et al. (2021) show a few other examples of MMS reconnection events, including magnetosheath reconnection in thin current sheets and electron-only reconnection to analyze the role of the pressure-strain. Examples from turbulent PIC simulations are also shown by Bandyopadhyay et al. (2021) for comparison with MMS data. Broadly speaking, the simulations and different kinds of reconnection events sampled by MMS do not show any systematic difference in the pressure-strain interaction in the diffusion region. However, $-(\mathbf{P}_\alpha \cdot \nabla) \cdot \mathbf{u}_\alpha$ can be negative in some cases at the reconnecting X-lines, indicating that internal energy is locally being converted into kinetic energy. This contrasts with the electromagnetic energy conversion rate (as measured by $\mathbf{j} \cdot \mathbf{E}'$), which is positive for most of the cases. Understanding the ratio of ion to electron conversion between kinetic and internal energy also poses an intriguing challenge. Like the example shown here, most reconnection cases show that the electron energy conversion rate (as measured by $-(\mathbf{P}_\alpha \cdot \nabla) \cdot \mathbf{u}_\alpha$) is larger than the ion energy conversion rate. This is in contrast to the global energy conversion, which is dominated by the ions in the magnetosheath. A recent work by Barbhuiya and Cassak (2022) provided an explanation of this based on scaling analysis. Further statistical studies are required to fully understand the role of pressure-strain in heating due to magnetic reconnection.

Finally, it should be noted that the above works focused mainly on the $-(\mathbf{P}_\alpha \cdot \nabla) \cdot \mathbf{u}_\alpha$ term that relates the transfer of bulk kinetic energy to internal energy. However, as shown in Fadanelli et al. (2020), it is possible to look at all the terms that compose the electromagnetic, kinetic, and internal energy equations. Such a description allows to perform a point-by-point analysis of all energy conversion channels. While the study of how these various terms compare with and balance each other in the context of reconnection was done using a Hybrid-Vlasov simulation, this exercise remains to be done with spacecraft observations.

3.1.4 Electron Vorticity Indicative of the Electron Diffusion Region

The electron vorticity ($\mathbf{\Omega}_\mathrm{e} = \nabla \times \mathrm{u}_\mathrm{e}$) can be used as a proxy for delineating the EDR of magnetic reconnection. Figure 15a-d show the 11 July 2017 event (Torbert et al. 2018), during which MMS traversed a magnetotail current sheet along the trajectory shown in Fig. 15e: (a) the four-spacecraft tetrahedral-averaged magnetic field components in boundary normal coordinates (LMN), (b) the current density calculated from the curlometer technique (Dunlop et al. 2002), (c and d) electron vorticity and its magnitude (black profile) compared to ω_ce (the electron cyclotron angular frequency; blue). The electron vorticity is enhanced around 'b' marked by the vertical dashed red line in Figs. 15a-d and red arrow in Fig. 15e.

Electron velocity vectors and electron distribution functions measured at the four spacecraft (not shown) demonstrate that the enhanced electron vorticity is due to the intense shear of the velocity mostly along $-\hat{\mathbf{e}}_M$, which originates from the variation along $\pm\hat{\mathbf{e}}_N$ of the meandering electrons' velocity (Fig. 15f). Since the meandering electrons carry the out-of-plane current (j_M in Fig. 15b), the electron vorticity enhancement should coincide with the strong gradient of the current density. Indeed, $\mathbf{\Omega}_\mathrm{e}$ peaks (blue arrow in Fig. 15c) are located at the edges of the current density (j_M) profile (Fig. 15b).

In these observations, the largest component of electron vorticity, i.e., $\Omega_{\mathrm{e},L}$ (Fig. 15c), can be approximately written as $\Omega_{\mathrm{e},L} \sim -\partial u_{\mathrm{e},M}/\partial N \sim (1/(en_\mathrm{e}))\,\partial j_M/\partial N \sim$

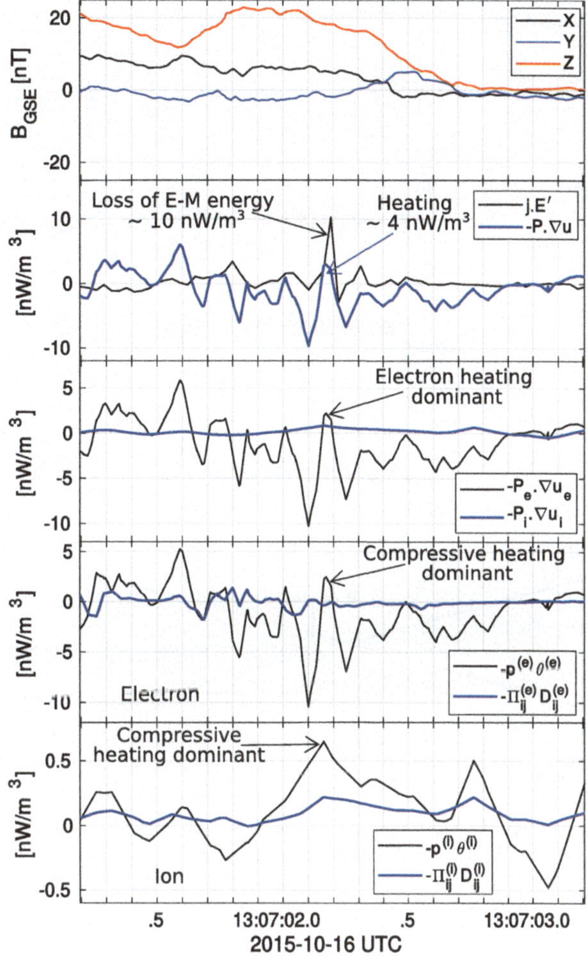

Fig. 14 MMS data in the magnetopause reconnection event on 16 October 2015. The X-line crossing was around 13:07:02.4 UTC. From top the plotted quantities are the magnetic field in GSE coordinates ($\mathbf{B}_{\mathrm{GSE}}$), the Zenitani measure ($\mathbf{j} \cdot \mathbf{E}'$) and pressure-strain rate ($-(\mathbf{P} \cdot \nabla) \cdot \mathbf{u}$), electron ($-(\mathbf{P}_\mathrm{e} \cdot \nabla) \cdot \mathbf{u}_\mathrm{e}$) and ion ($-(\mathbf{P}_\mathrm{i} \cdot \nabla) \cdot \mathbf{u}_\mathrm{i}$) pressure-strain rates, compressive ($-p^{(e)}\theta^{(e)}$) and incompressive ($-\Pi_{ij}^{(e)} D_{ij}^{(e)}$) pressure-strain rates for electrons, and compressive ($-p^{(i)}\theta^{(i)}$) and incompressive ($-\Pi_{ij}^{(i)} D_{ij}^{(i)}$) pressure-strain rates for ions

$(1/(en_\mathrm{e}\mu_0))\, \partial^2 B_L/\partial N^2$. If B_L changes from 0 at the neutral sheet to B_{edge} at the southern/northern edge of an EDR with a thickness of d_e, $\Omega_{\mathrm{e},L} \sim (1/(en_\mathrm{e}\mu_0))\, B_{edge}/d_\mathrm{e}^2 \sim \omega_{\mathrm{ce}}$. Thus, a peak of $\Omega_{\mathrm{e},L}$ that is comparable to or larger than ω_{ce} (Fig. 15d) delineates the N-directional edge of the EDR of a reconnecting current sheets on the d_e scale. This demonstrates why $|\mathbf{\Omega}_\mathrm{e}|$ compared with ω_{ce} can be a physical measure for EDR identification (Hwang et al. 2019).

3.1.5 Magnetic Flux Transport Method

The magnetic flux transport (MFT) method represents a novel way of detecting diffusion regions in situ. It is based on the definition of reconnection as the transport of magnetic flux across magnetic separatrices that intersect at an X-line (Vasyliunas 1975). This method measures signatures of active reconnection in the in-plane velocity of magnetic flux, $\mathbf{U}_\psi$, and its divergence, $\nabla \cdot \mathbf{U}_\psi$. Previously derived in 2D (Liu et al. 2018a; Liu and Hesse 2016) from Faraday's law and the advection equation of magnetic flux, $\partial\psi/\partial t + \mathbf{U}_\psi \cdot \nabla_\perp \psi = 0$,

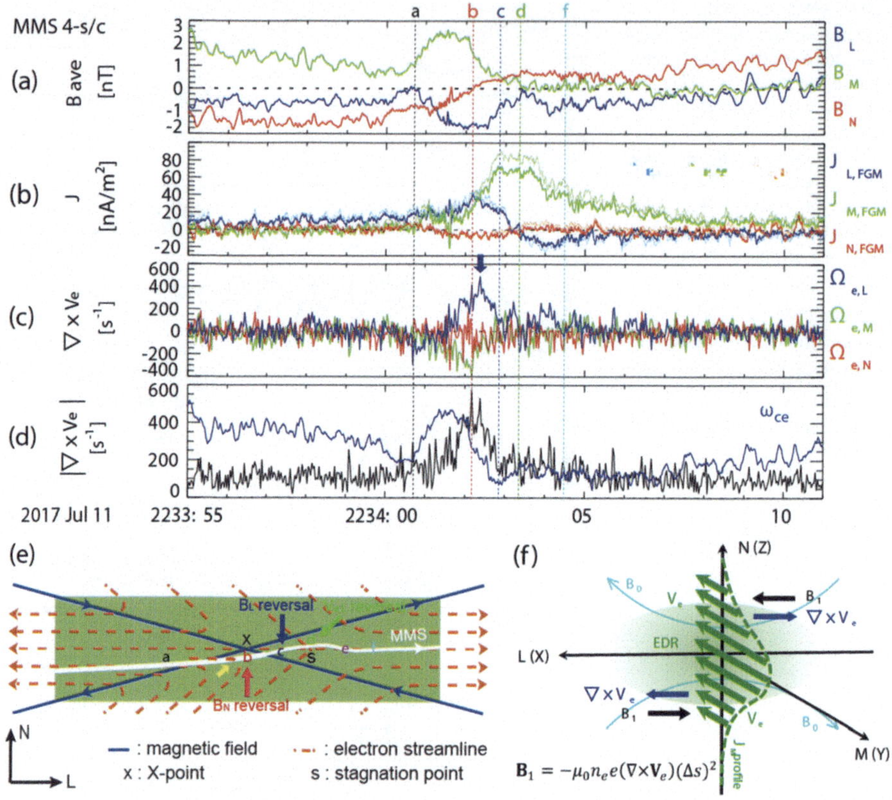

Fig. 15 MMS observation of a magnetotail current sheet crossing along the trajectory shown in white in panel (**e**). (**a**) The tetrahedral-averaged magnetic field, (**b**) current density calculated from the curlometer technique, (**c**, **d**) electron vorticity and its magnitude compared to ω_{ce}. Panel (**f**) illustrates the origin of the enhanced electron vorticity near the northern/southern edge of the EDR. Adapted from Hwang et al. (2019)

$\mathbf{U}_\psi$ was simplified and adapted for application in 3D (Li et al. 2021, 2023) as

$$\mathbf{U}_\psi = (E_M/B_{LN})(\hat{\mathbf{e}}_M \times \hat{\mathbf{b}}_{LN}), \tag{39}$$

where E_M is the out-of-plane (M) component of the electric field in LMN coordinates, $B_{LN} = \sqrt{B_L^2 + B_N^2}$ is the magnetic field component in the 2D reconnection (LN) plane, $\hat{\mathbf{e}}_M$is the unit vector in the M direction, and $\hat{\mathbf{b}}_{LN} \equiv \mathbf{B}_{LN}/B_{LN}$ the unit vector of the in-plane magnetic field $\mathbf{B}_{LN}$. The underlying assumptions are that the advection equation does not have a source or loss term, i.e., no magnetic field generation or diffusion occurs, and that $k_M \ll k_\perp$, where k_M and $k_\perp$ are the wavenumbers corresponding to the length scales of the magnetic field variation parallel and perpendicular to the M direction, respectively (Li et al. 2023, 2021). The latter essentially means that the scale of variation in the out-of-plane (M) direction is much larger than the current sheet thickness. Physically, it represents quasi-2D reconnection (Liu et al. 2018b, 2019; Li et al. 2020). The LMN coordinates can be determined by methods such as minimum variance analysis (Sonnerup and Scheible 1998), maximum directional derivative (Shi et al. 2019), or a combination of methods (Genestreti et al. 2018) (see Sect. 2.2.1 for more details). In simulations, $\mathbf{U}_\psi$ can be calculated if the guide field (M) direction is known (e.g., Li et al. 2023). Based on measured electromagnetic fields, MFT locates reconnection sites in diffusion regions without using information on plasma flows (Qi et al. 2022). This is ideal for identifying diffusion regions where ion and/or electron outflow jets are not well developed.

The MFT method has been demonstrated to accurately identify reconnection in 2D and 3D kinetic turbulence (Li et al. 2021, 2023) and 3D shock turbulence (Ng et al. 2022) simulations. Recent MMS observations further demonstrated the capability and accuracy of MFT statistically, by directly measuring MFT signatures for active reconnection throughout Earth's magnetosphere (Qi et al. 2022). Reconnection signatures in MFT are (i) co-existing Alfvénic inflow and outflow magnetic flux ($\mathbf{U}_\psi$) jets, and (ii) a significantly enhanced divergence of flux transport ($\nabla \cdot \mathbf{U}_\psi$) at an X-line exceeding the threshold of order $0.1\omega_{ce}$. We note that the first signature should be observed in a proper frame in which the corresponding X-line is seen to be quasi-stationary.

Here we show an example of application of MFT to the Eriksson event (Eriksson et al. 2018) observed in the magnetosheath on 25 October 2015 by MMS in Fig. 16. Panels (e,f) show the MFT quantities: $\mathbf{U}_\psi$ reveals bi-directional inflow MFT jets in the N direction (blue) and a super-Alfvénic outflow jet in the L direction (red), as a signature of reconnection; $\nabla \cdot \mathbf{U}_\psi$ is on the order of the electron gyro-frequency $f_{ce} = \omega_{ce}/(2\pi)$, exceeding the threshold of $0.1 f_{ce}$ for identification. The MFT signatures are clear despite the fact that no ion jets were detected along the spacecraft trajectory in this event, interpreted as reconnection in an extended current sheet. A total of 37 previously reported EDR or reconnection-line crossing events were analyzed, including well-known MMS events (e.g., Burch et al. 2016a; Torbert et al. 2018) and electron-only events (e.g., Phan et al. 2018). Almost all ($\geq$95%) of the events can be identified through either of the MFT signatures (Qi et al. 2022). The range of the observed $\mathbf{U}_\psi$ is on the order of ion to electron Alfvén speeds, and $\nabla \cdot \mathbf{U}_\psi$ is of order $0.1 f_{ce}$ or higher. This order of magnitude is consistent with simulations (Li et al. 2021, 2023). The MFT method can thus provide a clear identification of reconnection in diffusion regions in space.

3.1.6 Electron Diffusion Region Detection with Machine Learning Methods

Machine learning methods recently became a useful tool for space physics data analysis and were employed for a variety of tasks, including classification, event detection, and prediction (Sects. A.3-A.5). In the following, we present some recent developments which apply this approach to the detection of EDRs. This type of event is "rare": Webster et al. (2018) reported 32 events, Lenouvel et al. (2021) identified 18 new events, while Fuselier et al. (2016) estimated 56 events for the first 2.5 years of the nominal MMS mission. Therefore, this, at first glance, does not argue for the use of machine learning for EDR detection. However, by using special features of the MMS measurements, namely the details of the electron distribution function, it is possible to extract valuable information which can be processed by machine learning algorithms.

The first algorithm is detailed in Lenouvel et al. (2021) and is only summarized here. It is a rather classical feed-forward MultiLayer Perceptron (e.g., Rumelhart et al. 1986), i.e., a neural network with multiple layers of neurons connected to each other, using MMS observations as features and 4 classes as outputs. Notably one of the key features is a scalar parameter specifically characterizing the asymmetry observed in the crescent-shaped electron distribution functions (hereafter referred to as "electron crescents" or "crescents" for notational simplicity) on the dayside magnetopause (Hesse et al. 2014; Bessho et al. 2016; see also Sect. 3.3.1). The main drawback of this early model was the large number of false

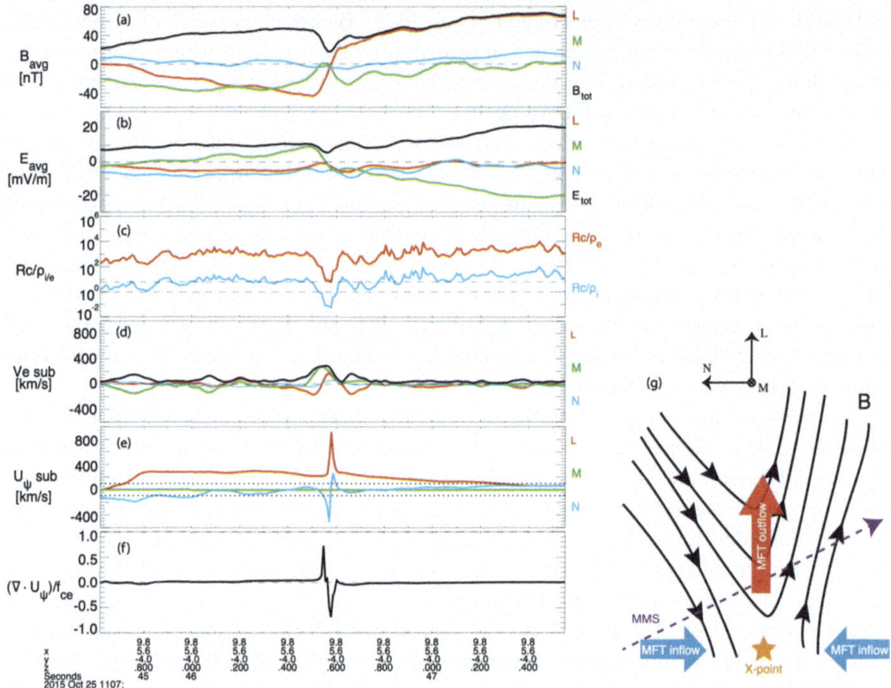

Fig. 16 MMS observations of the Eriksson et al. event (adapted from Qi et al. 2022). (**a**) Magnetic field and (**b**) electric field averaged over four spacecraft. (**c**) Radius of curvature R_C normalized to the electron (red) and ion (blue) gyro-radius. (**d**) Electron bulk flow velocity and (**e**) MFT velocity $\mathbf{U}_\psi$, where the ion bulk flow velocity is subtracted. Dotted lines denote the upstream Alfvén speed. (**f**) $\nabla \cdot \mathbf{U}_\psi$ normalized to f_{ce}. (**g**) Sketch of the trajectory of MMS and expected MFT inflows and outflow, adapted from Eriksson et al. (2018)

positives which needed to be visually analyzed to extract the best EDR candidates. It nevertheless enabled an increase in the number of EDR events and eventually led to a statistical analysis of the sign of the energy conversion rate $\mathbf{j} \cdot \mathbf{E}'$, and a discussion on the distinction between inner and outer EDRs (Lenouvel et al. 2021).

The second algorithm (Lenouvel 2022) has a different structure and is based on a Convolutional Neural Network (CNN) (Lecun et al. 1998), i.e., a deep learning algorithm specifically adapted to image recognition and classification. The idea of the architecture builds on the characteristic feature of electron crescents seen in the full distribution functions rather than reducing them to a scalar as in the first algorithm. The training set is based on all 50 events described in Webster et al. (2018) and a total of 214 crescents yielded by Lenouvel et al. (2021) using data from the four MMS spacecraft. The distribution functions are transformed into 32 by 32 pixel images where each pixel holds the value of the electron phase space density (PSD) for a given angle ($\theta = \arctan(v_{\perp 2}/v_{\perp 1})$) and energy ranges (see Fig. 17 for details). The range of PSD on log scale (min-max for all images) is coded on 256 levels. Data augmentation is then used to increase the number of training samples, by first extracting 112 most clear crescents (exhibiting a clear left-right asymmetry in the $v_{\perp 1}$-$v_{\perp 2}$ plane) and then combining them with each other (by averaging two images after random small rotation and adding logarithmic noise), which produces a dataset of ${}_{112}C_2 = 112\,!/(110\,!2\,!) = 6126$ new synthetic crescents from all possible combinations. The CNN architecture is formed by a succession of dedicated layers which aim at extracting features or patterns in input data.

Finally, applying this algorithm on the full MMS phase 1b of the prime mission resulted in the discovery of 17 new events (from the analysis of MMS 1 and 2 data only). For future studies, a list of events combining those obtained from both algorithms, with the addition of individual events analyzed in the literature during phases 1a and 1b, is available at Zenodo (https://zenodo.org/record/8319481).

To demonstrate how the detection works and performs, a model was trained after removing the distribution functions (736 distribution functions including real crescents, synthetic crescents and random distribution functions) from the EDR event reported by Burch et al. (2016a) that took place on 16 October 2015 at 13:07:02 UTC. The model could then be applied to the whole event (from 13:05:25 UTC to 13:07:44 UTC) without any bias that could be due to data leakage between the training dataset and the data from this example. In Fig. 17, the red vertical lines correspond to times of the distribution functions labeled as "crescents" by the model; 26 are visible on the plot and, after visual inspection, 18 of them were considered as correctly identified by the algorithm (true positives).

In this application, only the distribution function is used to make a prediction and no post-processing is applied, so false positives are to be expected. Removing these false positives would require the use of additional parameters to be associated with the distribution functions, such as the electron density (low densities are known to produce incomplete distribution functions that are interpreted as asymmetric by the model) or other key EDR parameters including the electron-frame electric field $\mathbf{E}'$ and $\mathbf{j} \cdot \mathbf{E}'$. One needs to bear in mind that peaks in the last two parameters may not occur at the same time as the presence of electron crescents (the separation time can go up to a few hundred milliseconds), so the automatic removal of false detections is not an easy matter and will be the topic of future work with a more advanced automatic EDR detection model. Nonetheless, the ability of these automatic EDR detection methods shows that they are valuable to identify and analyze rare and complex physical plasma processes.

3.1.7 Magnetic Nulls

Magnetic nulls are singularities or critical points where the magnetic field vanishes, and can be essential for characterizing 3D magnetic topology and understanding magnetic reconnection in 3D (Pontin and Priest 2022 and references therein). See recent works (e.g., Fu et al. 2015; Olshevsky et al. 2020; Guo et al. 2022; Ekawati and Cai 2023) for details about the methods to identify and analyze the magnetic nulls.

3.2 Analysis Methods for the Electron Diffusion Region

3.2.1 Estimation of Anomalous Resistivity, Viscosity, and Diffusion

In Graham et al. (2022) anomalous terms associated with lower hybrid waves were estimated from direct spacecraft observations in magnetopause reconnection events, and an assessment was performed to see whether the anomalous terms could contribute to the reconnection electric field E_M. The anomalous terms are based on expansions of the electron continuity and momentum equations:

$$\frac{\partial n}{\partial t} + \nabla \cdot (n\mathbf{u}) = 0, \tag{40}$$

$$m\frac{\partial (n\mathbf{u})}{\partial t} + m\nabla \cdot (n\mathbf{u}\mathbf{u}) + \nabla \cdot \mathbf{P} + en\,(\mathbf{E} + \mathbf{u} \times \mathbf{B}) = 0, \tag{41}$$

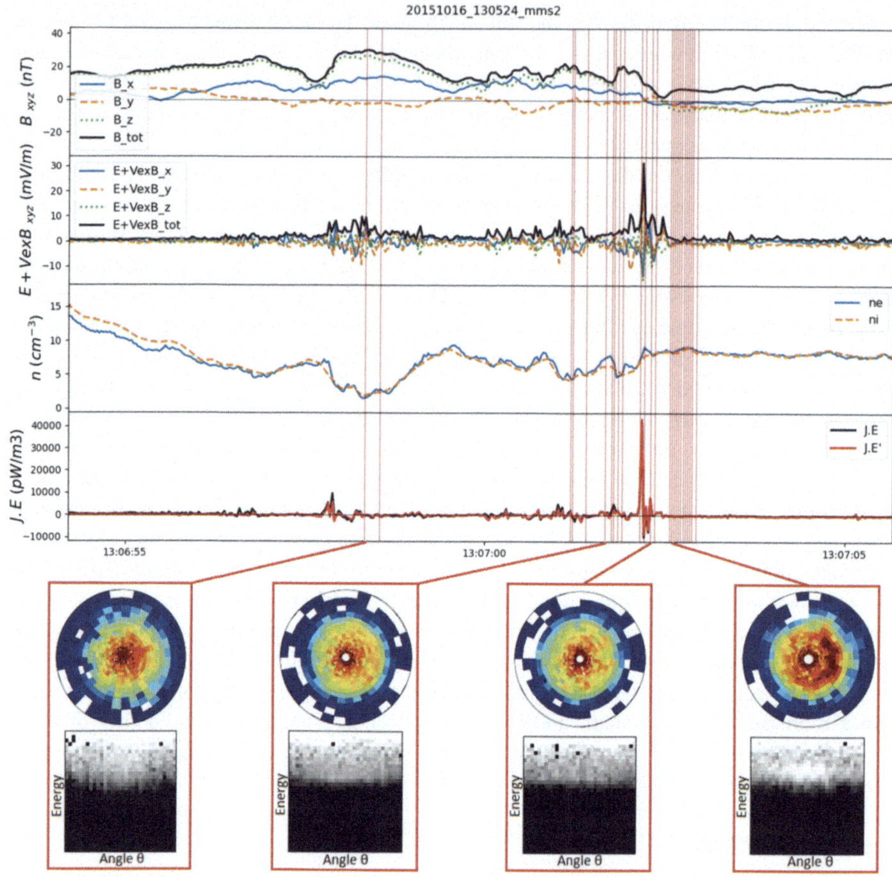

Fig. 17 EDR event from Burch et al. (2016a) with detected crescent distribution functions. The top four panels show MMS2 observations of the magnetic field $\mathbf{B}$, electron-frame electric field $\mathbf{E}'$ (showing the departure from the ideal conditions), electron and ion densities, and the energy conversion rate $\mathbf{j} \cdot \mathbf{E}'$. Four of the identified "crescents" are shown below the time series plots, both in classical phase space density units (top) and in transformed 32×32 images (bottom). The two-dimensional electron distribution slices are displayed in the $v_{\perp 1}$-$v_{\perp 2}$ plane where $v_{\perp 1}$ is directed along $(-\mathbf{u}_e \times \mathbf{B}) \times \mathbf{B}$ (approximately the $\mathbf{E} \times \mathbf{B}$ direction, where $\mathbf{E}$ is the electric field, and $\mathbf{u}_e$ is the electron bulk velocity), and $v_{\perp 2}$ is directed along $-\mathbf{u}_e \times \mathbf{B}$ (approximately the direction of $\mathbf{E}$)

where m is the electron mass, e is the unit charge, n is the number density, $\mathbf{u}$ is the bulk velocity, $\mathbf{P}$ is the pressure tensor, $\mathbf{E}$ is the electric field, and $\mathbf{B}$ is the magnetic field. To derive the anomalous terms the quantities in these equations are separated into fluctuating and quasi-stationary components: $\mathrm{Q} = \delta\mathrm{Q} + \langle \mathrm{Q} \rangle$, where $\delta\mathrm{Q}$ is the fluctuating component due to waves, and $\langle \mathrm{Q} \rangle$ is an ensemble average over Q, and $\langle \delta\mathrm{Q} \rangle = 0$. The anomalous terms are obtained by taking the ensemble average of the momentum equation. The ensemble average of the product of two quantities is $\langle \mathrm{QR} \rangle = \langle \mathrm{Q} \rangle \langle \mathrm{R} \rangle + \langle \delta\mathrm{Q}\delta\mathrm{R} \rangle$. From the continuity equation (40) a cross-field diffusion coefficient can be defined as

$$D_{\perp} = -\frac{\langle \delta n \delta u_N \rangle}{\nabla \langle n \rangle_N}, \tag{42}$$

where **N** is the direction normal to the local boundary. From the momentum equation (41) we obtain:

$$\langle \mathbf{E} \rangle + \langle \mathbf{u} \rangle \times \langle \mathbf{B} \rangle = -\frac{\nabla \cdot \langle \mathbf{P} \rangle}{\langle n \rangle e} - \frac{m}{\langle n \rangle e} \nabla \cdot (\langle n \rangle \langle \mathbf{u} \rangle \langle \mathbf{u} \rangle) + \mathbf{D} + \mathbf{T} + \mathbf{I}. \tag{43}$$

The time derivative term in Eq. (41) is assumed to be small, so it is neglected in Eq. (43). Here **D**, **T**, and **I** are the anomalous resistivity, anomalous viscosity, and anomalous inertial terms, which are given by

$$\mathbf{D} = -\frac{\langle \delta n \delta \mathbf{E} \rangle}{\langle n \rangle}, \tag{44}$$

$$\mathbf{T} = -\frac{\langle n \mathbf{u} \times \boldsymbol{B} \rangle}{\langle n \rangle} + \langle \mathbf{u} \rangle \times \langle \mathbf{B} \rangle, \tag{45}$$

$$\mathbf{I} = -\frac{m}{e \langle n \rangle} \left[\nabla \cdot (\langle n \mathbf{u} \mathbf{u} \rangle) - \nabla \cdot (\langle n \rangle \langle \mathbf{u} \rangle \langle \mathbf{u} \rangle) \right]. \tag{46}$$

This approach to calculate the anomalous terms corresponds to Reynolds averaging, which is often used to study fluid turbulence. The above terms were derived in Graham et al. (2022). Similar definitions are used in numerical simulations (Che et al. 2011; Price et al. 2016; Le et al. 2018; Price et al. 2020), although often the particle fluxes are treated as a single quantity, which can modify the contributions from the anomalous terms (Price et al. 2020).

The anomalous terms result from the correlations between fluctuating quantities associated with the waves. In numerical simulations the average is taken over the reconnection out-of-plane (M) direction (e.g., Price et al. 2016, 2020; Le et al. 2017), although the ensemble average can be performed over time (Le et al. 2018). With MMS an approximate average can also be obtained by averaging over the four spacecraft.

To calculate the anomalous contributions requires fields and particle measurements that can resolve the lower hybrid wave fluctuations. This is possible for the electron particle data using the highest time resolution electron distributions and moments, which can be sampled every 7.5 ms, rather than the nominal 30 ms sampling rate during burst mode (Pollock et al. 2016). Resolution of 7.5 ms is achieved by reducing the azimuthal resolution in the spacecraft spin plane of the electron measurements (Rager et al. 2018; Appendix A). Since we are interested in the bulk changes in the distributions rather than fine structures in computing the lower order moments, this reduced azimuthal resolution does not present a major problem.

The anomalous terms were calculated in Graham et al. (2022) and Fig. 18 shows the steps to calculate the M component of **D**. We summarize the steps they used to calculate the anomalous terms:

(1) The vector quantities are converted to the LMN coordinate system.

(2) All electric and magnetic field data are resampled to the sampling frequency of the 7.5 ms electron moments.

(3) Four spacecraft timing analysis on B_L at the current sheet to determine the velocity of the boundary in the normal direction and the time delays between the spacecraft.

(4) The time delays are used to offset the spacecraft times so all spacecraft cross the current sheet at the same time as MMS1.

(5) Quasi-stationary quantities $\langle Q \rangle$ are obtained by bandpass filtering the signals below 5 Hz, and averaging the data over the four spacecraft. For the magnetopause, as shown in

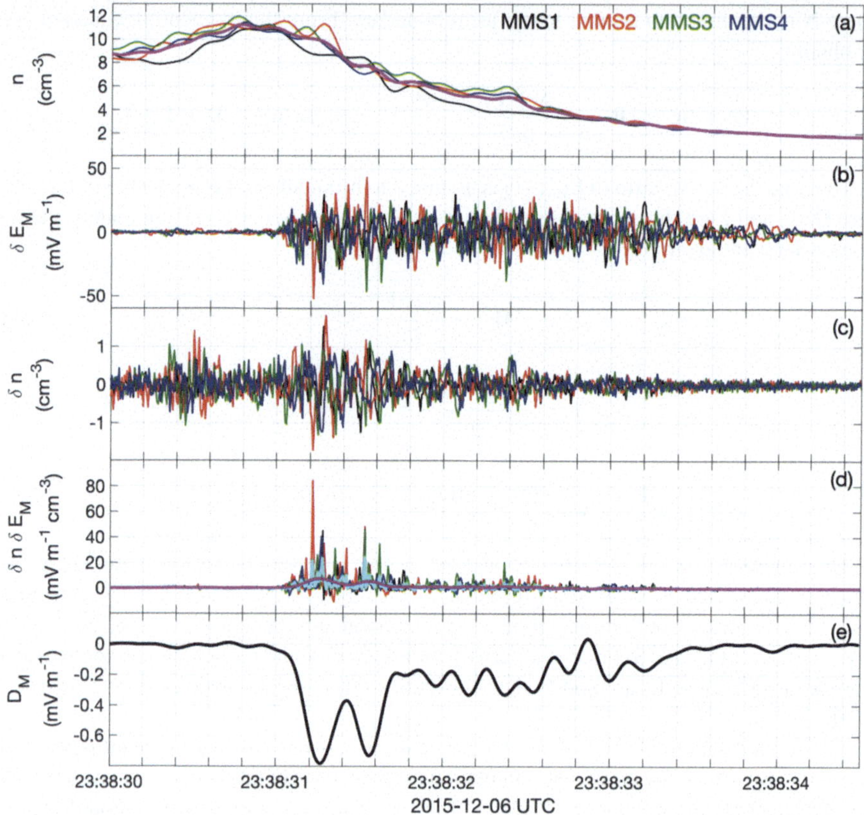

Fig. 18 Example of the calculation of the M component of the anomalous resistivity $\mathbf{D}$. (**a**) Background (lowpass filtered) components of the electron number density n (MMS1-MMS4 data are plotted in black, red, green, and blue, respectively) and $\langle n \rangle$ calculated by averaging over the four spacecraft (magenta line). (**b**) and (**c**) δE_M and δn for the four spacecraft. (**d**) $\delta n \delta E_M$ calculated for the four spacecraft, and the four-spacecraft averaged $\delta n \delta E_M$ (cyan), and $\langle \delta n \delta E_M \rangle$ (magenta) obtained by low-pass filtering the averaged $\delta n \delta E_M$. I M component of $\mathbf{D}$ calculated from $\langle \delta n \delta E_M \rangle$ and $\langle n \rangle$. In all panels (**a**)-(**d**) the quantities from MMS2-MMS4 have been time shifted so they cross the current sheet at the same time

Fig. 18, the frequency of the lower hybrid waves is typically comparable to or above 10 Hz, so this removes most of the fluctuations due to these waves.

(6) The fluctuating quantities δQ associated with lower hybrid waves are obtained by band pass filtering the data above 5 Hz (Figs. 18b,c).

(7) The correlations between fluctuating quantities $\langle \delta Q \delta R \rangle$ are obtained by averaging $\delta Q \delta R$ over the four spacecraft; then low-pass filtering the product below 5 Hz. This removes most of the remaining higher-frequency components from these terms (Fig. 18d). See Graham et al. (2022) for further details on the calculation of the anomalous terms (see also Table 5 in Appendix C).

3.2.2 Evaluation of Terms in the Electron Vlasov Equation

A wide variety of collisionless plasma phenomena have been studied and understood by utilizing the kinetic description provided by the Vlasov equation (e.g., Nicholson 1983;

Califano et al. 2016; Gershman et al. 2017). Here, we summarize a methodology for utilizing the MMS Fast Plasma Investigation (FPI) Dual Electron Spectrometer (DES) data (Pollock et al. 2016) in order to compute each derivative of the electron PSD f_e that appears in the electron Vlasov equation, given by

$$\frac{df_e}{dt} = \frac{\partial f_e}{\partial t} + \mathbf{v} \cdot \nabla f_e - \frac{e}{m_e} (\mathbf{E} + \mathbf{v} \times \mathbf{B}) \cdot \nabla_{\mathbf{v}} f_e = 0. \tag{47}$$

For example, since the generalized Ohm's law (electron momentum equation (41)) is derived from the moments of the Vlasov equation (47), the method, as explained below, allows us to discuss how and which part of the velocity-space distribution of each term of Eq. (47) contributes to each term of Eq. (41).

Figure 19 demonstrates the computation techniques needed for each Vlasov equation term in the context of an electron spatial-scale current sheet encountered by MMS on 23 December 2016 (2019, Shuster et al. 2021a,b). This event is discussed in more detail in Norgren et al. 2024, this collection). The Vlasov equation $df_e/dt = 0$ is a statement indicating that PSD is conserved along a particle's Lagrangian trajectory through phase space. In the Eulerian frame of the MMS spacecraft, it is necessary to consider and evaluate each term of the Vlasov equation (47) (see Shuster et al. (2019, 2023) for a more thorough discussion concerning the computation methods outlined here). Figure 19a-d, Figs. 19e-h, and Figs. 19i-m present a high-level, visual comparison of the three distinct methods for evaluating the terms $\partial f_e/\partial t$, $\mathbf{v} \cdot \nabla f_e$, and $-(e/m_e)(\mathbf{E} + \mathbf{v} \times \mathbf{B}) \cdot \nabla_{\mathbf{v}} f_e$, respectively.

For the event shown in Fig. 19, the current layer's thickness was about 3 to 5 d_e, where the local electron skin depth d_e was about 1.5 km. The normal velocity of the structure, V_N, was roughly 50 km/s, where N indicates the direction normal to the current layer. The current layer passed by each MMS spacecraft in about a tenth of a second. Thus, the spatial thickness was roughly (50 km/s)·(0.1 s) = 5 km, comparable to the inter-spacecraft spacing of the four MMS spacecraft.

Shuster et al. (2023) explain how to use higher order finite difference approximations to obtain more accurate measures of the temporal and velocity-space derivative terms. Furthermore, Shuster et al. (2023) show how $\partial f_e/\partial t$ may provide a useful estimate for ∇f_e in situations where the plasma is believed to be quasi-steady state:

$$\frac{Df_e}{Dt} \equiv \frac{\partial f_e}{\partial t} + \mathbf{V}_{\text{str}} \cdot \nabla f_e \approx 0 \qquad \rightarrow \qquad \frac{\partial f_e}{\partial N} \approx \left(-\frac{1}{V_N} \right) \frac{\partial f_e}{\partial t}, \tag{48}$$

where $\partial f_e/\partial t$ is computed in the frame of the spacecraft. The temporal derivative notation Df_e/Dt is used to indicate a time derivative taken in a frame moving in position space with the velocity of the structure, $\mathbf{V}_{\text{str}} = V_N \hat{\mathbf{e}}_N$. Here the structure is assumed to be planar, with spatial variations only in the N direction (Fig. 19). As noted by Shuster et al. (2019), we point out the connection between the spatial gradient term ∇f_e and the bulk electron pressure divergence term $\nabla \cdot \mathbf{P}_e$ via the integral identity utilized when deriving the electron momentum equation from the electron Vlasov equation (47):

$$m_e \int \mathbf{v} (\mathbf{v} \cdot \nabla f_e) d^3 v = \nabla \cdot \mathbf{P}_e + \nabla \cdot (m_e n_e \mathbf{u}_e \mathbf{u}_e). \tag{49}$$

For certain environments, such as magnetopause reconnection sites, the inertial term on the right-hand side of Eq. (49) is commonly negligible compared to $\nabla \cdot \mathbf{P}_e$. Thus, we can understand how velocity-space structures of each term of the Vlasov equation (47), as shown

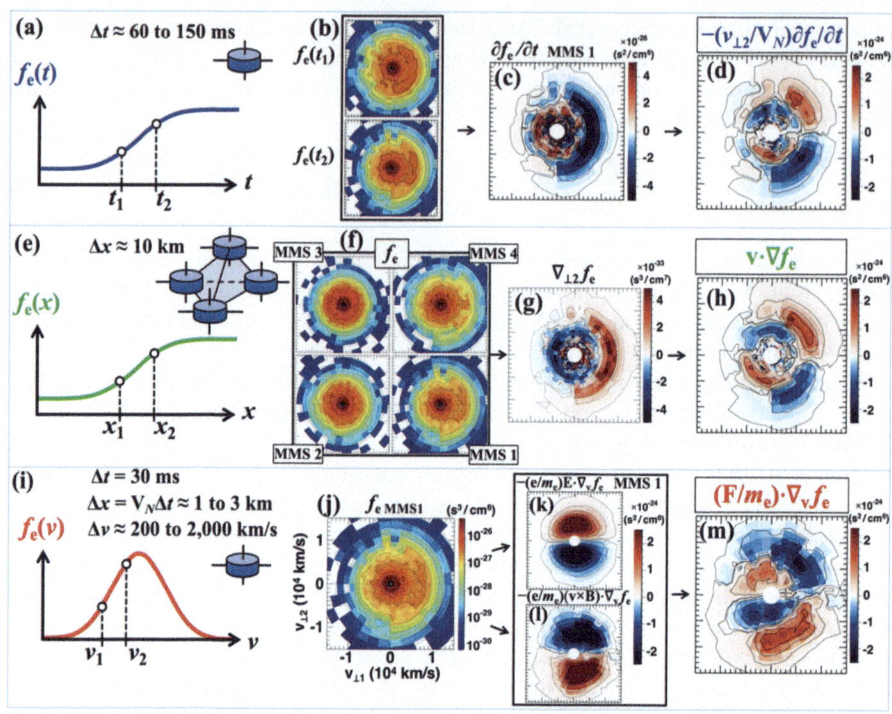

Fig. 19 This figure presents a visual summary of the velocity-space structure and qualitative balance of the three Vlasov equation terms for an electron-scale current sheet. The computation technique for **(a-d)** $\partial f_e/\partial t$, **(e-h)** $\mathbf{v}\cdot\nabla f_e$, and **(i-m)** $(\mathbf{F}/m_e)\cdot\nabla_\mathbf{v} f_e$ is shown schematically. Each velocity-space panel represents a slice taken in the $v_{\perp 1}$-$v_{\perp 2}$ plane. In this example, $\hat{\mathbf{e}}_N$ points roughly along $\hat{\mathbf{e}}_{\perp 2}$, so that the quantities shown in panels (d) and (h) are roughly equivalent. Comparing panels (h) and (m), one can see a notable quadrupolar pattern but with a polarity difference, suggesting that Eq. (47) is roughly satisfied in velocity space under the assumption of negligible $\partial f_e/\partial t$ (panel (c)). Adapted from Shuster et al. (2023)

in Fig. 19, contribute to collisionless plasma processes including the generation of a nonideal electric field (here, through the interrelationship among $\mathbf{v}\cdot\nabla f_e$, $\nabla\cdot\mathbf{P}_e$, and the electric field, based on Eq. (49) and the generalized Ohm's law) and field-to-electron energy conversion, which are key features of the EDR.

Regarding the velocity-space gradient term on the righthand side of Eq. (47), one may perform the derivative computations in the $\{E, \theta, \phi\}$ coordinates native to the FPI detectors, or one may choose to first interpolate the distribution to a Cartesian grid with $\{v_x, v_y, v_z\}$ coordinates before calculating the derivatives (see Shuster et al. (2023) for more details). As a consistency check, the results of each approach ought to be qualitatively consistent. We note also that the electric and magnetic fields may be averaged to the DES 30 ms cadence to obtain a measurement of the full velocity-space gradient term that appears in the Vlasov equation.

3.2.3 Non-Maxwellianity

In many plasmas, particle distributions can deviate significantly from thermal equilibrium, namely a Maxwellian distribution. Non-Maxwellian distributions develop during magnetic reconnection and can be unstable to a range of instabilities. Several scalar parameters have been defined to quantify the deviation of the observed distributions from a Maxwellian or bi-Maxwellian distribution function (Greco et al. 2012; Servidio et al. 2017; Liang et al. 2020; Graham et al. 2021; Argall et al. 2022; see also Sect. 3.2.4 for other versions of non-Maxwellianity not mentioned in this section). Here we outline the one developed in Graham et al. (2021). In Greco et al. (2012), Servidio et al. (2017), and Graham et al. (2021) the definitions of non-Maxwelliantity are based on the magnitude of the differences between the observed distribution and a model Maxwellian distribution. In Liang et al. (2020) and Argall et al. (2022) the definition of non-Maxwellianity is based on the increase in kinetic entropy from a Maxwellian distribution. In Graham et al. (2021) the non-Maxwellianity parameter was defined as

$$\epsilon = \frac{1}{2n} \int_{v,\theta,\phi} |f(v,\theta,\phi) - f_{model}(v,\theta,\phi)| \, v^2 sin\theta dv d\theta d\phi, \tag{50}$$

where n is the number density, f is the observed particle distribution function, f_{model} is the model particle distribution function, v is the speed, θ is the polar angle, and ϕ is the azimuthal angle in velocity space. The model distribution can be either a Maxwellian or bi-Maxwellian distribution with the same density, bulk velocity, and temperature as the observed distribution. The integral is performed in the same way as the particle moments calculations. The $1/(2n)$ factor normalizes ε to a dimensionless quantity with values between 0 and 1. A value of 0 indicates no deviation from the model distribution, while 1 corresponds to complete deviation from the model distribution. In Graham et al. (2021) a bi-Maxwellian distribution was used as the model distribution, given by:

$$f_{model}(\mathbf{v}) = \frac{n}{\pi^{3/2} v_{th,||}^3} \frac{T_{||}}{T_{\perp}} exp\left(-\frac{(v_{||} - V_{||})^2}{v_{th,||}^2} - \frac{(v_{\perp,1} - u_{\perp})^2 + v_{\perp,2}^2}{v_{th,||}^2 (T_{\perp}/T_{||})} \right), \tag{51}$$

where $T_{||}$ and $T_{\perp}$ are the parallel and perpendicular temperatures, $v_{th,||} = \sqrt{2k_B T_{||}/m}$ is the thermal speed, k_B is Boltzman's constant, m is the particle mass, and $u_{\perp}$ is the magnitude of the perpendicular bulk velocity. The velocity coordinates are defined such that $v_{||}$ is aligned with the magnetic field, $v_{\perp,1}$ is aligned with the component of the bulk velocity perpendicular to the magnetic field, and $v_{\perp,2}$ is orthogonal to $v_{||}$ and $v_{\perp,1}$. The calculation of ε corresponds to a zeroth order moment calculation, so the largest contributions to ε typically occur in the thermal energy range. The parameters used to calculate f_{model} are obtained from the observed particle moments, so no fitting to the observed distribution is required. In Graham et al., the bi-Maxwellian distribution function was used rather than an isotropic Maxwellian so temperature anisotropies, which are simple to identifFy in the particle moments, are not the cause of non-Maxwellianity.

Figure 20 shows an example of ε calculated for an EDR observed on 22 October 2015. The figure shows a reconnection event, as indicated by the reversal in the direction of the magnetic field (panel a), an increase in density (panel b), and a northward ion outflow (panel c).

The EDR was observed at the time indicated by the yellow-shaded region (e.g., Phan et al. 2016; Toledo-Redondo et al. 2016). Figure 20d shows ε calculated from the above equations. There is a significant increase in ε, which peaks in the EDR, indicating that non-Maxwellian electron distributions develop there. Additionally, large ε occurs in the outflow region, indicating that non-Maxwellian electron distributions are not limited to the EDR. More generally, all the observed EDRs in the first phase of the MMS mission exhibited enhanced electron non-Maxwellianities (Graham et al. 2021). In Graham et al. (2021) a

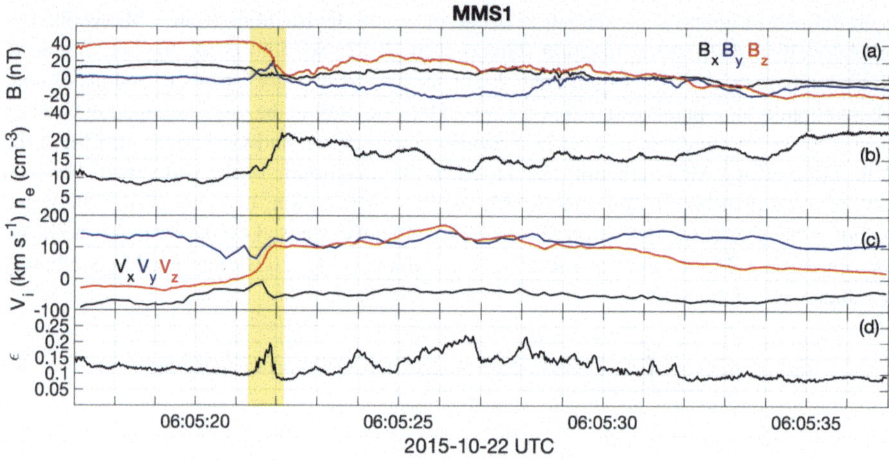

Fig. 20 Example of electron non-Maxwellianity for the electron diffusion region crossing observed by MMS1 on 22 October 2015. (**a**) Magnetic field. (**b**) electron number density. (**c**) Ion bulk velocity. (**d**) Non-Maxwellianity ε of electrons. The EDR is indicated by the yellow-shaded region

statistical analysis of electron distributions was performed on six months of data to determine which values of ε corresponded to enhanced non-Maxwellian electron distributions compared to typical values.

It should be noted that ε can be artificially large due to the low counting statistics from the particle detectors when the density is low, such as in Earth's magnetotail. This problem can be mitigated by averaging multiple particle distributions in such cases.

3.2.4 Kinetic Entropy

Due to the fact that collisions in space plasmas are weak, processes that involve kinetic physics often distort the distribution function so that it is no longer in equilibrium. Structures in the distribution function are linked to specific energy conversion processes (Egedal et al. 2010b, 2010a; Hoshino et al. 2001) and are used to create maps of the reconnection diffusion region (Chen et al. 2008, 2016) and study the spatio-temporal evolution of reconnection (Shuster et al. 2014, 2015; Egedal et al. 2016; Barbhuiya et al. 2022). Inhomogeneities in the distribution were quantified to provide indicators of when kinetic physics is important (Scudder and Daughton 2008; Aunai et al. 2013; Swisdak 2016; Sect. 3.1.2). The measured distribution was compared to an equivalent distribution in equilibrium to quantify the amount of free energy available to be dissipated (Greco et al. 2012; Servidio et al. 2017; Graham et al. 2021; Lindberg et al. 2022; Sect. 3.2.3). Similar considerations were made with respect to kinetic entropy (Boltzmann 1877), noting that the measured distribution will be non-Maxwellian if its entropy is less than that of an equilibrium Maxwellian distribution with the same density and effective temperature (Kaufmann and Paterson 2009; Liang et al. 2019, 2020). The theoretical development of non-Maxwellianity (Liang et al. 2020) was adapted to the non-uniform velocity space grid of MMS plasma measurements, and then was applied to a magnetotail reconnection event to show that non-Maxwellianity is indeed linked to kinetic processes in the EDR (Argall et al. 2022).

To develop a theory of kinetic entropy for a distribution function in velocity space, we start with Boltzmann's equation $S = k_B \ln \Omega$, where k_B is Boltzmann's constant and $\Omega =$

$N_{tot}!/\prod_{j,k} N_{jk}!$ is the total number of microstates that correspond to a given macrostate, N_{tot} is the total number of particles in the system and is assumed constant, N_{jk} is the number of particles in the jth position-space and kth velocity-space cell of phase space, and the product over j and k is over all position- and velocity-space cells, respectively. We then break phase space up into discrete bins and separate the total entropy into position and velocity space entropy (Mouhot and Villani 2011) to obtain a value for entropy that is local in position space and hence is measurable by a single spacecraft (Liang et al. 2019),

$$S = S_r + S_V, \tag{52}$$

where

$$S_r = k_B \left\{ N_{tot} \ln\left(\frac{N_{tot}}{\Delta^3 r}\right) - \int d^3 r n(\mathbf{r}) \ln[n(\mathbf{r})] \right\}, \tag{53}$$

$$S_V = \int d^3 r s_V(\mathbf{r}), \tag{54}$$

$$s_V(\mathbf{r}) = k_B \left\{ n(\mathbf{r}) \ln\left[\frac{n(\mathbf{r})}{\Delta^3 v}\right] - \int d^3 v f(\mathbf{r}, \mathbf{v}) \ln[f(\mathbf{r}, \mathbf{v})] \right\}. \tag{55}$$

Here, $\Delta^3 r$ and $\Delta^3 v$ are the position and velocity space volume elements of phase space, $f(\mathbf{r}, \mathbf{v})$ is the distribution function, and $n(\mathbf{r})$ is the number density. Equation (55) is the velocity-space entropy density and its last term is often referred to as the total kinetic entropy density, $s = -k_B \int d^3 v\, f(\mathbf{r}, \mathbf{v}) \ln[f(\mathbf{r}, \mathbf{v})]$ (Kaufmann and Paterson 2009).

The entropy of a measured distribution can be compared with that of an equivalent drifting Maxwellian distribution described by (omitting the dependence on $\mathbf{r}$)

$$f_M(\mathbf{v}) = n \left(\frac{m}{2\pi k_B T}\right)^{3/2} e^{\left[-m(\mathbf{v}-\mathbf{u})^2/(2k_B T)\right]}, \tag{56}$$

where m is the particle mass, $\mathbf{u}$ is the bulk flow velocity, and T is the effective temperature. Combining Eq. (56) with Eq. (55) to calculate the total and velocity-space entropy densities of a Maxwellian distribution (s_M and $s_{M,V}$, respectively) yields

$$s_M = \frac{3}{2} k_B n \left[1 + \ln\left(\frac{2\pi k_B T}{m n^{3/2}}\right)\right], \tag{57}$$

$$s_{M,V} = \frac{3}{2} k_B n \left[1 + \ln\left(\frac{2\pi k_B T}{m\left(\Delta^3 v\right)^{3/2}}\right)\right]. \tag{58}$$

The difference between the equivalent Maxwellian and measured total and velocity-space entropy densities defines two non-Maxwellianity parameters

$$\overline{M}_{KP} = \frac{s_M - s}{(3/2) k_B T}, \tag{59}$$

$$\overline{M} = \frac{s_{M,V} - s_V}{s_{M,V}}, \tag{60}$$

where the Kaufmann and Paterson non-Maxwellianity, $\overline{M}_{KP}$, is normalized by the internal energy per particle of an ideal gas (Kaufmann and Paterson 2009), and $\overline{M}$ is normalized by the velocity space Maxwellian entropy density (Liang et al. 2020). While $\overline{M}$ is bounded between 0 and 1 and so provides a better measure for making comparisons of non-Maxwellianity between distributions, the normalization of $\overline{M}_{KP}$ better relates entropy to the energetics of the system (Cassak et al. 2023).

The development so far has considered uniform velocity space grids typical of theory and simulations. Particle detectors, on the other hand, have logarithmically spaced energy bins. Because of this, the velocity space volume element depends on velocity: $d^3v\,(\mathbf{v})$. Revisiting the above derivations in spherical velocity space coordinates leads to (Argall et al. 2022)

$$s_V = s + k_B n \ln n - k_B \int d^3v\,(\mathbf{v})\, f\,(\mathbf{v}) \ln\left[d^3v\,(\mathbf{v})\right], \tag{61}$$

$$\overline{M} = \frac{s_M - s - k_B \int d^3v\,(\mathbf{v}) \ln\left[d^3v\,(\mathbf{v})\right]\left[f_M\,(\mathbf{v}) - f\,(\mathbf{v})\right]}{1 + k_B n \ln n - k_B \int d^3v\,(\mathbf{v}) \ln\left[d^3v\,(\mathbf{v})\right] f_M\,(\mathbf{v})}. \tag{62}$$

A number of corrections are required when integrating the measured particle distribution functions. These include corrections for photoelectrons and spacecraft potential, and normalization of the energy and look-angle space of the instrument (FPI in the case of MMS). The subsequent spherical, normalized energy space is obtained by applying the transformations outlined in Moseev and Salewski (2019), and Argall et al. (2022) and its Supplemental Material. In addition, to calculate the equivalent Maxwellian distribution, it is essential to create a look-up table to minimize numerical errors (Argall et al. 2022).

Figure 21 shows calculations of non-Maxwellianity from MMS observations and PIC simulations within the EDR of a magnetotail reconnection event. The 2D profile of $\overline{M}$ (panel a) shows significant departures from Maxwellianity in the vicinity of the EDR, indicating that important kinetic effects are taking place. The trajectory (magenta dashed line) indicates the path that MMS took through the EDR and the corresponding data is shown in the following panels. Three different measures of non-Maxwellianity, including $\overline{M}$ and $\overline{M}_{KP}$ show qualitative agreement between observations and simulations.

Distribution functions observed by MMS (Fig. 21f,g,h) and in PIC (Fig. 21k) are taken at the black vertical dashed lines (and the "x" in panel a). PIC distributions (Fig. 21i,j) outside the area shown in Fig. 21a were taken from regions representative of the MMS locations within the reconnection domain. The electron distribution functions are from the upstream (Fig. 21f,i), inflow (Fig. 21g,j), and X-line (Fig. 21h,k) regions. Upstream, the distributions are nearly Maxwellian and have the lowest non-Maxwellianity of the three regions sampled. The inflow distributions are elongated in the direction parallel to the magnetic field due to their bouncing within a parallel potential well (Egedal et al. 2010b), and X-line distributions are striated due to their meandering current sheet motion (Ng et al. 2011). This shows that entropy measures of non-Maxwellianity are able to identify regions where important kinetic effects are taking place.

Unfortunately, by breaking phase space up into discrete bins (i.e., by considering a distribution of particles instead of combinations of individual particles), we lose information about the system; $\overline{M}$ depends on the scale of the velocity space grids and their non-uniformity and thus is not equal to $\overline{M}_{KP}$. For details about how the difference between $\overline{M}$ and $\overline{M}_{KP}$ allows us to quantify the amount of information loss incurred by discretizing velocity space, see Argall et al. (2022). Implications include considering the thermal velocity of the target environment when designing plasma instruments and quantifying limits to observations of dissipation.

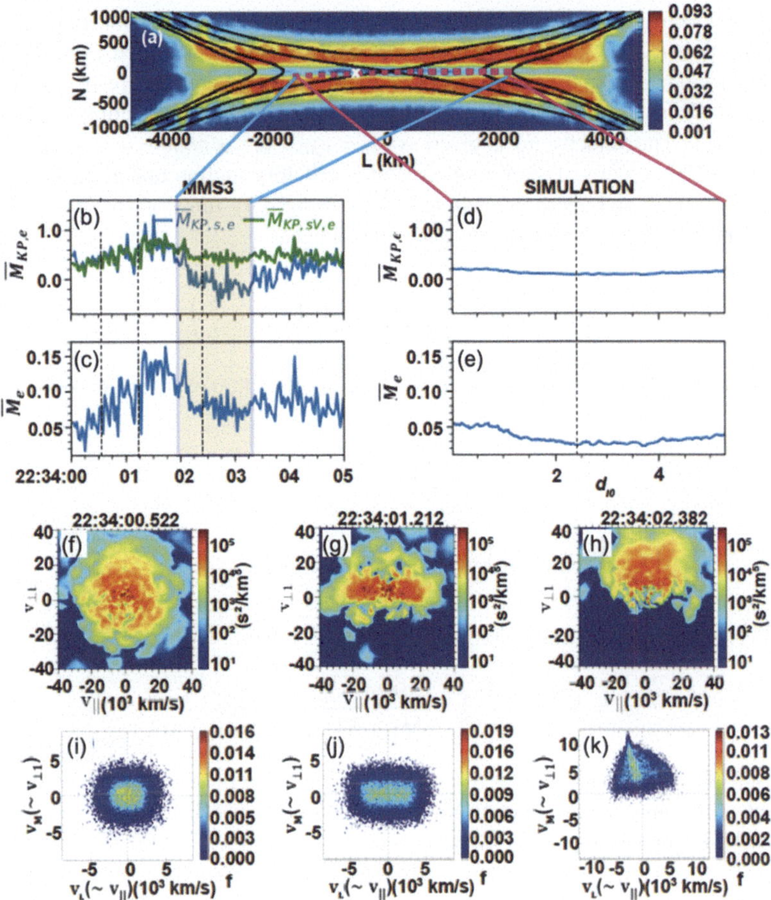

Fig. 21 MMS and PIC simulation comparisons of non-Maxwellianity in a magnetotail electron diffusion region. (**a**) The 2D profile of non-Maxwellianity surrounding the EDR with the MMS trajectory shown (dotted magenta line). Different electron non-Maxwellianity quantities as observed (**b**,**c**) by MMS and (**d**,**e**) in PIC, where $\overline{M}_{KP}$ (Eq. (59)) is computed using s (blue) and s_V (Eq. 61) (green). The MMS panels show a larger view than is depicted in the simulation for comparison with other published studies. Vertical dashed lines (and the "x" in panel **a**) indicate where electron distributions were obtained (**f**,**g**,**h**) by MMS and (**k**) in PIC simulation. PIC distributions in (**i**,**j**) were taken outside of the bounds of panel (**a**) from locations representative of the regions MMS sampled. Adapted from Argall et al. (Phys. Plasmas, 29, 022902, 2022; licensed under a Creative Commons Attribution (CC BY) license)

3.3 Reconnection Electric Field Estimation

3.3.1 Extracting E_M from Multiple Crescents in Electron Velocity Distributions

In the event on 11 July 2017 in the Earth's magnetotail (Torbert et al. 2018), MMS detected multi-crescent electron velocity distribution functions (VDFs) in the EDR. Multi-crescent VDFs are produced because the reconnection electric field E_M accelerates electrons while they are meandering across the reconnecting current sheet. In the following, we explain how to extract the information about E_M from a multi-crescent VDF.

Suppose electrons are meandering in a current sheet with a magnetic field $B_L = bN$ and a Hall electric field $E_N = -kN$, where N is the position measured from the current sheet center ($B_L = 0$) plane, and b and k are the slope of B_L and E_N, respectively, in the N direction. Consider an electron moving back and forth in the N direction, starting from $N = 0$ at $t = 0$. The velocity v_M is given as $v_M = v_{M0} - (eE_M/m_e)t - (eb/2m_e)N^2$, where v_{M0} is the initial velocity. The N motion is an oscillatory motion under $E_N = -kN$ and the magnetic force, and approximately described using Airy functions as $N = c_1 \mathrm{Ai}(r) + c_2 \mathrm{Bi}(r)$, where c_1 and c_2 are constants, and r is defined as

$$r = -\left(\frac{e^2 E_M b}{m_e^2}\right)^{1/3} \left[t - \frac{m_e}{eE_M}\left(v_{M0} + \frac{k}{b}\right)\right]. \tag{63}$$

Bessho et al. (2018) derived the equation of v_N as a function of v_M, as follows:

$$v_N = N\left(\frac{eb}{m_e}\right)^{1/2}\left(-v_M - \frac{eb}{2m_e}N^2 - \frac{k}{b}\right)^{1/2} \times \cot\left\{\frac{2}{3}\left(\frac{eb}{m_e}\right)^{1/2}\frac{m_e}{eE_M}\left[\left(-v_M - \frac{eb}{2m_e}N^2 - \frac{k}{b}\right)^{3/2} - \left(-v_{M0} - \frac{k}{b}\right)^{3/2}\right]\right\}, \tag{64}$$

which is for a case where $v_{M0} < -k/b$. In the other case, where $v_{M0} > -k/b$, we have

$$v_N = N\left(\frac{e^2 E_y b}{m_e^2}\right)^{1/3} \frac{\mathrm{Bi}(r_0)\frac{d\mathrm{Ai}(r)}{dr} - \mathrm{Ai}(r_0)\frac{d\mathrm{Bi}(r)}{dr}}{\mathrm{Ai}(r_0)\,\mathrm{Bi}(r) - \mathrm{Bi}(r_0)\,\mathrm{Ai}(r)}, \tag{65}$$

where r_0 is the initial value of r at $N = 0$. Both equations represent multiple curves in the v_M-v_N plane, because the cot function and Airy functions Ai and Bi are oscillatory functions.

In the MMS observation of the Torbert event, the electron VDF at the neutral line shows $v_{M0} > -k/b$; therefore, in Bessho et al. (2018) (see Fig. 22), to compare the observed multi-crescent VDF with the theory, Eq. (65) was used. In the following, we will explain how to extract the reconnection electric field E_M by comparing the theory and the observed VDF.

Procedure 1: Deriving the field quantities from observed field data

From the magnetic field and electric field data, we obtain the slope b for B_L and the slope k for E_N. Figure 22a shows magnetic fields (top), electric fields (middle), and the distance from the neutral line (bottom), obtained by MMS2 and MMS3. The *LMN* coordinates were obtained by a hybrid method (Denton et al. 2016) of MDD (Shi et al. 2005; Sect. 2.2.1) and MVA (Sonnerup and Scheible 1998). The N distance was obtained by a method similar to Denton et al. (2016), from the time integral of $V_N = (\mathrm{d}B_L/\mathrm{d}t)/(\partial B_L/\partial N)$, which represents the MMS barycenter velocity relative to the current sheet. The values of b and k were obtained as $b = 9.0\times10^{-2}$ nT/km (from MMS3 data), and $k = 1.2\times10^3$ mV/km^2 (from MMS2 data).

Procedure 2: Obtaining v_{M0} from the VDF at the neutral line $N = 0$

There are two equations, Eq. (64) and Eq. (65), depending on the value of v_{M0}. Using the VDF on the neutral line, we identify the population of electrons that have just arrived at the neutral line and started the meandering motion. Figure 22b shows the VDF at the neutral line $N = 0$ ($B_L = 0$ line) by MMS3, and the population near the white vertical line shows that $v_{M0} = -0.7\times10^4$ km/s. In this event, the $E_N \times B_L$ drift velocity, $-k/b$, is -1.3×10^4 km/s; therefore, the condition $v_{M0} > -k/b$ needs to be used, and we will use Eq. (65).

Procedure 3: Compare the theory and the multi-crescent VDF, and determine E_M

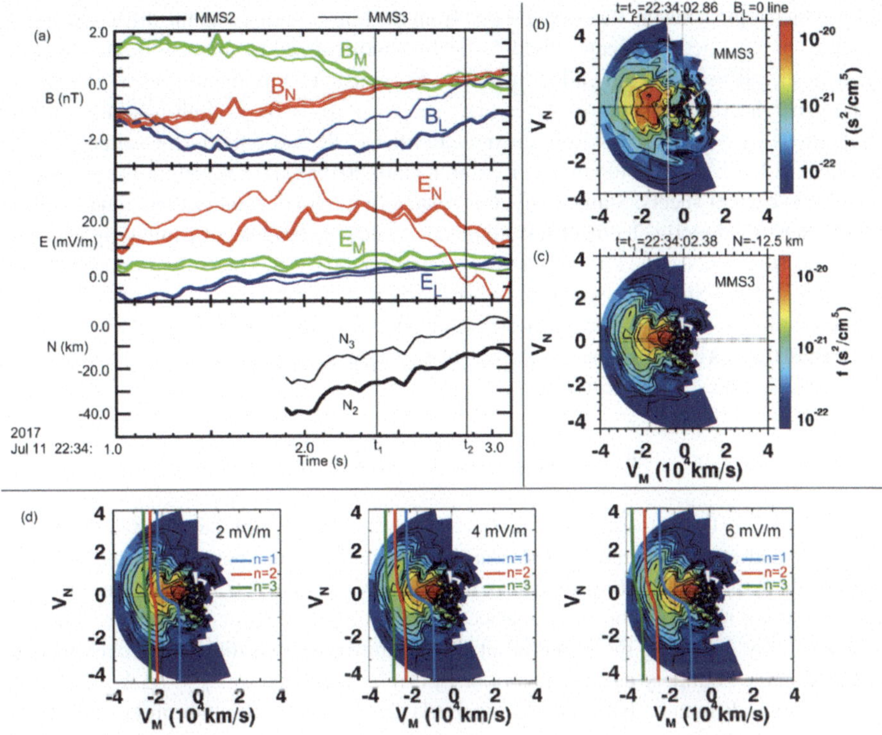

Fig. 22 MMS data in the event on 11 July 2017. (**a**) MMS data for the field quantities and the distance N from the neutral line. (**b**) Electron VDF at the neutral line. (**c**) VDF at t=t_1. (**d**) Comparison between the theory (blue, red, green curves by Eq. (65)) and the multi-crescent VDF. Adapted from Bessho et al. (2018)

Now, we have obtained all the required parameters, b, k, N, and v_{M0} to draw the theoretical curves on the multi-crescent VDF, except for an undetermined parameter, E_M. In this event, Fig. 22c is the VDF that was compared with the theory. In the VDF, there are three crescent-like stripes. Figure 22d shows three plots with different E_M values, 2 mV/m, 4 mV/m, and 6 mV/m. Values of $E_M = 2$ mV/m and $E_M = 6$ mV/m are excluded because the separations between theoretical curves based on Eq. (65) do not match the separations in the observed VDF. In contrast, $E_M = 4$ mV/m is consistent with the observed VDF. In this way, we can determine the reconnection electric field from a multi-crescent VDF. Note that this value, $E_M = 4$ mV/m is close to the value of $E_M = 3.2$ mV/m determined by Genestreti et al. (2018), if we allow for an uncertainty of 1-2 mV/m in the MMS measurements.

3.3.2 Remote Sensing at the Reconnection Separatrix Boundary

Background: The change of magnetic field connectivity during magnetic reconnection occurs within the micro-scale diffusion region. Assuming that reconnection develops in the x-z plane (Fig. 23), the y-component of the electric field E_y around the diffusion region matches the convectional electric field at the inflow and outflow regions to conserve the magnetic flux, $E_y \sim -u_{\text{in}} B_{\text{in}} \sim -u_{\text{out}} B_{\text{out}}$ where u_{in}, u_{out} and B_{in}, B_{out} are the plasma flow speeds and the magnetic field strengths, respectively, at the boundaries of the diffusion region and subscripts "in" and "out" mean the inflow and outflow boundaries, respectively. The magnitude of E_y, which represents the flux transfer rate from the inflow region into the diffusion region, is called the reconnection electric field E_r or the unnormalized reconnection rate. Near the center of the diffusion region called the EDR, E_r is sustained by the non-ideal component of the generalized Ohm's law as $E_r \sim |E'_y| = |E_y + (\mathbf{u}_e \times \mathbf{B})_y|$. High-resolution in-situ observations by MMS successfully encountered the EDR and directly measured E'_y near the EDRs in Earth's magnetopause (e.g., Burch et al. 2016a) and magnetotail (e.g., Torbert et al. 2018) current sheets. Statistically, it is relatively rare to encounter these small regions and measure E_r directly. To meet this challenge, a new technique for estimating E_r using in-situ measurements at the reconnection separatrix boundary was recently proposed (Nakamura et al. 2018a). Since the extent of the separatrix is longer than the micro-scale EDR, the probability for encountering the separatrix is much greater than that for the EDR.

Methods: Ignoring variations in the out-of-plane direction, then E_r, which corresponds to E_y at the X-line, can be written as

$$E_r = -E_{y\_x\text{-}line} = \frac{\partial A_{y,\mathrm{x-line}}}{\partial t} \sim \frac{\partial A_{ys}}{\partial t}, \tag{66}$$

where $A_{y,\mathrm{x-line}}$ is the out-of-plane component of the vector potential at the X-line and A_{ys} is the potential at the separatrix. Note that in the 2-D limit, the potential at the X-line is constant along the reconnection separatrix boundary (Vasyliunas 1975). Since the separatrix may be moving relative to observing spacecraft because of structure motion (see Fig. 23), by sequentially obtaining the potential at the separatrix by two different probes that are separated in the boundary normal direction, E_r can be estimated as $E_r \sim \Delta A_{ys}/\Delta t$. Here Δt is the time difference between the separatrix detections by the two probes and ΔA_{ys} (corresponding to A_3-A_2 in Fig. 23) is the difference of A_{ys} between the probes during this separatrix crossing. Defining the separation of the two probes as $\Delta\mathbf{x}$=(Δx, Δz), as illustrated in Fig. 23, and assuming: (i) a constant reconnection rate during Δt, and (ii) the uniform electric and magnetic fields between the two probes, then ΔA_{ys} can be estimated as $\Delta A_{ys} \sim (\Delta\mathbf{X} \times \mathbf{B})_y$, where $\Delta\mathbf{X} = \Delta\mathbf{x} - \mathbf{V}_c \Delta t$ takes out the effect of background structural motion in the $\mathbf{E} \times \mathbf{B}$ drift velocity, $\mathbf{V}_c = (\mathbf{E} \times \mathbf{B})/B^2$. Then, from Eq. (66), E_r can be described as,

$$\begin{aligned} E_r &\sim \frac{(\Delta\mathbf{X} \times \mathbf{B})_y}{\Delta t} = [-(\mathbf{V}_{\mathrm{tim}} - \mathbf{V}_c) \times \mathbf{B}]_y \\ &= -B_x \frac{\Delta z}{\Delta t} + B_z \frac{\Delta x}{\Delta t} + B_x \left(\frac{\mathbf{E} \times \mathbf{B}}{B^2}\right)_z - B_z \left(\frac{\mathbf{E} \times \mathbf{B}}{B^2}\right)_x, \end{aligned} \tag{67}$$

where $\mathbf{V}_{\mathrm{tim}} = \Delta\mathbf{x}/\Delta$t is the separatrix velocity from the timing analysis. Equation (67) suggests that if two probes that are separated in the normal direction sequentially detect the separatrix signatures, the reconnection electric field E_r can be remotely estimated (Nakamura et al. 2018a).

Applications: In Nakamura et al. (2018a), the proposed technique was first tested using virtual satellites in a 2-D fully kinetic PIC simulation of reconnection without a guide field. To identify separatrix boundaries from observation data, Nakamura et al. (2018a) proposed to detect high-energy parallel electron beams that stream away from the X-line along the separatrix and resulting enhancements of the B_y component (Hall fields). The remotely estimated E_r from virtual observations for both electron beams and Hall fields indeed agree well with the directly obtained E_r at the X-line, indicating the adequacy of this remote sensing technique.

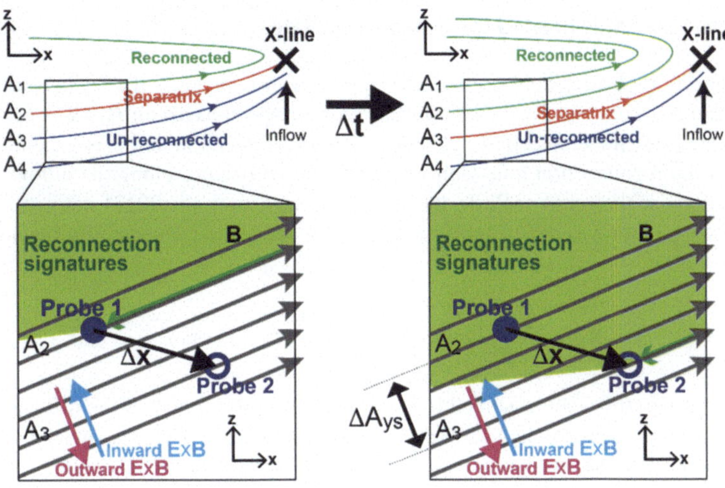

Fig. 23 Schematic of a reconnecting current sheet in motion in the x-z plane during Δt, focusing on a region near the separatrix boundary where two probes separated by $\Delta\mathbf{x}$ sequentially detect the separatrix signatures (adapted from Nakamura et al. 2018a). The field line motion (E×B drift velocity) can be inward (cyan arrow) or outward (magenta arrow), depending on the magnitude of the reconnection electric field and the magnitude and direction of the structure velocity

In real in-situ observations, this technique requires sufficiently high-cadence magnetic and electric field data under an assumption that temporal and spatial variations of the fields while obtaining $\mathbf{V}_{\text{tim}}$ (i.e., while multi-probes sequentially detect the separatrix) are negligible. The MMS mission satisfies this requirement with its burst mode magnetic (Russell et al. 2016) and electric (Ergun et al. 2016; Lindqvist et al. 2016; Torbert et al. 2016a) field measurements and its small inter-spacecraft separation (10^{1-2} km). The plasma particle measurements at burst mode cadences (Pollock et al. 2016) are also useful to detect signatures of the separatrix such as parallel electron beams (Varsani et al. 2017). Nakamura et al. (2018a) first applied this technique to an MMS observation event on 10 August 2016 of the plasma sheet crossing in the near-Earth region accompanied by a strong substorm, which was initially reported by Nakamura et al. (2017). By identifying the separatrix boundary from the Hall field enhancements, they estimated the reconnection electric field as $E_{\text{r}} \sim 15\pm5$ mV/m. Wellenzohn et al. (2021) applied this technique to another MMS plasma sheet crossing event accompanied by a small substorm on 12 July 2018. By identifying the separatrix boundary from high-energy parallel electron beams and resulting electric field disturbances, they estimated $E_{\text{r}} \sim 2\pm0.5$ mV/m. In another example, Nakamura et al. (2018b) focused on the magnetotail EDR crossing event by MMS on 11 July 2017 accompanied by a small substorm, which was initially reported by Torbert et al. (2018). In this event, E_{r} can be obtained not only from the remote sensing technique at the separatrix near the EDR, but also from the direct measurement within the EDR. The obtained E_{r} values from remote and direct methods both are in the range 2.5 ± 0.5 mV/m. These initial results of the remote sensing technique performed using MMS observations suggest a positive correlation between E_{r} and the intensity of substorms. A future statistical approach is required to comprehensively establish the relation between the local reconnection E_{r} and the geomagnetic disturbances.

3.3.3 Separatrix Angle Related to Reconnection Rate

The reconnection electric field normalized by $V_{A0}B_0$, where B_0 is the background reconnecting magnetic field and V_{A0} is the upstream Alfvén speed based on B_0 ($R \sim E_r/V_{A0}B_0$), measures how fast the connectivity changes during reconnection and it is generally called the normalized reconnection rate. Considering the force balance along the inflow and outflow directions at the meso-scale in the $\beta \ll 1$ limit, Liu et al. (2017) derived a general theory showing that for symmetric reconnection R is related to the separatrix opening angle θ near the IDR as,

$$R \sim \frac{E_r}{V_{A0}B_0} \sim \tan\theta \left(\frac{1-\tan^2\theta}{1+\tan^2\theta}\right)^2 \sqrt{1-\tan^2\theta}. \tag{68}$$

Here the separatrices are assumed to be straight from the center of the diffusion region through the observation points. Given that the adequacy of Eq. (68) was indeed confirmed by fully kinetic simulations (e.g., Liu et al. 2017, 2018a; Nakamura et al. 2018b), this theory implies that the reconnection rate R can be obtained by measuring the opening angle θ at an ion-scale distance from the X-line.

As sketched in Fig. 24a, the separatrix opening angle θ just outside the IDR needed in Eq. (68) matches the flaring angle $\tan^{-1}(|B_{N\mathrm{s}}|/|B_{L\mathrm{s}}|)$ made by the magnetic fields adjacent to the separatrix ($\mathbf{B}_\mathrm{s}$), which also matches the flaring angle $\tan^{-1}(|B_{N\mathrm{d}}|/|B_{L\mathrm{d}}|)$ made by the magnetic fields at the upstream/downstream edge of the diffusion region ($\mathbf{B}_\mathrm{d}$). In light of these relations, Nakamura et al. (2018b) introduced the following quantity f_r, that is a function of $|B_N|/|B_L|$ measured along the spacecraft path,

$$f_r\left(\frac{|B_N|}{|B_L|}\right) \sim \frac{|B_N|}{|B_L|}\left(\frac{1-\left(\frac{|B_N|}{|B_L|}\right)^2}{1+\left(\frac{|B_N|}{|B_L|}\right)^2}\right)^2 \sqrt{1-\left(\frac{|B_N|}{|B_L|}\right)^2}. \tag{69}$$

They then applied this function to estimate the reconnection rate R in the magnetotail EDR crossing event by MMS on 11 July 2017 (Torbert et al. 2018). This function gives R whenever the spacecraft crosses the separatrix near the diffusion region; i.e., where $|B_N|/|B_L| = |B_{N\mathrm{s}}|/|B_{L\mathrm{s}}|$. Thus, R can be estimated by detecting the separatrix signature and computing f_r at that time. In Nakamura et al. (2018b), the separatrix boundary was accurately deduced using close comparisons with a fully kinetic simulation of this MMS event. The reconnection rate R was then successfully obtained as $R \sim$0.15-0.2 (Fig. 24b), which indeed agrees well with the rate directly observed within the EDR (Genestreti et al. 2018).

Note that this technique of estimating the normalized reconnection rate R would be applicable only in a limited region where the separatrix opening angle is sufficiently close to the exhaust opening angle critical to the rate. This condition would be satisfied just outside the edge of the IDR as sketched in Fig. 24a. In addition, since the separatrix line in the LN plane is nearly straight from outside the IDR toward the corner of the EDR (Nakamura et al. 2018b), this condition would be satisfied even within the IDR.

4 Summary and Outlook

Most of the methods discussed in the present review have not yet been extensively used in the analysis of in-situ data, including from the MMS mission. Thus, our hope is that the use of

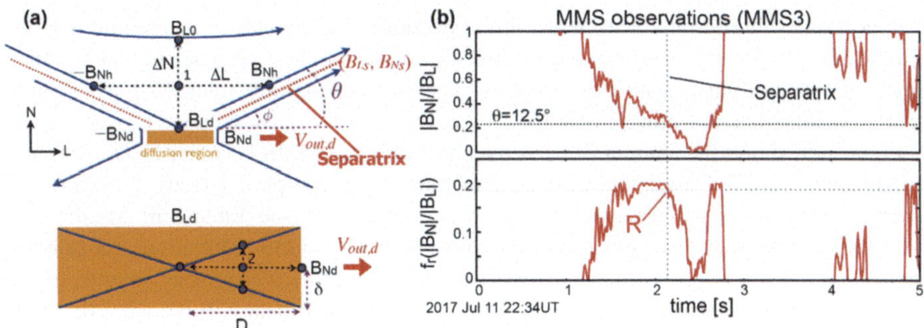

Fig. 24 (**a**) Sketch of the reconnection geometry around the diffusion region (adapted from Liu et al. 2017). (**b**) MMS observations of $|B_N|/|B_L|$ and f_r (Eq. (69)) during the separatrix crossing on 11 July 2017 (adapted from Nakamura et al. 2018b)

these methods will help advance our understanding of magnetic reconnection and associated processes in space, including wave excitation and particle acceleration, their feedback on the reconnection process, and coupling to macro-scale phenomena. Further improvement of the methodology would also be expected. In particular, since the MMS mission has focused essentially on electron-scale processes, cross-scale aspects of magnetic reconnection have not been well explored to date. Thus, there is vast room for the future development of data analysis methods to understand these multi-scale processes. See, for example, Broeren et al. (2021) for a potentially multi-scale method that can reconstruct a 3D magnetic field when more than four spacecraft are available, and Bard and Dorelli (2021) for a new type of reconstruction techniques based on Physics-Informed Neural Networks (Raissi et al. 2019).

Appendix A: Region and Current Sheet Identification

Magnetic reconnection is known to occur in thin current sheets often located at transitions between different plasma regions, as seen at the dayside magnetopause and in the nightside magnetotail. This has led to the assignment of regions of interest that define where the MMS spacecraft may encounter reconnection, parameters that identify where reconnection is likely to occur, and data management systems [the Automated Burst System (ABS), the Scientist-in-the-Loop (SITL), and the Ground Loop System (GLS)] to capture the right type of data. This appendix reviews the ways in which MMS locates, identifies, and captures reconnecting current sheets. A few methods based on machine learning for automated identification of regions in and around the magnetosphere are also summarized.

A.1 Automated Burst System

Prior to MMS, observations of the EDR, where electrons are demagnetized and magnetic energy is converted to electron kinetic energy, had been enigmatic, with few direct observations (Nagai et al. 2011, 2013; Scudder et al. 2012; Tang et al. 2013; Oka et al. 2016). This was because spacecraft lacked the spatial and temporal resolution to resolve electron-scale dynamics. MMS has overcome these limitations by having four spacecraft in a tetrahedron configuration at unmatched spatial scales and sampling rates. Since launch, MMS has identified more than 50 EDRs (see Webster et al. 2018, Lenouvel et al. 2021, and Genestreti et al. 2022 for partial lists) and greatly expanded our knowledge of what catalyzes the global reconnection cycle (Fuselier et al. 2024, this collection). The amount of data required to obtain this success greatly exceeds the downlink allocations of the deep space network, which means that the satellites require an automated way of selecting time intervals for high time resolution burst data downlink, the ABS.

Figure 25a shows the amount of burst data held onboard MMS between 17 October and 15 November, 2022. Data is categorized into Category 0 (purple), 1 (red), 2 (yellow), 3 (green), and 4 (blue) based on the active science objectives (see Table 1 in Argall et al. 2020 for an example). Category 0 and Category 1 correspond to calibration and primary science data, respectively, while Category 2-4 correspond to lower-priority science goals. The orange dashed line indicates the threshold beyond which data will be overwritten, as the remaining buffers are reserved for new regions of interest. Figure 25b shows the amount of overwritten data in each category. Data classified into the lowest-priority category (Category 4) are overwritten first, unless some memory cleanup activity takes place, as on 15 November 2022. This figure demonstrates that MMS is able to store and downlink all of its highest priority data and effectively manage low-priority data to both achieve tertiary science objectives and make room for new data. Such categorization is important for the burst memory management system to be effective.

In an ABS, trigger data numbers (TDNs) are calculated by applying a look-up table of gains G_i and offsets O_i to each measured data quantity x_i to create a scalar value that is a linear combination of the x_i (Baker et al. 2016; Fuselier et al. 2016):

$$\mathrm{TDN} = \sum_{i=0}^{N-1} G_i x_i + O_i . \tag{70}$$

When the TDNs exceed a threshold value, either the data are marked for downlink or the satellite is triggered into burst mode. Missions such as WIND, THEMIS, Cluster, STEREO have burst mode schemes that operate only when triggered. Some WIND and THEMIS triggers used to detect plasma boundaries such as the magnetopause are documented by Phan et al. (2015). Triggers used on STEREO for shock detection, their evolution, and their efficacy are described by Jian et al. (2013). MMS has enough memory to capture burst mode data at all times within its science region of interest. This is so that the Scientist-in-the-Loop has a chance to review the low time resolution data and make their own selections before the unselected burst data is erased from memory. The TDNs are available publicly and the look-up tables of gains and offsets can be updated depending on the active science objectives.

A.2 Scientist-in-the-Loop

The SITL is a role that is passed among the personnel on the MMS science team. The SITL scientist is responsible for manually selecting time intervals for burst data downlink by examining low time resolution data that is recorded simultaneously with the burst data and is downlinked at the end of each orbit. The SITL is guided by the mission-level science objectives, which are assigned a priority of 1-4 (Category 1-4) and a range of Figure-of-Merit values. The SITL assigns a Figure-of-Merit value to time intervals that fit the science objectives and the corresponding burst data is downlinked in priority-order. The SITL process has been successful and is being used as the primary method for burst data selections even during extended mission phases, despite the automated processes, as discussed in Sects. A.1 and A.3, having being operated.

Figure 26 shows a typical view of the data that the SITL scientist sees when using an interactive tool called "EVA" to make burst data selections (see Argall et al. 2020 for more

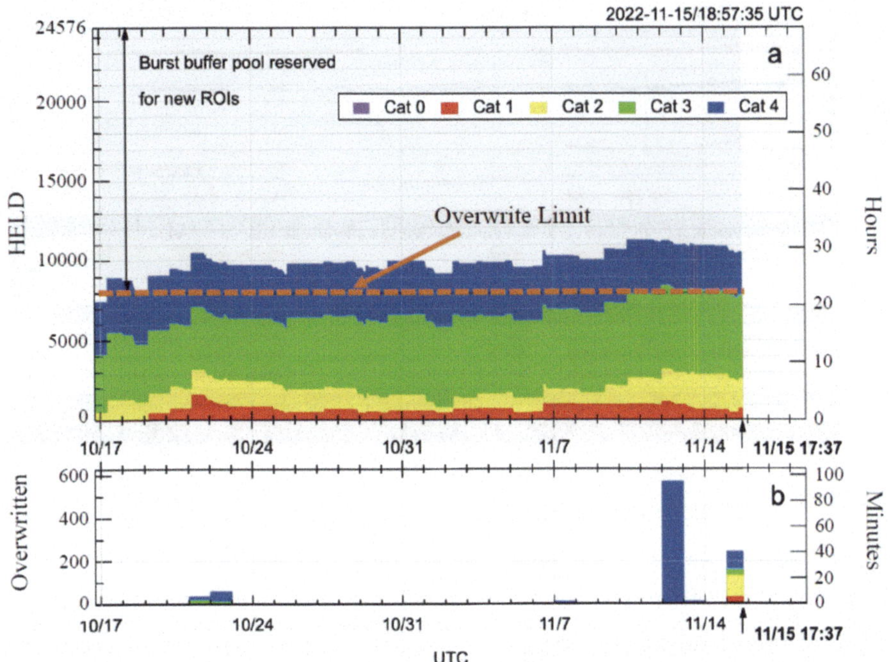

Fig. 25 Burst memory buffers at risk of being overwritten. (**a**) Amount of Category 0 (purple), 1 (red), 2 (yellow), 3 (green), and 4 (blue) stored or HELD onboard MMS, shown in terms of the number of burst buffers (left axis) and the corresponding hours of data (right axis). (**b**) The amount of overwritten data in terms of Category (color), memory buffers (left axis) and total time (right axis). Category 0 data are for special operations (e.g., calibration) or critical science data; selections are rare and are transmitted to ground as soon as the downlink is available, so no Category 0 buffers are present in the figure

details). Labels and scales have been removed on purpose except for the bottom two panels, which show the ABS and SITL selections, respectively. Note that while the SITL scientist selects most of the intervals selected by the ABS, the SITL makes more selections (vertical bars in the bottom panels) – a tedious process that could be alleviated, in part, by automated selection models. This is where the Ground Loop System (GLS) comes into play.

A.3 Ground Loop System

The GLS is designed to be a system of machine learning (ML) or empirical models that automate the event classification process using all of the data available to the SITL (much more than what is available to the ABS). Data available to the SITL is of restricted use because its quality is lower than of the science-quality (Level-2) data freely available to the public (Baker et al. 2016). Thus, ML models trained on SITL data may not perform as well when applied to Level-2 data, and vice versa. Argall et al. (2020) implemented the first GLS ML model for automated burst selections. Their model aimed to automate the SITL scientist's top priority – to select magnetopause crossings. Selecting magnetopause crossings was important because EDRs are not resolved in the low-resolution data available to the SITL; however, EDRs occur in reconnection events at the magnetopause and the magnetopause is easily identifiable in the SITL data.

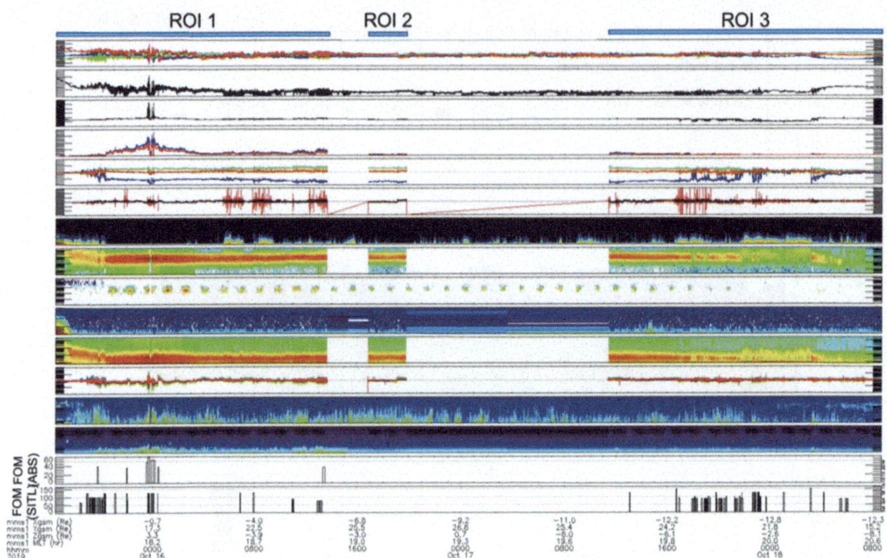

Fig. 26 Data that the SITL uses to make selections. The SITL views data from the regions of interest, ROI 1, 2, and 3, as it becomes available, and uses a tool to interact with the plot, make selections, and submit them to the science data center. Quantities and scales are purposefully not shown except for the ABS (second to last panel) and SITL (last panel) selections. Note that while the SITL basically selected the same intervals (vertical bars) as the ABS, the SITL selected many more

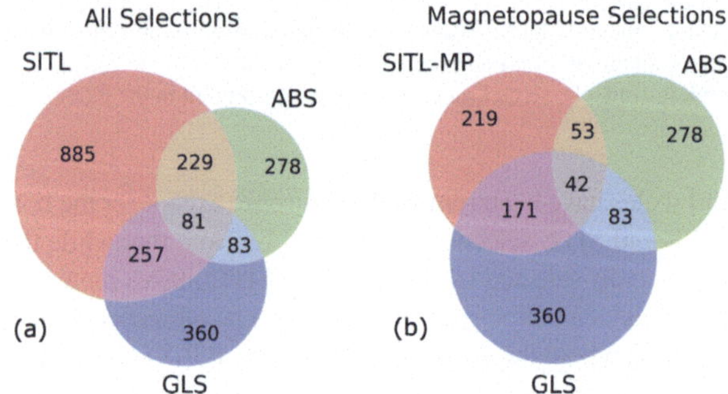

Fig. 27 The GLS and ABS are complementary systems that make selections of high interest to the SITL. (a) Comparison of all selections; the SITL selects a significant number of selections made by both the GLS (71%) and ABS (64%). (b) Comparison of selections designated as magnetopause (MP) crossings by the SITL. In both cases, there is little overlap between the GLS and ABS. Adapted from Argall et al. (2020)

Figure 27a is a Venn diagram showing the overlap of SITL, ABS, and GLS selections using all selections made between 19 October 2019 and 25 March 2020. The SITL selects 71% of all GLS selections and 64% of all ABS selections, indicating that the ABS and GLS are making good selections. Figure 27b is the same format, but includes only those selections that were designated magnetopause crossings by the SITL. The GLS selects 78% (171/219) of all magnetopause crossings selected by the SITL. The other 22% of selections include intervals that were magnetopause-like (i.e., flux transfer events, reconnection jets), but were not specifically called magnetopause crossings by the SITL. This indicates that the

GLS was effective at what it was designed to do. In both cases, the ABS and GLS had little overlap in selections – they are complementary systems that provide helpful information to the SITL.

The GLS was designed in such a way that new supervised ML models could be trained easily by searching and parsing the human-readable string SITL scientists assign to each of their selections. (A suggestion to future missions that wish to implement a similar system is to have a defined set of keywords to identify similar events.) These ML models could be implemented simultaneously in a hierarchical structure to automate science campaigns (Fig. 8 in Argall et al. 2020). The ground-level for the hierarchical campaigns is to implement region identifiers.

A.4 Machine Learning-Based Region Identification

The selection of intervals of interest can be facilitated by the ability to identify key regions of the near-Earth environment. While historically identification of regions and boundaries was done through visual inspection of data, recent work has shown that modern methods relying on ML are an alternative method that can be more efficient. Novel methods are increasingly needed as the amount of data available nowadays precludes scientists from visually mining numerous and huge datasets accumulated over many different missions, sometimes over decades. Such a method was recently implemented by Nguyen et al. (2022a) and applied to statistical analysis and interpretation of the location and shape of the Earth's magnetopause (Nguyen et al. 2022b,c,d).

The identification of three key near-Earth regions, the magnetosphere, the magnetosheath, and the solar wind, was made based on an automatic classification method that uses in situ data (magnetic field and ion moments) from multiple spacecraft. The classification was performed with the so-called Gradient Boosting algorithm (Friedman 2001). While not offering as much flexibility as deep learning methods, this algorithm, based on the iterative fit of the residuals obtained by the successive training and predictions made by decision trees, has been recognized to perform well on complex, eventually imbalanced classification problems (Brown and Mues 2012). Furthermore, it typically needs less labeled data and is lighter to train than deep neutral networks. The prediction of the algorithm is shown to outperform routines based on manually set thresholds. Data from 11 different spacecraft (THEMIS, ARTEMIS, Cluster, one Double Star (TC-1), and one MMS spacecraft) were analyzed for a total of 83 cumulated years. A total of 15,062 magnetopause crossings and 17,227 bow shock crossings were identified. An example automated identification is illustrated in Fig. 28 for an outbound orbit of the MMS mission. It highlights excellent agreement between label data and the method identification of the main regions and boundaries during this pass, except for magnetopause boundary layers (transition regions). The code is available online (Table 7 in Appendix C) and the datasets can easily be enhanced for future use by the community.

A.5 Automated Region Classification Based on 3D Particle Velocity Distributions

Vast amounts of data produced by space missions make it difficult for scientists to identify interesting events manually, and automatic data classification methods can be convenient. Below we describe one approach that allows us to classify three-dimensional (3D) ion energy distribution samples according to the plasma regions (Olshevsky et al. 2021). The classified data can then be used for statistical studies and identifying boundaries, such as bow shocks and the magnetopause, where reconnection may occur. We will focus on the dayside magnetosphere and will define the following regions: solar wind (SW), ion foreshock (IF),

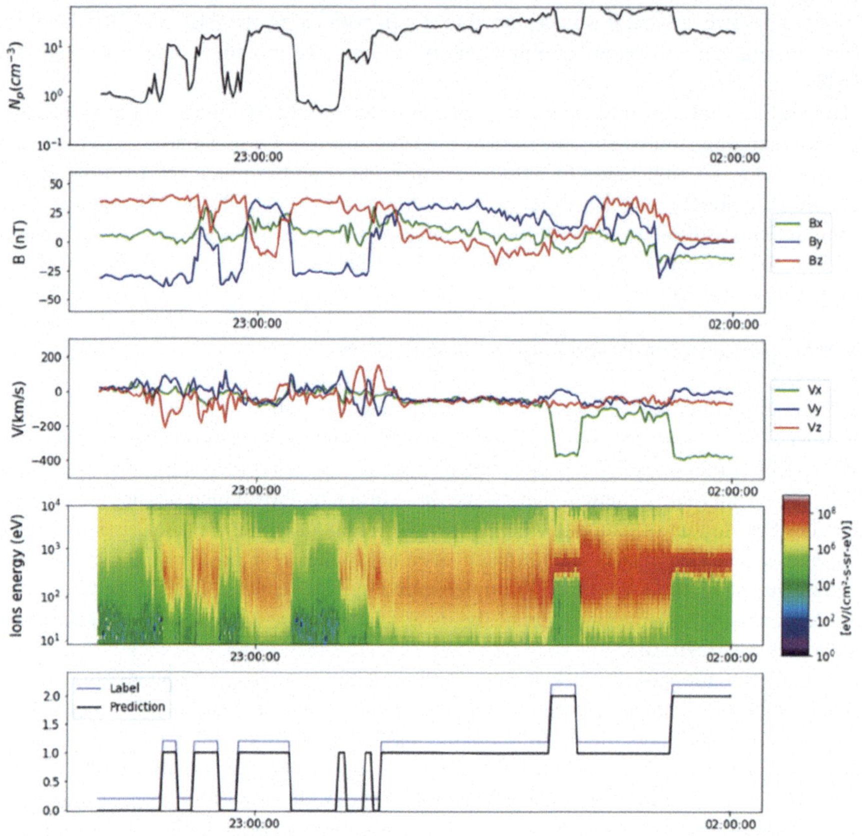

Fig. 28 MMS observations on 31 December 2015 (adapted from Nguyen et al. 2022a). The top to fourth panels show the ion density, magnetic field, ion velocity in GSM coordinates, and omnidirectional differential energy fluxes of ions. The bottom panel shows the evolution of the label (blue), intentionally shifted for visual inspection, and the prediction made by the ML algorithm (black), in which "0" means the magnetosphere, "1" the magnetosheath, and "2" the solar wind

magnetosheath (MSH), and magnetosphere (MSP), all of which have a distinct signature in ion VDFs. A training dataset representing "clean" samples (i.e., excluding boundaries) of the ion distributions was selected from the above four regions, and was used to train a 3D convolutional neural network classifier (Maturana and Scherer 2015) by a supervised ML approach. The classifier is then applied to ion distributions (other than the training dataset) and assigns a probability of each distribution belonging to one of the four classes.

To illustrate the method, we show in Fig. 29 MMS observations during a 6-hour interval on 16 October 2015 that contains multiple magnetopause crossings (transitions between MSH and MSP). The crossing at 13:06 UT contains an encounter with the EDR at the site of magnetic reconnection (Burch et al. 2016a). The MSP intervals are characterized by the northward magnetic field (B_z dominant), low density and plasma flow, and increased ion flux at energies above 10 keV. The MSH intervals are, on the contrary, characterized by high density, fast plasma flow ($V_y \sim 150$ km/s), and ion flux peaking at ~0.3 keV. The bottom panel shows the results of the neural network classifier; one can see that both the

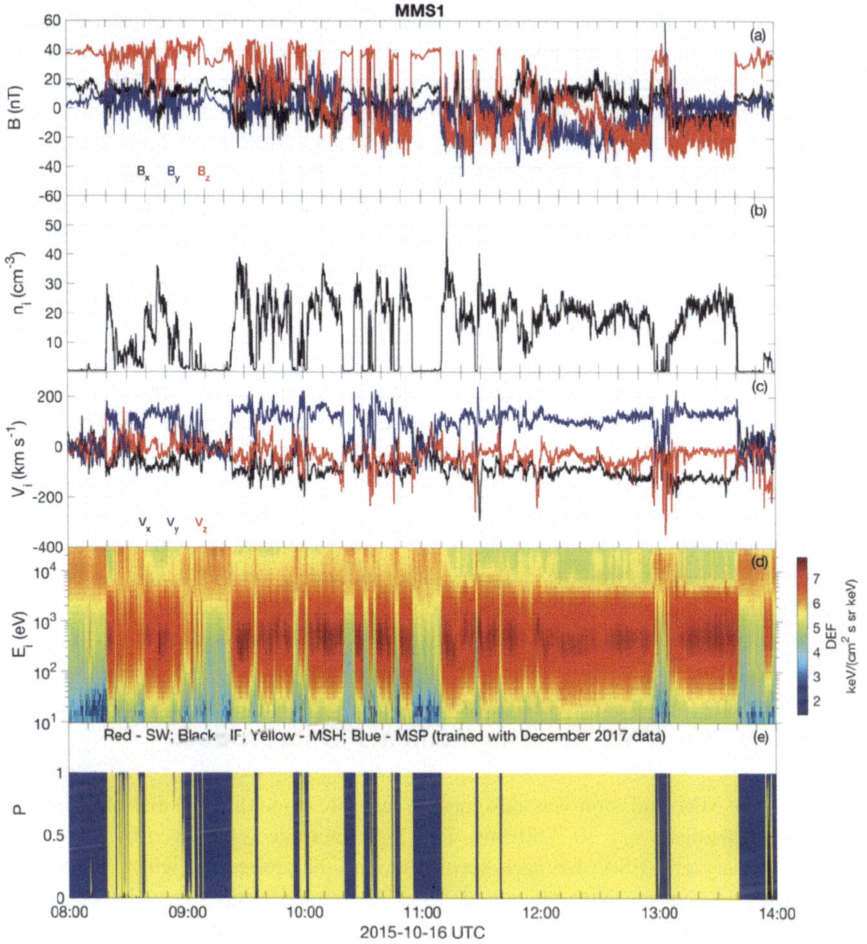

Fig. 29 Example of plasma region classification of MMS data on 16 October 2015. The panels from top to bottom show (**a**) the magnetic field, (**b**) plasma density, (**c**) ion velocity, (**d**) ion energy spectrum, and (**e**) the probabilities provided by the classifier

MSP (blue) and MSH (yellow) regions have been correctly classified except for boundary layers between the two regions.

Plasma regions can be identified using plasma moments and electromagnetic fields, as illustrated above and in several other ML approaches including the one introduced in Sect. A.4 (Nguyen et al. 2022a; Cheng et al. 2022). However, ion VDFs contain more information than the moments of the distribution, having a unique footprint for each plasma region and thus possibly improving the reliability of classification. Trained scientists use plots of ion data for visual region classification. As it is difficult to visualize the time series of 3D VDFs, they would normally reduce the full 3D measurements in some way, for example, to omnidirectional energy-time spectrograms (e.g., Fig. 29d) where all look directions are summed up.

Another helpful representation is the angle-angle (azimuth-polar angle) plots of PSDs for a specific energy channel. The approach to the classification by Olshevsky et al. (2021), in its essence, is based on image recognition of such plots. One recognizes each of specific footprints in such images corresponding to different regions. During a human inspection, only a small number of energies can generally be analyzed simultaneously. ML has no such limitation, and can analyze all the data simultaneously. Thus, a 3D image recognition is performed, in which a data cube composed of a stack of 32 (number of energy channels for MMS FPI Dual Ion Spectrometers (DIS) (Pollock et al. 2016)) angle-angle plots is analyzed. This approach is enabled by the homogeneous dataset provided by DIS in fast mode, i.e., the numbers of angular bins and the energy ranges are fixed (except for the special solar wind mode), and thus there is no need to reduce or resample the data.

The results for classification can be used in many different ways. One use is to define times when MMS is in a particular plasma region, for example, in a pristine magnetosheath (e.g., Svenningsson et al. 2023). Another use is to analyze the probabilities' time series to identify the boundaries between the different plasma regions. For example, by analyzing the transition between the SW/IF and MSH, Lalti et al. (2022) identified $\sim$3000 bow shock crossings by MMS. Similarly, magnetopause crossings can be identified from the transitions between the MSH and MSP classes. Another potential application is to use the probabilities (Fig. 29e) to quantify plasma mixing in boundary layers, which, for example, can occur in Kelvin-Helmholtz vortices (Settino et al. 2022). As a final remark, we note that the method as described is based solely on the classification of individual ion VDFs, and can be further extended to ingest the information on time evolution and/or other data (e.g., magnetic field, spacecraft position, etc.), which can enable an even more robust classification.

Appendix B: The MMS "Quarter Moments" Data Product

The FPI on the MMS mission was designed to measure three-dimensional electron (ion) phase space densities every 30 (150) ms. This high data rate is achieved by arranging 8 electrostatic analyzers (ESA) for each species around the spacecraft, where the field-of-view of each ESA can be electrostatically deflected to four uniformly spaced look directions spanning 45 degrees. The high voltage power supply executes a full energy sweep (from $\sim$10 eV to $\sim$30 keV) for each of the four deflection states, so that the suite of analyzers simultaneously samples 8 uniformly spaced azimuthal angles for each energy sweep. This means that FPI samples a $32\times16\times8$ regular (energy, zenith, azimuth) array every 7.5 (37.5) ms for electrons (ions), allowing for the possibility of recovering plasma moments a factor of 4 times faster than the nominal 30 (150) ms cadence. Figure 30 illustrates how the four deflection states are combined into the final FPI Level 2 phase space density "skymaps."

To recover 3D plasma moments at 7.5 (37.5) ms for electrons (ions), we use cubic spline interpolation to reconstruct 32 azimuthal samples – independently for each energy and zenith – from the 8 azimuthal samples corresponding to a single deflection state. To mitigate spline boundary condition issues, we interpolate an augmented set of data over the domain $[-2\pi, 3\pi]$ in which data from $[0, 2\pi]$ is copied to $[-2\pi, 0]$ and $[2\pi, 3\pi]$. The interpolated data is then passed to the production moments algorithm as a (32 energy) $\times$ (16 zenith) $\times$ (32 azimuth) array.

Figure 31 shows a validation test in which 7.5 ms perpendicular electron bulk velocity data is compared to the $E\times B$ drift velocity (from the electric field and magnetometer experiments averaged down to 7.5 ms). The agreement is excellent, demonstrating that this is a simple, robust method for extracting accurate plasma moments. These "quarter moment" products have been used in many MMS publications (e.g., Phan et al. 2018)

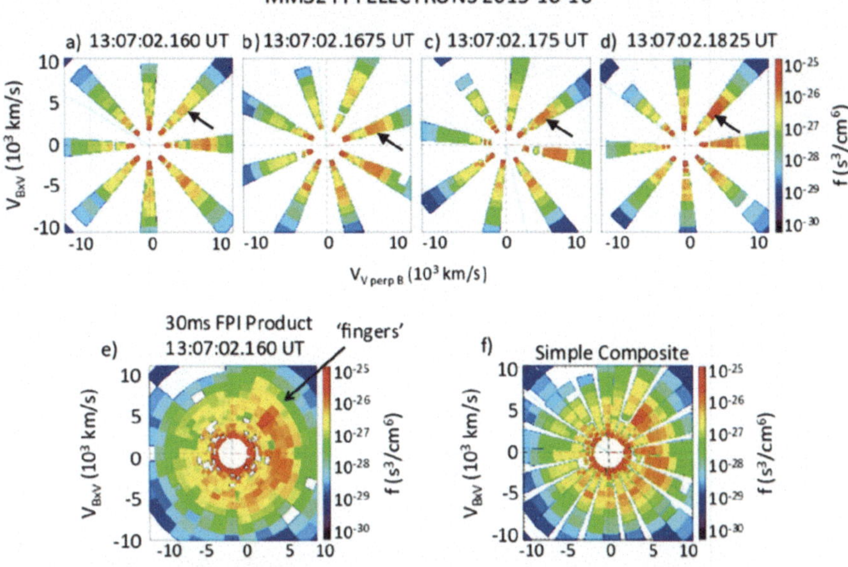

Fig. 30 The four FPI electrostatic deflection states (top four images), each composed of a full 32-step energy sweep and obtained at 7.5 ms cadence for electrons, are combined to produce a full Level 2 "skymap" (bottom left images) every 30 ms (adapted from Rager et al. 2018)

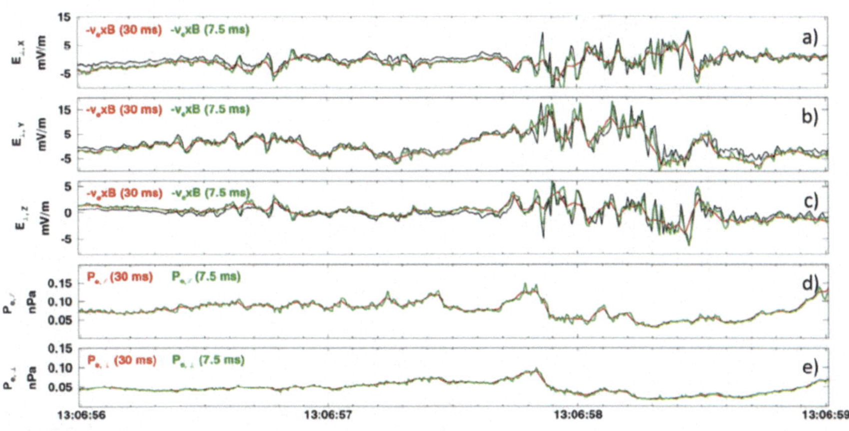

Fig. 31 The spline interpolation technique does an excellent job of recovering electron bulk velocity at 7.5 ms resolution (green line) (adapted from Rager et al. 2018). The top three panels show three components of the electron convective electric field at 30 ms (red) and 7.5 ms (green) resolutions. The perpendicular electric field measured by the double probe instruments (Ergun et al. 2016; Lindqvist et al. 2016), averaged down to 7.5 ms, is shown in black. The bottom two panels show the electron pressures in the directions parallel and perpendicular to the magnetic field

Appendix C: Tables Summarizing Methods Reviewed in the Paper

Table 1 Methods for large-scale context

Method	# of spacecraft (SC) needed	Underlying theory, concept, or name of data mining method	Input	Output	Assumptions	Spatial or temporal scale of interest	Other requirements if any	References
Maximum magnetic shear model	1-SC in solar wind	Time-of-flight analysis	ion moments (n, $\mathbf{u}_\mathrm{i}$), $\mathbf{B}$ in solar wind	Magnetopause magnetic shear plot; dayside reconnection location		Entire magnetopause	T96 & IMF draping models	Trattner et al. (2007); Kobel and Flückiger (1994); Tsyganenko (1995)
Global MHD model[1]	1-SC in solar wind	MHD equations	MHD parameters in solar wind	MHD quantities everywhere in space and time	MHD	MHD		Tóth et al. (2005); see the CCMC website[1] for other codes
Data mining (DM) reconstruction[2]	Multi-mission	DM using distance weighted kNN method and basis function magnetic field architectures	Magnetometer archives; Solar wind $u_{sw} B_z$; geomagnetic indices SML, SMR & their time derivatives	3D magnetic field parametrized by $\sim 10^3$ parameters derived from data	Magnetostatic	>1 R_E and >5 min	Global activity indices availability	Stephens et al. (2019); Sitnov et al. (2008); Tsyganenko and Sitnov (2007)
Global in situ data reconstruction	Multi-mission	kNN statistical method	Solar wind parameters	3D global reconstruction of outer magnetosphere and magnetosheath properties	SW propagation from L1, magnetopause and bow shock models for normalizing data location	Dayside magnetosphere, magnetopause, and magnetosheath		Michotte de Welle et al. (2022)

[1] The code can be run at CCMC (https://ccmc.gsfc.nasa.gov/)

[2] u_{sw}: solar wind speed; B_z: north-south component of the interplanetary magnetic field in geocentric solar magnetospheric (GSM) coordinates

Table 2 Methods for coordinate systems, frame velocity, and spacecraft trajectory estimation

Method	# of spacecraft (SC) needed	Underlying theory, concept	Input	Output	Assumptions	Spatial or temporal scale of interest	References
MVAB	1	$\nabla \cdot \mathbf{B} = 0$	$\mathbf{B}$	Variance directions; normal direction; LMN coordinates	1D structure to get a good estimate for the normal ($\mathbf{e}_N$) direction	N/A	Sonnerup and Scheible (1998)
MVAVe	1	None	$\mathbf{u}_e$	Variance directions	Presence of electron outflow jets	Sub-ion scale	Genestreti et al. (2018)
MDDB	4	None	$\mathbf{B}$, SC position	Gradient directions; LMN coordinates	Steady structure	>SC separation	Shi et al. (2019)
Dimensionality	4	MDDB	MDDB eigenvalues	Dimensionality indices	Steady structure	>SC separation	Rezeau et al. (2018)
Hybrid method	4	MDDB + MVAB or MVAVe	$\mathbf{B}$, SC position (and $\mathbf{u}_e$)	LMN coordinates	Large maximum eigenvalues	>SC separation	Denton et al. (2018); Heuer et al. (2022)
HT analysis	1	$\partial \mathbf{B}/\partial t = 0$	$\mathbf{B}$, $\mathbf{u}$ (or $\mathbf{E}$)	Structure velocity	Not in diffusion regions	Ion or MHD scale	De Hoffmann and Teller (1950); Khrabrov and Sonnerup (1998)
MFR	1	Faraday's law, $\partial \mathbf{B}/\partial t = 0$	$\mathbf{E}$ (or $\mathbf{u}$), $\mathbf{B}$	Normal direction and velocity	1D structure, $d/dt = 0$, constant velocity	>SC separation	Dunlop and Woodward (1998)
4SC timing	4	None	Various	Normal direction and velocity	1D structure, $d/dt = 0$, constant velocity	>SC separation	Dunlop and Woodward (1998)
STD	4	B changes from convection	$\mathbf{B}$(t), SC position	Structure velocity	Partial time derivative=0	>SC separation	Shi et al. (2006, 2019)
Reconstructed X-line motion	4	$\nabla \cdot \mathbf{B} = 0$; Ampère's law	$\mathbf{B}$, $\mathbf{j}$, SC position	Velocity in reconnection plane	Various depending on model	Spacecraft spacing	Denton et al. (2021)
Comparison with simulation	1 or more	PIC code applicable	Various	Path through simulation domain	Simulation is realistic	Simulation scale	Shuster et al. (2017); Schroeder et al. (2022)
MCA	4	None	$\mathbf{B}$, SC position	B-variance eigenvectors; planarity; elongation; dimensionality	Steady structure	>SC separation	Fadanelli et al. (2019)

Table 3 Methods for reconstructing 2D/3D structures

Method	# of spacecraft (SC) needed	Underlying theory, concept	Input	Output	Assumptions	Spatial or temporal scale of interest	Other requirements if any	References
3D B-field quadratic reconstruction	4	Maxwell equations	**B**, **j** at 4 SC locations	Full 3D vector B-field near tetrahedron	No quadratic terms in minimum variance direction	Spacecraft spacing		Torbert et al. (2020)
3D polynomial reconstruction[1]	4	$\nabla \cdot \mathbf{B} = 0$; Ampère's law	**B**, **j** at 4 SC locations	Full 3D vector B-field near tetrahedron	Various depending on model	Spacecraft spacing		Denton et al. (2020)
3D Polynomial Reconstruction with multiple input times[2]	4	$\nabla \cdot \mathbf{B} = 0$; Ampère's law	**B**, **j** at 4 SC locations	Full 3D vector B-field near tetrahedron	Various depending on model	Spacecraft spacing		Denton et al. (2022)
3D empirical reconstruction using stochastic optimization method	4	Simultaneous perturbation stochastic approximation	**B**, **j** at 4 SC locations	Full 3D vector B-field near tetrahedron	$\nabla \cdot \mathbf{B} = 0$; Ampère's law	Spacecraft spacing		Zhu et al. (2022)
Simulation-assisted 2D reconstruction of an EDR	Multi-SC	Comparison with kinetic simulation	Various	2D fields of all fluid quantities	Simulation is realistic	Ion skin depth		Schroeder et al. (2022)
Grad-Shafranov reconstruction	1	Static magnetic field, $\nabla \cdot \mathbf{B} = 0$, Ampère's law	**B**, pressure	2D maps of **B** & pressure	2D, MHD equilibrium	MHD scale	Predefined structure velocity	Sonnerup et al. (2006a,b)
MHD reconstruction	1	MHD equations	MHD parameters (occasionally incl. P-anisotropy)	2D maps of MHD quantities	2D, steady structure; various depending on model	MHD scale	Predefined structure velocity	Sonnerup et al. (2006a,b); Chen and Hau (2018); Teh (2019); Tian et al. (2020)

Table 3 *(Continued)*

Method	# of spacecraft (SC) needed	Underlying theory, concept	Input	Output	Assumptions	Spatial or temporal scale of interest	Other requirements if any	References
EMHD reconstruction[3]	1	EMHD equations	**B**, **E**, electron moments (n, $\mathbf{u}_e$, T_e)	2D maps of B, E, and electron moments	2D, steady structure; various depending on model	Electron scale	Predefined structure velocity	Sonnerup et al. (2016); Hasegawa et al. (2021); Korovinskiy et al. (2021, 2023)
3D RBF-based reconstruction	Multi-SC	Euler potentials expressed by modified radial basis functions	**B**	Full 3D vector B-field near tetrahedron	Steady structure; $\nabla \cdot \mathbf{B} = 0$	Spacecraft spacing		Chen et al. (2019)

[1] The code is available at Zenodo (https://doi.org/10.5281/zenodo.3906853)

[2] The code is available at Zenodo (https://doi.org/10.5281/zenodo.6395044)

[3] The code is available at Zenodo (https://doi.org/10.5281/zenodo.5144478)

Table 4 Methods for diffusion region identification

Method	# of spacecraft (SC) needed	Underlying theory, concept, or name of machine learning method	Input	Output	Assumptions	Spatial or temporal scale of interest	Other requirements if any	References
Electron-frame dissipation measure[1]	1	$\mathbf{j}\cdot(\mathbf{E}+\mathbf{u}_e\times\mathbf{B}) - \rho_c\,(\mathbf{u}_e\cdot\mathbf{E})$	$\mathbf{j}$, $\mathbf{E}$, $\mathbf{B}$, $\mathbf{u}_e$ $[\rho_c]$	Scalar in units of $[\mathrm{W/m^3}]$	Maxwell's equations			Zenitani et al. (2011a)
Agyrotropy	1	Particle distribution function	$\mathbf{P}$, $\hat{\mathbf{b}}$	Scalar	None	Diffusion region		Swisdak (2016)
Pressure strain	4	$-(\mathbf{P}\cdot\nabla)\cdot\mathbf{u}$	$\mathbf{P}$, $\mathbf{u}$	Ion and electron dissipation rate in scalar	Differentiable velocity; curlometer			Yang et al. (2017); Bandyopadhyay et al. (2021)
Electron vorticity	4	$\Omega_e \sim \omega_{ce}$ (electron gyrofrequency) defines the characteristic frequency of the EDR	$\mathbf{v}_e$	Electron vorticity ($\mathbf{\Omega}_e = \nabla\times\mathrm{u}_e$)	None	EDR		Hwang et al. (2019)
Magnetic flux transport (MFT)[2]	1 or 4	$\partial_t\psi + \mathbf{U}_\psi\cdot\nabla_\perp\psi = 0$ (Advection equation of magnetic flux)	$\mathbf{B}$, $\mathbf{E}$, SC position	MFT velocity ($\mathbf{U}_\psi$) or its divergence ($\nabla\cdot\mathbf{U}_\psi$)	Quasi-2D reconnection ($k_M \ll k_\perp$); no magnetic field diffusion or generation	Diffusion region	Predefined LMN coordinates and frame velocity	Li et al. (2021); Qi et al. (2022)

Table 4 (*Continued*)

Method	# of spacecraft (SC) needed	Underlying theory, concept	Input	Output	Assumptions	Spatial or temporal scale of interest	Other requirements if any	References
Machine learning 1 to classify regions	1	feed-forward multilayer perceptron	Plasma and field quantities, reduced electron distribution functions	4 classes (IDR, EDR, boundary layers, magnetosphere)	None	IDR, EDR, boundary layers, magnetosphere		Lenouvel et al. (2021)
Machine learning 2 to identify EDRs	1	Convolutional Neural Network	Electron distribution functions	EDR / no EDR	None	Diffusion regions		Lenouvel (2022)

[1] ρ_c: electric change density

[2] k_M: wave number corresponding to the scale of the spatial variation in the direction of the reconnection X-line; $k_\perp$: wave number corresponding to the scale of the spatial variation in the plane transverse to the X-line

Table 5 Methods for analyzing the electron diffusion region

Method	# of spacecraft (SC) needed	Underlying theory, concept	Input	Output	Assumptions	Spatial or temporal scale of interest	Other requirements if any	References
Anomalous transport terms estimation from lower hybrid waves[1]	4	Fluid equations, wave-particle interactions	**B**, **E**, particle moments (n, $\mathbf{u}$, $\mathbf{P}$)	Anomalous transport terms	None	Sub-ion scales	Particle moments that resolve the waves	Graham et al. (2022)
Computing $\partial f_e/\partial t$ in the spacecraft frame	1	Electron Vlasov equation	Electron distribution function f_e at multiple times	$\partial f_e/\partial t$ (sometimes an approximation for $\partial f_e/\partial N$)	Cadence of electron measurements is sufficient for $\partial f_e/\partial t$ estimation	Electron scale for EDR	Consult with the FPI team for proper implementation	(2019, Shuster et al. 2021a,b
Computing the spatial gradient $\mathbf{v} \cdot \nabla f_e$ term	4	Electron Vlasov equation	f_e, velocity-space coordinates for f_e, SC position	∇f_e and $\mathbf{v} \cdot \nabla f_e$	SC separation is sufficient for estimating ∇f_e	Electron scale for EDR	Consult with the FPI team for proper implementation	(2019, Shuster et al. 2021a,b
Computing the $\mathbf{F} \cdot \nabla_\mathbf{v} f_e$ term in the spacecraft frame	1	Electron Vlasov equation	f_e, velocity-space coordinates for f_e, **E**, **B**	$\nabla_\mathbf{v} f_e$ and each of the 3rd term of the Vlasov equation	Velocity-space resolution is sufficient for estimating $\nabla_\mathbf{v} f_e$	Electron scale for EDR	Consult with the FPI team for proper implementation	(2019, Shuster et al. 2021a,b
Non-Maxwellianity[1]	1	Particle distribution function	**B**, particle distribution, particle moments	Scalar non-Maxwellianity value	None	Kinetic scale of particles of interest	None	Greco et al. (2012); Servidio et al. (2017); Graham et al. (2021)
Kinetic entropy	1	Boltzmann's entropy theory	Particle distribution, particle moments	Entropy densities, scalar non-Maxwellianity value	None	Kinetic scale of particles of interest	Velocity space is properly resolved by particle measurements	Argall et al. (2022)

[1] The scripts and data required to reproduce the figures in Graham et al. (2022) can be found at https://zenodo.org/records/6370048, and the routines require the irf-matlab software package: https://github.com/irfu/irfu-matlab

[1] The routines to compute non-Maxwellianity can be found at https://github.com/irfu/irfu-matlab

Table 6 Methods for reconnection electric field (E_M) estimation

Method	# of spacecraft (SC) needed	Underlying theory or concept	Input	Output	Assumptions	Spatial or temporal scale of interest	Other requirements if any	References
Direct measurement	1	Faraday's law	**E**	E_M	2D, steady	N/A	Accurate LMN	Genestreti et al. (2018); Burch et al. (2020)
Inflow velocity	1		$\mathbf{u}_e$, n, **B**	Inflow velocity; normalized reconnection rate	2D, steady	EDR	Accurate LMN	Burch et al. (2020)
Multi-crescent VDF[1]	1	Equations of motion	B_L, E_N, N, u_{eM}	E_M	B_L=bN, E_N=-kN, No guide field	Electron scale	N=0 crossing	Bessho et al. (2018)
Remote sensing at separatrix	More than 2	Magnetic flux conservation	Timing velocity at separatrix, **E**, **B**	E_M	2D, constant reconnection rate, uniform field between multi-probes	Ion to MHD scale	Separatrix identification	Nakamura et al. (2018a)
Separatrix angle	1 or 4 (depending on how separatrix is detected)	Flux and force balance along inflow and outflow direction	B_N, B_M (or B_z, B_x for tail)	Normalized reconnection rate	2-D, low beta, separatrix angle=exhaust opening angle	Electron to ion scale	Separatrix identification	Liu et al. (2017); Nakamura et al. (2018b)

[1] N: distance in the normal (N) direction from the current sheet center (B_L=0) plane; see Sect. 3.3.1 for the definitions of b and k

Table 7 Methods for region and current sheet identification

Method	# of spacecraft (SC) needed	Underlying theory, concept, or name of machine learning method	Input	Output	Assumptions	Spatial or temporal scale of interest	Other requirements if any	References
ML-based region identification[1]	1	Gradient boosting/machine learning	**B**, ion moments (n, $\mathbf{u}_\mathrm{i}$, T_i)	Near-Earth region labelling, magnetopause and bow shock crossings	None	MHD[2] and larger	N/A	Nguyen et al. (2022a)
Region classification using 3D particle distributions	1	Convolutional Neural Network/supervised machine learning	3D ion velocity distributions	Near-Earth region labelling, magnetopause and bow shock crossings	None	MHD and larger	N/A	Olshevsky et al. (2021)

[1] The code is available at github (https://github.com/gautiernguyen/in-situ_Events_lists)

[1] Magnetohydrodynamics

Acknowledgements We thank the reviewers whose comments helped improve this manuscript. The work by H.H. was supported by JSPS Grant-in-aid for Scientific Research KAKENHI 21K03504. Part of the work done while H.H. was at SwRI was supported by NASA Contract No. NNG04EB99C at SwRI. M.R.A. acknowledges support from NASA grant 80NSSC23K0409 and NSF award 2308670. R.E.D. was supported by a NASA grant 80NSSC22K1109. The work by D.B.K. has been supported by the Austrian Science Fund (FWF): I 3506-N27 and Austrian FFG project ASAP15/873685. T.C.L. was supported by NSF award AGS-2000222. T.K.M.N. was supported by the Austrian Research Fund (FWF) P32175-N27. S.M.P was supported by Contract 499935Q. Y.Q. was supported by NASA MMS mission NNG04EB99C. The authors acknowledge the International Space Science Institute in Bern for supporting the international team of the workshop "Magnetic Reconnection: Explosive Energy Conversion in Space Plasmas".

Funding Open access funding provided by Uppsala University.

Declarations

Competing Interests The authors declare they have no conflicts of interest.

Publisher's Note Springer Nature remains neutral with regard to jurisdictional claims in published maps and institutional affiliations.

References

Andreeva VA, Tsyganenko NA (2016) J Geophys Res Space Phys 121:2249–2263. https://doi.org/10.1002/2015JA022242

Argall MR, Small CR, Piatt S, et al (2020) Front Astron Space Sci 7:54. https://doi.org/10.3389/fspas.2020.00054

Argall MR, Barbhuiya MH, Cassak PA, et al (2022) Phys Plasmas 29:022902. https://doi.org/10.1063/5.0073248

Aunai N, Hesse M, Kuznetsova M (2013) Phys Plasmas 20:092903. https://doi.org/10.1063/1.4820953

Baker DN, Riesberg L, Pankratz CK, Panneton RS, Giles BL, Wilder FD, Ergun RE (2016) Space Sci Rev 199:545–575. https://doi.org/10.1007/s11214-014-0128-5

Bandyopadhyay R, Chasapis A, Matthaeus WH, et al (2021) Phys Plasmas 28:112305. https://doi.org/10.1063/5.0071015

Barbhuiya MH, Cassak PA (2022) Phys Plasmas 29:122308. https://doi.org/10.1063/5.0125256

Barbhuiya MH, Cassak PA, Shay MA, et al (2022) J Geophys Res Space Phys 127:e2022JA030610. https://doi.org/10.1029/2022JA030610

Bard C, Dorelli J (2021) Front Astron Space Sci 8:732275. https://doi.org/10.3389/fspas.2021.732275

Bessho N, Chen L-J, Hesse M (2016) Geophys Res Lett 43(5):1828–1836. https://doi.org/10.1002/2016gl067886

Bessho N, Chen L-J, Wang S, Hesse M (2018) Geophys Res Lett 45:12142–12152. https://doi.org/10.1029/2018GL081216

Boltzmann L (1877) Wiener Ber 76:373–435

Borovsky JE, Yakymenko K (2017) J Geophys Res Space Phys 122(3):2973–2998. https://doi.org/10.1002/2016JA023625

Broeren T, Klein KG, TenBarge JM, Dors I, Roberts OW, Verscharen D (2021) Front Astron Space Sci 8:727076. https://doi.org/10.3389/fspas.2021.727076

Brown I, Mues C (2012) Expert systems with applications. Expert Syst Appl 39:3446–3453. https://doi.org/10.1016/j.eswa.2011.09.033

Buhmann M (2003) Radial basis functions: theory and implementations. Cambridge University Press, Cambridge

Burch JL, Torbert RB, Phan TD, et al (2016a) Science 352(6290):aaf2939. https://doi.org/10.1126/science.aaf2939

Burch JL, Moore TE, Torbert RB, Giles BL (2016b) Space Sci Rev 199(1–4):5–21. https://doi.org/10.1007/s11214-015-0164-9

Burch JL, Webster JM, Hesse M, et al (2020) Geophys Res Lett 47:e2020GL089082. https://doi.org/10.1029/2020GL089082

Califano F, Manfredi G, Valentini F (2016) J Plasma Phys 82:701820603. https://doi.org/10.1017/S002237781600115X

Cassak PA, Barbhuiya MH (2022) Phys Plasmas 29:122306. https://doi.org/10.1063/5.0125248

Cassak PA, Barbhuiya MH, Liang H, Argall MR (2023) Phys Rev Lett 130:085201. https://doi.org/10.1103/PhysRevLett.130.085201

Cerri SS (2016) Plasma turbulence in the dissipation range-theory and simulations. PhD thesis. Universität Ulm. https://doi.org/10.18725/OPARU-3355

Che H, Drake JF, Swisdak M (2011) Nature 474:184–187. https://doi.org/10.1038/nature10091

Chen G-W, Hau L-N (2018) J Geophys Res Space Phys 123:7358–7369. https://doi.org/10.1029/2018JA025842

Chen L-J, Bessho N, Lefebvre B, et al (2008) J Geophys Res 113:A12213. https://doi.org/10.1029/2008JA013385

Chen L-J, Hesse M, Wang S, Bessho N, Daughton W (2016) Geophys Res Lett 43:452–461. https://doi.org/10.1002/2016GL068243

Chen W, Wang X, Tsyganenko NA, Andreeva VA, Semenov VS (2019) J Geophys Res Space Phys 124:10141–10152. https://doi.org/10.1029/2019JA027078

Chen Y, Tóth G, Hietala H, et al (2020) Earth Space Sci 7:e2020EA001331. https://doi.org/10.1029/2020EA001331

Cheng IK, Achilleos N, Smith A (2022) Front Astron Space Sci 9:1016453. https://doi.org/10.3389/fspas.2022.1016453

Chin DC (1999) Opt Eng 38:606–611. https://doi.org/10.1117/1.602104

Daughton W, Scudder J, Karimabadi H (2006) Phys Plasmas 13:072101. https://doi.org/10.1063/1.2218817

De Hoffmann F, Teller E (1950) Phys Rev 80(4):692–703. https://doi.org/10.1103/physrev.80.692

Denton RE, Sonnerup BUÖ, Hasegawa H, et al (2016) Geophys Res Lett 43:5589–5596. https://doi.org/10.1002/2016GL069214

Denton RE, Sonnerup BUÖ, Russell CT, et al (2018) J Geophys Res Space Phys 123:2274–2295. https://doi.org/10.1002/2017JA024619

Denton RE, Torbert RB, Hasegawa H, et al (2020) J Geophys Res Space Phys 125:e2019JA027481. https://doi.org/10.1029/2019JA027481

Denton RE, Torbert RB, Hasegawa H, et al (2021) J Geophys Res Space Phys 126:e2020JA028705. https://doi.org/10.1029/2020JA028705

Denton RE, Liu Y-H, Hasegawa H, Torbert RB, Li W, Fuselier SA, Burch JL (2022) J Geophys Res Space Phys 127:e2022JA030512. https://doi.org/10.1029/2022JA030512

Dunlop MW, Woodward TI (1998) Multi-spacecraft discontinuity analysis: orientation and motion. In: Paschmann G, Daly P (eds) Analysis methods for multi-spacecraft data. International Space Science Institute, SR-001, Switzerland, pp 271–306

Dunlop MW, Southwood DJ, Glassmeier K-H, Neubauer FM (1988) Adv Space Res 8:273–277. https://doi.org/10.1016/0273-1177(88)90141-X

Dunlop MW, Balogh A, Glassmeier K-H, Robert P (2002) J Geophys Res 107:1384. https://doi.org/10.1029/2001JA005088

Dunlop MW, Zhang Q-H, Bogdanova YV, et al (2011) Ann Geophys 29:1683–1697. https://doi.org/10.5194/angeo-29-1683-2011

Egedal J, Lê A, Katz N, Chen L-J, Lefebvre B, Daughton W, Fazakerley A (2010a) J Geophys Res Space Phys 115:A03214. https://doi.org/10.1029/2009JA014650

Egedal J, Lê A, Zhu Y, et al (2010b) Geophys Res Lett 37:L10102. https://doi.org/10.1029/2010GL043487

Egedal J, Le A, Daughton W, et al (2016) Phys Rev Lett 117:185101. https://doi.org/10.1103/PhysRevLett.117.185101

Egedal J, Ng J, Le A, et al (2019) Phys Rev Lett 123:225101. https://doi.org/10.1103/PhysRevLett.123.225101

Ekawati S, Cai D (2023) J Geophys Res Space Phys 128:e2021JA029571. https://doi.org/10.1029/2021JA029571

Ergun RE, Tucker S, Westfall J, et al (2016) Space Sci Rev 199(1–4):167–188. https://doi.org/10.1007/s11214-014-0115-x

Eriksson E, Vaivads A, Graham DB, et al (2018) Geophys Res Lett 45:8081–8090. https://doi.org/10.1029/2018GL078660
Escoubet CP, Schmidt R, Goldstein ML (1997) Space Sci Rev 79:11–32. https://doi.org/10.1007/978-94-011-5666-0_1
Fadanelli S, Lavraud B, Califano F, et al (2019) J Geophys Res Space Phys 124:6850–6868. https://doi.org/10.1029/2019JA026747
Fadanelli S, Lavraud B, Califano F, et al (2020) J Geophys Res Space Phys 125:e2020JA028333. https://doi.org/10.1029/2020JA028333
Friedman JH (2001) Ann Stat 29(5):1189–1232. https://doi.org/10.1214/aos/1013203451
Fu HS, Vaivads A, Khotyaintsev YV, Olshevsky V, André M, Cao JB, Huang SY, Retinò A, Lapenta G (2015) J Geophys Res Space Phys 120:3758–3782. https://doi.org/10.1002/2015JA021082
Fuselier SA, Petrinec SM, Trattner KJ (2000) Geophys Res Lett 27:473–476. https://doi.org/10.1029/1999GL003706
Fuselier SA, Frey HU, Trattner KJ, Mende SB, Burch JL (2002) J Geophys Res 107(A7):1111. https://doi.org/10.1029/2001JA900165
Fuselier SA, Lewis WS, Schiff C, et al (2016) Space Sci Rev 199(1–4):77–103. https://doi.org/10.1007/s11214-014-0087-x
Fuselier SA, Petrinec SM, Reiff PH, et al (2024) Space Sci Rev 220:34. https://doi.org/10.1007/s11214-024-01067-0
Genestreti KJ, Nakamura TKM, Nakamura R, et al (2018) J Geophys Res Space Phys 123:9130–9149. https://doi.org/10.1029/2018JA025711
Genestreti KJ, Li X, Liu Y-H, et al (2022) Phys Plasmas 29:082107. https://doi.org/10.1063/5.0090275
Gershman DJ, F-Viñas A, Dorelli JC, et al (2017) Wave-particle energy exchange directly observed in a kinetic Alfvén-branch wave. Nat Commun 8:14719. https://doi.org/10.1038/ncomms14719
Gjerloev JW (2012) J Geophys Res Space Phys 117:A09213. https://doi.org/10.1029/2012JA017683
Gosling JT, Thomsen MF, Bame SJ, Elphic RC, Russell CT (1990) J Geophys Res 95:8073. https://doi.org/10.1029/JA095iA06p08073
Graham DB, Khotyaintsev YV, André M, et al (2021) J Geophys Res Space Phys 126:e2021JA029260. https://doi.org/10.1029/2021JA029260
Graham DB, Cozzani G, Khotyaintsev YV, et al (2024) Space Sci Rev 220
Graham DB, Khotyaintsev YV, André M, et al (2022) Nat Commun 13(2954). https://doi.org/10.1038/s41467-022-30561-8
Greco A, Valentini F, Servidio S, Matthaeus WH (2012) Phys Rev E 86(6):066405. https://doi.org/10.1103/PhysRevE.86.066405
Guo R, Pu Z, Wang X, Xiao C, He J (2022) J Geophys Res Space Phys 127:e2021JA030248. https://doi.org/10.1029/2021JA030248
Hasegawa H (2012) Monogr Environ Earth Planets 1(2):71–119. https://doi.org/10.5047/meep.2012.00102.0071
Hasegawa H, Sonnerup BUÖ, Dunlop MW, et al (2004) Ann Geophys 22:1251–1266. https://doi.org/10.5194/angeo-22-1251-2004
Hasegawa H, Sonnerup BUÖ, Denton RE, et al (2017) Geophys Res Lett 44:4566–4574. https://doi.org/10.1002/2017GL073163
Hasegawa H, Denton RE, Nakamura R, et al (2019) J Geophys Res Space Phys 124:122–138. https://doi.org/10.1029/2018JA026051
Hasegawa H, Nakamura TKM, Denton RE (2021) J Geophys Res Space Phys 126:e2021JA029841. https://doi.org/10.1029/2021JA029841
Hasegawa H, Denton RE, Dokgo K, et al (2023) J Geophys Res Space Phys 128:e2022JA031092. https://doi.org/10.1029/2022JA031092
Hau L-N, Phan T-D, Sonnerup BUÖ, Paschmann G (1993) Geophys Res Lett 20:2255–2258. https://doi.org/10.1029/93GL02491
Hau L-N, Chang C-K, Chen G-W (2020) Astrophys J 900:97. https://doi.org/10.3847/1538-4357/aba2d0
Hesse M, Schindler K, Birn J, Kuznetsova M (1999) Phys Plasmas 6:1781–1795. https://doi.org/10.1063/1.873436
Hesse M, Neukirch T, Schindler K, Kuznetsova M, Zenitani S (2011) Space Sci Rev 160:3–23. https://doi.org/10.1007/s11214-010-9740-1
Hesse M, Aunai N, Sibeck DG, Birn J (2014) Geophys Res Lett 41:8673–8680. https://doi.org/10.1002/2014GL061586
Heuer SV, Genestreti KJ, Nakamura TKM, Torbert RB, Burch JL, Nakamura R (2022) Geophys Res Lett 49:e2022GL100652. https://doi.org/10.1029/2022GL100652
Hoshino M, Mukai T, Terasawa T, Shinohara I (2001) J Geophys Res Space Phys 106:25979–25997. https://doi.org/10.1029/2001JA900052

Hu Q, Sonnerup BUÖ (2002) J Geophys Res 107(A7). https://doi.org/10.1029/2001JA000293
Hwang K-J, Choi E, Dokgo K, et al (2019) Geophys Res Lett 46:6287–6296. https://doi.org/10.1029/2019GL082710
Hwang K-J, Nakamura R, Eastwood JP, et al (2023) Space Sci Rev 219:71. https://doi.org/10.1007/s11214-023-01010-9
Jian LK, Russell CT, Luhmann JG, Curtis D, Schroeder P (2013) AIP Conf Proc 1539:195–198. https://doi.org/10.1063/1.4811021
Karimabadi H, Daughton W, Scudder J (2007) Geophys Res Lett 34:L13104. https://doi.org/10.1029/2007GL030306
Kaufmann RL, Paterson WR (2009) J Geophys Res Space Phys 114:A00D04. https://doi.org/10.1029/2008JA014030
Khrabrov AV, Sonnerup BUÖ, (1998) DeHoffmann-Teller analysis. In: Paschmann G, Daly P (eds) Analysis Methods for Multi-Spacecraft Data. International Space Science Institute, SR-001, Bern, pp 221–248
Kingsep AS, Chukbar KV, Yan'kov VV (1990) In: Kadomtsev BB (ed) Reviews of plasma physics, vol 16. Consultants Bureau, New York, pp 243–288
Kobel E, Flückiger EO (1994) J Geophys Res 99:23617–23622. https://doi.org/10.1029/94JA01778
Korovinskiy DB, Divin AV, Semenov VS, Erkaev NV, Kiehas SA, Kubyshkin IV (2020) Phys Plasmas 27:082905. https://doi.org/10.1063/5.0015240
Korovinskiy DB, Kiehas SA, Panov EV, Semenov VS, Erkaev NV, Divin AV, Kubyshkin IV (2021) J Geophys Res Space Phys 126:e2020JA029045. https://doi.org/10.1029/2020JA029045
Korovinskiy D, Panov E, Nakamura R, Kiehas S, Hosner M, Schmid D, Ivanov I (2023) Front Astron Space Sci 10:1069888. https://doi.org/10.3389/fspas.2023.1069888
Lalti A, Khotyaintsev YV, Dimmock AP, Johlander A, Graham DB, Olshevsky V (2022) J Geophys Res Space Phys 127:e2022JA030454. https://doi.org/10.1029/2022JA030454
Le A, Daughton W, Chen L-J, Egedal J (2017) Geophys Res Lett 44:2096–2104. https://doi.org/10.1002/2017GL072522
Le A, Daughton W, Ohia O, Chen L-J, Liu Y-H, Wang S, Nystrom WD, Bird R (2018) Phys Plasmas 25:062103. https://doi.org/10.1063/1.5027086
Lecun Y, Bottou L, Bengio Y, Haffner P (1998) In: Proceedings of the IEEE, vol 86, pp 2278–2324. https://doi.org/10.1109/5.726791
Lenouvel Q (2022) Identification by machine learning and analysis of electron diffusion regions at the Earth's magnetopause observed by MMS. PhD thesis. https://theses.hal.science/tel-04075287v1
Lenouvel Q, Génot V, Garnier P, et al (2021) Earth Space Sci 8:e2020EA001530. https://doi.org/10.1029/2020EA001530
Li TC, Liu Y-H, Hesse M, Zou Y (2020) J Geophys Res 125:e2019JA027094. https://doi.org/10.1029/2019JA027094
Li TC, Liu Y-H, Qi Y (2021) Astrophys J Lett 909:L28. https://doi.org/10.3847/2041-8213/abea0b
Li TC, Liu Y-H, Qi Y, Zhou M (2023) Phys Rev Lett 131:085201. https://doi.org/10.1103/PhysRevLett.131.085201
Liang H, Cassak PA, Servidio S, et al (2019) Phys Plasmas 26:082903. https://doi.org/10.1063/1.5098888
Liang H, Barbhuiya MH, Cassak PA, Pezzi O, Servidio S, Valentini F, Zank GP (2020) J Plasma Phys 86:825860502. https://doi.org/10.1017/S0022377820001270
Lindberg M, Vaivads A, Raptis S, Lindqvist P-A, Giles BL, Gershman DJ (2022) Entropy 24:745. https://doi.org/10.3390/e24060745
Lindqvist P-A, Olsson G, Torbert RB, et al (2016) Space Sci Rev 199(1–4):137–165. https://doi.org/10.1007/s11214-014-0116-9
Liu Y-H, Hesse M (2016) Phys Plasmas 23:060704. https://doi.org/10.1063/1.4954818
Liu Y-H, Hesse M, Guo F, Daughton W, Li H, Cassak PA, Shay MA (2017) Phys Rev Lett 118(8):085101. https://doi.org/10.1103/PhysRevLett.118.085101
Liu Y-H, Hesse M, Guo F, Li H, Nakamura TKM (2018a) Phys Plasmas 25:080701. https://doi.org/10.1063/1.5042539
Liu Y-H, Hesse M, Li TC, Kuznetsova M, Le A (2018b) J Geophys Res 123:4908. https://doi.org/10.1029/2018JA025410
Liu Y-H, Li TC, Hesse M, Sun WJ, Liu J, Burch JL, Slavin JA, Huang K (2019) J Geophys Res 124:2819. https://doi.org/10.1029/2019JA026539
Liu Y-H, Hesse H, Genestreti K, et al (2024) Space Sci Rev (in press). arXiv:2406.00875
Lyon JG, Fedder JA, Mobarry CM (2004) J Atmos Sol-Terr Phys 66:1333–1350. https://doi.org/10.1016/j.jastp.2004.03.020
Marshall AT, Burch JL, Reiff PH, Webster JM, Torbert RB, Ergun RE, et al (2020) J Geophys Res Space Phys 125:e2019JA027296. https://doi.org/10.1029/2019JA027296
Marshall AT, Burch JL, Reiff PH, et al (2022) Phys Plasmas 29:012905. https://doi.org/10.1063/5.0071159

Maturana D, Scherer S (2015) In: IEEE/RSJ international conference on intelligent robots and systems (IROS), pp 922–928. https://doi.org/10.1109/IROS.2015.7353481
McPherron RL, Russell CT, Aubry MP (1973) J Geophys Res 78(16):3131–3149. https://doi.org/10.1029/JA078i016p03131
Michotte de Welle B, Aunai N, Nguyen G, Lavraud B, Génot V, Jeandet A, Smets R (2022) J Geophys Res Space Phys 127:e2022JA030996. https://doi.org/10.1029/2022JA030996
Moseev D, Salewski M (2019) Phys Plasmas 26:020901. https://doi.org/10.1063/1.5085429
Mouhot C, Villani C (2011) Acta Math 207:29–201. https://doi.org/10.1007/s11511-011-0068-9
Nagai T, Shinohara I, Fujimoto M, Matsuoka A, Saito Y, Mukai T (2011) J Geophys Res 116:A04222. https://doi.org/10.1029/2010JA016283
Nagai T, Zenitani S, Shinohara I, Nakamura R, Fujimoto M, Saito Y, Mukai T (2013) J Geophys Res Space Phys 118:7703–7713. https://doi.org/10.1002/2013JA019135
Nakamura R, Nagai T, Birn J, Sergeev VA, Le Contel O, Varsani V, et al (2017) Earth Planets Space 69(1):129. https://doi.org/10.1186/s40623-017-0707-2
Nakamura TKM, Genestreti KJ, Liu Y-H, et al (2018b) J Geophys Res Space Phys 123:9150–9168. https://doi.org/10.1029/2018JA025713
Nakamura TKM, Nakamura R, Varsani A, Genestreti KJ, Baumjohann W, Liu Y-H (2018a) Geophys Res Lett 45:3829–3837. https://doi.org/10.1029/2018GL078340
Narita Y (2017) Nonlinear Process Geophys 24:203–214. https://doi.org/10.5194/npg-24-203-2017
Ng J, Egedal J, Le A, Daughton W, Chen L-J (2011) Phys Rev Lett 106:065002. https://doi.org/10.1103/PhysRevLett.106.065002
Ng J, Chen L-J, Bessho N, Shuster J, Burkholder B, Yoo J (2022) Geophys Res Lett 49:e2022GL099544. https://doi.org/10.1029/2022GL099544
Nguyen G, Aunai N, Michotte de Welle B, Jeandet A, Lavraud B, Fontaine D (2022a) J Geophys Res Space Phys 127:e2021JA029773. https://doi.org/10.1029/2021JA029773
Nguyen G, Aunai N, Michotte de Welle B, Jeandet A, Lavraud B, Fontaine D (2022b) J Geophys Res Space Phys 127:e2021JA029774. https://doi.org/10.1029/2021JA029774
Nguyen G, Aunai N, Michotte de Welle B, Jeandet A, Lavraud B, Fontaine D (2022c) J Geophys Res Space Phys 127:e2021JA030112. https://doi.org/10.1029/2021JA030112
Nguyen G, Aunai N, Michotte de Welle B, Jeandet A, Lavraud B, Fontaine D (2022d) J Geophys Res Space Phys 127:e2021JA029776. https://doi.org/10.1029/2021JA029776
Nicholson DR (1983) Introduction to plasma theory. Wiley, New York
Norgren C, Chen L-J, Graham DB, et al (2024) Space Sci Rev 220
Oka M, Phan T-D, Øieroset M, Angelopoulos V (2016) J Geophys Res Space Phys 121:1955–1968. https://doi.org/10.1002/2015JA022040
Oka M, Birn J, Egedal J, et al (2023) Space Sci Rev 219:75. https://doi.org/10.1007/s11214-023-01011-8
Olshevsky V, Pontin DI, Williams B, et al (2020) Astron Astrophys 644:A150. https://doi.org/10.1051/0004-6361/202039182
Olshevsky V, Khotyaintsev YV, Lalti A, et al (2021) J Geophys Res Space Phys 126:e2021JA029620. https://doi.org/10.1029/2021JA029620
Onsager TG, Thomsen MF, Gosling JT, Bame SJ (1990) Geophys Res Lett 17(11):1837–1840. https://doi.org/10.1029/GL017i011p01837
Onsager TG, Thomsen MF, Elphic RC, Gosling JT (1991) J Geophys Res 96:20999–21011. https://doi.org/10.1029/91JA01983
Paschmann G, Daly PW (1998) Analysis methods for multi-spacecraft data. ISSI Scientific Report Series, vol SR-001. ESA Publ. Div, Noordwijk
Paschmann G, Daly PW (2008) Multi-spacecraft analysis methods revisited. ISSI Scientific Report Series, vol SR-008. ESA Publ. Div, Noordwijk
Paschmann G, Schwartz S, Escoubet CP, Haaland S (2005) Outer magnetospheric boundaries: cluster results. Space sciences series of ISSI, vol 20. Springer, Dordrecht. https://doi.org/10.1007/1-4020-4582-4
Paschmann G, Øieroset M, Phan TD (2013) Space Sci Rev 47:309–341. https://doi.org/10.1007/978-1-4899-7413-6_12
Payne DS, Farrugia CJ, Torbert RB, Germaschewski K, Rogers AR, Argall MR (2021) Phys Plasmas 28:112901. https://doi.org/10.1063/5.0068317
Phan TD, Escoubet CP, Rezeau L, et al (2005) Space Sci Rev 118:367–424. https://doi.org/10.1007/s11214-005-3836-z
Phan TD, Hasegawa H, Fujimoto M, Øieroset M, Mukai T, Lin RP, Paterson W (2006) Geophys Res Lett 33:L09104. https://doi.org/10.1029/2006GL025756
Phan TD, Shay MA, Eastwood JP, Angelopoulos V, Øieroset M, Oka M, Fujimoto M (2015) Space Sci Rev 199:631–650. https://doi.org/10.1007/s11214-015-0150-2

Phan TD, Eastwood JP, Cassak PA, et al (2016) Geophys Res Lett 43(12):6060–6069. https://doi.org/10.1002/2016GL069212
Phan TD, Eastwood JP, Shay MA, et al (2018) Nature 557:202–206. https://doi.org/10.1038/s41586-018-0091-5
Pollock C, Moore TE, Jacques A, et al (2016) Space Sci Rev 199(1–4):331–406. https://doi.org/10.1007/s11214-016-0245-4
Pontin DI, Priest ER (2022) Living Rev Sol Phys 19:1. https://doi.org/10.1007/s41116-022-00032-9
Price L, Swisdak M, Drake JF, Cassak PA, Dahlin JT, Ergun RE (2016) Geophys Res Lett 43:6020–6027. https://doi.org/10.1002/2016GL069578
Price L, Swisdak M, Drake JF, Graham DB (2020) J Geophys Res Space Phys 125(4). https://doi.org/10.1029/2019JA027498
Pritchett PL (2001) J Geophys Res 106:3783–3798. https://doi.org/10.1029/1999JA001006
Qi Y, Li TC, Russell CT, Ergun RE, Jia Y-D, Hubbert M (2022) Astrophys J Lett 926:L34. https://doi.org/10.3847/2041-8213/ac5181
Qiu J, Hu Q, Howard TA, et al (2007) Astrophys J 659(1):758–772. https://doi.org/10.1086/512060
Raeder J, Cramer WD, Germaschewski K, Jensen J (2017) Space Sci Rev 206:601–620. https://doi.org/10.1007/s11214-016-0304-x
Rager AC, Dorelli JC, Gershman DJ, et al (2018) Geophys Res Lett 45:578–584. https://doi.org/10.1002/2017GL076260
Raissi M, Perdikaris P, Karniadakis GE (2019) J Comput Phys 378:686–707. https://doi.org/10.1016/j.jcp.2018.10.045
Reiff PH, Daou AG, Sazykin SY, et al (2016) Geophys Res Lett 43:7311–7318. https://doi.org/10.1002/2016GL069154
Reiff PH, Webster JM, Daou AG, et al (2017) CCMC modeling of magnetic reconnection in electron diffusion regions. In: Foullon C, Malandraki O (eds) Space weather of the heliosphere: processes and forecasts. Proceedings IAU Symposium, vol 335, pp 142–146. https://doi.org/10.1017/S1743921317010845
Reiff PH, Marshall A, Webster J, Sazykin S, Russell CT, Rastaetter L (2018) MMS observations and CCMC modeling of field line stretching at separator lines. Fall AGU e-Lightning poster. https://agu2018fallmeeting-agu.ipostersessions.com/default.aspx?s=B2-10-20-70-BD-2D-A2-4E-35-27-A4-FE-DC-C0-6D-DA. https://doi.org/10.1002/essoar.10502075.1
Reyes PI, Pinto VA, Moya PS (2021) Space Weather 19(9):e2021SW002766. https://doi.org/10.1029/2021SW002766
Rezeau L, Belmont G, Manuzzo R, Aunai N, Dargent J (2018) J Geophys Res Space Phys 123(1):227–241. https://doi.org/10.1002/2017ja024526
Rumelhart D, Hinton G, Williams R (1986) Nature 323:533–536. https://doi.org/10.1038/323533a0
Russell CT, Anderson BJ, Baumjohann W, et al (2016) Space Sci Rev 199(1–4):189–256. https://doi.org/10.1007/s11214-014-0057-3
Safrankova J, Nemecek Z, Dusik S, Prech L, Sibeck DG, Borodkova NN (2002) Ann Geophys 20:301–309. https://doi.org/10.5194/angeo-20-301-2002
Schlichting H (1979) Boundary layer theory. McGraw-Hill, New York
Schroeder JM, Egedal J, Cozzani G, Khotyaintsev YV, Daughton W, Denton RE, Burch JL (2022) Geophys Res Lett 49:e2022GL100384. https://doi.org/10.1029/2022GL100384
Scudder J, Daughton W (2008a) J Geophys Res 113:A06222. https://doi.org/10.1029/2008JA013035
Scudder JD, Holdaway RD, Daughton WS, Karimabadi H, Roytershteyn V, Russell CT, Lopez JY (2012) Phys Rev Lett 108:225005. https://doi.org/10.1103/PhysRevLett.108.225005
Sergeev V, Angelopoulos V, Kubyshkina M, et al (2011) J Geophys Res Space Phys 116:A00I26. https://doi.org/10.1029/2010JA015689
Servidio S, Chasapis A, Matthaeus WH, et al (2017) Phys Rev Lett 119:205101. https://doi.org/10.1103/PhysRevLett.119.205101
Settino A, Khotyaintsev YV, Graham DB, Perrone D, Valentini F (2022) J Geophys Res Space Phys 127:e2021JA029758. https://doi.org/10.1029/2021JA029758
Shay MA, Drake JF, Swisdak M (2007) Phys Rev Lett 99:155002. https://doi.org/10.1103/PhysRevLett.99.155002
Shelley EG, Ghielmetti AG, Balsiger H, et al (1995) Space Sci Rev 71:497–530. https://doi.org/10.1007/BF00751339
Shen C, Li X, Dunlop M, Shi QQ, Liu ZX, Lucek E, Chen ZQ (2007) J Geophys Res 112:A06211. https://doi.org/10.1029/2005JA011584
Shi QQ, Shen C, Pu ZY, et al (2005) Geophys Res Lett 32:L12105. https://doi.org/10.1029/2005GL022454
Shi QQ, Shen C, Dunlop MW, et al (2006) Geophys Res Lett 33:L08109. https://doi.org/10.1029/2005GL025073
Shi QQ, Tian AM, Bai SC, et al (2019) Space Sci Rev 215(4):35. https://doi.org/10.1007/s11214-019-0601-2

Shuster JR, Chen L-J, Daughton W, et al (2014) Geophys Res Lett 41:5389–5395. https://doi.org/10.1002/2014GL060608

Shuster JR, Chen L-J, Hesse M, Argall MR, Daughton W, Torbert RB, Bessho N (2015) Geophys Res Lett 42:2586–2593. https://doi.org/10.1002/2015GL063601

Shuster JR, Argall MR, Torbert RB, et al (2017) Geophys Res Lett 44:1625–1633. https://doi.org/10.1002/2017GL072570

Shuster JR, Gershman DJ, Chen L-J, et al (2019) Geophys Res Lett 46:7862–7872. https://doi.org/10.1029/2019GL083549

Shuster JR, Bessho N, Wang S, Ng J (2021b) Phys Plasmas 28:122902. https://doi.org/10.1063/5.0069559

Shuster JR, Gershman DJ, Dorelli JC, et al (2021a) Nat Phys 17:1056–1065. https://doi.org/10.1038/s41567-021-01280-6

Shuster JR, Gershman DJ, Giles BL, et al (2023) J Geophys Res Space Phys 128:e2022JA030949. https://doi.org/10.1029/2022JA030949

Sibeck DG, Lopez RE, Roelof EC (1991) J Geophys Res 96:5489–5495. https://doi.org/10.1029/90JA02464

Sitnov MI, Sharma AS, Papadopoulos K, Vassiliadis D (2001) Phys Rev E 65:016116. https://doi.org/10.1103/PhysRevE.65.016116

Sitnov MI, Tsyganenko NA, Ukhorskiy AY, Brandt PC (2008) J Geophys Res Space Phys 113(A7):A07218. https://doi.org/10.1029/2007JA013003

Sitnov MI, Stephens GK, Tsyganenko NA, Ukhorskiy AY, Wing S, Korth H, Anderson BJ (2017) Spatial structure and asymmetries of magnetospheric currents inferred from high-resolution empirical geomagnetic field models. In: Haaland S, Runov A, Forsyth C (eds) Dawn-dusk asymmetries in planetary plasma environments. American Geophysical Union (AGU), pp 199–212. https://doi.org/10.1002/9781119216346.ch15

Sitnov MI, Stephens GK, Tsyganenko NA, et al (2019) J Geophys Res Space Phys 124(11):8427–8456. https://doi.org/10.1029/2019JA027037

Sitnov MI, Stephens GK, Tsyganenko NA, et al (2020) Space Weather 18:e2020SW002561. https://doi.org/10.1029/2020SW002561

Sitnov M, Stephens G, Motoba T, Swisdak M (2021) Front Phys 9. https://doi.org/10.3389/fphy.2021.644884

Sonnerup BUÖ, Cahill LJ Jr (1967) J Geophys Res 72(1):171–183. https://doi.org/10.1029/JZ072i001p00171

Sonnerup BUÖ, Guo M (1996) Geophys Res Lett 23(25):3679–3682

Sonnerup BUÖ, Scheible M (1998) Minimum and maximum variance analysis. In: Paschmann G, Daly PW (eds) Analysis methods for multi-spacecraft data. ISSI Scientific Report Series, vol SR-001. ESA Publ., Noordwijk, pp 185–220

Sonnerup BUÖ, Teh W-L (2008) J Geophys Res 113(A5):A05202. https://doi.org/10.1029/2007JA012718

Sonnerup BUÖ, Teh W-L (2009) J Geophys Res 114:A04206. https://doi.org/10.1029/2008JA013897

Sonnerup BUÖ, Haaland S, Paschmann G, Dunlop MW, Rème H, Balogh A (2006a) J Geophys Res 111:A05203. https://doi.org/10.1029/2005JA011538

Sonnerup BUÖ, Hasegawa H, Teh W-L, Hau L-N (2006b) J Geophys Res Space Phys 111:A09204. https://doi.org/10.1029/2006JA011717

Sonnerup BUÖ, Teh W-L, Hasegawa H (2008) Grad-Shafranov and MHD reconstructions. In: Paschmann G, Daly PW (eds) ISSI Scientific Report Series, vol SR-008. ESA Publications Division, pp 81–90

Sonnerup BUÖ, Hasegawa H, Denton RE, Nakamura TKM (2016) J Geophys Res Space Phys 121(5):4279–4290. https://doi.org/10.1002/2016ja022430

Spall JC (1998) Johns Hopkins APL Tech Dig 19:482–492

Spall JC (2003) Introduction to stochastic search and optimization: estimation, simulation, and control. Wiley, New York

Stephens GK, Sitnov MI (2021) Front Phys 9:653111. https://doi.org/10.3389/fphy.2021.653111

Stephens GK, Sitnov MI, Korth H, Tsyganenko NA, Ohtani S, Gkioulidou M, Ukhorskiy AY (2019) J Geophys Res Space Phys 124(2):1085–1110. https://doi.org/10.1029/2018JA025843

Stephens GK, Bingham ST, Sitnov MI, et al (2020) Space Weather 18(12):e2020SW002583. https://doi.org/10.1029/2020SW002583

Stephens GK, Sitnov MI, Weigel RS, et al (2023) J Geophys Res Space Phys 128:e2022JA031066. https://doi.org/10.1029/2022JA031066

Stern DP (1976) Rev Geophys Space Phys 14(2):200. https://doi.org/10.1029/RG014i002p00199

Svenningsson I, Yodradnova E, Khotyaintsev YV, André M, Cozzani G (2023) In: EGU general assembly 2023, pp EGU23–EGU8664. https://doi.org/10.5194/egusphere-egu23-8664

Swisdak M (2016) Geophys Res Lett 43:43–49. https://doi.org/10.1002/2015GL066980

Tang X, Cattell C, Dombeck J, et al (2013) Geophys Res Lett 40:2884–2890. https://doi.org/10.1002/grl.50565

Tang B-B, Li WY, Graham DB, et al (2019) Geophys Res Lett 46:3024–3032. https://doi.org/10.1029/2019GL082231

Teh W-L (2019) J Geophys Res Space Phys 124:1644–1650. https://doi.org/10.1029/2018JA026416

Teh W-L, Zenitani S (2020) Earth Space Sci 7:e2020EA001449. https://doi.org/10.1029/2020EA001449

Terasawa T, Kawano H, Shinohara I, et al (1996) J Geomagn Geoelectr 48(5–6):603–614

Tian AM, Xiao K, Degeling AW, Shi QQ, Park J-S, Nowada M, Pitkanen T (2020) Astrophys J 889:35. https://doi.org/10.3847/1538-4357/ab6296

Toledo-Redondo S, André M, Khotyaintsev YV, et al (2016) Geophys Res Lett 43(13):6759–6767. https://doi.org/10.1002/2016gl069877

Torbert RB, Russell CT, Magnes W, et al (2016a) Space Sci Rev 199(1–4):105–135. https://doi.org/10.1007/s11214-014-0109-8

Torbert RB, Vaith H, Granoff M, et al (2016b) Space Sci Rev 199:285–305. https://doi.org/10.1007/s11214-015-0182-7

Torbert RB, Burch JL, Argall MR, et al (2017) J Geophys Res Space Phys 122:11901–11916. https://doi.org/10.1002/2017JA024579

Torbert RB, Burch JL, Phan TD, et al (2018) Science 362(6421):1391–1395. https://doi.org/10.1126/science.aat2998

Torbert RB, Dors I, Argall MR, et al (2020) Geophys Res Lett 47:e2019GL085542. https://doi.org/10.1029/2019GL085542

Tóth G, Sokolov IV, Gombosi T, et al (2005) J Geophys Res 110:A12226. https://doi.org/10.1029/2005JA011126

Trattner KJ, Fuselier SA, Petrinec SM, Yeoman TK, Mouikis C, Kucharek H, Rème H (2005) J Geophys Res 110:A04207. https://doi.org/10.1029/2004JA010722

Trattner KJ, Mulcock JS, Petrinec SM, Fuselier SA (2007) J Geophys Res 112:A08210. https://doi.org/10.1029/2007JA012270

Trattner KJ, Petrinec SM, Fuselier SA, Omidi N, Sibeck DG (2012) J Geophys Res 117:A01213. https://doi.org/10.1029/2011JA017080

Trattner KJ, Burch JL, Ergun RE, et al (2017) J Geophys Res 122:11991–12005. https://doi.org/10.1002/2017JA024488

Trattner KJ, Burch JL, Ergun RE, et al (2018) J Geophys Res Space Phys 123:10177–10188. https://doi.org/10.1029/2018JA026081

Trattner KJ, Petrinec SM, Fuselier SA (2021) Space Sci Rev 217:41. https://doi.org/10.1007/s11214-021-00817-8

Trenchi L, Marcucci MF, Pallocchia G, et al (2008) J Geophys Res 113:A07S10. https://doi.org/10.1029/2007JA012774

Trenchi L, Marcucci MF, Pallocchia G, et al (2009) Mem Soc Astron Ital 80:287

Tsyganenko N (1991) Planet Space Sci 39(4):641–654. https://doi.org/10.1016/0032-0633(91)90058-I

Tsyganenko NA (1995) J Geophys Res 100:5599–5612

Tsyganenko NA, Andreeva VA (2016) J Geophys Res Space Phys 121:10,786–10,802. https://doi.org/10.1002/2016JA023217

Tsyganenko NA, Sitnov MI (2005) J Geophys Res Space Phys 110:A03208. https://doi.org/10.1029/2004JA010798

Tsyganenko NA, Sitnov MI (2007) J Geophys Res Space Phys 112:A06225. https://doi.org/10.1029/2007JA012260

Varsani A, Nakamura R, Sergeev VA, et al (2017) J Geophys Res Space Phys 122:10891–10909. https://doi.org/10.1002/2017JA024547

Vasyliunas VM (1975) Rev Geophys 13:303–336. https://doi.org/10.1029/RG013i001p00303

Webster JM, Burch JL, Reiff PH, et al (2018) J Geophys Res Space Phys 123(6):4858–4878. https://doi.org/10.1029/2018JA025245

Wellenzohn S, Nakamura R, Nakamura TKM, et al (2021) J Geophys Res Space Phys 126:e2020JA028917. https://doi.org/10.1029/2020JA028917

Wettschereck D, Aha DW, Mohri T (1997) Artif Intell Rev 11(1):273–314. https://doi.org/10.1023/A:1006593614256

Wilder FD, Ergun RE, Burch JL, et al (2018) J Geophys Res Space Phys 123:6533–6547. https://doi.org/10.1029/2018JA025529

Yang Y, Matthaeus WH, Parashar TN, et al (2017) Phys Rev E 95:061201(R). https://doi.org/10.1103/PhysRevE.95.061201

Zenitani S, Hesse M, Klimas A, Black C, Kuznetsova M (2011b) Phys Plasmas 18:122108. https://doi.org/10.1063/1.3662430

Zenitani S, Hesse M, Klimas A, Kuznetsova M (2011a) Phys Rev Lett 106:195003. https://doi.org/10.1103/PhysRevLett.106.195003

Zenitani S, Shinohara I, Nagai T (2012) Geophys Res Lett 39:L11102. https://doi.org/10.1029/2012GL051938

Zhu X, Spall JC (2002) Int J Adapt Control Signal Process 16:397–409. https://doi.org/10.1002/acs.715

Zhang H, Fu S, Pu Z, et al (2019) Astrophysical Journal 880:122. https://doi.org/10.3847/1538-4357/ab290e

Zhu X, Cohen IJ, Mauk BH, Nikoukar R, Turner DL, Torbert RB (2022) Front Astron Space Sci 9:878403. https://doi.org/10.3389/fspas.2022.878403

Authors and Affiliations

H. Hasegawa[1] · M.R. Argall[2] · N. Aunai[3] · R. Bandyopadhyay[4] · N. Bessho[5,6] · I.J. Cohen[7] · R.E. Denton[8] · J.C. Dorelli[6] · J. Egedal[9] · S.A. Fuselier[10,11] · P. Garnier[12] · V. Génot[12] · D.B. Graham[13] · K.J. Hwang[10] · Y.V. Khotyaintsev[13,14] · D.B. Korovinskiy[15] · B. Lavraud[12,16] · Q. Lenouvel[12] · T.C. Li[8] · Y.-H. Liu[8] · B. Michotte de Welle[3] · T.K.M. Nakamura[15,17] · D.S. Payne[18] · S.M. Petrinec[19] · Y. Qi[20] · A.C. Rager[6] · P.H. Reiff[21] · J.M. Schroeder[9] · J.R. Shuster[2] · M.I. Sitnov[7] · G.K. Stephens[7] · M. Swisdak[18] · A.M. Tian[22] · R.B. Torbert[23,24] · K.J. Trattner[20] · S. Zenitani[15,25]

✉ H. Hasegawa
hase@stp.isas.jaxa.jp

✉ D.B. Graham
dgraham@irfu.se

✉ Y.V. Khotyaintsev
Yuri.Khotyaintsev@rymdfysik.uu.se

1 Institute of Space and Astronautical Science, Japan Aerospace Exploration Agency, Sagamihara, Kanagawa 252-5210, Japan

2 Space Science Center, Institute for the Study of Earth, Oceans, and Space, University of New Hampshire, Durham, NH 03824, USA

3 CNRS, Ecole polytechnique, Sorbonne Université, Université Paris Sud, Observatoire de Paris, Institut Polytechnique de Paris, Université Paris-Saclay, PSL Research Univsersity, Laboratoire de Physique des Plasmas, Palaiseau, France

4 Department of Astrophysical Sciences, Princeton University, Princeton, NJ 08544, USA

5 Department of Astronomy, University of Maryland, College Park, MD 20742, USA

6 Heliophysics Science Division, NASA Goddard Space Flight Center, Greenbelt, MD 20771, USA

7 Applied Physics Laboratory, The Johns Hopkins University, Laurel, MD, USA

8 Department of Physics and Astronomy, Dartmouth College, Hanover, NH, USA

9 Department of Physics, University of Wisconsin-Madison, Madison, WI 53706, USA

10 Southwest Research Institute, San Antonio, TX, USA

11 University of Texas at San Antonio, San Antonio, TX, USA

12 Institut de Recherche en Astrophysique et Planétologie, CNRS, Université Paul Sabatier, CNES, Toulouse, France

13 Swedish Institute of Space Physics, Uppsala, Sweden

14 Department of Physics and Astronomy, Uppsala University, Uppsala, Sweden

15 Space Research Institute, Austrian Academy of Sciences, Graz, Austria

[16] Laboratoire d'Astrophysique de Bordeaux, Université Bordeaux, CNRS, Pessac, France

[17] Krimgen LLC, Hiroshima 732-0828, Japan

[18] Institute for Research in Electronics and Applied Physics, University of Maryland, College Park, MD, USA

[19] Lockheed Martin ATC, Palo Alto, CA, USA

[20] Laboratory for Atmospheric and Space Physics, University of Colorado, Boulder, CO, USA

[21] Rice Space Institute, Rice University, Houston, TX, USA

[22] Shandong Key Laboratory of Optical Astronomy and Solar-Terrestrial Environment, School of Space Science and Physics, Institute of Space Sciences, Shandong University, Weihai, Shandong 264209, People's Republic of China

[23] Southwest Research Institute, Durham, NH, USA

[24] Physics Department, University of New Hampshire, Durham, NH, USA

[25] Research Center for Urban Safety and Security, Kobe University, Kobe 657-8501, Japan

Space Science Reviews (2025) 221:81
https://doi.org/10.1007/s11214-025-01210-5

Simulation Models for Exploring Magnetic Reconnection

Michael Shay[1] · Subash Adhikari[2,1] · Naoki Beesho[3] · Joachim Birn[4] · Jörg Büchner[5] · Paul Cassak[2] · Li-Jen Chen[6] · Yuxi Chen[7] · Giulia Cozzani[8,9] · James Drake[10,11] · Fan Guo[12] · Michael Hesse[13] · Neeraj Jain[14] · Yann Pfau-Kempf[8,15] · Yu Lin[16] · Yi-Hsin Liu[17] · Mitsuo Oka[18] · Yuri Omelchenko[4,19] · Minna Palmroth[8] · Oreste Pezzi[20] · Patricia H. Reiff[21] · Marc Swisdak[11] · Frank Toffoletto[21] · Gabor Toth[7] · Richard A. Wolf[21]

Received: 11 July 2024 / Accepted: 6 August 2025 / Published online: 9 September 2025

Abstract
Simulations have played a critical role in the advancement of our knowledge of magnetic reconnection. However, due to the inherently multiscale nature of reconnection, it is impossible to simulate all physics at all scales. For this reason, a wide range of simulation methods have been crafted to study particular aspects and consequences of magnetic reconnection. This article reviews many of these methods, laying out critical assumptions, numerical techniques, and giving examples of scientific results. Plasma models described include magnetohydrodynamics (MHD), Hall MHD, Hybrid, kinetic particle-in-cell (PIC), kinetic Vlasov, Fluid models with embedded PIC, Fluid models with direct feedback from energetic populations, and the Rice Convection Model (RCM).

Keywords Plasma simulation · Magnetic reconnection · Plasma physics · Magnetosphere · Solar corona · Turbulence · Numerical methods

1 Introduction

Numerical computation has always played an important role in science. The term "computer" was used during the Renaissance to describe a person who performed mathematical calculations, and such computers were used extensively to calculate the positions of the planets. However, with the advent of digital computers last century, the role of such computation has exploded and revolutionized science in general. The study of magnetic reconnection has seen such a revolution in the last several decades as both numerical power has increased and numerical techniques have become more sophisticated.

Magnetic reconnection is considered a multiscale process because it allows physics that emerges at very small length and time scales to have global consequences in the system. A straightforward example of this large separation of scales is magnetic reconnection on the sun side of Earth's magnetosphere. In this region, magnetic field lines are finally broken on a length scale of the order of 5 km which is the electron inertial length $d_e \equiv c/\omega_{pe}$. However, the dynamical effects of this breaking of field lines include driving global convection of the magnetosphere, a system spanning 100s of Earth radii (R_E) which is hundreds of thousands of d_e. A grid scale of about a d_e over 100 Earth Radii requires about 100,000 spatial grid

Extended author information available on the last page of the article

points in only 1 dimension. Clearly, accurately resolving the physics breaking the frozen-in constraint while simulating global scales is impossible.

The impossibility of globally simulating the whole 3D system and resolving all scales has led to the generation of a wide range of simulation models, each of which has its own strengths and weaknesses. Through many decades of research, scientists have carefully crafted these models for the particular application or applications they are studying. Typically, the more realistic physics that is included in the simulation, the more computationally expensive it is. Studies of the basic physics of magnetic reconnection (Biskamp 1996) have very often used kinetic PIC simulations which include all relevant physics, but require a simplified geometry and boundary conditions. Global magnetospheric simulations include the complex boundaries associated with the solar wind and the ionosphere, but until recently were required to be fluid models due to the cost of including kinetic effects.

In this paper we will provide an overview of the primary simulation models that are currently being used to study magnetic reconnection. Please note, however, that the topic of plasma simulation is extremely complex and detailed and cannot be fully covered in a single book, much less a single article. To assist the reader, we have included a table of information for representative simulations codes associated with the types of models described in this article (See Table 1 at the end of this section). If the reader wishes to dive even deeper into a particular model, there are many references available, many of which are cited in the individual sections of this paper and in the table. There are also excellent books devoted to the subject (e.g., Büchner et al. 2003; Büchner 2023).

In the field of magnetic reconnection research, more than one system of units is used. As of this writing, one can generally say that scientists specializing in theory/simulation primarily use cgs units and scientist specializing in observational analysis use SI units. We have chosen as much as possible to use cgs units in this paper, although the section on the Rice Convection Model has been left in SI units. For an excellent description on how to convert units between cgs and SI, please see the *NRL Plasma Formulary* (Huba 1998).

In magnetic reconnection the diffusion region occurs in thin boundary layers where the physics changes, ultimately allowing magnetic topology to change (Liu et al. 2025a). Although not exhaustive, the new physics which emerges in the diffusion region can be characterized by examining Ohm's law, which comes from the fluid electron momentum equation.

$$\mathbf{E} = -\frac{\mathbf{u}\times\mathbf{B}}{c} + \eta\mathbf{J} + \frac{\mathbf{J}\times\mathbf{B}}{n_e e c} + \frac{m_e}{e^2}\frac{\partial}{\partial t}\left(\frac{\mathbf{J}}{n_e}\right) - \frac{m_e}{e^2}\left(\frac{\mathbf{J}}{n_e e}\right)\cdot\nabla\left(\frac{\mathbf{J}}{n_e}\right) - \frac{1}{n_e e}\nabla\cdot\mathbf{p}_e, \tag{1}$$

where $\mathbf{E}$ and $\mathbf{B}$ are the electric and magnetic fields, $\mathbf{J}$ is the current density, $\mathbf{u}$ is the single fluid bulk flow velocity, η is the resistivity, $\mathbf{p}_e$ is the electron pressure tensor, n_e is the number density, e is the proton charge, m_e is the electron mass and c is the speed of light. The first term on the right hand side of Eq. (1) is the ideal term, the second term is the resistive term, the third term is the hall term, the fourth and fifth terms collectively represent electron inertia and the final term is due to the electron pressure tensor.

For the organization of the paper, we choose to move generally from fluid models to kinetic models as exemplified by terms on the right hand side of Eq. (1). we start with magnetohydrodynamics (MHD) and gradually increase in physical complexity until reaching fully kinetic simulations. We end with the Rice Convection Model (RCM), a widely used model for the inner magnetosphere, which acts as an inner boundary for magnetic field lines which are reconnecting in the magnetosphere. Section 2 describes MHD – first two terms. Section 3 describes Hall MHD including electron inertia – third and fourth terms. Section 4 describes Hybrid Simulations – generally also third and fourth terms. Section 5 describes

Table 1 List of representative simulation codes discussed in this manuscript

Code	Language	Parallelization	Reference	Public
	URL			
Regional MHD	Fortran 77	None	Birn et al. (2025)	No
	https://doi.org/10.1029/2024JA033648			
Global MHD (BATSRUS)	Fortran 90	MPI, OpenMP, OpenACC	Tóth et al. (2012)	Yes
	https://github.com/SWMFsoftware/BATSRUS			
Test particle in MHD	Fortran 77	None	Birn et al. (2022)	No
	https://doi.org/10.3389/fspas.2022.908730			
Hall MHD/EMHD (F3D)	Fortran 90	MPI	Shay et al. (2004)	No
	https://doi.org/10.1063/1.1705650			
Hybrid PIC/EMHD (CHIEF)	Fortran 90, C++	MPI	Muñoz et al. (2018)	No
	-			
Global Hybrid PIC (HYPERS)	Fortran 77, C++	MPI	Omelchenko et al. (2021a)	Yes[1]
	https://ccmc.gsfc.nasa.gov/models/HYPERS-Global~2021/			
Kinetic PIC (P3D)	Fortran 90	MPI	Zeiler et al. (2002)	Yes
	https://github.com/spudam/P3D-PLASMA-PIC			
Kinetic PIC (VPIC)	C++	MPI / OpenMP / Kokkos	Bowers et al. (2008), Bird et al. (2021)	Yes
	https://github.com/lanl/vpic / https://github.com/lanl/vpic-kokkos			
Kinetic PIC (ExPIC)	Fortran 77	MPI	Bessho et al. (2019)	No
	https://doi.org/10.1029/2019GL083397			
Embedded Kinetic PIC (FLEKS)	C++	MPI	Chen et al. (2023)	Yes
	https://github.com/SWMFsoftware/FLEKS			
Kglobal	Fortran 90	MPI	Drake et al. (2019)	No
	-			
Hybrid Vlasov (HVM)	Fortran 90, Fortran 77	MPI, OpenMP, OpenACC, CUDA	Valentini et al. (2007)	No
	https://doi.org/10.1016/j.jcp.2007.01.001			
Global Hybrid Vlasov (Vlasiator)	C++	MPI, OpenMP	Palmroth et al. (2018a), Ganse et al. (2023)	Yes
	https://github.com/fmihpc/vlasiator			
Inner Magnetosphere (RCM)	Fortran 95	MPI	Toffoletto et al. (2003)	Yes[2]
	-			

[1] Available at NASA CCMC

[2] By request

kinetic particle-in-cell simulations – the physics of the fifth term is added, although it arises from the collective effects of many individual electrons. Section 6 describes embedding PIC codes into fluid models like MHD. Section 7 describes Kglobal, an MHD model which self consistently evolves energetic particles. Section 8 describes kinetic Vlasov models. And Section 9 describes the Rice Convection Model.

2 MHD

2.1 Equations of MHD

Magnetohydrodynamics (MHD) is the simplest fluid model used to study large scale plasma dynamics. MHD is based on the assumption that the characteristic length and time scales of the system under study are much larger than the length and time scales of the plasma species, usually Debye length (λ_D) or gyroradius and gyroperiod. Therefore, MHD represents the slow evolution of plasmas, often electrons and ions as a single fluid. The macroscopic behavior of the fluid in presence of a magnetic field is described by MHD using hydrodynamics and Maxwell's equations.

Let us consider a fluid (in this case a plasma with ions and electrons), moving with a flow velocity $\mathbf{u}$, characterized by a mass density $\rho = m_i n_i$ where m_i is the mass of protons ($m_i \gg m_e$) and n_i is the number density of protons (with quasi-neutrality $n_i = n_e = n$), thermal pressure p, and a magnetic field $\mathbf{B}$. The evolution of these fields in space and time are governed by the MHD equations given by

$$\frac{\partial \rho}{\partial t} = -\nabla \cdot (\rho \mathbf{u}), \tag{2}$$

$$\rho \frac{\partial \mathbf{u}}{\partial t} + \rho(\mathbf{u} \cdot \nabla)\mathbf{u} = -\nabla p + \frac{1}{4\pi}(\nabla \times \mathbf{B}) \times \mathbf{B} + \nu \nabla^2 \mathbf{u}, \tag{3}$$

$$\frac{d}{dt}\left(\frac{p}{\rho^\gamma}\right) = 0, \tag{4}$$

$$\frac{\partial \mathbf{B}}{\partial t} = \nabla \times (\mathbf{u} \times \mathbf{B}) + \frac{\eta c^2}{4\pi} \nabla^2 \mathbf{B}, \tag{5}$$

where γ is the adiabatic index (usually 5/3), ν is the dynamic viscosity, η is the resistivity and $\eta c^2/4\pi$ collectively is known as the magnetic diffusivity. Here Eq. (2) is the continuity equation representing conservation of mass density, Eq. (3) is the momentum conservation equation, Eq. (4) is the simple adiabatic gas equation representing the conservation of energy and Eq. (5) is the induction equation, where the first term on the right is the advection term and the second one represents diffusion. Note, as a result of MHD approximations, the displacement current term is omitted in the induction equation. In an ideal situation, there are no dissipative processes and therefore ν and $\eta = 0$ gives ideal MHD equations. Studies have shown that ideal MHD description is a very good approximation to study dynamical properties of strongly magnetized plasmas.

2.2 Regional MHD Simulations

Several large-scale MHD approaches do not model the entire magnetosphere but only sections of it, such as certain magnetopause regions (dealt with elsewhere in this volume) or the magnetotail. Here we focus particularly on the magnetotail. The basic numerical approach used in regional MHD simulations is essentially similar to that used in (some) global simulations. It is typically based on explicit finite difference methods to solve the MHD equations. Minor differences might exist in adding resistive terms, which are usually necessary in local MHD to initiate reconnection. On the other hand, many global models use finite volume approaches in part due to the complex geometries involved (e.g., Powell et al. 1999).

The main difference, however, consists of the setup or initialization. Whereas global simulations typically involve of period of interaction with the solar wind to create a realistic magnetotail, local MHD simulations generally start from some equilibrium or near-equilibrium that models the stretched magnetotail (e.g., Schindler 1972). This approach provides more flexibility in treating different scenarios, for instance, varying the tail flaring between y and z (Birn and Hesse 2000) or including a local B_z hump (Merkin and Sitnov 2016; Birn et al. 2018). This flexibility has also proven useful in PIC simulations that go beyond the commonly used initial 1D Harris sheet, most notably in addressing the holy grail of reconnection onset (e.g., Liu et al. 2014; see also Liu et al. 2025a).

On the other side, interactions with the ionosphere or the solar wind are incorporated only in some ad hoc fashion, if at all. Regional magnetotail MHD simulations therefore have been most successful in treating dynamic tail phenomena on relatively short time scales that are typically substorm related. The successes include

1. The demonstration that X-line formation and plasmoid ejection can be part of a 2D or 3D tearing-type instability of the tail (e.g., Birn and Hones 1981).
2. The demonstration that the build-up of the substorm current wedge (SCW), involving dipolarization and Region-1-type field-aligned currents (McPherron et al. 1973), can be due to the braking and azimuthal diversion of earthward flow from a near-tail reconnection site (Birn and Hesse 1991; Scholer and Otto 1991). This basic picture has been modified more recently, most notably by the addition of a Region-2 current system connecting to the ionosphere at lower latitude (Birn and Hesse 2014; Kepko et al. 2015), in agreement with observations (Sergeev et al. 2014). While the buildup of the SCW in the simulations is based on the shear and vorticity of the earthward flow, the persistence of the currents relies on the changes of the magnetic flux and pressure patterns brought about by the severance of a plasmoid and the resulting los of entropy and redistribution of the pressure. It is noteworthy that these features can be, and have been, found also in global simulations. In the regional simulations, however, they arise as consequences of an instability without involvement of external driving or feedback from the ionosphere. This would be harder to extract from the global simulations.
3. Regional MHD simulations have also been used to address the evolution prior to the onset of reconnection in the tail, demonstrating, specifically, the formation of a thin concentrated current sheet embedded in the near-tail plasma sheet. These approaches have included interaction with the solar wind in two complementary, ad-hoc ways. In one approach magnetic flux is added to the tail lobes (Birn and Schindler 2002), the other is based on low-latitude magnetic flux reduction from convection around the Earth toward the dayside (Hsieh and Otto 2014, 2015). Both mechanisms are expected from solar wind interaction. For more details, see the review by Sitnov et al. (2019).
4. Recently, regional MHD simulations have also demonstrated that the magnetotail may become unstable even under ideal 2D MHD constraints, when it includes a region of

inverse (i.e. tailward) gradient of the normal magnetic field B_z (denoted 'B_z hump' instability; (Merkin and Sitnov 2016; Birn et al. 2018)).

2.3 Global MHD

Global MHD models representing the (outer) magnetosphere of Earth typically extend around $100 - 200\,R_E$ in the flank and tail directions, and around $30\,R_E$ (beyond the bow shock) towards the Sun, where the solar wind is coming from. When, occasionally, the solar wind becomes sub-Alfvenic, the bow shock disappears and Alfvén wings form. In this case the upstream boundary has to be moved much further to minimize the boundary effects. It is computationally very demanding to obtain an accurate solution in such a large domain while resolving various structures such as the current sheets at the dayside magnetopause and in the tail. There are various approaches to overcome this difficulty, including the block-adaptive grid of the BATSRUS code (Powell et al. 1999; Tóth et al. 2012), the stretched Cartesian grid of OpenGGCM (Raeder et al. 1997) or the stretched spherical grid of the Lyon-Fedder-Mobarry (LFM) code (Lyon et al. 2004).

A general issue associated with all MHD simulations is preserving $\nabla \cdot \mathbf{B} = 0$ (e.g., Tóth et al. 2012). This issue also creates issues associated with the upstream boundary condition. Typically, we have observations at a single point near L1, a Lagrange point in between the Sun and the Earth, and assume that the solar wind and interplanetary magnetic field (IMF) have no variation in the transverse direction. If B_x varies, and it certainly does, these assumptions lead to a finite $\nabla \cdot \mathbf{B} = \partial B_x / \partial x$ propagating into the domain. There are various approaches to handle this situation. One is to ignore the problem and propagate the finite $\nabla \cdot \mathbf{B}$ with the flow using some variation of the 8-wave scheme (Powell 1994). Another common approach is to set B_x to a constant value, for example 0. Finally, one can relax the condition that the transverse gradients in the y and z directions are zero, and guess those gradients from the temporal evolution of B_x. Unfortunately this is an underspecified problem, so there is no unique solution. A typical approach is to smooth B_x in time, and apply some minimum variance constraint. While theoretically nice, in practice the minimum variance approach does not work great. The likely reason is that the magnetic field is turbulent, so local changes in B_x are not representative of the large scale tilt of the propagation plain.

The inner boundary conditions are usually applied at a sphere of radius 1.5 R_E to 3 R_E surrounding the Earth. One can use a semi-empirical electrodynamic solver (Ridley et al. 2004), or a fully empirical model (Weimer 1996, 2001) to calculate the $\mathbf{E} \times \mathbf{B}$ drift velocity at the inner boundary. Another important use of an electrodynamic solver is that it can also provide the $\mathbf{E} \times \mathbf{B}$ drift to an inner magnetosphere model (Wolf et al. 1982; Toffoletto et al. 2003; Buzulukova et al. 2010; Liemohn et al. 2001; Jordanova et al. 1994; Zaharia et al. 2006), which can calculate realistic ring current and associated pressure (and density) in the closed field line region during geomagnetic storms (Liemohn et al. 2018). The global MHD model can then relax its pressure (and density) towards the values supplied by the inner magnetosphere model. In return, the global MHD model can supply the plasma boundary conditions for the inner magnetosphere model at the edge of the closed field region as well as the magnetic field configuration (De Zeeuw et al. 2004; Meng et al. 2013).

Global models can properly represent the overall dynamics of the interaction of the solar wind with the magnetosphere, including the formation of the bow shock and the magnetopause, as well as the main current sheet in the magnetotail. Magnetic reconnection will happen on the dayside magnetopause and in the magnetotail in agreement with the theory of the Dungey cycle (Dungey 1961). The magnetic reconnection in the MHD simulation is not represented by the actual kinetic physics but it is approximated by numerical diffusion or artificial resistivity. Despite these caveats, global MHD models generate reconnection sites where the magnetic field changes sign and the reconnection rate is approximately correct (see Appendix A of Wang et al. (2022a)).

When the grid is relatively coarse, the numerical diffusion will easily adjust to reconnect the incoming magnetic flux carried by the solar wind. For a constant solar wind and IMF driving the simulation will settle to a steady state solution. Using fine grids in combination with low dissipation numerical methods can lead to a more dynamic reconnection process in the model. On the dayside, simulations can produce Flux Transfer Events (Raeder 2006) and in the tail flux ropes can be produced even by ideal MHD simulations. If the flux ropes in the tail are triggered by sign changes of the IMF B_z, the MHD simulations can match observations very well. A more challenging problem is reproducing substorms and sawtooth events. MHD models cannot do this well (Haiducek et al. 2020), and one needs to add either ionospheric outflow to regulate the reconnection rate (Brambles et al. 2013; Zhang et al. 2020), or kinetic reconnection physics (Wang et al. 2022a) to produce the typical spatial and temporal scales of sawtooth oscillations.

2.4 Test Particles in MHD Simulations

Test particle approaches consist of tracing charged particle orbits in electromagnetic fields that are either prescribed in some plausible fashion or obtained from a simulation that typically does not contain individual particle information, most commonly based on MHD. The approach bridges the gap between large-scale MHD and small-scale particle simulations. In contrast to the latter, it can treat together realistic 3D space and large evolution time scales and realistic electron mass. However, it is not self-consistent and relies on whether the MHD model or the postulated **E**, **B** fields capture the main physics. But that may also be considered an advantage as it permits studying the effects of large-scale fields in isolation.

In the magnetospheric context ions are usually treated by integration of the full orbit

$$\frac{D\mathbf{u}}{Dt} = \frac{e}{m}\left(\mathbf{E} + \frac{1}{\gamma c}\mathbf{u} \times \mathbf{B}\right). \tag{6}$$

Here γ is the relativistic factor, which may be more relevant for electrons, $\mathbf{u} = \gamma \mathbf{w}$, where $\mathbf{w}$ is the particle velocity, c is the speed of light and $D/Dt = \partial/\partial t + \mathbf{w} \cdot \nabla$ denotes the derivative along the full orbit. This equation is typically evolved using a high order Runge-Kutta (Press et al. 1992) or the Boris method (Birdsall and Langdon 1991). Full integration of Equation (6) over extended orbits is not practical for electrons, as it is more time consuming and might accumulate too large errors. Also, the adiabatic drift approximation, based on conservation of the magnetic moment μ, is valid over larger areas in the magnetosphere and can be adequate for identifying typical acceleration mechanisms (e.g., Delcourt and Sauvaud 1994; Li et al. 1998; Zaharia et al. 2000; Gabrielse et al. 2012).

However, when the full history of electron orbits is considered; this may involve encounters of the reconnection site and low magnetic field, or high curvature regions, where the conservation of μ breaks down, and full orbit integration is required. Consequently, several codes have been developed that involve a transition between full orbits, integrated by Eq. (6), and drift orbits (e.g., Birn et al. 2004; Schriver et al. 2005; Ashour-Abdalla et al. 2011; Sorathia et al. 2017). The drift is described by the guiding center drift velocity (e.g., Birn et al. 2004)

$$\mathbf{v}_d = \mathbf{v}_E - \frac{\mu c}{\gamma e}\frac{\mathbf{B} \times \nabla B}{B^2} - \frac{\gamma m_e c v_\parallel}{e}\frac{\mathbf{B}}{B^2} \times \frac{d\mathbf{b}}{dt} - \frac{m_e c}{e}\frac{\mathbf{B}}{B^2} \times \frac{d(\gamma \mathbf{v}_E)}{dt} \tag{7}$$

where μ is the (relativistic) magnetic moment, $\mathbf{v}_E = \mathbf{E} \times \mathbf{B}/B^2$ and $\mathbf{b} = \mathbf{B}/B$. In addition, the field-aligned velocity is advanced by

$$\frac{du_{\parallel}}{dt} = -\frac{e}{m_e} E_{\parallel} - \frac{\mu}{\gamma m_e} \frac{\partial B}{\partial s} - \left(\mathbf{u}_E + \mathbf{u}_{\nabla B}\right) \cdot \frac{d\mathbf{b}}{\partial t} \tag{8}$$

where now $\mathbf{u} = \gamma \mathbf{v}$ and $\mathbf{v} = \mathbf{v}_E + \mathbf{v}_d + \mathbf{v}_{\parallel}$ describes the guiding center velocity, $\mathbf{v}_{\nabla B}$ is the grad B drift, given by the second term on the right side of Eq. (7), and d/dt is the derivative along the guiding center path. The transition between full orbit and drift orbit is typically determined from an adiabaticity criterion that is based on the ratio between the local field line curvature radius and the gyro radius based on the local magnetic field strength (e.g., Buchner and Zelenyi 1989). On the switch from drift to full orbit a phase has to be generated, which is typically chosen randomly. Although this can alter individual orbits and make them not reversible, it was found to have no significant effect on general conclusions about sources and properties of distributions (e.g., Birn et al. 2004).

Two different techniques are used, tracing particle motion either forward or backward in time. Forward tracing requires larger numbers of particles, sometimes comparable to those in PIC simulations, to obtain sufficient numbers at the points of interest (e.g., Scholer and Jamitzky 1987; Sachsenweger et al. 1989; Peroomian and El-Alaoui 2008; Ukhorskiy et al. 2017, 2018). However, since the particles are not interacting this approach is even more suitable for parallel processing than full particle simulations. In principle, this approach can also include wave scattering and collisions (albeit in an ad-hoc non self-consistent manner) to add to the simple collisionless advance.

Backward tracing is generally based on Liouville's theorem of the conservation of phase space density f to map f from source locations to the final location of interest (e.g., Curran and Goertz 1989; Birn and Hesse 1994; Birn et al. 2004). It requires fewer orbits to identify properties at selected final locations, but relies on the validity of Liouville's theorem, i.e. the absence of collisions. Backward tracing permits an easier identification of different sources contributing to the final population. Thus, sometimes a combination of both techniques is employed (e.g., Ashour-Abdalla et al. 2011).

Further complications are related to the use of MHD simulation results. Since the fields are given only on a finite grid, they have to be interpolated in space and time.

1. The advance of the drift equations (7) and (8) requires a third order spatial interpolation in B for continuous transition between grid cells, which could lead to spurious maxima or minima. This can be avoided, however, by employing a monotonicity algorithm (Hyman 1983; Birn et al. 2004).
2. Simple interpolation of the electric field could also yield spurious parallel components. This can be avoided, however, by various techniques, for instance, by interpolating $E_{\parallel}$ and $E_{\perp}$ separately (e.g., Birn et al. 2022).

3 Hall MHD

3.1 Introduction

The ideal-MHD model, as discussed in Sect. 2.1, is well-suited for magnetized plasmas when the dynamics is slow compared to the gyration time of charged particles around the magnetic fields and the length scales over which quantities vary is much larger than the gyroradius of the charged particles. However, going back many years in the study of neutral fluids, fluid models can lead to incorrect and paradoxical results at boundaries layers [e.g., d'Alembert's paradox (Sect. 4.7 of Choudhuri 1998)]. In a magnetized plasma, these problems can occur where plasmas of two different origins abut against each other (such as at Earth's magnetopause), at shocks and discontinuities such as Earth's bow shock, and at localized regions where the magnetic field goes to zero, such as in the solar corona near sunspots.

Magnetic reconnection, in particular, occurs at a boundary layer at a region where at least two components of the magnetic field go to zero, so it is a key example of a physical process that cannot be faithfully modeled by ideal-MHD. Often in numerical simulations, ideal-MHD is used anyway, with numerical dissipation allowing reconnection to occur with the hope that it mimics the actual process. Another approach employs resistivity to model the effect of collisions; this is a useful approach in systems for which collisions are dynamically relevant, but many settings where reconnection occurs – especially in space and the solar corona – are weakly collisional or effectively collisionless (Priest and Forbes 2000; Cassak and Shay 2012). There are examples where either approach can be good enough for the questions being asked. For other questions that rely on a faithful representation of the physics in the regions where ideal- and resistive-MHD break down, a new model is necessary. In this section, we discuss a number of approaches within the fluid description that are used to go beyond ideal- and resistive-MHD simulations. In later sections, simulation techniques using the kinetic theory of gases are discussed. There are previous review papers discussing Hall-MHD and numerical approaches (Vasyliunas 1975; Huba 1995, 2003; Gómez 2006).

3.2 The Hall-MHD Model

The equations of Hall-MHD are similar to those of MHD with one key difference. In resistive-MHD, Ohm's law is given (in cgs units) by

$$\mathbf{E}+\frac{\mathbf{u}\times\mathbf{B}}{c}=\eta\mathbf{J}, \tag{9}$$

where $\mathbf{E}$ and $\mathbf{B}$ are the electric and magnetic fields, $\mathbf{u}$ is the single fluid bulk flow velocity, η is the resistivity, and $\mathbf{J}=(c/4\pi)\nabla\times\mathbf{B}$ is the current density; in ideal-MHD, η is set to zero. To go beyond this model, we revisit where Ohm's law comes from.

The equation of motion of an electron fluid (i.e., Newton's 2nd law) in a fully ionized plasma (in cgs units) is (Braginskii 1965)

$$m_e\frac{d\mathbf{u}_e}{dt}=-e\left(\mathbf{E}+\frac{\mathbf{u}_e\times\mathbf{B}}{c}\right)-\frac{1}{n_e}\nabla\cdot\mathbf{p}_e+\mathbf{R}_e, \tag{10}$$

where m_e is the electron mass, $\mathbf{u}_e$ is the electron bulk flow velocity, $-e$ is the electron charge, n_e is the electron density, $\mathbf{p}_e$ is the electron pressure which we write more generally as a tensor for now, and $\mathbf{R}_e$ represents the rate of change of momentum resulting from collisions between electrons and other electrons or other charged or neutral particles in the plasma. There are rigorous ways to determine the role of collisions (e.g., Braginskii 1965) that we do not employ here. Instead, we use the often used simpler approach that assumes $\mathbf{R}_e=m_e\nu_{ei}(\mathbf{u}_i-\mathbf{u}_e)$, where ν_{ei} is the electron-ion collision frequency, and $\mathbf{u}_i$ is the ion bulk flow velocity, and for simplicity we assume the plasma has only electrons and ions (it is fully ionized).

To recover the resistive-MHD Ohm's law from this equation, first the "electron inertia term" $m_e d\mathbf{u}_e/dt$ and the "electron pressure gradient term" $-(\nabla \cdot \mathbf{p}_e)/n_e$ are ignored for reasons we return to in Sect. 3.6. Second, the single fluid bulk flow velocity used in MHD is $\mathbf{u} = (m_i n_i \mathbf{u}_i + m_e n_e \mathbf{u}_e)/(m_i n_i + m_e n_e)$ and the current density $\mathbf{J} = n_i q_i \mathbf{u}_i - n_e q_e \mathbf{u}_e \simeq n_e q_e (\mathbf{u}_i - \mathbf{u}_e)$, where the latter form uses the assumption of quasi-neutrality $n_i q_i - n_e q_e \simeq 0$ with q_i and q_e being the charge of ions and electrons respectively. Using these expressions to write $\mathbf{u}_e$ in terms of $\mathbf{u}$ and $\mathbf{J}$ gives $\mathbf{u}_e = \mathbf{u} - (\mathbf{J}/ne)[m_i n_i/(m_i n_i + m_e n_e)] \simeq \mathbf{u} - (\mathbf{J}/ne)$, where in the latter form, we use the approximation that $m_i \gg m_e$, since it is at least 1836 in an electron-ion plasma. Using these approximations in Eq. (10) and dividing by e gives

$$\mathbf{E} + \frac{\mathbf{u} \times \mathbf{B}}{c} = \frac{\mathbf{J} \times \mathbf{B}}{n_e e c} + \eta \mathbf{J}, \tag{11}$$

where the resistivity η is defined as $m_e \nu_{ei}/n_e e^2$. If one additionally ignores the $\mathbf{J} \times \mathbf{B}/n_e e c$ term, what remains is the resistive-MHD Ohm's law in Eq. (9). If instead, one ignores the resistive term $\eta \mathbf{J}$, the result is

$$\mathbf{E} + \frac{\mathbf{u} \times \mathbf{B}}{c} = \frac{\mathbf{J} \times \mathbf{B}}{n_e e c}. \tag{12}$$

The term on the right is called the "Hall electric field" $\mathbf{E}_H$ (or simply the "Hall term"), and Eq. (12) is called the "Hall-MHD Ohm's law". Simply coupling this equation to the rest of the ideal-MHD equations gives the Hall-MHD model:

$$\frac{\partial \rho}{\partial t} + \nabla \cdot (\rho \mathbf{u}) = 0, \tag{13}$$

$$\rho \left[\frac{\partial \mathbf{u}}{\partial t} + (\mathbf{u} \cdot \nabla) \mathbf{u} \right] = -\nabla p + \frac{\mathbf{J} \times \mathbf{B}}{c}, \tag{14}$$

$$\frac{\partial p}{\partial t} + (\mathbf{u} \cdot \nabla) p = -\gamma p (\nabla \cdot \mathbf{u}) \tag{15}$$

$$\frac{\partial \mathbf{B}}{\partial t} = -c \nabla \times \mathbf{E}, \tag{16}$$

$$\mathbf{E} + \frac{\mathbf{u} \times \mathbf{B}}{c} = \frac{\mathbf{J} \times \mathbf{B}}{n_e e c} \tag{17}$$

$$\nabla \times \mathbf{B} = \frac{4\pi \mathbf{J}}{c} \tag{18}$$

with the auxiliary equation $\nabla \cdot \mathbf{B} = 0$, and where γ is the single fluid ratio of specific heats, typically taken to be 5/3. Note n_e in Eq. (17) is related to the MHD mass density via $\rho = m_i n_i + m_e n_e$, so $n_e = \rho/(m_i e/q_i + m_e) \simeq Z\rho/m_i$, where we again assume $m_i \gg m_e$, we use quasi-neutrality to write $q_i n_i \simeq e n_e$, and define $Z = q_i/e$ as the degree of ionization. With these assumptions, Eqs. (13) - (18) form a closed set of equations, and therefore can be used to model physical systems. Technically, these equations actually give the "ideal Hall-MHD model" since resistivity is not retained here.

It is important to note a confusing aspect of these equations. The Hall term is proportional to $\mathbf{J} \times \mathbf{B}$, and Eq. (14) also contains a term including $\mathbf{J} \times \mathbf{B}$. It is tempting to draw relations between the two terms because of this outward similarity, but this should not be done. The two terms have different dimensions: $\mathbf{J} \times \mathbf{B}/c$ is a force density and $\mathbf{J} \times \mathbf{B}/n_e e c$ is an electric field. They have completely different manifestations and impacts on the physics, and therefore actually are not related despite their similar forms.

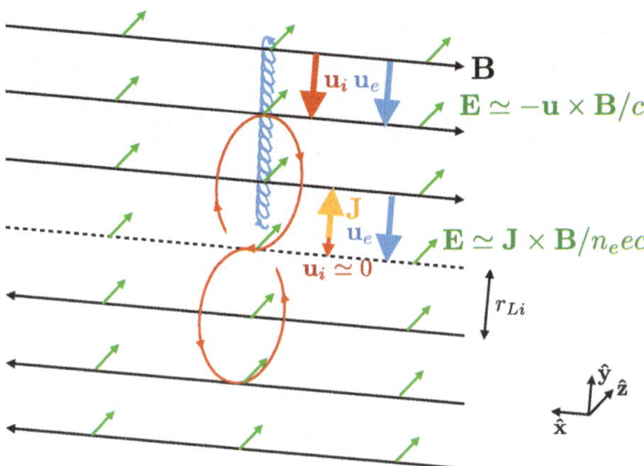

Fig. 1 **Physics of the Hall effect in reversing magnetic fields.** A reversing anti-parallel magnetic field in the $\pm\hat{\mathbf{x}}$ direction is drawn using black arrows, with the neutral line as the dashed line. A uniform electric field in the $\hat{\mathbf{z}}$ direction is drawn in green. The smaller blue trajectory is that of an electron $E \times B$ drifting towards the neutral line with bulk velocity $\mathbf{u}_e$. An ion $E \times B$ drifting far from the neutral line has bulk ion velocity $\mathbf{u}_i$ identical to $\mathbf{u}_e$, so the electric field in this region is given by $\mathbf{E} = -\mathbf{u} \times \mathbf{B}/c$, where $\mathbf{u}$ is the single fluid (MHD) bulk flow velocity. Within an ion gyroradius r_{Li} of the neutral line, the electron continues with bulk flow $\mathbf{u}_e$, while the ion demagnetizes (in the red trajectory) and $\mathbf{u}_i$ becomes small. In this region, there is a non-zero current density $\mathbf{J}$ (orange), and the electric field in this region is predominantly given by the Hall electric field $\mathbf{E} = \mathbf{J} \times \mathbf{B}/n_e ec$

3.3 Hall-MHD Physics

The only difference between the ideal-MHD and Hall-MHD models is the Hall electric field, and here we investigate the physics introduced by this term. Let $\tilde{L}$ be a characteristic length scale over which the plasma properties vary, and let $\tilde{r}_{Li}$ be the characteristic (Larmor) radius of the ions as they gyrate around a magnetic field of characteristic strength $\tilde{B}$. The Hall electric field is small and can be neglected if $\tilde{L} \gg \tilde{r}_{Li}$, and doing so brings us back to ideal-MHD. It is important to retain the Hall electric field, and therefore use Hall-MHD, when $\tilde{L} \lesssim \tilde{r}_{Li}$.

To see why structure below the ion gyroscale gives rise to the Hall effect, consider a magnetic field $\mathbf{B}$ that reverses direction over a scale $\tilde{r}_{Li}$ or less, but on larger scales than the characteristic electron gyroradius $\tilde{r}_{Le}$. For specificity, consider a magnetic field pointing in the $\pm\hat{\mathbf{x}}$ direction, as sketched as the black arrows in Fig. 1. The neutral line, where the magnetic field strength vanishes, is the dashed black line. Suppose also there is a uniform electric field $\mathbf{E}$ pointing everywhere in the $\hat{\mathbf{z}}$ direction, as sketched as the green arrows. (We use a reversed magnetic field and uniform electric field for illustrative purposes due to its relation to the reconnection process, but the Hall effect is important for any magnetic field configuration that varies on $\tilde{r}_{Li}$ scales.)

At distances from the magnetic field reversal that greatly exceed $\tilde{r}_{Li}$, ions and electrons undergo the $E \times B$ drift that gives rise to a bulk flow towards the neutral line; the bulk flow velocity of ions and electrons, $\mathbf{u}_i$ and $\mathbf{u}_e$, respectively, are identical. The $E \times B$ drift of the electrons above the neutral line is sketched as the blue curve. In the region farther from the neutral line than $\tilde{r}_{Li}$, the electric field is given by $\mathbf{E} = -\mathbf{u} \times \mathbf{B}/c$, where $\mathbf{u} = \mathbf{u}_i = \mathbf{u}_e$ is the MHD bulk flow velocity.

As the ions reach a distance from the neutral line that is equal to its gyroradius, the ions cross the neutral line and are immersed in a magnetic field pointing in the opposite direction.

Their gyromotion changes direction, and they make figure 8 orbits, sketched as the red curve. (They also accelerate in the $\hat{\mathbf{z}}$ direction due to the electric field, but this is omitted from the sketch.) Consequently, their bulk velocity becomes small, $\mathbf{u}_i \simeq 0$.

Since we assumed $\tilde{L} > \tilde{r}_{Le}$, the electrons have a smaller gyroradius and do not see the magnetic field reversal, so they continue to undergo the $E \times B$ drift towards the neutral line. The key is that the ions and electrons are undergoing different dynamics between distances from the neutral line of $\tilde{r}_{Li}$ and $\tilde{r}_{Le}$!

In this region, the difference in the bulk motion between ions and electrons implies there is a net current density $\mathbf{J}$, sketched as the orange arrow, and called the Hall current. The current density is perpendicular to the magnetic field $\mathbf{B}$. Then, between $\tilde{r}_{Li}$ and $\tilde{r}_{Le}$ from the neutral line, the electric field is given by the Hall electric field $\mathbf{E}_H = \mathbf{J} \times \mathbf{B}/n_e ec$. This exemplifies why the Hall electric field is important between ion and electron gyroscales.

This situation in a plasma is analogous to the Hall effect in condensed matter physics, where it was originally discovered by Edwin Hall (a graduate student) in 1879 (Hall et al. 1879). It has extensive applications in that field of study. The derivation generalizing shear Alfvén waves to include the Hall effect happened as early as 1954 by Jim Dungey (Dungey 1954), the same person who first understood magnetic reconnection and gave the process its name.

3.4 Linear Waves in Ideal Hall-MHD

It would take us too far afield to elucidate how the Hall electric field modifies all the physics of ideal-MHD. Rather, we highlight one important example – linear waves. In ideal-MHD, there are three propagating linear waves available to a uniform plasma: the shear Alfvén wave and the fast and slow magnetosonic waves. The Hall electric field introduces two wave modes that become important between ion and electron gyroscales – the whistler wave and the kinetic Alfvén wave.

3.4.1 The Whistler Wave

First, we consider transverse waves propagating along a uniform background magnetic field of strength B_0, which without loss of generality we take to be in the y direction, in a plasma of equilibrium mass density $\rho_0 \simeq m_i n_i$ and with no equilibrium bulk flow or current $\mathbf{u}_0 = 0$ and $\mathbf{J}_0 = 0$. The wave vector $\mathbf{k}$ is also in the y direction. Linearizing the Hall-MHD equations (13)-(18) about the given equilibrium with $\mathbf{k}$ parallel to $\mathbf{B}_0$ and solving for the dispersion relation gives

$$\omega^2 = k^2 v_{A0}^2 \left(1 + \frac{k^2 d_{i0}^2}{2} + \sqrt{k^2 d_{i0}^2 + \frac{k^4 d_{i0}^4}{4}} \right), \tag{19}$$

where ω is the wave frequency, $v_{A0} = B_0/(4\pi\rho_0)^{1/2}$ is the Alfvén speed, and $d_{i0} = (m_i c^2/4\pi n_0 q_i^2)^{1/2}$ is the ion inertial length.

In the limit of $kd_{i0} \to 0$, this dispersion relation reduces to $\omega^2 \to k^2 v_{A0}^2$, which is simply the shear Alfvén wave from ideal-MHD. In the other extreme, consider the limit $kd_{i0} \to \infty$. The 1 in Eq. (19) becomes negligible compared to $k^2 d_{i0}^2/2$ outside the square root, and the $k^2 d_{i0}^2$ is negligible compared to $k^4 d_{i0}^4/4$ inside the square root, so to low order the dispersion relation becomes

$$\omega^2 \to k^4 v_{A0}^2 d_{i0}^2 \qquad (\text{as } kd_{i0} \to \infty). \tag{20}$$

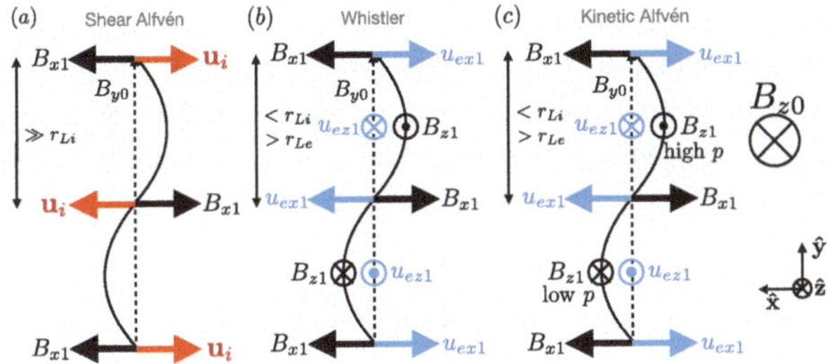

Fig. 2 Sketch of how the Hall effect impacts linear waves in MHD. The wave structure for (a) shear Alfvén waves, (b) parallel propagating whistler waves, and (c) kinetic Alfvén waves. The equilibrium magnetic field B_{y0} is the dashed black arrow. The perturbed magnetic field $\mathbf{B}_1$ are the black arrows. The perturbed ion bulk flow $\mathbf{u}_i$ are the red arrows, which occur in (a) because the wavelength is much larger than the ion gyroradius. In (b) and (c), the wavelength is at or below ion gyroscales, so the bulk flow is due to electron motion in the blue arrows. In (c), there is a large out-of-plane equilibrium magnetic field B_{z0}, so the magnetic perturbation B_{z1} changes the magnetic pressure which to first order and requires a change to the gas pressure p

This is the dispersion relation for the so-called "whistler wave."

Physically, the whistler wave is the sub-ion gyroscale counterpart of the shear Alfvén wave. The plasma properties in a shear Alfvén wave are sketched in Fig. 2(a). The equilibrium magnetic field B_{y0} is the dashed black arrow. A magnetic perturbation transverse to the equilibrium B_{x1} is the large black arrow. Since the size of the wave is far larger than the ion gyroradius, the frozen-in ions and electrons feel a restoring force analogous to plucking a guitar string, generating bulk flows in the x direction (shown for ions as the red arrows). The shear Alfvén wave is a linearly polarized wave.

The whistler wave, sketched in panel (b), occurs when the magnetic field varies between ion and electron gyroscales. Then, the ions are not frozen-in and do not respond to the plucked magnetic field line, but the electrons are frozen in and feel a restoring force. The x component of the perturbed electron flow u_{ex1} is sketched as the blue arrows. Since the electrons have a bulk velocity but the ions do not, this means there is a net current density, which induces a oscillating magnetic field out of the plane, labeled as B_{z1}. Similarly, the varying B_{x1} requires a u_{ez1} to sustain the current. This turns the wave into a circularly polarized wave. The polarization is right handed, *i.e.,* a receiver sees the magnetic field rotate in a counterclockwise direction.

There is a crucial difference between the shear Alfvén wave and the whistler wave. The sheer Alfvén wave has a phase speed $\omega/k = v_{A0}$, which is a constant independent of ω and k, so a shear Alfvén wave travels at the same speed regardless of its wavelength. It is an example of a non-dispersive wave that retains its waveform as it propagates (like a light wave in vacuum or a sound wave in a neutral fluid). In contrast, the phase speed for the whistler wave, from Eq. (20), is $\omega/k = kv_{A0}d_{i0}$. Thus, the phase speed is faster for shorter wavelength waves. This is an example of a dispersive wave, since a wave packet does not retain its shape. The dispersive nature is what gives the wave its name; it makes a characteristic "whistle" from high frequencies descending to low frequencies when it is detected at a location away from where it was generated. For example, when a lightning strike in one hemisphere excites a whistler wave on Earth's magnetic field, the high frequencies arrive before the low frequencies in the other hemisphere.

This analysis reveals the critical scale at which the Hall effect becomes important for transverse bending of the magnetic field. The three terms in Eq. (19) proportional to d_{i0} are absent if the Hall term is left out of the governing equations (*i.e.,* in ideal-MHD). By comparing the first term to the other terms, we see they become important when kd_{i0} is on the order of 1. This means that the Hall term becomes important at length scales below d_{i0}.

The ion inertial scale is related to a form of the ion gyroradius. To see this, note that an ion moving at the Alfvén speed v_{A0} that gyrates around a magnetic field of strength B_0 has a gyroradius r_{Li} given by

$$r_{Li} = \frac{v_{A0}}{\Omega_{ci0}} = d_{i0}, \tag{21}$$

where $\Omega_{ci0} = q_i B_0/m_i c$ is the ion gyrofrequency. Thus, the scale at which the Hall effect becomes important for transverse perturbations is when gradient scales are comparable to the ion inertial scale. Using characteristic scales for Earth's dayside magnetosheath, $d_{i0} \simeq$ 70 km; for the solar corona, $d_{i0} \simeq 2$ m. Thus, we have a feel for length scales over which we need to use Hall-MHD instead of ideal-MHD in two important space applications.

3.4.2 The Kinetic Alfvén Wave

Now, we consider the kinetic Alfvén wave. These waves are almost completely longitudinal, but not perfectly longitudinal. To allow for a strong analogy to the whistler wave, consider a uniform magnetic field that has a very large z component $B_{z0} \gg 0$ and very small y component B_{y0}, as sketched in Fig. 2(c). As in panels (a) and (b), the wave propagates in the y direction. If B_{z0} were zero, it would be the whistler wave. Physically, if the wavelength is between ion and electron gyroscales, this perturbation reacts similar to a whistler in that it sets up an electron flow that bends the magnetic field out of the plane. The perturbed magnetic field therefore has a component in the same or opposite direction as B_{z0}. Where B_{z0} and the perturbed magnetic field are parallel, the magnetic pressure $(B_{z0} + B_{z1})^2/8\pi$ is greater than the initial pressure $B_{z0}^2/8\pi$ to first order in the perturbed field. (Note, for the whistler wave, the pressure difference $B_{z1}^2/8\pi$ is second order in the perturbed magnetic field, so it is negligible.) Similarly, where B_{z1} opposes B_{z0}, the magnetic pressure decreases to first order. Now, if the plasma is overall low β (since we took B_{z0} to be large), this magnetic pressure imbalance cannot be maintained, so the gas pressure has to change in order to balance pressure. Electrons move along the magnetic field and ions move across the magnetic field to move from the high magnetic pressure region to the low magnetic pressure region, setting up a high and low gas pressure region as denoted in panel (c). This describes the physics of the kinetic Alfvén wave.

To find the dispersion relation of the kinetic Alfvén wave, we back up to the full dispersion relation of any wave in Hall-MHD. Linearizing Eqs. (13) - (18) around a uniform magnetic field, density, and pressure with an arbitrary linear perturbation and solving gives the following dispersion relation (Rogers et al. 2001):

$$\begin{aligned} \omega^6 &- (c_{ms}^2 + v_{Ak}^2 + k^2 v_{Ak}^2 d_{i0}^2) k^2 \omega^4 + \\ &[c_{ms}^2 + c_s^2(1 + k^2 d_{i0}^2)] k^4 v_{Ak}^2 \omega^2 - k^6 v_{Ak}^4 c_s^2 = 0, \end{aligned} \tag{22}$$

where $v_A^2 = B_0^2/(4\pi\rho_0)$ is the total Alfvén speed, $c_s^2 = \gamma p_0/\rho_0$ is the sound speed, $c_{ms}^2 = c_s^2 + v_A^2$ is the total fast magnetosonic speed, and $v_{Ak}^2 = B_{y0}^2/4\pi\rho_0$ is the Alfvén speed based only on B_{y0}. In the $k^2 d_{i0} \to 0$ long wavelength limit, this equation reduces to the dispersion relation for ideal-MHD waves. In the limit of large kd_{i0}, one solution is a high frequency solution which is approximately given by balancing the first two terms in Eq. (22), which gives

$\omega^2 \simeq k^4 v_{Ak}^4 d_{i0}^2$, the whistler wave dispersion relation in Eq. (20). There is also a medium frequency solution which arises from balancing the middle two terms in Eq. (22). This ratio in general is

$$\omega^2 \simeq \frac{c_{ms}^2 + c_s^2(1 + k^2 d_{i0}^2)}{c_{ms}^2 + v_{Ak}^2 + k^2 v_{Ak}^2 d_{i0}^2} k^2 v_{Ak}^2. \tag{23}$$

In the limit in which $c_{ms}^2 \ll c_s^2 k^2 d_{i0}^2$ and $c_{ms}^2 \gg v_{Ak}^2 + k^2 v_{Ak}^2 d_{i0}^2$, the resulting dispersion relation is $\omega^2 \simeq (c_s^2/c_{ms}^2) k^4 d_{i0}^2 v_{Ak}^2$, which in the $v_A^2 \gg c_s^2$ (low β) limit gives $\omega^2 \simeq (c_s^2/v_A^2) k^4 d_{i0}^2 v_{Ak}^2$. Since $k^2 v_{Ak}^2 = k_\parallel^2 v_A^2$, this becomes

$$\omega^2 \simeq k_\parallel^2 k^2 v_A^2 \rho_s^2, \tag{24}$$

where $\rho_s^2 = c_s^2/\Omega_{ci}^2$ is the ion Larmor radius based on the sound speed. This is the dispersion relation for the kinetic Alfvén wave. As with the whistler wave, the kinetic Alfvén wave is dispersive with $\omega/k \propto k_\parallel$, so it gets faster for smaller wavelengths. The length scale at which the Hall term becomes important for the kinetic Alfvén wave is ρ_s (as opposed to d_{i0} for the whistler wave).

3.5 A Numerical Algorithm for Hall-MHD

Including the Hall electric field has a significant impact on numerical simulations relative to MHD simulations. One way to think of why this is the case is that the waves in ideal-MHD are non-dispersive, so waves at any scale from the large scale down to the computational grid scale travel at the same speed. In Hall-MHD, as discussed in the previous section, both whistler and kinetic Alfvén waves are dispersive, so waves at the grid scale (which needs to be sub-ion gyroscale to capture the Hall electric field) are considerably faster than waves at the large scale. This makes the Hall-MHD equations "stiff" – one must use a much smaller time step in Hall-MHD than in ideal MHD, leading to a significant increase in the run time and expense of the simulation.

There are numerous algorithms that can be used to numerically evolve the equations in time. We provide one in detail, and mention references to other algorithms that have been used. We highlight the F3D code (Shay et al. 2004), which has been used for many years to study magnetic reconnection. First, the evolution equations are written in conservative form as

$$\frac{\partial n}{\partial t} + \nabla \cdot \mathbf{J}_i = 0, \tag{25}$$

$$\frac{\partial \mathbf{J}_i}{\partial t} + \nabla \cdot \left(\frac{\mathbf{J}_i \mathbf{J}_i}{n} + \frac{p\mathbf{I}}{m_i} + \frac{\mathbf{BB}}{4\pi m_i} - \frac{B^2 \mathbf{I}}{8\pi m_i} \right) = 0, \tag{26}$$

$$\frac{\partial p}{\partial t} + \nabla \cdot (\mathbf{u}p) + (\gamma - 1) p (\nabla \cdot \mathbf{u}) = 0, \tag{27}$$

$$\frac{\partial \mathbf{B}}{\partial t} + c \nabla \times \mathbf{E} = 0, \tag{28}$$

with auxiliary equations $\mathbf{E} = \mathbf{J} \times \mathbf{B}/nec - \mathbf{J}_i \times \mathbf{B}/nc$ and $\mathbf{J} = (c/4\pi)\nabla \times \mathbf{B}$. Here, $n \simeq \rho/m_i$ is the number density of ions (approximately equal to the number density of electrons due to quasi-neutrality) and $\mathbf{J}_i = n\mathbf{u}$ is the ion flux density.

In F3D, these equations are stepped forward using the trapezoidal leapfrog technique, a predictor-corrector method that is well-equipped to handle conservative equations (Zalesak 1979, 1981). Each of the above equations can be written as a conservative equation of the form (Guzdar et al. 1993)

$$\frac{\partial \psi}{\partial t} + \nabla \cdot \mathcal{F} - D\nabla^2 \psi + F(\xi) = 0, \tag{29}$$

where ψ is the plasma variable in question, $\mathcal{F}$ is a suitably defined flux, D is a second order diffusion coefficient which can be added to the equations to represent resistivity, viscosity, or a numerical dissipation to improve code stability, and $F(\xi)$ is a suitably defined sink/source term in terms of any other plasma variables ξ. (Equation (28) is not exactly in this form, but an analogous expression holds.)

To write the numerical algorithm, we use the standard notation where a superscript n on a plasma variable refers to the time step in question, so the initial values are set at $n = 0$, the first time step is $n = 1$, and so on. To evolve ψ^n to ψ^{n+1} in a time step Δt, the trapezoidal leapfrog algorithm is

$$\begin{aligned} \psi^{n+1/2} &= \frac{\psi^{n-1} + \psi^n}{2} + \Delta t \big[-\nabla \cdot \mathcal{F}^n \\ &\quad + D\nabla^2 \psi^{n-1} - F(\xi^n) \big] \end{aligned} \tag{30}$$

$$\begin{aligned} \psi^{n+1} &= \psi^n \Delta t \big[-\nabla \cdot \mathcal{F}^{n+1/2} \\ &\quad + D\nabla^2 \psi^n - F(\xi^{n+1/2}) \big]. \end{aligned} \tag{31}$$

The first equation uses the data at the n'th time step and the data at the previous $n - 1$'st time step to "predict" ψ half a time step in the future. Then, the flux and source terms are evaluated at this intermediate time step to evolve ψ the next half time step to the desired step $n + 1$. To go from $n = 0$ to $n = 1$, data is needed at $n = -1$, which is simply taken to be the same as the data at $n = 0$. This algorithm is second order in the time step Δt, meaning that the error from the algorithm is approximately a coefficient times the square of the time step. This is an example of an "explicit" time stepping algorithm because the data at the future time step is found completely using known data at the current or previous time steps.

The above shows how the equation is stepped forward temporally, but one also needs to calculate spatial derivatives. The approach F3D uses to calculate spatial derivatives is with a finite difference technique, which means the spatial derivatives are simply approximated by the derivative over the size of a grid scale instead of over an infinitesimal distance. For example, one approximation for the partial derivative in the x direction on a grid with grid scale Δx is $\partial \psi_j / \partial x \simeq (\psi_{j+1} - \psi_{j-1})/2\Delta x$, where the j subscript refers to the index of the spatial cell for which the derivative is desired. This is a second-order scheme because the error relative to the exact derivative scales like Δx^2. The F3D code uses the approximation

$$\frac{\partial \psi_j}{\partial x} \simeq \frac{2}{3\Delta x}(\psi_{j+1} - \psi_{j-1}) - \frac{1}{12\Delta x}(\psi_{j+2} - \psi_{j-2}), \tag{32}$$

which is fourth order (the error scales like Δx^4), and the cost for this higher order derivative is that it requires data from two adjacent cells on each side rather than one. Analogous expressions hold for spatial derivatives in the y and z directions. Second order derivatives are given by the associated fourth order accurate approximation $\partial^2 \psi_j / \partial x^2 \simeq [-(2/3)(\psi_{j+1} - \psi_{j-1}) - (1/12)(\psi_{j+2} + \psi_{j-2})]/(\Delta x)^2$.

The F3D code is written in Fortran 90 and is parallelized using Message Passing Interface (MPI) for use on supercomputers. The computational domain is rectangular with a fixed, regular grid. It can be run in two or three dimensions. When in two dimensions, the vectors can have an out of plane component even though all quantities are invariant in the out-of-plane direction; this is often referred to in the literature as "2.5 dimensional". The F3D code does not explicitly enforce that $\nabla \cdot \mathbf{B} = 0$, but it has been demonstrated that the value is small when the initial magnetic field is divergence free. Numerous aspects of the F3D code are on user-controlled switches that can turn terms or effects on or off, and the initial plasma variable profiles and values are controlled by the user. Other features of F3D that are used to go beyond the Hall-MHD model will be treated in Sect. 3.6.

To run the F3D code, the user chooses the simulation domain size and desired grid scale to resolve the relevant physics, which is typically at least 5 times smaller than the relevant ion gyroradius. As F3D is an explicit finite difference code, the time step Δt can be no larger than allowed by the so-called Courant-Friedrichs-Lewy (CFL) condition (Press et al. 1992), which requires $\Delta t \leq \Delta x / v_{\text{fastest}}$, where v_{fastest} is the fastest speed that can occur in the system. For Hall-MHD, the fastest speed is typically the whistler or kinetic Alfvèn wave speed at or near the grid scale, but can be the fast magnetosonic speed for some ambient plasma conditions. 2D reconnection simulations with F3D are typically performed with time step about 40% of the CFL condition.

Any employed diffusion coefficients then need to be chosen. Often, the resistivity is not used for Hall-MHD, but a fourth-order diffusion numerical dissipation of the form $-D_4(\partial^4\psi/\partial x^4 + \partial^4\psi/\partial y^4 + \partial^4\psi/\partial z^4)$ on the right hand side of Eq. (29) is included in F3D to damp structures at the grid scale while minimally affecting larger scale structures. In order to preserve the five point stencil used for the other finite differences, second order accuracy in the fourth order derivatives is employed, with $\partial^4\psi_j/\partial x^4 \simeq [\psi_{j+2} - 4\psi_{j+1} + 6\psi_j - 4\psi_{j-1} + \psi_{j-2}]/(\Delta x)^4$. The appropriate diffusion coefficient scales with $D_4 \sim v_{\text{fastest}}[\pi/(\Delta x)]^3$. An appropriate value for this coefficient is when it is large enough to control numerical issues at the grid scale while not impacting the large scale physics. A good approach to optimize this value is to run multiple simulations with only varying D_4, and finding a range of values for which the numerics are good and the large scale features are only weakly dependent on D_4; it is typically within an order of magnitude of the scaling prediction.

We have focused on F3D as an example of a Hall-MHD code because of its algorithmic simplicity and because it has long been used to study reconnection. There are drawbacks to the code. As a finite difference code, it does not capture shocks, and therefore if the number density gets fairly small in any given simulation the code typically crashes. The code performs well up to about 1000-2000 processors on high powered supercomputers; MHD codes without the Hall effect can be made to scale much better to 10s of thousands of processors. It is also restricted to a rectangular geometry with a regular grid.

There are a number of other Hall-MHD codes that have been used to study magnetic reconnection, some of which we gather here. Some codes used to study Hall-MHD reconnection in a rectangular domain have included VOODOO (Huba 2003), HMHD (Lottermoser and Scholar 1997), another code called HMHD (Huang et al. 2011), and the UI Hall-MHD code (Ma and Bhattacharjee 2001). Codes that have been used to study Hall-MHD in the context of planetary magnetospheres (including Earth's) are a multi-fluid code (Winglee 2004), Block Adaptive Tree Solar-wind Roe Upwind Scheme (BATS-R-US) (Tóth et al. 2008), and Gkyell (Wang et al. 2018). Codes used in the tokamak geometry include NIMROD (Glasser et al. 1999) and M3D-C1 (Jardin et al. 2008).

3.6 Further Extensions of Hall-MHD

Here, we briefly discuss fluid model extensions beyond ideal-MHD that contain the Hall electric field and other terms. We start with terms that were dropped from Eq. (10).

3.6.1 Hall-MHD with Electron Inertia

One extension of Hall-MHD is to retain the electron inertia term $m_e d\mathbf{u}_e/dt$ from Eq. (10). We call this model "Hall-MHD with electron inertia;" it is often called the "two-fluid model" in the literature, but we refrain from this nomenclature since a completely different set of equations is also typically given the same name.

Using the electron inertia term as is would lead to a new variable $\mathbf{u}_e$ with a time derivative in the model. Instead, the standard approach is to recognize that the prefactor m_e is small for electron-ion plasmas, so the only way this term important is if the electrons are moving very fast. The ions would be too slow to keep up in such a case, so on time scales where this term is important, we can treat the ions as approximately stationary. Then, the electron bulk flow velocity $\mathbf{u}_e$ is related to the current density via $\mathbf{u}_e = -\mathbf{J}/n_e e$. Since $\mathbf{J}$ and n_e are already included in the Hall-MHD description, the set of equations remains closed. Such a simplification is convenient but not absolutely necessary. For example, Sect. 4.4 describes a hybrid code in which the ion flows and density effects in the electron inertia term are included.

Analytically, the inertial electric field from dividing Eq. (10) by $-e$ is given by $-(m_e/e)d\mathbf{u}_e/dt$, and replacing $\mathbf{u}_e$ by $-\mathbf{J}/n_e e$ gives $(m_e/e^2)d(\mathbf{J}/n_e)/dt$. Then, Ohm's law in Hall-MHD with electron inertia becomes

$$\mathbf{E} + \frac{\mathbf{u} \times \mathbf{B}}{c} = \frac{\mathbf{J} \times \mathbf{B}}{n_e e c} + \frac{m_e}{e^2} \frac{d}{dt} \left(\frac{\mathbf{J}}{n_e} \right). \tag{33}$$

It is important to note that the same approximation $\mathbf{u}_e \simeq -\mathbf{J}/n_e e$ is used in the convective derivative term, so that $d/dt \simeq \partial/\partial t - (1/n_e e)\mathbf{J} \cdot \nabla$.

One needs to treat the $\partial/\partial t$ term on the right hand side of Eq. (33). To do so, we eliminate $\mathbf{E}$ in Faraday's law using Eq. (33); some algebra reveals that the equation becomes

$$\frac{\partial \mathbf{B}'}{\partial t} = -c\nabla \times \mathbf{E}', \tag{34}$$

where $\mathbf{B}' = (1 - d_e^2 \nabla^2)\mathbf{B}$ is an auxiliary magnetic field and $\mathbf{E}' = \mathbf{J} \times \mathbf{B}'/nec - \mathbf{J}_i \times \mathbf{B}/nec$ is an auxiliary electric field, and the factor of n in the inertia term is treated as a constant on the time scales of interest. Coupling this equation with Eqs. (25)-(27) gives a closed set of equations.

To solve these equations, we note that Eq. (34) looks just like the usual Faraday's law except for the primes, so the same numerical technique can be used to solve for $\mathbf{B}'$. Once $\mathbf{B}'$ is found, one uses that variable as the known source term in the equation $(1 - d_e^2 \nabla^2)\mathbf{B} = \mathbf{B}'$ to solve for $\mathbf{B}$. This is an elliptic differential equation with many algorithms that can be used to solve it. The simplest may be a relaxation technique (Press et al. 1992), but it is relatively slow, especially when used for codes that have been parallelized for use on a supercomputer. A faster version of relaxation is called multigrid (Trottenberg et al. 2000). F3D employs the Fast Fourier Transform approach.

We now briefly discuss the physics introduced by the electron inertia term and the advantages for including it in simulations. By comparing the electron inertia term to the Hall term, we determine the condition under which it is important to retain the electron inertia term. We know the Hall electric field is important at scales below the ion inertial scale d_i, but vanishes at the X-line where $\mathbf{B} = 0$. The electron inertia term can be important at small scales, so we seek the scale at which it becomes comparable to the Hall electric field. Setting them equal gives $J_y B_x/nec \sim (m_e/e^2)[(\mathbf{J}/ne) \cdot \nabla](J_z/n)$. In the scaling sense, we use $B_x \sim B_{up}$, $J_z \sim cB_{up}/4\pi\delta$ where δ is the scale at which the two terms are equal, so $J_y B_{up}/nec \sim (m_e/e^2)(J_y/ne\delta)(cB_{up}/4\pi\delta n)$, which simplifies to $\delta^2 \sim d_e^2$, where $d_e^2 = m_e c^2/4\pi n e^2$ is the electron inertial scale. Thus, at length scales below d_e, the electron inertia term can be important.

Consequently, one reason researchers include the electron inertia term into their Hall-MHD model is to aim to capture electron scale physics more accurately than without it. We point out, however, that the MHD model itself was derived with the assumption that m_e/m_i is small, and therefore including the electron inertia term as we have done does not actually provide a self-consistent treatment of sub-d_e scale physics. The two fluid model or the kinetic approach is needed to more accurately capture electron scale physics.

Then why include electron inertia? The answer is that it helps with the numerics. To see this, we consider waves in the Hall-MHD with electron inertia system, generalizing the treatment in Sects. 3.4.1 and 3.4.2. The dispersion relation for perfectly parallel propagating waves in Hall-MHD with electron inertia, generalizing Eq. (19), becomes

$$\omega^2 = \frac{k^2 v_{A0}^2}{D_e}\left(1 + \frac{k^2 d_{i0}^2}{2D_e} + \sqrt{\frac{k^2 d_{i0}^2}{D_e} + \frac{k^4 d_{i0}^4}{4D_e^2}}\right), \tag{35}$$

where $D_e = 1 + k^2 d_e^2$. When kd_e is negligible, this dispersion relation reduces to the Hall-MHD result in Eq. (19). When $kd_e \gg 1$, the waves become electron cyclotron waves with $\omega^2 = \Omega_{ce}^2$, where $\Omega_{ce} = eB_0/m_e c$. The reason this is useful numerically is that the whistler wave is dispersive, and in Hall-MHD the phase speed goes to infinity as the wavelength goes to zero. In Hall-MHD with electron inertia, the whistler rolls over to the electron cyclotron wave which does not propagate, so there is a maximum speed of the waves. This means that the time step required to run the simulation does not go to zero, and the simulations with very high resolution are less stiff and therefore cheaper to carry out.

3.6.2 Electron Magnetohydrodynamics (EMHD)

The electron-MHD (EMHD) model (Kingsep et al. 1990) is used when the large (MHD) scales are not of interest, and only the scales between electron and ion scales are of interest. In such a limit, the MHD terms concerning ion velocity are dropped. Because of this, the density and pressure can no longer change (on the time scales of interest), so the only remaining governing equation is Faraday's law, with the electric field given solely by the Hall electric field:

$$\frac{\partial \mathbf{B}}{\partial t} = -c\nabla \times \mathbf{E}, \tag{36}$$

$$\mathbf{E} = \frac{\mathbf{J} \times \mathbf{B}}{nec}. \tag{37}$$

Using Ampère's law $\mathbf{J} = (c/4\pi)\nabla \times \mathbf{B}$, it is common to combine these equations into a single equation for $\mathbf{B}$,

$$\frac{\partial \mathbf{B}}{\partial t} = -\frac{c}{4\pi n e}\nabla \times [(\nabla \times \mathbf{B}) \times \mathbf{B}]. \tag{38}$$

This closed vector equation comprises the EMHD model. One can include electron inertia in a manner analogous to the full Hall-MHD model: $\partial \mathbf{B}'/\partial t = -(1/n_e e)\nabla \times (\mathbf{J} \times \mathbf{B}')$, where $\mathbf{J} = (c/4\pi)\nabla \times \mathbf{B}$ and $\mathbf{B}' = (1 - d_e^2\nabla^2)\mathbf{B}$.

3.6.3 The Electron Pressure Term in Hall-MHD

We now consider the electron pressure term in Eq. (10). There, electron pressure is written as a tensor $\mathbf{p}_e$, which is the most general form following directly from the Vlasov/Boltzmann equation in kinetic theory. Retaining it in Ohm's law would give a term on the right hand side of Eq. (33) of the form $-(1/n_e e)\nabla \cdot \mathbf{p}_e$. We now consider a few commonly used approximations to simplify this.

Isotropic Electron Pressure In MHD, the total pressure is assumed to be isotropic, so that $\mathbf{p} = p\mathbf{I}$. In this case, the term entering Ohm's law is $-(1/n_e e)\nabla p_e$. When substituted into Faraday's law [Eq. (16)], this term gives a contribution of $(c/e)\nabla \times (\nabla p_e/n) = (c/e)\nabla(1/n_e) \times \nabla p_e = -(c/en_e^2)\nabla n_e \times \nabla p_e$. This contribution to the electric field is called the "Biermann battery" (Biermann 1950) and it is often of great importance in reconnection in high energy density plasmas (Fox et al. 2012). In most space applications that employ a fluid model, the electrons are assumed to be adiabatic with p_e/n_e^γ equal to a constant or isothermal with p_e/n_e equal to a constant. In either limit, the Biermann battery term vanishes identically.

One may presume from this result that the electron pressure gradient term does not have any contribution to Hall-MHD, but this is not true. Including the scalar electron pressure gradient in the Hall-MHD Ohm's law gives

$$\mathbf{E} + \frac{\mathbf{u}_e \times \mathbf{B}}{c} = -\frac{1}{n_e e}\nabla p_e, \tag{39}$$

where we write $\mathbf{u}_e \simeq \mathbf{u} - \mathbf{J}/n_e e$ for the bulk electron velocity. Taking the cross product of this equation with $\mathbf{B}$ and solving for the bulk flow velocity $\mathbf{u}_{e,\perp}$ perpendicular to the magnetic field $\mathbf{B}$ gives

$$\mathbf{u}_{e\perp} = c\frac{\mathbf{E} \times \mathbf{B}}{B^2} + \frac{c}{n_e e B^2}\nabla p_e \times \mathbf{B}. \tag{40}$$

This equation implies the electron perpendicular bulk flow velocity is a sum of the $E \times B$ drift $c\mathbf{E} \times \mathbf{B}/B^2$ and the electron diamagnetic drift speed $\mathbf{u}_{*e} = c\nabla p_e \times \mathbf{B}/n_e e B^2$. This important result shows that the electron diamagnetic drift is captured in Hall-MHD provided the electron pressure is non-zero. An important implication of this is that magnetic flux convects at the sum of the $E \times B$ and electron diamagnetic drift speed when the electron pressure is retained in Ohm's law.

Gyrotropic Electron Pressure The next level of approximation for the electron pressure tensor is motivated by the fact that electrons in a magnetic field often have a different temperature in the directions parallel and perpendicular to the magnetic field, *i.e.*, the electrons are gyrotropic. The electron pressure parallel to the magnetic field is $p_{e,\parallel}$ and perpendicular to the magnetic field is $p_{e,\perp}$. The general way to write a gyrotropic electron pressure tensor $\mathbf{p}_{e,g}$ for a magnetic field in an arbitrary direction $\hat{\mathbf{b}} = \mathbf{B}/B$ is (Parker 1957a)

$$\mathbf{p}_{e,g} = p_{e,\perp}\mathbf{I} + (p_{e,\parallel} - p_{e,\perp})\hat{\mathbf{b}}\hat{\mathbf{b}}. \tag{41}$$

Using this form in the generalized Ohm's law for the electron pressure, and an analogous term in the pressure gradient force in the momentum equation allows for the modeling of a plasma with a gyrotropic pressure.

Having a closed set of equations requires closures on the parallel and perpendicular pressures. The most widely known closure is the Chew-Goldberger-Low (CGL) closure, which assumes that there is no heat flux and the plasma is magnetized (Chew et al. 1956). A direct calculation using the Vlasov equation and these assumptions gives the CGL "double adiabatic laws"

$$\frac{d}{dt}\left(\frac{p_\perp}{nB}\right) = 0, \qquad \frac{d}{dt}\left(\frac{p_\parallel B^2}{n^3}\right) = 0, \tag{42}$$

A second model of great importance to reconnection is the Egedal closure (Egedal et al. 2013). This arises in the upstream and downstream regions of magnetic reconnection regions. The key physics is that the magnetic fields in this region are mirror fields that can trap electrons. The presence of an electric field heats them parallel to the electric field, leading to elongated gyrotropic distributions in the parallel direction. For large magnetic field strength, the equations reduce to the isothermal equation of state. For small magnetic field strength, the equations reduce to the CGL equations. The two limits were interpolated to find a closure that could be implemented into a fluid code; see Egedal et al. (2013) for details.

3.7 Examples of Reconnection Simulation Results with the Hall Electric Field

3.7.1 Ion-Coupled Reconnection

The Hall-MHD model holds a significant place of importance in the history of reconnection simulations. From the earliest days of reconnection research, it was known that the Sweet-Parker model (Sweet 1958; Parker 1957b) was too slow to explain solar flares (Parker 1963), and it is also too slow to explain magnetotail reconnection (Parker 1973) and the sawtooth crash in tokamaks (Edwards et al. 1986; Yamada et al. 1994). It was discovered that using a localized resistivity leads to reconnection fast enough to explain the observed reconnection rates (Ugai and Tsuda 1977; Sato and Hayashi 1979), but the resistivity model was not derived from first principles and it remains inconclusive whether it can be. Then, reconnection with the Hall electric field was found to also produce rates comparable to observed values without relying on ad hoc terms (Aydemir 1992; Wang and Bhattacharjee 1993; Kleva et al. 1995; Ma and Bhattacharjee 1996). The "GEM Challenge" study ((Birn et al. 2001) and references therein), one of the most cited papers ever about reconnection, compared comparable simulations with different simulation models, and all models containing the Hall electric field led to fast reconnection with a reconnection rate approximately 0.1, as shown in Fig. 3. We note that this figure contains results from Hybrid and PIC simulations, which are described in Chaps. 4 and 5 respectively. It was only recently that an explanation of why the Hall electric field contributes to make the reconnection rate 0.1 was presented (Liu et al. 2022). Thus, Hall-MHD represents the minimal first-principles physics model that reproduces the reconnection rate achieved in kinetic models.

Another characteristic feature of collisionless reconnection is the quadrupolar structure of the out-of-plane magnetic field. For anti-parallel reconnection, it was posited that the Hall electric field would cause there to be a quadrupolar out-of-plane magnetic field due to the in-plane currents within the ion diffusion region (where the ions decouple from the magnetic field but the electrons remain frozen-in) (Sonnerup 1979). Although this is a completely nonlinear effect, it is analogous to the out-of-plane magnetic field generation in the

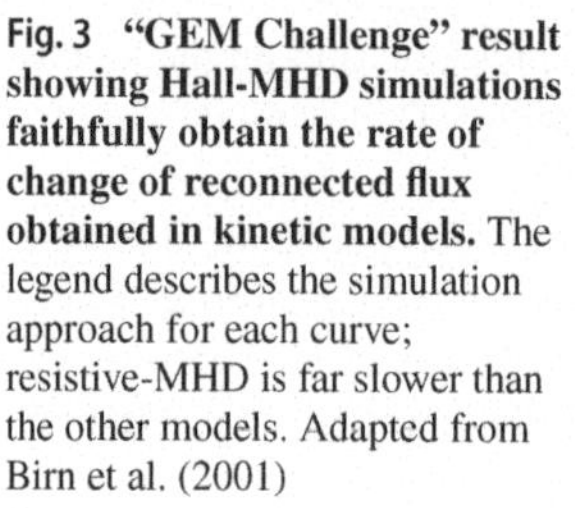

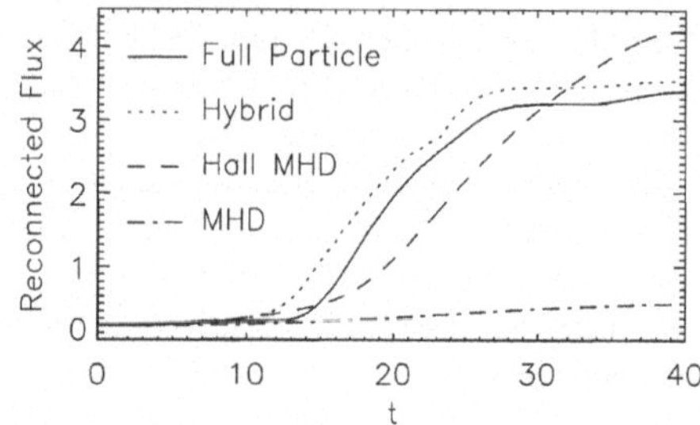

Fig. 3 "GEM Challenge" result showing Hall-MHD simulations faithfully obtain the rate of change of reconnected flux obtained in kinetic models. The legend describes the simulation approach for each curve; resistive-MHD is far slower than the other models. Adapted from Birn et al. (2001)

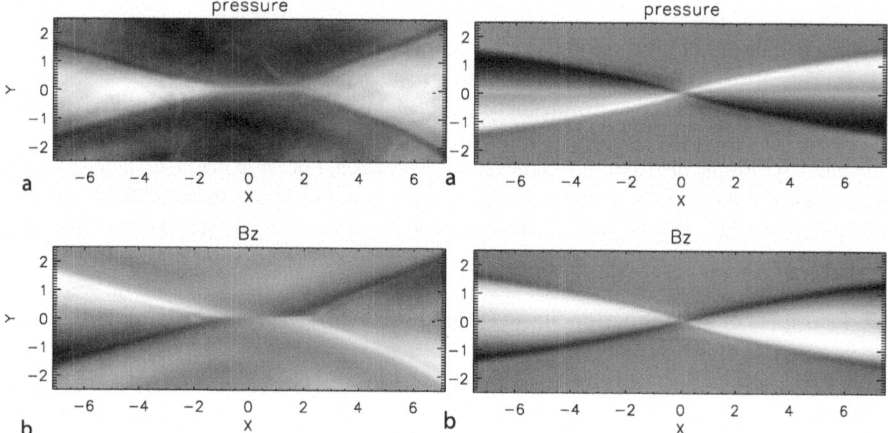

Fig. 4 Pressure and out-of-plane magnetic field in Hall MHD reconnection simulations. The legend describes the simulation approach for each curve; resistive-MHD is far slower than the other models. Adapted from Rogers et al. (2003)

whistler wave shown in Fig. 2(b) (Drake and Shay 2007). It can also be thought of as the out-of-plane current dragging the reconnecting field out of the reconnection plane (Mandt et al. 1994). Hall-MHD simulations of reconnection produce a quadrupolar out-of-plane magnetic field (Huba and Rudakov 2002), as shown in the lower left plot in Fig. 4 from simulations in Rogers et al. (2003). When there is a strong out-of-plane (guide) magnetic field, the quadrupolar structure persists, but a quadrupolar structure in the gas pressure also arises with opposite polarity (Kleva et al. 1995). The physical reason is analogous to the gas pressure perturbation formation in kinetic Alfvén waves as shown in Fig. 2(c). These quadrupolar structures arise in Hall-MHD simulations of reconnection, as shown in the right two panels of Fig. 4 from a simulation with guide field three times as strong as the reconnecting magnetic field. Despite the ability to produce quadrupolar structure in these quantities, it is now known that the detailed structure of the quadrupolar structures is not precisely the same as observed in kinetic simulations (Shay et al. 2007; Karimabadi et al. 2007) or magnetospheric observations (Phan et al. 2007). It was shown that the inclusion of an electron pressure anisotropy leads to better agreement with kinetic modeling (Ohia et al. 2012). An electron pressure anisotropy with the CGL relations and without the Hall term can also produce reconnection rates comparable to Hall reconnection (Cassak et al. 2015).

Another aspect of Hall-MHD reconnection that is not captured in collisional (Sweet-Parker) reconnection is how reconnection that is localized in the out-of-plane direction

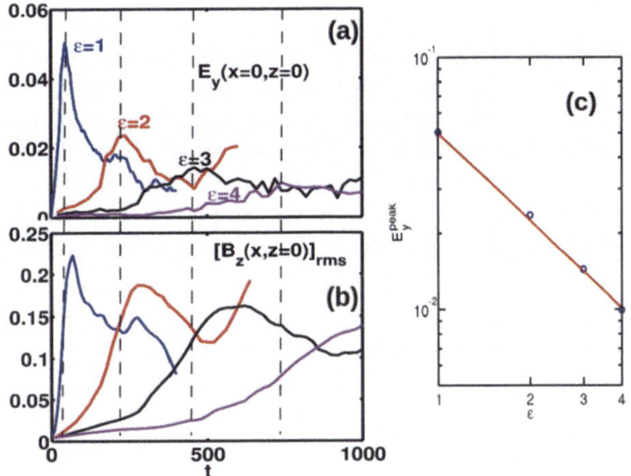

Fig. 5 Evolution of (a) out-of-plane electric field E_y at the X-point ($x = 0$, $z = 0$) and (b) root-mean-square value of the normal component of magnetic field B_z evaluated over the length of the reconnecting current sheet (between two outflow regions) Vertical dashed lines mark the times at which E_y attains its peak value E_y^{peak} for different values of the current sheet half-thickness ϵ. (c) Scaling of E_y^{peak} with ϵ (in log -scale) from simulations (blue circles) and a fit $E_y^{peak} = 0.05/\epsilon^{1.15}$ (red line). Adapted from Jain and Sharma (2015b)

spreads in that direction. Spreading has been studied in a number of Hall-MHD studies (Huba and Rudakov 2002; Shay et al. 2003; Huba and Rudakov 2003; Karimabadi et al. 2004; Nakamura et al. 2012; Shepherd and Cassak 2012; Arencibia et al. 2021, 2023) and EMHD studies (Jain et al. 2013) and they agree with kinetic simulations of reconnection spreading (Lapenta et al. 2006). In particular, it was shown that anti-parallel reconnection does not spread in resistive-MHD, but it does spread in Hall-MHD and kinetic models (Nakamura et al. 2012; Arencibia et al. 2021).

3.7.2 Electron-Only Reconnection

Collisionless magnetic reconnection in electron scale current sheets has been investigated using fixed ion models (both fluid and kinetic) for several decades before the observational discovery of electron-only reconnection by MMS (Bulanov et al. 1992; Mandt et al. 1994; Drake et al. 1994, 1997; Shay and Drake 1998; Attico et al. 2000; Chacón et al. 2007; Zocco et al. 2009; Jain and Sharma 2009; Jain et al. 2012; Jain and Büchner 2015; Jain and Sharma 2015a,b). It was, however, not termed as electron only reconnection as it was not conceived at that time that reconnection without ion participation is possible. It was rather termed as "early phase of reconnection" (Jain and Sharma 2009; Jain et al. 2012; Jain and Sharma 2015a,b). Later, electron only reconnection in Earth's magnetotail was also interpreted as an early phase of the standard reconnection (Hubbert et al. 2022; Farrugia et al. 2021) and there is ongoing discussion if the observed reconnection events without ion coupling are electron-only reconnection events or just an early phase of standard reconnection (Lu et al. 2020, 2022; Wang et al. 2020b; Yi et al. 2022).

Jain and Sharma (2009) were the first to propose that the early phase of reconnection after its onset in electron scale current sheets will be dominated by electron dynamics without coupling to ion dynamics and carried out EMHD simulations to study the physics of the early phase. This study was followed by further theoretical and simulation studies, both in 2-D and 3-D, using EMHD model and comparison with space observations by Cluster spacecraft (Jain et al. 2012; Jain and Sharma 2015a,b; Jain and Büchner 2014a,b, 2015; Jain et al. 2017b,a). These and other EMHD studies are also relevant for electron only reconnection.

Figure 5 shows results from the 2-D EMHD (x-z plane) simulations of magnetic reconnection in electron scale current sheets of different half-thicknesses (ϵ) (Jain and Sharma 2015b). In these simulations, the equilibrium current density is $\mathbf{J} = -n_0 e u_{ey0} \hat{y} =$

$(B_0 c/4\pi\epsilon)\,\mathrm{sech}^2(z/\epsilon)\hat{y}$ corresponding to the equilibrium anti-parallel magnetic field $\mathbf{B} = B_0 \tanh(z/\epsilon)\hat{x}$. The density n_0 is uniform and thus the current is due to the electron flow u_{ey0}. The results are shown in the normalized units: length by electron inertial length $d_e = c/\omega_{pe} = c/(4\pi n_0 e^2/m_e)^{1/2}$, magnetic field by B_0 and time by $\omega_{ce}^{-1} = (eB_0/m_e c)^{-1}$. Reconnection rate, measured by out-of-plane electric field E_y at the X-point and shown in Fig. 5a, reaches its peak value when the growth of the rms values of B_z begins to slow down, consistent with the Faraday law which gives $\partial E_y/\partial x = -1/c\partial B_z/\partial t$. The peak reconnection rate E_y^{peak} scales with ϵ as $E_y^{peak} = 0.05/\epsilon^{1.15}$ and drops from $E_y^{peak} = 0.05\, v_{Ae} B_0/c$ to $E_y^{peak} = 0.01\, v_{Ae} B_0/c$ as the ϵ increases from $\epsilon = d_e$ to $\epsilon = 4d_e$. This range of reconnection rate in units based on ion Alfvén speed is $E_y^{peak} = 0.43 - 2.15\, v_{Ai} B_0/c$ (using ion $m_i/m_e = 1836$) which is much larger than the value of the reconnection rate ($0.1\, v_{Ai} B_0/c$) for the standard ion-coupled reconnection. Reconnection rates for the electron only reconnection have been reported in the similar range by MMS observations (Burch et al. 2020) and PIC simulations (Sharma Pyakurel et al. 2019). A recent analytical model (Liu et al. 2025b) that incorporates the dispersive properties of Alfvén waves (Sect. 3.4) was proposed to explain these higher electron-only reconnection rates.

Note that the EMHD simulation results in Fig. 5 are independent of the strength of the guide magnetic field because a uniform guide field does not appear in 2-D EMHD equations (Jain and Büchner 2015). However, in 3-D, guide field can introduce current aligned instabilities in addition to the tearing instability as has been predicted by 3-D EMHD eigen value analysis (Jain and Büchner 2015) and simulations (Jain et al. 2017a). In 3-D, current aligned electron Kelvin-Helmholtz instabilities can grow in electron scale current sheets even in the absence of guide magnetic field (Jain and Büchner 2014a,b; Greess et al. 2021). These EMHD studies are relevant for the MMS observations of the electron shear flow generated electron Kelvin-Helmholtz vortices within the diffusion region of collisionless magnetic reconnection (Zhong et al. 2022, 2018; Hwang et al. 2019).

3.8 The Future of Hall-MHD

Interestingly, many computational plasma physicists in the reconnection community are moving away from the Hall-MHD model and its fluid extensions to study the small-scale properties of reconnection. The Hall-MHD model was crucial for understanding the minimal physics that gives rise to a 0.1 reconnection rate. As questions have moved to other aspects of reconnection including heating and particle acceleration, many researchers opt for kinetic models to more realistically capture the small scale physics than can be done with Hall-MHD. Treatments of particle acceleration and heating at large scales (Arnold et al. 2021) do not require the Hall electric field.

As example of an avenue of modern reconnection research and modeling where the Hall-MHD model remains highly beneficial is in the MHD-EPIC approach to global magnetospheric modeling (Daldorff et al. 2014), as is discussed more fully in Sect. 6. In this approach, the fluid model is used in regions of the magnetosphere where no small important scale physics takes place, which allows for faster run times. In regions where small scale physics does take place, the code couples to a particle-in-cell code that captures this physics. The numerical results between the two models are passed back and forth to each other across their boundaries. The Hall-MHD model is well suited to be used in the transition region between the PIC and MHD models to facilitate a more accurate transition between the two models. The Hall-MHD model has been used to study global magnetospheric systems for Earth and other planets and moons to great effect (Paty and Winglee 2004; Dorelli et al. 2015; Dong et al. 2019; Li et al. 2023b), and it is anticipated that further advances will continue to be made with the Hall-MHD approach for systems too large to employ global kinetic codes.

4 Hybrid Simulations

The Earth's magnetosphere is a complex plasma system characterized by a multitude of multiscale processes governing the interaction of the solar wind with the Earth's dipole magnetic field. Modeling small-scale turbulent processes in the foreshock and magnetosheath requires inclusion of ion kinetic effects and Hall physics (Karimabadi et al. 2014; Omelchenko et al. 2021a). Furthermore, kinetic treatment of hot and cold ion populations is greatly needed for improved modeling of ionospheric outflows and their impact on the magnetopause and magnetotail. To study magnetic reconnection, one also needs to incorporate finite electron-mass effects (e.g. Biskamp 2000; Birn and Priest 2007; Gonzalez and Parker 2016), or mimic these effects with *ad hoc* (resistivity) models.

The necessity to account for smaller and faster scales in kinetic simulations in a manner that would guarantee their numerical accuracy and computational efficiency creates challenges in global modeling of the Earth's magnetosphere. Since describing plasma kinetics with pure "first-principles" models is still not feasible, various approximations have been developed as candidates for future "beyond MHD" operational modeling, with multiple levels of physical fidelity included. However, to what degree kinetic processes may influence the "fluid-like" behavior of the magnetosphere on global scales still remains an open question. As we argue below, many of these challenges can be addressed by self-consistent hybrid modeling, where Maxwell's equations are solved in the quasi-neutral Darwin limit, ion species are treated kinetically, and the plasma electrons are approximated as an inertia-less fluid. These hybrid models can be broken into two categories, according to the computational techniques used to represent kinetic ions: Particle-in-Cell (PIC) models (Winske et al. 2003; Lipatov 2002) and Vlasov models (von Alfthan et al. 2014). In what follows we discuss only the hybrid-PIC approach because it has already been applied successfully to perform three-dimensional (3D) simulations of global plasma systems that range from the Earth's magnetosphere (e.g. Lin and Wang 2005; Omelchenko et al. 2021a), planets (Herčík et al. 2013; Jarvinen et al. 2020), and small space bodies (Fatemi et al. 2017; Kallio et al. 2019) to compact laboratory plasmas (Omelchenko and Sudan 1997; Lin et al. 2008; Thoma et al. 2013; Omelchenko 2015; Omelchenko and Karimabadi 2022). The hybrid-Vlasov approach (von Alfthan et al. 2014) is relatively new and considerably more computationally expensive, with production runs being still restricted to quasi-3D setups (Pfau-Kempf et al. 2020).

An important issue to grasp reconnection correctly is the consideration of the finite electron inertia, as it has been shown by EMHD simulations. This could reveal not only the properties of electron-only reconnection, proposed by (Jain and Sharma 2009) and recently discovered by MMS in the magnetsheath (Phan et al. 2018), but also the transition from electron- to ion reconnection. These effects can be treated by hybrid-approach with kinetic ions and an inertial electron fluid (Sect. 4.4).

4.1 Model Equations: Massless Electrons

The standard hybrid model (Winske et al. 2003) assumes plasma quasi-neutrality, neglects the displacement current in Maxwell's equations, and treats ions as full-orbit macro-particles (in the PIC approach) moving in self-consistent electric and magnetic fields. The plasma electrons are approximated as an inertialess fluid with scalar pressure described by either an adiabatic law or evolution equation. Together with a self-consistent PIC method for the ion components, this leads to a set of hybrid equations that include Ampere's law in the magnetostatic limit, Faraday's law, and an algebraic expression for electric field (generalized Ohm's law) with the Hall, electron pressure gradient, and resistive terms (e.g. Omelchenko and Karimabadi 2012):

$$\frac{d\mathbf{x}_i}{dt} = \mathbf{v}_i, \tag{43}$$

$$m_i \frac{d\mathbf{v}_i}{dt} = q_i \left(\mathbf{E} + \frac{\mathbf{v}_i \times \mathbf{B}}{c}\right), \tag{44}$$

$$\nabla \times \mathbf{B} = \frac{4\pi}{c}\mathbf{J}, \ \mathbf{J} = \mathbf{J}_e + \mathbf{J}_i, \tag{45}$$

$$\frac{\partial \mathbf{B}}{\partial t} = -c\nabla \times \mathbf{E}, \tag{46}$$

$$\mathbf{E} = \frac{\mathbf{J}_e \times \mathbf{B}_t}{en_e c} - \frac{\nabla p_e}{en_e} + \eta \mathbf{J}, \ \mathbf{B}_t = \mathbf{B} + \mathbf{B}_{ext}, \tag{47}$$

$$en_e = \rho_i, \tag{48}$$

$$p_e = n_e T_e \sim n_e^{\gamma}. \tag{49}$$

Eqs. ((43) and (44)) are the equations of motion for each ion particle. q_i is the charge on each ion, and in these equations the electric and magnetic fields are interpolated from the grid onto the each particle's position. In Eqs. (47)-(49) n_e, $\mathbf{J}_e$ are the electron number and current density, respectively; p_e is the electron pressure, here assumed to governed by Eq. (49)) with an adiabatic constant of γ; T_e is the electron temperature; ρ_i, $\mathbf{J}_i$ are the total ion charge density and current density (found by summing up individual particles around each grid point); $\mathbf{E}$ is the electric field; $\mathbf{B}$, $\mathbf{B}_{ext}$ are the "self-generated" ($\mathbf{B}|_{t=0} = 0$) and "external" (steady state) magnetic fields, respectively.

The applied plasma resistivity, η is either constant or chosen to be a function of plasma parameters (e.g. Lin et al. 2007; Omelchenko et al. 2021a). The resistive term in the generalized Ohm's law (Eq. (47)) may (i) describe finite conductivity of plasma or space bodies (e.g., the Moon (Fatemi et al. 2017; Omelchenko et al. 2021b)), (ii) imitate finite electron inertia effects in magnetic reconnection events, and (iii) enable fast magnetic field diffusion at low-density ("vacuum") cells, $n_e \leq n_{min}$, where n_{min} is a small density cutoff value (Omelchenko et al. 2021b). Failure to properly treat low-density regions in hybrid simulations may lead to non-physical results (Omelchenko et al. 2021b; Poppe 2019).

4.2 Key Physics Beyond MHD in Hybrid Models

The "mesoscale" hybrid model occupies the middle ground between the "large-scale" fluid and "micro-scale" first-principles modeling paradigms. For global magnetospheric simulations, the hybrid model enables a number of "beyond MHD" capabilities, as explained below.

Modeling turbulent processes in the foreshock and magnetosheath. Unless large resistive damping (or smoothing) is applied, the hybrid model accurately captures the Hall physics for mesh cell sizes, $\Delta \leq d_i$, where $d_i = c/\omega_{pi}$ and ω_{pi} are the local ion inertial length and plasma frequency, respectively. The Hall effects phase out on coarser meshes, $\Delta \gg d_i$, where the Alfvén term becomes greater than the Hall term in Eq. (47) and the whistler mode frequency, $\propto \Delta^{-2}$ becomes lower than the Alfvén mode frequency, $\propto \Delta^{-1}$. In fact, in this case the Hall term can completely be removed from the electric field in Faraday's law (Eq. (46)) and kept only in the equations of ion motion (Karimabadi et al. 2004). The ability of a hybrid code with the Hall term to run stably on coarser meshes ($\Delta \gtrsim d_i$) may depend on the numerical implementation of Faraday's law (Omelchenko and Karimabadi 2012).

Global hybrid codes have been used to address the ultra-low frequency (ULF) physics of the curved bow shock on the ion inertial/Larmor radius scales as the physics of the bow shock is predominantly determined by kinetic physics associated with charged particles from the solar wind. Of particular interest are the foreshock waves and diffuse ion distributions (Wang et al. 2009) and transient perturbations originating from the wave-particle processes in the quasi-parallel shock or due to the shock interaction with incoming solar wind discontinuities, including hot flow anomalies (Lin 2002; Lin et al. 2022b), foreshock bubbles (Omidi et al. 2010; Wang et al. 2020a), foreshock cavities (Lin and Wang 2005; Blanco-Cano et al. 2011), and high-speed jets (Omelchenko et al. 2021a; Palmroth et al. 2018b). The 3D hybrid simulations with ANGIE3D (AuburN Global hybrId CodE in 3-D) link the foreshock perturbations to the surface perturbations and kinetic-scale shear Alfvén waves (KAWs) at the magnetopause through mode conversion from the incoming compressional waves (Lin and Wang 2005; Shi et al. 2013), as well as the subsequent excitation of toroidal-mode field line resonances in the magnetosphere (Shi et al. 2021). It has also been shown that 3D models are essential for addressing the nonlinear physics of mode coupling and ion diffusion at the magnetopause (Lin et al. 2012).

The whistler mode plays a significant role in regulating turbulence in the magnetosheath and mediating magnetic reconnection in the Earth's magnetosphere (e.g. Dorelli and Birn 2003; Drake et al. 2008). Hybrid simulations generally have to resolve the quadratic dispersion of this mode, $\omega \propto k^2$. This requirement may create computational bottlenecks in simulations of strongly inhomogeneous magnetospheric and laboratory plasmas, where whistler timescales typically span several orders of magnitude (Omelchenko and Karimabadi 2012, 2022). If not accurately integrated in time (or resistively damped), the spurious short-wavelength oscillations may grow explosively unstable from noise and terminate simulation (Lin et al. 2008). It should also be noted that although particle noise in hybrid-PIC simulations typically degrades their physical resolution, the Lagrangian (particle) approach enables transport of ion species with less numerical diffusion compared to the Eulerian approach to solving the Vlasov equation on velocity meshes (von Alfthan et al. 2014).

Collisionless reconnection at the magnetopause and in the tail plasma sheet. The physics of magnetic reconnection in the magnetosphere can be investigated by carrying out global hybrid simulations with an ad-hoc current-dependent resistivity. For the dayside magnetopause, the modeling topics include the structures of ion diffusion region and outflow regions (Tan et al. 2011), global evolution of flux transfer events (FTEs) and magnetic flux ropes (Omidi and Sibeck 2007; Guo et al. 2020, 2021d), propagation of kinetic Alfvén waves and Poynting flux from reconnection (Wang et al. 2019), ion cusp precipitation and energy spectrum (Omidi and Sibeck 2007; Tan et al. 2012), the triggering of reconnection by solar wind discontinuities (Omidi et al. 2009; Pang et al. 2010; Guo et al. 2021c), and magnetosheath turbulence (Chen et al. 2021; Ng et al. 2021). Likewise, high-latitude reconnection tailward of the cusp under northward IMF has also been simulated (Lin and Wang 2006; Guo et al. 2021e). The 3D global physics of storm-time magnetotail reconnection, fast flow and entropy bubbles, the Hall-effects control of dawn-dusk asymmetry (Lin et al. 2014; Lu et al. 2016; Lin et al. 2017), and the associated global Alfvénic coupling between the magnetotail and the ionosphere under southward IMF have been simulated using the ANGIE3D code (Cheng et al. 2020). An Attempt has also been made to investigate the subsequent connection of fast flows to the ring current and radiation belt by combining ANGIE3D with the Comprehensive Inner Magnetosphere-Ionosphere (CIMI) model (Lin et al. 2021).

Inclusion of multi-species plasma ion populations of solar wind origin and improved representation of ionospheric outflow. In general, global hybrid models may include multiple ion species for representing solar wind and ionospheric outflow plasmas. Ionospheric outflow ions should be treated kinetically and self-consistently in order to properly account for their impact on the Earth's magnetosphere. Multi-fluid MHD models do not account for ion resonance acceleration and cyclotron effects, especially for heavy ions (Toledo-Redondo et al. 2021). Self-consistent 3D hybrid simulations of the impact of oxygen outflow on the magnetotail configuration and stability have recently been performed with the HYPERS code (Mouikis et al. 2021; Omelchenko et al. 2023).

Modeling local reconnection and electron scale physics. In MHD simulations, magnetic reconnection is often a result of mesh-dependent diffusion that is difficult to control numerically. The hybrid-PIC model is inherently more robust in this regard because ions are modeled as Lagrangian particles. As a result, reconnection onset and dynamics are controlled by the Hall physics and parameter-dependent resistivity. The hybrid-PIC model is also known to accurately reproduce reconnection rate when the ion inertial and cyclotron scales are properly resolved (Stanier et al. 2015). Further modifications of the hybrid model, which for instance may incorporate finite electron mass effects (e.g. Omelchenko et al. 2021c), could increase physical fidelity of reconnection modeling in the future.

Modeling non-MHD waves in a global context. The standard hybrid model supports ion cyclotron, whistler, and kinetic-Alfvén wave modes, which play an important role in regulating plasma turbulence in the Earth's magnetosphere and impact its global behavior. Modern observations report streams of non-Maxwellian ions that excite plasma turbulence through numerous kinetic instabilities that cannot be modeled within MHD. Global hybrid modeling naturally incorporates the ion kinetic effects into global models of the Earth's magnetosphere, which helps advance our understanding of the effects of turbulent plasma dynamics on global physical processes.

Other applications. 3D hybrid codes in space physics have been used to simulate shock-driven ion acceleration (Caprioli 2014; Guo and Giacalone 2013), solar wind turbulence (Franci et al. 2018; Roytershteyn et al. 2015), the Moon's wake (Fatemi et al. 2017; Kallio et al. 2019; Omelchenko et al. 2021b), planetary magnetospheres (Jarvinen et al. 2020), and small space bodies (Alho et al. 2019; Delamere 2009). Comparing results from full-scale 3D hybrid simulations of small space bodies with satellite observations provides yet another important route for validation and further extension of the hybrid approach to plasma modeling. Importantly, the recent advances in ionosphere-magnetosphere coupling (Lin et al. 2022b), code optimization (Dong et al. 2021) and multiscale computing (Omelchenko et al. 2021a) have greatly improved the prospects for hybrid simulations to become a key factor to consider in the overall theory of global solar wind-magnetosphere-ionosphere interactions.

4.3 Numerics

Equations (47)-(49), together with the self-consistent equations of motion for the ion macroparticles, may present computational challenges when used for modeling complex 3D plasma systems, such as the Earth's magnetosphere. Below we discuss some recent computational advances aimed at overcoming these problems.

4.3.1 Spatial Scales

Hybrid-PIC simulations typically intend to resolve the spatial scales of the order of the ion inertial length, d_i and ion cyclotron radius, r_{ci}. The physical validity regime of the hybrid model ranges from large MHD scales down to $kr_{ci} \sim 1$ and $\omega t \sim 1$. The actual physical resolution of a hybrid simulation is largely determined by (i) how well these characteristic lengths are resolved on a mesh, (ii) how many particles are used. For the typical solar wind proton inertial length, $d_i \sim 100\ km$, the Earth's radius, $R_E \sim 60d_i$. To encompass the whole magnetosphere, the computational mesh in a global simulation should cover the magnetopause with a typical standoff distance, $R_{MP} \sim 10R_E \sim 600d_i$ and the magnetotail stretching from the Earth to far distances, $R \sim 100R_E \sim 6000d_i$. The need to accurately account for the "far-field" inflow and outflow boundary conditions in the presence of a magnetic dipole may additionally require multiplying these dimensions by a factor of 2-3. Approximating such large 3D domains with uniform meshes with cell sizes of the order of $\sim 1d_i$ is prohibitively expensive for the hybrid model because of the need to advance ions and fields at all cells on kinetic scales.

To overcome these restrictions, several options are available. First, one may increase the cell size beyond $1d_i$ at the expense of lower accuracy in resolving the Hall physics (Omelchenko and Karimabadi 2012). Second, one may artificially increase the ratio between the solar wind ion inertial length d_i ($d_i \sim r_{ci}$ for the outer magnetosphere regions with ion $\beta \sim 1$) and the magnetopause distance R_{MP}, in order to better accommodate the available computation resources while still choosing a sufficiently large value of R_{MP}/d_i for assuring the separation between the global and local-kinetic scales (Omidi et al. 2004). Both approaches efficiently "downscale" the Earth's magnetosphere. For instance, ANGIE3D (Lin et al. 2014; Lin and Wang 2005) does it by artificially reducing the solar wind plasma density, i.e. by inflating the characteristic inertial ion length and proportionally increasing the Alfvén speed. Alternatively, H3D (Karimabadi et al. 2014), hybrid-VPIC (Dong et al. 2021), and HYPERS (Omelchenko and Karimabadi 2012) employ the physical ion inertial length but scale the realistic magnetopause standoff distance down by a factor of 4-6 by using a weaker magnetic dipole.

Regardless of a chosen magnetosphere scaling method, present-day 3D hybrid codes typically use $R_{MP}/d_i \geq 100$. One of the largest 3D hybrid simulations to date was performed with HYPERS for $R_{MP}/d_i \simeq 160$ on a uniform mesh with approximately $1000 \times 2000 \times 2000$ cells (Omelchenko et al. 2021b). To speed up global 3D simulations, hybrid codes may employ nonuniform meshes, among which 'stretched" (logically mapped) Cartesian meshes are the simplest. Nonuniform meshes typically maintain high resolution in a central domain of interest, while expanding cells towards domain boundaries (Omelchenko et al. 2021a; Lin and Wang 2005) to guarantee that the dipole field vanishes at the inflow/outflow (GSM X) and lateral (GSM Y and Z) boundaries so that robust local boundary conditions can be implemented. Sometimes, for simplicity, the lateral domain boundaries may be assumed to be periodic (e.g. Turc et al. 2015; Müller et al. 2011). This simplification, however, makes a global simulation valid for shorter simulation periods, until reflected particles or electromagnetic perturbations reach the periodic boundaries. To improve the counting statistics for macro-particles, splitting techniques may be used to enhance energetic particle distributions in the dayside magnetosphere (Omelchenko et al. 2021a) and magnetotail (Lin et al. 2007; Wang et al. 2009).

To further reduce the number of computational cells in global simulations, one may employ curvilinear (e.g., spherical) meshes (e.g. Dyadechkin et al. 2013; Guo et al. 2021b).

For example, to capture the short-wavelength physics along the shock normal and the magnetopause, early hybrid simulations, focusing on the dayside regions, employed cylindrical (2D) (Swift 1996; Lin et al. 1996; Lin 2002) and spherical (3D) (Lin and Wang 2005) coordinate systems. The spherical coordinate lines, however, have a singularity on the polar axis, which was handled by rotating the polar coordinates to the equator and omitting a conic region around them, while keeping the physical polar regions inside the domain (Lin and Wang 2005). A 2D hybrid simulation of the magnetotail also used curvilinear coordinates to accommodate the tail geometry (Swift and Lin 2001; Lin 2002). Similar to the need of assuring proper numerical resolution for resolving the kinetic scales along the curved or oblique boundary surfaces on the Cartesian meshes, special care is also necessary for the curvilinear meshes, especially when their coordinate lines are not orthogonal (Swift and Lin 2001). Compared to Cartesian meshes, curvilinear meshes may introduce additional discretization errors due to their (i) typically lower orders of numerical approximation, (ii) anisotropic particle-mesh weighting. These errors lead to various numerical artefacts and non-conservation of particle momentum ("self-forces"). As a result, it is necessary to benchmark results from simulations obtained with curvilinear meshes with similar simulations performed with Cartesian meshes (Dyadechkin et al. 2013).

Some global hybrid codes employ adaptive mesh refinement (AMR) (Leclercq et al. 2016; Müller et al. 2011). Hybrid AMR simulations, however, may suffer from spurious particle "self-forces" and wave reflections that occur at the mesh refinement interfaces. To mitigate these artefacts, AMR algorithms are typically complemented with smoothing procedures, which, however, should be performed with caution in order to avoid affecting underlying physics in the regions of interest.

4.3.2 Temporal Scales

In addition to the "slow" Alfvénic (MHD) timescales, hybrid codes need to follow the "fast" ion kinetic, ion cyclotron and whistler timescales. This requirement typically makes global hybrid simulations numerically "stiff" in the near-Earth space, where timesteps, required for numerical accuracy and stability, may become prohibitively small (Omelchenko and Karimabadi 2012). To partially mitigate these effects, the kinetic ions may be replaced in this region by a dense fluid (Swift 1996; Lin et al. 2021). Hybrid-PIC simulations inherently generate spurious oscillations with large wave numbers, $k \sim 1/\Delta$ and high frequencies, $\omega \sim k^2 \sim 1/\Delta^2$. If not properly integrated or resistively damped, these noisy oscillations may explosively grow and abort simulation (Lin et al. 2008). Deleterious instabilities may be avoided by applying "noise filtering" (smoothing) or/and various "flux-limiting" techniques for electric and magnetic fields. These modifications, however, need to be implemented with caution, as the may produce artificial solutions not supported by the hybrid model.

For accuracy, typical full-orbit particle solvers ("pushers") require that time steps, Δt_p should be small enough that $\Omega \Delta t_p \ll 1$, where Ω is the local ion gyro-frequency. Using the same time step for all particles may create another numerical bottleneck in global hybrid simulations of the Earth's magnetosphere. For instance, particle time steps of the order of $\Omega_0 \Delta t_p \sim 0.05$ (where Ω_0 is the ion gyro-frequency computed with respect to the IMF strength, B_{IMF}) fairly well describe ion gyro-motion in the solar wind (e.g. Turc et al. 2015; Guo et al. 2021b). At the same time, gyro-orbits and drifts of ions with $\Omega \gtrsim 10\ \Omega_0$ (e.g. found in the cusp or some parts of the magnetosheath), will not be reproduced with accuracy. As a remedy, in some simulations, sub gyro-orbit time steps may be employed in these (large magnetic field) regions of the magnetosphere (Lin et al. 2014).

To summarize, predicting optimum global time steps for the particles and fields in global hybrid simulations is difficult in practice. This challenge has been addressed by replacing time stepping with an asynchronous approach to time integration, which combines discrete-event simulation (DES) with elements of artificial intelligence: Event-driven Multi-Agent Planning System (EMAPS) (Omelchenko and Karimabadi 2006b, 2022). EMAPS enables time advance of individual particles and local fields on meshes of arbitrary topology by integrating them on their self-adaptive timescales, similar to Conway's Game of Life, where simulation elements evolve asynchronously based on a set of local interaction rules, rather than being updated synchronously at global time steps (Gardner 1970). Thus, EMAPS effectively performs the role of an intelligent "simulation time operating system". This approach was first applied to model 1D collisionless plasma shocks (Omelchenko and Karimabadi 2006a) and fluids (Omelchenko and Karimabadi 2006b, 2007). Implemented in HYPERS (Omelchenko and Karimabadi 2012), EMAPS has enabled efficient and accurate global 3D hybrid simulations of the Earth's magnetosphere (Omelchenko et al. 2021a,b, 2023).

Global 3D hybrid simulations of the Earth's magnetosphere are typically performed for simulation periods, $\Omega_0 t \sim 100 - 500$, where Ω_0 is the IMF based proton cyclotron frequency. Assuming $B_{IMF} = 5\ nT$, these simulations formally span relatively short (compared to MHD) magnetospheric times, $t < 20$ min. For convenience, in order to present physical results in "magnetospheric hours", some modelers multiply this simulation time by a model scaling factor (Lin et al. 2022b). Although this scaling is useful for comparing "macro-scale" simulation phenomena with observations, it is not appropriate for describing ion kinetic effects, e.g. those that drive the "magnetokinetic" formation of high-speed jets (Omelchenko et al. 2021a).

4.3.3 Plasmasphere and Ionosphere

Currently, hybrid codes cannot model global magnetospheric convection lasting many hours or days, e.g. a steady-state process of magnetotail loading and unloading. Therefore, global hybrid models typically assume a simple perfectly conducting or resistive ionosphere, where dipole magnetic field lines are "tied up" to the inner boundary (zero electric field) or allowed to diffuse due to its finite resistivity, respectively. model (e.g. Lin et al. 2021),

To avoid computing fast kinetic timescales, a cold, incompressible, dense ion fluid may be assumed to co-exist together with low-density particle ions in the inner magnetosphere within the distance of plasmasphere, where the plasma density is high (Swift 1996; Lin and Wang 2005; Lin et al. 2014). In ANGIE3D, this region is bounded by the near-Earth (inner) boundary, which is located at a radial distance at $r \simeq 3.5\ R_E$ in the inner magnetosphere. The field-aligned currents, calculated near this inner boundary and mapped along the geomagnetic dipole field lines down to the ionospheric altitude (1000 km), are used as input to the ionospheric potential equation solved on a sphere (Lin et al. 2014, 2021):

$$\nabla \cdot (-\boldsymbol{\Sigma} \cdot \nabla \Phi) = J_{\parallel} sin I, \tag{50}$$

where Σ is the conductance tensor, Φ is the electric potential, $J_{\parallel}$ is the mapped field-aligned current density, and I is the inclination of the dipole field at the ionosphere. The static analytical model of Hall and Pederson conductance that accounts for EUV and diffuse auroral contributions can be used for the conductance tensor. Similarly to the global MHD models, the ionospheric electric field is mapped along the dipole lines back to the inner magnetospheric boundary, to serve as a boundary condition for the cold ion fluid (Lin et al. 2021).

4.4 Hybrid Simulations with Inertial Electrons: Hybrid-PIC Code CHIEF

The MMS mission investigates physical processes like magnetic reconnection, shock waves and turbulence, which span from ion to electron scales. Simulation studies of these processes should ideally cover full kinetic physics from ion to electron scales for which the necessary present and near-future computational resources are prohibitively expensive. Therefore simulation models, which cover different scale ranges and physical phenomena, are used.

Hybrid-kinetic plasma simulation model, introduced in the previous Sect. 4.1, treats ions as kinetic species and electrons as a massless fluid. This restricts their applicability to physical processes in which not only electron kinetic effects are not important but also to the scales exceeding by far the electron scales. Hybrid-kinetic codes with inertia-less electrons, discussed in Sect. 4.1, can, therefore, be used to simulate global phenomena and in some cases for specifically limited physics studies of magnetic reconnection, plasma turbulence and shock waves.

The validity of hybrid-kinetic model can, however, be extended down to electron length scales, viz., to electron inertial length by considering electrons as an inertial fluid (Jain et al. 2023). Since the electron kinetic physics is still ignored such plasma model might computationally be more feasible than the fully kinetic model and describe larger scale phenomena and plasma process like magnetic reconnection, plasma turbulence and shock formation in collisionless plasmas, in which electron scale structures develop.

4.4.1 Model Equations: Inertial Electrons

Hybrid-kinetic model treats ions as kinetic species and electrons as an inertial fluid. In hybrid-kinetic simulation codes, ion dynamics can be described by solving either the ion's Vlasov equation using Eulerian methods or the equations of motion for ion macro-particles using semi-Lagrangian Particle-in-Cell (PIC) method. Solving Vlasov equation is computationally more expensive. Here we discuss the hybrid-PIC codes which treat ions as Lagrangian macro-particles modelled via the PIC method. Following are the governing equations of hybrid-PIC model.

$$\frac{d\mathbf{x}_i}{dt} = \mathbf{v}_i, \tag{51}$$

$$m_i \frac{d\mathbf{v}_i}{dt} = e(\mathbf{E} + \frac{\mathbf{v}_i \times \mathbf{B}}{c}), \tag{52}$$

$$\nabla \times \mathbf{E} = -\frac{1}{c}\frac{\partial \mathbf{B}}{\partial t}, \tag{53}$$

$$\nabla \times \mathbf{B} = \frac{4\pi}{c}\mathbf{J}, \tag{54}$$

$$\mathbf{J} = e(n_i \mathbf{u}_i - n_e \mathbf{u}_e), \tag{55}$$

$$n_i = n_e \tag{56}$$

$$\mathbf{E} = -\frac{\mathbf{u}_e \times \mathbf{B}}{c} - \frac{1}{en}\nabla p_e - \frac{m_e}{e}\left(\frac{\partial \mathbf{u}_e}{\partial t} + (\mathbf{u}_e \cdot \nabla)\mathbf{u}_e\right) + \eta \mathbf{J}, \tag{57}$$

$$p_e = C n_e^{\gamma} \tag{58}$$

The electric and magnetic fields ($\mathbf{E}$ and $\mathbf{B}$ respectively) in Maxwell's equations, Eqs. (53) and (54), are coupled to the plasma dynamics via the total current density $\mathbf{J} = e(n_i \mathbf{u}_i - n_e \mathbf{u}_e)$ resulting from the bulk motion of ions (number density n_i, bulk velocity $\mathbf{u}_i$) and electrons (number density n_e, bulk velocity $\mathbf{u}_e$). Ion's number density n_i and the bulk velocity $\mathbf{u}_i$ is obtained from their positions $\mathbf{x}_i$ and velocities $\mathbf{v}_i$ governed by Eqs. (51) and (52). Electron dynamics is governed by quasi-neutrality condition, Eq. (56), momentum equation of the inertial electron fluid, Eq. (57), and equation of state relating electron scalar pressure p_e with electron number density n_e, Eq. (58). Here, e is the fundamental charge, m_i ion mass, m_e electron mass, η collisional resistivity, γ the adiabatic constant and C is a constant (to be determined from initial conditions). Equations (51)-(58) are the fundamental equations of the hybrid-kinetic model with inertial electron fluid (Jain et al. 2023). These equations differ from the equations of hybrid-kinetic model with inertia-less electron fluid only by the electron inertial terms proportional to m_e on the RHS of Eq. (57). Addition of electron inertial terms in Eq. (57) makes the numerical solution of these equations much more involved in comparison to the case of inertia-less electron fluid. The algebraic calculation of electric field from Eq. (57) is not as straightforward as in the case of the inertia-less electron fluid. One needs to now calculate time derivative of $\mathbf{u}_e$ or find some other way to obtain electric field. The calculation of magnetic field also now requires numerical solution of additional elliptic partial differential equations arising because of the finite electron inertia.

In majority of the hybrid-kinetic codes with inertial electrons, evolution of magnetic field is followed by solving an evolution equation for the generalized vorticity $\mathbf{W} = \nabla \times \mathbf{u}_e - e\mathbf{B}/m_e c$ obtained by taking curl of Eq. (57) and using Eq. (53). This equation is,

$$\frac{\partial \mathbf{W}}{\partial t} = \nabla \times [\mathbf{u}_e \times \mathbf{W}] - \nabla \times \left(\frac{\nabla p_e}{m_e n} \right) - \nabla \times \left(\frac{e\eta}{m_e} \mathbf{J} \right). \tag{59}$$

The magnetic field is then calculated by solving an elliptic partial differential equation (PDE) which is obtained by substituting for $\mathbf{u}_e$ from Eq. (54) and (55), $\mathbf{u}_e = \mathbf{u}_i - c\nabla \times \mathbf{B}/(4\pi e n)$, in the expression for $\mathbf{W} = \nabla \times \mathbf{u}_e - e\mathbf{B}/m_e c$.

$$\frac{c}{4\pi e} \nabla \times \left(\frac{\nabla \times \mathbf{B}}{n} \right) + \frac{e\mathbf{B}}{m_e c} = \nabla \times \mathbf{u}_i - \mathbf{W}. \tag{60}$$

Some of the hybrid-kinetic codes make approximations of electron inertial terms in Eq. (57) and (60) to simplify their numerical solutions (Lipatov 2002; Shay et al. 1998; Kuznetsova et al. 1998). Spatial density variations are neglected in Eq. (60) (Shay et al. 1998; Kuznetsova et al. 1998). The electric field was then calculated from the generalized Ohm's law by neglecting the electron inertial term with time derivatives of the electron fluid velocity (Kuznetsova et al. 1998). These approximations are valid when length scale of variations is much larger than the electron inertial length. For a detailed discussion of these approximations, see Muñoz et al. (2018). These hybrid-kinetic codes which partially included electron inertial effects have mainly been used to study collisionless magnetic reconnection (Shay et al. 1999; Kuznetsova 2000; Kuznetsova et al. 2001). In particular Shay et al. (1998) used an evolution equation for a scalar electron pressure while Kuznetsova et al. (1998) included the full electron pressure tensor to take into account the non-gyrotropic effects.

More recently a hybrid-kinetic code CHIEF (Code Hybrid With Inertial Electron Fluid) was developed (Muñoz et al. 2018). This code solves Eqs. (57) and (60) without making any of the electron inertia related approximations used by other codes. The details of the numerical algorithm to solve Eqs. (51)-(58) are discussed by Muñoz et al. (2018). CHIEF was used to simulate kinetic plasma turbulence and it was found that the electron inertia related approximations are not valid in electron scale current sheets formed in the turbulence (Jain et al. 2022; Muñoz et al. 2023).

Some hybrid-kinetic codes with electron inertia calculate electric field from an elliptic PDE instead of Eq. (57) (Amano et al. 2014; Valentini et al. 2007). The elliptic PDE for the electric field is obtained by taking curl of Faraday's law, Eq. (53), and utilizing Eqs. (54) and (57). Electron inertia effects were considered in the elliptic equation for the electric field while, still, the electron inertia term was ignored that contains the divergence of the electric field. Two dimensional simulations of kinetic plasma turbulence have shown that this approximation is not valid from ion to electron scales (Jain et al. 2022).

4.4.2 Hybrid-PIC Code CHIEF for Magnetic Reconnection Studies

Hybrid-PIC code CHIEF can be used to study magnetic reconnection with a large guide magnetic field in which case electron non-gyrotropy is weak and reconnection electric field is expected to be balanced by the electron inertial terms in generalized Ohm's law. Indeed, PIC simulations of magnetic reconnection have shown that electron inertial terms are significant to balance the reconnection electric field when guide field is large (Hesse and Winske 1998; Hesse et al. 2016; Liu et al. 2013; Pritchett 2005).

A particular reconnection study for which CHIEF code can be employed is electron-only reconnection in which ions do not couple to electrons (Sharma Pyakurel et al. 2019), examples of which are discussed in Sect. 3.7.2. Electron-only reconnection has recently been discovered in space observations (Phan et al. 2018). In many of electron only reconnection events observed by MMS, guide magnetic field is significantly larger than the asymptotic value of the reconnecting component magnetic field (Phan et al. 2018; Zhou et al. 2021; Man et al. 2020; Stawarz et al. 2022). In the statistical survey of electron only reconnection events in Earth's magnetosheath, guide field in majority of the events is 1 to 10 times larger than the reconnecting component of the magnetic field (Stawarz et al. 2022). MMS observations of electron scale reconnection in Earth's magnetosheath show that the electron non-gyrotropy, which balances the reconnection electric field in weak or zero guide field case, reduces with the increasing strength of the guide magnetic field (Wilder et al. 2018). No evidence of agyrotropy was found in another MMS observations of large guide field reconnection (Eriksson et al. 2016). For the electron only events observed with large guide magnetic field and/or absence of non-gyrotropy, electrons can be modeled as an inertial electron fluid. Hybrid-PIC code CHIEF treats electrons as inertial fluid without any approximations and can be used to study electron-only reconnection.

4.4.3 Outlook

Hybrid-kinetic simulations with electron inertia provide a computationally less expensive (in comparison to fully kinetic simulations) tool to study magnetic reconnection with guide field in which bulk electron inertia is the dominant mechanism breaking the frozen-in condition of magnetic field. There have been some hybrid-kinetic simulations studies with electron inertia of guide field magnetic reconnection (Kuznetsova et al. 1998; Kuznetsova 2000; Shay et al. 1998; Califano et al. 2020; Muñoz et al. 2023). More studies are, however, required to address still many open questions about the guide field magnetic reconnection. The EMHD limit (stationary ions) of the hybrid-kinetic model with electron inertia is particularly useful to study the nature of reconnection in electron scale current sheets and is relevant for the recently discovered electron-only reconnection (Phan et al. 2018). At the same time, simulations of kinetic plasma turbulence from ion to electron scales using hybrid-kinetic model with electron inertia will shed light on the conditions under which electron scale current sheets form and reconnect with or without ion coupling in the turbulence.

5 Fully Kinetic Particle-in-Cell Simulations

5.1 Introduction

The particle-in-cell (PIC) method is a simulation method in which the plasma is treated as a collection of particles (electrons and ions), where each species is typically composed of up to 10^{12} particles in 3D cases. In PIC simulations, the motions of individual particles and the evolution of electric and magnetic fields are solved self-consistently. The electric and magnetic fields are defined on discrete grid points. In this subsection, we will explain an explicit PIC simulation, where all the quantities are updated based on the quantities obtained in the previous time step. Let us assume that the total particle number in a simulation is N_p (in other words, $N_p/2$ for ions, and $N_p/2$ for electrons). The equation of motion for the j-th particle's position $\boldsymbol{x}_j(t)$ and momentum $\boldsymbol{p}_j(t)$, where j represents an integer between 1 to N_p, is discretized in time, while Maxwell's equations for electromagnetic fields $\boldsymbol{E}(\boldsymbol{x},t)$ and $\boldsymbol{B}(\boldsymbol{x},t)$ are discretized in both space and time, using a grid spacing Δx (assuming that the grids are uniform in all the coordinates, i.e. $\Delta x = \Delta y = \Delta z$) and a time step Δt, respectively. They are given as

$$\frac{\boldsymbol{x}_j(t_n) - \boldsymbol{x}_j(t_{n-1})}{\Delta t} = \frac{\boldsymbol{p}_j(t_{n-1/2})}{m_j \gamma_j(t_{n-1/2})}, \tag{61}$$

$$\frac{\boldsymbol{p}_j(t_{n+1/2}) - \boldsymbol{p}_j(t_{n-1/2})}{\Delta t} = q_j \left[\boldsymbol{E}(\boldsymbol{x}_j(t_n), t_n) + \frac{\boldsymbol{p}_j(t_n)}{c m_j \gamma_j(t_n)} \times \boldsymbol{B}(\boldsymbol{x}_j(t_n), t_n) \right], \tag{62}$$

$$\frac{\boldsymbol{E}(\boldsymbol{x}, t_{n+1}) - \boldsymbol{E}(\boldsymbol{x}, t_n)}{\Delta t} = -4\pi \boldsymbol{J}(\boldsymbol{x}, t_{n+1/2}) + c\nabla_f \times \boldsymbol{B}(\boldsymbol{x}, t_{n+1/2}), \tag{63}$$

$$\frac{\boldsymbol{B}(\boldsymbol{x}, t_{n+3/2}) - \boldsymbol{B}(\boldsymbol{x}, t_{n+1/2})}{\Delta t} = -c\nabla_f \times \boldsymbol{E}(\boldsymbol{x}, t_{n+1}), \tag{64}$$

where m_j is a mass, γ_j is the Lorentz factor, q_j is a charge, c is the speed of light, and the operator $\nabla_f \times$ represents the finite difference version of the curl operation. The time is discretized to be $t_a = a\Delta t$, where a represents an integer n or a half integer $n + 1/2$. Field quantities such as $\boldsymbol{B}(\boldsymbol{x}, t_n)$ are defined at a grid position $\boldsymbol{x}$, while quantities such as $\boldsymbol{B}(\boldsymbol{x}_j(t_n), t_n)$ are defined at the position of a particle $\boldsymbol{x}(t_n)$. Note that $\boldsymbol{B}(\boldsymbol{x}_j(t_n), t_n)$ in the right-hand side of Eq. (62) represents the mean of $\boldsymbol{B}(\boldsymbol{x}_j(t_n), t_{n+1/2})$ and $\boldsymbol{B}(\boldsymbol{x}_j(t_n), t_{n-1/2})$. As seen in Eqs. (61) and (62), the position $\boldsymbol{x}_j$ is computed at integer time, $t = t_n$, while the momentum $\boldsymbol{p}_j$ is computed at half-integer time, $t = t_{n+1/2}$. This time staggering gives the second-order accuracy, i.e. the error is $O(\Delta t^2)$. In the same way, for the spatial discretization for $\boldsymbol{E}(\boldsymbol{x}, t)$ and $\boldsymbol{B}(\boldsymbol{x}, t)$, the Yee lattice (Yee 1966) is used, in which each component of electric and magnetic fields is defined as in Fig. 6. Also, each component of the current density $\boldsymbol{J}(\boldsymbol{x}, t)$ is defined at the same position as $\boldsymbol{E}(\boldsymbol{x}, t)$. These fields, which are spatially and temporarily staggered, are advanced using Eqs. (63) and (64), keeping the second-order accuracy in space and time. Equation (64) guarantees that the Gauss's law for the magnetic field, $\nabla_f \cdot \boldsymbol{B} = 0$, where $\nabla_f \cdot$ represents the finite difference version of the divergence, is satisfied when it is satisfied at $t = 0$.

To solve Eq. (62), the Boris method (Boris 1970) is commonly used. This method has three steps: (1) The momentum is updated from $\boldsymbol{p}_j(t_{n-1/2})$ to $\boldsymbol{p}_j(t_n)^*$, using only the electric field $\boldsymbol{E}(\boldsymbol{x}_j(t_n), t_n)$ for a half time step $\Delta t/2$. (2) The momentum vector $\boldsymbol{p}_j(t_n)^*$ is rotated to be $\boldsymbol{p}_j(t_n)^{**}$ using only the magnetic field $\boldsymbol{B}(\boldsymbol{x}_j(t_n), t_n)$ for a full time step Δt. (3) The

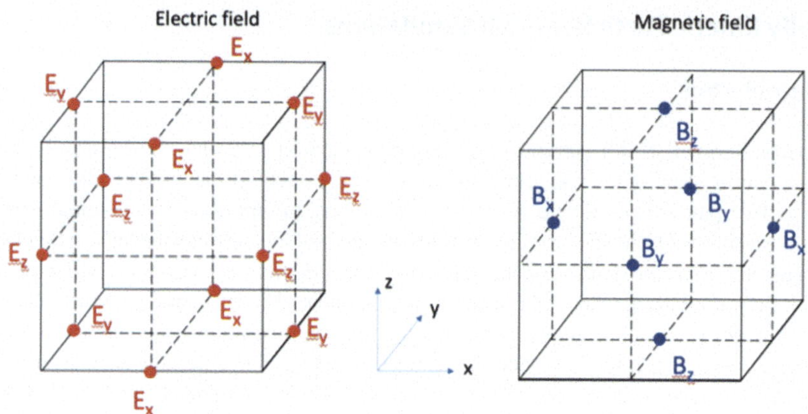

Fig. 6 Yee lattice and electric and magnetic fields in a cell in a 3D case, where the length of each side is Δx. Electric fields $\boldsymbol{E}$ are defined at the midpoint of each side of the cube, while magnetic fields $\boldsymbol{B}$ are defined at the center of each face of the cube. In a 2D case, all the quantities are defined in the x-y plane, projecting each position onto the cell in the x-y plane

rotated momentum is further updated from $\boldsymbol{p}_j(t_n)^{**}$ to $\boldsymbol{p}_j(t_{n+1/2})$, using the electric field $\boldsymbol{E}(\boldsymbol{x}_j(t_n), t_n)$ for another half time step $\Delta t/2$.

In the PIC method, each particle is not a point particle, but it has a finite size to reduce noise. The shape of a particle depends on simulation codes, but the most-commonly used shape function is a triangular function, $S_x(x - x_j) = (1- \mid x - x_j \mid /\Delta x)$ when $|x - x_j|/\Delta x < 1$ and zero otherwise, where only the x component is considered. In 2D and 3D simulations, the y and z components of the shape functions are multiplied, as $S(\boldsymbol{x} - \boldsymbol{x}_j) = S_x(x - x_j)S_y(y - y_j)$ for 2D and $S(\boldsymbol{x} - \boldsymbol{x}_j) = S_x(x - x_j)S_y(y - y_j)S_z(z - z_j)$ for 3D. Using these shape functions, the charge density $\rho(\boldsymbol{x}, t)$ is computed as $\rho(\boldsymbol{x}, t) = \sum_j q_j S(\boldsymbol{x} - \boldsymbol{x}_j)$. This way of charge assignment is reversed to compute the electric field exerted from each grid point to a particle's position. To avoid the self-force (the force due to the electric field generated by the particle itself), we must first average the electric fields defined on half-integer grids to obtain the mean electric field on each integer grid, before assigning the electric fields to the particle. The magnetic fields are assigned from grids to the particle position in the same way.

The current density can also be calculated using $\boldsymbol{J}(\boldsymbol{x}, t) = \sum_j q_j \boldsymbol{v}_j S(\boldsymbol{x} - \boldsymbol{x}_j)$, where $\boldsymbol{v}_j$ is the velocity, but the calculated $\boldsymbol{J}(\boldsymbol{x}, t)$ does not satisfy the continuum equation, $[\rho(\boldsymbol{x}, t_{n+1}) - \rho(\boldsymbol{x}, t_n)]/\Delta t + \nabla_f \cdot \boldsymbol{J}(\boldsymbol{x}, t_{n+1/2}) = 0$; therefore, the electric field calculated using Eq. (63) with this $\boldsymbol{J}(\boldsymbol{x}, t)$ does not satisfy the Gauss's law, $\nabla_f \cdot \boldsymbol{E}(\boldsymbol{x}, t_{n+1}) = 4\pi\rho(\boldsymbol{x}, t_{n+1})$. This means that we must either correct the electric field $\boldsymbol{E}(\boldsymbol{x}, t_{n+1})$ to satisfy the Gauss's law, or use another method to compute $\boldsymbol{J}(\boldsymbol{x}, t)$. A technique for the former is explained in Birdsall and Langdon (1991). For the latter, for example, Villasenor and Bunemann (1992) developed a rigorous charge conservation method for 2D and 3D PIC simulations, which guarantees that both the Gauss's law and the continuum equation are satisfied at the same time, when they are satisfied at $t = 0$.

The time step Δt must satisfy the Courant–Friedrichs–Lewy condition, $\Delta t < \Delta x/(cN_d^{1/2})$, where N_d represents the dimensionality ($N_d = 1, 2$, or 3). Also, the grid spacing Δx should be close to the Debye length λ_D, otherwise strong numerical heating occurs. Even when those conditions are satisfied, if particles are relativistic, a numerical Cherenkov instability can occur and the noise field becomes extremely large. When this occurs, a noise reduction method such as by Godfrey (1980) can be used.

To study magnetic reconnection, it is important to separate the spatiotemporal scale of protons and that of electrons, which is controlled by the ion to electron mass ratio, m_i/m_e. In general, to reduce the computing time, full PIC simulations use an artificial mass ratio, such as $m_i/m_e = 25$ and 100, which is smaller than the real mass ratio $m_i/m_e = 1836$. There are two effects if we use a smaller mass ratio. One is that the thickness of the electron-scale current layer near the reconnection X-line, which is of the order of the electron skin depth $d_e = c/\omega_{pe}$, becomes thicker than that for a realistic case. The other is that the time scale of the electron physics becomes longer than the realistic case. For example, if we choose the mass ratio 25, the spatial scale separation between the ion scale (d_i) and the electron scale (d_e) is 5 times, and the temporal scale separation between the ion scale (Ω_i^{-1}) and the electron scale (Ω_e^{-1}) is 25 times. Even though these are much smaller than those in the realistic case, the choice of the mass ratio 25 is acceptable in order to see the scale separation physics that occurs in a study of magnetic reconnection.

Various boundary conditions can be implemented including periodic, conducting wall, and open boundaries (Daughton and Scudder 2006; Ohtani and Horiuchi 2009). For open boundaries, particles that reach the boundaries are removed, and new particles are injected into the simulation box at each time. How to inject particles depends on simulation codes. For example, in Daughton and Scudder (2006), the flux of injecting particles is calculated assuming that the spatial derivative of the distribution function at the boundary is zero in the normal direction. Also, electromagnetic fluctuations can pass through the boundaries.

Some research groups use PIC codes with the adaptive mech refinement (AMR) technique (Fujimoto 2011; Innocenti et al. 2013). In AMR PIC simulations, the simulation region is subdivided based on the required spatial resolution: the regions where the small-scale physics becomes important are solved using fine grids, while the other regions outside the fine regions are solved using coarse grids. If more smaller-scale resolution is required, further finer levels of grids are produced. This technique can reduce the required particle number for fully kinetic simulations.

5.2 Magnetotail Reconnection

In this section, we introduce simulation results of magnetic reconnection obtained by the standard fully kinetic PIC simulations with uniform grids. In the Earth's magnetotail, the strength of magnetic field across the current sheet is symmetric, and the guide field (B_y field in the GSM coordinates) is small in many reconnection events. Many authors have been studying symmetric magnetic reconnection with zero guide-field, using the Earth's magnetotail parameters (Hoshino 1987; Pritchett et al. 1991; Horiuchi and Sato 1994; Dreher et al. 1996; Zhu and Winglee 1996; Hesse et al. 1996). In these simulations, the initial plasma is set up based on a Harris equilibrium (Harris 1962a): the magnetic field $B_x = B_0\tanh(z/w)$, and the density $n = n_0\text{sech}^2(z/w) + n_b$, where B_0 is the asymptotic magnetic field, w is the sheet thickness, n_0 is the peak density of the current sheet, and n_b is the background density. Also the conditions for the Harris equilibrium, $B_0^2/(8\pi) = n_0(T_i + T_e)$ and $|V_{di} - V_{de}| = [2c/(weB_0)](T_i + T_e)$, and $V_{di}/V_{de} = -T_i/T_e$, are satisfied, where T_i and T_e are the ion and electron temperatures, respectively, e is the elementary charge, and V_{di} and V_{de} are the y-directional drift velocity in the current sheet component of ions and electrons, respectively.

The following describes an example of a 2D PIC simulation of magnetotail reconnection, Hesse et al. (2018a). The system size is $L_x \times L_z = 102.4d_i \times 51.2d_i$, where

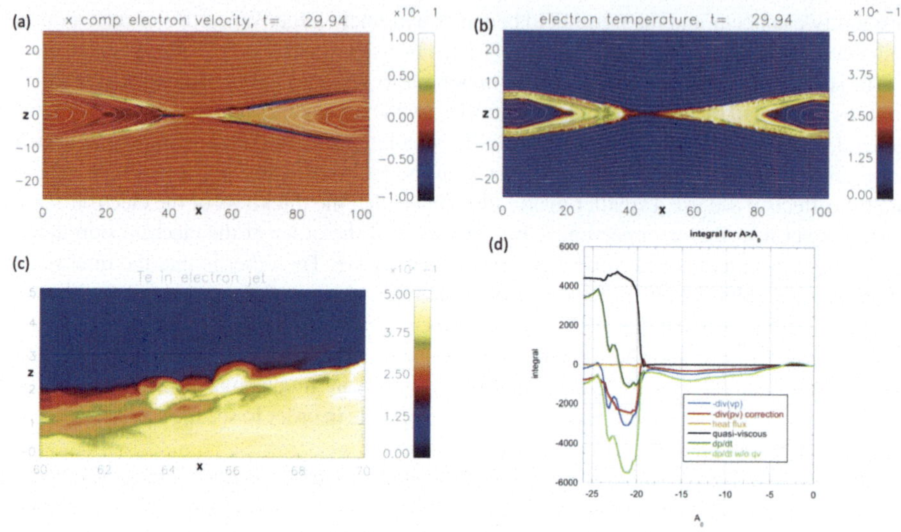

Fig. 7 2D PIC simulation result for magnetotail reconnection. (a) Electron fluid velocity V_{ex}, (b) electron temperature T_e, (c) zoom-in view of T_e, and (d) the heating term in Eq. (65). Adapted from Hesse et al. (2018a, 2019)

d_i is the ion skin depth, c/ω_{pi} with ω_{pi} being the ion plasma frequency based on n_0 ($\omega_{pi} = (4\pi n_0 e^2/m_i)^{1/2}$), and 3200 × 3200 grids are used. The mass ratio is $m_i/m_e = 100$, the sheet thickness is $w = 0.5d_i$, the temperature ratio is $T_i/T_e = 5$, the density ratio $n_b/n_0 = 0.2$, and the ratio of the plasma frequency (based on n_0) to the electron cyclotron frequency (based on B_0) is $\omega_{pe}/\Omega_e = (4\pi n_0 e^2/m_e)^{1/2}/[eB_0/(m_e c)] = 2.0$, which gives the ratio of the light speed to the Alfvén speed (based on B_0 and n_0, $v_{A0} = B_0/(4\pi m_i n_0)^{1/2}$) to be $c/v_{A0} = 20.0$. The x boundaries are periodic, and the z boundaries are conducting walls. To initiate magnetic reconnection, a perturbation is added to the magnetic field as $\delta B_x = (a_0\pi/L_z)\cos(2\pi x/L_x)\sin(\pi z/L_z)$ and $\delta B_z = -(a_0 2\pi/L_x)\sin(2\pi x/L_x)\cos(\pi z/L_z)$, which gives a reconnection X-line at the origin $x = 0$ and $z = 0$. Here a_0 is the amplitude of the perturbation. The total number of particles used in the simulation is 7×10^{10}.

After the simulation starts, the reconnection electric field E_y is generated in the diffusion region near the X-line, where both the ion and electron motions are decoupled from the magnetic field line motion, which allows a pair of magnetic field lines across the current sheet, one is in the positive z region ($B_x > 0$) and the other is in the negative z direction ($B_x < 0$), are reconnected and energy conversion occurs from the magnetic energy to the kinetic and thermal energies of ions and electrons. The reconnection rate is measured as $E_y/(B_0 v_{A0}/c)$, and in this simulation it is around 0.2 (Hesse et al. 2018b). Both ion and electron outflows are produced from the X-line toward the positive and negative x directions. Figure 7(a) shows the electron fluid velocity V_{ex}, where the bipolar positive and negative V_{ex} peaks appear along the $z = 0$ in $40 < x/d_i < 60$, and each peak value reaches near the electron Alfvén speed, $v_{Ae} = B_0/(4\pi m_e n_0)^{1/2} = (m_i/m_e)^{1/2} v_{A0}$. Outside the region of $40 < x/d_i < 60$, there are strong inflows, which also reach near v_{Ae}, toward the X-line along the separatrices. Because of these strong counter streaming electron flows (outflows and inflows), an electrostatic instability occurs that produces waves propagating along the separatrices toward the X-line, and electrons are heated due to wave-particle interactions.

Figure 7(b) and (c) show the electron temperature in the entire box and a zoom-in view that includes a separatrix. Electrons are heated inside the separatrices (panel (b)), and the zoom-in view (panel (c)) shows that there are two solitary structures due to the nonlinear evolution of the wave (at $x = 64d_i$ and $66d_i$ along the separatrix) where electron temperature significantly enhances.

The locations of the instability, along the separatrices, correspond to the boundary of the high electron temperature, which suggests the importance of the electrostatic instability to heat electrons. To understand the effect of the instability on the heating, Hesse et al. (2018a) analyzed the pressure equation:

$$\frac{\partial p}{\partial t} = -\nabla \cdot (\boldsymbol{V} p) - \frac{2}{3}\sum_{l} P_{ll}\frac{\partial}{\partial x_l}V_l - \frac{1}{3}\sum_{l,i}\frac{\partial}{\partial x_i}Q_{lii} - \frac{2}{3}\sum_{l,i(l\neq i)} P_{li}\frac{\partial}{\partial x_i}V_l, \tag{65}$$

where all the quantities are for electrons (the subscript e is omitted): p is the scalar pressure, $\boldsymbol{V}$ is the fluid velocity, and P_{ij} and Q_{ijk} are the pressure tensor and the heat tensor, respectively. The first two terms represent the compression effect, the third term is due to the heat flux, and the fourth term represents the quasi-viscous effect due to the off-diagonal components of the pressure tensor. Figure 7(d) shows the contribution of each term in Eq. (65), integrated over the region of $A > A_0$, where A is the y component of the mangnetic flux function (i.e., $B_x = \partial A/\partial z$ and $B_z = -\partial A/\partial x$). Note that $A = 0$ at the outermost z boundaries, and A is decreasing as we approach $z = 0$. Figure 7(d) indicates that the quasi-viscous term (the fourth term) is the dominat term to provide the pressure increase, leading heating in the reconnection region.

5.3 Magnetopause Reconnection

PIC simulations of magnetopause reconnection can include certain challenges due to asymmetries in the densities, temperatures and magnetic field strengths of the abutting plasmas (Sonnerup et al. 1986; Cassak and Shay 2007). Specifically, magnetospheric plasma is usually relatively sparse, hot, and threaded by a strong magnetic field, while magnetosheath plasma (which arises from shocked solar wind) is denser, cooler, and includes a somewhat weaker field.

From a simulation perspective, the density asymmetry – which can exceed an order of magnitude – can be particularly problematic. PIC simulations are inherently noisy. The random fluctuations tend to follow Poissonian statistics with an amplitude scaling as $1/\sqrt{N_{pc}}$, with N_{pc} the number of (macro) particles per computational cell. If variations in the number of macroparticles directly translate to variations in density, a $16:1$ ratio between the magneosheath and magnetospheric plasma densities will produce noise levels an unacceptable four times larger in the latter than the former. One obvious approach – throwing more particles at the problem – can quickly become computationally burdensome. An alternative is the use of particle weighting, in which each particle is assigned a weight that determines its significance in the calculation of particle moments (e.g., charge and current density). Doing so allows for an initially uniform distribution of particles with a roughly constant noise level. More sophisticated algorithms allow for the splitting and joining of particles as the simulation progresses to account for the development of density variations and to address computational load imbalances.

A 2D simulation of the magnetopause with p3d, a PIC code employing weighted particles (Zeiler et al. 2002), was presented in Swisdak et al. (2018). In its normalization, a reference magnetic field strength B_0 and density n_0 define the velocity unit

$v_{A0} = B_0/(4\pi m_i n_0)^{1/2}$. Times are normalized to the inverse ion cyclotron frequency $\Omega_{i0}^{-1} = m_i c/(eB_0)$, lengths to the ion inertial length $d_{i0} = c/\omega_{pi0}$ (where $\omega_{pi0} = (4\pi n_0 e^2/m_i)^{1/2}$ is the ion plasma frequency), electric fields to $v_{A0}B_0/c$, and temperatures to $m_i v_{A0}^2$.

The initial conditions closely mimic those observed during the diffusion region encounter described in Burch et al. (2016). In the system considered here, B_0 and n_0 correspond to their asymptotic magnetosheath values: $B_0 = 23$ nT and $n_0 = 11.3$ cm^{-3}. The simulation uses an LMN coordinate system in which the reconnecting field parallels the L axis (roughly north-south), the M axis runs roughly east-west, with dawnward positive, and the N axis points radially away from Earth and completes the right-handed triad. The computational domain has dimensions $(L_L, L_N) = (40.96, 20.48)$ with periodic boundary conditions used in all directions. While particles can move in the M direction, variations in physical quantities are not permitted: $\partial/\partial M = 0$.

The reconnecting component of the field B_L and the ion and electron temperatures, T_i and T_e, vary as functions of N with hyperbolic tangent profiles of width 1. The asymptotic values of n, B_L, T_i, and T_e in code units are 1.0, 1.0, 1.37, and 0.12 in the magnetosheath and 0.06, 1.70, 7.73, and 1.28 in the magnetosphere. Pressure balance determines the initial density profile. The guide field $B_M = 0.099$ is much smaller than B_L (i.e., the reconnection is nearly anti-parallel) and initially uniform. While not an exact kinetic equilibrium, the unperturbed configuration is in force balance and would not undergo significant evolution during the timescales of interest. Instead, a small initial perturbation is introduced to trigger reconnection onset.

The ion-to-electron mass ratio is chosen to be 100, which is sufficient to separate the electron and ion scales (the electron inertial length $d_{e0} = 0.1 d_{i0}$). The normalized speed of light is $c = 15$ so that $\omega_{pe}/\Omega_e = 1.5$ in the asymptotic magnetosheath and ≈ 0.2 in the asymptotic magnetosphere; the observed ratios are larger, ≈ 46 and 7, and as a consequence the simulation's Debye length is larger than in the real system. However, since the development of reconnection does not appreciably depend on physical effects at the Debye scale the expected impact is minimal. The spatial grid has resolution $\Delta = 0.01$ in normalized units while the Debye length in the simulation's magnetosheath, ≈ 0.03, is the smallest physical scale. To ameliorate numerical noise, particularly in the low-density magnetosphere, each grid cell initially contains 3000 weighted macroparticles.

Figure 8 shows results. The asymmetry in the field strength is apparent in the distribution of the field lines, with the separatrices extending much farther (in the N direction) into the plasma of the magnetosheath (top) than the magnetosphere (bottom). The Hall electric and magnetic fields (panels c and e) differ significantly from the case of symmetric reconnection, with the former concentrated almost exclusively on the magnetospheric side while the latter is almost completely dipolar rather than quadrupolar. Due to the use of weighted particles, the numerical noise is similar on both sides.

5.4 The 3D Nature of Magnetic Reconnection

The third dimension pointing out of the 2D reconnection plane introduces numerous plasma instabilities (e.g. Daughton et al. 2011; Graham et al. 2025). In addition to these secondary instabilities, inherent 3D nature of reconnection X-line is also omitted in 2D pictures, which includes the effect of limited X-line extent (Shay et al. 2003; Liu et al. 2019; Huang et al. 2020; Pyakurel et al. 2021; Huang et al. 2024), its tendency of spreading (Huba 2003; Shay et al. 2003; Lapenta et al. 2006; Nakamura et al. 2012; Shepherd and Cassak 2012; Li et al. 2020; Arencibia et al. 2023; Li et al. 2023c; Lin et al. 2025), and its orientation preference (Sonnerup 1974; Swisdak and Drake 2007; Hesse et al. 2013; Aunai et al. 2016; Liu et al. 2013, 2018).

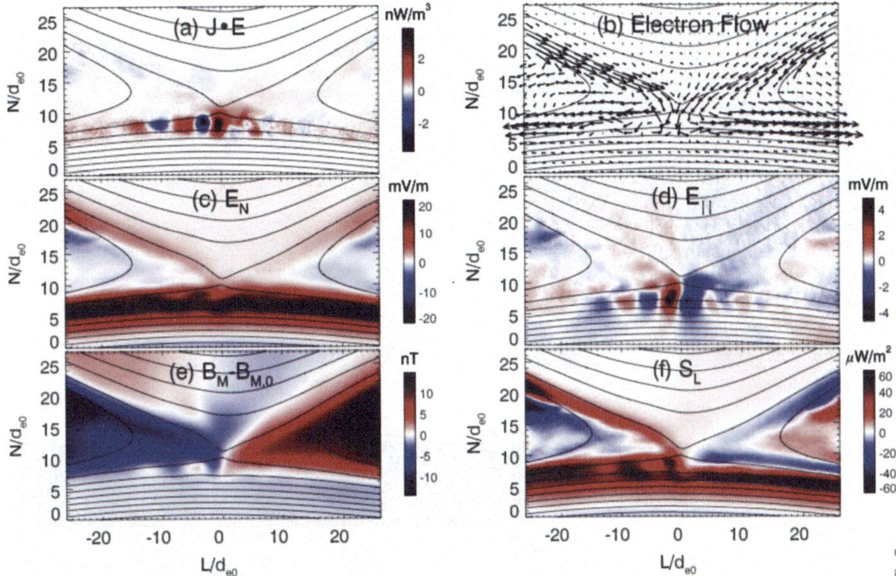

Fig. 8 Simulation results overplotted with magnetic field lines. (a) The $\mathbf{J} \cdot \mathbf{E}$ term from Poynting's theorem; (b) In-plane electron flow field; (c) E_N, the normal component of the electric field; (d) $E_{\|}$, the component of the electric field parallel to the magnetic field; (e) $B_M - B_{M,0}$, the change in the out-of-plane component of the magnetic field from its (spatially constant) initial value; (f) S_L, the horizontal component of the Poynting flux. Adapted from Swisdak et al. (2018)

We briefly highlight the property of reconnection X-lines with a short extent here. To sustain a current sheet, electrons and ions drift in opposite directions. This fact introduces the asymmetry along the X-line (current) direction. To reveal this effect, Liu et al. (2019) and Huang et al. (2020) studied magnetic reconnection with the X-line spatially confined in the current direction. They included thick current layers to prevent the reconnection from spreading out of the two ends of a thin current sheet that has a thickness on an ion inertial (d_i) scale. The x component of the magnetic field is given as $B_x = B_0 \tanh[z/L(y)]$, where the half-thickness $L(y) = L_{min} + (L_{max} - L_{min})[1 - f(y)]$ and $f(y) = [\tanh((y + w_0)/S) - \tanh((y - w_0)/S)]/[2\tanh(w_0/S)]$, $L_{min} = 0.5d_i$, $L_{max} = 4d_i$, and $S = 5d_i$. The parameter w_0, which controls the y-extent of the thin current region (L_{y-thin}), is varied from $w_0 = 2d_i$ to $20d_i$, corresponding to the y-extent of the thin current region from $L_{y-thin} \sim 4d_i$ ($w_0 = 2d_i$) to $L_{y-thin} \sim 30d_i$ ($w_0 = 20d_i$). The density is $n = n_0 \mathrm{sech}^2[z/L(y)] + n_b$, and $n_b = 0.3n_0$. The size of the system is $L_x \times L_y \times L_z = 32d_i \times 64d_i \times 16d_i$. The periodic boundary condition is used for x and y, and the conducting walls are placed in the z boundaries. Over 2.6×10^{10} particles for each species are used.

The resulting reconnection is shown in Fig. 9, which is for $L_{y-thin} \sim 30d_i$. Liu et al. (2019) found that the reconnection rate and the outflow speed drop significantly when the extent of the thin current sheet, L_{y-thin}, is less than $\mathcal{O}(10d_i)$. When the thin current sheet extent is long enough, it consists of two distinct regions: a suppressed reconnecting region (on the ion-drifting side, marked in Fig. 9(b)) exists adjacent to the active region where reconnection proceeds normally as in a 2D case with a typical fast rate value ≈ 0.1. The extent of this suppression region is $\mathcal{O}(10d_i)$, and it suppresses reconnection when L_{y-thin} is comparable or shorter. The time scale of current sheet thinning toward fast reconnection can be translated into the spatial scale of this suppression region, because the electron drifts

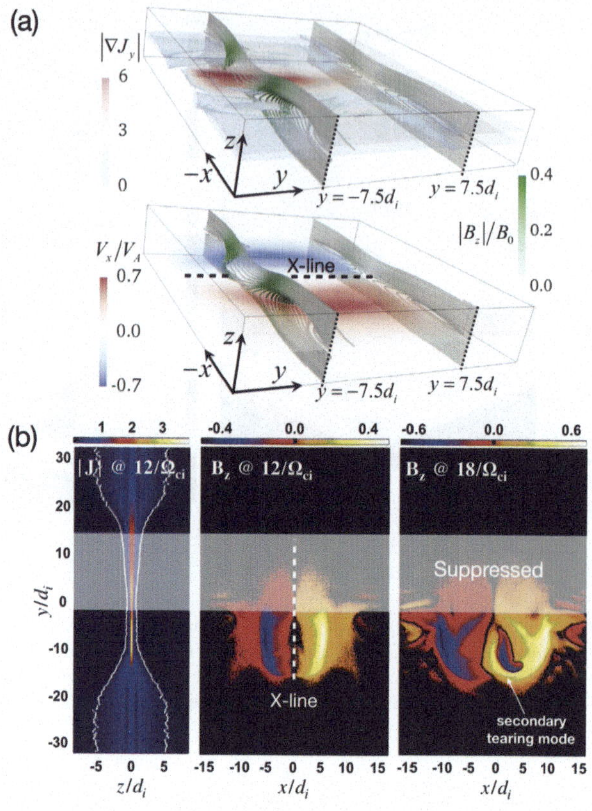

Fig. 9 3D PIC simulation results. (a) 3D view of reconnection with a limited X-line extent, where the thin current sheet region extends $30d_i$ in y (Huang et al. 2020). The mass ratio in the simulation is 25. (b) The current density on the $x = 0$ plane (left) and magnetic field B_z on the $z = 0$ plane (middle and right) (Liu et al. 2019). The mass ratio is 75. The gray shaded area represents the "suppressed reconnecting region". Adapted from Huang et al. (2020) and Liu et al. (2019)

inside the ion diffusion region transport the reconnected magnetic flux (that is critical in driving outflows and furthers the current sheet thinning) away from this region. This is a consequence of the Hall effect in 3D.

Huang et al. (2020) incorporated the length scale of this suppression region $\mathcal{O}(10d_i)$ to quantitatively model the reduction of the reconnection rate and the maximum outflow speed observed in the short X-line limit. The average reconnection rate drops because of the limited active region (where the current sheet thins down to the electron inertial scale) within the X-line. The outflow speed reduction correlates with the decrease of the $J \times B$ force, which can be modeled by the phase shift between the J and B profiles, also as a consequence of the flux transport out of the reconnection plane.

While the existence of this suppression region may explain the shortest possible azimuthal extent of dipolarizing flux bundles at Earth (Liu et al. 2015), it may also explain the dawn-dusk asymmetry observed at the magnetotail of Mercury (Sun et al. 2016, 2022), which has a global dawn-dusk extent much shorter than that of Earth.

5.5 Particle Acceleration

There have been quite remarkable advances in using PIC simulations to understand particle acceleration processes in magnetic reconnection, discussed in Oka et al. (2023), Drake et al. (2025), and Guo et al. (2023) of this collection. The simulation provided energetic particle flux, spectra and even detailed distributions that can be compared with in situ ob-

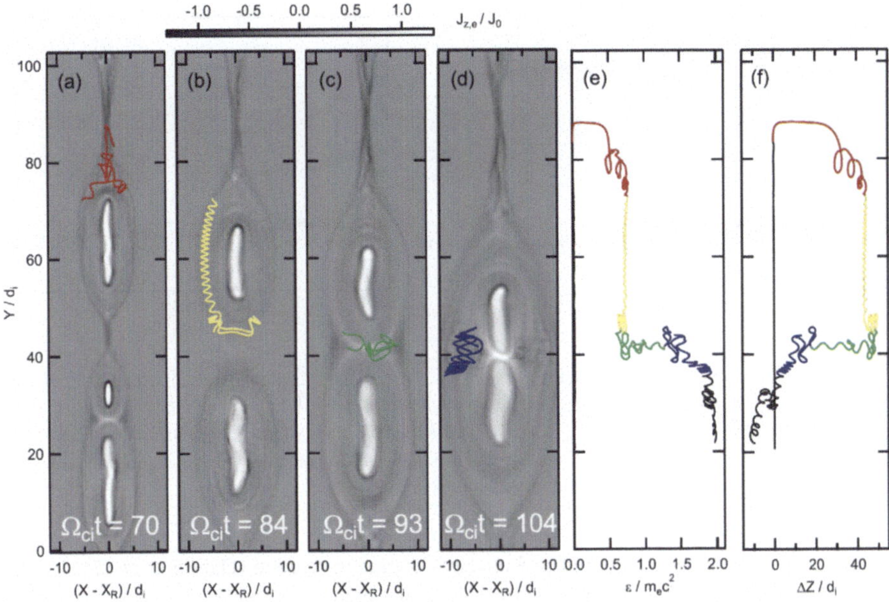

Fig. 10 An example of particle trajectory analysis. Adapted from Oka et al. (2010)

servations. We introduce several key diagnostics recently used for gaining insight in particle energization.

First, it has been a common practice to output particle trajectories to study the acceleration process (e.g., Hoshino et al. 2001; Drake et al. 2006; Fu et al. 2006; Oka et al. 2010; Guo et al. 2015). These have led to the identification of different acceleration mechanisms, as discussed in Oka et al. (2023) and Guo et al. (2023) of this collection. Figure 10 shows a representative particle trajectory adapted from Oka et al. (2010). This particle is first accelerated by an X-line (a), then further energized due to electric field during anti-reconnection between two merging island (c). The acceleration persists after the particle is ejected out of the X-line region. In addition, one can output the electric and magnetic fields and other quantities associated with particles, to complement the understanding of acceleration mechanisms.

The limitation of just showing several particle trajectories, even with the best effort, is that the "representative" examples are usually cherry-picking results and it is difficult to evaluate the relative importance of each mechanism. There has been recent effort to evaluate the acceleration processes over a large number of trajectories, (Guo et al. 2019, 2021a; Kilian et al. 2020; French et al. 2022; Li et al. 2023a).

Another method, developed and widely used over the last decade, is to study the collective energy gain, such as guiding center and pressure-restrained terms (discussed in Oka et al. 2023, this collection) using ensemble averaged moments (Dahlin et al. 2014, 2015; Li et al. 2015, 2017, 2018, 2019b; Du et al. 2018). For example, the acceleration due to the curvature (Fermi) and gradient (betatron) drifts can be evaluated under guiding center approximation. Figure 11 shows an example under the guiding-center approximation, and shows the curvature drift term is the main acceleration term. Moreover, it is possible to collect the energy dependent acceleration rates by considering particles with different energy,

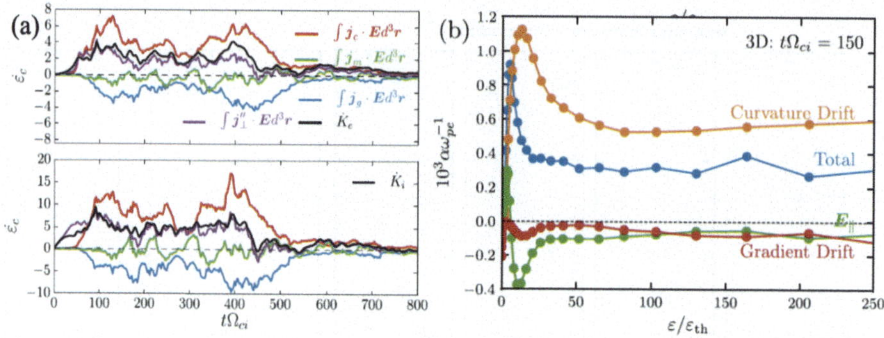

Fig. 11 (a) An example of guiding-center drift analysis. Particle energization due to different drift currents for electrons (top) and ions (bottom). j_c is due to particle curvature drift. j_g is due to particle grad-B drift. j_m is due to magnetization. $j_{//} = j_c + j_g + j_m$. $\dot{K}_e$ and $\dot{K}_i$ are the energy change rates for electrons and ions, respectively. They are all normalized by $m_e c^2 \omega_{pe}$. (b) Similar to (a) but shows energy dependent values (Li et al. 2019b) at a given time. Adapted from Li et al. (2017)

so the energization can be studied in a energy-dependent fashion (Dahlin et al. 2017; Guo et al. 2014; Li et al. 2018, 2019b). Figure 11b shows such an example.

5.6 Simulations of Magnetic Reconnection in Shock Waves

Magnetic reconnection can occur in current sheets generated in plasma turbulence, and PIC simulations have also been applied to turbulent environments (Wu et al. 2013; Matthaeus et al. 2016; Haggerty et al. 2017; Shay et al. 2018; Vega et al. 2020; Adhikari et al. 2021; Rueda et al. 2021), including turbulence in Kelvin Helmholtz vortices in the magnetopause flank region (Nakamura and Daughton 2014; Nakamura et al. 2017, 2022), and the transition region in shock waves (Matsumoto et al. 2015; Bohdan et al. 2017, 2020; Bessho et al. 2019, 2020, 2022, 2023; Ng et al. 2022).

Here, let us review 2D and 3D PIC simulation studies of magnetic reconnection in the shock turbulence. Bessho et al. (2019) used a 2D domain to study a quasi-parallel shock under the parameters in the Earth's bow shock. The size of the simulation domain is $L_x \times L_y = 375 d_i \times 51.2 d_i$, where the ion skin depth d_i has 40 grids. The plasma is uniform at $t = 0$, both ions and electrons are Maxwellian with their temperatures T_i and T_e, respectively, and the magnetic field is given as $\boldsymbol{B} = [B_0 \cos\theta, B_0 \sin\theta, 0]$, where θ is the shock angle with respect to the x axis. Periodic boundaries are used in the y direction, and conducting walls are placed in the x direction. To all the plasma particles, a negative drift speed, $-v_d$, in the x direction is given, and a uniform positive z component of electric field, as $E_z = v_d B_0 \sin\theta / c$, is set in the domain. At the right boundary, $x = L_x$, new particles for both ions and electrons are injected, using the same temperatures as the initially loaded particles, with the negative drift speed $-v_d$. At the left boundary, $x = 0$, all the particles are specularly reflected, and the incident particles and the reflected particles generate counter-streaming beams, which cause a beam instability. As a result, a non-linear wave grows near the left boundary, and a wave steepening occurs. Eventually, a shock wave forms, propagating toward the positive x direction.

In the 2D simulation, the following parameters are used: the electron and ion beta $\beta_e = \beta_i = 1$, the ratio of the plasma frequency to the electron cyclotron frequency $\omega_{pe}/\Omega_e = 4$, the shock angle $\theta = 25°$, and the mass ratio $m_i/m_e = 200$. With these parameters, the

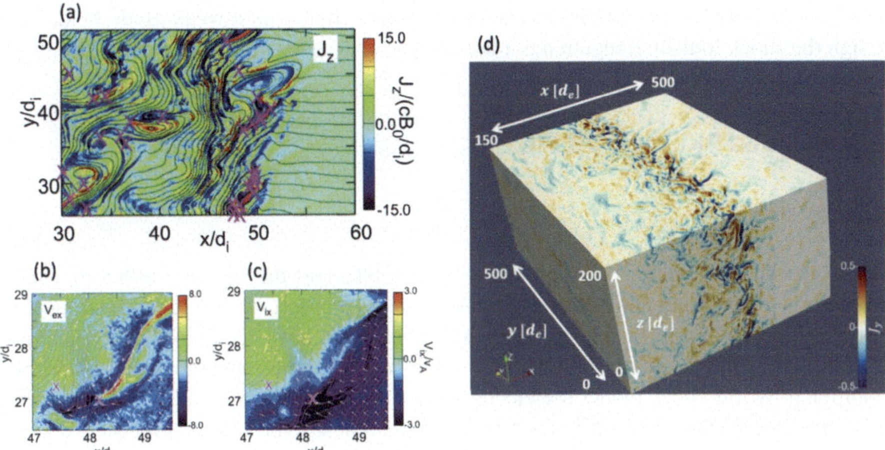

Fig. 12 PIC simulations of reconnection in shocks. (a) 2D simulation domain and the current density J_z. (b) Electron fluid velocity V_{ex}. (c) Ion fluid velocity V_{ix}. (d) 3D simulation domain and the current density J_z. Adapted from Bessho et al. (2019) and Ng et al. (2022)

electron thermal speed becomes $v_{Te} = 14.4v_A$. The drift speed is set to be $v_d = 9v_A$. In the simulation (the downstream rest frame), the shock speed is $2.4v_A$, which corresponds to the Alfvén Mach number of the shock wave $M_A = 11.4$. In other words, the shock speed in the laboratory frame is $11.4v_A$, which is less than v_{Te}, consistent with the Earth's bow shock.

In the simulation, the shock transition region shows a non-resonant ion-ion beam instability due to the interactions between the ions reflected by the shock and the incident ions, and many current sheets are generated, some of which show signatures of magnetic reconnection. In Fig. 12(a), the color shows the current density J_z in the 2D simulation domain, where the black curves are magnetic field lines projected onto the x-y plane, and magenta X marks represent the positions of reconnection X-lines. One of the reconnecting current sheet is zoomed up in Fig. 12(b) and (c), where the electron fluid velocity V_{ex} and the ion fluid velocity V_{ix} are shown. There is one magnetic island above the current sheet, and there are bipolar electron jets generated from the X-line. In contrast, the ion velocity plot does not show ion jet structures, and the ions are passing through the reconnection region with a negative V_{ix}. Therefore, this region is a site of electron-only reconnection, where only electrons are participating in reconnection, while ions cannot respond to the strong gradient of magnetic fields in the thin current sheet, whose thickness is less than the ion skin depth d_i. Electron-only reconnection has been observed in the Earth's magnetosheath (Phan et al. 2018; Gingell et al. 2021; Stawarz et al. 2022) and the transition region of the Earth's bow shock (Wang et al. 2019; Gingell et al. 2019, 2020). Note that the shock transition region has a negative B_z magnetic field, $B_z \sim -4B_0$ (not shown), and the reconnecting magnetic field is the same order. Therefore, in the 2D simulation, reconnection in the shock transition region is guide-field reconnection.

Ng et al. (2022) performed a 3D PIC simulation to study reconnection in the shock transition region. The simulation parameters are: $\beta_e = \beta_i = 1.41$, $\omega_{pe}/\Omega_e = 4$, $m_i/m_e = 100$, $\theta = 30°$, $v_d = 10v_A$, and the system size $L_x \times L_y \times L_z = 200d_i \times 50d_i \times 20d_i$. The z direction is set to be a periodic boundary. Figure 12(d) shows the current density J_z. In the 3D simulation, the current direction can be not only in the z direction, but also in the y direction; therefore, the reconnection plane does not have to be in the x-y plane as in the

2D simulation, and some current sheets show reconnection with a weak guide field, even though the shock transition region has a large negative B_z.

6 Embedded PIC: MHD-AEPIC

6.1 Overview

Due to the large separation between the kinetic scales and the size of Earth's magnetosphere, it is highly computationally expensive to apply a purely kinetic code for simulating global magnetospheric dynamics. Various hybrid methods have been proposed to incorporate kinetic physics into global simulations while keeping the computational costs feasible. Traditional hybrid codes model the electron species as a fluid and simulate the ions with either macro-particles or a grid-based Vlasov solver. These hybrid models reduce the separation between the kinetic scales and the global scale by removing the electron kinetic scales from the model so that it becomes feasible to apply them to Earth's magnetosphere.

The Magnetohydrodynamic with Adaptively Embedded Particle-in-Cell (MHD-AEPIC) model represents another type of hybrid approach to incorporate kinetic effects into global models (Daldorff et al. 2014; Shou et al. 2021; Chen et al. 2023). This type of hybrid model couples a kinetic code with a global fluid model, and only applies the kinetic code to simulate part of the simulation domain, where kinetic physics is crucial while using the fluid model to simulate the rest of the domain. Compared to a purely kinetic model, this type of hybrid model reduces the computational cost by reducing the domain size for the kinetic code, and it is best suited for applications where the important kinetic physics is localized. The MHD-AEPIC model, and its precursor, the Magnetohydrodynamic with Embedded Particle-in-Cell (MHD-EPIC) model, are the first two-way coupled models that work for global applications. Since then, similar coupled models have been developed by different independent teams. For example, Makwana et al. (2017) also developed a model that couples a PIC code with an MHD code, and Rieke et al. (2015) tried to couple a Vlasov solver with a two-fluid code.

6.2 Methodology

6.2.1 Development History

The original MHD-EPIC model was developed by Daldorff et al. (2014), in which the semi-implicit particle-in-cell code iPIC3D (Markidis et al. 2010) is coupled with the global fluid model BATS-R-US (Powell et al. 1999) through the Space Weather Modeling Framework (SWMF) (Tóth et al. 2005). The model has been successfully applied to study magnetic reconnections in the magnetospheres of Ganymede (Tóth et al. 2016; Zhou et al. 2019, 2020), Earth (Chen et al. 2017, 2020; Wang et al. 2022a,b), Mercury (Chen et al. 2019) and Mars (Ma et al. 2018). A PIC region has to be a box in the MHD-EPIC model. To cover the kinetic regions of interest, the MHD-EPIC model supports applying multiple independent kinetic regions (Tóth et al. 2016) in the same simulation domain, and it also allows rotating a box so that the corresponding PIC region does not have to be aligned with the global grid (Chen et al. 2020). These two features expand the capabilities of the MHD-EPIC model. However, not all the kinetic regions of interest can be covered by one or a few boxes. If the kinetic region moves at the global spatial scale during a simulation, the PIC box has to be very large to cover the whole region of interest, which is computationally expensive.

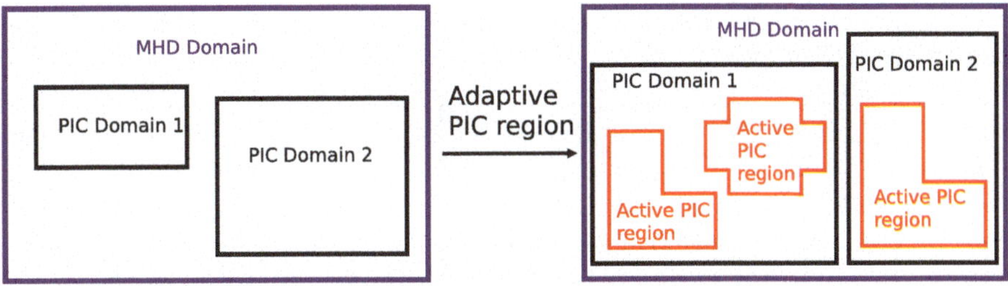

Fig. 13 A schematic shows the improvement of the MHD-AEPIC (right) model from the MHD-EPIC (left) model. Adapted from Chen et al. (2023)

To overcome these difficulties, the MHD-AEPIC model has been developed, which allows a dynamic PIC region of any shape (see Fig. 13).

To support dynamic PIC regions, two different PIC codes, the Adaptive Mesh Particle Simulator (AMPS) (Shou et al. 2021) and the FLexible Exascale Kinetic Simulator (FLEKS) (Chen et al. 2023), have been developed as the PIC component of the MHD-AEPIC model. Since FLEKS is more widely used for MHD-AEPIC simulations, we focus on the FLEKS code in this section.

6.2.2 Coupling Algorithm

MHD-AEPIC supports coupling with single-fluid MHD, multi-species MHD, multi-ion MHD (Glocer et al. 2009b), and five- and six-moment multi-fluid models (Huang et al. 2019). The most widely used fluid component is the single-fluid Hall MHD with a separate electron pressure equation, and we briefly describe the coupling algorithm for this case here.

In an MHD-AEPIC simulation, the PIC code covers part of the whole simulation domain. The MHD model provides the initial conditions for the PIC code at the beginning of the coupled simulation. Once the initialization is done, both the PIC and MHD models update independently for one or a few time steps until the next coupling time point is reached. During the coupling, the MHD model provides the boundary conditions for the PIC code, and the PIC code provides the updated magnetic field and plasma quantities to overwrite the overlapped MHD region. Since the MHD and PIC codes solve different sets of equations, conversion between the MHD and PIC variables is needed. When calculating PIC variables from MHD variables, we need densities, velocities, and pressures for both electron and ion species, and they are calculated as follows:

- Charge neutrality is assumed, so both the electron and ion densities can be easily obtained from total MHD density.
- From the MHD magnetic field, the current density can be calculated. Since the sum of electron and ion momentum is the total MHD momentum, and the velocity difference between electrons and ions produces the current, the electron and ion velocities can be obtained.
- Since we usually solve both ion and electron pressure equations on the MHD side, the electron and ion pressures can be obtained directly to initialize thermal PIC macro-particles.

Once the electron velocity is obtained, it is used to calculate the electric field **E** for PIC from the generalized Ohm's law:

$$\mathbf{E} = -\frac{\mathbf{U}_e \times \mathbf{B}}{c}, \tag{66}$$

where **B** is the MHD magnetic field and $\mathbf{U}_e$ is the electron bulk velocity including the Hall term. We note that no matter Hall physics is included or not into the MHD model, the equation above is applied to calculate the initial and boundary electric field for PIC.

Calculating MHD variables from PIC variables is more straightforward: we simply sum up the mass, momentum and energy of the electron and ion macro-particles to obtain the plasma variables required. We refer the readers to Daldorff et al. (2014) for more details. Currently, the PIC codes used for MHD-EPIC/MHD-AEPIC coupling have to use a Cartesian mesh, but the MHD model BATSU-R-US can use non-uniform Cartesian or non-Cartesian grids. The interpolation between the PIC and MHD grids is done by a second-order linear interpolation.

6.2.3 Particle-in-Cell Algorithm

The embedded PIC model is a particular version of the PIC models discussed in detail in Sect. 5. The original MHD-EPIC implementation used the iPIC3D model while MHD-AEPIC uses FLEKS. Both MHD-EPIC and FLEKS are semi-implicit (Brackbill and Lapenta 2008; Lapenta 2017; Chen and Tóth 2019), meaning that the electric field is solved for by an implicit scheme. We choose the semi-implicit PIC algorithm because it has a relaxed stability constraint so that the Debye length does not have to be resolved and the stability constraint for the time step is based on the thermal speed instead of the speed of light. Based on our numerical experiments, we found the stability of the PIC code is extremely important for a successful MHD-EPIC/MHD-AEPIC simulation. To improve the stability, we designed the Gauss's Law satisfying Energy-Conserving Semi-Implicit Method (GL-ECSIM) (Chen and Tóth 2019), which is based on the Energy-Conserving Semi-Implicit Method (ECSIM) by Lapenta (2017). GL-ECSIM shares the same energy conservation property as ECSIM, i.e., the total energy of the system can be exactly conserved with proper parameters. In practice, we found the code is more stable with parameters that slowly dissipate the total energy numerically. In addition, satisfying Gauss's law (charge conservation) is also crucial for the stability and accuracy of the PIC code. GL-ECSIM applies a novel method to satisfy Gauss's lay by adjusting particle positions at the end of each cycle. The details of GL-ECSIM can be found in Chen and Tóth (2019).

In a long MHD-AEPIC simulation, the macro-particle number per cell may vary significantly due to the transport of particles. The uneven distribution of particle numbers can cause load imbalance and reduce computational efficiency. To alleviate this problem, we designed particle splitting and merging algorithms for FLEKS. A particle splitting (merging) algorithm is applied to split (merge) particles when the number of particles per cell is below (above) a threshold (Chen et al. 2023).

6.2.4 Kinetic Region Adaptation

The most important improvement of MHD-AEPIC over MHD-EPIC is the adaptive PIC region. Although the PIC grid is still Cartesian, its cells can be switched on or off so that the active cells can fit any shape of kinetic regions. We note that the PIC cells can be activated or deactivated dynamically during a simulation. The active PIC region can be defined either based on geometric or physical criteria. For physics-based adaptation, BATS-R-US calculates the physical criteria and sends the corresponding grid information to FLEKS to turn on or turn off cells.

6.2.5 Kinetic Scaling

In some applications, the difference between kinetic and global spatial and temporal scales makes it difficult, if not impossible, to resolve the kinetic scales in an MHD-EPIC, or even MHD-AEPIC simulation. Fortunately, the large separation of scales can be exploited, and the kinetic scales can be increased by changing the mass per charge ratio without affecting the global dynamics (Tóth et al. 2017). This technique is not needed or even applicable for Ganymede and Mercury simulations, where the kinetic and global scales are not very different. On the other hand, kinetic scaling is applicable and extremely useful for modeling Earth's magnetosphere. We typically increase the kinetic scales by a factor of 4 to 16. See Tóth et al. (2017) for more detail.

6.3 Applications

The MHD-EPIC/MHD-AEPIC model has been applied to investigate the physical processes and consequences of both magnetopause and magnetotail reconnection. In these simulations, the PIC code is usually used to cover either the magnetopause or the magnetotail current sheet, where reconnection happens. Since the initial and boundary conditions of the PIC code are obtained from the MHD model, the physical parameters inside the PIC region, such as the plasma quantities and the shape of the current sheet, are more realistic than those in a standalone PIC simulation. On the other hand, the information from the PIC code is also fed back to the MHD model so that we can evaluate the global consequences of the kinetic magnetic reconnection.

Here we briefly describe a few applications of the MHD-EPIC/MHD-AEPIC model to study Earth's magnetosphere. Chen et al. (2017) studied both the kinetic features of magnetopause reconnection and the evolution of flux transfer events (FTEs) show in Fig. 14. Near the reconnection site, the simulation successfully produced key kinetic features of asymmetric magnetic reconnection, such as the crescent electron phase space distribution and the lower hybrid drift instability. Due to the multiple X-line reconnections inside the PIC code, FTEs are generated quasi-periodically at low latitudes, then propagate toward the cusps. We briefly describe the evolution of the FTEs here:

- During the growth of an FTE, its cross-section increase, and its length extends along the dawn-dusk direction (from $t = 100$ s to $t = 150$ s in Fig. 14).
- Since its ambient plasma flow speed varies, an FTE may become tilted ($t = 240$ s in Fig. 14).
- There may be multiple FTEs on the magnetopause, and a few FTEs can merge into one (from $t = 320$ to $t = 660$ s in Fig. 14).
- FTEs can be dissipated at high latitudes due to the reconnection between the FTE magnetic field and the cusp field lines (Fig. 5 of Chen et al. 2017).

Chen et al. (2020) simulated the GEM dayside kinetic processes challenge event, and compared simulation results with both MMS observations and ground-based SuperDARN observations. The MHD-EPIC simulation shows there are usually multiple X-lines at the magnetopause, and the expanding speed of the X-line endpoints is comparable with SuperDARN observations.

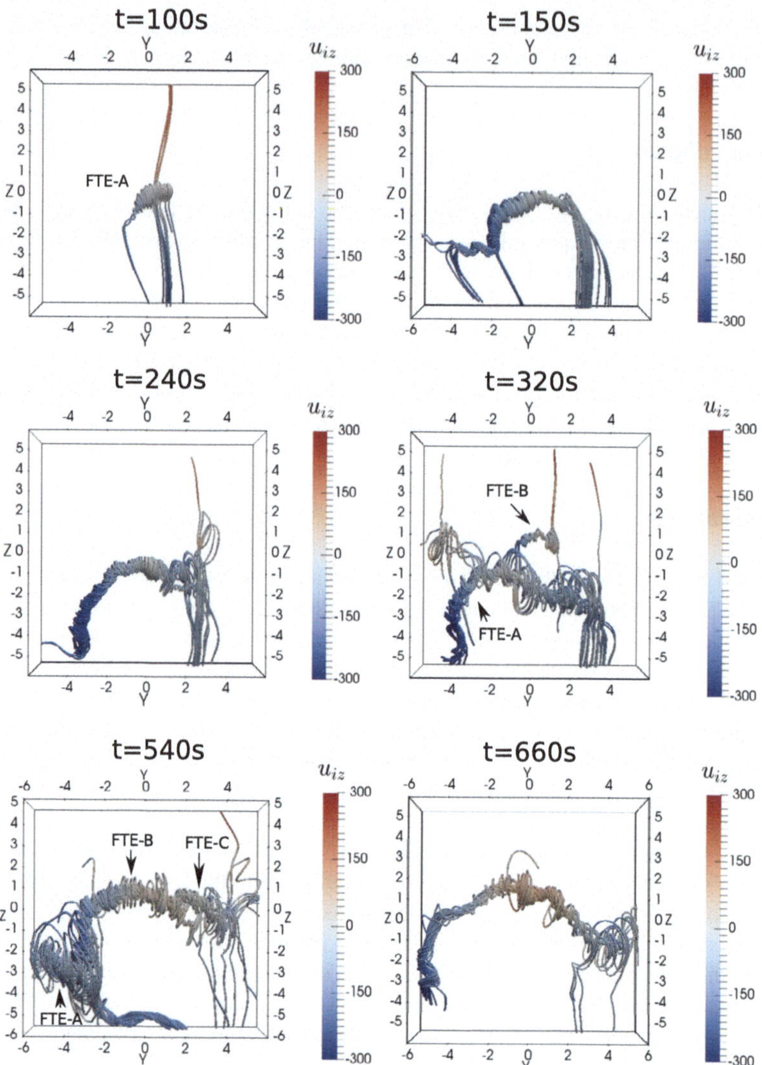

Fig. 14 Evolution of FTEs. Viewed from the Sun, a series of snapshots are shown with magnetic field lines colored by ion velocity u_{iz}[km/s]. Adapted from Chen et al. (2017)

The MHD-EPIC model has also been applied to study magnetotail reconnection. Wang et al. (2022b) found the MHD-EPIC simulation can produce global-scale magnetospheric sawtooth-like oscillations periodically even under steady solar wind conditions, while the ideal- and Hall-MHD simulations do not produce such variations. It suggests that kinetic reconnection physics may play an important role in driving sawtooth oscillations. Recently, the development of the MHD-AEPIC model has enabled us to simulate storm events with a dynamic PIC region that covers the highly dynamic magnetotail reconnection sites.

Currently, research employing MHD-AEPIC focuses on modeling extreme geomagnetic storm events. Extreme events occur infrequently, which makes it difficult to validate MHD models employing simple numerical diffusion to approximate reconnection physics. Using a higher-fidelity model, such as MHD-AEPIC, can improve the reliability of simulations of extreme events.

7 Kglobal: Particle Acceleration Self-Consistently Embedded in a Fluid Model

Solar flares convert magnetic energy into particle energy via magnetic reconnection. Observations of power-law tails in particle distribution functions imply that a large fraction of the released energy goes to energetic (i.e., non-thermal) electrons and ions (Warmuth and Mann 2016). However, the particle spectra found in particle-in-cell (PIC) simulations of reconnection in the relevant regime typically do not form power-laws, except in the limit of extremely low upstream plasma β (Dahlin et al. 2015, 2017; Zhang et al. 2021). Why? With structures extending $\sim 10^4$ km and a Debye length of ~ 1 cm (for $n \sim 10^{10}$ cm^{-3} and $T_e \sim 100$ eV), the corona spans ten orders of magnitude in physical scale. Explicit PIC models must resolve kinetic scales and hence can only simulate a tiny fraction of the macroscopic domain. The dependence of the Larmor radius on energy means nonthermal particles can quickly acquire orbits that approach the size of the simulation domain, halting further energy gains.

In contrast, MHD simulations study macroscopic domains with a fluid description that averages over small spatial and temporal scales. Following test particles in the MHD fields produces information about how particles gain energy but, without feedback coupling the particles and the fields, runaway energy gain can occur so that the system as a whole does not conserve total energy. It is possible to embed PIC models into MHD descriptions at selected locations, but such models presume that particle energy gain occurs in the vicinity of magnetic nulls, which is not consistent with the development and interaction of macroscale magnetic islands or the development of turbulence in large-scale current layers.

The *kglobal* model incorporates the physics necessary to explore particle energization from both the PIC and MHD descriptions (Drake et al. 2019; Arnold et al. 2019). The fundamental question is whether kinetic-scale boundary layers play an essential role in particle energy gain – or if they can be ordered out of the equations to facilitate simulations of macroscale systems. Kinetic boundary layers control the regions where $E_{\|}$, the component of the electric field parallel to the magnetic field, is non-zero. However, Fermi reflection rather than $E_{\|}$ is the dominant driver of energetic particles (Dahlin et al. 2016; Li et al. 2019a). Particle energy gain from Fermi reflection takes place over macro-scale regions and occurs even where $E_{\|} = 0$. As a consequence, kinetic-scale boundary layers are not required to describe the non-thermal energization in macroscale systems.

In order to keep the physics most important for describing particle energization while still being able to model macroscale systems, *kglobal* combines aspects of these descriptions. Guiding center particles move through a computational domain that includes a grid for fluid quantities. These particles feed back on the fluid through their gyrotropic pressure tensor. They can be small in number density but can contribute a pressure comparable to the pressure of the reconnecting magnetic field. The entire system conserves total energy.

The basic version of *kglobal* includes three species: fluid ions, fluid electrons, and particle electrons (the latter of which form the nonthermal population). We note that in a very strict sense the model may be viewed as multifluid plus guiding center methods. However, the motion of the fluid electrons is severely constrained: their density is determined by quasi-neutrality, their parallel velocity from the requirement that parallel currents vanish, and their perpendicular velocity equal to the $E \times B$ flow (as it is for every species). The electromagnetic fields follow the usual MHD equations

$$\frac{\partial \mathbf{B}}{\partial t} = -c\, \nabla \times \mathbf{E}_\perp \qquad \mathbf{E}_\perp = -\frac{1}{c}\mathbf{v}_i \times \mathbf{B} \tag{67}$$

and the ion fluid satisfies the usual MHD continuity equation

$$\frac{\partial n_i}{\partial t} + \nabla \cdot n_i \mathbf{v}_i = 0 \tag{68}$$

and energy equation

$$\frac{d}{dt}\left(\frac{P_i}{n_i^\gamma}\right) = 0 \tag{69}$$

The ion momentum equation takes the form

$$\rho_i \frac{d\mathbf{v}_i}{dt} = \frac{1}{c}\mathbf{J} \times \mathbf{B} - \nabla P_i - \nabla_\perp P_{ef} - m_e n_{ef} v_{\parallel ef}^2 \boldsymbol{\kappa} + e n_i E_\parallel \mathbf{b} - (\nabla \cdot \mathrm{T}_{ep})_\perp \tag{70}$$

in which the left-hand side and first terms on the right-hand side are the same as in MHD (P_i and P_{ef} are the ion and fluid electron pressure, respectively). However the final terms on the right-hand side include the curvature $\boldsymbol{\kappa} = \mathbf{b} \cdot \nabla\, \mathbf{b}$, large-scale parallel electric field $E_\parallel$, and particle electron stress tensor T_{ep} and quantify the self-consistent back-reaction of the particles on the system.

The perpendicular motion of the particle electrons is given by the conservation of the first adiabatic invariant

$$\mu_{ep} = \frac{p_{ep\perp}^2}{2B} = \text{const.} \tag{71}$$

while the parallel motion satisfies

$$\frac{d}{dt} p_{e\parallel} = p_{e\parallel} \mathbf{v_E} \cdot \boldsymbol{\kappa} - \frac{\mu_e}{\gamma_e} \mathbf{b} \cdot \nabla\, B - eE_\parallel \tag{72}$$

This equation includes a contribution from the parallel electric field given by

$$E_\parallel = -\frac{1}{n_i e}\left(\mathbf{B} \cdot \nabla \left(\frac{m_e n_{ef} v_{ef\parallel}^2}{B}\right) + \mathbf{b} \cdot \nabla\, P_c + \mathbf{b} \cdot \nabla \cdot \mathrm{T}_{ep}\right) \tag{73}$$

Finally, the fluid electron density enforces quasi-neutrality

$$n_{ef} = n_i - n_{ep} \tag{74}$$

the parallel flow eliminates parallel currents

$$n_{ef} v_{ef\parallel} = n_i v_{i\parallel} - n_{ep} v_{ep\parallel} \tag{75}$$

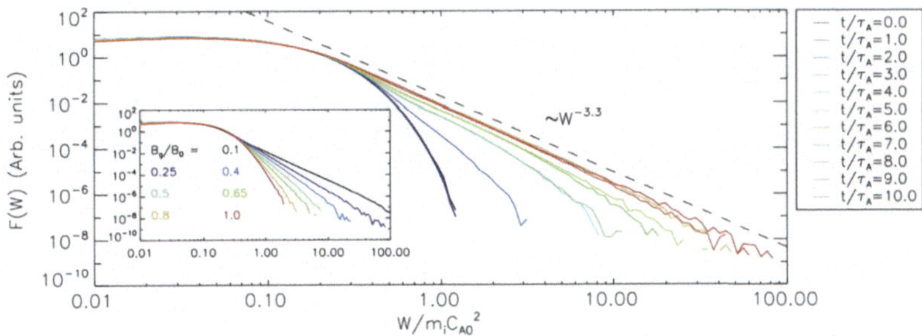

Fig. 15 Energetic electron spectra from *kglobal*. A log-log plot of the electron differential density F(W) versus energy (W) at multiple times from a reconnection simulation with a guide field $B_g/B_0 = 0.25$. A power-law develops after $t/\tau_A \sim 3-5$. Inset: The late-time F(W) for several guide fields, illustrating the dependence on the *ratio* of the guide-to-ambient magnetic field. Adapted from Arnold et al. (2021)

and the pressure equation takes the usual form

$$\frac{d}{dt}\left(\frac{P_{ef}}{n_{ef}^{\gamma}}\right) = 0 \tag{76}$$

A full derivation of these equations is given in Drake et al. (2019) and Arnold et al. (2019).

Simulations with these equations pass several tests. They describe the linear propagation of stable, circularly polarized Alfvén waves and the linear growth of firehose modes. The latter plays an important role in controlling the feedback of energetic particles during magnetic reconnection since magnetic tension is suppressed on the approach to firehose marginal stability. In addition, they accurately capture the dynamics of electron acoustic waves and describe the suppression of transport of hot electrons parallel to the ambient magnetic field. The inclusion of the large scale $E_\parallel$ is important in describing the development of return currents that form as hot electrons escape from regions of electron acceleration in macroscale energy release events such as flares (Egedal et al. 2012).

Reconnection simulations with *kglobal* have produced power-law spectra of energetic electrons that extend nearly three decades in energy, while simultaneously generating the super-hot thermal electrons characteristic of flare observations (Arnold et al. 2021). Figure 15 shows the electron energy spectrum for a typical simulation. Electrons in the initial Maxwellian distribution (black curve) transform into a nonthermal spectrum in a few Alfvén crossing times (τ_A). Consistent with observations, the total energy content of the nonthermal electrons can exceed that of the hot thermal electrons even though the number density does not. The strength of the ambient out-of-plane guide field strongly impacts the energy content and power-law index of the nonthermal electrons (see inset of Fig. 15): the guide field increases the radius of curvature of a reconnected field line, thereby weakening Fermi reflection (Drake et al. 2006). In contrast, the size of the global system has relatively little influence.

The governing equations of *kglobal* can be extended to include the contributions of nonthermal (particle) ions. Unlike for electrons, whose small mass can be used to simplify the equations, the ion inertia can not be neglected and must be included. Recent work has incorporated these equations into the computational model and early results reveal the simultaneous development and evolution of extended electron and proton power law distributions (Yin et al. 2024a,b).

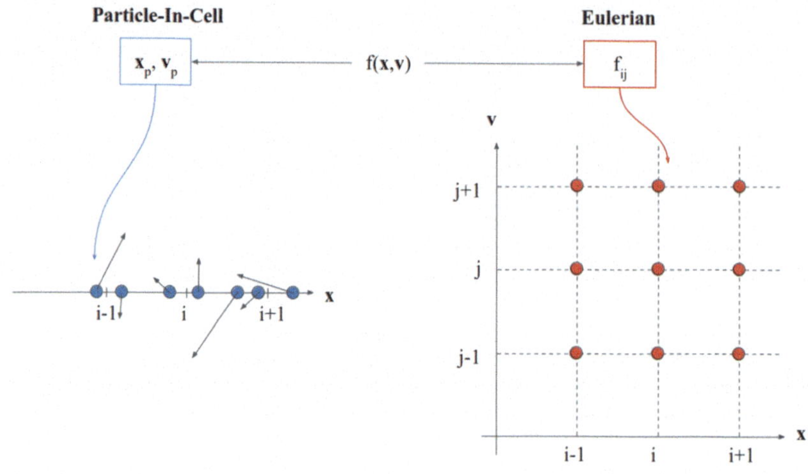

Fig. 16 Sketch of the typical sampling of the plasma distribution function $f(\mathbf{x}, \mathbf{v})$ adopted in PIC (left) and Eulerian (methods). Adapted from Finelli (2022)

8 Vlasov

8.1 Overview

Eulerian Vlasov-Maxwell numerical simulations are a useful tool for investigating fundamental kinetic-scale plasma processes, such as turbulence and magnetic reconnection, as well as the interaction between the solar wind and planetary magnetospheres.

Thanks to the clean description of the plasma dynamics in the entire phase space at the expense of a larger computational cost, Eulerian algorithms complement well Particle-In-Cell (PIC) codes. The almost noise-free description of velocity space is generally guaranteed by the discretization of the plasma distribution function on a six-dimensional phase-space grid characterized by collocation points in both physical and velocity space. On the other hand, PIC methods suffer from the intrinsic stochastic shot noise which becomes especially relevant at small scales and in cases where the number of particles per cell is not large. However, in Eulerian methods, setting a six-dimensional grid in the entire phase space dramatically increases the computational cost. The bottleneck is generally constituted by the memory necessary to store the plasma distribution function, as shown in the following simple example.

The main difference when sampling the plasma distribution function in PIC and Eulerian approaches is depicted in Fig. 16. In PIC methods, the grid is defined only on the physical space $\mathbf{x}$, and the distribution function is sampled through macroparticles (blue circles), each one representative of a large number of effective plasma particles. In the Eulerian approach, the grid is defined on the entire phase space $(\mathbf{x}, \mathbf{v})$ and the distribution function is known on this ensemble of grid points (red circles). In both PIC and Eulerian methods, the electromagnetic fields and the moments of the distribution function (e.g., density, bulk speed, etc.) are defined on the physical-space grid $\mathbf{x}$.

Let us imagine a generic phenomenon occurring in a plasma composed of protons and electrons that requires a physical-space grid discretized with 512^3 points. Since the electromagnetic fields have the same memory requirements in both PIC and Eulerian methods, we neglect them here for simplicity. PIC simulations indicate that the plasma dynamics is overall well described with ~ 1000 particles per cell. The memory (assuming double-precision variables) required to store particles positions and velocities is ~ 6 TB. Similarly, Eulerian Vlasov simulations with a homogeneous Cartesian grid in velocity space, generally adopt at least $\sim 51^3$ velocity-space points to well describe the fine details of the velocity plasma distribution function and preserve mass and entropy conservation, thus requiring about ~ 260 TB of memory. Hence, PIC simulations can usually be performed in larger physical-space computational boxes compared to Eulerian ones. Moreover, 3D simulations are more easily achievable with PIC methods while they remain often prohibitive for Eulerian codes, or require more advanced methods such as adaptive mesh refinement or sparse velocity space techniques to become tractable.

In the following, we will introduce two Vlasov-Maxwell algorithms that have been intensively used by the scientific community — namely the Vlasiator code (von Alfthan et al. 2014; Palmroth et al. 2018a) and the Hybrid Vlasov-Maxwell HVM code (Valentini et al. 2007) — and, then, discuss the results of global and local simulations of magnetic reconnection based on these two codes.

8.2 Models and Algorithms

Picturing a virtual journey from large to small scales, the first relevant Vlasov-Maxwell model widely adopted for performing Eulerian simulations is the hybrid Vlasov-Maxwell model. The hybrid model considers protons as a kinetic species, while electrons are a background fluid. It is a low-frequency approximation of the full Vlasov-Maxwell system of equations that assumes quasi-neutrality and neglects the displacement current (Mangeney et al. 2002). Faraday's law is used to evolve the magnetic field, while the electric field is provided by the generalized Ohm's law which includes the Hall term, the electron pressure gradient term, and possibly the terms related to electron inertia (see, e.g., Valentini et al. 2007 for further details about the generalized Ohm's law).

The following paragraphs present Vlasiator (von Alfthan et al. 2014; Palmroth et al. 2018a) and the Hybrid Vlasov-Maxwell (HVM) (Valentini et al. 2007) codes as particular examples of algorithms based on the Vlasov approach. The reader is invited to refer to recent review papers (Califano and Cerri 2022; Palmroth et al. 2018a) for more details on how to solve the Vlasov equation numerically.

The hybrid Vlasov-Maxwell model has been adopted for global simulations of the interaction between the solar wind and the Earth's magnetosphere through the Vlasiator algorithm (von Alfthan et al. 2014; Palmroth et al. 2018a). The Vlasiator code adopts a splitting algorithm to decompose the six-dimensional Vlasov equation into a set of two three-dimensional advection equations (Strang 1968) in physical and velocity space, respectively. The solution of each advection equation is obtained by a semi-Lagrangian method (Zerroukat and Allen 2012), which relieves from the strict limitations to the time step length posed by the CFL condition[1] in velocity space acceleration in regions of strong magnetic field. Position space is discretized on a cell-adaptive Cartesian grid (Honkonen et al. 2013; Ganse et al. 2023) and at each position in space the velocity-space grid is stored on a uniform, Cartesian grid. Vlasiator developed a sparse velocity-space method in which only regions of the velocity distribution function above a set phase-space density are stored and propagated (von Alfthan et al. 2014), yielding a gain of two orders of magnitude in terms of memory and computations. This technique made two-dimensional (2D position space periodic in the third dimension, 3D velocity space) magnetospheric simulations possible,

[1] The CFL condition dictates that an element of the distribution function f may not cross more than one velocity space grid spacing in one time step under the acceleration due to the Lorentz force.

as well as quasi-three dimensional simulations with a very limited extent in the third dimension (Pfau-Kempf et al. 2020). The implementation of adaptive mesh refinement in position space, allowing to focus resolution on regions of interest while saving computations in less-resolved regions, is what made full three-dimensional, global magnetospheric simulations achievable with Vlasiator on modern, bleeding-edge supercomputers (Grandin et al. 2023; Ganse et al. 2023; Palmroth et al. 2023). The electric and magnetic fields are propagated using an upwind constrained transport method (Londrillo and Del Zanna 2004) with divergence-free magnetic field reconstruction (Balsara 2009) on a uniform Cartesian grid matching the finest refinement level of the Vlasov spatial grid, requiring a dedicated coupling scheme between the grids (Papadakis et al. 2022). Note that, despite the sparse velocity-space grid, Vlasiator also requires a huge amount of RAM memory similar, as order of magnitude, with the simple estimate discussed above.

The hybrid Vlasov-Maxwell model has also been adopted for local simulations of plasma turbulence at sub-proton scales using the HVM code (Valentini et al. 2007) (see Califano and Cerri 2022 for a recent review). The HVM code reduces the six-dimensional Vlasov equation to a set of six one-dimensional advection equations. Each equation is then solved through the van Leer method (van Leer 1977). Fields are computed through the Current-Advance Method (CAM). The grid is homogeneous in both physical and velocity space. Periodic boundary conditions are implemented in physical space. In velocity space, the proton distribution function is set to zero after a large number of thermal speeds, thus ensuring that mass conservation is preserved. Numerical resolution in physical space is usually about 4-5 gridpoints per ion skin depth, while velocity-space grid resolution is of about $0.2 - 0.25$ ion thermal speed. The latter condition is an optimal compromise to (i) guarantee that velocity-space filamentation and distortion of the ion distribution function are well observed in velocity space, thus allowing to investigate energy conversion in the entire phase space, (ii) preserve the entropy conservation to an excellent value ($\lesssim 0.1\%$), and (iii) maintain simulations numerically feasible.

Moving towards smaller scales, fully-kinetic Vlasov-Maxwell Eulerian algorithms have been recently implemented to describe electron-scale dynamics. Given the larger computational cost of Eulerian simulations with respect to PIC methods, the former are generally more recent than the latter and possibly implement different assumptions to simplify the Maxwell equations. The full Maxwell system has been retained in several codes (Umeda et al. 2009, 2010; Delzanno 2015; Ghizzo et al. 2017; Juno et al. 2018; Pezzi et al. 2019a; Allmann-Rahn et al. 2022). However, different approximations of the Maxwell equations have been proposed to alleviate the CFL constraint which sets a very small time step when the wave phase speed approaches the speed of light. In this regard, Wiegelmann and Büchner (2001) neglect the displacement current while allowing for charge separation, while Tronci and Camporeale (2015) ignore both the displacement current and the charge separation. Yet a different approach neglects only the transverse part of the displacement current responsible for ordinary mode propagating at the speed of lights (Schmitz and Grauer 2006a; Pezzi et al. 2019a; Shiroto 2023). Finally, an original approach has been developed based on the Vlasiator model, whereby a small section of interest from an ion-hybrid Vlasiator run is used to initialize an electron-hybrid setup in which the ions are kept static while the electron distribution function evolves (Battarbee et al. 2021). This so-called eVlasiator approach has successfully reproduced properties of electron distributions observed in the vicinity of reconnection diffusion regions (Alho et al. 2022).

8.3 Examples of Applications Focused on the Study of Magnetic Reconnection

8.3.1 Local Simulations

In this section, we will present key results relevant to magnetic reconnection that have been obtained with the Hybrid Vlasov Maxwell code (HVM). Further treatments, based on fully-kinetic Eulerian Vlasov-Maxwell simulations, will not be covered in detail in this article, but the reader is referred to the works of Schmitz and Grauer (2006b), Inglebert et al. (2011), Zenitani and Umeda (2014), Sarrat et al. (2017), Pezzi et al. (2019a), as well as Table 2 in the review by Palmroth et al. (2018a) which lists works using Vlasov-based methods in space and astrophysics.

The HVM code, which retains alpha particles (Perrone et al. 2012; Valentini et al. 2016) and inter-particle collisions (Pezzi et al. 2019b), has been used for years to investigate plasma processes occurring at ion kinetic scales. It has been massively employed to study the properties of plasma turbulence (Valentini et al. 2010; Servidio et al. 2012, 2014, 2015; Cerri et al. 2017), showing that turbulent fluctuations generate manifestly non-Maxwellian proton distribution functions (Greco et al. 2012). This emergent velocity-space complexity has been envisioned as a cascade process occurring in velocity space (e.g., Tatsuno et al. 2009; Schekochihin et al. 2016; Servidio et al. 2017): HVM results have allowed to characterize it in a full Vlasov system rather than in the gyrokinetic approximation (Cerri et al. 2018; Pezzi et al. 2018, 2021b). Characterizing non-equilibrium plasma distribution functions is significant to understanding energy transfer and dissipation processes occurring at ion kinetic scales in nearly-collisionless plasmas such as the solar wind (Matthaeus et al. 2020; Cassak et al. 2023), as also reported in different studies based on the HVM code (Sorriso-Valvo et al. 2018; Pezzi et al. 2019c, 2021a; Fadanelli et al. 2021). In the perspective of the Holloway and Dorning (1991) work showing that non-Maxwellian plasmas can support the propagation of undamped plasma waves, the HVM code has been adopted to study the onset of a novel type of electrostatic fluctuations triggered by trapped ions (Valentini et al. 2011a,b, 2014).

One of the first studies employing the HVM code for investigating magnetic reconnection reported the onset of a fast reconnection process obtained as a result of magnetic islands developed by the electromagnetic current filamentation (Califano et al. 2001). In the following years, despite the large number of studies adopting the HVM code, the vast majority of the research work mostly focused on the investigation of fully developed plasma turbulence. More recently, Finelli et al. (2021) investigated the magnetic reconnection in a similar manner as PIC-based studies discussed in Sect. 5, that is, modeling an isolated Harris-like current sheet (Harris 1962b) in equilibrium or pressure balance. Such a current sheet, usually doubled in the physical-space domain to accommodate for periodic boundary conditions, quickly starts reconnecting thanks to an initial perturbation of proton density and/or current (thus magnetic field).

In particular, Finelli et al. (2021) compared results from three different models (i) the HVM model with (isotropic) isothermal electrons including finite electron-inertia, (ii) a modified HVM model, called hybrid-Vlasov-Landau-fluid (HVLF); (iii) a fully-kinetic PIC code (iPIC3D Markidis et al. 2010). The HVLF model is equipped to include anisotropies of the gyrotropic electron pressure with a Landau-fluid (LF) closure for the transport of the gyrotropic electron thermal energy along magnetic field lines (Sulem and Passot 2015). Using these three models, Finelli et al. (2021) performed 2D-3V magnetic reconnection simulations with moderate guide field ($B_g = 0.25\ B_0$, where B_0 is the asymptotic magnetic field) and with reduced mass ratio $m_p/m_e = 100$. The initial setup consists of a double Harris

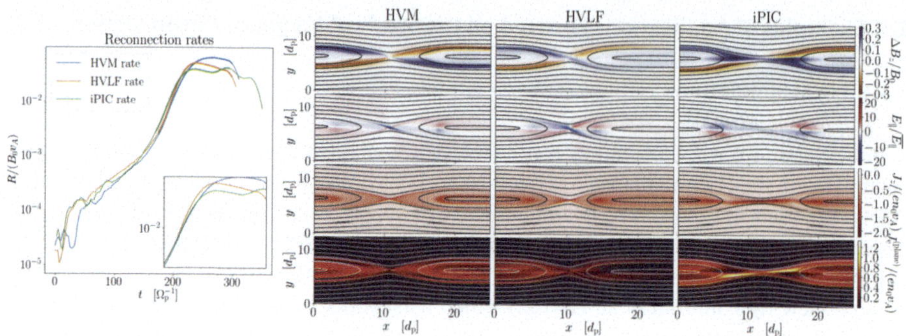

Fig. 17 Magnetic reconnection modeled using three different codes (HVM, HVLF and PIC). Left: normalized reconnection rate $R/B_0 v_\mathrm{A}$. The inset shows the time interval corresponding to the ticker curves in the main plot. To ease the comparison, the curves in the inset are shifted in time. Right: (first row) out-of-reconnection-plane magnetic field $\Delta B_z/B_0 = (B_z - B_z(t=0))/B_0$ showing the expected Hall quadrupolar pattern; (second row) electric field parallel to the ambient magnetic field $E_\parallel/\overline{E_\parallel}$, where $\overline{E_\parallel}$ is the root mean square of $E_\parallel$ in the shown region; (third row) current density in the out-of-plane direction J_z; (fourth row) electron current density in the plane $J_\mathrm{e}^{(\mathrm{in-plane})}$. The superposed black or white curves are the magnetic field lines. The three columns show results from the three different models. The left column show results from HVM at the simulation time $t = 237.5\ \Omega_\mathrm{cp}^{-1}$; the center column show results from the HVLF code at the simulation time $t = 232.5\ \Omega_\mathrm{cp}^{-1}$; the right column show results from the PIC code at the simulation time $t = 235.0\ \Omega_\mathrm{cp}^{-1}$. Ω_cp is the proton cyclotron frequency. Adapted from Finelli et al. (2021)

current sheet (Harris 1962b) perturbed by long wavelength magnetic field fluctuations with random phase (with $1 < |\mathbf{k}| d_\mathrm{p} < 9$, where $\mathbf{k}$ is the wave vector of the fluctuations and d_p is the proton inertial length). The size of the simulations domains is $L_x \times L_y = 24\pi d_\mathrm{p} \times 12\pi d_\mathrm{p}$ discretized with $N_x \times N_y = 1024 \times 512$ grid points. In the HVM and HVLF simulations, the velocity space domain in each direction (x, y and z) is $[-6.4, +6.4]\ v_{th,p}$, where $v_{th,p}$ is the proton thermal speed, and it is discretized by 51^3 grid points. Figure 17 shows results comparing the three models.

While the reconnection linear phase evolution, as well as the overall reconnection signatures and patterns, are quite similar for all three models, Finelli et al. (2021) report that the region of intense current at the centre of the current sheet is more elongated in the case of the HVLF and PIC simulations than in the HVM run. Also, the normalized reconnection rate $R/B_0 v_A$ computed in the quasi-steady state is higher for the HVM ($R/B_0 v_A \sim 0.06$, where B_0 is the upstream magnetic field and v_A is the Alfvén speed) than for the HVLF and PIC simulations ($R/B_0 v_A \sim 0.04$). Despite these differences, the results of all three simulations agree qualitatively. In terms of electron dynamics, which is not captured by the HVM code, the HVLF model reproduces the main features obtained with the fully kinetic treatment of the PIC code.

Magnetic reconnection and turbulence are intricately coupled in plasmas (Stawarz et al. 2024, this collection), where coherent structures such as magnetic holes, magnetic islands, and current sheets naturally develop (see, e.g., Matthaeus et al. 2015). Then, current sheet widths tend to approach the kinetic scale, leading to magnetic reconnection. Hybrid-Vlasov simulations of turbulent plasmas have been successful in modelling both turbulence-induced "standard" reconnection with ion-coupling and electron-only magnetic reconnection (Califano et al. 2020; Arrò et al. 2020). A key result in the context of the interplay between reconnection and turbulence is the fact that turbulence is mediated by magnetic reconnection. More specifically, reconnection plays a key role in driving the onset of sub-ion turbulent cascade (Franci et al. 2017; Cerri et al. 2017; Manzini et al. 2023; Adhikari et al. 2024). Hybrid-Vlasov simulations with the HVM code have played a crucial role in providing evidence for the role played by reconnection in this context.

8.3.2 Global Simulations

The main goal of Vlasiator is to model the solar wind–magnetosphere interaction with a hybrid-Vlasov approach. For this reason, the computational efforts have been mainly directed toward performing global simulations of the entire magnetosphere. As a consequence, a broad variety of magnetospheric plasma phenomena have been investigated using Vlasiator (notably collisionless shock (e.g., Johlander et al. 2022) and foreshock physics (e.g. Turc et al. 2023), magnetosheath jets (e.g. Suni et al. 2021), auroral proton precipitation (e.g. Grandin et al. 2020, 2023) to mention a few). As magnetic reconnection plays a key role in magnetosphere dynamics, it has been investigated in several Vlasiator studies, both at the magnetopause (Pfau-Kempf et al. 2016; Hoilijoki et al. 2017, 2019; Akhavan-Tafti et al. 2020; Pfau-Kempf et al. 2020) and in the magnetotail (Palmroth et al. 2017; Juusola et al. 2018; Runov et al. 2021; Palmroth et al. 2023). Since Vlasiator does not include an explicit resistive term, magnetic reconnection is enabled by the numerical diffusivity or resistivity.

In this section, we present key results from selected Vlasiator studies in two subsections, one devoted to magnetopause reconnection and the other focusing on magnetotail reconnection. We start with discussing 2D-3V simulations, and then present 3D-3V simulations since recent algorithmic improvements allowed running global, three-dimensional (3D-3V) hybrid-Vlasov simulations of Earth's magnetosphere (Ganse et al. 2023).

Magnetopause Reconnection: Global simulations allow us to study the interaction between the solar wind and the magnetosphere, including how and to which extent dayside magnetic reconnection is affected by the solar wind and magnetosheath dynamics. Hoilijoki et al. (2017) investigate this topic by using a 2D-3V Vlasiator global simulation in the GSE polar xz plane, focusing in particular on the laminar or bursty nature of magnetic reconnection during steady solar wind conditions (Fig. 18).

The simulation domain covers $x = [-94, \ +48]\,R_\mathrm{E}$ and $z = [-56, \ +56]\,R_\mathrm{E}$ ($R_\mathrm{E} = 6371\,\mathrm{km}$ is the Earth's radius) and it features a 2D line dipole centered at the origin modelling the Earth's magnetosphere dipole and scaled to match the geomagnetic dipole strength. The steady solar wind has a density $n = 1\,\mathrm{cm}^{-3}$, a constant velocity $\mathbf{v}_{SW} = -750\,\mathrm{km/s}\,\hat{\mathbf{x}}$ and a proton temperature of 0.5 MK. The interplanetary magnetic field (IMF) is directed purely southward and it has a magnitude of 5 nT. The resolution is uniform in the simulation domain; the spatial resolution is 300 km ($\sim 0.047\,R_\mathrm{E} \sim 1.3\,d_{p,SW}$, where $d_{p,SW}$ is the proton inertial length in the solar wind) and the velocity space resolution is 30 km/s ($\sim 0.33\,v_{\mathrm{th,p,SW}}$, where $v_{\mathrm{th,p,SW}}$ is the solar wind proton thermal speed). The solar wind flows into the simulation domain from the boundary at $x = +48\,R_\mathrm{E}$ with constant parameters. The boundaries of the simulation box are periodic in the out-of-plane y direction while the $-x$ and $\pm z$ boundaries apply copy boundary conditions. The inner boundary of the magnetosphere is a circle of radius $4.7\,R_\mathrm{E}$ centered at the origin and it enforces a static Maxwellian proton velocity distribution and perfect conductor field boundary conditions.

Hoilijoki et al. (2017) reported that, despite the steady solar wind conditions, magnetic reconnection at the subsolar magnetopause does not reach a steady state and it is very dynamic. Indeed, magnetic islands are constantly produced and the presence of multiple X-points is observed as is evident in Fig. 18b. The motion of the X-points appears to be mostly dictated by the outflow produced by the neighboring X-points. Hoilijoki et al. (2017) suggest that including the ion kinetic physics in the model promotes the development of a dynamic and bursty reconnection process at the dayside.

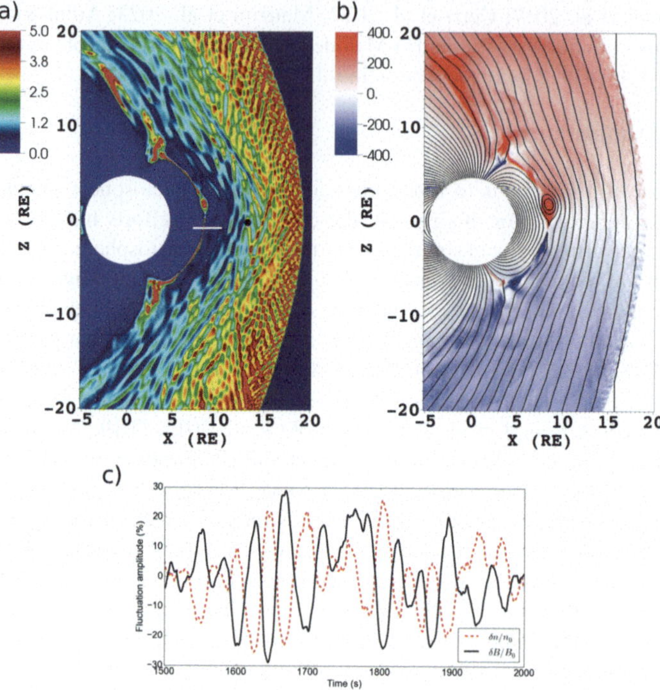

Fig. 18 (a) Plasma β; (b) proton V_Z. The black lines show the magnetic field lines. (c) Magnetic field strength (black solid) and plasma density (red dashed) fluctuations from the virtual spacecraft location indicated with the black dot in panel (a). The anticorrelation between magnetic field and density fluctuations is compatible with mirror mode waves. Adapted from Hoilijoki et al. (2017)

This study investigates also the reconnection rate at the multiple simultaneous X-points and how the rate is affected by the local plasma conditions near the X-point. In particular, the presence of mirror modes (Fig. 18c) in the magnetosheath appears to affect the reconnection rate, in agreement with spacecraft observations (Laitinen et al. 2010). The dependence of the magnetopause reconnection rate upon the IMF direction is further investigated in global Vlasiator simulation by Hoilijoki et al. (2019). In particular, the run presented in Hoilijoki et al. (2017) is compared to a run with similar parameters but with a positive component of the IMF ($B_{IMF} = [3.54,\ 0,\ -3.54]\ nT$). The Sun-ward tilt of the IMF results in a smaller tangential field at the magnetopause, leading to a reduction of the reconnection rate with respect to the purely southward-directed IMF case. The presence of a non-zero $B_{x,IMF}$ introduces an asymmetry that impacts the reconnection process in terms of flux transfer events (FTE) size, speed and occurrence rate. In particular, FTEs are observed more frequently in the Northern Hemisphere and they are smaller in size with respect to the Southern Hemisphere.

The findings of Hoilijoki et al. (2017) have been confirmed by Pfau-Kempf et al. (2020), who analyze the reconnection process in a three-dimensional setup reproducing the magnetopause surface. In particular, Pfau-Kempf et al. (2020) report a Vlasiator simulation of the noon–midnight meridional plane which is extended to cover 7 R_{E} in the dawn–dusk direction. The study by Pfau-Kempf et al. (2020) is the first example of a 3D-3V Vlasiator simulation of a cylindrical geometry mimicking the subsolar dayside magnetosphere. While the dimensionality is increased, the cylindrical geometry and the limited extent in the

y direction allow keeping the computational cost affordable and much lower than a global 3D-3V simulation modeling the entire magnetosphere.

The simulation domain covers $x = [-16, +31]\,R_E$, $y = [-3.5, +3.5]\,R_E$ and $z = [-35, +35]\,R_E$ and the solar wind and IMF parameters are the same used for the 2D-3V run reported in Hoilijoki et al. (2017) and discussed above. The spatial resolution is $\sim 0.24\,R_E$, which is larger than the resolution of 2D-3V Vlasiator simulations because of the increased computational cost of 3D-3V runs.

Identifying the magnetic reconnection site in 3D settings is not as straightforward as in 2D-3V simulations, where the local behaviour of the flux function allows to identify saddle points corresponding to X-points that are associated with reconnection sites. Hence, in this study the X-line location is estimated by combining the four-field junction method (Laitinen et al. 2006) with the identification of the locations exhibiting a flow reversal in the z direction. However, the four-field junction method is insufficient for identifying multiple reconnection X-lines. Pfau-Kempf et al. (2020) find that, despite the uniform initial condition and the cylindrical symmetry in y, the X-line is not a straight line and it exhibits variations along the y direction. It is suggested that structures in the magnetosheath break the translation symmetry along y. Analogously to Hoilijoki et al. (2017), Pfau-Kempf et al. (2020) point out that reconnection is bursty and patchy, with multiple reconnection sites being present at the various z and y locations across the magnetopause, despite the homogeneous and steady-state solar wind conditions.

Magnetotail Reconnection: Recently, magnetotail reconnection has been investigated in the context of a 3D-3V Vlasiator simulation investigating the dynamics of plasma eruptions (Palmroth et al. 2023). The three-dimensional simulation in both ordinary and velocity space is made possible by technological advances, notably by enabling static adaptive mesh refinement (AMR) for ordinary space (Ganse et al. 2023). With AMR, regions of high scientific interest such as the magnetotail plasma sheet are sampled with higher resolution ($0.16\ R_E$) with respect to other regions in the simulations, the coarser resolution is $1.26\ R_E$.

The 3D-3V simulation domain covers $x = [-111, +50]\,R_E$ and $y, z = [-58, +58]\,R_E$. The simulation parameters and initial conditions (IMF, solar wind density and speed) are the same adopted in (Hoilijoki et al. 2017). However, since this is a 3D-3V run, the Earth's dipole is 3D and the inner boundary is a sphere of radius $4.7\ R_E$, while the $\pm y$ boundaries apply copy boundary conditions, as the $\pm z$ boundaries. Differently from (Hoilijoki et al. 2017; Palmroth et al. 2018a; Juusola et al. 2018), where Ohm's law included only the Hall term, this run includes the electron pressure gradient term as well. A polytropic closure is adopted for electrons, $\mathbf{P}_e = p_e\mathbf{I}$ and $p_e = n^\gamma T_e$, where $\mathbf{P}_e$ is the electron pressure tensor, p_e is the scalar pressure, T_e is the electron temperature and n is the density. The polytropic index γ is set to $5/3$ (adiabatic).

Palmroth et al. (2023) focus on the investigation of magnetotail, revealing complex dynamics where magnetic reconnection and kinking instability co-exist in the magnetotail current sheet. In particular, in Fig. 19 it is shown that both processes are required to induce a global topological reconfiguration of the magnetotail, with the formation of a tail-wide plasmoid which is released and rapidly moves tailward. As mentioned above, the identification of the reconnection sites is challenging in 3D systems since we cannot rely on the identification based on the flux function. The reconnection site (X-line) in the magnetotail is identified by a combination of magnetic field and velocity proxies. X-lines and O-lines are identified as the locations where both $B_r = B_z = 0$, where B_r is the radial magnetic field component. The quantity $\partial B_z/\partial r$ allows us to distinguish between X-lines ($\partial B_z/\partial r > 0$ in the magnetotail) and O-lines ($\partial B_z/\partial r < 0$ in the magnetotail). The locations where an X-line is co-located with a v_x reversal (diverging plasma flow) are identified as reconnection sites.

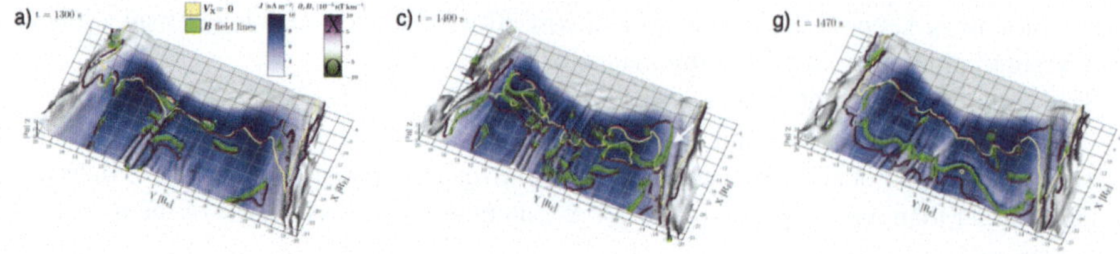

Fig. 19 Evolution of the magnetotail current sheet in a 3D-3V Vlasiator simulation. The panels show the current sheet surface (defined as $B_r = 0$) at different times, $t = 1300$ s (a), $t = 1400$ s (c), $t = 1470$ s (g). The color of the surface corresponds to the current density J. The yellow line indicates the flow reversal between the Earthward and tailward reconnection outflow. The magenta and green lines are locations where $B_r = 0$ and $B_z = 0$ and correspond to X-lines and O-lines (differentiated using the sign of $\partial B_z/\partial r$, which is positive at the X-lines and negative at the O-lines). The primary reconnection line is where the X-line (magenta) and flow reversal (yellow) contours are approximately co-located. The background grid shows the coordinates but also the magnetic-field topology: the black grid shows areas where the magnetic field is directed northward, and the white grid shows the areas where it is southward-directed. Adapted from Palmroth et al. (2023)

Palmroth et al. (2023) further confirm that reconnection is ongoing at those locations by showing reconnection signatures such as the Hall electric field and ion demagnetization. In the 3D-3V run, a dominant tail-wide reconnection X-line is found at $X \sim -15\ R_E$ throughout the simulation. The dominant X-line is very dynamic and new X-lines and O-lines with limited extent in the Y-direction are constantly formed.

9 The Rice Convection Model

9.1 Introduction:

As a reminder, the equations in this section are written in SI units. The Rice Convection Model (RCM), by definition, never includes the physics of reconnection. However, reconnection is a microscale/mesoscale process. Only a small fraction of the magnetic flux in the magnetosphere is included in reconnection at any given time. Including the RCM allows us to discuss how the reconnection process impacts the rest of the magnetosphere, specifically the inner magnetosphere, where the magnetic field lines are closed. Since the RCM's outer boundary condition comes from the plasmasheet, any process in the tail, such as reconnection, can impact the inner magnetosphere. Specifically, reconnection can generate low entropy bubbles that move toward the Earth at high speeds and can be a significant source of transport of plasma and magnetic field from the tail to the inner magnetosphere. In addition, the RCM helps quantitatively predict the plasmas from the inner magnetosphere which can reconnect on the dayside.

9.2 Assumptions and Equations

The physics behind the RCM can be found in detail in Toffoletto et al. (2003) and Wolf (1983), and a detailed discussion of the use of the RCM can be found in Wolf et al. (2016) and Toffoletto et al. (2003), and ring current models are described by Toffoletto (2020). In the RCM, the distribution of magnetospheric particles is assumed to be isotropic, that is divided up into multiple energy channels. For each channel, a key variable is the isotropic energy invariant

$$|\lambda_k| = W_k\,(\mathbf{x}, t)\, V^{2/3} \tag{77}$$

where W_k is the particle kinetic energy, including bounce and gyro motion, k is the energy channel label, and the sign of λ_k is positive for positive ions and negative for electrons. The flux tube volume is

$$V = \int_{sh}^{nh} \frac{ds}{B(\mathbf{x}, t)} \tag{78}$$

where the integral extends along the field line from the southern to the northern ionosphere.

The motions of magnetospheric particles in the inner magnetosphere are assumed to be governed by

$$\text{Drift velocity} \ll \text{bounce motion} \ll \text{gyro motion}$$

The RCM calculates the bounce-averaged drift velocity, including gradient, curvature, and $E \times B$ drifts, i.e.,

$$\mathbf{v}_k = \frac{\left(\mathbf{E} - \frac{1}{q_k}\nabla W_k(\mathbf{x}, t)\right) \times \mathbf{B}(\mathbf{x}, t)}{B(\mathbf{x}, t)^2} \tag{79}$$

where q_k is the charge of a particle of species k. Inertial drift is assumed to be negligible. The quantity $\eta_k(\mathbf{x}, t)$ is defined as the number of particles per unit magnetic flux for particles of a specific chemical species and a specific value of energy invariant. It follows a conservation law (Wolf 1983)

$$\left[\frac{\partial}{\partial t} + \mathbf{v}_k(\lambda_k, \mathbf{x}, t) \cdot \nabla\right]\eta_k = -L(\eta_k) \tag{80}$$

where L is the loss rate of particles due to precipitation and charge exchange. The classic RCM neglects particles flowing up from the ionosphere to the magnetosphere. In the early 2000's Stan Sazykin and Darren DeZeeuw implemented a grid-based scheme using the CLAWPAK package (De Zeeuw et al. 2004; Mandli et al. 2016) that was more robust but a bit more diffusive than the earlier Lagrangian scheme.

The flux tube content η_k is related to the thermodynamic pressure P

$$PV^{5/3} = \frac{2}{3}\sum_k \eta_k|\lambda_k| \tag{81}$$

while the flux tube content is related to the plasma distribution function $f_k(\lambda)$ as

$$\eta_k = \frac{4\pi 2^{1/2}}{m_k^{3/2}} \int_{\lambda_{min}}^{\lambda_{max}} |\lambda|^{1/2} f_k(\lambda) d\lambda \tag{82}$$

where $(\lambda_{max} - \lambda_{min})$ is the width of the invariant energy channel associated with species k. The species k is defined for a given chemical species (usually e^-, H^+, O^+), and the specific value of the energy invariant.

The electric field can be expressed as the sum of a potential component and an inductive component

$$\mathbf{E} = -\nabla\Phi - \mathbf{v}_{induction} \times \mathbf{B} \tag{83}$$

In the RCM, the inductive electric field is included implicitly through time-dependent magnetic field mappings. The inductive magnetic field in the ionosphere is assumed to be zero there; however, it is not zero in the magnetosphere.

There are two more complications in the electric field:

1. In the classic RCM, we assume the electric field is perpendicular to the magnetic field.
2. There are two coordinate systems used in the classic RCM. One moves with the Earth as it rotates, and the potential in that system is labeled Φ_i. The other does not rotate with the Earth, is approximately an inertial system, and is labeled Φ.

We can translate from one system to the other in the ionosphere using the formula

$$\Phi = \Phi_i - \frac{\omega_E B_0 R_E^3 \sin^2(\theta_i)}{R_i} \tag{84}$$

where ω_E is the angular rotation rate of the Earth, B_0 is the magnetic field at the Earth's equator, R_E is the radius of the Earth, θ_i is the colatitude, and R_i is the radius of the ionosphere. Equation (84) applies to the equatorial plane, but it applies only to a dipole magnetic field. To compute Φ in the equatorial plane of the magnetosphere, the RCM calculates Φ by mapping between the ionosphere to the equatorial plane, assuming Φ is constant along each field line.

In the thin-shell approximation, the equation for the conservation of current is $(\nabla \cdot \mathbf{J} = 0)$ can be written

$$\nabla_i \cdot \left[\overleftrightarrow{\Sigma} \cdot \left(\nabla_i \Phi_i \right) \right] = \left(J_{\|nh} - J_{\|sh} \right) \sin(I) \tag{85}$$

where $\overleftrightarrow{\Sigma}$ is the field-line integrated conductivity tensor due to both hemispheres, I is the dip angle of the magnetic field in the ionosphere, and $J_{\|nh} - J_{\|sh}$ is the ionospheric field-aligned current density.

The Vasyliunas (1970) equation, which is based on force balance

$$\mathbf{J} \times \mathbf{B} - \nabla P = 0 \tag{86}$$

is given by

$$\frac{J_{\|nh} - J_{\|sh}}{B_i} = \frac{\hat{b}}{B} \cdot \nabla V \times \nabla P \tag{87}$$

which relates field-aligned currents in the ionosphere to pressure gradients in the magnetosphere, and B_i is the magnetic field at the southern- and northern-ionospheric footprints of the field line (assumed the same). The derivation makes use of the fact the right-hand side of Eq. (87) can be evaluated anywhere along the field line. The RCM equations are solved on a fixed ionospheric grid that has variable grid spacing in latitude to better resolve the auroral zone. The RCM grid is time-dependent in the equatorial plane, ranging from just inside the magnetopause on the dayside to $10 - 20 R_E$ on the night side.

9.3 Inputs, Boundary, and Initial Conditions and Outputs

The magnetic field model: For many years, the RCM assumed a constant magnetic field, but, beginning about 2000 the RCM used a time-dependent semi-empirical model such as the Tsgyanenko models (1989, 1995, 2003). The RCM can also use the Hilmer and Voigt

(1995) magnetic field model. In classic RCM runs, the magnetic field is not designed to be consistent with Ampere's law and equation (86). Ways of including Eq. (86) consistently in the RCM are described in Sect. 9.4. In the ionosphere, where the magnetic field is assumed to be dipolar, the magnetic field is

$$\mathbf{B}_i = -\frac{\mu_0}{4\pi}\frac{\mathbf{M}_E - 3\hat{r}(\mathbf{M}_E \cdot \hat{r})}{R_i^3} \tag{88}$$

where $\hat{r}$ is the radial unit vector in the ionosphere, and $\mathbf{M}_E$ is the magnetic moment, which is in the southern direction, and

$$B_i = \hat{r} \cdot \mathbf{B}_i \tag{89}$$

Ionospheric conductance: Ionospheric conductance has two major drivers: Solar heating (e.g., the Sheffield University Ionosphere Plasmasphere (SUPIM) model, Bailey et al. 1997). The second is Auroral heating. The standard treatment uses the electron precipitating energy flux and average energy (Robinson et al. 1987).

Loss models: Separate models are needed for electrons and ions. The simplest electron loss model assumes a fixed fraction of strong pitch-angle scattering, often between 33% and 67% (Schumaker et al. 1989). That procedure is reasonable for the plasma sheet but over-estimates the electron loss rate in the inner magnetosphere. A slightly more sophisticated model (Chen and Schulz 2001) is somewhat more realistic. For the Ion loss model, there are many theoretical models of ion charge exchange. The overall ion loss rate for an energy and L value is calculated using an algorithm developed by James Bishop (Freeman et al. 1993; Bishop 1996).

The RCM needs boundary conditions both at its outer (large L) boundary and at its lowest (low L) boundary. The large$-L$ boundary depends on MLT as well as UT. Note that the large$-L$ boundary can't be aligned with the grid, except in a few cases. At this boundary, the number η_s which is the number of particles for a given type s per weber of magnetic flux, is needed as well as the potential distribution, which can be a simple function of solar wind conditions or an empirical model such as the Weimer model (1985).

The low-L boundary (low-latitude) is set at least a few degrees latitude from the equator. Given the aligned-dipole assumption, the latitudinal current density should, in principle, be zero at the equator. The RCM and many other models use a thin-wire approximation to represent the region near the equator, to provide a boundary condition

$$\frac{\partial J_\theta}{\partial \theta} + S\frac{\partial J_\phi}{\partial \phi} = 0 \tag{90}$$

which was derived by Blanc and Richmond (1980). Here J is the current density, θ is the colatitude coordinate, and ϕ is the longitudinal coordinate. S is a function of ϕ that was defined by Blanc and Richmond (1980).

Initial conditions: These are needed to provide the value of the initial value of η_k, which is a function of grid location and time. The RCM can be initialized with an empty value and run for a period of time to fill in the inner magnetosphere. Alternatively, the RCM uses the Spence et al. (1989) model for the initial pressure distribution. Earlier versions of the RCM assumed a Maxwellian distribution, but more recent versions have the option to assume a kappa distribution (e.g., Yang et al. 2015).

RCM outputs: The main RCM outputs are the electric potential (Φ), field-aligned currents ($J_{\|nh} - J_{\|sh}$), the distribution function (η_s) and moments (pressure and density) within the RCM modeling region in the ionosphere and the magnetospheric equatorial plane.

9.4 Generalizations of the RCM

There have been many modifications to the RCM, particularly since 2000.

Other planets: Tom Hill and several of his students modified the RCM to be appropriate for Jupiter and Saturn. Jupiter's moon Io has volcanoes that loft neutrals and positive ions into the inner magnetosphere, and there is also a similar effect at Saturn's moon Enceladus. Centrifugal force is stronger than gravity near Io, and the region beyond Io is consequently interchange unstable. In the simulations, plasma develops finger-like structures, moving outward because of centrifugal force and azimuthally because of Coriolis force. The clearest magnetospheric signature of interchange transport occurs in the inner magnetosphere of Saturn, where the hot plasma injection-dispersion structures are evident (Hill et al. 2005 and references therein).

CRCM (Comprehensive Ring Current Model): This model (Fok et al. 2001) was similar to the classic RCM, except that it used a much more complete equation for the distribution function. Whereas the classic RCM assumed an isotropic pitch-angle distribution, CRCM assumes conservation of the first and second invariant. Additional terms account for precipitation and charge-exchange losses and pitch-angle scattering. CIMI (Comprehensive Inner Magnetosphere Model) includes radiation belt electrons and the plasmasphere (Fok et al. 2014).

RCM-E: In the classic RCM, the magnetic field was not required to satisfy the force balance equation $\mathbf{J} \times \mathbf{B} = \nabla P$ but $P = (2/3)V^{-\gamma} \sum_k |\lambda_k| \eta_k$ and η_k was based on the theoretical equation (80), with $\gamma = 5/3$. The RCM-E (equilibrium) is run for a small time step (typically $1-5$ minutes), and a modified MHD code, called the "friction code", recalculates the magnetic field in order to make it approximately consistent with $\mathbf{J} \times \mathbf{B} = \nabla P$ (Lemon et al. 2003). For conditions of strong convection, the time development of the RCM-E would cause the magnetic field to be tail-like and more like a substorm growth phase. If PV^γ was constant on the nightside large L-boundary, it became difficult to form a realistic strong ring current (Lemon et al. 2004). The RCM-E usually exhibited the pressure balance inconsistency (Erickson and Wolf 1980). In other words, the more theoretically consistent model became a less realistic representation of observations. Other modelers have developed models that are variants of the RCM-E — e.g., Chen et al. (2012) or the RAM-SCB model (Zaharia et al. 2006) — that use an alternative ring current and force balance model. The solution to the pressure balance inconsistency turned out to be bursty bulk flows (BBFs), which are localized regions of the inner and middle plasma sheet (Angelopoulos et al. 1992) that flow rapidly earthward. The flow bursts, which are also often called "bubbles", often move very fast (typically 400 km/s). These flows correspond to regions of low PV^γ (Pontius and Wolf 1990). Angelopoulos et al. (1992) found that BBFs account for a large fraction of the total earthward flow in the plasma sheet. BBFs usually terminate about the inner edge of the plasma sheet, although some of the fast flows penetrate the ring current (Gkioulidou et al. 2014; Yang et al. 2016). Lemon et al. (2004) produced a substantial ring current injection by reducing the PV^γ at the RCM's outer boundary over a limited region of local time, simulating a ring current injection during a storm. Yang et al. (2014a) showed a possible relation to the streamers observed in the polar cap and bubbles in the plasma sheet, and Yang et al. (2014b) argued that this effect could account for pressure balance inconsistency and that during storms low entropy flux tubes could account for up to 60% of the ring current (Yang et al. 2015). See also the Sect. 9.5 below.

RCM-I (RCM-Inertial): The main problem with using RCM-E to represent BBFs is they move so fast that the assumption of force balance is not valid. Yang et al. (2019) developed a more complex version of the RCM that includes inertial effects in a very approximate way. Equation (85) is replaced by a much more complex expression that involves a $\partial\Phi/\partial t$ term.

9.5 Large, Coupled Models That Include RCM

Single fluid, global MHD models have become powerful tools in recent years (e.g., Lyon et al. 2004; Raeder et al. 2001, Zhang et al. 2019). However, these models do not capture all the important physics in the inner magnetosphere where gradient and curvature drifts become important but are neglected in MHD. Coupling these models is a daunting task as the modeling regions overlap in space and information is fed back and forth between them. There are different physics assumptions associated with each model: the RCM model assumes slow flow and force balance and neglects waves, while MHD does not. However, MHD does not include energy-dependent drifts that are important in the inner magnetosphere. There have been several successful efforts to couple the RCM with global MHD, which provides many of the inputs used by the RCM (boundary and initial conditions) such as the magnetic field, plasma density, and pressure as well as the ionospheric potential. In return, the RCM provides the density and pressure that are derived from computing the moments from the RCM distribution function that have been subject to energy-dependent drifts.

SMWF: The earliest successful coupling effort was De Zeeuw et al. (2004) which merged the BATS-R-US Global MHD (Tóth et al. 2005) code with the RCM. The RCM was embedded in the Global MHD code as a subroutine that later became part of the Space Weather Modeling Framework (SWMF) (Tóth et al. 2005). Each coupling exchange requires computing many field-line integrals to obtain the flux tube volume (Eq. (78)), which is used by the RCM, and the mapping of the 2D RCM quantities into the 3D domain of the MHD code, which is then used to update the MHD. This field line tracing requires using a parallelized and efficient field line tracer that exploits the nested adaptive grid used in the MHD code (e.g., De Zeeuw et al. 2004. This version of the model demonstrated that including the RCM increased the pressure in the inner magnetosphere ring current region as compared to standalone MHD. The RCM also was able to model the Region-2 currents in the ionosphere. Later versions of the SWMF also include other inner magnetosphere models such as the Comprehensive Ring current model (CRCM) (Glocer et al. 2013) and other models (Tóth et al. 2012), becoming the first coupled magnetosphere model to be used in the NASA's Community Coordinated Modeling Center (CCMC).

LFM-RCM: Pembroke et al. (2012) coupled the RCM to the Lyon Fedder Mobary (LFM) global MHD code (Lyon et al. 2004) that included the MIX ionosphere model (Merkin and Lyon 2010). This approach used a loose coupling scheme where the models (LFM, MIX, and RCM) ran independently as separate processes and used the InterComm software package to exchange information at pre-set intervals (Lee and Sussman 2004). In this model, RCM returned both pressure and density to the MHD code and included a simple static plasmasphere based on the Gallagher et al. (2000) empirical model. Since the RCM was only tracking the distribution function and not computing the potential as in the standalone RCM, the assumption of zero dipole tilt could be relaxed in the coupled model, allowing for more realistic simulations. The resulting coupled model was very dynamic, especially during geomagnetic storm simulations, and significantly impacted the ring current region (e.g., Wiltberger et al. 2017). To keep the code stable, the RCM boundary was restricted to regions where the field line average plasma beta was less than 1. With moderately strong solar wind driving, the coupled model produced a strong ring current and Region-2 currents.

OpenGGCM-RCM: Hu et al. (2010) and Cramer et al. (2017) coupled the OpenGGCM global MHD code to RCM (Raeder et al. 2001) that also includes the Coupled Thermosphere-Ionosphere Model (CTIM) (Fuller-Rowell et al. 1996). The RCM is embedded within the MHD code, where the feedback to the MHD code used a configurable ramp-up region based on the strength of the magnetic field for numerical stability. Cramer et al. (2017) found that most of the transport of plasma to the inner magnetosphere is via low entropy bubbles, consistent with Yang et al. (2015). Raeder et al. (2016) also used the coupled model to simulate a geomagnetic storm and showed that it developed subauroral polarization streams (SAPS) from electron precipitation computed from the MHD code. Hu et al. (2011) examined the entropy profile in an idealized Open GGCM simulation and found that violations of the frozen-in-flux in MHD could lead to an entropy profile that produced a low entropy bubble that was earthward of an entropy enhancement. Such a configuration causes the bubble/blob pair to move earthward/tailward, which thins the current sheet in the region between them and can ultimately result in tearing or other configuration changes.

MAGE: The newest version of a global magnetosphere model is the Multiscale Atmosphere Geospace Environment Model (MAGE) that couples the RCM to the Grid Agnostic MHD for Extended Research Application (GAMERA) global MHD code (Zhang et al. 2019; Sorathia et al. 2021), the ReMIX ionosphere model (Merkin and Lyon 2010) which is a revised version of the MIX solver, and the NCAR Thermosphere-Ionosphere-Electrodynamics General Circulation Model (TIE-GCM) (Roble et al. 1988). The GAMERA MHD model is derived from the LFM model but with improved numerical algorithms and updated software designed for efficient use on modern supercomputers. The coupling to the RCM has also been significantly improved and modernized, for example, it uses a highly configurable and customizable parallel field line tracer. Other improvements include moving the RCM boundary in MAGE further from the Earth compared to the coupled LFM-RCM code, allowing more plasma from the plasmasheet to move into the RCM modeling region, and the option of a Maxwellian distribution to compute RCM distribution functions replaced with a Kappa distribution (Sciola et al. 2023). The new model also includes improved loss rate mechanisms for electrons (Bao et al. 2023) where the electron precipitation model is based on RCM-computed electron energy fluxes that are used to modify the ionospheric conductances (Lin et al. 2022a). The use of this conductance model was found to influence the formation of the SAPS channel (Lin et al. 2021). The model also includes a dynamic plasmasphere density that is tracked using a zero-energy channel in the RCM that is fed back to the MHD model (Bao et al. 2023). Pham et al. (2022) used the coupled MAGE model to investigate the impact on thermospheric density perturbations produced by traveling ionospheric disturbances. Sciola et al. (2023) found that in the MAGE model, fast magnetospheric flows associated with low entropy channels can contribute over 50% of the ring current population, consistent with Yang et al. (2015) and Cramer et al. (2017).

9.6 Summary

While the RCM does not model reconnection, it is impacted by it. Reconnection in the tail produces low entropy flux tubes that rapidly interchange their way toward the Earth (e.g., Wiltberger et al. 2015; Sorathia et al. 2021). Some of these flux tubes make it into the inner magnetosphere and play an important role in the formation and structure of the ring current region. It can also have ionospheric effects such as the formation of streamers. Over the years, the RCM has helped illuminate the impact of processes in the tail on the inner magnetosphere. This is especially true using the new generation coupled models that have been developed in recent years. However, there are several limitations in the models that present challenges. One is the modeling of the region in the tail where the magnetic field is transitioning from a stretched tail-like configuration to a dipole. When fast flows appear in this region, as they often do, neither MHD, which neglects gradient/curvature drifts, nor the RCM, which assumes slow flow, are applicable.

Furthermore, including the RCM in MHD models can also affect the location and effectiveness of dayside reconnection as well. Since the Region-2 FACs modeled by the RCM shields the inner magnetosphere from convection, it also strongly affects the distribution of return flow back to the dayside boundary. The RCM has been successful in modeling the plasmaspheric plumes that can bring dense plasmas to the dayside reconnection regions (Goldstein et al. 2002; Huba and Sazykin 2014; Bao et al. 2023). Dayside SWMF runs that include RCM all successfully place MMS within 1 R_E of at an X-line (or separatrix) whenever clear EDRs are observed (e.g. Reiff et al. 2017).

The next generation coupled models will need to add the effect of ionospheric plasma sources (e.g., Glocer et al. 2009a; Varney et al. 2016), which will ultimately require a multi-fluid MHD model coupled to a ring current model and includes a model of ionospheric outflow to track all the species in the magnetosphere.

10 Conclusions

In this paper, we have presented a "brief" overview of the large collection of computational methods that are used to study magnetic reconnection. It should be clear to the reader of this text, that simulating magnetic reconnection is nearly a separate field of study in and of itself. It should be and has been (Büchner et al. 2003) the topic of entire books. The single element to take away from this paper is that simulating a multifaceted and multiscale problem like magnetic reconnection is not a simple endeavor. Both plasma models and simulation initial conditions must be tuned carefully to match the goals of the study.

We have reviewed simulation methods for magnetic reconnection in space plasmas, from macroscopic MHD scales to microscopic kinetic scales. Basically, macroscopic plasma behaviors can be simulated based on fluid modes, and as we resolve smaller scale physics, kinetic models need to be implemented in simulations. In addition, we have reviewed novel approaches to incorporate multi-scale physics.

MHD simulations are useful to study large scale physics, including performing global simulations for planetary magnetospheres. We have discussed the basic algorithm for MHD simulations, and also how to implement test particles that follow MHD fields. Hall MHD simulations contain Hall physics, which allows kinetic scale waves to propagate, mediating fast reconnection. We have reviewed recent progresses of Hall MHD studies, and also the effect of the electron inertia term and EMHD.

Hybrid PIC simulations and full PIC simulations include particle kinetic physics, where particle motions are directly solved by equations of motion of particles. For hybrid simulations, we have reviewed techniques to overcome the limitation of spatiotemporal resolution in global models, and also the implementation of electron kinetic physics into hybrid simulations. For full PIC simulations, we have reviewed simulation studies of magnetic reconnection in the magnetotail and magnetopause, particle acceleration, and shock driven reconnection.

Next, we have reviewed two novel approaches to address multi-scale physics in magnetic reconnection: embedded PIC, and kglobal. In the embedded PIC approach, the macro-scale region is solved using MHD equations, and local kinetic domains are embedded in the MHD domain, where full PIC techniques are employed. In kglobal, on the other hand, equations for the ion and electron fluids are combined with the particle equations for electrons, and the macro-scale evolution is modified by the kinetic physics.

Finally, we have reviewed two types of other simulation techniques: Vlasov simulations and the Rice convection model. In Vlasov simulations, kinetic effects are implemented in simulations by solving 2D-3V or 3D-3V Vlasov equations. We have discussed recent progresses of studies by Vlasov models from local reconnection simulations to global simulations. The Rice convection model is a kinetic approach to simulate physics of the inner magnetosphere, where bounce averaged particle drift motion is taken into account.

10.1 Outlook

We first wish to stress that even relatively simple numerical models continue to shed light on the essential physics of magnetic reconnection, even though some of these models have been around for decades. A good example of this is the explanation of fast magnetic reconnection in collisionless plasmas–the longstanding Reconnection Rate Problem. Only in the last few years, built upon the endeavors of previous theoretical efforts, has a convincing theory that passed the cross-examination of all these numerical models (i.e., PIC, hybrid, Hall-MHD, EMHD, resistive-MHD, pair plasmas) been developed to offer the rate prediction (Liu et al. 2022). Comparing and contrasting "numerical experiments" of reconnection in these models provides invaluable, rigorous constraints to a theory. Such constraints from different models will continue to play a pivotal role in the theory development of any nonlinear phenomenon in plasmas, especially when one attempts to discern the "cause" from the "consequences".

A major driver on the progress of simulations of reconnection has been the uncanny steady exponential increase in computing power known as Moore's law (Schaller 1997). If technology manages to continue Moore's law, current simulations will undoubtedly continue to yield major insights. For example, fully kinetic particle in cell simulations are only just now reaching around 100 ion inertial scale sizes in three dimensions. However, such massive simulations require extremely large data storage, with significant environmental implications.

Beyond increasing computational power, more efficient algorithms continue to be a source of study. For example, GPUs have allowed a significant increase in computational speed for modest sized systems (Bard and Dorelli 2014). For global hybrid models, novel algorithms that drive particles based on "events" rather than time steps are another new direction. The HYPERS global hybrid simulations (Omelchenko and Karimabadi 2012) described in Sect. 4 is one such example.

A major thrust in the coming decade is expected to be the feedback between the multiple disparate scales associated with reconnection. One such multi-scale example is the connection between microscales and mesoscales during reconnection at the dayside of Earth's magnetosphere, where the formation of x-lines depends strongly on magnetic geometry. Coupling different models, as described in Sect. 6, is a promising way forward to address these questions, but there are significant challenges to overcome, especially in regards to boundary conditions and disparate length and time scales.

Another example of such multi-scale physics is reconnection particle energization throughout the heliosphere. A major challenge in the effort to understand the energization of electrons and ions has been the inability to explore the kinetic dynamics of particles in very large systems. Because particle-in-cell (PIC) and hybrid models have to resolve kinetic scales, which are a small fraction of typical macroscales (a factor of around 10^{-10} in the case of solar flares). The consequence is that although PIC simulations have been successful in identifying some of the dominant acceleration mechanisms of electrons and ions (Drake et al. 2006; Dahlin et al. 2014; Guo et al. 2014; Li et al. 2019a), their success in producing the extended powerlaws seen in flare and magnetotail observations has been limited. The largest PIC simulations have revealed electron powerlaws that extend only a single decade in energy (Li et al. 2019a) while observations reveal that powerlaws in flares extend across many decades in energy (Lin et al. 2003; Vilmer 2012). The production of powerlaw distributions of ions has been an even greater challenge (Zhang et al. 2021, 2024). However, PIC modeling revealed that Fermi reflection during the growth and merger of large-scale flux ropes dominates particle acceleration (Dahlin et al. 2016; Guo et al. 2014; Li et al. 2019a), a result that led to a new computational model, kglobal, that combined MHD fluid and particle descriptions while eliminating the kinetic scales that constrained the macro-scale modeling of flares (Drake et al. 2019; Arnold et al. 2019). A major accomplishment was the first exploration of magnetic reconnection-driven electron acceleration in macro-systems which produced electron powerlaw distributions extending nearly three decades in energy and revealed that the ambient guide field is the dominant control factor of the powerlaw index (Arnold et al. 2021). The success of the kglobal model motivates two major extensions: to include particle ions so that the partitioning of energy between the two species can be explored; and to incorporate the kglobal particle algorithm into a global simulation code such as the Adaptively Refined MHD Solver (ARMS) (DeVore 1991) flare simulation code or the Block Adaptive Tree Solar-wind Roe Upwind Scheme (BATS-R-US) (Gombosi et al. 2002). The resulting model would be able to simulate macroscale energy release during magnetic reconnection with energetic electrons and produce synthetic photon spectra for comparison with remote observations of solar flares observations.

Funding Information We acknowledge support from NASA grant 80NSSC20K0198 (MAS), NASA 80NSSC20K1813 (MAS), NASA 80NSSC24K0172 (MAS and PAC), NSF grant AGS-2024198 (MAS). NSF PHY-2308669 (PAC); DE-SC0020294 (PAC); NASA 80NSSC22K0323 (PAC); NASA 80NSSC23K0409 (PAC); NASA 80NSSC18K0834 (JB) and 80NSSC24K1452 (JB). We acknowledge high-performance computing support from the Derecho system (https://doi.org/10.5065/qx9a-pg09) provided by the NSF National Center for Atmospheric Research (NCAR), sponsored by the National Science Foundation. We also acknowledge the National Energy Research Scientific Computing Center (NERSC), a U.S. Department of Energy Office of Science User Facility operated under Contract No. DE-AC02-05CH11231.

Declarations

Competing Interests The authors have no competing interests to declare that are relevant to the content of this article.

References

Adhikari S, Parashar TN, Shay MA, Matthaeus WH, Pyakurel PS, Fordin S, Stawarz JE, Eastwood JP (2021) Energy transfer in reconnection and turbulence. Phys Rev E 104:065206

Adhikari S, Yang Y, Matthaeus WH, Cassak PA, Parashar TN, Shay MA (2024) Scale filtering analysis of kinetic reconnection and its associated turbulence. Phys Plasmas 31(2):020701

Akhavan-Tafti M, Palmroth M, Slavin JA, Battarbee M, Ganse U, Grandin M, Le G, Gershman D, Eastwood JP, Stawarz J (2020) Comparative analysis of the Vlasiator simulations and MMS observations of multiple X-line reconnection and flux transfer events. J Geophys Res Space Phys 125(7):e2019JA027410. https://doi.org/10.1029/2019JA027410

Alho M, Wedlund CS, Nilsson H, Kallio E, Jarvinen R, Pulkkinen T (2019) Hybrid modeling of cometary plasma environments. Astron Astrophys 630:A45. https://doi.org/10.1051/0004-6361/201834863

Alho M, Battarbee M, Pfau-Kempf Y, Khotyaintsev YV, Nakamura R, Cozzani G, Ganse U, Turc L, Johlander A, Horaites K, Tarvus V, Zhou H, Grandin M, Dubart M, Papadakis K, Suni J, George H, Bussov M, Palmroth M (2022) Electron signatures of reconnection in a global evlasiator simulation. Geophys Res Lett 49(14):e2022GL098329. https://doi.org/10.1029/2022GL098329

Allmann-Rahn F, Grauer R, Kormann K (2022) A parallel low-rank solver for the six-dimensional Vlasov–Maxwell equations. J Comput Phys 469:111562. https://doi.org/10.1016/j.jcp.2022.111562

Amano T, Higashimori K, Shirakawa K (2014) A robust method for handling low density regions in hybrid simulations for collisionless plasmas. J Comput Phys 275:197–212

Angelopoulos V, Baumjohann W, Kennel C, Coroniti FV, Kivelson M, Pellat R, Walker R, Lühr H, Paschmann G (1992) Bursty bulk flows in the inner central plasma sheet. J Geophys Res Space Phys 97(A4):4027–4039

Arencibia M, Cassak PA, Shay MA, Priest ER (2021) Scaling theory of three-dimensional magnetic reconnection spreading. Phys Plasmas 28(8):082104

Arencibia M, Cassak PA, Shay MA, Qiu J, Petrinec SM, Liang H (2023) Three-dimensional magnetic reconnection spreading in current sheets of non-uniform thickness. J Geophys Res 128:e2022JA030999

Arnold H, Drake J, Swisdak M, Dahlin J (2019) Large-scale parallel electric fields and return currents in a global simulation model. Phys Plasmas 26(10):102903

Arnold H, Drake JF, Swisdak M, Guo F, Dahlin JT, Chen B, Fleishman G, Glesener L, Kontar E, Phan T, et al (2021) Electron acceleration during macroscale magnetic reconnection. Phys Rev Lett 126(13):135101

Arrò G, Califano F, Lapenta G (2020) Statistical properties of turbulent fluctuations associated with electron-only magnetic reconnection. Astron Astrophys 642:A45. https://doi.org/10.1051/0004-6361/202038696

Ashour-Abdalla M, El-Alaoui M, Goldstein ML, Zhou M, Schriver D, Richard R, Walker R, Kivelson MG, Hwang KJ (2011) Observations and simulations of non-local acceleration of electrons in magnetotail magnetic reconnection events. Nat Phys 7(4):360–365

Attico N, Califano F, Pegoraro F (2000) Fast collisionless reconnection in the whistler frequency range. Phys Plasmas 7(6):2381–2387. https://doi.org/10.1063/1.874076

Aunai N, Hesse M, Lavraud B, Dargent J, Smets R (2016) Orientation of the x-line in asymmetric magnetic reconnection. J Plasma Phys 82:535820401

Aydemir AY (1992) Nonlinear studies of $m = 1$ modes in high-temperature plasmas. Phys Fluids B 4:3469

Bailey G, Balan N, Su Y (1997) The Sheffield university plasmasphere ionosphere model—a review. J Atmos Sol-Terr Phys 59(13):1541–1552

Balsara DS (2009) Divergence-free reconstruction of magnetic fields and WENO schemes for magnetohydrodynamics. J Comput Phys 228(14):5040–5056. https://doi.org/10.1016/j.jcp.2009.03.038

Bao S, Wang W, Sorathia K, Merkin V, Toffoletto F, Lin D, Pham K, Garretson J, Wiltberger M, Lyon J, Michael A (2023) The relation among the ring current, subauroral polarization stream, and the geospace plume: MAGE simulation of the 31 March 2001 super storm. J Geophys Res Space Phys 128(12):e2023JA031923. https://doi.org/10.1029/2023JA031923

Bard CM, Dorelli JC (2014) A simple GPU-accelerated two-dimensional MUSCL-Hancock solver for ideal magnetohydrodynamics. J Comput Phys 259:444–460

Battarbee M, Brito T, Alho M, Pfau-Kempf Y, Grandin M, Ganse U, Papadakis K, Johlander A, Turc L, Dubart M, Palmroth M (2021) Vlasov simulation of electrons in the context of hybrid global models: an evlasiator approach. Ann Geophys 39(1):85–103. https://doi.org/10.5194/angeo-39-85-2021

Bessho N, Chen LJ, Wang S, Hesse M, Wilson III LB (2019) Magnetic reconnection in a quasi-parallel shock. Geophys Res Lett 46:9352

Bessho N, Chen LJ, Wang S, Hesse M, Wilson III LB, Ng J (2020) Magnetic reconnection and kinetic waves generated in the Earth's quasi-parallel bow shock. Phys Plasmas 27:092901

Bessho N, Chen LJ, Stawarz JE, Wang S, Hesse M, Wilson III LB, Ng J (2022) Strong reconnection electric fields in shock-driven turbulence. Phys Plasmas 29:042304

Bessho N, Chen LJ, Hesse M, Ng J, Wilson III LB, Stawarz JE (2023) Electron acceleration and heating during magnetic reconnection in the Earth's quasi-parallel bow shock. Astrophys J 954:25

Biermann L (1950) Über den Ursprung der Magnetfelder auf Sternen und im interstellaren Raum (mit einem Anhang von A. Schlüter). Z Naturforsch Teil A 5:65

Bird R, Tan N, Luedtke SV, Harrell SL, Taufer M, Albright B (2021) Vpic 2.0: next generation particle-in-cell simulations. IEEE Trans Parallel Distrib Syst 33(4):952–963. https://doi.org/10.1109/TPDS.2021.3084795

Birdsall CK, Langdon AB (1991) Plasma physics via computer simulation. Adam Hilger, Bristol

Birn J, Hesse M (1991) The substorm current wedge and field-aligned currents in MHD simulations of magnetotail reconnection. J Geophys Res Space Phys 96(A2):1611–1618

Birn J, Hesse M (1994) Particle acceleration in the dynamic magnetotail: orbits in self-consistent three-dimensional mhd fields. J Geophys Res Space Phys 99(A1):109–119

Birn J, Hesse M (2000) Large-scale stability of the magnetotail. In: Fifth international conference on substorms, vol 443, p 15

Birn J, Hesse M (2014) The substorm current wedge: further insights from mhd simulations. J Geophys Res Space Phys 119(5):3503–3513

Birn J, Hones EW Jr (1981) Three-dimensional computer modeling of dynamic reconnection in the geomagnetic tail. J Geophys Res Space Phys 86(A8):6802–6808

Birn J, Priest ER (eds) (2007) Reconnection of magnetic fields Cambridge University Press, Cambridge. https://doi.org/10.1017/CBO9780511536151

Birn J, Schindler K (2002) Thin current sheets in the magnetotail and the loss of equilibrium. J Geophys Res Space Phys 107(A7):SMP–18

Birn J, Drake JF, Shay MA, Rogers BN, Denton RE, Hesse M, Kuznetsova M, Ma ZW, Bhattacharjee A, Otto A, Pritchett PL (2001) GEM magnetic reconnection challenge. J Geophys Res 106:3715

Birn J, Thomsen M, Hesse M (2004) Electron acceleration in the dynamic magnetotail: test particle orbits in three-dimensional magnetohydrodynamic simulation fields. Phys Plasmas 11(5):1825–1833

Birn J, Merkin V, Sitnov M, Otto A (2018) MHD stability of magnetotail configurations with a B_Z hump. J Geophys Res Space Phys 123(5):3477–3492

Birn J, Hesse M, Runov A (2022) Electron anisotropies in magnetotail dipolarization events. Front Astron Space Sci 9:908730

Birn J, Merkin V, Sitnov M, Hesse M (2025) 3-D MHD stability of magnetotail configurations with a B_Z hump. J Geophys Res Space Phys 130(6):e2024JA033648

Bishop J (1996) Multiple charge exchange and ionization collisions within the ring current-geocorona-plasmasphere system: generation of a secondary ring current on inner L shells. J Geophys Res Space Phys 101(A8):17325–17336

Biskamp D (1996) Magnetic reconnection in plasmas. Astrophys Space Sci 242:165–207

Biskamp D (2000) Magnetic reconnection in plasmas. Cambridge University Press, Cambridge. https://doi.org/10.1017/CBO9780511599958

Blanc M, Richmond A (1980) The ionospheric disturbance dynamo. J Geophys Res Space Phys 85(A4):1669–1686

Blanco-Cano X, Kajdič P, Omidi N, Russell C (2011) Foreshock cavitons for different interplanetary magnetic field geometries: simulations and observations. J Geophys Res Space Phys 116(A9):A09101

Bohdan A, Niemiec J, Kobzar O, Pohl M (2017) Electron pre-acceleration at nonrelativistic high-Mach-number perpendicular shocks. Astrophys J 847:71

Bohdan A, Pohl M, Niemiec J, Vafin S, Matsumoto Y, Amano T, Hoshino M (2020) Kinetic simulations of nonrelativistic perpendicular shocks of young supernova remnants. Astrophys J 893:6

Boris JP (1970) Relativistic plasma simulation-optimization of a hybrid code. In: Proceeding of the 4th conference on numerical simulation of plasmas

Bowers KJ, Albright BJ, Yin L, Bergen B, Kwan TJT (2008) Ultrahigh performance three-dimensional electromagnetic relativistic kinetic plasma simulation. Phys Plasmas 15(5):055703. https://doi.org/10.1063/1.2840133

Brackbill JU, Lapenta G (2008) Magnetohydrodynamics with implicit plasma simulation. Commun Comput Phys 4:433–456

Braginskii SI (1965) Transport processes in a plasma. In: Leontovich MA (ed) Reviews of plasma physics, vol 1. Consultants Bureau, New York, pp 205–311

Brambles O, Lotko W, Zhang B, Ouellette J, Lyon J, Wiltberger M (2013) The effects of ionospheric outflow on icme and sir driven sawtooth events. J Geophys Res Space Phys 118(10):6026–6041

Büchner J (ed) (2023) Space and astrophysical plasma simulation: methods, algorithms, and applications. Springer, Cham. https://doi.org/10.1007/978-3-031-11870-8

Buchner J, Zelenyi L (1989) Regular and chaotic charged-particle motion in magnetotail-like field reversals 1. Basic theory of trapped motion. J Geophys Res 94(A9):11821–11842

Büchner J, Dum C, Scholer M (2003) Space plasma simulation. Lecture Notes in Physics, vol 615. Springer, Berlin. https://doi.org/10.1007/3-540-36530-3

Bulanov SV, Pegoraro F, Sakharov AS (1992) Magnetic reconnection in electron magnetohydrodynamics. Phys Fluids, B Plasma Phys 4(8):2499–2508. https://doi.org/10.1063/1.860467

Burch JL, Torbert RB, Phan TD, Chen LJ, Moore TE, Ergun RE, Eastwood JP, Gershman DJ, Cassak PA, Argal MR, Wang S, Hesse M, Pollock CJ, Giles BL, Nakamura R, Mauk BH, Fuselier SA, Russell CT, Strangeway RJ, Drake JF, Shay MA, Khotyaintsev YV, Lindqvist PA, Marklund G, Wilder FD, Young DT, Torkar K, Goldstein J, Dorelli JC, Avanov LA, Oka M, Baker DN, Jaynes AN, Goodrich KA, Cohen IJ, Turner DL, Fennell JF, Blake JB, Clemmons J, Goldman M, Newman D, Petrinec SM, Trattner K, Lavraud B, Reiff PH, Baumjohann W, Magnes W, Steller M, Lewis W, Saito Y, Coffey V, Chandler M (2016) Electron-scale measurements of magnetic reconnection in space. Science 352(6290):aaf2939

Burch JL, Webster JM, Hesse M, Genestreti KJ, Denton RE, Phan TD, Hasegawa H, Cassak PA, Torbert RB, Giles BL, Gershman DJ, Ergun RE, Russell CT, Strangeway RJ, Le Contel O, Pritchard KR, Marshall AT, Hwang KJ, Dokgo K, Fuselier SA, Chen LJ, Wang S, Swisdak M, Drake JF, Argall MR, Trattner KJ, Yamada M, Paschmann G (2020) Electron inflow velocities and reconnection rates at Earth's magnetopause and magnetosheath. Geophys Res Lett 47(17):e2020GL089082. https://doi.org/10.1029/2020GL089082

Buzulukova N, Fok MC, Pulkkinen A, Kuznetsova M, Moore TE, Glocer A, Brandt PC, Tóth G, Rastatter L (2010) Dynamics of ring current and electric fields in the inner magnetosphere during disturbed periods: CRCM–BATS-R-US coupled model. J Geophys Res 115:A05210. https://doi.org/10.1029/2009JA014621

Califano F, Cerri SS (2022) Eulerian approach to solve the Vlasov equation and hybrid-Vlasov simulations. In: Space and astrophysical plasma simulation: methods, algorithms, and applications. Springer, Berlin, pp 123–161

Califano F, Attico N, Pegoraro F, Bertin G, Bulanov SV (2001) Fast formation of magnetic islands in a plasma in the presence of counterstreaming electrons. Phys Rev Lett 86:5293–5296. https://doi.org/10.1103/PhysRevLett.86.5293

Califano F, Cerri SS, Faganello M, Laveder D, Sisti M, Kunz MW (2020) Electron-only reconnection in plasma turbulence. Front Phys 8:317

Caprioli D (2014) Hybrid simulations of particle acceleration at shocks. Nucl Phys B, Proc Suppl 256–257:48–55. https://doi.org/10.1016/j.nuclphysbps.2014.10.005

Cassak PA, Shay MA (2007) Scaling of asymmetric magnetic reconnection: general theory and collisional simulations. Phys Plasmas 14:102114. https://doi.org/10.1063/1.2795630

Cassak PA, Shay MA (2012) Magnetic reconnection for coronal conditions: reconnection rates, secondary islands and onset. Space Sci Rev 172:283–302. https://doi.org/10.1007/s11214-011-9755-2

Cassak PA, Baylor RN, Fermo RL, Beidler MT, Shay MA, Swisdak M, Drake JF, Karimabadi H (2015) Fast magnetic reconnection due to anisotropic electron pressure. Phys Plasmas 22:020705

Cassak PA, Barbhuiya MH, Liang H, Argall MR (2023) Quantifying energy conversion in higher-order phase space density moments in plasmas. Phys Rev Lett 130(8):085201. https://doi.org/10.1103/PhysRevLett.130.085201. arXiv:2306.01106

Cerri SS, Servidio S, Califano F (2017) Kinetic cascade in solar-wind turbulence: 3D3V hybrid-kinetic simulations with electron inertia. Astrophys J Lett 846(2):L18. https://doi.org/10.3847/2041-8213/aa87b0

Cerri SS, Kunz MW, Califano F (2018) Dual phase-space cascades in 3D hybrid-Vlasov-Maxwell turbulence. Astrophys J Lett 856(1):L13. https://doi.org/10.3847/2041-8213/aab557. arXiv:1802.06133

Chacón L, Simakov AN, Zocco A (2007) Steady-state properties of driven magnetic reconnection in 2D electron magnetohydrodynamics. Phys Rev Lett 99(23):235001. https://doi.org/10.1103/PhysRevLett.99.235001

Chen MW, Schulz M (2001) Simulations of diffuse aurora with plasma sheet electrons in pitch angle diffusion less than everywhere strong. J Geophys Res Space Phys 106(A12):28949–28966

Chen Y, Tóth G (2019) Gauss's law satisfying energy-conserving semi-implicit particle-in-cell method. J Comput Phys 386:632. https://doi.org/10.1016/j.jcp.2019.02.032

Chen MW, Lemon CL, Guild TB, Schulz M, Roeder JL, Le G (2012) Comparison of self-consistent simulations with observed magnetic field and ion plasma parameters in the ring current during the 10 August 2000 magnetic storm. J Geophys Res Space Phys 117(A9):A09232

Chen Y, Tóth G, Cassak P, Jia X, Gombosi TI, Slavin J, Markidis S, Peng B (2017) Global three-dimensional simulation of Earth's dayside reconnection using a two-way coupled magnetohydrodynamics with embedded particle-in-cell model: initial results. J Geophys Res 122:10318. https://doi.org/10.1002/2017JA024186

Chen Y, Tóth G, Jia X, Slavin J, Sun W, Markidis S, Gombosi T, Raines J (2019) Studying dawn-dusk asymmetries of Mercury's magnetotail using MHD-EPIC simulations. J Geophys Res 124:8954. https://doi.org/10.1029/2019JA026840

Chen Y, Tóth G, Hietala H, Vines SK, Zou Y, Nishimura Y, Silveira MV, Guo Z, Lin Y, Markidis S (2020) Magnetohydrodynamic with embedded particle-in-cell simulation of the geospace environment modeling dayside kinetic processes challenge event. Earth Space Sci. https://doi.org/10.1029/2020ea001331

Chen LJ, Ng J, Omelchenko Y, Wang S (2021) Magnetopause reconnection and indents induced by foreshock turbulence. Geophys Res Lett 48(11):e2021GL093029

Chen Y, Tóth G, Zhou H, Wang X (2023) FLEKS: a flexible particle-in-cell code for multi-scale plasma simulations. Comput Phys Commun 287:108714. https://doi.org/10.1016/j.cpc.2023.108714

Cheng L, Lin Y, Perez J, Johnson JR, Wang X (2020) Kinetic Alfvén waves from magnetotail to the ionosphere in global hybrid simulation associated with fast flows. J Geophys Res Space Phys 125(2):e2019JA027062

Chew GF, Goldberger ML, Low FE (1956) Boltzmann equation and the one-fluid hydrodynamic equations in the absence of particle collisions. Proc R Soc Lond A 236:112

Choudhuri AR (1998) The physics of fluids and plasmas. Cambridge University Press, Cambridge

Cramer WD, Raeder J, Toffoletto F, Gilson M, Hu B (2017) Plasma sheet injections into the inner magnetosphere: two-way coupled openggcm-rcm model results. J Geophys Res Space Phys 122(5):5077–5091

Curran DB, Goertz C (1989) Particle distributions in a two-dimensional reconnection field geometry. J Geophys Res Space Phys 94(A1):272–286

Dahlin JT, Drake JF, Swisdak M (2014) The mechanisms of electron heating and acceleration during magnetic reconnection. Phys Plasmas 21(9):092304. https://doi.org/10.1063/1.4894484. arXiv:1406.0831

Dahlin JT, Drake JF, Swisdak M (2015) Electron acceleration in three-dimensional magnetic reconnection with a guide field. Phys Plasmas 22(10):100704. https://doi.org/10.1063/1.4933212. arXiv:1503.02218

Dahlin J, Drake J, Swisdak M (2016) Parallel electric fields are inefficient drivers of energetic electrons in magnetic reconnection. Phys Plasmas 23(12):120704

Dahlin J, Drake J, Swisdak M (2017) The role of three-dimensional transport in driving enhanced electron acceleration during magnetic reconnection. Phys Plasmas 24(9):092110

Daldorff LKS, Tóth G, Gombosi TI, Lapenta G, Amaya J, Markidis S, Brackbill JU (2014) Two-way coupling of a global Hall magnetohydrodynamics model with a local implicit particle-in-cell model. J Comput Phys 268:236. https://doi.org/10.1016/j.jcp.2014.03.009

Daughton W, Scudder J (2006) Fully kinetic simulations of undriven magnetic reconnection with open boundary conditions. Phys Plasmas 13:072101

Daughton W, Roytershteyn V, Karimabadi H, Yin L, Albright BJ, Bergen B, Bowers KJ (2011) Role of electron physics in the development of turbulent magnetic reconnection in collisionless plasmas. Nat Phys 7:539–542. https://doi.org/10.1038/nphys1965

De Zeeuw D, Sazykin S, Wolf R, Gombosi T, Ridley A, Tóth G (2004) Coupling of a global MHD code and an inner magnetosphere model: initial results. J Geophys Res 109(A12):219. https://doi.org/10.1029/2003JA010366

Delamere PA (2009) Hybrid code simulations of the solar wind interaction with Pluto. J Geophys Res 114:A03220. https://doi.org/10.1029/2008JA013756

Delcourt D, Sauvaud J (1994) Plasma sheet ion energization during dipolarization events. J Geophys Res Space Phys 99(A1):97–108

Delzanno G (2015) Multi-dimensional, fully-implicit, spectral method for the Vlasov–Maxwell equations with exact conservation laws in discrete form. J Comput Phys 301:338–356. https://doi.org/10.1016/j.jcp.2015.07.028

DeVore CR (1991) Flux-corrected transport techniques for multidimensional compressible magnetohydrodynamics. J Comput Phys 92(1):142–160

Dong C, Wang L, Hakim A, Bhattacharjee A, Slavin JA, DiBraccio GA, Germaschewski K (2019) Global ten-moment multifluid simulations of the solar wind interaction with Mercury: from the planetary conducting core to the dynamic magnetosphere. Geophys Res Lett 46(21):11584–11596

Dong C, Le A, Wang L, Stanier A, Wetherton B, Daughton W, Bhattacharjee A, Slavin J, DiBraccio G (2021) Global hybrid-VPIC simulations of the solar wind interaction with Mercury's dynamic magnetosphere: reconnection and foreshock. In: EGU general assembly, 19–30 Apr. 2021, vol EGU-12954. https://doi.org/10.5194/egusphere-egu21-12954

Dorelli J, Birn J (2003) Whistler-mediated magnetic reconnection in large systems: magnetic flux pileup and the formation of thin current sheets. J Geophys Res Space Phys 108(A3):1133. https://doi.org/10.1029/2001JA009180

Dorelli JC, Glocer A, Collinson G, Tóth G (2015) The role of the Hall effect in the global structure and dynamics of planetary magnetospheres: Ganymede as a case study. J Geophys Res 120:5377

Drake JF, Shay MA (2007) The fundamentals of collisionless reconnection. In: Birn J, Priest E (eds) Reconnection of magnetic fields: magnetohydrodynamics and collisionless theory and observations. Cambridge University Press, Cambridge

Drake JF, Kleva RG, Mandt ME (1994) Structure of thin current layers: implications for magnetic reconnection. Phys Rev Lett 73(9):1251–1254. https://doi.org/10.1103/PhysRevLett.73.1251

Drake JF, Biskamp D, Zeiler A (1997) Breakup of the electron current layer during 3-d collisionless magnetic reconnection. Geophys Res Lett 24(22):2921–2924. https://doi.org/10.1029/97GL52961

Drake JF, Swisdak M, Che H, Shay MA (2006) Electron acceleration from contracting magnetic islands during reconnection. Nature 443(7111):553–556. https://doi.org/10.1038/nature05116

Drake JF, Shay MA, Swisdak M (2008) The Hall fields and fast magnetic reconnection. Phys Plasmas 15(4):042306. https://doi.org/10.1063/1.2901194

Drake J, Arnold H, Swisdak M, Dahlin J (2019) A computational model for exploring particle acceleration during reconnection in macroscale systems. Phys Plasmas 26(1):012901

Drake J, Antiochos S, Bale S, Chen B, Cohen C, Dahlin J, Glesener L, Guo F, Hoshino M, Imada S, et al (2025) Magnetic reconnection in solar flares and the near-sun solar wind. Space Sci Rev 221:27. https://doi.org/10.1007/s11214-025-01153-x

Dreher J, Arendt U, Schindler K (1996) Particle simulations of collisionless reconnection in magnetotail configuration including electron dynamics. J Geophys Res Space Phys 101(A12):27375–27381

Du S, Guo F, Zank GP, Li X, Stanier A (2018) Plasma energization in colliding magnetic flux ropes. Astrophys J 867(1):16. https://doi.org/10.3847/1538-4357/aae30e. arXiv:1809.08357

Dungey JW (1954) The attenuation of Alfvén waves. J Geophys Res 59(3):323–328. https://doi.org/10.1029/JZ059i003p00323

Dungey J (1961) Interplanetary magnetic field and the auroral zones. Phys Rev Lett 93:47. https://doi.org/10.1103/PhysRevLett.6.47

Dyadechkin S, Kallio E, Jarvinen R (2013) A new 3-d spherical hybrid model for solar wind interaction studies. J Geophys Res Space Phys 118:5157–5168. https://doi.org/10.1002/jgra.50497

Edwards AW, Campbell DJ, Engelhardt WW, Farhbach HU, Gill RD, Granetz RS, Tsuji S, Tubbing BJD, Weller A, Wesson J, Zasche D (1986) Rapid collapse of a plasma sawtooth oscillation in the jet tokamak. Phys Rev Lett 57:210–213

Egedal J, Daughton W, Le A (2012) Large-scale electron acceleration by parallel electric fields during magnetic reconnection. Nat Phys 8(4):321–324

Egedal J, Le A, Daughton W (2013) A review of pressure anisotropy caused by electron trapping in collisionless plasma, and its implications for magnetic reconnection. Phys Plasmas 20:061201

Erickson GM, Wolf R (1980) Is steady convection possible in the Earth's magnetotail? Geophys Res Lett 7(11):897–900

Eriksson S, Wilder FD, Ergun RE, Schwartz SJ, Cassak PA, Burch JL, Chen LJ, Torbert RB, Phan TD, Lavraud B, Goodrich KA, Holmes JC, Stawarz JE, Sturner AP, Malaspina DM, Usanova ME, Trattner KJ, Strangeway RJ, Russell CT, Pollock CJ, Giles BL, Hesse M, Lindqvist PA, Drake JF, Shay MA, Nakamura R, Marklund GT (2016) Magnetospheric multiscale observations of the electron diffusion region of large guide field magnetic reconnection. Phys Rev Lett 117(1):015001. https://doi.org/10.1103/PhysRevLett.117.015001

Fadanelli S, Lavraud B, Califano F, Cozzani G, Finelli F, Sisti M (2021) Energy conversions associated with magnetic reconnection. J Geophys Res Space Phys 126(1):e2020JA028333. https://doi.org/10.1029/2020JA028333

Farrugia CJ, Rogers AJ, Torbert RB, Genestreti KJ, Nakamura TKM, Lavraud B, Montag P, Egedal J, Payne D, Keesee A, Ahmadi N, Ergun R, Reiff P, Argall M, Matsui H, Wilson III LB, Lugaz N, Lugaz N, Burch JL, Russell CT, Fuselier SA, Dors I (2021) An encounter with the ion and electron diffusion regions at a flapping and twisted tail current sheet. J Geophys Res Space Phys 126(3):e2020JA028903. https://doi.org/10.1029/2020JA028903

Fatemi S, Poppe AR, Delory GT, Farrell WM (2017) AMITIS: a 3D GPU-based hybrid-PIC model for space and plasma physics. J Phys Conf Ser 837:012017. https://doi.org/10.1088/1742-6596/837/1/012017

Finelli F (2022) Magnetic reconnection in space plasmas: advanced numerical models and detection in turbulence. Phd thesis, University of Pisa, Pisa, Italy. https://etd.adm.unipi.it/t/etd-06062022-180752/

Finelli F, Cerri SS, Califano F, Pucci F, Laveder D, Lapenta G, Passot T (2021) Bridging hybrid- and full-kinetic models with Landau-fluid electrons. Astron Astrophys 653:A156. https://doi.org/10.1051/0004-6361/202140279

Fok MC, Wolf R, Spiro R, Moore T (2001) Comprehensive computational model of Earth's ring current. J Geophys Res Space Phys 106(A5):8417–8424

Fok MC, Buzulukova N, Chen SH, Glocer A, Nagai T, Valek P, Perez J (2014) The comprehensive inner magnetosphere-ionosphere model. J Geophys Res Space Phys 119(9):7522–7540

Fox W, Bhattacharjee A, Germaschewski K (2012) Magnetic reconnection in high-energy-density laser-produced plasmas. Phys Plasmas 19(5):056309

Franci L, Cerri SS, Califano F, Landi S, Papini E, Verdini A, Matteini L, Jenko F, Hellinger P (2017) Magnetic reconnection as a driver for a sub-ion-scale cascade in plasma turbulence. Astrophys J Lett 850(1):L16. https://doi.org/10.3847/2041-8213/aa93fb

Franci L, Hellinger P, Guarrasi M, Chen CH, Papini E, Verdini A, Matteini L, Landi S (2018) Three-dimensional simulations of solar wind turbulence with the hybrid code CAMELIA. J Phys Conf Ser 1031(1):012002. https://doi.org/10.1088/1742-6596/1031/1/012002

Freeman J, Wolf R, Spiro R, Hausman B, Bales B, Hilmer R, Nagai A, Lambour R (1993) Magnetospheric specification model development code documentation, scientific description, and software documentation. Contract F19628-90-K-0012, Rice Univ for Air Force Geophys Lab, Hanscom Air Force Base, Mass, July

French O, Guo F, Zhang Q, Uzdensky D (2022) Particle Injection and Nonthermal Particle Acceleration in Relativistic Magnetic Reconnection. arXiv e-prints. arXiv:2210.08358

Fu XR, Lu QM, Wang S (2006) The process of electron acceleration during collisionless magnetic reconnection. Phys Plasmas 13(1):012309. https://doi.org/10.1063/1.2164808

Fujimoto K (2011) A new electromagnetic particle-in-cell model with adaptive mesh refinement for high-performance parallel computation. J Comput Phys 230(23):8508–8526. https://doi.org/10.1016/j.jcp.2011.08.002

Fuller-Rowell T, Rees D, Quegan S, Moffett R, Codrescu M, Millward G (1996) A coupled thermosphere-ionosphere model (ctim). STEP report 239(4)

Gabrielse C, Angelopoulos V, Runov A, Turner D (2012) The effects of transient, localized electric fields on equatorial electron acceleration and transport toward the inner magnetosphere. J Geophys Res Space Phys 117(A10):A10213

Gallagher DL, Craven PD, Comfort RH (2000) Global core plasma model. J Geophys Res Space Phys 105(A8):18819–18833

Ganse U, Koskela T, Battarbee M, Pfau-Kempf Y, Papadakis K, Alho M, Bussov M, Cozzani G, Dubart M, George H, Gordeev E, Grandin M, Horaites K, Suni J, Tarvus V, Kebede FT, Turc L, Zhou H, Palmroth M (2023) Enabling technology for global 3D + 3V hybrid-Vlasov simulations of near-Earth space. Phys Plasmas 30(4):042902

Gardner M (1970) Mathematical games. Sci Am 222(6):132–140

Ghizzo A, Sarrat M, Del Sarto D (2017) Vlasov models for kinetic Weibel-type instabilities. J Plasma Phys 83(1):705830101. https://doi.org/10.1017/S0022377816001215

Gingell I, Schwartz SJ, Eastwood JP, Burch JL, Ergun RE, Fuselier S, Gershman DJ, Giles BL, Khotyaintsev YV, Lavraud B, Lindqvist PA, Paterson WR, Phan TD, Russell CT, Stawarz JE, Strangeway RJ, Torbert RB, Wilder F (2019) Observations of magnetic reconnection in the transition region of quasi-parallel shocks. Geophys Res Lett 46(3):1177–1184. https://doi.org/10.1029/2018GL081804

Gingell I, Schwartz SJ, Eastwood JP, Stawarz JE, Burch JL, Ergun RE, Fuselier SA, Gershman DJ, Giles BL, Khotyaintsev YV, Lavraud B, Lindqvist PA, Paterson WR, Phan TD, Russell CT, Strangeway RJ, Torbert RB, Wilder F (2020) Statistics of reconnecting current sheets in the transition region of Earth's bow shock. J Geophys Res 125:e2019JA027119

Gingell I, Schwartz SJ, Kucharek H, Farrugia CJ, Trattner KJ (2021) Observing the prevalence of thin current sheets downstream of Earth's bow shock. Phys Plasmas 28:102902

Gkioulidou M, Ukhorskiy A, Mitchell D, Sotirelis T, Mauk B, Lanzerotti L (2014) The role of small-scale ion injections in the buildup of Earth's ring current pressure: Van Allen probes observations of the 17 March 2013 storm. J Geophys Res Space Phys 119(9):7327–7342

Glasser AH, Sovinec CR, Nebel RA, Gianakon TA, Plimpton SJ, Chu MS, Schnack DD (the NIMROD Team) (1999) The nimrod code: a new approach to numerical plasma physics. Plasma Phys Control Fusion 41(3A):A747. https://doi.org/10.1088/0741-3335/41/3A/067

Glocer A, Tóth G, Gombosi T, Welling D (2009a) Modeling ionospheric outflows and their impact on the magnetosphere, initial results. J Geophys Res Space Phys 114(A5):A05216

Glocer A, Tóth G, Ma YJ, Gombosi T, Zhang JC, Kistler LM (2009b) Multifluid block-adaptive-tree solar wind Roe-type upwind scheme: magnetospheric composition and dynamics during geomagnetic storms – initial results. J Geophys Res 114:A12203. https://doi.org/10.1029/2009JA014418

Glocer A, Fok M, Meng X, Toth G, Buzulukova N, Chen S, Lin K (2013) Crcm+ bats–r–us two–way coupling. J Geophys Res Space Phys 118(4):1635–1650

Godfrey BB (1980) Time-biased field solver for electromagnetic PIC code. In: Proceedings of the ninth conference on numerical simulation of plasmas, p OD–4

Goldstein J, Spiro RW, Reiff PH, Wolf RA, Sandel BR, Freeman JW, Lambour RL (2002) IMF-driven overshielding electric field and the origin of the plasmaspheric shoulder of May 24, 2000. Geophys Res Lett 29(16):1819. https://doi.org/10.1029/2001GL014534

Gombosi TI, Tóth G, De Zeeuw DL, Hansen KC, Kabin K, Powell KG (2002) Semirelativistic magnetohydrodynamics and physics-based convergence acceleration. J Comput Phys 177:176–205. https://doi.org/10.1006/jcph.2002.7009

Gómez D (2006) Parallel simulations of Hall-MHD plasmas. Space Sci Rev 122:231–238. https://doi.org/10.1007/s11214-006-7287-y

Gonzalez W, Parker E (eds) (2016) Magnetic reconnection: concepts and applications. Astrophysics and Space Science Library, vol 427. Springer, Cham. https://doi.org/10.1007/978-3-319-26432-5

Graham D, Cozzani G, Khotyaintsev YV, et al (2025) The role of kinetic instabilities and waves in collisionless magnetic reconnection. Space Sci Rev 221:20. https://doi.org/10.1007/s11214-024-01133-7

Grandin M, Turc L, Battarbee M, Ganse U, Johlander A, Pfau-Kempf Y, Dubart M, Palmroth M (2020) Hybrid-Vlasov simulation of auroral proton precipitation in the cusps: comparison of northward and southward interplanetary magnetic field driving. J Space Weather Space Clim 10:51. https://doi.org/10.1051/swsc/2020053

Grandin M, Luttikhuis T, Battarbee M, Cozzani G, Zhou H, Turc L, Pfau-Kempf Y, George H, Horaites K, Gordeev E, Ganse U, Papadakis K, Alho M, Tesema F, Suni J, Dubart M, Tarvus V, Palmroth M (2023) First 3d hybrid-Vlasov global simulation of auroral proton precipitation and comparison with satellite observations. J Space Weather Space Clim. https://doi.org/10.1051/swsc/2023017

Greco A, Valentini F, Servidio S, Matthaeus WH (2012) Inhomogeneous kinetic effects related to intermittent magnetic discontinuities. Phys Rev E 86(6):066405. https://doi.org/10.1103/PhysRevE.86.066405

Greess S, Egedal J, Stanier A, Daughton W, Olson J, Lê A, Myers R, Millet-Ayala A, Clark M, Wallace J, Endrizzi D, Forest C (2021) Laboratory verification of electron-scale reconnection regions modulated by a three-dimensional instability. J Geophys Res Space Phys 126(7):e2021JA029316. https://doi.org/10.1029/2021JA029316

Guo F, Giacalone J (2013) The acceleration of thermal protons at parallel collisionless shocks: three-dimensional hybrid simulations. Astrophys J 773:158. https://doi.org/10.1088/0004-637X/773/2/158

Guo F, Li H, Daughton W, Liu YH (2014) Formation of hard power laws in the energetic particle spectra resulting from relativistic magnetic reconnection. Phys Rev Lett 113(15):155005. https://doi.org/10.1103/PhysRevLett.113.155005. arXiv:1405.4040

Guo F, Liu YH, Daughton W, Li H (2015) Particle acceleration and plasma dynamics during magnetic reconnection in the magnetically dominated regime. Astrophys J 806(2):167. https://doi.org/10.1088/0004-637X/806/2/167. arXiv:1504.02193

Guo F, Li X, Daughton W, Kilian P, Li H, Liu YH, Yan W, Ma D (2019) Determining the dominant acceleration mechanism during relativistic magnetic reconnection in large-scale systems. Astrophys J Lett 879(2):L23. https://doi.org/10.3847/2041-8213/ab2a15. arXiv:1901.08308

Guo Z, Lin Y, Wang X, Vines SK, Lee S, Chen Y (2020) Magnetopause reconnection as influenced by the dipole tilt under southward imf conditions: hybrid simulation and mms observation. J Geophys Res Space Phys 125(9):e2020JA027795

Guo F, Li X, Daughton W, Li H, Kilian P, Liu YH, Zhang Q, Zhang H (2021a) Magnetic energy release, plasma dynamics, and particle acceleration in relativistic turbulent magnetic reconnection. Astrophys J 919(2):111. https://doi.org/10.3847/1538-4357/ac0918. arXiv:2008.02743

Guo Z, Lin Y, Wang X (2021b) Investigation of the interaction between magnetosheath reconnection and magnetopause reconnection driven by oblique interplanetary tangential discontinuity using three-dimensional global hybrid simulation. J Geophys Res Space Phys 126:e2020JA028558. https://doi.org/10.1029/2020JA028558

Guo Z, Lin Y, Wang X, Du A (2021c) Magnetic reconnection inside solar wind rotational discontinuity during its interaction with the quasi-perpendicular bow shock and magnetosheath. J Geophys Res Space Phys 126(12):e2021JA029979

Guo J, Lu S, Lu Q, Lin Y, Wang X, Huang K, Wang R, Wang S (2021d) Re-reconnection processes of magnetopause flux ropes: three-dimensional global hybrid simulations. J Geophys Res Space Phys 126(6):e2021JA029388

Guo J, Lu S, Lu Q, Lin Y, Wang X, Zhang Q, Xing Z, Huang K, Wang R, Wang S (2021e) Three-dimensional global hybrid simulations of high latitude magnetopause reconnection and flux ropes during the northward imf. Geophys Res Lett 48(21):e2021GL095003

Guo F, Liu YH, Zenitani S, Hoshino M (2023) Magnetic Reconnection and Associated Particle Acceleration in High-energy Astrophysics. arXiv e-prints. https://doi.org/10.48550/arXiv.2309.13382. arXiv:2309.13382

Guzdar PN, Drake JF, McCarthy D, Hassam AB, Liu CS (1993) Three-dimensional fluid simulations of the nonlinear drift-resistive ballooning modes in tokamak edge plasmas. Phys Fluids B 5(10):3712–3727

Haggerty CC, Parashar TN, Matthaeus WH, Shay MA, Yang Y, Wan M, Wu P, Servidio S (2017) Exploring the statistics of magnetic reconnection x-points in kinetic particle-in-cell turbulence. Phys Plasmas 24:102308

Haiducek JD, Welling DT, Morley SK, Gañushkina NY, Chu X (2020) Using multiple signatures to improve accuracy of substorm identification. J Geophys Res 125(4):e2019JA027559. https://doi.org/10.1029/2019JA027559

Hall EH, et al (1879) On a new action of the magnet on electric currents. Am J Math 2(3):287–292

Harris EG (1962a) The equilibrium of oppositely directed magnetic fields. Nuovo Cimento 23:115–121

Harris EG (1962b) On a plasma sheath separating regions of oppositely directed magnetic field. Nuovo Cimento 23(1):115–121

Herčík D, Trávníček PM, Johnson R, Kim EH, Hellinger P (2013) Mirror mode structures in the asymmetric Hermean magnetosheath: hybrid simulations. J Geophys Res Space Phys 118:405–417. https://doi.org/10.1029/2012JA018083

Hesse M, Winske D (1998) Electron dissipation in collisionless magnetic reconnection. J Geophys Res Space Phys 103(A11):26479–26486. https://doi.org/10.1029/98JA01570

Hesse M, Birn J, Baker DN, Slavin JA (1996) Mhd simulations of the transition of magnetic reconnection from closed to open field lines. J Geophys Res Space Phys 101(A5):10805–10816. https://doi.org/10.1029/95JA02857
Hesse M, Aunai N, Zenitani S, Kuznetsova M, Birn J (2013) Aspects of collisionless magnetic reconnection in asymmetric systems. Phys Plasmas 20:061210
Hesse M, Liu YH, Chen LJ, Bessho N, Kuznetsova M, Birn J, Burch JL (2016) On the electron diffusion region in asymmetric reconnection with a guide magnetic field. Geophys Res Lett 43(6):2359–2364. https://doi.org/10.1002/2016GL068373
Hesse M, Liu YH, Chen LJ, Bessho N, Wang S, Burch J, Moretto T, Norgren C, Genestreti K, Phan T, et al (2018b) The physical foundation of the reconnection electric field. Phys Plasmas 25(3):032901
Hesse M, Norgren C, Tenfjord P, Burch JL, Liu YH, Chen LJ, Bessho N, Wang S, Nakamura R, Eastwood JP, Hoshino M, Torbert RB, Ergun RE (2018a) On the role of separatrix instabilities in heating the reconnection outflow region. Phys Plasmas 25:122902
Hesse M, Norgren C, Tenfjord P, Burch JL, Liu YH, Chen LJ, Bessho N, Wang S, Nakamura R, Eastwood JP, Hoshino M, Torbert RB, Ergun RE (2019) Erratum: “on the role of separatrix instabilities in heating the reconnection outflow region”. Phys Plasmas 26:049901
Hill T, Rymer A, Burch J, Crary F, Young D, Thomsen M, Delapp D, André N, Coates A, Lewis G (2005) Evidence for rotationally driven plasma transport in Saturn’s magnetosphere. Geophys Res Lett 32(14):L14S10
Hilmer RV, Voigt GH (1995) A magnetospheric magnetic field model with flexible current systems driven by independent physical parameters. J Geophys Res Space Phys 100(A4):5613–5626
Hoilijoki S, Ganse U, Pfau-Kempf Y, Cassak PA, Walsh BM, Hietala H, von Alfthan S, Palmroth M (2017) Reconnection rates and x line motion at the magnetopause: global 2d-3v hybrid-Vlasov simulation results. J Geophys Res Space Phys 122(3):2877–2888. https://doi.org/10.1002/2016JA023709
Hoilijoki S, Ganse U, Sibeck DG, Cassak PA, Turc L, Battarbee M, Fear RC, Blanco-Cano X, Dimmock AP, Kilpua EKJ, Jarvinen R, Juusola L, Pfau-Kempf Y, Palmroth M (2019) Properties of magnetic reconnection and ftes on the dayside magnetopause with and without positive imf bx component during southward imf. J Geophys Res Space Phys 124(6):4037–4048. https://doi.org/10.1029/2019JA026821
Holloway JP, Dorning JJ (1991) Undamped plasma waves. Phys Rev A 44(6):3856–3868. https://doi.org/10.1103/PhysRevA.44.3856
Honkonen I, von Alfthan S, Sandroos A, Janhunen P, Palmroth M (2013) Parallel grid library for rapid and flexible simulation development. Comput Phys Commun 184(4):1297–1309. https://doi.org/10.1016/j.cpc.2012.12.017
Horiuchi R, Sato T (1994) Particle simulation study of driven magnetic reconnection in a collisionless plasma. Phys Plasmas 1(11):3587–3597
Hoshino M (1987) The electrostatic effect for the collisionless tearing mode. J Geophys Res Space Phys 1:7368–7380
Hoshino M, Mukai T, Terasawa T, Shinohara I (2001) Suprathermal electron acceleration in magnetic reconnection. J Geophys Res 106(A11):25979–25998. https://doi.org/10.1029/2001JA900052
Hsieh MS, Otto A (2014) The influence of magnetic flux depletion on the magnetotail and auroral morphology during the substorm growth phase. J Geophys Res Space Phys 119(5):3430–3443
Hsieh MS, Otto A (2015) Thin current sheet formation in response to the loading and the depletion of magnetic flux during the substorm growth phase. J Geophys Res Space Phys 120(6):4264–4278
Hu B, Toffoletto F, Wolf R, Sazykin S, Raeder J, Larson D, Vapirev A (2010) One-way coupled OpenGGCM/RCM simulation of the 23 March 2007 substorm event. J Geophys Res Space Phys 115(A12):A12205
Hu B, Wolf R, Toffoletto F, Yang J, Raeder J (2011) Consequences of violation of frozen-in-flux: evidence from openggcm simulations. J Geophys Res Space Phys 116(A6):A06223
Huang YM, Bhattacharjee A, Sullivan BP (2011) Onset of fast reconnection in Hall magnetohydrodynamics mediated by the plasmoid instability. Phys Plasmas 18:072109
Huang Z, Tóth G, van der Holst B, Chen Y, Gombosi T (2019) A six-moment multi-fluid plasma model. J Comput Phys 387:134. https://doi.org/10.1016/j.jcp.2019.02.023
Huang K, Liu YH, Lu Q, Hesse M (2020) Scaling of magnetic reconnection with a limited x-line extent. Geophys Res Lett 47:e2020GL088147
Huang K, Lu Q, Liu YH, Lu S, Li X, Tang H, Peng E (2024) Secondary reconnection between interlinked flux tubes driven by magnetic reconnection with a short x-line. Geophys Res Lett 51:e2024GL111812
Huba JD (1995) Hall magnetohydrodynamics in space and laboratory plasmas. Phys Plasmas 2(6):2504–2513
Huba JD (1998) NRL plasma formulary. 98-358, Naval Research Laboratory
Huba JD (2003) Hall magnetohydrodynamics – a tutorial. In: Büchner J, Dum CT, Scholer M (eds) Space plasma simulation. Lecture Notes in Physics, vol 615. Springer, Berlin, pp 166–192. https://doi.org/10.1007/3-540-36530-3_9

Huba JD, Rudakov LI (2002) Three-dimensional Hall magnetic reconnection. Phys Plasmas 9:4435

Huba JD, Rudakov LI (2003) Hall magnetohydrodynamics of neutral layers. Phys Plasmas 10:3139

Huba JD, Sazykin S (2014) Storm time ionosphere and plasmasphere structuring: SAMI3-RCM simulation of the 31 March 2001 geomagnetic storm. Geophys Res Lett 41(23):8208–8214. https://doi.org/10.1002/2014GL062110

Hubbert M, Russell CT, Qi Y, Lu S, Burch JL, Giles BL, Moore TE (2022) Electron-only reconnection as a transition phase from quiet magnetotail current sheets to traditional magnetotail reconnection. J Geophys Res Space Phys 127(3):e2021JA029584. https://doi.org/10.1029/2021JA029584

Hwang KJ, Choi E, Dokgo K, Burch JL, Sibeck DG, Giles BL, Goldstein ML, Paterson WR, Pollock CJ, Shi QQ, Fu H, Hasegawa H, Gershman DJ, Khotyaintsev Y, Torbert RB, Ergun RE, Dorelli JC, Avanov L, Russell CT, Strangeway RJ (2019) Electron vorticity indicative of the electron diffusion region of magnetic reconnection. Geophys Res Lett 46(12):6287–6296. https://doi.org/10.1029/2019GL082710

Hyman JM (1983) Accurate monotonicity preserving cubic interpolation. SIAM J Sci Stat Comput 4(4):645–654

Inglebert A, Ghizzo A, Reveille T, Sarto DD, Bertrand P, Califano F (2011) A multi-stream Vlasov modeling unifying relativistic Weibel-type instabilities. Europhys Lett 95(4):45002. https://doi.org/10.1209/0295-5075/95/45002

Innocenti ME, Lapenta G, Markidis S, Beck A, Vapirev A (2013) A multi level multi domain method for particle in cell plasma simulations. J Comput Phys 238:115

Jain N, Büchner J (2014a) Nonlinear evolution of three-dimensional instabilities of thin and thick electron scale current sheets: plasmoid formation and current filamentation. Phys Plasmas 21(7):072306. https://doi.org/10.1063/1.4887279

Jain N, Büchner J (2014b) Three dimensional instabilities of an electron scale current sheet in collisionless magnetic reconnection. Phys Plasmas 21(6):062116. https://doi.org/10.1063/1.4885636

Jain N, Büchner J (2015) Effect of guide field on three-dimensional electron shear flow instabilities in electron current sheets. J Plasma Phys 81(6):905810606. https://doi.org/10.1017/S0022377815001257

Jain N, Sharma AS (2009) Electron scale structures in collisionless magnetic reconnection. Phys Plasmas 16(5):050704. https://doi.org/10.1063/1.3134045

Jain N, Sharma AS (2015a) Electron-scale nested quadrupole Hall field in cluster observations of magnetic reconnection. Ann Geophys 33(6):719–724. https://doi.org/10.5194/angeo-33-719-2015

Jain N, Sharma AS (2015b) Evolution of electron current sheets in collisionless magnetic reconnection. Phys Plasmas 22(10):102110. https://doi.org/10.1063/1.4933120

Jain N, Sharma AS, Zelenyi LM, Malova HV (2012) Electron scale structures of thin current sheets in magnetic reconnection. Ann Geophys 30(4):661–666. https://doi.org/10.5194/angeo-30-661-2012

Jain N, Büchner J, Dorfman S, Ji H, Surjalal Sharma A (2013) Current disruption and its spreading in collisionless magnetic reconnection. Phys Plasmas 20(11):112101

Jain N, Büchner J, Muñoz PA (2017a) Nonlinear evolution of electron shear flow instabilities in the presence of an external guide magnetic field. Phys Plasmas 24(3):032303. https://doi.org/10.1063/1.4977528

Jain N, von Stechow A, Muñoz PA, Büchner J, Grulke O, Klinger T (2017b) Electron-magnetohydrodynamic simulations of electron scale current sheet dynamics in the Vineta.II guide field reconnection experiment. Phys Plasmas 24(9):092312. https://doi.org/10.1063/1.5004564

Jain N, Muñoz PA, Farzalipour Tabriz M, Rampp M, Büchner J (2022) Importance of accurate consideration of the electron inertia in hybrid-kinetic simulations of collisionless plasma turbulence: the 2D limit. Phys Plasmas 29(5):053902. https://doi.org/10.1063/5.0087103

Jain N, Muñoz PA, Büchner J (2023) Hybrid-kinetic approach: inertial electrons. In: Büchner J (ed) Space and astrophysical plasma simulation: methods, algorithms, and applications. Springer, Cham, pp 283–311. https://doi.org/10.1007/978-3-031-11870-8_9

Jardin S, Ferraro N, Luo X, Chen J, Breslau J, Jansen K, Shephard M (2008) The M3D-C1 approach to simulating 3D 2-fluid magnetohydrodynamics in magnetic fusion experiments. J Phys Conf Ser 125:012044

Jarvinen R, Alho M, Kallio E, Pulkkinen TI (2020) Ultra-low-frequency waves in the foreschock of Mercury: a global hybrid modelling study. Mon Not R Astron Soc 491:4147–4161. https://doi.org/10.1093/mnras/stz3257

Johlander A, Battarbee M, Turc L, Ganse U, Pfau-Kempf Y, Grandin M, Suni J, Tarvus V, Bussov M, Zhou H, Alho M, Dubart M, George H, Papadakis K, Palmroth M (2022) Quasi-parallel shock reformation seen by magnetospheric multiscale and ion-kinetic simulations. Geophys Res Lett 49(2):e2021GL096335. https://doi.org/10.1029/2021GL096335

Jordanova VK, Kozyra JU, Khazanov GV, Nagy AF, Rasmussen CE, Fok MC (1994) A bounce-averaged kinetic model of the ring current ion population. Geophys Res Lett 21:2785

Juno J, Hakim A, TenBarge J, Shi E, Dorland W (2018) Discontinuous Galerkin algorithms for fully kinetic plasmas. J Comput Phys 353:110–147. https://doi.org/10.1016/j.jcp.2017.10.009

Juusola L, Hoilijoki S, Pfau-Kempf Y, Ganse U, Jarvinen R, Battarbee M, Kilpua E, Turc L, Palmroth M (2018) Fast plasma sheet flows and x line motion in the Earth's magnetotail: results from a global hybrid-Vlasov simulation. Ann Geophys 36(5):1183–1199. https://doi.org/10.5194/angeo-36-1183-2018

Kallio E, Dyadechkin S, Wurz P, Khodachenko M (2019) Space weathering on the Moon: farside-nearside solar wind precipitation asymmetry. Planet Space Sci 166:9–22. https://doi.org/10.1016/j.pss.2018.07.013

Karimabadi H, Krauss-Varban D, Huba JD, Vu HX (2004) On magnetic reconnection regimes and associated three-dimensional asymmetries: hybrid, Hall-less hybrid, and Hall-MHD simulations. J Geophys Res 109:A09205

Karimabadi H, Daughton W, Scudder J (2007) Multi-scale structure of the electron diffusion region. Geophys Res Lett 34:L13104

Karimabadi H, Roytershteyn V, Vu HX, Omelchenko YA, Scudder J, Daughton W, Dimmock A, Nykyri K, Wan M, Sibeck D, Tatineni M, Majumdar A, Loring B, Geveci B (2014) The link between shocks, turbulence, and magnetic reconnection in collisionless plasmas. Phys Plasmas 21(6):062308. https://doi.org/10.1063/1.4882875

Kepko L, McPherron R, Amm O, Apatenkov S, Baumjohann W, Birn J, Lester M, Nakamura R, Pulkkinen TI, Sergeev V (2015) Substorm current wedge revisited. Space Sci Rev 190:1–46. https://doi.org/10.1007/s11214-014-0124-9

Kilian P, Li X, Guo F, Li H (2020) Exploring the acceleration mechanisms for particle injection and power-law formation during transrelativistic magnetic reconnection. Astrophys J 899(2):151. https://doi.org/10.3847/1538-4357/aba1e9. arXiv:2001.02732

Kingsep AS, Chukbar KV, Yan'kov YY (1990) Electron magnetohydrodynamics. In: Reviews of plasma physics, vol 16. Consultants Bureau, New York

Kleva R, Drake J, Waelbroeck F (1995) Fast reconnection in high temperature plasma. Phys Plasmas 2:23

Kuznetsova M (2000) Toward a transport model of collisionless magnetic reconnection. J Geophys Res Space Phys 105(A4):7601–7616. https://doi.org/10.1029/1999JA900396

Kuznetsova M, Hesse M, Winske D (1998) Kinetic quasi-viscous and bulk flow inertia effects in collisionless magnetotail reconnection. J Geophys Res 103:199–213

Kuznetsova MM, Hesse M, Winske D (2001) Collisionless reconnection supported by nongyrotropic pressure effects in hybrid and particle simulations. J Geophys Res Space Phys 106(A3):3799–3810. https://doi.org/10.1029/1999JA001003

Laitinen TV, Janhunen P, Pulkkinen TI, Palmroth M, Koskinen HEJ (2006) On the characterization of magnetic reconnection in global MHD simulations. Ann Geophys 24(11):3059–3069. https://doi.org/10.5194/angeo-24-3059-2006

Laitinen TV, Khotyaintsev YV, André M, Vaivads A, Reme H (2010) Local influence of the magnetosheath plasma beta fluctuations on magnetopause reconnection. Ann Geophys 28:1053

Lapenta G (2017) Exactly energy conserving semi-implicit particle in cell formulation. J Comput Phys 334:349. https://doi.org/10.1016/j.jcp.2017.01.002

Lapenta G, Krauss-Varban D, Karimabadi H, Huba JD, Rudakov LI, Ricci P (2006) Kinetic simulations of x-line expansion in 3D reconnection. Geophys Res Lett 33:L10102

Leclercq L, Modolo R, Leblanc F, Hess S, Mancini M (2016) 3D magnetospheric parallel hybrid multi-grid method applied to planet–plasma interactions. J Comput Phys 309:295–313. https://doi.org/10.1016/j.jcp.2016.01.005

Lee JY, Sussman A (2004) Efficient communication between parallel programs with intercomm. Tech. Rep., Citeseer

Lemon C, Toffoletto F, Hesse M, Birn J (2003) Computing magnetospheric force equilibria. J Geophys Res Space Phys 108(A6):1237

Lemon C, Wolf R, Hill T, Sazykin S, Spiro R, Toffoletto F, Birn J, Hesse M (2004) Magnetic storm ring current injection modeled with the rice convection model and a self-consistent magnetic field. Geophys Res Lett 31(21):L21801

Li X, Baker D, Temerin M, Reeves G, Belian R (1998) Simulation of dispersionless injections and drift echoes of energetic electrons associated with substorms. Geophys Res Lett 25(20):3763–3766

Li X, Guo F, Li H, Li G (2015) Nonthermally dominated electron acceleration during magnetic reconnection in a low-β plasma. Astrophys J Lett 811(2):L24. https://doi.org/10.1088/2041-8205/811/2/L24. arXiv:1505.02166

Li X, Guo F, Li H, Li G (2017) Particle acceleration during magnetic reconnection in a low-beta plasma. Astrophys J 843(1):21. https://doi.org/10.3847/1538-4357/aa745e

Li X, Guo F, Li H, Birn J (2018) The roles of fluid compression and shear in electron energization during magnetic reconnection. Astrophys J 855(2):80. https://doi.org/10.3847/1538-4357/aaacd5. arXiv:1801.02255

Li X, Guo F, Li H (2019a) Particle acceleration in kinetic simulations of nonrelativistic magnetic reconnection with different ion–electron mass ratios. Astrophys J 879(1):5

Li X, Guo F, Li H, Stanier A, Kilian P (2019b) Formation of power-law electron energy spectra in three-dimensional low-β magnetic reconnection. Astrophys J 884(2):118. https://doi.org/10.3847/1538-4357/ab4268. arXiv:1909.01911

Li TC, Liu YH, Hesse M, Zou Y (2020) Three-dimensional X-line spreading in asymmetric magnetic reconnection. J Geophys Res Space Phys 125(2):e27094. https://doi.org/10.1029/2019JA027094. arXiv:1907.02025

Li X, Guo F, Liu YH, Li H (2023a) A model for nonthermal particle acceleration in relativistic magnetic reconnection. Astrophys J Lett 954(2):L37. https://doi.org/10.3847/2041-8213/acf135. arXiv:2302.12737

Li C, Jia X, Chen Y, Toth G, Zhou H, Slavin JA, Sun W, Poh G (2023b) Global Hall mhd simulations of Mercury's magnetopause dynamics and ftes under different solar wind and imf conditions. J Geophys Res Space Phys 128(5):e2022JA031206

Li TC, Liu YH, Qi Y, Zhou M (2023c) Extended magnetic reconnection in kinetic plasma turbulence. Phys Rev Lett 131:085201

Liemohn MW, Kozyra JU, Clauer CR, Ridley AJ (2001) Computational analysis of the near-Earth magnetospheric current system during two-phase decay storms. J Geophys Res 106:29,531

Liemohn M, Ganushkina N, Zeeuw DD, Rastaetter L, Kuznetsova M, Welling D, Tóth G, Ilie R, Gombosi T, van der Holst B (2018) Real-time SWMF at CCMC: assessing the Dst output from continuous operational simulations. Space Weather 16:1583. https://doi.org/10.1029/2018SW001953

Lin Y (2002) Global hybrid simulation of hot flow anomalies near the bow shock and in the magnetosheath. Planet Space Sci 50(5–6):577–591

Lin Y, Wang XY (2005) Three-dimensional global hybrid simulation of dayside dynamics associated with the quasi-parallel bow shock. J Geophys Res Space Phys 110(A12):1–13. https://doi.org/10.1029/2005JA011243

Lin Y, Wang X (2006) Formation of dayside low-latitude boundary layer under northward interplanetary magnetic field. Geophys Res Lett 33(21):L21104

Lin Y, Lee L, Yan M (1996) Generation of dynamic pressure pulses downstream of the bow shock by variations in the interplanetary magnetic field orientation. J Geophys Res Space Phys 101(A1):479–493

Lin RP, Krucker S, Hurford GJ, Smidth DM, Hudson HS (2003) RHESSI observations of particle acceleration and energy release in an intense solar gamma-ray line flare. Astrophys J 595:L69–L76

Lin Y, Wang XY, Chang SW (2007) Connection between bow shock and cusp energetic ions. Geophys Res Lett 34:L11107. https://doi.org/10.1029/2007GL030038

Lin Y, Wang XY, Brown MR, Schaffer MJ, Cothran CD (2008) Modeling Swarthmore spheromak reconnection experiment using hybrid code. Plasma Phys Control Fusion 50:074012. https://doi.org/10.1088/0741-3335/50/7/074012

Lin Y, Johnson JR, Wang X (2012) Three-dimensional mode conversion associated with kinetic Alfvén waves. Phys Rev Lett 109(12):125003

Lin Y, Wang XY, Lu S, Perez JD, Lu Q (2014) Investigation of storm time magnetotail and ion injection using three-dimensional global hybrid simulation. J Geophys Res Space Phys 119(9):7413–7432. https://doi.org/10.1002/2014JA020005

Lin Y, Wing S, Johnson JR, Wang X, Perez JD, Cheng L (2017) Formation and transport of entropy structures in the magnetotail simulated with a 3-d global hybrid code. Geophys Res Lett 44(12):5892–5899

Lin Y, Wang X, Fok MC, Buzulukova N, Perez JD, Cheng L, Chen LJ (2021) Magnetotail-inner magnetosphere transport associated with fast flows based on combined global-hybrid and CIMI simulation. J Geophys Res Space Phys 126:e2020JA028405. https://doi.org/10.1029/2020JA028405

Lin D, Wang W, Merkin VG, Huang C, Oppenheim MM, Sorathia K, Pham KH, Michael A, Bao S, Wu Q, et al (2022a) Ionospheric dawnside subauroral polarization streams: a unique feature of major geomagnetic storms. Authorea Preprints

Lin Y, Wang X, Sibeck DG, Wang CP, Lee SH (2022b) Global asymmetries of hot flow anomalies. Geophys Res Lett 49(4):e2021GL096970

Lin SC, Liu YH, Li X (2025) The spreading of magnetic reconnection x-line in particle-in-cell simulations–mechanism and the effect of drift-kink instability. J Geophys Res 130:e2024JA033494

Lipatov AS (2002) The hybrid multiscale simulation technology. Scientific computation. Springer, Berlin. https://doi.org/10.1007/978-3-662-05012-5

Liu YH, Daughton W, Karimabadi H, Li H, Roytershteyn V (2013) Bifurcated structure of the electron diffusion region in three-dimensional magnetic reconnection. Phys Rev Lett 110(26):265004. https://doi.org/10.1103/PhysRevLett.110.265004

Liu YH, Birn J, Daughton W, Hesse M, Schindler K (2014) Onset of reconnection in the near magnetotail: pic simulations. J Geophys Res Space Phys 119(12):9773–9789

Liu J, Angelopoulos V, Zhou X-Z, Yao ZH, Runov A (2015) Cross-tail expansion of dipolarizing flux bundles. J Geophys Res 120:2516

Liu YH, Hesse M, Li TC, Kuznetsova M, Le A (2018) Orientation and stability of asymmetric magnetic reconnection x line. J Geophys Res Space Phys 123:4908–4920

Liu YH, Li TC, Hesse M, Huang K (2019) Three-dimensional magnetic reconnection with a spatially confined x-line extent: implications for dipolarizing flux bundles and the dawn-dusk asymmetry. J Geophys Res Space Phys 128:2819–2830

Liu YH, Cassak P, Li X, Hesse M, Lin SC, Genestreti K (2022) First-principles theory of the rate of magnetic reconnection in magnetospheric and solar plasmas. Commun Phys 5:97. https://doi.org/10.1038/s42005-022-00854-x

Liu YH, Hesse M, Genestreti K, Nakamura R, Burch JL, Cassak PA, Bessho N, Eastwood JP, Phan T, Swisdak M, et al (2025a) Ohm's law, the reconnection rate, and energy conversion in collisionless magnetic reconnection. Space Sci Rev 221:16. https://doi.org/10.1007/s11214-025-01142-0

Liu YH, Pyakurel P, Li X, Hesse M, Bessho N, Genestreti K, Thapa SB (2025b) An analytical model of "electron-only" magnetic reconnection rates. Commun Phys 8(1):128. https://doi.org/10.1038/s42005-025-02034-z

Londrillo P, Del Zanna L (2004) On the divergence-free condition in Godunov-type schemes for ideal magnetohydrodynamics: the upwind constrained transport method. J Comput Phys 195(1):17–48. https://doi.org/10.1016/j.jcp.2003.09.016

Lottermoser RF, Scholar M (1997) Undriven magnetic reconnection in magnetohydrodynamics and Hall magnetohydrodynamics. J Geophys Res Space Phys 102(A3):4875–4892. https://doi.org/10.1029/96JA03634

Lu S, Lin Y, Angelopoulos V, Artemyev A, Pritchett P, Lu Q, Wang X (2016) Hall effect control of magnetotail dawn-dusk asymmetry: a three-dimensional global hybrid simulation. J Geophys Res Space Phys 121(12):11–882

Lu S, Wang R, Lu Q, Angelopoulos V, Nakamura R, Artemyev AV, Pritchett PL, Liu TZ, Zhang XJ, Baumjohann W, Gonzalez W, Rager AC, Torbert RB, Giles BL, Gershman DJ, Russell CT, Strangeway RJ, Qi Y, Ergun RE, Lindqvist PA, Burch JL, Wang S (2020) Magnetotail reconnection onset caused by electron kinetics with a strong external driver. Nat Commun 11(1):5049. https://doi.org/10.1038/s41467-020-18787-w

Lu S, Lu Q, Wang R, Pritchett PL, Hubbert M, Qi Y, Huang K, Li X, Russell CT (2022) Electron-only reconnection as a transition from quiet current sheet to standard reconnection in Earth's magnetotail: particle-in-cell simulation and application to MMS data. Geophys Res Lett 49(11):e2022GL098547. https://doi.org/10.1029/2022GL098547

Lyon J, Fedder J, Mobarry C (2004) The Lyon–Fedder–Mobarry (lfm) global mhd magnetospheric simulation code. J Atmos Sol-Terr Phys 66(15–16):1333–1350

Ma ZW, Bhattacharjee A (1996) Fast impulsive reconnection and current sheet intensfication due to electron pressure gradients in semi-collisional plasmas. Geophys Res Lett 23:1673

Ma ZW, Bhattacharjee A (2001) Hall magnetohydrodynamic reconnection: the geospace environment modeling challenge. J Geophys Res Space Phys 106(A3):3773–3782. https://doi.org/10.1029/1999JA001004

Ma Y, Russell C, Tóth G, Chen Y, Nagy A, Harada Y, McFadden J, Halekas J, Lillis R, Connerney J, Espley J, DiBraccio G, Markidis S, Peng IB, Fang X, Jakosky B (2018) Reconnection in the Martian magnetotail: Hall-mhd with embedded particle-in-cell simulations. J Geophys Res 123:3742. https://doi.org/10.1029/2017JA024729

Makwana K, Keppens R, Lapenta G (2017) Two-way coupling of magnetohydrodynamic simulations with embedded particle-in-cell simulations. Comput Phys Commun 221:81. https://doi.org/10.1016/j.cpc.2017.08.003

Man HY, Zhou M, Yi YY, Zhong ZH, Tian AM, Deng XH, Khotyaintsev Y, Russell CT, Giles BL (2020) Observations of electron-only magnetic reconnection associated with macroscopic magnetic flux ropes. Geophys Res Lett 47(19):e2020GL089659. https://doi.org/10.1029/2020GL089659

Mandli KT, Ahmadia AJ, Berger M, Calhoun D, George DL, Hadjimichael Y, Ketcheson DI, Lemoine GI, LeVeque RJ (2016) Clawpack: building an open source ecosystem for solving hyperbolic pdes. PeerJ Comput Sci 2:e68

Mandt ME, Denton RE, Drake JF (1994) Transition to whistler mediated magnetic reconnection. Geophys Res Lett 21(1):73–76. https://doi.org/10.1029/93GL03382

Mangeney A, Califano F, Cavazzoni C, Travnicek P (2002) A numerical scheme for the integration of the Vlasov–Maxwell system of equations. J Comput Phys 179(2):495–538. https://doi.org/10.1006/jcph.2002.7071

Manzini D, Sahraoui F, Califano F (2023) Subion-scale turbulence driven by magnetic reconnection. Phys Rev Lett 130:205201. https://doi.org/10.1103/PhysRevLett.130.205201

Markidis S, Lapenta G, Rizwan-Uddin (2010) Multi-scale simulations of plasma with ipic3d. Math Comput Simul 80:1509–1519. https://doi.org/10.1016/j.matcom.2009.08.038

Matsumoto Y, Amano T, Kato TN, Hoshino M (2015) Stochastic electron acceleration during spontaneous turbulent reconnection in a strong shock wave. Science 347:974

Matthaeus WH, Wan M, Servidio S, Greco A, Osman KT, Oughton S, Dmitruk P (2015) Intermittency, nonlinear dynamics and dissipation in the solar wind and astrophysical plasmas. Philos Trans R Soc Lond Ser A 373(2041):20140154. https://doi.org/10.1098/rsta.2014.0154

Matthaeus WH, Parashar TN, Wan M, Wu P (2016) Turbulence and proton–electron heating in kinetic plasma. Astrophys J Lett 827:L7

Matthaeus WH, Yang Y, Wan M, Parashar TN, Bandyopadhyay R, Chasapis A, Pezzi O, Valentini F (2020) Pathways to dissipation in weakly collisional plasmas. Astrophys J 891(1):101. https://doi.org/10.3847/1538-4357/ab6d6a

McPherron RL, Russell CT, Aubry MP (1973) Satellite studies of magnetospheric substorms on August 15, 1968: 9. Phenomenological model for substorms. J Geophys Res 78(16):3131–3149

Meng X, Tóth G, Glocer A, Fok MC, Gombosi TI (2013) Pressure anisotropy in global magnetospheric simulations: coupling with ring current models. J Geophys Res 118:5639. https://doi.org/10.1002/jgra.50539

Merkin V, Lyon J (2010) Effects of the low-latitude ionospheric boundary condition on the global magnetosphere. J Geophys Res Space Phys 115(A10):A10202

Merkin V, Sitnov M (2016) Stability of magnetotail equilibria with a tailward bz gradient. J Geophys Res Space Phys 121(10):9411–9426

Mouikis C, Omelchenko Y, Roytershteyn V (2021) Comparative study of the magnetotail loading of ionospheric O+ using 3D global hybrid simulations. In: AGU Fall Meeting 2021, held in New Orleans, LA, 13-17 December 2021, vol SH35G-02

Müller J, Simon S, Motschmann U, Scüller J, Glassmeier KH (2011) A.I.K.E.F.: adaptive hybrid model for space plasma simulations. Comput Phys Commun 182:946–966

Muñoz PA, Jain N, Kilian P, Büchner J (2018) A new hybrid code (CHIEF) implementing the inertial electron fluid equation without approximation. Comput Phys Commun 224:245–264. https://doi.org/10.1016/j.cpc.2017.10.012

Muñoz PA, Jain N, Farzalipour Tabriz M, Rampp M, Büchner J (2023) Electron inertia effects in 3D hybrid-kinetic collisionless plasma turbulence. Phys Plasmas 30(9):092302. https://doi.org/10.1063/5.0148818

Nakamura TKM, Daughton W (2014) Turbulent plasma transport across the Earth's low-latitude boundary layer. Geophys Res Lett 41:8704

Nakamura TKM, Nakamura R, Alexandrova A, Kubota Y, Nagai T (2012) Hall magnetohydrodynamic effects for three-dimensional magnetic reconnection with finite width along the direction of the current. J Geophys Res 117:03220

Nakamura TKM, Hasegawa H, Daughton W, Eriksson S, Li W, Nakamura R (2017) Turbulent mass transfer caused by vortex induced reconnection in collisionless magnetospheric plasmas. Nat Commun 8:1582

Nakamura TKM, Blasl KA, Hasegawa H, Umeda T, Liu YH, Peery SA, Plaschke F, Nakamura R, Holmes JC, Stawarz JE, Nystrom WD (2022) Multi-scale evolution of Kelvin–Helmholtz waves at the Earth's magnetopause during southward imf periods. Phys Plasmas 29:012901

Ng J, Chen LJ, Omelchenko YA (2021) Bursty magnetic reconnection at the Earth's magnetopause triggered by high-speed jets. Phys Plasmas 28:092902. https://doi.org/10.1063/5.0054394

Ng J, Chen LJ, Bessho N, Shuster J, Burkholder B, Yoo J (2022) Electron-scale reconnection in three-dimensional shock turbulence. Geophys Res Lett 49:e2022GL099544

Ohia O, Egedal J, Lukin VS, Daughton W, Le A (2012) Demonstration of anisotropic fluid closure capturing the kinetic structure of magnetic reconnection. Phys Rev Lett 109:115004. https://doi.org/10.1103/PhysRevLett.109.115004

Ohtani H, Horiuchi R (2009) Open boundary condition for particle simulation in magnetic reconnection research. J Plasma Fusion Res 4:24

Oka M, Phan TD, Krucker S, Fujimoto M, Shinohara I (2010) Electron acceleration by multi-island coalescence. Astrophys J 714(1):915–926. https://doi.org/10.1088/0004-637X/714/1/915. arXiv:1004.1154

Oka M, Birn J, Egedal J, Guo F, Ergun RE, Turner DL, Khotyaintsev Y, Hwang KJ, Cohen IJ, Drake JF (2023) Particle acceleration by magnetic reconnection in geospace. Space Sci Rev 219(8):75. https://doi.org/10.1007/s11214-023-01011-8. arXiv:2307.01376

Omelchenko YA (2015) Formation, spin-up, and stability of field-reversed configurations. Phys Rev E 92:023105. https://doi.org/10.1103/PhysRevE.92.023105

Omelchenko Y, Karimabadi H (2006a) Event-driven, hybrid particle-in-cell simulation: a new paradigm for multi-scale plasma modeling. J Comput Phys 216(1):153–178. https://doi.org/10.1016/j.jcp.2005.11.029

Omelchenko Y, Karimabadi H (2006b) Self-adaptive time integration of flux-conservative equations with sources. J Comput Phys 216(1):179–194. https://doi.org/10.1016/j.jcp.2005.12.008

Omelchenko Y, Karimabadi H (2007) A time-accurate explicit multi-scale technique for gas dynamics. J Comput Phys 226(1):282–300. https://doi.org/10.1016/j.jcp.2007.04.010

Omelchenko Y, Karimabadi H (2012) HYPERS: a unidimensional asynchronous framework for multiscale hybrid simulations. J Comput Phys 231(4):1766–1780. https://doi.org/10.1016/j.jcp.2011.11.004

Omelchenko YA, Karimabadi H (2022) EMAPS - an intelligent agent-based technology for simulation of multiscale systems. In: Büchner J (ed) Space and astrophysical plasma simulation. Springer, Berlin. https://doi.org/10.1007/978-3-031-11870-8_13

Omelchenko YA, Sudan R (1997) A 3-d Darwin-EM hybrid, PIC code for ion ring studies. J Comput Phys 133:146–159. https://doi.org/10.1006/jcph.1997.5670

Omelchenko YA, Chen LJ, Ng J (2021a) 3D space-time adaptive hybrid simulations of magnetosheath high-speed jets. J Geophys Res Space Phys 126:e2020JA029035. https://doi.org/10.1029/2020JA029035

Omelchenko YA, Roytershtyen V, Chen LJ, Ng J, Hietala H (2021b) HYPERS simulations of solar wind interactions with the Earth's magnetosphere and the Moon. J Atmos Sol-Terr Phys 215:105581. https://doi.org/10.1016/j.jastp.2021.105581

Omelchenko YA, Rudakov LI, Ng J, Crabtree C, Ganguli G (2021c) On the rate of energy deposition by an ion ring velocity ring. Phys Plasmas 28:052102. https://doi.org/10.1063/5.0046309

Omelchenko YA, Mouikis C, Ng J, Roytershteyn V, Chen L-J (2023) Multiscale hybrid modeling of the impact response of the Earth's magnetotail to ionospheric O+ outflow. Front Astron Space Sci 10:33. https://doi.org/10.3389/fspas.2023.1056497. https://ui.adsabs.harvard.edu/abs/2023FrASS..1056497O

Omidi N, Sibeck D (2007) Flux transfer events in the cusp. Geophys Res Lett 34(4):L04106

Omidi N, Blanco-Cano X, Russell C, Karimabadi H (2004) Dipolar magnetospheres and their characterization as a function of magnetic moment. Adv Space Res 33:1996–2003. https://doi.org/10.1016/j.asr.2003.08.041

Omidi N, Phan T, Sibeck D (2009) Hybrid simulations of magnetic reconnection initiated in the magnetosheath. J Geophys Res Space Phys 114(A2):A02222

Omidi N, Eastwood J, Sibeck D (2010) Foreshock bubbles and their global magnetospheric impacts. J Geophys Res Space Phys 115(A6):A06204

Palmroth M, Hoilijoki S, Juusola L, Pulkkinen TI, Hietala H, Pfau-Kempf Y, Ganse U, von Alfthan S, Vainio R, Hesse M (2017) Tail reconnection in the global magnetospheric context: Vlasiator first results. Ann Geophys 35(6):1269–1274. https://doi.org/10.5194/angeo-35-1269-2017

Palmroth M, Ganse U, Pfau-Kempf Y, Battarbee M, Turc L, Brito T, Grandin M, Hoilijoki S, Sandroos A, von Alfthan S (2018a) Vlasov methods in space physics and astrophysics. Living Rev Comput Astrophys 4:1. https://doi.org/10.1007/s41115-018-0003-2

Palmroth M, Hietala H, Plaschke F, Archer M, Karlsson T, Blanco-Cano X, Sibeck D, Kajdič P, Ganse U, Pfau-Kempf Y, Battarbee M, Turc L (2018b) Magnetosheath jet properties and evolution as determined by a global hybrid-Vlasov simulation. Ann Geophys 36:1171–1182. https://doi.org/10.5194/angeo-36-1171-2018

Palmroth M, Pulkkinen TI, Ganse U, Pfau-Kempf Y, Koskela T, Zaitsev I, Alho M, Cozzani G, Turc L, Battarbee M, Dubart M, George H, Gordeev E, Grandin M, Horaites K, Osmane A, Papadakis K, Suni J, Tarvus V, Zhou H, Nakamura R (2023) Magnetotail plasma eruptions driven by magnetic reconnection and kinetic instabilities. Nat Geosci 16(7):570–576

Pang Y, Lin Y, Deng X, Wang X, Tan B (2010) Three-dimensional hybrid simulation of magnetosheath reconnection under northward and southward interplanetary magnetic field. J Geophys Res Space Phys 115(A3):A03203

Papadakis K, Pfau-Kempf Y, Ganse U, Battarbee M, Alho M, Grandin M, Dubart M, Turc L, Zhou H, Horaites K, Zaitsev I, Cozzani G, Bussov M, Gordeev E, Tesema F, George H, Suni J, Tarvus V, Palmroth M (2022) Spatial filtering in a 6D hybrid-Vlasov scheme to alleviate adaptive mesh refinement artifacts: a case study with Vlasiator (versions 5.0, 5.1, and 5.2.1). Geosci Model Dev 15(20):7903–7912. https://doi.org/10.5194/gmd-15-7903-2022

Parker EN (1957a) Newtonian development of the dynamical properties of ionized gases of low density. Phys Rev 107:924–933. https://doi.org/10.1103/PhysRev.107.924

Parker EN (1957b) Sweet's mechanism for merging magnetic fields in conducting fluids. J Geophys Res 62:509

Parker EN (1963) The solar-flare phenomenon and the theory of reconnection and annihilation of magnetic fields. Astrophys J 8:177

Parker EN (1973) The reconnection rate of magnetic fields. Astrophys J 180:247

Paty C, Winglee R (2004) Multi-fluid simulations of Ganymede's magnetosphere. Geophys Res Lett 31:L24806

Pembroke A, Toffoletto F, Sazykin S, Wiltberger M, Lyon J, Merkin V, Schmitt P (2012) Initial results from a dynamic coupled magnetosphere-ionosphere-ring current model. J Geophys Res Space Phys 117(A2):A02211

Peroomian V, El-Alaoui M (2008) The storm-time access of solar wind ions to the nightside ring current and plasma sheet. J Geophys Res Space Phys 113(A6):A06215

Perrone D, Valentini F, Servidio S, Dalena S, Veltri P (2012) Vlasov simulations of multi-ion plasma turbulence in the solar wind. Astrophys J 762(2):99. https://doi.org/10.1088/0004-637X/762/2/99

Pezzi O, Servidio S, Perrone D, Valentini F, Sorriso-Valvo L, Greco A, Matthaeus WH, Veltri P (2018) Velocity-space cascade in magnetized plasmas: numerical simulations. Phys Plasmas 25(6):060704. https://doi.org/10.1063/1.5027685. arXiv:1803.01633

Pezzi O, Cozzani G, Califano F, Valentini F, Guarrasi M, Camporeale E, Brunetti G, Retinò A, Veltri P (2019a) Vida: a Vlasov–Darwin solver for plasma physics at electron scales. J Plasma Phys 85(5):905850506. https://doi.org/10.1017/S0022377819000631

Pezzi O, Perrone D, Servidio S, Valentini F, Sorriso-Valvo L, Veltri P (2019b) Proton-proton collisions in the turbulent solar wind: hybrid Boltzmann-Maxwell simulations. Astrophys J 887(2):208. https://doi.org/10.3847/1538-4357/ab5285. arXiv:1903.03398

Pezzi O, Yang Y, Valentini F, Servidio S, Chasapis A, Matthaeus WH, Veltri P (2019c) Energy conversion in turbulent weakly collisional plasmas: Eulerian hybrid Vlasov-Maxwell simulations. Phys Plasmas 26(7):072301. https://doi.org/10.1063/1.5100125

Pezzi O, Liang H, Juno JL, Cassak PA, Vásconez CL, Sorriso-Valvo L, Perrone D, Servidio S, Roytershteyn V, TenBarge JM, Matthaeus WH (2021a) Dissipation measures in weakly collisional plasmas. Mon Not R Astron Soc 505(4):4857–4873. https://doi.org/10.1093/mnras/stab1516. arXiv:2101.00722

Pezzi O, Pecora F, Le Roux J, Engelbrecht NE, Greco A, Servidio S, Malova HV, Khabarova OV, Malandraki O, Bruno R, Matthaeus WH, Li G, Zelenyi LM, Kislov RA, Obridko VN, Kuznetsov VD (2021b) Current sheets, plasmoids and flux ropes in the heliosphere. Part II: theoretical aspects. Space Sci Rev 217(3):39. https://doi.org/10.1007/s11214-021-00799-7. arXiv:2101.05007

Pfau-Kempf Y, Hietala H, Milan SE, Juusola L, Hoilijoki S, Ganse U, von Alfthan S, Palmroth M (2016) Evidence for transient, local ion foreshocks caused by dayside magnetopause reconnection. Ann Geophys 34(11):943–959. https://doi.org/10.5194/angeo-34-943-2016

Pfau-Kempf Y, Palmroth M, Johlander A, Turc L, Alho M, Battarbee M, Dubart M, Grandin M, Ganse U (2020) Hybrid-Vlasov modeling of three-dimensional dayside magnetopause reconnection. Phys Plasmas 27:092903. https://doi.org/10.1063/5.0020685

Pham KH, Zhang B, Sorathia K, Dang T, Wang W, Merkin V, Liu H, Lin D, Wiltberger M, Lei J, et al (2022) Thermospheric density perturbations produced by traveling atmospheric disturbances during August 2005 storm. J Geophys Res Space Phys 127(2):e2021JA030071

Phan TD, Drake JF, Shay MA, Mozer FS, Eastwood JP (2007) Evidence for an elongated (>60 ion skin depths) electron diffusion region during fast magnetic reconnection. Phys Rev Lett 99:255002

Phan TD, Eastwood JP, Shay MA, Drake JF, Sonnerup BUÖ, Fujimoto M, Cassak PA, Øieroset M, Burch JL, Torbert RB, Rager AC, Dorelli JC, Gershman DJ, Pollock C, Pyakurel PS, Haggerty CC, Khotyaintsev Y, Lavraud B, Saito Y, Oka M, Ergun RE, Retino A, Le Contel O, Argall MR, Giles BL, Moore TE, Wilder FD, Strangeway RJ, Russell CT, Lindqvist PA, Magnes W (2018) Electron magnetic reconnection without ion coupling in Earth's turbulent magnetosheath. Nature 557(7704):202–206. https://doi.org/10.1038/s41586-018-0091-5

Pontius D Jr, Wolf R (1990) Transient flux tubes in the terrestrial magnetosphere. Geophys Res Lett 17(1):49–52

Poppe AR (2019) Comment on "the dominant role of energetic ions in solar wind interaction with the moon" by Omidi et al. J Geophys Res Space Phys 124:6927–6932. https://doi.org/10.1029/2019JA026692

Powell KG (1994) An approximate Riemann solver for magnetohydrodynamics (that works in more than one dimension). Tech. Rep. 94-24, Inst. for Comput. Appl. in Sci. and Eng., NASA Langley Space Flight Center, Hampton, Va

Powell K, Roe P, Linde T, Gombosi T, De Zeeuw DL (1999) A solution-adaptive upwind scheme for ideal magnetohydrodynamics. J Comput Phys 154:284–309. https://doi.org/10.1006/jcph.1999.6299

Press WH, Teukolsky SA, Vetterling WT, Flannery BP (1992) Numerical recipes in Fortran 77. Cambridge University Press, Cambridge, pp 281–286. Chap. 7.3

Priest E, Forbes T (2000) Magnetic reconnection. Cambridge University Press, Cambridge

Pritchett PL (2005) Onset and saturation of guide-field magnetic reconnection. Phys Plasmas 12(6):062301. https://doi.org/10.1063/1.1914309

Pritchett P, Coroniti FV, Pella R (1991) Collisionless reconnection in two-dimension magnetotail equilibria. J Geophys Res Space Phys 96(A7):11523–11538

Pyakurel PS, Shay MA, Drake JF, Phan TD, Cassak PA, Verniero JL (2021) Faster form of electron magnetic reconnection with a finite length X-line. Phys Rev Lett 127(15):155101. https://doi.org/10.1103/PhysRevLett.127.155101

Raeder J (2006) Flux transfer events: 1. Generation mechanism for strong southward IMF. Ann Geophys 24(1):381–392. https://doi.org/10.5194/angeo-24-381-2006

Raeder J, Berchem J, Ashour-Abdalla M, Frank L, Paterson W, Ackerson K, Kokubun S, Yamamoto T, Slavin J (1997) Boundary layer formation in the magnetotail: geotail observations and comparisons with a global MHD simulation. Geophys Res Lett 24:951. https://doi.org/10.1029/97GL00218

Raeder J, McPherron R, Frank L, Kokubun S, Lu G, Mukai T, Paterson W, Sigwarth J, Singer H, Slavin J (2001) Global simulation of the geospace environment modeling substorm challenge event. J Geophys Res Space Phys 106(A1):381–395

Raeder J, Cramer WD, Jensen J, Fuller-Rowell T, Maruyama N, Toffoletto F, Vo H (2016) Sub-auroral polarization streams: a complex interaction between the magnetosphere, ionosphere, and thermosphere. J Phys Conf Ser 767:012021

Reiff PH, Webster JM, Daou AG, Marshall A, Sazykin SY, Rastaetter L, Welling DT, DeZeeuw D, Kuznetsova MM, Glocer A, Russell CT (2017) CCMC modeling of magnetic reconnection in electron diffusion region events. In: Foullon C, Malandraki OE (eds) Proceedings of the international astronomical union, vol 13, pp 142–146. https://doi.org/10.1017/S1743921317010845

Ridley A, Gombosi T, Dezeeuw D (2004) Ionospheric control of the magnetosphere: conductance. Ann Geophys 22:567–584. https://doi.org/10.5194/angeo-22-567-2004

Rieke M, Trost T, Grauer R (2015) Coupled Vlasov and two-fluid codes on GPUs. J Comput Phys 283:436–452. https://doi.org/10.1016/j.jcp.2014.12.016

Robinson R, Vondrak R, Miller K, Dabbs T, Hardy D (1987) On calculating ionospheric conductances from the flux and energy of precipitating electrons. J Geophys Res Space Phys 92(A3):2565–2569

Roble R, Ridley EC, Richmond A, Dickinson R (1988) A coupled thermosphere/ionosphere general circulation model. Geophys Res Lett 15(12):1325–1328

Rogers BN, Denton RE, Drake JF, Shay MA (2001) Role of dispersive waves in collisionless magnetic reconnection. Phys Rev Lett 87(19):195004

Rogers BN, Denton RE, Drake JF (2003) Signatures of collisionless magnetic reconnection. J Geophys Res 108(A3).1111. https://doi.org/10.1029/2002JA009699

Roytershteyn V, Karimabadi H, Omelchenko Y, Germaschewski K (2015) Kinetic simulations of collisionless turbulence across scales. In: Solar heliospheric and interplanetary environment (SHINE 2016), the SHINE conference held 5-10 July, 2015 at the Stoweflake resort in Stowe, VT

Rueda JAA, Verscharen D, Wicks RT, Owen CJ, Nicolaou G, Walsh AP, Zouganelis I, Germaschewski K, Domínguez SV (2021) Three-dimensional magnetic reconnection in particle-in-cell simulations of anisotropic plasma turbulence. J Plasma Phys 87:905870228

Runov A, Grandin M, Palmroth M, Battarbee M, Ganse U, Hietala H, Hoilijoki S, Kilpua E, Pfau-Kempf Y, Toledo-Redondo S, Turc L, Turner D (2021) Ion distribution functions in magnetotail reconnection: global hybrid-Vlasov simulation results. Ann Geophys 39(4):599–612. https://doi.org/10.5194/angeo-39-599-2021

Sachsenweger D, Scholer M, Möbius E (1989) Test particle acceleration in a magnetotail reconnection configuration. Geophys Res Lett 16(9):1027–1030

Sarrat M, Ghizzo A, Del Sarto D, Serrat L (2017) Parallel implementation of a relativistic semi-Lagrangian Vlasov–Maxwell solver. Eur Phys J D 71(11):271. https://doi.org/10.1140/epjd/e2017-80188-4

Sato T, Hayashi T (1979) Externally driven magnetic reconnection and a powerful magnetic energy converter. Phys Fluids 22:1189

Schaller RR (1997) Moore's law: past, present and future. IEEE Spectr 34(6):52–59

Schekochihin AA, Parker JT, Highcock EG, Dellar PJ, Dorland W, Hammett GW (2016) Phase mixing versus nonlinear advection in drift-kinetic plasma turbulence. J Plasma Phys 82(2):905820212. https://doi.org/10.1017/S0022377816000374. arXiv:1508.05988

Schindler K (1972) A self-consistent theory of the tail of the magnetosphere. In: Earth's magnetospheric processes: proceedings of a symposium organized by the summer advanced study institute and ninth ESRO summer school, held in Cortina, Italy, August 30-September 10, 1971. Springer, Berlin, pp 200–209

Schmitz H, Grauer R (2006a) Darwin–Vlasov simulations of magnetised plasmas. J Comput Phys 214(2):738–756. https://doi.org/10.1016/j.jcp.2005.10.013

Schmitz H, Grauer R (2006b) Kinetic Vlasov simulations of collisionless magnetic reconnection. Phys Plasmas 13(9):092309. https://doi.org/10.1063/1.2347101

Scholer M, Jamitzky F (1987) Particle orbits during the development of plasmoids. J Geophys Res Space Phys 92(A11):12181–12186

Scholer M, Otto A (1991) Magnetotail reconnection: current diversion and field-aligned currents. Geophys Res Lett 18(4):733–736

Schriver D, Ashour-Abdalla M, Zelenyi L, Gombosi T, Ridley A, De Zeeuw D, Toth G, Monostori G (2005) Modeling the kinetic transport of electrons through the Earth's global magnetosphere. In: Proceedings of the 7th international school/symposium for space simulations. Research institute for sustainable humanosphere, Kyoto university, Kyoto, Japan, pp 345–346

Schumaker TL, Gussenhoven MS, Hardy DA, Carovillano RL (1989) The relationship between diffuse auroral and plasma sheet electron distributions near local midnight. J Geophys Res Space Phys 94(A8):10061–10078

Sciola A, Merkin VG, Sorathia K, Gkioulidou M, Bao S, Toffoletto F, Pham K, Lin D, Michael A, Wiltberger M, Ukhorskiy A (2023) The contribution of plasma sheet bubbles to stormtime ring current buildup and evolution of its energy composition. J Geophys Res Space Phys 128(1):e2023JA031693. https://doi.org/10.1029/2023JA03169310.22541/essoar.168677230.00750251/v1. https://ui.adsabs.harvard.edu/abs/2023JGRA..12831693S

Sergeev V, Nikolaev A, Tsyganenko N, Angelopoulos V, Runov A, Singer H, Yang J (2014) Testing a two-loop pattern of the substorm current wedge (scw2l). J Geophys Res Space Phys 119(2):947–963

Servidio S, Valentini F, Califano F, Veltri P (2012) Local kinetic effects in two-dimensional plasma turbulence. Phys Rev Lett 108:045001. https://doi.org/10.1103/PhysRevLett.108.045001

Servidio S, Osman KT, Valentini F, Perrone D, Califano F, Chapman S, Matthaeus WH, Veltri P (2014) Proton kinetic effects in Vlasov and solar wind turbulence. Astrophys J Lett 781(2):L27. https://doi.org/10.1088/2041-8205/781/2/L27

Servidio S, Valentini F, Perrone D, Greco A, Califano F, Matthaeus WH, Veltri P (2015) A kinetic model of plasma turbulence. J Plasma Phys 81(1):325810107. https://doi.org/10.1017/S0022377814000841

Servidio S, Chasapis A, Matthaeus WH, Perrone D, Valentini F, Parashar TN, Veltri P, Gershman D, Russell CT, Giles B, Fuselier SA, Phan TD, Burch J (2017) Magnetospheric multiscale observation of plasma velocity-space cascade: Hermite representation and theory. Phys Rev Lett 119(20):205101. https://doi.org/10.1103/PhysRevLett.119.205101. arXiv:1707.08180

Sharma Pyakurel P, Shay MA, Phan TD, Matthaeus WH, Drake JF, TenBarge JM, Haggerty CC, Klein KG, Cassak PA, Parashar TN, Swisdak M, Chasapis A (2019) Transition from ion-coupled to electron-only reconnection: basic physics and implications for plasma turbulence. Phys Plasmas 26(8):082307. https://doi.org/10.1063/1.5090403

Shay MA, Drake JF (1998) The role of electron dissipation on the rate of collisionless magnetic reconnection. Geophys Res Lett 25:3759

Shay MA, Drake JF, Denton RE, Biskamp D (1998) Structure of the dissipation region during collisionless magnetic reconnection. J Geophys Res Space Phys 103(A5):9165–9176. https://doi.org/10.1029/97JA03528

Shay MA, Drake JF, Rogers BN, Denton RE (1999) The scaling of collisionless, magnetic reconnection for large systems. Geophys Res Lett 26(14):2163–2166. https://doi.org/10.1029/1999GL900481

Shay MA, Drake JF, Swisdak M, Dorland W, Rogers BN (2003) Inherently three-dimensional magnetic reconnection: a mechanism for bursty bulk flows? Geophys Res Lett 30:1345

Shay MA, Drake JF, Swisdak M, Rogers BN (2004) The scaling of embedded collisionless reconnection. Phys Plasmas 11:2199. https://doi.org/10.1063/1.1705650

Shay MA, Drake JF, Swisdak M (2007) Two-scale structure of the electron dissipation region during collisionless magnetic reconnection. Phys Rev Lett 99:155002. https://doi.org/10.1103/PhysRevLett.99.155002

Shay MA, Haggerty CC, Matthaeus WH, Parashar TN, Wan M, Wu P (2018) Turbulent heating due to magnetic reconnection. Phys Plasmas 25:012304

Shepherd LS, Cassak PA (2012) Guide field dependence of 3D X-line spreading during collisionless magnetic reconnection. J Geophys Res 117:A10101

Shi F, Lin Y, Wang X (2013) Global hybrid simulation of mode conversion at the dayside magnetopause. J Geophys Res Space Phys 118(10):6176–6187

Shi F, Lin Y, Wang X, Wang B, Nishimura Y (2021) 3-d global hybrid simulations of magnetospheric response to foreshock processes. Earth Planets Space 73(1):1–17

Shiroto T (2023) An improved Darwin approximation in the classical electromagnetism. Phys Plasmas 30(4):044501. https://doi.org/10.1063/5.0138048

Shou Y, Tenishev V, Chen Y, Tóth G, Ganushkina N (2021) Magnetohydrodynamic with adaptively embedded particle-in-cell model: MHD-AEPIC. J Comput Phys 446:110656. https://doi.org/10.1016/j.jcp.2021.110656

Sitnov M, Birn J, Ferdousi B, Gordeev E, Khotyaintsev Y, Merkin V, Motoba T, Otto A, Panov E, Pritchett P, et al (2019) Explosive magnetotail activity. Space Sci Rev 215:31. https://doi.org/10.1007/s11214-019-0599-5

Sonnerup BUÖ (1974) Magnetopause reconnection rate. J Geophys Res 79(10):1546–1549. https://doi.org/10.1029/JA079i010p01546

Sonnerup BUÖ (1979) Magnetic field reconnection. In: Lanzerotti LJ, Kennel CF, Parker EN (eds) Solar System plasma physics, vol 3. North-Holland, Amsterdam, p 46

Sonnerup BUÖ, Paschmann G, Papamastorakis I, Sckopke N, Haerendel G, Bame SJ, Asbridge JR, Gosling JT, Russell CT (1986) Evidence for magnetic field reconnection at the Earth's magnetopause. J Geophys Res Space Phys 86(A12):10049–10067

Sorathia K, Merkin V, Ukhorskiy A, Mauk B, Sibeck D (2017) Energetic particle loss through the magnetopause: a combined global mhd and test-particle study. J Geophys Res Space Phys 122(9):9329–9343
Sorathia K, Michael A, Merkin VG, Ukhorskiy AY, Turner DL, Lyon J, Garretson J, Gkioulidou M, Toffoletto F (2021) The role of mesoscale plasma sheet dynamics in ring current formation. Front Astron Space Sci 8:761875
Sorriso-Valvo L, Perrone D, Pezzi O, Valentini F, Servidio S, Zouganelis I, Veltri P (2018) Local energy transfer rate and kinetic processes: the fate of turbulent energy in two-dimensional hybrid Vlasov–Maxwell numerical simulations. J Plasma Phys 84(2):725840201. https://doi.org/10.1017/S0022377818000302
Spence HE, Kivelson MG, Walker RJ, McComas DJ (1989) Magnetospheric plasma pressures in the midnight meridian: observations from 2.5 to 35 re. J Geophys Res Space Phys 94(A5):5264–5272
Stanier A, Daughton W, Chacón L, Karimabadi H, Ng J, Huang YM, Hakim A, Bhattacharjee A (2015) Role of ion kinetic physics in the interaction of magnetic flux ropes. Phys Rev Lett 115(17):175004. https://doi.org/10.1103/PhysRevLett.115.175004
Stawarz JE, Eastwood JP, Phan TD, Gingell IL, Pyakurel PS, Shay MA, Robertson SL, Russell CT, Le Contel O (2022) Turbulence-driven magnetic reconnection and the magnetic correlation length: observations from magnetospheric multiscale in Earth's magnetosheath. Phys Plasmas 29(1):012302. https://doi.org/10.1063/5.0071106
Stawarz J, Muñoz P, Bessho N, Bandyopadhyay R, Nakamura T, Eriksson S, Büchner J, Chasapis A, Drake J, et al (2024) The interplay between collisionless magnetic reconnection and turbulence. Space Sci Rev 220:90. https://doi.org/10.1007/s11214-024-01124-8
Strang G (1968) On the construction and comparison of difference schemes. SIAM J Numer Anal 5(3):506–517. https://doi.org/10.1137/0705041
Sulem PL, Passot T (2015) Landau fluid closures with nonlinear large-scale finite Larmor radius corrections for collisionless plasmas. J Plasma Phys 81(1):325810103. https://doi.org/10.1017/S0022377814000671
Sun WJ, Fu SY, Slavin JA, Raines JM, Zong QG, Poh GK, Zurbuchen TH (2016) Spatial distribution of Mercury's flux ropes and reconnection fronts: Messenger observations. J Geophys Res 121:7590
Sun W, Dewey RM, Aizawa S, Huang J, Slavin JA, Fu S, Wei Y, Bowers CF (2022) Review of Mercury's dynamic magnetosphere: post-messenger era and comparative magnetospheres. Sci China Earth Sci 65:25
Suni J, Palmroth M, Turc L, Battarbee M, Johlander A, Tarvus V, Alho M, Bussov M, Dubart M, Ganse U, Grandin M, Horaites K, Manglayev T, Papadakis K, Pfau-Kempf Y, Zhou H (2021) Connection between foreshock structures and the generation of magnetosheath jets: Vlasiator results. Geophys Res Lett 48(20):e2021GL095655. https://doi.org/10.1029/2021GL095655
Sweet PA (1958) The neutral point theory of solar flares. In: Lehnert B (ed) Electromagnetic phenomena in cosmical physics, vol 123. Cambridge University Press, New York
Swift DW (1996) Use of a hybrid code for global-scale plasma simulation. J Comput Phys 126:109–121
Swift DW, Lin Y (2001) Substorm onset viewed by a two-dimensional, global-scale hybrid code. J Atmos Sol-Terr Phys 63(7):683–704
Swisdak M, Drake JF (2007) Orientation of the reconnection X-line. Geophys Res Lett 34:L11106. https://doi.org/10.1029/2007GL029815
Swisdak M, Drake JF, Price L, Burch JL, Cassak PA, Phan TD (2018) Localized and intense energy conversion in the diffusion region of asymmetric magnetic reconnection. Geophys Res Lett 45(11):5260–5267
Tan B, Lin Y, Perez J, Wang X (2011) Global-scale hybrid simulation of dayside magnetic reconnection under southward imf: structure and evolution of reconnection. J Geophys Res Space Phys 116(A2):A02206
Tan B, Lin Y, Perez J, Wang X (2012) Global-scale hybrid simulation of cusp precipitating ions associated with magnetopause reconnection under southward imf. J Geophys Res Space Phys 117(A3):A03217
Tatsuno T, Dorland W, Schekochihin AA, Plunk GG, Barnes M, Cowley SC, Howes GG (2009) Nonlinear phase mixing and phase-space cascade of entropy in gyrokinetic plasma turbulence. Phys Rev Lett 103(1):015003. https://doi.org/10.1103/PhysRevLett.103.015003. arXiv:0811.2538
Thoma C, Welch DR, Hsu SC (2013) Particle-in-cell simulations of collisionless shock formation via head-on merging of two laboratory supersonic plasma jets. Phys Plasmas 20:082128. https://doi.org/10.1063/1.4819063
Toffoletto F (2020) Modelling techniques. In: Jordanova VK, Ilie R, Chen MW (eds) Ring current investigations: the quest for space weather prediction. Elsevier, Amsterdam
Toffoletto F, Sazykin S, Spiro R, Wolf R (2003) Inner magnetospheric modeling with the rice convection model. Space Sci Rev 107:175–196. https://doi.org/10.1023/A:1025532008047
Toledo-Redondo S, André M, Aunai N, Chappell CR, Dargent J, Fuselier SA, Glocer A, Graham DB, Haaland S, Hesse M, Kistler LM, Lavraud B, Li W, Moore TE, Tenfjord P, Vines SK (2021) Impacts of ionospheric ions on magnetic reconnection and Earth's magnetosphere dynamics. Rev Geophys 59:e2020RG000707. https://doi.org/10.1029/2020RG000707

Tóth G, Sokolov IV, Gombosi TI, Chesney DR, Clauer C, Zeeuw DLD, Hansen KC, Kane KJ, Manchester WB, Powell KG, Ridley AJ, Roussev II, Stout QF, Volberg O, Wolf RA, Sazykin S, Chan A, Yu B, Kóta J (2005) Space weather modeling framework: a new tool for the space science community. J Geophys Res 110:A12226. https://doi.org/10.1029/2005JA011126

Tóth G, Ma YJ, Gombosi TI (2008) Hall magnetohydrodynamics on block adaptive grids. J Comput Phys 227:6967

Tóth G, Van der Holst B, Sokolov IV, De Zeeuw DL, Gombosi TI, Fang F, Manchester WB, Meng X, Najib D, Powell KG, et al (2012) Adaptive numerical algorithms in space weather modeling. J Comput Phys 231(3):870–903

Tóth G, Jia X, Markidis S, Peng B, Chen Y, Daldorff L, Tenishev V, Borovikov D, Haiducek J, Gombosi T, Glocer A, Dorelli J (2016) Extended magnetohydrodynamics with embedded particle-in-cell simulation of Ganymede's magnetosphere. J Geophys Res 121:1273–1293. https://doi.org/10.1002/2015JA021997

Tóth G, Chen Y, Gombosi TI, Cassak P, Markidis S, Peng B (2017) Scaling the ion inertial length and its implications for modeling reconnection in global simulations. J Geophys Res 122:10336. https://doi.org/10.1002/2017JA024189

Tronci C, Camporeale E (2015) Neutral Vlasov kinetic theory of magnetized plasmas. Phys Plasmas 22(2):020704. https://doi.org/10.1063/1.4907665

Trottenberg U, Oosterlee CW, Schuller A (2000) Multigrid. Academic Press, San Diego

Turc L, Fontaine D, Savoini P, Modolo R (2015) 3d hybrid simulations of the interaction of a magnetic cloud with a bow shock. J Geophys Res Space Phys 120:6133–6151. https://doi.org/10.1002/2015JA021318

Turc L, Roberts OW, Verscharen D, Dimmock AP, Kajdič P, Palmroth M, Pfau-Kempf Y, Johlander A, Dubart M, Kilpua EKJ, Soucek J, Takahashi K, Takahashi N, Battarbee M, Ganse U (2023) Transmission of foreshock waves through Earth's bow shock. Nat Phys 19(1):78–86

Ugai M, Tsuda T (1977) Magnetic field line reconnexion by localized enhancement of resistivity, 1, evolution in a compressible mhd fluid. J Plasma Phys 17:337

Ukhorskiy A, Sitnov M, Merkin V, Gkioulidou M, Mitchell D (2017) Ion acceleration at dipolarization fronts in the inner magnetosphere. J Geophys Res Space Phys 122(3):3040–3054

Ukhorskiy AY, Sorathia KA, Merkin VG, Sitnov MI, Mitchell DG, Gkioulidou M (2018) Ion trapping and acceleration at dipolarization fronts: high-resolution mhd and test-particle simulations. J Geophys Res Space Phys 123(7):5580–5589

Umeda T, Togano K, Ogino T (2009) Two-dimensional full-electromagnetic Vlasov code with conservative scheme and its application to magnetic reconnection. Comput Phys Commun 180(3):365–374. https://doi.org/10.1016/j.cpc.2008.11.001

Umeda T, Miwa JI, Matsumoto Y, Nakamura TKM, Togano K, Fukazawa K, Shinohara I (2010) Full electromagnetic Vlasov code simulation of the Kelvin-Helmholtz instability. Phys Plasmas 17(5):052311. https://doi.org/10.1063/1.3422547

Valentini F, Trávníček P, Califano F, Hellinger P, Mangeney A (2007) A hybrid-Vlasov model based on the current advance method for the simulation of collisionless magnetized plasma. J Comput Phys 225(1):753–770. https://doi.org/10.1016/j.jcp.2007.01.001

Valentini F, Califano F, Veltri P (2010) Two-dimensional kinetic turbulence in the solar wind. Phys Rev Lett 104:205002. https://doi.org/10.1103/PhysRevLett.104.205002

Valentini F, Califano F, Perrone D, Pegoraro F, Veltri P (2011a) New ion-wave path in the energy cascade. Phys Rev Lett 106(16):165002

Valentini F, Perrone D, Veltri P (2011b) Short-wavelength electrostatic fluctuations in the solar wind. Astrophys J 739(1):54. https://doi.org/10.1088/0004-637X/739/1/54

Valentini F, Vecchio A, Donato S, Carbone V, Briand C, Bougeret J, Veltri P (2014) The nonlinear and nonlocal link between macroscopic Alfvénic and microscopic electrostatic scales in the solar wind. Astrophys J Lett 788(1):L16. https://doi.org/10.1088/2041-8205/788/1/L16

Valentini F, Perrone D, Stabile S, Pezzi O, Servidio S, De Marco R, Marcucci F, Bruno R, Lavraud B, De Keyser J, Consolini G, Brienza D, Sorriso-Valvo L, Retinò A, Vaivads A, Salatti M, Veltri P (2016) Differential kinetic dynamics and heating of ions in the turbulent solar wind. New J Phys 18(12):125001. https://doi.org/10.1088/1367-2630/18/12/125001. arXiv:1611.04802

van Leer B (1977) Towards the ultimate conservative difference scheme. IV. A new approach to numerical convection. J Comput Phys 23:276. https://doi.org/10.1016/0021-9991(77)90095-X

Varney R, Wiltberger M, Zhang B, Lotko W, Lyon J (2016) Influence of ion outflow in coupled geospace simulations: 1. Physics-based ion outflow model development and sensitivity study. J Geophys Res Space Phys 121(10):9671–9687

Vasyliunas VM (1970) Mathematical models of magnetospheric convection and its coupling to the ionosphere. In: Particles and fields in the magnetosphere: proceedings of a symposium organized by the summer advanced study institute, held at the university of California, Santa Barbara, Calif., August 4–15, 1969. Springer, Berlin, pp 60–71

Vasyliunas VM (1975) Theoretical models of magnetic field line merging, 1. Rev Geophys 13(1):303

Vega C, Roytershteyn V, Delzanno GL, Boldyrev S (2020) Electron-only reconnection in kinetic-Alfvén turbulence. Astrophys J Lett 893(1):L10. https://doi.org/10.3847/2041-8213/ab7eba

Villasenor J, Bunemann O (1992) Rigorous charge conservation for local electromagnetic field solvers. Comput Phys Commun 63:306

Vilmer N (2012) Solar flares and energetic particles. Philos Trans R Soc A, Math Phys Eng Sci 370(1970):3241–3268

von Alfthan S, Pokhotelov D, Kempf Y, Hoilijoki S, Honkonen I, Sandroos A, Palmroth M (2014) Vlasiator: first global hybrid-Vlasov simulations of Earth's foreshock and magnetosheath. J Atmos Sol-Terr Phys 120:24–35. https://doi.org/10.1016/j.jastp.2014.08.012

Wang X, Bhattacharjee A (1993) Nonlinear dynamics of the $m = 1$ instability and fast sawtooth collapse in high-temperature plasmas. Phys Rev Lett 70(11):1627–1630

Wang X, Lin Y, Chang SW (2009) Hybrid simulation of foreshock waves and ion spectra and their linkage to cusp energetic ions. J Geophys Res Space Phys 114(A6):A06203

Wang L, Germaschewski K, Hakim A, Dong C, Raeder J, Bhattacharjee A (2018) Electron physics in 3-d two-fluid 10-moment modeling of Ganymede's magnetosphere. J Geophys Res Space Phys 123(4):2815–2830. https://doi.org/10.1002/2017JA024761

Wang H, Lin Y, Wang X, Guo Z (2019) Generation of kinetic Alfvén waves in dayside magnetopause reconnection: a 3-d global-scale hybrid simulation. Phys Plasmas 26(7):072102

Wang R, Lu Q, Lu S, Russell CT, Burch JL, Gershman DJ, Gonzalez W, Wang S (2020b) Physical implication of two types of reconnection electron diffusion regions with and without ion-coupling in the magnetotail current sheet. Geophys Res Lett 47(21):e2020GL088761. https://doi.org/10.1029/2020GL088761

Wang CP, Wang X, Liu TZ, Lin Y (2020a) Evolution of a foreshock bubble in the midtail foreshock and impact on the magnetopause: 3-d global hybrid simulation. Geophys Res Lett 47(22):e2020GL089844

Wang X, Chen Y, Toth G (2022a) Global magnetohydrodynamic magnetosphere simulation with an adaptively embedded particle-in-cell model. J Geophys Res 127:e2021JA030091. https://doi.org/10.1029/2021JA030091

Wang X, Chen Y, Tóth G (2022b) Simulation of magnetospheric sawtooth oscillations: the role of kinetic reconnection in the magnetotail. Geophys Res Lett 49(15):e2022GL099638. https://doi.org/10.1029/2022GL099638

Warmuth A, Mann G (2016) Constraints on energy release in solar flares from RHESSI and goes X-ray observations-ii. Energetics and energy partition. Astron Astrophys 588:A116

Weimer D (1996) A flexible, IMF dependent model of high-latitude electric potential having "space weather" applications. Geophys Res Lett 23:2549

Weimer D (2001) An improved model of ionospheric electric potentials including substorm perturbations and application to the geosphace environment modeling November 24, 1996, event. J Geophys Res 106:407

Wiegelmann T, Büchner J (2001) Evolution of magnetic helicity in the course of kinetic magnetic reconnection. Nonlinear Process Geophys 8(3):127–140. https://doi.org/10.5194/npg-8-127-2001

Wilder FD, Ergun RE, Burch JL, Ahmadi N, Eriksson S, Phan TD, Goodrich KA, Shuster J, Rager AC, Torbert RB, Giles BL, Strangeway RJ, Plaschke F, Magnes W, Lindqvist PA, Khotyaintsev YV (2018) The role of the parallel electric field in electron-scale dissipation at reconnecting currents in the magnetosheath. J Geophys Res Space Phys 123(8):6533–6547. https://doi.org/10.1029/2018JA025529

Wiltberger M, Merkin V, Lyon J, Ohtani S (2015) High-resolution global magnetohydrodynamic simulation of bursty bulk flows. J Geophys Res Space Phys 120(6):4555–4566

Wiltberger M, Merkin V, Zhang B, Toffoletto F, Oppenheim M, Wang W, Lyon J, Liu J, Dimant Y, Sitnov M, et al (2017) Effects of electrojet turbulence on a magnetosphere-ionosphere simulation of a geomagnetic storm. J Geophys Res Space Phys 122(5):5008–5027

Winglee RM (2004) Ion cyclotron and heavy ion effects on reconnection in a global magnetotail. J Geophys Res 109:A09206

Winske D, Yin L, Omidi N, Karimabadi H, Quest K (2003) Hybrid simulation codes: past, present and future - a tutorial. In: Büchner J, Scholer M, Dum CT (eds) Space plasma simulation, lecture notes in physics, vol 615. Springer, Berlin, pp 136–165. https://doi.org/10.1007/3-540-36530-3_8

Wolf R (1983) The quasi-static (slow-flow) region of the magnetosphere. In: Solar-terrestrial physics: principles and theoretical foundations based upon the proceedings of the theory institute held at Boston college, August 9–26, 1982. Springer, Berlin, pp 303–368

Wolf RA, Harel M, Spiro RW, Voigt G, Reiff PH, Chen CK (1982) Computer simulation of inner magnetospheric dynamics for the magnetic storm of July 29, 1977. J Geophys Res 87:5949–5962. https://doi.org/10.1029/JA087iA08p05949

Wolf RA, Spiro RW, Sazykin S, Toffoletto FR, Yang J (2016) Forty-seven years of the rice convection model. In: Magnetosphere-Ionosphere Coupling in the Solar System. American Geophysical Union (AGU), chapter 17, pp 215–225. https://doi.org/10.1002/9781119066880.ch17

Wu P, Wan M, Matthaeus WH, Shay MA, Swisdak M (2013) Von Kármán energy decay and heating of protons and electrons in a kinetic turbulent plasma. Phys Rev Lett 111:121105
Yamada M, Levinton FM, Pomphrey N, Budny R, Manickam J, Nagayama Y (1994) Investigation of magnetic reconnection during a sawtooth crash in a high-temperature tokamak plasma. Phys Plasmas 1:3269–3276
Yang J, Toffoletto FR, Wolf RA (2014a) RCM-E simulation of a thin arc preceded by a north-south-aligned auroral streamer. Geophys Res Lett 41(8):2695–2701
Yang J, Wolf RA, Toffoletto FR, Sazykin S, Wang CP (2014b) Rcm-e simulation of bimodal transport in the plasma sheet. Geophys Res Lett 41(6):1817–1822
Yang J, Toffoletto FR, Wolf RA, Sazykin S (2015) On the contribution of plasma sheet bubbles to the storm time ring current. J Geophys Res Space Phys 120(9):7416–7432
Yang J, Toffoletto FR, Wolf RA (2016) Comparison study of ring current simulations with and without bubble injections. J Geophys Res Space Phys 121(1):374–379
Yang J, Wolf R, Toffoletto F, Sazykin S, Wang W, Cui J (2019) The inertialized rice convection model. J Geophys Res Space Phys 124(12):10294–10317
Yee K (1966) Numerical solution of initial boundary value problems involving Maxwell's equations in isotropic media. IEEE Trans Antennas Propag 14:302
Yi Y, Zhou M, Song L, Pang Y, Deng X (2022) Electron-only magnetic reconnection: lessons learned from magnetic island coalescence. Geophys Res Lett 49(6):e2022GL098124. https://doi.org/10.1029/2022GL098124
Yin Z, Drake JF, Swisdak M (2024a) A computational model for ion and electron energization during macroscale magnetic reconnection. Phys Plasmas 31(6):062901. https://doi.org/10.1063/5.0199679
Yin Z, Drake JF, Swisdak M (2024b) Simultaneous proton and electron energization during macroscale magnetic reconnection. Astrophys J 974(1):74. https://doi.org/10.3847/1538-4357/ad7131
Zaharia S, Cheng C, Johnson JR (2000) Particle transport and energization associated with substorms. J Geophys Res Space Phys 105(A8):18741–18752
Zaharia S, Jordanova VK, Thomsen MF, Reeves GD (2006) Self-consistent modeling of magnetic fields and plasmas in the inner magnetosphere: application to a geomagnetic storm. J Geophys Res 111:A11S14. https://doi.org/10.1029/2006JA011619
Zalesak ST (1979) Fully multidimensional flux-corrected transport algorithms for fluids. J Comput Phys 31:335
Zalesak ST (1981) High order "zip" differencing of convective terms. J Comput Phys 40(2):497–508. https://doi.org/10.1016/0021-9991(81)90225-4
Zeiler A, Biskamp D, Drake JF, Rogers BN, Shay MA, Scholer M (2002) Three-dimensional particle simulations of collisionless magnetic reconnection. J Geophys Res 107(A9):1230. https://doi.org/10.1029/2001JA000287
Zenitani S, Umeda T (2014) Some remarks on the diffusion regions in magnetic reconnection. Phys Plasmas 21(3):034503. https://doi.org/10.1063/1.4869717
Zerroukat M, Allen T (2012) A three-dimensional monotone and conservative semi-Lagrangian scheme (SLICE-3D) for transport problems. Q J R Meteorol Soc 138(667):1640–1651. https://doi.org/10.1002/qj.1902
Zhang B, Sorathia KA, Lyon JG, Merkin VG, Garretson JS, Wiltberger M (2019) Gamera: a three-dimensional finite-volume mhd solver for non-orthogonal curvilinear geometries. Astrophys J Suppl Ser 244(1):20
Zhang B, Brambles OJ, Lotko W, Lyon JG (2020) Is nightside outflow required to induce magnetospheric sawtooth oscillations. Geophys Res Lett 47(6):e2019GL086419
Zhang Q, Guo F, Daughton W, Li H, Li X (2021) Efficient nonthermal ion and electron acceleration enabled by the flux-rope kink instability in 3d nonrelativistic magnetic reconnection. Phys Rev Lett 127(18):185101
Zhang Q, Guo F, Daughton W, Li X, Li H (2024) Plasma dynamics and nonthermal particle acceleration in 3d nonrelativistic magnetic reconnection. Astrophys J 974(1):47
Zhong ZH, Tang RX, Zhou M, Deng XH, Pang Y, Paterson WR, Giles BL, Burch JL, Tobert RB, Ergun RE, Khotyaintsev YV, Lindquist PA (2018) Evidence for secondary flux rope generated by the electron Kelvin-Helmholtz instability in a magnetic reconnection diffusion region. Phys Rev Lett 120(7):075101. https://doi.org/10.1103/PhysRevLett.120.075101
Zhong ZH, Zhou M, Liu YH, Deng XH, Tang RX, Graham DB, Song LJ, Man HY, Pang Y, Khotyaintsev YV (2022) Stacked electron diffusion regions and electron Kelvin–Helmholtz vortices within the ion diffusion region of collisionless magnetic reconnection. Astrophys J Lett 926(2):L27. https://doi.org/10.3847/2041-8213/ac4dee
Zhou H, Tóth G, Jia X, Chen Y, Markidis S (2019) Embedded kinetic simulation of Ganymede's magnetosphere: improvements and inferences. J Geophys Res Space Phys 124(7):5441–5460. https://doi.org/10.1029/2019JA026643

Zhou H, Tóth G, Jia X, Chen Y (2020) Reconnection-driven dynamics at Ganymede's upstream magnetosphere: 3-d global Hall MHD and MHD-EPIC simulations. J Geophys Res Space Phys 125(8):e2020JA028162. https://doi.org/10.1029/2020JA028162
Zhou M, Man HY, Deng XH, Pang Y, Khotyaintsev Y, Lapenta G, Yi YY, Zhong ZH, Ma WQ (2021) Observations of secondary magnetic reconnection in the turbulent reconnection outflow. Geophys Res Lett 48(4):e2020GL091215. https://doi.org/10.1029/2020GL091215
Zhu Z, Winglee RM (1996) Tearing instability, flux ropes, and the kinetic current sheet kink instability in the Earth's magnetotail: a three-dimensional perspective from particle simulations. J Geophys Res Space Phys 101(A12):4885–4897
Zocco A, Chacón L, Simakov AN (2009) Current sheet bifurcation and collapse in electron magnetohydrodynamics. Phys Plasmas 16(11):110703. https://doi.org/10.1063/1.3264102

Publisher's Note Springer Nature remains neutral with regard to jurisdictional claims in published maps and institutional affiliations.

Authors and Affiliations

Michael Shay[1] · Subash Adhikari[2,1] · Naoki Beesho[3] · Joachim Birn[4] · Jörg Büchner[5] · Paul Cassak[2] · Li-Jen Chen[6] · Yuxi Chen[7] · Giulia Cozzani[8,9] · James Drake[10,11] · Fan Guo[12] · Michael Hesse[13] · Neeraj Jain[14] · Yann Pfau-Kempf[8,15] · Yu Lin[16] · Yi-Hsin Liu[17] · Mitsuo Oka[18] · Yuri Omelchenko[4,19] · Minna Palmroth[8] · Oreste Pezzi[20] · Patricia H. Reiff[21] · Marc Swisdak[11] · Frank Toffoletto[21] · Gabor Toth[7] · Richard A. Wolf[21]

✉ M. Shay
shay@udel.edu

✉ G. Cozzani
giulia.cozzani@cnrs-orleans.fr

1 Bartol Research Institute, Department of Physics and Astronomy, University of Delaware, Newark, 19716, DE, USA

2 Department of Physics and Astronomy, West Virginia University, Morgantown, 26506, WV, USA

3 Department of Astronomy, University of Maryland, College Park, 20742, MD, USA

4 Center for Space Plasma Physics, Space Science Institute, Boulder, 80301, CO, USA

5 Max Planck Institute for Solar System Research, Göttingen, 27077, Germany

6 NASA Goddard Space Flight Center, Greenbelt, 20771, MD, USA

7 University of Michigan, Ann Arbor, 48109, MI, USA

8 Department of Physics, University of Helsinki, P.O. Box 68, 00014, Uusimaa, Finland

9 LPC2E, OSUC, Univ Orleans, CNRS, CNES, Orleans, F-45071, France

10 Department of Physics, University of Maryland, College Park, 20740, MD, USA

11 Institute for Research in Electronics and Applied Physics, University of Maryland, College Park, 20740, MD, USA

12 Los Alamos National Laboratory, Los Alamos, 87545, NM, USA

13 NASA Ames Research Center, Moffett Field, 94035, CA, USA

14 Center for Astronomy and Astrophysics, Technical University Berlin, Berlin, 10623, Germany

15 Now at Advanced Computing Facility, CSC - IT Center for Science, Espoo, 02101, Uusimaa, Finland

[16] Physics Department, Auburn University, Auburn, 36832, AL, USA

[17] Department of Physics and Astronomy, Dartmouth College, Hanover, 03750, NH, USA

[18] Space Sciences Laboratory, University of California, Berkeley, 94720, CA, USA

[19] Trinum Research Inc., San Diego, 92126, CA, USA

[20] Istituto per la Scienza e Tecnologia dei Plasmi (ISTP), Consiglio Nazionale delle Ricerche, Bari, I-70126, Italy

[21] Department of Physics and Astronomy, Rice University, Houston, 77005, TX, USA

VII. Summary and Outlook

Space Science Reviews (2025) 221:17
https://doi.org/10.1007/s11214-025-01143-z

Outstanding Questions and Future Research on Magnetic Reconnection

R. Nakamura[1,2] · J.L. Burch[3] · J. Birn[4] · L.-J. Chen[5] · D.B. Graham[6] · F. Guo[7] · K.-J. Hwang[3] · H. Ji[8] · Y.V. Khotyaintsev[6] · Y.-H. Liu[9] · M. Oka[10] · D. Payne[11] · M.I. Sitnov[12] · M. Swisdak[11] · S. Zenitani[1] · J.F. Drake[11] · S.A. Fuselier[3,13] · K.J. Genestreti[3] · D.J. Gershman[5] · H. Hasegawa[14] · M. Hoshino[15] · C. Norgren[6] · M.A. Shay[16] · J.R. Shuster[17] · J.E. Stawarz[18]

Received: 7 June 2024 / Accepted: 16 January 2025 / Published online: 11 February 2025

Abstract
This short article highlights unsolved problems of magnetic reconnection in collisionless plasma. Advanced in-situ plasma measurements and simulations have enabled scientists to gain a novel understanding of magnetic reconnection. Nevertheless, outstanding questions remain concerning the complex dynamics and structures in the diffusion region, cross-scale and regional couplings, the onset of magnetic reconnection, and the details of particle energization. We discuss future directions for magnetic reconnection research, including new observations, new simulations, and interdisciplinary approaches.

Keywords Magnetic reconnection · Magnetospheric Multiscale (MMS) mission · Diffusion region · Onset · Cross-scale · Energetics

1 Introduction

Magnetic reconnection is a fundamental energy conversion process in plasmas. While changes in the topology of the magnetic field take place inside a small region, acceleration and heating of the plasma are distributed over larger scales. Acceleration and heating drive plasma transport and lead to explosive magnetic energy release likewise on large scales during phenomena such as substorms, solar flares and gamma ray bursts. With modern space technology, geospace is an ideal plasma laboratory for studying how collisionless magnetic reconnection operates in nature since plasmas and fields in action can be directly measured at high cadence. With the advanced in-situ measurement capability to resolve electron-scale physics, the four Magnetospheric Multiscale (MMS) spacecraft (Burch et al. 2016) have significantly advanced the study of magnetic reconnection and relevant plasma processes. The rich studies conducted in the MMS era motivated us to summarize the current understanding of magnetic reconnection that arises from new observations mainly in geospace and in other environments as well as from theoretical studies (Burch and Nakamura 2025, this collection).

Extended author information available on the last page of the article

Studies based on in-situ observations from MMS and numerical simulations confirmed some theoretical predictions and led to a number of new discoveries on dynamics of reconnection at smallest scale: the electron kinetic scale (Genestreti et al. 2025, this collection). In particular, progress has been made in observations and theories related to the reconnection rate and energy conversion processes (Liu et al. 2025, this collection), and in the kinetic behavior of both electrons and ions in the vicinity of the diffusion region (Norgren et al. 2025, this collection). The diverse roles of waves and turbulence in magnetic reconnection are also among the important discoveries from the MMS observations (Graham et al. 2025, this collection; Stawarz et al. 2024, this collection). Some of these features were not predicted or not the focus of theory or numerical simulations before the MMS era.

MMS, combined with other spacecraft and empirical and/or theoretical modeling, has allowed us to gain new insights into the macroscale consequences of reconnection. These include the large-scale consequences of solar-wind magnetospheric interactions (Fuselier et al. 2024, this collection) and particle acceleration (Oka et al. 2023b, this collection), as well as the coupling among magnetic reconnection-related processes at different scales (Hwang et al. 2023, this collection). All these studies benefited from the development of new data analysis techniques (Hasegawa et al. 2024, this collection) and simulation/modeling schemes (Shay et al. 2025, this collection), which allow direct comparisons between the observed and simulated velocity distributions of particles and electromagnetic signatures.

Recent observations throughout the different environment in the solar system (Drake et al. 2025, this collection; Gershman et al. 2024, this collection) and advanced laboratory experiments (Ji et al. 2023, this collection) enabled us to study different scales of magnetic reconnection in different parameter regimes and deepen our understanding of the process. New kinetic and fluid simulations have also significantly contributed to understanding magnetic reconnection in both collisionless and collisional astrophysical plasmas (Guo et al. 2024, this collection).

While significant advancements in magnetic reconnection research have been made with these endeavors, there remain several unsolved questions. These questions relate to kinetic physics and macroscale consequences in different environments, both within and beyond geospace. In this short paper, we highlight several unsolved questions and propose future research directions in the short term (years) using MMS as well as in the long term (decades).

2 Unsolved Problems

2.1 Complex Dynamics and Structures in the Diffusion Region

Substantial progress has been made in understanding the relationship between magnetic reconnection and kinetic plasma waves (e.g., Graham et al. 2025, this collection). These include specification of the types and locations of the waves that can develop during reconnection and identification of particle distributions that can excite the waves (e.g., Burch et al. 2018). However, much less is known about the effects of these waves on the plasma and it is likewise not well known how these waves can affect reconnection. In particular, an ongoing question is whether anomalous resistivity due to wave-particle interactions contributes to magnetic reconnection, for example by modifying the reconnection electric field (e.g., Yoo et al. 2024). MMS was able to directly quantify the anomalous resistivity associated with reconnection by resolving the changes in electron distributions and moments associated with lower hybrid waves around the X-line (Graham et al. 2022). The results revealed that the anomalous resistivity balances with anomalous viscosity so that its contributions to the reconnection electric fields were small, which is consistent with the findings of previous theoretical and observational studies. However, these waves contribute to significant cross-field diffusion that can develop and thereby broaden narrow boundary layers and facilitate electron mixing. Further work can be done with MMS to answer the question on the role of waves in reconnection by examining the interactions between electron and higher-frequency waves. While the current direct investigation of wave-particle interaction using the highest-resolution electron distributions is limited to the lower hybrid frequency range, the wave-particle correlator technique can be applied to reconnection current sheets to study higher-frequency wave-particle interactions. This technique has been used to compute the energy transfer between waves and particles for whistler waves in the magnetosheath (Kitamura et al. 2022).

Furthermore, MMS has produced discoveries that have not been predicted by theory or numerical simulations. MMS observations have shown that the agyrotropic electron distributions found in the electron diffusion region (EDR) can become unstable to large-amplitude waves (Graham et al. 2025, this collection), such as upper hybrid waves and electron Bernstein waves, due to beam-plasma interactions. These waves provide potential sources of radio emission and can modify the electron distributions in the EDR, but their overall impact on reconnection remains to be quantified. These observations also clearly demonstrate the presence of physical processes at scales below the electron gyroscale, i.e. down to the Debye scale, inside the EDR. The proper description of EDR physics must therefore include Debye-scale processes, which are often only marginally resolved in typical simulations (see Sect. 3.3).

MMS observations have also shown that some EDRs exhibit turbulent structures (Khotyaintsev et al. 2020) or strong oscillations (Cozzani et al. 2021) in and around EDRs. The oscillations were attributed to kinking of the current sheet by an electromagnetic drift wave propagating in the out-of-plane direction, suggesting that magnetic reconnection needs to be considered in three dimensions. Kinetic simulations have shown that EDRs can become structured and turbulent when there is scale separation between the electron Debye length and the electron inertial length (Jara-Almonte et al. 2014). More generally, MMS observations have reported both turbulent and more laminar EDRs at the magnetopause and in the magnetotail (Liu et al. 2025, this collection; Graham et al. 2025, this collection). At present, the underlying processes that determine whether an EDR behaves in a laminar or turbulent manner are not fully understood. This raises the important question of whether more complicated EDRs are missed or overlooked in observations. Although many EDRs have been identified by MMS, their identification has generally relied on predictions from kinetic simulations of laminar reconnection. Further work is needed to identify more complex EDRs. Methods such as tunable algorithms (e.g., Bergstedt et al. 2020) or machine-learning techniques (e.g., Argall et al. 2020; Hasegawa et al. 2024, this collection; Bergstedt and Ji 2024) can be applied to identify relevant magnetic structures from observations. With more EDRs, case studies, which have dominated the research so far, can give way to statistical studies. This transition leads to a more comprehensive understanding of complex EDR dynamics.

At present, guide-field reconnection is not as well understood as antiparallel reconnection. Electrons in the EDR tend to remain strongly magnetized in the presence of a strong guid field. When the electrons are magnetized, the off-diagonal pressure terms play a reduced role in supporting the reconnection electric field. This is in contrast to antiparallel reconnection when the reconnection electric field is supported by the off-diagonal pressure terms generated by electron agyrotropy, which is often used to identify EDRs. Kinetic simulations demonstrate the formation of a narrow sublayer (of intensified current density) embedded within the broader, electron inertia-scale EDR (Liu et al. 2014b). The off-diagonal pressure term only becomes significant within this sublayer, which is on the electron gyroscale (Genestreti et al. 2025, this collection). Additionally, a strong guide field creates out-of-plane field-aligned electron flow around the X-line. This electron flow is free energy for the development of electrostatic waves and turbulence in the EDR. The reduced role of agyrotropy and the role of electrostatic turbulence in guide-field reconnection requires further investigation. Interestingly, the same out-of-plane electron flow from magnetic reconnection in the strong guide field limit may explain some features of electron precipitation associated with the auroral spiral structure (Huang et al. 2022).

2.2 Cross-Scale Dynamics and Regional Coupling

Magnetic reconnection operates in the presence of a diffusion region with dissipative electric fields which are generated in the EDR. Electron-kinetic physics prevails in the EDR, whereas Hall physics becomes significant in the ion diffusion region (IDR). The influence of magnetic reconnection further extends to macroscopic systems, such as magnetospheric boundaries and mesoscale plasma structures in geospace, for which ideal magnetohydrodynamics (MHD) provides a good overall description. These nested reconnection regions around the X-line are interconnected via the exchange and transport of particles, momentum, energy and Poynting flux. Thus, reconnection is intrinsically a multiscale and cross-scale process all the way up to the macroscale. In-situ observations and state-of-the-art numerical simulations have significantly advanced our understanding of the multiscale aspects of reconnection (Hwang et al. 2023, this collection) occurring throughout geospace, as highlighted in Fig. 1. They also revealed new questions, as discussed in the following sub-sections. By answering these questions, they may change the current understanding, leading to a paradigm shift.

2.2.1 Electron-Only to Ion-Coupled Reconnection

MMS data-model analyses have shown that reconnection is ubiquitous in the shock transition region, the foreshock, and the magnetosheath downstream of both quasi-parallel and quasi-perpendicular shocks (Fig. 1b, adapted from Bessho et al. 2022). Of particular interest in this region is the electron-only reconnection, newly discovered in observations (Phan et al. 2018), which has stimulated new theoretical studies (Liu et al. 2025, this collection) and new investigations on the interplay between turbulence and reconnection (Stawarz et al. 2024, this collection). In turbulent systems, electron-only reconnection is considered to occur mainly because the scale of the turbulent fluctuations limits the maximum size of the reconnection region, particular along the reconnection outflow. Alternatively, it has been suggested that electron-only reconnection might represent the early stage of regular reconnection before the reconnection exhaust becomes large enough to involve ions (e.g. Hubbert et al. 2022). Such finite lifetime effects may be relevant for magnetotail reconnection. However, confirming such a scenario is challenging. In the simulations, electron-only reconnection was shown to have faster reconnection rate than for regular reconnection (Sharma Pyakurel et al. 2019). It is uncertain observationally whether the transition from electron-only to ion-coupled reconnection is regulated by the reduction in the reconnection rate. Further investigations and observations are needed to gain a complete understanding of electron-only reconnection, its role in cross-scale reconnection dynamics, and the scale-dependent energy conversion.

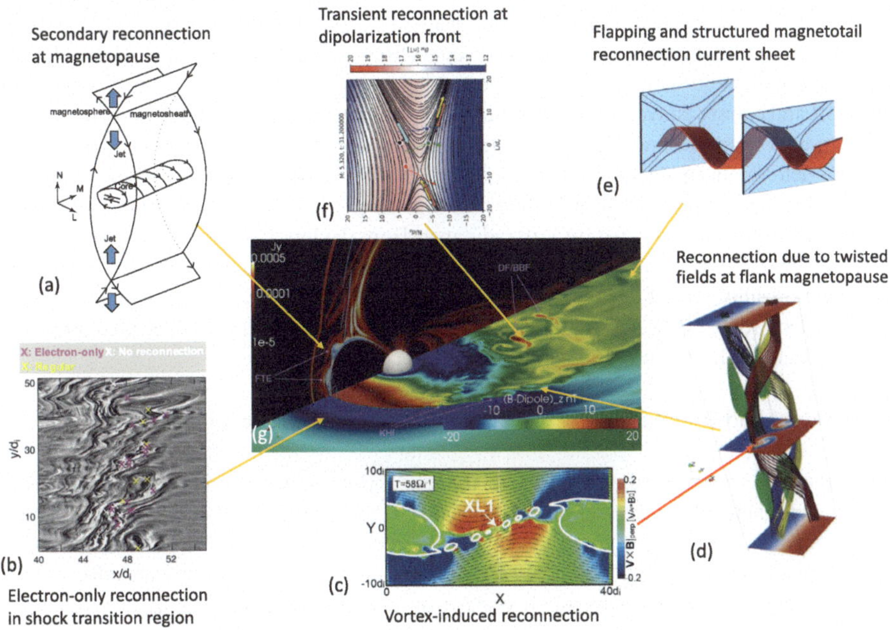

Fig. 1 **Reconnection in geospace**. In addition to global dayside and nightside magnetic reconnection, recent in-situ measurements and simulations have revealed 3D, complex, and localized reconnection features throughout geospace. (a-f) Examples of different types of reconnection that are actively studied in the MMS era. (g) 3D view of the magnetosphere from an MHD model (Credit: V. G. Merkin) adapted from Sitnov et al. (2016), where several key mesoscale processes KHI, BBF/DF and FTE related to localized reconnections are indicated. The highlighted reconnection features are: (a) secondary and/or multiple reconnection at the magnetopause (adapted from Øieroset et al. 2016), (b) turbulent reconnection in the shock transition region (adapted from Bessho et al. 2022), (c-d) 3D multiscale KHI/KHV induced reconnection (adapted from (c) Nakamura et al. 2011 and (d) Faganello et al. 2012), (e) structured and disturbed EDR in the magnetotail current sheet (adapted from Cozzani et al. 2021) and (f) transient and localized reconnection at the dipolarization front (adapted from Hosner et al. 2024)

2.2.2 Velocity Shear-Driven Asymmetric Reconnection

Asymmetries in density and magnetic shear are important factors in different regimes of reconnection (Genestreti et al. 2025, this collection). These effects as well as the effects of flow shear are prominent at the flank-side magnetopause. In this region, flow shear drives the complex multiscale evolution of the magnetopause current sheet, which can develop into a turbulent layer depending on the ambient conditions (Hwang et al. 2023, this collection; Stawarz et al. 2024, this collection). When the interplanetary field (IMF) is northward, the unperturbed low-latitude magnetopause on the flank is stable to the formation of extended reconnection diffusion regions, but unstable to the large-scale Kelvin-Helmholtz instability (KHI) driven by shear flows. Under super-Alfvénic conditions the vortex flow produced by the nonlinear growth of the KHI can locally compress the magnetic shear layer (current sheet), forcing the onset of vortex-induced reconnection (VIR) (Nakamura et al. 2017). Multiple reconnection regions appear in the current sheet as shown in Fig. 1c and can result in a complex turbulent boundary layer. When the IMF is southward, meaning a strong magnetic shear at the magnetopause favorable for reconnection, the evolution of the current sheet varies depending on the initial condition (magnetic shear vs. flow shear). However, the two modes can interact with each other, leading to complex and intercorrelated dynamics. Understanding the interplay between reconnection and the KHI (and/or Rayleigh-Taylor instability associated with density asymmetry) is important, as it would control solar wind transport and energy conversion across the flankside magnetopause. Furthermore, reconnection can also occur around the flow-shear plane due to a 3-D twist of the magnetospheric and magnetosheath magnetic fields induced by Kelvin-Helmholtz vortices (Faganello et al. 2012). This type of reconnection is shown in Fig. 1d and is called "mid-latitude reconnection" (MIR). MIR occurs several Earth radii apart from the low-latitude VIR location, while being magnetically connected in 3D. Hence, the potential "communication" between the two reconnection sites can affect solar wind transport in a complex way. These examples show that magnetic reconnection at the flank-magnetopause provides an excellent laboratory for studying multiscale (forced) 3D reconnection.

2.2.3 Extent and Orientation of X-Lines; Primary and Secondary X-Lines

While magnetic reconnection at the magnetopause and magnetotail is considered the driver of global magnetosphere circulation, reconnection in these large-scale currents has variability in space and time and signatures of multiple reconnection sites (Fuselier et al. 2024, this collection; Hwang et al. 2023, this collection). Interpreting in-situ reconnection events is often complicated as the large-scale context of reconnection cannot be ascertained from observations with limited coverage. There remain unsolved questions regarding the temporal and spatial scales of reconnection in mesoscale and large-scale contexts for both the magnetopause and the magnetotail.

At the magnetopause, the location and extent of the primary X-line are considered to be determined by the global solar wind-magnetosphere interaction, enabling us to predict this interaction via the maximum magnetic shear model (Trattner et al. 2007; Hasegawa et al. 2024, this collection, and reference therein). This is an empirical model that uses upstream conditions and global parameters. However, observations have also revealed more complicated structures, with localized and transient behavior of multiple reconnection sites at the magnetopause. An example is magnetic reconnection at the center of a magnetic flux rope where the jets from the adjacent two reconnection sites collide and form a compressed thin current sheet (Fig. 1a, adapted from Øieroset et al. 2016). Some simulations suggest that local physics can influence the orientation and variation of the X-line (e.g., Liu et al. 2018). The relationships between the primary and secondary X-lines are yet unsolved problems. Are secondary reconnection sites created by turbulence or external (e.g., magnetosheath) conditions near the primary X-line? Alternatively, is the migration of the primary X-line initiated by the local physics of the secondary reconnection sites? Lastly, is it possible that multiple X-lines develop simultaneously, with the roles of primary vs. secondary being later established? The evolutionary paths of plasmoids and flux ropes commonly generated on the dayside magnetopause via secondary/multiple X-lines are also yet to be understood.

The configuration of the magnetotail current sheet is typically quasi-2D and symmetric, so that the formation of a large-scale extended X-line is expected; however, the statistical distribution of observed reconnection events suggests that the near-Earth magnetotail reconnection is localized within $\sim$5 Earth radii (Nagai and Shinohara 2021). One of the major challenges with observations is determining the extent of the reconnection region in the out-of-plane direction (as reviewed in Hwang et al. 2023, this collection). The dawn-dusk extent of bursty bulk flows (BBFs), associated localized dipolarization fronts (DFs) and localized thin current sheets can be more easily detected due to their larger cross-section (relative to the diffusion region). The finite extent of these transients may indicate a finite dimension of the source, i.e., the magnetic reconnection region. Alternatively, their size may be decoupled from that of their source as (a) a ballooning/interchange instability may break up a wider flow into localized channels as it penetrates into the inner magnetosphere or (b) the structured flows/DFs may be created by the interchange instability itself. Furthermore, transient localized reconnection can also take place at a DF (Fig. 1f adapted from Hosner et al. 2024) so that the DF is modified as it propagates Earthward from the source region. Understanding the extent of the reconnection region is crucial, as it affects large-scale dynamics, e.g., magnetic flux and mass transport, as well as particle acceleration in the magnetosphere. Recently, the application of data mining tools has provided some insight into the extents of X-lines (Stephens et al. 2023). The larger spacecraft separations along the MMS spacecraft orbit planned in 2024 may enable new studies of 3D nature of X-lines including the out-of-plane direction in the Earth's magnetotail. Furthermore, as discussed in Sect. 2.1, MMS reported strong oscillation along the X-line inside EDRs (Fig. 1e adapted from Cozzani et al. 2021). Such complex dynamics in the diffusion region may affect the extent of the X-line and the local reconnection rate (Liu et al. 2024).

2.3 Onset of Reconnection

While the free energy of reconnection is determined by the large-scale background plasma conditions and has large-scale consequences, the dissipation of the tearing mode occurs at the ion or electron gyro-scale/gyroradii. The onset problem is therefore multiscale in nature, and an under-explored topic in reconnection physics. Limited coverage of all necessary scales by in-situ plasma observatories makes it very difficult to compare with theoretical/numerical descriptions. Here we highlight the onset problems of different types of current sheets including magnetotail, solar flares, magnetopause and other forced current sheets.

2.3.1 Reconnection Onset in Earth's Magnetotail

To understand the onset of near-Earth magnetotail reconnection one needs to understand both the formation of the thin current sheet and the instability leading to explosive energy release. The observed thin current sheets are generally embedded in a thicker plasma sheet with anisotropy and agyrotropy in both ions and electrons and contain radial or azimuthal gradients (Runov et al. 2021, and references therein). Detection of the formation and evolution of thin current sheets from in-situ observations is limited because of the sparse dataset. The current best approach to resolve multiscale current sheet structures is the data-mining method (Sitnov et al. 2019b), which helped resolve the location of the X-lines (Stephens et al. 2023) detected by the MMS mission in the form of IDRs.

MHD models suggest that thin current sheets are created because of the deformation of the high-latitude magnetopause boundary by the reconnected and transported magnetic flux from the dayside (Birn and Schindler 2002) or because of depletion of the closed magnetic flux at the near-Earth current sheet transported toward the dayside (Hsieh and Otto 2014). The basic concept of the former effect was obtained in the isotropic plasma description of MHD models and was also verified via 2D PIC simulations (Hesse and Schindler 2001). However, modeling of the onset current sheet with very small, but still finite B_Z (normal component to the current sheet), where the anisotropic and agyrotropic pressure contributions play a role, is challenging. The mechanism leading to the onset of magnetotail reconnection with finite B_Z has been extensively studied via simulations, which revealed two primary onset scenarios (Sitnov et al. 2019a, and references therein). The first is the electron tearing instability preceded by an external driving of the current sheet as described above to form an electron scale current sheet (e.g. Hesse and Schindler 2001; Liu et al. 2014a). The second is a magnetic flux release instability in an ion-scale current sheet with a B_Z hump (Sitnov and Schindler 2010). It may develop in the ideal-MHD regime and in the form of the kinetic ion tearing instability. The problem is, however, that simulations of both the ion and electron tearing instability reveal that the new X-lines form just $\sim 15\ d_i$ ($< 2R_E$) from the left boundary of the simulation box, which is much closer to Earth than nearly all observed X-lines in the magnetotail. Recently a new class of current sheets has been explored (Sitnov and Arnold 2022). In these current sheets, weak anisotropy in the ion species extends them much farther than corresponding isotropic (Harris-like) current sheets. The new "overstretched ion-scale current sheets" are agyrotropic and are supported by the off-diagonal pressure gradient terms originating from ion Speiser motions (Arnold and Sitnov 2023). However, comprehensive stability theories for these new current sheets have yet to be developed and simulations of reconnection onset are still an active area of research.

Using in-situ observations to detect reconnection onset is another challenge. A recent particle-in-cell (PIC) simulation suggested that a possible observable onset feature is a slightly agyrotropic electron distribution (Spinnangr et al. 2022). However, to date, there is no identified MMS observations within less than 10 ion gyroperiods from onset in the vicinity of the EDR to confirm such predictions. Nevertheless, several observations of MMS electrons near current sheets are suggested to indicate precursors of larger scale reconnection onset and these observations are consistent with predictions from simulations. These include: observations of thin electron-scale current sheets with slow electron flows (Wang et al. 2018), divergent electron velocity flow observations without magnetic topology change (Motoba et al. 2022), and observations of electron-scale islands and Hall currents in the vicinity (or as a consequence of the formation) of a major X-line (Genestreti et al. 2023). However, all these observations are snapshots of some stages of reconnection evolution predicted by some simulations. Multiscale observations, which monitor both the ion- and electron-scale evolution of the current sheet simultaneously, are essential for confirming the different onset mechanisms of fast reconnection in the magnetotail current sheet.

2.3.2 Reconnection Onset in Solar Flares

The mechanisms of flare onset and associated particle accelerations are also a research area with outstanding questions (Drake et al. 2025, this collection). Similar to magnetotail reconnection, both the build up and trigger for the sudden release of magnetic energy need to be explained to understand the mechanism of solar flare onset. As in the magnetotail, the large-scale accumulation of energy preceding reconnection onset and its transport down to kinetic length scales are important for solar flares in coronal loops; hence it is again a multiscale problem. While solar kinetic scales are inaccessible from observations, the complex 3D evolution of solar flares has been extensively studied through multiwavelength observations as well as in-situ measurements of accelerated particles that propagate away from coronal loops. Theories for magnetic reconnection onset in flares, such as breakout (Antiochos et al. 1999) and tether cutting (Jiang et al. 2021), have been successful in producing standard eruptive morphologies such as a twisted CME flux rope escaping at high speed and fast reconnection in the flare current sheet below the flux rope. The kink instability of flux ropes in the solar corona (Török and Kliem 2005), on the other hand, has also been suggested to be important for flare onset. However, it has not been established definitively from observations or simulations whether Alfvénic motions cause the onset and drive reconnection or vice versa (Drake et al. 2025, this collection). Furthermore, observed local precursors such as preflare-heating and its role in subsequent eruption, remain to be understood (e.g. Battaglia et al. 2019; Hudson et al. 2021).

In contrast to the near antiparallel geometry of the magnetotail current sheet, the guide field plays a crucial role in the evolution of the reconnection current sheet in solar flares. In the presence of a strong guide field, the thermal pressure of the current sheet cannot play a major role in the force balance, since the guide field contributes to magnetic pressure at the center of the reversal and prevents the collapse of the converging fields (Leake et al. 2020; Dahlin et al. 2022). It is also possible for a current sheet with a small finite guide field to evolve toward a "mixed" equilibrium, where the current sheet relaxation process leads to local guide field amplification (Yoon et al. 2023). The amplification of the guide field enhances the previously negligible magnetic pressure, and creates a condition where both the thermal pressure and the magnetic pressure play a significant role in stabilizing the current sheet (Yoon et al. 2023). A similar guide field amplification process has been reported in a 3D MHD simulation study (Dahlin et al. 2022). Here, a local accumulation of magnetic shear followed by outward expansion to form a thin current sheet was shown immediately before flare onset, after which the guide field decreased precipitously. Strong magnetic shear has also been associated with larger and more rapid increases in ion kinetic and thermal energy after reconnection onset in the corona, making it a potential candidate to explain the switch-on nature of solar flares (Leake et al. 2020). The role of other instabilities, such as the kink instability, in flare onset is still an open question (Drake et al. 2025, this collection).

The dynamics of reconnection in flare current sheets span an enormous range of scales in a much more complex geometry than in the magnetotail. In a collisional plasma with high Lundquist number ($\sim 10^{14}$) such as the solar corona, the Sweet-Parker current layers are highly unstable to the plasmoid instability (Shibata and Tanuma 2001; Loureiro et al. 2007; Bhattacharjee et al. 2009). These layers are modified well before they can reach kinetic scales and current sheet breakups have been successfully simulated with fluid models (Daldorff et al. 2022). In thin currents that form between flux-ropes, on the other hand, the super-Dreicer fields induce a transition to kinetic reconnection (Stanier et al. 2019), which cannot be detected from observations. How the dynamics of reconnection current layers at kinetic scales couple to energy release at the macroscale is still an open question (Drake et al. 2025, this collection).

2.3.3 Reconnection Onset in Different Forced Current Sheets

The onset problems of magnetic reconnection in the magnetotail and solar flares discussed in the previous sections are about a slow build-up of a thin current sheet followed by a sudden onset. Magnetic reconnection can also be forced to take place by external drivings. In this subsection, we discuss how such forced magnetic reconnection takes place. We highlight magnetopause reconnection and reconnection at the other transient forced current sheets. Forced reconnection has been intensively studied in the MMS era. In forced current sheets, the reconnection onset problem is less related to "when" but is more related to "where" and "under what conditions".

The magnetopause current sheet is continuously driven by the solar wind. Magnetopause reconnection is enabled or disabled depending on the asymmetries in the density and the magnetic and flow shear across the magnetopause current sheet. These factores are reviewed in Hwang et al. (2023, this collection) and Fuselier et al. (2024, this collection) for geospace and in Gershman et al. (2024, this collection) for planetary magnetosphere and heliopause. The diamagnetic drift stabilization (Swisdak et al. 2010) or shear flow-based suppression (Cassak 2011) provide sufficient but not necessary conditions for determining where reconnection is suppressed. The suppression conditions have been successfully tested at Earth and planetary magnetospheres. However, the Earth's magnetopause does not fulfill typically the diamagnetic-drift stabilization condition, i.e., reconnection is 'possible' for the typical range of changes in plasma β across the terrestrial magnetopause over large range of magnetic shear angles (Cassak and Fuselier 2016). The mechanism of determining the location of magnetopause reconnection as well as the multiple and transient nature of the magnetopause reconnection is therefore not fully understood (see also Sect. 2.2).

Numerous studies have shown that local, transient thin current sheets form and reconnect as a consequence of reconnection (or non-reconnection) related flows or field disturbances (Hwang et al. 2023, this collection; Stawarz et al. 2024, this collection), as discussed in Sect. 2.2. Examples of such current sheets are highlighted in Fig. 1. Unlike large-scale magnetopause or magnetotail current sheets, these current sheets can be localized and/or transient within complex dynamic processes. These include flow shear (Kelvin-Helmholtz instability) driven reconnection at the flank magnetopause (Nakamura et al. 2017), shown in Fig. 1c, and shock- and turbulent-driven reconnection in the magnetosheath or foreshock region (Bessho et al. 2022), shown in Fig. 1b. Open questions remain on interplay between reconnection and turbulence such as: how often reconnection can be generated by turbulence, how a turbulence-generated current sheet is forced by fluctuation, and what impact magnetic reconnection has on turbulence dissipation and nonlinear interactions (Stawarz et al. 2024, this collection). Furthermore, the reconnection jet itself can also be a driver of secondary reconnection due to the collision of reconnection jets from multiple X-lines (Øieroset et al. 2016), shown in Fig. 1a. In the near-Earth magnetotail transition region, reconnection events have been reported where a flux rope interacted with the dipole field (Poh et al. 2019), or at the dipolarization front in the flow braking region (Marshall et al. 2020; Hosner et al. 2024), shown in Fig. 1f. These types of reconnection events are usually forced by some primary processes. Important questions are: how these primary processes create such current sheets and how these reconnection events subsequently affect the overall system. Exploring different regions in space with dedicated in-situ measurements may lead to the further discoveries of different types of thin current sheets throughout the solar system.

2.4 Energetics, Acceleration, and Heating

The energy explosively released through magnetic reconnection goes into plasma flow energy, heating, and nonthermal particle acceleration in systems ranging from electron-scale current sheets to magnetospheres of accreting black holes. The nature and controlling factors of energetics in the vast array of reconnection systems are among the most compelling questions in reconnection research. Recent developments in laboratory (Ji et al. 2023, this collection), geospatial (Oka et al. 2023b, this collection), solar (Drake et al. 2025, this collection) and astrophysical (Guo et al. 2024, this collection) investigations present an unprecedented opportunity to establish a common framework for energetics across different systems. In the following sections, we list long-standing open questions, and in particular, highlight how the released magnetic energy is partitioned between thermal and nonthermal components and between electrons and ions in the realms of magnetotail observations, solar flares, astrophysical systems, and laboratory experiments.

2.4.1 Magnetotail Observations

In-situ observations in the magnetotail enable the study of particle acceleration at various regions related to reconnection, e.g., the diffusion region, separatrix, magnetic islands or flux ropes, outflows and the dipolarization front (Oka et al. 2023b, this collection). Distinct power law spectra for both electrons and protons are associated with reconnection. The partition between the non-thermal and thermal populations varies for different magnetic reconnection events. A puzzle is that a significant increase of the thermal electron population is not always associated with a hard non-thermal tail (Oka et al. 2022). One should note that a nonthermal population is also observed in a quiet plasma sheet. How particles are heated and accelerated to non-thermal energies in the magnetotail remains to be understood. For ions, there are fewer studies on the energy partition between thermal and nonthermal components. A recent study suggests that ion energization is dominated by electric field fluctuations near the ion cyclotron frequency (Ergun et al. 2020). How energies are partitioned between ions and electrons is also an important unsolved problem. When the ion and electron energy fluxes were compared in the ion diffusion region of magnetotail reconnection, they were dominated by the ion enthalpy flux, with smaller contributions from the electron enthalpy flux and the heat flux and the ion kinetic energy flux (Eastwood et al. 2013).

The important role of turbulence in particle acceleration was identified in low-β magnetotail reconnection events for both ions and electrons (Ergun et al. 2020). While the formation of the nonthermal tail distribution is generally considered on the basis of the guiding-center approximation, how particles interact with turbulence/waves and how they receive energization "kicks" from fluctuations, which are inherently nonadiabatic interactions, remain open questions. It is also interesting to learn how turbulence regulates the repartitioning of energy released by reconnection as a function of distance from the X-line, since energy may be transferred from the bulk outflow into the particle thermal energy or kinetic energy of energetic particles over some distance. Another factor affecting the energization processes in magnetotail reconnection is the finite extent of the reconnection regions and their multiplicity, as discussed in Sect. 2.2. Electrons and low-energy ions have gyroradii smaller than the typical size of the reconnection outflows and can be confined within the reconnection region. However, heavier or energetic ions can have a gyroradius comparable to the transverse scale of the reconnection outflow and thus can no longer be trapped within the outflow, and their acceleration may stop. For such ions to gain further increase in the energy, they need to interact with multiple reconnection events. However, such structures and the evolution of multiple reconnection sites in the magnetotail are difficult to identify from observations. A further caveat that must also be considered in magnetotail events is that the particle distribution observed from a spacecraft prior to an event is generally not identical to the source population observed afterward. Understanding the energetics of reconnection in the magnetotail via simultaneous coverage of the acceleration regions in a larger context, i.e. from the X-line to the outflow regions, is essential.

2.4.2 Solar Flares

Macroscale energy release from magnetic reconnection has been extensively observed through remote sensing of solar flares, and also via recent in-situ measurements of the near-sun solar wind related to interchange reconnection within the coronal holes and reconnection in the heliospheric current sheet (Drake et al. 2025, this collection). Solar flare observations first suggested that the magnetic energy released during reconnection is partitioned into nonthermal and thermal components of electrons and ions. In contrast to magnetotail reconnection, data suggest that the contributions of nonthermal electrons are comparable to or exceed those of thermal electrons. Significant ion energy gains are detected in the emission, although the observed emission is limited in energy range. Combining in-situ observations of flare ejecta from the Parker Solar Probe and Solar Orbiter is expected to improve our understanding of ion energetics.

Theory and modeling efforts have significantly advanced our understanding of the macroscale particle acceleration mechanisms related to reconnection, as summarized in Drake et al. (2025, this collection). Different models that integrate MHD with particle descriptions have been shown to be effective in producing power-law spectra (Arnold et al. 2021; Li et al. 2022; Yin et al. 2024a,b; Seo et al. 2024). These models, as they cover kinetic to large-scale MHD regimes, make it possible to compare and predict imaging spectroscopy observations of solar flares and the highest energy particle acceleration in astrophysical objects. To improve understanding of the energetics of flare reconnection, it is essential to compare observations and model predictions of the role of the guide field or location of the acceleration sites.

2.4.3 Astrophysical Systems

In astrophysical systems, magnetic reconnection has been proposed as a mechanism to explain high-energy phenomena and radiation signatures in systems such as pulsar wind nebulae, pulsar magnetospheres, relativistic jets, gamma-ray bursts, accretion disks, and magnetars (Uzdensky 2011; Hoshino and Lyubarsky 2012; Arons 2012; Guo et al. 2020). Magnetic reconnection can take place in relativistic magnetically dominated regions in these systems. High-energy emissions are observed during reconnection. Heating versus acceleration is one of the key issues in the reconnection studies as discussed in the review by Guo et al. (2024, this collection). Relativistic reconnection events trigger acceleration in various regimes where the power-law tail slope can approach unity (Sironi and Spitkovsky 2014; Guo et al. 2014; Werner et al. 2016; Li et al. 2023). Direct acceleration due to the reconnection electric field can also lead to power-law spectra (Zenitani and Hoshino 2001) in addition to the more common Fermi/betatron processes among different systems (Guo et al. 2015, 2019). However, the overall framework of the energy partition problem is similar to those of other systems. Treating the large-scale fluid behavior and the basic particle acceleration process simultaneously is a challenging problem as in other systems, considering the enormous ratio between the system size and the plasma kinetic scales. Different theories have successfully explained magnetic reconnection as a source of nonthermal particles. Many unresolved questions (e.g., how much energy goes to thermal and nonthermal) are also relevant to space plasmas, but the phenomena exist at much more varied scales, including those observed surrounding black holes in the Event Horizon Telescope.

2.4.4 Laboratory Reconnection Energetics

With the advantage of being able to systematically quantify reconnection energetics, laboratory experiments have made substantial progress on this topic (Ji et al. 2023, this collection), particularly when combined with numerical simulations and space observations. As the magnetic energy is converted into flows, thermal and nonthermal energization takes place at the X line, separatrices, exhausts, and far downstream. Consistent with space observations and fully kinetic simulations, the ion energy gain exceeded that of the electrons in laboratory experiments of reconnection (Yamada et al. 2018). Recent experiments detected directly accelerated electrons via reconnection electric fields and nonthermal electrons for anti-parallel reconnection in low-β plasmas (Chien et al. 2023). These new experiments are expected to enable new comparative studies with space-based reconnection studies. The range of system sizes achievable in laboratory experiments is thus far within 10 ion-inertial lengths from the X-line; hence, the aspects of dynamics and energy conversion at global scales are open challenges. The effects of plasma collisions need to be carefully handled for comparative studies with space plasma. Future experiments in new facilities such as FLARE (Ji et al. 2022) will access both the collisional and collisionless regimes, promising fruitful comparisons with magnetic reconnection in space and astrophysical systems.

3 Future Research

The outstanding questions reviewed in the previous section motivate us to advance the current observation and computing capabilities, and to think beyond existing capabilities. In this section, we discuss new research aspects that can increase our understanding of magnetic reconnection in space plasma environments.

3.1 Interdisciplinary Studies

The recent developments in magnetic reconnection in astrophysical systems has strong connections with reconnection in space, solar and laboratory environments and these connections can be extended further in the future. The development of collisionless magnetic reconnection and kinetic simulations, starting in the 1990s in the space plasma community, laid the solid ground for studying relativistic magnetic reconnection in the astrophysics community. It has become common knowledge that kinetic physics supports fast magnetic reconnection and that magnetic reconnection likely leads to plasma heating and particle acceleration (Birn et al. 2012, and references therein). The other way around, the development of relativistic magnetic reconnection led to new knowledge and motivations for reconnection physics and particle acceleration mechanisms applicable to the nonrelativistic regime. For example, recent progress of theories of reconnection rate was initiated by studies of relativistic magnetic reconnection (Liu et al. 2017). The development of nonthermal powerlaws in simulations of relativistic magnetic reconnection removed doubts surrounding the development of such spectral forms in the nonrelativistic case (Guo et al. 2024, this collection). Motivated by these advances, particle power-law distributions have recently been achieved in nonrelativistic studies (Arnold et al. 2021; Li et al. 2019; Zhang et al. 2021, 2024). Such connections and communication between different communities should continue, and discussions should be strongly encouraged.

Through the common framework of theory and simulations, processes occurring in solar and astrophysical systems captured with large-scale remote-sensing images can be bridged to those in space and laboratory environments where plasma is "directly" measured. The understanding and knowledge gained from in-situ kinetic-scale measurements in geospace and in the laboratory can be applied to other planetary environments. This knowledge also serves as a foundation for understanding larger scale systems such as solar flares and astrophysical phenomena (e.g., relativistic jets in quasars). Direct comparison of the energy spectra between solar flare and magnetotail reconnection has proven to be a successful scheme for studying particle acceleration in magnetic reconnection (Oka et al. 2023a; Drake et al. 2025, this collection). The efficiency of reconnection in the solar wind-planetary interaction uses the common framework observed throughout the solar system (Fuselier et al. 2024, this collection; Gershman et al. 2024, this collection) and serves as a reference for other stellar systems. The 3D dynamics and evolution of the reconnection current sheet detected from in-situ measurements (Hwang et al. 2023, this collection) as well as in controlled laboratory settings (Ji et al. 2023, this collection) benefits from the knowledge of larger scale context gained from solar flare studies (Drake et al. 2025, this collection) and vice versa. That is, to identify the energy conversion site and its dynamics in the solar context one can take into account knowledge from in-situ observations such as those made by MMS (Genestreti et al. 2025; Liu et al. 2025; Norgren et al. 2025; Graham et al. 2025; Stawarz et al. 2024, all in this collection). Communications between communities with different skill sets and bases of knowledge are essential.

3.2 Multiscale Observations

As outlined in Sect. 2.2, cross-scale dynamics and regional coupling remain challenging, unsolved problems. Ion-scale and electron-scale physics have been studied by multi-spacecraft missions such as Cluster and MMS, respectively, and THEMIS has enabled studying larger scale evolution. However, it is necessary to have a larger number of spacecraft covering a wide range of scales simultaneously. In order to realize observations for answering the cross-scale science questions, various future mission concepts have been proposed. These include *Plasma Observatory* (Retinò et al. 2022), which would cover simultaneously the ion and fluid scales at different magnetosphere boundaries, and multipoint observations with sufficient energy range to study terrestrial magnetotail reconnection, including the larger context, such as *MagneToRE* (Maruca et al. 2021), *MagCon* (Kepko et al. 2023), *WEDGE* (Turner et al. 2023), and a CubeSat constellation mission *AME* (Dai et al. 2020).

It would also be interesting to study the distant magnetotail, as reconnection signatures have been identified in the far downtail region ($X \sim -100$ to -200 R_E). While the future multi-spacecraft mission *HelioSwarm* (Klein et al. 2023), is designed to study solar wind turbulence, and also crosses the magnetotail down to $X \sim -60$ R_E, it is important to push further downtail beyond this distance. Such an extension in observational capabilities would allow us to study larger-scale reconnection signatures, including chains of plasmoids, and enable some comparisons with solar flares. These comparions are possible despite the fact that the ion kinetic scale in the magnetotail is on the order of 100–1000 km, whereas it is only 1 m in the solar corona.

Improving solar flare observations is also crucial for facilitating interdisciplinary and comparative studies. In the next few years, the *Solar-C* (Shimizu et al. 2020) and *MUSE* (Cheung et al. 2022) missions will be launched. These missions will study reconnection-related phenomena by conducting spectroscopic observations at EUV wavelengths with wide and seamless temperature coverage (1–1000 eV) and high temporal and spatial resolutions. However, to understand energetics and fast-varying plasma processes such as shocks and reconnection, it is also important to conduct imaging spectroscopy via X-rays (e.g. Oka et al. 2023a; Glesener et al. 2023). Unlike EUV emissions, which can be delayed due to ionization and recombination processes (e.g. Imada et al. 2011; Shen et al. 2013), X-ray continua are produced via bremsstrahlung emission without delay. Recent advancements in photon-counting techniques and improved focusing optics are likely to cover large dynamic ranges at high temporal and spatial resolutions. Therefore, high-precision imaging-spectroscopy of reconnection-related phenomena is expected to be realized. The energy spectrum is obtained seamlessly from thermal to nonthermal energy ranges, which is a crucial step toward a better comparative study between solar and space plasmas. Currently, mission concepts such as *PhoENiX* (Narukage et al. 2020) and *FIERCE* (Shih et al. 2023) are being developed to achieve such imaging spectroscopy via X-rays.

3.3 Future Numerical Simulations and Modeling

Currently, modeling of magnetic reconnection largely relies on numerical simulations, as presented in Shay et al. (2025, this collection). In PIC simulations, various parameters such as the mass ratio (m_i/m_e) and the ratio of the plasma frequency to the electron cyclotron frequency (ω_{pe}/ω_{ce}), are often not realistic to reduce the computational cost. There is currently no consensus on how realistic these parameters should be to provide physically meaningful results. These parameters need to be chosen carefully since the artificial mass ratio controls the separation between ion-scale and electron-scale physics and modifies plasma wave properties. Interestingly, Debye-scale turbulence was reported to alter electron-scale dynamics (Jara-Almonte et al. 2014) when a realistic frequency-ratio parameter is used.

A major unsolved area of research is the interaction of magnetic reconnection with both mesoscale and global-scale dynamics. By "mesoscale," we mean that the length scales are much larger than the ion diffusion region but still smaller than the global magnetospheric scales. Examples of such multiscale interactions are the generation and dynamics of bursty bulk flows in the magnetotail as well as the interaction of reconnection and turbulence both in the magnetosheath and upstream of the Earth's bow shock. A major issue with studying these multiscale interactions is that PIC simulations are too computationally expensive to include meso- and global-scales. To capture the multiscale nature of magnetic reconnection and its interaction with global-scale dynamics, several novel numerical schemes such as interlocking PIC and MHD models (Daldorff et al. 2014; Tóth et al. 2016) have been developed. Additionally, a recently proposed hybrid simulation model, *kglobal*, couples particle gyrokinetics with MHD simulations for particle acceleration studies (Shay et al. 2025, this collection).

Several new directions are emerging, both in software and hardware. Owing to the strong requirements for electric power, recent supercomputers have begun to use "accelerators" such as graphic processing units (GPUs). Since the programming model is different, it is often necessary to develop GPU variants of simulation codes. A growing number of simulation codes have been recently developed for GPUs to overcome the issue of yet-to-be-improved software development environments. Another new direction is machine learning (ML) or artificial intelligence (AI) technologies (Camporeale et al. 2024). ML/AI is useful not only for postprocessing the simulation data but also for predicting solutions for physics problems (Raissi et al. 2019; Karniadakis et al. 2021). Furthermore, quantum computers could be game changers (Grumbling and Horowitz 2019), although the timeline for the creation of practical hardware for simulations is still unknown. They may allow us to calculate a far larger number of variables than classical computers do. However, since basic principles and logic circuits are very different, the development of algorithms for new simuations is required to start from scratch. In the next decade, when algorithms and hardware are further advanced, the role of quantum computing in plasma simulations is expected to become clearer.

4 Conclusions

Recent advancements in in-situ plasma measurements, which have enabled the study of collisionless magnetic reconnection physics, including kinetic physics, led to new discoveries as well as many of the open questions discussed in the previous sections. While they address examples mostly from geospace, many of these open questions are also applicable to other systems including other planets, astrophysical systems, and laboratories. However, in-situ measurements are limited by a specific range of observed plasma parameters and specific scales. Remote observations, on the other hand, usually cover the large-scale context of magnetic reconnection but have a limited energy range and limited resolution that does not cover the microscale. Future observational capabilities addressing the multiscale problems of magnetic reconnection are desired. In the MMS era, the advancement of simulations has also opened new possibilities for close comparisons between observations and simulations at different scales. Applying these simulations, which are "validated" by comparison with in-situ measurements, to other systems in different parameter regimes via next-generation computing techniques is expected to further advance our understanding of the physics of magnetic reconnection.

Acknowledgements The authors gratefully acknowledge all the contributions from the participants of the International Space Science Institute (ISSI) Workshop on *Magnetic Reconnection: Explosive Energy Conversion in Space Plasmas*, Bern, Switzerland, June 27 - July 1, 2022. We thank the ISSI and its staff for hosting and supporting the workshop.

Funding Open access funding provided by Österreichische Akademie der Wissenschaften. This research was funded in part by the Austrian Science Fund (FWF) [10.55776/P32175]. For open access purposes, the author has applied a CC BY public copyright license to any author accepted manuscript version arising from this submission. JES is supported by the Royal Society University Research Fellowship URF/R1/201286. The work at UMD is supported by NASA 80NSSC22K0352 and NSF PHY2109083.

Declarations

Competing Interests The authors declare no competing interests.

References

Antiochos SK, DeVore CR, Klimchuk JA (1999) A model for solar coronal mass ejections. Astrophys J 510(1):485–493. https://doi.org/10.1086/306563. arXiv:astro-ph/9807220

Argall MR, Small CR, Piatt S, et al (2020) MMS SITL ground loop: automating the burst data selection process. Front Astron Space Sci 7:54. https://doi.org/10.3389/fspas.2020.00054. arXiv:2004.07199

Arnold H, Sitnov MI (2023) PIC simulations of overstretched ion-scale current sheets in the magnetotail. Geophys Res Lett 50(15):e2023GL104534. https://doi.org/10.1029/2023GL104534

Arnold H, Drake JF, Swisdak M, et al (2021) Electron acceleration during macroscale magnetic reconnection. Phys Rev Lett 126(13):135101. https://doi.org/10.1103/PhysRevLett.126.135101. arXiv:2011.01147

Arons J (2012) Pulsar wind nebulae as cosmic pevatrons: a current sheet's tale. Space Sci Rev 173(1–4):341–367. https://doi.org/10.1007/s11214-012-9885-1. arXiv:1208.5787

Battaglia M, Kontar EP, Motorina G (2019) Electron distribution and energy release in magnetic reconnection outflow regions during the pre-impulsive phase of a solar flare. Astrophys J 872(2):204. https://doi.org/10.3847/1538-4357/ab01c9

Bergstedt K, Ji H (2024) A novel method to train classification models for structure detection in in-situ spacecraft data. Earth Space Sci 11:e2023EA002965. https://doi.org/10.1029/2023EA002965

Bergstedt K, Ji H, Jara-Almonte J, et al (2020) Statistical properties of magnetic structures and energy dissipation during turbulent reconnection in the Earth's magnetotail. Geophys Res Lett 47(19):e88540. https://doi.org/10.1029/2020GL088540

Bessho N, Chen LJ, Stawarz JE, et al (2022) Strong reconnection electric fields in shock-driven turbulence. Phys Plasmas 29(4):042304. https://doi.org/10.1063/5.0077529

Bhattacharjee A, Huang YM, Yang H, et al (2009) Fast reconnection in high-Lundquist-number plasmas due to the plasmoid instability. Phys Plasmas 16(11):112102. https://doi.org/10.1063/1.3264103. arXiv:0906.5599

Birn J, Schindler K (2002) Thin current sheets in the magnetotail and the loss of equilibrium. J Geophys Res Space Phys 107(A7):1117. https://doi.org/10.1029/2001JA000291

Birn J, Artemyev AV, Baker DN, et al (2012) Particle acceleration in the magnetotail and aurora. Space Sci Rev 173(1–4):49–102. https://doi.org/10.1007/s11214-012-9874-4

Burch JL, Nakamura R (2025) Magnetic reconnection in space: an introduction. Space Sci Rev 221

Burch JL, Moore TE, Torbert RB, et al (2016) Magnetospheric multiscale overview and science objectives. Space Sci Rev 199(1–4):5–21. https://doi.org/10.1007/s11214-015-0164-9

Burch JL, Webster JM, Genestreti KJ, et al (2018) Wave phenomena and beam-plasma interactions at the magnetopause reconnection region. J Geophys Res Space Phys 123(2):1118–1133. https://doi.org/10.1002/2017JA024789

Camporeale E, Marino R, the Editorial Board (2024) Our vision for JGR: Machine Learning and Computation. J Geophys Res, Mach Learn Comput 1:e2024JH000184. https://doi.org/10.1029/2024JH000184

Cassak PA (2011) Theory and simulations of the scaling of magnetic reconnection with symmetric shear flow. Phys Plasmas 18(7):072106. https://doi.org/10.1063/1.3602859

Cassak PA, Fuselier SA (2016) Reconnection at Earth's dayside magnetopause. In: Gonzalez W, Parker E (eds) Magnetic reconnection: concepts and applications. Astrophysics and Space Science Library, vol 427. Springer, Cham, p 213–276. https://doi.org/10.1007/978-3-319-26432-5_6

Cheung MCM, Martínez-Sykora J, Testa P, et al (2022) Probing the physics of the solar atmosphere with the Multi-slit Solar Explorer (MUSE). II. Flares and eruptions. Astrophys J 926(1):53. https://doi.org/10.3847/1538-4357/ac4223

Chien A, Gao L, Zhang S, et al (2023) Non-thermal electron acceleration from magnetically driven reconnection in a laboratory plasma. Nat Phys 19(2):254–262. https://doi.org/10.1038/s41567-022-01839-x. arXiv:2201.10052

Cozzani G, Khotyaintsev YV, Graham DB, et al (2021) Structure of a perturbed magnetic reconnection electron diffusion region in the Earth's magnetotail. Phys Rev Lett 127(21):215101. https://doi.org/10.1103/PhysRevLett.127.215101. arXiv:2103.12527

Dahlin JT, Antiochos SK, Qiu J, et al (2022) Variability of the reconnection guide field in solar flares. Astrophys J 932(2):94. https://doi.org/10.3847/1538-4357/ac6e3d. arXiv:2110.04132

Dai L, Wang C, Cai Z, et al (2020) AME: a cross-scale constellation of CubeSats to explore magnetic reconnection in the solar-terrestrial relation. Front Phys 8:89. https://doi.org/10.3389/fphy.2020.00089

Daldorff LKS, Tóth G, Gombosi TI, et al (2014) Two-way coupling of a global Hall magnetohydrodynamics model with a local implicit particle-in-cell model. J Comput Phys 268:236–254. https://doi.org/10.1016/j.jcp.2014.03.009

Daldorff LKS, Leake JE, Klimchuk JA (2022) Impact of 3D structure on magnetic reconnection. Astrophys J 927(2):196. https://doi.org/10.3847/1538-4357/ac532d. arXiv:2202.04761

Drake JF, Antiochos SK, Bale SD, et al (2025) Magnetic reconnection in solar flares and the near-Sun solar wind. Space Sci Rev 221

Eastwood JP, Phan TD, Drake JF, et al (2013) Energy partition in magnetic reconnection in Earth's magnetotail. Phys Rev Lett 110(22):225001. https://doi.org/10.1103/PhysRevLett.110.225001

Ergun RE, Ahmadi N, Kromyda L, et al (2020) Particle acceleration in strong turbulence in the Earth's magnetotail. Astrophys J 898(2):153. https://doi.org/10.3847/1538-4357/ab9ab5

Faganello M, Califano F, Pegoraro F, et al (2012) Double mid-latitude dynamical reconnection at the magnetopause: an efficient mechanism allowing solar wind to enter the Earth's magnetosphere. Europhys Lett 100(6):69001. https://doi.org/10.1209/0295-5075/100/69001

Fuselier SA, Petrinec SM, Reiff PH, et al (2024) Global-scale processes and effects of magnetic reconnection on the geospace environment. Space Sci Rev 220(4):34. https://doi.org/10.1007/s11214-024-01067-0

Genestreti KJ, Farrugia CJ, Lu S, et al (2023) Multi-scale observation of magnetotail reconnection onset: 2. Microscopic dynamics. J Geophys Res Space Phys 128(11):e2023JA031760. https://doi.org/10.1029/2023JA031760. arXiv:2311.05411

Genestreti K, Nakamura R, Liu YH, et al (2025) Structure of the electron diffusion region during magnetic reconnection. Space Sci Rev

Gershman DJ, Fuselier SA, Cohen IJ, et al (2024) Magnetic reconnection at planetary bodies and astrospheres. Space Sci Rev 220(1):7. https://doi.org/10.1007/s11214-023-01017-2

Glesener L, Albert CA, et al (2023) The need for focused, hard X-ray investigations of the Sun. Bull Am Astron Soc 55(3). https://doi.org/10.3847/25c2cfeb.78fa7c49. arXiv:2306.05447

Graham DB, Khotyaintsev YV, André M, et al (2022) Direct observations of anomalous resistivity and diffusion in collisionless plasma. Nat Commun 13:2954. https://doi.org/10.1038/s41467-022-30561-8

Graham DB, Cozzani G, Khotyanitsev YV, et al (2025) The role of kinetic instabilities and waves in collisionless magnetic reconnection. Space Sci Rev 221

Grumbling E, Horowitz M (eds) (2019) Quantum computing: progress and prospects. The National Academies Press, Washington, DC. https://doi.org/10.17226/25196

Guo F, Li H, Daughton W, et al (2014) Formation of hard power laws in the energetic particle spectra resulting from relativistic magnetic reconnection. Phys Rev Lett 113(15):155005. https://doi.org/10.1103/PhysRevLett.113.155005. arXiv:1405.4040

Guo F, Liu YH, Daughton W, et al (2015) Particle acceleration and plasma dynamics during magnetic reconnection in the magnetically dominated regime. Astrophys J 806(2):167. https://doi.org/10.1088/0004-637X/806/2/167. arXiv:1504.02193

Guo F, Li X, Daughton W, et al (2019) Determining the dominant acceleration mechanism during relativistic magnetic reconnection in large-scale systems. Astrophys J Lett 879(2):L23. https://doi.org/10.3847/2041-8213/ab2a15. arXiv:1901.08308

Guo F, Liu YH, Li X, et al (2020) Recent progress on particle acceleration and reconnection physics during magnetic reconnection in the magnetically-dominated relativistic regime. Phys Plasmas 27(8):080501. https://doi.org/10.1063/5.0012094. arXiv:2006.15288

Guo F, Liu YH, Zenitani S, et al (2024) Magnetic reconnection and associated particle acceleration in high-energy astrophysics. Space Sci Rev 220:43. https://doi.org/10.1007/s11214-024-01073-2. arXiv:2309.13382

Hasegawa H, Argall MR, Aunai N, et al (2024) Advanced methods for analyzing in-situ observations of magnetic reconnection. Space Sci Rev 220:68. https://doi.org/10.1007/s11214-024-01095-w. arXiv:2307.05867

Hesse M, Schindler K (2001) The onset of magnetic reconnection in the magnetotail. Earth Planets Space 53:645–653. https://doi.org/10.1186/BF03353284

Hoshino M, Lyubarsky Y (2012) Relativistic reconnection and particle acceleration. Space Sci Rev 173(1–4):521–533. https://doi.org/10.1007/s11214-012-9931-z

Hosner M, Nakamura R, Schmid D, et al (2024) Reconnection inside a dipolarization front of a diverging earthward fast flow. J Geophys Res Space Phys 129(1):e2023JA031976. https://doi.org/10.1029/2023JA031976

Hsieh MS, Otto A (2014) The influence of magnetic flux depletion on the magnetotail and auroral morphology during the substorm growth phase. J Geophys Res Space Phys 119(5):3430–3443. https://doi.org/10.1002/2013JA019459

Huang K, Liu YH, Lu Q, et al (2022) Auroral spiral structure formation through magnetic reconnection in the aurora acceleration region. Geophys Res Lett 49:e2022GL100466

Hubbert M, Russell CT, Qi Y, et al (2022) Electron-only reconnection as a transition phase from quiet magnetotail current sheets to traditional magnetotail reconnection. J Geophys Res Space Phys 127(3):e29584. https://doi.org/10.1029/2021JA029584

Hudson HS, Simões PJA, Fletcher L, et al (2021) Hot X-ray onsets of solar flares. Mon Not R Astron Soc 501(1):1273–1281. https://doi.org/10.1093/mnras/staa3664

Hwang KJ, Nakamura R, Eastwood JP, et al (2023) Cross-scale processes of magnetic reconnection. Space Sci Rev 219(8):71. https://doi.org/10.1007/s11214-023-01010-9

Imada S, Murakami I, Watanabe T, et al (2011) Magnetic reconnection in non-equilibrium ionization plasma. Astrophys J 742(2):1–11. https://doi.org/10.1088/0004-637x/742/2/70

Jara-Almonte J, Daughton W, Ji H (2014) Debye scale turbulence within the electron diffusion layer during magnetic reconnection. Phys Plasmas 21(3):032114. https://doi.org/10.1063/1.4867868

Ji H, Daughton W, Jara-Almonte J, et al (2022) Magnetic reconnection in the era of exascale computing and multiscale experiments. Nat Rev Phys 4:263–282. https://doi.org/10.1038/s42254-021-00419-x

Ji H, Yoo J, Fox W, et al (2023) Laboratory study of collisionless magnetic reconnection. Space Sci Rev 219:76. https://doi.org/10.1007/s11214-023-01024-3. arXiv:2307.07109

Jiang C, Feng X, Liu R, et al (2021) A fundamental mechanism of solar eruption initiation. Nat Astron 5:1126–1138. https://doi.org/10.1038/s41550-021-01414-z. arXiv:2107.08204

Karniadakis GE, Kevrekidis IG, Lu L, et al (2021) Physics-informed machine learning. Nat Rev Phys 3(6):422–440. https://doi.org/10.1038/s42254-021-00314-5

Kepko L, Gabrielse C, Gkioulidou M, et al (2023) Magnetospheric Constellation (MagCon). Bull Am Astron Soc 55(3). https://doi.org/10.3847/25c2cfeb.0e470159

Khotyaintsev YV, Graham DB, Steinvall K, et al (2020) Electron heating by Debye-scale turbulence in guide-field reconnection. Phys Rev Lett 124(4):045101. https://doi.org/10.1103/PhysRevLett.124.045101. arXiv:1908.09724

Kitamura N, Amano T, Omura Y, et al (2022) Direct observations of energy transfer from resonant electrons to whistler-mode waves in magnetosheath of Earth. Nat Commun 13:6259. https://doi.org/10.1038/s41467-022-33604-2

Klein KG, Spence H, Alexandrova O, et al (2023) HelioSwarm: a multipoint, multiscale mission to characterize turbulence. Space Sci Rev 219(8):74. https://doi.org/10.1007/s11214-023-01019-0. arXiv:2306.06537

Leake JE, Daldorff LKS, Klimchuk JA (2020) The onset of 3D magnetic reconnection and heating in the solar corona. Astrophys J 891(1):62. https://doi.org/10.3847/1538-4357/ab7193. arXiv:2001.02971

Li X, Guo F, Li H, et al (2019) Formation of power-law electron energy spectra in three-dimensional low-β magnetic reconnection. Astrophys J 884(2):118. https://doi.org/10.3847/1538-4357/ab4268. arXiv:1909.01911

Li X, Guo F, Chen B, et al (2022) Modeling electron acceleration and transport in the early impulsive phase of the 2017 September 10th solar flare. Astrophys J 932(2):92. https://doi.org/10.3847/1538-4357/ac6efe. arXiv:2205.04946

Li X, Guo F, Liu YH, et al (2023) A model for nonthermal particle acceleration in relativistic magnetic reconnection. Astrophys J Lett 954(2):L37. https://doi.org/10.3847/2041-8213/acf135. arXiv:2302.12737

Liu YH, Birn J, Daughton W, et al (2014a) Onset of reconnection in the near magnetotail: PIC simulations. J Geophys Res Space Phys 119(12):9773–9789. https://doi.org/10.1002/2014JA020492

Liu YH, Daughton W, Karimabadi H, et al (2014b) Do dispersive waves play a role in collisionless magnetic reconnection? Phys Plasmas 21:022113

Liu YH, Hesse M, Guo F, et al (2017) Why does steady-state magnetic reconnection have a maximum local rate of order 0.1? Phys Rev Lett 118(8):085101. https://doi.org/10.1103/PhysRevLett.118.085101. arXiv:1611.07859

Liu YH, Hesse M, Li TC, et al (2018) Orientation and stability of asymmetric magnetic reconnection X line. J Geophys Res Space Phys 123(6):4908–4920. https://doi.org/10.1029/2018JA025410. arXiv:1805.07774

Liu YN, Fujimoto K, Cao JB (2024) Intense magnetic reconnection process embedded in three-dimensional turbulent current sheet. Geophys Res Lett 51(1):e2023GL106466. https://doi.org/10.1029/2023GL106466

Liu YH, Hesse M, Genestreti K, et al (2025) Ohm's law, the reconnection rate, and energy conversion in collisionless magnetic reconnection. Space Sci Rev 221. https://doi.org/10.1007/s11214-025-01142-0. arXiv:2406.00875

Loureiro NF, Schekochihin AA, Cowley SC (2007) Instability of current sheets and formation of plasmoid chains. Phys Plasmas 14(10):100703. https://doi.org/10.1063/1.2783986. arXiv:astro-ph/0703631

Marshall AT, Burch JL, Reiff PH, et al (2020) Asymmetric reconnection within a flux rope-type dipolarization front. J Geophys Res Space Phys 125(1):e27296. https://doi.org/10.1029/2019JA027296

Maruca BA, Agudelo Rueda JA, Bandyopadhyay R, et al (2021) MagneToRE: mapping the 3-d magnetic structure of the solar wind using a large constellation of nanosatellites. Front Astron Space Sci 8:108. https://doi.org/10.3389/fspas.2021.665885

Motoba T, Sitnov MI, Stephens GK, et al (2022) A new perspective on magnetotail electron and ion divergent flows: MMS observations. J Geophys Res Space Phys 127(10):e2022JA030514. https://doi.org/10.1029/2022JA030514

Nagai T, Shinohara I (2021) Dawn-dusk confinement of magnetic reconnection site in the near-Earth magnetotail and its implication for dipolarization and substorm current system. J Geophys Res Space Phys 126(11):e29691. https://doi.org/10.1029/2021JA02969110.1002/essoar.10507373.1

Nakamura TKM, Hasegawa H, Shinohara I, et al (2011) Evolution of an MHD-scale Kelvin-Helmholtz vortex accompanied by magnetic reconnection: two-dimensional particle simulations. J Geophys Res Space Phys 116(A3):A03227. https://doi.org/10.1029/2010JA016046

Nakamura TKM, Haswgawa H, Daughton W, et al (2017) Turbulent mass transfer caused by vortex induced reconnection in collisionless magnetospheric plasmas. Nat Commun 8:1582. https://doi.org/10.1038/s41467-017-01579-0

Narukage N, Oka M, Fukazawa Y, et al (2020) Satellite mission: PhoENiX (Physics of Energetic and Non-thermal plasmas in the X (= magnetic reconnection) region). Proc SPIE 11444, Space Telescopes and Instrumentation 2020: Ultraviolet to Gamma Ray, 1144429. https://doi.org/10.1117/12.2561341

Norgren C, Chen LJ, Graham DB, et al (2025) Electron and ion dynamics in reconnection diffusion regions. Space Sci Rev 221

Øieroset M, Phan TD, Haggerty C, et al (2016) MMS observations of large guide field symmetric reconnection between colliding reconnection jets at the center of a magnetic flux rope at the magnetopause. Geophys Res Lett 43(11):5536–5544. https://doi.org/10.1002/2016GL069166

Oka M, Phan T, Øieroset M, et al (2022) Electron energization and thermal to non-thermal energy partition during Earth's magnetotail reconnection. Phys Plasmas 29(5):052904. https://doi.org/10.1063/5.0085647

Oka M, Birn J, Egedal J, et al (2023b) Particle acceleration by magnetic reconnection in geospace. Space Sci Rev 219(8):75. https://doi.org/10.1007/s11214-023-01011-8. arXiv:2307.01376

Oka M, Caspi A, Chen B, et al (2023a) Particle acceleration in solar flares with imaging-spectroscopy in soft X-rays. Bull Am Astron Soc 55(3). https://doi.org/10.3847/25c2cfeb.c1b1eb07. arXiv:2306.04909

Phan TD, Eastwood JP, Shay MA, et al (2018) Electron magnetic reconnection without ion coupling in Earth's turbulent magnetosheath. Nature 557(7704):202–206. https://doi.org/10.1038/s41586-018-0091-5

Poh G, Slavin JA, Lu S, et al (2019) Dissipation of earthward propagating flux rope through re-reconnection with geomagnetic field: an MMS case study. J Geophys Res Space Phys 124(9):7477–7493. https://doi.org/10.1029/2018JA026451

Raissi M, Perdikaris P, Karniadakis GE (2019) Physics-informed neural networks: a deep learning framework for solving forward and inverse problems involving nonlinear partial differential equations. J Comput Phys 378:686–707. https://doi.org/10.1016/j.jcp.2018.10.045

Retinò A, Khotyaintsev Y, Le Contel O, et al (2022) Particle energization in space plasmas: towards a multi-point, multi-scale plasma observatory. Exp Astron 54(2–3):427–471. https://doi.org/10.1007/s10686-021-09797-7

Runov A, Angelopoulos V, Artemyev AV, et al (2021) Global and local processes of thin current sheet formation during substorm growth phase. J Atmos Sol-Terr Phys 220:105671. https://doi.org/10.1016/j.jastp.2021.105671

Seo J, Guo F, Li X, et al (2024) Proton acceleration in low-β magnetic reconnection with energetic particle feedback. Astrophys J 977:146. https://doi.org/10.3847/1538-4357/ad8e64. arXiv:2404.12276

Sharma Pyakurel P, Shay MA, Phan TD, et al (2019) Transition from ion-coupled to electron-only reconnection: basic physics and implications for plasma turbulence. Phys Plasmas 26(8):082307. https://doi.org/10.1063/1.5090403. arXiv:1901.09484

Shay M, Adhikari S, Bessho N, et al (2025) Simulation models for exploring magnetic reconnection. Space Sci Rev 221 arXiv:2406.05901

Shen C, Reeves KK, Raymond JC, et al (2013) Non-equilibrium ionization modeling of the current sheet in a simulated solar eruption. Astrophys J 773(2):110. https://doi.org/10.1088/0004-637x/773/2/110

Shibata K, Tanuma S (2001) Plasmoid-induced-reconnection and fractal reconnection. Earth Planets Space 53(6):473–482. https://doi.org/10.1186/BF03353258. arXiv:astro-ph/0101008

Shih AY, Glesener L, Krucker S, et al (2023) Fundamentals of impulsive energy release in the corona. Bull Am Astron Soc 55(3). https://doi.org/10.3847/25c2cfeb.14f5155c

Shimizu T, Imada S, Kawate T, et al (2020) The solar-C (EUVST) mission: the latest status. In: den Herder JWA, Nikzad S, Nakazawa K (eds) Proc SPIE 11444, Space Telescopes and Instrumentation 2020: Ultraviolet to Gamma Ray, 114440N. https://doi.org/10.1117/12.2560887

Sironi L, Spitkovsky A (2014) Relativistic reconnection: an efficient source of non-thermal particles. Astrophys J Lett 783(1):L21. https://doi.org/10.1088/2041-8205/783/1/L21. arXiv:1401.5471

Sitnov MI, Arnold H (2022) Equilibrium kinetic theory of weakly anisotropic embedded thin current sheets. J Geophys Res Space Phys 127(11):e2022JA030945. https://doi.org/10.1029/2022JA030945

Sitnov MI, Schindler K (2010) Tearing stability of a multiscale magnetotail current sheet. Geophys Res Lett 37(8):L08102. https://doi.org/10.1029/2010GL042961

Sitnov MI, Merkin VG, Raeder J (2016) Great mysteries of the Earth's magnetotail. Eos 97 https://doi.org/10.1029/2016EO048185

Sitnov M, Birn J, Ferdousi B, et al (2019a) Explosive magnetotail activity. Space Sci Rev 215(4):31. https://doi.org/10.1007/s11214-019-0599-5

Sitnov MI, Stephens GK, Tsyganenko NA, et al (2019b) Signatures of nonideal plasma evolution during substorms obtained by mining multimission magnetometer data. J Geophys Res Space Phys 124(11):8427–8456

Spinnangr SF, Hesse M, Tenfjord P, et al (2022) Electron behavior around the onset of magnetic reconnection. Geophys Res Lett 49(23):e2022GL102209. https://doi.org/10.1029/2022GL102209

Stanier A, Daughton W, Le A, et al (2019) Influence of 3D plasmoid dynamics on the transition from collisional to kinetic reconnection. Phys Plasmas 26(7):072121. https://doi.org/10.1063/1.5100737. arXiv:1906.04867

Stawarz JE, Muñoz PA, Bessho N, et al (2024) The interplay between collisionless magnetic reconnection and turbulence. Space Sci Rev 220:90. https://doi.org/10.1007/s11214-024-01124-8

Stephens GK, Sitnov MI, Weigel RS, et al (2023) Global structure of magnetotail reconnection revealed by mining space magnetometer data. J Geophys Res Space Phys 128(2):e2022JA031066. https://doi.org/10.1029/2022JA031066

Swisdak M, Opher M, Drake JF, et al (2010) The vector direction of the interstellar magnetic field outside the heliosphere. Astrophys J 710(2):1769. https://doi.org/10.1088/0004-637X/710/2/1769

Török T, Kliem B (2005) Confined and ejective eruptions of kink-unstable flux ropes. Astrophys J Lett 630(1):L97–L100. https://doi.org/10.1086/462412

Tóth G, Jia X, Markidis S, et al (2016) Extended magnetohydrodynamics with embedded particle-in-cell simulation of Ganymede's magnetosphere. J Geophys Res Space Phys 121(2):1273–1293. https://doi.org/10.1002/2015JA021997

Trattner KJ, Mulcock JS, Petrinec SM, et al (2007) Probing the boundary between antiparallel and component reconnection during southward interplanetary magnetic field conditions. J Geophys Res Space Phys 112(A8):A08210. https://doi.org/10.1029/2007JA012270
Turner DL, Genestreti K, Argall M, et al (2023) Cross-scale physics and the acceleration of particles in collisionless plasmas throughout the heliosphere and beyond: II. Magnetic reconnection. Bull Am Astron Soc 55(3). https://doi.org/10.3847/25c2cfeb.cf78b56a
Uzdensky DA (2011) Magnetic reconnection in extreme astrophysical environments. Space Sci Rev 160(1–4):45–71. https://doi.org/10.1007/s11214-011-9744-5. arXiv:1101.2472
Wang R, Lu Q, Nakamura R, et al (2018) An electron-scale current sheet without bursty reconnection signatures observed in the near-Earth tail. Geophys Res Lett 45(10):4542–4549. https://doi.org/10.1002/2017GL076330
Werner GR, Uzdensky DA, Cerutti B, et al (2016) The extent of power-law energy spectra in collisionless relativistic magnetic reconnection in pair plasmas. Astrophys J Lett 816(1):L8. https://doi.org/10.3847/2041-8205/816/1/L8. arXiv:1409.8262
Yamada M, Chen LJ, Yoo J, et al (2018) The two-fluid dynamics and energetics of the asymmetric magnetic reconnection in laboratory and space plasmas. Nat Commun 9:5223. https://doi.org/10.1038/s41467-018-07680-2
Yin Z, Drake JF, Swisdak M (2024a) A computational model for ion and electron energization during macroscale magnetic reconnection. Phys Plasmas 31:062901. https://doi.org/10.1063/5.0199679. arXiv:2401.14500
Yin Z, Drake JF, Swisdak M (2024b) Simultaneous proton and electron energization during macroscale magnetic reconnection. Astrophys J 974(1):74. https://doi.org/10.3847/1538-4357/ad7131. arXiv:2407.10933
Yoo J, Ng J, Ji H, et al (2024) Anomalous resistivity and electron heating by lower hybrid drift waves during magnetic reconnection with a guide field. Phys Rev Lett 132:145101. https://doi.org/10.1103/PhysRevLett.132.145101
Yoon YD, Wendel DE, Yun GS (2023) Equilibrium selection via current sheet relaxation and guide field amplification. Nat Commun 14:139. https://doi.org/10.1038/s41467-023-35821-9
Zenitani S, Hoshino M (2001) The generation of nonthermal particles in the relativistic magnetic reconnection of pair plasmas. Astrophys J Lett 562(1):L63–L66. https://doi.org/10.1086/337972. arXiv:1402.7139
Zhang Q, Guo F, Daughton W, et al (2021) Efficient nonthermal ion and electron acceleration enabled by the flux-rope kink instability in 3D nonrelativistic magnetic reconnection. Phys Rev Lett 127(18):185101. https://doi.org/10.1103/PhysRevLett.127.185101. arXiv:2105.04521
Zhang Q, Guo F, Daughton W, et al (2024) Multispecies ion acceleration in 3D magnetic reconnection with hybrid-kinetic simulations. Phys Rev Lett 132(11):115201. https://doi.org/10.1103/PhysRevLett.132.115201. arXiv:2210.04113

Authors and Affiliations

R. Nakamura[1,2] · J.L. Burch[3] · J. Birn[4] · L.-J. Chen[5] · D.B. Graham[6] · F. Guo[7] · K.-J. Hwang[3] · H. Ji[8] · Y.V. Khotyaintsev[6] · Y.-H. Liu[9] · M. Oka[10] · D. Payne[11] · M.I. Sitnov[12] · M. Swisdak[11] · S. Zenitani[1] · J.F. Drake[11] · S.A. Fuselier[3,13] · K.J. Genestreti[3] · D.J. Gershman[5] · H. Hasegawa[14] · M. Hoshino[15] · C. Norgren[6] · M.A. Shay[16] · J.R. Shuster[17] · J.E. Stawarz[18]

✉ R. Nakamura
rumi.nakamura@oeaw.ac.at

1 Space Research Institute, Austrian Academy of Sciences, Schmiedlstraße 6, 8042 Graz, Austria

2 International Space Science Institute, Bern, Switzerland

3 Southwest Research Institute, San Antonio, TX 78238, USA

4 Center for Space Plasma Physics, Space Science Institute, Boulder, CO 80301, USA

[5] Goddard Space Flight Center, NASA, Greenbelt, MD 20771, USA

[6] Swedish Institute of Space Physics, Uppsala, Sweden

[7] Los Alamos National Laboratory, Los Alamos, NM 87545, USA

[8] Department of Astrophysical Sciences, Princeton University, Princeton, NJ 08544, USA

[9] Department of Physics and Astronomy, Dartmouth College, Hanover, NH 03750, USA

[10] Space Science Laboratory, UC Berkeley, Berkeley, CA 94720, USA

[11] University of Maryland, College Park, MD 20742, USA

[12] The Johns Hopkins University Applied Physics Laboratory, Laurel, MD, 20723, USA

[13] The University of Texas at San Antonio, San Antonio, TX, 78249, USA

[14] Institute of Space and Astronautical Science, JAXA, Sagamihara, Japan

[15] Department of Earth and Planetary Science, The University of Tokyo, Tokyo, 113-0033, Japan

[16] Department of Physics and Astronomy, University of Delaware, Newark, DE 19716, USA

[17] Space Science Center, University of New Hampshire, Durham, NH 03824, USA

[18] Department of Mathematics, Physics, and Electrical Engineering, Northumbria University, Newcastle upon Tyne, UK

The manufacturer's authorised representative in the EU is Springer Nature Customer Service Centre GmbH, Europaplatz 3, 69115 Heidelberg, Germany. If you have any concerns regarding our products, please contact ProductSafety@springernature.com

Printed and bound by CPI Group (UK) Ltd, Croydon, CR0 4YY
07/07/2026
02160918-0004